Lipid Mediators

THE HANDBOOK OF IMMUNOPHARMACOLOGY

Series Editor: Clive Page
King's College London, UK

Titles in this series

Cells and Mediators

Immunopharmacology of
Eosinophils
(edited by H. Smith and R. Cook)

The Immunopharmacology of Mast
Cells and Basophils
(edited by J.C. Foreman)

Lipid Mediators
(edited by F. Cunningham)

Adhesion Molecules
(edited by C.D. Wegner,
forthcoming)

Immunopharmacology of
Lymphocytes
(edited by M. Rola-Pleszczynski,
forthcoming)

Immunopharmacology of Platelets
(edited by M. Joseph,
forthcoming)

Immunopharmacology of
Neutrophils
(edited by P.G. Hellewell and
T.J. Williams, forthcoming)

Systems

Immunopharmacology of the
Gastrointestinal System
(edited by J.L. Wallace)

Immunopharmacology of Joints
and Connective Tissue
(edited by J. Dingle and M.E.
Davies, forthcoming)

Immunopharmacology of the Heart
(edited by M.J. Curtis,
forthcoming)

Immunopharmacology of Epithelial
Barriers
(edited by R. Goldie, forthcoming)

Immunopharmacology of the Renal
System
(edited by C. Tetta)

Immunopharmacology of the
Microvasculature
(edited by S. Brain, forthcoming)

Drugs

Immunotherapy for Immune-
related Diseases
(edited by W.J. Metzger,
forthcoming)

Immunopharmacology of AIDS
(forthcoming)

Immunosuppressive Drugs
(forthcoming)

Glucocorticosteroids
(forthcoming)

Angiogenesis
(forthcoming)

Immunopharmacology of Free
Radical Species
(forthcoming)

Lipid Mediators

edited by

Fiona M. Cunningham
Department of Veterinary Basic Sciences,
The Royal Veterinary College,
University of London, UK

ACADEMIC PRESS

Harcourt Brace & Company, Publishers
London   San Diego   New York
Boston   Sydney   Tokyo   Toronto

ACADEMIC PRESS LIMITED
24/28 Oval Road
London NW1 7DX

United States Edition published by
ACADEMIC PRESS INC.
San Diego, CA 92101

This book is printed on acid-free paper

A catalogue record for this book
is available from the British Library

ISBN 0–12–198875–9

Typeset by J&L Composition Ltd, Filey, North Yorkshire
Printed and bound in Great Britain by The Bath Press, Avon

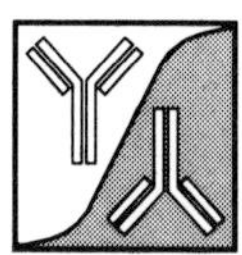

Contents

3. *Biological Properties of Cyclooxygenase Products* 61

J.R. Vane and R.M. Botting

4. *Other Essential Fatty Acids, Their Metabolites and Cell–Cell Interactions* 99

M.R. Buchanan and S.J. Brister

5. *Inhibitors of Fatty Acid-Derived Mediators* 117

D.W.P. Hay and D.E. Griswold

6.   Biosynthesis and Catabolism of Platelet-Activating Factor   181

F. Snyder

7.   Cellular Sources of Platelet-Activating Factor and Related Lipids   193

D.L. Bratton, K.L. Clay and P.M. Henson

8.   Biological Properties of Platelet-Activating Factor   221

A.G. Stewart

9.   *Platelet-Activating Factor: Receptors and Receptor Antagonists*   297

S.-B. Hwang

Colour plate 1a, b referring to Chapter 4 appears between pages 110–111

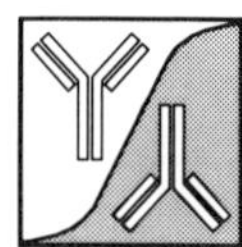

Contributors

R. Botting
The William Harvey Research Institute,
St. Bartholomew's Hospital Medical School,
Charterhouse Square,
London EC1M 6BQ, UK

Donna L. Bratton
Department of Paediatrics,
National Jewish Center for Immunology and Respiratory
Medicine,
1400 Jackson Street,
Denver,
Colorado 80206, USA

S. Brister
Department of Surgery,
McMaster University,
Hamilton, Ontario,
Canada

Michael R. Buchanan
Director of Surgical Research,
H.G.D. – McMaster Clinic,
Hamilton Civic Hospitals,
Hamilton General Division,
237 Barton Street East,
Hamilton, Ontario L8L 2X2,
Canada

K. Clay
Department of Paediatrics,
National Jewish Center for Immunology and Respiratory
Medicine,
1400 Jackson Street,
Denver,
Colorado 80206, USA

Dr. D. Griswold
Department of Pharmacology,
SmithKline Beecham Pharmaceuticals,
709 Swedeland Road, P.O. Box 1539,
King of Prussia,
PA 19406–0939,
USA

Douglas W.P. Hay
Department of Pharmacology,
SmithKline Beecham Pharmaceuticals,
709 Swedeland Road, P.O. Box 1539,
King of Prussia,
PA 19406–0939,
USA

P.M. Henson
Department of Paediatrics,
National Jewish Center for Immunology and Respiratory
Medicine,
1400 Jackson Street,
Denver,
Colorado 80206, USA

San-Bao Hwang
Alteon Inc.
165 Ludlow Avenue,
Northvale,
NJ 07647,
USA

D.L. Smith
Syntex Research,
Palo Alto,
CA 94304, USA

Fred Snyder
Vice Chairman,
Medical Sciences Division,
Oak Ridge Associated Universities,
Post Office Box 117,
Oak Ridge,
Tennessee 37831–0117,
USA

A.G. Stewart
Microsurgery Research Centre,
St. Vincent's Hospital,
41, Victoria Parade,
Fitzroy 3065,
Victoria,
Australia

G. Taylor
Department of Clinical Pharmacology,
Royal Postgraduate Medical School,
Hammersmith Hospital,
Du Cane Road,
London W12 0NN, UK

J.R. Vane
The William Harvey Research Institute,
St. Bartholomew's Hospital Medical School,
Charterhouse Square,
London EC1M 6BQ, UK

R. Wellings
Department of Clinical Pharmacology,
Royal Postgraduate Medical School,
Hammersmith Hospital,
Du Cane Road,
London W12 0NN, UK

A.L. Willis
Palo Alto Institute for Molecular Biology,
2462 Wyandotte Street,
Mountainview,
CA 94043, USA

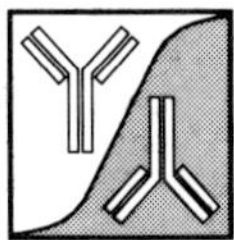

Series Preface

The consequences of diseases involving the immune system such as AIDS, and chronic inflammatory diseases such as bronchial asthma, rheumatoid arthritis and atherosclerosis, now account for a considerable economic burden to governments worldwide. In response to this, there has been a massive research effort investigating the basic mechanisms underlying such diseases, and a tremendous drive to identify novel therapeutic applications for the prevention and treatment of such diseases. Despite this effort, however, much of it within the pharmaceutical industries, this area of medical research has not gained the prominence of cardiovascular pharmacology or neuropharmacology. Over the last decade there has been a plethora of research papers and publications on immunology, but comparatively little written about the implications of such research for drug development. There is also no focal information source for pharmacologists with an interest in diseases affecting the immune system or the inflammatory response to consult, whether as a teaching aid or as a research reference. The main impetus behind the creation of this series was to provide such a source by commissioning a comprehensive collection of volumes on all aspects of immunopharmacology. It has been a deliberate policy to seek editors for each volume who are not only active in their respective areas of expertise, but who also have a distinctly *pharmacological* bias to their research. My hope is that *The Handbook of Immunopharmacology* will become indispensable to researchers and teachers for many years to come, with volumes being regularly updated.

The series follows three main themes, each theme represented by volumes on individual component topics.

The first covers each of the major cell types and classes of inflammatory mediators. The second covers each of the major organ systems and the diseases involving the immune and inflammatory responses that can affect them. The series will thus include clinical aspects along with basic science. The third covers different classes of drugs that are currently being used to treat inflammatory disease or diseases involving the immune system, as well as novel classes of drugs under development for the treatment of such diseases.

To enhance the usefulness of the series as a reference and teaching aid, a standardized artwork policy has been adopted. A particular cell type, for instance, is represented identically throughout the series. An appendix of these standard drawings is published in each volume. Likewise, a standardized system of abbreviations of terms has been implemented and will be developed by the editors involved in individual volumes as the series grows. A glossary of abbreviated terms is also published in each volume. This should facilitate cross-referencing between volumes. In time, it is hoped that the glossary will be regarded as a source of standard terms.

While the series has been developed to be an integrated whole, each volume is complete in itself and may be used as an authoritative review of its designated topic.

I am extremely grateful to the officers of Academic Press, and in particular to Dr Carey Chapman, for their vision in agreeing to collaborate on such a venture, and greatly hope that the series does indeed prove to be invaluable to the medical and scientific community.

C.P. Page

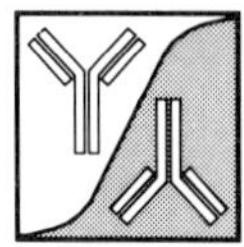

Preface

The structural identification of lipid mediators such as platelet-activating factor and the leukotrienes has led, over the past 10–15 years, to an explosion of information on the physiological and pathophysiological properties of these compounds. Of even greater significance is the application of this knowledge to the rational design of novel therapeutic agents for treating a range of diseases. This is well illustrated by the development in recent years of inhibitors of the formation and actions of leukotrienes and PAF receptor antagonists and the evaluation of such compounds in the clinic. Inhibitors of prostaglandin formation and, more recently, prostaglandin analogues have already established their place in the physician's armamentarium. As more becomes known about synthetic pathways, receptors and receptor–effector coupling mechanisms for the lipid mediators, the opportunities for pharmacological intervention will, no doubt, increase.

In compiling this volume on lipid mediators for the *Handbook of Immunopharmacology* series I have been fortunate that such a number of distinguished scientists, whose research endeavours have greatly advanced our understanding of this field, agreed to lend their support to the venture. Pooling the contributions of these authors, each with their different specialized interests, has, I believe, produced a comprehensive overview of lipid mediators, from synthesis through to inhibition. It could be argued that any book on this subject will be out of date almost as soon as it is written. However, I believe that this volume will prove to be an invaluable reference text for both those familiar with and those new to the exciting, and ever changing, world of lipid mediators.

F.M. Cunningham

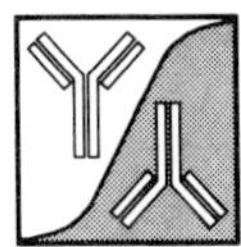

1. Metabolism of Arachidonic Acid: An Overview

A.L. Willis and D.L. Smith

Lipid Mediators
ISBN 0–12–198875–9

1. Introduction

Arachidonic acid ($C_{20:4}$ ω-6) is a member of the ω-6 series of EFAs. It is an example of a "derived" EFA in that it can be derived biochemically from dietary linoleic acid ($C_{18:2}$ ω-6) by alternate steps of desaturation and chain elongation (Fig. 1.1). This interrelationship between EFA, discovered by Mead and others (see Mead and Willis, 1987) was the first major discovery concerning arachidonic acid metabolism, but its implications were not realized until the mid-1960s, when Van Dorp *et al.* (1964) and Bergstrom *et al.* (1964) independently discovered that arachidonic acid was the biochemical precursor for the prostaglandins (PGE_2 and $PGF_{2\alpha}$), whose structure had been elucidated by Bergstrom's group some 10 years earlier. Thus, the earlier pioneering discoveries of the EFAs by Burr and Burr (1930) and of the prostaglandins by von Euler (1936) were at last linked (see von Euler, 1982).

Undoubtedly the development that triggered an explosive interest in arachidonic acid metabolism was the recognition that aspirin-type drugs owed their mode of action to inhibition of the enzyme that produced prostaglandins from arachidonic acid (Ferreira *et al.*, 1971; Smith and Willis, 1971; Vane, 1971). The mechanisms involved in biosynthesis of the prostaglandins and other eicosanoids is outlined below. For detailed structures of most eicosanoids (up to 1987) and metabolites, the reader is referred to the reviews of Willis (1987), Smith (1987) and Smith *et al.* (1987). Nomenclature of the various eicosanoid structures, enzymes and proposed receptor types is also reviewed in detail elsewhere (Willis *et al.*, 1990).

2. Mechanisms Involved in Bioavailability of Arachidonic Acid for Eicosanoid Production

It is commonly considered that arachidonic acid has to be "released" from its stores in tissue lipids in order to be converted into eicosanoids, although recent studies by Kuhn *et al.* (1990) indicated that arachidonic acid can also be oxidized *in situ* and then released as a preformed or partially preformed eicosanoid.

The principal source of arachidonic acid for eicosanoid biosynthesis is thought to be phospholipids of cell membranes where arachidonic acid is located mainly in the 2 position (Fig. 1.2). PLA, PLC and PLD have all been implicated in release of arachidonic acid.

2.1   PHOSPHOLIPASES A_2

Consequently, PLA_2s (which cleave fatty acids from the 2 position of the phospholipid) are probably the most important group of enzymes involved in arachidonic acid

release, although definitive proof has been largely lacking (see Smith, 1992). These phospholipases have been subdivided into three main classes.

2.1.1   Secretory (Pancreatic) Type I Phospholipases A_2

Isolated from human and bovine pancreas, this is secreted as a proenzyme and is activated by cleavage of a heptapeptide from the N terminus. Enzymes of this class are closely related to the cobra venom enzymes (Davidson and Dennis, 1990). They are optimally active at alkaline pH and require over 1 mM Ca^{2+} for activity. These enzymes show no specificity towards the acyl group at the 2 position of the phospholipid and do not discriminate between PC and phosphatidylethanolamine (PE) (see Smith, 1992).

2.1.2   Secretory (Non-pancreatic) Type II Phospholipases A_2

This enzyme class has been isolated from platelets, synovial fluid and liver and the protein sequences have been determined. Unlike the type I class of enzymes they are not produced as proenzymes and lack a cysteine at residue 11. The pH optimum and specificity are similar to those for the type I PLA_2s (see Smith, 1992).

Although not strictly a secreted enzyme (only partially secreted in platelets; Kramer *et al.*, 1989), there is some evidence implicating activity of this enzyme class with prostanoid production. Thus, manoalide and its analogues inhibited a membrane-associated type II PLA_2 with corresponding diminution of prostanoid production (Lister *et al.*, 1989; Reynolds *et al.*, 1991).

2.1.3   Cytosolic Phospholipases A_2

Three types have been isolated and characterized and two of them are unusual in not requiring calcium for their activity. One of these is activated during myocardial ischaemia and has activity specific to certain phospholipids (see Smith, 1992). The third PLA_2 of this class is a large (110 kDa) cytosolic protein that undergoes a Ca^{2+}-dependent activation. There is considerable evidence that this enzyme is involved in prostanoid production in the kidney (see Smith, 1992), in addition to providing a substrate for the epoxygenase pathway (Force *et al.*, 1991).

2.2   PHOSPHOLIPASE C

As first proposed by Bell *et al.* (1979), this phospholipase can cleave the phosphatidyl moiety from PI and then, subsequently, diglyceride lipase cleaves arachidonic acid from the 2 position of the glyceryl moiety. The differential release of fatty acids from different phospholipid precursors was further described as being attributable to differential action of PLA_2 and PLC, the latter acting principally on PI and the former acting mainly on PC and PE (Bills *et al.*, 1977). For a more complete review, see Holmsen (1987) and Smith (1992).

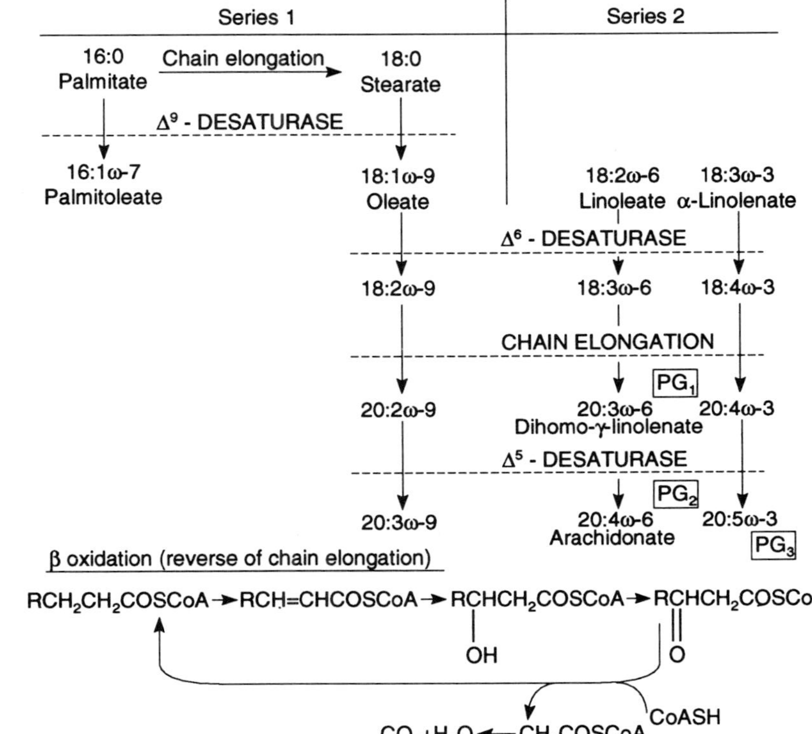

Figure 1.1 EFA metabolism. Redrawn from Willis (1987).

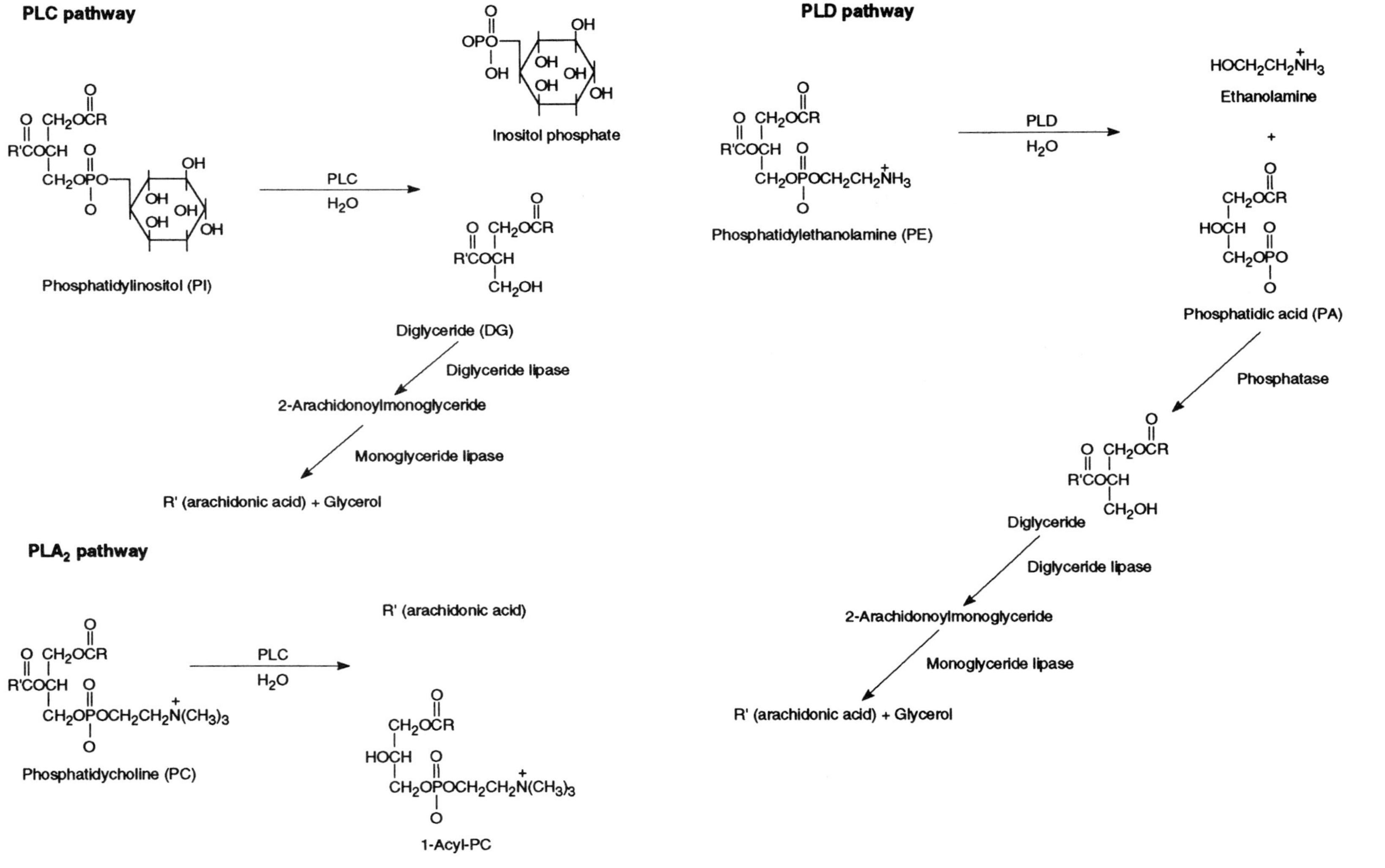

Figure 1.2 PLC/glyceride lipase, PLD/glyceride lipase and PLA₂ pathways that may be involved in release of arachidonic acid from phospholipids for prostanoid synthesis. For purposes of illustration, different phospholipids are shown functioning with different phospholipases; however, each phospholipase has its own range of substrate specificities. Redrawn from Smith (1992).

2.3 PHOSPHOLIPASE D

More recently, evidence has been available for a role for PLD in the release of arachidonate from cell membrane phospholipids (see Smith, 1992). This mechanism involves initial conversion by PLD of PE or PC to phosphatidic acid followed by formation of diglyceride and monoglyceride (see Smith, 1992).

2.4 OTHER LIPASES

Theoretically, any esterified source for arachidonic acid can serve as a source for eicosanoid production. Reported sources are cholesteryl esters of the adrenal cortex (Vahouny *et al.*, 1982) and low-density lipoproteins (Habenicht *et al.*, 1990). According to these authors, an important but overlooked source of arachidonic acid release is lipoprotein lipase that cleaves fatty acids from the triglycerides of chylomicrons. This could explain the short peak in prostaglandin production following oral administration of DGLA to humans (Stone *et al.*, 1979) and arachidonic acid to rats (Lands and Kulmacz, 1986).

2.5 METABOLIC POOLS OF PROSTAGLANDIN PRECURSORS

In rabbits maintained on an EFA-deficient diet, reduced levels of E- and F-type prostaglandins in brain and other tissues were seen that were not accompanied by reduced phospholipid levels of arachidonate but were associated with reduced linoleate content (Hassam *et al.*, 1979; Willis *et al.*, 1982). Consequently, a concept was developed that "spillover" from metabolic pools of eicosanoid precursor produced from the alternate steps of desaturation/elongation (Fig. 1.2) may be associated with steady state "basal" levels of PG production. This concept (see Willis, 1987) is also supported by the findings that PGE_1 is produced by unstimulated human platelets. This is not related to platelet phospholipid arachidonic acid or DGLA levels, but is related to small amounts of free DGLA (the PGE_1 precursor) probably spilling over from metabolic pools formed via linoleic acid (see Lagarde *et al.*, 1982; Willis and Smith, 1982).

2.6 TRIGLYCERIDE AND FREE ACID STORES

In theory, arachidonic acid and other precursor fatty acids released from adipose tissue during lipolysis could be a source of eicosanoids. Certainly, triglycerides have been reported to be a source of prostaglandins in the thyroid (Haye *et al.*, 1974). Even more intriguing are the interstitial lipid droplets in the kidney. These contain both arachidonic acid and adrenic acid, both incorporated into triglycerides and also present as free acids (Comai *et al.*, 1975). Interestingly, administration of indomethacin to rabbits increases both the physical size of the lipid droplets and their prostaglandin precursor composition (Comai *et al.*, 1974) indicating that the free acid form of the precursors (derived from the triglycerides?) might function in the normal endogenous turnover of renal prostaglandins.

3. Enzymic Conversion of Arachidonic Acid to Prostaglandins and Related "Prostanoids" (Fig. 1.3)

3.1 PROSTAGLANDIN-H SYNTHASE

This could be considered as the world's most important enzyme, since it is the target site for inhibition by non-steroidal anti-inflammatory drugs of the aspirin type. These currently account for US $3000–5000 million in annual revenue in the USA (Smith and Marnett, 1991). This enzyme, first called "prostaglandin synthetase" then "cyclooxygenase", is now officially designated "PGH (endoperoxide) synthase" (see Willis *et al.*, 1990), and has now been completely sequenced (Smith and Marnett, 1991; Smith, 1992). Mechanisms of its regulation, enzyme expression and inhibition have been extensively studied, as reviewed in more detail elsewhere (DeWitt, 1991; Smith and Marnett, 1991; Smith, 1992). The area of prostaglandin synthesis inhibitors has been reviewed by Nelson (1989) and by Smith (1992).

In brief, PGH (endoperoxide) synthase activity consists of two related catalytic functions. There is both a cyclooxygenase activity that catalyses the formation of PGG_2 and a peroxidase activity catalysing a two-electron reduction of PGG_2 to PGH_2. PGH synthase is an integral membrane protein found mainly in microsomal membranes, although it has also been found on the nuclear envelope and often on the plasma membrane (see Smith and Marnett, 1991; Smith, 1992).

The initial cyclooxygenase reaction supports the originally proposed mechanism of Samuelsson (1965) on suggested incorporation of molecular oxygen into the PGH_2 molecule, which is unable to occur under anaerobic conditions. This fact was once used to help determine the structure (as PGH_2) of the "labile aggregation-stimulating substance" (LASS) formed from arachidonic acid (Willis, 1974b) as PGH_2. As first proposed by Lands *et al.* (1973), this initial reaction is catalysed by low amounts of fatty acid hydroperoxide ("peroxide tone"), and is "suicidal" in that incorporation of molecular oxygen by the enzyme results in inactivation of the enzyme activity. This suicide inactivation seems associated with peroxidase activity, since it is partially prevented by peroxidase-reducing substrates (see Smith and Marnett, 1991). Furthermore, Mn^{2+}-PPIX-reconstituted PGH synthase (which lacks appreciable peroxidase activity) undergoes auto-inactivation at a slower rate than the native enzyme (see Smith and Marnett, 1991).

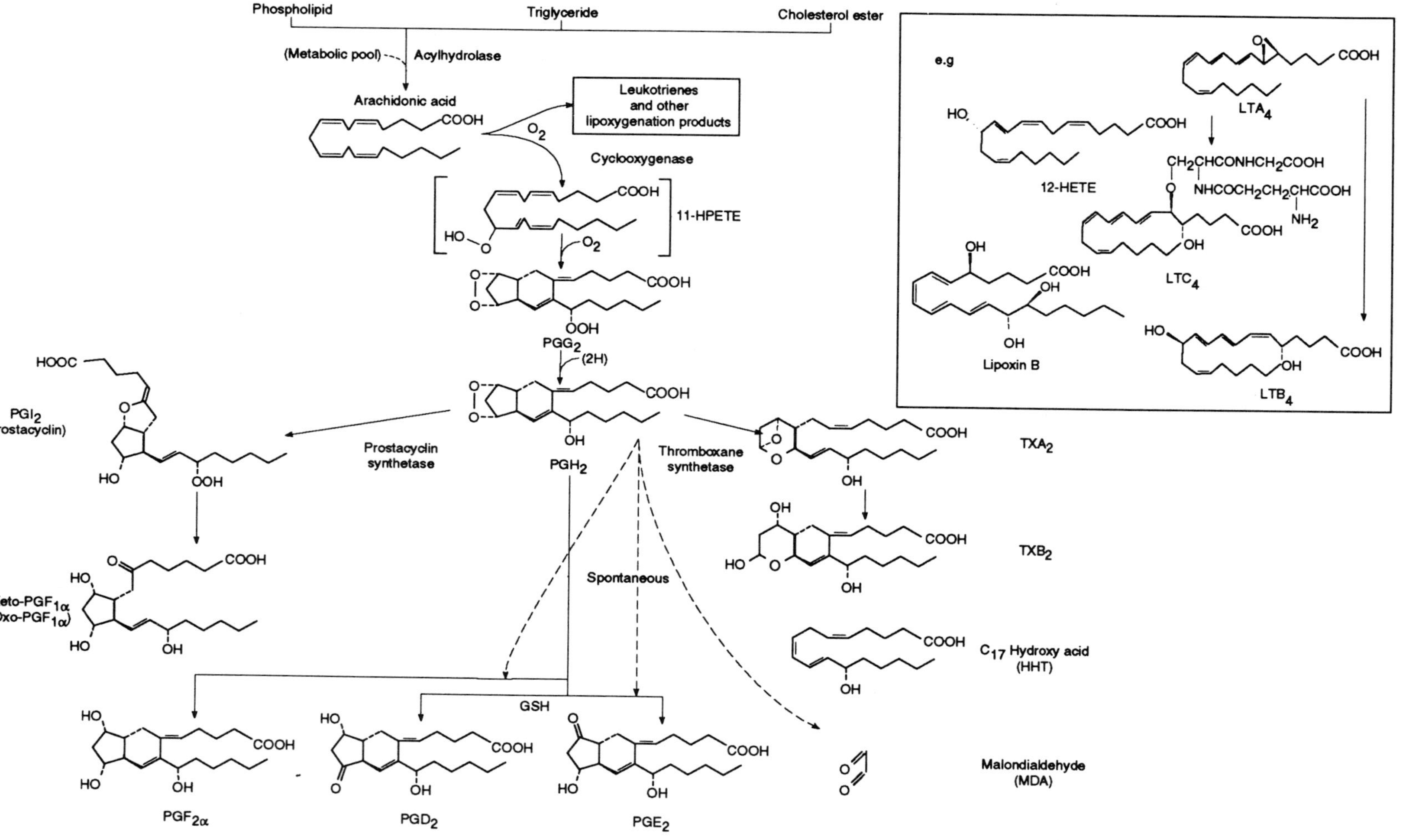

Figure 1.3 Principal products of cyclooxygenase (prostaglandins, thromboxanes and prostacyclin). It has been reported that 5(6)-epoxides of arachidonic acid (formed by cytochrome P-450) can be converted by cyclooxygenase to 5-hydroxy-PGI$_1$. Redrawn from Willis (1987).

The second important step in conversion of arachidonic acid to PGH_2 by the PGH synthase is the peroxidase activity that catalyses the reduction of a variety of hydroperoxides to alcohols at the expense of a reducing co-substrate. The peroxidase of PGH synthase preferentially reduces fatty acid hydroperoxides, including PGG_2 (see Smith and Marnett, 1991).

Association of prostaglandin synthesis at more than one subcellular location suggests the possibility that prostaglandin synthesis at different subcellular sites may be elicited by different stimuli. In turn, this could be associated with differences in mechanisms of release of arachidonate or (perhaps) other eicosanoid precursors (Smith and Marnett, 1991; Smith, 1992). The possibility thus arises that there could be differential production of prostaglandins and lipoxygenase products. Such considerations might also explain reported differential release of PGE_1 and PGE_2 (see Horrobin, 1980).

3.2   FURTHER TRANSFORMATION OF PROSTAGLANDIN H_2 INTO BIOLOGICALLY ACTIVE EICOSANOIDS

Although the endoperoxide PGH_2 has important biological effects (notably aggregation of platelets), the principal biologically active prostanoids are formed enzymatically and/or non-enzymatically from PGH_2 (see Smith and Marnett, 1991; Smith, 1992). PGD_2 is formed from PGH_2 by PGD synthase, PGE by PGE synthase (once called "isomerase") and PGF by PGF synthase. The unstable metabolite prostacyclin (PGI_2) is produced from PGH_2 by PGI synthase, and the even more highly unstable metabolite TXA_2 is formed by the action of TXA synthase. Although most prostanoid-producing cells form only one or a few of these products because of the predominance of a single PGH-metabolizing enzyme (e.g. thromboxane production by platelets) non-enzymatic processes can result in formation of PGE_2 and PGD_2 from endoperoxide that escapes the cell; non-enzymatic formation of PGD_2 can be catalysed by serum albumin (Hamberg and Fredholm, 1976). Cooperative formation of prostanoids by two or more adjacent cell types can also occur, as exemplified by the formation of PGI_2 in aspirin-treated vascular endothelium by endoperoxide "stolen" from adjacent clumping platelets (Marcus et al., 1980). Cooperative cellular interactions involved in the production of some lipoxygenase products is discussed later in this chapter.

Non-prostanoid by-products of PGH_2 breakdown (including non-enzymatic breakdown) are 12-HHTrE (previously called "12-HHT") and MDA, which have therefore been used as indirect measurements of prostanoid production. By some unexplained mechanism, MDA formation is largely catalysed by thromboxane rather than prostanoid synthesis (see Willis, 1987).

3.3   ENDOGENOUS FACTORS CONTROLLING CELLULAR LEVELS OF PROSTAGLANDIN-H SYNTHASE

This topic has been extensively reviewed by DeWitt (1991) and Smith (1992). Most studies on mechanisms of PGH synthase levels have involved the use of cultured cells, especially fibroblasts and monocytes/macrophages. Interestingly, many of the agents that stimulate arachidonate release (the initial trigger to eicosanoid biosynthesis) also increase the cellular synthesis of PGH synthase, probably by stimulating transcription of the PGH synthase gene. Thus, after treatment with aspirin, enzyme replacement generally takes more than 12 h (Bailey et al., 1985; Habenicht et al., 1985) but this recovery time can be shortened to as little as 1–4 h by treatment with agents such as PDGF and IL-1 (Bailey et al., 1985; Habenicht et al., 1985; Burch et al., 1989; Lin et al., 1989), an action additional to their ability to release arachidonate (Habenicht et al., 1981; Burch et al., 1989). In addition, arachidonate release may be initiated, albeit indirectly, by PGH synthase expression (Lister et al., 1988, 1989; Ulevitch et al., 1988).

In model systems, PGH synthase seems regulated by two mechanisms, viz. those associated with basal (steady state) prostaglandin synthesis and those associated with evoked prostaglandin synthesis that is typically several-fold higher than that of basal synthesis. As reviewed above (Section 2.6), arachidonic acid may be provided from "metabolic pools" related to non-oxidative EFA metabolism rather than the explosive burst of arachidonic acid release produced by phospholipase activation.

3.3.1   Regulation of Steady State Prostaglandin Synthesis

This type of regulation is characteristic of fibroblasts and vascular endothelial cells. Prostaglandin synthesis typically can be increased five- to twenty-fold but is accompanied by small (one- to three-fold) or undetectable increases in PGH synthase protein levels. In these cells increased prostaglandin synthesis appears to result from increased transcription of the PGH synthase gene, increased turnover of PGH synthase protein, and sustained availability, presumably via release of arachidonic acid substrate (see DeWitt, 1991).

3.3.2   Newly Recognized Isoenzymes 1 and 2 of Prostaglandin-H Synthase

Recently, it has become apparent that there are two isoenzymes of PGH synthase, designated PGH synthases-1 and -2 (in some publications designated COX-1 and COX-2).

PGH synthase-1 is the enzyme that was originally purified from ovine and bovine vesicular gland and platelets (see Smith et al., 1991). The deduced amino

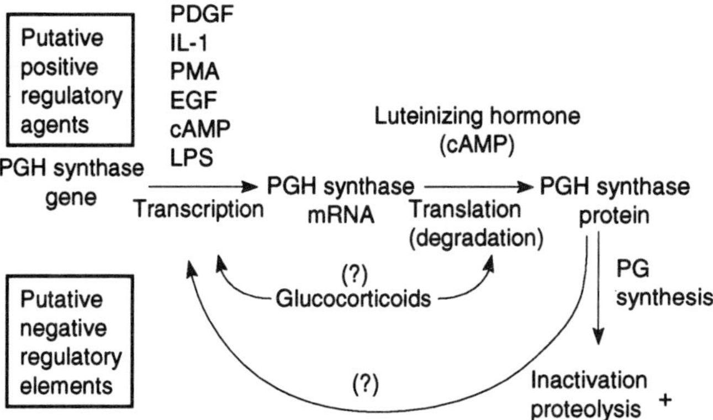

Figure 1.4 Putative scheme for regulation of PGH synthase expression (see text for details). Factors that increase prostaglandin synthesis by increasing PGH synthase synthesis appear to enhance transcription of the gene, although in some instances increased PGH synthase expression may result from increased translation of extant PGH synthase mRNA. During translation PGH synthase mRNA may be degraded, limiting synthesis. PGH synthase protein could attenuate transcription of the gene or translation of the mRNA. Glucocorticoids appear to decrease PGH synthase expression, although whether it is at the level of transcription or translation is not known. Redrawn from DeWitt (1991).

acid sequences of PGH synthases-1 from sheep, human and mouse are approximately 90% identical at the amino acid level, with the major differences being found in the signal peptide (between amino acid residues 26 and 27) and also the 12 amino acids immediately preceding an endoplasmic reticulum retention signal at the C terminus (see Smith, 1992).

PGH synthase-2 was originally isolated as a v-*src*-inducible gene product in chicken fibroblasts (Simmons *et al.*, 1989; Xie *et al.*, 1991) and as a phorbol ester-inducible immediate–early gene product (called T1S10) in murine 3T3 cells (Kujubu *et al.*, 1991).

Enzymatic activity of the murine PGH synthase-1 has been shown to exhibit both the cyclooxygenase and peroxidase activities characteristic of PGH synthase-2 (Kujubu and Herschman, 1992).

Although the amino acid sequence of PGH synthase-1 is about 75% homologous with that of PGH synthase-2, there are significant differences, including an 18-amino acid insert very near the C terminus of the enzyme.

Unlike the case with PGH synthase-1, the mRNA for PGH synthase-2 is much less ubiquitous, being detectable in prostate, brain, testis and lung but is barely detected or undetectable in kidney and most other organs (Simmons *et al.*, 1989). Mouse 3T3 fibroblasts are one of the few common cell lines with PGH synthase-2 mRNA (Kujubu and Herschman, 1992), although it has now also been shown to be present in vascular endothelial cells (Hla and Neilson, 1992; Holtzman *et al.*, 1992). By contrast, PGH synthase-1 appears to be (as mRNA) distributed in many cell lines and in extracts of virtually all mammalian tissues (see Smith, 1992). Existence and

a role for PGH synthase-2 has also been proposed in alveolar macrophages (O'Sullivan *et al.*, 1992). The cDNA for PGH synthase-2 has also been isolated (Hla and Neilson, 1992).

3.3.3   Acute Expression of Prostaglandin-H Synthase (Fig. 1.4)

Levels of mRNA for PGH synthase-2 are up-regulated much more dramatically than the mRNA for PGH synthase-1, and this elevation (at least in serum-stimulated mouse 3T3 cells) is both more rapid and of shorter duration than for the PGH synthase-1 mRNA. Stimuli include both serum and growth factors (Lin *et al.*, 1989; DeWitt *et al.*, 1991; Kujubu *et al.*, 1991; Simmons *et al.* 1991). In addition, lipopolysaccharide can prime alveolar macrophages for enhanced synthesis of prostanoids by a mechanism that apparently involves induction of PGH synthase-2 (O'Sullivan *et al.*, 1992).

These data tend to indicate that PGH synthase-2 acts more as an induced/regulated enzyme than PGH synthase-1. This concept is further supported by the work of O'Banion *et al.* (1991, 1992) that serum (in mouse fibroblasts) or IL-1β (in human monocytes) specifically induces a 4.1 kb mRNA for PGH synthase(-2), whose induction is decreased by anti-inflammatory glucocorticoids, while levels of a 2.8 kb mRNA for PGH synthase(-1), are not. This confirmed an even earlier report (Masferrer *et al.*, 1990) that dexamethasone inhibits induction of a PGH synthase enzyme (determined both by enzyme activity and mass) in mouse peritoneal macrophages that is different from the PGH synthase in non-adherent cells or renal medulla. Such findings were confirmed *in vivo* and related to lethality of endotoxin (Masferrer *et al.*, 1992).

Evidence for involvement of inducible PGH synthase-2 in reproductive physiology is provided by the fact that human chorionic gonadotrophin stimulates the expression of PGH synthase-2 in rat ovarian follicles in association with ovulation (Sirois and Richards, 1992; Sirois *et al.*, 1992).

Given the differences in tissue sensitivity between some inhibitors of PGH synthase (e.g. paracetamol, Section 7.1) it would be of great interest to examine whether this could be due to differences in PGH synthase isoenzymes as suggested from the preliminary data of Meade *et al.* (unpublished, quoted by Smith, 1992). There is currently some speculation that anti-inflammatory drugs that selectively inhibit induction and/or activity of PGH synthase-2 may have fewer side-effects such as ulcerogenicity (Xie *et al.*, 1992).

Further discussion of such regulatory mechanisms will be restricted to cells involved in immunology, viz. fibroblasts, vascular endothelial cells and macrophages/monocytes.

3.3.3.1 Stimulation of Prostaglandin Synthesis in Fibroblasts

Mitogens (e.g. PDGF, IL-1) derived from aggregating platelets and monocytes stimulate fibroblasts to divide and to produce secondary factors such as GM-CSFs and prostaglandins (Albrightson *et al.*, 1985; Zucali *et al.*, 1986) which have a modulatory role in inflammation and wound repair (Korn *et al.*, 1980).

PDGF and IL-1 are particularly powerful in their ability to induce prostaglandin synthesis with a concomitant increase in PGH synthase levels. Summarizing the current literature on fibroblast prostaglandin synthesis, DeWitt (1991) stated that quiescent fibroblasts contain sufficient PGH synthase to account for both the early (complete by 10 min) and late phases (2–6 h) of PGE_2 production that are stimulated by PDGF. The initial burst resulted primarily from PDGF-stimulated phospholipid hydrolysis, whereas the second phase utilized arachidonate derived from LDLs, secondary to the ability of PDGF to stimulate expression of LDL receptors. Hence, pretreatment with cycloheximide (which blocks protein synthesis) and actinomycin (which inhibits transcription) blocked the second wave of PGE_2 production, but did not alter levels of PGH synthase protein. DeWitt (1991) points out that minor experimental discrepancies between workers may be due to differences in cell line and amounts of LDL in the serum of the culture medium.

That is not to say that PGH synthase activity cannot be modulated in fibroblasts. Thus, microsomal PGH synthase activity was increased two-fold after 3 h of stimulation with medium containing 10% serum. Consequently, it seems likely that PDGF can stimulate *de novo* PGH synthase without increasing levels of PGH synthase protein *per se*. Presumably, inactive enzyme is replaced with active enzyme (i.e. turnover is increased). This turnover of PGH synthase changes depending upon the growth state of the cell, with minimal turnover in quiescent cells and rapid turnover in growing or mitogen-stimulated cells. This situation was demonstrated by showing that aspirin-treated quiescent cells do not recover their ability to synthesize PGE_2 for up to 12 h (Habenicht *et al.*, 1981, 1985), although recovery can be as rapid as 3–4 h after a 1 h lag in cells stimulated with PDGF (Habenicht *et al.*, 1985). Furthermore, this PDGF-stimulated increase in PGE_2 synthesis was blocked by cycloheximide (Lin *et al.*, 1989).

Studies with actinomycin and measurements of mRNA have shown that transcriptional activation for PGH synthase occurs in 3T3 fibroblasts stimulated both with serum and PDGF (Habenicht *et al.*, 1985; DeWitt *et al.*, 1991) which can slightly precede the increase in PGH synthase (DeWitt, 1991).

The half-life of PGH synthase in cycling mouse fibroblast 3T3 cells can be tested (Habenicht *et al.*, 1985). Thus, cells stimulated for 6 h with PDGF were tested for their continued ability to synthesize PGE_2 from added arachidonate in the presence of cycloheximide. Although PGE_2 synthesis was reduced by 80% within 4 h, approximately 20% of the PGE_2 synthetic capacity remained even after 10 h. Similar fast and slow turnover pools of PGH synthase have been seen in vascular endothelial cells (see below).

In contrast to PDGF, IL-1 stimulates prostaglandin synthesis in fibroblasts by several mechanisms, including increases in PLA_2, increased PGH synthase activity and increased signal transduction efficiency (e.g. increased G protein coupling). While metabolic labelling of PGH synthase with $[^{35}S]$methionine increased in concert with PGH synthase activity, it was not clear whether the actual mass of PGH synthase protein increased. Experiments with staurosporine and another inhibitor of PKC suggested that PKC was involved in mediating the response to IL-1, although studies with more specific stimulators of PKC activation indicate that PKC activation alone is insufficient to account for the total IL-1-mediated induction. Overall, it appears that both transcription and translation are necessary for the increase in production of PGH synthase, with possibly distinct "transcriptional" (0–4 h) and "translational" (4–8 h) phases (Raz *et al.*, 1990; DeWitt, 1991).

3.3.3.2 Stimulation of Prostaglandin Production in Vascular Endothelial Cells

The net effect of low-dose aspirin therapy is to inhibit permanently the platelet production of the pro-aggregatory endoperoxide PGH_2 and its amplifying metabolite TXA_2 until new platelets are generated in the bone marrow (about 10 days).

By contrast, studies on cultured vascular endothelial cells showed that they begin to recover from aspirin inhibition of PGH synthase in 2–24 h (depending upon conditions; Weksler, 1987), data complementary to those obtained in *ex vivo* samples of saphenous veins removed from patients who received 20 mg of aspirin per day for 7 days (Heavey *et al.*, 1985).

As for fibroblasts, there appear to be different pools of PGH synthase with different rates of turnover. Thus, when the degradation rate of [^{32}S]methionine-labelled PGH synthase was measured in human vascular endothelial cells (Wu *et al.*, 1989; Tsai *et al.*, 1991) the data suggested that there were two pools of PGH synthase turning over with half-lives of less than 10 min or more than 150 min, respectively.

A variety of stimuli (histamine, thrombin, bradykinin, ionophore A23187, IL-1 and phorbol esters) induce production of prostaglandins (mainly PGI$_2$) in vascular endothelial cells (see DeWitt, 1991). This production of PGI$_2$ seems coupled through some PKC mechanism to production of vascular nitric oxide (De Nucci *et al.*, 1988). These stimuli appear to be of two types: those that induce an acute (10–30 min) release of prostanoids (histamine, thrombin, bradykinin, A23187) and those that elicit a sustained (> 1 h) release of prostanoids (IL-1, IL-2 and phorbol esters). It is thought that the former group acts via stimulation of arachidonic acid release, while the latter group both release arachidonic acid *and* increase the expression of PGH synthase (DeWitt, 1991).

IL-1 is important immunologically since it is a product of activated monocytes and a potent modulator of endothelial cell growth. IL-1 stimulates mRNA for PGH synthase and increases both PGH synthase protein levels and PGI$_2$ synthesis. This effect persists for up to 24 h. Since cycloheximide stimulates the increase in mRNA levels (Maier *et al.*, 1990), superinduction of PGH synthase mRNA seems apparent. This raises the possibility that PGH synthase may regulate its own transcription, as seen analogously in fibroblasts (DeWitt, 1991; DeWitt *et al.*, 1991). IL-2 and PMA appear to act similarly to IL-1, except that IL-2 also stimulates (some 30-fold) levels of PGI synthase (DeWitt, 1991). These increases in PGH synthase levels appear to be solely the result of increased production of enzyme, since the half-life of the enzyme (about 10 min) is unaltered by stimuli that induce prostaglandin production.

3.3.3.3   *Stimulation of Prostaglandin Production in Macrophages/ Monocytes*

Prostaglandins, particularly those of the E type, are pro-inflammatory and fever-producing and cause hyperalgesia (Willis *et al.*, 1972; Higgs *et al.*, 1984). Prostaglandins can also attenuate the immune response via inhibition of B and T cell proliferation and function (Kato and Askenase, 1984; Thompson *et al.*, 1984; Makoul *et al.*, 1985; Morgan *et al.*, 1985; Chouaib *et al.*, 1987) and inhibit accessory monocyte/macrophage cell action and function (Taffet and Russell, 1981; Snyder *et al.*, 1982;

Boraschi *et al.*, 1984). Monocytes/macrophages can be an important source of PGE$_2$, TXA$_2$ and LTB$_4$. Macrophages can be stimulated to produce these eicosanoids by lipopolysaccharide, zymosan, immune complexes or complement (Rouzer *et al.*, 1982; Hansch *et al.*, 1984; Morgan *et al.*, 1985; Zoeller *et al.*, 1987; Fu *et al.*, 1990; Raz *et al.*, 1990; Pueringer and Hunninghake, 1992). PGE$_2$ inhibits macrophage proliferation (Kurland *et al.*, 1978) and cytokine expression (Taffet and Russell, 1981) and further macrophage accessory function, including the ability to stimulate B and T cell proliferation and differentiation (see DeWitt, 1991).

The macrophage precursor cell lines HL-60 and U937 have been used to study the regulation of PGH synthase (Harris and Ralph, 1985; Hass *et al.*, 1989). Both cell lines can be differentiated into macrophage-like cells by treatment with PMA, diacylglycerol or vitamin D$_3$. In addition, HL-60 cells can differentiate into neutrophil-like cells with the addition of dimethylsulphoxide (Harris and Ralph, 1985).

Upon differentiation, prostaglandin synthesis increases as much as 50- to 100-fold (Goerig *et al.*, 1987; Koehler *et al.*, 1990; Spotts and Hann, 1990). This increase is due partially to increased PGH synthase expression, as determined by various methods (e.g. prostaglandin production in response to exogenous arachidonate or immunoquantification of PGH synthase protein levels; Goerig *et al.*, 1987; Spotts and Hann, 1990).

3.4   METABOLIC INACTIVATION OF EICOSANOIDS (Fig. 1.5)

Prostaglandins have usually been considered to be local hormones that act at or near their site of synthesis. This concept derived from early findings that prostaglandin metabolites had no significant biological activity (Anggard, 1966) and the observation that PGE and PGF are almost completely inactivated by a single passage through the pulmonary circulation (Ferreira and Vane, 1967). This concept was further supported by the low circulating levels of prostaglandins in blood and by the availability of prostaglandins and the principal metabolizing enzyme, 15-hydroxyprostaglandin dehydrogenase, at multiple organ sites (Smith *et al.*, 1987). In addition to the effects of metabolism, the actions of some eicosanoids (notably PGI$_2$ and TXA$_2$) are limited by their chemical instability.

The local-hormone concept was later modified to account for the findings that PGA and PGC are not inactivated by a single passage through the pulmonary circulation (McGiff *et al.*, 1969; Jones, 1972). Thus (it was once proposed), these hypotensive prostaglandins might be circulating hormones that exert their activity in blood vessels and other tissues (Vane, 1969).

As detailed below, common metabolic pathways for PGE, PGF, PGI$_2$ and TXB$_2$ include 15-hydroxy-dehydrogenase, Δ^{13}-reductase (and Δ^{5}-reductase for TXB$_2$),

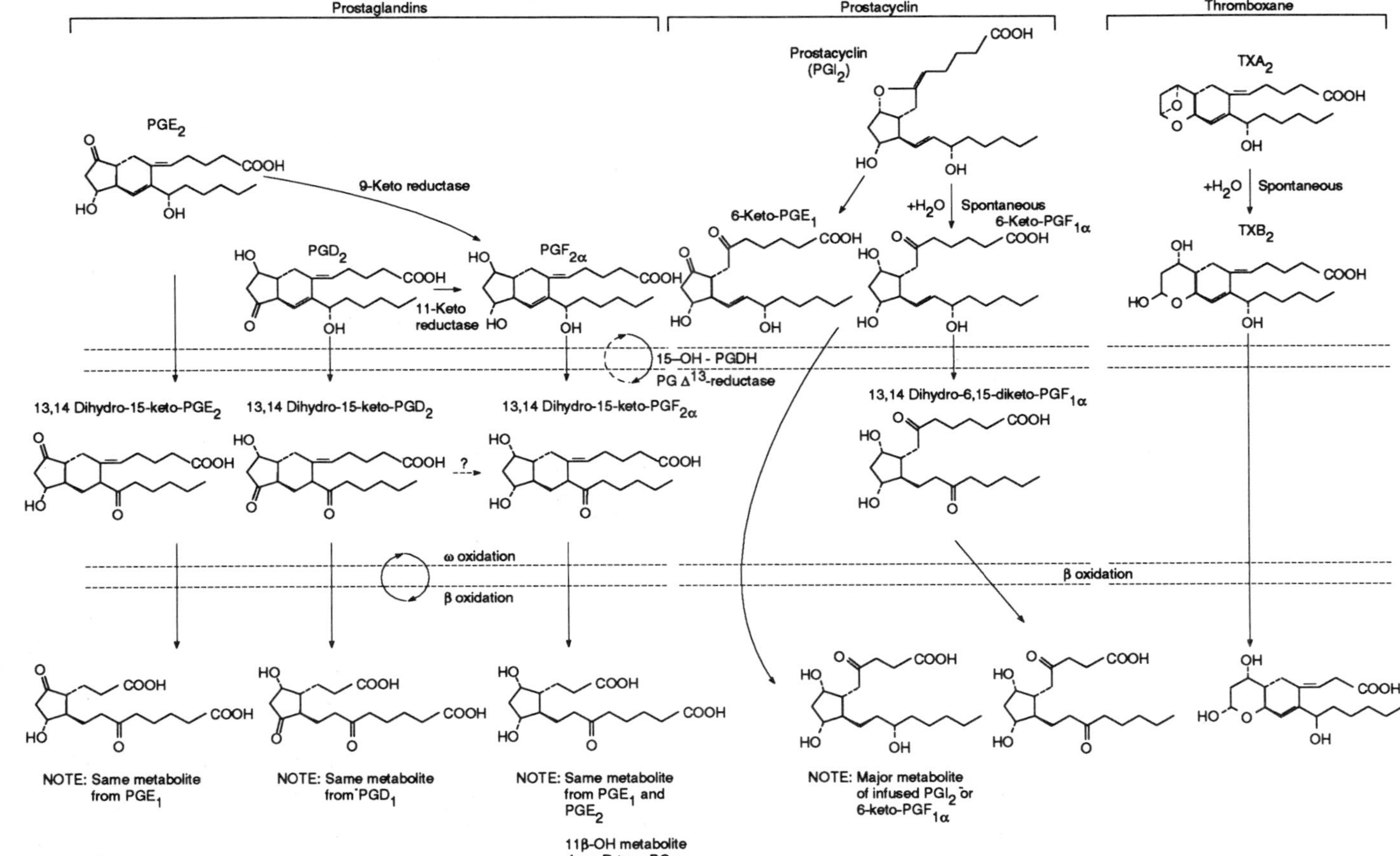

Figure 1.5 Main pathways in the metabolism and urinary excretion of primary prostaglandins, prostacyclin, and thromboxanes. *Primary prostaglandins*: These compounds can (after uptake into cells) be rapidly metabolized. The principal pathway is dehydrogenation of the 15-hydroxyl group, followed by reduction of the double bond (although these events can occur, to some extent, in the opposite order). Oxidation of upper (β oxidation) and lower (ω oxidation) side-chain then occurs. An estimate of whole body PGE turnover can be obtained by measurement of the major (PGEM) tetranor metabolite. Likewise for the tetranor metabolite of PGD compounds. The largest amounts of tetranor metabolite have the PGF ring structure (PGFM). However, occurrence of this compound reflects not only turnover of PGF compounds but also PGE and PGD compounds that can be enzymatically converted to F-type prostaglandins. Indeed, *in vivo* metabolic studies in the monkey have shown that most polar metabolites of PGD_2 found in urine have a PGF ring structure. *Prostacyclin*: prostacyclin (PGI_2) spontaneously breaks down to 6-keto-PGF_{1α}. Both PGI_2 and 6-keto-PGF_{1α} can be acted upon by prostaglandin 15-dehydrogenase and Δ^{13}-reductase, although PGI_2 seems to be a much better substrate for the 15-hydroxy dehydrogenase. However, dinor-6-keto-PGF_{1α} seems to be the main urinary metabolite of both prostacyclin and 6-keto-PGF_{1α} in humans, accounting for about 20% of the material infused intravenously. Another important exception for PGI metabolism may be its localized conversion to 6-keto-PGE_1 which has similar biological effects on platelets. *Thromboxanes*: The hydrolytic conversion of TXA_2 into TXB_2 is non-enzymatic but very rapid. In both monkeys and humans, the major urinary metabolite of TXB_2 is a dinor product formed by a single step of β oxidation. The thromboxane produced seemed to be derived mainly from platelets. Thus, in normal donors, urinary thromboxane B_2 metabolites were markedly reduced by doses of aspirin that were (presumably) only sufficient to affect platelets. Redrawn from Willis (1987). See Smith *et al.* (1987) for more details.

and β and ω oxidation. These pathways are also involved in PGD_2 metabolism, either before or after conversion to an F ring isomer via 11-ketoreductase. β Oxidation is an important pathway for TXB_2 metabolism, and TXB_2 is also uniquely metabolized via 11-hydroxydehydrogenase.

The key metabolizing enzyme for PGE and $PGF_α$ prostaglandis in several tissues (especially kidney, placenta and lung) and for PGI_2 in liver and kidney (but not lung) is 15-hydroxyprostaglandin dehydrogenase (Jones, 1972; Schlegel et al., 1974; Lee and Levine, 1975; Sun et al., 1976; Wong et al., 1980). Followed in sequence by action of $Δ^{13}$-reductase, this results in 15-keto-13,14-dihydro derivatives of these prostaglandins (Hamberg and Samuelsson, 1971; Granstrom, 1972; Wong et al., 1980), which for $PGF_{2α}$ and PGE_2 are the major circulating metabolites (Hamberg and Israelsson, 1970; Granstrom, 1971). Further metabolism via β and ω oxidation of the upper side-chain can also occur (although infrequently for PGI_2), resulting in 2,3-dinor- or 2,3,4,5-tetranor-15-keto-13,14-dihydro-20-carboxyl derivatives, the latter of which are the major urinary metabolites of PGE_2 and $PGF_{2α}$ (Granstrom and Samuelsson, 1971; Hamberg and Samuelsson, 1971). In addition to utilizing these pathways, PGI_2 can also be hydrolysed to the inactive derivative 6-keto-$PGF_{1α}$, the major circulating metabolite of PGI_2 (Rosenkranz et al., 1981); 6-keto-$PGF_{1α}$ can also be further metabolized by the enzymes described above. The major urinary metabolite of PGI_2 is 2,3-dinor-6-keto-$PGF_{1α}$ (Rosenkranz et al., 1980).

PGD_2 and TXB_2 are poor substrates for 15-hydroxy-prostaglandin dehydrogenase (Sun et al., 1976). They are β oxidized, although not as extensively as PGE_2 and $PGF_{2α}$, and the major urinary metabolites of PGD_2 and TXB_2 are 2,3-dinor compounds (Roberts et al., 1977a,b; Ellis et al., 1979). For TXB_2 the major urinary metabolite is 2,3-dinor-TXB_2, and, for PGD_2, the major D ring metabolite is 2,3-dinor-13,14-dihydro-PGD_2. As with PGI_2, ω oxidation is relatively unimportant for PGD_2 or TXB_2.

Another important pathway for TXB_2 metabolism is 11-hydroxydehydrogenase, which converts TXB_2 to 11-dehydro-TXB_2 (Roberts et al., 1978). This is further metabolized via β and ω oxidation (Roberts et al., 1978). Additional pathways also exist for metabolism of PGD_2, based on conversion to a PGF ring via 11-ketoreductase (Hensby, 1974; Ellis et al., 1979). In monkeys and humans, this is the predominant metabolic pathway. 2,3-Dinor-9α,11β-PGF_2 is an important metabolite of PGD_2 in human plasma and urine (Ellis et al., 1979; Liston and Roberts, 1985; Pugliese et al., 1985; Roberts and Sweetman, 1985). Metabolism to PGF_2 isomers may be followed by other enzymatic metabolism (via β and ω oxidation, 15-hydroxyprostaglandin dehydrogenase, and $Δ^{13}$-reductase). Several derivatives of 9α,11β-PGF_2 (such as ω-carboxyl-2,3-dinor-15-keto-9α,11β-PGF_2) are present in human plasma and urine.

Several metabolites resulting from routes other than those described above are observed (for example, β oxidation of the lower side-chain, C-19 or C-20 hydroxylation, and interconversion of the E and F rings). In most cases, these are relatively minor. A more comprehensive description and listing of prostaglandin metabolites is presented elsewhere (Roberts, 1987; Smith et al., 1987).

Metabolism of prostanoids does not in every case result in inactivation or greatly diminished activity; in some cases, activity is enhanced or made more selective. For example, 14,15-dihydro-PGE_2 has about half the activity of PGE_2 on uterine smooth muscle but is inactive in other tissues (Crutchley and Piper, 1975), suggesting a mechanism for generating a uterine stimulation-selective compound. By contrast, 15-keto-$PGF_{2α}$ is twice as potent as $PGF_{2α}$ in producing bronchial muscle contractility (Dawson et al., 1974). A further example of amplification is the metabolism of PGD_2 to PGJ_2, which is more potent as an inhibitor of platelet aggregation and cell proliferation (Fukushima et al., 1982; Mahmud et al., 1984); however, it is not known to what extent this is an enzymatic conversion.

9α,11β-PGF_2, a biologically active major metabolite of PGD_2, inhibits platelet aggregation and cell proliferation but elevates rat blood pressure and constricts canine coronary strips and human bronchial smooth muscle (Pugliese et al., 1985; Roberts and Sweetman, 1985; Roberts, 1987; Seibert et al., 1987). In addition, 6-keto-PGE_1, a metabolite of prostacyclin (PGI_2), is more stable in vivo than the parent compound but strongly inhibits the aggregation of blood platelets (Wong et al., 1980) although, unlike PGI_2, it is a constrictor of coronary vasculature (see Willis, 1987). In a related fashion, PGE_1 is metabolized to 13,14-dihydro-PGE_1, which also inhibits platelet aggregation (Westwick, 1976) but has a more prolonged and selective antiplatelet activity in the circulation (Ney et al., 1991).

Some metabolites sensitize smooth muscle to stimulation by other agonists (Boot et al., 1976), suggesting that prostaglandin metabolites may be involved in producing hyperalgesia during inflammation (Willis and Cornelson, 1973); PGE-induced hyperalgesia becomes maximal only after prolonged infusion (Ferreira, 1972) or after elimination of PGE_2 from the tissue (Kuhn and Willis, 1973).

In general, metabolism of lipoxygenase products produces alteration, rather than inactivation, of the primary product. In addition, it is also difficult to determine what the primary product is. Metabolites of lipoxygenase products and their biological activity are described by Smith et al. (1987), and this subject is further discussed in Section 4.

4. *Lipoxygenase Pathways* (Fig. 1.6)

Unlike the "prostanoids" the lipoxygenase products do not have a "prostanoic acid" backbone or (in the case of

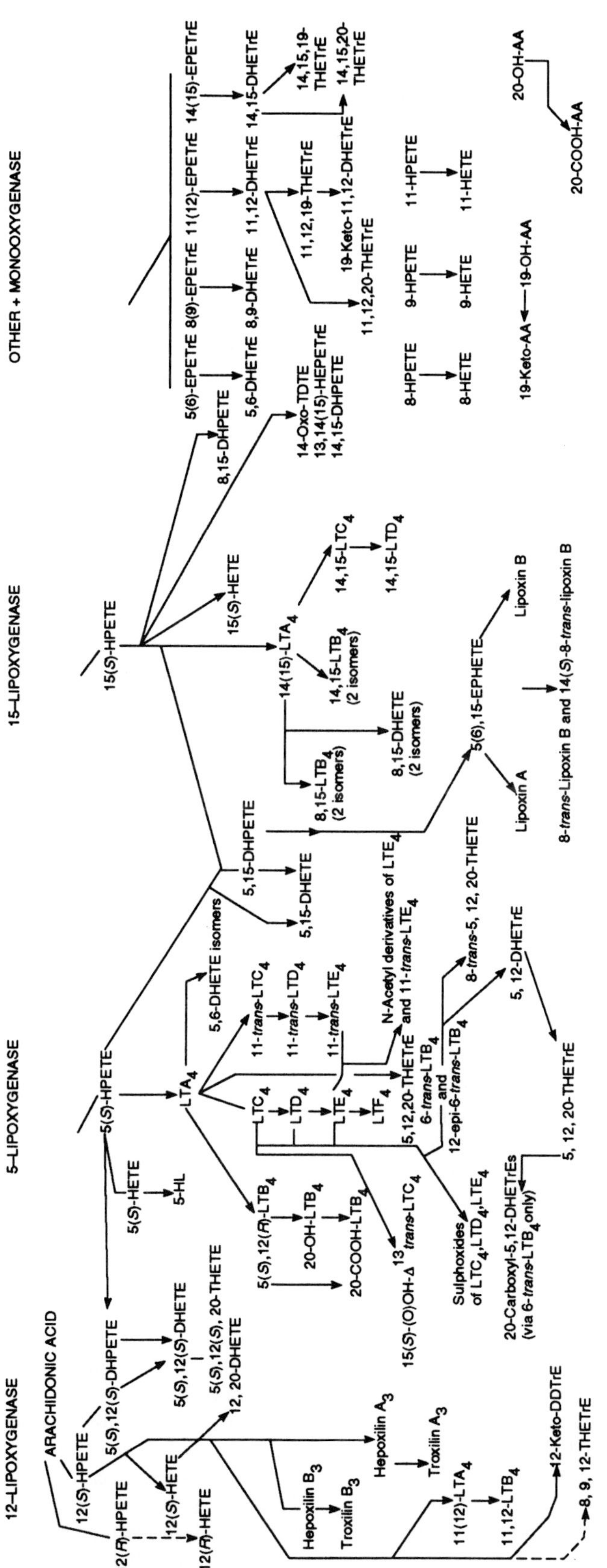

Figure 1.6 Overview of lipoxygenase pathways. Redrawn from Willis (1987).

the thromboxanes) a closely related structure. The first animal lipoxygenase to be described was that converting arachidonic acid into 12-lipoxygenase products (Hamberg and Samuelsson, 1974; Nugteren, 1975).

Discovery of the 5-lipoxygenase pathway leading to formation of the leukotrienes (see Samuelsson, 1982; Hammarstrom, 1983) caused an explosion of interest in this area, largely displacing interest in the classic prostaglandins (see Fig. 1 of Willis, 1987). Subsequently, interest in the 15-lipoxygenase area led to discovery of the lipoxins (Serhan *et al.*, 1984). These and other more complex pathways are described below.

4.1 12-Lipoxygenase Products

12-Lipoxygenase in plants and animals converts arachidonic acid into 12(S)-HPETE and subsequently into the hydroxy compound 12-HETE. Found in blood platelets in 1975, it was the first animal lipoxygenase to be described (Hamberg and Samuelsson, 1974; Nugteren, 1975). The specificity in conversion of other polyunsaturated fatty acids was reported by Nugteren (1977), who showed that unsaturation at the ω-6 position was obligatory. Consequently, the PGE_1 precursor DGLA is converted into 12-lipoxygenase products, even though its metabolism differs from arachidonic acid in many other ways (Willis and Smith, 1989a; Section 6).

4.1.1 Further Metabolism of 12-Lipoxygenase Products

12-HPETE can be further metabolized in leucocytes or platelets via epoxidation to LTA- and LTB-like compounds (11(12)-LTA_4 and 11,12-LTB_4) (Hammarstrom, 1983; Westlund, 1987); or in lung, platelets, vascular tissue and pancreatic islets to hydroxylated epoxides and their derived 8,11,12- and 10,11,12-trihydroxy compounds (Jones *et al.*, 1978; Bryant and Bailey, 1979; Pace-Asciak *et al.*, 1981, 1983). These latter compounds have been called hepoxilins and trioxilins A_3 and B_3 (Pace-Asciak *et al.*, 1983) since they are thought to have a role in controlling insulin secretion (Pace Asciak and Martin, 1984; Pace-Asciak *et al.*, 1985). Aortic hepoxilin A_3 can be further metabolized into hepoxilin A_3-C, a glutathione conjugate that, like hepoxilin A_3, can potentiate the vasoconstrictor response to adrenaline (Laneuville *et al.*, 1991).

12-Lipoxygenases from platelets have been shown to be distinct from those of leucocytes in terms of substrate specificity (e.g. the platelet enzyme is a poor substrate for linoleic and linolenic acids), and also differs in immunogenicity (Takahashi *et al.*, 1988; Hada *et al.*, 1991). Also, the leucocyte enzyme but not the platelet enzyme can efficiently utilize LTA_4 as a substrate for formation of lipoxins (Hada *et al.*, 1991).

Other known metabolites of 12-HPETE include, in platelets, the trihydroxy compound 8,9,12-THETrE (Jones *et al.*, 1978; Bryant and Bailey, 1980), which has

no known epoxy intermediate, and in leucocytes a 12-carbon compound with a terminal oxo group (12-oxo-DDTrE) (Glasgow *et al.*, 1986).

It is now known that two or more lipoxygenases can act sequentially on a substrate, sometimes in adjacent cells of a different type. Thus, 12-HPETE (from platelets) can be further metabolized in leucocytes, via 5-lipoxygenase, to 5,12-diHPETE. Conversely, in leucocytes and platelets, 5,12-diHETEs can be formed by action of 12-lipoxygenase on 5-HETE (Borgeat *et al.*, 1981; Lindgren *et al.*, 1981; Hammarstrom, 1983; Ueda *et al.*, 1986; Yokoyama *et al.*, 1986).

12-HETE from platelets can also be metabolized directly. It can be terminally hydroxylated in neutrophils via an ω-hydroxylase to form 12,20-diHETE (Marcus *et al.*, 1984).

In contrast to 12(S)-HPETE formed in platelets, arachidonic acid has been shown to be metabolized to 12(R)-HPETE in psoriatic lesions (Woollard, 1986) and via cytochrome P-450-dependent monooxygenase in the cornea (Schwartzman *et al.*, 1987; Masferrer *et al.*, 1989).

4.1.2 Possible Biological Role of 12-Lipoxygenase Products

12-HPETE and 12-HETE may act as mediators in several biological systems, including leucocyte chemotaxis, and migration of vascular smooth muscle cells during atherosclerosis (see Willis, 1987; Willis and Smith, 1989b). Lesser known actions include the reported ability of 12-HPETE to act as a second messenger for presynaptic inhibition of sensory cells in the marine mollusc *Aplysia californica*, by inhibiting the action of FMRFamide (Phe-Met-Arg-Phe-NH_2) on serotonin-induced depolarizaton (Piomelli *et al.*, 1987). 12-HETE and 15-HETE (see below) have recently been implicated as mediators of the action of erythropoietin in stimulating the growth and differentiation of erythroid progenitor cell-derived colonies (Beckman *et al.*, 1987). In addition, 12-HETE stimulates pituitary cell and ovarian prolactin release (Kiesel *et al.*, 1987; Wang *et al.*, 1989); it also stimulates ovarian PGE_2 release and appears to augment such stimulatory effects of LHRH (Wang *et al.*, 1989). Among additional biological activities, 12-HETE also stimulates bone resorption (Meghji *et al.*, 1988).

12(R)-HPETE formed in psoriatic lesions and in cornea can inhibit Na^+/K^+-ATPase (Schwartzman *et al.*, 1987) and appears to mediate ocular inflammation (Masferrer *et al.*, 1989).

In addition to having biological actions in its own right, 12-HPETE has been reported to inhibit platelet thromboxane synthesis (Hammarstrom and Falardeau, 1977) and 12-HETE inhibits 5-lipoxygenase (Borgeat *et al.*, 1982). If such effects occur in the intact animal, these effects have obvious implications in the inhibition of thromboxane-mediated platelet activation and to modulation of leukotriene-mediated effects.

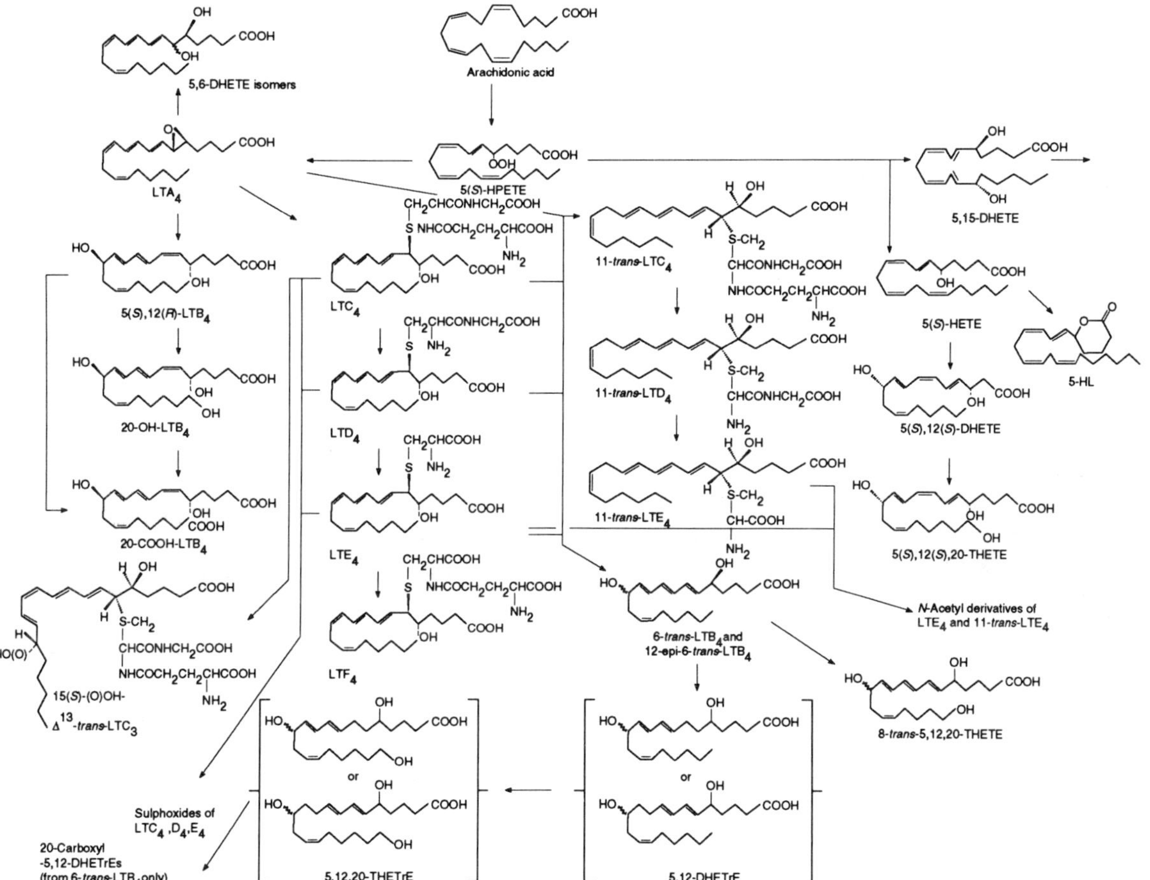

Figure 1.7 *5-Lipoxygenase products*: 5(*S*)-HPETE, 5(*S*)-hydroperoxy-6-*trans*,8-*cis*,11-*cis*, 14-*cis*-eicosatetraenoic acid; 5(*S*)-HETE, 5(*S*)-Hydroxy-6-*trans*,8-*cis*,11-*cis*,14-*cis*-eicosatetraenoic acid; 5(*S*),12(*S*)-DHETE, 5(*S*),12(*S*)-dihydroxy-6-*trans*,8-*cis*,10-*trans*,14-*cis*-eicosatetraenoic acid; 5(*S*),12(*S*),20-THETE, 5(*S*),12(*S*),20-Trihydroxy-6-*trans*,8-*cis*,10-*trans*,14-*cis*-eicosatetraenoic acid; 5,15-DHETE (previously called 5,15-diHETE), 5(*S*),15(*S*)-dihydroxy-6-*trans*,8-*cis*,11-*cis*,13-*trans*-eicosatetraenoic acid; LTA$_4$, 5(*S*),6(*S*)-5(6)-epoxy-7-*trans*,9-*trans*,11-*cis*-14-*cis*-eicosatetraenoic acid; LTC$_4$, 5(*S*)-hydroxy-6(*R*)-*S*-glutathionyl-7-*trans*, 9-*trans*,11-*cis*,14-*cis*-eicosatetraenoic acid; LTD$_4$, 5(*S*)-hydroxy-6(*R*)-S-cysteinylglycyl-7-*trans*,9-*trans*,11-*cis*,14-*cis*-eicosatetraenoic acid; LTE$_4$, 5(*S*)-hydroxy-6(*R*)-*S*-cysteinyl-7-*trans*,9-*trans*,11-*cis*,14-*cis*-eicosatetraenoic acid; LTF$_4$, 5(*S*)-hydroxy-6(*R*)-*S*-cysteinyl-γ-glutamyl-7-*trans*,9-*trans*,11-*cis*,14-*cis*-eicosatetraenoic acid; 5(*S*),12(*R*)-LTB$_4$ (commonly designated LTB$_4$), 5(*S*),12(*R*)-dihydroxy-6-*cis*,8-*trans*,10-*trans*,14-*cis*-eicosatetraenoic acid; 20-hydroxy-LTB$_4$ (ω-hydroxy-LTB$_4$), 5(*S*),12(*R*),20-trihydroxy-6-*cis*,8-*trans*,10-*trans*,14-*cis*-eicosatetraenoic acid; 20-carboxy-LTB$_4$ (20-COOH-LTB$_4$), 5(*S*),12(*R*)-dihydroxy-6-*cis*,8-*trans*,10-*trans*,14-*cis*-eicosatetraen-1,20-dioic acid; 6-*trans*-LTB$_4$ and 12-epi-6-*trans*-LTB$_4$, 5(*S*),12(*R*)- and 5(*S*),12(*S*)-dihydroxy-6-*trans*,8-*trans*,10-*trans*,14-*cis*-eicosatetraenoic acid; 11-*trans*-LTC$_4$, 5(*S*)-hydroxy-6(*R*)-S-glutathionyl-7-*trans*,9-*trans*,11-*trans*,14-*cis*-eicosatetraenoic acid; 11-*trans*-LTD$_4$, 5(*S*)-hydroxy-6(*R*)-*S*-cysteinylglycyl-7-*trans*,9-*trans*,11-*trans*,14-*cis*-eicosatetraenoic acid; 11-*trans*-LTE$_4$, 5(*S*)-hydroxy-6(*R*)-*S*-cysteinyl-7-*trans*,9-*trans*,11-*trans*,14-*cis*-eicosatetraenoic acid; *Sulphoxides of LTC$_4$, LTD$_4$, LTE$_4$*. 5,6-DHETE (previously called 5,6-diHETE), 5,6-dihydroxy-7-*trans*,9-*trans*,11-*cis*,14-*cis*-eicosatetraenoic acid; 8-*trans*-5,12,20-THETE, 5,12,20-trihydroxy-6-*trans*,8-*trans*,10-*trans*,14-*cis*-eicosatetraenoic acid; 5,12-DHETrE, 5,12-dihydroxy-6-*trans*,8-*trans*,14-*cis*-eicosatrienoic acid or 5,12,-dihydroxy-8-*trans*,10-*trans*,14-*cis*-eicosatrienoic acid; 5,12,20-THETrE, 5,12,20-trihydroxy-6-*trans*-8-*trans*-14-*cis*-eicosatrienoic acid or 5,12,20-trihydroxy-8-*trans*,10-*trans*,14-*cis*-eicosatrienoic acid; 20-COOH-5,12-DHETrE (20-Carboxy-5, 12-DHETrE), 5,12-dihydroxy-6-*trans*,8-*trans*,14-*cis*-eicosatrien-1,20-dioic acid or 5,12-dihydroxy-8-*trans*,10-*trans*,14-*cis*-eicosatrien-1,20-dioic acid; 5-HL (5-HETE lactone), 5-hydroxy-6-*trans*,8-*cis*,11-*cis*,14-*cis*-eicosatetraenoic acid-delta lactone; 15(*S*)-hydroxy-Δ^{13}-*trans*-LTC$_3$ (15(*S*)-OH-Δ^{13}-*trans*-LTC$_3$), 5(*S*),15(*S*)-dihydroxy-6(*R*)-*S*-glutathionyl-7-*trans*,9-*trans*,11-*cis*,13-*trans*-eicosatetraenoic acid; *N*-Acetyl-LTE$_4$, 5-hydroxy-6-*S*-(2-acetamido-3-thiopropionyl)-7-*trans*,9-*trans*,11-*cis*,14-*cis*-eicosatetraenoic acid; *N*-Acetyl-11-*trans*-LTE$_4$, 5-hydroxy-6-*S*-(2-acetamido-3-thiopropionyl)-7-*trans*,9-*trans*,11-*trans*,14-*cis*-eicosatetraenoic acid. Redrawn from Willis (1987).

4.2 5-LIPOXYGENASE PRODUCTS (INCLUDING THE LEUKOTRIENES) (Fig. 1.7)

The principal 5-lipoxygenase metabolites of arachidonic acid, especially in leucocytes but also in lung and many other cells and tissues, are the leukotrienes, so named for their occurrence in leucocytes and a characteristic conjugated triene structure (see Hammarstrom, 1983). The leukotrienes formed from arachidonic acid are derived through a 5-hydroperoxy intermediate (5-HPETE) (Borgeat *et al.*, 1976; Borgeat and Samuelsson, 1979a,b; Hammarstrom, 1982, 1983).

The leukotrienes derived from arachidonic acid have four double bonds and are thus called the 4-series, using subscript nomenclature similar to that for the prostanoids. LTA_4 is formed first and has an epoxide moiety bridging the carbons at positions 5 and 6 (Borgeat and Samuelsson, 1979a,b; Hammarstrom, 1982, 1983; Shimizu *et al.*, 1984; Yokoyama *et al.*, 1986; Borgeat, 1987). LTB_4 is a 5,12-hydroxy compound derived from LTA_4 (Borgeat and Samuelsson, 1979a; Samuelsson, 1982; Hammarstrom, 1982, 1983; Fitzpatrick *et al.*, 1983; Haeggstrom *et al.*, 1985, 1986; Borgeat, 1987) that has important chemotactic and other biological activities thought to be involved in inflammatory disease (Section 4.2.1). Also derived from LTA_4 are, in biosynthetic sequence, LTC, LTD, LTE and LTF (Lindgren *et al.*, 1981; Samuelsson, 1982, 1983; Hammarstrom, 1983; Feinmark and Cannon, 1986; Pace-Asciak *et al.*, 1986; Claesson and Haeggstrom, 1988). These are peptido-lipids, linked via a sulphidopeptide group at position 5. The peptide group for LTC, LTD, LTE and LTF are, respectively, glutathionyl, cysteinyl-glycyl, cysteinyl, and cysteinyl-γ-glutamyl.

4.2.1 Biological Role of the 5-Lipoxygenase Products

LTB_4 is a potent chemotactic agent and inflammatory mediator (Ford-Hutchinson *et al.*, 1980; Samuelsson, 1983a,b) and can enhance NK cell cytotoxicity (Chang *et al.*, 1989). Other biological implications of 5-lipoxygenase products and drugs that inhibit their formation or actions are reviewed extensively by Musser and Kreft (1992).

LTC_4 through to LTE_4 are potent vasoconstrictors and bronchoconstrictors (see Samuelsson, 1983a,b), and have several other biological actions, including ability to increase vascular permeability and produce negative inotropic effects on cardiac contractions (Drazen *et al.*, 1980; Samuelsson, 1983a,b; Samuelsson, 1985). These leukotrienes are components (LTD_4 is the major one in human lung) of SRS-A (Lewis *et al.*, 1980a,b,c; Morris *et al.*, 1980), which had been long thought to have a major role in human allergic asthma (see Brocklehurst, 1982). As a bronchoconstrictor, LTB_4 is some 100-fold less potent than LTC_4 (Samuelsson, 1983a,b, 1985).

Other biological effects of leukotrienes have also been demonstrated. For example, leukotrienes might be second messengers of the somatostatin-induced increase in neuronal M current (a time-and-voltage-dependent K^+ current that persists at slightly depolarized membrane potentials) (Schweitzer *et al.*, 1990). Also, LTB_4 and LTC_4 (but not LTD_4 or LTE_4) and cyclooxygenase inhibitors stimulate the release of MDGF from activated macrophages (Phan *et al.*, 1987), whereas PGE_2 (Phan *et al.*, 1987), prostacyclins (Willis *et al.*, 1986) and lipoxygenase inhibitors (Phan *et al.*, 1987) inhibit MDGF release. It is therefore possible that cyclooxygenase products are important in controlling excessive fibro-proliferative states, while lipoxygenase products are important in wound healing and tissue repair (see Phan *et al.*, 1987).

An often overlooked non-leukotriene metabolite of the 5-lipoxygenase pathway is 5-HETE (Borgeat and Samuelsson, 1979b; Samuelsson, 1982; Hammarstrom 1982, 1983; Borgeat, 1987). This shares with 5-HPETE, 12-HETE, 15-HPETE, and 15-HETE an ability to stimulate ovarian progesterone and PGE_2 production (Wang *et al.*, 1989). 5-HETE can also, like $PGF_{2\alpha}$ (but less potently) increase contractility of human myometrium (Bennett *et al.*, 1987), and, like 12-HETE and the leukotrienes, stimulate bone resorption (Meghji *et al.*, 1988). 5-HETE can also enhance NK cell cytotoxicity (Chang *et al.*, 1989).

4.2.2 Further Metabolism of Leukotrienes and Effects on Biological Activity

LTA_4 can also be metabolized to isomers of 5,6-diHETE (Borgeat and Samuelsson, 1979a; Ueda *et al.*, 1986).

LTB_4 can be ω oxidized to form 20-hydroxy- and 20-carboxyl-LTB_4. The latter compound has similar bronchoconstrictor activity to LTB_4 but is less potent as a chemotactic agent (Hansson *et al.*, 1981; Jubiz *et al.*, 1982; Hammarstrom, 1983; Powell, 1984). In monocytes, LTB_4 is principally metabolized to two dihydro metabolites (Fauler *et al.*, 1989).

LTC_4 can be metabolized to 15(S)-hydroperoxy-Δ^{13}-*trans*-LTC_3 (which contracts guinea-pig ileum, though less potently than LTC_4 (Orning and Hammarstrom, 1983).

LTA_4 and LTC_4 through to LTE_4 can also be isomerized to 11-*trans*-LTC_4 via LTE_4, with a resulting decreased potency in ability to induce contraction of smooth muscle or to increase vascular permeability (Lewis *et al.*, 1980c, 1983; Bernstrom and Hammarstrom, 1981; Samuelsson, 1982). They can also be metabolized to two 12-epimeric forms of 6-*trans*-LTB_4 that have no spasmogenic or immunoreactive activity (Hansson *et al.*, 1981; Lee *et al.*, 1982, 1983; Borgeat, 1987). These can subsequently be metabolized to various other di- and triHETrEs (some of which can be ω carboxylated), and at least one triHETE (Powell, 1986).

Sulphoxides of LTC_4 through to LTE_4 can be formed,

and these possess little or no spasmogenic activity. Also formed are N-acetyl derivatives of LTE_4 and 11-*trans*-LTE_4, which also have much less potent vascular activity than LTE_4 (Lee *et al.*, 1983; Lewis *et al.*, 1983; Bernstrom and Hammarstrom, 1986; Orning *et al.*, 1986). ω Oxidation of LTC_4 and LTE_4 can also occur, forming N-acetyl-ω-carboxyl-leukotrienes (Orning *et al.*, 1988; Perrin *et al.*, 1989). In addition, LTE_4 can be sequentially β oxidized to form 16-carboxyl-17,18,19, 20-tetranor-14,15-dihydro-N-acetyl-LTE_4 (Delorme *et al.*, 1988; Perrin *et al.*, 1989) and 18-carboxyl-19,20-dinor-N-acetyl-LTE_4 (Delorme *et al.*, 1988).

As discussed above in Section 4.1.1, 5-HETE can enter the 12-lipoxygenase pathway to form $5(S),12(S)$-diHETE (Borgeat *et al.*, 1981; Lindgren *et al.*, 1981; Samuelsson, 1982; Hammarstrom, 1983; Ueda *et al.*, 1986). 5-HETE can also enter the 15-lipoxygenase pathway, to form 5,15-diHETE (Turk *et al.*, 1982; Ueda *et al.*, 1986; Borgeat, 1987). It may also enter both pathways, to form 5,12,20-triHETE (Hansson *et al.*, 1981; Lindgren *et al.*, 1981; Samuelsson, 1982).

4.3   15-LIPOXYGENASE PRODUCTS AND THEIR ACTIVITY

The action of 15-lipoxygenase in polymorphonuclear leucocytes, lymphocytes and reticulocytes (Borgeat *et al.*, 1976; Goetzl, 1981; Bryant *et al.*, 1982; Borgeat, 1987) produces 15-HPETE, which can be metabolized to 15-HETE in several tissues (Borgeat *et al.*, 1976; Goetzl, 1981; Byrant *et al.*, 1982; Hunter *et al.*, 1985; Borgeat, 1987; Kumlin *et al.*, 1990) and several hydroxylated or leukotriene like derivatives (see below).

15-HPETE can inhibit prostacyclin synthesis (see Dusting *et al.*, 1982); 15-HETE can activate a basophil 5-lipoxygenase and can also inhibit platelet 12-lipoxygenase and leucocyte 5-lipoxygenase (Vanderhoek *et al.*, 1980, 1982). Like 5-HPETE, 15-HPETE can also enhance NK cell cytotoxicity (Chang *et al.*, 1989) and, like 5-HETE and the leukotrienes, 15-HETE can stimulate prolactin release in superfused pituitary cells (Kiesel *et al.*, 1987). In addition, 15-HETE has recently been found to be the major arachidonic acid metabolite in human bronchi (Kumlin *et al.*, 1990).

4.3.1   Lipoxins and Related Compounds

15-HPETE can be metabolized by 5-lipoxygenase to produce 5,15-diHPETE or 5,15-diHETE (Sok *et al.*, 1982; Turk *et al.*, 1982; Ueda *et al.*, 1986; Borgeat, 1987); 5,15-diHETE can also be produced by action of 15-lipoxygenase on 5-HETE. In leucocytes and reticulocytes, these dihydroxy or dihydroperoxy compounds can produce, via a 5(6)-epoxy intermediate (5(6),15-EPHETE), the conjugated trihydroxy compounds known as lipoxin A and lipoxin B (Kuhn *et al.*, 1984; Serhan *et al.*, 1984, 1985; Puustinen *et al.*, 1986).

Several biological actions have been reported for the lipoxins, including stimulation of neutrophil degranulation and superoxide anion generation in neutrophils, contraction of lung strips, and inhibition of NK cell activity (Serhan *et al.*, 1985). 8-*trans*-Lipoxin B and $14(S)$-8-*trans*-lipoxin B can also be formed (Serhan *et al.*, 1986).

Other active metabolites of 15-HPETE, via LTA_4 (Sok *et al.*, 1982; Hammerstrom, 1983; Bryant *et al.*, 1985; Borgeat, 1987), include 14,15-LTC_4 and 14,15-LTD_4 from human leucocytes (Hammarstrom, 1983), which can contract guinea-pig ileum (Sok *et al.*, 1982), and 14,15-LTB_4 (from leucocytes, platelets and other cells; Lundberg *et al.*, 1981; Maas *et al.*, 1981; Turk *et al.*, 1982; Hammarstrom, 1983; Hunter *et al.*, 1985; Wong *et al.*, 1985; Wetterholm *et al.*, 1988), which inhibits NK cell cytotoxicity and superoxide generation (Samuelsson, 1985).

Other metabolites of 15-HPETE from various sources (leucocytes, reticulocytes, platelets, etc.) include: 8,15-diHPETE (Bild *et al.*, 1978; Schewe *et al.*, 1986; Yokoyama *et al.*, 1986); 8,15-diHETE (Maas *et al.*, 1981; Hopkins *et al.*, 1984; Hunter *et al.*, 1985; Schewe *et al.*, 1986); 14,15-diHPETE (Yokoyama *et al.*, 1986); 8,15-LTB_4 (Lundberg *et al.*, 1981; Maas *et al.*, 1981; Turk *et al.*, 1982; Hammerstrom, 1983; Hopkins *et al.*, 1984; Hunter *et al.*, 1985; Wong *et al.*, 1985; Yokoyama *et al.*, 1986); 13,14(15)-HEPETrE (Bryant *et al.*, 1985); and a 14-carbon aldehyde (Glasgow *et al.*, 1986). There is at least one report that some 8,15-diHETEs have hyperalgesic properties (Levine *et al.*, 1986).

A unique non-cyclooxygenase biosynthetic route to prostaglandins has recently been described, involving lipoxygenase and allene-oxide synthase enzymes. An example of this was demonstrated using 8-hydroxy-$15(S)$-hydroperoxy eicosanoids from arachidonic acid and eicosapentaenoic acid as substrates. A hydroxylated allene oxide intermediate was cyclized to form prostacyclin A analogues (two from arachidonic acid, four from eicosapentaenoic acid; Brash *et al.*, 1990).

4.4   MONOOXYGENASE PRODUCTS
(Fig 1.8)

Metabolism of arachidonic acid can occur via NADPH-dependent cytochrome P-450 monooxidase (epoxygenase) catalysis. The result is four *cis*-epoxytrienoic acids, each with an epoxide group bridging the positions of what was one of the four double bonds in arachidonic acid: 5(6)-, 8(9)-, 11(12)- and 14(15)-EPETrE (Chacos *et al.*, 1982; Falck and Manna, 1982). These compounds produce several biological effects: they affect sodium transport in the kidney; they may be involved in blood pressure regulation and other vasoactivity, stimulation of steroid and peptide hormone release, inhibition of leucocyte and platelet aggregation, and in regulation of tissue blood flow. Several of these effects may be mediated through enhancement of calcium transport in various

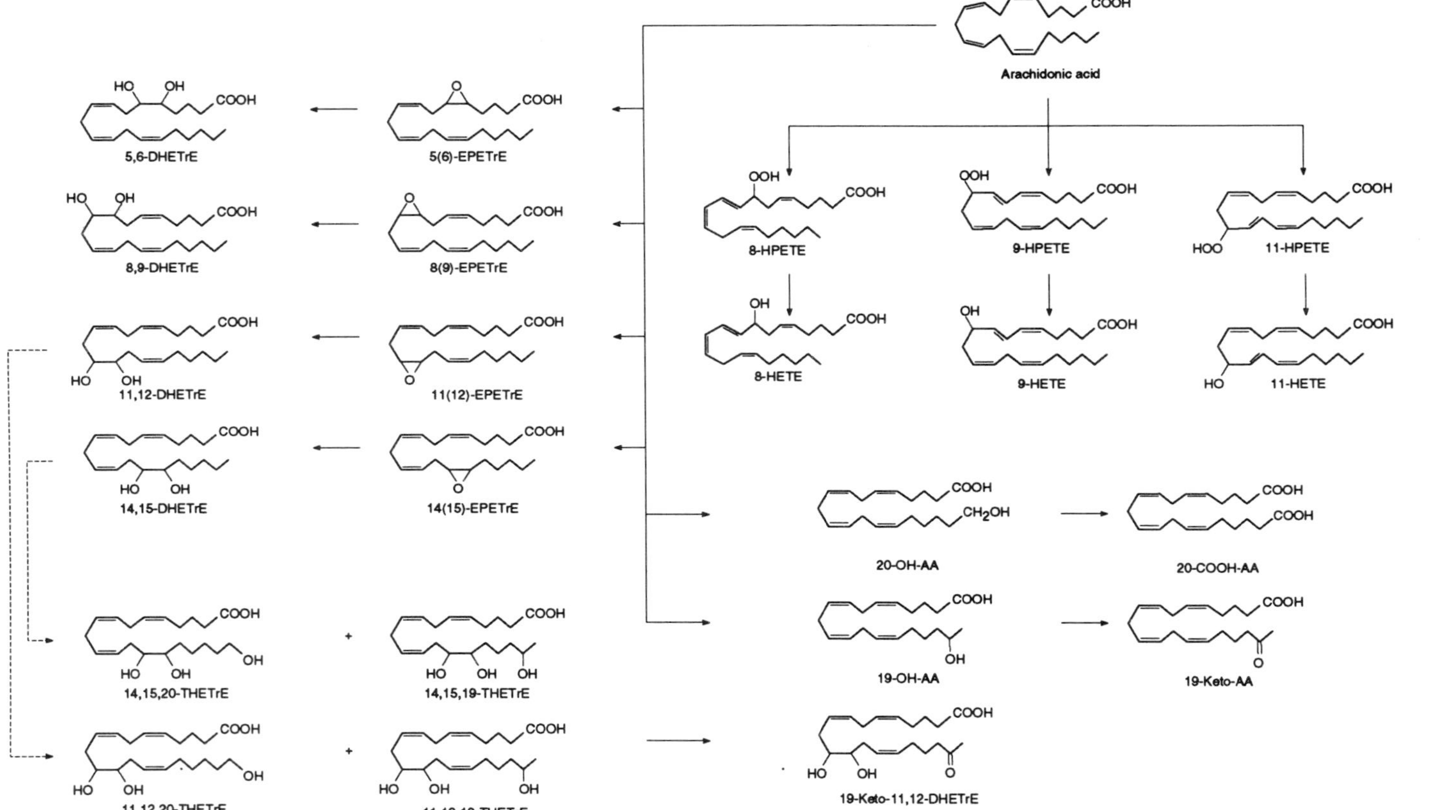

Figure 1.8 Lipoxygenase products of arachidonic acid. Monooxygenase, 8-, 9- and 11-lipoxygenase products: 5(6)-EPETrE (5(6)-oxido-20:3), 5(6)-epoxy-8-*cis*-11-*cis*,14-*cis*-eicosatrienoic acid; 8(9)-EPETrE (8(9)-oxido-20:3), 8(9)-epoxy-5-*cis*,11-*cis*,14-*cis*-eicosatrienoic acid; 11(12)-EPETrE (11(12)-oxido-20:3), 11(12)-epoxy-5-*cis*,8-*cis*,14-*cis*-eicosatrienoic acid; 14(15)-EPETrE (14(15)-oxido-20:3), 14(15)-epoxy-5-*cis*,8-*cis*,11-*cis*-eicosatrienoic acid; 8(*S*)-HPETE, 8(*S*)-hydroperoxy-5-*cis*,9-*trans*,11-*cis*,14-*cis*-eicosatetraenoic acid; 8(*S*)-HETE, 8(*S*)-hydroxy-5-*cis*,9-*trans*,11-*cis*,14-*cis*-eicosatetraenoic acid; 9-HPETE, 9-hydroperoxy-5-*cis*,7-*trans*,11-*cis*,14-*cis*-eicosatetraenoic acid; 9-HETE, 9-hydroxy-5-*cis*,7-*trans*,11-*cis*-14-*cis*-eicosatetraenoic acid; 11-HPETE, 11-hydroperoxy-5-*cis*,8-*cis*,12-*trans*,14-*cis*-eicosatetraenoic acid; 11-HETE, 11-hydroxy-5-*cis*,8-*cis*,12-*trans*,14-*cis*-eicosatetraenoic acid; 20-OH-AA (20-hydroxy-arachidonic acid), 20-hydroxy-5-*cis*,8-*cis*,11-*cis*,14-*cis*-eicosatetraenoic acid; 20 COOH-AA (20-carboxy-arachidonic acid), 5-*cis*-8-*cis*,11-*cis*,14-*cis*-eicosatetraen-1,20-dioic acid; 19-OH-AA (19-hydroxy-arachidonic acid), 19-hydroxy-5-*cis*,8-*cis*,11-*cis*,14-*cis*-eicosatetraenoic acid; 19-keto-AA (19-keto-arachidonic acid), 19-keto-5-*cis*,8-*cis*,11-*cis*,14-*cis*-eicosatetraenoic acid; 5,6-DHETrE, 5,6-dihydroxy-8-*cis*,11-*cis*,14-*cis*-eicosatrienoic acid; 8,9-DHETrE, 8,9-dihydroxy-5-*cis*,11-*cis*,14-*cis*-eicosatrienoic acid; 11,12-DHETrE, 11,12-dihydroxy-5-*cis*,8-*cis*,14-*cis*-eicosatrienoic acid; 14,15-DHETrE, 14,15-dihydroxy-5-*cis*,8-*cis*,11-*cis*-eicosatrienoic acid; 14,15,20-THETrE, 14,15-20-trihydroxy-5-*cis*,8-*cis*,11-*cis*-eicosatrienoic acid; 14,15,19-THETrE, 14,15,19-trihydroxy-5-*cis*,8-*cis*,11-*cis*-eicosatrienoic acid; 11,12,20-THETrE, 11,12,20-trihydroxy-5-*cis*,8-*cis*,14-*cis*-eicosatrienoic acid; 11,12,19-THETrE, 11,12,19-trihydroxy-5-*cis*,8-*cis*,14-*cis*-eicosatrienoic acid; 19-keto-11,12-DHETrE, 11,12-dihydroxy-19-keto-5-*cis*,8-*cis*,14-*cis*-eicosatrienoic acid. Redrawn from Willis (1987).

tissues (Snyder *et al.*, 1986; Proctor *et al.*, 1987, 1989; Nishimura *et al.*, 1989; Ellis *et al.*, 1990a,b; Hirai *et al.*, 1990; Force *et al.*, 1991; McGiff and Carroll, 1991).

It has recently been found that vasodilatory activity of the 5(6)-EPETrE in rat tail and rabbit brain requires conversion by PGH synthase, as further discussed in the next section (Carroll *et al.*, 1990; Ellis *et al.*, 1990).

Further metabolism of all of the epoxides can result in formation of the corresponding dihydroxy and trihydroxy compounds oxidized at position 19 or 20 (Oliw and Oates, 1981).

Cytochrome-P-450 monooxygenase can also oxidize arachidonic acid at the 20 (terminal) or 19 position, to form the corresponding hydroxyl, carboxyl or oxo compounds (Morrison and Pascoe, 1981; Oliw and Oates, 1981; Oliw *et al.*, 1982; Nishimura *et al.*, 1989).

Cyclooxygenase-catalysed endothelium-dependent vasoconstrictor metabolites of 20-hydroxy-arachidonic acid include 20-hydroxy-PGG_2 and 20-hydroxy-PGH_2 (Schwartzman *et al.*, 1989). Like the monooxygenase-derived EPETrEs, the effects of 20-HETE and 20-carboxy-arachidonic acid may involve ion transport (Escalante *et al.*, 1991; McGiff and Carroll, 1991).

4.5   OTHER LIPOXYGENASE PRODUCTS

8-HETE, 9-HETE and 11-HETE (Goetzl and Sun, 1979; Hunter *et al.*, 1985; Lianos *et al.*, 1985; Gschwendt *et al.*, 1986) have been described and can presumably be formed via the action of lipoxygenases. An 8(*S*)-lipoxygenase has been found in mouse skin treated topically with phorbol ester or calcium ionophore A23187; thus, 8(*S*)-HETE may have a role in the cellular inflammatory response to TPA (Hughes and Brash, 1991). It has been reported that 11-HETE, 15-HPETE and 15-HETE are produced by cyclooxygenase pathways (Bailey *et al.*, 1983; Hecker *et al.*, 1987a,b).

5. *"Crossover" Products*

Recently, instances of conversion of arachionic acid to prostaglandins have been demonstrated that require the prior formation of lipoxygenase or monooxygenase products. At least two PGA_2 analogues with a *trans* configuration of the side-chain are produced from the lipoxygenase product 8-hydroxy-15(*S*)-HPETE (Brash *et al.*, 1990). It has been speculated that similar reactions are involved in the formation of PGA analogues in corals.

As a second example, the monooxygenase product 5(6)-EPETrE can be converted to 5-hydroxy-$PGI_{1\alpha}$ and 5-hydroxy-$PGI_{1\beta}$ (Oliw, 1984). This is consistent with the PGH synthase dependence of the vasodilator activity of 5(6)-EPETrE in rat tail and rabbit brain (Carroll *et al.*, 1990; Ellis *et al.*, 1990).

As a final example, novel endothelium-dependent vasoconstrictor metabolites of the monooxygenase product 20-hydroxy-arachidonic acid include 20-hydroxy-PGG_2 and PGH_2 (Schwartzman *et al.*, 1989).

An additional type of "crossover" product has also been described, involving PGH synthase-catalysed conversion of arachidonic acid to various HETE and HPETE compounds: 11-HETE, 15-HETE and 15-HPETE (Bailey *et al.*, 1983; Hecker *et al.*, 1987a,b).

6. *Alternative Substrates to Arachidonic Acid in the Formation of Eicosanoids*

Although the focus of this chapter is on the metabolism of arachidonic acid, other polyunsaturated precursors for eicosanoids exist and may, on occasion (especially after dietary manipulation), be the preferred substrate and/or influence the metabolism of arachidonic acid or its eicosanoid pathways.

DGLA is precursor for prostaglandins (notably PGE_1). However, PGE_1 is normally present in large amounts only in seminal plasma, probably because of the large amount of DGLA in the phospholipids of vesicular gland. DGLA is poorly converted to 1 series thromboxanes that (in any case) seem lacking in appreciable biological effects and there is some evidence that DGLA may inhibit formation of TXA_2. DGLA is also unique in that it lacks Δ^5 unsaturation and therefore cannot be converted into prostacyclins or conventional leukotrienes. However, it is converted into 12-lipoxygenase and 15-lipoxygenase products, including 15-HETrE that can inhibit conversion of arachidonic acid into leukotrienes. It is therefore the opinon of these authors that DGLA is uniquely suited to exert biochemical "leverage" in improving the balance of advantageous to deleterious eicosanoids. Evidence for the above concepts, and the effects and claimed therapeutic benefit of administering the immediate precursor of DGLA, γ-linolenic acid (from seed oils) are reviewed in detail in an extensively referenced book chapter (Willis and Smith, 1989a), and the cardiovascular aspects are reviewed by Willis and Smith (1989b).

EPA is a principal constituent of marine-lipid diets that appear to confer protection against vascular occlusive disease in Eskimos and the Japanese (see Willis and Smith, 1989b). Like DGLA, EPA can, at least in part, displace arachidonic acid from its principal role as an eicosanoid precursor. EPA can form TXA_3 that is virtually inactive as a prothrombotic agent. It can also form the prostacyclin PGI_3, which has antithrombotic properties. Moreover, it reduces formation of active leukotrienes derived from arachidonic acid (because LTA_5 competes with LTA_4 for enzymatic conversion to further leukotriene metabolites); LTB_5 also appears much less active than LTB_4 as a pro-inflammatory mediator. Finally, EPA and, especially, docosahexanoic acid (DHA), another component of marine-lipid diets, tend to reduce conversion of arachidonic acid to PGH and its prostanoid metabolites. These complex biochemical aspects of EPA biochemistry are extensively reviewed by Salem (1989), and the biological significance is amplified in the review by Willis and Smith (1989b).

Adrenic acid ($C_{22:4}$ ω-6), found in the lipids of the adrenal and kidney interstitial lipid droplets, is a chain extension derivative of arachidonic acid whose fatty acid chain is elongated by two carbons at the carboxyl (α) end of the molecule (see Smith, 1987; Willis, 1987). It can serve as a substrate for PGE, PGF and thromboxanes (Tobias *et al.*, 1975; Sprecher *et al.*, 1982) and (theoretically) other eicosanoids as well that are elongated by two carbon atoms (see Smith, 1987). A similarly elongated version of DGLA (docosatrienoic acid, $C_{22:3}$ ω-6) that is also present in adrenal cortex (Takayasu *et al.*, 1970) could lead to formation of dihomo-PGE_1, an inhibitor of platelet aggregation (Kloeze, 1970; Van Dorp, 1971). In both cases, however, endogenous roles for these precursors have not been established.

Finally, linoleic acid ($C_{18:2}$ ω-6) is a precursor of 13-HODE, an eicosanoid that may possess a role in reducing platelet adhesion to vascular endothelium (Buchanan *et al.*, 1985a,b). This is formed by the same enzyme that produces 15-lipoxygenation of arachidonic acid, DGLA and EPA. Indeed, linoleate may be the preferred substrate for this enzyme, at least in polymorphonuclear leucocytes (Soberman *et al.*, 1985).

7. Mode of Action of Anti-Inflammatory Drugs

7.1 NON-STEROIDAL ("ASPIRIN-TYPE") DRUGS

It is now some 20 years since evidence was produced that PGE_2 was a mediator of inflammation and fever (Crunkhorn and Willis, 1969, 1971; Juhlin and Michaelsson, 1969; Willis, 1969, 1970; Milton and Wendlandt, 1971). However, the three simultaneous reports that aspirin-type drugs could inhibit prostaglandin production (Ferreira *et al.*, 1971; Smith and Willis, 1971; Vane, 1971) by inhibition of the conversion of arachidonic acid to prostaglandins (Vane, 1971) provided further stimulus for investigating the role of prostaglandins in inflammatory disease. This resulted in further definition of the role of prostaglandins in inflammation and fever (see Willis *et al.*, 1982; Higgs *et al.*, 1984), including pain production (Ferreira, 1972; Willis and Cornelsen, 1973). Indeed, the "endogenous pyrogen" released from macrophages during inflammatory conditions has now been identified as IL-1, which acts to release pro-inflammatory and pyrogenic E-type prostaglandins (Dinarello and Meir, 1986).

However, the study of platelets has yielded the most important advances in the understanding of aspirin-like drugs. The pioneering work of Weiss and Aledort (1967) and of O'Brien (1968) first demonstrated the classic inhibition of "secondary aggregation" and release of platelet granule contents by drugs of the aspirin type. The findings of O'Brien (1968) were of particular note since it was reported that sodium salicylate and 4-acetomidophenol (paracetamol) were virtually inactive compared to aspirin. This finding runs parallel both to effects of these drugs on bleeding time in humans (Sutor *et al.*, 1971) and ability to inhibit platelet production (Willis and Smith, 1982).

The fact that classic prostaglandins did not induce platelet aggregation or reverse the effects of aspirin led to the discovery of the role of the unstable endoperoxides (Willis and Kuhn, 1973; Hamberg *et al.*, 1974; Willis, 1974a,b; Willis *et al.*, 1974) and subsequent discovery of their conversion into the thromboxanes (Hamberg *et al.*, 1975), which act upon the same receptor to induce platelet activation. The platelet then became a useful tool with which to elucidate the enzymatic mechanism of action of aspirin. It was known many years ago that aspirin irreversibly acetylated some platelet protein (Al-Mondhiry *et al.*, 1970) and the platelet, unlike nucleated cells, cannot generate proteins, including fresh PGH synthase. Consequently, the platelet provided an ideal system in which to study the mode of action of aspirin. It was an obvious next step to acetylate platelets with radiolabelled aspirin and then use detection of the label as a marker to isolate the PGH synthase protein (Roth *et al.*, 1975).

PGH synthase has also been purified from bovine and ovine seminal vesicles (Miyamoto *et al.*, 1976; Van der Ouderaa *et al.*, 1977, 1979) with an MW of approximately 72 kDa, as determined by SDS–PAGE, corresponding well with the MW calculated from the primary sequence of 65 kDa (DeWitt and Smith, 1988; Merlie *et al.*, 1988; Yokoyama *et al.*, 1988). Such minor differences in molecular mass are due to attachment of 2–3 oligosaccharides per protein subunit (Van der Ouderaa *et al.*, 1977; Mutsaers *et al.*, 1985). From cDNA studies of sheep, mouse and human PGH synthase, the amino acid sequences of the proteins were deduced and a sequence homology of about 90% noted. Aspirin acts by acetylating a single serine residue (Ser530) and the irreversible activity of aspirin is attributable to a lack of enzymatic activity capable of hydrolysing the acetyl-Ser530 ester (Roth *et al.*, 1983; DeWitt and Smith, 1988; Merlie *et al.*, 1988; Yokoyama *et al.*, 1988; Shimokawa and Smith, 1990). Indeed, if this serine residue at position 530 is replaced by an alanine residue that lacks the hydroxyl side-chain for acetylation, then aspirin inhibits PGH synthase in a reversible manner (Shimokawa and Smith, 1992; Smith 1992). Other non-steroidal anti-inflammatory drugs generally lack this irreversible effect, as exemplified by the short-lived (about 24 h) ability of indomethacin to inhibit platelet aggregation and prostanoid production (Kocsis *et al.*, 1973).

Although non-steroidal anti-inflammatory drugs in

general inhibit prostaglandin synthesis in all tissues regardless of species, paracetamol reportedly inhibits prostaglandin biosynthesis only in brain (and possibly other neural tissue), not in sites of inflammation (Willis *et al.* 1972) or in spleen (see Flower and Vane, 1972) or platelets (Willis and Smith, 1982). This explains why paracetamol has antipyretic and analgesic properties but is not anti-inflammatory and does not interfere with platelet aggregation (O'Brien 1968) or bleeding time (Sutor *et al.*, 1971). To our knowledge, the reasons for this selective effect are still unexplored, although recent advances in the molecular biology of prostaglandin endoperoxide synthase and the discovery of two isoforms (see Smith, 1992; Section 3.3.2) may lead to such an explanation.

Another anomalous finding that (until very recently) has defied adequate explanation pertains to sodium salicylate. This compound is an effective anti-inflammatory agent, but a very poor inhibitor of PGH synthase *in vitro*. However, treatment with anti-inflammatory doses of sodium salicylate does reduce the content of PGE_2 in rat inflammatory exudate or mouse and gerbil brain (Willis *et al.*, 1972) or of the PGE tetranor metabolite in human urine (Hamberg, 1972). Thus, the compound clearly does act, albeit indirectly via inhibition of prostaglandin production. Although the salicylate metabolite gentisic acid inhibits PGH synthase (Flower, 1974), the action of this compound is unlikely to explain why salicylate treatment of donor animals did not inhibit spontaneous or arachidonic acid-evoked prostaglandin synthesis in inflammatory exudate or brain (see Willis, 1989). A possible explanation may stem from the recent observation that very low concentrations of salicylate (approximately 20 nM) block the *de novo* production of PGH synthase in cultured HUVECs (Sanduja *et al.*, 1991). Similar effects have been reported for aspirin and salicylate (Wu *et al.*, 1991) and for naproxen (Zvelewska *et al.*, 1992), a mechanism apparently due to reduced gene expression since it was accompanied by reduced PGH synthase mRNA. Whether this effect is generally produced in other cell types remains to be established, but an adequate mechanism for the apparent anomaly of sodium salicylate now seems established. The inhibition of "LOX" (now identified as 13-HODE) production by salicylate (Buchanan *et al.*, 1985a,b) would also seem worth exploring again since this material is now known to be produced by the same enzyme system as that producing 15-lipoxygenase products of arachidonic acid (Soberman *et al.*, 1985).

7.2 STEROIDAL ANTI-INFLAMMATORY DRUGS

Acute treatment with hydrocortisone did not inhibit prostaglandin production by platelets (Smith and Willis, 1971; Willis and Smith, 1982). Such findings may be expected from a present knowledge of its mode of action (via inhibition of specific protein synthesis) since platelets are anucleated and therefore cannot synthesize proteins. Acute treatment with hydrocortisone also did not inhibit prostaglandin production in spleen (Ferreira *et al.*, 1971), although chronic treatment with hydrocortisone could do so (Grodzinska and Dembinska-Kiec, 1977).

Subsequently, it was reported that anti-inflammatory steroids could inhibit production of prostaglandins by homogenates (not microsomes) of human skin (Greaves and McDonald Gibson, 1972) and in mouse fibroblasts in culture, an effect attributable to diminished arachidonate release (Hong and Levine, 1976). Similar effects were seen in perfused lungs of guinea-pigs (Blackwell *et al.*, 1978; Gryglewski *et al.*, 1978), and the latter system has since been used extensively by Flower's group to examine the mechanisms involved.

The finding that RNA and protein synthesis were necessary for corticosteroid inhibition of prostaglandin production (Danon and Assouline, 1978) first pointed to the possibility of induction of some protein that inhibited prostaglandin production, perhaps via inhibition of arachidonate release. The existence of such a peptide factor produced by macrophages (and hence named "macrocortin") was first reported by Blackwell *et al.* (1980). Subsequently, similar material was reported to be produced in response to corticosteroids from a variety of sources, including monocytes, neutrophils, renal medullary cells, endometrium, stomach tissue and skin fibroblasts (see the review by Flower, 1988; Flower *et al.*, 1989; D.L. Smith, 1989). It is thought that these factors (now collectively named lipocortins) are subsets of the glycoprotein lipocortin I. Lipocortins I through VI have been identified and, through molecular biological approaches to sequence determination, are now classified as part of the annexin family (see Bailey, 1991; Rothwell and Flower, 1992). The predominantly active form (lipocortin I) has an MW of 40 kDa (See D.L. Smith, 1989).

Lipocortins I and II are homologous with the calcium-regulated phospholipid and actin-binding proteins termed calpactins II and I, respectively (see Flower *et al.*, 1989; D.L. Smith, 1989). With the availability of sequence data, it is apparent that many different names were being used for identical proteins. A new terminology has thus been suggested, based upon membrane-binding properties of the proteins and the extant lipocortin numerical sequence; thus, lipocortin I may also be termed annexin I (Crumpton and Dedman, 1990).

The original hypothesis that lipocortin mediates the ability of corticosteroids to reduce eicosanoid production seems well established (Flower, 1988; Flower *et al.*, 1989; Browning *et al.*, 1990; Bailey, 1991; Rothwell and Flower, 1992). For example, lipocortin I is produced in peripheral blood leucocytes in response to hydrocortisone (Goulding *et al.*, 1990). Also, the PGE/eicosanoid-mediated growth arrest of a lung adenocarcinoma cell line is prevented by both dexamethasone and lipocortin I. Furthermore, the effect of dexamethasone is blocked by a monoclonal antibody that neutralizes lipocortin I (Croxtall and Flower, 1992). Most convincing is *in vivo* data showing that recombinant lipocortin I mimics the effects of the anti-inflammatory glucocorticoids. Furthermore, the neutralizing anti-lipocortin I antibodies reverse the anti-inflammatory and antipyretic actions of glucocorticoids *in vivo*, indicating that endogenous lipocortin I mediates these effects. However, controversy still surrounds the exact mechanisms through which lipocortin mediates its anti-inflammatory effects. Work indicating that production of lipocortin I (or its mRNA) could not be induced in cultured cells has been countered by findings that it is produced *in vivo* or in cells or tissues examined immediately after harvesting from animals or humans that have been treated with anti-inflammatory corticosteroids (see Bailey, 1991). Lipocortin appears to exist primarily intracellularly but can be detected on the cell surface. Although lipocortin would appear to be ill suited to serve as a released humoral factor, it is found in exceptionally high concentrations in human seminal fluid (see Croxtall and Flower, 1992).

In vitro studies indicate that the original assumption that lipocortin is a direct inhibitor of PLA_2 is probably untenable since it appears to merely bind to the substrate phospholipid (Davidson *et al.*, 1987). Alternative mechanisms for lipocortin I, such as the induction of protein phosphatases that dephosphorylate and thus inactivate membrane-bound PLA_2 are possible. In addition, there is evidence that reduced cellular levels of PGH synthase (see below) may involve conversion of PGH synthase mRNA into an inactive cryptic form (Bailey, 1991).

An alternative (or additional) mode of action of corticosteroids has been suggested by Lee *et al.* (1988) through inhibition of the production of IL-1, which is a known potent stimulator of eicosanoid release (see previous discussion in this review).

Finally, glucocorticoids have also been reported to inhibit production of a PGH synthase in dermal fibroblasts and monocytes/macrophages (Raz *et al.*, 1990). This activity may be limited to a specially regulated PGH synthase (O'Banion *et al.*, 1992). In addition, dexamethasone inhibits mitogen induction of the *TIS10* PGH synthase gene (Kujubu and Herschman, 1992). Masferrer *et al.* (1990) also showed that dexamethasone inhibits induction of the PGH synthase produced selectively in response to lipopolysaccharide. This protective effect versus endotoxin-induced overproduction of prostaglandins could also be seen *in vivo* with correspondingly reduced lethality of the endotoxin (Masferrer *et al.*, 1992). According to Bailey (1991), this protective effect of dexamethasone (at least in vascular smooth muscle cells) is due to inhibition of translational control of mRNA for the PGH synthase and is therefore mimicked by cycloheximide (which inhibits translation) but not by actinomycin (which inhibits transcription). It has yet to be established with certainty whether this effect

of corticosteroids on PGH synthase levels is selective to the newly described type 2 form of the enzyme. Regardless, in at least one cell type (bone marrow-derived macrophages) about 70% of the observed inhibition of prostaglandin production by glucocorticoids was attributable to reduced PGH synthase and only about one-third due to reduced PLA_2 activity (Goppelt-Struebe *et al.*, 1989)

8. Concluding Remarks

The vast complexity of the eicosanoid field in general and the recent advances applying molecular biological techniques to the study of individual enzymes and their control has made detailed review of each aspect of arachidonate metabolism impossible. However, the reader is referred to several recent detailed reviews and compendia (Borgeat, 1987; Flower *et al.*, 1989; Holmsen, 1987; Mead and Willis, 1987; Roberts, 1987; D.L. Smith, 1987, 1989; Smith *et al.*, 1987, 1991; Willis, 1987, 1989; Flower, 1988; Nelson, 1989; Salem, 1989; W.L. Smith, 1989; Willis and Smith, 1989a,b; Willis *et al.*, 1990; Bailey, 1991; Smith and Marnett, 1991; Rothwell and Flower, 1992). In addition, only passing mention can be given to the enormous literature on utilization of polyunsaturated fatty acid substrates other than arachidonic acid. This is justified on the basis that only with change in the dietary fatty acids would this likely to be a major factor.

It is a sobering thought that rational design of therapeutic agents via selective inhibition of one or other eicosanoid type has not yet materialized. However, we now are beginning to realize the almost impossible overall complexity of arachidonate metabolism and the complexity of biological effects of the eicosanoids produced via various receptor subtypes, cooperative and interactive effects and so on. Perhaps, then, it is time that the field returned to its roots: biological assays and good old-fashioned "phenomenology", which is surely a prerequisite in clinical trials!

9. References

Albrightson, C.R., Baenziger, N.L. and Needleman, P. (1985). Exaggerated human vascular cell prostaglandin biosynthesis mediated by monocytes. Role of monokines and interleukin 1. J. Immunol. 135, 1872–1877.

Al Mondhiry, H., Marcus, A.J. and Spaet, T.H. (1970). On the mechanisms of platelet function inhibition by acetyl salicylic acid. Proc. Soc. Exp. Biol. Med. 133, 632–636.

Anggard, E. (1966). The biological activities of three metabolites of prostaglandin E_1. Acta Physiol. Scand. 66, 509–510.

Bailey, J.M. (1991). New mechanisms for effects of anti-inflammatory glucocorticoids. Biofactors 13, 97–102.

Bailey, J.M., Bryant, R.W., Whiting, J. and Salata, K. (1983).

Characterization of 11-HETE and 15-HETE, together with prostacyclin, as major products of the cyclo-oxygenase pathway in cultured rat aorta smooth muscle cells. J. Lipid Res. 24, 1419–1428.

Bailey, J.M., Muza, B., Hla, T. and Salata, K. (1985). Restoration of prostacyclin synthase in vascular smooth muscle cells after aspirin treatment. Regulation by epidermal growth factor. J. Lipid Res. 26, 54–61.

Beckman, B.S., Mason-Garcia, M., Nystuen, L., King, L. and Fisher, J.W. (1987). The action of erythropoietin is mediated by lipoxygenase metabolites in murine fetal liver cells. Biochem. Biophys. Res. Commun. 147, 392–398.

Bell, R.L., Donald, A.K., Stanford, N. and Majerus, P.W. (1979). Diglyceride lipase: a pathway for arachidonate release from human platelets. Proc. Natl Acad. Sci. USA 76, 3238–3241.

Bennett, P.R., Elder, M.G. and Myatt, L. (1987). The effects of lipoxygenase metabolites of arachidonic acid on human myometrial contractibility. Prostaglandins 33, 837–844.

Bergstrom, S., Danielsson, H. and Samuelsson, B. (1964). The enzymatic formation of prostaglandin E_2 from arachidonic acid. Biochim. Biophys. Acta 90, 207–210

Bernstrom, K. and Hammarstrom, S. (1981). Metabolism of leukotriene D by porcine kidney. J. Biol. Chem. 256, 9579–9582.

Bernstrom, K. and Hammarstrom, S. (1986). Metabolism of leukotriene E_4 by rat tissues: formation of N-acetylleukotriene E_4. Arch. Biochem. Biophys. 244, 486–491.

Bild, G.S., Bhat, S.G., Axelrod, B. and Iatridis, P.G. (1978). Inhibition of aggregation of human platelets by 8,15-dihydroperoxides of 5,9,11,13-eicosatetraenoic and 9,11,13-eicosatrienoic acids. Prostaglandins 16, 795–801.

Bills, T.K., Smith, J.B. and Silver, M.J. (1977). Selective release of arachidonic acid from the phospholipids of human platelets in response to thrombin. J. Clin. Invest. 60, 1–6

Blackwell, G.J., Flower, R.J., Nijkamp, F.P. and Vane, J.R. (1978). Phospholipase A_2 activity of guinea-pig isolated perfused lungs. Stimulation and inhibition by antiinflammatory steroids. Br. J. Pharmacol. 62, 79–89.

Blackwell, G.J., Carnuccio, R., DiRosa, M., Flower, R.J., Parente, L. and Persico, P. (1980). Macrocortin: a polypeptide causing the antiphospholipase effect of glucocorticoids. Nature 287, 147–149

Boot, J.R., Dawson, W. and Harvey, J. (1976). Comparative biological activity of prostaglandin E_2 and its C20 metabolites on smooth muscle preparations. Adv. Prostaglandin Thromboxane Res. 2, 958 (Abstr.).

Boraschi, D., Censini, C. and Tagliabue, A. (1984). Interferon-gamma reduces macrophage-suppressive activity by inhibiting prostaglandin E_2 release and inducing interleukin 1 production. J. Immunol. 133, 764–768.

Borgeat, P. (1987). In "CRC Handbook of Eicosanoids: Prostaglandins and Related Lipids", Vol I (ed A.L. Willis), pp 193–210. CRC Press, Boca Raton

Borgeat, P. and Samuelsson, B. (1979a). Arachidonic acid metabolism in polymorphonuclear leukocytes: unstable intermediate in formation of dihydroxy acids. Proc. Natl Acad. Sci. USA 76, 3213–3217

Borgeat, P. and Samuelsson, B. (1979b). Transformation of arachidonic acid by rabbit polymorphonuclear leukocytes. J. Biol. Chem. 254, 2643–2646.

Borgeat, P., Hamberg, M. and Samuelsson, B. (1976). Transformation of arachidonic acid and homo-γ-linolenic acid by rabbit polymorphonuclear leukocytes. J. Biol. Chem. 251, 7816–7820.

Borgeat, P., Picard, S., Vallerand, P. and Sirois, S. (1981). Transformation of arachidonic acid in leukocytes. Isolation and structural analysis of a novel dihydroxy derivative. Prostaglandins Med. 6, 557–570.

Borgeat, P., Fruteau De Laclos, B., Picard, S., Vallerand, P. and Sirois, P. (1982). In "Leukotrienes and Other Lipoxygenase Products" (eds B. Samuelsson and R. Paoletti), pp 45–51. Raven Press, New York.

Brash, A.R., Baertschi, S.W. and Harris, T.M. (1990). Formation of prostaglandin A analogues via an allene oxide. J. Biol. Chem. 265, 6705–6712.

Brocklehurst, W.E. (1982). The forty year quest of "slow reacting substance of anaphylaxis". Prog. Lipid Res. 20, 709–712.

Browning, J.L., Ward, M.P., Wallner, B.P. and Pepinski, R.B. (1990). Studies on the structural properties of lipocortin-1 and the regulation of its synthesis by steroids. Prog. Clin. Biol. Res. 349, 27–45.

Bryant, R.W. and Bailey, J.M. (1979). Isolation of a new lipoxygenase metabolite of arachidonic acid-8,11,12-trihydroxy-5,9,14-eicosatrienoic from human platelets. Prostaglandins 17, 9–18.

Bryant, R.W. and Bailey, J.M. (1980). Isolation of glucose-sensitive platelet lipoxygenase products from arachidonic acid. Adv. Prostaglandin Thromboxane Res. 6, 95–99.

Bryant, R.W., Bailey, J.M., Schewe, T. and Rapoport, S.M. (1982). Positional specificity of a reticulocyte lipoxygenase: conversion of arachidonic acid to 15-S-hydroperoxyeicosatetraenoic acid. J. Biol. Chem. 257, 6050–6055.

Bryant, R.W., Schewe, J., Rapoport, S.M. and Bailey, J.M. (1985). Leukotriene formation by a purified reticulocyte lipoxygenase enzyme. Conversion of arachidonic acid and 15-hydroxyeicosatetraenoic acid to 14,15-leukotriene A_4. J. Biol. Chem. 260, 3548–3555.

Buchanan, M.R., Butt, R.W., Magas, Z., Ryn, J.V., Hirsch, J. and Nazir, D.J. (1985a). Endothelial cells produce a lipoxygenase derived chemorepellant which influences platelet/ endothelial cell interactions – Effects of aspirin and salicylate. Thromb. Haemost. 53, 306–311.

Buchanan, M.R., Haas, T.A, Lagarde, M. and Guichardant, M. (1985b). 13-Hydroxyoctadecadienoic acid is the vessel wall chemorepellant factor, LOX. J. Biol. Chem. 260, 16056–16059.

Burch, R.M., White, M.F. and Connor, J.R. (1989). Interleukin 1 stimulates prostaglandin synthesis and cyclic AMP accumulation in Swiss 3T3 fibroblasts. Interactions between two second messenger systems. J. Cell. Physiol. 139, 29–33.

Burr, G.O. and Burr, M.M. (1930). On the nature and role for the fatty acids essential to nutrition. J. Biol. Chem. 86, 587–621.

Carroll, M.A., Garcia, M.P., Falck, J.R. and McGiff, J.C. (1990). 5,6-Epoxyeicosatrienoic acid, a novel arachidonate metabolite. Mechanism of vasoactivity in the rat. Circ. Res. 67, 1082–1088.

Chacos, N., Falck, J.R., Wixtrom, C. and Capdevila, J. (1982). Novel epoxides formed during the liver cytochrome P-450 oxidation of arachidonic acid. Biochem. Biophys. Res. Commun. 104, 916–922.

Chang, K.J., Saito, H., Tatsuno, I., Tamura, Y., Watanabe,

K. and Yoshida, S. (1989). Comparison of the effect of lipoxygenase metabolites of arachidonic acid and eicosapentaenoic acid on human natural killer cell cytotoxity. Prostaglandins Leukot. Essent. Fatty Acids 38, 87–90.

Chouaib, S., Robb, R.J., Welte, K. and Dupont, B. (1987). Analysis of prostaglandin E_2 effect on T lymphocyte activation. Abrogation of prostaglandin E_2 inhibitory effect by the tumor promotor 12.0 tetradecanoyl phorbol-13 acetate. J. Clin. Invest. 80, 333–340.

Claesson, H.E. and Haeggstrom, J. (1988). Human endothelial cells stimulate leukotriene synthesis and convert granulocyte-released leukotriene A_4 into leukotrienes B_4, C_4, D_4, and E_4. Eur. J. Biochem. 173, 93–100.

Comai, K., Prose, P., Farber, S.J. and Paulsrud, J.R. (1974). Correlation of renal medullary prostaglandin content and renal interstitial cell lipid droplets. Prostaglandins 6, 375–379.

Comai, K., Farber, S.J. and Paulsrud, J.R. (1975). Analyses of renal medullary lipid droplets from normal, hydronephrotic, and indomethacin treated rabbits. Lipids 10, 555–561.

Croxtall, J.D. and Flower, R.J. (1992). Lipocortin 1 mediates dexamethasone-induced growth arrest of the A549 lung adenocarcinoma cell line. Proc. Natl Acad. Sci. USA 89, 3571–3575.

Crumpton, M.J. and Dedman, J.R. (1990). Protein terminology tangle. Nature 345, 212.

Crunkhorn, P. and Willis, A.L. (1969). Actions and interactions of prostaglandins administered intradermally in rat and man. Br. J. Pharmacol. 36, 216P.

Crunkhorn, P. and Willis, A.L. (1971). Cutaneous reactions to intradermal prostaglandins. Br. J. Pharmacol. 41, 49–56.

Crutchley, D.J. and Piper, P.J. (1975). Comparative bioassay of prostaglandin E_2 and its three pulmonary metabolites. Br. J. Pharmacol. 54, 397–399.

Danon, A. and Assouline, G. (1978). Inhibition of prostaglandin biosynthesis by corticosteroids requires RNA and protein synthesis. Nature 273, 552–554.

Davidson, F.F. and Dennis, E.A. (1990). Evolutionary relationships and implications for the regulation of phospholipase A_2 from snake venom to human secretory forms. J. Mol. Evol. 31, 228–238.

Davidson, F.F., Dennis, E.A., Powell, M. and Glenney, J.R. Jr (1987). Inhibition of phospholipase A_2 by "lipocortins" and calpactins. J. Biol. Chem. 262, 1698–1705.

Dawson, W., Lewis, R.L., McMahon, R.E. and Sweatman, W.J.F. (1974). Potent bronchoconstrictor activity of 15-keto prostaglandin $F_{2\alpha}$. Nature 250, 331–332.

Delorme, D., Foster, A., Girard, Y. and Rokach, J. (1988). Synthesis of beta-oxidation products as potential leukotriene metabolites and their detection in bile of anesthetized rat. Prostaglandins 36, 291–302.

De Nucci, G., Gryglewski, R.J., Warner, J.C. and Vane, J.R. (1988). Receptor-mediated release of endothelium-derived relaxing factor and prostacyclin from bovine aortic endothelial cells is coupled. Proc. Natl. Acad. Sci. USA 85, 2334–2338.

DeWitt, D.L. (1991). Prostaglandin endoperoxide synthase: regulation of enzyme expression. Biochim. Biophys. Acta 1083, 121–134.

DeWitt, D.L. and Smith, W. (1988). Primary structure of the prostaglandin G/H synthase from sheep vesicular gland determined from the complementary DNA sequence. Proc. Natl Acad. Sci. USA 85, 1412–1416.

DeWitt, D.L., Kraemer, S.A. and Meade, E.A. (1991). Serum induction and superinduction of PGG/H synthase mRNA levels in 3T3 fibroblasts. Adv. Prostaglandin Thromboxane Leukotriene Res. 21A, 65–68.

Dinarello, C.A. and Meir, J.W. (1986). Interleukins. Annu. Rev. Med. 37, 173–178.

Drazen, J.M., Austen, K.F., Lewis, R.A., Clark, D.A., Goto, G., Marfat, A. and Corey, E.J. 1980. Comparative airway and vascular activities of leukotrienes C-1 and D *in vivo* and *in vitro*. Proc. Natl Acad. Sci. USA 77, 4354–4358.

Dusting, G.J., Moncada, S. and Vane, J.R. (1982). Prostacyclin: its biosynthesis, actions and clinical potential. Adv. Prostaglandin Thromboxane Leukotriene Res. 10, 59–106.

Ellis, C.K., Smigel, M.D., Oates, J.A., Oelz, O. and Sweetman, B.J. (1979). Metabolism of prostaglandin D_2 in the monkey. J. Biol. Chem. 254, 4152–4163.

Ellis, E.F., Police, R.J. Yancey, L., McKinney, J.S. and Amruthesh, S.C (1990a). Dilation of cerebral arterioles by cytochrome P-450 metabolites of arachidonic acid. Am. J. Physiol. 259, H1171–H1177.

Ellis, E.F., Amruthesh, S.C., Police, R.J. and Yancey, L.M. (1990b). Brain synthesis and cerebrovascular action of cytochrome P-450/monoxygenase metabolites of arachidonic acid. Adv. Prostaglandin Thromboxane Leukotriene Res. 21, 201–304.

Escalante, B., Erlij, D., Falck, J.R. and McGiff, J.C. (1991). Effect of cytochrome P450 arachidonate metabolites on ion transport in rabbit kidney loop of Henle. Science 251, 799–802.

Falck, J.R. and Manna, S. (1982). 8,9-Epoxyarachidonic acid: a cytochrome P-450 metabolite. Tetrahedron Lett. 23, 1755–1756.

Fauler, J., Marx, K.H., Kaever, V. and Frolich, J.C. (1989). Human monocytes convert leukotriene B_4 to two dihydro-leukotriene B_4 metabolites. Prostaglandins Leukot. Essent. Fatty Acids 38, 413–430.

Feinmark, S.J. and Cannon, P.J. (1986). Endothelial cell leukotriene C_4 synthesis results from intracellular transfer of leukotriene C_4 synthesized by polymorphonuclear leukoycytes. J. Biol. Chem. 261, 16466–16472.

Ferreira, S.H. (1972). Prostaglandins, aspirin-like drugs, and analgesia. Nature (New Biol.) 240, 200–203

Ferreira, S.H. and Vane, J.R. (1967). Prostaglandins: their disappearance from and release into the circulation. Nature 217, 868–873

Ferreira, S.H., Moncada, S. and Vane, J.R. (1971). Indomethacin and aspirin abolish prostaglandin release from the spleen. Nature (New Biol.) 231, 237–239.

Fitzpatrick, F.A., Haeggstrom, J., Granstrom, E. and Samuelsson, B. (1983). Metabolism of leukotriene A_4 by an enzyme in blood plasma: a possible leukotactic mechanism. Proc. Natl Acad. Sci. USA 80, 5425–5429.

Flower, R.J. (1974). Drugs that inhibit prostaglandin biosynthesis. Pharmacol. Rev. 26, 33–67.

Flower, R.J. (1988). Eleventh Gaddum Memorial Lecture: Lipocortin and the mechanism of action of glucocorticoids. Br. J. Pharmacol. 94, 987–1015.

Flower, R.J. and Vane, J.R. (1972). Inhibition of prostaglandin synthetase in brain explains the anti-pyretic activity of paracetamol (4-acetamidophenol). Nature 240, 410–411.

Flower, R.J., Blackwell, G.J. and Smith, D.L. (1989). In "Handbook of Eicosanoids: Prostaglandins and Related Lipids", Vol II (ed A.L. Willis), pp 35–46. CRC Press, Boca Raton

Force, T., Hyman, G., Hajjar, R., Sellmayer, A. and Bonventre,

J.V. (1991). Noncyclooxygenase metabolites of arachidonic acid amplify the vasopressin-induced Ca^{2+} signal in glomerular mesangial cells by releasing Ca^{2+} from intracellular stores. J. Biol. Chem. 266, 4295–4302.

Ford-Hutchinson, A.W., Bray, M.A., Doig, M.V., Shipley, M.E. and Smith, M.J.H. (1980). Leukotriene B, a potent chemokinetic and aggregating substance released from polymorphonuclear leukocytes. Nature 286, 264–265

Fu, J.Y., Masferrer, J.L., Seibert, K., Raz, A. and Needleman, P. (1990). The induction and suppression of prostaglandin H_2 synthase (cyclooxygenase) in human monocytes. J. Biol. Chem. 265, 16737–16740.

Fukushima, M., Kato, K., Ota, K., Arai, Y. and Narumiya, S. (1982). 9-Deoxy-Δ^9-prostaglandin D_2: a prostaglandin D_2 derivative with potent antineoplastic and weak smooth muscle contracting activities. Biochem. Biophys. Res. Commun. 109, 626–633.

Glasgow, W.C., Harris, T.M. and Brash, A.R. (1986). A short-chain aldehyde is a major lipooxygenase product in arachidonic acid stimulated porcine leukocytes. J. Biol. Chem. 261, 200–204.

Goerig, M., Habenicht, A.J.R., Heitz, R., Zeh, W., Katus, H., Kommerell, B., Ziegler, R. and Glomset, J.A. (1987). sn-1,2-Diacylglycerols and phorbol diester stimulate thromboxane synthesis by de novo synthesis of prostaglandin H synthase in human promyelocytic leukemia cells. J. Clin. Invest. 79, 903–907.

Goetzl, E.J. (1981). Selective feed-back inhibition of the 5-lipoxygenase of arachidonic acid in human T-lymphocytes. Biochem. Biophys. Res. Commun. 101, 344–350.

Goetzl, E.J. and Sun, F.F. (1979). Generation of unique monohydroxyeicosatetraenoic acids from arachidonic acid by human neutrophils. J. Exp. Med. 150, 406–411.

Goppelt-Struebe, M., Wolter, D. and Resch, K. (1989). Glucocorticoids inhibit prostaglandin synthesis not only at the level of phospholipase A_2 but also at the level of cyclooxygenase PGE isomerase. Br. J. Pharmacol. 98, 1287–1295.

Goulding, N.J., Godolphin, J.L., Sharland, P.R., Peers, S.H., Sampson, M., Maddison, P.J. and Flower, R.J. (1990). Anti-inflammatory lipocortin 1 production by peripheral blood leukocytes in response to hydrocortisone. Lancet i, 1416–1418.

Granstrom, E. (1971). Metabolism of prostaglandin $F_{2\alpha}$ in guinea pig lung. Eur. J. Biochem. 20, 451–458.

Granstrom, E. (1972). On the metabolism of prostaglandin $F_{2\alpha}$ in female subjects: structures of two metabolites in blood. Eur. J. Biochem. 27, 462–469.

Granstrom, E. and Samuelsson, B. (1971). On the metabolism of prostaglandin $F_{2\alpha}$ in female subjects. J. Biol. Chem. 246, 5254–5263.

Greaves, M.W. and McDonald-Gibson, W. (1972). Prostaglandin biosynthesis by human skin and its inhibition by corticosteroids. Br. J. Pharmacol. 46, 172–175.

Grodzinska, L. and Dembinska-Kiec, A. (1977). Hydrocortisone and the release of prostaglandins from the spleen. Prostaglandins 13, 125–129.

Gryglewski, R.J., Panczenko, B., Korbut, R., Grodzinska, L. and Ocetkiewicz, A. (1978). Corticosteroids inhibit prostaglandin release from perfused mesenteric blood vessels of rabbit and from perfused lungs of sensitized guinea pig. Prostaglandins 10, 343–355.

Gschwendt, M., Furstenberger, G., Kittstein, W., Besemfelder, E., Hull, W.E., Hagedorn, H., Opferkuch, H.J. and Marks, F. (1986). Generation of the arachidonic acid metabolite 8-HETE by extracts of mouse skin treated with phorbol ester

in vivo. Identification by ^{1}H-NMR spectroscopy. Carcinogenesis 7, 449–455.

Habenicht, A.J., Glomset, J.A., King, W.C., Nist, C., Mitchell, C.D. and Ross, R. (1981). Early changes in phosphatidylinositol and arachidonic acid metabolism in quiescent 3T3 cells stimulated to divide by platelet-derived growth factor. J. Biol. Chem. 256, 12329–12335.

Habenicht, A.J., Goerig, M., Grulich, J., Rothe, D., Gronwald, R. Loth, U., Schettler, G., Kommerell, B. and Ross, R. (1985). Human platelet-derived growth factor stimulates prostaglandin synthesis by activation and rapid de novo synthesis of cyclooxygenase. J. Clin. Invest. 75, 1381–1387.

Habenicht, A.J., Salbach, P., Goerig, M., Zeh, W., Hanssen-Timmen, U., Blattner, C., King, W.C. and Glomset, J.A., (1990). The LDL receptor pathway delivers arachidonic acid for eicosanoid formation in cells stimulated by platelet-derived growth factor. Nature 345, 634–636.

Hada, T., Ueda, N., Takahashi, Y. and Yamamoto, S. (1991). Catalytic properties of human platelet 12-lipoxygenase as compared with the enzymes of other origins. Biochim. Biophys. Acta 1083, 89–93.

Haeggstrom, J., Radmark, O. and Fitzpatrick, F.A. (1985). Leukotriene A_4-hydrolase activity in guinea pig and human liver. Biochim. Biophys. Acta 835, 378–384.

Haeggstrom, J., Meijer, J. and Radmark, O. (1986). Leukotriene A_4. Enzymatic conversion into 5,6-dihydroxy-7,9,11,14-eicosatetraenoic acid by mouse liver cytosolic epoxide hydrolase. J. Biol. Chem. 261, 6332–6337.

Hamberg, M. (1972). Inhibition of prostaglandin synthesis in man. Biochim. Biophys. Res. Commun. 49, 720–726.

Hamberg, M. and Fredholm, B.B. (1976). Isomerization of prostaglandin H_2 into prostaglandin D_2 in the presence of serum albumin. Biochim. Biophys. Acta 431, 189–193.

Hamberg, M. and Israelsson, U. (1970). Metabolism of prostaglandin E_2 in guinea pig liver. J. Biol. Chem. 245, 5107–5114.

Hamberg, M. and Samuelsson, B. (1971). On the metabolism of prostaglandins E_1 and E_2 in man. J. Biol. Chem. 246, 6713–6721.

Hamberg, M. and Samuelsson, B.M. (1974). Prostaglandin endoperoxides. Novel transformations of arachidonic acid in human platelets. Proc. Natl Acad. Sci. USA 71, 3400–3404.

Hamberg, M., Svensson, J. and Samuelsson, B. (1974). Prostaglandin endoperoxides. A new concept concerning the mode of action and release of prostaglandins. Proc. Natl Acad. Sci. USA 71, 3824–3828.

Hamberg, M., Svensson, J. and Samuelsson, B. (1975). Thromboxanes. A new group of biologically active compounds derived from prostaglandin endoperoxides. Proc. Natl Acad. Sci. USA 72, 2994–2998.

Hammarstrom, S. (1982). Leukotriene formation by mastocytoma and basophilic leukemia cells. Prog. Lipid Res. 20, 89–95.

Hammarstrom, S. (1983). Leukotrienes. Annu. Rev. Biochem. 52, 355–377.

Hammarstrom, S. and Falardeau, P. (1977). Resolution of prostaglandin endoperoxide synthase and thromboxane synthase of human platelets. Proc. Natl Acad. Sci. USA 74, 3691–3695.

Hansch, G.M., Seitz, M., Martinotti, G., Betz, M., Rauterberg, E.W. and Gemsa, D. (1984). Macrophages release arachidonic acid, prostaglandin E_2, and thromboxane in response to late complement components. J. Immunol. 133, 2145–2150.

Hansson, G., Lindgren, J.A., Dahlen, S.E., Hedqvist, P. and Samuelsson, B. (1981). Identification and biological activity of novel ω-oxidized metabolites of leukotriene B_4 from human leukocytes. FEBS Lett. 130, 107–112.

Harris, P. and Ralph, P. (1985). Human leukemic models of myelomonocytic development. A review of the HL-60 and U937 cell lines. J. Leukocyte Biol. 37, 407–422.

Hass, R., Bartels, H., Topley, N., Hadam, M., Kohler, L., Goppelt-Struebe, M. and Resch, K. (1989). TPA-induced differentiation and adhesion of U937 cells: changes in ultrastructure, cytoskeletal organization, and expression of cell surface antigens. Eur. J. Cell Biol. 48, 282–293.

Hassam, A.G., Willis, A.L., Denton, J.P., Stevens, P. and Crawford, M.A. (1979). The effect of essential fatty acid-deficient diet on the levels of prostaglandins and their fatty acid precursors in the rabbit brain. Lipids 14, 78–80.

Haye, B., Champion, S. and Jacuemin, C. (1974). Existence of two pools of prostaglandins during stimulation of the thyroid by TSH. FEBS Lett. 41, 89–93.

Heavey, D.J., Barrow, S.E., Hickling, N.E. and Ritter, J.M. (1985). Aspirin causes short-lived inhibition of bradykinin-stimulated prostacyclin production in man. Nature 318, 186–188.

Hecker, M., Hatzelmann, A. and Ullrich, V. (1987a). Preparative HPLC purification of prostaglandin endoperoxides and isolation of novel cyclooxygenase-derived arachidonate metabolites. Biochem. Pharmacol. 36, 851–855.

Hecker, M., Ullrich, V., Fischer, C. and Meese, C.O. (1987b). Identification of novel arachidonic acid metabolites formed by prostaglandin H synthase. Eur. J. Biochem. 169, 113–123.

Hensby, C.N. (1974). The enzymatic conversion of prostaglandin D_2 to prostaglandin $F_{2\alpha}$. Prostaglandins 8, 369–375.

Higgs, G.A., Moncada, S. and Vane, J.R. (1984). Eicosanoids in inflammation. Ann. Clin. Res. 16, 287–299.

Hirai, A., Yoshida, S., Nishimura, M., Seki, K., Tamura, Y.U. and Yoshida, S. (1990). Role of expoxygenase metabolites of arachidonic acid in intracellular signal transduction. Adv. Prostaglandin Thromboxane Leukotriene Res. 21, 827–830.

Hla, T. and Neilson, K. (1992). Human cyclooxygenase-2 cDNA. Proc. Natl Acad. Sci. USA 89, 7384–7388.

Holmsen, H. (1987). In "Handbook of the Eicosanoids: Prostaglandins and Related Lipids", Vol I, Part A (ed A.L. Willis), pp 119–131. CRC Press, Boca Raton

Holtzman, M.J., Turk, J. and Shornick, L.P. (1992). Identification of a pharmacologically distinct prostaglandin H synthase in cultured epithelial cells. J. Biol Chem. 267, 21438–21445.

Hong, S.L.C. and Levine, L. (1976). Inhibition of arachidonic acid release from cells as the biochemical action of anti-inflammatory corticosteroids. Proc. Natl Acad. Sci. USA 73, 1730–1734.

Hopkins, N.K., Oglesby, T.D., Bundy, G.L. and Gorman, R.R. (1984). Biosynthesis and metabolism of 15-hydroperoxy-5,8,11,13-eicosatetraenoic acid by human umbilical vein endothelial cells. J. Biol. Chem. 259, 14048–14053.

Horrobin, D.F. (1980). The regulation of prostaglandin biosynthesis: negative feedback mechanics and the selective control of formation of 1 and 2 series prostaglandins. Relevance to inflammation and immunity. Med. Hypotheses 6, 687–709.

Hughes, M.A. and Brash, A.R. (1991). Investigation of the mechanism of biosynthesis of 8-hydroxyeicosatetraenoic acid in mouse skin. Biochim. Biophys. Acta 1081, 347–354.

Hunter, J.A., Finkbeiner, W.E., Nadel, J.A., Goetzl, E.J. and Holtzman, M.J. (1985). Predominant generation of 15-lipoxygenase metabolites of arachidonic acid by epithelial cells from human trachea. Proc. Natl Acad. Sci. USA 82, 4633–4637.

Jones, R.L. (1972). 15-Hydroxy-9-oxoprosta-11,13-dienoic acid as the product of a prostaglandin isomerase. J. Lipid Res. 13, 511–518.

Jones, R.L., Kerry, P.J., Poyser, N.L., Walker, I.C. and Wilson, N.H. (1978). The identification of triydroxyeicosatrienoic acids as products from the incubation of arachidonic acid with washed blood platelets. Prostaglandins 16, 583–589.

Jubiz, W., Radmark, O., Malmsten, C., Hansson, G.H., Lindgren, J. A., Palmblad, J., Uden, A.M. and Samuelsson, B. (1982). A novel leukotriene produced by stimulation of leukocytes with formylmethionylleucylphenylalanine. J. Biol. Chem. 257, 6106–6110.

Juhlin, L. and Michaelsson, G. (1969). Cutaneous vascular reactions to prostaglandins in healthy subjects and in patients with urticaria and atopic dermatitis. Acta Dermato-Venereol. 49, 251–261.

Kato, K. and Askenase, P.W. (1984). Reconsitution of an inactive antigen-specific T-cell suppressor factor by incubation of the factor with prostaglandins. J. Immunol. 133, 2025–2031.

Kiesel, L., Przylipiak, A., Rabe, T., Przylipiak, M. and Runnebaum, B. (1987). Arachidonic acid and its lipoxygenase metabolites stimulate prolactin release in superfused pituitary cells. Hum. Reprod. 2, 281–285.

Kloeze, J. (1970). Prostaglandins and platelet aggregation in vivo. I. Influence of PGE_1 and omega-homo-PGE_1 on transient thrombocytopenia and PGE_1 on the LD_{50} of ADP. Thromb. Diath. Haemorrh. 23, 286–292.

Kocsis, J.J. Hernandovich, J., Silver, M.J., Smith, J.B. and Ingerman, C. (1973). Duration of inhibition of platelet prostaglandin formation and aggregation by ingested aspirin or indomethacin. Prostaglandins 3, 141–144.

Koehler, L., Hass, R., Wessel, K., DeWitt, D.L., Kaever, V., Resch, K. and Goppelt-Struebe, M. (1990). Altered arachidonic acid metabolism during differentiation of the human monoblastoid cell line U937. Biochim. Biophys. Acta 1042, 395–403.

Korn, J.H., Halushka, P.V. and LeRoy, E.C. (1980). Mononuclear cell modulation of connective tissue function. Suppression of fibroblast growth by stimulation of endogenous prostaglandin production. J. Clin. Invest. 65, 543–554.

Kramer, R.M., Hession, C., Johansen, B., Hayes, G., McGray, P., Chow, E.P., Tizard, R. and Pepinsky, R.B. (1989). Structure and properties of a human non-pancreatic phospholipase A_2. J. Biol. Chem. 264, 5768–5775.

Kuhn, D.C. and Willis, A.L. (1973). Prostaglandin E_2, inflammation, and pain threshold in rat paws. Br. J. Pharmacol. 49, 183P–184P.

Kuhn, H., Wiesner, R. and Stender, H. (1984). The formation of products containing a conjugated tetraenoic system by pure reticulocyte lipoxygenase. FEBS Lett. 177, 255–259.

Kuhn, H., Belkner, J., Wiesner, R. and Brash, A.R. (1990). Oxygenation of biological membranes by the pure reticulocyte lipoxygenase. J. Biol. Chem. 265, 18351–18361.

Kujubu, D.A. and Herschman, H.A. (1992). Dexamethasone inhibits mitogen induction of the TIS10 prostaglandin synthase cyclooxygenase gene. J. Biol. Chem. 267, 7991–7994.

Kujubu, C.A., Fletcher, B.S., Varnum, B.C., Lim, R.W. and Herschman, H.R. (1991). TIS10, a phorbol ester tumor promoter-inducible mRNA from Swiss 3T3 cells, encodes a novel prostaglandin and synthase/cyclooxygenase homologue. J. Biol. Chem. 266, 12866–12872.

Kumlin, E., Ohlson, E., Bjorck, T., Hamberg, M., Granstrom, E., Dahlen, B., Zetterstrom, O. and Dahlen, S.-E. (1990). 15(S)-hydroxyeicosatetraenoic acid (15-HETE) is the major arachidonic acid metabolite in human bronchi. Adv. Prostaglandin Thomboxane Leukotriene Res. 21, 441–444.

Kurland, J.I., Bockman, R.S., Broxmeyer, H.E. and Moore, M.A.S. (1978). Limitation of excessive myelopoiesis by the intrinsic modulation of macrophage-derived prostaglandin E. Science 199, 552–555.

Lagarde, M., Guichardant, M. and Dechavanne, M. (1982). Human platelet PGE_1 and dihomogammalinolenic acid. Comparison to PGE_2 and arachidonic acid. Prog. Lipid Res. 20, 439–443.

Lands, W.E. and Kulmacz, R.J. (1986). The regulation of the biosynthesis of prostglandins and leukotrienes. Prog. Lipid Res. 25, 105–109.

Lands, W.E.M., Le Tellier, P.R., Rome, L.H. and Vanderhoek, J.Y. (1973). Inhibition of prostglandin biosynthesis. Adv. Biosci. 9, 15–28.

Laneuville, O., Corey, E.J., Couture, R. and Pace-Asciak, C.R. (1991). Hepoxilin A_3 (HxA_3) is formed by the rat aorta and is metabolized into HXA_3-C, a glutathione conjugate. Biochim. Biophys. Acta 1084, 60–88.

Lee, C.W., Lewis, R.A., Corey, E.J., Barton, A., Oh, H., Tauber, A.I. and Austen, K.F. (1982). Oxidative inactivation of leukotriene C_4 stimulated human polymorphonuclear leukocytes. Proc. Natl Acad. Sci. USA 79, 4166–4170.

Lee, C.W., Lewis, R.A., Tauber, A.I., Mehrota, M., Corey, E.J. and Austen, K.F. (1983). The myeloperoxidase-dependent metabolism of leukotrienes C_4, D_4, and E_4 to 6-*trans*-leukotriene B_4 diastereomers and the subclass-specific S-diastereomeric sulfoxides. J. Biol. Chem. 258, 15004–15010.

Lee, S.C. and Levine, L. (1975). Prostaglandin metabolism. II. Identification of two 15-hydroxyprostaglandin dehydrogenase types. J. Biol. Chem. 250, 548–552.

Lee, S.W., Tsou, A.P., Chan, H., Thomas, J., Petrie, K., Eugie, E.M. and Allison, A.C. (1988). Glucocorticoids selectively inhibit the transcription of the interleukin 1-beta genes and decrease the stability of interleukin 1-beta mRNA. Proc. Natl Acad. Sci. USA 85, 1204–1208.

Levine, J.D., Lam, D., Taiwo, Y.O., Donatoni, P. and Goetzl, E.J. (1986). Hyperalgesic properties of 15-lipoxygenase products of arachidonic acid. Proc. Natl Acad. Sci. USA 83, 5331–5334.

Lewis, R.A., Austen, K.F., Drazen, J.M., Clark, D.A. and Corey, E.J. (1980a). Slow reacting substance of anaphylaxis: identification of leukotrienes C-1 and D *in vivo* and *in vitro*. Proc. Natl Acad. Sci. USA 77, 4354–4358.

Lewis, R.A., Austen, K.F., Drazen, J.M., Clark, D.A., Marfat, A. and Corey, E.J. (1980b). Slow reacting substances of anaphylaxis: identification of leukotriene C-1 and D from human and rat sources. Proc. Natl Acad. Sci. USA 77, 3710–3714.

Lewis, R.A., Drazen, J.M., Austen, K.F., Clark, D.A. and Corey, E.J. (1980c). Identification of the C(6)-S-conjugate of leukotriene A with cysteine as a naturally-occurring slow reacting substance of anaphylaxis (SRS-A). Importance of the 11-*cis* geometry for biological activity. Biochem. Biophys. Res. Commun. 96, 271–277.

Lewis, R.A., Lee, C.W., Levine, L., Morgan, R.A., Weiss, J.W., Drazen, J.M., Oh, M., Hoover, D., Corey, E.J. and Austen, K.F. (1983). Biology of the C-6 sulfidopeptide leukotrienes. Adv. Prostaglandin Thomboxane Leukotriene Res. 11, 15–25.

Lianos, E.A., Rahman, M.A. and Dunn, M.J. (1985). Glomerular arachidonate lipoxygenation in rat nephrotoxic serum nephritis. J. Clin. Invest. 76, 1355–1359.

Lin, A.H., Bienkowski, M.J. and Gorman, R.R. (1989). Regulation of prostaglandin H synthase on RNA levels and prostaglandin biosynthesis by platelet-derived growth factor. J. Biol. Chem. 264, 17379–17383.

Lindgren, J.A., Hansson, G. and Samuelsson, B. (1981). Formation of hydroxylated eicosatetraenoic acids in preparations of human polymorphonuclear leukocytes. FEBS Lett. 128, 329–335.

Lister, M.D., Deems, R.A., Watanabe, Y., Ulevitch, R.J. and Dennis, E.A. (1988). Kinetic analysis of the Ca^{2+}-dependent, membrane-bound, macrophage phospholipase A_2 and the effects of arachidonic acid. J. Biol. Chem. 263, 7506–7513.

Lister, M.D., Glaser, K.B., Ulevitch, R.J. and Dennis, E.A. (1989). Inhibition studies on the membrane-associated phospholipase A_2 *in vitro* and prostaglandin E_2 production *in vivo* of the macrophage-like P388D1 cell. Effects of manoalide, 7,7-dimethyl-5,8-eicosadienoic acid, and p-bromophenacyl bromide. J. Biol. Chem. 264, 8520–8528.

Liston, T.E. and Roberts, L.J. II (1985). Metabolic fate of radiolabelled prostaglandin D_2 in a normal human male volunteer. J. Biol. Chem. 260, 13172–13180.

Lundberg, U., Radmark, O., Malmsten, C. and Samuelsson, B. (1981). Transformation of 15-hydroperoxy-5,9,11,13-eicosatetraenoic acid into novel leukotrienes. FEBS Lett. 126, 127–132.

McGiff, J.C., and Carroll, M.A. (1991). Cytochrome P450-dependent arachidonate metabolites, renal function and blood pressure regulation. Adv. Prostaglandin Thomboxane Leukotriene Res. 21, 675–682.

McGiff, J.C., Terragno, N.A., Strand, J.C., Lee, J.B., Lonigro, A.J. and Ng, K.K.F. (1969). Selective passage of prostaglandins across the lung. Nature 223, 742–745.

Maas, R.L., Brash, A.R. and Oates, J.A. (1981). A second pathway of leukotriene biosynthesis in porcine leukocytes. Proc. Natl Acad. Sci. USA 78, 5523–5527.

Mahmud, I., Smith, D.L., Willis, A.L., Whyte, M.A., Nelson, J.T., Cho, D., Tokes, L.G. and Alvarez, R. (1984). On the identification and biological properties of prostaglandin J_2. Prostaglandins Leukotrienes Med. 16, 131–146.

Maier, J.A., Hla, T. and Maciag, T. (1990). Cyclooxygenase is an immediate–early gene induced by interleukin-1 in human endothelial cells. J. Biol. Chem. 265, 10805–10808.

Makoul, G.T., Robinson, D.R., Bhalla, A.K. and Glimcher, L.H. (1985). Prostaglandin E_2 inhibits the activation of cloned T cell hybridomas. J. Immunol. 134, 2645–2650.

Marcus, A.J., Weksler, B.B., Jaffe, E.A. and Broekman, M.J. (1980). Synthesis of prostacyclin from platelet-derived endoperoxides by cultured human endothelial cells. J. Clin. Invest. 66, 979–986.

Marcus, A.J., Safier, L.B., Ullman, H.L., Broekman, M.J., Islam, N., Oglesby, T.D. and Gorman, R.R. (1984). 12S,20-Dihydroxyeicosatetraenoic acid: a new eicosanoid synthesized by neutrophils from 12S-hydroxyeicosatetraenoic acid produced by thrombin- or collagen-stimulated platelets. Proc. Natl Acad. Sci. USA 81, 903–907.

Masferrer, J.L., Murphy, R.C., Pagano, P.J., Dunn, M.W. and Lanaido-Schwartzman, M. (1989). Ocular effects of a novel cytochrome P-450-dependent arachidonic acid metabolite. Invest. Ophthalmol. Vis. Sci. 30, 451–460.

Masferrer, J., Zweifel, B.S., Selbert, K. and Needleman, P. (1990). Selective regulation of cellular cyclooxygenase by dexamethasone and endotoxin in mice. J. Clin. Invest. 86, 1375–1379.

Masferrer, J.L., Seibert, K., Zweifel, B. and Needleman, P. (1992). Endogenous glucocorticoids regulate an inducible cyclo-oxygenase enzyme. Proc. Natl Acad. Sci. USA 89, 3917–3921.

Mead, J.F. and Willis, A.L. (1987). In "Handbook of the Eicosanoids: Prostaglandins and Related Lipids", Vol I, Part A (ed A.L. Willis), pp 85–98. CRC Press, Boca Raton

Meghji, S., Sandy, J.R., Scutt, A.M., Harvey, W. and Harris, M. (1988). Stimulation of bone resorption by lipoxygenase metabolites of arachidonic acid. Prostaglandins 36, 139–149.

Merlie, J.P., Fagan, D., Mudd, J. and Needleman, P. (1988). Isolation and characterization of the complementary DNA for sheep seminal vesicle prostaglandin endoperoxide synthase (cyclooxygenase). J. Biol. Chem. 263, 3550–3553.

Milton, A.S. and Wendlandt, S. (1971). Effects on body temperature of prostaglandins of the A, E, and F series on injection into the third ventricle of unanaesthetized cats and rabbits. J. Physiol. 218, 325–326.

Miyamoto, T., Ogino, N., Yamamoto, S. and Hayaishi, O. (1976). Purification of prostaglandin endoperoxide synthetase from bovine vesicular gland microsomes. J. Biol. Chem. 251, 2629–2636.

Morgan, E.L., Hobbs, M.V. and Weigle, W.O. (1985). Lymphocyte activation by the F_c region of immunoglobulin. I. Role of prostaglandins in the down regulation of F_c fragment-induced polyclonal antibody production. J. Immunol. 134, 2247–2253.

Morris, H.A., Taylor, G.W., Piper, P.J. and Tippins, J.R. (1980). Structure of slow-reacting substance of anaphylaxis from guinea-pig lung. Nature 285, 104–105.

Morrison, A.R. and Pascoe, N. (1981). Metabolism of arachidonate through NADPH-dependent oxygenase of renal cortex. Proc. Natl Acad. Sci. USA 78, 7375–7378.

Musser, J.H. and Kreft, A.F. (1992). 5-Lipoxygenase: properties, pharmacology, and the quinolinyl (bridged) aryl class of inhibitors. J. Med. Chem. 35, 2501–2524.

Mutsaers, J.H.G.M., Van Halbeek, H., Kamerling, J.P. and Vliegenthart, J.F.G. (1985). Determination of the structure of the chains of prostaglandin endoperoxide synthase from sheep. Eur. J. Biochem. 147, 569–574.

Nelson, P.H. (1989). In "Handbook of the Eicosanoids: Prostaglandins and Related Lipids", Vol II (ed A.L. Willis), pp 59–133. CRC Press, Boca Raton.

Ney, P., Braun, M., Szymanski, C., Bruch, L. and Schror, K. (1991). Antiplatelet, antineutrophil and vasodilating properties of 13,14-dihydro-PGE$_1$ (PGE0) – an *in vivo* metabolite of PGE$_1$ in man. Eicosanoids 4, 177–184.

Nishimura, M., Hirai, A., Omura, M., Tamura, Y. and Yoshida, S. (1989). Arachidonic acid metabolites by cytochrome P-450 dependent monooxygenase pathway in bovine adrenal fasciculata cells. Prostaglandins 38, 413–430.

Nugteren, D.H. (1975). Arachidonate lipoxygenase in blood platelets. Biochim. Biophys. Acta 380, 299–307.

Nugteren, D.H. (1977). In "Prostaglandins in Hematology" (eds M.J. Silver, J.B. Smith and J.J. Kocsis), pp 11–25. Spectrum Publications, Jamaica, NY.

O'Banion, M.K., Sadowski, H.B., Winn, V. and Young, D.A. (1991). A serum- and glucocorticoid-regulated 4-kilobase mRNA encodes a cyclooycgenase-related protein. J. Biol. Chem. 266, 23261–23267.

O'Banion, M.K., Winn, V.D. and Young, D.A. (1992). cDNA cloning and functional activity of a glucocorticoid-regulated inflammatory cyclooxygenase. Proc. Natl Acad. Sci. USA 89, 4888–4892.

O'Brien, J.R. (1968). Effect of anti-inflammatory agents on platelets. Lancet i, 894–895.

O'Sullivan, M.G., Chilton, F.H., Huggins, E.M. and McCall, C.E. (1992). Lipopolysaccharide priming of alveolar macrophages for enhanced synthesis of prostanoids involves induction of a novel prostaglandin H synthase. J. Biol. Chem. 267, 14547–14550.

Oliw, E.H. (1984). Metabolism of 5(6)-oxidoeicosatrienoic acid by ram seminal vessels. Formation of two stereoisomers of 5-hydroxyprostaglandin I$_1$. J. Biol. Chem. 259, 2716–2721.

Oliw, E.H. and Oates, J.A. (1981). Oxygenation of arachidonic acid by hepatic microsomes of the rabbit. Biochim. Biophys. Acta 666, 327–340.

Oliw, E.H., Guengerich, F.P. and Oates, J.A. (1982). Oxygenation of arachidonic acid by hepatic monooxygenases. Isolation and metabolism of four epoxide intermediates. J. Biol. Chem. 257, 3771–3781.

Orning, L. and Hammarstrom, S. (1983). Isolation and characterization of 15-hydroxylated metabolites of leukotriene C$_4$. FEBS Lett. 153, 253–256.

Orning, L., Norin, E., Gustafsson, B. and Hammarstrom, S. (1986). *In vivo* metabolism of leukotriene C$_4$ in germ free and conventional rats. Fecal excretion of N-acetylleukotriene E$_4$. J. Biol. Chem. 261, 766–771.

Orning, L., Keppler, A., Midtvedt, T. and Hammarstrom, S. (1988). *In vivo* formation of omega-oxidized metabolite of leukotriene C$_4$ in the rat. Prostaglandins 35, 493–501.

Pace-Asciak, C.R. and Martin, J.M. (1984). Hepoxilin, a new family of insulin secretagogues formed by intact pancreatic islets. Prostaglandins Leukotrienes Med. 16, 173–180.

Pace-Asciak, C.R., Mizuno, K. and Yamamoto, S. (1981). Formation of 8,11,12-trihydroxyeicosatrienoic acid by a rat lung high speed supernatant fraction. Biochim. Biophys. Acta 665, 352–354.

Pace-Asciak, C.R., Granstrom, E. and Samuelsson, B. (1983). Arachidonic acid epoxides. Isolation and structure of two hydroxy epoxide intermediates in the formation of 8,11,12- and 10,11,12-trihydroxyeicosatrienoic acids. J. Biol. Chem. 258, 6835–6840.

Pace-Asciak, C.R., Martin, J.M., Corey, E.J. and Su, W.-G. (1985). Endogenous release of hepoxilin A, from isolated perfused pancreatic islets of langerhans. Biochem. Biophys. Res. Commun. 128, 942–946.

Pace-Asciak, C.R., Klein, J. and Speilberg, S.P. (1986). Metabolism of leukotriene A$_4$ into C$_4$ by human platelets. Biochim. Biophys. Acta 877, 68–74.

Perrin, P., Zirolli, J., Stene, D.O., Lellouche, J.P., Beaucourt, J.P. and Murphy, R.C. (1989). *In vivo* formation of beta-oxidized metabolites of leukotriene E$_4$ in the rat. Prostaglandins 37, 53–60.

Phan, S.H., McGarry, B.M., Loeffler, K.M. and Kunkel, S.L. (1987). Regulation of macrophage-derived fibroblast growth factor released by arachidonate metabolites. J. Leukocyte Biol. 41, 106–113.

Piomelli, D., Volterra, A., Dale, N., Siegelbaum, S.A., Kandel,

E.R., Schwartz, J.H. and Belardetti, F. (1987). Lipoxygenase metabolites of arachidonic acid as second messengers for presynaptic inhibition of *Aplysia* sensory cells. Nature 328, 38–43.

Powell, W.S. (1984). Properties of leukotriene B_4 20-hydroxylase from polymorphonuclear leukocytes. J. Biol. Chem. 259, 3082–3089.

Powell, W.S. (1986). Novel pathway for the metabolism of 6-*trans*-leukotriene B_4 by human polymorphonuclear leukocytes. Biochem. Biophys. Res. Commun. 136, 707–712.

Proctor, K.G., Falck, J.R. and Capdevila, J. (1987). Intestinal validation by epoxyeicosatrienoic acids: arachidonic acid metabolites produced by a cytochrome P450 monooxygenase. Circ. Res. 60, 50–59.

Proctor, K.G., Capdevila, J.H., Falck, J.R., Fitzpatrick, F.A., Mullane, R.M. and McGiff, J.C. (1989). Cardiovascular and renal actions of cytochrome P-450 metabolite of arachidonic acid. Blood Vessels 26, 53–64.

Pueringer, R.J. and Hunninghake, G.W. (1992). Lipopolysaccharide stimulates *de novo* synthesis of PGH synthase in human alveolar macrophages. Am. J. Physiol. 262, L78–L85.

Pugliese, G., Spokas, E.G., Marcinkiewicz, E. and Wong, P.Y.-K. (1985). Hepatic transformation of prostaglandin D_2 to a new prostanoid, $9_\alpha,11\beta$-prostaglandin F_2, that inhibits platelet aggregation and constricts blood vessels. J. Biol. Chem. 260, 14621–14625.

Puustinen, T., Webber, S.E., Nicolaou, K.C., Haeggstrom, J., Serhan, C.N. and Samuelsson, B. (1986). Evidence for a 5(6)-epoxytetraene intermediate in the biosynthesis of lipoxins in human leukocytes. FEBS Lett. 207, 127–132.

Raz, A., Wyche, A., Fu, J., Seibert, K. and Needleman, P. (1990). Regulation of prostanoid synthesis in human fibroblasts and human blood monocytes by interleukin-1, endotoxin, and glucocorticoids. Adv. Prostaglandin Thomboxane Leukotriene Res. 20, 22–27.

Reynolds, L.J., Hihelich, E.D. and Dennis, E.A. (1991). Inhibition of venom phospholipases A_2 by manoalide and manoalogue. Stoichiometry of incorporation. J. Biol. Chem. 266, 16512–16517.

Roberts, L.J. II (1987). In "CRC Handbook of the Eicosanoids: Prostaglandins and Related Lipids", Vol 1 (ed A.L. Willis), pp 233–244. CRC Press, Boca Raton.

Roberts, L.J. II and Sweetman, B.J. (1985). Metabolic fate of endogenously synthesized prostaglandin D_2 in a human female with mastocytosis. Prostaglandins 30, 383–400.

Roberts, L.J. II, Sweetman, B.J., Morgan, J.L., Payne, N.A. and Oates, J.A. (1977a). Identification of the major urinary metabolite of thromboxane B_2 in the monkey. Prostaglandins 13, 631–647.

Roberts, L.J. II, Sweetman, B.J., Payne, N.A. and Oates, J.A. (1977b). Metabolism of thromboxane B_2 in man. Identification of the major urinary metabolite. J. Biol. Chem. 252, 7415–7417.

Roberts, L.J. II, Sweetman, B.J. and Oates, J.A. (1978). Metabolism of thromboxane B_2 in the monkey. J. Biol. Chem. 253, 5305–5318.

Rosenkranz, B., Fischer, C., Reiman, I., Weimer, K.E., Beck, G. and Frolich, J.C. (1980). Identification of the major metabolite of prostacyclin and 6-ketoprostaglandin $F_{1\alpha}$ in man. Biochim. Biophys. Acta 619, 207–213.

Rosenkranz, B., Fischer, C. and Frolich, J.C. (1981). Prostacyclin metabolites in human plasma. Clin. Pharmacol. Ther. 29, 420–424.

Roth, G.J., Stanford, N. and Majerus, P.W. (1975). Acetylation of prostaglandin synthase. Proc Natl Acad. Sci. USA 72, 3073–3076

Roth, G.J., Machuga, E.T. and Ozols, J. (1983). Isolation and covalent structure of the aspirin-modified, active-site region of prostaglandin synthetase. Biochemistry 22, 4672–4675.

Rothwell, N.J. and Flower, R. (1992). Lipocortin-1 exhibits novel actions providing clinical opportunities. Trends Pharmacol. Sci. 13, 45–46.

Rouzer, C.A., Scott, W.A., Hamill, A.L., Liu, F.T., Katz, D.H. and Cohn, Z.A. (1982). Secretion of leukotriene C and other arachidonic acid metabolites by macrophages challenged with immunoglobulin E immune complexes. J. Exp. Med. 156, 1077–1086.

Salem, N. Jr (1989). In "New Protective Roles for Selected Nutrients" (eds G.A. Spiller and J. Scala), pp 39–108. Alan R. Liss, New York.

Samuelsson, B. (1965). On the incorporation of oxygen in the conversion of 8,11,14-eicosatrienoic acid into prostaglandin E_1. J. Am. Chem. Soc. 87, 3011–3013.

Samuelsson, B. (1982). Leukotrienes: a novel group of compounds including SRS-A. Prog. Lipid Res. 20, 23–30.

Samuelsson, B. (1983a). Leukotrienes: a new class of mediators of immediate hypersensitivity reaction and inflammation. Adv. Prostaglandin Thromboxane Leukotriene Res. 11, 1–26.

Samuelsson, B. (1983b). Leukotrienes: mediators of immediate hypersensitivity reactions and inflammation. Science 220, 568–575.

Samuelsson, B. (1985). Leukotrienes and related compounds. Adv. Prostaglandin Thomboxane Leukotriene Res. 15, 1–9.

Sanduja, R., Loose-Mitchell, D. and Wu, K.K. (1991). Inhibition of the *de novo* synthesis and message expression of prostaglandin H synthase by salicylates. Adv. Prostaglandin Thomboxane Leukotriene Res. 21A, 149–156.

Sano, H., Hla, T., Maier, J.A.M., Crofford, L., Case, J.P., Maciag, T. and Wilder, R.L. (1992). *In vivo* cyclooxygenase expression of synovial tissues of patients with rheumatoid arthritis and osteoarthritis and rats with adjuvant and streptoccocal cell wall arthritis. J. Clin. Invest. 89, 97–108.

Schewe, T., Kuhn, M. and Rapoport, S.M. (1986). Positional specificity of lipoxygenases and their suitability for testing potential drugs. Prostaglandins Leukotrienes Med. 23, 155–160.

Schlegel, W., Demers, L.M., Hildebrandt, S.T. and Behrman, H.R. (1974). Partial purification of human placental 15-hydroxyprostaglandin dehydrogenase: kinetic properties. Prostaglandins 5, 417–433.

Schwartzman, M.L., Balazy, M., Masferrer, J., Abraham, N.G., McGiff, J.C. and Murphy, R.C. (1987). 12(*R*)-Hydroxyeicosatetraenoic acid: a cytochrome-P-450-dependent arachidonate metabolite that inhibits Na, K^+-ATPase in the cornea. Proc. Natl Acad. Sci. USA 84, 8125–8129.

Schwartzman, M.L., Falck, J.R., Yadagiri, P. and Escalante, B. (1989). Metabolism of 20-hydroxyeicosatetraenoic acid by cyclooxygenase. Formation and identification of novel endothelium-dependent vasoconstrictor metabolites. J. Biol. Chem. 264, 11658–11662.

Schweitzer, P., Madamba, S. and Siggins, G.R. (1990). Arachidonic acid metabolites as mediators of somatostatin-induced increase of neuronal M-current. Nature 346, 464–467.

Seibert, K., Sheller, J.R. and Roberts, L.J. II (1987). $9_\alpha,11\beta$-Prostaglandin F_2: formation and metabolism by human lung and contractile effects on human bronchial smooth muscle. Proc. Natl Acad. Sci. USA 84, 256–270.

Serhan, C.N., Hamberg, M. and Samuelsson, B. (1984). Lipoxins: novel series of biologically active compounds formed from arachidonic acid in human leukocytes. Proc. Natl Acad. Sci. USA 81, 5335–5339.

Serhan, C., Fahlstadius, P., Dahlen, S.E., Hamberg, M. and Samuelsson, B. (1985). Biosynthesis and biological activities of lipoxins. Adv. Prostaglandin Thomboxane Leukotriene Res. 15, 163–166.

Serhan, C.N., Hamberg, M., Samuelsson, B., Morris, J. and Wishka, D.G. (1986). On the stereochemistry and biosynthesis of lipoxin B. Proc. Natl Acad. Sci. USA 83, 1983–1987.

Shimizu, T., Radmark, O. and Samuelsson, B. (1984). Enzyme with dual lipoxygenase activities catalyzes leukotriene A_4 synthesis from arachidonic acid. Proc. Natl Acad. Sci. USA 81, 689–693.

Shimokawa, T. and Smith, W.L. (1992). Prostaglandin endoperoxide synthase. The aspirin acetylation region. J. Biol. Chem. 267, 12387–12392.

Simmons, D.L., Levy, D.B., Yannoni, Y. and Erikson, R.L. (1989). Identification of a phorbol ester-repressible v-src-inducible gene. Proc. Natl Acad. Sci. USA 86, 1178–1182.

Sirois, J. and Richards, J.S. (1992). Purification and characterization of a novel, distinct isoform of prostaglandin endoperoxide synthase induced by human chorionic gonadotropin in granulosa cells of rat preovulatory follicles. J. Biol. Chem. 267, 6382–6388.

Sirois, J., Simmons, D.L. and Richards, J.S. (1992). Hormonal regulation of messenger ribonucleic acid encoding a novel isoform of prostaglandin endoperoxide H synthase in rat preovulatory follicles. Induction *in vivo* and *in vitro*. J. Biol. Chem. 267, 11586–11592.

Smith, D.L. (1987). In "Handbook of the Eicosanoids: Prostaglandins and Related Lipids", Vol I (ed A.L. Willis), pp 47–83. CRC Press, Boca Raton.

Smith, D.L. (1989). In "Handbook of Eicosanoids: Prostaglandins and Related Lipids", Vol II (ed A.L. Willis), pp 27–34. CRC Press, Boca Raton

Smith, D.L., Stone, K.J. and Willis, A.L. (1987). In "Handbook of Eicosanoids: Prostaglandins and Related Compounds", Vol I (ed A.L. Willis), pp 245–301. CRC Press, Boca Raton.

Smith, J.B. and Willis, A.L. (1971). Aspirin selectively inhibits prostaglandin production in human platelets. Nature (New Biol.) 231, 235–237.

Smith, W.L. (1989). The eicosanoids and their biochemical mechanisms of action. Biochem. J. 259, 315–324.

Smith, W.L. (1992). Prostanoid biosynthesis and mechanisms of action. Am. J. Physiol. 263, F181–F190.

Smith, W.L. and Marnett, L.J. (1991). Prostaglandin endoperoxide synthase: structure and catalysis. Biochim. Biophys. Acta 1083, 1–17.

Smith, W.L., Marnett, L.J. and DeWitt, D.L. (1991). Prostaglandin and thromboxane biosynthesis. Pharmacacol. Ther. 49, 153–179.

Snyder, D.S., Beller, D.I. and Unanue, E.R. (1982). Prostaglandins modulate macrophage Ia expression. Nature 299, 163–165.

Snyder, G., Lattanzio, F., Yadagiri, P., Falck, J.R. and Capdevila, J. (1986). 5,6-Epoxyeicosatrienoic acid mobilizes Ca^{2+} in anterior pituitary cells. Biochem. Biophys. Res. Commun. 139, 1188–1194.

Soberman, R.J., Harper, T.W., Betteridge, D., Lewis, R.A. and Austen, K.F. (1985). Characterization and separation of the arachidonic acid 5-lipoxygenase and linoleic acid ω-6 lipoxygenase (arachidonic acid 15-lipoxygenase) of human polymorphonuclear leukocytes. J. Biol. Chem. 260, 4508–4515.

Sok, D.E., Han, C.O., Shieh, W.-R., Zhou, B.-N. and Sih, C.J. (1982). Enzymatic formation of 14,15-leukotriene A and C(14)-sulfur-linked peptides. Biochem. Biophys. Res. Commun. 104, 1363–1370.

Spotts, G.D. and Hann, S.R. (1990). Enhanced translation and increased turnover of c-myc proteins occur during differentiation of murine erythroleukemia cells. Mol. Cell Biol. 10, 3952–3964.

Sprecher, H., Van Rollins, M., Sun, F., Wyche, A. and Needleman, P. (1982). Dihomo-prostaglandins and thromboxane. A prostaglandin family from adrenic acid that may be preferentially synthesized in the kidney. J. Biol. Chem. 257, 3912–3918.

Stone, K.J., Willis, A.L., Hart, W.M., Kirtkand, S.J., Kernoff, P.B.A. and McNicol, G.P. (1979). The metabolism of dihomo-gamma-linolenic acid in man. Lipids 14, 174–180.

Sun, F.F., Armour, S.B., Bockstanz, V.R. and McGuire, J.C. (1976). Studies on 15-hydroxyprostaglandin dehydrogenase from monkey lung. Adv. Prostaglandin Thomboxane Res. 1, 163–169.

Sutor, A.H., Bowie, E.J. and Owen, C.A. Jr (1971). Effect of aspirin, sodium salicylate, and acetaminophen on bleeding. Mayo Clinic Proc. 46, 178–181.

Taffet, S.M. and Russell, S.W. (1981). Macrophage-mediated tumor cell killing. Regulation of expression of cytolytic activity by prostaglandin E. J. Immunol. 126, 424–427.

Takahashi, Y., Ueda, N. and Yamamoto, S. (1988). Two immunologically and catalytically distinct arachidonate 12-lipoxygenases of bovine platelets and leukocytes. Arch. Biochem. Biophys. 266, 613–621.

Takayasu, K., Okuda, K. and Yoshikawa, I. (1970). Fatty acid composition of human and rat adrenal lipids: occurrence of ω6 docosatrienoic acid in human adrenal cholesterol ester. Lipids 5, 743–750.

Thompson, P.A., Jelinek, D.F. and Lipsky, P.E. (1984). Regulation of human B cell proliferation by prostaglandin E_2. J. Immunol. 133, 2446–2453.

Tobias, L.D., Vane, F.M. and Paulsrud, J.R. (1975). The biosynthesis of 1a,1b-dihomo-PGE_2 and 1a,1b-dihomo-$PGF_{2\alpha}$ from acetone–pentane powder of sheep vesicular gland microsomes. Prostaglandins 10, 443–468.

Tsai, A.L., Sanduja, R. and Wu, K.K. (1991). Evidence for two pools of prostaglandin H synthase in human endothelial cells. Adv. Prostaglandin Thomboxane Leukotriene Res. 21A, 141–144

Turk, J., Maas, R.L., Brash, A.R., Roberts, L.J. and Oates, J.A. (1982). Arachidonic acid 15-lipoxygenase products from human eosinophils. J. Biol. Chem. 257, 7068–7076.

Ueda, N., Kaneko, S., Yoshimoto, T. and Yamamoto, S. (1986). Purification of arachidonate 5-lipoxygenase from porcine leukocytes and its reactivity with hydroperoxyeicosatetraenoic acids. J. Biol. Chem. 261, 7982–7988.

Ulevitch, R.J., Watanabe, Y., Sano, M., Lister, M.D., Deems, R.A. and Dennis, E.A. (1988). Solubilization, purification, and characterization of a membrane-bound phospholipase A_2 from the P388D1 macrophage-like cell line. J. Biol. Chem. 263, 3079–3085.

Vahouny, G.V., Chanderbhan, R., Bisgaier, C., Hodges, V.A. and Naghshineh, S. (1982). Essential fatty acids and adrenal steroidogenesis. Prog. Lipid Res. 20, 233–240.

Van der Ouderaa, F.J., Buytenhek, M., Nugteren, D.H. and

Van Dorp, D.A. (1977). Purification and characterization of prostaglandin endoperoxide synthetase from sheep vesicular glands. Biochim. Biophys. Acta 487, 315–331.

Van der Ouderaa, F.J., Buytenhek, M., Slikkerveer, F.J. and Van Dorp, D.A. (1979). On the haemoprotein character of prostaglandin endoperoxide synthetase. Biochim. Biophys. Acta 572, 29–41.

Vanderhoek, J.Y., Bryant, R.W. and Bailey, J.W. (1980a). 15-Hydroxy-5,8,11,13-eicosatetraenoic acid: a potent and selective inhibitor of platelet lipoxygenase. J. Biol. Chem. 255, 5996–5998.

Vanderhoek, J.Y., Bryant, R.W. and Bailey, J.M. (1980b). Inhibition of leukotriene biosynthesis by the leukocyte product 15-hydroxy-5,8,11,13-eicosatetraenoic acid. J. Biol. Chem. 255, 10064–10066.

Vanderhoek, J.Y., Tare, N.S., Bailey, J.M., Goldstein, A.L. and Pluznic, D. (1982). New role for 15-hydroxyeicosatetraenoic acid: activator of leukotriene biosyntheis in PT18 mast/basophil cells. J. Biol. Chem. 257, 12191–12195.

Van Dorp, D.A. (1971). Enzymatic conversion of all-cis-polyunsaturated fatty acids into prostaglandins. Ann. N. Y. Acad. Sci. 180, 181–199.

Van Dorp, D.A. Beerthuis, R.K., Nugteren, D.H. and Vonkeman, H. (1964). The biosynthesis of prostaglandins. Biochim. Biophys. Acta 90, 207–210.

Vane, J.R. (1969). The release and fate of vasoactive hormones in the circulation. Br. J. Pharmacol. 35, 209–242.

Vane, J.R. (1971). Inhibition of prostaglandin synthesis as a mechanism of action for aspirin-like drugs. Nature (New Biol.) 231, 232–235.

von Euler, U.S. (1936). On the specific vasodilating and plain muscle stimulating substances from accessory genital glands in man and certain animals (prostaglandins and vesiglandins). J. Physiol. (London) 88, 213–234.

von Euler, U.S. (1981). Prostaglandins, historical remarks. Prog. Lipid Res. 20, xxxi–xxxv.

Wang, J., Yuen, B.H. and Leung, P.C. (1989). Stimulation of progesterone and prostaglandin E_2 production by lipoxygenase metabolites of arachidonic acid. FEBS Lett. 244, 154–158.

Weiss, H.J. and Aledort, L.M. (1967). Impaired platelet-connective tissue reaction in man after aspirin ingestion. Lancet ii, 495–497.

Weksler, B.B. (1987). Regulation of prostaglandin synthesis in human vascular cells. Ann. N.Y. Acad. Sci. 509, 142–148.

Westlund, P. (1987). Formation of novel dihydroxy eicosatetraenoic acids in human platelets. Identification of 11,12-DHETEs and 5,12-DHETEs. Adv. Prostaglandin Thromboxane Leukotriene Res. 17A, 99–104.

Westwick, J. (1976). The effect of pulmonary metabolites of prostaglandins E_1, E_2, and $F_{2\alpha}$ on ADP-induced aggregation of human and rabbit platelets. Br. J. Pharmacol. 58, 297P–298P.

Wetterholm, A., Haeggstrom, J., Hamberg, M., Meijer, J. and Radmark, O. (1988). 14,15-Dihydroxy-5,8,10,12-eicosatetraenoic acid. Enzymatic formation from 14,15-leukotriene A_4 by cytosolic epoxide hydrolase. Eur. J. Biochem. 173, 531–536.

Willis, A.L. (1969). Parallel assay of prostaglandin-like activity in rat inflammatory exudate by means of cascade superfusion. J. Pharm. Pharmacol. 21, 126–128.

Willis, A.L. (1970). A method for studying release of prostaglandins from superfused serum of isolated spleen. Br. J. Pharmacol. 38, 470P–472P.

Willis, A.L. (1974a). An enzymatic mechanism for the anti-thrombotic and anti-hemostatic actions of aspirin. Science 183, 325–327.

Willis, A.L. (1974b). Isolation of chemical trigger for thrombosis. Prostaglandins 5, 1–25.

Willis, A.L. (1987). In "Handbook of the Eicosanoids: Prostaglandins and Related Lipids", Vol I (ed A.L. Willis), pp 3–46. CRC Press, Boca Raton.

Willis, A.L. (1989). In "Handbook of Eicosanoids: Prostaglandins and Related Lipids", Vol. II (ed A.L. Willis), pp 3–24. CRC Press, Boca Raton.

Willis, A.L. and Cornelsen, M. (1973). Repeated injection of prostaglandin E_2 in rat paws induced chronic swelling and a marked decrease in pain threshold. Prostaglandins 3, 353–357.

Willis, A.L. and Kuhn, D.C. (1973). A new potential mediator of arterial thrombosis whose biosynthesis is inhibited by aspirin. Prostaglandins 4, 127–130.

Willis, A.L. and Smith, J.B. (1982). Some perspectives on platelets and prostaglandins. Prog. Lipid Res. 20, 387–406.

Willis, A.L. and Smith, D.L. (1989a). In "New Protective Roles for Selected Nutrients" (eds G.A. Spiller and J. Scala), pp 39–108. Alan R. Liss, New York.

Willis, A.L. and Smith, D.L. (1989b). Therapeutic impact of eicosanoids in atherosclerotic disease. Eicosanoids 2, 69–99.

Willis, A.L., Davison, P., Ramwell, P.W., Brocklehurst, W.E. and Smith, B. (1972). In "Prostaglandins in Cellular Biology" (eds P.W. Ramwell and B.B. Pharriss), pp 227–268. Plenum Press, New York.

Willis, A.L., Vane, F.M., Kuhn, D.C., Scott, C.G. and Petrin, M. (1974). An endoperoxide aggregator (LASS), formed in platelets in response to thrombotic stimuli. Purification, identification, and unique biological significance. Prostaglandins 8, 453–507.

Willis, A.L., Hassam, A.G., Crawford, M.A., Stevens, P. and Denton, J.P. (1982). Relationships between prostaglandins, prostacyclin and EFA precursors in rabbits maintained on EFA-deficient diets. Prog. Lipid Res. 20, 161–167.

Willis, A.L., Smith, D.L. and Vigo, C. (1986). Suppression of principal atherosclerotic mechanisms by prostacyclins and other eicosanoids. Prog. Lipid Res. 25, 645–666.

Willis, A.L., Smith, W.L., Smith, D.L. and Eglen, R. (1990). In "The Terminology of Biotechnology: A Multidisciplinary Problem" (ed. K. Loening), pp 93–108. Springer-Verlag, Berlin.

Wong, P.Y.-K., Malik, K.U., Desiderio, D.M., McGiff, J.C. and Sun, F.F. (1980). Hepatic metabolism of prostacyclin (PGI_2) in the rabbit: formation of a potent novel inhibitor of platelet aggregation. Biochem. Biophys. Res. Commun. 93, 486–494.

Wong, P.Y.-K., Westlund, P., Hamberg, M., Granstrom, E., Chao, P.H.-W. and Samuelsson, B. (1985). 15-Lipoxygenase in human platelets. J. Biol. Chem. 260, 9162–9165.

Woollard, P.M. (1986). Stereochemical difference between 12-hydroxy-5, 8,10,14-eicosatetraenoic acid in platelets and psoriatic lesions. Biochem. Biophys. Res. Commun. 136, 169–176.

Wu, K.K., Lo, S.S., Papp, A.C. and Vijjeswarapu, H.N. (1989). Stimulation of de $novo$ synthesis of prostaglandin G/H synthase in endothelial cells. Adv. Prostaglandin Thromboxane Leukotriene Res. 19, 237–241.

Wu, K.K., Sanduja, R., Tsai, A.L., Ferhanoglu, B. and Loose-Mitchell, D.S. (1991). Aspirin inhibits interleukin l-induced prostaglandin H synthase expression in cultured endothelial cells. Proc. Natl Acad. Sci. USA 88, 2384–2387.

Xie, W., Chipman, J.G., Robertson, D.L., Erikson, R.L. and

Simmons, D.L. (1991). Expression of a mitogen-response gene encoding prostaglandin synthase is regulated by mRNA splicing. Proc. Natl Acad. Sci. USA 88, 2692–2696.

Xie, W., Robertson, D.L. and Simmons, D.L. (1992). Mitogen-inducible prostaglandin G/H synthase: a new target for nonsteroidal antiinflammatory drugs. Drug Development Res. 25, 249–265.

Yokoyama, C., Shinjo, F., Yoshimoto, T., Yamamoto, S., Oates, J.A. and Brash, A.R. (1986). Arachidonate 12-lipoxygenase purified from porcine leukocytes by affinity chromatography and its reactivity with hydroperoxyeicosatetraenoic acids. J. Biol. Chem. 261, 16714–16721.

Yokoyama, C., Takai, T. and Tanabe, T. (1988). Primary structure of sheep prostaglandin endoperoxide synthase deduced from cDNA sequence. FEBS Lett. 231, 347–351.

Zoeller, R.A., Wightman, P.D., Anderson, M.S. and Raetz, C.R. (1987). Accumulation of lysophosphatidylinositol in RAW 264.7 macrophage tumor cells stimulated by lipid A precursors. J. Biol. Chem. 262, 17212–17220.

Zucali, J.R., Dinarello, C.A., Oblon, D.J., Gross, M.A., Anderson, L. and Weiner, R.S. (1986). Interleukin 1 stimulates fibroblasts to produce granulocyte–macrophage colony-stimulating activity and prostaglandin E_2. J. Clin. Invest. 77, 1857–1863.

Zvelewska, T., Sanduja, R., Ohashi, K., Loose-Mitchell, D.S., Sanduja, S.K. and Wu, K.K. (1992). Inhibition of endothelial cell prostaglandin H synthase gene expression by naproxen. Biochim. Biophys. Acta 1131, 78–82.

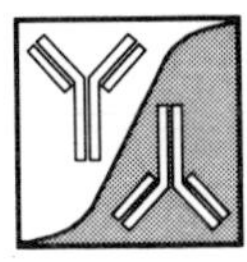

2. *Measurement of Fatty Acids and Their Metabolites*

G.W. Taylor and R. Wellings

1. *Introduction*

Measurements of fatty acids and the associated work-up procedures of extraction and purification were for many years the preserve of the analytical chemist. It was with the discovery that lipids such as the prostanoids and leukotrienes were key elements of the inflammatory process in humans that every laboratory wanted to

Lipid Mediators
ISBN 0–12–198875–9

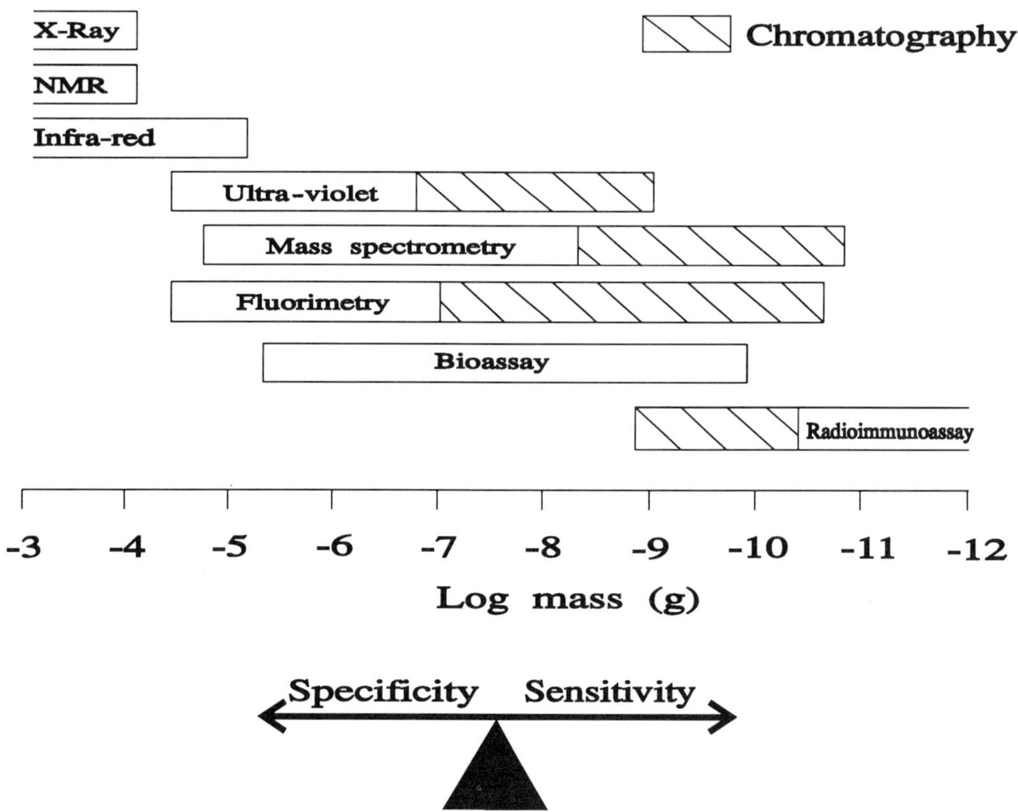

Figure 2.1 Assay specificity versus sensitivity. In an assay, there is a balance between the needs of sensitivity and those of specificity. Using a chromatographic step before the assay increases the confidence in the specificity of the assay and can also facilitate the detection of many compounds present in mixtures. Sample losses on any handling step will work in the other direction to reduce the sensitivity of the assay as a whole.

measure these substances. Although the assay methodology has kept up with (and often led) the needs of research workers to measure trace amounts of lipids in biological fluids, these developments have not always been accompanied by a similar level of understanding in biomedical laboratories of the limitations and potential drawbacks of these sensitive assays. There is, however, nothing difficult or mysterious about measuring lipid mediators, and, as long as the limitations of the methods are understood, there is no reason why most assay procedures cannot be carried out in any reasonably equipped laboratory. There are a number of reviews on lipid mediator analysis (Frolich *et al.*, 1984; Taylor *et al.*, 1986a; Granstrom and Kindahl, 1990) and two volumes of *Methods of Enzymology* have been dedicated to these compounds (Lands and Smith, 1982; Murphy and Fitzpatrick, 1990). In this chapter, we will outline a number of areas of concern that we have encountered when measuring lipid mediators in our laboratory.

2. *Approaches to Assays*

Assay kits are now readily available for a number of mediators, and this is where many problems start: when

it says "specific" on the side of the box, many people believe that this is true and act accordingly.

2.1 SPECIFICITY VERSUS SENSITIVITY

It is taken for granted that all assays are designed to give quantitative data. There is, however, an equally important facet to any assay (and one that is not always considered): a quantitative result assumes the concomitant qualitative identification of the analyte under study. For example, a report of circulating levels of 6-oxo-PGF$_{1\alpha}$ carries with it the implied statement that 6-oxo-PGF$_{1\alpha}$ has been correctly identified in the first place. Unless you know what you are measuring, there is no point in carrying out the assay. A full structural identification (including stereochemistry) of an analyte can be obtained using techniques such as nuclear magnetic resonance spectroscopy and X ray spectroscopy, but these methods clearly lack the required sensitivity to detect mediators present at normal physiological levels (often $<10^{-9}$ g/ml). Detection at these low levels can be carried out with antibody-based techniques such as radioimmunoassay

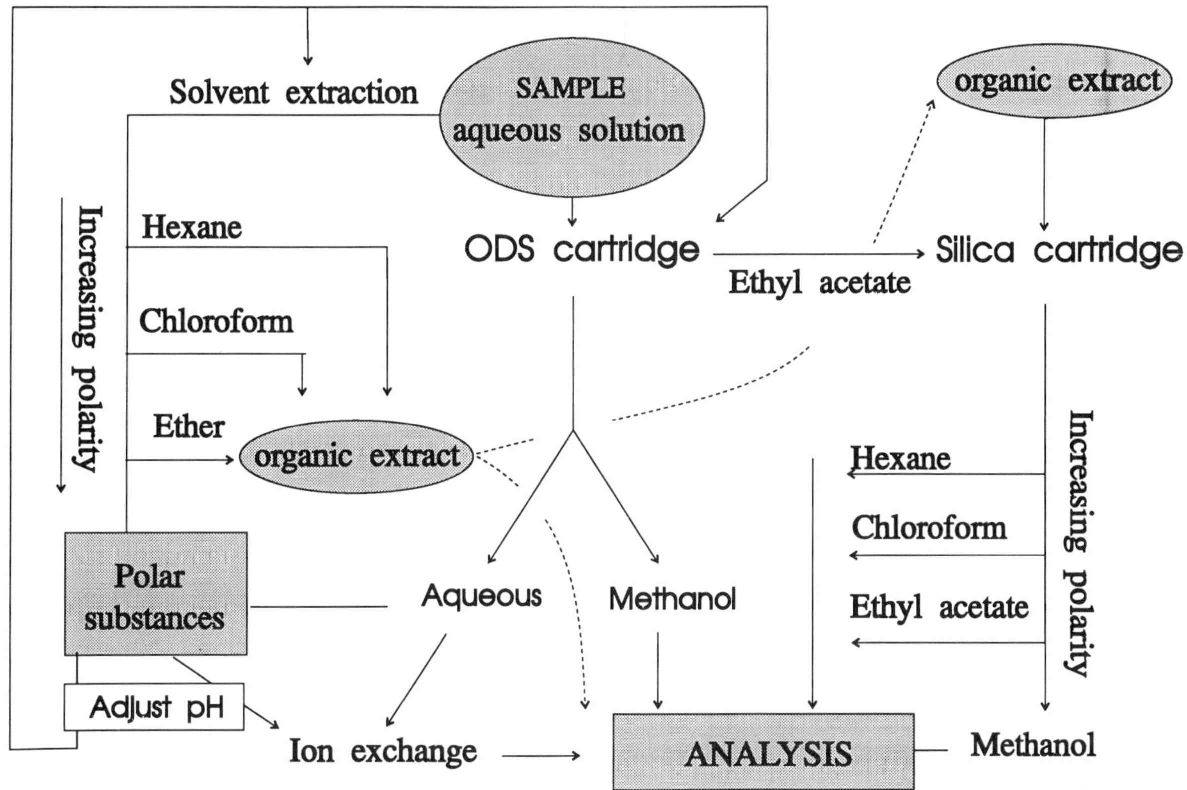

Figure 2.2 Development of extraction procedures. Extraction of analytes can be carried out by either solvent or solid phase extraction. Depending on the polarity of the analyte, it will partition between the aqueous solution (e.g. plasma) and the organic solvent/solid phase cartridge.

(RIA); the specificity of RIA relies on the specificity of the antibody for the analyte (and nothing else), and this is where many problems may arise. There is thus a balance between the requirements for sensitivity and those of specificity, and this requires a judgement to be made about the overall confidence of both qualitative and quantitative aspects of any assay (Fig. 2.1). A chromatography or extraction step can be used before the assay to afford an extra degree of confidence to the specificity of the assay.

2.2   EXTRACTION

If extraction is required before the assay to remove contaminating impurities, the method used should be simple, reproducible and rapid, and with a good extraction yield (Fig. 2.2). It is essential that the extraction procedure does not lead to a modification of the analyte. Extremes of pH should be avoided to minimize hydrolysis of labile bonds, the elimination of water from alcohols or isomerization. Solvents should be removed at low temperatures (<30°C), using a stream of nitrogen wherever possible to prevent oxidation. The possibility of chemical interactions of the solvents with the analyte should also be considered: methanol in the presence of strong acids can result in some esterification. At every

stage of an extraction, some impurities are added to the sample (such as phthalate plasticizers and silicone grease impurities) and these may interfere with the final result. The most common methods for extracting lipids are organic-solvent extraction and solid phase extraction (with cartridges such as Sep-Paks or Bond Eluts) or by chromatographic methods such as HPLC.

2.2.1   Solvent Extraction

Historically, solvent extraction was the method of choice for extracting lipids. With a wide variety of solvents available, it was generally possible to partition lipids into the organic phase, leaving salts and other impurities in the aqueous phase. Different substances will dissolve in solvents of different polarity: in general, substances with a greater proportion of polar functional groups (amino, hydroxyl, carboxyl) will dissolve in more polar solvents such as methanol, whereas hydrophobic neutral lipids are more soluble in non-polar solvents such as hexane or benzene. Clearly to be useful in an extraction from biological fluids, the organic solvent should be immiscible with water. A suitable series of solvents (in order of increasing polarity) which can be tested when developing assays is hexane<chloroform<diethyl ether<ethyl acetate. If solids are to be extracted, then mixtures

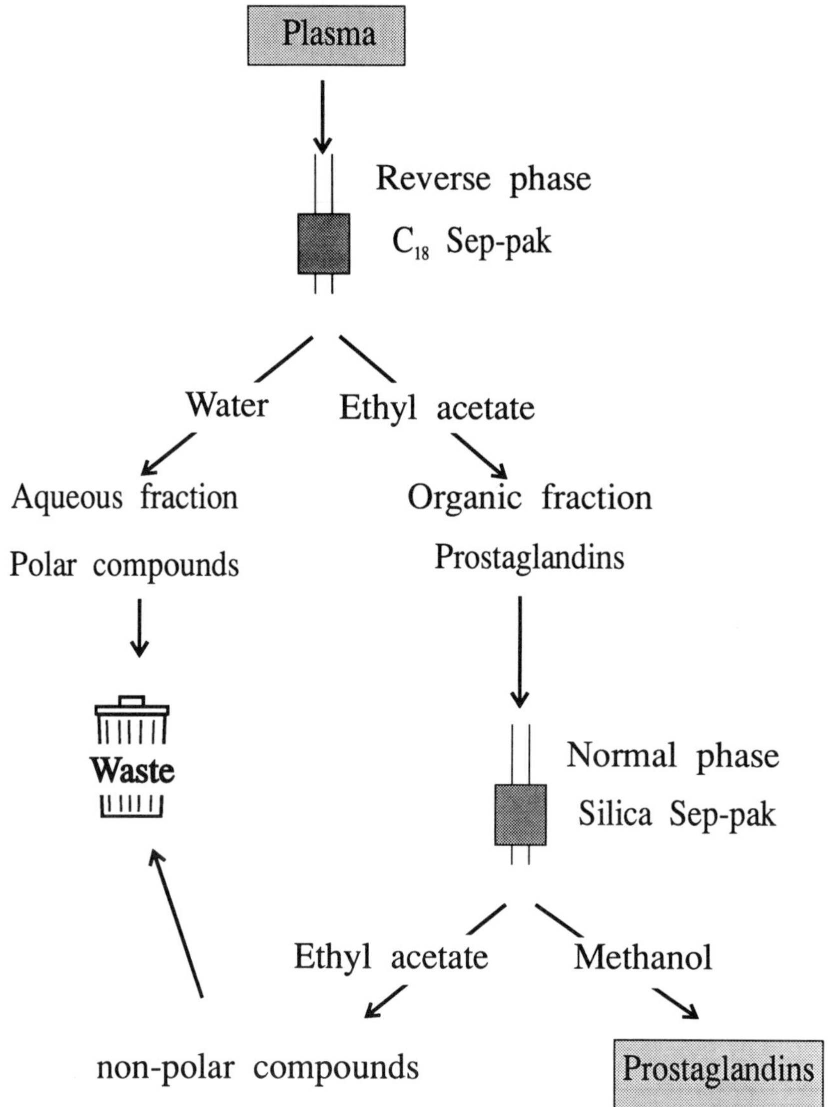

Figure 2.3 Extraction of prostanoids. Solid phase extraction of prostanoids makes use of the selectivity of different phases. Reverse-phase cartridges are used as the first step to extract trace amounts of prostanoids from large volumes of plasma. Salts and polar impurities are removed with a water wash and the less polar prostaglandins eluted with ethyl acetate. The silica (normal-phase) cartridge binds prostaglandins and allows the less polar lipids to pass through. Prostaglandins are eluted with a more polar solvent, methanol. *Conditions*: Reverse-phase cartridges (C18 Sep Paks, Waters Ass) were pre-washed with methanol (5 ml) and water (5 ml) before use. Silica cartridges were preconditioned in methanol (5 ml) and then ethyl acetate (5 ml).

such as chloroform/methanol or ether/ethanol may be employed. The efficiency of extraction from aqueous solutions can be altered by adding salt to the solution ("salting out") or by changing the pH of the solution to suppress or increase the amount of ionization of the analyte/impurities. For example, simple carboxylic acids have a pK_a of 4.5. Thus, below pH 3 a carboxylic acid is in the protonated form (-COOH) and soluble in ether, while above pH 6 the acid is deprotonated (COO^-) and is water-soluble. There are a number of points to bear in mind when using solvent extraction:

1. Many solvents are flammable.
2. Toxicity must be considered, particularly when large volumes are used.
3. Multiple extractions using smaller volumes of solvent are more efficient than a single extraction using a large volume.
4. Solvents such as ether will dissolve up to 10% of its weight in water during extraction, and drying over anhydrous sodium sulphate is recommended.
5. Ether contains small amounts of peroxides which may result in oxidation of unsaturated analytes. Peroxides

are less volatile than ether and, during removal of the solvent, they may concentrate in the flask and cause explosions.

6. Many extractions produce large volumes of waste solvents, with the accompanying problems of disposal.

2.2.2 Solid Phase Extraction

Increased selectivity can often be achieved with solid phase extraction cartridges (Powell, 1980; Metz *et al.*, 1982). Large volumes of biological fluids can be concentrated on the cartridges, but only small volumes of solvents are required for their extraction. The solid phases in general use include normal-phase silica, reverse-phase ODS (C_{18}) and, less usually, ion exchange. In general, analytes elute from reverse-phase cartridges in order of increasing hydrophobicity by decreasing the polarity of the solvent, and from normal-phase columns in order of decreasing hydrophobicity as the polarity of the solvent is lowered. Ion exchange cartridges, which make use of charged functional groups within the analyte ($-COOH/COO^-$, $-NH_2/NH_3^+$), have also found some use in lipid analysis. The selectivity that can be obtained with solid phase cartridges can be seen in the extraction of prostaglandins from plasma using both ODS and silica cartridges (Fig. 2.3). Prostanoids in plasma will bind to ODS cartridges, allowing salts and polar impurities to pass through with a water wash. Prostanoids and other less polar substances are eluted from the ODS cartridge with ethyl acetate. On applying the ethyl acetate solution to a silica cartridge, the prostanoids again bind, but neutral lipids and other non-polar substances are washed through. Prostanoids are then eluted in a more polar solvent, methanol. In many cases, greater discrimination between analyte and impurities may be obtained by

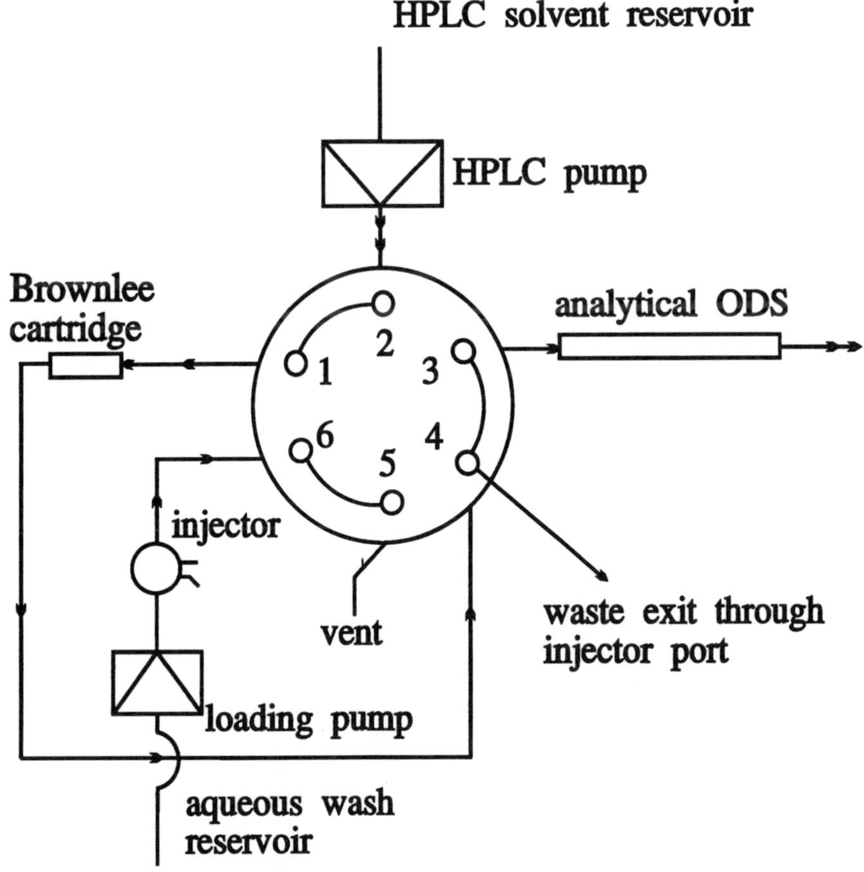

Figure 2.4 On-line extraction. LTE$_4$ is extracted from urine with a one-step extraction/HPLC system using a rheodyne injector. Initially, the ports are connected 1–6, 4–5 and 2–3. Urine is pumped through port 6 and out through port 1 on to the Brownlee extraction cartridge; LTE$_4$ binds, but salts are removed with a water wash via port 5 and to waste. The Brownlee cartridge is switched into the HPLC system with the following connections: ports 1–2, 3–4 and 5–6 (shown in the figure). LTE$_4$ is eluted with the HPLC solvent from the Brownlee cartridge through ports 4 and 3 and on to the HPLC column. On-line extraction minimizes the handling steps and reduces handling losses. *Conditions:* The Brownlee cartridge contains MCH10 packing (Varian). HPLC (of LTE$_4$) is carried out at 2 ml/min on a Hypersil ODS column (25×0.5 cm, 5 m, Shandon) eluting with a 35 min linear gradient of methanol/water (57:43 to 83:17, buffered to pH 3.8 with acetic acid/ammonia) in the presence of 0.5 mM oxalic acid (see Fig. 2.13) (Richmond *et al.*, 1987).

including further washes with solvents of intermediate polarity (such as methyl formate), or by changing the pH to alter the ionic state of analyte.

Occasionally, it is possible to exploit the unique nature of an analyte to modify the analyte and simplify the extraction procedure. For example, 6-oxo-PGF$_{1\alpha}$ exists as a polar hydroxy acid; in the presence of acid, lactonization occurs to form a less polar lactone which can be extracted (Barrow and Taylor, 1987). The *cis*-diol present in PGF analogues will react with butylboronic acid to generate a cyclic boronate compound (Liston and Roberts, 1985). This work has been extended by incorporating the boronic acid on to a solid phase cartridge for the extraction of PGF compounds (Lawson *et al.*, 1985) and, following methoximation, TXB$_2$ (Catella and Fitzgerald, 1990). Perhaps the most specific form of solid phase extraction involves binding of the analyte to antibodies linked to a sepharose matrix (Chiabrando *et al.*, 1987; Vrbanac *et al.*, 1990); only the analyte and other cross-reacting species bind to the antibody. The analyte can then be eluted with a solvent such as acetone. Separate antibodies are generally required for each prostanoid to be extracted, although the inherent cross-reactivity of antibodies can often be exploited to extract both the parent prostanoid and any structurally similar metabolites (e.g. TXB$_2$ and 2,3-dinor-TXB$_2$). In our laboratory, we routinely use antibody extraction to extract TXB$_2$, 6-oxo-PGF$_{1\alpha}$ and their 2,3-dinor metabolites from urine prior to analysis by gas chromatography–mass spectrometry (GC–MS) (Barrow *et al.*, 1989; Moore *et al.*, 1991; Taylor *et al.*, 1991).

2.2.3   On-Line Extraction for High-Performance Liquid Chromatography

In many cases, purification by HPLC is preceded by extraction with solid phase cartridges. By coupling the extraction cartridge directly to HPLC, the number of separate handling steps (each with some loss of sample) is minimized (Pickett and Douglas, 1985; Powell, 1987; Richmond, *et al.*, 1987). In our laboratory, we use on-line extraction (shown in Fig. 2.4) to purify LTE$_4$ before immunoassay. A rheodyne injector is connected to a solid phase Brownlee cartridge containing MCH10 packing (Varian) and also to an HPLC column. Urine (or other fluid) is pumped on to the Brownlee cartridge and washed with water; the cartridge is switched on-line and the HPLC solvent elutes the leukotrienes directly on to the HPLC column, where chromatography is carried out. HPLC fractions are then assayed by RIA. With this system there is a single stage where solvent is removed before the sample is resuspended in RIA buffer. Extraction yields are good, and the chances of sample deterioration are minimized.

2.3   INTERNAL AND EXTERNAL STANDARDS

It is well recognized that the extraction of subnanogram quantities of inflammatory mediators from complex biological matrices rarely results in 100% yields, particularly when multiple steps are used. This is exemplified in Table 2.1, which shows the recovery of radiolabelled [^{3}H]LTE$_4$ post-HPLC from human urine. In this experiment, carried out over 1 week, the recovery of [^{3}H]LTE$_4$ varied from 24.1 to 67.8%, a 2.5-fold difference. If no correction is made for this variation in the recovery of the analyte, then the final "recovery-corrected" data may be out by a similar factor. This problem is exacerbated if the overall recovery is low: differences in recovery of 5–15% between samples will have a far greater effect on the final data than differences of 45–60%. Clearly, some estimate of extraction losses must be made, and this is normally carried out by using either external or internal standards. With an external standard, the analyte is taken through the extraction protocol a number of times and the mean recovery determined. All analyses are then corrected for this mean recovery. This option relies on tightly controlled recoveries: a maximum of 10% variation in recovery (i.e. mean ± 5% of mean) is unlikely to reduce confidence in the quantitative data. If, as in the case of LTE$_4$ noted above, the variation is greater, using the mean recovery for all samples is no longer acceptable. For example, at the two extremes shown in Table 2.1, a sample of 1 ng of LTE$_4$ would be measured as 241 pg (minimum recovery), 678 pg (maximum recovery) or 403 pg (mean recovery). The data from such measurements could not be trusted.

Table 2.1 Percentage of [^{3}H]LTE$_4$ from 14 human urine samples. These data are taken from a routine analysis of urine for LTE$_4$ and are based on the individual recoveries post-HPLC as determined from the radiolabelled internal standard

47.2	34.3	67.8	51.7	45.9
32.0	34.8	30.1	41.6	42.8
45.9	42.8	42.1	24.1	39.2
30.9				

Mean recovery = 40.3

The most accurate method of accounting for variable extraction yields is to include an internal standard, which confers a sample-dependent marker of extraction yield. The internal standard should be closely related to the analyte, with similar stability, interactions with the matrix (including protein binding) and extraction efficiencies. Although compounds of similar chemical structure may be used, with, for example, 19-hydroxy-PGB$_2$ acting as the internal standard in an HPLC–RIA analysis of lipoxygenase products (Borgeat and Picard, 1988), the

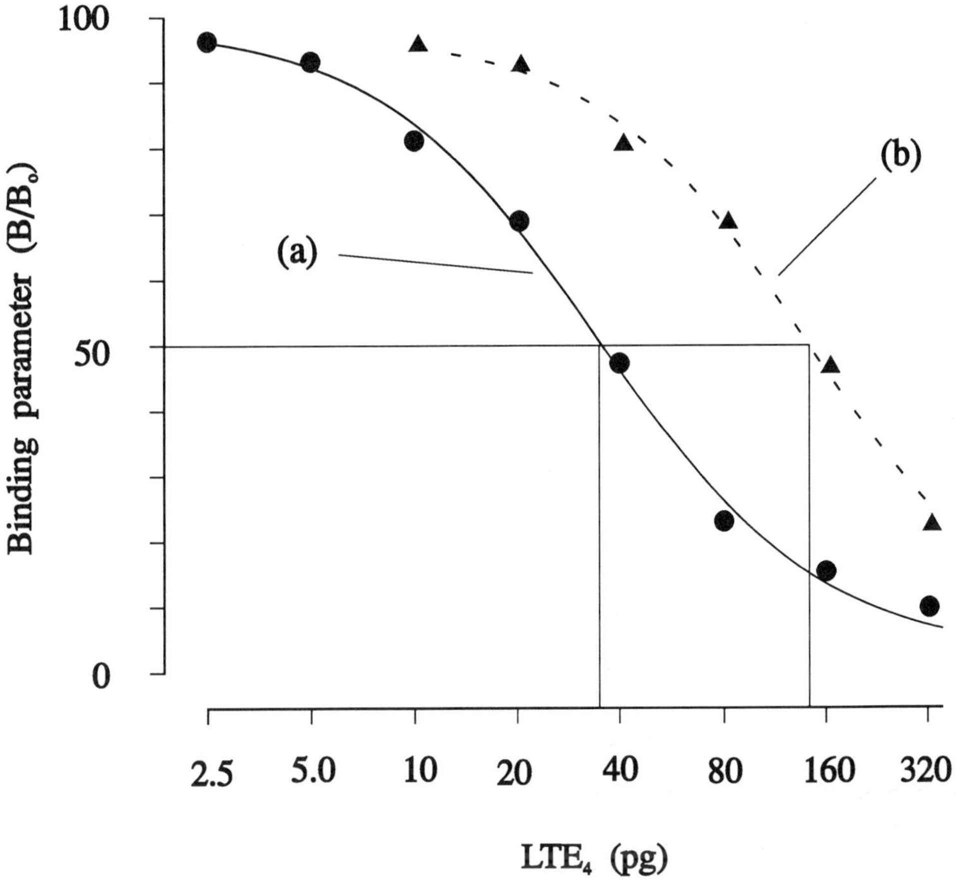

Figure 2.5 Antibody-binding curve. (a) An antibody-binding curve for LTE₄ using fresh ligand. The accuracy of the assay is dependent on the accuracy of this curve. The usable limits of this assay are between 5 and 160 pg. (b) LTE₄ will decompose on storage, and thus more ligand is apparently required to generate the binding curve, which is shifted to the right. The 50% binding value (B_{50}) is shown for both curves. *Conditions*: The antibody was raised in rabbits against LTE₄ and is available from Cascade Biochem Ltd, University of Reading, PO Box 68, Reading, Berks, UK.

ideal internal standard is one that is essentially of identical chemical structure – those labelled with heavy isotopes (e.g. ^{2}H, ^{3}H, ^{14}C). Internal standards should be added at the time of sample collection (to account for any analyte changes on storage) and must be treated in an identical manner to the sample throughout all stages of the assay, including detection and quantification. In our laboratory, trace amounts of [^{3}H]LTE₄ are used as internal standards in HPLC–RIA assays of LTE₄, and ^{2}H-labelled analogues used for GC–MS assays of prostanoids.

Radiolabelled lipids for use as internal standards are available from Amersham International and New England Nuclear. Some deuterated eicosanoids are available from Cayman Chemicals, Ann Arbor, MI, USA (or from their agent, Cascade Biochem, University of Reading, PO Box 68, Reading, Berks, UK). Other ^{2}H-labelled lipids may be prepared from [^{2}H$_8$]arachidonic acid.

2.4 ACCURACY AND PRECISION

Even when the problems of extraction recoveries have been taken into account there are still a number of sources of error associated with the quantitative aspects of any assay.

2.4.1 Errors

There is an inherent error in all assays and this will affect the confidence in the final result. Such errors arise from the cumulative effect of each of the small errors which occur at every stage of the assay. The sum of these errors can be reported as the coefficient of variation (CV) of the assay. The CV can be obtained by measuring the same sample a number of times (n) and determining the mean ($\bar{x}$) and the standard deviation (σ_{n-1}) of the data:

$$CV = (\sigma_{n-1}/\bar{x}) \times 100\%.$$

Both intra-assay and inter-assay CVs should be determined. A reliable assay will have CVs of less than 5%, although as the assay becomes more complex, and includes multiple extraction and purification steps, it is not unusual for values of over 10% to be recorded. It is not uncommon for CVs determined in day-to-day assays (without the knowledge of the operator) to exceed those

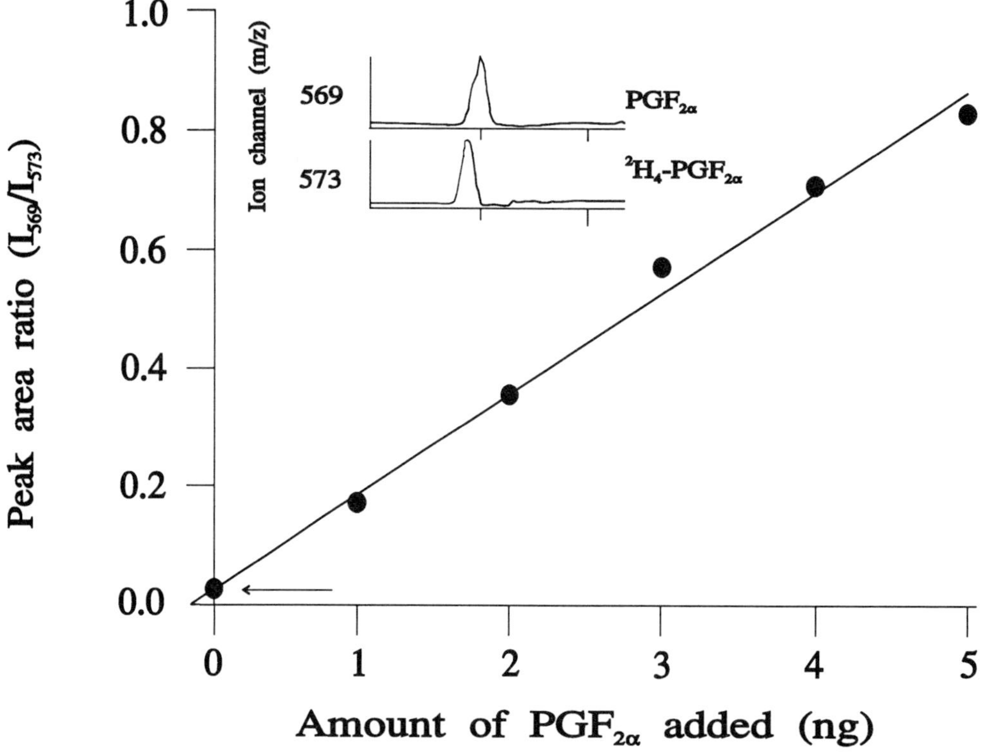

Figure 2.6 Extracted standard curve. Part of an extracted standard curve from a GC–MS assay for PGF$_{2\alpha}$ is shown. The intensity ratio of PGF$_{2\alpha}$/[^{2}H$_{4}$]PGF$_{2\alpha}$ (ion channels at *m/z* 569 and 573, inset) is plotted against the amount of standard used. The amount of endogenous PGF$_{2\alpha}$ present (present naturally in plasma or arising from an impurity in the internal standard) can be determined from the intercept on the *x* axis.

reported when the assay was originally developed. This is simply because routine assays are rarely conducted with the same care and attention to detail as during the initial development.

2.4.2   Standard Curves

Standards are needed in order to use a quantitative assay, and any errors in the measurement of the standards will appear in the final data. These errors can be minimized by using a number of standards of known concentration (to encompass the expected range of the unknowns to be determined) and preparing a standard curve. The sigmoidal binding curve from an assay for LTE$_4$ is such a curve where the effect of authentic LTE$_4$ on the binding of [^{3}H]LTE$_4$ is plotted against the binding parameter B/B_0 (Fig. 2.5a). The amount of analyte present can then be determined from interpolation between points on the curve: the more points on the curve, the more accurate the interpolation. If the standards are made up incorrectly or have decomposed on storage, then any data based on such standards will be incorrect. The LTE$_4$-binding curve exhibits a B_{50} value of approximately 40 pg. When an old sample of LTE$_4$ was used as the ligand, the binding curve shifted to the right, with a B_{50} value of 160 pg (Fig. 2.5b). This effect was due to the slow decomposition of LTE$_4$ to weakly immunoreactive species such as the

sulphoxide (relative immunoreactivity of 24%). Thus, four-fold more LTE$_4$ needed to be used to prepare the standard curve, with the consequent apparent increase in LTE$_4$ values determined in the assay.

If an internal standard has been included in the assay, the standard curve can be set up to take this into account. Part of an extracted curve for a GC–MS assay for TXB$_2$ is shown in Fig. 2.6. Here, 5 ng of the standard ([^{2}H$_4$]TXB$_2$) has been included with five separate TXB$_2$ standards (0, 1, 2, 3, 4 and 5 ng) and the mixtures extracted as normal. The ratio [^{2}H$_4$]TXB$_2$/TXB$_2$ (determined in the mass spectrometer) is plotted against the amount of TXB$_2$ used to generate the standard curve (from which the amount of analyte may be determined). The overall assay error will be determined by the "goodness of fit" of the standard curve and, as long as the same amount of internal standard has been used for both the samples and the standard curve, then handling and extraction losses are taken into account. The standard curve should pass through the origin. There are two explanations if this does not happen: either there is contamination by unlabelled TXB$_2$ in the ^{2}H-labelled internal standard (easily confirmed by MS) or there is a basal level of TXB$_2$ present in the matrix used to prepare the standard curve (in this case, pooled plasma). A non-zero intercept will not affect the quantification except at low levels.

2.4.3 Limit of Detection

Every assay has a limit of detection and a usable range, and this is exemplified by the RIA for LTE_4 (see Fig. 2.5a). In principle, less than 5 pg of LTE_4 can be measured with this assay. At this level, however, the binding curve flattens out and small difference in B/B_0 can represent a relatively large difference in measured LTE_4. The error in the measurement may become significant. The usable range of the assay is between 5 and 160 pg. Over 160 pg, the curve again flattens out, and the error once more becomes significant. In a linear standard curve (as generated in a GC–MS assay, Fig. 2.6), it cannot be assumed that linearity will continue beyond the range plotted.

There is a further consideration that determines the true detection limit of the assay. When small amounts of analyte are present in complex matrices, the effects of non-specific binding to glass surfaces during extraction can result in a complete loss of material. Thus, an assay with a theoretical detection limit of 5 pg (added from a stock solution) may not detect 5 pg of the analyte present in urine.

3. *Methodology*

Having considered the limitations associated with assays, we will now consider the methodology in current use.

3.1 BIOASSAYS

Bioassays, which use the biological response of a tissue or cell to an agonist, are perhaps one of the most important, and least utilized, methods of measurement and have been instrumental in the development of the eicosanoid field (Vane, 1983). Although a good bioassay can offer both sensitivity and specificity, the requirement for animal tissues, interference by other compounds, the lack of sensitivity to biologically inactive metabolites and the need for trained staff to carry out routine assays has limited its use in recent years. Although there are a number of *in vivo* models for studying the properties of bioactive lipids (for example, measuring changes in vascular permeability in the guinea-pig (Peck *et al.*, 1981)), these are rarely used in day-to-day assays. *In vitro* methods have found a much wider application.

3.1.1 Tissue Bioassays

The response of tissues to agonists can be determined either in an organ bath or in a superfusion cascade (Vane, 1983). As long as the response is dose-dependent, then it can be used as the basis of a bioassay. Different tissues respond in different ways to agonists and this can be used to increase the specificity of the assay and to differentiate a number of agonists. For example, PGE_2 can be distinguished from $PGF_{2\alpha}$ by its action on rabbit coeliac artery, and from PGI_2 on bovine coronary artery. Both TXA_2 and PGG_2 contract rabbit aorta but differ in their action on rabbit coeliac artery. PGI_2 has no effect on rabbit aorta but relaxes rabbit coeliac artery. The combined use of isolated strips of guinea-pig lung parenchyma and ileum have also been used as a sensitive and selective bioassay for LTB_4 (Samhoun and Piper, 1984). An extra degree of confidence in a bioassay can be obtained by using the specific antagonists. In the bioassay for LTC_4, LTD_4 or LTE_4 on guinea-pig ileum smooth muscle, the contraction is abolished by the LTD_4 antagonist FPL 55712 (Parker *et al.*, 1982; Piper, 1986). In this bioassay, the specific activities of the cysteinyl leukotrienes differ ($LTD_4>LTC_4>LTE_4$) and it is not possible to quantify LTC_4 in the presence of LTD_4 or LTE_4. The assay is not responsive to the urinary oxidative metabolites of LTE_4.

3.1.2 Cell-Based Assays

Blood-derived cells (in particular neutrophils and platelets) are a readily available source of bioassay tissue. Chemotaxis (to neutrophils) has been used to measure LTB_4 and some other diHETEs (Camp *et al.*, 1982; Bray, 1983; Czarnetzki and Rosenbach, 1986) but has now been superseded by RIA and GC–MS methods. Chemotaxis is still useful when identifying novel compounds; for example, 8,15-diHETE has been isolated from guinea-pig lung by monitoring its action on eosinophils. Chemotaxins generally exhibit a bell-shaped dose–response curve, and it is important to carry out the assay at multiple dilutions.

PAF is one substance that is still measured with a bioassay: its effect on the aggregation of platelets (Moon *et al.*, 1990). The assay can be extended by preloading the platelets with [³H]serotonin, and measuring the release of radioactivity on stimulation (Henson, 1990). The bioassay of PAF has several complications: the presence of other substances in the sample which will elicit the same biological action as PAF, may complicate the quantification and require the use of antagonists, and there may also be endogenous inhibitors present in the biological sample.

Although bioassays are invaluable in the discovery and isolation of biologically active species nowadays, they are mainly used only when a more convenient procedure is not available, for example when measuring endothelium-derived relaxing factor.

3.2 RADIOIMMUNOASSAY AND RELATED ASSAYS

There are a number of assays which use the specificity of an antibody to quantify a metabolite. Polyclonal antibodies are generally raised in rabbits to the analyte (conjugated to proteins such as albumin or keyhole limpet haemocyanin), although there is an increasing tendency to use monoclonal antibodies.

3.2.1 Principles

RIA is probably the most common method employed for the measurement of eicosanoids, and several types have been developed (Granstrom and Kindahl, 1978; Salmon, 1978; Salmon *et al.*, 1982; Hayes *et al.*, 1983). A number of kits are commercially available for measuring eicosanoids and other lipid mediators. RIA is based on the competition between the analyte and a radiolabelled analogue for an antibody (see Fig. 2.5a); as the amount of analyte increases, binding of the radioactive tracer to the antibody decreases. Analytes can thus be quantified by comparison with a standard binding curve. As part of the work-up procedure, free (unbound) radioactivity must be separated using dextran-coated charcoal (leaving antibody-bound radioactivity behind), and it is this step that slows down the assay. The scintillation proximity assay is a modification of RIA which employs antibody-coated fluoromicrospheres which contain scintillant. Again, competition between the analyte and its radio-labelled analogue is set up. However, only those β particles which are bound to the antibody (and thus in close proximity) will cause excitation and be detected; unbound radioactivity is not detected. The time-consuming process of centrifugation and decanting of charcoal and the requirement for scintillant are eliminated. This type of assay has been developed for the measurement of PAF (Sugatani *et al.*, 1990). ELISA is also a development of RIA. Here, competition for an antibody occurs between an analyte in solution (to be measured) and an immobilized analyte (normally bound to an ELISA plate). The amount of this (first) antibody bound to the plate is inversely proportional to the amount of the analyte in solution, and is determined by interaction with a second enzyme-linked antibody. The final conversion of a colourless substrate to a coloured product (by the now-bound enzyme) is directly proportional to the initial analyte concentration. This method, which has been applied to eicosanoids (Reincke *et al.*, 1989; Lellouche *et al.*, 1990), has similar sensitivity to RIA with the inherent advantage that radioactivity is not required.

3.2.2 Confidence in Immunoassays

The specificity of any antibody-based assay is only as good as the specificity of the antibody. In our laboratory, we have produced antibodies to LTE_4 (Table 2.2). The antibody will detect LTE_4 and, to a lesser extent, LTC_4 and LTD_4 but does not cross-react to any extent with arachidonic acid, LTB_4 or the prostanoids.

There are three areas of concern when using the antibody to LTE_4. First, the relative biological activity of the cysteinyl leukotrienes generally follows the order $LTD_4 > LTC_4 > LTE_4$ (Piper, 1986) and, thus, except in samples where only LTE_4 is present, the overall "immuno-reactivity" of the sample will not bear any relation to the biological activity of the cysteinyl leukotrienes present. The true level of each of the leukotrienes can be determined following purification, or by enzymatic

Table 2.2 Cross-reactivity of a LTE_4 antibody

Analyte	Cross-reactivity (%)
LTE_4	100
LTC_4	16
LTD_4	25
LTB_4	<0.2
Arachidonic acid	<0.1
PGE_2, $PGF_{2\alpha}$, PGD_2, prostacyclin, TXA_2	<0.1
Unknowns	?

conversion of LTC_4/LTD_4 present to LTE_4 (Heavey *et al.*, 1987), thus giving a sum of the leukotriene load. Secondly, although the cross-reactivity of arachidonic acid with this antibody is low (<0.1%), there are many experiments where there is a vast excess of this substrate over product. In such cases, the excess substrate may be identified incorrectly as "immunoreactive" LTE_4. Thirdly, the cross-reactivity with unknowns present in the sample can be a particular problem, especially in complex biological matrices such as urine and plasma, where the presence of other unrelated immunoreactive species preclude the use of direct RIA for LTE_4. A well-known example where direct RIA has led to problems is that of circulating levels of $6\text{-oxo-PGF}_{1\alpha}$, the stable hydrolysis product of prostacyclin. Early RIAs gave plasma levels of $6\text{-oxo-PGF}_{1\alpha}$ of 10–20 pg/ml (Mitchell, 1978; Vermylen *et al.*, 1981); however, the true circulating level of this prostanoid is 1–2 pg/ml (Fitzgerald *et al.*, 1981; Blair *et al.*, 1982). Thus, over 80% of the immunoreactive $6\text{-oxo-PGF}_{1\alpha}$ was due to impurities: a five-fold increase in true $6\text{-oxo-PGF}_{1\alpha}$ levels from 1–2 to 5 pg/ml would appear as a small increase (20–25 pg/ml) in immuno-reactive $6\text{-oxo-PGF}_{1\alpha}$ (Fig. 2.7). The effect of background impurities is often negligible when high levels of prostanoids are measured; there are no difficulties in using RIA directly when determining serum TXB_2 (at 100–200 ng/ml). When problems of background inter-ference arise, they may be minimized by using more specific antibodies and by employing suitable extraction and purification protocols such as TLC (Barrow and Taylor, 1987) or, more recently, HPLC to increase the confidence in the final assay result.

3.3 HIGH-PERFORMANCE LIQUID CHROMATOGRAPHY

HPLC is a high-resolution separative technique which became generally available in the mid-1970s. HPLC can be used to separate and purify compounds of similar structure, and, when coupled to a suitable detection system such as RIA, will offer a further degree of confidence in the qualitative aspects of an assay. To be useful as part of an assay, the chromatography, retention time and recovery must be reproducible and independent of the amount or purity of the sample. The solvents

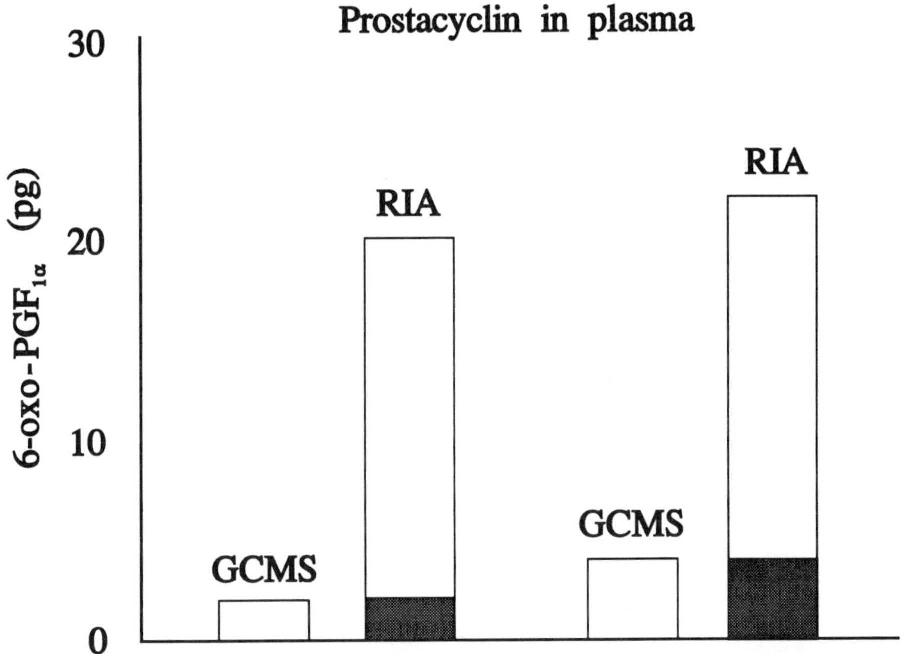

Figure 2.7 RIA versus GC–MS for plasma 6-oxo-PGF₁ₐ. Many RIAs of PGI₂ in plasma give values of 20 pg/ml against the correct level of 1–2 pg/ml determined by GC–MS (left panels). Not only are the immunoreactive basal levels incorrect, but a (possibly significant) two-fold increase in the true plasma 6-oxo-PGF₁ₐ levels would only appear as a minor perturbation in the immunoreactivity (right panels).

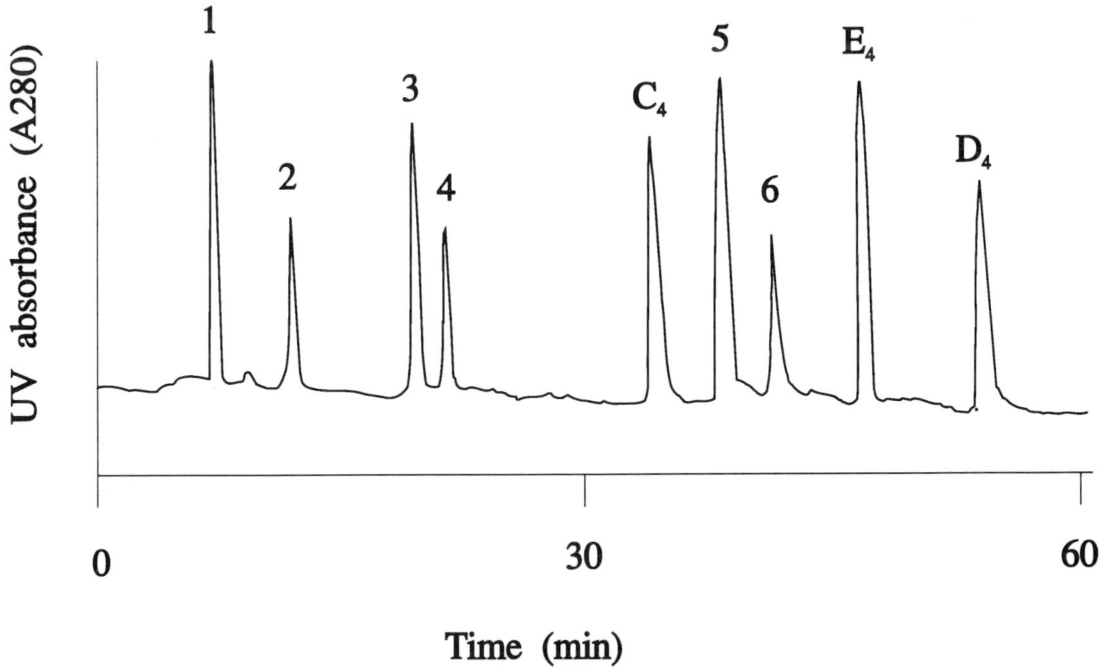

Figure 2.8 Reverse-phase HPLC of leukotrienes. The high-resolution separative capability of reverse-phase HPLC allows the separation of LTC₄, LTD₄ and LTE₄ from a number of their metabolites: 1, ω-tetranorcarboxy-LTE₄; 2, ω-dinorcarboxy-LTE₄; 3, ω-carboxy-LTE₄; 4, ω-hydroxy-LTE₄; 5, N-acetyl-LTE₄; 6, LTE₄ sulphoxide. _Conditions_: Chromatography was carried out on a Waters HPLC column using a Hypersil ODS column at 1.5 ml/min in a multistep gradient of water/methanol (pH 4, adjusted with acetic acid/ammonia and containing 0.5 mM oxalic acid): 80:20 to 60:40 (20 min, Waters gradient 7), 60:40 to 43:57 (8 min, gradient 5), 43:57 to 17:83 (45 min, linear gradient), 17:83 to 0:100 (5 min, linear gradient). Detection was by UV absorbance at 280 nm.

must be compatible with the detection method (UV absorbance, RIA). HPLC can best be thought of as a highly selective method of partitioning an analyte between a solid (stationary) phase normally contained within a steel column and a mobile solvent phase. The affinity of an analyte for the stationary phase depends on its structure, and it is this property which is exploited in developing new separation techniques by HPLC. There are two basic forms of solid phase used in lipid analysis: normal phase and reverse phase (cf. solid phase extraction cartridges). In each case, the composition of the mobile phase can be varied.

3.3.1 Normal Phase

Normal-phase HPLC is an extension of classical open-column chromatography: the solid phase is polar (often silica) and analytes are eluted with non-polar solvents such as hexane and chloroform. Non-polar (hydrophobic) compounds elute first, and, as the polarity of the solvent is increased (for example by increasing the concentration of methanol), more polar compounds will elute. There are a number of normal-phase separations used in the

fatty acids/metabolite field (Boeynams *et al.*, 1980; Powell, 1980).

3.3.2 Reverse Phase

The vast majority of HPLC separations of lipid mediators are now carried out in the reverse-phase mode. The excellent resolving power of the technique can be seen in the separation of a number of closely related leukotriene metabolites (Fig. 2.8) (Maltby *et al.*, 1990).

In reverse-phase HPLC, the stationary phase is hydrophobic and is prepared by coupling an alkyl chain to an inert bed of silica or a similar polymeric material. Elution is carried out with polar solvents such as water, methanol or acetonitrile (i.e. the "reverse" of classical normal-phase chromatography). In general, polar solutes elute first, with less polar substances retained in the column, eluting as the content of the organic modifier in the mobile phase is increased. There are a number of commercially available reverse-phase columns with stationary phases of differing polarity. These include C_{18}, C_8, C_4, C_2, CN and NH_2. Greater retention of non-polar solutes occurs with the longer alkyl chain phases. The most popular reverse phase

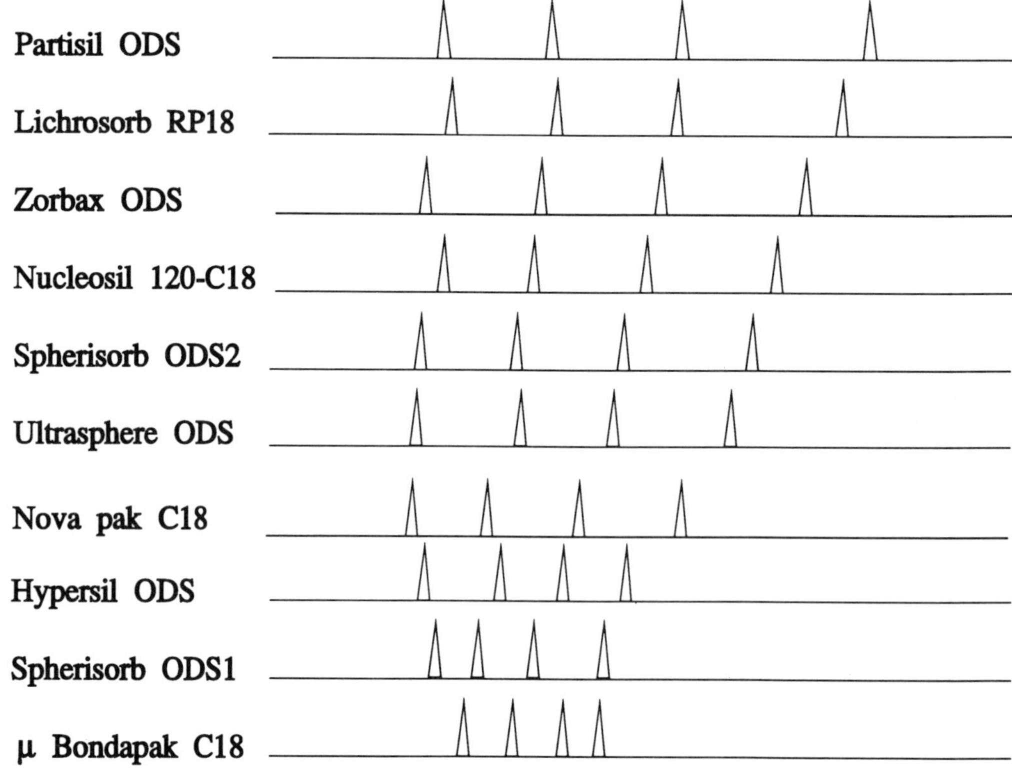

Relative retention time

Figure 2.9 Comparison of ODS reverse-phase columns. ODS columns from different manufacturers vary in their characteristics, and this will affect the retention time and resolution of analytes. This is exemplified in the separation of dimethyl phthalate, toluene, biphenyl and phenanthrene on a number of ODS columns in a methanol/water system. Data were kindly supplied by Hichrom Ltd, 1 The Markham Centre, Station Road, Theale, Reading, Berks, UK.

Table 2.3 Absorbance maxima of common lipids

Analyte	Chromophore	λ_{max} (nm)	ε_{max}
Saturated fatty acids, glycerides, PAF	None	—	—
Arachidonic acid,	Single double bond	<220	—
PGD_2, PGE_2, $PGF_{2\alpha}$, 6-oxo-$PGF_{1\alpha}$, TXA_2, EETs			
PGA_2	Enone	220	10 000
PGB_2	Dienone	280	30 000
HETEs	Conjugated diene (E,Z)	234	20 000
LTB_4	Conjugated triene	270	50 000
DiHETEs	Conjugated triene	268–272	40 000
LTC_4, LTD_4, LTE_4	Conjugated triene	280	40 000
Lipoxins	Conjugated tetraene	302	50 000

Note that simple carbonyl groups (e.g. in 6-oxo-$PGF_{1\alpha}$) absorb weakly in the UV at approximately 280 nm (ε=200).

is ODS. These ODS columns are available from a number of manufacturers, but they can vary markedly in their characteristics and this can radically affect the separations that may be achieved (Fig. 2.9). The stationary phase particles may be either spherical or irregular: as with any chromatographic technique, better resolution can be obtained with small (<5 μm) spherical particles; however, large (10 μm) irregular particles, such as used in the μBondapak columns, allow greater loading capacities. The amount of reverse-phase packing attached to the silica (the "carbon load") can also vary between 3 and 15%. The retention of non-polar solutes normally increases with carbon load. One further consideration is that of end-capping. The assay can be affected if the coverage of the bonded phase is non-uniform, leaving a number of surface polar groups. Most modern phases now cap the free polar groups with methylsilyl derivatives.

3.3.3 High-Performance Liquid Chromatography–Ultraviolet Absorbance

The simplest method of detecting analytes post-HPLC is to make use of their natural UV absorbance (Table 2.3). UV detection is suitable for analytes with chromophores which absorb strongly (molar extinction coefficient (ε)>10 000) and with absorbance maxima (λ_{max})well away from that of the HPLC solvents in general use (λ_{max}>220 nm). HPLC–UV detection has been used to detect HETEs, leukotrienes and lipoxins (Jackson et al., 1984; Müller and Sorrell, 1985; Powell, 1983, 1985, 1987; Serhan, 1989). Most prostaglandins of biological interest possess single non-conjugated double bonds and only absorb strongly at low wavelengths (<200 nm) (Terragno et al., 1981). Simple saturated fatty acids and glycerides do not absorb in the UV region. Even with analytes having intense UV absorbances such as the leukotrienes there are dangers in using UV detection for trace amounts in complex biological matrices where there may be many UV-absorbing impurities present, and this is exemplified in Fig. 2.10. Here, plasma containing LTC_4 (10 ng/ml) was extracted and subjected to reverse-phase

HPLC: there is a UV-absorbing species with the retention time of LTC_4, but the absorbance is equivalent to a concentration of over 5 μg/ml. Clearly, any LTC_4 present was hidden by the presence of a co-eluting impurity. This problem can, in part, be addressed by using multiwavelength monitoring. For a given analyte, the ratio of two (or more) wavelengths is constant. If the wavelength ratio of an HPLC peak is different from that of the authentic material then, even if it has the correct retention time, it is not the correct analyte. Multiwavelength monitoring can also help to distinguish compound classes; for example, the A_{280}/A_{254} ratio for LTC_4 (2.6:1) differs from that of LTB_4 and the other diHETEs ($\approx$ 1.3:1) (Fig. 2.11).

3.3.3.1 Derivatization

Derivatization may be used to detect "UV-invisible" compounds by adding a strongly UV-absorbing chromophore with high values of ε, and values of λ_{max} in the 250–270 nm range. Derivatization is normally carried out before HPLC. For example, $PGF_{2\alpha}$, PGE_2 and PGD_2 have been converted to p-bromophenacyl ester derivatives which absorb at 254 nm (Morozowich and Douglas, 1975). The limit of detection of this derivative is approximately 1 ng. Increased sensitivity can be obtained with anthroyloxyphenacyl derivatives (Watkins and Peterson, 1982; Salari et al., 1987); also, these derivatives fluoresce, allowing the detection limit to be reduced to 50 pg.

3.3.4 High-Performance Liquid Chromatography–Radioimmunoassay

HPLC–RIA is an excellent way of overcoming the problems of immunoreactive impurities (Section 3.3.2) and increasing the specificity of RIA (Alam et al., 1979; Richmond et al., 1987). The value of HPLC to identify the analyte and remove impurities can be seen in the analysis of human urine. Urine contains both TXB_2 and 6-oxo-$PGF_{1\alpha}$ eluting at 24–26 and 10–12 ml, respectively (Fig. 2.12). There are, however, two other TXB_2 immunoreactive species and one other 6-oxo-$PGF_{1\alpha}$-like

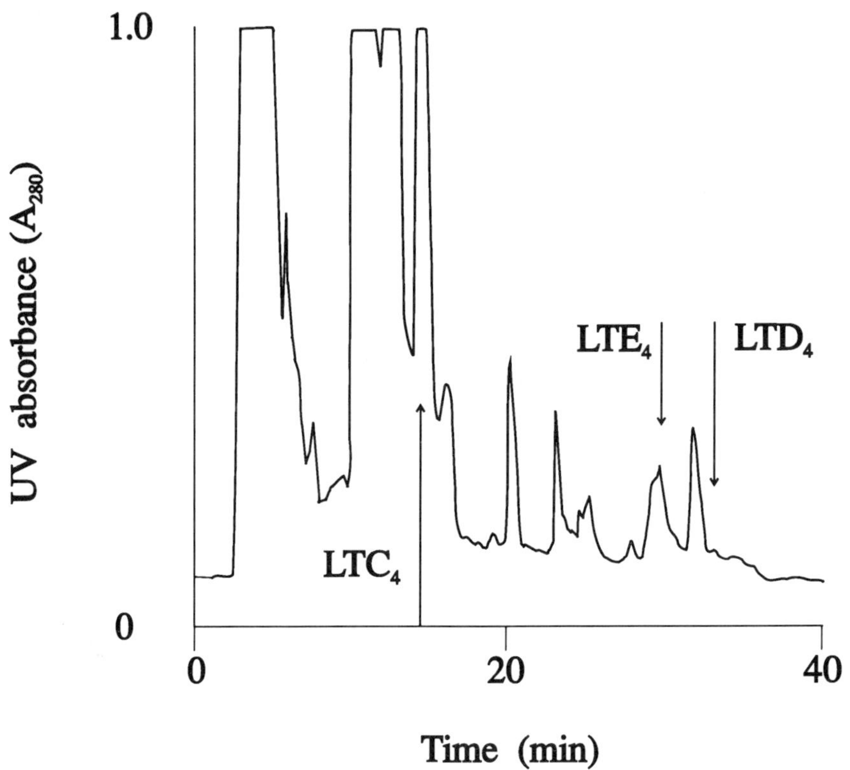

Figure 2.10 HPLC–UV detection in plasma. Human plasma was spiked with 10 ng of LTC$_4$/LTD$_4$/LTE$_4$ analysed by HPLC with UV (A_{280}) detection. Although there are UV-absorbing peaks at the retention times of LTC$_4$, LTD$_4$ and LTE$_4$, they are associated with impurities and not the leukotrienes. *Conditions*: Plasma was passed through a Hypersil ODS column eluting at 2 ml/min with a methanol/water gradient (see Fig. 2.4).

material in urine, which may be accounted for partly by the presence of lactones and partly by immunoreactive impurities. Direct RIA of urine for either of these prostanoids would clearly overestimate the amounts of TXB$_2$ and 6-oxo-PGF$_{1\alpha}$ present. More recently, Mucha and colleagues (1991) demonstrated that urine contains substances which interfere with the assay of 11-dehydro-TXB$_2$, and recommended HPLC purification.

If HPLC–RIA is to be used as a routine assay, it is important to ascertain the HPLC retention time of the analyte, its recovery and the variable background immunoreactivity. This is where internal standards come into play. In the assay for urinary LTE$_4$ used in our department (Richmond *et al.*, 1987; Taylor *et al.*, 1989), trace amounts of [^{3}H]LTE$_4$ are included in the urine on collection. The urine is purified by on-line extraction/HPLC and 10 HPLC fractions are analysed for radio-activity and immunoreactivity (Fig. 2.13). The retention time of LTE$_4$ is defined by the radiolabelled internal standard. Although LTE$_4$ is well separated from other immunoreactive species such as *N*-acetyl-LTE$_4$, it is only accepted when there is a concomitant rise and fall of immunoreactivity (internal standard) and radioactivity. The internal standard also allows us to determine the recovery in each sample (which, as we have already seen

in Table 2.1, is variable) and estimate the background immunoreactivity. HPLC–RIA is now the method of choice for measuring the cysteinyl leukotrienes (Huber *et al.*, 1989; Taylor *et al.*, 1989; Manning *et al.*, 1990; Sampson *et al.*, 1990; Tagari *et al.*, 1990).

3.3.5 Other High-Performance Liquid Chromatography Detection Systems

A number of other detection systems are available for HPLC, including electrochemical detection of LTB$_4$ and the lipoxins (Herrmann *et al.*, 1987, 1988) and radiometric detection of PAF (Wientzek *et al.*, 1985). Both require the sample to be suitably derivatized before analysis and neither method has found widespread support in this area of research. HPLC–thermospray MS is considered in Section 3.6.

3.4 GAS CHROMATOGRAPHY

GC, originally developed in the 1950s, is still widely used today for the routine separation of volatile fatty acid derivatives. The technique is based on the partition of the analyte between a stationary phase and a gaseous phase. Helium and hydrogen are commonly used as carrier gases, but the attendant fire risks with hydrogen

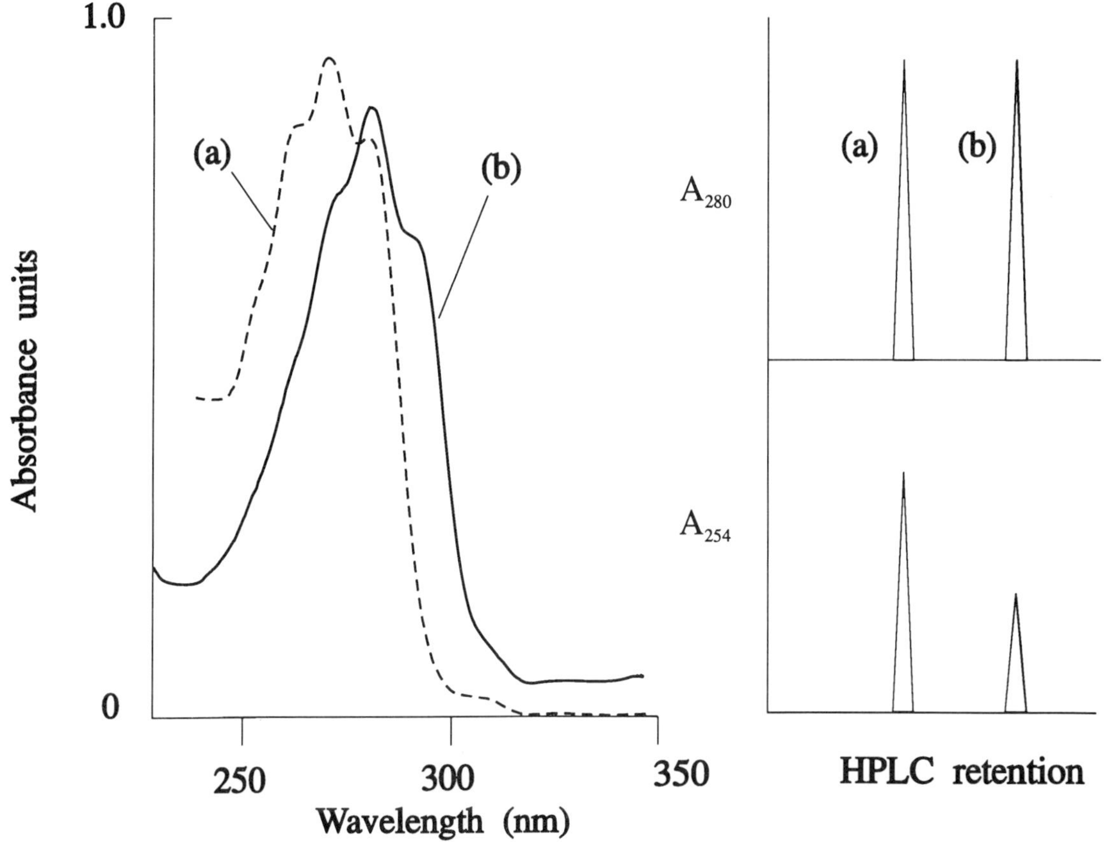

Figure 2.11 Multiple-wavelength monitoring. The left panel shows the ultraviolet spectrum of (a) LTB$_4$ (λ_{max}=270 nm) and (b) LTC$_4$ (λ_{max}=280 nm). The absorbance ratios (A_{280}/A_{254}) for the two compounds are 1.3:1 and 2.6:1, respectively. The two leukotrienes can be distinguished by monitoring these two wavelengths post-HPLC (right panel).

has limited its use in routine GC analysis. As with HPLC, the choice of the GC column is important: modern columns consist of a fine capillary of fused silica to which is bonded the stationary (liquid) phase. These columns can be used at high temperatures (>300°C). There is a wide variety of GC columns available for fatty acid analysis, and these have been reviewed by Ackmann (1984) and Body (1984). The excellent resolution of GC columns and the invariant retention times of analytes offers a high degree of specificity when GC is coupled to a suitable detector.

3.4.1 Detectors

There are a number of GC detectors in general use, of which the FID and ECD are the most common for lipid analysis. With flame ionization, the sample eluting from the GC column is burnt in a hydrogen flame, producing ions which are detected as an electric current; the intensity of the current is dependent both on the type and quantity of analyte. FID is a universal detector which can be used down to 10^{-11}g, but it lacks selectivity: any compound eluting from the GC column will be detected. The ECD is somewhat more selective: only those compounds

containing an electron-capturing group will be detected (Fig. 2.14). For GC–ECD analysis, analytes are converted to electron-capturing derivatives; these include the O-trifluoroacetate, heptafluorobutyrate and pentafluorobenzyl esters (Gyllenhaal *et al.*, 1976; Sonesson *et al.*, 1987). Perhaps the most selective and sensitive detection technique is the mass spectrometer.

3.5 GAS CHROMATOGRAPHY–MASS SPECTROMETRY

GC–MS offers the dual benefits of high-resolution capillary GC coupled with the specificity of mass detection, and has been used successfully for many years for the analysis and identification of prostanoids and other lipid mediators. MS is independent of compound class and offers unrivalled specificity for both qualitative and quantitative analysis. In particular, GC–MS allows detection of multiple analytes in one assay. For analysis by GC and GC–MS, there is an absolute requirement that the samples are volatile and also thermally stable at the operating temperature of the column. This is achieved by converting the analytes into suitable derivatives.

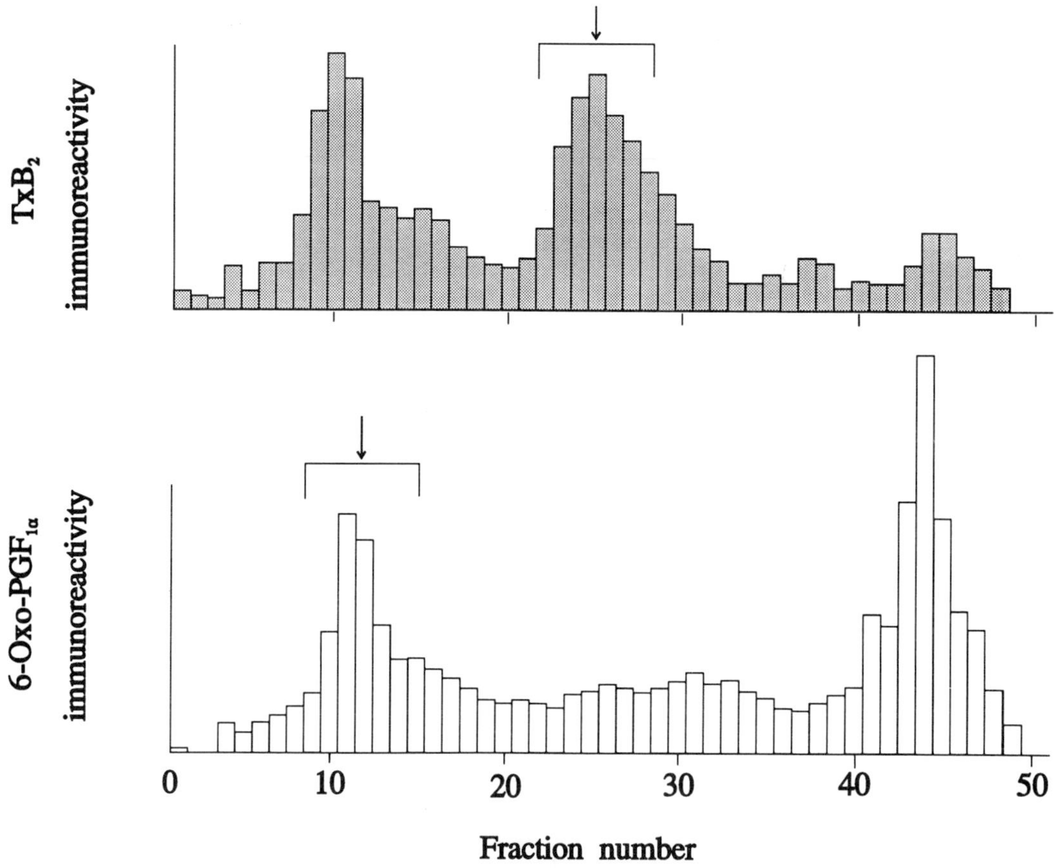

Figure 2.12 HPLC–RIA analysis of TXB$_2$ and 6-oxo-PGF$_{1\alpha}$ in human urine. The elution positions of authentic TXB$_2$ and 6-oxo-PGF$_{1\alpha}$ are arrowed. Multiple immunoreactive peaks are present. *Conditions*: HPLC was carried out at 1 ml/min on a Nova Pak ODS column (Waters Ass) using a 30 min gradient of acetonitrile/water (28:72 to 60:40). RIAs were performed with commercially available kits from Amersham International (K. Moore, unpublished results).

3.5.1   Derivatization

Many fatty acids and their metabolites are neither volatile nor thermally stable because of the presence of polar functional groups such as hydroxyl and carboxyl moieties. Volatility is low because of the strong intermolecular interactions (particularly hydrogen bonding) between analyte molecules, consequently more energy (and a higher temperature) is required to break these intermolecular bonds and vaporize the sample. To reduce these intermolecular interactions, the polar functions must be converted to less polar derivatives (Knapp, 1979) (Table 2.4). Derivatization has two further advantages: it confers stability on many lipids (the thermal stability of methyl esters is much greater than that of the parent acids), and also leads to improved resolution on the GC column, which in turn enhances both sensitivity and specificity. *N*-Acylation, methoxime formation, esterification and ether formation are the most appropriate methods for derivatizing lipids (Barrow and Taylor, 1987). Specific derivatization with fluorinated reagents such as penta-

fluorobenzyl bromide or bistrifluoromethylbenzyl bromide are used for EC detection and GC–ECMS (Fitzgerald *et al.*, 1981; Blair *et al.*, 1982; Strife and Murphy, 1984; Barrow and Taylor, 1987; Catella and Fitzgerald, 1990). For polar compounds such as phosphates, sulphates, glucuronides or phosphatidylcholine derivatives (e.g. PAF), hydrolysis to the hydroxyl group is recommended. This method has been used to convert PAF to a compound suitable for GC analysis (Murphy and Clay, 1987; Haroldsen *et al.*, 1991).

3.5.2   Ionization mode

Once the sample has been derivatized and has passed through the GC column, it enters the mass spectrometer, where it is ionized in either the EI, CI or EC mode (Barrow and Taylor, 1987). Under EI ionization, the analyte is bombarded with a beam of high-energy electrons (70 eV). Collisions occur, and the analyte molecule is converted to a radical cation, the molecular ion

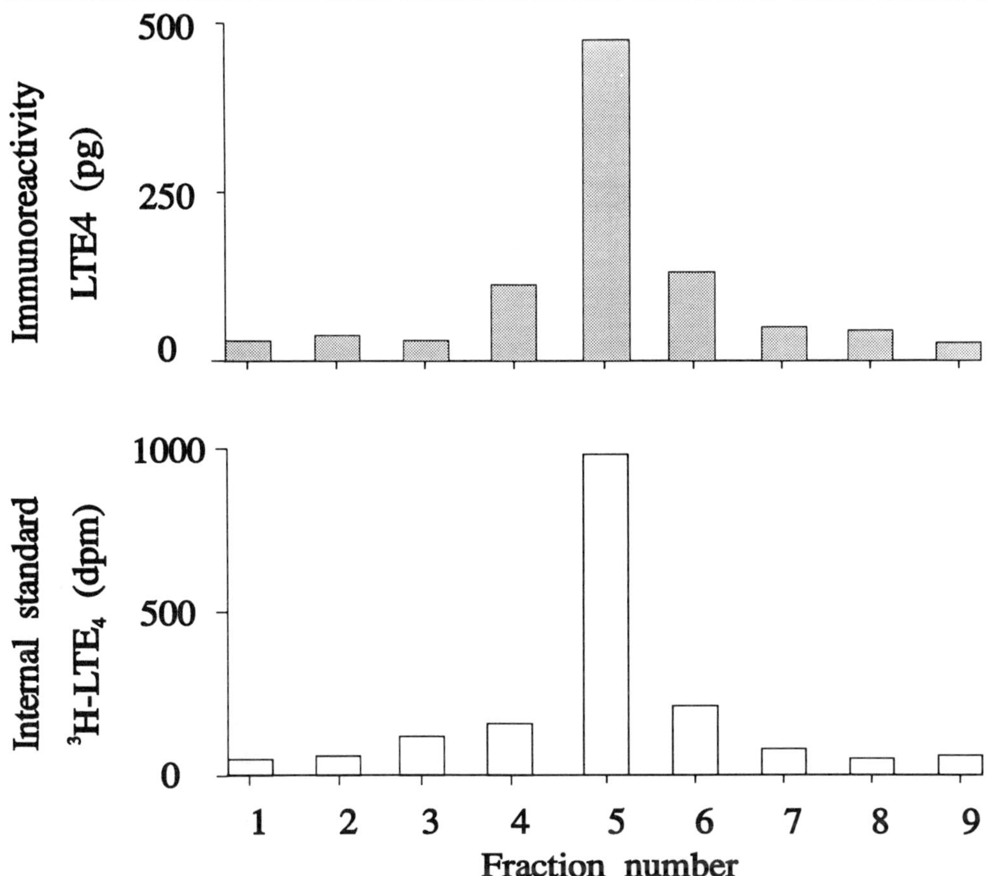

Figure 2.13 HPLC–RIA analysis of urinary LTE$_4$ in the presence of [^{3}H]LTE$_4$ as an internal standard. The elution position and recovery of LTE$_4$ in each sample is obtained from the radioprofile of the internal standard. The presence of LTE$_4$ is only accepted when there is a concomitant rise and fall of radioactivity (internal standard, bottom panel) and immunoreactivity (top panel). Analysis of 10 fractions also allows the true background immunoreactivity to be determined. *Conditions*: HPLC conditions are outlined in Fig. 2.4 (Taylor *et al.*, 1989). LTE$_4$ RIAs were carried out with either a commercial kit (Amersham International) or using an antibody raised in this department (see Fig. 2.5).

$M^{\cdot+}$, with the concomitant formation of a low-energy thermal electron.

$$M + \beta^- \ (70 \ eV) \rightarrow M^{\cdot+} + \beta^- + \beta^- \ (thermal)$$

The molecular ion then breaks down to a number of fragments. The fragmentation pattern can be used to characterize the analyte (Fig. 2.15a). The proportion of the total ion current which is carried by any fragment is low, and thus the sensitivity of an EI-based assay is reduced (Vesterqvist and Green, 1984). Under CI conditions, the sample is surrounded by a reagent gas such as methane. It is this gas which is initially ionized by the electron beam, initially to $CH_4^{\cdot+}$ and then to the secondary ionizing agent, CH_5^+.

$$CH_4 + \beta^- \ (70 \ eV) \rightarrow CH_4^{\cdot+} \ \beta^- + \beta^- \ (thermal)$$
$$CH_4^{\cdot+} + CH_4 \rightarrow CH_5^+ + CH_3^{\cdot}$$

This in turn ionizes the analyte (M) to the protonated species $(M+H^+)$.

$$CH_5^+ + M \rightarrow (M+H^+) + CH_4$$

Fragmentation is reduced, and the intensity of the protonated ion species (or a simple fragment) is often high; however, the sensitivity of the method rarely exceeds that of EIMS (Barrow and Taylor, 1987). By far the most sensitive ionization method is EC (often called negative-ion CI). The initial step is EI ionization of a reagent gas such as methane or ammonia; however, the secondary ionizing species is the low-energy thermal electron formed by this process. Ionization of the analyte occurs by capture of a low-energy thermal electron to form a radical molecular anion ($M^{\overline{\cdot}}$).

$$M + \beta^- \rightarrow (M^{\overline{\cdot}})$$

As few compounds readily capture electrons, derivatization with a fluorinated aromatic group (such as pentafluorobenzyl or, more recently, with the bistrifluoromethylbenzyl group) is required. Fragmentation is very

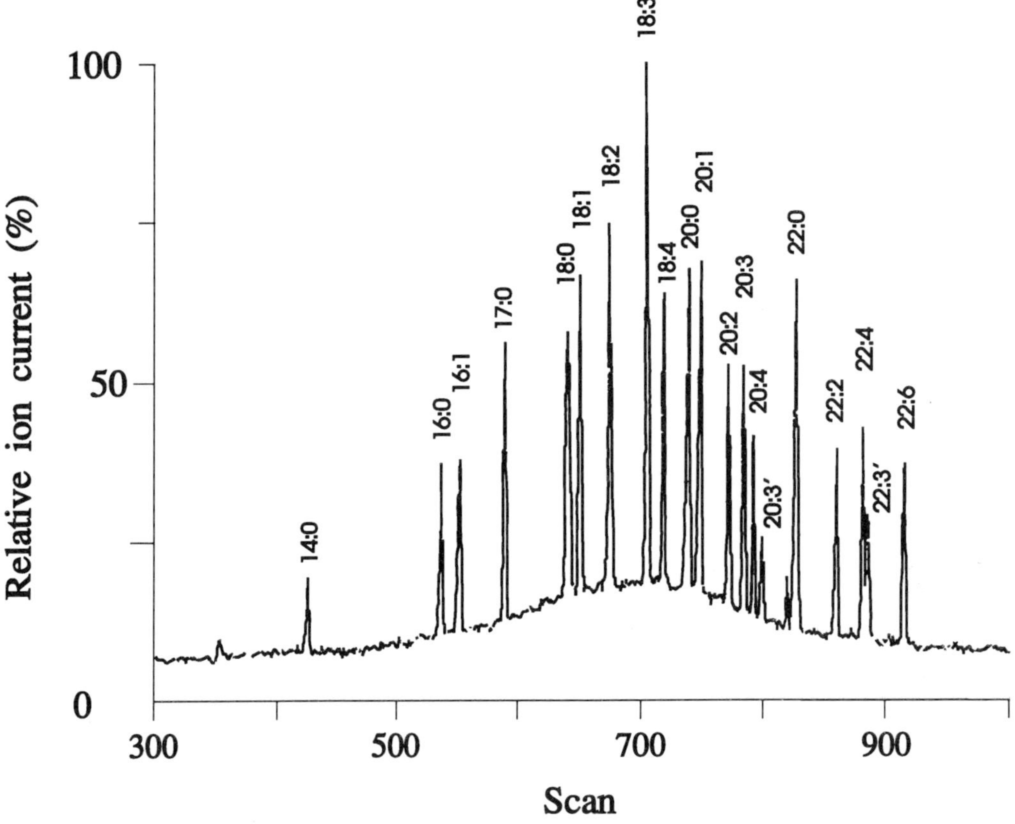

Figure 2.14 GC–EIMS analysis of fatty acid methyl esters. GC can be used to separate a mixture of fatty acids as their methyl esters. The retention time is dependent on chain length ($C_{14} < C_{16} < C_{18} < C_{20} < C_{22}$) and the degree of unsaturation ($C_{18:0} < C_{18:1} < C_{18:2} < C_{18:3} < C_{18:4}$). All unsaturated fatty acids were in the *cis* configuration: isomerization of one double bond to the *trans* configuration results in an increased retention time ($C_{20:3}$, and $C_{22:3}$,). Detection was by EIMS with a single peak observed for each ester. The specificity relies on the high resolution capability of the GC column. Increased specificity can be obtained by monitoring the molecular ion or a suitable fragment ion. *Conditions*: Samples were injected in octane through a Grob injector at 210°C on to a DB5 capillary column. The column was maintained at 100°C for 1 min, followed by a gradient of 10°C/min to 325°C. Helium was used as the carrier gas. MS was undertaken in the EI mode.

simple, with the molecular anion (or a simple fragment) carrying the majority of the ion current (Fig. 2.15b). The sensitivity of the technique is 10–100 times that of either EI or positive-ion CI, and is the method of choice for GC–MS assays where sensitivity is at a premium.

3.5.3   Selected Ion Monitoring

To maximize sensitivity, the mass spectrometer is normally set to detect a small number of ions in the selected ion mode. Using ECMS, as little as 1 pg of a (suitably derivatized) analyte can be detected. As with all assays, the balance between sensitivity and specificity must be considered: a full EI mass spectrum requires more material but offers great specificity, whereas sensitivity is obtained with ECMS and selected ion monitoring, but confidence in the structural identification is reduced. In general, this is offset by the high-resolution capability offered by GC. Operators must always be aware that a

sample with the correct mass and eluting at the correct time can still be associated with impurities.

3.5.4   Prostanoid Analysis by Gas Chromatography–Electron Capture Mass Spectrometry

The value of GC–ECMS can best be illustrated by considering a multi-prostaglandin analysis; PGE_2, PGD_2, $PGF_{2\alpha}$, TXB_2 and 6-oxo-$PGF_{1\alpha}$ can all be quantified in a single assay. The prostanoids (together with their ^{2}H-labelled internal standards) are extracted and converted to their methoxime, *O*-trimethylsilyl ether–pentafluorobenzyl (or bistrifluoromethylbenzyl) derivatives. These derivatives perform well on GC: $PGF_{2\alpha}$, TXB_2 and 6-oxo-$PGF_{1\alpha}$ elute as single peaks and PGE_2 and PGD_2 elute as pairs of *syn* and *anti*-methoxime isomers (Fig. 2.16). Under EC conditions, the fluorinated benzyl group is lost and each prostanoid generates an intense carboxylate anion

Table 2.4 Derivatization of polar functional groups for MS/GC–MS

Functional group	Derivative	Method
Amine	Acetyl $-NH.COCH_3$	Treat with 0.5 ml of methanol/acetic anhydride (3:1 v/v, 3 h). Some lactonization can occur with leukotrienes (Taylor *et al.*, 1983)
		N-Acetylation occurs readily with acetic anhydride in the presence of a basic catalyst such as triethylamine (Hagmann *et al.*, 1986). *N*-Acetyl derivatives are stable
	Heptafluorobutyl/ trifluoracetyl $NH.CO.C_nF_{n+1}$	Suitable for ECMS. Amines react readily with the acid anhydride in dry acetonitrile (1:10 v/v) at room temperature (Blau and King, 1979)
	Bistrifluoromethyl benzoate $-NH.COC_6H_3.(CF_3)_2$	Suitable for ECMS. Dissolve the sample in dry ethyl acetate (50 µl) and add bistrifluoromethylbenzoyl chloride (0.5 µl) and diisopropylethylamine (0.5 µl). Stand at room temperature for 30 min (Murray *et al.*, 1989)
	Bistrifluoromethyl benzylamine $-NH.CH_2C_6H_3.(CF_3)_2$ (or the pentafluorobenzylamine)	Suitable for ECMS. React the amine with bistrifluoromethylbenzyl bromide (5 µl) in dry acetontrile (40 µl) in the presence of diisopropylethylamine (10 µl) for 30 min at room temperature (Murray *et al.*, 1988). Both mono- and di-derivatives are formed.
	Trimethylsilylylamine $-NH.Si(CH_3)_3$	Amines react readily with most silylating reagents to form the trimethylsilyl derivative but are normally protected with other functional groups
Carbonyl	Methoxime $-C=N.OCH_3$	Reaction with 100 µl of a solution of methoxyamine hydrochloride in pyridine (10 mg/ml) for 18 h produces two isomers. Excess reagent can be removed by solvent or solid phase extraction (Taylor and Barrow, 1987)
Carboxylic acid	Methyl ester $-COO.CH_3$	Add fresh ethereal diazomethane (0.5 ml) to a solution of the analyte in methanol and stand for 30 min. The solution should still be yellow at 30 min. Note that diazomethane is explosive, and side-reactions can occur (Blau and King, 1979)
		Prepare methanolic HCl by bubbling HCl gas into 1 ml of dry methanol for 1 min. Reaction with fatty acids occurs generally within 3 h, although benzoic acid reacts more slowly (Taylor *et al.*, 1983)
		Methyl esters can also be prepared with BF_3/methanol or methyl anilinium hydrochloride (Blau and King, 1979)
	Isopropyl ester $-COO.C_3H_7$	Dissolve the substrate in propan-2-ol (100 µl), add acetyl chloride (20 µl) and heat at 70°C for 90 min (Murray *et al.*, 1989). This reaction is not suitable in cases where acetylation may also occur (e.g. with amines)
	Bistrifluoromethyl benzyl ester $-COO.CH_2C_6H_3.(CF_3)_2$	Reaction conditions as for amines, but with an increased reaction time (1 h)
Hydroxyl	*O*-Acetate	*O*-Acetylation requires stronger conditions than *N*-acetylation and can be achieved in pyridine/acetic anhydride (1:10 v/v) or trifluoroacetic anhydride/acetic acid (2:1 v/v) for 1 h at room temperature. Many *O*-acetylated compounds will slowly hydrolyse in solution

Table 2.4 Continued

Functional group	Derivative	Method
	Bistrifluoromethyl benzoate $-OCOC_6H_3(CF_3)_2$	Alcohols can be benzoylated under similar conditions used for amines, but with an overnight reaction.
	O-Trimethylsilyl ether $-O-Si.(CH_3)_3$	There are a number of trimethylsilylating reagents available, with varying strengths as trimethylsilyl donors. In our experience, most fatty acid hydroxyl functions are readily trimethylsilylated by treating with 100 µl of bistrimethylsilyl trifluoroacetamide (BSTFA) at room temperature for 18 h. The reaction can also be carried out at 60°C for 1 h. For hindered hydroxyl groups, a catalyst such as trimethylchlorosilane can be included with BSTFA. Pierce (1979) has published a monograph on silylation of organic compounds. In general, it is best to use trimethylsilylation as the final step in any derivatization process as trimethylsilyl derivatives (particularly those of carboxylic acids) are readily hydrolysed in water. If trimethylsilyl derivatives cannot be analysed immediately on preparation, they should be stored in the silylating reagent at $-20°C$ to prevent hydrolysis
Phospholipids, phosphate, sulphates and glucuronides	Hydrolyse chemically or enzymatically and treat as hydroxyl group	

Unless stated, excess reagents may be removed under a stream of nitrogen. It should be noted that fluorinated benzyl bromides are powerful lachrymators, and suitable precautions must be taken.

Isotopic labelling can conveniently be carried out by using [^{2}H$_3$]methanol (Sigma), [^{2}H$_6$]acetic anhydride (Sigma), [^{2}H$_3$]methoxylamine hydrochloride (Pierce and Warriner) and [^{2}H$_9$]bistrimethylsilyltrifluoroacetamide (MDS Isotopes). [^{2}H]Diazomethane can conveniently be prepared from deuterodiazald (Aldrich).

((M-PFBz)⁻ in Fig. 2.15) which is detected in the selected-ion monitoring mode for increased sensitivity. The ^{2}H-labelled internal standards elute immediately before the ^{1}H species, confirming the retention time of each prostanoid. Using a separate standard curve, these prostanoids can be detected at concentrations as low as 1 pg/ml. The mass-selective benefits of GC–MS also allow the dehydro (PG$_3$ series) analogues of the prostanoids to be assayed (Braden *et al.*, 1990). GC–ECMS using selected-ion monitoring is now used routinely in this and other laboratories where sensitivity is at a premium (Barrow *et al.*, 1989; Catella and Fitzgerald, 1990; Morrow *et al.*, 1991; Taylor *et al.*, 1991).

At this point we must mention the debt owed by many of us to Dr John Pike (Upjohn, Kalamazoo), who for many years was the only source of ^{2}H-labelled prostanoids as internal standards.

3.5.5 Gas Chromatography–Mass Spectrometry Analysis of Oxygenated Lipids and Leukotrienes

As with prostanoids, ECMS offers greater sensitivity over EIMS methods for the analysis of oxygenated arachidonic acid metabolites, and the method is finding wide application. There are a number of EETs which are amenable to GC–MS analysis (Turk *et al.*, 1985). There are some difficulties in achieving good-quality separations and a number of groups have suggested converting the epoxide into its saturated analogue (Toto *et al.*, 1987) or the corresponding diol (Catella *et al.*, 1990) before further derivatization and assay by GC–ECMS. Oliw (1985) also suggested trapping these epoxides as chlorhydrin adducts. The HETEs and diHETEs (including LTB$_4$) can also be measured by GC–ECMS (Murphy, 1984; Mathews *et al.*, 1988; Turk *et al.*, 1988; Holtzman *et al.*, 1989). Under the high-temperature conditions of GC–MS, some of the diHETEs (particularly those with a *trans–cis–trans* orientation) undergo isomerization and rearrangement (Borgeat and Pilote, 1988). This leads to poor chromatography with subsequent reduction in resolution (specificity) and sensitivity.

The cysteinyl leukotrienes LTC$_4$, LTD$_4$ and LTE$_4$ are too polar and thermally labile to pass through a GC column. To overcome this restriction, Balazy and Murphy (1986) used catalytic hydrogenation to convert these compounds to 5-hydroxyeicosanoic acid, which can be measured by GC–MS. As each of the leukotrienes generated the same final product, HPLC separation was

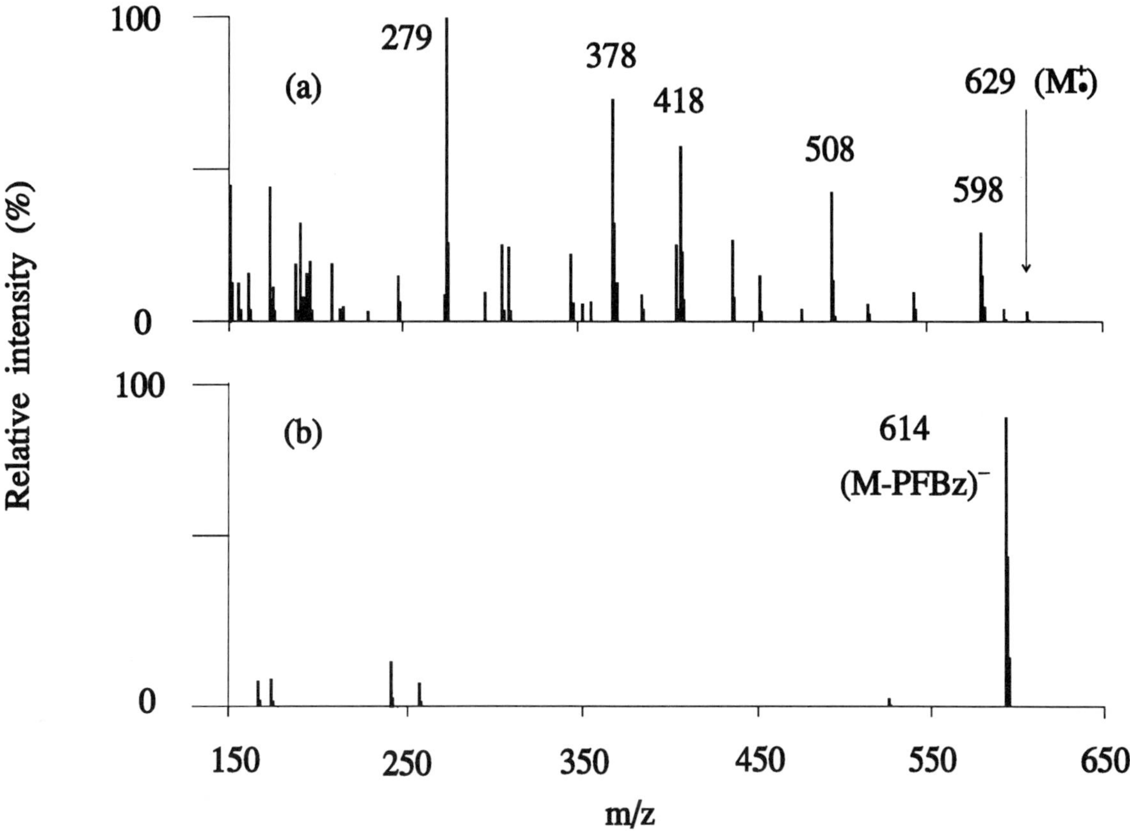

Figure 2.15 EIMS and ECMS spectra of 6-oxo-PGF$_{1\alpha}$. (a) EI mass spectrum (post-GC) of the methoxime, O-trimethylsilyl ether–methyl ester derivative of 6-oxo-PGF$_{1\alpha}$. The intensity of the molecular ion (M$^{+\cdot}$, m/z 629) is low, and fragmentation is extensive. The sensitivity of a GC–EIMS assay is poor (>200 pg). (b) EC mass spectrum (post-GC) of the methoxime, O-trimethylsilyl ether–pentafluorobenzyl (PFBz) ester derivative. The base peak in the spectrum is (M-PFBz)$^{-}$ at m/z 614, corresponding to loss of the PFBz radical. There are few other ions present. Sensitivity in the ECMS mode is 20–50 times that of EIMS, with detection limits at 10 pg on the column. Identical spectra are obtained with the analogous bistrifluoromethylbenzyl esters. Conditions: MS was carried out on a Finnigan 4500 instrument scanning from mass 100–700 over 1 s. The electron energy was set at 70 eV for EIMS and 100 eV for ECMS. Ammonia (0.4 torr) was used as the reagent gas for ECMS.

required before reductive derivatization, and this has limited the use of this method.

3.6   HIGH-PERFORMANCE LIQUID CHROMATOGRAPHY–THERMOSPRAY MASS SECTROMETRY

For GC analysis, there is an absolute requirement for sample volatility, which is usually achieved by the formation of suitable derivatives. Derivatization may, however, be accompanied by unexpected side-reactions, and is not always successful. There are no such requirements for volatility if HPLC is used as the inlet system for a mass spectrometer via the thermospray interface. Thermospray MS is a "soft" form of ionization and generates molecular ion species with little fragmentation

directly post-HPLC and has been applied to a number of fatty acid metabolites including HETEs, prostanoids and leukotrienes (Richmond et al., 1986; Taylor et al., 1986b; Yergey et al., 1986; Kim and Salem, 1987). Care must be taken as some compounds such as LTC$_4$ ionize poorly under thermospray conditions and others decompose. The major limitation of the thermospray technique at present is that of sensitivity: although full spectra can be obtained at the submicrogram level, the limits of detection for assay purposes are insufficient (at 1–5 ng) for the technique to compete effectively with HPLC–RIA or GC–MS for fatty acid metabolites.

3.7   OTHER MASS SPECTROMETRY METHODS

There are a number of other MS methods that have been occasionally used for lipid analysis. Complex mixtures can

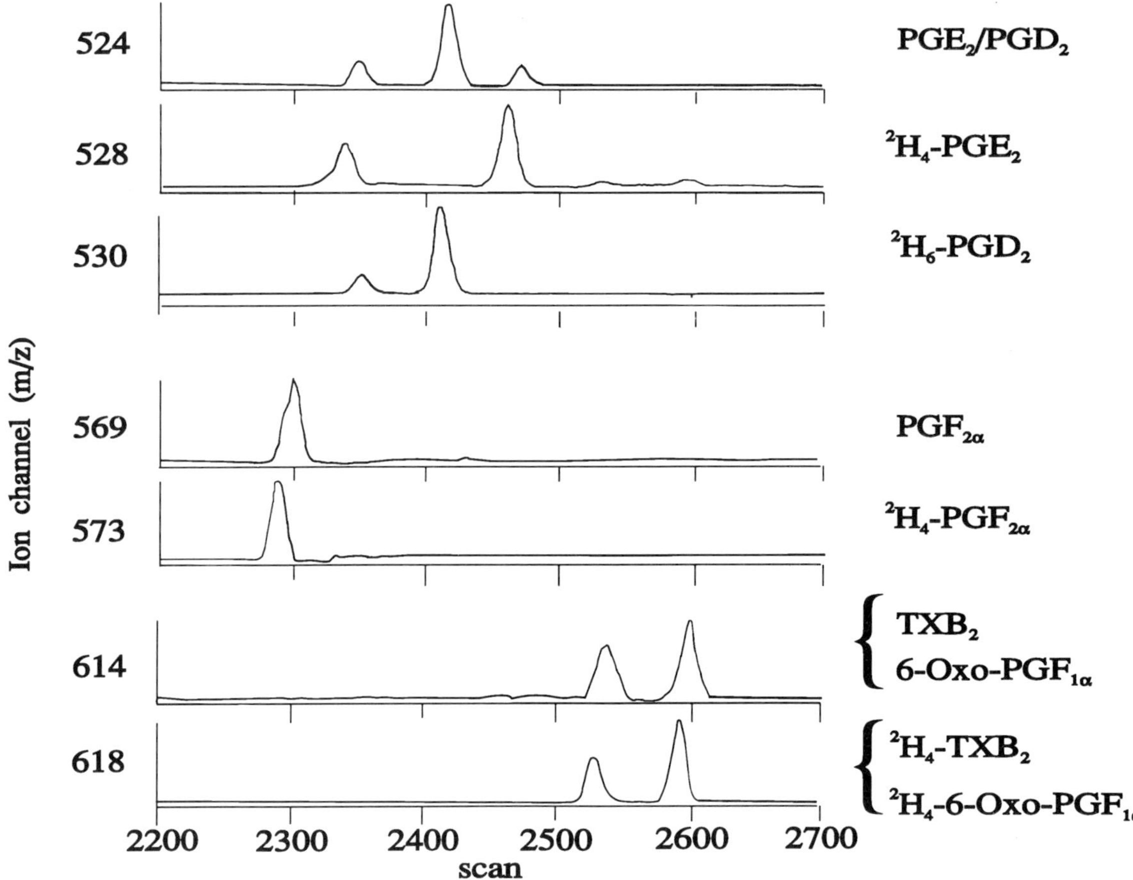

Figure 2.16 GC–ECMS of prostanoids. PGE$_2$, PGD$_2$, PGF$_{2\alpha}$, 6-oxo-PGF$_{1\alpha}$ and TXB$_2$ are analysed as their methoxime, *O*-trimethylsilyl ether–pentafluorobenzyl (or bistrifluoromethylbenzyl) ester derivatives. Each prostanoid generates an intense (M-PFBz)$^-$ ion (see Fig. 2.15). Their ^{2}H-labelled internal standards elute fractionally earlier and are used to define the retention time of each prostanoid, and using a suitable standard curve (see Fig. 2.6) allow accurate quantification. PGE$_2$ and PGD$_2$ are monitored in the *m/z* 524 ion channel and elute from the GC column as *syn* and *anti* isomers (with one set of isomers co-eluting). PGF$_{2\alpha}$ (*m/z* 569), TXB$_2$ and 6-oxo-PGF$_{1\alpha}$ (*m/z* 614) each give a single GC peak. *Conditions*: Derivatized samples were injected in dodecane at 250°C via a Grob injector on to a DB5 capillary GC column. The column was kept at 200°C for 1 min, followed by a gradient of 20°C/min to 325°C. Helium was the carrier gas. The mass spectrometer was operated in the EC mode using ammonia as the reagent gas. ^{2}H-labelled standards were a kind gift from Dr J Pike, Upjohn, Kalamazoo, USA.

be handled with the recently developed method of GC–MS–MS. Here, the specificity of GC–MS is enhanced by using an extra mass spectrometer as a filtering device to remove unwanted ions (Schweer *et al.*, 1990). The technique works well, and is only limited by the technology: not all laboratories have access to GC–MS–MS. Supercritical flow chromatography–MS is a modification of GC–MS which uses carbon dioxide at low temperatures as the eluant. It has been used in the analysis of HETEs (Crabtree and Adler, 1991). Probe techniques are also proving to be of value in lipid analysis. One useful probe technique is fast-atom bombardment (FAB) MS, which enables molecular weight information to be generated from most lipid mediators (Murphy *et al.*, 1982; Taylor *et al.*, 1983). FAB has been used to assay PAF; although it can discriminate between the various

PAF analogues present naturally (Weintraub *et al.*, 1985), it lacks sensitivity and requires relatively pure samples.

4. Group Analysis

Having discussed the various methods available for measuring lipid mediators, the question arises as to which method is most suitable in a particular case. Table 2.5 summarizes the methods for the analysis of a number of lipid groups. The choice of method will depend on the amount and purity of the analyte. Trace amounts of analytes in complex matrices require more complex techniques.

Assay methodology moves forward rapidly and each year sees a number of "novel" assay methods published

Table 2.5 Assay methodology

Method	Analyte	Comments
Bioassay (tissue)	Prostaglandins, LTC, LTD, LTE	A number of tissues respond to these substances. The routine use of bioassays has diminished, but is still valuable in detecting novel substances or when other assays are not available. Specificity can be achieved with the use of antagonists or HPLC
Bioassay (cells)	PAF	Platelet aggregation (in the presence of antagonists of other bioactive substances) is still a convenient way of detecting PAF
	LTB_4 and other diHETEs	Chemotaxis. Blood-derived cells (neutrophils) are readily available. The response of many chemotaxins is bell shaped: multiple dilutions may need to be assayed
RIA (direct)	Prostanoids	RIA kits are readily available. In general, direct RIA suffers from the problems of interfering immunoreactive substances (particularly in blood and urine). Extraction can help to overcome this problem
	Serum TXB_2	The high levels generated when blood clots makes RIA ideal for serum TXB_2 measurements
	LTC, LTD, LTE	Not applicable because of the number of immunoreactive impurities. HPLC should be used
	LTB_4 (from ionophore treatment of cells)	The large amounts generated make the assay more suitable. HPLC should be used occasionally to confirm the identity of LTB_4
	PAF	Immunoassays are now being developed for PAF. There are multiple forms of PAF which may complicate the assay
HPLC–RIA	Prostanoids	Laborious, with the possibility of isomerization and lactonization on the column. Not recommended if GC–MS is available
	Leukotrienes	The method of choice for leukotrienes. Ideally, a radiolabelled internal standard should be used
HPLC–UV	Fatty acids, PAF and glycerides, most prostaglandins	Generally not used because of the lack of a good chromophore ($\lambda_{max} < 220$ nm). Derivatization with chromophores or fluorophors has extended the value of the method
	PGB_2	PGB_2 has been used as a UV-absorbing internal standard for leukotriene analysis
	HETEs	These can be detected at 234 nm. This is a useful method for estimating radical-induced polyunsaturated fatty acid breakdown
	Leukotrienes	Useful for measurements in simple matrices such as cell supernatants. Dual-wavelength monitoring can be used to distinguish between $LTC_4/LTD_4/LTE_4$ and LTB_4
GC	Fatty acids	Samples must be derivatized. Electron-capturing derivatives offer the greatest sensitivity. Specificity is poor
GC–MS	Fatty acids and simple analogues	Analysis in the EI mode (structure) or EC mode (sensitivity) are suitable

Table 2.5 Continued

Method	Analyte	Comments
	PAF	PAF cannot be analysed directly, but must first be chemically converted to the diacylglycerol. This entails a loss of specificity
	Prostaglandins	Recommended method for prostaglandin analysis. A number of prostaglandins can be assayed at one time (with suitable internal standards)
	LTB_4 and other diHETEs	GC–MS is generally suitable. Some diHETEs isomerize and decompose on the GC column
	LTC_4, LTD_4, LTE_4	Not directly amenable to GC–MS. One suitable method (although not in general use) involves converting the leukotrienes to 5-hydroxyeicosanoic acid before GC–MS analysis
	EETs	GC resolution of the various EETs is difficult. These compounds may be reduced and/or hydrolysed to their dihydroxy derivatives before analysis
Other MS methods	Prostaglandins	GC–MS–MS offers further enhancements in specificity (and some increase in sensitivity) over GC–MS
	All lipids	All lipids will generate thermospray mass spectra. The method lacks sensitivity for routine use
	PAF	Probe methods such as FAB–MS can be used. FAB will discriminate the naturally occurring PAF analogues but lacks sensitivity

in specialist journals. Unfortunately, many of these methods are never heard of again. The test of a good assay is that it is recommended by others, is used routinely and the data are reported in "main-line" biological or medical journals. In our laboratory, we use GC–MS for the analysis of prostanoids and HPLC–RIA for leukotrienes, and would recommend these techniques.

5. Concluding Remarks

Even when the assay methodology has been worked out, there are still a number of traps for the unwary. The dangers of *ex vivo* generation of the analyte on collection must be considered. For example, the basal level of TXB_2 (<5 pg/ml) can be increased markedly by platelet activation on venepuncture. This may be overcome by measuring a circulating metabolite of TXB_2 (e.g. 11-dehydro-TXB_2) or urinary metabolites (2,3-dinor-TXB_2) (Patrono, 1989). If urinary metabolites are to be measured, the effect of renal and hepatic function on excretion rate is important and also must be considered (Huber *et al.*, 1989; Moore *et al.*, 1990, 1991).

Although we have discussed the problems of sample deterioration on storage (which may be overcome by inclusion of internal standards), there is a potentially more serious complication: that of analyte formation. Morrow and colleagues (1990) have recently demonstrated that PGF compounds were formed in plasma on storage by non-enzymatic processes.

The analytical data must make biological sense. Plasma measurements which appear accurate in every way but which correlate with impossibly high circulating levels of bioactive lipids must be treated with caution: either the chemistry or the biology is wrong.

It is essential to understand the limitations of any assay and to realize that there is always a balance between the needs of sensitivity and those of specificity. In our view, it is more important to know what you are measuring than to have an "accurate" quantitative result on an unknown.

6. References

Ackman, R.G. (1984). In "CRC Handbook of Chromatography: Lipids", Vol 1 (ed H.K. Mangold), pp 95–240. CRC Press, Boca Raton.

Alam, I., Ohuchi, K. and Levine, L. (1979). Determination of cyclooxygenase products and prostaglandin metabolites using high pressure liquid chromatography and radioimmunoassay. Anal. Biochem. 93, 339–345.

Balazy, M. and Murphy, R.C. (1986). Determination of sulfidopeptide leukotrienes in biological fluids by gas chromatography/mass spectrometry. Anal. Chem. 58, 1098–1101.

Barrow, S.E. and Taylor, G.W. (1987). In "Prostaglandins and Related Substances: A Practical Approach" (eds S. Nigam and T. Slater), pp 99–141. IRL Press, Oxford.

Barrow, S.E., Ward, P.S., Sleightholm, M., Ritter, J.M. and Dollery, C.T. (1989). Cigarette smoking: profiles of thromboxane and prostacyclin products in human urine. Biochim. Biophys. Acta 993, 121–127.

Blair, I.A., Barrow, S.E., Waddell, K.A., Lewis, P.J. and Dollery, C.T. (1982). Prostacyclin is not a circulating hormone in man. Prostaglandins 23, 579–588.

Blau, K. and King, G.K. (1978). "The Handbook of Derivatives for Chromatography". Heyden, London.

Body, D.R. (1984). In "CRC Handbook of Chromatography: Lipids", Vol 1 (ed H.K. Mangold), pp 241–275. CRC Press, Boca Raton.

Boeynams, J.M. Brash, A.R., Oates, J.A. and Hubbard, W.C. (1980). Preparation and assay of monohydroxyeicosatetraenoic acids. Anal. Biochem. 104, 259–267.

Borgeat, P. and Picard, S. (1988). 19-Hydroxyprostaglandin B_2 as an internal standard for on-line extraction–high performance liquid chromatography analysis of lipoxygenase products. Anal. Biochem. 171, 283–289.

Borgeat, P. and Pilote, S. (1988). Rearrangement of 5S,12S-dihydroxy-6,8,10,14-(E,Z,E,Z) eicosatetraenoic acid during gas chromatography: formation of a cyclohexadiene derivative. Prostaglandins 35, 723–731.

Braden, G.A., Knapp, H.R., Fitzgerald, D.J. and FitzGerald, G.A. (1990). Dietary fish oil accelerates the response to coronary thrombolysis with tissue-type plasminogen activator. Evidence for a modest platelet inhibitory effect *in vivo*. Circulation 82, 178–187.

Bray, M. (1983). Pharmacology and pathophysiology of leukotriene B_4. Br. Med. Bull. 39, 249–254.

Camp, R.D.R., Woollard, P.M., Mallet, A.I. and Fincham, N.J. (1982). Neutrophil aggregating and chemokinetic properties of a 5,12,20-trihydroxy-6,8,10,14-eicosatetraenoic acid isolated from human leukocytes. Prostaglandins 23, 631–640.

Catella, F. and Fitzgerald, G.A. (1990). Methods Enzymol. 86, 42–50.

Catella, F., Lawson, J.A., Fitzgerald, D.J. and Fitzgerald, G.A. (1990). Endogenous biosynthesis of arachidonic acid epoxides in humans: increased formation in pregnancy-induced hypertension. Proc. Natl Acad. Sci. USA 87, 5893–5897.

Chiabrando, C., Begnini, A., Picinnilli, A., Carminati, C., Cozzi, E., Remuzzi, G. and Fanelli, R. (1987). Antibody mediated extraction/negative ion mass spectrometric measurement of thromboxane A_2 in human and rat urine. Anal. Biochem. 163, 255–262.

Crabtree, D.V. and Adler, A.J. (1991). Derivatisation of hydroxyeicosatetraenoic acid methyl esters with pentafluorobenzoic anhydride and analysis with supercritical flow chromatography chemical ionisation mass spectrometry. J. Chromatogr. 543, 405–411.

Czarnetzki, B.M. and Rosenbach, T. (1986). Chemotaxis of human neutrophils and eosinophils towards leukotriene B_4 and its 20-ω-oxidation products *in vitro*. Prostaglandins 31, 851–857.

Fitzgerald, G.A., Brash, A.R., Falardeau, P. and Oates, J.A. (1981). Estimated rate of prostacyclin secretion into the circulation of normal man. J. Clin. Invest. 68, 1272–1275.

Frolich, J.C. and the Group of Standardisation of Methods in Eicosanoid Research (1984). Measurement of eicosanoids. Prostaglandins 27, 349–368.

Granstrom, E.E. and Kindahl, H. (1978). Radioimmunoassay of prostaglandins and thromboxanes. Adv. Prostaglandin Thromboxane Res. 5, 119–210.

Granstrom, E.E. and Kindahl, H. (1990). A critical approach to eicosanoid assay. Adv. Prostaglandin Thromboxane Res. 5, 295–302.

Gyllenhaal, O., Brottel, H. and Harvig, P. (1976). Determination of free fatty acids as pentafluorobenzyl esters by electron capture gas chromatography. J. Chromatogr. 129, 295–302.

Hagmann, W., Denzingler, C., Rapp, S., Weckbecker, G. and Keppler, D. (1986). Identification of the major endogenous leukotriene metabolite in the bile of rats as N-acetyl leukotriene E_4. Prostaglandins 31, 239–251.

Haroldsen, P.E., Gaskell, S.J., Weintraub, S.T. and Pinckard, R.N. (1991). Isotopic exchange during derivatisation of platelet activating factor for gas chromatography–mass spectrometry. J. Lipid Res. 32, 723–729.

Hayes, E.C., Lombardo, D.L., Girard, Y., Maycock, A., Rokach, J., Rosenthal, A.S., Young, R.N., Egan, R.W. and Zweerink, H.J. (1983). Measuring leukotrienes of slow reacting substance of anaphylaxis: development of a specific radioimmunoassay. J. Immunol. 131, 429–433.

Heavey, D.J., Soberman, R.J., Lewis, R., Spur, B. and Austen, K.F. (1987). Critical considerations in the development of an assay for sulfidopeptide leukotrienes in plasma. Prostaglandins 33, 693–708.

Henson, P.M. (1990). Methods Enzymol. 187, (eds R.C. Murphy and F.A. Fitzpatrick) 130–134.

Herrmann, T., Steinhilber, D. and Roth, H.J. (1987). Determination of leukotriene B_4 by high-performance liquid chromatography with electrochemical detection. J. Chromatogr. 416, 170–175.

Herrman, T., Steinhilber, D., Knospe, J. and Roth, H.J. (1988). Determination of picogram amounts of lipoxin A_4 and lipoxin B_4 by high-performance liquid chromatography with electrochemical detection. J. Chromatogr. 428, 237–245.

Holtzman, M.J., Turk, J. and Pentland, A.A. (1989). Regiospecific monooxygenase with novel stereopreference is the major pathway for arachidonic acid oxygenation in isolated epidermal cells. J. Clin. Invest. 84, 1446–1453.

Huber, M., Kastner, S., Scholmerich, J., Gerok, W. and Keppler, D. (1989). Analysis of cysteinyl leukotrienes in human urine: enhanced excretion in patients with liver cirrhosis and hepatorenal syndrome. Eur. J. Clin. Invest. 19, 53–60.

Jackson, E.M., Mott, G.E., Hoppens, C., McManus, L.M., Weintraub, S.T., Ludwig, J.C. and Pinckard, R.N. (1984). High performance liquid chromatography of platelet-activating factors. J. Lipid Res. 25, 753–756.

Kim, H.Y. and Salem, N. Jr (1987). Application of thermospray high-performance liquid chromatography/mass spectrometry for the determination of phospholipids and related compounds. Anal. Chem. 59, 722–726.

Knapp, D. (1979). "The Handbook of Analytical Derivatisation Reactions". Wiley, Chichester.

Lands, W.E. and Smith, W.L. (eds) (1982). Prostaglandins and arachidonate metabolites. Methods Enzymol. 86.

Lawson, J.A., Brash, A.R., Doran, J. and Fitzgerald, G.A.

(1985). Measurement of urinary 2,3 dinorthromboxane B_2 and thromboxane B_2 using bonded-phase phenylboronic acid columns and capillary gas chromatography negative ion chemical ionisation mass spectrometry. Anal. Biochem. 150, 463–470.

Lellouche, F., Fradin, A., Fitzgerald, D.J. and Maclouf, J. (1990). Enzyme immunoassay measurement of the urinary metabolites of thromboxane A_2 and prostacyclin. Prostaglandins 40, 297–310.

Liston, T.E. and Roberts, L.J.II (1985). Transformation of prostaglandin D_2 to $9\alpha,11\beta$ (15S)-trihydroxyprosta-(5Z,13E)-dien-1-oic acid ($9\alpha,11\beta$ prostaglandin F_2): a unique biologically active prostaglandin produced enzymatically *in vivo* by humans. Proc. Natl Acad. Sci. USA 82, 6030-6034.

Maltby, N.H., Taylor, G.W., Ritter, J.M., Moore, K., Fuller, R.W. and Dollery, C.T. (1990). The metabolism of leukotriene C_4 in man. J. Allergy Clin. Immunol. 85, 3–9.

Manning, P.J., Rokach, J., Malo, J.L., Ethier, D., Cartier, A., Girard, Y., Charleson, S. and O'Byrne, P.M. (1990). Urinary leukotriene E_4 levels during early and late asthmatic responses. J. Allergy Clin. Immunol. 86, 211–220.

Mathews, W.R., Bundy, G.L., Wynalda, M.A., Guido, D.M., Schneider, W.P. and Fitzpatrick, F.A. (1988). Development and comparative evaluation of radioimmunoassay and gas chromatographic/mass spectrometric procedures for determination of leukotriene B_4. Anal. Chem. 60, 349–353.

Metz, S.A., Hall, M.E., Harper, T.W. and Murphy, R.C. (1982). Rapid extraction of leukotrienes from biological fluids and quantitation by high-performance liquid chromatography. J. Chromatogr. 233, 193–201.

Mitchell, M.D. (1978). A sensitive radioimmunoassay for 6-oxo-$PGF_{1\alpha}$: preliminary observations on circulating concentrations Prostaglandins Med. 1, 13–21.

Moon, D.G., Van der Zee, H., Morton, K.D., Krasodomski, J.A., Kaplan, J.E. and Fenton, J.W.II (1990). Platelet activating factor and sheep platelets: a sensitive new bioassay. Thromb. Res. 57, 551–554.

Moore, K.P., Taylor, G.W., Maltby, N.H., Siegers, D., Fuller, R.W., Dollery, C.T. and Williams, R. (1990). Increased production of cysteinyl leukotrienes in hepatorenal syndrome. J. Hepatol. 11, 263–271.

Moore, K.P., Ward, P.S., Taylor, G.W. and Williams, R. (1991). Systemic and renal production of thromboxane A_2 and prostacyclin in hepatorenal syndrome. Gastroenterology 100, 1069–1077.

Morozowich, W. and Douglas, S.L. (1975). Resolution of prostaglandin *p*-nitrophenacyl esters by liquid chromatography and conditions for rapid *p*-nitrophenacylation. Prostaglandins 10, 19–40.

Morrow, J.D., Harris, T.M. and Roberts, L.J. II (1990). Noncyclooxygenase formation of a series of novel prostaglandins: analytical ramifications for measurement of eicosanoids. Anal. Biochem. 184, 1–10.

Morrow, J.D., Prakash, C., Awad, J.A., Duckworth, T.A., Zackert, W.E., Blair, I.A., Oates, J.A. and Jackson Roberts, L. (1991). Quantification of the major urinary metabolite of prostaglandin D_2 by a stable isotope dilution mass spectrometric assay. Anal. Biochem. 193, 142–148.

Mucha, I., Riutta, A. and Vapaatalo, H. (1991). Factors affecting the reliability of direct radioimmunoassay for urinary 11-dehydrothromboxane B_2 with ^{125}I-labelled ligand: albumin and heterogeneous immunoreactivity. Eicosanoids 4, 1–7.

Müller, M. and Sorrell, T.C. (1985). Quantitation of sulfido-peptide leukotrienes by reversed-phase high-performance liquid chromatography. J. Chromatogr. 343, 213–218.

Murphy, R.C. (1984). Mass spectrometric quantitation and analysis of leukotrienes and other 5-lipoxygenase metabolites. Prostaglandins 28, 597–601.

Murphy, R.C. and Clay, K.L. (1987). Measurement of platelet activating factor by physicochemical techniques. Am. Rev. Resp. Dis. 136, 207–210.

Murphy, R.C. and Fitzpatrick, F.A. (eds) (1990). Arachidonic acid and related lipid mediators. Methods Enzymol. 187.

Murphy, R.C., Mathews, W.R., Rokach, J. and Fenselau, C. (1982). Comparison of biological-derived and synthetic leukotriene C_4 by fast atom bombardment mass spectrometry. Prostaglandins 23, 201–206.

Murray, S., Gooderham, N.J., Boobis, A.R. and Davies, D.S. (1988). Measurement of MeIQx and DiMeIQx in fried beef by capillary gas chromatography electron capture negative ion chemical ionisation mass spectometry. Carcinogenesis 9, 321–325.

Murray, S., O'Malley, G., Taylor, I.K., Mallet, A.I. and Taylor, G.W. (1989). Assay for N^τ-methylimidazoleacetic acid, a major metabolite of histamine, in urine and plasma using capillary column gas chromatography-negative ion mass spectrometry. J. Chromatogr. 491, 15–25.

Oliw, E.H. (1985). Analysis of epoxyeicosatrienoic acids by gas chromatography–mass spectrometry using chlorohydrin adducts. J. Chromatogr. 339, 175–181.

Parker, C.W., Huber, M.M. and Falkenheim, S.F. (1982). Pharmacological characteristics of slow reacting substances. Methods Enzymol. 86, (eds W.E. Lands and W.L. Smith) 655–667.

Patrono, C. (1989). Measurement of thromboxane synthesis in man. Eicosanoids 2, 249–251.

Peck, M.J., Piper, P.J. and Williams, T.J. (1981). The effect of leukotrienes C_4 and D_4 on the microvasculature of guinea-pig skin. Prostaglandins 21, 315–321.

Pickett, W.C. and Douglas, M.B. (1985). Automated extraction and HPLC resolution of lipoxygenase and cyclooxygenase products utilising high pressure column switching. Prostaglandins 29, 83–90.

Pierce, A.E. (1979). "Silylation of Organic Compounds". Pierce Chemical Co., Rockford, IL.

Piper, P.J. (1986). In "Leukotrienes: Their Biological Significance" (ed P.J. Piper), pp 59–66. Raven Press, New York.

Powell, W.S. (1980). Rapid extraction of oxygenated metabolites of arachidonic acid from biological samples using octadecyl silica. Prostaglandins 20, 947–957.

Powell, W.S. (1983). Separation of unlabelled metabolites of arachidonic acid from their deuterium and tritium labelled analogs by argentation high pressure liquid chromatography. Anal. Biochem. 128, 93–103.

Powell, W.S. (1985). Reversed phase high pressure liquid chromatography of arachidonic acid metabolites formed by cyclooxygenases and lipoxygenases. Anal. Biochem. 148, 59–69.

Powell, W.S. (1987). Precolumn extraction and reverse phase high pressure liquid chromatography of prostaglandins and leukotrienes. Anal. Biochem. 164, 117–131.

Reinke, M., Piller, M. and Brune, K. (1989). Development of an enzyme-linked immunosorbent assay of thromboxane A_2 using a monoclonal antibody. Prostaglandins 37, 577–586.

Richmond, R., Clarke, S.R., Watson, D., Chappell, G.C., Dollery, C.T. and Taylor, G.W. (1986). Generation of hydroxyeicosatetraenoic acids by human inflammatory cells: analysis by thermospray liquid chromatography mass spectrometry. Biochim. Biophys. Acta 881, 159–166.

Richmond, R., Turner, N.C., Maltby, N., Heavey, D., Vial, J., Dollery, C.T. and Taylor, G.W. (1987). A single step procedure for the extraction and purification of leukotrienes C_4, D_4 and B_4. J. Chromatogr. 417, 241–251.

Salari, H., Yeung, M., Douglas, S. and Morozowich, W. (1987). Detection of prostaglandins by high performance liquid chromatography after conversion to p-(9-anthroxyloxy) phenacyl esters. Anal. Biochem. 165, 220–229.

Salmon, J.A. (1978). A radioimmunoassay for 6-keto prostaglandin $F_{1\alpha}$. Prostaglandins 15, 383–395.

Salmon, J.A., Simmons, P.M. and Palmer, R.M.J. (1982). A radioimmunoassay for leukotriene B_4. Prostaglandins 24, 225–235.

Samhoun, M.N. and Piper, P.J. (1984). The combined use of isolated strips of guinea-pig lung parenchyma and ileum as a sensitive and selective bioassay for leukotriene B4. Prostaglandins 27, 711–724.

Sampson, A.P., Spencer, D.A., Green, C.P., Piper, P.J. and Price, J.F. (1990). Leukotrienes in the sputum and urine of cystic fibrosis children. Br. J. Clin. Pharmacol. 30, 861–869.

Schweer, H., Meese, C.O. and Seyberth, H.W. (1990). Determination of 11α-hydroxy-9,15-dioxo-2,3,4,5,20 pentanor-19-carboxyprostanoic acid and 9α,11α-dihydroxy-15-dioxo-2,3,4,5,20 pentanor-19-carboxyprostanoic acid by gas chromatography negative ion chemical ionisation triple stage quadrupole mass spectrometry. Anal. Biochem. 189, 54–58.

Serhan, C.N. (1989). On the relationship between leukotriene and lipoxin production by human neutrophils: evidence for differential metabolism of 15-HETE and 5-HETE. Biochim. Biophys. Acta 1004, 158–168.

Sonesson, A., Larson, L. and Jiminez, J. (1987). Use of pentafluorobenzyl and pentafluoropropionyl–pentafluorobenzyl esters of bacterial fatty acids for gas chromatographic analysis with electron capture detection. J. Chromatogr. 417, 366–380.

Strife, R.J. and Murphy, R.C. (1984). Preparation of pentafluorobenzyl esters of arachidonic acid lipoxygenase metabolites: analysis by gas chromatography negative ion chemical ionisation mass spectrometry. J. Chromatogr. 305, 3–12.

Sugatani, J., Lee, D.Y., Hughes, K.T. and Saito, K. (1990). Development of a novel scintillation proximity assay for platelet activating factor measurement: comparison with bioassay and GC–MS techniques. Life Sci. 46, 1443–1450.

Tagari, P., Rasmussen, J.B., Delorme, D., Girard, Y., Erikson, L.O., Charleson, S. and Ford-Hutchinson, A.W. (1990). Comparison of urinary leukotriene E_4 and 16-carboxytetranordihydro leukotriene LTE_4 excretion in allergic asthmatics after inhaled antigen. Eicosanoids 3, 75–80.

Taylor, G.W., Morris, H.R., Beaubien, B. and Clinton, P.M. (1983). In "Leukotrienes and Other Lipoxygenase Products" (ed P.J. Piper), pp 277–282. Research Studies Press/Wiley, Chichester.

Taylor, G.W., Chappell, G.C., Clarke, S.R., Heavey, D.H., Turner, N.C., Watson, D. and Dollery, C.T. (1986a). In "Leukotrienes: Their Biological Significance" (ed P.J. Piper), pp 67–90. Raven Press, New York.

Taylor, G.W., Watson, D., Dollery, C.T. and Richmond, R. (1986b). In "Topics in Lipid Research: From Structural Elucidation to Biological Function" (eds R.A. Klein and B. Schmitz), pp 85–93. Royal Society of Chemistry, London.

Taylor, G.W., Taylor, I.K., Maltby, N.H., Black, P., Turner, N., Fuller, R.W. and Dollery, C.T. (1989). Urinary leukotriene E_4 following allergen challenge and in patients with acute asthma and allergic rhinitis. Lancet i, 584–588.

Taylor, I.K., Ward, P.S., O'Shaughnessy, K.M., Dollery, C.T., Black, P., Barrow, S.E., Taylor, G.W. and Fuller, R.W. (1991). Thromboxane A_2 biosynthesis in acute asthma and following antigen challenge. Am. Rev. Resp. Dis. 143, 119–125.

Terragno, A., Rdzik, R. and Terragno, N.A. (1981). High performance liquid chromatography and UV detection for the separation and quantitation of prostaglandins. Prostaglandins 21, 101–112.

Toto, R., Siddhanta, A., Manna, S., Pramanik, B., Falck, J.R. and Capdevilla, J. (1987). Arachidonic acid epoxygenase: detection of epoxyeicosatrienoic acids in human urine. Biochim. Biophys. Acta 919, 132–139.

Turk, J., Wolf, B.A., Comens, P.G., Colca, J., Jakschik, B. and McDaniel, M.L. (1985). Arachidonic acid metabolism in isolated pancreatic islets. Negative ion mass spectrometric quantitation of monoxygenase product synthesis by liver and islets. Biochim. Biophys. Acta 835, 1–7.

Turk, J., Stump, W.T., Wolf, B.A., Easom, R.A. and McDaniel, M.L. (1988). Quantitative stereochemical analysis of subnanogram amounts of 12-hydroxy-(5,8,10,14)-eicosatetraenoic acid by sequential chiral phase liquid chromatography and stable isotope dilution mass spectrometry. Anal. Biochem. 174, 580–588.

Vane, J.R. (1983). Adventures and excursions in bioassay – the stepping stones to prostacyclin. Br. J. Pharmacol. 79, 821–838.

Vermylen, J., Defreyn, G., Carreras, L.F., Machin, S.J., Schaeren, J.V. and Verstraete, M. (1981). Thromboxane synthetase inhibition as antithrombotic strategy. Lancet ii, 1073–1075.

Vesterqvist, O. and Green, K. (1984). Development of a GC–MS method for quantitation of 2,3 dinor-6-keto-$PGF_{1\alpha}$ and determination of the urinary excretion rates in healthy humans under normal conditions and following drugs. Prostaglandins 28, 139–154.

Vrbanac, J.J., Cox, J.W., Eller, T.D. and Knapp, D.R. (1990). Immunoaffinity purification-chromatographic analysis of arachidonic acid metabolites Methods Enzymol. 187, (eds R.C. Murphy and F.A. Fitzpatrick) 62–70.

Watkins, W.D. and Peterson, M.B. (1982). Fluorescent/ ultraviolet absorbing ester derivative formation and analysis of eicosanoids by high-pressure liquid chromatography. Anal. Biochem. 125, 30–40.

Weintraub, S.T., Ludwig, J.C., Mott, G.E., McManus, L.M., Lear, C. and Pinckard, R.N. (1985). Fast atom bombardment identification of molecular species of platelet activating factor produced by stimulated human polymorphonuclear leukocytes. Biochem. Biophys. Res. Commun. 129, 868–875.

Wientzek, M., Arthur, G., Man, R.Y.K. and Choy, P.C. (1985). A sensitive method for the quantitation of lysophosphatidylcholine in canine heart. J. Lipid Res. 26, 1166–1169.

Yergey, J.A., Kim, H.Y. and Salem, N. Jr (1986). High-performance liquid chromatography/thermospray mass spectrometry of eicosanoids and novel oxygenated metabolites of docosahexaenoic acid. Anal. Chem. 58, 1344–1348.

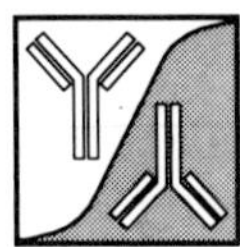

3. Biological Properties of Cyclooxygenase Products

J.R. Vane and R.M. Botting

Lipid Mediators
ISBN 0–12–198875–9

1. *In the Cardiovascular System*

The major prostanoid products involved in maintaining homeostasis in the cardiovascular system are TXA_2 formed by platelets and PGI_2 released by the vascular endothelium, although small amounts of other prostaglandins are also generated.

1.1   THROMBOXANE A_2

Blood platelets convert arachidonic acid through the cyclooxygenase pathway to the potent pro-aggregatory and vasoconstrictor eicosanoid TXA_2, which is the major cyclooxygenase product formed by platelets, along with small amounts of PGE_2 and $PGF_{2\alpha}$.

TXA_2 has a chemical half-life at body pH and temperature of 30 s, breaking down to the inactive TXB_2. The activity of the rabbit aorta-contracting substance released from sensitized guinea-pig lungs, first described by Piper and Vane (1969), can be accounted for by TXA_2. TXA_2 constricts large blood vessels, has variable vasoconstrictor activity in the microcirculation and is a potent stimulus for platelet aggregation. The enzyme that synthesizes TXA_2 from prostaglandin endoperoxides was first localized in the high-speed particulate fraction of human and horse blood platelets (Moncada et al., 1976a; Needleman et al., 1976). The enzyme has been solubilized and separated from the cyclooxygenase (Diczfalusy et al., 1977; Hammarstrom and Falardeau, 1977; Yoshimoto et al., 1977). Thromboxane synthase from human platelet "microsomes" was isolated and characterized as a cytochrome P-450-containing protein with a molecular mass of 58 800 kDa (Haurand and Ullrich, 1985). This enzyme promotes conversion of PGH_2 to TXB_2 (derived from TXA_2) and to HHT (with an unknown biological function) in a 1:1 ratio.

Purification of TXA_2 synthase from human platelets (Nüssing et al., 1990) led to the cloning and sequencing of its cDNA (Yokoyama et al., 1991) and expression in insect cells using a recombinant baculovirus (Yokoyama et al., 1993). TXA_2 synthase from human lung has also been cloned and sequenced (Ohashi et al., 1992).

Using polyclonal and monoclonal antibodies directed against purified human platelet thromboxane synthase and an ELISA, the enzyme content of different human cells and tissues has been measured (Nüsing and Ullrich, 1990). After platelets, blood monocytes have the highest content of thromboxane synthase, whereas lung fibroblasts and promyelocytic cells have only very low levels. Contrary to previous reports, no enzyme was found in purified human polymorphonuclear leucocytes (Goldstein et al., 1978; Conti et al., 1987). Of the tissues examined, lung and liver have the highest content, with low levels of enzyme in spleen, kidney, brain, lymph nodes and gallbladder. However, it may also be that, in some instances, the production of TXA_2 by tissues is due to the presence of migratory cells such as histiocytes or macrophages. High TXB_2 production by some organs such as the kidney may be the result of an inflammatory reaction, with monocytes migrating into the tissue and differentiating into macrophages (Morrison et al., 1978). Certain vascular tissues can also produce TXA_2, including human umbilical artery, rabbit pulmonary artery and cultured bovine endothelial cells (Tuvemo et al., 1976; Salzman et al., 1980; Ingerman-Wojenski et al., 1981).

TXA_2 has been synthesized and its structure confirmed (Bhagwat et al., 1985). Its potent pro-aggregatory effects are due to receptor-activated phosphoinositide metabolism (Siess et al., 1983). The platelet receptors for TXA_2 can be additionally stimulated by the endoperoxide, PGH_2. Platelet responses to both PGH_2 and TXA_2 are blocked by the same antagonists and the putative receptors have been designated TXA_2/PGH_2 receptors (Burch et al., 1985). Stimulation of these receptors on platelets leads to activation of a PLC followed by liberation of DAG and activation of PKC. The resultant influx of extracellular calcium, together with mobilization of calcium from intracellular stores, leads to shape change and aggregation of the platelets (Takahara et al., 1990).

1.2   PROSTACYCLIN

In 1976, Vane and his colleagues discovered that blood vessels made a previously unknown prostanoid, which they called PGX (Moncada et al., 1976b), and later renamed prostacyclin (Johnson et al., 1976). This is a bicyclic eicosanoid with a short chemical half-life at physiological pH of approximately 3 min, which degrades to the relatively inactive substance 6-oxo-$PGF_{1\alpha}$. As for TXA_2 in platelets (see earlier), synthesis of prostacyclin begins with the release of arachidonic acid from cell membrane phospholipids by PLA_2 or indirectly by PLC. Arachidonic acid is the substrate for PGH synthase, a 71 kDa haem-containing enzyme (Markey et al., 1987) which transforms arachidonic acid to the unstable prostaglandin PGG_2 (cyclooxygenase activity) and PGG_2 to the unstable prostaglandin PGH_2 (peroxidase activity). PGH_2 is then acted upon by microsomal prostacyclin synthase to form prostacyclin (Gryglewski et al., 1976). Cyclooxygenase is inhibited by aspirin and other aspirin-like drugs (Vane, 1971; Flower and Vane, 1972) and its activity is modulated by hydroperoxides, including 15-HPETE, hydrogen peroxide and PGG_2. These peroxides also affect the activity of prostacyclin synthase. For example, the activity of cyclooxygenase is stimulated by low (10^{-10}–10^{-7} M) and inhibited by high ($>10^{-6}$ M) concentrations of lipid hydroperoxides (Warso and Lands, 1983). Lipid hydroperoxides, e.g. HPETEs, are always inhibitory for prostacyclin synthase (Gryglewski et al., 1976; DeWitt and Smith, 1983). This inhibitory action of hydroperoxides may be mediated by oxygen-derived free radicals such as the hydroxyl radical OH·, which are formed during the reduction of HPETEs to HETEs (Egan et al., 1976; Sagone et al., 1980). Indeed, the inhibition of cyclooxygenase from bull seminal vesicles by high concentrations of hydrogen peroxide is reversed by uric acid or mannitol, which are known OH· scavengers (Deby *et al.*, 1984). Hydroperoxides have no effect on thromboxane synthase activity (Ham *et al.*, 1979).

Prostacyclin is the main product of arachidonic acid metabolism by endothelial cells of the larger arteries and veins but it is also formed by endothelial cells of microvessels (Gerritsen, 1987). The ability of the large vessel wall to synthesize prostacyclin is greatest at the intimal surface and progressively decreases towards the adventitia (Moncada *et al.*, 1977). Production of prostacyclin by cultured cells from vessel walls also shows that endothelial cells are the most active producers of prostacyclin (Weksler *et al.*, 1977; MacIntyre *et al.*, 1978), although smooth muscle cells also make considerable amounts. Stripping of the endothelium from a rabbit aorta *in vivo* removed virtually all ability of the luminal surface to produce prostacyclin from exogenously added arachidonic acid. Recovery of prostacyclin production to 60% of that of uninjured vessels took about 35 days and still had not returned completely after 70 days (Weksler *et al.*, 1982).

A number of endogenous substances promote the release of prostacyclin from endothelial cells. These include thrombin, arachidonic acid, PGH_2, adenine nucleotides, trypsin, substance P and bradykinin (for a review, see Gryglewski *et al.*, 1988). In addition, pulsatile pressure releases prostacyclin from isolated arteries (Quadt *et al.*, 1982; Pohl *et al.*, 1987), which could be an important release mechanism for prostacyclin *in vivo*. However, circulating levels of prostacyclin in plasma are low (3 pg/ml plasma; Blair *et al.*, 1982), indicating that prostacyclin released by the endothelial cells probably acts as a local mediator on the underlying smooth muscle.

The release of prostacyclin following receptor activation is initiated by a rapid increase in intracellular calcium above a threshold level (>0.8 mM), which then declines but remains above resting levels for several minutes (Pearson *et al.*, 1983). The initial increase in intracellular calcium is due to the release of calcium by IP_3, itself released as a consequence of stimulation of PLC, whereas the sustained elevation of calcium is due to an entry of calcium from extracellular sources. The nature of this influx has not yet been elucidated but it does not occur through voltage-operated calcium channels since these do not exist in endothelial cells (Hallam and Pearson, 1986). The increase in intracellular calcium then stimulates PLA_2 and thus the release of arachidonic acid.

The release of prostacyclin is dependent on intracellular but not extracellular calcium, is transient and decreases upon repeated stimulation with the same agonist (Pearson *et al.*, 1983; Luckhoff and Busse, 1986; White and Martin, 1989). This desensitization does not depend upon a feedback effect of prostacyclin through activation of adenylate cyclase (Czervionke *et al.*, 1979; Brotherton *et al.*, 1982).

Prostacyclin potently inhibits aggregation of platelets, an effect mediated by stimulation of platelet adenylate cyclase which results in accumulation of cAMP (Tateson *et al.*, 1977) and stimulation of PKA. Thus, prostacyclin initiates the phosphorylation of several proteins which are essential for platelet function. These include the 22 000 kDa polypeptide on the dense tubular system, the myosin light-chain kinase and the glycoprotein Ib (GPIb) receptor involved in the binding of von Willebrand factor to the platelet. Furthermore, by inhibiting the activity of PLA_2 and PLC, prostacyclin decreases the production of DAG, phosphatidic acid, arachidonic acid and IP_3 (Siess, 1989). Prostacyclin only weakly inhibits adhesion of platelets to thrombogenic surfaces. Its anti-adhesive efficacy varies according to the nature of the exposed surface (Adelman *et al.*, 1981). Inhibition by aspirin of endothelial cell cyclooxygenase, and therefore of prostacyclin release, does not increase the adhesion of platelets to monolayers of endothelial cells (Hoak *et al.*, 1985); however, it does decrease the thromboresistance of rabbit aortic endothelial cells to whole citrated blood (Gryglewski, 1990). The pro-aggregatory effect induced by inhibition of cyclooxygenase in endothelial cells is explained by removal of the synergism between prostacyclin and EDRF, derived either from endothelial or circulating blood cells, which on its own is insufficient to inhibit platelet aggregation (Radomski *et al.*, 1987).

Apart from its platelet suppressant and vasodilator actions (for a review, see Dusting *et al.*, 1986), prostacyclin has other actions which may be protective for the endothelial lining. For example, it promotes the outflow of free cholesterol from endothelial cells, suppresses the accumulation of cholesterol esters by macrophages, inhibits the release of growth factors from endothelial cells, platelets and macrophages (Willis *et al.*, 1986, 1987; Willis and Smith, 1989), and attenuates the release of free radicals and mediators from white blood cells (Kainoh *et al.*, 1990). The fibrinolytic (Dembinska-Kiec *et al.*, 1982; Szczeklik *et al.*, 1983) and cytoprotective actions of prostacyclin are well documented but the mechanisms underlying these effects are poorly understood (see Gryglewski *et al.*, 1988). A deficiency in prostacyclin formation by blood vessel walls occurs in atherosclerosis and diabetes, whereas its overproduction is associated with endotoxic shock (Nawroth *et al.*, 1984).

1.3 PROSTACYCLIN AND THROMBOXANE BALANCE

Prostacyclin and TXA_2 are both formed from the endoperoxide PGH_2, derived from arachidonic acid freed from the phospholipids of cell membranes. TXA_2 is an unstable ($t_{1/2} = 30$ s at 37°C), powerful vasoconstrictor agent and aggregator of platelets. Prostacyclin is also unstable ($t_{1/2} = 3$ min at 37°C) but it induces vasodilatation and inhibits platelet aggregation (Vane, 1982).

It was proposed in 1976 that TXA_2 and prostacyclin

represent the opposite poles of a homeostatic mechanism for regulation of platelet aggregation *in vivo* (Moncada *et al.*, 1976a). Evidence has been accumulating in recent years that EDRF (now identified as nitric oxide) produced by endothelial cells and circulating neutrophils may additionally be part of this regulating system (for a review, see Vane *et al.*, 1990a).

Prostacyclin disperses platelet aggregates *in vitro* (Moncada *et al.*, 1976b; Ubatuba *et al.*, 1979) and in the circulation of humans (Szczeklik *et al.*, 1978a). Moreover, it inhibits thrombus formation in models using the carotid artery of the rabbit (Ubatuba *et al.*, 1979) and the coronary artery of the dog (Aiken *et al.*, 1979), protects against sudden death (thought to be due to platelet clumping) induced by intravenous arachidonic acid in rabbits (Bayer *et al.*, 1979) and inhibits platelet aggregation in pial venules of the mouse when applied locally (Rosenblum and El Sabban, 1979).

As discussed earlier, prostacyclin inhibits platelet aggregation by stimulating adenylate cyclase, leading to an increase in platelet cAMP levels (Tateson *et al.*, 1977). In this respect prostacyclin is much more potent than either PGE_1 or PGD_2 and its effect is longer-lasting. Prostacyclin inhibits platelet aggregation (platelet–platelet interaction) at much lower concentrations than those needed to inhibit adhesion (platelet–collagen interaction) (Higgs *et al.*, 1978a). Thus, the loss of EDRF/nitric oxide generation by removal of or damage to endothelial cells plus prostacyclin formation by deeper tissues will permit platelets to stick as a monolayer to vascular tissue and to interact with it. This will allow platelets to participate in the repair of a damaged vessel wall while at the same time preventing or limiting thrombus formation. In addition, platelets adhering to a site where prostacyclin synthase is present could well feed the enzyme with endoperoxide, thereby producing prostacyclin and preventing other platelets from clumping on to the adherent platelets, reinforcing the limitation of the cells to a monolayer (Marcus *et al.*, 1980).

Shortly after balloon de-endothelialization of the aortae of rabbits there appeared a closely adherent layer of spread platelets. A small reduction of adherent platelets was observed in animals receiving prostacyclin at 50–100 ng/kg/min). Only high infusion rates of 650–850 ng/kg/min) inhibited this platelet adhesion (Adelman *et al.*, 1981). Eldor *et al.* (1981) have demonstrated that balloon catheter de-endothelialization of the rabbit aorta abolishes the capacity for generation of prostacyclin in the luminal surface, which then recovers slowly over a period of 70 days, concomitant with the appearance of neo-intimal cells on the vessel surface. The authors also observed in the de-endothelialized areas a "carpet of platelets" which slowly disappeared during re-endothelialization. Thus, the loss of EDRF/nitric oxide production from endothelial cells allowed platelet adhesion, but the generation of prostacyclin by the subendothelium prevented platelet aggregates from forming, since prostacyclin potently inhibits aggregation but not adhesion.

All this work suggests that prostacyclin, although not responsible for all the thromboresistant properties of the vascular endothelium, plays a very important part in limiting the deposition of platelets to a monolayer, without the formation of aggregates.

1.4   PROSTACYCLIN AND THROMBOXANE A_2 IN DISEASE

A number of cardiovascular, thrombotic diseases have been associated with an imbalance in the prostacyclin–thromboxane system. Platelets from patients with arterial thrombosis, deep venous thrombosis, or recurrent venous thrombosis produce more prostaglandin endoperoxides and TXA_2 than normal platelets and have a shortened survival time (Lagarde and Dechavanne, 1977). Platelets from rabbits made atherosclerotic by a high-fat diet (Shimamoto *et al.*, 1978) or from patients who have survived myocardial infarction (Szczeklik *et al.*, 1978b; Henriksson *et al.*, 1986) are abnormally sensitive to aggregating agents and produce more TXA_2 than controls. Elevated TXB_2 levels have been demonstrated in the blood of patients with Prinzmetal's angina (Lewy *et al.*, 1979) and vasotonic angina (Robertson *et al.*, 1981). Hirsh and colleagues (1981) also studied TXB_2 levels in coronary sinus blood of patients with unstable angina. They concluded that local TXA_2 release is associated with recent episodes of angina but were unable to distinguish whether the release was the cause or an effect. Patients with arteriosclerosis obliterans and diabetes mellitus had more TXB_2 and less 6-oxo-$PGF_{1\alpha}$ in their plasma than control patients without these diseases (Udvardy *et al.*, 1987).

TXA_2 produced during ligation of the coronary artery of the dog causes arrhythmias (Coker *et al.*, 1981) and vasoconstriction induced by TXA_2 in the gastric mucosa of the dog results in gastric ulceration (Whittle *et al.*, 1981). During transplant rejection there is an increased level of TXB_2 excreted in the urine preceding the acute crisis (Foegh *et al.*, 1981); furthermore, there are high levels of TXB_2 in the venous blood of patients suffering from endotoxin shock (Reines *et al.*, 1982).

However, measurements of TXA_2 and prostacyclin concentrations in peripheral blood are subject to sampling errors (FitzGerald *et al.*, 1983a). Assays of the 2,3-dinor metabolites of these eicosanoids in the urine provide a more accurate estimate of their true *in vivo* rate of synthesis (Lawson *et al.*, 1985; Patrono *et al.*, 1986). By measuring urinary 2,3-dinor-TXB_2 and 2,3-dinor-6-oxo-$PGF_{1\alpha}$, increased synthesis of TXA_2 and prostacyclin was shown in patients during the development of acute myocardial infarction and in unstable angina (Fitzgerald *et al.*, 1986; Henriksson *et al.*, 1986). The largest rise in TXA_2 formation occurred in patients with unstable angina, indicating a high rate of platelet activation (Fitzgerald *et al.*, 1986). Administration of aspirin only reduced the urinary concentrations of 2,3-dinor-TXB$_2$ in a proportion of the unstable angina patients, highlighting the problems in assessing the fluctuating TXA$_2$ production in this disease (Henriksson *et al.*, 1986). The increased prostacyclin synthesis during these cardiovascular disorders may reflect a compensatory response of the vascular endothelium.

In general, it seems that in diseases where there is a tendency for thrombosis to develop, TXA$_2$ production is elevated whereas prostacyclin production may be either elevated or reduced. The opposite is found in some diseases associated with an increased bleeding tendency.

1.5 ANTITHROMBOTIC ACTION OF ASPIRIN

1.5.1 Mechanism of Action

The studies of Vane (1971) and Smith and Willis (1971) demonstrated that aspirin irreversibly inhibits the cyclooxygenase enzyme of guinea-pig lung homogenates and of human platelets, thus preventing the synthesis of all arachidonic acid metabolites. This inhibition results from the irreversible acetylation of the cyclooxygenase component of prostaglandin-endoperoxide synthase, leaving the peroxidase activity of the enzyme unaffected (Van der Ouderaa *et al.*, 1980). Acetylation of the residue Ser530 in the platelet enzyme occurs (Roth and Majerus, 1975; Roth *et al.*, 1975; Roth and Siok, 1978). This places a bulky group close to the active site of the enzyme and hinders access of the substrate to the active site. Ser530 is not itself part of the active site since its replacement by an alanine residue does not affect cyclooxygenase activity. However, the mutant enzyme can no longer be acetylated and inactivated by aspirin (DeWitt *et al.*, 1990). Ram seminal vesicle cyclooxygenase has been cloned (DeWitt and Smith, 1988) and only a single copy of the cyclooxygenase gene was found in the sheep. However, some biochemical and pharmacological studies suggest the existence of isoenzymes and this has been confirmed (Raz *et al.*, 1989; Xie *et al.*, 1991). There are at least two forms of cyclooxygenase, a constitutive enzyme (COX-1) and a structurally different enzyme (COX-2) which is induced by inflammatory stimuli. The expression of COX-2 is inhibited by glucocorticoids (Masferrer *et al.*, 1990; Kujubu and Herschman, 1992), suggesting another important way in which these steroids have an anti-inflammatory action.

In contrast to the irreversible action of aspirin, other NSAIDs such as indomethacin or ibuprofen produce reversible cyclooxygenase inhibition by competing with the substrate, arachidonic acid, for the active site of the enzyme (Vane *et al.*, 1990b).

1.5.2 Clinical Studies

During the past two decades, numerous randomized clinical trials have taken place assessing the efficacy of aspirin in the prevention of thrombosis. These have covered a wide spectrum of thrombotic conditions encompassing coronary artery and cerebrovascular occlusion, peripheral vascular disease, postoperative deep-vein thrombosis, pulmonary embolism and occlusion of physiological or prosthetic vascular grafts (for reviews, see De Gaetano *et al.*, 1986; Reilly and FitzGerald, 1988). Although these trials have evaluated aspirin for the prevention of arterial occlusion, a possible fibrinolytic action may contribute to the overall antithrombotic effect (Gryglewski, 1970).

1.5.3 Selectivity of Low Doses of Aspirin

Once inhibition of platelet cyclooxygenase by more than 95% has been achieved, a further increase in the dose of aspirin will not produce any increase in antithrombotic effect (Patrono, 1989). Daily dosing with 324 mg of aspirin will therefore not achieve any more suppression of platelet function than daily administration of 50 mg. Both doses inhibited *ex vivo* thromboxane-dependent platelet aggregation and prolonged *in vivo* bleeding time to the same extent (De Caterina *et al.*, 1985).

However, administration of higher doses of aspirin increases the likelihood of side-effects, particularly gastrointestinal toxicity. Evidence from large-scale clinical trials indicates that the occurrence of side-effects is dose-related in the range of doses between 160 and 1500 mg daily, whereas antithrombotic efficacy is not (Antiplatelet Trialists' Collaboration, 1988).

Likewise, Ritter *et al.* (1989) have shown a clear separation between the inhibition of cyclooxygenase of the platelets and that of the blood vessel wall. After treatment with 600 mg of aspirin, the urinary excretion of thromboxane metabolites dropped to zero and did not recover within the experimental period of 6 h. The levels of prostacyclin derivatives also dropped to almost zero, but recovered completely within 3–5 h. During daily administration of 20–60 mg of aspirin, measurements of 6-oxo-PGF$_{1\alpha}$ and 2,3-dinor-6-oxo-PGF$_{1\alpha}$ in urine have indicated less than 30% inhibition of prostacyclin production (Patrignani *et al.*, 1982; FitzGerald *et al.*, 1983b; Benigni *et al.*, 1989).

Thus, aspirin potently inhibits TXA$_2$ generation by platelets in doses which largely spare prostacyclin synthesis in the vessel wall. This is attributed partly to the recovery of cyclooxygenase in endothelial cells and partly to the pharmacokinetics of aspirin after absorption. Platelets encounter orally administered aspirin in the presystemic circulation before it is deacetylated in the liver and diluted by venous blood (Pedersen and FitzGerald, 1984). Since they cannot synthesize fresh cyclooxygenase enzyme, TXA$_2$ production is irreversibly abolished. On the basis of these observations, slow-release aspirin preparations are being formulated to "dribble" aspirin into the presystemic circulation at a slow rate over many hours with the aim of ablating TXA$_2$ formation in platelets without disturbing prostacyclin biosynthesis in the vessel wall (Charman *et al.*, 1992).

1.6 CONCLUSIONS

The most convincing argument for the involvement of TXA_2 in thrombotic conditions is that aspirin, which potently inhibits its synthesis in platelets, also significantly reduces the risk of formation of intravascular thrombi. This is supported by evidence from measurements of TXB_2 levels in plasma and 2,3-dinor-TXB_2 concentrations in urine, which are raised in many thrombotic conditions and suppressed by aspirin. It is also significant that subthreshold concentrations of pro-aggregatory agents, such as TXA_2, can synergize with other endogenous aggregating substances and provide an amplifying signal for these other platelet agonists (Murray and FitzGerald, 1989). Prevention of TXA_2 synthesis by aspirin can, therefore, exert a substantial influence on the formation of intra-arterial thrombi.

The chronic administration of small doses of aspirin to produce a selective inhibition of platelet cyclooxygenase while largely leaving prostacyclin synthesis by the vessel wall unaffected is an aim which is actively being pursued.

2. In Inflammation

Prostaglandins are produced in pathological processes and have biological activities which contribute to the pathology. The most convincing evidence that they are mediators of disease, however, comes not only from their ability to mimic the effects of the disease but also from the demonstration that compounds which selectively interfere with their synthesis have therapeutic effects. For many years the mechanism of action of aspirin and other NSAIDs was ill-defined. The discovery that individual and chemically diverse members of this large group of drugs all act by inhibiting cyclooxygenase (Vane, 1971) has provided a unifying explanation of their therapeutic actions and has firmly established certain prostaglandins as important mediators of inflammatory disease (for reviews, see Higgs *et al.*, 1984; Vane and Botting, 1990).

2.1 INFLAMMATORY PROPERTIES OF CYCLOOXYGENASE PRODUCTS

Phospholipase is activated when tissues are subjected to mechanical, chemical or immunological stimulation. This liberates arachidonic acid as a substrate for cyclooxygenase and, since inflammation is the response of living tissue to injury, prostaglandin production always accompanies the inflammatory response. PGE_2 is the predominant eicosanoid detected in inflammatory conditions ranging from experimental acute oedema and sunburn through to chronic arthritis in humans. Because inflammation is one of the few conditions in which PGE_2 is a major product of cyclooxygenase, it is possible that the process of inflammation directs the enzymatic pathway towards this product.

PGE_2 is a potent dilator of vascular smooth muscle,

accounting for the characteristic vasodilatation and erythema (redness) seen in acute inflammation (Solomon *et al.*, 1968). The effect of vasodilatation is to increase the flow of blood through inflamed tissues and this augments the extravasation of fluid (oedema) caused by agents which increase vascular permeability such as bradykinin and histamine (Williams and Peck, 1977). PGE_2 also acts synergistically with other mediators to produce inflammatory pain. Without having any direct pain-producing activity, PGE_2 sensitizes receptors on afferent nerve endings to the actions of bradykinin and histamine (Ferreira, 1972). Finally, PGE_2 is a potent pyretic agent and its production, probably stimulated by the release of IL-1, in bacterial and viral infections contributes to the fever associated with these diseases (Saxena *et al.*, 1979).

Many other cyclooxygenase products have been detected in inflammatory lesions. These include $PGF_{2\alpha}$, PGD_2, prostacyclin (as 6-oxo-$PGF_{1\alpha}$) and TXA_2 (as TXB_2), but usually they are present at less than a quarter of the concentrations of PGE_2 (Trang *et al.*, 1977; Sturge *et al.*, 1978; Bombardieri *et al.*, 1981; Egg, 1984). Of these products, prostacyclin is probably the most important in terms of inflammatory signs. Prostacyclin has a similar vasodilator potency to that of PGE_2 and is a more potent hyperalgesic agent than PGE_2. It is likely, therefore, that both PGE_2 and prostacyclin contribute to the development of inflammatory erythema and pain (Higgs *et al.*, 1978b).

The concentration of a PGE_2-like substance was about 20 ng/ml in the synovial fluid of patients with rheumatoid arthritis. This decreased to zero in patients taking aspirin, a good demonstration of its effect on prostaglandin synthesis clinically (Higgs *et al.*, 1974). Carrageenin-impregnated polyester sponges implanted subcutaneously in rats were used to induce experimental inflammation (Simmonds *et al.*, 1983). Periodic examination of the inflammatory exudate contained within the sponges showed that the concentration of PGE_2 increased throughout the 24 h experiment. In addition, the output of TXA_2 and LTB_4 increased to a peak after 4–6 h and then declined over the remaining duration of the experiment. PGE_2 causes vasodilatation and hyperalgesia, and the chemotactic property of LTB_4 probably attracts PMNs into the region (Ford-Hutchinson *et al.*, 1984). However, the role (if any) of TXA_2 in the inflammatory response is not understood.

Evidence supporting the role of prostaglandins in the inflammatory reaction was obtained by using carrageenin to induce inflammation in the rat paw. The release of endogenous prostaglandins was eliminated by aspirin, and subsequent administration of low doses of exogenous PGE_2 (1.0 ng) or prostacyclin (10 ng) caused an increase in oedema (Moncada *et al.*, 1973). Moreover, the synovial fluid from rabbits with antigen-induced arthritis contained significantly more PGE_2 than fluid from control joints. The high levels of PGE_2 could be detected for up to 46 days after joint challenge (Henderson *et al.*, 1985).

2.2 INHIBITION OF CYCLOOXYGENASE

The inhibition of prostaglandin synthesis by NSAIDs has been demonstrated in a wide variety of cell types and tissues, ranging from whole animals and humans to microsomal enzyme preparations. Within 2 years of the discovery that these drugs inhibit prostaglandin synthesis, several classes of inhibitors had been identified (Flower, 1974). A more recent reviewer (Shen, 1979) lists no fewer than 12 major chemical series known directly to affect prostaglandin production and concludes that "it is remarkable that in the short span of time since the first observations with aspirin and indomethacin, such a variety of chemical structures have been identified as inhibitors of prostaglandin synthesis".

This diverse group of chemicals has been broadly classified to exhibit three types of inhibition: reversible competitive, irreversible and reversible non-competitive (Lands, 1981). Examples of reversible competitive inhibitors are fatty acids, closely related to the substrate arachidonic acid, which have a comparable affinity for the enzyme but are not converted to oxygenated products. The anti-inflammatory drugs ibuprofen, flufenamic acid and sulindac are reversible enzyme inhibitors which are competitive inhibitors of arachidonate binding (Rome and Lands, 1975). Aspirin selectively acetylates the hydroxyl group of a single serine residue located 70 amino acids away from the C terminus of the enzyme (Roth and Majerus, 1975). Acetylation leads to irreversible cyclooxygenase inhibition and thus new enzyme has to be synthesized before more prostanoids are produced. When the purified enzyme is acetylated, only the cyclooxygenase but not the peroxidase activity is inhibited. The stoichiometry of this reaction is 1:1, with one acetyl group transferred per enzyme monomer of this dimeric protein (Kulmacz and Lands, 1985). At low concentrations, aspirin acetylates prostaglandin-endoperoxide synthase rapidly (within minutes) and selectively. At high concentrations, over longer time periods, aspirin will also acetylate non-specifically a variety of proteins and nucleic acids (Roth *et al.*, 1975).

The possibility that aspirin-like drugs influence the release of other substances, such as histamine and bradykinin, was experimentally discounted and further studies were designed to show that the anti-enzyme effect of aspirin-like drugs correlated with their anti-inflammatory effects. Comparing the effects of two optical isomers of naproxen, Tomlinson and Ringold (Tomlinson *et al.*, 1972) showed that the one which possesses anti-inflammatory properties (in adjuvant arthritis and carrageenin oedema) was also a potent inhibitor of PGE_2 synthesis. The other isomer was much less active in all the tests. In a survey of a whole range of NSAIDs, at therapeutic doses the peak plasma concentrations, even allowing for protein binding, were more than sufficient to inhibit prostaglandin formation in an isolated enzyme preparation (Flower, 1974).

The degree to which microsomal prostaglandin synthetase preparations from different tissues are inhibited by the aspirin-like drugs varies considerably, and it is possible that the synthetase system (or at least one component protein) exists in multiple forms within the organism and that each has its own drug specificity. Results with paracetamol encourage this idea. Paracetamol has analgesic and antipyretic effects but little anti-inflammatory activity. It has only weak activity against most cyclooxygenase preparations but is considerably more active in diminishing prostaglandin synthesis within the CNS (Flower and Vane, 1972).

2.3 DIFFERENCES BETWEEN ASPIRIN AND SALICYLATE

Aspirin and salicylate are considered to be equally potent as anti-inflammatory agents but salicylate is less potent than aspirin in inhibiting prostaglandin synthesis by a crude enzyme preparation from guinea-pig lung (Vane, 1971). In contrast, an anti-inflammatory dose (3 g) of aspirin or salicylate reduced the urinary output of prostaglandin metabolites in humans by 85–95% (Hamberg, 1972).

In rats, the concentrations of aspirin and salicylate were measured in inflammatory exudate using carrageenin-impregnated polyester sponge implants (Higgs *et al.*, 1987). The findings paralleled the previously described pharmacokinetic study in humans. After oral administration of aspirin (200 mg/kg) to these rats, the peak concentration of aspirin (about 1.5 mg/ml) in inflammatory exudate was considerably lower than that of salicylate. Furthermore, administration of aspirin or salicylate equally reduced the concentrations of PGE_2 and TXB_2 in inflammatory exudate. Finally, the effects of aspirin and salicylate on the inhibition of cyclooxygenase in explants of inflamed tissue were examined. Comparison of the potencies with the relative concentrations of aspirin and salicylate measured in inflammatory exudates after oral administration showed that, although aspirin did not reach high enough concentrations to inhibit cyclo-oxygenase to any great extent, there was a high enough concentration of salicylate to inhibit prostaglandin synthesis considerably.

However, in carrageenin-induced pleurisy in rats, (Chiabrando *et al.*, 1989) aspirin was four times more potent than salicylate in reducing exudation and cell migration into the pleural cavity. An anti-inflammatory dose of salicylate of 100 mg/kg did not reduce the prostaglandin or TXA_2 content of the inflammatory exudate and higher doses of 200 mg/kg only reduced 6-oxo-$PGF_{1\alpha}$ levels. Equi-active doses of aspirin lowered the content of 6-oxo-$PGF_{1\alpha}$, TXB_2, PGD_2 but not PGE_2 in the exudate. Thus, inhibition of cyclooxygenase is not a likely mechanism of anti-inflammatory action of salicylate in this model.

Studies in humans have also led to the same conclusions (Ritter *et al.*, 1989). Intravenous sodium salicylate (600 mg) did not inhibit synthesis of prostanoids whereas the same dose of aspirin prevented Bk-stimulated prostaglandin and platelet TXA_2 synthesis as measured by serum levels of 6-oxo-$PGF_{1\alpha}$, 13,14-dihydro-15-oxo-$PGF_{2\alpha}$ and TXB_2. Moreover, the potent anti-inflammatory drug choline magnesium trisalicylate (Danesh *et al.*, 1989) did not affect TXA_2 synthesis in platelets, which was decreased by equivalent doses of aspirin. Since aspirin and salicylate appear to manifest anti-inflammatory effects by different mechanisms, it seems unlikely that the action of aspirin can be explained by its conversion to salicylate *in vivo*. However, it has been suggested that both aspirin and salicylate have a common action in inhibiting the expression of cyclooxygenase, as the concentration of salicylate required for this is at least two orders of magnitude lower than that for inhibiting cyclooxygenase activity (Sanduja *et al.*, 1991).

2.4 THROMBOXANE A_2 IN INFLAMMATION

Various preparations of PMNs make predominantly TXA_2 (Morley *et al.*, 1979). The suggestion that TXA_2 production in PMN preparations comes from contaminating platelets is not well based, for in one series of experiments the participation of different cells in TXA_2 production has been clearly elucidated. Thromboxane concentrations were measured in the serum from clotted blood and in inflammatory exudates taken from the same animals (Higgs *et al.*, 1983). Following the depletion of circulating platelets, TXB_2 was undetectable in the serum but unchanged in the exudates. Conversely, when the animals were made neutropenic, serum TXB_2 levels were not reduced whereas the TXB_2 concentration in the exudates was reduced to less than 10% of control values. These observations indicate that platelets are the source of TXA_2 in clotting blood but do not contribute to TXA_2 production in inflammation, while PMNs appear to be the major source of TXA_2 in acute inflammation. However, TXA_2 does not have a pivotal role in the inflammatory response.

3. *In the Immune System*

Most cells of the immune system make prostaglandins (Roberts *et al.*, 1979; Stenson and Parker, 1980; Pawlowski *et al.*, 1982; Scott *et al.*, 1982; Parker, 1986). Monocytes and macrophages produce the largest amounts while lymphocytes make very little (Goldyne and Stobo, 1979; Kennedy *et al.*, 1980). Mouse macrophages stimulated with zymosan produce mainly PGE_2 and PGI_2 whereas stimulated human monocytes and macrophages

secrete large amounts of TXA_2 as well as PGE_2 (Rouzer *et al.*, 1982; Tripp *et al.*, 1985; Fels *et al.*, 1986; Laviolette *et al.*, 1988; Pawlowski, 1989). Polymorphonuclear leucocytes make moderate amounts of PGE_2 while mast cells produce almost exclusively PGD_2 (Stenson and Parker, 1983).

PGE_2 has the best documented regulatory effects on immune function, manifested mainly as immunosuppression. Inhibition of immune function by endogenous prostaglandins after *in vivo* antigen challenge was first demonstrated by Webb and Osheroff (1976). Their study showed that raised levels of immunoreactive prostaglandin could be detected in the spleen of mice immediately after challenge with specific antigen. Treatment of the mice with indomethacin both prevented the increase in prostaglandin levels and increased the number of antigen-specific plaque-forming cells in the spleen, providing evidence for an *in vivo* modulation of the response to antigen by endogenously generated prostaglandins. Gordon *et al.* (1976) suggested that PGE_2 released by activated macrophages blocks the secretion of lymphokines from lymphocytes, thus exerting a negative-feedback control on macrophage function and reducing further activation of macrophages. Subsequently, PGE_2 was shown to inhibit [³H]thymidine incorporation into cultured human peripheral blood mononuclear cells stimulated with concanavalin A (Goodwin *et al.*, 1977). Conversely, treatment of phytohaemagglutinin-stimulated cultures of human mononuclear cells with indomethacin increased [³H]thymidine uptake into these cells (Goodwin *et al.*, 1978), indicating control by endogenous prostaglandins in this system. Concentrations of PGE_2 similar to those found at inflammatory sites (3–300 nmol/l) suppress *in vitro* lymphocyte functions such as responsiveness to mitogens (Goodwin *et al.*, 1978; Novagrodsky *et al.*, 1979), clonal proliferation (Gordon *et al.*, 1979; Eckles and Gershwin, 1981), lymphocyte migration (Van Epps, 1981), cell-mediated cytotoxicity (Schultz *et al.*, 1979; Darrow and Tomar, 1980; Meerpohl and Bauknecht, 1986) and the release of lymphokines (Gordon *et al.*, 1976; Venza-Teti *et al.*, 1980; Baker *et al.*, 1981; Rappaport and Dodge, 1982; Soppi *et al.*, 1982). PGE_2 inhibits IL-2 and IFNγ production from T lymphocytes (Betz and Fox, 1991) as well as IL-1 (Kunkel *et al.*, 1986a) and tumour necrosis factor release from macrophages (Kunkel *et al.*, 1986b, 1988).

However, immature cells of the immune system are stimulated by PGE_2. For example, PGE_2 induces immature thymocytes and B lymphocytes to differentiate and acquire the functional characteristics of mature cells (Parker, 1986). Some of the stimulatory effects of PGE_2 are mediated by the inhibition of suppressor T lymphocytes (Staite and Panayi, 1982).

The immunoregulatory role of PGE_2 is not essential for the normal functioning of the organism, since chronic administration of cyclooxygenase inhibitors to humans or experimental animals does not have a deleterious effect on the function of the immune system (Goodwin and Ceuppens, 1983). It has been suggested, however, that PGE_2 produced by tumour cells could account for the depression of the immune system associated with cancer. Tumour cells produce large amounts of prostaglandins (Sykes and Maddox, 1972; Bennett *et al.*, 1977) which induce a generalized state of immunodeficiency (Plescia *et al.*, 1975; Jaffe and Santoro, 1987; Plescia and Racis, 1988). This immunosuppression was prevented in tumour-bearing mice with inhibitors of prostaglandin synthesis such as indomethacin (Hial *et al.*, 1976; Pelus and Strausser, 1976; Pollard and Luckert, 1981). The circulating monocytes of patients with Hodgkin's disease, a lymphoproliferative disorder, and other lymphomas secrete more PGE_2 *in vitro* than normal and are especially effective in suppressing T lymphocyte responses. This suppression can be reduced by indomethacin *in vitro* and to some extent *in vivo*, thus confirming the importance of prostaglandins in the immunodeficiency of patients with Hodgkin's disease (Bockman, 1980; Bockman *et al.*, 1987).

Inhibition of prostaglandin synthesis with cyclo-oxygenase inhibitors can have important consequences for the immune system in chronic inflammatory conditions. Treatment of rheumatoid arthritis with aspirin-like drugs leads to inhibition of prostaglandin formation and thus to removal of the immunosuppressant effects of these eicosanoids. The question therefore arises whether it is beneficial to inhibit prostaglandin production in chronic inflammatory states. In an *in vitro* study in which macrophages and lymphocytes were incubated together, indomethacin reduced PGE_2 production by macrophages and increased the release of IL-2, which resulted in proliferation of the lymphocytes. Addition of exogenous PGE_2 reversed this action of indomethacin (Lewis and Barrett, 1986).

Another consequence of removing the suppression of immune processes by prostaglandins may be the enhancement of cartilage breakdown seen with NSAIDs *in vitro* and *in vivo* (De Brito *et al.*, 1987; Bottomley *et al.*, 1988; Desa *et al.*, 1988; Pettipher *et al.*, 1988). Indomethacin treatment increased lymphocyte infiltration, cartilage degradation and loss of proteoglycan from the knee joints of rabbits with adjuvant-induced arthritis (Pettipher *et al.*, 1988). This could have been the result of increased IL-1 production by macrophages because of the removal of PGE_2, which normally inhibits IL-1 production (Bahl *et al.*, 1991).

These experiments in rabbits have been confirmed in patients with osteoarthritis of the hip who were awaiting arthroplasty. When treated with indomethacin, a potent inhibitor of prostaglandin synthesis, these patients lost joint space more rapidly than when treated with azapropazone, an analgesic with weak cyclooxygenase inhibitory activity. The cartilage proteoglycan content remained higher in the azapropazone-treated group and their synovium contained significantly more PGE_2, TXB_2 and 5-HETE than did those in the indomethacin-treated group (Rashad *et al.*, 1989). Thus, indomethacin

appeared to accelerate cartilage breakdown in these patients. Prostaglandins may therefore acquire an importance in pathological states to maintain control over the immune system and dampen down the activity of overactive cells involved in the immune response.

Almost all of the actions of PGE_2 on the cells of the immune system are brought about through stimulation of adenylate cyclase and changes in intracellular cAMP concentrations (Smith *et al.*, 1971; Melmon *et al.*, 1974). These changes are mediated through high-affinity prostaglandin receptors, which have been described on rat thymocytes, human peripheral blood mononuclear cells, purified monocytes and macrophages (Goodwin *et al.*, 1979; Meerpohl and Bauknecht, 1986).

4. In the Central Nervous System

Until recently, very little was known about the function of prostaglandins in the CNS. However, during the past 10 years several laboratories discovered the presence of high concentrations of PGD_2 in the brains of rats (Narumiya *et al.*, 1982; Hiroshima *et al.*, 1986), monkeys (Onoe *et al.*, 1988) and humans (Ogorochi *et al.*, 1984). This made PGD_2 unusual among the prostanoids in that it was present in large amounts in the brains of mammals. In addition, PGD synthase (Urade *et al.*, 1985) and 15-hydroxy-PGD_2 dehydrogenase from rat and pig brains (Watanabe *et al.*, 1980; Tokumoto *et al.*, 1982) were identified and characterized. Using immunohistochemical techniques, it was then possible to measure the distribution of PGD synthase-like immunoreactivity in brain tissue. In young rats, PGD synthase activity was localized mainly inside neurones, whereas in adult rats it was found mainly in oligodendrocytes, although some of the activity was still associated with neurones (Urade *et al.*, 1987). PGE synthase has also been purified from human brain and identified as an anionic form of glutathione *S*-transferase (Ogorochi *et al.*, 1987), and $PGF_{2\alpha}$ was identified in postmortem human brains (Ogorochi *et al.*, 1984).

The intracerebral distribution of the PGD_2-binding protein, the putative receptor for this eicosanoid, was studied by Watanabe and his colleagues (Watanabe *et al.*, 1983, 1989; Yamashita *et al.*, 1983). The binding protein was highly concentrated in the olfactory bulb, cingulate cortex, occipital cortex, cerebellar cortex, hippocampus, hypothalamus and the preoptic area. There was less binding in the brain stem, midbrain and in white matter. A similar distribution of PGE_2 receptors was found in the monkey brain (Watanabe *et al.*, 1989).

4.1 PROSTAGLANDIN D_2 IN PHYSIOLOGICAL SLEEP

The preoptic area of the brain is known to be concerned with the control of sleep and with temperature regulation. Lesions of the preoptic area of the rat resulted in sleep disturbances (Nauta, 1946) whereas electrical stimulation of this area in the cat induced slow-wave sleep (Sterman and Clemente, 1962). While investigating the effects of PGD_2 on body temperature, Ueno *et al.* (1982a,b) accidentally discovered that nanomolar amounts of PGD_2 microinjected into the preoptic area could induce sleep. When saline was injected into the preoptic area of control rats placed on an unsteady platform, the rats remained awake. However, when several nanomoles of PGD_2 were microinjected into the preoptic area, the rats' awake period was decreased by almost 50% and the amount of sleep was increased by more than five times that of control rats receiving saline. The action of PGD_2 was confined to the preoptic area as injections into other areas, including the posterior hypothalamus, were ineffective. The effect was dose-dependent and specific for PGD_2, since other prostaglandins such as PGE_2 or $PGF_{2\alpha}$ were much less active.

Continuous infusions of small amounts of PGD_2 into the third ventricle of rats produced typical physiological sleep patterns. Rats are nocturnal animals and remain awake most of the night. When PGD_2 was infused at a rate of 0.6 pmol/min during the night, slow-wave sleep and rapid eye movement (REM) sleep increased by 33 and 56%, respectively (Ueno *et al.*, 1983). The effect was dose-dependent and specific for PGD_2, since other prostaglandins were inactive. The sleep induced by PGD_2 could not be distinguished from physiological sleep when compared using EEGs and electromyograms and by the general behaviour of the rats. The rats could be easily aroused by the sound of hand clapping, their sleeping postures remained normal and their sleep was episodic, which is typical of the normal sleep of rats. Moreover, the synthesis of endogenous PGD_2 showed a circadian fluctuation in parallel with the sleep–awake cycle. In the light period, when rats mainly sleep, PGD synthase activity was higher than in the dark period (Ueno *et al.*, 1985). Inhibition of PGD_2 synthesis with intravenous infusions of the prostaglandin synthase inhibitor diclofenac sodium dose-dependently decreased the amount of slow-wave sleep and REM sleep in these rats (Naito *et al.*, 1988).

Quadrivalent, inorganic selenium compounds were found to be potent and specific inhibitors of PGD synthase without affecting other enzymes of the arachidonic acid cascade (Islam *et al.*, 1991). When selenium chloride was infused continuously into the third ventricle of rats through a microdialysis probe, sleep was inhibited time- and dose-dependently. The effect was reversed on termination of the infusion (Matsumara *et al.*, 1991). Additionally, Krueger (1985) observed that PGD_2 induced sleep in rabbits and Roberts and coworkers (1980) reported that patients with systemic mastocytosis went into a deep sleep after the periodic production of large amounts of PGD_2 by their mast cells. This provides circumstantial evidence that PGD_2 induces sleep in humans as well.

4.2   PROSTAGLANDIN E_2 IN WAKEFULNESS

PGD_2 and PGE_2 sometimes exhibit opposite biological activities. For instance, whereas PGD_2 lowers body temperature (Ueno *et al.*, 1982a), PGE_2 causes an increase (Laychock *et al.*, 1979). Furthermore, PGE_2 stimulates secretion of luteinizing hormone-releasing hormone (Harms *et al.*, 1973) while PGD_2 suppresses its release (Kinoshita *et al.*, 1982). Intracerebral injections of PGE_2 into the preoptic area of rats (Matsumura *et al.*, 1988) reduced the amount of diurnal sleep, and infusions of PGE_2 into the third cerebral ventricle of rats during the day reduced the amount of slow-wave sleep by 30% and that of REM sleep by 60% (Matsumura *et al.*, 1989a). To assess the involvement of endogenous PGE_2 in the sleep–awake cycle, AH6809 (6-isopropoxy-9-oxoxanthine-2-carboxylic acid), a specific antagonist of the effect of PGE_2 on smooth muscle and on hyperalgesia, was administered. AH6809 completely abolished the effect of injected PGE_2 and, when infused into the third ventricle of rats during the night, increased the amount of slow-wave sleep by 22% and REM sleep by 89% above the controls (Matsumura *et al.*, 1989b). This infusion of AH6809 had no effect on body temperature by itself or on the pyrogenic action of PGE_2, indicating that the PGE_2 receptors for sleep and temperature regulation are probably different and the awaking effects of PGE_2 are independent of the pyrogenic effect.

These studies were repeated in rhesus monkeys and confirmed that PGD_2 induces sleep which is indistinguishable from normal sleep by an action in, or near to, the preoptic area (Onoe *et al.*, 1988, 1990). However, in the monkey the effect of PGE_2 on wakefulness was much greater when it was infused into the posterior hypothalamus than when it was infused into the preoptic area (Hayaishi *et al.*, 1990). PGE_2 has very little hyperthermic activity when administered into the posterior hypothalamus, thus confirming that the waking effect of PGE_2 is not due to its hyperthermic action (Matsumura *et al.*, 1989b).

4.3   CONCLUSIONS

The above studies are consistent with the hypothesis that PGD_2 induces sleep by an action on a sleep centre located in or near the preoptic area and that PGE_2 induces wakefulness by an action on an awake centre located in or near the posterior hypothalamus. These loci coincide with the sleep and awake centres postulated by Economo in 1917 (see Economo, 1926) and subsequently by Nauta (1946) and Moruzzi and Magoun (1949).

It is interesting to speculate that production of PGD_2 and PGE_2 could be influenced by cytokines generated under pathological conditions (Krueger *et al.*, 1990). Thus, the sleep-producing effects of muramyl dipeptide and of IL-1 may be mediated by the release of PGD_2 from astrocytes (Yamamoto *et al.*, 1988; for a recent review, see Hayaishi, 1991).

5. *In Pain Perception*

The discovery that aspirin-like drugs with their potent analgesic actions inhibit the biosynthesis of prostaglandins (Ferreira *et al.*, 1971; Smith and Willis, 1971; Vane, 1971) suggested that these eicosanoids were involved in the mediation of pain.

At the time of the forging of the link between aspirin and the prostaglandins, and for several years thereafter, the measurement of prostaglandin release in body fluids did not allow clear identification of which prostaglandin was involved. For instance, in the early days, relatively high levels of PGE_1 were reported in various tissues but we now know that in mammals (excluding seminal fluid) there is little or no release of PGE_1. It could be that 6-keto-PGE_1, a breakdown product of prostacyclin, was the actual culprit in these assays.

In general, we now know that:

1. Prostaglandins are not stored, so that concentrations found in tissues or exudates represent new synthesis and release. Since mechanical stimuli in themselves lead to prostaglandin synthesis, many of the early observations were confused by extra amounts being formed due to homogenization of tissues, venepuncture, centrifugation of samples, etc.
2. PGE_1 is largely limited in mammals to the seminal fluid.
3. Most tissues release PGE_2 and/or prostacyclin, although $PGF_{2\alpha}$ release is important in the uterus. Formation of PGD_2 is largely limited to mast cells and the brain.

In this section, the prostaglandins will be identified wherever the experimental results (with hindsight) allow it; otherwise, when the term "prostaglandins" is used, it most probably (again with hindsight) refers to PGE_2 and/or prostacyclin.

5.1 PROSTAGLANDINS AND HYPERALGESIA

In 1972, Ferreira observed that subdermal infusions in humans of PGE_1 or PGE_2 in low concentrations produced hyperalgesia. Infusions were used to mimic the continuous release of mediators at the site of an injury. The hyperalgesic effect of the prostaglandins was cumulative, since it depended not only on the concentration but also on the duration of the infusion. Neither Bk nor histamine showed this property.

During separate subdermal infusions of PGE_1, Bk or histamine (or a mixture of Bk and histamine) there was no pain; however, when PGE_1 was added to Bk or histamine, or a mixture of both, strong pain occurred. Furthermore, in areas made hyperalgesic by an infusion of PGE_1, a second infusion either of histamine or Bk caused pain which gradually increased in intensity. However, at the site where Bk or histamine had been previously infused (without producing hyperalgesia) an infusion of PGE_1 caused little or no pain.

Ferreira concluded that inflammatory mediators such as Bk or histamine had a direct pain-producing action only when the chemical receptors were sensitized by prostaglandins. Another possibility was an indirect action of the mediators on the sensitized receptors due to the oedema they produced, but this hypothesis was discounted.

5.1.1 Studies using Experimental Models

PGE_1 and PGE_2 also sensitized pain nerve endings in the spleen of lightly anaesthetized dogs (Ferreira *et al.*, 1973). In this preparation, sensitization to the reflex hypertension induced by Bk was observed when PGE_1 or PGE_2 was infused in low concentrations into the splenic artery. Interestingly, the spleen generates and releases a PGE_2-like prostaglandin continuously into the venous outflow, and this release is readily increased by intra-arterial Bk (Ferreira *et al.*, 1973). Thus, it is possible that there is already a background of sensitization, and injected pain-producing substances will always find an appropriate environment in which to act.

Bk injected in high doses into the knee joint cavity of anaesthetized dogs induces a reflex response compatible with stimulation of pain receptors (Phelps *et al.*, 1966; Melmon *et al.*, 1967). In this preparation, PGE_1 or PGE_2 also has the same sensitizing effect to the pain-producing effects of Bk (Moncada *et al.*, 1974, 1975). In contrast to the spleen, which has a continuous basal release of prostaglandins, normal synovial fluid does not contain prostaglandins (Herman and Moncada, 1975). This correlates well with the observation that the responses to Bk injected into the dog knee joint were more reproducible after prostaglandin release had been induced by a continuous infusion of saline into the joint (Moncada *et al.*, 1975). There is, however, a release of PGE_2 during carrageenin-induced inflammation in this model.

Results of electrophysiological experiments have confirmed the findings of these behavioural studies on the dog knee joint. Thus, a close-arterial injection of PGE_2 into the normal joint generated electrical discharges in sensory articular afferent nerves which were similar to those induced by an experimental inflammation (Schiable and Schmidt, 1988a,b).

Sensitization of pain receptors by PGE_2 has also been described by Kuhn and Willis (1973) and Willis and Cornelson (1973), who showed the development of hyperalgesia in the rat paw after single or repeated injections of PGE_2; Juan and Lembeck (1974) observed a similar effect of PGE_1 in the circulation of the rabbit ear, potentiating the pain response induced by several substances. Staszewska-Barczak *et al.* (1976) found that E-type prostaglandins sensitized the surface of the dog's heart to Bk topically applied to the surface (thereby causing a reflex rise in blood pressure) and Handwerker (1975) showed sensitization to the thermally induced discharge of sensory nerves by PGE_2.

The use of infusions (Ferreira, 1972; Ferreira *et al.*, 1973; Moncada *et al.*, 1975) allowed the study of the characteristics of PGE_2-induced hyperalgesia, two important features of which are its cumulative nature and the long duration of action. The cumulative nature of the PGE_2-induced hyperalgesia was defined by Ferreira (1972), who showed that the hyperalgesic effect of PGE_2 depended not only on the concentration infused but also on the duration of the infusion. Even very small amounts of PGE_2 infused into the knee joint of the dog were able to induce hyperalgesia, provided that the infusions were maintained for a long enough period (Moncada *et al.*, 1975).

Thus, PGE_2 causes long-lasting hyperalgesia (Juhlin and Michaelson, 1969; Ferreira, 1972; Moncada *et al.*, 1975) with a delayed onset lasting more than 3 h (Ferreira *et al.*, 1978a). This may explain why in various pathological conditions such as rheumatoid arthritis or special types of headache there is a delay of several hours before the antialgesic effect of cyclooxygenase inhibitors (such as NSAIDs) becomes apparent. However, in many clinical situations and experimental animal models, the analgesic effect of aspirin-like drugs is relatively rapid.

An explanation for this discrepancy was suggested when the unstable prostaglandin prostacyclin was discovered (Moncada *et al.*, 1976b). Thus, prostacyclin rather than PGE_2 may be involved for it was more potent than PGE_2 in producing hyperalgesia in both the rat and dog models (Ferreira *et al.*, 1978a). Furthermore, hyperalgesia induced by prostacyclin reached a plateau within 15–30 min and was relatively short lived, disappearing within 1 h. The presence of 6-keto-$PGF_{1\alpha}$, the non-enzymatic breakdown product of prostacyclin, was demonstrated in inflammatory exudates (Chang *et al.*, 1976; Doherty *et al.*, 1987; Berkenkopf and Weichman, 1988). The immediate and short-lasting inflammatory effect of prostacyclin was confirmed by Higgs *et al.* (1978b).

The difference in the duration of effects of PGE_2 and prostacyclin might explain the effectiveness of cyclooxygenase inhibitors in clinical situations. In some headaches, the prompt action of aspirin-like drugs would indicate the participation of prostacyclin, probably generated by the affected vessels. Processes in which the hyperalgesia terminates only after successive administrations of the analgesic drug suggest a predominant release of PGE_2 (as in sunburn or back pain, for example).

5.1.2 Mechanism of Hyperalgesia

The mechanism by which prostaglandins sensitize sensory nerve endings is not fully understood; however, electrophysiological techniques have shown an increased response frequency of the receptors to control stimuli, probably because of a lowering of the normally high threshold of polymodal nociceptors associated with sensory C fibres (Chahl and Iggo, 1977; Martin *et al.*, 1987; Cohen and Perl, 1990).

Stimulation of prostaglandin receptors (Coleman *et al.*, 1985) generally regulates the synthesis of cAMP by activating or inhibiting adenylate cyclase (Brunton *et al.*, 1976). PGE_2 and prostacyclin stimulate specific receptors thought to be located on nociceptive primary afferent nerves (Taiwo and Levine, 1989a), which are linked to adenylate cyclase through a stimulatory guanine nucleotide regulatory protein (G_s; Taiwo and Levine, 1989b). Activation of this system increases the level of intracellular cAMP (Hamprecht and Schultz, 1973; Collier and Roy, 1974a,b; Dismukes and Daly, 1975) and sensitizes the primary afferent nociceptors (Ferreira and Nakamura, 1979). The increased levels of cAMP may be associated with an increased intracellular calcium concentration (Ferreira and Nakamura, 1979; Greengard, 1979; Taiwo *et al.*, 1989).

5.1.3 Chronic Pain

Prolonged periods of pain stimulation generate a state of persistent hyperalgesia, as experienced in the initial stages of chronic pain (Ferreira *et al.*, 1990). Daily injections of PGE_2 or prostacyclin for a minimum of 14 days caused the development of a persistent hyperalgesic state which lasted up to a month. In contrast, rats receiving injections of dibutyryl cAMP for 14 days – at doses sufficient to cause an intense, acute and hyperalgesic response – did not show any persistent hyperalgesia. This persistent hyperalgesia depends on *de novo* protein synthesis, because treatment with cycloheximide reduced the intensity of the persistent hyperalgesic state by about 40%. Thus, the formation of a regulatory protein, which controls adenylate cyclase activation, might be important in the persistence of chronic pain. When persistent hyperalgesia was blocked with a cyclooxygenase inhibitor, it could be fully restored by a single injection of PGE_2 at concentrations which normally produced a mild and short-lived hyperalgesia. It appears, therefore, that nociceptor up-regulation can be memorized by nociceptors. The mechanism of persistent hyperalgesia may be complex and could additionally involve the release of substance P (Nakamura-Craig and Smith, 1989) and the activation of PKC (Follenfant *et al.*, 1990).

5.1.4 Conclusions

From all these results, it is clear that PGE_2 and prostacyclin, released in almost any form of tissue damage, sensitize the pain receptors to different types of stimulation (chemical, thermal, mechanical). Other mediators released during the inflammatory process, such as IL-1β (Ferreira *et al.*, 1988), interact to stimulate the pain receptors, but, at the concentrations present, their activity is effective only against a background of sensitization induced by the presence of prostanoids.

5.2 ANALGESIA WITH ASPIRIN-LIKE DRUGS

Aspirin-like drugs are weak analgesics in contrast to the "strong" narcotic analgesics like morphine. Several differences separate the groups. Strong analgesics induce tolerance and addiction. They block inhibitory pathways controlling muscle tone, giving rise to the phenomenon described as "plastic rigidity" or "Straub tail" in mice and rats. They relieve pain primarily through an effect on the CNS independent of the aetiology of pain. Aspirin-like drugs, on the other hand, do not induce tolerance and addiction. They are selective peripheral analgesics against pain produced in some clinical or experimental conditions, although a central component in the analgesic action of some NSAIDs has been postulated (Ferreira *et al.*, 1978b; Ferreira, 1979; Brune *et al.*, 1981; Catania *et al.*, 1991).

In clinical experience, aspirin-like drugs are effective as analgesics in pain of low and moderate intensity but not pain of high intensity (Lim, 1966; Insel, 1990). The pain is usually associated with inflammatory tissue damage, or processes in which the involvement of chemical mediators has been suggested, such as postoperative pain, osteoarthritis, rheumatoid arthritis, ankylosing spondylitis, cancer pain, dysmenorrhoea and some forms of headache (Beaver, 1988). Aspirin-like drugs have no measurable activity against pain of high intensity caused by muscular spasm, distension of a hollow viscera (DeLeo *et al.*, 1989) or acute noxious stimulation of the skin, or against pain in which nerve trunks are involved.

5.2.1   Studies in Experimental Models

Aspirin-like drugs are effective in experimental models involving the previous induction of an inflammatory state (Hesse *et al.*, 1930; Randall and Selitto, 1957; Gilfoil *et al.*, 1963; Margolin, 1965; Winter, 1965; Winter and Flataker, 1965; Pardo and Rodriguez, 1966). When there is no previous inflammation, aspirin-like drugs are effective analgesics only when there is a delay between the application of the stimulus and the development of the "pain response". For instance, there is a delay between the intraperitoneal injection of phenylbenzoquinone and the development of the stretching response in rodents (Keith, 1960). Aspirin and other aspirin-type drugs block this delayed stretching response but hardly affect the almost immediate stretching response induced by hypertonic saline (Collier *et al.*, 1968). They have a low activity against the early stretching induced by Bk (Collier *et al.*, 1968) but are very active against the delayed response to Bk (Emele and Shanaman, 1963). Aspirin-like drugs are not effective against nociception of short duration induced by pinching or stimulating the tail or toes of the mouse, rat or guinea-pig (Collier and Chesher, 1956; Winter and Flataker, 1965).

5.2.2   Mechanism of Analgesia

The site of action and the mechanism by which aspirin-like drugs produce analgesia has been the subject of much discussion. The observation that anti-inflammatory and analgesic activities run parallel in many cases (Randall, 1963) led several authors to believe that at least part of

the analgesic activity was peripheral (Randall and Selitto, 1957; Siegmund *et al.*, 1957; Winder, 1959). Finally, Guzman *et al.* (1964) and then Lim *et al.* (1964) by a cross-circulation experiment in anaesthetized dogs provided definitive evidence which showed that aspirin-like drugs had a peripheral analgesic action. Other authors (Whittle, 1964; Scott, 1968; Takesue *et al.*, 1976; Milne and Twomey, 1980; Rooks *et al.*, 1985; Schweizer and Brom, 1985; Böttcher *et al.*, 1987) using alternative models have confirmed this concept.

As a mechanism for this peripheral activity, Lim *et al.* (1964) and Lim (1968) claimed a direct antagonism at a receptor level between algesic substances and aspirin-like drugs. However, the repeated observation that some sort of inflammation or damage of the tissue needs to be present in order to observe the analgesic effect of aspirin-like drugs (Winder, 1959; Randall, 1963; Collier, 1969a) led several authors to suggest that analgesia was produced as a result of an action against pain-producing substances released during inflammation (e.g. Bk, as is inferred from the work of Guzman *et al.* (1962) and Lim *et al.* (1964)) or against the release of an intermediary substance responsible in the final step for the action of several mediators in inflammation (Collier, 1969a,b) or as Winder (1959) put it, the suppression of some "preinflammatory process" in the course of the reaction by tissues to injury. The same process could lead both to stimulation of pain endings and eventually to frank inflammation.

Three main findings allowed the development of the concept of the mechanism of analgesia induced by aspirin-like drugs. First there was the observation that prostaglandin-like activity is released during inflammation (Willis, 1969). Next came the discovery that aspirin-like drugs inhibit the biosynthesis of prostaglandins (Vane, 1971) and, thirdly, the finding that PGE_2 and prostacyclin induce hyperalgesia rather than pain in concentrations likely to be present in inflammatory exudates (Ferreira, 1972; Ferreira *et al.*, 1973, 1978a; Moncada *et al.*, 1975).

It was, therefore, proposed that prostaglandins sensitize rather than stimulate sensory nerve endings to the pain-producing activity of other stimuli. The pain-producing activity of Bk in the presence of exogenous PGE_2 is unaffected by aspirin-like drugs (Moncada *et al.*, 1975) while analgesia with the cyclooxygenase inhibitors can be reversed by PGE_2 or prostacyclin (Berkenkopf and Weichman, 1988). PGE_2, prostacyclin and carbacyclin are all potent inducers of the stretching response in rodents which is not prevented by cyclooxygenase inhibitors (Collier and Schneider, 1972; Smith *et al.*, 1985). Moreover, both PGE_2 and 6-keto-$PGF_{1\alpha}$ (the stable product of prostacyclin hydrolysis) have been identified in the pleural cavity during the stretching response to zymosan in mice (Doherty *et al.*, 1987).

Aspirin-like drugs increase the pain threshold by no more than 50% (Wolff *et al.*, 1941; Winder *et al.*, 1946;

Winder, 1947; Deneau *et al.*, 1953) in experimental models, and in clinical conditions they are ineffective against pain of high intensity (see above). This fits with the proposed mechanism of action, for removal of sensitization would explain why this analgesic activity can be overcome by increasing the original stimulus and would explain why direct immediate damage to the nerve ending or trunks or, on the other hand, extensive damage leading to a massive concentration of pain-producing substances, is unlikely to be affected by aspirin-like drugs.

There are, then, three possibilities. First, as in the spleen model (Guzman *et al.*, 1962; Ferreira *et al.*, 1973), the increased basal release and any stimulated release of prostaglandins is readily blocked by aspirin-like drugs, thus allowing an immediate observation of their "antihyperalgesic" effect. Secondly, there are models in which the stimulus leads to prostaglandin release; an example would be the delayed stretching response induced by phenylbenzoquinone or Bk (Keith, 1960; Emele and Shanaman, 1963). Thirdly, in models like the skin or knee joint, an inflammatory state has to develop, leading to PGE_2 and/or prostacyclin release, and only then can the antihyperalgesic effects of aspirin-like drugs be observed.

The knee joint of the dog becomes sensitive to the analgesic effects of aspirin-like drugs when inflammation is induced (Pardo and Rodriguez, 1966; Rosenthale *et al.*, 1966). There is a release of PGE_2 during carrageenin-induced inflammation in this model and a sensitization by PGE_1 or PGE_2 to the pain induced by Bk. Herman and Moncada (1975) observed a close correlation between the concentration of a prostaglandin-like activity and degree of incapacitation during endotoxin-induced inflammation in the knee joint of the dog.

Experimental models in which prostanoids do not seem to be involved like dextran-induced rat paw oedema or PCA develop much less hyperalgesia (Van Arman *et al.*, 1968; Moncada *et al.*, 1978). Conversely, sunburn hyperalgesia (an example used by Lewis in 1942 to suggest the presence of a factor specifically sensitizing nerve endings) is effectively inhibited by aspirin-like drugs (Hsia *et al.*, 1974) and is a model in which the presence of prostaglandins has been firmly established (Greaves and Sondergaard, 1970).

Patients suffering from chronic headaches are hypersensitive to the pain-producing effects of injections of Bk, and the intracranial pain receptors of rabbits display high sensitivity to Bk (Sicuteri *et al.*, 1966). Interestingly enough, indomethacin is an effective inhibitor of this pain-producing activity of Bk in both humans and rabbits. The blockade indicates an involvement of the prostaglandin system, probably in the same way as in the dog spleen (Ferreira *et al.*, 1973). Thus, the local release of PGE_2 by injected Bk sensitizes the pain receptors to the pain-producing activity of Bk. Certainly, both *in vitro* and *in vivo*, Bk is an effective stimulus for PGE_2 release

(Piper and Vane, 1971; McGiff *et al.*, 1972; Moncada *et al.*, 1972; Ferreira *et al.*, 1973).

5.2.3   Conclusion

We can conclude that the aspirin-like drugs do not reduce the effect of prostanoids but reduce only those effects caused by inflammatory substances or stimuli that induce their generation.

6. *In the Gastrointestinal Tract*

Prostaglandin synthesis can be demonstrated in every part of the gastrointestinal tract. In the rat, using vortex generation, the rank order of prostacyclin synthesis as determined by bioassay techniques was greatest in gastric muscle and forestomach, followed in rank order by gastric mucosa, colon, rectum, ileum, caecum, duodenum, jejunum and oesophagus (Whittle and Salmon, 1983). In contrast to gastric mucosal tissue, the extracts of ileal tissue from rats and rabbits contained high levels of PGE_2, which exceeded those of 6-oxo-$PGF_{1\alpha}$ (Whittle and Salmon, 1983) as found in another study, using "whole-cell" preparations of rat jejunum (Peskar *et al.*, 1981). Studies on the cellular distribution of PGE_2 in the small intestine of the rat suggest that synthesis is located predominantly in the subepithelium (Smith *et al.*, 1982).

6.1   GASTRIC ANTISECRETORY ACTIONS

The first report of the potent antisecretory actions of PGE_1 and PGE_2 came from studies in the dog (Robert *et al.*, 1967), which have subsequently been extensively confirmed. Gastric output of acid, fluid volume, and pepsin secretion induced by secretagogues such as histamine or pentagastrin, or by vagal stimulation (using a 2-deoxyglucose), as well as after feeding, were inhibited during intravenous infusion of either innervated or denervated gastric pouches (Nezamis *et al.*, 1971; Wilson and Levine, 1972; Dajani *et al.*, 1975; Robert, 1977). Likewise, pentagastrin- or histamine-stimulated acid output was inhibited by local intra-arterial infusion of PGE_2 in the anaesthetized dog (Gerkens *et al.*, 1978; Walus *et al.*, 1980a).

In anaesthetized rats (Whittle *et al.*, 1985) and conscious dogs (Kauffman *et al.*, 1979; Konturek *et al.*, 1980), intravenous infusion of prostacyclin was more active than PGE_2 in reducing pentagastrin- or histamine-stimulated acid secretion. However, when administered by close-arterial infusion in the anaesthetized dog, PGE_2 was the more potent secretory inhibitor (Gerkens *et al.*, 1978; Walus *et al.*, 1980a). In the monkey, intravenous infusion of prostacyclin inhibited acid output while having no effect on non-parietal secretion (Shea-Donohue *et al.*, 1982).

In canine isolated parietal cells obtained through

density gradients, PGE_2 and prostacyclin likewise inhibited histamine-stimulated acid secretion (Skoglund *et al.*, 1982). These findings clearly demonstrate that the inhibitory action is directly on the secretory apparatus within the parietal cell. There is good biochemical evidence, at least *in vitro*, that prostaglandins exert their antisecretory action on adenylate cyclase in the parietal cell (Levine *et al.*, 1982; Ruoff and Becker, 1982; Fiorucci and McArthur, 1991).

6.2   GASTRIC VASCULAR ACTIONS

Under resting conditions, intravenous infusions of PGE_1, PGE_2 or prostacyclin in the rat increased mucosal blood flow while reducing systemic arterial blood pressure, thus indicating a direct vasodilator action on the gastric mucosa (Main and Whittle, 1973; Whittle *et al.*, 1978; Konturek *et al.*, 1980). Local intra-arterial administration of PGE_1 and PGE_2 also reduced gastric vascular resistance in the dog stomach (Nakano and Prancan, 1972; Walus *et al.*, 1980b; Kauffman and Whittle, 1982). Using radiolabelled microspheres in the dog, infusion of low doses of prostacyclin into the left atrium or directly into the gastric artery increased blood flow in fundic (Gerber and Nies, 1982) and antral (Einzig *et al.*, 1980) mucosa, as was also found following intravenous infusion in the pig (Gaskill *et al.*, 1982).

In further studies in the dog, prostacyclin and PGE_2 were equipotent vasodilators in the blood-perfused gastric circulation *in situ* (Gerkens *et al.*, 1978; Spenny and Barton, 1981; Kauffman and Whittle, 1982) following close-arterial infusion or bolus injection. However, in the gastric circulation of the isolated stomach of the rat and rabbit perfused with Krebs' solution, prostacyclin was some 30 times more potent than PGE_2 as a vasodilator (Salvati and Whittle, 1981). These findings may reflect a species variation of differences in binding, breakdown or availability of the two prostanoids in blood.

Studies in the dog and rat using clearance techniques have shown that parenteral administration of indomethacin in doses sufficient to inhibit mucosal prostaglandin biosynthesis reduced gastric mucosal blood flow (Main and Whittle, 1975a; Gallavan and Jacobson, 1982), while direct microscopic visualization showed a reduction in arteriolar diameter in the rat gastric submucosa (Guth and Moler, 1982). Indomethacin increased gastric vascular resistance in a blood-perfused canine fundic segment *in situ*, as did indomethacin perfusion in the rabbit isolated gastric circulation (Salvati and Whittle, 1981). Indomethacin also reduced blood flow in the cat and rabbit stomach, as determined by radiolabelled microsphere techniques (Hierton, 1981). These results in several species thus support a vasodilator role for endogenous cyclooxygenase products in the mucosal circulation, at least under resting conditions.

6.3   GASTRIC ANTI-ULCER AND PROTECTIVE ACTIONS

PGE_1, PGE_2 and prostacyclin inhibit the formation of gastric erosions and ulcers induced by a wide variety of techniques, including pyloric ligation, steroid administration, restraint or cold stress, treatment with reserpine, serotonin, bile salts or NSAIDs, while the methylated analogues of PGE_2 and other prostaglandin analogues likewise have potent anti-ulcer activity (Ferguson *et al.*, 1973; Lee *et al.*, 1973; Lippman, 1974; Kawarada *et al.*, 1975; Mann, 1976; Whittle, 1976; Whittle *et al.*, 1978). As an acidic environment in the gastric lumen is a necessary prerequisite in the development of many forms of gastric damage, it is important to distinguish an antisecretory action from any other protective mechanism. In an early survey of many synthetic and naturally occurring prostanoids, a disassociation between antisecretory and anti-ulcer activity was proposed by Robert, who used the all-embracing term "cytoprotection" (Robert *et al.*, 1979a).

Further early evidence for a direct protective mechanism of the prostanoids came from studies in which methyl analogues of PGE_2 reduced the gross gastric damage induced by acidified bile salts, either alone or in combination with indomethacin (Whittle, 1976; Carmichael *et al.*, 1978). A striking demonstration of the cytoprotective phenomenon was the finding that gastric damage assessed macroscopically, induced by topical application of strong acids, base hypertonic solutions, or ethanol in rats, could be reduced by co-administration of various prostaglandins (Miller, 1983). The ability of various prostanoids to protect the gastric mucosa from the macroscopically apparent damage induced by ethanol is well documented (Robert *et al.*, 1979a) and has become a standard experimental model for the investigation of the protective mechanisms.

The use of the term "cytoprotection" as an all-embracing description of the anti-ulcer properties of various prostaglandins has become controversial. Although many studies have shown that PGE_2 and prostacyclin and their analogues can reduce the formation of macroscopically observed mucosal erosions and deep necrotic lesions induced by local irritants and anti-inflammatory agents, detailed histological studies have only more recently been reported.

Although prostanoids may not directly enhance cellular migration (Svanes *et al.*, 1984), protection of the basal lamina and migrating cells from damage would allow rapid epithelial restitution (Wallace and Whittle, 1985) and restore the disrupted surface cells and normal barrier function, whereas preservation of the mucosal proliferative zone would allow the slower migration of these deeper cells (Tarnawski *et al.*, 1985). Such rapid restitution (Morris and Wallace, 1981; Morris and Harding, 1984; Svanes *et al.*, 1984) of the surface epithelial cell layer may contribute to the overall tissue protection from ethanol damage when evaluated 30 min or longer after challenge. It may also contribute to the apparent cytological preservation of the epithelial layer continuity by higher doses of the prostaglandin analogue, as estimated by histological techniques as well as by the release of the cytoplasmic enzyme marker lactate dehydrogenase, and lysosomal enzymes (Whittle and Steel, 1985). Other studies have shown that the fall in gastric DNA and RNA synthesis that accompanies mucosal damage by ethanol or acidified aspirin is prevented by administration of prostacyclin or PGE_2, or an analogue (Konturek et al., 1981a; Miller et al., 1982). Indomethacin delays the healing of experimental ulcers in rats, and in patients taking NSAIDs the rate of mitosis at the edge of ulcer sites was significantly reduced (Inauen et al., 1988; Levi et al., 1990).

6.3.1 Anti-ulcer and Protective Actions in Humans

Studies in humans have confirmed the experimental findings that the prostaglandins can protect against gastric damage induced by anti-inflammatory agents. Oral administration of liquid formulations of PGE_2 three times a day significantly reduced faecal blood loss in rheumatic patients taking indomethacin, as determined by the output of radiolabelled erythrocytes, over a 1 week treatment period (Johansson et al., 1980). Using comparable techniques, aspirin-induced faecal blood loss in healthy volunteers was significantly reduced by PGE_2 administered orally four times a day (Cohen and Pollett, 1976; Cohen et al., 1980; Cohen, 1981) and by a PGE_1 analogue in non-antisecretory doses using a similar protocol (Cohen et al., 1985). Oral ingestion of tablets of PGE_2 also reduced the aspirin-induced faecal blood loss over a 5 day period (Yik et al., 1982). In another clinical study, the elevated blood loss induced by aspirin, which was chemically determined in serial gastric washings, was reduced on the second day of oral treatment with PGE_2 (Hunt and Franz, 1981). Likewise, oral administration of PGE_2 reduced both microbleeding and DNA output into the gastric lumen following aspirin ingestion (Konturek et al., 1983).

Treatment with the PGE_1 analogue misoprostol significantly reduced the incidence of gastric ulcers in a clinical study of 420 osteoarthritic patients receiving NSAIDs (Graham et al., 1988).

6.3.2 Mechanisms of Gastric Mucosal Protection

The mechanisms underlying the anti-ulcer and protective properties of exogenously administered prostaglandins are not clearly defined, apart from their ability to inhibit acid secretion. Several important concepts, based on changes in the major gastric parameters, have been proposed to explain the "cytoprotective" process.

6.3.2.1 Gastric Barrier Changes

An increased diffusion of luminal hydrogen ions back into the mucosal tissue, which is accompanied by a fall in transmucosal potential difference and efflux of potassium, has been implicated in the gastric damage induced by topical irritants such as bile salts, aspirin, salicylates, other anti-inflammatory agents and ethanol, the so-called "barrier breakers" (Davenport, 1964, 1965). Thus, one potential mechanism by which prostaglandins could protect the gastric mucosa from damage would be by reducing such acid back-diffusion.

6.3.2.2 Gastric Vascular Actions

Since many of the protective prostaglandins of the E and I series and their synthetic analogues are potent vaso-dilators in the gastric circulation, increases in gastric mucosal blood flow could be beneficial in maintaining the functional integrity of the gastric tissue (McGreevy and Moody, 1977; Moody et al., 1978; Whittle, 1980). This would be especially important when the mucosa is under attack from noxious elements, both by enhancing the supply of essential nutrients and by preventing accumulation of hydrogen ions and other potentially damaging products. Indeed, the ability of the gastric mucosa to withstand or clear such back-diffusing acid may well be of greater importance than the absolute magnitude of the acid back-diffusion (Kivilaakso et al., 1978), since only with acid accumulation within the tissue will the intramucosal pH fall and lead to tissue necrosis.

6.3.2.3 Mucus, Bicarbonate and Fluid Secretion

Gastric mucus is a viscoelastic polymeric gel that is secreted from surface epithelial cells (Allen and Garner, 1980; Allen, 1981; Morris et al., 1984). It both lubricates the mucosal surface to allow free passage of solid particles and protects it from mechanical damage from such material.

Intragastric administration of PGE_2 to humans stimu-lated the release of viscous mucus, judged visually and by chemical methods (Johansson and Kollberg, 1979). In the rat, PGE_2 and 16,16-dimethyl-PGE_2 increased the soluble mucus in the gastric lumen but did not alter the glycoprotein content of the adherent mucus (Bolton et al., 1978). Gastric mucus could play a defensive role against mucosal injury (Allen and Garner, 1980; Allen, 1981). Other than acting as a physical barrier, mucus may act to create an unstirred layer of secreted bicarbonate (Bahari et al., 1982) and hence help to neutralize hydrogen ions diffusing back from the lumen into the mucosa.

Stimulation of duodenal bicarbonate secretion by prostanoids of the E and I series has also been described (Flemstrom and Garner, 1982; Heylings et al., 1985). The regulation of duodenal bicarbonate secretion by an endogenous prostanoid may play a physiological role in the protection of the duodenum from excess acidity derived from the gastric lumen (Ahlquist et al., 1983).

Apart from the potential local buffering action of the alkaline secretion, the net stimulation of fluid output into the gastric lumen may also contribute to mucosal protection. The stimulation of gastric fluid secretion induced by various prostanoids including 16,16-dimethyl-PGE_2 has been demonstrated in the rat, cat and dog (Bolton and Cohen, 1978; Bolton *et al.*, 1978; Gascoigne and Hirst, 1981), and this net flow of secretion may help to enhance the disposal of acid or to act as a diluent for any topical irritant (McGreevy and Moody, 1980; Whittle and Steel, 1985).

The stimulation of mucus, bicarbonate and fluid secretion by prostanoids could also be important in the process of epithelial restitution that follows mucosal damage (Morris and Harding, 1984). Thus, the formation of a mucus cap, rich in bicarbonate and plasma exudate, over areas of mucosal damage would create a local environment suitable for cellular migration and subsequent re-epithelialization of the damaged mucosa (Morris and Harding, 1984; Morris *et al.*, 1984; Wallace and Whittle, 1985). Such encouragement of the rapid repair process by prostanoids could thus contribute to their overall mucosal protective processes.

6.3.2.4 *Conclusions*

It can be seen that many divergent mechanisms evoking changes in a wide range of gastric parameters have been proposed to explain the potent protective actions of various prostanoids on the mucosa. Most proposals have been supported by convincing experimental evidence, yet, as discussed above, objections to each can be raised, based on data from alternative experimental models. It is likely, therefore, that no single action can account for "cytoprotective" processes, and it is possible that many of the properties act in concert or synergistically to produce the overall phenomenon. Indeed, the contribution of each may well depend on the type of gastric damage under investigation and the nature of the prostaglandin being investigated. The observations *in vitro* that prostaglandins can prevent the ulceration of bullfrog gastric mucosa and the aspirin-induced disruption of rat gastric epithelial cells in culture (Barzilai *et al.*, 1980; Terano *et al.*, 1984) may point to fundamental alterations in phospholipid biochemistry of cell membranes.

Endogenous prostaglandins may interact with other protective mediators in the preservation of the integrity of the gastric mucosa. For example, gastric mucosal damage induced by indomethacin is augmented when synthesis of nitric oxide or release of sensory local neuropeptides is suppressed (Whittle *et al.*, 1990).

6.4 ACTIONS OF NON-STEROIDAL ANTI-INFLAMMATORY AGENTS

An important concept concerning the damaging effects of aspirin and other salicylates was developed by Davenport and his colleagues, who proposed that the normal resistance of the gastric mucosa to the back-diffusion of gastric acid from the lumen into the mucosal tissues can be disrupted by the topical administration of aspirin and similar compounds (Davenport, 1964, 1965). Such a topical irritant action, although clearly of importance, cannot be the sole mechanism, for several NSAIDs cause gastric damage when administered parenterally in both experimental animals and humans (Grossman *et al.*, 1961; Bugat *et al.*, 1976; Whittle, 1977; Kauffman and Grossman, 1978; Konturek *et al.*, 1981b; Whittle, 1983; Whittle *et al.*, 1985).

Since PGE_2 and prostacyclin exert potent protective actions on the gastric mucosa and may also be involved in the local regulation of gastric function, inhibition of prostaglandin biosynthesis by NSAIDs could be expected to disrupt the functional integrity of the gastric mucosa. Such effects would occur whether the aspirin-like drugs were administered orally or parenterally in therapeutic doses, and thus are a likely biochemical basis of their gastric toxicity.

Early studies in the rat indicated that indomethacin at ulcerogenic doses reduced the level of prostaglandins in gastric mucosal homogenates (Main and Whittle, 1975a), bioassayed as PGE_2, and the levels of both PGE_2 and $PGF_{2\alpha}$ in the stomach, assayed by RIA (Fitzpatrick and Wynalda, 1976). The clinically used compounds indomethacin, aspirin, naproxen, flurbiprofen and ketoprofen all caused a dose-related inhibition of gastric muscosal prostaglandin generation determined *ex vivo* following both parenteral and oral administration (Whittle *et al.*, 1980). Using similar techniques, others have confirmed that intragastric or intravenous administration of aspirin inhibits both PGE_2 and prostacyclin generation by the rat gastric mucosa *ex vivo* (Konturek *et al.*, 1982).

When administered in divided doses over 24 h, indomethacin, flurbiprofen and naproxen all induced a comparable degree of gastric erosion formation, which paralleled the inhibition of mucosal prostacyclin production (Whittle *et al.*, 1980). In contrast, sodium salicylate did not inhibit mucosal cyclooxygenase or cause macroscopic damage (Whittle *et al.*, 1980). Such findings support the association between these two events. Aspirin, however, induced a greater degree of gastric damage in doses causing a similar mucosal cyclooxygenase inhibition to the other compounds, which may reflect additional local irritant actions (for a review, see Whittle, 1992).

Further evidence that prostaglandins maintain gastric mucosal integrity comes from the use of specific antibodies directed towards PGE_2 and prostacyclin. Both in the dog and the rabbit, immunization against endogenous prostanoids resulted in the development of chronic lesions in the stomach and small intestine (Redfern *et al.*, 1987; Redfern and Feldman, 1989). Thus, the physiological actions of PGE_2 and prostacyclin were prevented by the binding of the specific antibodies to their endogenous ligands.

6.5 PROSTANOID ACTIONS ON INTESTINAL FUNCTION

Many early studies demonstrated that PGE_1, PGE_2 and $PGF_{2\alpha}$ affected the tone of isolated strips of intestinal smooth muscle of all species studied (Bennett *et al.*, 1968a,b, 1975). This property has been used extensively in the detection and estimation of these prostaglandins by biological assay. Longitudinal smooth muscle from segments of rat, guinea-pig and human small intestine was contracted by PGE_1 and $PGF_{2\alpha}$, whereas circular smooth muscle was relaxed by the E series prostanoids (Bennett *et al.*, 1968a).

6.5.1 Gastrointestinal Motility

PGE_1, PGE_2 and $PGF_{2\alpha}$ stimulated gastrointestinal motility *in vivo* in several species including the guinea-pig, dog, rat and human following oral or intravenous administration (Bennett *et al.*, 1968b; Horton *et al.*, 1968; Cummings *et al.*, 1973; Newman *et al.*, 1975; Dajani *et al.*, 1979). The actions of these prostaglandins and their analogues on small intestinal transit in the rat, like those on gastric emptying, are complex and depend not only on the type of prostanoid, but also on the dose and route of administration (Ruwart and Rush, 1984). In humans, oral administration of PGE_1 accelerated the transit rate through the small intestine and colon, accompanied by abdominal colic (Horton *et al.*, 1968; Misiewicz *et al.*, 1969).

6.5.2 Fluid Transport

Oral, intrajejunal and parenteral administration of prostaglandins of the E and F series and their analogues potently alters fluid and electrolyte transport across the intestinal mucosa and induces diarrhoea (Horton *et al.*, 1968; Main and Whittle, 1975b; Bennett, 1976; Robert, 1976). Diarrhoea often accompanies the therapeutic use of these prostaglandins and their methylated analogues in obstetrics (Karim and Fung, 1976) and is a major side-effect associated with their more chronic use as anti-ulcer agents (Vantrappen *et al.*, 1982). Several of the non-osmotic laxatives may also exert their actions in part by the stimulation of intestinal prostaglandin formation (Beubler and Juan, 1979; Rachmilewitz *et al.*, 1980; Smith *et al.*, 1982).

In contrast to prostaglandins of the E and F series, prostacyclin and its stable analogues have limited spasmogenic action on isolated gastrointestinal tissue (Whittle and Boughton-Smith, 1979). Prostacyclin relaxed longitudinal muscle from segments of human intestine *in vitro* and antagonized the contractions induced by PGE_2 and $PGF_{2\alpha}$ (Bennett *et al.*, 1981). Prostacyclin and its analogues do not induce the typical profuse mucoid diarrhoea seen with PGE_2 and its

methylated analogues following parenteral or oral administration in animals (Robert *et al.*, 1979b; Whittle and Boughton-Smith, 1979).

6.5.3 Intestinal Damage and Protection

Many NSAIDs induce damage to the small intestine in several species, with extensive effects being seen in the rat. Intestinal damage, consisting of palpable nodules with adhesions between adjacent loops of ileum and jejunum, becomes macroscopically apparent some 24 h after a single dose of an agent such as indomethacin (Robert, 1975, 1976; Fang *et al.*, 1977; Whittle, 1981). This type of intestinal damage can be inhibited by oral or parenteral administration of several types of prostaglandin, including analogues of PGE_2 (Dajani *et al.*, 1976), the prostacyclin analogue 6β-PGI_1 and a 16-phenoxy derivative (Whittle and Boughton-Smith, 1979).

7. In Fever

One of the earliest events which takes place following infection, inflammation or tissue damage is the development of fever. It is now generally accepted that fever results from the action of exogenous pyrogens such as lipopolysaccharide or muramyl dipeptide derived from invading bacteria, which then elicit a response in the host. A number of cell types, particularly monocytes and macrophages, synthesize a low molecular-weight protein, namely IL-1, in response to stimulation with bacterial pyrogens (Dinarello, 1984). Human IL-1 exists in two forms, IL-1α and IL-1β, both of which raise the body temperature when injected intravenously into conscious rabbits. IL-1β is approximately five times as active as IL-1α, but the response to both can be completely abolished by an injection of an NSAID given 30 min before that of IL-1 (Davidson *et al.*, 1989, 1990).

The development of fever is thus dependent on stimulation of arachidonic acid metabolism by IL-1 in various cells such as macrophages and fibroblasts (Kunkel *et al.*, 1986a; Newton and Covington, 1987) and on increased production of PGE_2 by these cells.

7.1 EXPERIMENTAL STUDIES IN ANIMALS

In the 1970s, Milton and Wendlandt discovered that PGE_1 and PGE_2 (but not PGA_1, $PGF_{1\alpha}$ or $PGF_{2\alpha}$) were hyperthermic when injected into cats, rabbits and rats (Milton and Wendlandt, 1971a,b). Furthermore, Feldberg and Saxena (1971) demonstrated that PGE_1 infused into the cerebral ventricles of the cat raised its body temperature by an action on the preoptic area of the anterior hypothalamus. These experiments indicated that the prostaglandins may be the ideal endogenous substances to modulate increases in body temperature. CSF collected from the third cerebral ventricle, or from the cisterna magna of the cat, normally contains very little prostaglandin-like activity. However, when bacterial pyrogens were injected into the cerebral ventricles or intravenously, they produced a febrile response and increased the PGE_2 content of the CSF. Both the rise in body temperature and the increase in PGE_2 were abolished by the administration of aspirin, paracetamol or indomethacin (Feldberg and Gupta, 1973; Feldberg *et al.*, 1973). The conclusion was that PGE_2, produced by exogenous pyrogens, was the mediator of hyperpyrexia and the connection between prostaglandins and thermoregulation was thereby established.

It has already been noted that stimuli which produce fever cause the synthesis of an endogenous pyrogenic substance now called IL-1. In 1975, Harvey and Milton prepared endogenous pyrogen from cells of cat peritoneal exudate and infused or injected it intravenously into cats. A dose-dependent rise in body temperature followed and the levels of PGE_2 in the CSF increased markedly. Parenteral injection or infusion of bacterial endotoxin to the cats produced similar rises in body temperature and raised PGE_2 concentrations in the CSF. Interestingly, CSF taken from patients suffering from high fever has been shown to contain measurable amounts of PGE-like activity (Saxena *et al.*, 1979), while that of afebrile patients had no such activity.

The mechanism by which prostaglandins raise the body temperature was investigated in the Welsh Mountain sheep (Bligh and Milton, 1973). This animal has a particularly efficient mechanism for maintaining a constant body temperature by regulating both heat loss and heat gain mechanisms. PGE_1 (which has similar actions to PGE_2) was infused into the lateral ventricles of the sheep at different ambient temperatues (10, 18 or 45°C). The results showed that PGE_1 increased the body temperature by inhibiting evaporative heat loss mechanisms such as panting and by inhibiting surface heat loss through vasoconstriction. In addition, heat gain mechanisms were stimulated to increase metabolic heat production through shivering. The interpretation of these results was that PGE_1 had an action on the pathway between cold sensors and the mechanisms of heat production and on effectors in the pathway between warm sensors and heat loss mechanisms.

Prostaglandins do not appear to be involved in the regulation of a normal body temperature, which would explain why NSAIDs have no effect on the normal temperature in the afebrile animal (Cammock *et al.*, 1976; Bernheim *et al.*, 1980).

the CNS or formed by peripheral cells or tissues and transported to the preoptic area of the hypothalamus. This would also answer the question of whether NSAIDs exert their antipyretic action by inhibiting prostaglandin production at a peripheral or central site.

Studies with radiolabelled endotoxin have shown that the radioactivity does not penetrate into the brain (Dascombe and Milton, 1979). Similarly, there is no convincing evidence that IL-1 can cross the blood–brain barrier and become localized in the preoptic area of the hypothalamus (Dinarello *et al.*, 1978). More recently, Stitt (1985) suggested that IL-1 could act on an area of the brain which is effectively located outside the blood–brain barrier, the *organum vasculosum laminae terminalis*, in which the synthesis of modulators of fever could be initiated. However, Blatteis *et al.* (1989) gave ^{125}I-labelled IL-1 systemically and found no radioactive label in the whole brains of the animals, whereas it was possible to detect ample radioactivity in blood, lungs, liver, kidney and urine. Administration of IL-1 increases the circulating levels of PGE_2 (Rotondo *et al.*, 1988) as well as PGE_2 levels in the CSF (Abdul and Milton, 1989), in parallel with increases in body temperature. Both effects of IL-1 were abolished by the NSAID ketoprofen.

There are a number of indications that PGE_2 synthesized peripherally crosses into the brain and raises the body temperature by an action on the hypothalamus. For example, IL-1 stimulates PLA_2 activity in monocytes, but not in neuronal membranes, probably by an effect on the PLA_2 enzyme itself (Milton and Rotondo, 1987). In addition, both lipopolysaccharide and IL-1 release PGE_2 from rabbit monocytes, an effect which can be inhibited by ketoprofen (Abdul and Milton, 1989). A peripheral site of action for antipyretic agents would explain why drugs such as indomethacin, which only penetrate slowly into the CNS, are effective antipyretics when given peripherally (for a review, see Milton, 1992).

7.3   CONCLUSIONS

There is no doubt that pyrogens can cause fever when injected directly into the CNS and that antipyretic drugs such as aspirin or paracetamol, which penetrate easily into the CNS, reduce this fever. However, the time-course of this pyrogenic action is different from when pyrogens are given peripherally, and Dascombe and Milton (1975) suggested that pyrogens injected directly into the brain set up a local inflammatory response which leads to production of PGE_2 and can be blocked by antipyretic drugs.

7.2   MECHANISMS OF HYPERPYREXIA

It has therefore been established that increased synthesis of PGE_2 is responsible for the hyperpyrexia resulting from infection with bacterial pyrogens. What is not clear is whether PGE_2 is synthesized close to its site of action in

8. *In the Kidney*

The kidney has the capacity to form large amounts of prostaglandins. Different regions of the nephron produce different proportions of various eicosanoids which appear to act locally. Thus, prostaglandins synthesized in cortical structures such as the glomeruli and cortical arterioles do not enter the renal medullary compartment, and vice versa. The medulla has a higher biosynthetic capacity to form prostaglandins than cortical structures and high cyclooxygenase activity is found in the endoplasmic reticulum of collecting tubule cells, medullary interstitial cells and vascular endothelial cells. The major prostaglandins formed by the kidney are PGE_2, PGI_2, $PGF_{2\alpha}$ and TXA_2 (for reviews, see Dunn, 1983; Schlondorff and Ardaillou, 1986).

8.1   PROSTAGLANDIN PRODUCTION BY THE KIDNEY

Cortical structures of normal kidneys produce mainly PGE_2 and PGI_2 and very small amounts of TXA_2. This has been demonstrated using isolated nephron segments (Farman *et al.*, 1987), isolated glomeruli (Sraer *et al.*, 1979) and cultured cell preparations (Petrulis *et al.*, 1981). Human glomeruli synthesize mainly PGI_2 whereas cortical collecting tubules make predominantly PGE_2. Glomerular epithelial and mesangial cells have the synthetic capacity to form both PGI_2 and PGE_2. These prostanoids are therefore uniquely situated to influence RBF, GFR and the release of renin.

The renal medulla produces mostly PGE_2, for which it has a synthetic capacity approximately 20 times that of the cortex (Zusman and Keiser, 1977), Here, PGE_2 may have a role in regulating medullary blood flow, sodium and chloride reabsorption and the action of arginine vasopressin. Conversely, arginine vasopressin, angiotensin II, catecholamines and sympathetic nerve stimulation all have the ability to stimulate renal prostaglandin synthesis.

Prostaglandin levels in the urine are generally regarded as reflecting production of prostaglandins by the kidneys (Patrono and Dunn, 1987). However, PGE_2 and PGI_2 (as its breakdown product 6-oxo-$PGF_{1\alpha}$) enter the kidney from the arterial circulation. More than 80% of extrarenal PGE_2 is metabolized by the kidney and therefore does not appear in the urine, but only about 25% of PGI_2 is degraded, so that a substantial proportion of PGI_2 metabolites excreted originates from extrarenal sources (Bugge *et al.*, 1987). Some of the PGE_2 and PGI_2 generated by the kidney can be released into the renal circulation rather than into kidney tubules and does not appear in the urine (Boyd *et al.*, 1986). It is difficult to estimate the proportion of the prostaglandins originating in the renal cortex compared to medullary tissues; however, chemical medullectomy of the kidney can reduce excretion of PGE_2 by 75%, suggesting that the PGE_2 excreted may represent PGE_2 mostly of medullary origin (Bing *et al.*, 1983). Nevertheless, prostaglandin excretion is considered to provide a measure of intrarenal prostaglandin synthesis which can be reduced by the administration of NSAIDs.

8.2   EFFECTS OF PROSTAGLANDINS ON KIDNEY FUNCTION

PGE_2 and PGI_2 are vasodilators in the kidney whereas TXA_2 is a vasoconstrictor. Intrarenal infusions of PGE_2 or arachidonic acid increase renal blood flow. The prostaglandins are also natriuretic, inhibiting tubular sodium reabsorption and, in the thick ascending limb of the loop of Henle, they reduce chloride transport. Prostaglandins modulate vasoconstriction brought about by angiotensin II, vasopressin or nerve stimulation, thus counteracting the reduction of RBF and GFR, and contraction of glomerular mesangial cells induced by these treatments (Scharschmidt and Dunn, 1983). Prostaglandins also attenuate the osmotic action of vasopressin on the mammalian nephron, thereby increasing urine flow (Orloff and Zusman, 1978). Thus, prostaglandins have the capacity to influence RBF, GFR, tubular electrolyte transport, water reabsorption and the release of renal hormones. However, in normal animals and in healthy humans, prostaglandins play only a minor role in the control of renal function. Their function is mainly to modulate the effects of nervous and endocrine control systems operating in the kidney (Dunn and Zambraski, 1980). This modulation achieves particular importance during disturbances of fluid and electrolyte balance, activation of other hormone systems or in various disease states such as liver cirrhosis or heart failure.

PGI_2 and PGE_2 synthesized in the renal cortex are important stimulators of renin release (Osborn *et al.*, 1984). Several mechanisms of renin release are dependent on intact prostaglandin synthesis, since administration of inhibitors of prostaglandin synthesis abolishes the renin release induced by activation of these mechanisms (Bugge *et al.*, 1988, 1990). Prostaglandins also interact with the renal kallikrein–kinin system (McGiff, 1980), but do not affect the natriuretic action of atrial natriuretic peptide (Salazar *et al.*, 1988).

Aspirin and other NSAIDs inhibit renal cyclooxygenase (Caterson *et al.*, 1978) and decrease renal prostaglandin production. PGE_2 excretion is decreased (Oliw *et al.*, 1978) and renal vasodilatation, produced by an intrarenal infusion of arachidonic acid, is abolished (Anderson *et al.*, 1983).

8.3   INHIBITION OF CYCLOOXYGENASE IN ANIMAL STUDIES

Acute administration of NSAIDs to normal animals has no effect on the kidney, implying that renal prostaglandins do not play an important role in normal kidney physiology. However, inhibition of cyclooxygenase in sodium-depleted animals decreased RBF and GFR while raising plasma renin levels (Blasingham and Nasjletti, 1980; De Forrest *et al.*, 1980; Izumi *et al.*, 1985). While sodium depletion, by activating the renin–angiotensin system, sensitized the kidneys to the deleterious effects of NSAIDs, there was no clear evidence that sodium depletion actually increased renal prostaglandin production. Similarly, aspirin caused marked renal vasoconstriction in animals with a low cardiac output, presumably by eliminating the modulating influence of prostaglandins on sympathetic and angiotensin-mediated vasoconstrictor activity (Reigger *et al.*, 1989).

Maintenance of normal kidney function is also dependent on prostaglandins in animal models of various abnormal and disease states such as liver cirrhosis, heart failure, diabetes and experimental renal damage. Cyclooxygenase inhibition reduced RBF and GFR in bile duct-ligated cirrhotic animals in which renal PGE_2 synthesis was increased (Zambraski and Dunn, 1984). During reduced renal perfusion pressure, increased prostaglandin production by the kidney counteracted the vasoconstrictor activity of angiotensin II and protected against acute tubular necrosis (Goto *et al.*, 1987; Kaufman *et al.*, 1987). In streptozocin-induced diabetic rats, glomerular PGE_2 synthesis was increased (Barnett *et al.*, 1987) and control of GFR was dependent on prostaglandin production (Jensen *et al.*, 1986; Kirschenbaum and Chaudhari, 1986). In experimental models of renal disease, synthesis of both vasodilator PGE_2 and vasoconstrictor TXA_2 was increased. In these conditions, such as experimental glomerulonephritis (Kaizu *et al.*, 1985) or experimental reduction of renal mass (Stahl *et al.*, 1986), administration of NSAIDs decreased GFR.

8.4 INHIBITION OF CYCLOOXYGENASE IN HUMAN STUDIES

The majority of studies in which aspirin was given to normal subjects (Robert *et al.*, 1972; Reimann *et al.*, 1985; Patrono and Dunn, 1987) did not show any changes in renal function, even though PGE_2 excretion was reduced. However, in a number of disease states, including congestive heart failure, cirrhosis and renal insufficiency, the local release of a vasodilator prostaglandin helped to maintain RBF. Patients are therefore at risk of renal ischaemia when prostaglandin synthesis is diminished by NSAIDs. An antiproteinuric effect of aspirin-like drugs is easily demonstrable when there is disturbed renal function. In such instances, indomethacin, for example, decreases GFR and the effective RBF, possibly via inhibition of prostaglandin synthesis (Arisz *et al.*, 1976; Donker *et al.*, 1976). Recent clinical studies found that, in patients with mild chronic renal failure, even a brief course of ibuprofen may precipitate acute renal failure (Whelton *et al.*, 1990). Aspirin substantially reduced urine volume and sodium excretion as well as urinary PGE_2 in patients with liver disease (Arroyo *et al.*, 1983, 1986), while the deleterious effect of indomethacin

was even more pronounced than that of aspirin. Patients with cardiovascular disease, particularly with congestive heart failure, also showed a large decrease in GFR, and 50–80% reduction in water and sodium secretion after 1 g of aspirin (Bock *et al.*, 1986). The effects of NSAIDs on sodium and water excretion generally paralleled those on the changes in RBF and GFR. The reduction of electrolyte excretion must result from both a reduction in filtered sodium and water combined with enhanced tubular reabsorption, secondary to the inhibition of renal PGE_2 synthesis (Dunn and Zambraski, 1980; Zambraski and Dunn, 1992).

9. *In Reproduction and Parturition*

There has been much interest in the possible involvement of prostaglandins in reproductive physiology.

9.1 PROSTAGLANDINS IN SEMINAL FLUID

Human seminal fluid contains a number of prostaglandins, namely PGE_1, PGE_2, PGE_3, $PGF_{1\alpha}$, $PGF_{2\alpha}$ (Samuelsson, 1963), 19-hydroxy-PGE_1, 19-hydroxy-PGE_2, 19-hydroxy-$PGF_{1\alpha}$ and 19-hydroxy-$PGF_{2\alpha}$ (Taylor and Kelly, 1974, 1975). The 8β isomers of these prostaglandins have also been isolated from semen (Taylor, 1979) and small amounts of 18,19-dehydro-PGE_1 and 18,19-dehydro-PGE_2 have been identified (Oliw *et al.*, 1986). The total amount of prostaglandins in semen is very high, with approximately 1 mg in one ejaculate; this comprises a high proportion of PGEs and 19-hydroxy-PGEs (Templeton *et al.*, 1978; Bendvold *et al.*, 1987). The origin of these prostaglandins has been unclear, but the results of recent studies support the assumption that prostaglandins present in semen originate from the seminal vesicles. This is based on evidence obtained from measurements of prostaglandin concentrations in semen before and after vasectomy (Bendvold *et al.*, 1985), in fluid obtained from the vas deferens (Bendvold *et al.*, 1985) and in semen obtained from men with cystic fibrosis and lacking seminal vesicles (Bendvold *et al.*, 1986).

Von Euler suggested over 50 years ago (von Euler, 1936) that prostaglandins may be involved in the regulation of ejaculation. More recently, a relaxant effect of PGE_2 has been observed on the smooth muscle of the epididymis *in vitro* in the rat (Consentino *et al.*, 1984). There seems to be a correlation between lowered concentrations of prostaglandins in semen and some cases of male infertility. In some men, injection of PGE_1 into the corpus cavernosum is an effective treatment for impotence (Keogh and Earle, 1989; Williams, 1991), presumably owing to its ability to relax corporeal smooth muscle.

There is also some evidence that sperm density and

motility are related to the PGE concentration of seminal fluid (Kelly *et al.*, 1979; Bendvold *et al.*, 1984). The addition of physiological amounts of 19-hydroxy-PGEs to seminal fluid significantly improved sperm motility, whereas 19-hydroxy-PGFs had the opposite effect (Aitken and Kelly, 1985; Bygdeman *et al.*, 1985). The functional performance of the sperm was also affected. The addition of 19-hydroxy-PGE_2 significantly improved and the addition of 19-hydroxy-PGFs reduced the capacity of the sperm to penetrate cervical mucus *in vitro*. The mechanism by which 19-hydroxy-PGEs and 19-hydroxy-PGFs influence sperm motility may involve an effect on the sperm content by ATP, since 19-hydroxy-PGFs decreased and 19-hydroxy-PGEs increased the ATP content of semen (Comhaire *et al.*, 1983; Bygdeman *et al.*, 1987).

It is, therefore, possible that the physiological role of prostaglandins in human seminal fluid is to regulate the functional capacity of spermatozoa. However, the substantial absorption of prostaglandins by the vagina has encouraged speculation that prostaglandins deposited during coitus may facilitate conception by stimulating the contractile activity of the cervix, uterine body and Fallopian tubes. This would result in an enhanced transport of semen.

9.2 REGULATION OF UTERINE CONTRACTIONS

PGE_2 and $PGF_{2\alpha}$ are synthesized by decidua and myometrium, fetal membranes and umbilical cord. The capacity of these tissues to elaborate prostaglandins rises progessively during pregnancy. Concentrations of prostaglandins in blood and amniotic fluid are elevated during labour, correlating with increased cervical dilatation (Keirse, 1979; Mitchell, 1981, 1984; Okazaki *et al.*, 1981). It is not certain whether this is a major determinant of the onset of labour or only serves to sustain uterine contractions that have been initiated by oxytocin. In any event, inhibitors of cyclooxygenase such as aspirin can increase the length of gestation, prolong the duration of spontaneous labour and interrupt premature labour (Keirse, 1979; Wiqvist, 1979). The last-mentioned effect has prompted clinical investigation of these agents for the prevention of premature delivery. While effective, their potential impact on fetal development such as the premature closure of the ductus arteriosus (Heyman *et al.*, 1976), together with the availability of other tocolytic agents, has limited the use of cyclooxygenase inhibitors for this purpose.

It has been suggested that activity of the pregnant human uterus is regulated by a balance between the intrinsic uterine stimulant $PGF_{2\alpha}$ and the inhibitor of uterine motility progesterone (Csapo, 1973). Studies with the antiprogestin steroid RU486, which acts as an antagonist at the progesterone receptor (Baulieu, 1985), have confirmed this hypothesis. The human uterus is virtually inactive in early pregnancy. However, if patients are treated with RU486, spontaneous regular uterine contractions can be recorded. Moreover, such treatment also increases the sensitivity of the myometrium to the PGE_2 analogue 16-phenoxytetranor-PGE_2 methylsulphonylamide (Sulprostone). A combination of RU486 and Sulprostone may be an effective method of terminating early pregnancies and would be associated with a lower frequency of side-effects than the use of Sulprostone alone (Bygdeman and Swahn, 1985). Administration of PGE_2 or $PGF_{2\alpha}$ will induce labour and cervical ripening at any stage of pregnancy (Embrey, 1971; Keirse, 1979; Thiery, 1979), thus they are useful for the treatment of most cases of missed abortion, late intrauterine death, molar gestation and premature rupture of the amniotic membranes (Thiery and Amy, 1977; Toppozada, 1985).

The most common reason for using prostaglandin analogues such as Sulprostone is termination of pregnancy. Induction of early abortion with prostaglandins has a success rate approaching 90%, but the side-effects, especially uterine pain, vomiting, diarrhoea and uterine bleeding, are a disadvantage (Bygdeman *et al.*, 1984).

There is evidence implicating PGE_2 in prelabour physiological ripening of the cervix (Wingerup *et al.*, 1978). Local administration of PGE_2 is used clinically, particularly in gel form, to ripen the cervix prior to spontaneous or induced labour (Lange *et al.*, 1984; Brindley and Sokol, 1988). 15-Methyl-$PGF_{2\alpha}$ can also be used for this purpose, and other suitable prostaglandin preparations are under development.

9.3 LUTEOLYTIC ACTION

$PGF_{2\alpha}$ produced in the uterus is a luteolytic hormone in some subprimate species. This knowledge has led to the development of prostaglandin analogues for veterinary use in synchronizing oestrus in farm animals such as sheep, cattle, pigs and horses. This method has simplified breeding procedures and is used to provide safe, early abortions before the animals are sent to market. However, a luteolytic prostaglandin analogue for human use has not been developed. The possible roles of prostaglandins in reproductive processes have been reviewed by several authors (Goldberg and Ramwell, 1975; Horton and Poyser, 1976; Karim and Hillier, 1979).

10. *Acknowledgements*

The authors would like to thank Joanne Robinson for excellent secretarial assistance. The William Harvey Research Institute is supported by Glaxo Group Research Ltd, the Parke Davis Division of Warner Lambert and the ONO Pharmaceutical Company.

11. References

Abdul, H.T. and Milton, A.S. (1989). The effects of dexamethasone on prostaglandin E_2 release into rabbit CSF during fever and on the pyrogen stimulated release of prostaglandin E_2 from rabbit monocytes. J. Physiol. (Lond.) 409, 60P.

Adelman, B., Stemerman, M.B., Merrell, D. and Handin, R. (1981). The interaction of platelets with aortic subendothelium: inhibition of adhesion and secretion by prostaglandin I_2. Blood 58, 198–205.

Ahlquist, D.A., Dozois, R.R., Zinsmeister, A.R. and Malagelada, J.-R. (1983). Duodenal prostaglandin synthesis and acid load in health and duodenal ulcer disease. Gastroenterology 85, 522–528.

Aiken, J.W., Gorman, R.R. and Shebuski, R.J. (1979). Prevention of blockage of partially obstructed coronary arteries with prostacyclin correlates with inhibition of platelet aggregation. Prostaglandins 17, 483–494.

Aitken, R.J. and Kelly, R.W. (1985). Analysis of the direct effects of prostaglandins on human sperm function. J. Reprod. Fertil. 73, 139–146.

Allen, A. (1981). In "Basic Mechanisms of Gastrointestinal Mucosal Cell Injury and Protection" (ed J.W. Harmon), pp 351–367. Williams and Wilkins, Baltimore.

Allen, A. and Garner, A. (1980). Mucus and bicarbonate secretion in the stomach and their possible role in mucosal protection. Gut 21, 249–262.

Anderson, W.P., Bartley, P.J., Casley, D.J. and Selig, S.E. (1983). Comparison of aspirin and indomethacin pretreatments on responses to reduced renal artery pressure in conscious dogs. J. Physiol. 336, 101–112.

Antiplatelet Trialists' Collaboration (1988). Secondary prevention of vascular disease by prolonged antiplatelet treatment. Br. Med. J. 296, 320–331.

Arisz, L., Donker, A.J.M., Brentjens, J.R.H. and van der Hem, G.K. (1976). The effect of indomethacin on proteinuria and kidney function in the nephrotic syndrome. Acta Med. Scand. 199, 121–125.

Arroyo, V., Planas, R., Gaya, J., Deulofeu, R., Rimola, A., Perez-Ayuso, R.M., Rivera, F. and Rodes, J. (1983). Sympathetic nervous activity, renin–angiotensin system and renal excretion of prostaglandin E_2 in cirrhosis. Relationship to functional renal failure and sodium and water excretion. Eur. J. Clin. Invest. 13, 271–278.

Arroyo, V., Gines, P., Rimola, A. and Gaya, J. (1986). Renal function abnormalities, prostaglandins, and effects of non-steroidal anti-inflammatory drugs in cirrhosis with ascites. An overview with emphasis on pathogenesis. Am. J. Med. 81 (Suppl. 2B), 104–122.

Bahari, H.M.M., Ross, I.N. and Turnberg, L.A. (1982). Demonstration of a pH gradient across the mucus layer on the surface of human gastric mucosa in vitro. Gut 23, 513–516.

Bahl, A.K., Foreman, J.C. and Dale, M.M. (1991). The effect of prostaglandin E_2 and non-steroidal anti-inflammatory drugs on cell-associated interleukin one. Adv. Prostaglandin Thromboxane Leukotriene Res. 21, 513–515.

Baker, P.E., Fahey, J.V. and Munck, A. (1981). Prostaglandin inhibition of T-cell proliferation is mediated at two levels. Cell. Immunol. 61, 52–57.

Barnett, R., Scharschmidt, L., Ko, Y.H. and Schlondorff, D. (1987). Comparison of glomerular and mesangial prostaglandin synthesis and glomerular contraction in two rat models of diabetes mellitus. Diabetes 36, 1468–1475.

Barzilai, A., Schiessel, R., Kivilaakso, E., Matthews, J.B., Fleischer, L.A., Bartzokis, G. and Silen, W. (1980). Effect of 16,16-dimethyl prostaglandin E_2 on ulceration of isolated amphibian gastric mucosa. Gastroenterology 78, 1508–1512.

Baulieu, E.E. (1985). In "The Antiprogestin Steroid RU 486 and Human Fertility Control" (eds E.E. Baulieu and S.J. Segal), pp 1–25. Plenum Press, New York.

Bayer, B.-L., Blass, K.-E. and Förster, W. (1979). Anti-aggregatory effect of prostacyclin (PGI$_2$) in vivo. Br. J. Pharmacol. 66, 10–12.

Beaver, W.T. (1988). Impact of non-narcotic oral analgesics on pain management. Am. J. Med. 84(Suppl. 5A), 3–15.

Bendvold, E., Svanborg, K., Eneroth, P., Gottlieb, C. and Bygdeman, M. (1984). The natural variations in prostaglandin concentration in human seminal fluid and its relation to sperm quality. Fertil. Steril. 41, 743–747.

Bendvold, E., Svanborg, K., Bygdeman, M. and Noren, S. (1985). On the origin of prostaglandins in human seminal fluid. Int. J. Androl. 8, 37–43.

Bendvold, E., Gottlieb, C., Svanborg, K., Gilljam, H., Strandvik, B., Bygdeman, M. and Eneroth, P. (1986). Absence of prostaglandins in semen of men with cystic fibrosis is an indication of the contribution of the seminal vesicles. J. Reprod. Fertil. 78, 311–314.

Bendvold, E., Gottlieb, C., Svanborg, K., Bygdeman, M. and Eneroth, P. (1987). Concentration of prostaglandins in seminal fluid of fertile men. Int. J. Androl. 10, 463–469.

Benigni, A., Gregorini, G., Frusca, T., Chiabrando, C., Ballerini, S., Valcamonico, A., Orisio, S., Picinelli, A., Pinciroli, V., Fanelli, R., Gastaldi, A. and Remuzzi, G. (1989). Effect of low-dose aspirin on fetal and maternal generation of thromboxane by platelets in women at risk for pregnancy-induced hypertension. N. Engl. J. Med. 321, 357–362.

Bennett, A. (1976). Prostaglandins as factors in diseases of the alimentary tract. Adv. Prostaglandin Thromboxane Res. 2, 547–555.

Bennett, A., Eley, K.G. and Scholes, G.B. (1968a). Effects of prostaglandins E_1 and E_2 on human, guinea pig and rat isolated small intestine. Br. J. Pharmacol. 34, 630–638.

Bennett, A., Eley, K.G. and Scholes, G.B. (1968b). Effect of prostaglandins E_1 and E_2 on intestinal motility in the guinea pig and rat. Br. J. Pharmacol. 34, 639–647.

Bennett, A., Eley, K.G. and Scholes, G.B. (1975). The effects of prostaglandins on guinea pig isolated intestine and their possible contribution to muscle activity and tone. Br. J. Pharmacol. 54, 197–204.

Bennett, A., DelTacca, M., Stamford, I.F. and Zebro, T. (1977). Prostaglandins from tumors of human large bowel. Br. J. Cancer 35, 881–884.

Bennett, A., Hensby, C.N., Sanger, G.J. and Stamford, I.F. (1981). Metabolites of arachidonic acid formed by human gastrointestinal tissue and their actions on the muscle layers. Br. J. Pharmacol. 74, 434–444.

Berkenkopf, J.W. and Weichman, B.M. (1988). Production of prostacyclin in mice following injection of acetic acid, phenylbenzoquinone and zymosan: its role in the writhing response. Prostaglandins 36, 693–709.

Bernheim, H.A., Gilbert, T.M. and Stitt, J.T. (1980). Prostaglandin E levels in the third ventricular cerebrospinal fluid of rabbits during fever and changes in body temperature. J. Physiol. (Lond.) 301, 69–78.

Betz, M. and Fox, B.S. (1991). Prostaglandin E_2 inhibits the production of Th1 lymphokines but not of Th2 lymphokines. J. Immunol. 146, 108–113.

Beubler, E. and Juan, H. (1979). Effect of ricinoleic acid and other laxatives on net water flux and prostaglandin E release by the rat colon. J. Pharm. Pharmacol. 31, 681–685.

Bhagwat, S.S., Hamann, P.R., Still, W.C., Bunting S. and Fitzpatrick, F.A. (1985). Synthesis and structure of the platelet aggregation factor thromboxane A_2. Nature 315, 511–513.

Bing, R.F., Russel, G.I., Thurston, H., Swales, J.D., Godfrey, N., Lazarus, Y. and Jackson, J. (1983). Chemical renal medullectomy. Effect on urinary prostaglandin E_2 and plasma renin in response to variations in sodium intake and in relation to blood pressure. Hypertension 5, 951–957.

Blair, I.A., Barrow, S.E., Waddell, K.A., Lewis, P.J. and Dollery, C.T. (1982). Prostacyclin is not a circulating hormone in man. Prostaglandins 23, 579–589.

Blasingham, M.C. and Nasjletti, A. (1980). Differential renal effects of cyclooxygenase inhibition in sodium-replete and sodium-deprived dog. Am. J. Physiol. 239, F360–F365.

Blatteis, C.M., Dinarello, C.A., Shibata, M., Llanos-Q, J. Quan, N. and Busija, D.W. (1989). In "Thermal Physiology Nineteen Eighty-Nine", Excerpta Medica International Congress Series 871 (ed. J.B. Mercer), pp 385–390. Elsevier, Amsterdam.

Bligh, J. and Milton, A.S. (1973). The thermoregulatory effects of prostaglandin E_1 when infused into a lateral cerebral ventricle of the Welsh Mountain sheep at different ambient temperatures. J. Physiol. (Lond.) 229, 30–31P.

Bock, H.A., Frölich, J.C., Ritz, R. and Brunner, F.P. (1986). Effects of intravenous aspirin on prostaglandin synthesis and kidney function in intensive care patients. Nephrol. Dial. Transplant 1, 164–169.

Bockman, R.S. (1980). Stage-dependent reduction in T colony formation in Hodgkin's disease. Coincidence with monocyte synthesis of prostaglandins. J. Clin. Invest. 66, 523–531.

Bockman, R.S., Hickok, N. and Rapuano, B. (1987). In "Prostaglandins in Cancer Research" (eds E. Garachi, R. Paoletti and M.G. Santoro), pp 77–85. Springer-Verlag, Berlin.

Bolton, J.P. and Cohen, M.M. (1978). Stimulation of non-parietal cell secretion in canine Heidenhain pouches by 16,16-dimethyl prostaglandin E_2. Digestion 17, 291–299.

Bolton, J.P., Palmer, D. and Cohen, M.M. (1978). Stimulation of mucus and non-parietal cell secretion by the E_2 prostaglandins. Am. J. Dig. Dis. 23, 359–364.

Bombardieri, S., Cattani, P., Ciabattoni, G., Di Munno, O., Pasero, G., Patrono, C., Pinca, E. and Pugliese, F. (1981). The synovial prostaglandin system in chronic inflammatory arthritis: differential effect of steroidal and non-steroidal anti-inflammatory drugs. Br. J. Pharmacol. 73, 893–901.

Böttcher, I., Schweizer, A., Glatt, M. and Werner, H. (1987). A sulphonamide-indanone derivative cGP 28237 (ZK 34228), a novel non-steroidal anti-inflammatory agent without gastro-intestinal ulcerogenicity in rats. Drugs Exp. Clin. Res. 13, 237–245.

Bottomley, K.M.K., Griffiths, R.J., Rising, T.J. and Steward, A. (1988). A modified mouse air pouch model for evaluating the effects of compounds on granuloma-induced cartilage degradation. Br. J. Pharmacol. 93, 627–635.

Boyd, R.M., Nasjletti, A., Heerdt, P.M. and Baer, P.G. (1986). PGI_2 synthesis and excretion in dog kidney: evidence for renal PG compartmentalization. Am. J. Physiol. 250, F58–F65.

Brindley, B.A. and Sokol, R.J. (1988). Induction and augmentation of labor: basis and methods for current practice. Obstet. Gynecol. Surv. 43, 730–743.

Brotherton, A.F.A., MacFarlane, D.E. and Hoak, J.C. (1982). Prostacyclin biosynthesis in vascular endothelium is not inhibited by cyclic AMP. Studies with 3-isobutyl-1-methylxanthine and forskolin. Thromb. Res. 28, 637–647.

Brune, K., Rainsford, K.D., Wagner, K. and Peskar, B.A. (1981). Inhibition by anti-inflammatory drugs of prosta-glandin production in cultured macrophages. Naunyn-Schmiedebergs Arch. Pharmakol. 315, 269–276.

Brunton, L.I., Wiklund, R.A., Van Arsdale, P.M. and Gilman, A.G. (1976). Binding of (^{3}H) prostaglandin E_1 to putative receptors linked to adenylate cyclase of cultured cell clones. J. Biol. Chem. 251, 3037–3044.

Bugat, R., Thompson, M.R., Aures, D. and Grossman, M.I. (1976). Gastric mucosal lesions produced by intravenous infusion of aspirin in cats. Gastroenterology 71, 754–759.

Bugge, J.F., Vikse, A., Dahl, E. and Kiil, F. (1987). Renal degradation and distribution between urinary and venous output of prostaglandins E_2 and I_2. Acta Physiol. Scand. 130, 467–474.

Bugge, J.F., Stokke, E.S. and Kiil, F. (1988). Properties of the macula densa mechanism for renin release in the dog. Acta Physiol. Scand. 132, 401–412.

Bugge, J.F., Stokke, E.S., Vikse, A. and Kiil, F. (1990). Stimulation of renin release by PGE_2 and PGI_2 infusion in the dog: enhancing effect of ureteral occlusion or administra-tion of ethacrynic acid. Acta Physiol. Scand. 138, 193–201.

Burch, R.M., Mais, D.E., Saussy, D.L. Jr and Halushka, P.V. (1985). Solubilization of a thromboxane A_2/prostaglandin H_2 antagonist binding site from human platelets. Proc. Natl Acad. Sci. USA 82, 7434–7438.

Bygdeman, M. and Swahn, M.L. (1985). Progesterone receptor blockage. Effect on uterine contractility and early pregnancy. Contraception 32, 45–51.

Bygdeman, M., Christensen, N.J., Green, K. and Vesterqvist, O. (1984). In "Prostaglandins and Fertility Regulation" (eds M. Toppozada, M. Bygdeman and E.S.E. Hafez), pp 83–90. MTP Press, Lancaster.

Bygdeman, M., Bendvold, E., Gottlieb, C., Svanborg, K. and Eneroth, P. (1985). In "Prostaglandins, Leukotrienes and Lipoxins" (ed J.M. Bailey), pp 423–431. Plenum Press, New York.

Bygdeman, M., Gottlieb, K., Svanborg, K. and Swahn, M.L. (1987). Role of prostaglandins in human reproduction: recent advances. Adv. Prostaglandin Thromboxane Leukotriene Res. 17, 1112–1116.

Cammock, S., Dascombe, M.J. and Milton, A.S. (1976). Prostaglandins in thermoregulation. Adv. Prostaglandin Thromboxane Res. 1, 375–380.

Carmichael, H.A., Nelson, L.M. and Russell, R.J. (1978). Cimetidine and prostaglandin: evidence for different modes of action on the rat gastric mucosa. Gastroenterology 74, 1229–1232.

Catania, A., Arnold, J., Macaluso, A., Hiltz, M.E. and Lipton, J.M. (1991). Inhibition of acute inflammation in the periphery by central action of salicylates. Proc. Natl Acad. Sci. USA 88, 8544–8547.

Caterson, R.J., Duggin, G.G., Horvath, J., Mohandas, J. and Tiller, D. (1978). Aspirin, protein transacetylation and inhibition of prostaglandin synthetase in the kidney. Br. J. Pharmacol. 64, 353–358.

Chahl, L.A. and Iggo, A. (1977). The effects of bradykinin and prostaglandin E_1 on rat cutaneous afferent nerve activity. Br. J. Pharmacol. 59, 343–347.

Chang, W.C., Murota, S. and Tsurufugi, S. (1976). A new prostaglandin transformed from arachidonic acid in carrageenin-induced granuloma. Biochem. Biophys. Res. Commun. 72, 1259–1264.

Charman, W.N., Kerins, D.M. and FitzGerald, G.A. (1992). In "Aspirin and Other Salicylates" (eds J.R. Vane and R.M. Botting), pp 73–106. Chapman and Hall, London.

Chiabrando, C., Castelli, M.G. and Cozzi, E. (1989). Antiinflammatory actions of salicylates: aspirin is not a prodrug for salicylate against rat carrageenin pleurisy. Eur. J. Pharmacol. 159, 257–264.

Cohen, M.M. (1981). Prevention of aspirin-induced faecal blood loss with oral prostaglandin E_2: dose–response studies in man. Prostaglandins 21(Suppl.), 155–160.

Cohen, M.M. and Pollet, J.M. (1976). Prostaglandin E_2 prevents aspirin and indomethacin damage to human gastric mucosa. Surg. Forum 27, 400–401.

Cohen, M.M., Cheung, G. and Lyster, D.M. (1980). Prevention of aspirin-induced faecal blood loss by prostaglandin E_2. Gut 21, 602–606.

Cohen, M.M., Clark, L., Armstrong, L. and D'Souza, J. (1985). Reduction in aspirin-induced fecal blood loss with low-dose misoprostol tablets in man. Dig. Dis. Sci. 30, 605–611.

Cohen, R.H. and Perl, E.R. (1990). Contributions of arachidonic acid derivatives and substance P to the sensitization of cutaneous nociceptors. J. Neurophysiol. 64, 457–464.

Coker, S.J., Parratt, J.R., Ledingham, I.M. and Zeitlin, I.J. (1981). Thromboxane and prostacyclin release from ischaemic myocardium in relation to arrhythmias. Nature 291, 323–324.

Coleman, R.A., Humphrey, P.P.A. and Kennedy, I. (1985). In "Trends in Autonomic Pharmacology", Vol 3 (ed S. Kalsner), p 35. Taylor and Francis, Philadelphia.

Collier, H.O.J. (1969a). A pharmacological analysis of aspirin. Adv. Pharmacol. Chemother. 7, 333–405.

Collier, H.O.J. (1969b). New light on how aspirin works. Nature 223, 35–37.

Collier, H.O.J. and Chesher, G.B. (1956). Antipyretic and analgesic properties of 2-hydroxyisophthalic acids. Br. J. Pharmacol. 11, 20–26.

Collier, H.O.J. and Roy, A.C. (1974a). Morphine-like drugs inhibit the stimulation by E prostaglandins of cyclic AMP formation by rat brain homogenates. Nature 248, 24–25.

Collier, H.O.J. and Roy, A.C. (1974b). Inhibition of E prostaglandin-sensitive adenyl cyclase as the mechanism of morphine analgesia. Prostaglandins 7, 361–376.

Collier, H.O.J. and Schneider, C. (1972). Nociceptive response to prostaglandins and analgesic actions of aspirin and morphine. Nature (New Biol.) 236, 141–143.

Collier, H.O.J., Dineen, L.C., Johnson, C.A. and Schneider, C. (1968). The abdominal constriction response and its suppression by analgesic drugs in the mouse. Br. J. Pharmacol. 32, 295–310.

Comhaire, F., Vermeulen, L., Ghedira, K., Mas, J., Irvine, S. and Callipolidis, G. (1983). Adenosine triphosphate in human semen: a quantitative estimate of fertilizing potential. Fertil. Steril. 40, 500–504.

Consentino, M.J., Emilson, L.B.W. and Cockett, A.T.K. (1984). Prostaglandins in semen and their relationship to male fertility: a study of 145 men. Fertil. Steril. 41, 88–94.

Conti, P., Reale, M., Cancelli, A. and Angeletti, P.U. (1987). Lipoxin A augments release of thromboxane from human polymorphonuclear leukocyte suspensions. FEBS Lett. 225, 103–108.

Csapo, A.I. (1973). The prospects of PGs in postconceptional therapy. Prostaglandins 3, 245–289.

Cummings, J.H., Newman, A., Misiewicz, J.J., Milton-Thompson, G.J. and Billings, J.A. (1973). Effect of intravenous prostaglandin $F_{2\alpha}$ on small intestinal function in man. Nature 243, 169–171.

Czervionke, R.L., Smith, J.B., Hoak, J.C., Fry, G.L. and Haycroft, D.L. (1979). Use of a radioimmunoassay to study thrombin-induced release of PGI_2 from cultured endothelium. Thromb. Res. 14, 781–786.

Dajani, E.Z., Driskill, D.R., Bianchi, R.G. and Collins, P.W. (1975). Comparative gastric antisecretory and antiulcer effects of prostaglandin E_1 and its methyl ester in animals. Prostaglandins 10, 205–215.

Dajani, E.Z., Callison, D.A., Bianchi, R.G. and Driskill, D.R. (1976). Gastric antisecretory effects of E prostaglandins in rhesus monkeys. Am. J. Dig. Dis. 21, 1020–1028.

Dajani, E.Z., Bertermann, R.E., Roge, E.A.W., Schweingruber, F.L. and Woods, E.M. (1979). Canine gastrointestinal motility effects of prostaglandin $F_{2\alpha}$. Arch. Int. Pharmacodyn. 237, 15–24.

Danesh, B.J., McLaren, M., Russel, R.I., Lowe, G.D.O. and Forbes, C.D. (1989). Comparison of the effect of aspirin and choline magnesium trisalicylate on thromboxane biosynthesis in human platelets: role of the acetyl moiety. Haemostasis 19, 169–173.

Darrow, T.L. and Tomar, R.H. (1980). Prostaglandin-mediated regulation of the mixed lymphocyte culture and generation of cytotoxic cells. Cell. Immunol. 65, 172–178.

Dascombe, M.J. and Milton, A.S. (1975). The effects of cyclic adenosine $3',5'$-monophosphate and other adenine nucleotides on body temperature. J. Physiol. (Lond.) 250, 143–160.

Dascombe, M.J. and Milton, A.S. (1979). Study on the possible entry of bacterial endotoxin and prostaglandin E_2 into the central nervous system from the blood. Br. J. Pharmacol. 66, 565–572.

Davenport, H.W. (1964). Gastric mucosal injury by fatty and acetylsalicyclic acids. Gastroenterology 46, 245–253.

Davenport, H.W. (1965). Damage to the gastric mucosa: effects of salicylates and stimulation. Gastroenterology 49, 184–196.

Davidson, J., Milton, A.S. and Rotondo, D. (1989). Effect of dexamethasone and ketoprofen on the pyrogenic response to human recombinant interleukin-1α (IL-1α) and interleukin-1β (IL-1β). Br. J. Pharmacol. 97, 429P.

Davidson, J., Milton, A.S. and Rotondo, D. (1990). A study of the pyrogenic actions of interleukin-1α and interleukin-1β: interactions with a steroidal and a non-steroidal anti-inflammatory agent. Br. J. Pharmacol. 100, 542–546.

De Brito, F.B., Holmes, M.J.G., Carney, S.L. and Willoughby, D.A. (1987). Drug effects on a novel model of connective tissue breakdown. Agents Actions 21, 287–290.

Deby, C., Pincemail, J., Deby-Dupont, G., Braquet, P. and Goutier, R. (1984). In "Icosanoids and Cancer" (eds R. Paoletti and P. Gnastes de Paulet), pp 31–40. Raven Press, New York.

De Caterina, R., Giannessi, D. and Boem, A. (1985). Equal antiplatelet effects of aspirin 50 or 324 mg/day in patients after acute myocardial infarction. Thromb. Haemost. 54, 528–532.

De Forrest, J.M., Davies, J.O., Freeman, R.H., Seymour, A.A., Rowe, B.P., Williams, G.M. and Davis, T.P. (1980). Effects of indomethacin and meclofenamate on renin release and renal haemodynamic function during chronic sodium depletion in conscious dogs. Circ. Res. 47, 99–107.

De Gaetano, G., Cerletti, C., Dejana, E. and Vermylen, J. (1986). Current issues in thrombosis prevention with antiplatelet drugs. Drugs 31, 517–549.

DeLeo, J.A., Colburn, R.W., Coombs, D.W. and Ellis, M.A. (1989). The differentiation of NSAIDs and prostaglandin action using a mechanical visceral pain model in the rat. Pharmacol. Biochem. Behav. 33, 253–255.

Dembinska-Kiec, A., Kostka-Trabka, E. and Gryglewski, R.J. (1982). Effect of prostacyclin on fibrinolytic activity in patients with atherosclerosis obliterans. Thromb. Haemost. 47, 190.

Deneau, G.A., Waud, R.A. and Gowdey, C.W. (1953). A method for the determination of the effects of drugs on the pain threshold of human subjects. Can. J. Med. Sci. 31, 387–393.

Desa, F.M., Chander, C.L., Howat, D.W., Moore, A.R. and Willoughby, D.A. (1988). Indomethacin and cartilage breakdown. J. Pharm. Pharmacol. 40, 667.

DeWitt, D.L. and Smith, W.L. (1983). Purification of prostacyclin synthase from bovine aorta by immunoaffinity chromatography. Evidence that the enzyme is a hemoprotein. J. Biol. Chem. 258, 3285–3293.

DeWitt, D.L. and Smith, W.L. (1988). Primary structure of prostaglandin G/H synthase from sheep vesicular gland determined from the complementary DNA sequence. Proc. Natl Acad. Sci. USA 85, 1412–1416.

DeWitt, D.L., El-Harith, E.A., Kraemer, S.A., Yao, E.F., Armstrong, R.L. and Smith, W.L. (1990). The aspirin and heme-binding sites of ovine and murine prostaglandin endoperoxide synthases. J. Biol. Chem. 265, 5192–5198.

Diczfalusy, U., Falardeau, P. and Hammarstrom, S. (1977). Conversion of prostaglandin endoperoxides to C_{17}-hydroxy acids catalyzed by human platelet thromboxane synthase. FEBS Lett. 84, 271–274.

Dinarello, C.A. (1984). Interleukin-1. Rev. Infect. Dis. 6, 51–95.

Dinarello, C.A., Weiner, P. and Wolff, S.M. (1978). Radiolabelling and disposition in rabbits of purified human leukocytic pyrogen. Inflammation 2, 179–189.

Dismukes, K. and Daly, J.W. (1975). Accumulation of adenosine 3',5'-monophosphate in rat brain slices: effects of prostaglandins. Life Sci. 17, 199–210.

Doherty, N.S., Beaver, T.H., Chan, K.Y., Coutant, J.E. and Westrich, G.L. (1987). The role of prostaglandins in the nociceptive response induced by intraperitoneal injection of zymosan in mice. Br. J. Pharmacol. 91, 39–47.

Donker, A.J.M., Arisz, L., Brentjens, J.R.H., van der Hem, G.K. and Hollemans, H.J. (1976). The effect of indomethacin on kidney function and plasma renin activity in man. Nephron 17, 288–296.

Dunn, M.J. (1983). In "Renal Endocrinology", pp 1–74. Williams and Wilkins, Baltimore.

Dunn, M.J. and Zambraski, E.J. (1980). Renal effects of drugs that inhibit prostaglandin synthesis. Kidney Int. 18, 609–622.

Dusting, G.J., Mullane, K.M. and Moncada, S. (1986). In "Handbook of Hypertension", Vol 7 (eds A. Zanchetti and R.C. Tarazi), pp 408–426. Elsevier, Amsterdam.

Eckles, D.D. and Gershwin, M.E. (1981). Pharmacologic and biochemical production of human T-lymphocyte colony formation: Hormonal influences. Immunopharmacology 3, 259–274.

Economo, C.V. (1926). In "Handbuch der Normalen und Pathologischen Physiologie", Vol 17 (eds A. Bethe, G.V. Bergmann, G. Embden and A. Ellinger), pp 591–610. Julius Springer, Berlin.

Egan, R.W., Paxton, J. and Keuhl, F.A. Jr (1976). Mechanism for irreversible self-deactivation of prostaglandin synthetase. J. Biol. Chem. 251, 7329–7335.

Egg, D. (1984). Concentrations of prostaglandin D_2, E_2, $F_{2\alpha}$, 6-keto-$F_{1\alpha}$ and thromboxane B_2 in synovial fluid from patients with inflammatory joint disorders and osteoarthritis. Z. Rheumatol. 43, 89–96.

Einzig, S., Rao, G.H.R. and White, J.G. (1980). Differential sensitivity of regional vascular beds in the dog to low-dose prostacyclin infusion. Can. J. Physiol. Pharmacol. 58, 940–946.

Eldor, A., Falcone, D., Hajjar, D., Minick, C. and Weksler, B. (1981). Recovery of prostacyclin production by de-endothelialized rabbit aorta. Critical role of neointimal smooth muscle cells. J. Clin. Invest. 67, 735–741.

Embrey, M.P. (1971). PGE compounds for induction of labour and abortion. Ann. N.Y. Acad. Sci. 180, 518–523.

Emele, J.F. and Shanaman, J. (1963). Bradykinin writhing: method for measuring analgesia. Proc. Soc. Exp. Biol. Med. 114, 680–682.

Fang, W.F., Broughton, A. and Jacobson, E.D. (1977). Indomethacin-induced intestinal inflammation. Am. J. Dig. Dis. 22, 749–760.

Farman, N., Pradelles, P. and Bonvalet, J.P. (1987). PGE_2, $PGF_{2\alpha}$, 6-keto-$PGF_{1\alpha}$, and TXB_2 synthesis along the rabbit nephron. Am. J. Physiol. 252, F53–F59.

Feldberg, W. and Gupta, K.P. (1973). Pyrogen fever and prostaglandin-like activity in cerebrospinal fluid. J. Physiol. (Lond.) 228, 41–53.

Feldberg, W. and Saxena, P.N. (1971). Further studies on prostaglandin E_1 fever in cats. J. Physiol. (Lond.) 219, 739–745.

Feldberg, W., Gupta, K.P., Milton, A.S. and Wendlandt, S. (1973). The effect of pyrogen and antipyretics on prostaglandin activity in cisternal CSF of unanaesthetised cats. J. Physiol. (Lond.) 234, 279–303.

Fels, A.O.S., Pawlowski, N.A., Abraham, E.L. and Cohn, Z.A. (1986). Compartmentalized regulation of macrophage arachidonic acid metabolism. J. Exp. Med. 163, 752–757.

Ferguson, W.W., Edmonds, A.W., Starling, J.R. and Wangensteen, S.L. (1973). Protective effect of prostaglandin E_1 (PGE_1) on lysosomal enzyme release in serotonin-induced gastric ulceration. Ann. Surg. 177, 648–654.

Ferreira, S.H. (1972). Prostaglandins, aspirin-like drugs and analgesia. Nature 240, 200–203.

Ferreira, S.H. (1979). In "Mechanisms of Pain and Analgesic Compounds" (eds R.F. Beers and E.G. Bassett), pp 309–321. Raven Press, New York.

Ferreira, S.H. and Nakamura, M. (1979). I-prostaglandin hyperalgesia. A cAMP/Ca^{2+} dependent process. Prostaglandins 18, 1790.

Ferreira, S.H., Moncada, S. and Vane, J.R. (1971). Indomethacin and aspirin abolish prostaglandin release from spleen. Nature 231, 237–239.

Ferreira, S.H., Moncada, S. and Vane, J.R. (1973). Prostaglandins and the mechanism of analgesia produced by aspirin-like drugs. Br. J. Pharmacol. 49, 86–97.

Ferreira, S.H., Nakamura, M. and Abreu Castro, M.S. (1978a). The hyperalgesic effects of prostacyclin and PGE_2. Prostaglandins 16, 31–37.

Ferreira, S.H., Lorenzetti, B.B. and Correa, F.M.A. (1978b). Central and peripheral antialgesic action of aspirin-like drugs. Eur. J. Pharmacol. 53, 39–48.

Ferreira, S.H., Lorenzetti, B.B., Bristow, A.F. and Poole, S. (1988). Interleukin-1β or a potent hyperalgeric agent antagonized by a tripeptide analogue. *Nature* 334, 698–700.

Ferreira, S.H., Lorenzetti, B.B. and De Campos, D.I. (1990). Induction, blockade and restoration of a persistent hypersensitive state. Pain 42, 365–371.

Fiorucci, S. and McArthur, K.E. (1991). Prostaglandin E_2 desensitizes cAMP-mediated pepsinogen secretion in chief cells. Am. J. Physiol. 261, G858–G865.

Fitzgerald, D.J., Roy, L., Catella, F. and FitzGerald, G.A. (1986). Platelet activation in unstable coronary disease. N. Engl. J. Med. 315, 983–989.

FitzGerald, G.A., Pedersen, A.K. and Patrono, C. (1983a). Analysis of prostacyclin and thromboxane biosynthesis in cardiovascular disease. Circulation 67, 1174–1177.

FitzGerald, G.A., Oates, J.A. and Hawiger, J. (1983b). Endogenous biosynthesis of prostacyclin and thromboxane and platelet function during chronic administration of aspirin in man. J. Clin. Invest. 71, 676–683.

Fitzpatrick, F.A. and Wynalda, M.A. (1976). In vivo suppression of prostaglandin biosynthesis by non-steroidal anti-inflammatory agents. Prostaglandins 12, 1037–1051.

Flemstrom, G. and Garner, A. (1982). Gastroduodenal HCO_3 transport: characteristics and proposed role in acidity regulation and mucosal protection. Am. J. Physiol. 242, G183–G193.

Flower, R.J. (1974). Drugs which inhibit prostaglandin biosynthesis. Pharmacol. Rev. 26, 33–67.

Flower, R.J. and Vane, J.R. (1972). Inhibition of prostaglandin synthetase in brain explains the antipyretic activity of paracetamol (4-acetamidophenol). Nature 240, 410.

Foegh, M.L., Winchester, J.F. and Zmudka, M. (1981). Urine i-TXB_2 in allograft rejection. Lancet ii, 431–434.

Follenfant, R.L., Nakamura-Craig, M. and Garland, L.G. (1990). Sustained hyperalgesia in rats evoked by 15-hydroperoxyeicosatetraenoic acid is attenuated by the protein kinase inhibitor H-7. Br. J. Pharmacol. 99, 289P.

Ford-Hutchinson, A.W., Brunet, G., Savard, P. and Charleson, S. (1984). Leukotriene B_4, polymorphonuclear leukocytes and inflammatory exudates in the rat. Prostaglandins 28, 13–27.

Gallavan, R.J. and Jacobson, E.D. (1982). Prostaglandins and the splanchnic circulation. Proc. Soc. Exp. Biol. Med. 170, 391–397.

Gascoigne, A.D. and Hirst, B.H. (1981). Prostaglandins alter the relationship between gastric hydrogen ion concentration and flow: evidence for the stimulation of non-parietal secretion in the cat. J. Physiol. (Lond.) 316, 427–438.

Gaskill, H.V., Sirinek, K.R. and Levine, B.A. (1982). Prostacyclin-mediated gastric cytoprotection is dependent on mucosal blood flow. Surgery 92, 220–225.

Gerber, J.G. and Nies, A.S. (1982). Canine gastric mucosal vasodilation with prostaglandins and histamine analogs. Dig. Dis. Sci. 27, 870–874.

Gerkens, J.F., Gerber, J.C., Shand, D.G. and Branch, R.A. (1978). Effect of PGI_2, PGE_2 and 6-keto-$PGF_{1\alpha}$ on canine gastric blood flow and acid secretion. Prostaglandins 16, 815–823.

Gerritsen, M.E. (1987). Functional heterogeneity of vascular endothelial cells. Biochem. Pharmacol. 35, 2701–2711.

Gilfoil, T.M., Klavins, I. and Grumbach, L. (1963). Effects of acetylsalicylic acid on the oedema and hyperesthesia of the experimentally inflamed rat's paw. J. Pharmacol. Exp. Ther. 142, 1–5.

Goldberg, V.J. and Ramwell, P.W. (1975). Role of prostaglandins in reproduction. Physiol. Rev. 55, 325–351.

Goldstein, J.M., Mahnsten, C.L. and Kindahl, H. (1978).

Thromboxane generation by human peripheral blood polymorphonuclear leukocytes. J. Exp. Med. 148, 787–792.

Goldyne, M.E. and Stobo, J.D. (1979). Synthesis of prostaglandins by subpopulations of human peripheral blood monocytes. Prostaglandins 18, 687–695.

Goodwin, J.S. and Ceuppens, J. (1983). Regulation of the immune response by prostaglandins. J. Clin. Immunol. 3, 295–315.

Goodwin, J.S., Bankhurst, A.D. and Messner, R.P. (1977). Suppression of human T-cell mitogenesis by prostaglandin. Existence of a prostaglandin-producing suppressor cell. J. Exp. Med. 146, 1719.

Goodwin, J.S., Messner, R.P. and Peake, G.T. (1978). Prostaglandin suppression of mitogen-stimulated lymphocytes in vitro. J. Clin. Invest. 62, 753–760.

Goodwin, J.S., Wiik, A., Lewis, M., Bankhurst, A.D. and Williams, R.C. (1979). High affinity binding sites for prostaglandin E on human lymphocytes. Cell. Immunol. 43, 150–159.

Gordon, D., Bray, M. and Morley, J. (1976). Control of lymphokine secretion by prostaglandins. Nature 262, 401–407.

Gordon, D., Henderson, D.C. and Westwick, J. (1979). Effects of prostaglandin E_2 and I_2 on human lymphocyte transformation in the absence and presence of inhibitors of prostaglandin biosynthesis. Br. J. Pharmacol. 67, 17–22.

Goto, F., Jackson, E.K., Ohnishi, A., Herzer, W. and Branch, R.A. (1987). Effect of cyclooxygenase and thromboxane synthase inhibition on the response to angiotensin II in the hypoperfused canine kidney. J. Pharmacol. Exp. Ther. 243, 799–803.

Graham, D., Agrawal, N. and Roth, S. (1988). Prevention of NSAID-induced gastric ulcer with the synthetic prostaglandin, misoprostol: a multicentre, double-blind, placebo-controlled trial. Lancet ii, 1277–1280.

Greaves, M.W. and Sondergaard, J. (1970). Urticaria pigmentosa and factitious urticaria. Direct evidence for release of histamine and other smooth muscle contracting agents in dermographic skin. Arch. Dermatol. 101, 418–425.

Greengard, P. (1979). Some chemical aspects of neurotransmitter action. Trends Pharmacol. Sci. 1, 27–29.

Grossman, M.I., Matsumoto, K.K. and Lichter, R.J. (1961). Fecal blood loss by oral and intravenous administration of various salicylates. Gastroenterology 40, 383–388.

Gryglewski, R.J. (1970) In "Chemical Control of Fibrinolysis – Thrombolysis" (ed J.M. Schrör), pp 44–72. Wiley Interscience, New York.

Gryglewski, R.J. (1990). In "Endogenous Factors of Cardiovascular Regulation and Protection" (eds M. Cantin, R. Paoletti, P. Braquet and Y. Christen), pp 21–59. Excerpta Medica, Amsterdam.

Gryglewski, R.J., Bunting, S., Moncada, S., Flower, R.J. and Vane, J.R. (1976). Arterial walls are protected against deposition of platelet thrombi by a substance (prostaglandin X) which they make from prostaglandin endoperoxides. Prostaglandins 12, 685–713.

Gryglewski, R.J., Botting, R.M. and Vane, J.R. (1988). Mediators produced by the endothelial cell. Hypertension 12, 530–548.

Guth, P.H. and Moler, T.L. (1982). The role of endogenous prostanoids in the response of the rat gastric microcirculation to vasoactive agents. Microvasc. Res. 23, 336–346.

Guzman, F., Braun, C. and Lim, R.K.S. (1962). Visceral pain

and the pseudoaffective response to intraarterial injection of bradykinin and other algesic agents. Arch. Int. Pharmacodyn. Ther. 136, 353–384.

Guzman, F., Braun, C. and Lim, R.K.S. (1964). Narcotic and non-narcotic analgesics which block visceral pain evoked by intra-arterial injections of bradykinin and other algesic agents. Arch. Int. Pharmacodyn. Ther. 149, 571–588.

Hallam, T.J. and Pearson, J.D. (1986). Exogenous ATP raises cytoplasmic free calcium in fura-2 loaded piglet aortic endothelial cells. FEBS Lett. 207, 95–99.

Ham, E.A., Egan, R.W., Soderman, D.D., Gale, P.H. and Kuehl, F.A. Jr (1979). Peroxidase-dependent deactivation of prostacylin synthetase. J. Biol. Chem. 254, 2191–2194.

Hamberg, M. (1972). Inhibition of prostaglandin synthesis in man. Biochem. Biophys. Res. Commun. 49, 720–726.

Hammarstrom, S. and Falardeau, P. (1977). Resolution of prostaglandin endoperoxide synthase and thromboxane synthase of human platelets. Proc. Natl Acad. Sci. USA 74, 3691–3695.

Hamprecht, D. and Schultz, J. (1973). Stimulation by prostaglandin E_1 of adenosine $3',5'$-cyclic monophosphate formation in neuroblastoma cells in the presence of phosphodiesterase inhibitors. FEBS Lett. 34, 85–89.

Handwerker, H.O. (1975). Influence of prostaglandin E_2 on the discharge of cutaneous nociceptive C-fibres induced by radiant heat. Pfluegers Arch. Gesamte Physiol. Menschen Tiere 355, R116.

Harms, P.G., Ojeda, S.R. and McCann, S.M. (1973). Prostaglandin involvement in hypothalamic control of gonadotropin and prolactin release. Science 181, 760–761.

Harvey, C.A. and Milton, A.S. (1975). Endogenous pyrogen fever, prostaglandin release and prostaglandin synthetase inhibitors. J. Physiol. (Lond.) 250, 18–20P.

Haurand, M. and Ullrich, V. (1985). Isolation and characterisation of thromboxane synthase from human platelets as a cytochrome P-450 enzyme. J. Biol. Chem. 260, 15059–15067.

Hayaishi, O. (1991). Molecular mechanisms of sleep–wake regulation: roles of prostaglandins D_2 and E_2. FASEB J. 5, 2575–2581.

Hayaishi, O., Matsumura, H., Onoe, H., Koyama, Y. and Watanabe, Y. (1990). In "Sleep 90" (ed J. Horne), pp 405–408. Pontenagel Press, Bochum.

Henderson, B., Higgs, G.A., Moncada, S. and Salmon, J.A. (1985). Synthesis of eicosanoids by tissues of the synovial joint during the development of chronic erosive synovitis. Agents Actions 17, 360–362.

Henriksson, P., Wennmalm, A., Edhag, O., Vesterqvist, O. and Gréen, K. (1986). *In vivo* production of prostacyclin and thromboxane in patients with acute myocardial infarction. Br. Heart J. 55, 543–548.

Herman, A.G. and Moncada, S. (1975). Release of prostaglandins and incapacitation after injection of endotoxin in the knee joint of the dog. Br. J. Pharmacol. 53, 465P.

Hesse, E., Roesler, G. and Buhler, F. (1930). Zur biologischen wertbestimmung der analgetika und ihrer kombinationen. Naunyn-Schmiedebergs Arch. Exp. Pathol. Pharmakol. 158, 247–253.

Heylings, J.R., Hampson, S.E. and Garner, A. (1985). Endogenous E-type prostaglandins in regulation of basal alkaline secretion by amphibian duodenum *in vitro*. Gastroenterology 88, 290–294.

Heyman, M.A., Rudolph, A.M. and Silverman, W.H. (1976). Closure of the ductus arteriosus in premature infants by inhibition of prostaglandin synthesis. N. Engl. J. Med. 295, 530–533.

Hial, V., Horakova, Z., Shaff, F.E. and Beaven, M.A. (1976). Alteration of tumor growth by aspirin and indomethacin: studies with two transplanted tumours in mice. Eur. J. Pharmacol. 37, 367–376.

Hierton, C. (1981). Effects of indomethacin, naproxen and paracetamol in regional blood flow in rabbits: a microsphere study. Acta Pharmacol. Toxicol. 49, 327–333.

Higgs, E.A., Moncada, S., Vane, J.R., Caen, J., Michel, H. and Tobelem, G. (1978a). Effect of prostacyclin (PGI_2) on platelet adhesion to rabbit arterial subendothelium. Prostaglandins 16, 17–22.

Higgs, E.A., Moncada, S. and Vane, J.R. (1978b). Inflammatory effects of prostacyclin (PGI_2) and 6-oxo-$PGF_{1\alpha}$ in the rat paw. Prostaglandins 16, 153–162.

Higgs, G.A., Vane, J.R., Hart, F.D. and Wojtulewski, J.A. (1974). In "Prostaglandin Synthetase Inhibitors" (eds H.J. Robinson and J.R. Vane), pp 165–173. Raven Press, New York.

Higgs, G.A., Moncada, S., Salmon, J.A. and Seager, K. (1983). The source of prostaglandins and thromboxane in experimental inflammation. Br. J. Pharmacol. 79, 863–868.

Higgs, G.A., Moncada, S. and Vane, J.R. (1984). Eicosanoids in inflammation. Ann. Clin. Res. 16, 287–299.

Higgs, G.A., Salmon, J.A., Henderson, B. and Vane, J.R. (1987). Pharmacokinetics of aspirin and salicylate in relation to inhibition of arachidonate cyclo-oxygenase and anti-inflammatory activity. Proc. Natl Acad. Sci. USA 84, 1417–1420.

Hiroshima, O., Hayashi, H., Ito, S. and Hayaishi, O. (1986). Basel level of prostaglandin D_2 in rat brain by a solid-phase enzyme immunoassay. Prostaglandins 32, 63–80.

Hirsh, P., Hillis, L., Campbell, W., Firth, B. and Willerson, J. (1981). Release of prostaglandins and thromboxane into the coronary circulation in patients with ischemic heart disease. N. Engl. J. Med. 304, 685–691.

Hoak, J.C., Brotherton, A.A., Czervionke, R.L. and Fry, G.L. (1985). In "Interaction of Platelets with the Vessel Wall" (eds J.A. Oates, J. Hawiger and R. Ross), pp 117–124. American Physiological Society, Bethesda.

Horton, E.W. and Poyser, N.L. (1976). Uterine luteolytic hormone. A physiological role for prostaglandin $F_{2\alpha}$. Physiol. Rev. 56, 595–651.

Horton, E.W., Main, I.H.M., Thompson, C.J. and Wright, P.W. (1968). The effect of orally administered prostaglandin E_1 on gastric secretion and gastrointestinal motility in man. Gut 9, 655–658.

Hsia, S.L., Ziboh, V.A. and Snyder, D.S. (1974). In "Prostaglandin Synthetase Inhibitors" (eds H.J. Robinson and J.R. Vane), pp 353–361. Raven Press, New York.

Hunt, J.N. and Franz, D.F. (1981). Effect of prostaglandin E_2 on gastric mucosal bleeding caused by aspirin. Dig. Dis. Sci. 26, 301–305.

Inauen, W., Wyss, P.A., Kayser, S., Baumgartner, A., Schurer-Maly, C.C., Koelz, H.R. and Halter, F. (1988). Influence of prostaglandins, omprazole and indomethacin on healing of experimental gastric ulcers in the rat. Gastroenterology 95, 636–641.

Ingerman-Wojenski, C., Silver, M.J., Smith, J.B. and Macarak, E. (1981). Bovine endothelial cells in culture produce thromboxane as well as prostacyclin. J. Clin. Invest. 67, 1292–1296.

Insel, P.A. (1990). In "The Pharmacological Basis of Therapeutics" (eds A.C. Gilman, T.W. Rall, A.S. Nies and P. Taylor), pp 638–681. Pergamon Press, New York.

Islam, F., Watanabe, Y., Morii, H. and Hayaishi, O. (1991). Inhibition of rat brain prostaglandin D synthase by inorganic selenocompounds. Arch. Biochem. Biophys. 289, 161–166.

Izumi, Y., Franco-Saenz, R. and Mulrow, P.J. (1985). Effect of prostaglandin synthesis inhibitors on the renin–angiotensin system and renal function. Hypertension 7, 791–796.

Jaffe, B.M. and Santoro, M.G. (1987). In "Prostaglandins in Cancer Research" (eds E. Garaci, R. Paoletti and M.G. Santoro), pp 141–150. Springer-Verlag, Berlin.

Jensen, P.K., Steven, K., Blaehr, H., Christiansen, J.S. and Parving, H.H. (1986). Effects of indomethacin on glomerular hemodynamics in experimental diabetes. Kidney Int. 29, 490–495.

Johansson, C. and Kollberg, B. (1979). Stimulation by intragastrically administered E_2 prostaglandins of human gastric mucus output. Eur. J. Pharmacol. 9, 229–232.

Johansson, C., Kollberg, B., Nordemar, R., Samuelsson, K. and Bergstrom, S. (1980). Protective effect of prostaglandin E_2 in the gastrointestinal tract during indomethacin treatment of rheumatic diseases. Gastroenterology 78, 479–483.

Johnson, R.A., Morton, D.R. and Kinner, J.H. (1976). The chemical structure of prostaglandin X (prostacyclin). Prostaglandins 12, 915–928.

Juan, H. and Lembeck, F. (1974). Action of peptides and other algesic agents on paravascular pain receptors of the isolated perfused rabbit ear. Naunyn-Schmiedebergs Arch. Exp. Pathol. Pharmakol. 283, 151–164.

Juhlin, L. and Michaelson, G. (1969). Cutaneous vascular reactions to prostaglandins in healthy subjects and in patients with urticaria and atopic dermatitis. Acta Dermato-Venereol. (Stockh.) 49, 251–261.

Kainoh, M., Imai, R., Umetsu, T., Hattori, M. and Nishio, S. (1990). Prostacyclin and beraprost sodium as suppressors of activated rat polymorphonuclear leukocytes. Biochem. Pharmacol. 39, 477–484.

Kaizu, K., Marsh, D., Zipser, R. and Glassock, R.J. (1985). Role of prostaglandins and angiotensin II in experimental glomerulonephritis. Kidney Int. 28, 629–635.

Karim, S.M.M. and Fung, W.P. (1976). Effects of some naturally occurring prostaglandins and synthetic analogues on gastric secretion and ulcer healing in man. Adv. Prostaglandin Thromboxane Res. 2, 529–539.

Karim, S.M.M. and Hillier, K. (1979). Prostaglandins in the control of animal and human reproduction. Br. Med. Bull. 35, 173–180.

Kauffman, G.L. and Grossman, M.I. (1978). Prostaglandins and cimetidine inhibit formation of ulcers produced by parenteral salicylates. Gastroenterology 75, 1099–1102.

Kauffman, G.L. and Whittle, B.J.R. (1982). Gastric vascular actions of prostanoids and the dual effect of arachidonic acid. Am. J. Physiol. 242, G582–G587.

Kauffman, G.L., Whittle, B.J.R., Aures, D. and Grossman, M.I. (1979). Effects of prostacyclin and a stable analogue 6β-PGI_1 on gastric acid secretion, mucosal blood flow and blood pressure in conscious dogs. Gastroenterology 77, 1301–1306.

Kaufman, R.P., Anner, H., Kobzik, L., Valeri, C.R., Shepro, D. and Hechtman, H.B. (1987). Vasodilator prostaglandins (PG) prevent renal damage after ischemia. Ann. Surg. 205, 195–198.

Kawarada, Y., Lambek, J. and Matsumoto, T. (1975). Pathophysiology of stress ulcer and its prevention. II. Prostaglandin E_1 and microcirculatory responses in stress ulcer. Am. J. Surg. 129, 217–222.

Keirse, M. (1979). In "Human Parturition" (eds M. Keirse, A. Anderson and J. Bennebroek Gravenhorst), pp 101–142. Martinus Nijhoff, The Hague.

Keith, E.F. (1960). Evaluation of analgesic substances. Am. J. Pharm. 132, 202–230.

Kelly, R.W., Cooper, I. and Templeton, A.A. (1979). Reduced prostaglandin levels in the semen of men with very high sperm concentrations. J. Reprod. Fertil. 56, 195–199.

Kennedy, M.S., Stobo, J.D. and Goldyne, M.E. (1980). *In vitro* synthesis of prostaglandins and related lipids by populations of human peripheral blood mononuclear cells. Prostaglandins 20, 135–145.

Keogh, E.J. and Earle, C.M. (1989). Recent advances in the role of pharmaceutical agents in impotence. Reprod. Fertil. Dev. 1, 387–390.

Kinoshita, F., Nakai, Y., Katakami, H., Imura, H., Shimizu, T. and Hayaishi, O. (1982). Suppressive effect of prostaglandin D_2 on pulsatile luteinizing hormone release in conscious castrated rats. Endocrinology 110, 2207–2209.

Kirschenbaum, M.A. and Chaudhari, A. (1986). Effect of experimental diabetes on glomerular filtration rate and glomerular prostanoid production in the rat. Miner. Electrolyte Metab. 12, 352–355.

Kivilaakso, E., Fromm, D. and Silen, W. (1978). Relationship between ulceration and intramural pH of gastric mucosa during hemorrhagic shock. Surgery 84, 70–78.

Konturek, S.J., Robert, A., Hancher, A.J. and Nezamis, J.E. (1980). Comparison of prostacyclin and prostaglandin E_2 on gastric acid secretion, gastrin release and mucosal blood flow in dogs. Dig. Dis. Sci. 25, 673–679.

Konturek, S.J., Radecki, T., Brzozowski, T., Piastucki, I., Dembinski, A., Dembinska-Kiec, A., Zumuda, A., Gryglewski, R.J. and Gregory, H. (1981a). Gastric cytoprotection by epidermal growth factor: role of endogenous prostaglandins and DNA synthesis. Gastroenterology 81, 438–443.

Konturek, S.J., Radecki, T., Brzozowski, T., Piastucki, I., Dembinska-Kiec, A. and Zumuda, A. (1981b). Aspirin-induced gastric ulcers in cats. Prevention by prostacyclin. Dig. Dis. Sci. 26, 1003–1012.

Konturek, S.J., Piastucki, I., Brzozowski, T., Radecki, T., Dembinska-Kiec, A., Zumuda, A. and Gryglewski, R.J. (1982). Role of prostaglandins in the formation of aspirin-induced gastric ulcers. Gastroenterology 80, 4–9.

Konturek, S.J., Kwiecien, N., Obtulowicz, W., Polanski, M., Kopp, B. and Olesky, J. (1983). Comparison of prostaglandin E_2 and ranitidine in prevention of gastric bleeding by aspirin in man. Gut 24, 89–93.

Krueger, J.M. (1985). In "Endogenous Sleep Substances and Sleep Regulation" (eds S. Inoue and A.A. Borbely), pp 181–192. Japan Scientific Societies Press, Tokyo.

Krueger, J.M., Obal, F. Jr, Opp, M., Cady, A.B., Johannsen, L., Toth, L. and Majde, J. (1990). In "Sleep and Biological Rhythms" (ed. J. Montplaisir), pp 163–185. Oxford University Press, Oxford.

Kuhn, D.C. and Willis, A.L. (1973). Prostaglandin E_2 inflammation and pain threshold in rat paws. Br. J. Pharmacol. 49, 183P–184P.

Kujuba, D.A. and Herschman, H.R. (1992). Dexamethasone inhibits mitogen induction of the TIS10 prostaglandin synthase/cyclooxygenase gene. J. Biol. Chem. 267, 7991–7994.

Kulmacz, R.J. and Lands, W.E.M. (1985). Stoichiometry and kinetics of the interaction of prostaglandin H synthase with anti-inflammatory agents. J. Biol. Chem. 260, 12572–12578.

Kunkel, S.L., Chensue, S.W. and Phan, S.H. (1986a). Prostaglandins as endogenous mediators of interleukin 1 production. J. Immunol. 136, 186–192.

Kunkel, S.L., Wiggins, R.C., Chensue, S.W. and Larrick, J. (1986b). Regulation of macrophage tumour necrosis factor production by prostaglandin E_2. Biochem. Biophys. Res. Commun. 137, 404–410.

Kunkel, S.L., Spengler, M., May, M.A., Spengler, R., Larrick, J. and Remick, D. (1988). Prostaglandin E_2 regulates macrophage-derived tumor necrosis factor gene expression. J. Biol. Chem. 263, 5380–5384.

Lagarde, M. and Dechavanne, M. (1977). Increase of platelet prostaglandin cyclic endoperoxides in thrombosis. Lancet i, 88.

Lands, W.E.M. (1981). Actions of anti-inflammatory drugs. Trends Pharmacol. Sci. 2, 78–80.

Lange, I.R., Collister, C., Johnson, J., Cote, D., Torchia, M., Freund, G. and Manning, F.A. (1984). The effect of vaginal prostaglandin E_2 pessaries on induction of labor. Am. J. Obstet. Gynecol. 148, 621–629.

Laviolette, M., Carreau, M., Coulombe, R., Cloutier, D., Dupont, P., Rioux, J., Braquet, P. and Borgeat, P. (1988). Metabolism of arachidonic acid through the 5-lipoxygenase pathway in normal human peritoneal macrophages. J. Immunol. 141, 2104–2109.

Lawson, J.A., Brash, A.R., Doran, J. and FitzGerald, G.A. (1985). Measurement of urinary 2,3-dinor-thromboxane B_2 and thromboxane B_2 using bonded-phase phenylboronic acid columns and capillary gas chromatography–negative-ion chemical ionization mass spectrometry. Anal. Biochem. 150, 463–470.

Laychock, S.G., Johnson, D.N. and Harris, L.S. (1979). PGD_2 effects on rodent behavior and EEG patterns in cats. Pharmacol. Biochem. Behav. 12, 747–754.

Lee, Y.H., Cheng, W.D., Bianchi, R.G., Mollison, K. and Hansen, J. (1973). Effects of oral administration of PGE_2 on gastric secretion and experimental peptic ulcerations. Prostaglandins 3, 29–45.

Levi, S., Goodlad, R.A., Lee, C.Y., Stamp, G., Walport, M.J., Wright, N.A. and Hodgson, H.J.F. (1990). Inhibitory effect of non-steroidal anti-inflammatory drugs on mucosal cell proliferation associated with gastric ulcer healing. Lancet 336, 840–843.

Levine, R.A., Kohen, K.R., Schwartzel, E.H. and Ramsey, C.E. (1982). Prostaglandins E_2 histamine interactions on cAMP, cGMP and acid production in isolated fundic glands. Am. J. Physiol. 242, G21–G26.

Lewis, G.P. and Barrett, M.L. (1986). Immunosuppressive actions of prostaglandins and the possible increase in chronic inflammation after cyclooxygenase inhibitors. Agents Actions 19, 59–65.

Lewis, T. (1942). "Pain". MacMillan, New York.

Lewy, R., Smith, J., Silver, M., Saia, J., Walinsky, P. and Wiener, L. (1979). Detection of thromboxane B_2 in peripheral blood of patients with Prinzmetal's angina. Prostaglandins Med. 2, 243–248.

Lim, R.K.S. (1966). In "The Salicylates" (eds M.J.H. Smith and P.K. Smith), pp 155–202. Interscience, New York.

Lim, R.K.S. (1968). In "Pharmacology of Pain" (eds R.K.S. Lim, D. Armstrong and E.G. Pardo), pp 169–217. Pergamon Press, Oxford.

Lim, R.K.S., Guzman, F., Rodgers, D.W., Goto, K., Braun, C., Dickerson, G.D. and Engle, R.J. (1964). Site of action of narcotic and non-narcotic analgesics determined by blocking bradykinin-evoked visceral pain. Arch. Int. Pharmacodyn. Ther. 152, 25–58.

Lippman, W. (1974). Inhibition of indomethacin-induced gastric ulceration in the rat by perorally administered synthetic and natural prostaglandin analogues. Prostaglandins 7, 1–10.

Luckhoff, A. and Busse, R. (1986). Increased free calcium in endothelial cells under stimulation with adenine nucleotides. J. Cell. Physiol. 126, 414–420.

McGiff, J.C., Terragno, N.A, Malik, K.U. and Lonigro, A.J. (1972). Release of prostaglandin E like substance from canine kidney by bradykinin. Circ. Res. 31, 36–43.

McGiff, J.F. (1980). Interactions of prostaglandins with the kallikrein–kinin and renin–angiotensin systems. Clin. Sci. 59, 105S–116S.

McGreevy, J.M. and Moody, F.G. (1977). Protection of gastric mucosa against aspirin-induced erosions by enhanced blood flow. Surg. Forum 28, 357–359.

McGreevy, J.M. and Moody, F.G. (1980). A mechanism for prostaglandin cytoprotection. Br. J. Surg. 67, 873–876.

MacIntyre, D.E., Pearson, J.D. and Gordon, J.L. (1978). Localisation and stimulation of prostacyclin production in vascular cells. Nature 271, 549–551.

Main, I.H.M. and Whittle, B.J.R. (1973). The effects of E and A prostaglandins on gastric mucosal blood flow and acid secretion in the rat. Br. J. Pharmacol. 49, 428–436.

Main, I.H.M. and Whittle, B.J.R. (1975a). Investigation of the vasodilator and antisecretory role of prostaglandins in the rat gastric mucosa by use of non-steroidal anti-inflammatory drugs. Br. J. Pharmacol. 53, 217–224.

Main, I.H.M. and Whittle, B.J.R. (1975b). Potency and selectivity of methyl analogues of prostaglandin E_2 on rat gastrointestinal function. Br. J. Pharmacol. 54, 309–317.

Mann, N.S. (1976). Bile-induced acute erosive gastritis: its prevention by antacid, cholestyramine and prostaglandin E_2. Am. J. Dig. Dis. 21, 89–92.

Marcus, A.J., Weksler, B.B., Jaffe, E.A. and Broekman, M.J. (1980). Synthesis of prostacyclin from platelet-derived endoperoxides by cultured human endothelial cells. J. Clin. Invest. 66, 979–986.

Margolin, S. (1965). In "Non-Steroidal Antiinflammatory Drugs" (eds S. Garattini and M.N.G. Dukes), pp 214–217. Excerpta Medica, Amsterdam.

Markey, C.M., Alward, A., Weller, P.E. and Marnett, L.J. (1987). Quantitative studies of hydroperoxide reduction by prostaglandin H synthase. J. Biol. Chem. 262, 6266–6279.

Martin, H.A., Basbaum, A.I., Kwiat, G.C., Goetzl, E.J. and Levine, J.D. (1987). Leukotriene and prostaglandin sensitization of cutaneous high-threshold C- and A-delta mechanonociceptors in the hairy skin of rat hindlimbs. Neuroscience 22, 651–659.

Masferrer, J.L., Zweifel, B.S., Seibert, K. and Needleman, P. (1990). Selective regulation of cellular cyclooxygenase by dexamethasone and endotoxin in mice. J. Clin. Invest. 86, 1375–1379.

Matsumura, H., Goh, Y., Ueno, R., Sakai, T. and Hayaishi, O. (1988). Awaking effect of PGE_2 microinjected into the preoptic area of rats. Brain Res. 444, 254–272.

Matsumura, H., Honda, K., Goh, Y., Ueno, R., Sakai, T., Inoué, S. and Hayaishi, O. (1989a). Awaking effect of prostaglandin E_2 in freely moving rats. Brain Res. 481, 242–249.

Matsumura, H., Honda, K., Choi, W.S., Inoué, S., Sakai, T. and Hayaishi, O. (1989b). Evidence that brain prostaglandin E_2 is involved in physiological sleep–wake regulation in rats. Proc. Natl Acad. Sci. USA 86, 5666–5669.

Matsumura, H., Takahata, R. and Hayaishi, O. (1991). Inhibition of sleep by inorganic selenium compounds, inhibitors of prostaglandin D synthase. Proc. Natl Acad. Sci. USA 88, 9046–9050.

Meerpohl, H.G. and Bauknecht, T. (1986). Role of prostaglandins in the regulation of macrophage proliferation and cytotoxic functions. Prostaglandins 31, 961–972.

Melmon, K.L., Webster, M.E., Goldfiner, S.E. and Seegmiller, J.E. (1967). The presence of a kinin in inflammatory synovial effusion from arthritides of varying etiologies. Arthritis Rheum. 10, 13–20.

Melmon, K.L., Bourne, H.R., Weinstein, Y., Shearer, G.M., Kram, J. and Bauminger, S. (1974). Separation of specific antibody-forming mouse cells by their adherence to insolubilised endogenous hormones. J. Clin. Invest. 53, 22–27.

Miller, T.A. (1983). Protective effects of prostaglandins against gastric mucosal damage: current knowledge and proposed mechanisms. Am. J. Physiol. 245, G601–G623.

Miller, T.A., Gum, E.T., Gunn, E.J. and Henagan, J.M. (1982). Prostaglandin prevents alterations in DNA, RNA, and protein in damaged gastric mucosa. Dig. Dis. Sci. 27, 776–781.

Milne, G.M. and Twomey, T.M. (1980). The analgetic properties of piroxicam in animals and correlation with experimentally determined plasma levels. Agents Actions 10, 31–37.

Milton, A.S. (1992). In "Aspirin and Other Salicylates" (eds J.R. Vane and R.M. Botting), pp 213–244. Chapman and Hall, London.

Milton, A.S. and Rotondo, D. (1987). The effects of pyrogens on phospholipase A_2 (PLA_2) activity in the rabbit. J. Physiol. (Lond.) 391, 95P.

Milton, A.S. and Wendlandt, S. (1971a). Effects on body temperature of prostaglandins of the A, E and F series on injection into the third ventricle of unanaesthetised cats and rabbits. J. Physiol. (Lond.) 218, 325–326.

Milton, A.S. and Wendlandt, S. (1971b). The effects of 4-acetamidophenol (paracetamol) on the temperature response of the conscious rat to the intracerebral injection of prostaglandin E_1, adrenaline and pyrogen. J. Physiol. (Lond.) 217, 33P–34P.

Misiewicz, J.J., Waller, S.L., Kiley, N. and Horton, E.W. (1969). Effect of oral prostaglandin E_1 on intestinal transit in man. Lancet i, 648–651.

Mitchell, M.D. (1981). Prostaglandins during pregnancy and the perinatal period. J. Reprod. Fertil. 62, 305–315.

Mitchell, M.D. (1984). The mechanism(s) of human parturition. J. Dev. Physiol. 6, 107–118.

Moncada, S., Ferreira, S.H. and Vane, J.R. (1972). In "Abstracts of the V International Congress of Pharmacology", p 160. San Francisco.

Moncada, S., Ferreira, S.H. and Vane, J.R. (1973). Prostaglandins, aspirin-like drugs and oedema of inflammation. Nature 246, 217–219.

Moncada, S., Ferreira, S.H. and Vane, J.R. (1974). In "Prostaglandin Synthetase Inhibitors" (eds H.J. Robinson and J.R. Vane), pp 189–195. Raven Press, New York.

Moncada, S., Ferreira, S.H. and Vane, J.R. (1975). Inhibition of prostaglandin biosynthesis as the mechanism of analgesia of aspirin-like drugs in the dog knee joint. Eur. J. Pharmacol. 31, 250–260.

Moncada, S., Needleman, P., Bunting, S. and Vane, J.R. (1976a). Prostaglandin endoperoxide and thromboxane generating systems and their selective inhibition. Prostaglandins 12, 323–329.

Moncada, S., Gryglewski, R.J., Bunting, S. and Vane, J.R. (1976b). An enzyme isolated from arteries transforms prostaglandin endoperoxides to an unstable substance that inhibits platelet aggregation. Nature 263, 663–665.

Moncada, S., Herman, A.G., Higgs, E.A. and Vane, J.R. (1977). Differential formation of prostacyclin (PGX or PGI_2) by layers of the arterial wall. An explanation for the antithrombotic properties of vascular endothelium. Thromb. Res. 11, 323–344.

Moncada, S., Ferreira, S.H. and Vane, J.R. (1978). In "Inflammation" (eds J.R. Vane and S.H. Ferreira), pp 588–616. Springer-Verlag, Berlin.

Moody, F.G., McGreevy, J., Zalewsky, C., Cheung, L.Y. and Simons, M. (1978). The cytoprotective effect of mucosal blood flow in experimental erosive gastritis. Acta Physiol. Scand. (Suppl.), 35–43.

Morley, J., Bray, M.A., Jones, R.W., Nugteren, D.H. and Van Dorp, D.A. (1979). Prostaglandin and thromboxane production by human and guinea-pig macrophages and leukocytes. Prostaglandins 17, 730–736.

Morris, G.P. and Harding, P.L. (1984). In "Mechanisms of Mucosal Protection in the Upper Gastrointestinal Tract" (eds A. Allen, G. Flemstrom, A. Garner, W. Silen and L.A. Turnberg), pp 209–213. Raven Press, New York.

Morris, G.P. and Wallace, J.L. (1981). The roles of ethanol and of acid in the production of gastric mucosal erosions in rats. Virchows Arch. (Cell Pathol.) 38, 23–28.

Morris, G.P., Harding, R.K. and Wallace, J.L. (1984). A functional model for extracellular gastric mucus in the rat. Virchows Arch. (Cell Pathol.) 46, 239–251.

Morrison, A.R., Nishikawa, K. and Needleman, P. (1978). Thromboxane A_2 biosynthesis in the ureter obstructed isolated perfused kidney of the rabbit. J. Pharmacol. Exp. Ther. 205, 1–8.

Moruzzi, G. and Magoun, H.W. (1949). Brain stem reticular formation and activation of the EEG. Clin. Neurophysiol. 1, 455–473.

Murray, R. and FitzGerald, G.A. (1989). Regulation of thromboxane receptor activation in human platelets. Proc. Natl Acad. Sci. USA 86, 124–128.

Naito, K., Osama, H., Ueno, R., Hayaishi, O., Honda, K. and Inoué, S. (1988). Suppression of sleep by prostaglandin synthesis inhibitors in unrestrained rats. Brain Res. 453, 329–336.

Nakamura-Craig, M. and Smith, T.W. (1989). Substance P and peripheral hyperalgesia. Pain 38, 91–98.

Nakano, J. and Prancan, A.V. (1972). Effect of prostaglandins E_1 and A_1 on the gastric circulation in dogs. Proc. Soc. Exp. Biol. Med. 139, 1151–1154.

Narumiya, S., Ogorochi, T., Nakao, K. and Hayaishi, O. (1982). Prostaglandin D_2 in rat brain, spinal cord and pituitary: basal level and regional distribution. Life Sci. 31, 2093–2103.

Nauta, W.J.H. (1946). Hypothalamic regulation of sleep in rats. An experimental study. J. Neurophysiol. 9, 285–316.

Nawroth, P.P., Stern, D.M., Kaplan, K.L. and Nossel, H.L. (1984). Prostacyclin production by perturbed bovine aortic endothelial cells in culture. Blood 64, 801–806.

Needleman, P., Moncada, S., Bunting, S., Vane, J.R., Hamberg, M. and Samuelsson, B. (1976). Identification of an enzyme in platelet microsomes which generates thromboxane A_2 from prostaglandin endoperoxides. Nature 261, 558–560.

Newman, A., Prado, J., de Moraes-Fiho, P., Philippakos, P. and Misiewicz, J.J. (1975). The effect of intravenous infusions of prostaglandins E_2 and $F_{2\alpha}$ on human gastric function. Gut 16, 272–276.

Newton, R.C. and Covington, M. (1987). The activation of human fibroblast prostaglandin E production by interleukin 1. Cell. Immunol. 110, 338–349.

Nezamis, J.E., Robert, A. and Stowe, D.F. (1971). Inhibition by prostaglandin E_1 of gastric secretion in the dog. J. Physiol. (Lond.) 218, 369–383.

Novagrodsky, A., Rubin, A.L. and Stenzel, K.H. (1979). Selective suppression by adherent cells, prostaglandin, and cyclic AMP analogues of blastogenesis induced by different mitogens. J. Immunol. 122, 1–8.

Nüsing, R. and Ullrich, V. (1990). Immunoquantitation of thromboxane synthase in human tissues. Eicosanoids 3, 175–180.

Nüsing, R., Schneider-Voss, S. and Ullrich, V. (1990). Immuno-affinity purification of human thromboxane synthase. Arch. Biochem. Biophys. 280, 235–330.

Ogorochi, T., Narumiya, S., Mizuno, N., Yamashita, K., Miyazaki, H. and Hayaishi, O. (1984). Regional distribution of prostaglandins D_2, E_2 and $F_{2\alpha}$ and related enzymes in postmortem human brain. J. Neurochem. 43, 71–82.

Ogorochi, T., Ujihara, M. and Narumiya, S. (1987). Purification and properties of prostaglandin H–E isomerase from the cytosol of human brain: identification as anionic forms of glutathione S-transferase. J. Neurochem. 48, 900–909.

Ohashi, K., Ruan, K.-H., Kulmacz, R.J., Wu, K.K. and Wang, L.H. (1992). Primary structure of human thromboxane synthase determined from the cDNA sequence. J. Biol. Chem. 267, 789–793.

Okazaki, T., Casey, M.L., Okita, J.R., MacDonald, P.C. and Johnston, J.M. (1981). Initiation of human parturition. XII. Biosynthesis and metabolism of prostaglandins in human fetal membranes and uterine decidua. Am. J. Obstet. Gynecol. 139, 373–381.

Oliw, E., Lunden, I. and Änggård, E.E. (1978). *In vivo* inhibition of prostaglandin synthesis in rabbit kidney by non-steroidal anti-inflammatory drugs. Acta Pharmacol. Toxicol. 42, 179–184.

Oliw, E.H., Sprecher, H. and Hamberg, M. (1986). Isolation of two novel E prostaglandins in human seminal fluid. J. Biol. Chem. 261, 2675–2683.

Onoe, J., Ueno, R., Fujita, I., Nishino, H., Oomura, Y. and Hayaishi, O. (1988). Prostaglandin D_2, a cerebral sleep-inducing substance in monkeys. Proc. Natl Acad. Sci. USA 85, 4082–4086.

Onoe, H., Kim, K. and Hayaishi, O. (1990). In "Endogenous Sleep Factors" (eds S. Inoue and J.M. Krueger), pp 69–76. SPB Academic Publishing, The Hague.

Orloff, J. and Zusman, R. (1978). Role of prostaglandin E (PGE) in the modulation of action of vasopressin on water flow in the urinary bladder of the toad and mammalian kidney. J. Membr. Biol. 40, 297–304.

Osborn, J.L., Kopp, U.C., Thames, M.D. and DiBona, G.F. (1984). Interactions among renal nerves, prostaglandins and renal arterial pressure in the regulation of renin release. Am. J. Physiol. 247, F706–F713.

Pardo, E.G. and Rodriguez, R. (1966). Reversal by acetylsalicylic acid of pain induced functional impairment. Life Sci. 5, 775–781.

Parker, C.W. (1986). Leukotrienes and prostaglandins in the immune system. Adv. Prostaglandin Thromboxane Leukotriene Res. 16, 113–134.

Patrignani, P., Filabozzi, P. and Patrono, C. (1982). Selective cumulative inhibition of platelet thromboxane production by low-dose aspirin in healthy subjects. J. Clin. Invest. 69, 1366–1372.

Patrono, C. (1989). Aspirin and human platelets: from clinical trials to acetylation of cyclooxygenase and back. Trends Pharmacol. Sci. 10, 453–458.

Patrono, C. and Dunn, M.J. (1987). The clinical significance of inhibition of renal prostaglandin synthesis. Kidney Int. 32, 1–12.

Patrono, C., Ciabattoni, G., Pugliese, F., Pierucci, A., Blair, I.A. and FitzGerald, G.A. (1986). Estimated rate of thromboxane secretion into the circulation of normal humans. J. Clin. Invest. 77, 590–594.

Pawlowski, N.A. (1989). In "Human Monocytes" (eds M. Zembala and G.L. Asherson), pp 273–289. Academic Press, London.

Pawlowski, N.A., Scott, W.A., Andreach, M. and Cohn, Z.A. (1982). Uptake and metabolism of monohydroxyeicosatetraenoic acids by macrophages. J. Exp. Med. 155, 1653–1664.

Pearson, J.D., Shakey, L.L. and Gordon, J.L. (1983). Stimulation of prostaglandin production through purinoceptors of cultured porcine endothelial cells. Biochem. J. 214, 273–276.

Pedersen, A.K. and FitzGerald, G.A. (1984). Dose related kinetics of aspirin: presystemic acetylation of platelet cyclooxygenase in man. N. Engl. J. Med. 311, 1206–1211.

Pelus, L.M. and Strausser, H.R. (1976). Indomethacin enhancement of spleen-cell responsiveness to mitogen stimulation in tumorous mice. Int. J. Cancer 18, 653–660.

Peskar, B.M., Weiler, H., Kroner, E.E. and Peskar, B.A. (1981). Release of prostaglandins by small intestinal tissue of man and rat *in vitro* and the effect of endotoxin in the rat *in vivo*. Prostaglandins 21(Suppl.), 9–14.

Petrulis, A.S., Aikawa, M. and Dunn, M.J. (1981). Prostaglandin and thromboxane synthesis by rat glomerular epithelial cells. Kidney Int. 20, 469–474.

Pettipher, E.R., Henderson, B., Edwards, J.C.W. and Higgs, G.A. (1988). Indomethacin enhances proteoglycan loss from articular cartilage in antigen-induced arthritis. Br. J. Pharmacol. 94, 341P.

Phelps, P., Prockop, D.J. and McCarty, D.J. (1966). Crystal induced inflammation in canine joints. III. Evidence against bradykinin as a mediator of inflammation. J. Lab. Clin. Med. 68, 433–434.

Piper, P.J. and Vane, J.R. (1969). Release of additional factors in anaphylaxis and its antagonism by anti-inflammatory drugs. Nature 223, 29–35.

Piper, P.J. and Vane, J.R. (1971). The release of prostaglandins from lung and other tissues. Ann. N.Y. Acad. Sci. 180, 363–385.

Plescia, O.J. and Racis, S. (1988). Prostaglandins as physiological immunoregulators. Prog. Allergy 44, 153–171.

Plescia, O.J., Smith, A. and Grinwich, K. (1975). Subversion of the immune system by tumor cells and the role of prostaglandins. Proc. Natl Acad. Sci. USA 72, 1848–1852.

Pohl, U., Dezsi, L., Simon, B. and Busse, R. (1987). Selective inhibition of endothelium-dependent dilation in resistance-sized vessels *in vivo*. Am. J. Physiol. 253, H234–H239.

Pollard, M. and Luckert, P.H. (1981). Treatment of chemically induced intestinal cancers with indomethacin. Proc. Soc. Exp. Biol. Med. 167, 161–164.

Quadt, J.F.A., Voss, R. and Ten-Hoor, F. (1982). Prostacyclin production of the isolated pulsatingly perfused rat aorta. J. Pharmacol. Methods 7, 263–270.

Rachmilewitz, D., Karmeli, F. and Okon, E. (1980). Effects of bisacodyl on c-AMP and prostacyclin E_2 contents, (Na^+K^+) ATPase, adenyl cyclase and phosphodiesterase activities of rat intestine. Dig. Dis. Sci. 25, 602–608.

Radomski, M.W., Palmer, R.M.J. and Moncada, S. (1987). The anti-aggregating properties of vascular endothelium: interactions between prostacyclin and nitric oxide. Br. J. Pharmacol. 92, 639–646.

Randall, L.O. (1963). In "Physiological Pharmacology, The Nervous System", Vol 1, Part A (eds W.S. Root and F.G. Hofmann), pp 313–316. Academic Press, New York.

Randall, L.O. and Selitto, J.J. (1957). A method for measurement of analgesic activity on inflamed tissue. Arch. Int. Pharmacodyn. Ther. 11, 409–419.

Rappaport, R.S. and Dodge, G.R. (1982). Prostaglandin E inhibits the production of human interleukin 2. J. Exp. Med. 155, 943–948.

Rashad, S., Hemingway, A., Rainsford, K., Revell, P., Low, F. and Walker, F. (1989). Effect of non-steroidal anti-inflammatory drugs on the course of osteoarthritis. Lancet ii, 519–522.

Raz, A., Wyche, A. and Needleman, P. (1989). Temporal and pharmacological division of fibroblast cyclooxygenase expression into transcriptional and translational phases. Proc. Natl Acad. Sci. USA 86, 1657–1661.

Redfern, J.S. and Feldman, M. (1989). Role of endogenous prostaglandins in preventing gastrointestinal ulceration: Induction of ulcers by antibodies to prostaglandins. Gastroenterology 96, 596–605.

Redfern, J.S., Blair, A.J., Clubb, F.J., Lee, E. and Feldman, M. (1987). Gastroduodenal ulceration following active immunization with prostaglandin E_2 in dogs. Role of gastric acid secretion. Prostaglandins 34, 623–632.

Reigger, G.A., Eisner, D. and Kromer, E.P. (1989). Circulatory and renal control by prostaglandins and renin in low cardiac output in dogs. Am. J. Physiol. 256, H1079–H1086.

Reilly, I.A.G. and FitzGerald, G.A. (1988). Aspirin in cardiovascular disease. Drugs 35, 154–176.

Reimann, I.W., Golbs, E., Fischer, C. and Frölich, J.C. (1985). Influence of intravenous acetylsalicylic acid and sodium salicylate on human renal function and lithium clearance. Eur. J. Clin. Pharmacol. 29, 435–441.

Reines, H.D., Halushka, P.V., Cook, J.A., Wise, W.C. and Rambo, W. (1982). Plasma thromboxane concentrations are raised in patients dying with septic shock. Lancet ii, 174–175.

Ritter, J.M., Cockcroft, J.R., Doktor, H.S., Beacham, J. and Barrow, S.E. (1989). Differential effect of aspirin on thromboxane and prostaglandin biosynthesis in man. Br. J. Clin. Pharmacol. 28, 573–579.

Robert, A. (1975). An intestinal disease produced experimentally by a prostaglandin deficiency. Gastroenterology 69, 1045–1047.

Robert, A. (1976). Antisecretory, antiulcer, cytoprotective and diarrhoegenic properties of prostaglandins. Adv. Prostaglandin Thromboxane Res. 2, 507–520.

Robert, A. (1977). In "Progress in Gastroenterology", Vol. III (ed. G. Jerzy Glass), pp 777–801. Grune and Stratton, New York.

Robert, A., Nezamis, J.E. and Phillips, J.P. (1967). Inhibition of gastric secretion by prostaglandin. Am. J. Dig. Dis. 12, 1073–1076.

Robert, A., Nezamis, J.E., Lancaster, C. and Hanchar, A.J. (1979a). Cytoprotection by prostaglandins in rats – prevention of gastric necrosis produced by alcohol, HCl, NaOH, hypertonic NaCl and thermal injury. Gastroenterology 77, 433–443.

Robert, A., Hanchar, A.J., Lancaster, C. and Nezamis, J.E. (1979b). In "Prostacyclin" (eds J.R. Vane and S. Bergstrom), pp 147–157. Raven Press, New York.

Robert, M., Fellastre, J.P., Berger, H. and Malandain, H. (1972). Effects of intravenous infusion of acetyl salicylic acid on renal function. Br. Med. J. 2, 466–467.

Roberts, J.L. II, Lewis, R.A., Oates, J.A. and Austen, K.F. (1979). Prostaglandin, thromboxane and 12-hydroxy-5,8,10, 14-eicosatetraenoic acid production by ionophore-stimulated rat serosal mast cells. Biochim. Biophys. Acta 575, 185–192.

Roberts, J.L. II, Sweetman, B.J., Lewis, R.A., Austen, K.F. and Oates, J.A. (1980). Increased production of prostaglandin D_2 in patients with systemic mastocytosis. N. Engl. J. Med. 303, 1400.

Robertson, R., Robertson, D. and Roberts, J. (1981). Thromboxane A_2 in vasotonic angina pectoris. Evidence from direct measurements and inhibitor trials. N. Engl. J. Med. 304, 998–1003.

Rome, L.H. and Lands, W.E.M. (1975). Structural requirements for time-dependent inhibition of prostaglandin biosynthesis by anti-inflammatory drugs. Proc. Natl Acad. Sci. USA 72, 4863–4865.

Rooks, W.H. II, Maloney, P.J., Shott, L.D., Schuler, M.E. and Sevelius, H. (1985). The analgesic and anti-inflammatory profile of ketorolac and its tromethamine salt. Drugs Exp. Clin. Res. 11, 479–492.

Rosenblum, W.I. and El Sabban, F. (1979). Topical prostacyclin (PGI_2) inhibits platelet aggregation in pial venules of the mouse. Stroke 10, 399–401.

Rosenthale, M.E., Kassarich, J. and Schneider, F. Jr (1966). Effect of anti-inflammatory agents on acute experimental synovitis. Proc. Soc. Exp. Biol. Med. 122, 693–696.

Roth, G.J. and Majerus, P.W. (1975). The mechanism of the effect of aspirin on human platelets. I. Acetylation of a particulate fraction protein. J. Clin. Invest. 56, 624–632.

Roth, G.J. and Siok, C.J. (1978). Acetylation of the NH_2-terminal serine of prostaglandin synthase by aspirin. J. Biol. Chem. 253, 3782–3784.

Roth, G.J., Stanford, N. and Majerus, P.W. (1975). Acetylation of prostaglandin synthetase by aspirin. Proc. Natl Acad. Sci. USA 72, 3073–3076.

Rotondo, D., Abul, H.T., Milton, A.S. and Davidson, J. (1988). Pyrogenic immunomodulators increase the level of prostaglandin E_2 in the blood simultaneously with the onset of fever. Eur. J. Pharmacol. 154, 145–152.

Rouzer, C.A., Scott, W.A., Hamill, A.L. and Cohn, Z.A. (1982). Synthesis of leukotriene C and other arachidonic acid metabolites by mouse pulmonary macrophages. J. Exp. Med. 155, 720–733.

Ruoff, H.J. and Becker, M. (1982). Histamine-sensitive adenylate cyclase in human gastric mucosa: cellular localization and interaction by PGE_2 somatostatin and secretin. Agents Actions 12, 174–175.

Ruwart, J. and Rush, B.D. (1984). The effects of $PGF_{2\alpha}$, PGE_2 and 16,16-dimethyl PGE_2 on gastric emptying and small intestinal transit in rat. Prostaglandins 28, 915–928.

Sagone, A.L., Wells, R.M. and De Mocko, C. (1980). Evidence

that OH production by human PMNs is related to prostaglandin metabolism. Inflammation 4, 65–71.

Salazar, F.J., Bolterman, R. and Fiksen-Olsen, M.J. (1988). Role of prostaglandins in mediating the renal effects of atrial natriuretic factor. Hypertension 12, 274–278.

Salvati, P. and Whittle, B.J.R. (1981). Investigation of the vascular actions of arachidonate lipoxygenase and cyclooxygenase products on the isolated perfused stomach of rat and rabbit. Prostaglandins 22, 141–156.

Salzman, P.M., Salmon, J.A. and Moncada, S. (1980). Prostacyclin and thromboxane A_2 synthesis by rabbit pulmonary artery. J. Pharmacol. Exp. Ther. 215, 240–247.

Samuelsson, B. (1963). Isolation and identification of prostaglandins from human seminal plasma. 18. Prostaglandins and related factors. J. Biol. Chem. 238, 3229–3234.

Sanduja, R., Loose-Mitchell, D. and Wu, K.K. (1991). Inhibition of *de novo* synthesis and message expression of prostaglandin H synthase by salicylates. Adv. Prostaglandin Thromboxane Leukotriene Res. 21, 149–152.

Saxena, P.N., Beg, M.M.A., Singhal, K.C. and Ahmad, M. (1979). Prostaglandin-like activity in the cerebrospinal fluid of febrile patients. Indian J. Med. Res. 79, 495–498.

Scharschmidt, L.A. and Dunn, M.J. (1983). Prostaglandin synthesis by rat glomerular mesangial cells in culture. Effects of angiotensin II and arginine vasopressin. J. Clin. Invest. 71, 1756–1764.

Schiable, H.G. and Schmidt, R.F. (1988a). Excitation and sensitization of fine articular afferents from cat's knee joint by prostaglandin E_2. J. Physiol. 43, 91–104.

Schiable, H.G. and Schmidt, R.F. (1988b). Time course of mechanosensitivity changes in articular afferents during a developing experimental arthritis. J. Neurophysiol. 60, 2180–2195.

Schlondorff, D. and Ardaillou, R. (1986). Prostaglandins and other arachidonic acid metabolites in the kidney. Kidney Int. 29, 108–119.

Schultz, R.M., Stoychkov, J.N., Pavlidis, N., Chirigos, M.A. and Olkowski, Z.L. (1979). Role of E-type prostaglandin in the regulation of interferon-treated macrophage cytotoxic activity. J. Reticuloendothel. Soc. 26, 93–100.

Schweizer, A. and Brom, R. (1985). Differentiation of peripheral and central effects of analgesic drugs. Int. J. Tissue React. 7, 79–83.

Scott, D. Jr (1968). Aspirin action on receptor in the tooth. Science 161, 180–181.

Scott, W.A., Pawlowski, N.A., Murray, H.W., Andreach, M., Zrike, J. and Cohn, Z.A. (1982). Regulation of arachidonic acid metabolism by macrophage activation. J. Exp. Med. 155, 1148–1160.

Shea-Donohue, T., Nompleggi, D., Myers, L. and Dubois, A. (1982). A comparison of the effects of prostacyclin and the 15(S),15-methyl analogs of PGE_2 and $PGF_{2\alpha}$ on gastric parietal and non-parietal secretion. Dig. Dis. Sci. 27, 17–22.

Shen, T.Y. (1979). In "Antiinflammatory Drugs" (eds J.R. Vane and S.H. Ferreira), pp 305–347. Springer-Verlag, Berlin.

Shimamoto, T., Kobayashi, M., Takahashi, T., Takashima, Y., Sakamoto, M. and Morooka, S. (1978). An observation of thromboxane A_2 in arterial blood after cholesterol feeding in rabbits. Jpn. Heart J. 19, 748–753.

Sicuteri, F., Franchi, P.L., Del, B. and Franciullacci, M. (1966). In "International Symposium on Vasoactive Polypeptides, Bradykinin, and Related Kinins" (eds M. Rocha and H.A. Rothschild), pp 255–261. Edat, Sao Paulo.

Siegmund, E.A., Cadmus, R.A. and Lu, G. (1957). A method for evaluating both non-narcotic and narcotic analgesics. Proc. Soc. Exp. Biol. Med. 95, 729–731.

Siess, W. (1989). Molecular mechanisms of platelet activation. Physiol. Rev. 69, 58–178.

Siess, W., Siegel, F.L. and Lapetina, E.G. (1983). Arachidonic acid stimulates the formation of 1,2-diacylglycerol and phosphatidic acid in human platelets. Degree of phospholipase C activation correlates with protein phosphorylation, platelet shape change, serotonin release and aggregation. J. Biol. Chem. 258, 11236–11242.

Simmonds, P.M., Salmon, J.A. and Moncada, S. (1983). The release of leukotriene B_4 during experimental inflammation. Biochem. Pharmacol. 32, 1353–1359.

Skoglund, M.L., Nies, A.S. and Gerber, J.G. (1982). Inhibition of acid secretion in isolated canine parietal cells by prostaglandins. J. Pharmacol. Exp. Ther. 220, 371–374.

Smith, G.S., Warhurst, G. and Turnberg, L.A. (1982). Synthesis and degradation of prostaglandin E_2 in the epithelial and subepithelial layers of the rat intestine. Biochim. Biophys. Acta 713, 684–687.

Smith, J.H. and Willis, A.L. (1971). Aspirin selectively inhibits prostaglandin production in human platelets. Nature 231, 235–237.

Smith, J.W., Steiner, A.L. and Parker, C.W. (1971). Human lymphocytic metabolism. Effects of cyclic and noncyclic nucleotides on stimulation by phytohemagglutinin. J. Clin. Invest. 50, 442–449.

Smith, T.W., Follenfant, R.L. and Ferreira, S.H. (1985). Antinociceptive models displaying peripheral opioid activity. Int. J. Tissue React. 7, 61–67.

Solomon, L.M., Juhlin, L. and Kirchenbaum, M.E. (1968). Prostaglandins on cutaneous vasculature. J. Invest. Dermatol. 51, 280–282.

Soppi, E., Eskola, J. and Ruuskanen, O. (1982). Effects of indomethacin on lymphocyte proliferation, suppressor cell function, and leukocyte migration inhibitory factor (LMIF) production. Immunopharmacology 4, 236–242.

Spenny, J.G. and Barton, J.C. (1981). 15-hydroxyprostaglandin dehydrogenase and delta[13] reductase content of gastrointestinal organs of rabbits and rats. Prostaglandins 21(Suppl.), 15–23.

Sraer, J., Sraer, S.D., Chansel, D., Russo-Marie, F., Kouznetzova, B. and Ardaillou, R. (1979). Prostaglandin synthesis by isolated rat renal glomeruli. Mol. Cell. Endocrinol. 16, 29–37.

Stahl, R.A., Kudelka, S., Paravicini, M. and Schollmeyer, P. (1986). Prostaglandin and thromboxane formation in glomeruli from rats with reduced renal mass. Nephron 42, 252–257.

Staite, N.D. and Panayi, G.S. (1982). Regulation of human immunoglobulin production *in vitro* by prostaglandin E_2. Clin. Exp. Immunol. 49, 115–122.

Staszewska-Barczak, J., Ferreira, S.H. and Vane, J.R. (1976). An excitatory nociceptive cardiac reflex elicited by bradykinin and potentiated by prostaglandins and myocardial ischemia. Cardiovasc. Res. 10, 314–327.

Stenson, W.F. and Parker, C.W. (1980). Monohydroxyeicosatetraenoic acids (HETEs) induce degranulation of human neutrophils. J. Immunol. 124, 2100–2104.

Stenson, W.F. and Parker, C.W. (1983). Metabolites of arachidonic acid. Clin. Rev. Allergy 1, 369–384.

Sterman, M.B. and Clemente, C.D. (1962). Forebrain inhibitory mechanisms: sleep patterns induced by basal forebrain stimulation in the behaving cat. Exp. Neurol. 6, 103–117.

Stitt, J.T. (1985). Evidence for the involvement of the organum vasculosum laminae terminalis in the febrile response of rabbits and cats. J. Physiol. (Lond.) 368, 501–511.

Sturge, R.A., Yates, D.B., Gordon, D., Franco, M., Paul, W., Bray, A. and Morley, J. (1978). Prostaglandin production in arthritis. Ann. Rheum. Dis. 37, 315–320.

Svanes, K., Critchlow, J., Takeuchi, K., Magee, D., Ito, S. and Silen, W. (1984). In "Mechanisms of Mucosal Protection in the Upper Gastrointestinal Tract" (eds A. Allen, G. Flemstrom, A. Garner, W. Silen and L.A. Turnberg), pp 33–38. Raven Press, New York.

Sykes, J. and Maddox, J. (1972). Prostaglandin production by experimental tumors and effects of antiinflammatory compounds. Nature (New Biol.) 237, 59.

Szczeklik, A., Gryglewski, R.J., Nizankowski, R., Musial, J., Pieton, R. and Mruk, J. (1978a). Circulatory and anti-platelet effects of intravenous prostacyclin in healthy men. Pharmacol. Res. Commun. 10, 545–556.

Szczeklik, A., Gryglewski, R.J., Musial, J., Grodzinska, L., Serwonska, M. and Marcinkiewicz, E. (1978b). Thromboxane generation and platelet aggregation in survivors of myocardial infarction. Thromb. Diath. Haemorrh. 40, 66–74.

Szczeklik, A., Kopec, M. and Slade, K. (1983). Prostacyclin and the fibrinolytic system in ischaemic vascular disease. Thromb. Res. 29, 655–660.

Taiwo, Y.O. and Levine, J.D. (1989a). Prostaglandin effects after elimination of indirect hyperalgesic mechanisms in the skin of the rat. Brain Res. 492, 397–399.

Taiwo, Y.O. and Levine, J.D. (1989b). Contribution of guanine nucleotide regulatory proteins to prostaglandin hyperalgesia in the rat. Brain Res. 492, 400–403.

Taiwo, Y.O., Bjerknes, L.K., Goetzl, E.J. and Levine, J.D. (1989). Mediation of primary afferent peripheral hyperalgesia by the cAMP second messenger system. Neuroscience 32, 577–580.

Takahara, K., Murray, R., FitzGerald, G.A. and Fitzgerald, D.J. (1990). The response to thromboxane A_2 in human platelets: discrimination of two binding sites linked to distinct effector systems. J. Biol. Chem. 265, 6836–6844.

Takesue, E.I., Perrine, J.W. and Trapold, J.H. (1976). The anti-inflammatory profile of proquazone. Arch. Int. Pharmacodyn. Ther. 221, 122–131.

Tarnawski, A., Hollander, D., Stachura, J., Krause, W.J. and Gergely, H. (1985). Prostaglandin protection of the gastric mucosa against alcohol injury – a dynamic time-related process. Role of the mucosal proliferative zone. Gastroenterology 88, 334–352.

Tateson, J.E., Moncada, S. and Vane, J.R. (1977). Effects of prostacyclin (PGX) on cyclic AMP concentrations in human platelets. Prostaglandins 13, 389–397.

Taylor, P.L. (1979). The 8-isoprostaglandins: evidence for eight compounds in human semen. Prostaglandins 17, 259–267.

Taylor, P.L. and Kelly, R.W. (1974). 19-Hydroxylated E prostaglandins as the major prostaglandins of human semen. Nature 250, 665–667.

Taylor, P.L. and Kelly, R.W. (1975). The occurrence of 19-hydroxy F prostaglandins in human semen. FEBS Lett. 57, 22–25.

Templeton, A.A., Cooper, I. and Kelly, R.W. (1978). Prostaglandin concentrations in the semen of fertile men. J. Reprod. Fertil. 52, 147–150.

Terano, A., Mach, T., Stachura, J., Tarnawski, A. and Ivey, K.J. (1984). Effect of 16,16-dimethyl prostaglandin E_2 on aspirin-induced damage to rat gastric epithelial cells in tissue culture. Gut 25, 19–25.

Thiery, M. (1979). In "Human Parturition" (eds M. Keirse, A. Anderson and J. Bennebroek Gravenhorst), pp 155–164. Martinus Nijhoff, The Hague.

Thiery, M. and Amy, J. (1977). Spontaneous and induced labor: two roles for the prostaglandins. Obstet. Gynecol. Annu. 6, 127–171.

Tokumoto, H., Watanabe, K., Fukushima, D., Shimizu, T. and Hayaishi, O. (1982). An NADP-linked 15-hydroxyprosta-glandin dehydrogenase specific for prostaglandin D_2 from swine brain. J. Biol. Chem. 257, 13576–13580.

Tomlinson, R.V., Ringold, H.J., Qureshi, M.C. and Forchielli, E. (1972). Relationship between inhibition of prostaglandin synthesis and drug efficacy: support for the current theory on mode of action of aspirin-like drugs. Biochem. Biophys. Res. Commun. 46, 552–559.

Toppozada, M.K. (1985). Clinical application of prostaglandins in human reproduction. Adv. Prostaglandin Thromboxane Leukotriene Res. 15, 631–635.

Trang, L.E., Granstrom, E. and Lovgren, O. (1977). Levels of prostaglandins $F_{2\alpha}$ and E_2 and thromboxane B_2 in joint fluid in rheumatoid arthritis. Scand. J. Rheumatol. 6, 151–154.

Tripp, C.S., Leahy, K.M. and Needleman, P. (1985). Thromboxane synthase is preferentially conserved in activated mouse peritoneal macrophages. J. Clin. Invest. 76, 898–901.

Tuvemo, T., Strandberg, K., Hamberg, M. and Samuelsson, B. (1976). Maintenance of the tone of the human umbilical artery by prostaglandin and thromboxane formation. Adv. Prostaglandin Thromboxane Res. 1, 425–428.

Ubatuba, F.B., Moncada, S. and Vane, J.R. (1979). The effect of prostacyclin (PGI_2) on platelet behaviour, thrombus formation *in vivo* and bleeding time. Thromb. Haemost. 41, 425–435.

Udvardy, M., Török, I. and Rak, K. (1987). Plasma thromboxane and prostacyclin metabolite ratio in atherosclerosis and diabetes mellitus. Thromb. Res. 47, 479–484.

Ueno, R., Narumiya, S., Ogorochi, T., Nakayama, T., Ishikawa, Y. and Hayaishi, O. (1982a). Role of prostaglandin D_2 in the hyperthermia of rats caused by bacterial lipopolysaccharide. Proc. Natl Acad. Sci. USA 79, 6093–6097.

Ueno, R., Ishikawa, Y., Nakayama, T. and Hayaishi, O. (1982b). Prostaglandin D_2 induces sleep when microinjected into the preoptic area of conscious rats. Biochem. Biophys. Res. Commun. 109, 576–582.

Ueno, R., Honda, K., Inoue, S. and Hayaishi, O. (1983). Prostaglandin D_2, a cerebral sleep-inducing substance in rats. Proc. Natl Acad. Sci. USA 80, 1735–1737.

Ueno, R., Hayaishi, O., Osama, H., Honda, K., Inoue, S., Ishikawa, Y. and Kanayama, T. (1985). In "Endogenous Sleep Substances and Sleep Regulation" (eds S. Inoué and A.A. Borbely), pp 193–201. Japan Scientific Societies Press, Tokyo.

Urade, Y., Fujimoto, N. and Hayaishi, O. (1985). Purification and characterization of rat brain prostaglandin D synthetase. J. Biol. Chem. 260, 12410–12415.

Urade, Y., Fujimoto, N., Kaneko, T., Konisho, A., Mizuno, N. and Hayaishi, O. (1987). Postnatal changes in the localization of prostaglandin D synthetase from neurons to oligodendrocytes in the rat brain. J. Biol. Chem. 262, 15132–15136.

Van Arman, C.G., Nuss, G.W., Winter, C.A. and Flataker, L. (1968). In "Pharmacology of Pain" (eds R.K.S. Lim, D. Armstrong and E.G. Pardo), pp 25–32. Pergamon Press, Oxford.

Van der Ouderaa, F.J., Buytenhek, M., Nugteren, D.H. and van Dorp, D.A. (1980). Acetylation of prostaglandin endoperoxide synthetase with acetylsalicylic acid. Eur. J. Biochem. 109, 1–8.

Van Epps, D. (1981). Suppression of human lymphocyte migration by PGE_2. Inflammation 5, 81–87.

Vane, J.R. (1971). Inhibition of prostaglandin synthesis as a mechanism of action for the aspirin-like drugs. Nature 231, 232–235.

Vane, J.R. (1982). Prostacyclin: a hormone with a therapeutic potential. J. Endocrinol. 95, 3P–43P.

Vane, J.R. and Botting, R.M. (1990). The mode of action of anti-inflammatory drugs. Postgrad. Med. J. 66(Suppl. 4) S2–S17.

Vane, J.R., Änggård, E.E. and Botting, R.M. (1990a). Regulatory functions of the vascular endothelium. N. Engl. J. Med. 323, 27–36.

Vane, J.R., Flower, R.J. and Botting, R.M. (1990b). History of aspirin and its mechanism of action. Stroke 21(Suppl. IV), 12–23.

Vantrappen, G., Janssens, J., Popiela, T., Kulig, J., Tytgat, G.N.J., Huibregtse, K., Lambert, R., Pauchard, J.P. and Robert, A. (1982). Effect of $15(R)$-15-methyl prostaglandin E_2 (arbaprostil) on the healing of duodenal ulcer. Gastroenterology 83, 357–363.

Venza-Teti, D., Misefari, A., Sofo, V., Fimiani, V. and La Via, M.F. (1980). Interaction between prostaglandins and human T lymphocytes: effect of PGE_2 on E-receptor expression. Immunopharmacology 2, 165–171.

von Euler, U.S. (1936). On specific vasodilating and plain muscle stimulating substances from accessory genital glands in man and certain animals (prostaglandin and vesiglandin). J. Physiol. 88, 213–234.

Wallace, J.L. and Whittle, B.J.R. (1985). Acceleration of recovery of gastric epithelial integrity by 16,16-dimethyl prostaglandin E_2. Br. J. Pharmacol. 86, 837–842.

Walus, K.M., Pawlik, W. and Konturek, S.J. (1980a). Prostacyclin-induced gastric mucosal vasodilatation and inhibition of acid secretion in the dog. Proc. Soc. Exp. Biol. Med. 163, 228–232.

Walus, K.M., Gustaw, P. and Konturek, S.J. (1980b). Differential effects of prostaglandins and arachidonic acid on gastric circulation and oxygen consumption. Prostaglandins 20, 1089–1102.

Warso, M.A. and Lands, W.E.M. (1983). Lipid peroxidation in relation to prostacyclin and thromboxane physiology and pathophysiology. Br. Med. Bull. 39, 277–280.

Watanabe, K., Shimizu, T., Iguchi, S., Wakatsuka, H., Hayashi, M. and Hayaishi, O. (1980). An NADP-linked prostaglandin D dehydrogenase in swine brain. J. Biol. Chem. 225, 1779–1782.

Watanabe, Y., Yamashita, A., Tokumoto, H. and Hayaishi, O. (1983). Localization of prostaglandin D_2 binding protein and NADP-linked 15-hydroxyprostaglandin D_2 dehydrogenase in the Purkinje cells of miniature pig cerebellum. Proc. Natl Acad. Sci. USA 80, 4542–4545.

Watanabe, Ya, Watanabe, Yu, Hamada, K., Bommelaer-Bayt, M.-C., Dray, F., Kaneko, T., Yumoto, N. and Hayaishi, O. (1989). Distinct localization of prostaglandin D_2, E_2 and $F_{2\alpha}$ binding sites in monkey brain. Brain Res. 478, 143–148.

Webb, D.R. and Osheroff, P.L. (1976). Antigen stimulation of prostaglandin synthesis and control of immune responses. Proc. Natl Acad. Sci. USA 73, 1300–1309.

Weksler, B.B., Marcus, A.J. and Jaffe, E.A. (1977). Synthesis of prostaglandin I_2 (prostacyclin) by cultured human and bovine endothelial cells. Proc. Natl Acad. Sci. USA 74, 3922–3926.

Weksler, B.B., Eldor, A., Falcone, D., Levine, R.I., Jaffe, E.A. and Minick, C.R. (1982). In "Cardiovascular Pharmacology of the Prostaglandins" (eds A.G. Herman, P.M. Vanhoutte, H. Denolin and A. Goossens), pp 137–148. Raven Press, New York.

Whelton, A., Stout, R.L., Spilman, P.S. and Klassen, D.K. (1990). Renal effects of ibuprofen, piroxicam and sulindac in patients with symptomatic renal failure. Ann. Intern. Med. 112, 568–576.

White, D.G. and Martin, W. (1989). Differential control and calcium-dependence of production of endothelium-derived relaxing factor and prostacyclin by pig aortic endothelial cells. Br. J. Pharmacol. 97, 683–690.

Whittle, B.J.R. (1964). The use of changes in capillary permeability in mice to distinguish between narcotic and non-narcotic analgesics. Br. J. Pharmacol. 22, 246–253.

Whittle, B.J.R. (1976). Relationship between the prevention of rat gastric erosions and the inhibition of acid secretion by prostaglandins. Eur. J. Pharmacol. 40, 233–239.

Whittle, B.J.R. (1977). Mechanisms underlying gastric mucosal damage induced by indomethacin and bile salts, and the actions of prostaglandins. Br. J. Pharmacol. 60, 455–460.

Whittle, B.J.R. (1980). In "Gastro-Intestinal Mucosal Blood Flow" (ed L.P. Fielding), pp 180–191. Churchill Livingstone, Edinburgh.

Whittle, B.J.R. (1981). Temporal relationship between cyclo-oxygenase inhibition, as measured by prostacyclin biosynthesis, and the gastrointestinal damage induced by indomethacin in the rat. Gastroenterology 80, 94–98.

Whittle, B.J.R. (1983). The potentiation of taurocholate-induced rat gastric erosions following parenteral administration of cyclo-oxygenase inhibitors. Br. J. Pharmacol. 80, 545–551.

Whittle, B.J.R. (1992). In "Aspirin and Other Salicylates" (eds J.R. Vane and R.M. Botting), pp 464–509. Chapman and Hall, London.

Whittle, B.J.R. and Boughton-Smith, N.K. (1979). In "Prostacyclin" (eds J.R. Vane and S. Bergstrom), pp 159–171. Raven Press, New York.

Whittle, B.J.R. and Salmon, J.A. (1983). In "Intestinal Secretion" (ed L.A. Turnberg), pp 69–73. Smith Kline and French Publications, Welwyn Garden City.

Whittle, B.J.R. and Steel, G. (1985). Evaluation of the protection of rat gastric mucosa by a prostaglandin analogue using cellular enzyme marker and histologic techniques. Gastroenterology 88, 315–327.

Whittle, B.J.R., Boughton-Smith, N.K., Moncada, S. and Vane, J.R. (1978). Actions of prostacyclin (PGI_2) and its product 6-oxo-$PGF_{1\alpha}$ on the rat gastric mucosa *in vivo* and *in vitro*. Prostaglandins 15, 955–968.

Whittle, B.J.R., Higgs, G.A., Eakins, K.E., Moncada, S. and Vane, J.R. (1980). Selective inhibition of prostaglandin production in inflammatory exudates and gastric mucosa. Nature 284, 271–273.

Whittle, B.J.R., Kauffman, G. and Moncada, S. (1981). Ulceration of the gastric mucosa following vasoconstriction with thromboxane A_2. Nature 292, 472–474.

Whittle, B.J.R., Hansen, D. and Salmon, J.A. (1985). Gastric ulcer formation and cyclo-oxygenase inhibition in cat antrum follows parenteral administration of aspirin but not salicylate. Eur. J. Pharmacol. 115, 153–157.

Whittle, B.J.R., Lopez-Belmonte, J. and Moncada, S. (1990). Regulation of gastric mucosal integrity by endogenous nitric oxide: interactions with prostanoids and sensory neuropeptides in the rat. Br. J. Pharmacol. 99, 607–611.

Williams, G. (1991). Erectile dysfunction – advances. Practitioner, 235, 114–118.

Williams, T.J. and Peck, M.J. (1977). Role of prostaglandin-mediated vasodilatation in inflammation. Nature 270, 530–532.

Willis, A.L. (1969). In "Prostaglandins, Peptides, and Amines" (eds P. Mantegazza and E.W. Horton), pp 31–38. Academic Press, London.

Willis, A.L. and Cornelsen, M. (1973). Repeated injection of prostaglandin E_2 in rat paws induces chronic swelling and a marked decrease in pain threshold. Prostaglandins 3, 353–357.

Willis, A.L. and Smith, D.L. (1989). Therapeutic impact of eicosanoids in atherosclerotic disease. Eicosanoids 2, 69–99.

Willis, A.L., Smith, D.L. and Vigo, C. (1986). Suppression of principal atherosclerotic mechanisms by prostacyclins and other eicosanoids. Prog. Lipid Res. 25, 645–666.

Willis, A.L., Smith, D.L., Vigo, C. and Kluge, A.F. (1987). Effects of prostacyclin and orally active stable mimetic agent RS-93427-007 on basic mechanisms of atherogenesis. Lancet ii, 682–683.

Wilson, D.E. and Levine, R.A. (1972). The effect of prostaglandin E_1 on canine gastric acid secretion and gastric mucosal blood flow. Am. J. Dig. Dis. 17, 527–532.

Winder, C.V. (1947). A preliminary test for analgesic action in guinea-pigs. Arch. Int. Pharmacodyn. Ther. 74, 176–192.

Winder, C.V. (1959). Aspirin and algesimetry. Nature 184, 494–497.

Winder, C.V., Pfeiffer, C.C. and Maison, G.L. (1946). The nociceptive contraction of the cutaneous muscle of the guinea pig as elicited by radiant heat, with observations on mode of action of morphine. Arch. Int. Pharmacodyn. Ther. 72, 329–359.

Wingerup, L., Andersson, K.E. and Ulmsten, U. (1978). Ripening of the uterine cervix and induction of labour at term with prostaglandin E_2 in viscous gel. Acta Obstet. Gynecol. Scand. 57, 403–406.

Winter, C.A. (1965). In "Medicinal Chemistry: Analgesics" (ed G. Stevens), pp 9–74. Academic Press, New York, London.

Winter, C.A. and Flataker, L. (1965). Nociceptive thresholds as affected by parenteral administration of irritants and of various anti-nociceptive drugs. J. Pharmacol. Exp. Ther. 148, 373–379.

Wiqvist, N. (1979). In "Human Parturition" (eds M. Keirse, A. Anderson and J. Bennebroek Gravenhorst), pp 189–200. Martinus Nijhoff, The Hague.

Wolff, H.G., Hardy, J.D. and Goodell, H. (1941). Measurement of the effect on the pain threshold of acetylsalicylic acid, acetanilid, acetophenetidin, aminopyrine, ethylalcohol, trichlorethylene, a barbiturate, quinine, ergotamine tartrate and caffeine, an analysis of their relation to pain experience. J. Clin. Invest. 20, 63–80.

Xie, W., Chipman, J.G., Robertson, D.L., Erikson, R.L. and Simmons, D.L. (1991). Expression of a mitogen-responsive gene encoding prostaglandin synthase is regulated by mRNA splicing. Proc. Natl Acad. Sci. USA 88, 2692–2696.

Yamamoto, K., Miwa, T., Ueno, R. and Hayaishi, O. (1988). Muramyl dipeptide-elicited production of PGD_2 from astrocytes in culture. Biochem. Biophys. Res. Commun. 156, 882–888.

Yamashita, A., Watanabe, Y. and Hayaishi, O. (1983). Autoradiographic localization of a binding protein(s) specific for prostaglandin D_2 in rat brain. Proc. Natl Acad. Sci. USA 80, 6114–6118.

Yik, K.Y., Dreidger, A.A. and Watson, W.C. (1982). Prostaglandin E_2 tablets prevent aspirin-induced blood loss in man. Dig. Dis. Sci. 27, 972–975.

Yokoyama, C., Miyata, A., Ihara, H., Ullrich, V. and Tanabe, T. (1991). Molecular cloning of human platelet thromboxane A synthase. Biochem. Biophys. Res. Commun. 178, 1479–1484.

Yokoyama, C., Miyata, A., Suzuki, K., Nishikawa, Y., Yamamoto, S., Nüsing, R., Ullrich, V. and Tanabe, T. (1993). Expression of human thromboxane synthase using a baculovirus system. FEBS Lett. 318, 91–94.

Yoshimoto, T., Yamamoto, S., Okuma, M. and Hayaishi, O. (1977). Solubilization and resolution of thromboxane synthesizing system from microsomes of bovine blood platelets. J. Biol. Chem. 252, 5871–5874.

Zambraski, E.J. and Dunn, M.J. (1984). Importance of renal prostaglandin in control of renal function after chronic ligation of the common bile duct in dogs. J. Lab. Clin. Med. 103, 549–559.

Zambraski, E.J. and Dunn, M.J. (1992). In "Aspirin and Other Salicylates" (eds J.R. Vane and R.M. Botting), pp 510–530. Chapman and Hall, London.

Zusman, R.M. and Keiser, H.R. (1977). Prostaglandin biosynthesis by rabbit renomedullary interstitial cells in tissue culture. Stimulation by angiotensin II, bradykinin, and arginine vasopressin. J. Clin. Invest. 60, 215–223.

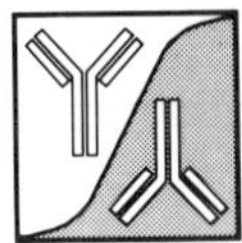

4. Other Essential Fatty Acids, Their Metabolites and Cell–Cell Interactions

M.R. Buchanan and S.J. Brister

1. Introduction

Fatty acid metabolites synthesized by blood and vascular wall cells are thought to play important roles in various cell–cell interactions which occur in response to injury, be it in inflammation, thrombosis or metastasis. There are many different steps in the genesis of these patho-physiological states, but one step which is common to all is the adhesion of circulating blood cells, e.g. platelets, leucocytes, macrophages and tumour cells, to the vascular endothelium and/or its underlying extracellular tissues. All of these cells synthesize a variety of fatty acid metabolites during these interactions. The preceding and following chapters provide us with significant insights into a number of these metabolites, particularly concerning those derived from arachidonic acid via the cyclooxygenase and lipoxygenase pathways, the regulation of their synthesis, and some of their putative actions. In this chapter we will focus only on a few of these fatty acid metabolites, in particular from the perspective of their role in cell–cell interactions, and how their effects may be altered by two other essential fatty acids, linoleic acid and eicosapentaenoic acid. We will highlight how cell–cell interactions influence their synthesis, which may suggest to the reader alternative mechanisms of actions for these metabolites other than those that have been mentioned. We will review how differences in the experimental design used to study their synthesis and their modes of action have influenced our interpretations of data and, consequently, the generation of certain hypotheses. For example, most studies which investigate the role(s) of the cyclooxygenase and lipoxygenase metabolites use specific enzyme inhibitors in combination with an exogenous substrate and/or specific metabolites, i.e. the add-back approach. These studies have generated an unprecedented abundance of data, which has lead to the generation of multiple hypotheses concerning the biological roles of these fatty

Lipid Mediators
ISBN 0–12–198875–9

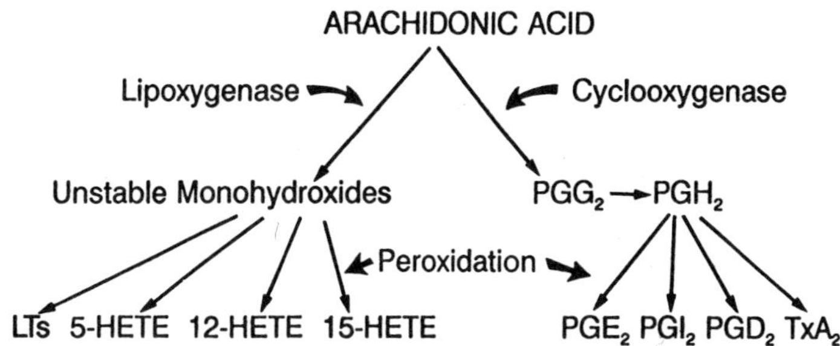

Figure 4.1 Metabolism of arachidonic acid via the cyclooxygenase and lipoxygenase enzyme pathways.

acid metabolites. We will review some of the possible pitfalls and limitations of that experimental design approach, which may impact on our basic understanding of fatty acid metabolism *per se* and its role in health and disease.

2. Prostaglandin I₂ Synthesis by Endothelial Cells

Numerous stimuli such as thrombin, endotoxin and cytokines induce the synthesis by, and subsequent release of, C_{20} molecules, i.e. arachidonic acid prostanoids from vessel wall cells, platelets, leucocytes, tumour cells and other circulating blood cells. (Hamberg *et al.*, 1974; Moncada and Vane, 1979; Defreyn *et al.*, 1982; Mehta *et al.*, 1985; Bastida *et al.*, 1989; Burrows *et al.*, 1991; Elliot and Van de Meent, 1991). These stimuli facilitate the release of arachidonic acid from the phospholipid stores by activating cell plasma membrane phospholipases. Free arachidonic acid rapidly isomerizes, resulting in a ring closure between C-8 and C-12 and the formation of an oxygen bridge between C-9 and C-11. A third oxygen molecule is then inserted at C-15, completing the formation of 15-hydroperoxyendoperoxide (PGG_2; Hamberg *et al.*, 1974). PGG_2 is then rapidly hydrolysed to PGH_2, which is also labile, and consequently is metabolized into a variety of eicosanoids (Fig. 4.1). In endothelial cells, PGH_2 is metabolized predominantly by prostaglandin synthase to form prostacyclin (PGI_2; Moncada and Vane, 1979).

PGI_2 is a potent inhibitor of platelet aggregation and a potent dilator of the vessel wall (Moncada and Vane, 1979). PGI_2 is thought to mediate its platelet effect by activating platelet adenylate cyclase, thereby elevating platelet cAMP levels and rendering the platelets hyporesponsive to subsequent pathogenic stimuli. The activation of adenylate cyclase by PGI_2 is thought to be mediated through a PGI_2 receptor and to be common to a variety of cells, (Moncada and Vane 1979; Hajjar *et al.*, 1982; Pohlman *et al.*, 1983). The PGI_2 effect is opposite to that of the platelet-derived cyclooxygenase arachidonic acid metabolite TXA_2, which acts on a platelet TXA_2

receptor to promote further platelet activation (Hamberg *et al.*, 1974; Hammarstrom and Falardeau, 1977). Consequently, Moncada and Vane (1979) postulated that the relative amounts of TXA_2 and PGI_2, produced by platelets and endothelial cells respectively, play important opposing roles in regulating platelet–vessel wall interactions following blood cell–vessel wall injury.

Platelets are also thought to play an important role in facilitating cancer cell activation and the subsequent escape of cancer cells from the intravascular space. Honn *et al.* (1981) suggested, therefore, that endothelial cell-derived PGI_2 also influences cancer cell–vessel wall interactions when impairing platelet function. PGI_2 has also been shown to have a direct effect on the cancer cells themselves, rendering them hyporesponsive. Thus, Honn and coworkers (1981) also postulated that, when PGI_2 renders cancer cells hyporesponsive to assorted pathophysiological stimuli, their vulnerability to NK cells and associated cell-mediated immunological responses may be enhanced.

The above hypotheses (Moncada and Vane, 1979; Honn *et al.*, 1981) assume that endothelial cell-derived PGI_2 *per se* is the most important naturally occurring prostanoid which down-regulates cell responsiveness.

3. Other Sources of Prostaglandin I₂

Other investigators have demonstrated that endothelial cells may also metabolize exogenous platelet-derived PGG_2 and PGH_2 to PGI_2. Marcus *et al.* (1982a) co-incubated [³H]arachidonic acid-labelled platelets and acetylsalicylic acid (ASA)-treated endothelial cells. (Roth *et al.* (1975) had previously demonstrated that ASA acetylates the cyclooxygenase, rendering it inactive.) Then Marcus *et al.* (1982a) stimulated the cell mixture with calcium ionophore (A23187) or thrombin, and detected both PGI_2 and TXA_2, measured as [³H]6-keto-PGF$_{1\alpha}$ and [³H]TXB₂, respectively. However, the generation of PGI_2 (i.e. [³H]6-keto-PGF$_{1\alpha}$), presumably by the ASA-treated endothelial cells, only occurred when: (1) there was a specific ratio of endothelial cells to platelets; (2) the platelets were in close proximity to the endothelial cells; and (3) the endothelial cells were stimulated

(Marcus *et al.*, 1982a,b). While the use of purified systems containing authentic substrates and single-cell preparation *in vitro* generates information concerning the possible pathways by which specific metabolites may be synthesized, Marcus's studies emphasize the need to examine how their synthesis may be altered in conditions which better mimic the *in vivo* setting. This is further emphasized by the following studies.

Defreyn *et al.* (1982) and Spitz *et al.* (1983) were the first to demonstrate that PGI_2 can also be generated in whole blood *in vitro* in the absence of endothelial cells. Further studies suggested that the generation of PGI_2 (measured as 6-keto-$PGF_{1\alpha}$) was the result of an enzymatic activity in leucocytes and/or monocytes on the endoperoxides (Deckmyn *et al.*, 1983). Deckmyn *et al.* (1983) also demonstrated that the PGI_2 synthesis in whole blood was increased in blood obtained from children with the haemolytic uraemic syndrome. These studies provided a direct correlation between changes in PGI_2 synthesis in whole blood and a pathological state, i.e. children with the haemolytic uraemic syndrome have an increased propensity for bleeding. These results therefore provide some biological relevance of the *ex vivo* observations to the clinical setting. Further studies confirmed that PGI_2 synthesis in whole blood involves leucocytes (Mehta *et al.*, 1985).

The above studies indicate that PGI_2 is derived not only from endothelial cells but also from circulating blood cells (Fig. 4.2). However, this latter point is completely ignored when developing strategies for antithrombotic treatment. For example, it has been suggested that ASA is a potentially useful antithrombotic agent because it acetylates cyclooxygenase, thereby blocking the ability of the enzyme to metabolize platelet arachidonic acid into the pro-aggregatory eicosanoid TXA_2 (Roth *et al.*, 1975). However, following the discovery of PGI_2, it was also recognized that ASA will also block vessel wall cyclooxygenase activity, thereby preventing the synthesis of PGI_2 by endothelial cells. Consequently, there has been an ongoing concern as to what dose of ASA should be used

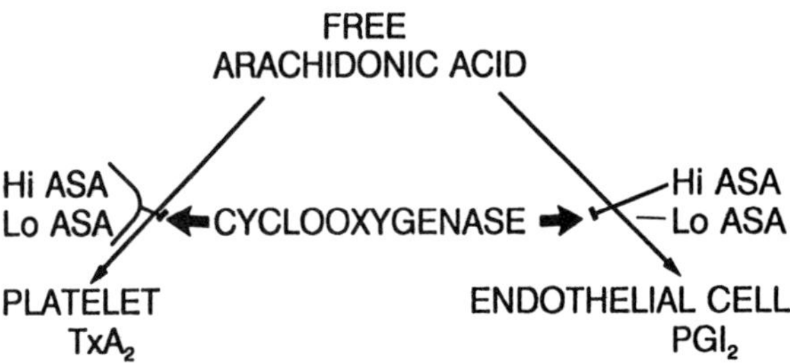

Figure 4.3 Differential inhibition of platelet and endothelial cell cyclooxygenase by low (Lo) and high (Hi) dose ASA.

as an antithrombotic, since any inhibition of PGI_2 may mask any beneficial effect achieved by inhibiting TXA_2. Some studies suggested that the platelet cyclooxygenase enzyme is more susceptible to inhibition by low-dose ASA than is the endothelial cell cyclooxygenase (Fig. 4.3; Baenziger *et al.*, 1979). It has also been established that, although inhibition of platelet cyclooxygenase activity is irreversible (since platelets are anucleated), the endothelial cells are able to synthesize cyclooxygenase *de novo*, and hence restore their capacity to synthesize PGI_2 after ASA treatment, albeit at different rates in arteries and veins (Buchanan *et al.*, 1980). Thus, it is currently argued that "low-dose" ASA may be better than "high-dose" ASA, since low-dose ASA will inhibit platelet cyclooxygenase activity selectively (Fig. 4.3). The above data support the concept that utilization of low-dose ASA (which only blocks platelet cyclooxygenase activity) may be useful as an antithrombotic; however, the ASA story does not recognize any contribution from PGI_2 derived from leucocytes in modulating platelet–vessel wall interactions, nor what the sensitivity of leucocyte cyclooxygenase to ASA is. It is possible that leucocyte-derived PGI_2 also plays a role in regulating thrombosis, but if so, this possibility should be confirmed. More importantly, it should be considered that PGI_2 synthesis (by both endothelial cells and leucocytes) and its subsequent degradation to 6-keto-$PGF_{1\alpha}$ may not be the end of the story, but rather the beginning.

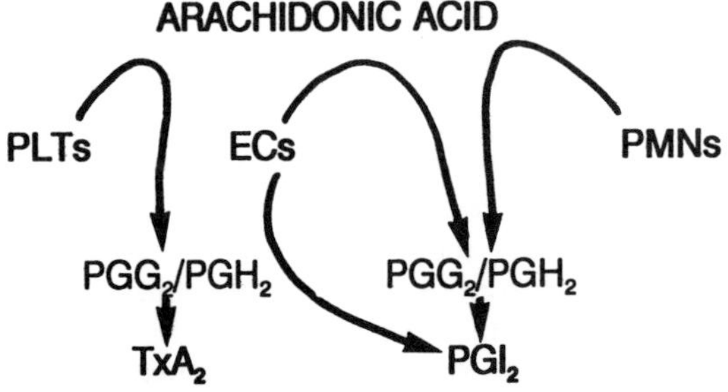

Figure 4.2 Synthesis of PGI₂ by endothelial cells (ECs) and leucocytes (PMNs). ECs not only metabolize their own endogenous arachidonic acid to PGI₂, but also are able to "steal" unstable endoperoxides from platelets (PLTs) to synthesize PGI₂. PMNs also metabolize endogenous arachidonic acid to PGI₂.

4. Conversion of Prostaglandin I_2 and 6-Ketoprostaglandin $F_{1\alpha}$ to 6-Ketoprostaglandin E_1

Wong and colleagues (Quilley *et al.*, 1979; Wong *et al.*, 1980a,b) demonstrated *in vitro* that the enzyme 9-hydroxyprostaglandin dehydrogenase (isolated from platelets) converts PGI_2 into 6-keto-PGE_1 in the presence of $NADP^+$ (Fig. 4.4.). Conversion of PGI_2 to 6-keto-PGE_1 by platelets has also been confirmed by other investigators (Griffiths and Moore, 1983). Others demonstrated that the leucocyte-derived enzyme 9-ketoreductase also converts the non-enzymatically

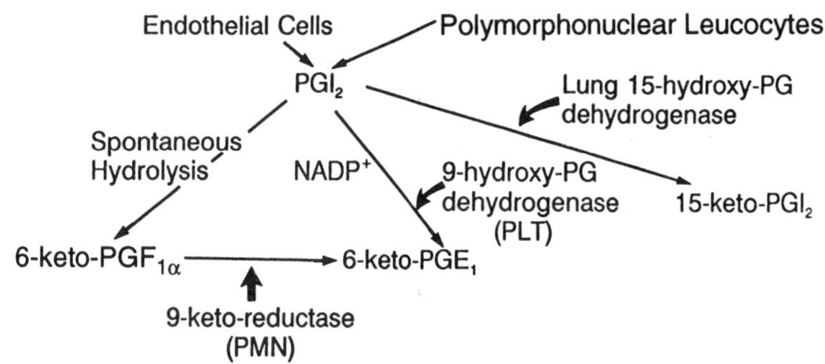

Figure 4.4 Metabolism of PGI$_2$ to active 6-keto-PGE$_1$ and other inactive products.

derived end-product of PGI$_2$, 6-keto-PGF$_{1\alpha}$, into 6-keto-PGE$_1$ (Fig. 4.4.). 6-Keto-PGE$_1$ is not only as stable as 6-keto-PGF$_{1\alpha}$, but it is also as potent as PGI$_2$ in increasing cAMP in a variety of cells. In addition, 6-keto-PGE$_1$, like PGI$_2$, is resistant to clearance in the lung (Baer *et al.*, 1986; Berry *et al.*, 1986). Unlike PGI$_2$, however, 6-keto-PGE$_1$ is a poor substitute for 15-hydroxyprostaglandin dehydrogenase degradation (Fig. 4.4; Berry and Houch, 1983; Moore and Griffiths, 1983; Berry *et al.*, 1984). These *in vitro* observations have led to the suggestion that, following vessel wall injury *in vivo*, i.e. when both platelets and leucocytes are in close proximity to a damaged vessel wall, the interacting cells will favour the formation of the more stable metabolite 6-keto-PGE$_1$ instead of the unstable product PGI$_2$. This, in effect, would result in generation of a more effective antiplatelet and/or vasoactive substance, thereby modulating thrombosis/haemostasis and/or inflammation, etc. Wong *et al.* (1980a) therefore postulated that 6-keto-PGE$_1$ was at least as important as PGI$_2$ *in vivo*.

The initial studies by Wong and colleagues, and their proposed hypothesis, i.e. that 6-keto-PGE$_1$ plays at least as important a role as PGI$_2$ in cell–cell interactions, sparked much interest for further experiments. However, the support for a biological role of 6-keto-PGE$_1$, particularly in modulating haemostasis/thrombosis, quickly dampened when the results of a number of studies were interpreted to suggest that 6-keto-PGE$_1$ was an *in vitro* artefact. Jackson *et al.* (1982) infused PGI$_2$ into healthy human volunteers, and then measured the time-dependent appearance of PGI$_2$-related metabolites in the urine. No 6-keto-PGE$_1$ was detected. Haslam and McCleneghan (1981) developed a sensitive assay to detect both PGI$_2$ and 6-keto-PGE$_1$ in human blood. They could not find any significant amounts of 6-keto-PGE$_1$. Pieroni *et al.* (1988) found that the major prostanoid produced by isolated perfused rabbit kidneys was PGI$_2$, and again was produced in significantly greater amounts than the minor amount of 6-keto-PGE$_1$. The conclusion, therefore, was that the results of Wong's studies were indeed an *in vitro* artefact. 6-Keto-PGE$_1$ was not detectable *in vivo*. However, our laboratory challenged these conclusions. We hypothesized that if 6-keto-PGE$_1$ is synthesized in response to "injury", as suggested by Wong *et al.* there should be no (or little) detectable 6-keto-PGE$_1$ in plasma obtained from healthy volunteers, but there should be detectable amounts of 6-keto-PGE$_1$ in plasma obtained from individuals in conditions in which cell–cell interactions were ongoing. Consistent with this hypothesis, we found that metabolism of both 6-keto PGF$_{1\alpha}$ and PGI$_2$ to 6-keto-PGE$_1$ only occurs in situations involving *cell–cell interactions*.

First, we demonstrated that single-cell preparations of intact and stimulated endothelial cells or leucocytes only metabolized arachidonic acid into PGI$_2$ and/or 6-keto-PGF$_{1\alpha}$ (Buchanan *et al.*, 1987b; Buchanan, 1989). Interestingly, leucocytes were as capable as endothelial cells in producing PGI$_2$, i.e. both leucocytes and endothelial cells produced similar amounts of PGI$_2$ ($\approx$ 300 pg/10^3 cells). Platelets did not synthesize either PGI$_2$ or 6-keto-PGF$_{1\alpha}$. However, when either leucocytes or platelets were co-incubated with endothelial cells, and then stimulated, significant amounts of PGI$_2$ and/or 6-keto-PGF$_{1\alpha}$ were metabolized further into 6-keto-PGE$_1$. Production of 6-keto-PGE$_1$ by leucocytes was 10-fold greater than the production by platelets (560 pg/10^3 PMNs versus 52 pg/10^3 platelets). These studies demonstrated that 6-keto-PGE$_1$ synthesis requires a cooperative effort between stimulated endothelial cells and platelets and/or leucocytes. Thus, both stimulation (possibly cell injury or cell perturbation) and subsequent cell–cell interactions are required for 6-keto-PGE$_1$ synthesis.

Second, similar results were seen *in vivo*. When rabbits were injected with increasing concentrations of thrombin, there was a dose-related decrease in the circulating platelet count over the subsequent 30 min after the thrombin injection. (Earlier studies demonstrated that decreased platelet recovery in response to thrombin and other prothrombotic stimuli reflects *in vivo* platelet aggregation (Buchanan and Hirsh, 1978).) The degree of platelet aggregation correlated with TXA$_2$ production (r=0.8421, P<0.001). The degree of platelet de-aggregation correlated with the generation of 6-keto-PGE$_1$ (r=0.806, P<0.001). There was no correlation between *in vivo* platelet aggregation and TXA$_2$ production, and PGI$_2$ synthesis *per se*.

These data demonstrate the importance of considering cell–cell interactions when determining which specific fatty acid metabolite(s) will be generated *in vivo*, and what their biological effects may be. Such considerations may impact on our understanding of the relationship between symptomatic events and underlying causes. For example, Clive *et al.* (1990) reported that the degree of platelet defect seen in patients with Bartter's syndrome correlates with the circulating plasma level of 6-keto-PGE_1. (Bartter's syndrome is associated with hypokalaemic/hypochloraemic alkalosis, increased renin, aldosterone and angiotensin II synthesis and decreased vascular sensitivity to the pressor effect of antiotensin II. The latter is associated with increased prostaglandin synthesis, as indicated by the marked increase in urinary excretion of prostaglandins.) Clive *et al.* (1990) have also found that the plasma levels of 6-keto-PGE_1 in these patients exceeded 130 pg/ml. These levels are markedly higher than those reported for healthy volunteers (Haslam and McCleneghan, 1981). Furthermore, the platelet defect in Bartter's syndrome patients (measured *ex vivo*) could be reversed by adding an antibody against 6-keto-PGE_1 to the plasma. These latter observations confirm that the platelet defect is contributed to, at least in part, by an inhibitory effect of 6-keto-PGE_1. These observations also imply that there are continuous ongoing platelet–leucocyte vessel wall interactions in these patients. In earlier studies, Gibson *et al.* (1987) reported that 6-keto-PGE_1 was detectable in plasma samples obtained from a patient with primary thrombocythaemia, but only in those plasma samples which were obtained from the patient during symptomatic and paradoxical haemorrhagic/thrombotic episodes. Both studies are consistent, therefore, with 6-keto-PGE_1 being produced in humans in response to injury involving cell–cell interactions.

Other studies suggest that 6-keto-PGE_1 is produced by specific endothelial cells within the vascular tree. Gerritsen and Cheli (1983) demonstrated that intact rabbit coronary artery microvessels *per se* produced 6-keto-PGE_1. These investigators concluded that cardiac tissue may have a specific ability to metabolize arachidonic acid to 6-keto-PGE_1. The possibility that the 6-keto-PGE_1 measured in their studies was generated by a vessel wall interaction with plateletes and/or leucocytes was excluded by their observations that isolated and cultured coronary artery microvascular endothelial cells also produced 6-keto-PGE_1. This raises the possibility that cardiac endothelium, unlike systemic or peripheral vascular endothelium, contains not only the cofactor to activate the platelet and/or leucocyte enzymes, but also contains the 9-hydroxyprostaglandin dehydrogenase or 9-ketoreductase enzyme itself.

In summary, highlighting the history of the 6-keto-PGE_1 studies is not meant to downplay the importance of PGI_2 in the regulation of platelet and/or vessel wall function; however, the observations that 6-keto-PGE_1 is formed from endogenous sources during pathophysiological events, particularly during those involving blood cell–vessel wall interactions, should not be ignored.

In addition, 6-keto-PGE_1 has been shown to have other effects *in vivo* (for reviews, see Adaikan *et al.*, 1984; Lewis *et al.*, 1984). For example, Miyamori *et al.* (1985) infused animals with either 6-keto-PGE_1 or PGI_2. They found that 15 ng/kg/min of 6-keto-PGE_1 inhibited platelet aggregation as effectively as 4 ng/kg/min of PGI_2. Interestingly, PGI_2 also had a marked hypotensive effect, whereas 6-keto-PGE_1 did not. This difference may have certain advantages in regulating platelet–vessel wall interactions. Other investigators have suggested that 6-keto-PGE_1 may also be involved in thrombosis. When Korbut *et al.* (1983) infused rabbits with an existing thrombus with either PGI_2 or 6-keto-PGE_1, both prostanoids triggered plasminogen proactivator to plasminogen activator inside the formed thrombus. In addition, the blood euglobulin clot lysis timed shortened. Similar results were seen *in vitro* when adding 6-keto-PGE_1 to blood in a test-tube; however, PGI_2 had no effect *in vitro*, but this may have been related to its instability in whole blood (Lucas *et al.*, 1986). Modulation of the vessel wall cAMP and cGMP pathways is also thought to be important in vessel wall homeostasis. Haas *et al.* (1990) have demonstrated that cAMP regulates triglyceride turnover in vessel wall cells, and consequently influences vessel wall linoleic acid metabolism. Kobayashi *et al.* (1991) demonstrated that increasing endothelial cell cGMP increases PGI_2 synthesis, and Ishii *et al.* (1990) have shown that increasing the endothelial cell nucleotides increases thrombomodulin expression on the endothelial cell surface. Thrombomodulin acts as a receptor for activated thrombin which, when bound to thrombomodulin, undergoes a conformational change, such that the thrombin is prevented from promoting further coagulation and, instead, activates protein C to facilitate an anticoagulant effect. Thus, a number of properties of the vessel wall are modulated by 6-keto-PGE_1, all of which are involved in the blood cell–vessel cell interactions which occur in response to injury.

5. *The Lipoxygenase Pathway*

It is also well recognized that arachidonic acid, once liberated from the plasma membrane phospholipid stores in response to stimuli or injury, is not only metabolized by the cyclooxygenase enzyme to a number of prostanoids, but is also metabolized by the lipoxygenase enzyme to the unstable hydroxide HPETE. HPETE, in turn, is hydrolysed rapidly by cytosolic-associated peroxidases into the monohydroxides HETE and leukotrienes (Fig. 4.5; Nugteren, 1975; Brash, 1985a). The details of the molecular and chemical metabolism of arachidonic acid by the lipoxygenase pathway are

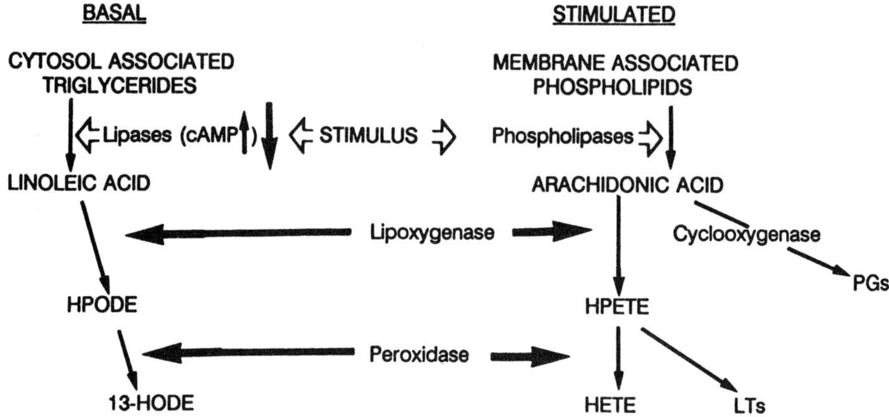

Figure 4.5 Metabolism of arachidonic and linoleic acids via the lipoxygenase enzyme. Under basal conditions, i.e. in unstimulated, but active endothelial cells, lipases, in the presence of elevated cAMP, facilitate linoleic acid release from the cytosolic triglyceride stores. Lipoxygenase, in turn, metabolizes the linoleic acid to 13-HODE. Upon stimulation, the cAMP levels drop, the membrane phospholipids become activated, and HETEs are metabolized via the lipoxygenase pathway instead of 13-HODE.

reviewed in other chapters. What we would like to emphasize here is that arachidonic acid is liberated from cell membrane-associated phospholipid stores by a number of phospholipases in response to *stimulation or injury* (Fig. 4.5). This applies to both circulating blood cells and to the stationary or "fixed" vascular wall cells. If these cells are not stimulated, they *do not* synthesize HETEs or leukotrienes. This implies that, in the unstimulated cells, the lipoxygenase enzyme is not active, i.e. the cells are in a "resting" state. However, studies in the last few years have clearly demonstrated that this is not so. Thus, neutrophils, tumour cells, macrophages, endothelial cells, smooth muscle cells, fibroblasts, epithelial cells and a variety of other cells continuously metabolize linoleic acid into the monohydroxide 13-HODE via the lipoxygenase pathway under resting conditions (Fig. 4.5; Boldingh,1976; Claeys *et al.*, 1982; Buchanan *et al.*, 1985b; Soberman *et al.*, 1985; Schade *et al.*, 1987; Daret *et al.*, 1989; Bastida *et al.*, 1990; Bull *et al.*, 1991). (The relevance of 13-HODE in cell pathophysiology will be discussed in detail in Section 7.)

The position of the first double bond in any HETE structure, e.g. 5-HETE, 12-HETE or 15-HETE, is specific to the cell. For example, 5-HETE is formed in leucocytes, 12-HETE in platelets and 15-HETE in other cells (Hamberg *et al.*, 1974; Lagarde *et al.*, 1984; Nadel *et al.*, 1991; Sloane *et al.*, 1991). In addition, cloning of cDNA for a variety of lipoxygenases indicates that the lipoxygenase for any one cell type can vary, depending upon the speies of origin (Izumi *et al.*, 1990a,b; Yoshimoto *et al.*, 1990a,b; Funk *et al.*, 1991). Furthermore, a genetic analysis of the lipoxygenase gene, for example in the pea, indicates that there are multiple loci (Domoney *et al.*, 1991). The latter study illustrates the complexity of correlating gene loci hybridization analysis and gene production with the possible synthesis of subsequent cDNAs. Thus, while a certain commonality exists in arachidonic acid metabolism via the lipoxygenase pathway for all cells, it is becoming increasingly apparent that the metabolite(s) generated by any one specific lipoxygenase in any one cell type cannot be extrapolated easily to the same cell type in other species or to other cell types in the same species. This is illustrated by the studies of Hada *et al.* (1991) in which the abilities of human, porcine and bovine 12-lipoxygenase to metabolize specific substrates were compared. Hada *et al.* found that each 12-lipoxygenase species has a similar high specificity for arachidonic acid. However, human 12-lipoxygenase does not metabolize linoleic acid. Bovine lipoxygenase has only 10% of the specificity for linoleic acid that it has for arachidonic acid. Porcine lipoxygenase has 70% of the specificity for linoleic acid that it has for arachidonic acid. In contrast, human 12-lipoxygenase has a greater affinity for EPA (a fatty acid in fish oil) than it has for arachidonic acid, whereas the porcine and bovine enzymes have a lower affinity for EPA than for arachidonic acid. Similar results have been obtained by others (Hawkins and Brash, 1987; Shannon *et al.*, 1990). In addition, the ratio of fatty acids, e.g. EPA and linoleic acid, will also influence which metabolites and how much are synthesized from other fatty acids (Horrobin, 1991). It is not surprising, therefore, that investigators may arrive at entirely different conclusions concerning what metabolites are generated by specific lipoxygenases if the cell source for the enzyme differs.

Marcus *et al.* (1987) reported that when intact human platelets (i.e. the 12-lipoxygenase) are incubated with arachidonic acid, arachidonic acid is metabolized to 12-HETE (and TXA_2 by the cyclooxygenase enzyme). However, when human leucocytes are added to the platelet suspension, 12,20-diHETE is formed. In addition, if LTB_4 synthesis is inhibited by a cytochrome

P-450 inhibitor, 12,20-diHETE synthesis is increased. They suggested that cytochrome P-450 monooxygenase (which activates cytochrome P-450) also contributes to the 2-hydroxylation of 12-HETE (Marcus *et al.*, 1987). These studies emphasize not only how the heterogeneity of any specific enzyme may influence what product is or is not generated, but also emphasize how cell–cell interactions influence what product is generated.

It should also be noted that the lipoxygenase enzyme is located in various sites within the cell. Platelet 12-lipoxygenase is concentrated in the supernatant (Nugteren, 1975; Lagarde *et al.*, 1984); however, Lagarde *et al.* (1984) have demonstrated that only 65% of 12-lipoxygenase activity is associated with the cytosol, while 20% of the enzyme is associated with the platelet plasma membranes and another 8% associated with the granules. It is therefore probable that these differences in enzyme location will influence not only substrate specificity but also the kinetics of their metabolism and where their biological effects are mediated. Brash (1985a,b) found that exogenous arachidonic acid is not metabolized as effectively as endogenous arachidonic acid by the lipoxygenases in intact platelets and leucocytes. In addition, free arachidonic acid is a poor substrate for the lipoxygenase enzymes in a purified system. Furthermore, the affinity of free arachidonic acid for the cyclooxygenase enzyme decreases as its affinity for the lipoxygenase enzyme increases, such as in the presence of albumin (Tremoli *et al.*, 1987; Broekman *et al.*, 1989). Yet, many studies use exogenous substrates in "purified systems" to determine the predominant metabolites synthesized.

To determine the role of HETEs in cell function, many investigators have used a traditional "pharamacological approach", that is, they measured a particular cell response, then inhibited the synthesis of the endogenous metabolite(s), and reassessed the cell responses or cell interactions, and then added a "purified metabolite" to determine if the normal cell responses were re-established. For example, Hadjiagapiou and Spector (1986) concluded that platelet 12-HETE blocks endothelial cell PGI_2 synthesis. If true, this could significantly influence our understanding of the mechanism underlying certain platelet–vessel wall interactions during haemostasis/thrombosis. They concluded that 12-HETE blocks PGI_2 synthesis on the basis that when endothelial cells were incubated for 1–2 h with HETE, washed to get rid of the HETE, and then incubated with arachidonic acid for 20 min, the cells were no longer able to metabolize arachidonic acid to PGI_2. However, Takayama (1989) found that if endothelial cells were incubated repeatedly with arachidonic acid the capacity of the endothelial cells to synthesize PGI_2 decreased for a finite duration, i.e. the cells became refractory. It should also be noted that decreasing amounts of PGI_2 were produced with each arachidonic acid incubation. This latter study demonstrates that exposure of the endothelial cells to excess substrate consumes the cyclooxygenase enzyme,

and thus the capacity of endothelial cells to synthesize more PGI_2 is dependent, in part, on *de novo* synthesis of more enzyme. Others have demonstrated that 5-HETE, 12-HETE and 15-HETE stimulate endothelial cell PGI_2 synthesis from *endogenous* arachidonic acid stores (Haas *et al.*, 1988). Others have shown that HETEs compete with arachidonic acid for either the cyclooxygenase or lipoxygenase enzyme, or both (Vanderhoek *et al.*, 1980; Camp and Fincham, 1985; Wilkenson *et al.*, 1985; Revtyak *et al.*, 1988; Grotemeyer, 1991). Thus, the apparent inability of endothelial cells to synthesize PGI_2 after HETE stimulation (as suggested by Hadjiagapiou and Spector, 1986) may simply reflect a depletion of the cyclooxygenase enzyme stores, not an inhibition of the enzyme *per se*. Consistent with this possibility, when rabbit or bovine vessel wall segments (prelabelled with [^{3}H]arachidonic acid) were incubated with increasing concentrations of 15-HETE, low concentrations of 15-HETE were shown to induce vasodilation, whereas high concentrations were shown to induce vasoconstriction (Lovelady *et al.*, 1988; Ma *et al.*, 1991; Takahashi *et al.*, 1991; Van Diest *et al.*, 1991). If the endothelial cells are removed from the vessel wall, the vasodilatory effect is abolished, consistent with the 15-HETE stimulating PGI_2 synthesis, not blocking it. The vasoconstriction associated with high concentrations of 15-HETE is thought to be due to an activation of the TXA_2 receptor expression on vascular smooth muscle cells, thereby facilitating vessel wall contraction. Humes *et al.* (1986) suggested that lipoxygenase metabolites may act as intracellular regulators of the cyclooxygenase pathway since 12-HETE blocks PGE_2 synthesis. Honn *et al.* (1989) has suggested that extracellular 12(*S*)-HETE induces reversible endothelial cell retraction, thereby facilitating the access of circulating tumour cells to the basement membrane.

Thus, there is little doubt that various HETEs when added exogenously have multiple effects on the vessel wall. However, the relevance of all of these observations has yet to be determined, since there is little or no evidence which demonstrates that any of these monohydroxides are present in the ambient surroundings of the vessel wall or concentrated locally at the site of blood cell–vessel wall cell interactions.

There are also two schools of thought concerning what effects HETEs have on platelet function. One school suggests that HETEs limit or down-regulate platelet function (Aharony *et al.*, 1982; Croset and Lagarde, 1983; Smith *et al.*, 1985; Nugteren, 1986; Setta and Stuart, 1986; Sekiya *et al.*, 1991). The other suggests that HETE is required for platelet function (Okuma and Uchino, 1979; Dutilh *et al.*, 1981; Buchanan *et al.*, 1986; Van Ryn-McKenna and Buchanan, 1989; Calzada *et al.*, 1991). It is interesting to note that those who argue that HETEs inhibit platelet function based their conclusions on studies using exogenous HETEs, whereas those who argue that HETEs facilitate platelet function base their conclusions on studies in which endogenous 12-HETE is modulated.

Exogenous HETEs have been shown to have both vasoconstrictive and vasodilatory effects on the vessel wall (Lovelady *et al.*, 1988; Ma *et al.*, 1991; Takahashi *et al.*, 1985, 1991; Van Diest *et al.*, 1991). Smith *et al.* (1985) found that 12-HETE also inhibits aortic smooth muscle cell proliferation more effectively than prostaglandins, but by what mechanism is not entirely clear. Nakao *et al.* (1984) found the opposite. These studies suggest that HETEs produced by a variety of circulating blood cells influence vessel wall physiology *per se*. Furthermore, it has been suggested that vessel wall-derived HETEs are also important. Thus, other investigators have reported that the effects of various HETEs and which HETE is produced by which vessel wall cell varies, depending upon whether the cells are derived from cardiac, cerebral or systemic vascular tissue (Lovelady *et al.*, 1988; Moore *et al.*, 1988, 1991; Revtyak *et al.*, 1988; Ma *et al.*, 1991). Moore *et al.* (1991) suggested that brain microvascular endothelial cells synthesize 12-HETE. This is based upon their results from incubation of pools of microvascular cells obtained from five to fifty rodent brains with 20 μM of [^{3}H]- or [^{14}C]arachidonic acid. Other investigators found that other vascular wall cells, e.g. smooth muscle cells, synthesize PGI$_2$ and 11-HETE and 15-HETE when an exogenous substrate, e.g. [^{14}C]arachidonic acid, is added, and when the cells are stimulated with increasing doses of thrombin (Bailey *et al.*, 1983). These data suggest that HETEs are important physiologically for vascular cell function and are produced by the vascular wall cells. Consistent with that possibility, Kuhn *et al.* (1985) and Hopkins *et al.* (1984) reported that the major lipoxygenase metabolite produced by cultured endothelial cells was 15-HETE. Kuhn *et al.* used calf aorta endothelial cells (10–14 subcultures) prelabelled with [^{14}C]arachidonic acid, such that the cells were loaded with the radiolabelled arachidonic acid in a 2:1 ratio of triglycerides:phospholipids. Hopkins *et al.* used human umbilical endothelial cells, prelabelled with [^{3}H]arachidonic acid for 2 h at 37°C. The concentrations added were 35–50 μM. It should be noted, however, that while these investigators demonstrated that a 15-lipoxygenase enzyme exists in vascular endothelial cell preparations, none of them demonstrated that the vascular wall cell enzyme synthesizes HETEs from *endogenous* arachidonic acid stores.

A number of investigators have suggested that arachidonic acid and its lipoxygenase derivatives act as secondary messengers to influence cell function (Abramson *et al.*, 1991; Chopra *et al.*, 1991; Fukuda *et al.*, 1991; Legrand *et al.*, 1991; Naor, 1991). Abramson *et al.* (1991) found that arachidonic acid binds to G proteins. Legrand *et al.* (1991) found that HETEs alter DAGs which, in turn, transduce protein kinase C by transmembranal signals. Eliott and van de Meent (1991) reported that IL-1α increases synthesis of PGE$_2$, PGI$_2$

and 15-HETE by human skin fibroblasts, and that pretreatment of the cells with cycloheximide blocked these effects, demonstrating that IL-1 stimulates *de novo* enzyme synthesis. Nadel *et al.* (1991) found that there was no intracellular 15-lipoxygenase in either rabbit or human erythrocytes, but there were significant amounts of the enzyme in the erythrocyte parent, the reticulocyte. They suggested that the intracellular 15-lipoxygenase pathway plays a role in selectively degrading mitochondria in the developing erythrocyte, particularly since the amount of 15-lipoxygenase varied significantly with reticulocyte/erythrocyte differentiation, the lipoxygenase activity being restricted to the developing cell. Eynard *et al.* (1986) demonstrated that exposure of platelets to increasing doses of ASA resulted in decreased endogenous platelet 12-HETE, which, in turn, was associated with decreased platelet function. Thus, four criteria which are required in order for a molecule to be considered a "secondary messenger" are satisfied by these studies: (1) the concentration of the molecule, i.e. HETE, can be altered in response to a stimulus, e.g. IL-1; (2) the cell responses are altered by the molecule in a manner consistent with those induced by the primary stimulus; (3) a mechanism for its removal or degradation is in place; and (4) causal relationships exist in the presence of the secondary messenger and cell responses (Robinson *et al.*, 1971). These data therefore support the concept that *endogenous* HETEs play a more important or complex role in cell physiology than merely competing extracellularly with other substrates for specific receptors or enzymes.

6. Vessel Wall Lipoxygenase Metabolism of Linoleic Acid

Endothelial cells are metabolically active under basal conditions. The triglycerides are continuously turned over in unstimulated cells. The turnover is dependent upon the level of cAMP (Denning *et al.*, 1983; Haas *et al.*, 1990). Linoleic acid is continuously released from the cytosolic triglyceride pool and metabolized by the endothelial cell enzyme 15-lipoxygenase into the mono-hydroxide 13-HODE (Fig. 4.5). 13-HODE is retained intracellularly and is not released into the ambient surroundings, unlike PGI$_2$ which is released when endothelial cells are stimulated. Furthermore, when endothelial cells are stimulated, 13-HODE synthesis rapidly ceases, such as following stimulation by thrombin, endotoxin or any of a number of cytokines (Buchanan *et al.*, 1985a,b, 1987b; Buchanan and Bastida, 1988; Bastida *et al.*, 1989, 1990; Haas *et al.*, 1990). The decrease in 13-HODE synthesis following endothelial cell stimulation is associated with a dramatic fall in intracellular cAMP levels (which is also a prerequisite for activation of the cell membrane phospholipases and subsequent release of arachidonic acid (Fig. 4.5)). The level of intracellular 13-HODE in vascular wall cells can be elevated by increasing the levels of cAMP both *in vitro* and *in vivo* (Haas *et al.*, 1990; Weber *et al.*, 1990). This effect appears to be associated with an increase in linoleic acid release from the triglyceride pool, thereby providing more substrate for 13-HODE synthesis via the lipoxygenase pathway.

Other investigators have also demonstrated that endothelial cells can metabolize exogenous linoleic acid into 13-HODE and 9-HODE via the cyclooxygenase pathway (Funk and Powell, 1983, 1985; Kaduce *et al.*, 1989; Kuhn *et al.*, 1990). It must be emphasized that these observations are again derived from studies in which the cells were exposed to high concentrations of exogenous substrate *in vitro*. This is an important point to consider when interpreting the above experiments. $C_{18:2}$ fatty acid substrates, such as linoleic acid, are poor substrates for cyclooxygenase (Nugteren, 1975; Hamberg and Samuelsson, 1980). As a consequence, high concentrations of linoleic acid are required to generate 9-HODE and 13-HODE via the cyclooxygenase pathway. Whether such concentrations exist free in the plasma in the biological setting *in vivo* is unknown, but unlikely. It is more likely that the generation of 9-HODE and 13-HODE via the cyclooxygenase pathway is an artefact due to the artificial conditions, particularly since none of the investigators who reported that endothelial cells synthesize 9-HODE and/or 13-HODE via the cyclooxygense pathway from exogenous linoleic acid provide any evidence, either *in vitro* or *in vivo*, that endothelial cells can synthesize 9-HODE and 13-HODE via the cyclooxygenase pathway from *endogenous* linoleic acid stores. We have never detected these metabolites from cultured cells or freshly isolated animal and human tissue.

7. Role of 13-Hydroxyoctadecadienoic Acid in Cell Adhesion Molecule Expression

It has been well established that the expression of adhesion molecules on the surface of blood and vessel wall cells is required for cell–cell adhesive interactions (Buchanan *et al.*, 1982, 1991; Gimbrone and Buchanan, 1982; Humphries *et al.*, 1986; Cheresh, 1987; Hynes, 1987; Grossi *et al.*, 1988, 1989; Singer *et al.*, 1989; Lafrenie *et al.*, 1990; Lauri *et al.*, 1990; Fonlupt *et al.*, 1991; Kishimoto *et al.*, 1991; Kuijpers *et al.*, 1991). The majority of these adhesion molecules comprise two glycoprotein chains and have been categorized into the superfamily of adhesion molecules called "integrins" (Cheresh, 1987; Hynes, 1987). More importantly, most of these integrins have a specific adhesive domain, which contains three specific peptides, aspartic acid, glycine and arginine (RGD) in their β chain. The RGD domain must be expressed in its adhesive form to facilitate receptor/ligand binding and subsequent cell–cell adhesion. However, what has not been well established is what

regulates the expression of these integrins in their adhesive form from within the cell?

A number of studies suggest that intracellular 13-HODE in endothelial cells and intracellular 13-HODE:15-HETE in malignant cells regulate adhesion molecule expression on the extracellular surface of cells (Buchanan *et al.*, 1985b; Buchanan and Bastida, 1988; Grossi *et al.*, 1988, 1989; Bastida *et al.*, 1989, 1990). When the intracellular 13-HODE levels are increased, there is a corresponding decrease in the ability of both cancer cells and endothelial cells to adhere to each other. In contrast, when endothelial cells are stimulated, such as by naturally occurring cytokines, endothelial cell 13-HODE synthesis decreases and cancer cell–endothelial cell adhesion increases. These observations prompted us to hypothesize that 13-HODE down-regulates vessel wall adhesivity (Buchanan and Bastida, 1988). Subsequently, a number of studies have supported this possibility (Grossi *et al.*, 1989; Simon *et al.*, 1989a; Bastida *et al.*, 1990; Haas *et al.*, 1990; Brister *et al.*, 1991; Chopra *et al.*, 1991), and suggest that specific adhesion molecules are involved. Stimulated endothelial cells express a specific integrin, the vitronectin receptor, on their apical cell surface, i.e. at a time when the endothelial cells are adhesive (Lafrenie *et al.*, 1990). (The vitronectin receptor is a two-way chain glycoprotein belonging to the adhesion molecule or "integrin" family (Cheresh, 1987; Hynes, 1987).) However, in unstimulated endothelial cells, i.e. in non-adhesive cells, the vitronectin receptor is colocalized with 13-HODE in vesicles lying immediately below the plasma membrane. This is illustrated in Plate 1 by histoimmunofluorescence photomicrographs. In Plate 1a the punctate pattern of histoimmunofluorescence of a human umbilical, unstimulated endothelial cell monolayer (using a rhodamine red-tagged antibody targeted at the RGD domain of the vitronectin receptor) shows that the vitronectin receptor is located in vesicles. The vesicles appear to be located approximately 3 μm below the endothelial cell apical surface, as determined by confocal laser microscopy. Plate 1b shows an identical pattern of histoimmunofluorescent staining when the same endothelial cell monolayer preparation is incubated with a fluorescent-tagged antibody against 13-HODE. The patterns of fluorescence were obtained using optics to differentiate rhodamine fluorescence (Plate 1a) and fluorescein fluorescence (Plate 1b). These studies indicate that both the vitronectin receptor and 13-HODE are colocalized inside vesicles of unstimulated endothelial cells.

When the endothelial cells are stimulated, however, the vitronectin receptor and 13-HODE dissociate, and the vitronectin receptor relocates in a cluster formation on the apical surface of the endothelial cells, i.e. at a time when endothelial cells are more reactive to circulating blood cells (Buchanan *et al.*, 1991). Cheresh (1987) reported similar results when endothelial cells are put into suspension, that is, they express an RGD adhesion molecule in cluster formation about their external surfaces, i.e. on both their apical and abluminal surfaces. We postulated, therefore, that 13-HODE interacts with lipophilic sites on the α chain of the vitronectin receptor, maintaining the receptor in a conformational presentation, such that its RGD-related adhesivity and/or expression on unstimulated endothelial cells is down-regulated, i.e maintaining the endothelial cells biocompatible or non-adhesive (Buchanan *et al.*, 1991). Consistent with the possibility that fatty acid metabolites *per se* can alter the ability of integrins to recognize their specific ligands, and hence alter their adhesivity, Conforti *et al.* (1990) demonstrated that the ability of the vitronectin receptor to recognize its specific ligands changes when its fatty acid environment is altered.

Other investigators have suggested that similar relationships exist between the linoleic acid and arachidonic acid-derived lipoxygenase metabolites 13-HODE and HETEs and the glycoprotein IIb/IIIa-dependent adhesion of tumour cells, platelets and endothelial cells (Grossi *et al.*, 1989; Bastida *et al.*, 1990). These phenomena appear to involve a reorganization of the cytoskeleton within the cells (Chopra *et al.*, 1991). Thus, a battery of evidence supports the concept that expression of cell surface molecules necessary for the adhesion of endothelial cells with platelets or with other circulating blood cells can be manipulated by altering the fatty acid environment, in particular by altering the relative amounts of lipoxygenase products derived from linoleic and arachidonic acids.

Finally, we found that when the endothelium is removed from the vessel wall the extracellular components which are exposed are initially not thrombogenic but, rather, are as thromboresistant as the intact endothelialized vessel wall (Buchanan *et al.*, 1987). At first, these results appeared to contradict many studies which concluded that an injured vessel wall was highly thrombogenic. However, with further experiments we found that the innate thrombogenic property of an injured vessel wall surface varies inversely with the amount of 13-HODE associated with it (Aznar-Salatti *et al.*, 1990, 1991; Bertomeu *et al.*, 1990; Weber *et al.*, 1990). These observations have recently been confirmed *ex vivo*, using freshly isolated human tissue (Brister *et al.*, 1991). If our initial hypothesis, namely that 13-HODE induces conformational changes to adhesion moieties, rendering them non-reactive, is correct, these latter studies suggest that 13-HODE not only regulates endothelial cell adhesivity but that 13-HODE may also be released unidirectionally from endothelial cells to their abluminal surface, thereby rendering basement membrane components non-reactive. Other investigators (de Graff *et al.*, 1989) disagree with the latter possibility since they found that exogenously added 13-HODE did not alter basement membrane thrombogenicity. However, in their studies, they added exogenous 13-HODE to endothelial cell basement membrane preparations which are isolated with ammonia hydroxide, i.e. with a chemical which not only destroys endothelial cell proteins and extracts lipids but which also "fixes" any remaining constituents. If 13-HODE actually does bind to glycoprotein substances or other adhesive components of the vascular basement membrane, rendering them in a non-adhesive conformation, it is unlikely that this would be possible with the fixed preparations used in the studies of de Graff *et al.* (1989). Thus, their conclusions must be interpreted with caution.

8. *13-Hydroxyoctadecadienoic Acid and Cell–Cell Interactions* in vivo

The concept of blocking pathogenic cell–cell interactions by agents which inhibit integrin expression or which compete with their binding of ligands has been considered for the treatment of metastasis (Humphries *et al.*, 1986) and thrombosis (Fitzgerald, 1989; Coller *et al.*, 1989; Roth, 1991). It has been suggested, for example, that an antithrombotic effect can be achieved by blocking adhesion molecules on platelets, such as with glycoprotein IIb/IIIa antibodies. What has been of concern, however, particularly with the antithrombotic treatment, is that blockage of the adhesion molecules themselves may be compounded by significant haemorrhagic side-effects (Fitzgerald, 1989). The recognition that adhesion molecules expressed on the vessel wall also contribute to some of these pathogenic adhesive events, and that their expression is related to the metabolism of 13-HODE, suggests that an alternative treatment of some of these events can be achieved by altering the vessel wall reactivity. We hypothesized that, by increasing vascular cAMP levels, endothelial cell triglyceride turnover would be increased. This, in turn, would result in increased amounts of free linoleic acid available for 13-HODE synthesis. We also hypothesized that salicylate, which blocks the peroxidation of the unstable precursor 13-HPODE to 13-HODE (Buchanan *et al.*, 1985a), would decrease 13-HODE synthesis, thereby increasing vessel wall adhesivity. If so, we should be able to either decrease or increase adhesivity of both the intact and the injured vessel wall. If the former occurred, fewer platelets should adhere to the vessel wall following injury, and thrombogenesis should be ameliorated. If the latter, thrombogenesis should be increased.

Rabbits were pretreated orally with salicylate or dipyridamole for at least 1 week. Dipyridamole was used to inhibit phosphodiesterase activity, thereby increasing vessel wall cAMP levels. After 1 week of the oral treatment with either drug (or an appropriate control), both carotid arteries in each animal were isolated and the endothelium was selectively removed using an air injury technique. Three hours later the animals were injected with radiolabelled platelets. The number of radiolabelled platelets adhering to the injured vessel wall surface 1 h later was determined as a measure of vessel wall thrombogenicity. Other vessel wall segments were removed and analysed for 13-HODE and cAMP levels.

Platelet adhesion to the injured carotid arteries *in vivo* increased two-fold in the salicylate-treated animals, as compared to platelet adhesion to injured carotid arteries in untreated rabbits. This increase in platelet adhesion was associated with a significant decrease in vessel wall 13-HODE. In contrast, platelet adhesion to the injured carotid arteries in the dipyridamole-treated animals decreased by 50%. This decrease was associated with an increase in 13-HODE synthesis, which correlated with both the dipyridamole plasma level and the vessel wall cAMP level. Neither the dipyridamole nor the salicylate treatment altered platelet function as assessed *ex vivo*. These studies are described in detail elsewhere (Weber *et al.*, 1990). Thus, this was the first study to provide direct evidence that the amount of 13-HODE associated with the vessel wall *in vivo* correlated with vessel wall thromboresistance. These observations support the possibility that manipulating vessel wall thromboresistance *per se* is an effective approach to inhibit platelet–vessel wall interactions *in vivo* following vessel wall injury.

More recently, we demonstrated that human blood vessels also synthesize 13-HODE. The levels of endogenous 13-HODE in both veins and arteries correlated inversely with vessel wall thrombogenicity (Brister *et al.*, 1991). Segments of internal mammary arteries and saphenous veins were obtained from patients undergoing elective coronary artery bypass grafting. The arteries produced six times more 13-HODE than veins but, more interestingly, 13-HODE synthesis in both arteries and veins decreased with age. This decrease was paralleled by an increase in vessel wall thrombogenicity measured *ex vivo*. Studies presently ongoing in our laboratory in a similar patient population are designed to determine whether increasing 13-HODE synthesis preoperatively in humans will also be associated with a corresponding decrease in vessel wall thrombogenicity at the time of surgery.

9. 13-Hydroxyoctadecadienoic Acid and the Atherosclerotic-Prone Watanabe Rabbit

The Watanabe rabbit, unlike other rabbit breeds, develops atherosclerotic plaques within 6–12 months, and consequently has been a useful animal model to study atherogenesis. The propensity for the Watanabe rabbit to develop atherosclerosis is thought to be due to an LDL receptor anomaly (Simon *et al.*, 1989b; Biancardi *et al.*, 1991). Recent studies, however, indicate that the Watanabe rabbit also has an age-dependent anomaly in vessel wall 15-lipoxygenase activity, that is, that 15-lipoxygenase activity increases with vessel wall thrombogenicity (Simon *et al.*, 1989a,b 1990). These investigators suggested that this increase in enzyme activity was a feedback mechanism to increase 13-HODE synthesis, thereby compensating for the pending thrombogenic state brought on presumably by the LDL receptor anomaly. However, more recently, a number of investigators have suggested a more complex link between the lipoxygenase pathway and LDL receptors in a variety of species. Rankin *et al.* (1991) found that lipoxygenases increase oxidation of the LDL receptor in murine peritoneal macrophages. Yla-Hertuala *et al.* (1990, 1991) found that lipoxygenase-associated oxidation of LDL in vessel wall fatty acid streaks was also associated with an invasion of macrophages in the blood vessels. In addition, they found that mRNA and protein for 15-lipoxygenase colocalized in the same macrophage-rich areas of human vessel fatty acid streaks, as did the mRNA for the acetylated LDL receptor.

In more recent studies, Simon *et al.* (1990) found that atherosclerotic plaque formation in the Watanabe rabbit could be inhibited without altering vessel wall 15-lipoxygenase activity. They concluded that there was no "causal" relationship between the lipoxygenase pathway and the apparent anomaly of increased atherosclerotic propensity in these animals. Thus, while a relationship exists between vessel wall 15-lipoxygenase activity, 15-lipoxygenase mRNA and LDL receptor mRNA expression, further experiments are required to confirm or refute a causal role of lipoxygenase in preventing atherosclerosis. However, it is difficult to disregard this animal model, since LDL receptor expression is fundamental for fatty acid uptake. It should also be noted that in the Watanabe rabbit, despite there being an increase in 15-lipoxygenase activity with age, the vessels of these animals do not produce significant amounts of 13-HODE as seen in non-atherosclerotic rabbits. This suggests that the uptake of the 13-HODE substrate (by the LDL receptor?) may be impaired because of the LDL receptor anomaly. It is also interesting to note that the increased propensity of the Watanabe rabbit to develop atherosclerosis, paralleled by a decrease in vessel wall 13-HODE synthesis, is similar to the observations seen in atherosclerotic patients undergoing cardiac surgery that human vessel wall 13-HODE levels vary inversely with vessel wall adhesivity (Brister *et al.*, 1991).

We have reviewed only a few of the studies investigating the role of cyclooxygenase and lipoxygenase metabolites, which are confounded by major differences in interpretation of what is metabolized under what conditions, and their roles in the pathophysiology of cells. For the most part, however, it becomes quite apparent that those studies in which endogenous metabolites are considered generate different results, and hence are interpreted differently, from those studies in which exogenous metabolites are used with an add-back approach. From the perspective of PGI_2 and 13-HODE synthesis by endothelial cells, we have modified the pathway of fatty acid metabolism to reflect only observations seen in studies using endogenous metabolites (Fig. 4.6). Under basal conditions, lipases remain active at a time when cAMP is elevated. Consequently, triglycerides are turned over, liberating linoleic acid. The linoleic acid either undergoes a series of desaturation and elongation reactions, or is also metabolized by lipoxygenase to 13-HODE, which in turn down-regulates adhesion molecule receptor expression. This maintains the endothelial cells biocompatible with the circulating blood cells to which they are constantly exposed. Upon stimulation or injury, however, cAMP levels drop, triglycerides turnover and 13-HODE synthesis ceases. Arachidonic acid is liberated from the phospholipid stores by a number of phospholipases. Cyclooxygenase metabolizes the free arachidonic acid to PGI_2, which undergoes further metabolism by either the platelet or the leucocyte to form a more stable product, 6-keto-PGE_1. 6-Keto-PGE_1, in turn, feeds back to re-elevate cAMP and re-establish homeostasis. Lipoxygenase products generated from arachidonic acid act intracellularly to up-regulate integrin expression, as suggested by Grossi *et al.* (1989), or as chemotactic agents to recruit more blood cells in the pathophysiological response.

Whether this is an oversimplification of the possible role of these metabolites or, in fact, adds a greater complexity has yet to be defined by further experiments. However, such a pathway is based upon studies in which only cell–cell interactions and modulating of endogenous stores metabolite synthesis are considered, i.e. without the addition of any exogenous metabolites.

10. Dietary Modification of Linoleic and Arachidonic Acid Metabolism in Endothelial Cells

Epidemiological evidence indicates that the prevalence of vascular diseases in those populations whose diets are enriched in fish, such as Greenland Eskimos and Japanese fishermen, or in populations who ingest significant amounts of vegetables containing linoleic acid, is significantly lower than in those populations whose diets are rich in animal fats (Nordoy, 1986; Woods *et al.*, 1987). The decreased prevalence in thrombosis in those populations ingesting fish is attributed to the presence of ω-3 fatty acids in the fish, such as EPA and docosahexaenoic acid (DHA). EPA and DHA compete with arachidonic acid for the cyclooxygenase and lipoxygenase enzymes (Powell and Gravelle, 1985). As a consequence, the synthesis of 2 series prostaglandins (e.g. TXA_2 and PGI_2) is blocked and the 3 series prostaglandins, PGI_3 from endothelial cells and TXA_3 from platelets, are produced. This change from the 2 to 3 series prostanoids in endothelial cells and platelets is thought to favour the down-regulation of thromboembolic events, since TXA_3 is thought to be less active than TXA_2, whereas PGI_3 and

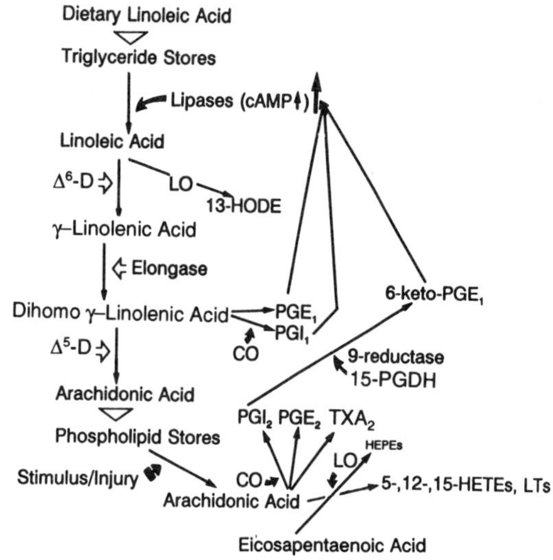

Figure 4.6 Negative/positive fatty acid feedback mechanism regulating linoleic/arachidonic acid metabolism. Δ^6–D, Δ^6-desaturase; Δ^5-D, Δ^5-desaturase; CO, cyclooxygenase; LO, lipoxygenase.

PGI_2 are thought to be equipotent (Nordoy, 1986). The most objective evidence to indicate enriched fish oil diets alters haemostasis is the observation that the bleeding time, a measure of platelet interactions with the vessel wall, is prolonged in those populations ingesting a fish diet. It is interesting to note, however, that this prolonged bleeding time is prolonged even further when these people take ASA. This suggests that the effect of the fish diet *per se* on prolonging the bleeding time is independent of a prostanoid effect. EPA also induces changes in the platelet lipid fatty acid composition, resulting in changes in platelet aggregation and bleeding. Croset and Lagarde (1985) demonstrated that the fatty acid composition of endothelial cell triglycerides and phospholipids, particularly with respect to linoleic and arachidonic acids, is markedly altered by the fatty acid composition of the incubating media used. Consequently, our understanding of the mechanism of the beneficial effect of EPA has yet to be clearly elucidated, but these and other studies clearly demonstrate that dietary fats significantly alter cell responses.

To date, the influence of changes in diet on endothelial cell 13-HODE synthesis *in vivo* has been limited to a few animal studies. Speizer *et al.* (1991) demonstrated that endothelial cells have a low Δ^6-desaturase activity, suggesting that a diet supplement of linoleic acid would increase the synthesis of linoleic acid metabolites, because linoleic acid was not effectively elongated into arachidonic acid in the vessel wall. If so, one might argue that an increase in linoleic acid in the vessel wall would enhance production of 13-HODE. Consistent with this possibility, we performed studies to determine the effects of semipurified synthetic diets on rabbit vessel wall and platelet fatty acid composition, and vessel wall and platelet

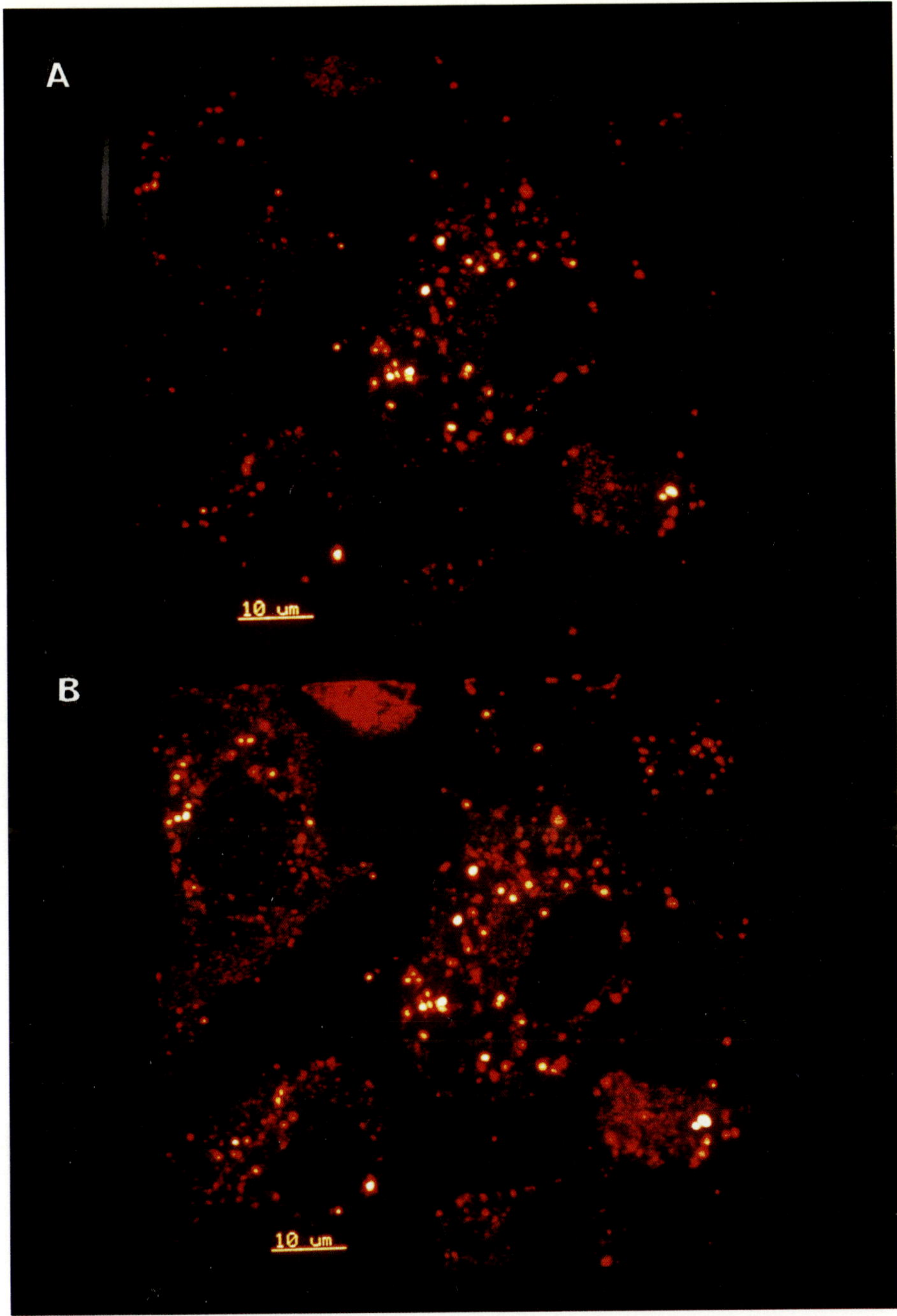

Plate 1 Histoimmunofluorescent localization of (a) the vitronectin receptor and (b) 13-HODE in an unstimulated endothelial cell monolayer, using confocal laser microscopy focused 3 μm below the apical surface of the cells. The patterns of fluorescence at two different wavelengths (rhodamine red, part a; fluorescein, part b) are identical, indicating colocalization of the vitronectin receptor and 13-HODE.

reactivity (Bertomeu *et al.*, 1990). Rabbits were fed diets rich in either linoleic acid (linoleic acid plus γ-linolenic acid) or oleic acid for 4 weeks. Both carotid arteries were injured, and then vessel wall thrombogenicity and 13-HODE synthesis were measured. As in the drug studies of Weber *et al.* (1990), when vessel wall 13-HODE increased two-fold (in those animals fed a diet rich in linoleic acid plus γ-linolenic acid), there was a significant decrease in vessel wall reactivity with platelets. This effect was specific for the vessel wall, since platelet adhesion and platelet aggregation *per se* were not altered. A fatty acid analysis of the vessel wall and the circulating blood cells demonstrated that linoleic acid and γ-linolenic acid were selectively incorporated into the vessel wall, not into circulating platelets nor leucocytes. This selective uptake in fatty acid provides an explanation as to why vessel wall thrombogenicity can be altered without interfering with the normal function of the circulating blood cells.

Other investigators demonstrated that the ratio or type of EPA and linoleic acid in the diet will significantly influence fatty acid metabolism *per se*. Wojenksi *et al.* (1991) fed animals diets rich in either fish oil (EPA plus DHA) or ethyl-EPA for 4 weeks. The ethyl-EPA was more effective in reducing plasma cholesterol, triglyceride and TXA_2 levels than the (EPA plus DHA) concentrate. This ethyl-EPA effect was associated with a prolonged bleeding time and decreased sensitivity of platelets to a variety of agonists. Other investigators have shown (in psoriasis rather than in thrombosis) that EPA is metabolized by lipoxygenase to the 5 series leukotrienes, which are thought to be less chemotactic than arachidonic acid-derived leukotrienes of the 4 series. Consequently, stimulation of keratinocyte proliferation and epithelial cell proliferation are decreased (Miller and Ziboh, 1987a; McCarthy, 1991). Whether these results can be extrapolated to vessel wall smooth muscle cell proliferation or chemotaxis during vessel wall inflammatory responses remains to be determined. However, there is some evidence to support this possibility. Fish oils are incorporated differently into diseased vessel wall plaques and healthy vessel walls following arterial endarterectomy (Rapp *et al.*, 1991). More importantly, however, the distribution of the fatty acids between the triglycerides and phospholipids differs, depending upon the presence and/or ratio of EPA and/or DHA in the diet. This has important implications concerning which fats may be metabolized under basal conditions (those released from triglycerides?) and following stimulation (those released from phospholipids?). Other investigators have also suggested that diets rich in EPA and linolenic acid influence vessel wall hypertension (Head *et al.*, 1991; Howe *et al.*, 1991). James *et al.* (1991) fed animals a diet containing 5% EPA with increasing concentrations of linoleic acid, and found that the subsequent uptake and elongation of fatty acids differed markedly as the linoleic acid concentration was increased; specifically, EPA uptake

decreased. Similar results have been obtained by other investigators (Horrobin, 1991).

11. Concluding Remarks

Many of the challenging frontiers of eicosanoid research have been reached, expanding the boundaries of our understanding of fatty acids metabolism by the cyclo-oxgenase and lipoxygenase enzymes and how these pathways influence cell–cell interactions. For the most part, however, our current concepts concerning which metabolites are important, and what their roles are in cell pathophysiology, are based upon the use of purified systems *in vitro*, and the modification of one specific metabolite in single-cell preparations *in vitro*. What is perhaps the challenge of the future, and which is undoubtedly more difficult, is to examine the roles of these metabolites in specific cell–cell responses, when modulating the synthesis and levels of endogenous metabolites, rather than masking their interrelationships with an exogenous substrate and/or authentic metabolites. In this chapter, we have tried to highlight how experimental design, cell–cell interactions and a study of intracellular metabolism may alter our understanding of fatty acid metabolism. While we have provided no definitive answers to either support or refute much of the current dogma, we hope that we have stimulated investigators to consider these parameters, which may result in the generation of completely different hypotheses than are presently being expounded.

12. Acknowledgements

We are grateful to Maureen Lantz for her secretarial assistance in preparing this manuscript. All of the studies performed by us and referred to in this chapter were supported, in part, by the Medical Research Council of Canada, and the Ontario Heart and Stroke Foundation.

13. References

Abramson, S.B., Leszczynska-Piziak, J. and Weissmann, G. (1991). Arachidonic acid as a second messenger. Interactions with a GTP-binding protein of human neutrophils. J. Immunol. 147, 231–236.

Adaikan, P.G., Tai, M.Y., Lau, L.C., Karim, S.M.M. and Kottegoda, S.R. (1984). A comparison of some pharmacological actions of prostaglandin E_1, 6-oxo-PGE_1 and PGI_2. Prostaglandins 27, 505–516.

Aharony, D., Smith, J.B. and Silver, M.J. (1982). Regulation of arachidonic-induced platelet aggregation by the lipoxygenase product, 12-hydroperoxyeicosatetraenoic acid. Biochim. Biophys. Acta 718, 193–200.

Aznar-Salatti, J., Bastida, E. Buchanan, M.R., Castillo, R., Ordinas, A., and Escolar, G. (1990). Differential localization of von Willebrand factor, fibronectin and 13-HODE in human endothelial cell cultures. Histochemistry 93, 507–511.

Aznar-Salatti, J., Bastida, E., Haas, T.A., Escolar, G., Ordinas, A., DeGroot, P.H.G. and Buchanan, M.R. (1991). Platelet adhesion to exposed endothelial cell extracellular matrixes is influenced by the method of preparation. Arteriosclerosis 11, 436–442.

Baenziger, N.L., Becherer, P.R. and Majerus, P.W. (1979). Characterization of prostacyclin synthesis in cultured human arterial smooth muscle cells, venous endothelial cells and skin fibroblasts. Cell 16, 967–974.

Baer, A.N., Costello, P.B., and Green, F.A. (1991). Stereo-specificity of the hydroxyeicosatetraenoic and hydroxyoctade-cadienoic acids produced by cultured bovine endothelial cells. Biochim. Biophys. Acta 1085, 45–52.

Baer, P.G., Cagen, L.M. and Nasjletti, A. (1986). 6-Keto-prostaglandin F1α is not added to urine during passage through the rat urinary bladder in vivo. Prostaglandins 32, 311–316.

Bailey, J.M., Bryant, R.W., Whiting, J. and Salata, K. (1983). Characterization of 11-HETE and 15-HETE together with prostacyclin, as major products of the cyclooxygenase pathway in cultured rat aorta smooth muscle cells. J. Lipid Res. 24, 1419–1428.

Bastida, E., Almirall, L., Ordinas, A., Buchanan, M.R., Bertomeu, M.C. and Haas, T.A. (1989). Effects of antiplatelet drugs and cytokines on endothelial cell 13-HODE synthesis and tumor cell/endothelial cell adhesion. Thromb. Haemost. 62, 443(Abstr.).

Bastida, E., Bertomeu, M.C., Haas, T.A., Almirall, L., Lauri, D., Orr, F.W. and Buchanan, M.R. (1990). Regulation of tumor cell adhesion by intracellular 13-HODE:15-HETE ratio. J. Lipid Med. 2, 281–293.

Berry, C.N. and Hoult, J.R.S. (1983). 6-Keto-prostaglandin E1: its formation by platelets from prostacyclin and resistance to pulmonary degradation. Pharmacology 26, 324–330.

Berry, C.N., Hoult, J.R.S., Griffiths, R.J. and Moore, P.K. (1984). Enzymatic inactivation of 6-keto-prostaglandin E1 in vitro: comparison with prostaglandin E1. Biochem. Pharmacol. 33, 1277–1284.

Berry, C.N., Griffiths, R.J., Hoult, J.R.S., Moore, P.K. and Taylor, G.W. (1986). Identification of 6-oxo-prostaglandin E1 as a naturally occurring prostanoid generated by rat lung. Br. J. Pharmacol. 87, 327–335.

Bertomeu, M.C., Crozier, G.L., Haas, T.A., Fleith, M. and Buchanan, M.R. (1990). Selective effects of dietary fats on vascular 13-HODE synthesis and platelet/vessel wall interactions. Thromb. Res. 59, 819–830.

Biancardi, G., Tanganelli, P., Palummo, N., Zatta, A., Prosdocimi, M. and Weber, G. (1991). Platelet function abnormalities in 6–12 month-old Watanabe heritable hyperlipidemic (WHHL) rabbits. Thromb. Res. 64, 109–115.

Boldingh, J. (1976). Reaction mechanisms and functions of plant lipoxygenase. Lipids 1, 147–156.

Brash, A.R. (1985a). Specific lipoxygenase attack on arachi-donate and linoleate esterified in phosphatidylcholine: precedent for an alternative mechanism in activation of eicosanoid biosynthesis. Adv. Prostaglandin Thromboxane Leukotriene Res. 15, 197–199.

Brash, A.R. (1985b). A review of possible roles of the platelet. Circulation 72, 702–707.

Brister, S.J., Haas, T.A., Bertomeu, M.C., Austin, J. and Buchanan, M.R. (1991). 13-HODE synthesis in internal mammary arteries and saphenous veins: implications in cardiovascular surgery. Adv. Prostaglandin Thromboxane Res. 21, 667–670.

Broekman, M.J., Eiroa, A.M. and Marcus, A.J. (1989). Albumin redirects platelet eicosanoid metabolism toward 12(S)-hydroxyeicosatetraenoic acid. J. Lipid Res. 30, 1925–1932.

Buchanan, M.R. (1989). In "Fondation Sanofi Thrombose Pour La Recherche" (ed H. Choay), pp 14–22. Choay, Paris.

Buchanan, M.R. and Bastida, E. (1988). Endothelium and underlying membrane reactivity with platelets, leukocytes and tumor cells regulation by the lipoxygenase-derived fatty acid metabolites, 13-HODE and HETEs. Med. Hypothesis 27, 317–325.

Buchanan, M.R. and Hirsh, J.A. (1978). Comparison of the effects of aspirin and dipyridamole on platelet aggregation in vivo and ex vivo. Thromb. Res. 13, 517–529.

Buchanan, M.R., Dejana, E., Cazenave, J.P., Richardson, M., Mustard, J.F. and Hirsh, J. (1980). Differences in inhibition of PGI2 production by aspirin in rabbit artery and vein segments. Thromb. Res. 20, 447–460.

Buchanan, M.R., Crowley, C.A., Rosin, R.E., Gimbrone, M.A. Jr and Babior, B.M. (1982). Studies on the interaction between GP-180-deficient neutrophils and vascular endothelium. Blood 60, 160–165.

Buchanan, M.R., Butt, R.W., Magas, Z., Ryn, J.V., Hirsh, J. and Nazir, D.J. (1985a). Endothelial cells produce a lipoxygenase derived chemorepellent which influences platelet/endothelial cell interactions: effects of aspirin and salicylate. Thromb. Haemost. 53, 306–311.

Buchanan, M.R., Haas, T.A., Lagarde, M. and Guichardant, M. (1985b). 13-Hydroxyoctadecadienoic acid is the vessel wall chemorepellent factor, LOX. J. Biol. Chem. 260, 16056–16059.

Buchanan, M.R., Butt, R.W., Hirsh, J., Markham, B.A. and Nazir, D.J. (1986). Role of lipoxygenase metabolism in platelet function: effect of aspirin and salicylate. Prostaglandins Leukotrienes Med. 21, 157–168.

Buchanan, M.R., Richardson, M., Haas, T.A., Hirsh, J. and Madri, J.A. (1987a). The basement membrane underlying the vascular endothelium is not thrombogenic: in vivo and in vitro studies with rabbit and human tissue. Thromb. Haemost. 58, 698–704.

Buchanan, M.R., Nakamura, K., Haas, T.A. and Hullin, F. (1987b). In "Biology of Icosanoids", (ed. M. Lagarde), pp 247–258. Colloque INSERM.

Buchanan, M.R., Bertomeu, M.C., Haas, T.A., Orr, F.W. and Eltringham-Smith, L.L. (1991). 13-HODE down-regulates the vitronectin receptor on human endothelial cells, thereby inhibiting endothelial cell/platelet interactions. Thromb. Haemost. 65, 1225.

Bull, A.W., Earles, S.M. and Bronstein, J.C. (1991). Metabolism of oxidized linoleic acid: distribution of activity for the enzymatic oxidation of 13-hydroxyoctadecadienoic acid to 13-oxooctadecadienoic acid in rat tissues. Prostaglandins 41, 43–50.

Burrows, F.J., Haskard, D.O., Hart, I.R., Marshall, J.F., Selkirk, S. and Poole, S. (1991). Influence of tumor-derived inter-leukin 1 on melanoma–endothelial cell interactions in vitro. Cancer Res. 51, 4768–4775.

Calzada, C., Vericel, E. and Lagarde, M. (1991). Decrease in platelet reduced glutathione increases lipoxygense activity and decreases vitamin E. Lipids 26, 696–699.

Camp, R.D.R. and Fincham, N.J. (1985). Inhibition of

ionophore-stimulated leukotriene B4 production in human leukocytes by monohydroxy fatty acids. Br. J. Pharmacol. 85, 837–841.

Cheresh, D.A. (1987). Human endothelial cells synthesize and express an Arg-Gly-Asp-directed adhesion receptor involved in attachment to fibrinogen and von Willebrand factor. Proc. Natl Acad. Sci. USA 84, 6471–6475.

Chopra, H., Timar, J., Chen, Y.Q., Rong, X.H., Grossi, I.M., Fitzgerald, L.A., Taylor, J.D. and Honn, K.V. (1991). The lipoxygenase metabolite 12(S)-HETE induces a cytoskeleton-dependent increase in surface expression of integrin $\alpha IIb\beta_3$ on melanoma cells. Int. J. Cancer 49, 774–786.

Claeys, M., Coene, M.C., Herman, A.G., Jouvenaz, G.H. and Nugteren, D.H. (1982). Characterization of monohydroxylated lipoxygenase metabolites of arachidonic and linoleic acid in rabbit peritoneal tissue. Biochim. Biophys. Acta 713, 160–169.

Clive, D.M., Stoff, J.S., Cardi, M., MacIntyre, D.E., Brown, R.S. and Salzman, E.W. (1990). Evidence that circulating 6-keto prostaglandin E_1 causes the platelet defect of Bartter's syndrome. Prostaglandins Leukot. Essent. Fatty Acids 41, 251–258.

Coller, B.S., Folts, J.D., Smith, S.R., Scudder, L.E. and Jordan, R. (1989). Abolition of in $vivo$ platelet thrombus formation in primates with monoclonal antibodies to the platelet GPIIb/IIIa receptor: correlation with bleeding time, platelet aggregation, and blockade of GPIIb/IIIa receptors. Circulation 80, 1766–1774.

Conforti, G., Zanetti, A., Pasquali-Ronchetti, I., Quaglino, D. Jr, Neyroz, P. and Dejana, E. (1990). Modulation of vitronectin receptor binding by membrane lipid composition. J. Biol. Chem. 265, 4011–4019.

Croset, M. and Lagarde, M. (1983). Stereospecific inhibition of PGH_2-induced aggregation by lipoxygenase products of eicosaenoic acids. Biochem. Biophys. Res. Commun. 1121, 878–883.

Croset, M. and Lagarde, M. (1985). Enhancement of eicosaenoic acid lipoxygenation in human platelets by 12-hydroperoxy derivative of arachidonic acid. Lipids 20, 743–750.

Daret, D., Blin, P. and Larrue, J. (1989). Synthesis of hydroxy fatty acids from linoleic acid by human blood platelets. Prostaglandins 38, 203–214.

Deckmyn, H., Proesmans, W. and Vermylen, J. (1983). Prostacyclin production by whole blood from children: impairment in the hemolytic uremic syndrome and excessive formation in chronic renal failure. Thromb. Res. 30, 13–18.

Defreyn, G., Deckmyn, H. and Vermylen, J. (1982). A thromboxane synthetase inhibitor agonist reorients endoperoxides metabolism in whole blood towards prostacyclin and prostaglandin E_2. Thromb. Res. 26, 389–400.

de Graaf, J.C., Bult, H., de Meyer, G.R., Sixma, J.J. and de Groot, P.G. (1989). Platelet adhesion to subendothelial structures under flow conditions: no effect of the lipoxygenase product 13-HODE. Thromb. Haemost. 62, 802–806.

Denning, G.M., Figard, P.H., Kaduce, T.L. and Spector, A.A. (1983). Role of triglycerides in endothelial cell arachidonic acid metabolism. J. Lipid Res. 24, 993–1001.

Domoney, C., Casey, R., Turner, L. and Ellis, N. (1991). Pisum lipoxygenase genes. Theor. Appl. Genet. 81, 800–805.

Dutilh, C.E., Haddeman, E., Don, J.A. and Hoor, F.T. (1981). The role of arachidonate lipoxygenase and fatty acids during irreversible blood platelet aggregation in $vitro$. Prostaglandins Med. 6, 111–126.

Elliot, G.R. and Van de Meent, D. (1991). Human recombinant interleukin-1α enhances basal and arachidonic acid stimulated synthesis of cyclooxygenase and lipoxygenase metabolites by human skin fibroblasts in $vitro$. Int. J. Immunopathol. Pharmacol. 4, 67–74.

Eynard, A.R., Galli, G., Tremoli, E., Maderna, P., Magni, F. and Paoletti, R. (1986). Aspirin inhibits platelet 12-hydroxy-eicosatetraenoic acid formation. J. Lab. Clin. Med. 107, 73–78.

Fitzgerald, D.J. (1989). Platelet inhibition with an antibody to glycoprotein IIb/IIIa. Circulation 80, 1766–1774.

Fonlupt, P., Croset, M. and Lagarde, M. (1991). 12-HETE inhibits the binding of PGH2/TXA2 receptor ligands in human platelets. Thromb. Res. 63, 239–248.

Fukuda, S., Sasaguri, Y., Yanagi, H., Ohuchida, M., Morimatsu, M. and Yagi, K. (1991). Effect of linoleic acid hydroperoxide on replication of adenovirus DNA in endothelial cells of bovine aorta. Atherosclerosis 89, 143–149.

Funk, C.D. and Powell, W.S. (1983). Metabolism of linoleic acid by prostaglandin endoperoxide synthase from adult and fetal blood vessels. Biochim. Biophys. Acta 754, 57–71.

Funk, C.D. and Powell, W.S. (1985). Release of prostaglandins and monohydroxy and trihydroxy metabolites of linoleic and arachidonic acids by adult and fetal aortae and ductus arteriosus. J. Biol. Chem. 260, 7481–7488.

Funk, C.D., Fucri, L. and FitzGerald, G.A. (1991). Molecular cloning, primary structure, and expression of the human platelet-erythroleukemia cell 12-lipoxygenase. Proc. Natl Acad. Sci. USA 87, 5638–5642.

Gerritsen, M.E. and Cheli, C.D. (1983). Arachidonic acid and prostaglandin endoperoxide metabolism in isolated rabbit coronary microvessels and isolated and cultivated coronary microvessel endothelial cells. J. Clin. Invest. 72, 1658–1671.

Gibson, B.E.S., Buchanan, M.R., Barr, R.D. and White, J.G. (1987). Primary thrombocythaemia in childhood: symptomatic episodes and their relationship to thromboxane A_2, 6-keto-PGE_1 and 12-hydroxyeicosatetraenoic acid production: a case report. Prostaglandins Leukotrienes Med. 26, 221–231.

Gimbrone, M.A. Jr and Buchanan, M.R. (1982). Interactions of platelets and leukocytes with vascular endothelium: in $vitro$ studies. Ann. N.Y. Acad. Sci. 401, 171–183.

Griffiths, R.J. and Moore, P.K. (1983). Conversion of prostacyclin to 6-oxo prostaglandin E_1 by rat, rabbit, guinea-pig and human platelets. Br. J. Pharmacol. 80, 395–402.

Grossi, I.M., Hatifield, J.S., Mosley, K.L., Taylor, J.D., FitzGerald, L.A. and Honn, K.V. (1988). Effects of TPA and 12-HETE on tumor cell expression of IRGpIIb/IIIa and tumor cell adhesion to subendothelial matrix. Proc. Am. Assoc. Cancer Res. 29, 248.

Grossi, I.M., FitzGerald, L.A., Umbarger, L.A., Nelson, K.K., Diglio, C.A., Taylor, J.D. and Honn, K.V. (1989). Bidirectional control of membrane expression and/or activation of tumor cell IRGpIIb/IIIa receptor and tumor cell adhesion by lipoxygenase products of arachidonic acid and linoleic acid. Cancer Res. 49, 1029–1037.

Grotemeyer, K.-H. (1991). Effects of acetylsalicylic acid in stroke patients. Evidence of non-responders in a sub-population of treated patients. Thromb. Res. 63, 587–593.

Haas, T.A., Bastida, E., Nakamura, K., Hullin, F., Admirall, L. and Buchanan, M.R. (1988). Binding of 13-HODE and 5-, 12- and 15-HETE to endothelial cells and subsequent platelet neutrophil and tumor cell adhesion. Biochim. Biophys. Acta 961, 153–159.

Haas, T.A., Bertomeu, M.C., Bastida, E. and Buchanan, M.R. (1990). Cyclic AMP regulation of endothelial cell triacylglycerol turnover, 13-hydroxyoctadecadienoic acid (13-HODE) synthesis and endothelial cell thrombogenicity. Biochim. Biophys. Acta 1051, 174–178.

Hada, T., Ueda, N., Takahashi, Y. and Yamamoto, S. (1991). Catalytic properties of human platelet 12-lipoxygenase as compared with the enzymes of other origins. Biochim. Biophys. Acta 1083, 89–93.

Hadjiagapiou, C. and Spector, A.A. (1986). 12-Hydroxyeicosatetraenoic acid reduces prostacyclin production by endothelial cells. Prostaglandins 31, 1135–1144.

Hajjar, D.P., Weksler, B.B., Falcone, D.J., Hefton, J.M., Taik-Goldman, K. and Minick, C.R. (1982). Prostacyclin modulates cholesteryl hydrolytic activity by its effects on cyclic adenosine monophosphate in rabbit aortic smooth muscle cell. J. Clin. Invest. 70, 479–488.

Hamberg, M. and Samuelsson, B. (1980). Stereochemistry in the formation of 9-hydroxy linoleic acid by fatty acid cyclooxgenase. Biochim. Biophys. Acta 617, 545–547.

Hamberg, M., Svensson, J. and Samuelsson, B. (1974). Prostaglandin endoperoxides: a new concept concerning the mode of action and release of prostaglandins. Proc. Natl Acad. Sci. USA 71, 3824–3828.

Hammarstrom, S. and Falardeau, P. (1977). Resolution of prostaglandins endoperoxide synthase and thromboxane synthase of human platelets. Proc. Natl Acad. Sci. USA 74, 3691–3695.

Haslam, R.J. and McClenaghan, M.D. (1981). Measurement of circulatory prostacyclin. Nature 292, 364–365.

Hawkins, D.J. and Brash, A.R. (1987). Eggs of the sea urchin: *Strongylocentrotus purpuratus* contain a prominent (11R) and (12R) lipoxygenase activity. J. Biol. Chem. 262, 7629–7639.

Head, R.J., Mano, M.T., Bexis, S., Howe, P.R.C. and Smith, R.M. (1991). Dietary fish oil administration retards the development of hypertension and influences vascular neuro-effector function in the stroke prone spontaneously hypertensive rat (SHRSP). Prostaglandins Leukot. Essent. Fatty Acids 44, 119–122.

Honn, K.V., Cicone, B. and Skoff, A. (1981). Prostacyclin: a potent antimetastic agent. Science 212, 1270–1272.

Honn, K.V., Grossi, I.M., Diglio, C.A., Wojtukiewicz, M. and Taylor, J.D. (1989). Enhanced tumor cell adhesion to the subendothelial matrix resulting from 12(S)-HETE-induced endothelial cell retraction. Fed. Proc. 3, 2285–2293.

Hopkins, N.K., Oglesby, T.D., Bundy, G.L. and Gorman, R.R. (1984). Biosynthesis and metabolism of 15-hydroperoxy-5,8,11,13-eicosatetraenoic acid by human umbilical vein endothelial cells. J. Biol. Chem. 259, 14048–14053.

Horrobin, D.F. (1991). Interactions between n-3 and n-6 essential fatty acids (EFAs) in the regulation of cardiovascular disorders and inflammation. Prostaglandins Leukot. Essent. Fatty Acids 44, 127–131.

Hoult, J.R.S. and Moore, P.K. (1986). 6-Keto-prostaglandin E_1: a naturally occurring stable prostacyclin-like mediator of high potency. TIPS, 197–200.

Howe, P.R.C., Lungerhauser, Y.K., Rogers, P.F., Gerkens, J.F., Head, R.J. and Smith, R.M. (1991). Effects of dietary sodium and fish oil on blood pressure development in stroke-prone spontaneously hypertensive rats. J. Hyperten. 9, 639–644.

Humes, J.L., Opas, E.E., Galavage, M., Soderman, D. and Bonney, R.J. (1986). Regulation of macrophage eicosanoid production by hydroperoxy- and hydroxy-eicosatetraenoic acids. Biochem. J. 233, 199–206.

Humphries, M.J., Olden, K. and Yamada, K.M. (1986). A synthetic peptide from fibronectin inhibits experimental metastasis of murine melanoma cells. Science 233, 467–469.

Hynes, R.O. (1987). Integrins: A family of cell surface receptors. Cell 48, 549–554.

Ishii, H., Kizaki, K., Uchiyama, H., Horie, S. and Kazama, M. (1990). Cyclic AMP increases thrombomodulin expression on membrane surface of cultured human umbilical vein endothelial cells. Thromb. Res. 59, 841–850.

Izumi, T., Radmark, O. and Samuelsson, B. (1990a). Purification of 15-lipoxygenase from human leukocytes, evidence for the presence of isozymes. Adv. Prostaglandin Thromboxane Leukotriene Res. 21, 101–104.

Izumi, T., Hoshiko, S., Radmark, O. and Samuelsson, B. (1990b). Cloning of the cDNA for human 12-lipoxygenase. Proc. Natl Acad. Sci. USA 87, 7477–7481.

Jackson, E.K., Goodman, R.P. Fitzgerald, G.A., Oates, J.A. and Branch, R.A. (1982). Assessment of the extent to which exogenous prostaglandin I_2 is converted to 6-keto-prostaglandin E_1 in human subjects. J. Pharmacol. Exp. Ther. 22, 183–196.

James, M.J., Cleland, L.G., Gibson, R.A. and Hawkes, J.S. (1991). Interaction between fish and vegetable oils in relation to rat leukocyte leukotriene production. J. Nutr. 121, 631–637.

Kaduce, T.L., Figard, P.H., Leifur, R. and Spector, A.A. (1989). Formation of 9-hydroxyoctadecadienoic acid from linoleic acid in endothelial cells. J. Biol. Chem. 264, 6823–6830.

Kishimoto, T.K., Warnock, R.A., Jutila, M.A., Butcher, E.C., Lane, C., Anderson, D.C. and Smith, C.W. (1991). Antibodies against human neutrophil LECAM-1 (LAM-1/Leu-8/DREG-56 antigen) and endothelial cell ELAM-1 inhibit a common CD18-independent adhesion pathway *in vitro*. Blood 78, 805–811.

Kobayashi, K., Toyoda, T. Sawada, S., Shirai, K., Yamomoto, K., Katoh, K. Kameda, M., Takada, O., Yamagami, M., Uno, M., Tsuji, H. and Nakagawa, M. (1991). Effect of cyclic GMP and sulfhydryl on prostacyclin production by human vascular endothelial cells. Jpn. Circ. J. 55, 643–647.

Korbut, R., Byrska-Danek, A.B. and Gryglewski, R.J. (1983). Fibrinolytic activity of 6-keto-prostaglandin E_1. Thromb. Haemost. 50, 893.

Kuhn, H., Ponicke, K., Halle, W., Wiesner, R., Schewe, T. and Forster, W. (1985). Metabolism of [1-^{14}C]-arachidonic acid by cultured calf aortic endothelial cells: evidence for the presence of a lipoxygense pathway. Prostaglandins Leukotrienes Med. 17, 291–303.

Kuhn, H., Belkner, J., Wiesner, R. and Alder, L. (1990). Occurrence of 9- and 13-ketooctadecadienoic acid in biological membranes oxygenated by the reticulocyte lipoxygenase. Arch. Biochem. Biophys. 279, 218–224.

Kuijpers, T.W., Hakkert, B.C., Hoogerwerf, M., Leeuwenberg, J.F.M. and Roos, D. (1991). Role of endothelial leukocyte adhesion molecule 1 and platelet activating factor in neutrophil adherence to IL-1 prestimulated endothelial cells. J. Immunol. 147, 1369–1376.

Lafrenie, R.M., Podor, T.J., Buchanan, M.R. and Orr, F.W. (1990). Interleukin-1a induced vitronectin receptor expression and tumor cell endothelial cell adhesion. FASEB J. 4, A1134 (Abstr.)

Lagarde, M., Croset, M., Auth, K.S. and Crawford, N. (1984). Subcellular localization and some properties of lipoxygenase activity in human blood platelets. Biochem. J. 222, 495–500.

Lauri, D., Bertomeu, M.C., Orr, F.W., Bastida, E., Sauder, D. and Buchanan, M.R. (1990). Interleukin-1 increases tumor cell adhesion to endothelial cells through an RGD dependent mechanism: *in vitro* and *in vivo* studies. Clin. Exp. Metastasis 8, 27–32.

Legrand, A.B., Lawson, J.A., Meyrick, B.O., Blair, I.A. and Oates, J.A. (1991). Substitution of 15-hydroxyeicosatetraenoic acid in the phosphoinositide signalling pathway. J. Biol. Chem. 266, 7570–7577.

Lewis, R., Scholkens, B.A. and Beck, G. (1984). Biological activities of 6-keto-prostaglandin E_1. Prostaglandins Leukotrienes Med. 16, 303–324.

Lovelady, G.K., Mirro, R., Armstead, W.M., Busija, D.W. and Leffler, C.W. (1988). Effect of 15-HETE on cerebral arterioles of newborn pigs. Prostaglandins 36, 507–513.

Lucas, F.V., Skrinska, V.A., Chisholm, G.M. and Hesse, B.L. (1986). Stability of prostacyclin in human and rabbit whole blood and plasma. Thromb. Res. 43, 379–387.

Ma, Y.H., Harder, D.R., Clark, J.E. and Roman, R.J. (1991). Effects of 12-HETE on isolated dog renal arcuate arteries. Am. J. Physiol. 261, H451–H456.

McCarthy, G. (1991). Fish oil and psoriasis. Lancet 338, 824.

Marcus, A.J., Broekman, M.J., Safier, L.B., Ullman, H.L., Islam, N., Serham, C.N., Rutherford, L.E., Korckhak, H.M. and Weissman, G. (1982a). Formation of leukotrienes and other hydroxy acids during platelet–neutrophil interactions *in vivo*. Biochem. Biophys. Res. Commun. 109, 130–137.

Marcus, A.J., Broekman, M.J., Weksler, B.B., Jaffe, E.A., Safier, L.B., Ullman, H.L., Islam, N. and Tack-Goldman, K. (1982b). Arachidonic and metabolism in endothelial cells and platelets. Ann. N.Y. Acad. Sci. 401, 195–202.

Marcus, A.J., Safier, L.B., Ullman, H.L., Islam, N., Broekman, M.J. and von Schacky, C. (1987). Studies on the mechanism of 2-hydroxylation of platelet 12-hydroxyeicosatetraenoic acid (12-HETE) by unstimulated neutrophils. J. Clin. Invest. 79, 179–187.

Mehta, J., Mehta, P., Lawson, D.L., Ostrowski, N. and Brigmon, L. (1985). Influence of selective thromboxane synthetase blocker CGS-13080 on thromboxane and prostacyclin biosynthesis in whole blood: evidence for synthesis of prostacyclin by leukocytes from platelet-derived endoperoxides. J. Lab. Clin. Med. 106, 246–252.

Miller, C.C. and Ziboh, V.A. (1989). Induction of epidermal hyper-proliferation by topical docosahexaenoic acid linked to decreased epidermal 13-hydroxyoctadecadienoic acid. Clin. Res. 37, 36.

Miyamori, I., Morise, T., Yasuhara, S., Takeda, Y., Koshida, H. and Takeda, R. (1985). Single-blind study of epoprostenol and 6-keto-prostaglandin E_1 in man: effects of platelet aggregation and plasma renin. Br. J. Clin. Pharmacol. 20, 681–683.

Moncada, S. and Vane, J.R. (1979). Arachidonic acid metabolites and the interactions between platelets and blood vessel walls. N. Engl. J. Med. 300, 1142–1147.

Moore, P.K. and Griffiths, R.J. (1983). 6-Keto-prostaglandin-E_1. Prostaglandins 26, 500–517.

Moore, P.K., Griffiths, R.J. and Lofts, F.T. (1983). The effect of some flavone drugs on the conversion of prostacyclin to 6-oxo-prostaglandin E_1. Biochem. Pharmacol. 32, 2813–2817.

Moore, S.A., Spector, A.A. and Hart, M.N. (1988). Eicosanoid metabolism in cerebromicrovascular endothelium. Cell Physiol. 23, 37–44.

Moore, S.A., Giordano, M.J., Kim, H.Y., Salem, N. and Spector, A.A. (1991). Brain microvessel 12-hydroxyeicosatetraenoic acid is the (S) enantiomer and is lipoxygenase derived. J. Neurochem. 922–929.

Morita, I. and Murota, S.I. (1987). Role of 12-lipoxygenase products of arachidonic acid on platelet aggregation. Adv. Prostaglandin Thromboxane Leukotriene Res. 17, 219–223.

Nadel, J.A., Conrad, D.J., Ueki, I.F., Schuster, A. and Sigal, E. (1991). Immunocytochemical localization of arachidonate 15-lipoxygenase in erythrocytes, leukocytes, and airway cells. J. Clin. Invest. 87, 1139–1145.

Nakao, J., Ito, H., Koshihara, Y. and Murota, S.I. (1984). Age-related increase in the migration of aortic smooth muscle cells induced by 12-L-hydroxy-5,8,10,14-eicosatetraenoic acid. Atherosclerosis 51, 179–187.

Naor, Z. (1991). Is arachidonic acid a second messenger in signal transduction? Mol. Cell. Endocrinol. 80, C181–C186.

Nordoy, A. (1986). The role of dietary fatty acids in thrombosis. Prog. Lipid Res. 25, 455–459.

Nugteren, D.H. (1975). Arachidonic lipoxygenase in blood platelets. Biochim. Biophys. Acta 380, 299–307.

Nugteren, D.H. (1986). Arachidonic acid-12-lipoxygenase from bovine platelets. Methods Enzymol. 86, 49–54.

Okuma, M. and Uchino, H. (1979). Altered arachidonate metabolism by platelets in patients with myeloproliferative disorders. Blood 54, 1258–1269.

Pieroni, J.P., Dray, F., Pace-Asciak, R. and McGiff, J.C. (1988). Identification of 6-keto-prostaglandin E_1 obtained from isolated perfused kidney of the rabbit. J. Pharmacol. Exp. Ther. 247, 63–68.

Pohlman, T., Yates, J., Needleman, P. and Klahr, S. (1983). Effects of prostacyclin on short circuit current and water flow in the toad urinary bladder. Am. J. Physiol. 244, 270–278.

Powell, W.S. and Gravelle, F. (1985). Metabolism of eicosapentaenoic acid by aorta: formation of a novel 13-hydroxylated prostaglandin. Biochim. Biophys. Acta 835, 201–211.

Quilley, C.P., Wong, P.Y.K. and McGiff, J.C. (1979). Hypotensive and renovascular actions of 6-keto-prostaglandin E_1: A metabolite of prostacyclin. Eur. J. Pharmacol. 57, 273–276.

Rankin, S.M., Parthasarathy, S. and Steinberg, D. (1991). Evidence for a dominant role of lipoxygenase(s) in the oxidation of LDL by mouse peritoneal macrophages. J. Lipid Res. 32, 449–456.

Rapp, J.H., Connor, W.E., Lin, D.S. and Porter, J.M. (1991). Dietary eicosapentaenoic acid and docosahexaenoic acid from fish oil: their incorporation into advanced human atherosclerotic plaques. Arteriosclerosis Thromb. 11, 903–911.

Revtyak, G.E., Johnson, A.R. and Campbell, W.B. (1988). Cultured bovine coronary arterial endothelial cells synthesize HETEs and prostacyclin. Am. Physiol. Soc. 254, C8–19.

Robinson, G.A., Butcher, R.W. and Sutherland, B.W. (1971). "Cyclic AMP." Academic Press, New York.

Roth, G.J. (1991). Developing relationships: arterial adhesion, glycoprotein Ib and leucine-rich glycoproteins. Blood 77, 5–19.

Roth, G.J., Stanford, N. and Majerus, P.W. (1975). Acetylation of prostaglandin synthase by aspirin. Proc. Natl Acad. Sci. USA 72, 3073–3076.

Schade, U.F., Burmeister, I. and Engel, R. (1987). Increased 13-hydroxyoctadecadienoic acid content in lipopolysaccharide stimulated macrophages. Biochem. Biophys. Res. Commun. 147, 695–700.

Sekiya, F., Takagi, J., Usui, T., Kawajiri, K., Kobayashi, Y., Sato, F. and Saito, Y. (1991). 12S-Hydroxyeicosatetraenoic acid plays a central role in the regulation of platelet activation. Biochem. Biophys. Res. Commun. 179, 345–351.

Setta, B.N.Y. and Stuart, M.J. (1986). 15-Hydroxy-5,8,11,13-eicosatetraenoic acid inhibits human vascular cyclooxygenase. J. Clin. Invest. 77, 202–211.

Shannon, V.R., Hansbrough, J.R., Takahashi, Y., Ueda, N., Yamamoto, S. and Holtzman, M.J. (1990). Biochemical and immunohistochemical evidence for selective expression of novel epithelial lipoxygenases. Adv. Prostaglandin Thromboxane Leukotriene Res. 21, 37–40.

Simon, T.C., Makheja, A.N. and Bailey, J.M. (1989a). The induced lipoxygenase in atherosclerotic aorta converts linoleic acid to the platelet chemorepellent factor 13-HODE. Thromb. Res. 55, 171–178.

Simon, T.C., Makheja, A.N. and Bailey, J.M. (1989b). Formation of 15-hydroxyeicosatetraenoic acid (15-HETE) as the predominant eicosanoid in aortas from Watanabe heritable hyperlipidemic and cholesterol-fed rabbits. Atherosclerosis 75, 31–38.

Simon, T.C., Makheja, A.N. and Bailey, J.M. (1990). Relationship of vascular 15-lipoxygenase induction to atherosclerotic plaque formation in rabbits. Prostaglandins Leukot. Essent. Fatty Acids 41, 273–278.

Singer, I.I., Scott, S., Kawka, W. and Kazazis, D.M. (1989). Adhesomes: specific granules containing receptors for laminin, C3bi-fibrinogen, fibronectin, and vitronectin in human polymorphonuclear leukocytes and monocytes. J. Cell. Biol. 109, 3169–3182.

Sloane, D.L., Leung, R., Craik, C.S. and Sigal, E. (1991). A primary determinant for lipoxygenase positional specificity. Nature 354, 149–152.

Smith, E.F. III, Darius, H., Ferber, H. and Schror, K. (1985). Inhibition of thromboxane and 12-HPETE formation by dazaxiben and its two thiopenic acid-substituted derivatives. Eur. J. Pharmacol. 112, 161–169.

Soberman, R.J., Harper, T.W., Betteridge, D., Lewis, R.A. and Austen, K.F. (1985). Characterization and separation of the arachidonic acid 5-lipoxygenase and linoleic acid ω-6 lipoxygenase (arachidonic acid 15-lipoxygenase) of human polymorphonuclear leukocytes. J. Biol. Chem. 260, 4508–4515.

Speizer, L.A., Watson, M.J. and Brunton, L.L. (1991). Differential effects of omega-3 fish oils on protein kinase activities *in vitro*. Am. J. Physiol. 261, E109–E114.

Spitz, B., Deckmyn, H., Assche, F.A.V. and Vermylen, J. (1983). Prostacyclin production in whole blood throughout normal pregnancy. Clin. Exp. Hypertens. 2, 191–202.

Takahashi, M., Ozaki, N., Kawakita, S., Nozaki, M. and Saeki, Y. (1985). Vascular effects of 15-hydroperoxidyeicosatetraenoic acid and 15-hydroxyeicosatetraenoic acid on canine arteries. Jpn. J. Pharmacol. 37, 325–334.

Takahashi, R., Begin, M.E., Ells, G. and Horrobin, D.F. (1991). Effects of eicosapentaenoic acid and arachidonic acid on incorporation and metabolism of radioactive linoleic acid in cultured human fibroblasts. Prostaglandins Leukot. Essent. Fatty Acids 42, 113–117

Takayama, H. (1989). Interactions via eicosanoids between vascular and blood cells. Acta Haematol. 52, 1322–1329.

Tremoli, E., Maderna, P., Caruso, D., Galli, G. and Paoletti, R. (1987). Platelet 12-hydroxyeicostetraenoic acid synthesis and aspirin: *in vitro* effects plasma and albumin. Adv. Prostaglandin Thromboxane Leukotriene Res. 17, 274–278.

Vanderhoek, J.Y., Brant, R.W. and Bailey, J.M. (1980). 15-Hydroxy-5,8,11,13-eicosatetraenoic acid: a potent and selective inhibitor of platelet lipoxygenase. J. Biol. Chem. 255, 5996–5998.

Van Diest, M.J., Verbeuren, T.J. and Herman, A.G. (1991). 15-Lipoxygenase metabolites or arachidonic acid evoke contractions and relaxations in isolated canine arteries: role of thromboxane receptors, endothelial cells and cyclooxygenase. J. Pharmacol. Exp. Ther. 256, 194–203.

Van Ryn-McKenna, J. and Buchanan, M.R. (1989). Relative effects of flurbiprofen on platelet 12-hydroxy-eicosatetraenoic acid and thromboxane A2 production: influence on collagen-induced platelet aggregation and adhesion. Prostaglandins Leukot. Essent. Fatty Acids 36, 171–174.

Weber, E., Haas, T.A., Muller, T.H., Eisert, W.G., Hirsh, J., Richardson, M. and Buchanan, M.R. (1990). Relationship between vessel wall 13-HODE synthesis and vessel wall thrombogenecity following injury: influences of salicylate and dipyridamole treatment. Thromb. Res. 57, 383–392.

Wilkenson, D., Hallam, C., Hainsley, P.E., Lord, G.H. and Mitchell, P.D. (1985). 5-L(S)-Hydroxy-5Z,8Z,11Z,13e-eicosatetraenoic acid is a substrate for 5-lipoxygenase. Biochem. Soc. Trans. 13, 180–182.

Wojenski, C.M., Silver, M.J. and Walker, J. (1991). Eicosapentaenoic acid ethyl ester as an antithrombotic agent: comparison to an extract of fish oil. Biochim. Biophys. Acta 1081, 33–38.

Wong, P.Y.-K., Lee, W.H., Chao, P.H.-W., Reiss, R.F. and McGiff, J.C. (1980a). Metabolism of prostacyclin by 9-hydroxyprostaglandin dehydrogenase in human platelets. J. Biol. Chem. 225, 9021–9024.

Wong, P.Y.K., Malik, K.U., Desiderio, D.M., McGiff, J.C. and Sun, F.F. (1980b). Hepatic metabolism of prostacyclin (PGI2) in the rabbit: formation of a potent novel inhibitor of platelet aggregation. Biochem. Biophys. Res. Commun. 93, 486–494.

Wood, D.A., Riemersma, R.A., Butler, S., Thomson, M., MacIntyre, C., Elton, R.D. and Oliver, M.F. (1987). Linoleic acid and eicosatetraenoic acids in adipose tissue and platelets and risk of coronary heart disease. Lancet i, 177–183.

Yla-Hertuala, S., Rosenfeld, M.E., Parthasarathy, S., Glass, C.K., Sigal, E., Witztum, J.L. and Steinberg, D. (1990). Colocalization of 15-lipoxygenase mRNA and protein with epitopes of oxidized low density lipoprotein in macrophage-rich areas of atherosclerotic lesions. Proc. Natl Acad. Sci. USA 87, 6959–6963.

Yla-Hertuala, S., Rosenfeld, M.E., Parthasarathy, S., Sigal, E., Sarkioja, T., Witztum, J.L. and Steinberg, D. (1991). Gene expression in macrophage-rich human atherosclerotic lesions. 15-Lipoxygenase acetyl low density lipoprotein receptor messenger RNA colocalize with oxidation specific lipid protein adducts. J. Clin. Invest. 87, 1146–1152.

Yoshimoto, T., Yamamoto, Y., Arakawa, T., Suzuki, H., Yamamoto, S., Yokoyama, C., Tanabe, T. and Toh, H. (1990a). Molecular cloning and expression of human arachidonate 12-lipoxygenase. Biochem. Biophys. Res. Commun. 172, 1230–1235.

Yoshimoto, T., Suzuki, H., Yamamoto, S., Takai, T., Yokoyama, C. and Tanabe, T. (1990b). Cloning and sequence analysis of the cDNA for arachidonate 12-lipoxygenase of porcine leukocytes. Proc. Natl Acad. Sci. USA 87, 2142–2146.

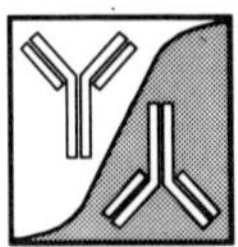

5. Inhibitors of Fatty Acid-Derived Mediators

D.W.P. Hay and D.E. Griswold

Lipid Mediators
ISBN 0–12–198875–9

1. General Introduction

The fatty acid-derived mediators of inflammation are, for the purposes of this chapter, defined as including arachidonic acid, its 5-, 12- and 15-lipoxygenase metabolites (e.g. leukotrienes, monoHETES and lipoxins), and its cyclooxygenase metabolites (e.g. PGD$_2$, PGE$_2$, PGF$_{2\alpha}$, PGI$_2$ and TXA$_2$).

As described comprehensively in other chapters, these molecules possess a plethora of activities in a variety of systems and have been implicated as critical mediators in the pathophysiology of several diseases, including asthma, arthritis, endotoxic shock and ARDS. Accordingly, with the ultimate goal of producing novel therapies for these and other diseases, the pharmaceutical industry and academic laboratories continue to commit a considerable amount of resources towards the development of potent and selective agents which will modulate their numerous actions.

As indicated in Fig. 5.1, there are several potential sites of pharmacological intervention in the arachidonic acid cascade. Thus far, the strategies that have been employed predominantly can be separated into two categories: (1) enzyme inhibitors, which control the release of the mediators of their precursors, and (2) receptor antagonists, which attenuate the effects of the released mediators at the target organ. We have utilized a hierarchical approach in this chapter to describe the status of research and clinical trials on modulators of the arachidonic acid cascade. Thus, initially there will be a description of inhibitors of the formation and release of arachidonic acid itself. This will be followed by a comprehensive discussion of inhibitors of the enzymes involved in the arachidonic acid metabolism cascade, predominantly cyclooxygenase and 5-lipoxygenase, and then a description of the receptor antagonists of various of the products produced by these enyzmes. To conclude, there will be a brief discussion of miscellaneous and novel classes of compounds that are beginning to emerge and a synopsis of our opinions of possible future directions in this field. As a reflection of the focus of research to date, the majority of the material describes inhibitors of the cyclooxygenase and 5-lipoxygenase enzymes and receptor antagonists of the peptidoleukotrienes and, to a lesser extent, LTB$_4$ and TXA$_2$. However, it is worth noting that an increasing amount of research is being conducted on other targets and strategies.

In each section attempts are made to present a historical perspective and a summary of the preclinical activity of representative members of each class of compounds. In addition, whenever available, a comprehensive discussion of the pivotal clinical data with compounds is given; from a therapeutic standpoint this is obviously the most important information.

As we hope will become apparent from the following discussion, considerable progress has been made with several of these strategies in the past several years and we are approaching a point where, at least with some diseases such as asthma, we will soon determine whether some of these classes of compounds will be of clinical benefit and perhaps the drugs of the future for several inflammatory disorders.

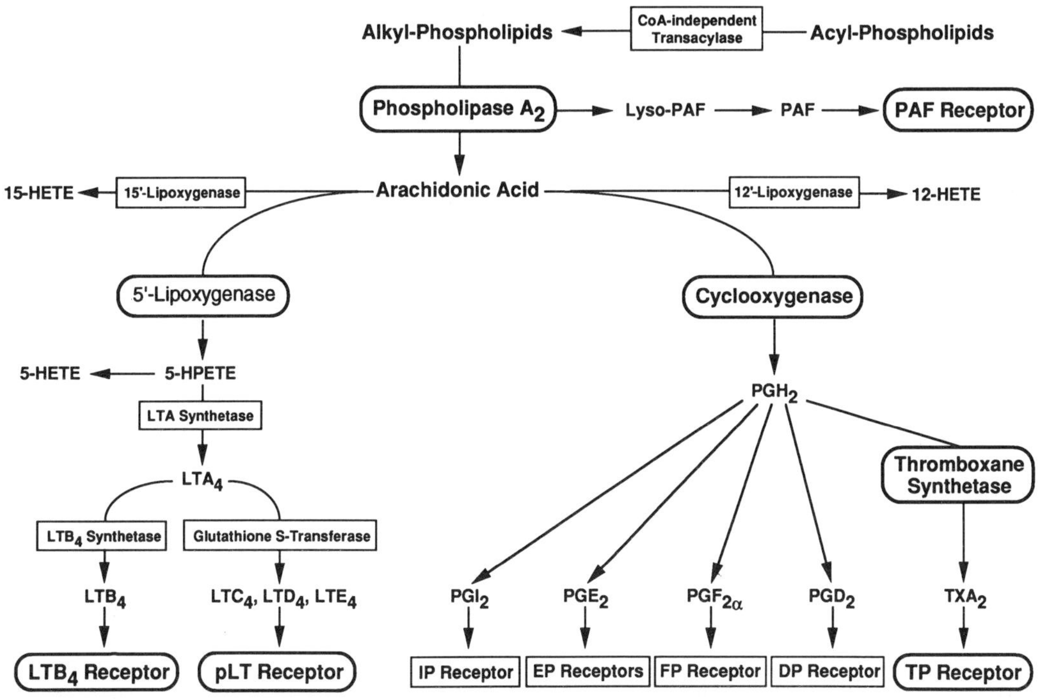

Figure 5.1 Some potential sites of pharmacological intervention in the arachidonic acid cascade. Present major targets for the pharmaceutical industry which are discussed in this chapter (except for PAF receptor antagonists) are indicated in the oval boxes.

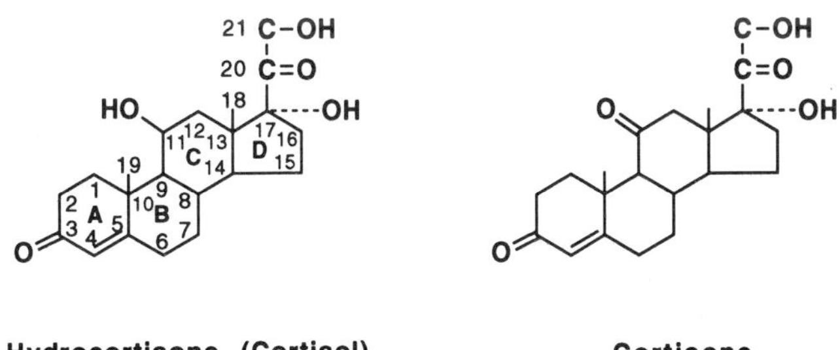

Hydrocortisone (Cortisol)   **Cortisone**

Figure 5.2 Chemical structures of hydrocortisone and cortisone.

2. *Inhibitors of Arachidonic Acid Release*

2.1 GLUCOCORTICOIDS

2.1.1 Introduction

Perhaps the most widely used and effective anti-inflammatory agents are represented by the glucocorticoids. The 42nd edition of *The Physician's Desk Reference* lists 124 glucocorticoid and combination products currently marketed in the USA. Historically, the first use of an adrenal glucocorticoid (17-hydroxy-11-dehydrocorticosterone) as an anti-inflammatory agent is credited to Hench and coworkers at the Mayo Clinic in the late 1940s (Hench *et al.*, 1949). Since it was known that the activities of the adrenal corticosteroids included profound effects on carbohydrate and protein metabolism, lipid metabolism in adipose tissue, electrolyte and water balance in addition to their anti-inflammatory activities, it was not

surprising that the therapeutic administration of glucocorticoids was attended by significant undesirable side-effects. They included sodium retention, fat mobilization, gluconeogenesis, muscle wasting and skin atrophy (when used topically) (Morris, 1978; Kirby, 1989). Since the glucocorticoids are clearly efficacious in a wide variety of inflammatory diseases, much of the work in the past 40 years or so has centred on elimination, or at least a marked reduction, of these side-effects.

2.1.2 Chemical/Pharmacological Classes

Most of the anti-inflammatory glucocorticoids share substitution of a hydroxyl group α to C-17 (see structures in Fig. 5.2). The natural steroids cortisol and cortisone possess this feature. One of the earliest improvements in the molecule was inclusion of an additional double bond in ring A (Δ^1) to yield prednisone and prednisolone (analogues of cortisone and cortisol, respectively). This change resulted in a four-fold increase in anti-inflammatory potency with about the same ability to induce sodium retention. Substitution in the B ring of prednisolone with a 6α methyl group produced a compound, methylprednisolone, which is five times more potent than cortisol with one-half of the sodium retention activity. Substitution of a methyl group at C-16 (ring D) eliminated sodium retention. This change coupled with substitution of a fluorine at C-6 of prednisolone yielded betamethasone (16β) and dexamethasone (16α), which are 25-fold more potent as anti-inflammatory agents with no sodium retention. However, none of these changes altered the metabolic or immunosuppressive effects of the resulting molecules (Morris, 1978; Haynes and Murad, 1985).

Recently, molecules have been designed to attempt to separate the metabolic effects from the anti-inflammatory/immunological effects. Deflazacort is a prednisolone analogue (oxazoline at C-16–C-17) which appears to have reduced metabolic effects while bestowing long-lasting immunosuppressive and anti-inflammatory activity (Scudeletti *et al.*, 1990). Other strategies have included design of metabolically labile compounds (e.g. androstene 17-thioketals; Woinar and Varma, 1983) and novel 12β-substituted betamethasone analogues, one of which, the tripropionate, notably had no atrophogenic activity in the skin (Avery *et al.*, 1990).

In addition, non-steroidal structures are being described with significant glucocorticoid receptor-binding affinities (e.g. phenylpyrazoles) (Schane *et al.*, 1985) and conformational relationships to glucocorticoids (e.g. xanthones) (Bianco *et al.*, 1989).

2.1.3 Mechanism of Action

While the precise molecular mechanism of action of glucocorticoids is still under intense scrutiny, the biological actions can be attributed to high-affinity binding to an intracellular receptor with subsequent acquisition (activation) of DNA-binding activity of the complex. The glucocorticoid–receptor complex is than translocated to the nucleus with subsequent binding to specific sequences of DNA (steroid regulatory elements), activation of transcription and synthesis of proteins (Danon and Assouline, 1978; Scudeletti *et al.*, 1990; Burnstein *et al.*, 1991). At least one of the proteins produced, lipocortin 1, has been shown to inhibit or interfere with PLA_2 activity (Cartwright *et al.*, 1989). This 37 kDa protein has been sequenced, cloned and expressed in sufficient quantities to allow *in vivo* experimentation. Lipocortin 1, which is present in lungs in high amounts compared with other organs (Pepinsky *et al.*, 1986), is anti-inflammatory and inhibits the release of arachidonic acid and formation of eicosanoids (Flower, 1988). Dexamethasone increased lipocortin 1 levels in human alveolar macrophages (Ambrose *et al.*, 1992). However, the biological relevance of lipocortin is controversial, and debate persists as to the validity of the proposed link

between lipocortin and the anti-inflammatory activity of glucocorticoids (Davidson and Dennis, 1989). Work is ongoing to ascribe the different facets of glucocorticoid action to these second-messenger proteins.

2.1.4 Preclinical Studies

The hallmark of glucocorticosteroid *in vivo* anti-inflammatory activity is their ability to inhibit both the fluid and cellular phases of the inflammatory response coupled with their effects on lymphocyte traffic and cytokine production. It seems clear that both leukotriene and prostanoid biosynthesis, as well as inflammatory cytokine production is inhibited by glucocorticoid administration (Morris, 1978; Haynes and Murad, 1985; Scudeletti *et al.*, 1990). As would be expected, glucocorticoids are active in most inflammatory models in experimental animals. Immunologically driven inflammatory models, such as dinitrofluorobenzene-induced contact sensitivity in the mouse, adjuvant-induced arthritis and experimental allergic encephalomyelitis, are exquisitely sensitive to glucocorticoids. In addition, classical acute inflammatory models such as carrageenan-induced paw oedema are also sensitive (Rosenthale, 1974; Swingle, 1974; D.E. Griswold, unpublished observations). Of interest is the relative insensitivity of arachidonic acid-induced inflammation to the action of glucocorticoids. This is particularly apparent in arachidonic acid-induced mouse ear oedema, where oral or topical administration gives only modest inhibition or is inactive (Carlson *et al.*, 1985). The explanation for such inactivity could be related to the fact that the substrate (arachidonic acid) is free and being utilized by oxygenases, therefore circumventing the need for PLA_2 action.

As mentioned above, non-steroidal glucocorticoid-like compounds have been investigated. A phenyl pyrazole, 3-(4-fluorophenyl)-4,5,8,9-tetrahydro-1,3′-dimethylspiro ([1]benzopyrano[6,5-*c*]pyrazole[7,5′]pyrimidine)-2′,4′, 6′-trione, synthesized by Winthrop, was found to bind to rat thymus glucocorticoid receptors, induce glycogen deposition, and thymolysis at oral doses of 5 mg/kg. Anti-inflammatory activity was observed against cotton pellet-induced granuloma formation in the rats (topically applied at 5 ng/kg). It was speculated that the activity was due to a conformational similarity to glucocorticoids (Schane *et al.*, 1985; Bianco *et al.*, 1989).

2.1.5 Clinical Studies

These pluripotent anti-inflammatory agents have proven to be particularly efficacious in immunologically driven inflammatory diseases and diseases with immune features. The use of glucocorticoids in the clinic has recently been reviewed (Kirby, 1989) and what follows is an outline of newer approaches and/or compounds which may emerge.

Rheumatoid arthritis is responsive to systemic and intra-articular glucocorticoids (Strandhoy, 1986). Unfortunately, it is clear that the side-effects are sufficiently overwhelming to make long-term systemic therapy inappropriate. These side-effects include suppression of the adrenal–pituitary axis, immunosuppression (recurrent infection), osteoporosis and inhibition of wound healing (Axelrod, 1989). As mentioned above, deflazacort (Lantadin, Flantadin) is a new systemic glucocorticoid which is marketed in various countries and is under development in several others. It appears to have less effect on bone resorption and provides long-term immunosuppressive activity (Scudeletti *et al.*, 1990). It will be of interest to see if this compound can deliver an improved risk-to-benefit ratio.

While controversial, intra-articular injection of long-lasting glucocorticoids (e.g. triamcinolone) into an affected joint has been highly effective when used infrequently. In this regard, THS-201 (Halpredone acetate) has been reported to be particularly effective intra-articularly in antigen-induced arthritis in mice (Kiniwa *et al.*, 1986).

Glucocorticoids have also been a mainstay in the treatment of psoriasis. Again, the use is not without consequences, with prolonged treatment leading to atrophy of the skin (Arnold *et al.*, 1990). As mentioned above, one of several recently synthesized betamethasone analogues unexpectedly did not cause atrophy of the skin while maintaining its anti-inflammatory activity (Avery *et al.*, 1990). This compound may prove very useful in the treatment of chronic inflammatory diseases of the skin.

A third area in which glucocorticoids have been widely used is in the treatment of asthma. In fact, oral glucocorticoids have been utilized since the 1950s (Aviado and Carrillo, 1970). However, their usage was associated with not unexpected significant side-effects. A major development in this area was the introduction of inhaled glucocorticoids, such as beclomethasone, triamcinolone and flunisolide, in the 1970s which maintained impressive clincal efficacy while possessing greatly diminished side-effects (Morris, 1978; Check and Kaliner, 1990). There continues to be a progressive increase in the use of aerosol glucocorticoids in asthma, in both adults and children, such that they are generally considered as a front-line therapy. In contrast, utilization of oral glucocorticoids has decreased substantially during this period. Glucocorticoids are efficacious via actions on multiple tissues and cells, suppressing many of the processes responsible for the pathophysiological manifestations of the disease (Kaliner, 1985). The concept of the use of a metabolically labile glucocorticoid was used by Jouveinal with the advent of tixocortol, a thiosteroid which is extremely labile. This drug has been shown to possess excellent local anti-inflammatory activity with minimal systemic action (Larochelle *et al.*, 1983).

2.1.6 Summary
Despite the clear efficacy of glucocorticoids, the problem of significant untoward side-effects has remained to this day. The challenge of developing less toxic compounds

is still being taken up and significant progress has been made. The growing understanding of the molecular events occurring upon exposure to glucocorticoids raises the potential for the design of molecules to take advantage of or to mimic specific second-messenger proteins. In addition, possibilities persist with regard to the synthesis of glucocorticoid analogues. With the advent of non-steroids which bind to glucocorticoid receptors comes the possibility of new, more specific, structure–activity relationships.

2.2   LIPOCORTINS

2.2.1   Introduction
As discussed above, the actual mechanism responsible for the impressive effects produced by glucocorticoids appears to involve the induction of second-messenger proteins which have been shown to possess anti-inflammatory activity; however, this remains a controversial area of research. These molecules, therefore, may be legitmate targets in their own right for drug development. Some work has been done to try to exploit this strategy.

2.2.2   Chemical/Pharmacological Classes
Lipocortin 1 has been sequenced and cloned and appears to be related to other membrane-associated proteins with Ca^{2+}- and lipid-binding properties. The EGF receptor protein-tyrosine kinase (p35), calpactin II and lipocortin 1 are all closely related proteins, while the retroviral protein-tyrosine kinase (p36) is closely related or identical to calpactin I and lipocortin 2 (Saris *et al.*, 1986; Flower, 1988). Recently, uteroglobin oligopeptides (antiflammins), which share homology with lipocortin 1, have also been found to possess PLA_2 inhibitory activity (Miele *et al.*, 1988).

2.2.3   Mechanism of Action
These are known to bind to lipids and Ca^{2+}, although the question as to whether the proteins act by directly inhibiting PLA_2 or by interfering with the substrate remains unresolved. The results of studies with recombinant lipocortin 1 (Flower, 1988), and also the antiflammins (Miele *et al.*, 1988), suggests potent inhibition of enzymatic activity.

2.2.4   Preclincal Studies
Recombinant lipocortin 1 has clearly been shown to possess many of the activities of glucocorticoids (*vide supra*). Inhibition of PLA_2 activity in normal and psoriatic skin was achieved by picomolar concentrations (Cartwright *et al.*, 1989). In addition, it was found to inhibit paw oedema induced by carrageenan and *Naja mocambique* venom PLA_2, but not oedema induced by bradykinin, dextran, serotonin or PAF (Cirino *et al.*, 1989). Antipyretic activity was also demonstrated in poly I:C-fevered rabbits (Davidson *et al.*, 1991). These activities, coupled with the fact that autoantibodies to lipocortin 1 have been

found in chronic inflammatory diseases (rheumatoid arthritis and systemic lupus erythematosus (Hirata *et al.*, 1981), suggest a role for lipocortin and related proteins as endogenous regulators of arachidonic acid release.

Of considerable interest is the finding that nonapeptides related to lipocortin 1 possess potent PLA_2 inhibitory and anti-inflammatory activity as reflected by inhibition of carrageenan-induced oedema (Miele *et al.*, 1988).

2.2.5   Clinical Studies

There are no reports of clinical studies with lipocortins.

2.2.6   Summary

While it is obvious that substantial work is necessary to clarify the role of endogenous PLA_2-inhibitory proteins, they may serve as the basis for design of potent alternatives to corticosteroids. The exact pharmacological profile that might be expected is at present unknown, but the risk-to-benefit ratio might be considerably improved.

2.3   PHOSPHOLIPASE A_2 INHIBITORS

2.3.1   Introduction

The PLA_2s, or phosphatidyl-glycerol 2-acylhydrolases (EC 3.1.1.4), form a diverse family of cell-associated and extracellular acylhydrolytic enzymes that catalyse the hydrolysis of the fatty acyl ester bond of membrane phosphoglycerides at the *sn*-2 position, resulting in the generation of free fatty acids, predominantly arachidonic acid, and lysophospholipids. These formed products are the precursors for the pro-inflammatory mediators, the eicosanoids (prostanoids, leukotrienes, lipoxins, hydroxy fatty acids) and PAF, respectively (Verheij *et al.*, 1981; Dennis, 1983; Vadas and Pruzanski, 1986; Chang *et al.*, 1987a).

There exist distinct classes of PLA_2s which were categorized initially as type (group) I and type (group) II (Heinrikson *et al.*, 1977; Johnson *et al.*, 1990) and which differ in their location, structure and pharmacological properties. Type I, which is a secreted, extracellular form, is exemplified by pancreatic lipases and also various snake venoms, but is also present in limited quantities in other tissues including spleen, kidney, lung and gastric mucosa (Dufton and Hider, 1983; Dufton *et al.*, 1983; Seilhamer *et al.*, 1986; Matsuda *et al.*, 1987; Tojo *et al.*, 1988a,b). In contrast, type II PLA_2s, which exist in membrane-bound (granular) and extracellular (secretory) forms, are more widespread and found in a variety of tissues and cell types including platelet, liver, kidney, placenta, spleen, intestine, in addition to synovial cells and fluid (Verger *et al.*, 1982; Forst *et al.*, 1986; Chang *et al.*, 1987b; Hayakawa *et al.*, 1987; Lai and Wada, 1988; Ono *et al.*, 1988; Aarsman *et al.*, 1989; Kramer *et al.*, 1989, 1990; Seilhamer *et al.*, 1989a,b).

The isolation, purification, cloning and sequencing of the human genes and cDNAs, and the mapping of the chromosomal location encoding the type I (pancreatic) and type II (non-pancreatic platelet and synovial fluid) forms of PLA_2 has been achieved (Hara *et al.*, 1988; Seilhamer *et al.*, 1988, 1989b; Kramer *et al.*, 1989, 1990; Johnson *et al.*, 1990; Wery *et al.*, 1991). The sequences of the secreted and membrane-bound forms of the mammalian type II PLA_2 enzyme from several species, including humans, are generally identical or very similar (see Johnson *et al.*, 1990).

Recent research indicates that there are at least two different forms of the mammalian non-pancreatic PLA_2, a high and a low molecular weight form. Thus, an alkaline-active, calcium-dependent (millimolar) form of PLA_2 containing 124 amino acid residues with a relatively low molecular mass ($\approx$ 14 kDa) has been described which is present in platelets and rheumatoid synovial fluid (Dennis, 1983; Seilhamer *et al.*, 1986, 1989a; Kramer *et al.*, 1989, 1990; Parks *et al.*, 1990). This soluble PLA_2, found extracellularly, does not demonstrate a preference for fatty acids in the *sn*-2 position of phospholipid substrates (Schalkwijk *et al.*, 1990) and it is inhibited by sulphydryl reducing agents such as dithiothreitol (DTT) (Vadas *et al.*, 1985).

The purification of a higher molecular-weight cytosolic PLA_2 was first described in sheep platelets (Loeb and Gross, 1986) and confirmed in many systems, including a human monocytic leukaemic cell (U937 cell) (Clark *et al.*, 1990 (110 kDa); Diez and Mong, 1990 (56 kDa); Kramer *et al.*, 1991 (100 kDa)) and a macrophage cell line (RAW 2647) (Leslie *et al.*, 1988 (60 kDa)). A more recent report by Clark and coworkers described the cloning and expression of a cDNA encoding a high molecular-weight (85 kDa) cytosolic PLA_2 whose amino acid sequence has no homology with the type I or low molecular-weight type II forms of PLA_2 and which is not inhibited by DTT; it contains a portion which is similar to the Ca^{2+}-binding domain of various protein kinase C enzymes (Clark *et al.*, 1990, 1991). Studies with this cytosolic high molecular-weight form of the type II PLA_2 indicates that, unlike the low molecular-weight form, it has selectivity for arachidonic acid at the *sn*-2 position of substrate phospholipids and, therefore, may be involved in the regulation of the release of arachidonic acid within the cell.

It is evident that several distinct forms of PLA_2 have already been identified and the likelihood is that this number will increase in the future. In relation to drug development and synthesis of specific PLA_2 inhibitors, a critical question is "Which of these different PLA_2s is (are) the most pathophysiologically relevant form(s) in the many diseases in which PLA_2 has been implicated?" Most of the information available to date would suggest that it is the low molecular-weight soluble form of PLA_2 (Johnson *et al.*, 1990; Vadas and Pruzanski, 1990); there is less information on the potential role of the 85 kDa enzyme.

Notwithstanding uncertainties regarding the type(s)

and sources of PLA_2s, there is growing evidence that these enzymes are involved in the pathogenesis of a variety of inflammatory disorders of the pulmonary, cardiovascular and gastrointestinal systems in addition to connective tissue and skin disorders, brain dysfunction and septic shock. For example, high levels of the low molecular-weight PLA_2 are detected in inflammatory exudates, including synovial fluid (Vadas and Pruzanski, 1984; Pruzanski *et al.*, 1985, 1989) and peritoneal fluids from arthritic patients and those with peritonitis (Vadas *et al.*, 1989b), and also in the plasma of patients with acute sepsis (Vadas *et al.*, 1988a). In addition, injection of purified synovial fluid and snake venom PLA_2 into knee joints of animals produces inflammatory and proliferative changes (Murakami *et al.*, 1990; Vadas *et al.*, 1989a) and administration of purified PLA_2 produces disease-like symptoms in animals (Vadas and Hay, 1983; Vadas and Pruzanski, 1986). The increased local and systemic levels of PLA_2 and its reactive products observed during infections, inflammatory diseases, tissue injury and brain dysfunction correlates with the severity and duration of disease-like symptoms (Vadas and Pruzanski, 1986; Chang *et al.*, 1987b; Gattaz *et al.*, 1987; Pruzanski *et al.*, 1988; Vadas *et al.*, 1988a,b). Recently, it was reported that injection of the recombinant human synovial fluid low molecular-weight PLA_2 into rabbit joints elicited a dramatic inflammatory and arthritogenic response (Bomalaski *et al.*, 1991c). It has been proposed that cytosolic PLA_2 activity may have relevance in affective disorders, including depression (Hibbeln *et al.*, 1989).

Accordingly, the development of PLA_2 inhibitors appears to be a logical and appropriate choice as a potential therapeutic strategy for a variety of common and serious disorders which are inflammatory in nature. The pursuit of type I or type II PLA_2 inhibitors has been conducted for approximately the past 20 years. However, either snake venom-derived or pancreatic lipases were utilized in drug design and compounds emanating from these efforts have generally not proven to be efficacious in animal inflammatory models. Thus, unfortunately, research on the regulation of human PLA_2 is in its relative infancy, and to date there are no data on the clinical effects of selective PLA_2 inhibitors. Outlined below is a summary of the preclinical effects of different classes of putative PLA_2 inhibitors, which have provided valuable information on the potential role of PLA_2 in disease states, and has fuelled the excitement regarding the potential therapeutic utility of this particular class of compounds.

2.3.2 Chemical/Pharmacological Classes

To date all reported efforts have focused on developing inhibitors of the low molecular-weight (14 kDa) PLA_2. There is a dearth of compounds that can be considered as *bona fide* potent and, in particular, selective PLA_2 inhibitors. This is in part reflective of the special difficulty in studying PLA_2 enzymes and putative inhibitors *in vitro*.

For example, the enzymes do not confirm to classical Michaelis–Menten kinetics with their activity modulated by a variety of cofactors, as well as being dependent on the substrate type and the concentration of the substrate and inhibitor in the lipid–water interface; for optimal activity of the enzyme the phospholipid substrate must be an aggregated form, of which there are several. Several adaptive kinetic models to study enzyme activity have been developed, although which is the most predictive and relevant one to evaluate the actions of compounds is unclear (Deems *et al.*, 1975; Jain and Berg, 1989). A critical issue is the source and purity of the PLA_2. The importance of this selection is highlighted dramatically in a study comparing the activity of manoalide as a PLA_2 inhibitor in which it was observed that there was an approximately 3 log unit difference in the IC_{50}, depending on the source of enzyme (Mayer *et al.*, 1988; Marshall and Chang, 1990). At present a significant hurdle in this research area is the lack of sources of significant quantities of mammalian PLA_2. The above and other issues and potential problems associated with the strategies for the *in vitro* and *in vivo* studies using putative PLA_2 inhibitors have been discussed in more depth elsewhere (Dennis, 1987; Mobilio and Marshall, 1989; Marshall and Chang, 1990).

Notwithstanding the difficulty in the identification and classification of PLA_2 inhibitors, a variety of compounds have been proposed to be PLA_2 inhibitors. The compounds have diverse purported mechanisms of action and come from an array of different chemical classes, strategies and sources. The different types of chemical classes include: (1) substrate or product analogues; (2) natural products and analogues; (3) covalent binding agents; (4) NSAIDs; (5) compounds that modulate Ca^{2+} levels; and (6) agents that affect substrate–enzyme interfacial binding or disrupt aggregated substrate. The main strategies employed are: (1) compounds derived from screening and (2) computer-assisted design primarily based on the inhibitor enzyme X ray structure (Chang *et al.*, 1987a; Mobilio and Marshall, 1989). Examples of most of these classes are given in Fig. 5.3. It should be emphasized that the ability of some of these agents, e.g. compounds in classes 4 and 5, to inhibit PLA_2 occurs only in much higher concentrations than are required for their recognized mechanism of action.

2.3.2.1 Substrate or Product Analogues

A number of phospholipid analogues possess activity as PLA_2 inhibitors (Chang *et al.*, 1987a; Mobilio and Marshall, 1989). For example, a group of fluorinated phospholipid analogues have been shown to be inhibitors of snake venom PLA_2, although they have limited potencies (the most effective had IC_{50}s of 70–1600 μM) (Yuan *et al.*, 1987). This approach was extended to synthesize phosphonate–phospholipid analogues that were designed to mimic the transition state of the phospholipid during catalysis. Some of these compounds

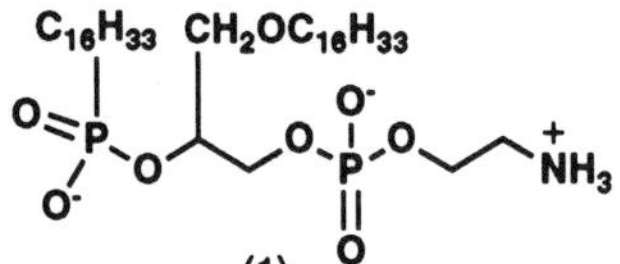

(1)

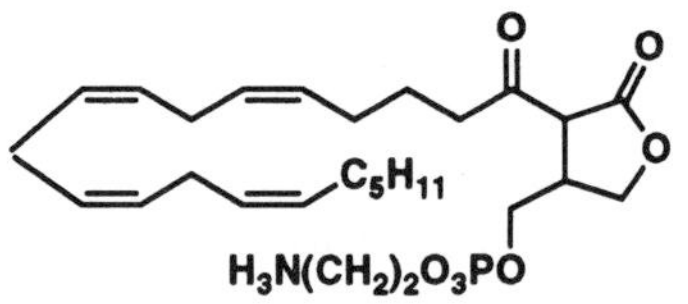

(2)

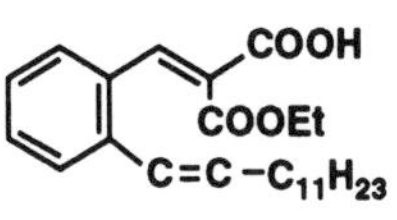

(3)

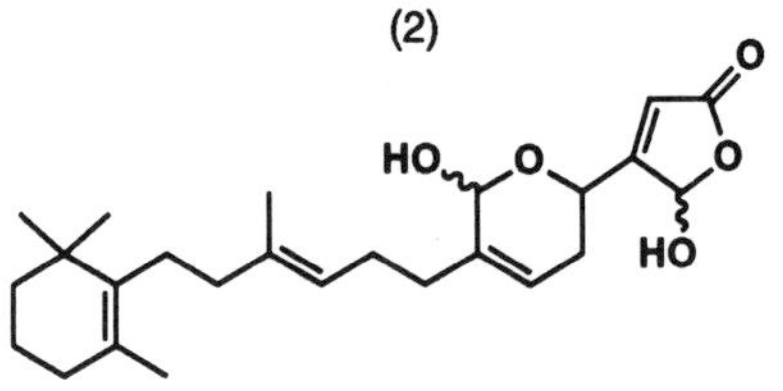

Manaolide

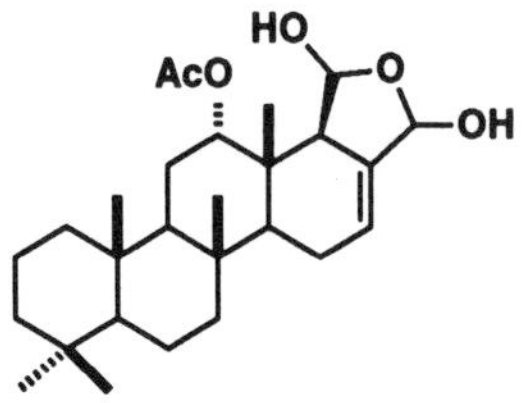

Scalaradial

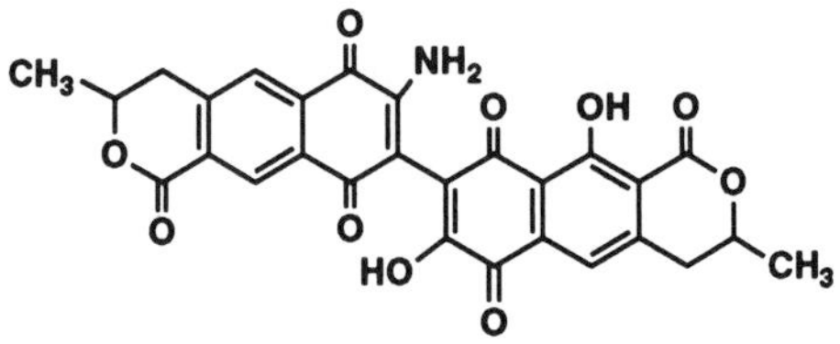

Plastatin

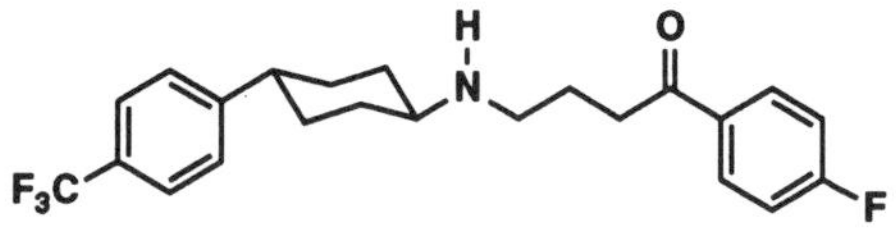

U-3538

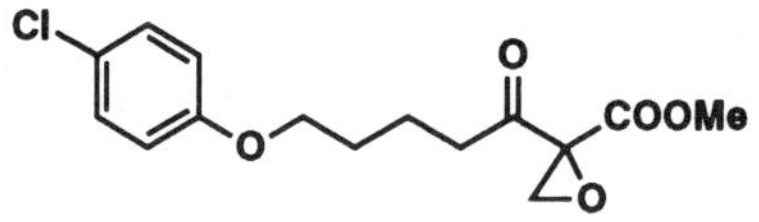

WY-49,422

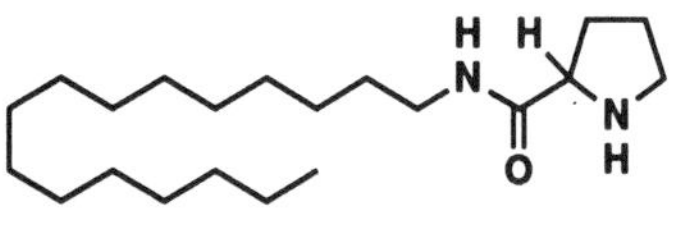

WY-48,489

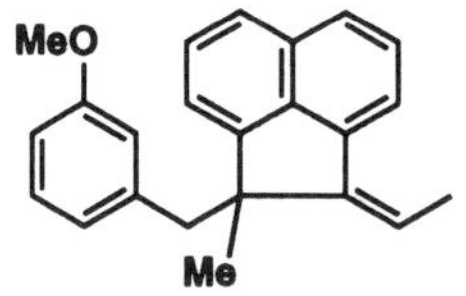

(4)

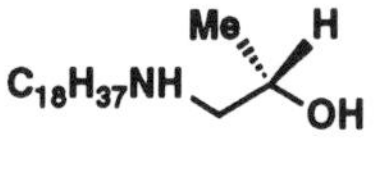

(5)

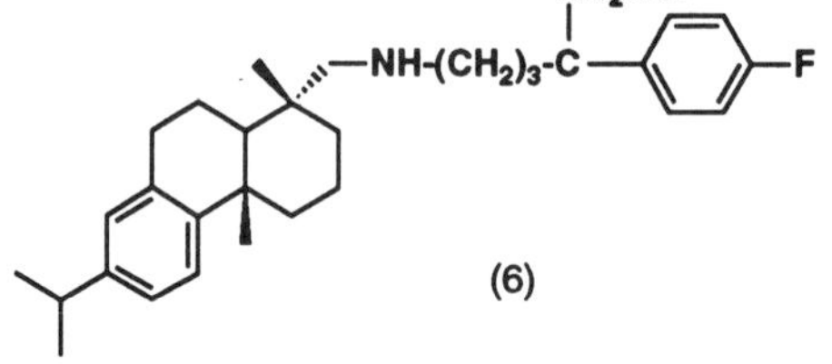

(6)

Figure 5.3 Names and chemical structures of purported PLA_2 inhibitors.

were significantly more potent than the fluorinated ketones; for example, compound (1) indicated in Fig. 5.3 had an IC_{50} of 5 μM (Yuan and Gelb, 1988). What were termed constrained phospholipid analogues also inhibited PLA_2 (Barlow *et al.*, 1988a,b; Gialih *et al.*, 1988; Lister and Hancock, 1988), with compound (2) (Fig. 5.3) possessing an IC_{50} of approximately 50 μM versus rat neutrophil or macrophage PLA_2 (Campbell *et al.*, 1988). Probably the most well-known product fatty acid analogue is the arachidonic acid analogue 5,8,11,14-ETYA, which inhibits rabbit peritoneal neutrophil PLA_2 with an IC_{50} of about 12–22 μM (Lanni and Becker, 1985).

Several other arachidonic acid analogues inhibit PLA$_2$, with compound (3) in Fig. 5.3 the most potent, possessing an IC$_{50}$ of 0.7 μM against *Naja naja* venom PLA$_2$ (Foster *et al.*, 1987).

Not surprisingly in view of their structures, these compounds generally lack selectivity and affect other enzymes involved in arachidonic acid metabolism. In addition, they lack potency and may have the potential for toxicity as a result of membrane-aggregating properties.

2.3.2.2 Natural Products

Manoalide, a sesterterpenoid isolated from the sponge *Luffariella variabilis*, irreversibly inhibits bee venom and snake venom PLA$_2$ by a covalent interaction with lysine residues; it reduces the hydrolysis of phosphatidylcholine by PLA$_2$ and enhances the hydrolysis of phosphatidyl-ethanolamine (Jacobs *et al.*, 1985; Lombardo and Dennis, 1985; Douglas *et al.*, 1986; Glaser and Jacobs, 1986).

Manoalide has recently been shown to be a potent inhibitor of PLA$_2$ from a variety of sources, with IC$_{50}$s in the range 0.008–10 μM (Marshall and Chang, 1990) and has been shown to possess anti-inflammatory activity in *in vivo* animal models (Jacobs *et al.*, 1985). A close relative of manoalide, luffariellolide, is a potent inhibitor of bee venom PLA$_2$ (IC$_{50}$ = 0.23 μM) via a mechanism which is partially reversible (Vainio *et al.*, 1985).

Scalaradial, a marine natural product isolated from the sponge *Cacospongia mollior* was determined to be a potent, irreversible inhibitor of bee venom PLA$_2$ (IC$_{50}$ = 0.07 μM), perhaps via an interaction close to the substrate binding site (De Carvalho and Jacobs, 1991).

The alkaloid aristololic acid inhibited *Vipera russelli* venom PLA$_2$ via a non-competitive action (IC$_{50}$ = 50 μM), and inhibited PLA$_2$ from another source (*Trimere-sunirus flavoviridis*) with a similar potency (IC$_{50}$ = 25 μM), but by a competitive mechanism (Glaser and Jacobs, 1986; Deems *et al.*, 1987).

Various other natural products have been proposed to inhibit PLA$_2$, including several retinoids (Hope *et al.*, 1988; Mihelich *et al.*, 1988), α-tocophenol, glycyrrhizin, polymyxin B (Okimasu *et al.*, 1983; Matsumoto and Saito, 1984; Douglas *et al.*, 1986), and the fungal metabolites plastatin (Fig. 5.3) and plipstatin (Fitzsimmons *et al.*, 1985; Singh *et al.*, 1985; Nishikiori *et al.*, 1986). Several of these substances are antioxidants, but it is not known if their ability to inhibit PLA$_2$ is related to this property.

There is obviously an array of diverse structures and chemical classes derived from natural products which are purported to inhibit PLA$_2$. Appropriate chemical modifications of these lead structures may lead to the synthesis of compounds of enchanced potency and with the necessary selectivity to be considered as useful experimental tools for preclinical research and perhaps clinical candidates.

2.3.2.3 Compounds Derived from Screening

About a decade ago a series of butyrophenone derivatives, including U-3538 (Fig. 5.3) were proposed to be PLA$_2$ inhibitors (Wallach and Brown, 1981). More recently, information was presented on the activity of representatives of two chemically distinct series of transition state inhibitors, the glycidic esters (WY-49,422) and alkyla-mines (WY-48,489) designed by Wyeth-Ayerst (Marshall and Chang, 1990) (Fig. 5.3). WY-49,422 and WY-48,489 inhibited PLA$_2$ activity from a variety of sources with potencies generally greater than other substrate analogues, such as the fluorinated phospholipid transition state or fatty acid analogues described above, but less than manoalide. WY-48,489 possessed selectivity for mammalian PLA$_2$s (Marshall and Chang, 1990).

2.3.2.4 Computer-Assisted Compound Design Based on Inhibitor Enzyme X Ray Structure

This approach has the potential to result in the design of both more potent and more selective PLA$_2$ inhibitors.

One of the most commonly used PLA$_2$ inhibitors is *p*-bromophenacyl bromide (BPB). BPB irreversibly inhibits bovine pancreatic and cobra venom PLA$_2$ (Roberts *et al.*, 1977) and the 2.5Å resolution, three-dimensional crystal structure of the inhibited enzyme was reported by Renetsder and coworkers (1988). BPB specifically and covalently binds to the catalytic His48 residue at the active site, which results in decreased enzymatic activity (Dennis, 1987; Renetsder *et al.*, 1988). Although BPB is a non-specific alkylating agent and obviously not a viable therapeutic candidate, the rational design of specific inhibitors of PLA$_2$ based on the three-dimensional structure of the inhibited enzyme was discussed by Renetsder *et al.* (1988).

Utilization of the structure of the bovine and porcine pancreatic PLA$_2$ led to the identification of two chemically distinct classes of inhibitors. Researchers from DuPont designed a series of conformationally rigid 1-*m*-hydroxybenzyl-2-substituted acenaphthenes directed against the active site of PLA$_2$; these were relatively potent possessing IC$_{50}$s of 0.2–1.4 μM; a representative example, compound (4), is given in Fig. 5.3 (Ripka *et al.*, 1987). A group of less potent long-chain alkylamine inhibitors (IC$_{50}$s = 10–30 μM) were identified, e.g. compound (5) in Fig. 5.3 (Davis *et al.*, 1988).

Using information obtained from examination of the X ray crystal structure of porcine pancreatic PLA$_2$, in addition to the reported actvity of butyrophenone derivatives such as U-3538 as PLA$_2$ inhibitors, there was a recent publication on a novel series of dehydroabietyla-mine derivatives which were shown to be active site inhibitors of porcine pancreatic and rat PMN low molecular-weight PLA$_2$; the most potent compounds, of which compound (6) in Fig. 5.3 is a representative, had IC$_{50}$s of 1.8–25 μM, respectively (Wilkerson *et al.*, 1991).

The hope is that recent elucidation of the three-dimensional crystal structure of the recombinant human rheumatoid arthritis synovial fluid PLA_2 (Scott *et al.*, 1991; Wery *et al.*, 1991) will assist greatly in the efforts to synthesize selective inhibitors of human PLA_2s.

2.3.3 Mechanism of Action

Not unexpectedly in view of the diversity of the compounds, and as indicated above, the numerous purported PLA_2 inhibitors have various proposed profiles and mechanisms of action. In many cases the precise molecular mechanism of action has not been elucidated.

2.3.4 Preclinical Studies

There is only limited information on the effects of PLA_2 inhibitors *in vitro* in cellular assays and, in particular, in *in vivo* models of inflammation; the latter is partly due to the limited potency and poor bioavailability of presently available compounds. Manoalide ($IC_{50} \approx 0.25$ μM) inhibited the release of LTC_4, PGE_2 or arachidonic acid from cultured mouse peritoneal macrophages induced by several, but not all, stimulants (Marshall, 1988; Mayer *et al.*, 1988), and attenuated the zymosan-induced release of LTC_4 and 6-keto-$PGF_{1\alpha}$ *in vivo* in mice when given via the intraperitoneal route ($IC_{50} \approx 0.2$ mg/kg) (Mayer *et al.*, 1988). In addition, topical application of manoalide inhibited phorbol ester-induced, but not arachidonic acid-induced, ear inflammation in the mouse (Jacobs *et al.*, 1985). The transition state inhibitor WY-48,489 inhibited the agonist-induced release of arachidonic acid and also its major products, PGE_2 and LTC_4, from mouse macrophages with IC_{50}s of about 4–20 μM (Marshall and Chang, 1990). This compound, and also WY-49,422, also exhibited some efficacy for inhibition of carrageenan-induced paw oedema in the rat and phorbol ester-induced ear oedema in mouse when given by oral and/or topical routes of administration. Several retinoids were reported to inhibit the release of arachidonic acid induced by various stimuli (Hope *et al.*, 1988; Mihelich *et al.*, 1988). Some of the recently identified novel series of dehydroabietylamine derivatives possessed inhibitory activity in standard *in vivo* rat and mouse anti-inflammatory assays and influenced intact cell arachidonic acid metabolism (Wilkerson *et al.*, 1991).

Therefore, different chemical classes of PLA_2 inhibitors have been shown to modify cellular arachidonic acid metabolism and eicosanoid synthesis, supporting a role for this enzyme in pro-inflammatory lipid mediator formation. Furthermore, in some cases this activity has been translated into efficacy in animal disease models.

2.3.5 Clinical Studies

No clinical studies have been reported with any PLA_2 inhibitors.

2.3.6 Summary

The PLA_2s are a diverse family of regulatory enzymes which play a pivotal role in a plethora of specialized cellular functions, most notably the regulation of the bio-synthesis of numerous lipid pro-inflammatory mediators. From a theoretical perspective, it is this latter attribute, in addition to the growing evidence that PLA_2, either directly or indirectly, appears to be involved in various inflammatory processes, which provides the strong rationale that potent and selective inhibitors of these enzymes may be powerful anti-inflammatory agents. No clinical information is yet available to substantiate this hypothesis and thus far the putative PLA_2 inhibitors examined preclinically have been limited by their lack of both selectivity and potency. To date, most of the PLA_2 inhibitors have come from natural product screening or are phospholipid analogues. However, more recently, computer-generated, molecular-modelling efforts have proven successful in designing novel, selective active site inhibitors.

The task of identifying and developing potent and selective PLA_2 inhibitors will not be facile as this area of research has many special issues which have to be addressed. Probably the most critical one is the elucidation of the relevant PLA_2(s) involved in a particular disease. This knowledge is central to the selection of the appropriate strategies and relevant *in vitro* and *in vivo* assays to test compounds of interest. The elucidation of the three-dimensional crystal structure of the various PLA_2s should assist greatly in the design of compounds which exhibit selectivity for inhibition of a particular form of the enzyme. Another major hurdle is the availability of sufficient quantities of the purified forms of the relevant PLA_2s; work is in progress towards this endeavour using modern recombinant technology.

Another issue is what will be the effects of long-term exposure to PLA_2 inhibitors other than on the regulation of the formation of pro-inflammatory mediators, notably potential effects on lipid metabolism and lipid content. In addition, will PLA_2 inhibitors possess similar gastro-intestinal liabilities to those that are associated with cyclo-oxygenase inhibitors? Furthermore, one has to consider the caveat that there are sources and mechanisms for the liberation of arachidonic acid other than hydrolysis by PLA_2 at the *sn*-2 position, for example the phosphatidylinositol pathway.

The above questions and issues can only be addressed satisfactorily by the examination of the effects of potent and select PLA_2 inhibitors in appropriate animal models and, eventually, in the clinic. Notwithstanding the many issues, the research area of PLA_2 inhibitors is an exciting, albeit problematic, one in which significant, recent advances have been made and which has the exciting potential of the development of a novel and powerful class of anti-inflammatory drugs.

3. *Cyclooxygenase Inhibitors*

3.1 Introduction

As detailed elsewhere, the metabolism of arachidonic acid takes place through the action of a variety of oxygenases,

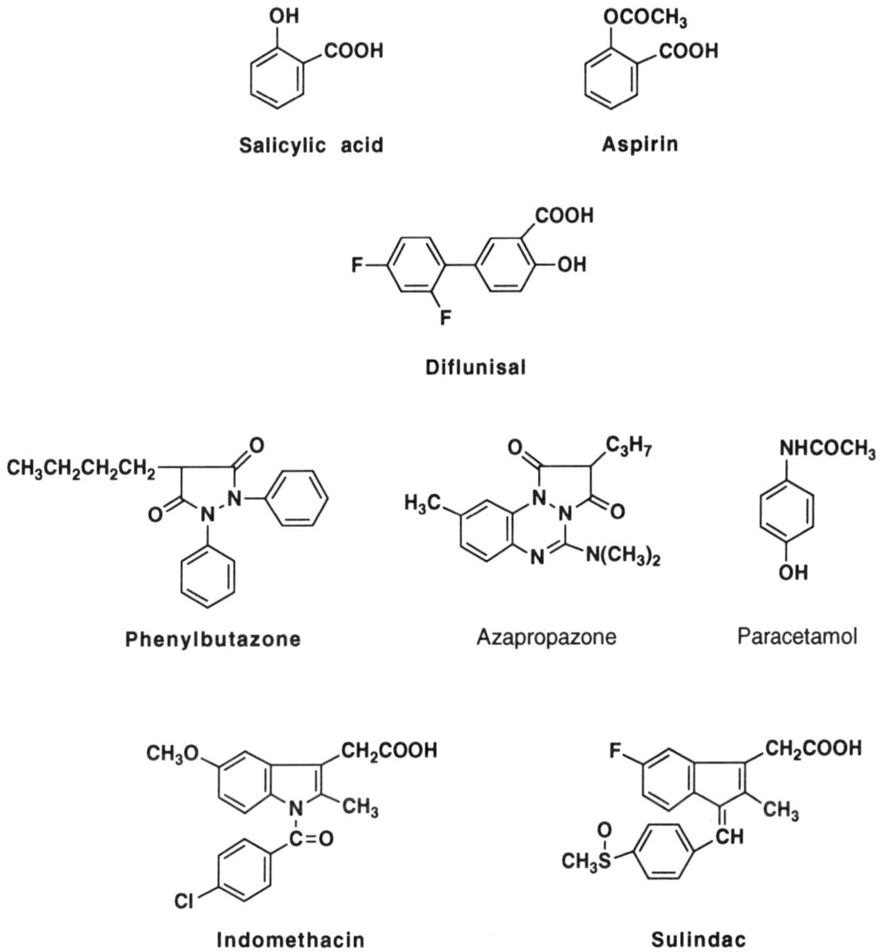

Figure 5.4 Names and chemical structures of representative traditional cyclooxygenase inhibitors.

including primarily cyclooxygenase (PGH synthase), 5-lipoxygenase, 12-lipoxygenase and 15-lipoxygenase. Selective inhibition of the cyclooxygenase-mediated metabolism of arachidonic acid, and thus inhibition of the production of prostanoids (PGH_2, prostacyclin, PGD_2, PGE_2, $PGF_{2\alpha}$ and TXA_2), appears to be the pharmacological mechanism of a still-growing class of agents known more commonly as NSAIDs. Technically, any anti-inflammatory agent that is significantly dissimilar structurally from the steroid nucleus would be an NSAID; however, the terminology has been applied to selective cyclooxygenase inhibitors for such a period of time that they have become synonymous.

The first such agent was aspirin, although its introduction in the 1870s predated any understanding of its mechanism of action. With the work of Vane and colleagues (Ferreira *et al.*, 1971; Vane, 1971) and Smith and Willis (Smith and Willis, 1971) showing that aspirin, sodium salicylate and indomethacin were inhibitors of prostaglandin biosynthesis (although sodium salicylate was only marginally active) came a concerted effort to design and discover newer compounds of improved potency. This has been a burgeoning effort and currently at least 16 compounds are widely used clinically. However, the well-known gastrointestinal and renal side-effects have dampened enthusiasm for additional NSAIDs without an improved risk-to-benefit ratio.

3.2 CHEMICAL/PHARMACOLOGICAL CLASSES

3.2.1 Traditional Cyclooxygenase Inhibitors (Fig. 5.4)

3.2.1.1 Salicylates

The simple *o*-hydroxybenzoic acid (salicylic acid) itself is anti-inflammatory, although it is too toxic for systemic use. Of the salicylates, the oldest member of this class – aspirin – is the most commonly used. In fact aspirin remains the most widely used anti-inflammatory, analgesic and antipyretic agent, with a world-wide yearly consumption estimated in the thousands of tonnes

(Flower *et al.*, 1985). Aspirin is used in a wide range of doses for different indications, from 300 mg/day for antiaggregatory action to 6 g/day, which is close to the dose which will have significant side-effects, for anti-arthritic activity (Flower *et al.*, 1985). Difluorophenyl salicylic acid (diflunisal) has also been utilized but appears to be essentially the same as aspirin in its actions (Rainsford, 1985).

3.2.1.2   *Phenylbutazone*

This compound and its metabolite oxyphenbutazone were among the first compounds introduced as NSAIDs (phenylbutazone was introduced in 1949). They are pyrazolons, and are substituted at N-1 and N-2 by phenyl groups. Oxyphenbutazone is *p*-hydroxylated at the N-1 phenyl group of phenylbutazone. These compounds are considered to be essentially equivalent both in pharmacology and toxicology. Azapropazone (apazone) is also a pyrazolon derivative, to which a benzotriazine has been fused to the pyrazolon ring. This derivative is claimed to have an equivalent anti-inflammatory activity to phenylbutazone but with less toxicity. Neither phenylbutazone nor its derivatives are currently used widely in human medicine (Flower *et al.*, 1985).

3.2.1.3   p-*Aminophenols*

This class includes phenacetin and paracetamol (acetaminophen). Both compounds are primarily analgesic/antipyretics with weaker anti-inflammatory activity. They were introduced in the late 1800s and paracetamol continues to be used. Paracetamol has been shown to inhibit PGI_2 production with concomitant inhibition of TXA_2 production (Drvota *et al.*, 1991). While it has no effect on platelet aggregation, the analgesic activity is equal to that of aspirin (Flower *et al.*, 1985). PGI_2 has been shown to be hyperalgesic in several systems (Doherty, 1987). If it spares pro-inflammatory prostanoids, this may explain the analgesic/antipyretic activity and the absence of anti-inflammatory activity of this class of compounds.

3.2.1.4   *Indomethacin and Sulindac*

Discovered as an anti-inflammatory agent, indomethacin is a methylated indole acetic acid derivative with potent cyclooxygenase inhibitory activity. The compound was introduced in 1963 and remains clinically useful today. It is a potent compound used at doses of 25 mg every 4–8 h. Sulindac, a related compound, is a methylsulphinylbenzylidene indene acetic acid derivative which is a prodrug for the sulphide active metabolite and is notable for its possible renal-sparing activity (Dunn, 1984). Sulindac has been shown to be effective in rheumatoid arthritis and osteoarthritis and is used at doses of 150–200 mg twice daily (Shen, 1985).

3.2.2   Newer Cyclooxygenase Inhibitors
(Fig 5.5)

3.2.2.1   *Fenamic Acids*

The members of this class are *N*-substituted phenylanthranilic acids. It includes meclofenamic acid, which is the most widely utilized member of the class and has *in vitro* activity against cyclooxygenase, and also 5-lipoxygenase (Boctor *et al.*, 1986). This activity has not, however, been apparent *in vivo* (D.E. Griswold, unpublished observations). The compound is effective in rheumatoid arthritis and osteoarthritis in doses of 200–300 mg/day. It is unlikely, therefore, that the weak 5-lipoxygenase activity ($IC_{50} = 48$ μM) could be achieved by these doses.

3.2.2.2   *Arylacetic Acids*

Diclofenac and fenclofenac are the major members of this class. Diclofenac is a potent, reversible inhibitor of cyclooxygenase and has been shown to be effective in rheumatoid arthritis in a dosage of 75–150 mg daily (Brooks *et al.*, 1986). It has been suggested to be an inhibitor of arachidonic acid release and/or a stimulator of re-acylation, and thus inhibition of 5-lipoxygenase products has been observed (Scholer *et al.*, 1986). This profile may explain the improved therapeutic index seen with diclofenac (Sengupta *et al.*, 1985).

3.2.2.3   *Arylpropionic acids*

While there is some confusion over the nomenclature in this area, the convention has been to categorize the α-methyl-substituted phenylacetic acids (e.g. ibuprofen, naproxen and fenoprofen) as arylpropionic acid derivatives. Ibuprofen (isobutylphenyl, α-methyl acetic acid) was described in 1969. At low doses (1200 mg/day) significant analgesia is apparent but only marginal anti-inflammatory activity was noted. Higher doses gave anti-inflammatory activity as well as gastrointestinal irritation (Roth, 1980). Naproxen is a methoxynapthalene derivative which was introduced in 1970. It is effective in rheumatoid arthritis and osteoarthritis and is given at doses of 400–750 mg/day. The side-effects observed are similar to those of other NSAIDs. Fenoprofen is the phenoxy derivative and was introduced in 1971. It is effective in the treatment of rheumatoid arthritis and osteoarthritis in doses of 1800–2400 mg daily (Strandhoy, 1986).

3.2.2.4   *Indole and Pyrrolealkanoic acids*

Tolmetin, introduced in 1976, was modelled after indomethacin and is a benzoylmethylpyrrole derivative. It has been shown to have an anti-inflammatory activity equivalent to aspirin in rheumatoid arthritis and osteoarthritis. It is used at doses of 1200–1500 mg daily (Brooks *et al.*, 1986).

Etodolac is an indole acetic acid whose structure is also reminiscent of indomethacin. Although the results are controversial, a 1 year trial versus aspirin indicated a

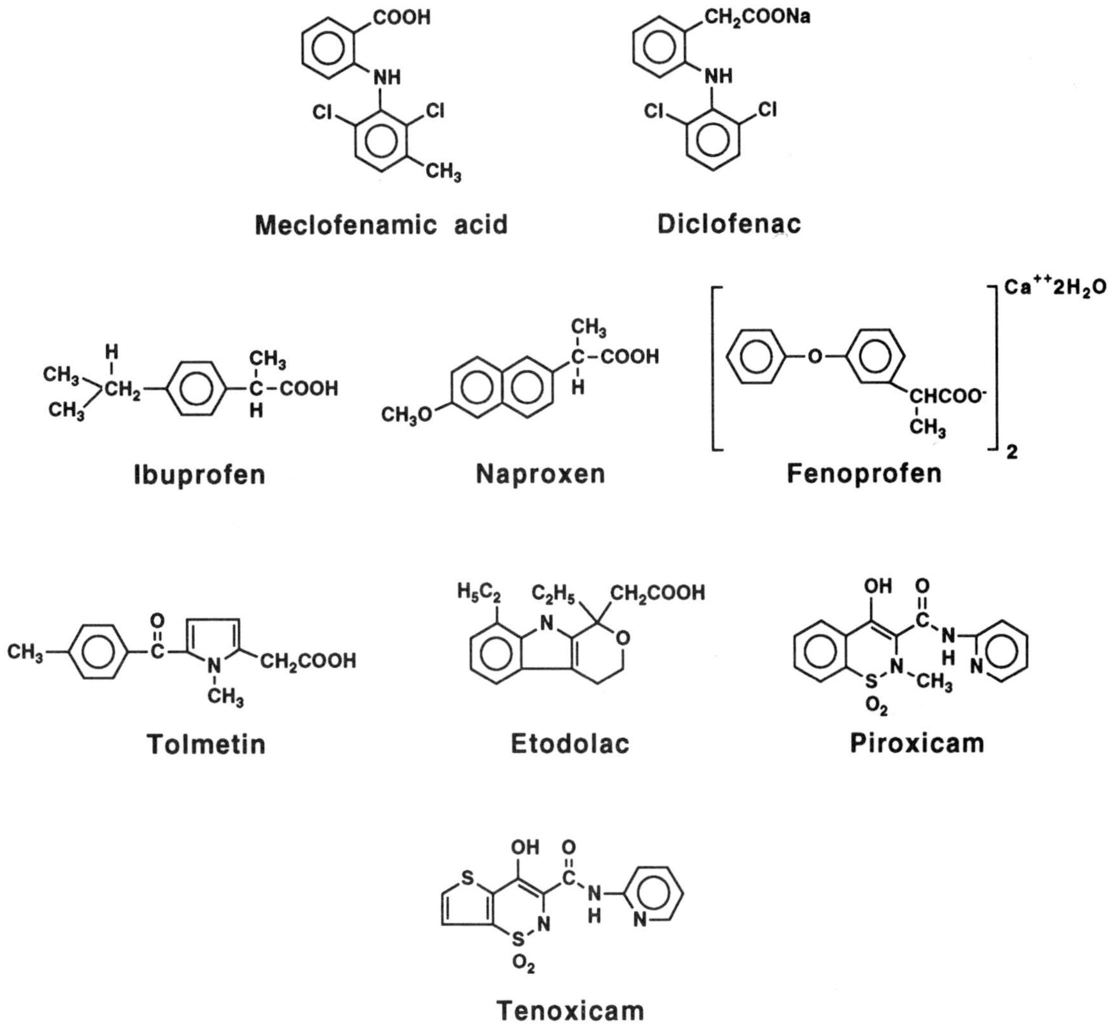

Figure 5.5 Names and chemical structures of representative newer cyclooxygenase inhibitors.

slowing of radiographic changes in rheumatoid arthritis (Paulus, 1989). The dose used is 200–600 mg/day. Etodolac appears to have less toxicity than indomethacin (Karbowski, 1991).

3.2.2.5   Oxicams

This new class includes piroxicam, isoxicam, sudoxicam and tenoxicam. Piroxicam was introduced in 1979 by Pfizer. It was the result of a design effort to achieve an extended plasma half-life, high potency and acceptable toxicity. Piroxicam has a long half-life (45 h) and is very potent (the usual dose is 20 mg/day). Tenoxicam, a thiophene analogue of piroxicam, was developed by Roche. It also has a long plasma half-life (60 h) and is used in a dosage of 20–40 mg/day. The oxicams appear to have less gastrointestinal irritation relative to indomethacin and aspirin (Brooks *et al.*, 1986; Bianchi *et al.*, 1991).

3.2.3   "Enhanced" Cyclooxygenase Inhibitors (Fig. 5.6)

3.2.3.1   Nabumetone

This new compound is a prodrug for 6-methoxy-2-naphthylacetic acid (6-MNA). The parent is a weak cyclooxygenase inhibitor while 6-MNA, which appears to be the active principle, is considerably more potent. The compound has been shown to possess markedly lower gastrointestinal irritancy coupled with excellent anti-inflammatory activity (Dandona and Jeremy, 1990; Melarange *et al.*, 1992).

3.2.3.2   Other agents

While little has been formally published, Pharmaprojects lists lornoxicam as a new oxicam derivative with extreme potency and it appears to be well tolerated. DUP-697 is a thiophene derivative which is suggested to be devoid

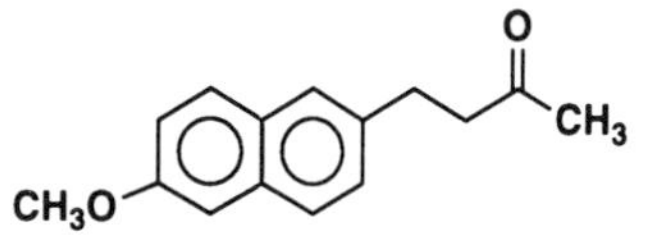

Nabumetone

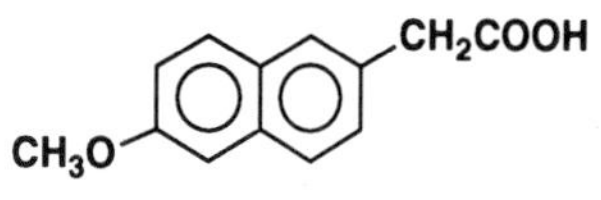

6-Methoxy-2-naphthyl acetic acid

Figure 5.6   Names and chemical structures of "enhanced" cyclooxygenase inhibitors.

of ulcerogenic liability. In addition, other prodrug and delivery vehicle strategies designed to lower the gastrointestinal toxicity of NSAIDs have been described. Incorporation of NSAIDs into β-cyclodextrins (Rainsford, 1990) has also been utilized with piroxicam (Passeri, 1990) as well as with a prodrug strategy (ampiroxicam). The above strategies are all directed towards providing an improved therapeutic index.

3.3   MECHANISM OF ACTION

Despite the intense competition and work in this area, the actual molecular mechanism of action remains vague. Aspirin has been shown to acetylate the enzyme and thus inactivate cyclooxygenase (Roth *et al.*, 1975). Indomethacin has recently been shown to bind to PGH synthase and was suggested to inhibit the enzyme by perturbation of tyrosyl residues involved in catalysis (Kulmacz *et al.*, 1991). Differential inhibition of platelet cyclooxygenase versus cyclooxygenases in other tissues with aspirin is explained by the irreversible nature of the interaction and the inability of the platelet to synthesize the enzyme (Fagan and Goldberg, 1986; Patrono, 1989; Drvota *et al.*, 1991). It is still not clear whether the entire spectrum of activity of salicylates is due exclusively to cyclooxygenase inhibition. It is interesting to note that sodium salicylate appeared to inhibit IL-1-stimulated PGH synthase mRNA synthesis in human umbilical vein endothelial cells in culture (Sanduja *et al.*, 1991). These data could explain the discrepancy between aspirin and salicylate regarding the ability of the former, but not the latter, to inhibit cyclooxygenase activity while having equivalent inhibitory activity on prostaglandin synthesis.

The recent finding of an inducible isoform of PGH synthase (PGH synthase 2) (Fletcher *et al.*, 1992) raises the possibility that this isoform may be strongly associated with the inflammatory response and that selective inhibition of this isoform or the induction of PGH synthase 2 might provide an anti-inflammatory agent without the

well-known side-effects of cyclooxygenase inhibitors, since the constitutive enzyme activity would be spared.

3.4   PRECLINICAL STUDIES

The profile of activity of this pharmacological class of agents is remarkably similar. With some differences in potency, bioavailability and half-life, they all inhibit carrageenan-induced rat paw oedema and adjuvant-induced arthritis paw swelling with $ED_{50}s$ consistent with their potency *in vitro* as inhibitors of cyclooxygenase activity. The spectrum of actions includes antioedematous, antipyretic, antiaggregatory and analgesic activities. Generally, cyclooxygenase inhibitors do not inhibit inflammatory cell infiltration; for those compounds which do, the activity is not correlated with *in vitro* or *in vivo* inhibition of cyclooxygenase products (Higgs *et al.*, 1980; Rainsford, 1985).

3.5   CLINICAL STUDIES

The most common use of NSAIDs is as analgesics. This is mainly accounted for by the over-the-counter use of paracetamol, aspirin and, more recently, ibuprofen. An associated usage is treatment of dysmenorrhoea, for which aspirin, ibuprofen and naproxen have been used, with the last two showing greater potency (Flower *et al.*, 1985).

The major prescription use of NSAIDs is in the treatment of arthritis, including rheumatoid, osteoarthritis and gout. The drug of choice for gout appears to be indomethacin while tolmetin has emerged as the drug of choice for ankylosing spondylitis (Brooks *et al.*, 1985). The preferred drug for osteoarthritis and rheumatoid arthritis is a matter of individual choice. Aspirin is often chosen initially, and is efficacious when used in sufficiently high doses. However, the required dosage is frequently in the toxic range and tinnitus can be used to judge the adequacy of the dose. If the response to aspirin is insufficient, other NSAIDs are recommended, with indomethacin often a second choice followed by naproxen. In comparative studies examining NSAID preference overall none was evident, although individual patients strongly preferred each of the agents tested (Paulus and Furst, 1989).

The overriding stimulus for further work in this area is not the pharmacological activity or potency of the inhibitors but their toxicity. It appears that all cyclooxygenase inhibitors, to one degree or another, cause gastrointestinal irritation often leading to frank ulceration (Day *et al.*, 1987). It has been postulated that PGE_2 plays a protective role in the gastrointestinal tract or, alternatively, that lipoxygenase products, which may predominate upon treatment with selective cyclooxygenase inhibitors, may be damaging to the gastrointestinal tract (Rainsford, 1987). Most of the effort currently expended in this area is geared towards the discovery and development of NSAIDs with greatly reduced gastrointestinal side-effects.

Nabumetone appears to have the above profile. This compound, as mentioned above, is a prodrug for the active metabolite 6-MNA. It appears that nabumetone may be the first member of a new class of "gastrointestinal-sparing NSAIDs". Studies on the production of protective prostanoids (epoprostenol and dinoprostone) by human and rat gastric mucosa revealed that nabumetone, in contrast to indomethacin and naproxen, had no effect; the active metabolite had modest inhibitory activity. It was suggested that this may explain the lack of direct local impact of orally administered nabumetone (Dandona and Jeremy, 1990). It is also possible that the improved side-effect profile for nabumetone stems from the recent observation that, relative to other cyclooxygenase inhibitors, it appears to more strongly inhibit PGH synthase 2 as opposed to PGH synthase 1 (Meade *et al.*, 1993).

3.6 SUMMARY

This class of compounds is clearly of importance therapeutically. They collectively fill an important niche in the therapy of acute and chronic inflammation. They are, by and large, flawed by the small risk-to-benefit ratio, and significant improvement in this area is being pursued. Time and experience will tell whether some of the newer compounds will have an improved therapeutic index.

4. *Selective 5-Lipoxygenase Inhibitors*

4.1 INTRODUCTION

The peptidoleukotrienes and LTB$_4$ have been implicated as being pivotal players in a variety of inflammatory disorders (*vide infra*). Accordingly, for more than a decade there has been a major effort from the pharmaceutical industry directed towards the development of potent and selective inhibitors of the formation of the peptidoleukotrienes and LTB$_4$, i.e. 5-lipoxygenase inhibitors, which would not interfere with the production of mediators from the cyclooxygenase arm of arachidonic acid metabolism. This research has had considerable success and there are several compounds which are about to enter, or are presently in, clinical trials.

Lipoxygenases are highly conserved and ubiquitous enzymes. Perhaps the most widely studied enzyme is soya bean lipoxygenase. Much of the early data and understanding of the mechanism and activity of 5-lipoxygenase came from this work. Mammalian 5-lipoxygenase has been cloned and the molecular mechanism elucidated (Dixon *et al.*, 1988; Rouzer *et al.*, 1988). The formation of LTA$_4$ from arachidonic acid involves a two-step catalysis involving formation of 5-HPETE by the removal of the 7-pro-5 hydrogen. This is concomitant with the transfer of an electron from the iron (III) atom in the active site (Yamamoto, 1989). The presence of an essential iron atom in the human enzyme has been recently demonstrated (Percival, 1991). Given these mechanisms,

several types of inhibitors have been studied, including antioxidant and radical trapping agents (e.g. Esculetin and 2-aminophenol). In addition, substrate analogues have been utilized (e.g. ETYA) (Papatheofanis and Lands, 1985). These experimental tools were useful in understanding the molecular biochemistry of lipoxygenases but were without significant utility as drugs for preclinical or clinical evaluation. Subsequently, several classes and drugs, discussed below, have, by virtue of their selective inhibition of 5-lipoxygenase, exhibited significant anti-inflammatory activity and some possess the appropriate profile for preclinical and clinical investigation.

4.2 CHEMICAL/PHARMACOLOGICAL CLASSES (Fig. 5.7)

4.2.1 Hydroxyureas

The initial interest in hydroxamates as 5-lipoxygenase inhibitors stemmed from a consideration of their iron-chelating capacity, since it was hypothesized that iron was present in human 5-lipoxygenase and that it was critical for 5-lipoxygenase activity. As reviewed recently, this postulate appears to be correct and hydroxamic acid derivatives are effective 5-lipoxygenase inhibitors (Summers, 1990; Garland and Salmon, 1991). Their metabolism and *in vivo* (*ex vivo*) activity have always been important in the characterization of these molecules, as the discovery of potent *in vitro* 5-lipoxygenase inhibitors which had poor *in vivo* activity plagued the early work. Summers and coworkers described the synthesis of "type B" hydroxamic acids which had improved *in vivo* potency (Summers *et al.*, 1988). These efforts effectively gave rise to A64077 (zileuton), currently the most advanced 5-lipoxygenase inhibitor in clinical evaluation; it is presently in phase II clinical trials for multiple indications. Zileuton is a potent 5-lipoxygenase inhibitor (IC$_{50}$ = 0.5 μM) which has been suggested to exhibit selectivity for inhibition of this enzyme (Carter *et al.*, 1989).

4.2.2 BWA4C and Wy-50,295

An additional example, BWA4C, was synthesized by Burroughs Wellcome. This compound was designed to address the issues of metabolism, clearance and short half-life associated with the hydroxyureas. While BWA4C is a potent *in vitro* inhibitor of 5-lipoxygenase (IC$_{50}$ = 0.05–0.2 μM), it was also found to inhibit 12- and 15-lipoxygenases at higher concentrations (3.3 μM). *Ex vivo* inhibition of LTB$_4$ production in whole blood was also observed (ED$_{50}$ ≈ 5 mg/kg p.o.) with a duration of action of up to 6 h in rats (Tateson *et al.*, 1988). BWA4C was reported to inhibit leucocyte migration and exhibit antipyretic activity, but not to suppress inflammatory oedema or pain, in several acute inflammatory models in rats and mice (Higgs *et al.*, 1988). This compound has also undergone clinical evaluation.

Another, orally active, 5-lipoxygenase inhibitor is

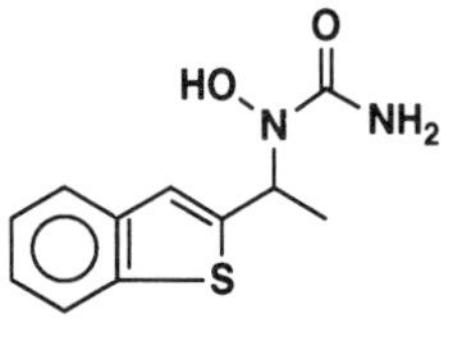

A64077 (Zileuton)

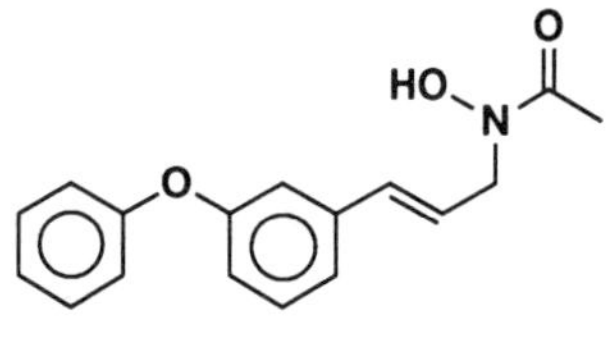

BWA4C

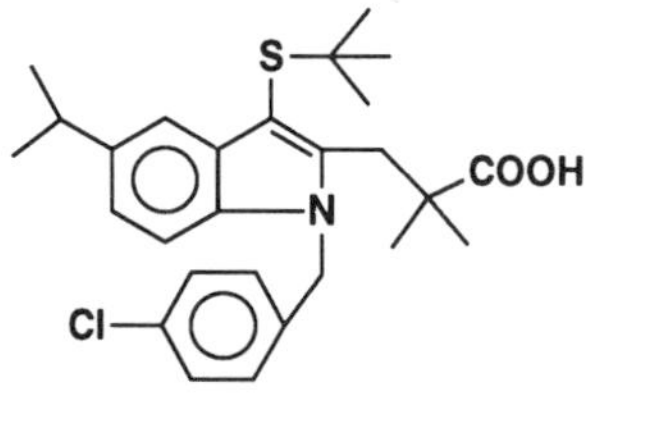

MK-886 (translocation inhibitor)

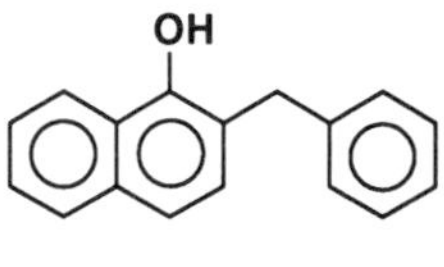

DUP 654

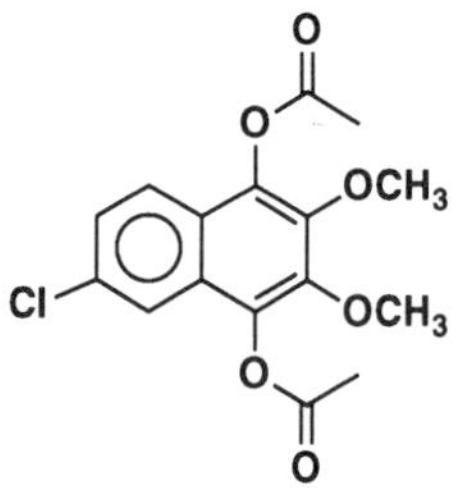

Lonapalene

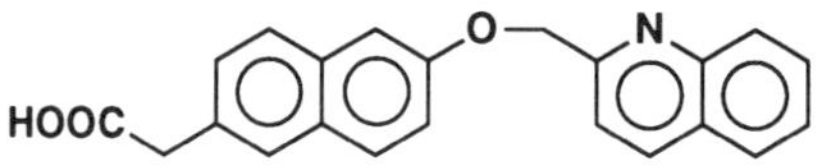

Wy-50,295

Figure 5.7 Names and chemical structures of representative selective 5-lipoxygenase inhibitors.

Wy-50,295, which is structurally related to arachidonic acid and which displays activity as an LTD_4 receptor antagonist. The compound potently inhibits 5-lipoxygenase activity (IC_{50} of 8 μM) and shows an impressive range of anti-inflammatory activity in both mouse ($ED_{50} \approx$ 13–56 mg/kg p.o.) and rat models ($ED_{35} \approx$ 75 mg/kg p.o.). This compound, however, lacks activity in whole human blood (> 200 μM), due ostensibly to a high binding affinity towards human plasma proteins (Carlson *et al.*, 1991). The development of this compound is not being pursued.

4.2.3   Translocation Inhibitor (MK-886)

An additional strategy for selective inhibition of 5-lipoxygenase product formation was provided by Merck in the form of MK-886, which was found to inhibit leukotriene production in the whole cell but had little or no activity on the isolated enzyme (Gillard *et al.*, 1989). Subsequent work revealed that MK-886 was inhibiting the translocation and docking of 5-lipoxygenase in the membrane. It was found that a FLAP was the molecular target for MK-886 (Evans *et al.*, 1991).

4.2.4   Other Compounds of Interest

DUP 654 is a phenolic structure which potently inhibits 5-lipoxygenase ($IC_{50} = 0.01$ μM). It appears to do so through an antioxidant mechanism (Summers, 1990). No selectivity data have been published but the topical anti-inflammatory activity reflects its 5-lipoxygenase potency. It potently inhibits arachidonic acid-induced inflammation in mouse skin ($ED_{50} = 11$ μg/ear), contact sensitivity ($ED_{50} = 0.8$ μg/ear) and phorbol ester-induced inflammation ($ED_{50} = 699$ μg/ear) (Ackerman *et al.*, 1989). No clinical data have been published.

REV-5901, an arylmethyl phenyl ether is one of the early compounds proposed as a 5-lipoxygenase inhibitor (Coutts *et al.*, 1985). It inhibited antigen- or A23187-induced release of leukotrienes from human lung tissue *in vitro* (Tennant *et al.*, 1987). REV-5901 also possessed LTD_4 receptor antagonist activity (Coutts *et al.*, 1985). REV-5901 reduced the increase in myocardial infarct size induced by coronary artery occlusion and reperfusion in dogs (Mullane *et al.*, 1987). The compound is no longer in clinical development.

Lonapalene   is   a   quinone   analogue   antioxidant

5-lipoxygenase inhibitor. It was being developed as a topical antipsoriatic agent and has a potent *in vitro* 5-lipoxygenase inhibitory activity (IC_{50} = 0.5 μM). The compound also inhibited arachidonic acid-induced skin inflammation, albeit with less potency than DUP 654 (ED_{50} = 1 mg/ear) (Young *et al.*, 1985).

A recent investigation of the effects of the enantiomers of an optically active methoxyalkylthiazole (ICI 216800) provided the first convincing evidence for a specific, chiral interaction with the 5-lipoxygenase enzyme; enantiomer-specific topical anti-inflammatory activity in rabbit skin was also demonstrated (McMillan *et al.*, 1990). The methoxyalkylthiazoles appear to represent a novel class of potent and selective 5-lipoxygenase inhibitors which act via a direct mechanism other than redox inhibition or iron chelation (competitive inhibition) (McMillan *et al.*, 1990; Bird *et al.*, 1991).

CGS 8515 (methyl 2-[(3,4-dihydro-3,4-dioxo-1-naphthalenyl)amino]benzoate) has been demonstrated to be a potent and selective 5-lipoxygenase inhibitor, attenuating the production of leukotrienes from a variety of cells and systems, including whole human blood (IC_{50} of 0.8 μM) (Ku *et al.*, 1988). Oral administration (2–50 mg/kg) significantly inhibited A23187-induced production of leukotrienes in rat whole blood *ex vivo* for at least 6 h, and attenuated leucocyte migration and exudate volume in carrageenan-induced pleurisy and sponge models in the rat (Ku *et al.*, 1988). It was also active in a rat endotoxic shock model (Matera *et al.*, 1988).

4.3   MECHANISM OF ACTION

There are several proposed mechanisms of action for this class of agents and they are discussed above for individual compounds. The majority of earlier 5-lipoxygenase inhibitors can best be classified as non-selective antioxidants, inhibiting 5-lipoxygenase by virtue of their ability to reduce the ferric enzyme to the inactive ferrous enzyme. Redox inhibitors may also intervene by decreasing the amount of lipid hydroperoxide, thereby preventing reactivation of 5-lipoxygenase. The recent description of the series of methoxyalkylthiazoles appears to be the first convincing example of compounds which may act as competitive inhibitors of 5-lipoxygenase. In addition, interruption of the presumed interaction of 5-lipoxygenase and the FLAP is an additional mechanism of intervention. Suicide inactivation (enzyme-activated irreversible inhibition) is another possible mechanism by which 5-lipoxygenase may be inhibited (Cashman, 1985; Summers, 1990).

4.4   PRECLINICAL STUDIES

The predominant model used to evaluate systemically active, selective 5-lipoxygenase inhibitors is *ex vivo* inhibition of leukotriene production, including in whole human blood. The successful compounds all perform well in this assay and generally have ED_{50}s of less than 10 mg/kg p.o. (*vide supra*). However, it has been more difficult to demonstrate clearly the anti-inflammatory activity of this class of compounds. They appear to be effective in arachidonic acid-induced inflammation in the skin (mouse ear) but only if low concentrations of arachidonic acid are used. All of these agents perform well when administered topically using this model and these data, plus the activity in phorbol ester-induced inflammation in the skin, have supported the development of these agents as therapy for inflammatory diseases of the skin. A variety of compounds have been demonstrated to inhibit antigen-induced bronchospasm in various species, both *in vitro* and *in vivo* (Hand *et al.*, 1986; McFarlane *et al.*, 1987; Piechuta *et al.*, 1987; Johnson and Stout, 1988; Payne *et al.*, 1988; Yamamura *et al.*, 1988). A variety of peritoneal and pleural cavity models have also been used with success (*vide supra*). In addition, several of these compounds have been shown to be effective in IBD models, and these results provided evidence for support of the clinical evaluation of zileuton in this disorder (Summers, 1990).

4.5   CLINICAL STUDIES

Zileuton is the most advanced selective 5-lipoxygenase inhibitor currently under clinical investigation. Zileuton was found to inhibit LTB_4 production by up to 82% at an oral dose of 800 mg in the phase I studies; the half-life was found to be approximately 3 h. No untoward side-effects were noted. In addition, there was no attenuation of *ex vivo* inhibition of LTB_4 production following 10 days of dosing (Rubin *et al.*, 1989).

Results of subsequent evaluation in IBD (ulcerative colitis) revealed that zileuton (800 mg) significantly reduced LTB_4 in rectal dialysates (Staerk Laursen *et al.*, 1990). A phase II, randomized, double-blind, placebo-controlled trial indicated a "trend towards true zileuton drug effect" in ulcerative colitis patients not receiving concomitant sulphasalazine (Roundtree and Calhoun, 1991). These promising preliminary results warrant further investigation. The effect of zileuton (800 mg) against early and late-phase bronchoconstrictor responses to allergen was also examined. The study showed a correlation between urinary LTE_4 levels and a trend towards inhibition of the antigen-induced fall in FEV_1, although it did not reach statistical significance. It was speculated that more complete inhibition of 5-lipoxygenase product formation may be needed (Hui *et al.*, 1991). Zileuton (800 mg) inhibited cold, dry air-induced bronchoconstriction in asthmatics, concomitant with substantial inhibition of A23187-induced production of LTB_4, but not TXB_2, in whole blood *ex vivo*; the compound was without effect on baseline lung function (Israel *et al.*, 1990).

The ability of BWA4C, the hydroxyurea 5-lipoxygenase inhibitor, to inhibit PAF-induced bronchospasm was recently examined. Although BWA4C (400 mg p.o.) significantly inhibited PAF-induced neutropenia, it was without effect on the changes in pulmonary function (sGaw and $p\dot{V}_{max\ 30}$) produced by PAF. *Ex vivo* LTB_4 production in whole blood was inhibited by 70–90%. It was speculated that pulmonary 5-lipoxygenase activity was not adequately inhibited because of an insufficient concentration of BWA4C in the lung (Spencer *et al.*, 1991b). The lack of effect against PAF-induced broncho-spasm contrasts with the inhibitory activity observed with SK&F 104353, the potent and selective peptidoleuko-triene receptor antagonist (Spencer *et al.*, 1991a). Oral AA-861 (1100 mg over 4 days) did not affect bronchial hyper-responsiveness to acetylcholine in 10 asthmatic patients (Fujimura *et al.*, 1986).

The other 5-lipoxygenase inhibitor which has been evaluated clinically is lonapalene (RS43179). In an excellent study in psoriatic patients, RS43179 (2% ointment) was shown to cause a 50% improvement in clinical signs (relative to the placebo vehicle). This was associated with a selective decrease in the levels of LTB_4-like chemotactic material in lesioned skin. No change in 12-HETE was observed (Black *et al.*, 1990). The data were taken as evidence for the role of LTB_4 in the development of a psoriatic lesion. Use of MK-886 (115 mg three times a day) orally in psoriasis gave different results in that, despite marked reductions of urinary LTE_4 levels, no improvement of the lesions or LTB_4-like material was observed. The results were suggested to be due to either a lack of sufficient drug in the skin or to the LTB_4-like material not being a 5-lipoxygenase product (Dejong *et al.*, 1991).

4.6   SUMMARY

While there is intense activity in this area in terms of both preclinical and clinical research, a clear picture of the utility of this class of compounds is not yet available. As more clinical experience is gained with zileuton, and with the anticipation of other compounds with improved potency and pharmacodynamics entering clinical develop-ment, these studies should assist greatly in the clarification of the involvement of leukotrienes in the disease states in which they have been implicated. The strongest rationale appears to be for therapeutic utility in IBD and asthma, but the participation of leukotrienes in multiple, reinforc-ing feedback loops involving cytokines, adhesion mole-cule expression and inflammatory activation makes other disease states, e.g. rheumatoid arthritis, attractive targets. This class of compounds has an attractive and poten-tially powerful pharmacological profile, which for some diseases may be an advantage over selective peptidoleuko-triene receptor antagonists, in that they will inhibit the release of not only the peptidoleukotrienes but also the pro-inflammatory LTB_4.

5.   *Dual Inhibitors of 5-Lipoxygenase and Cyclooxygenase*

5.1   INTRODUCTION

Interest in the ability of a compound to inhibit the generation of prostanoids and leukotrienes was stimulated by the realization that corticosteroids, by virtue of inhibition of the release of arachidonic acid, would achieve such an end. This, coupled with the growing dissatisfaction with the gastrointestinal toxicity of selec-tive cyclooxygenase inhibitors, provided great impetus to research in this area. The effects of early compounds in this class gave significant promise that control of eicosanoid biosynthesis would provide potent anti-inflammatory activity, and control both the fluid and cellular phases of the inflammatory response without sacrificing analgesic activity.

5.2   CHEMICAL/PHARMACOLOGICAL CLASSES   (Fig. 5.8)

5.2.1   Early Compounds

Benoxaprofen (2-(4-chlorophenyl)-a-methyl-5-benzoxa-zone acetic acid) was perhaps the best early example of a compound with *in vitro* lipoxygenase and cyclooxygenase inhibitory activity. Benoxaprofen inhibits lipoxygenase activity in rabbit ($IC_{50} = 25$ µg/ml) and human PMN (8–36 µM) (Walker and Dawson, 1979; Salmon *et al.*, 1985). In contrast, the compound only weakly inhibited bovine seminal vesicle PGH synthetase activity ($IC_{50} = 150$ µM) (Dawson, 1980). In addition, RBL-1 produc-tion of immunoreactive slow reacting substance of anaphylaxis and PGE_2 was also inhibited (0.4 and 1.3 µM, respectively) (Levine, 1983). Evidence of inhibition of lipoxygenase activity was not universal since mouse peritoneal macrophage production of LTC_4 was not inhibited while PGE_2 production was markedly reduced ($IC_{50} \approx 10$ µM) (Humes *et al.*, 1983). Inhibition of 5-lipoxygenase was also difficult to demonstrate *in vivo* (Salmon *et al.*, 1985), although benoxaprofen strongly inhibited monocyte and neutrophil functions (Anderson and Jooné, 1984) and inhibited inflammatory cell infiltra-tion (Dawson, 1980). Additionally, the monocyte–endothelial cell interaction was also inhibited (Brown *et al.*, 1984). Overall, the consensus is that benoxaprofen inhibits lipoxygenase and cyclooxygenase activity but it is not particularly potent. In addition, the compound appears capable of inhibiting the motility of inflammatory cells by an unknown mechanism.

A second early compound was BW-755C (3-amino-1-[*m*-(trifluoromethyl)phenyl]-2-pyrazoline). BW-755C is a widely studied compound that is structurally related to the photographic developer phenidone. Both BW-755C and benoxaprofen are antioxidants and appear to be general peroxidase-reducing cofactors. They inhibited a wide variety of oxidative enzymes, among them

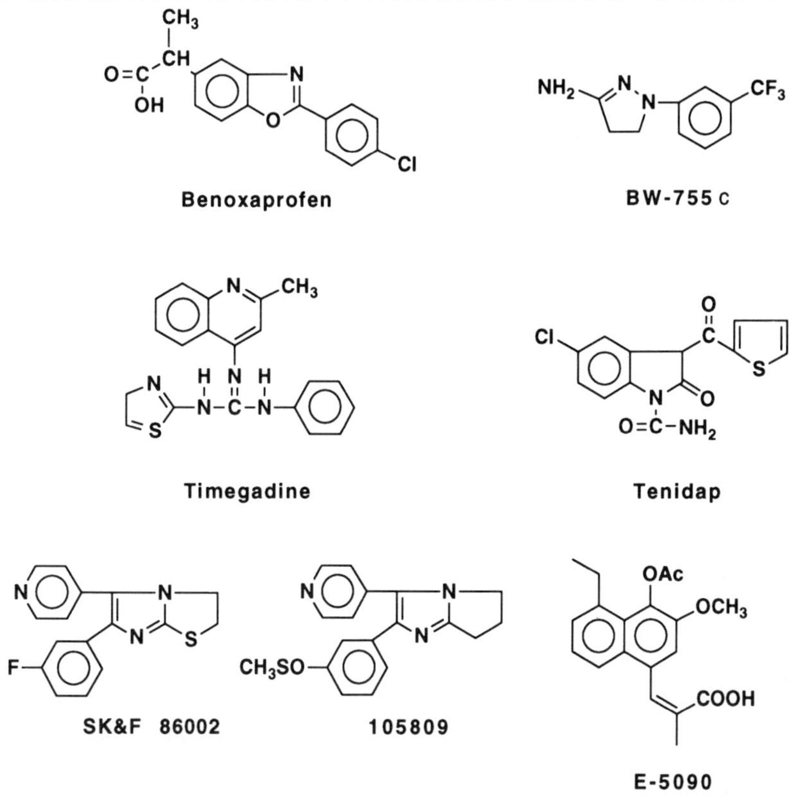

Benoxaprofen

BW-755 c

Timegadine

Tenidap

SK&F 86002

105809

E-5090

Figure 5.8 Names and chemical structures of representative dual cyclooxygenase and 5-lipoxygenase inhibitors.

5-lipoxygenase and cyclooxygenase (Marnett *et al.*, 1982). BW-755C is also effective in reducing the production of leukotrienes (IC_{50} = 9.2 μM) and prostaglandins (IC_{50} = 3.4 μM) from stimulated cells (Blackham *et al.*, 1985). This compound is notable for its effects on inflammatory cell infiltration and its general anti-inflammatory properties (Higgs and Mugridge, 1983; Salmon *et al.*, 1983). Both phenidone and BW-755c can induce methaemoglobinaemia and therefore have not been pursued clinically (Summers, 1990). However, these compounds have been extremely valuable experimental tools in discerning the potential therapeutic utility of dual inhibitors of 5-lipoxygenase and cyclooxygenase.

Timegadine (*N*-cyclohexyl-*N'*-4-(2-methylquinolyl)-*N'*-2-thiazolylguanidine) is an additional example of a dual inhibitor. This guanidine derivative has been demonstrated to inhibit 5-lipoxygenase activity (IC_{50} = 40 μM) and cyclooxygenase activity (IC_{50} = 11 μM). It also inhibited the production of LTB_4 from leucocytes (IC_{50} = 20 μM) and thromboxane from platelets (IC_{50} = 30 μM), and inhibited the release of arachidonic acid (IC_{50} = 27 μM). In addition, it has been shown to inhibit 12-HETE production (IC_{50} = 49 μM). This unique profile of activity suggests multiple molecular mechanisms (Ahnfelt-Rønne and Arrigoni-Martelli, 1982). As perhaps expected, timegadine also exhibits anti-inflammatory activity in a wide variety of model systems, e.g. carrageenan-induced inflammation and adjuvant-induced arthritis

(Bramm *et al.*, 1981). In addition, *in vitro* inhibition of granulocyte function was observed, including inhibition of lysosomal enzyme release and superoxide formation (Laghi Pasini *et al.*, 1984).

5.2.2 More Recent Compounds

Tenidap (CP-66,248, (*Z*)-5-chloro-2,3-dihydro-3-(hydroxy-2-thienylmethylene)-2-oxo-1*H*-indole-1-carboxamide) is perhaps a new member of a new class of NSAIDs which could more precisely be called CSAIDs. It is unclear as to exactly how tenidap provides anti-inflammatory activity. It was initially described as a dual inhibitor of 5-lipoxygenase and cyclooxygenase and has been demonstrated to inhibit the production of eicosanoids from A23187-stimulated human polymorphonuclear leucocytes (LTB_4, IC_{50} = 18 μM; PGE_2, IC_{50} = 32 μM) (Moilanen *et al.*, 1988). The compound was subsequently found to inhibit IL-1 production by mouse macrophages *in vitro* (IC_{50} = 8 μM) (Otterness *et al.*, 1991). Other activities include inhibition of the release of neutrophil collagenase (Blackburn *et al.*, 1991b). This promising agent is currently in phase III clinical trials for rheumatoid arthritis and osteoarthritis.

Bicyclic imidazoles such as SK&F 86002 and SK&F 105809 have been shown to be powerful anti-inflammatory agents. These compounds are inhibitors of 5-lipoxygenase (IC_{50} = 10 and 3 μM, respectively) and cyclooxygenase

(IC_{50} = 20 and 3 μM, respectively) (Griswold *et al.*, 1987; Marshall *et al.*, 1991). Perhaps more important was the profound inhibition of IL-1 and TNF production in human monocytes (IC_{50} = 1 and 2 μM, respectively) (Lee *et al.*, 1988; Marshall *et al.*, 1991). Not unexpectedly they are powerful anti-inflammatory analgesic and anti-arthritic agents in a wide spectrum of model systems (Griswold *et al.*, 1991). Notably these compounds are excellent in controlling the progression of collagen-induced arthritis in a mouse model system which is relatively insensitive to selective 5-lipoxygenase or selective cyclooxygenase inhibitors (Griswold *et al.*, 1988).

5.3 MECHANISM OF ACTION

The molecular mechanism of action of this class of compounds is unknown.

5.4 PRECLINICAL STUDIES

These compounds are characterized by their activity in a wide variety of inflammatory models (*vide supra*). The effects in prostanoid-dependent models, e.g. carrageenan-induced oedema and pleurisy and adjuvant-induced arthritis, appears to reflect the activity of the compounds as cyclooxygenase inhibitors. SK&F 105809, however, had only modest activity in adjuvant-induced arthritis and carrageenan-induced paw oedema (D. E. Griswold, unpublished observations). The impact of 5-lipoxygenase inhibition can be seen with these compounds using arachidonic acid-induced inflammation (mouse ear or peritoneal challenge). The influx of inflammatory cells is often the most sensitive end-point. The most outstanding feature of this class of compounds is the ability to inhibit both fluid and cellular phases of the inflammatory response. Another feature of many of these compounds is the inhibition of contact sensitivity to oxazolone or dinitrofluorobenzene, which implicates eicosanoids as being important in mounting a cell-mediated immune response where the end result is an inflammatory response.

5.5 CLINICAL STUDIES

Benoxaprofen has exhibited promise in the treatment of psoriasis, in that preliminary results indicated that six of 13 patients showed marked improvement. These results were taken to underscore the involvement of eicosanoids in the disease process (Allen and Littlewood, 1983). Other clinical reports suggested efficacy with this compound in rheumatoid arthritis (Huskisson and Scott, 1979) and perhaps ulcerative colitis (Hawkey and Rampton, 1983). These interesting but preliminary experiences set the tone and provided much of the early enthusiasm in the search for new enhanced NSAIDs and CSAIDs.

Probably the most well-characterized of the newer compounds is timegadine. A 24 week, double-blind controlled trial using 500 mg/day provided evidence that timegadine was superior to naproxen in controlling disease activity in rheumatoid arthritis, and the authors reported disease-modifying properties (Egsmose *et al.*, 1988). Further studies are required to clarify this latter property.

Tenidap is also of particular interest in that it appears to alter prostanoid and leukotriene levels in synovial fluid from patients with rheumatoid arthritis as well as reducing lysosomal enzyme release from neutrophils *ex vivo*. These results were correlated with clinical improvement (Blackburn *et al.*, 1991a). It will be important to determine whether cytokine production can also be reliably controlled.

5.6 SUMMARY

This class contains some of the most powerful anti-inflammatory agents examined to date, apart from corticosteroids. Preliminary information from clinical trials is exciting and these compounds offer the hope of profoundly altering the course of chronic inflammatory diseases such as rheumatoid arthritis. This may be particularly true for those compounds with the ability to inhibit cytokine production. Future clinical trials with diverse compounds should clarify the value of combined inhibition of eicosanoids and cytokines.

6. *Thromboxane Synthetase Inhibitors*
6.1 INTRODUCTION

In view of the considerable evidence implicating TXA_2 as an important mediator in the pathophysiology of several disorders, primarily of the cardiovascular and pulmonary systems (see Chapter 3 and *vide infra*), major efforts have been directed towards the development of agents which will inhibit its production (thromboxane synthetase inhibitors) or actions (receptor antagonists). The former strategy was the focus of the initial work in this area and the rationale was that, unlike cyclooxygenase inhibitors which will decrease the formation of all cyclooxygenase products, inhibitors of thromboxane synthetase will selectively attenuate the synthesis of TXA_2 but not other products of the pathway (Smith, 1989; Gresele *et al.*, 1991). This selective pharmacological manipulation may indirectly lead to an increase in the release of the beneficial products of the cyclooxygenase (Fitzgerald *et al.*, 1985; Smith, 1989; Patrono, 1990; Gresele *et al.*, 1991). In particular, prostaglandin endoperoxide substrate derived from platelets may be accumulated by closely associated endothelium, where it is converted by PGI_2 synthetase to PGI_2 (Fitzgerald *et al.*, 1985; Gresele *et al.*, 1991), which has potent anti-aggregatory and vasodilator properties (Dusting *et al.*, 1982; Fitzgerald *et al.*, 1985; Gresele *et al.*, 1991). This endoperoxide redirection or steal phenomenon, for which there is some, but not unequivocal, evidence (Defreyn *et al.*, 1982; Fitzgerald *et al.*, 1985; Carey and Haworth, 1986; Terashita *et al.*, 1986; Smith, 1987; Patrono, 1989; Vesterqvist *et al.*, 1991), has been speculated to be a significant homeostatic mechanism for controlling thrombus formation in damaged blood vessels (Patrono, 1989).

6.2   CHEMICAL/PHARMACOLOGICAL CLASSES

6.2.1   Low-Dose Aspirin

Over the past decade or so there has been considerable interest in the comparison of the biochemical and clinical effects of different dosing regimens of aspirin, particularly the low-dose regimen, to minimize gastrointestinal side-effects, to attempt to dissociate the inhibitory actions on TXA_2 formation from those on the production of the beneficial prostanoids (Antiplatelet Trialists' Collaboration, 1988; Patrono, 1989; Gresele *et al.*, 1991). The stimulus for these studies was probably the well-recognized ability of aspirin to inhibit platelet function at much lower doses (about 10- to 50-fold) than those necessary to produce its other actions (Patrono, 1989). In healthy volunteers daily dosing with 0.45 mg/kg of aspirin for 7 days produced a cumulative and essentially complete inhibition of platelet TXB_2 production without affecting renal synthesis of PGI_2, PGE_2 or $PGF_{2\alpha}$ (Patrignani *et al.*, 1982). In a study in healthy males examining the effects of various doses of aspirin, it was observed that PGI_2 formation was less sensitive to inhibition than thromboxane production, although inhibitory effects were apparent with most doses studied (Fitzgerald *et al.*, 1983a). It was concluded that selective and maximal inhibition of thromboxane synthesis without affecting PGI_2 production was unlikely with any dose of aspirin. When studying the effects of different dosing regimens of aspirin, and also selective thromboxane synthetase inhibitors, an important consideration is the evidence from the study by Reilly and Fitzgerald (1987) suggesting that over 95% inhibition of thromboxane generation is necessary to produce significant antiplatelet activity *in vivo*.

Administration of aspirin in a single oral dose of 500 mg inhibited thromboxane production for 2–3 days, with complete recovery of normal synthetic rates requiring 8–10 days. In contrast, the inhibition of PGI_2 synthesis was short lived (3–4 h) (Vesterqvist and Green, 1984; Vesterqvist, 1986). It was suggested that 500 mg aspirin every third day could be an alternative strategy to daily, low-dose aspirin for prolonged effects on thromboxane compared to PGI_2 production (Drovta *et al.*, 1991). Some of the findings regarding the extent and also the duration of the effects of different doses of aspirin on thromboxane and PGI_2 production in different sites may be attributed to the fact that endothelial cells, a major

source of PGI_2, are nucleated and can resynthesize cyclooxygenase, whereas platelets are non-nucleated and cannot manufacture new enzyme (Fagan and Goldberg, 1986; Patrono, 1989).

It is apparent that the data are equivocal, and substantial controversy exists regarding the optimal dosing regimen for aspirin as an antithrombotic agent which maintains the delicate balance between efficacy and side-effects. Accordingly, the focus of research efforts aimed at attenuating the production of TXA_2 has been directed to what can be classified as a more rational approach, namely selective and potent inhibitors of thromboxane synthetase. This class of compounds, by virtue of their more selective mechanism of action, should permit a more facile interpretation of the data from clinical trials and should provide a clearer picture of the true role of thromboxane, in addition to a greater understanding of the extent and significance of the endoperoxide redirection and sparing of the beneficial prostanoids, such as PGI_2.

6.2.2   Non-Aspirin Compounds

Over the past 15 years or so several potent and selective thromboxane synthetase inhibitors have been identified and studied comprehensively, and representatives of these compounds are indicated in Fig. 5.9.

Imidazole was the first compound which exhibited some selectivity for inhibition of thromboxane synthetase, albeit not very potently (Moncada *et al.*, 1977; Needleman *et al.*, 1977). For example, it had an IC_{50} of 22 μg/ml for inhibition of enzymatic conversion of the endoperoxides, PGG_2 and PGH_2 to TXA_2 by platelet microsomes; it inhibited cyclooxygenase only in much higher concentrations (Moncada *et al.*, 1977). Other purported thromboxane synthetase inhibitors, such as benzydamine, N-0164 and U-51506, are non-selective compounds (Moncada *et al.*, 1976; Needleman *et al.*, 1977).

In addition, 9,11-azoprosta-5,13-dienoic acid was reported to be a selective thromboxane synthetase inhibitor (Gorman *et al.*, 1977; Sun, 1977). Soon after these early observations, a series of 1-carboxyalkyl derivatives of imidazoles were reported to be more potent inhibitors of thromboxane synthetase, with the most potent compound, 1-carboxyheptylimidazole, possessing an IC_{50} of 0.14 μM against the enzyme derived from bovine platelets (Yoshimoto *et al.*, 1978). Subsequently, many of the compounds which have been identified as potent inhibitors of thromboxane synthetase are imidazole derivatives, specifically 1-substituted imidazoles; these include dazmagrel (UK-38,485), dazoxiben (UK-37,248) and OKY-046 (Fig. 5.9).

The other class of compounds to which most of the other potent thromboxane synthetase inhibitors identified belong are the substituted pyridines, and includes OKY-1581, CV 4151, furegrelate (U-63557A) and GGS 12970 (Fig. 5.9).

These compounds have undergone extensive preclinical

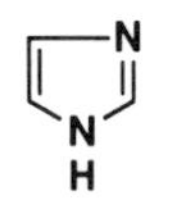

Imidazole

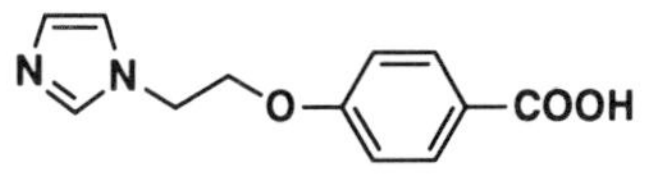

UK-37,248
(Dazoxiben)

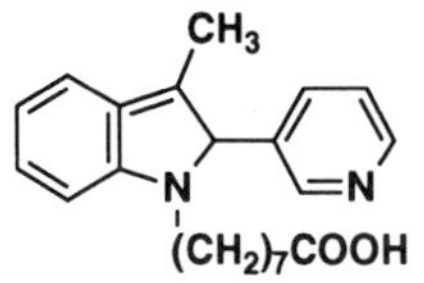

CGS 12970

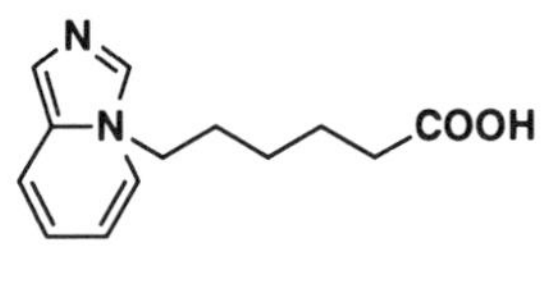

CGS 13080
(Pirmagrel)

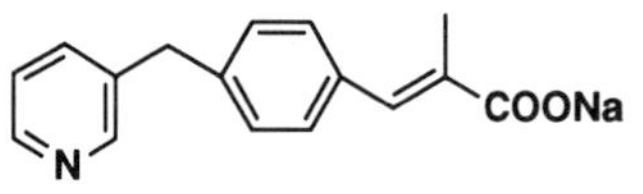

OKY-1581

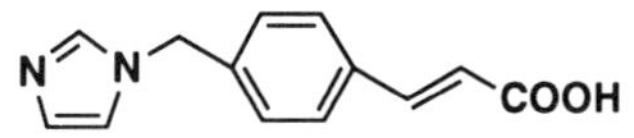

OKY-046

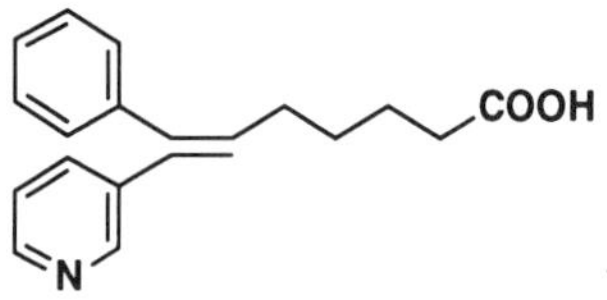

CV 4151

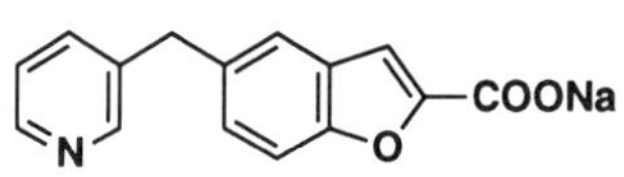

U-63557A
(Furegrelate)

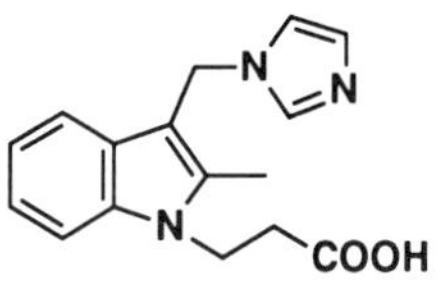

UK-38,485
(Dazmagrel)

Figure 5.9 Names and chemical structures of representative thromboxane synthetase inhibitors.

evaluation, both *in vitro* and *in vivo*, and some have been in several clinical trials. They are much more potent that imidazole, with IC_{50}s for inhibition of thromboxane synthetase in the 3–50 nM range (Randall *et al.*, 1981; Aiken, 1983; Gorman *et al.*, 1983; Ku *et al.*, 1983; Ambler *et al.*, 1985; Terashita *et al.*, 1986).

6.3   MECHANISM OF ACTION

The precise molecular mechanism of action of these agents is unknown.

6.4   PRECLINICAL STUDIES

The effects of thromboxane synthetase inhibitors have been examined in an array of *in vivo* animal models of

diseases of the cardiovascular, renal and pulmonary systems. In this section only some specific pertinent examples will be given as this topic has been reviewed elsewhere (Smith, 1989).

One of the earliest studies exploring the potential therapeutic utility of thromboxane synthetase inhibitors demonstrated that imidazole inhibited arachidonic acid-induced thrombosis in the rabbit (Puig-Parellada and Planas, 1977). Subsequently, dazoxiben (UK-37,248) was reported to inhibit thrombolytic sudden death elicited by arachidonic acid, but not the thromboxane mimetic U46619, in the same species (Lefer *et al.*, 1981; Darius and Lefer, 1985). In rabbits, acute thrombosis induced by local electrical stimulation in the carotid arteries was inhibited by dazoxiben, and also by aspirin (Randall and Wilding, 1982).

Dazoxiben has been observed to improve survival and other pathophysiological sequelae associated with experimental endotoxic shock in rats (Halushka *et al.*, 1983) and OKY-15181 prevented endotoxin-induced pulmonary hypertension in primates (Casey *et al.*, 1982). In sheep, furegrelate (U-63557A) attenuated the increase in pulmonary artery pressure and microvascular changes induced by endotoxaemia (Gunther *et al.*, 1984). There are several other reports of thromboxane synthesis inhibitors attenuating the pulmonary hypertension and the sequelae evident in various animal models of ARDS (Wise *et al.*, 1980; Watkins *et al.*, 1982; Ball *et al.*, 1983; Garcia-Szabo *et al.*, 1983, 1984; Winn *et al.*, 1983; Reines *et al.*, 1985).

Two studies indicated that intravenous furegrelate decreased myocardial infarct size following coronary ligation in rats (Hock *et al.*, 1985; Wargovich *et al.*, 1987). This was associated with a reduction in platelet aggregation and serum TXB_2 levels for up to 48 h, and also a decrease in neutrophil accumulation (Wargovich *et al.*, 1987). Increased plasma levels of TXB_2 have been detected in animals and humans soon after the start of myocardial ischaemia (Lewy *et al.*, 1979; Smith *et al.*, 1980; Walinsky *et al.*, 1984). In contrast to thromboxane synthetase inhibition (Smith *et al.*, 1980; Wargovich *et al.*, 1987), inhibition of cyclooxygenase has been reported to be without effect on ischaemic myocardial injury and infarct size (Ogletree and Lefer, 1976; Bonow *et al.*, 1981). There is conflicting data on the effectiveness of thromboxane synthetase inhibitors in models of myocardial reperfusion injury (Coker and Parratt, 1983; Huddleston *et al.*, 1983; Thiemermann and Schrör, 1985; Mullane and Fornabaio, 1988; Toki *et al.*, 1988). It is noteworthy that thromboxane receptor antagonists (Grover and Schumacher, 1988; Thiemermann *et al.*, 1988; Smith *et al.*, 1989) but not cyclooxygenase inhibitors (Mullane *et al.*, 1984; Grover and Schumacher, 1988; Mullane and Fornabaio, 1988), exerted a cardioprotective effect in similar studies.

In spontaneously hypertensive rats, 4-(imidazol-1-yl) acetophenone (Uderman *et al.*, 1982) or CV 4151 (Shibouta *et al.*, 1985) delayed, rather than inhibited, the age-related increases in blood pressure. Chronic administration of dazmegrel for about 2 weeks produced a significant decrease in systolic blood pressure (25–30 mmHg) although the rats were still hypertensive (Stier *et al.*, 1988); acute administration was without effect in animals with developed hypertension (Uderman *et al.*, 1984).

In a rat model of acute renal allograft rejection, intravenous infusion but not intraperitoneal administration of OKY-046 decreased renal TXB_2 production and improved renal function (Coffman *et al.*, 1989). In the same model, acute administration of dazoxiben improved glomerular filtration rate and renal blood flow (Coffman *et al.*, 1985). However, these effects were not maintained in the later stages of rejection and were not associated with significant inhibition of cellular infiltration or systemic cellular immunity. Increased urinary levels of TXB_2 are detected in renal transplant recipients and also in animal models of renal allograft transplantation (Foegh *et al.*, 1981; Tannenbaum *et al.*, 1984; Coffman *et al.*, 1985, 1986).

In the respiratory system, OKY-046 inhibited antigen-induced bronchoconstriction in guinea-pigs both *in vivo* and *in vitro* (Nagai *et al.*, 1987). Similarly, OKY-046 inhibited peptidoleukotriene-induced bronchoconstriction in isolated guinea-pig bronchus (Prié and Sirois, 1988) and OKY-1581 inhibited LTC_4 and LTD_4 bronchospasm in guinea-pigs *in vivo* (Ueno *et al.*, 1982). These data support evidence that antigen- and peptido-leukotriene-induced bronchoconstriction in guinea-pigs is mediated, in part, via the release of TXA_2 (Weichman *et al.*, 1982; Creese *et al.*, 1984; Cheng *et al.*, 1990). However, in human perfused lungs, injection of LTD_4 did not elicit release of thromboxanes (Chagnon *et al.*, 1985). Thromboxane synthetase inhibitors attenuated bronchospasm in cats elicited by arachidonic acid, PGH_2 or the calcium ionophore A23187 (Kriseman *et al.*, 1987; Tilden *et al.*, 1987). Intravenous infusion of OKY-046 inhibited antigen-induced airway hyper-responsiveness in ragweed-sensitized dogs; the compound was without effect on baseline pulmonary function, antigen-induced bronchoconstriction or associated neutrophil influx (Chung *et al.*, 1986).

The above studies represent a sample of the numerous preclinical studies conducted using the thromboxane synthetase inhibitors which have formed the basis for the interest in exploring the potential therapeutic utility of this class of compounds in a variety of disorders.

6.5 CLINICAL STUDIES

Several clinical trials have been conducted with various thromboxane synthetase inhibitors but, unfortunately, and in contrast to the tests from preclinical studies in animal models, the data have been unimpressive and disappointing.

The most widely studied compound clinically has been dazoxiben (UK-37,248), which was the first compound from this class to be administered to humans (Tyler *et al.*, 1981). In two studies, involving a total of 17 patients, intravenous administration of dazoxiben did not affect the haemodynamic and pulmonary changes associated with ARDS despite producing a significant and maintained decrease in plasma immunoreactive TXB_2 levels (Leeman *et al.*, 1985; Reines *et al.*, 1985). This contrasts with the impressive ability of thromboxane synthetase inhibitors, including dazoxiben, to attenuate the pulmonary hypertension, and many of pathophysiological sequelae manifest in several animal models of ARDS (*vide supra*). These data using a small group of patients may indicate that TXA_2 may be important early in the development of the disease, but not once it is fully developed. It is worth noting that in endotoxin models of ARDS it appears that thromboxane synthetase inhibitors, and also thromboxane receptor antagonists, are only effective when given prior to exposure to endotoxin (Wise *et al.*, 1980; Halushka *et al.*, 1983; Winn *et al.*, 1983; Reines *et al.*, 1985). In patients with primary pulmonary hypertension, 3 month treatment with CGS 13080 enhanced the decrease in pulmonary vascular resistance produced by the calcium channel inhibitor nifedipine (Rich *et al.*, 1987).

Four studies have been conducted in patients with Raynaud's phenomenon or syndrome. Six week treatment in 20 patients with 400 mg of dazoxiben per day produced an improvement in subjective clinical symptoms in some patients, although there was no effect on objective clinical parameters (Belch *et al.*, 1983). In the three additional studies dazoxiben failed to produce any beneficial effects on objective or subjective clinical assessments (Ettinger *et al.*, 1984; Luderer *et al.*, 1984; Rustin *et al.*, 1984). This lack of efficacy was observed despite a report describing increased levels of TXA_2 metabolites in patients with Raynaud's syndrome (Reilly *et al.*, 1986b).

In three small open clinical trials in patients with peripheral vascular disease, dazoxiben produced equivocal data which is difficult to interpret, in large part because of the limited number of patients studied (Ehrly, 1983; Raftery *et al.*, 1983; Vermylen and Deckmyn, 1983). In a double-blind, placebo-controlled study, albeit in only nine patients with severe peripheral disease, another thromboxane synthetase inhibitor, CGS 13080, administered at 100 or 200 mg q.i.d., was without effect on subjective or objective clinical parameters (Reilly *et al.*, 1986a). Using this dosing regimen there was only partial inhibition of thromboxane synthesis and there was evidence of continued platelet activation in the patients.

Probably the most promising, but preliminary, data emanate from trials examining the role of TXA_2 and the potential therapeutic utility of thromboxane synthetase inhibitors in asthma. Thus, oral administration of OKY-046 (200 mg q.i.d. for 6 days) inhibited exercise-induced asthma in seven of eleven subjects (Hoshino and Fukushima, 1991). These inhibitory effects were observed concomitantly with an inhibition of plasma TXB_2 levels and an elevation in the plasma levels of the PGI_2 metabolite 6-keto $PGF_{1\alpha}$. In a preliminary study in four atopic asthmatics, 200 mg of OKY-046, administered orally 1 h before and 1, 4 and 8 h after antigen challenge, was reported to inhibit the characteristic EAR and LAR by 56 and 39%, respectively (Iwamoto *et al.*, 1988). The same study revealed significant elevations in plasma TXB_2 levels during the EAR (approximately two-fold increase to 132 pg/ml) and LAR (approximately five-fold increase to 297 pg/ml); in the presence of OKY-046 the TXB_2 levels were less than 20 pg/ml. When given via the oral, but not the aerosol, route OKY-046 was also reported to inhibit the increased responsiveness to acetylcholine in asthmatic patients; the 5-lipoxygenase inhibitor AA-861,

given orally, was without effect (Fujimura *et al.*, 1986).

Several short-term trials in patients with stable angina have been performed with dazoxiben and, in general, there was no significant positive effect on haemodynamic parameters or clinical symptoms (McGibney *et al.*, 1982; Hendra *et al.*, 1983; Kiff *et al.*, 1983; Reuben *et al.*, 1983). Two studies suggested a marginal beneficial effect in pacing- or exercise-induced angina (Hutton *et al.*, 1983; Thaulow *et al.*, 1984). Similarly, studies in patients with vasospastic angina revealed no efficacy with either dazoxiben (Thieuleux *et al.*, 1986) or OKY-046 (Yui *et al.*, 1986). In contrast, in two open studies, one with OKY-046 in patients with angina (Tsuji *et al.*, 1985) and another with CV 4151 in patients with unstable angina (Kimura *et al.*, 1986), a beneficial effect was observed. The former observation is perhaps rather surprising in view of the evidence that the stable metabolites of TXA_2, 2,3-dinor-TXB_2 and 11-dehydro-TXB_2 are not increased during stable angina (Fitzgerald *et al.*, 1986). In contrast, levels of TXA_2 metabolites are elevated in patients with unstable angina (Tsuji *et al.*, 1985; Fitzgerald *et al.*, 1986).

Several studies with dazoxiben have demonstrated minimal or no beneficial effects in patients with various forms of renal diseases (Barnett *et al.*, 1984; Zipser *et al.*, 1984; Fitzgerald *et al.*, 1985; Patrono *et al.*, 1985). In a small uncontrolled study, which was not blind and which should be interpreted with caution, it was observed that six of seventeen kidney transplant patients treated with CGS 13080 did not experience rejection (Foegh *et al.*, 1988).

There are several potential reasons for the limited and disappointing clinical efficacy of thromboxane synthesis inhibitors (Fitzgerald *et al.*, 1985; Fiddler and Lumley, 1990). The most obvious one is that the data indicate that TXA_2 is not an important mediator in the pathophysiology of many of the diseases studied. It is of interest that in many of the disorders examined, the important rationale for exploring the efficacy of thromboxane synthetase inhibitors, namely evidence for increased production of TXA_2 and demonstrated effectiveness of aspirin, is not apparent (Fitzgerald *et al.*, 1985; Fiddler and Lumley, 1990).

The lack of clinical efficacy of the thromboxane synthesis inhibitors contrasts with their effectiveness in an array of animal models of diseases (*vide supra*) and accordingly calls into question the relevance of these models, at least with this class of compounds. It should be noted that, despite the lack of efficacy of cyclooxygenase inhibitors in many of the models of cardiovascular disease discussed above, evidence suggests that aspirin produces a significant reduction in cardiovascular disease and mortality in humans (Elwood, 1983; Mustard *et al.*, 1983; Antiplatelets Trialists' Collaboration, 1988).

Probably one of the main reasons for the disappointing clinical data is that it appears inappropriate and inadequate dosing regimens were utilized, especially in view of the pharmacodynamic profile of the compounds examined (Fitzgerald *et al.*, 1985; Fiddler and Lumley, 1990). It was proposed that the inhibition of serum TXB_2 levels must be essentially complete and maintained in order to significantly inhibit TXA_2-induced platelet activation and produce clinical effects (Fitzgerald *et al.*, 1985; Reilly and Fitzgerald, 1987). However, the majority of the thromboxane synthetase inhibitors only produce greater than 95% inhibition of TXB_2 production for periods of less than 5 h (see Fiddler and Lumley, 1990). Accordingly, to produce adequate and sustained inhibition of TXA_2-induced platelet activation in long-term clinical trials it is likely that dosing of patients every 4 or 5 h would be required; this was not employed. The above consideration may explain, at least in part, the discrepancies between the effectiveness of the thromboxane synthesis inhibitors in short-term studies in animal models, where measurements are taken close to or during maximal inhibition of thromboxane production, and the longer-term studies in clinical trials, in which there is likely to be intermittent and largely inadequate inhibition of thromboxane synthetase (Fitzgerald *et al.*, 1985).

Another consideration is the conflicting information on the ability of thromboxane synthetase inhibitors to increase the production and levels of PGI_2 and its metabolites (Fischer *et al.*, 1983; Fitzgerald *et al.*, 1983b; MacNab *et al.*, 1984; Patrignani *et al.*, 1984; Reilly and Fitzgerald, 1987; Vesterqvist *et al.*, 1991). The relevance of apparent differences in the sensitivity of thromboxane synthetase in different tissues, e.g. platelets versus kidney, to inhibitors (Patrignani *et al.*, 1984; Patrono *et al.*, 1985; Grone *et al.*, 1986; Foegh *et al.*, 1988) needs to be further explored and clarified.

By virtue of their mechanism of action in selectively inhibiting the production of TXA_2, thromboxane synthesis inhibitors should lead to the accumulation of the endoperoxide substrate PGH_2. This precursor may lead to enhanced production of PGI_2. However, PGH_2 may also be redirected to other prostanoids, including PGE_2, which produces aggregation of human platelets (Andersen *et al.*, 1980); PGE_2 levels have been shown to be elevated in humans after administration of thromboxane synthesis inhibitors (MacNab *et al.*, 1984; Patrignani *et al.*, 1984). Accordingly, the qualitative and quantitative nature of the final influence of the redirection of accumulated PGH_2 after thromboxane synthesis inhibitors will depend on the balance between the formation of pro-aggregatory and stimulatory prostanoids versus anti-aggregatory and inhibitory prostanoids. In addition, the true clinical significance of maintaining the production of beneficial prostanoids such as PGI_2 is unclear (Patrono, 1989).

Probably a more important consideration is the evidence that PGH_2 interacts with the same receptor and has the same pharmacological profile of activity as TXA_2 (Kennedy *et al.*, 1982; Halushka *et al.*, 1987, 1989a,b; Ogletree, 1987). Therefore, even with effective and well-maintained inhibition of thromboxane production clinic-ally, using an appropriate thromboxane synthesis inhibitor, PGH_2 may negate this by accumulating, substituting for TXA_2 and producing similar deleterious effects.

6.6   SUMMARY

Over the last 15 years or so, significant resources have been deployed in the search for a compound that is more selective than and clinically superior to aspirin, i.e. potent thromboxane synthetase inhibitors. However, and despite considerable evidence from animal models suggesting that TXA_2 may play a role in the pathophysiology of a variety of disorders, predominantly thrombolytic or cardiopulmonary in nature, clinical trials with potent and selective thromboxane synthetase inhibitors have been extremely disappointing. These trials have generally employed small groups of subjects. For the following three reasons, amongst others, the results are perhaps not surprising: (1) for many of the diseases examined clinically a strong rationale for a pathophysiological role of TXA_2 does not exist; (2) in general the compounds investigated have an inadequate pharmacodynamic profile to examine under the dosing regimens employed; (3) the substrate endoperoxide for thromboxane, PGH_2, can substitute for the biological activity of TXA_2 via an interaction with the same receptor. This last point brings forth the issue of whether the identification of compounds which have a superior pharmacodynamic profile to those presently tested will exhibit any greater clinical benefit. Overall, a superior clinical profile of the thromboxane synthesis inhibitors compared to aspirin has not been evident. In fact, an additional consideration is the continuing debate regarding whether there is an appropriate dosing regimen with aspirin which minimizes the inhibitory effects on PGI_2 production.

In view of the above serious questions and concerns, there has recently been a shift in emphasis and strategy towards the development of thromboxane receptor antagonists, and also consideration of the potential clinical utility of a combination therapy of a thromboxane synthetase inhibitor and a thromboxane receptor antagonist, either in one or two molecules. These are discussed below.

7. *Receptor Antagonists – Cyclooxygenase Pathway*

7.1   INTRODUCTION

As outlined comprehensively in Chapter 3, the various cyclooxygenase products exert a myriad of diverse biological effects and have been implicated in the pathophysiology of several diseases. The spectrum of actions of the cyclooxygenase products can be broadly categorized as stimulatory (pro-inflammatory) or inhibitory (anti-inflammatory) and are generally associated with

TXA$_2$ and PGF$_{2\alpha}$ or PGI$_2$ and PGE$_2$, respectively; PGD$_2$ has both stimulatory and inhibitory effects (Hanley, 1986; Konturek and Pawlik, 1986; Ogletree, 1987; Giles and Leff, 1988). Accordingly, the ultimate influence of products of the cyclooxygenase enzyme in a particular system will be the result of net effects of the opposing inputs. As an example, and to illustrate the potential therapeutic relevance of this delicate balance, it is recognized that, although cyclooxygenase inhibitors do not affect most asthmatics, approximately 4–10% of the subjects worsen (Szczeklik *et al.*, 1975; Harnett *et al.*, 1978; Szczeklik, 1990) and about 1% improve (Hume and Eddey, 1977; Lockey, 1978; Szczeklik and Nizankowska, 1983). The reason(s) for this clinical observation, designated aspirin-induced asthma (Samter and Beers, 1968; Szczeklik, 1986), remains uncertain but the most widely accepted theory is that it is due to an alteration in the balance of prostanoids in the airways with a shunting towards the 5-lipoxygenase pathway (Szczeklik, 1990). This aspirin-induced asthma also serves to highlight again the notion that, in specific diseases, inhibition of cyclooxygenase, with the resultant attenuation in the formation of both the stimulating and inhibitory products, may not be an effective strategy.

Rather than targeting the cyclooxygenase enzyme, a more appropriate strategy, which has been the cornerstone of significant efforts over several years, may be the development of agents that selectively interfere with the formation or actions of the individual stimulatory cyclooxygenase products, or alternatively compounds which are mimetics of the inhibitory members of the cascade. Building in inherent selectivity in these molecules for a particular cyclooxygenase product or receptor could markedly decrease their side-effect profile.

7.1.1   Prostanoid Receptors

It was originally hypothesized by Gardiner and Collier (1980) that, based on the activity of various prostaglandins and analogues in the respiratory tract, there were three distinct prostanoid receptors, designated contractant (χ), relaxant (ψ) and irritant (ω). However, using data from functional and binding studies in a variety of tissues with an array of agonists and antagonists this classification has been altered and expanded. The current evidence suggests that there are a multiplicity of prostanoid receptors, which have been divided into five basic groups: *DP*, *EP*, *FP*, *IP* and *TP* for which PGD$_2$, PGE$_2$, PGF$_{2\alpha}$, PGI$_2$ and TXA$_2$ are the most potent natural agonist, respectively (Coleman *et al.*, 1984, 1985a; Coleman, 1988; Gardiner, 1990). Subtypes of the EP receptor, designated EP$_1$, EP$_2$ and EP$_3$, have been proposed (Coleman, 1988; Gardiner, 1990). There is conflicting evidence for the existence of subtypes of the TP (or TXA$_2$/PGH$_2$) receptor (*vide infra*).

A significant hindrance in the classification of the prostanoid receptors and their relative tissue distributions, and also their potential role in the pathophysiology

of diseases, is that the individual prostanoids interact to some degree with the other prostanoid receptor subtypes. Furthermore, prostanoid receptors are present in such a large number of diverse tissues, making significant side-effects with agents that interact with them almost inevitable (Keen *et al.*, 1989; Gardiner, 1990). To date, this has probably been a major reason for the disappointing lack of success in the development of therapeutically useful prostanoid agonists or antagonists. This is despite an intensive research effort directed primarily towards the synthesis of stable analogues of PGI$_2$ and PGE$_2$ and antagonists of TXA$_2$. For the present discussion, information will be restricted to prostanoid receptor antagonists, particularly TXA$_2$ receptor antagonists which has been the most widely studied and most successful area of research in this field, and will not encompass the postanoid mimetics.

7.2   CHEMICAL/PHARMACOLOGICAL CLASSES

7.2.1   Thromboxane Receptor Antagonists

7.2.2.1   *Introduction*

TXA$_2$, first discovered and named in 1975 (Hamberg *et al.*, 1975b), is the potent and chemically unstable metabolite of PGH$_2$. Its synthesis and release is widespread in the body, particularly in platelets, where it is the major cyclooxygenase product (Patrono *et al.*, 1986). It possesses four major biological actions: (1) vasoconstriction via a direct mechanism (Ellis *et al.*, 1976; Lefer *et al.*, 1980); (2) platelet aggregation (Hamberg *et al.*, 1975b; Mais *et al.*, 1985b); (3) bronchoconstriction (Hamberg *et al.*, 1975a; Svensson *et al.*, 1977); and (4) enhanced membrane labilization and permeability (Schrör *et al.*, 1980; Brezinski *et al.*, 1986), probably the least recognized effect of TXA$_2$ (Lefer and Darius, 1987; Ogletree, 1987). In view of this profile of activity, in conjunction with other accumulating pieces of evidence, such as elevated levels of the mediator in body fluids, TXA$_2$ has been postulated to play a significant role in the pathophysiology of a diverse array of diseases, particularly thrombolytic and cardiopulmonary disorders such as myocardial ischaemia (Smith *et al.*, 1980; Burke *et al.*, 1983a,b), endotoxic or circulatory shock (Lefer *et al.*, 1979; Halushka *et al.*, 1983), sudden cardiopulmonary death (Lefer *et al.*, 1981), pulmonary hypertension, renal failure, coronary artery thrombosis, myocardial infarction and pregnancy-induced hypertension (Ogletree, 1987; Halushka, 1989; Smith, 1989), in addition to asthma (O'Byrne and Fuller, 1989).

The precursor for TXA$_2$, PGH$_2$, exhibits the same spectrum of pharmacological activity in various tissues as TXA$_2$, and the evidence suggests strongly that these two labile substances produce their effects via an interaction with a common receptor, the TXA$_2$/PGH$_2$ receptor (Kennedy *et al.*, 1982; Halushka *et al.*, 1987, 1989a,b;

Ogletree, 1987). Significant species differences are apparent in the TXA_2/PGH_2 receptor (Halushka *et al.*, 1989a,b). In addition, there is conflicting evidence as to the existence of subtypes of the TXA_2/PGH_2 receptor, specifically located on platelets and the vasculature (Needleman *et al.*, 1976; Fitzpatrick *et al.*, 1978; Lefer *et al.*, 1980; LeDuc *et al.*, 1981; Akbar *et al.*, 1985; Armstrong *et al.*, 1985; Mais *et al.*, 1985a,b; Swayne *et al.*, 1988). Preliminary studies suggest that the pulmonary (TXA_2/PGH_2) receptor may represent a unique receptor subclass (Brittain *et al.*, 1984; Steinbacher and Ogletree, 1985; McKenniff *et al.*, 1988). This may be a very important issue clinically, in that activation of a particular subtype of the TXA_2/PGH_2 receptor may be more prominent in a specific disorder.

As indicated above in Section 6, the thromboxane synthetase inhibitors have exhibited disappointing clinical efficacy in a variety of disorders. This has been partially attributed to the short duration of action of many of the compounds, but is probably largely due to some unfavourable consequences and limitations of the mechanism of action of this class of agents. Thus, from a theoretical standpoint, potent and selective TXA_2/PGH_2 receptor antagonists should possess the following advantages over TXA_2 synthetase inhibitors, and have an increased likelihood of clinical benefit: (1) receptor antagonists, but not synthetase inhibitors, will antagonize the actions of the precursor endoperoxides PGH_2 and PGG_2; (2) in the presence of synthetase inhibitors, but not receptor antagonists, there may be the accumulation of PGH_2 in sufficient quantities to activate the TXA_2/PGH_2 receptor and produce the same effects as TXA_2; (3) in the presence of synthetase inhibitors, but not receptor antagonists, there may be a reduction of endoperoxide metabolism towards the formation of detrimental (e.g. $PGF_{2\alpha}$) in addition to beneficial products (e.g. PGI_2), i.e. receptor antagonists will not interfere with eicosanoid metabolism; (4) receptor antagonists, but not synthetase inhibitors, will be very effective when given after the formation of thromboxane (Lefer and Darius, 1987; Smith, 1989; Patscheke, 1990b). In preclinical studies in some animal models, thromboxane receptor antagonists are more effective than thromboxane synthetase inhibitors (Lefer and Darius, 1987; Fitzgerald *et al.*, 1988; Quilley *et al.*, 1989; Smith, 1989).

In view of the above considerations the pharmaceutical industry has in recent years redirected their efforts in the area of thromboxane biology towards the synthesis and development of thromboxane receptor antagonists. Many structurally diverse, potent and selective compounds have emerged from this major effort, some of which are illustrated in Fig. 5.10.

7.2.1.2 *Mechanism of Action*

Thromboxane receptor antagonists act both competitively and non-competitively.

7.2.1.3 *Preclinical Studies*

The earliest reported thromboxane receptor antagonists were pinane TXA_2 (PTA_2) (Nicolaou *et al.*, 1979) and 13-azaprostanoic acid (13-APA) (LeBreton *et al.*, 1979), both of which have eicosanoid-like structures. 13-APA inhibited human platelet aggregation induced by arachidonic acid, PGH_2 and a stable endoperoxide analogue but did not inhibit aggregation induced by ADP or thrombin, and had no effect on thromboxane synthesis (LeBreton *et al.*, 1979). PTA_2 (0.1–1 µM) inhibited contraction of cat coronary artery induced by stable PGH_2 analogues, and higher concentrations (1–15 µM) antagonized platelet aggregation induced by the analogues but not by PGI_2 or PGD_2 (Nicolaou *et al.*, 1979). 13-APA prevented arachidonate-induced sudden death in rabbits (Burke *et al.*, 1983b) and PTA_2 (1 µM) antagonized the lysosomal labilizing action produced by thromboxane receptor activation (Lefer and Darius, 1987; Nicolaou *et al.*, 1979). PTA_2 gave protection in a model of acute myocardial ischaemia in cats (Schrör *et al.*, 1980).

PTA_2 and 13-APA have different profiles of activity; the former is a more potent antagonist of vascular versus platelet thromboxane receptors, whereas the latter compound does not discriminate between the two (Lefer and Darius, 1987). In addition, PTA_2 is a partial agonist at the thromboxane receptor in the rabbit, contracting blood vessels and inducing shape changes in platelets (Lefer and Darius, 1987), and it is not a very selective compound as it also inhibits thromboxane synthetase, albeit in higher concentrations (Nicolaou *et al.*, 1979).

Since the synthesis of 13-APA and PTA_2, various structurally diverse and more potent compounds have been reported and their *in vitro* and *in vivo* activities as thromboxane receptor antagonists have been comprehensively analysed. These compounds, which can be classified as eicosanoid-like and non-eicosanoid-like structures, include BM 13.177 (sulotroban) and BM 13.505 (daltroban), SQ 29,548 and SQ 28,668, SK&F 88046, AH 23848, L-655,240, ICI 185282, Bay U 3405, GR 32191 and ONO-3708 (Fig. 5.10). These and other compounds have been shown to potently and selectively antagonize thromboxane-mediated responses in a plethora of *in vitro* and *in vivo* systems (Weichman *et al.*, 1984a,b; Brittain *et al.*, 1985; Ogletree *et al.*, 1985a,b; Patscheke and Stegmeier, 1985; Patscheke *et al.*, 1986; Hall *et al.*, 1987; Kondo *et al.*, 1987; Birch *et al.*, 1988; Lumley *et al.*, 1989; McKenniff *et al.*, 1991; Norel *et al.*, 1991) and also to attenuate the sequelae in several animal models of diseases (Lefer, 1985; Lefer and Darius, 1987; Smith, 1989), predominantly cardiovascular, in which TXA_2 has been implicated, including acute myocardial ischaemia (Schrör *et al.*, 1980; Brezinski *et al.*, 1985), arterial thrombosis (Ashton *et al.*, 1986; Bush and Smith, 1987; Schumacher *et al.*, 1987), collagen-, U-46619- or arachidonic acid-induced sudden death (Burke *et al.*, 1983b; Darius *et al.*, 1985; Lefer *et al.*, 1987; Smith and McDonald, 1988; Perzborn *et al.*, 1990; Seuter *et al.*,

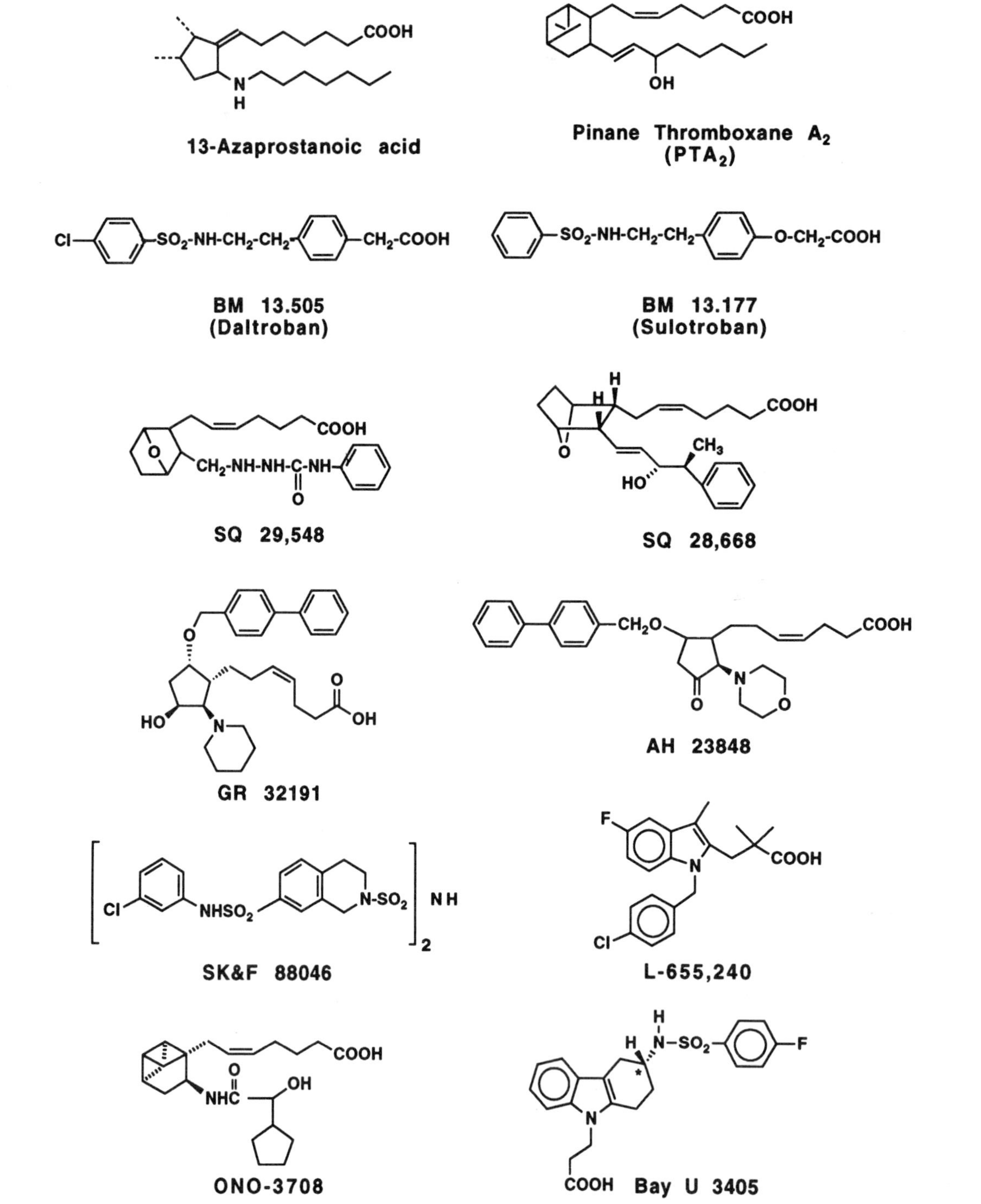

13-Azaprostanoic acid

Pinane Thromboxane A₂
(PTA₂)

BM 13.505
(Daltroban)

BM 13.177
(Sulotroban)

SQ 29,548

SQ 28,668

GR 32191

AH 23848

SK&F 88046

L-655,240

ONO-3708

Bay U 3405

Figure 5.10 Names and chemical structures of representative thromboxane receptor antagonists.

1990), cerebral haemorrhage (Perzborn *et al.*, 1990), reperfusion- or ischaemia-induced arrhythmias (Wainwright and Parratt, 1988) and myocardial infarction/reperfusion injury (Grover and Schumacher, 1988, 1989; Thiemermann *et al.*, 1988; Smith *et al.*, 1989). Some studies have revealed that a thromboxane receptor antagonist is more effective than a thromboxane synthetase inhibitor (Lefer and Darius, 1987; Fitzgerald *et al.*, 1988; Quilley *et al.*, 1989; Smith, 1989).

This area of research has been extensively reviewed in recent years and the reader is directed to some excellent publications for more information about the preclinical effects of the thromboxane receptor antagonists, particularly in *in vivo* animal models of disease (Lefer, 1985; Lefer and Darius, 1987; Smith, 1989).

7.2.1.4 Clinical Studies

It should be emphasized that, to date, the clinical studies with the thromboxane receptor antagonists have been rather disappointing and the hope of significantly increased efficacy relative to the thromboxane synthesis inhibitors has really not been fulfilled. The compounds examined in humans include AH 23848, GR 32191, BM 13.177, BM 13.505, and SQ 28,668. They were generally well tolerated and inhibited U46619-induced human platelet aggregation *ex vivo* following oral administration; there were differences in the potencies, efficacies and durations of action of the compounds (for a review, see Fiddler and Lumley, 1990).

Most of the reported studies were conducted with BM 13.177 (sulotroban), which was one of the first thromboxane receptor antagonists to be examined clinically. It has been shown to reduce the increased levels of PF_4 and β-thromboglobulin in patients with atherosclerosis (Riess *et al.*, 1984). In normal volunteers, BM 13.177 was shown to increase bleeding time by about 30% (Gresele *et al.*, 1984; Stegmeier *et al.*, 1984). Intravenous (7 days) followed by oral treatment (2 weeks) with BM 13.177 had no effect in advanced stages of peripheral obliterative arterial disease (Hossmann *et al.*, 1987). Continuous infusion of BM 13.177 for 48 h in 10 patients with diffuse proliferative lupus nephritis produced a significant increase in renal function (Pierucci *et al.*, 1989). The effects of BM 13.177 have been examined in patients undergoing percutaneous transluminal coronary angioplasty. In one study, administration of BM 13.177 (800 mg t.i.d.) to 20 patients for 3 months decreased the restenosis rate relative to aspirin (Ludwig *et al.*, 1987). In contrast, another study found no difference between patients treated with placebo and those treated with BM 13.177 (800 mg q.i.d.) (Hofling *et al.*, 1988). BM 13.177 significantly improved the graft occlusion rates (3.1 versus 11.5% in the placebo-treated group, $p < 0.02$) in patients undergoing coronary artery bypass grafting (Torka *et al.*, 1988). AH 23848, but not aspirin plus dipyridamole, significantly reduced platelet deposition on mature Dacron aortobifemoral grafts in patients with peripheral vascular disease (Lane *et al.*, 1984).

AH 23848 was without effect in clinical trials in stable angina pectoris (Debono *et al.*, 1986). A single dose of another Glaxo compound, GR 32191 (80 mg), produced greater than a 10-fold shift to the right in PGD_2 dose–response curves in asthmatic subjects and also produced a small (about 25%) but significant inhibition of the initial bronchoconstrictor response to allergen challenge; there was no effect on baseline lung function or on methacholine-induced bronchoconstriction (Beasley *et al.*, 1989). In addition, GR 32191 (120 mg) was without effect on exercise-induced asthma (Finnerty *et al.*, 1991). A novel thromboxane receptor antagonist, AA-2414, which inhibits contractions of guinea-pig trachea to U46619, PGD_2 and its metabolite $9\alpha,11\beta$-PGF_2, but has minimal effects on $PGF_{2\alpha}$-induced responses (Ashida *et al.*, 1989), was reported to attenuate bronchial hyper-responsiveness to methacholine in asthmatics when given orally (40 mg) for 4 days (Fujimura *et al.*, 1991); however, again the effect was small.

The disappointing data so far with the receptor antagonists, and also the thromboxane synthesis inhibitors, especially in relation to the impressive preclinical results in various animal models, calls into question the role of TXA_2 in the various pathophysiological conditions in which it has been implicated, in addition to the relevance of the various animal models of disease examined to the human conditions.

Recently, Patscheke (1990b) discussed the concept that non-competitive antagonists may possess greater potency and longer duration of action than competitive antagonists. Thus, the local release of high concentrations of TXA_2, e.g. from platelets, may normally be sufficient to overcome the effects of competitive antagonists of TXA_2/PGH_2-induced platelet aggregation such as BM 13.177 (Patscheke and Stegmeier, 1985) and SQ 29,548 (Ogletree *et al.*, 1985a). In contrast, other antagonists such as ER 32,191 (Lumley *et al.*, 1989), daltroban (Patscheke, 1990a), S-145 (Hanasaki *et al.*, 1989) and Bay U 3405 (Patsheke, 1990b), by virtue of their slow dissociation rates from the TXA_2/PGH_2 receptor, elicit an insurmountable inhibition in human platelets as the agonists cannot compete with the antagonist and bind to the receptor rapidly enough to trigger the response. Therefore, it is both the receptor affinity and dissociation rate which are important. It is of interest that it was calculated, using a structural analogue of SQ 29,548, SQ 28,688, that about 94% receptor occupation was needed to result in a small increase in template bleeding time in healthy subjects (Friedhoff *et al.*, 1986), suggesting that very high receptor occupancy is required to affect TXA_2-induced platelet activation. Most of the reported clinical studies to date have utilized sulotroban and it is not known if the above profile has limited its clinical efficacy. Accordingly, analysis of the therapeutic effects of some compounds producing insurmountable inhibition will clarify whether these interesting laboratory observations and the above postulate have any clinical relevance.

7.2.2 Dual Thromboxane Synthetase Inhibitors and Thromboxane Receptor Antagonists (Fig. 5.11)

There is increasing interest in the development of compounds which are combined thromboxane synthetase inhibitors and receptor antagonists. Theoretically, an agent with this profile should overcome the limitations of the individual compounds, particularly the synthetase inhibitors, with the receptor antagonists controlling the effects of the endoperoxides, including PGH_2, which

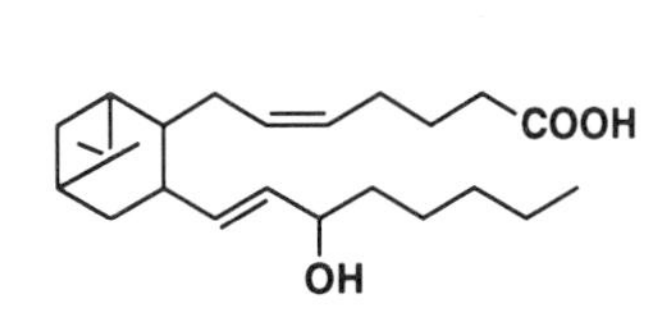

Figure 5.11 Names and chemical structures of purported dual thromboxane synthetase inhibitors and thromboxane receptor antagonists.

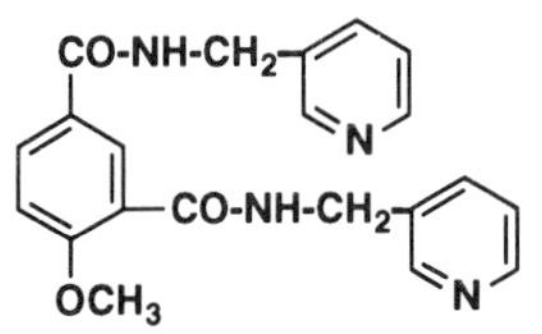

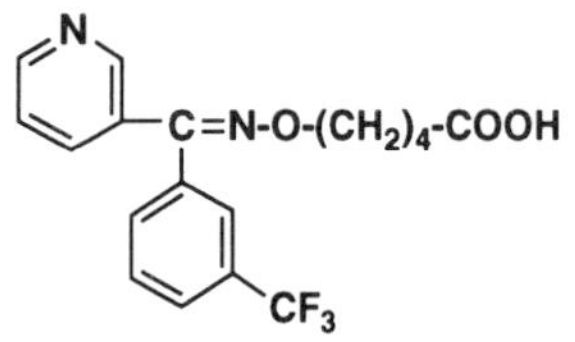

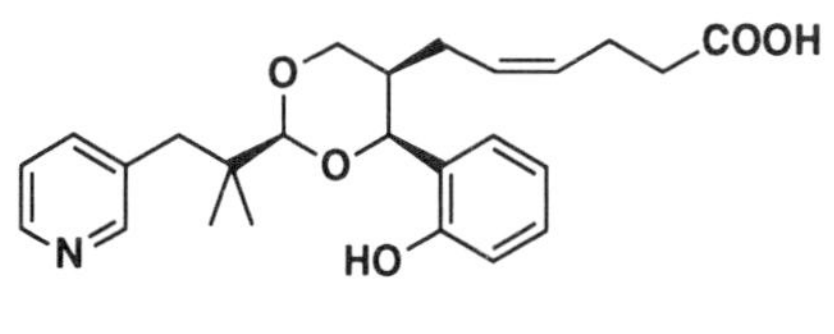

accumulate as a result of thromboxane synthetase inhibition (Patscheke, 1990b; Gresele *et al.*, 1991). This hypothesis is supported by the results of several preclinical studies which indicate an increased effectiveness *in vitro* and *in vivo* of the combination of a thromboxane receptor antagonist and a thromboxane synthetase inhibitor, than either agent alone (Smith, 1989; Patscheke 1990b; Gresele *et al.*, 1991). Furthermore, in healthy individuals, the combined administration of BM 13.177 and the synthetase inhibitor dazoxiben produced inhibition of platelet aggregation and increased the template bleeding time more than either drug alone; this effect was abolished by indomethacin (Gresele *et al.*, 1987).

PTA₂ acts both as a thromboxane receptor antagonist and a thromboxane synthetase inhibitor, although the latter requires about 100-fold higher concentrations (Nicolaou *et al.*, 1979). Ridogrel, a more recent dual inhibitor, has the opposite profile, being about 100-fold more potent as a thromboxane synthetase inhibitor ($IC_{50} \approx 10$ nM) than a thromboxane receptor antagonist ($IC_{50} \approx 2$ μM) (De Clerck *et al.*, 1989; Hoet *et al.*, 1990b). Ridogrel has been shown to prolong bleeding time more effectively than aspirin in healthy individuals (Hoet *et al.*, 1989) and reduces elevated levels of platelet-derived β-thromboglobulin in patients with peripheral arterial disease (Hoet *et al.*, 1990a). Unlike PTA₂ and ridogrel, picotamide, a non-prostanoid, has the more acceptable profile of possessing thromboxane synthethase inhibitory activity and thromboxane receptor antagonist effects at equivalent concentrations (Violi *et al.*, 1988; Gresele *et al.*, 1989). However, it is not a potent compound ($IC_{50} \approx 10^{-4}$M) and may not be very selective (Violi *et al.*, 1988; Gresele *et al.*, 1989, 1991). A recent preliminary report described some 2-(3-pyridylalkyl)-1,3-dioxane

derivatives which, by virtue of incorporating a 3-pyridyl moiety (representative of TXA₂ synthetase inhibitors) into a 1,3-dioxane-derived molecule (representative of TXA₂ receptor antagonists), were dual inhibitors. One of the derivatives, compound (1) in Fig. 5.11, 4(*Z*)-6-{(2*RS*,4*RS*, 5*SR*)-4-(2-hydroxyphenyl)-2-[2-(3-pyridyl)-1, 1-dimethylethyl]-1,3-dioxan-5-yl}hex-4-enoic acid, was a potent agent which exhibited TXA₂ synthetase inhibition and TXA₂ receptor antagonism at similar concentrations ($IC_{50} \approx 10$ nM) (Brewster *et al.*, 1990).

Among the different types of drugs that affect the biological activity of TXA₂, potent dual inhibitors may have the best potential to provide significant therapeutic benefit. A major issue with whichever of the different classes of the above compounds exhibits the greatest therapeutic benefit will be their ability to compete on the market place with the many cyclooxygenase inhibitors which possess a long track record of inexpensive, largely safe and efficacious usage. The new classes of compound are likely to have to demonstrate a considerable therapeutic and/or safety advantage over the cyclooxygenase inhibitors in addition to being competitively priced; this is a formidable task which may never be realized.

7.2.3 Other Prostanoid Receptor Antagonists

7.2.3.1 Prostaglandin D₂ Receptor Antagonists

There has been some interest in the development of specific PGD₂ receptor antagonists. A comprehensive review of the biology, pharmacology and potential physiological and pathophysiological significance of PGD₂ is given by Giles and Leff (1988). PGD₂ exerts a variety of what can be classified as stimulatory and inhibitory actions which appear to be mediated via an interaction with

PGD_2 (DP) and thromboxane (TP) and/or EP_1 receptors, respectively (Giles and Leff, 1988). For example, PGD_2 appears to inhibit platelet aggregation by acting through a different prostanoid receptor from that which mediates the inhibition elicited by PGI_2 and PGE_1 (MacIntyre and Gordon, 1977; Whittle *et al.*, 1978; Miller and Gorman, 1979; Tynan *et al.*, 1984). In addition, the relaxant effects of PGD_2 seem to be mediated by the DP receptor whereas the contractile effects are mediated by the TP or EP_1 receptor (Jones *et al.*, 1982; Coleman *et al.*, 1985a; Coleman and Kennedy, 1985; Ogletree *et al.*, 1985a; Giles and Leff, 1988; Ashida *et al.*, 1989; Beasley *et al.*, 1989). Thus, like other cyclooxygenase products, PGD_2 interacts with various prostanoid receptors.

Only one purported selective DP agonist has been proposed, BW 245C (Caldwell *et al.*, 1979), although it has been shown to also interact at EP and TP receptors (Giles *et al.*, 1989). Several putative PGD_2 receptor antagonists have been identified but they have limited potency and poor selectivity (Giles and Leff, 1988; Gardiner, 1990). For example, the first such compound N-0164 (Eakins *et al.*, 1976) antagonized, albeit very weakly, inhibition of platelet aggregation induced by PGD_2 or TXA_2/PGH_2 receptor activation but not that elicited by PGE_1 (MacIntyre and Gordon, 1977; Hamid-Bloomfield and Whittle, 1986). However, it also inhibited thromboxane synthesis (Kulkarni and Eakins, 1976). Another compound, AH6809 was reported to be a competitive antagonist ($pA_2 = 5.35$) at the platelet DP receptor (Coleman *et al.*, 1985b), but has been shown to interact at TP (Keery and Lumley, 1985) and EP_1 receptors (Coleman *et al.*, 1985b). Advances in our understanding of the receptors and mechanisms responsible for the biological effects of PGD_2 may become apparent with the utilization of the recently identified potent and apparently selective PGD_2 antagonist BW A868C (Giles *et al.*, 1989).

7.2.3.2 *Prostaglandin $F_{2\alpha}$ Receptor Antagonists*
The N-dimethylamine analogue of $PGF_{2\alpha}$ is the only reported antagonist of $PGF_{2\alpha}$ (Stinger *et al.*, 1982), although there is very little information on the compound. It was demonstrated to produce a dose-dependent shift to the right in $PGF_{2\alpha}$-induced pulmonary and systemic vascular effects in the rat, while having no effect on responses elicited by noradrenaline, U46619 or arachidonic acid (Stinger *et al.*, 1982). In contrast, a number of compounds – analogues of $PGF_{2\alpha}$ – are potent and selective agonists at the $PGF_{2\alpha}$ receptor (Coleman *et al.*, 1985a).

7.2.3.3 *Prostaglandin E Receptor Antagonists*
Agonists for the subtypes of the PGE receptor, EP_1, EP_2 and EP_3, have been discovered although they tend to interact with more than one of the three receptors (Gardiner, 1990). A selective, but rather weak, antagonist for the EP_1 receptor, SC 19220, was identified several years ago (Coleman *et al.*, 1980; Kennedy *et al.*, 1982; Gardiner, 1990), although no known selective antagonists of the two other EP receptor subtypes have been developed (Gardiner, 1990). AH 6809 is a more potent and selective EP_1 receptor antagonist (Coleman *et al.*, 1985a), although it has some DP and TP receptor-blocking activity (Keery and Lumley, 1985).

7.3 SUMMARY
Apart from cyclooxygenase inhibitors, the clinical trials of different classes of compounds which affect the prostanoid products of arachidonic acid metabolism, particularly the thromboxane receptor antagonists and thromboxane synthetase inhibitors, have proven to be disappointing. Despite significant effort over many years, there are still no potent, selective antagonists of many of the prostanoids. The hope is that with the identification of the more selective prostanoid agonists and antagonists, and progress in our knowledge of prostanoid receptor classification and biology, the anticipated promise of therapeutically beneficial effects of this class of compounds will be fulfilled.

8. *Receptor Antagonists – Lipoxygenase Pathway*
8.1 PEPTIDOLEUKOTRIENE (LEUKOTRIENE D_4) RECEPTOR ANTAGONISTS

8.1.1 Introduction
Over 50 years ago Kellaway and Trethewie published a manuscript entitled "The liberation of a slow reacting smooth muscle stimulatory substance of anaphylaxis" in which they described studies indicating that stimulated lung tissue releases a material, referred to as SRS, which contracted airway smooth muscle (Kellaway and Trethewie, 1940). In 1960, Brocklehurst first utilized the term "slow-reacting substance of anaphylaxis" (SRS-A) to describe this material released by human lungs following antigen challenge (Brocklehurst, 1960). Significant studies on SRS-A were conducted during the 1960s and 1970s, and research in this field was transformed by the key work of Samuelsson and coworkers, who identified and elucidated the structure and synthetic pathway of the peptidoleukotrienes LTC_4, LTD_4 and LTE_4, the arachidonic acid metabolites which were shown to account for the physicochemical and biological properties of SRS-A (Borgeat and Samuelsson, 1979; Murphy *et al.*, 1979; Corey *et al.*, 1980; Radmark *et al.*, 1980). In large part because of the availability of leukotrienes in significant quantities, there was a dramatic increase in the research on these compounds, particularly that which was directed towards investigation of their potential pathophysiological

(a)

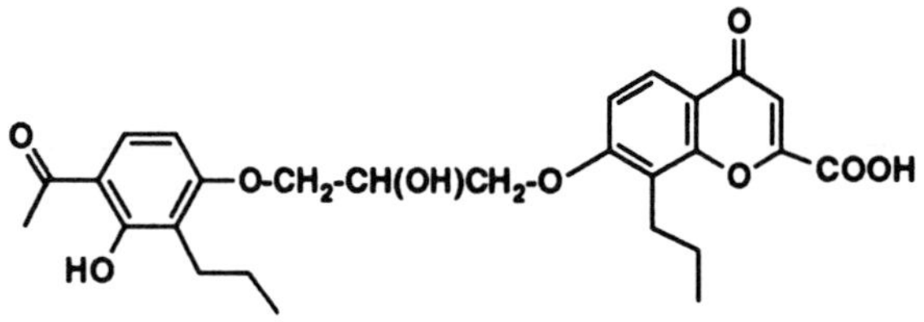

FPL-55712

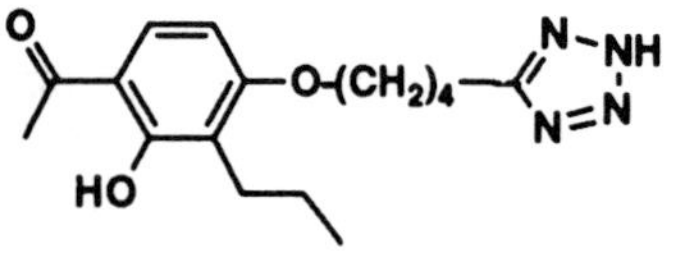

LY 171883

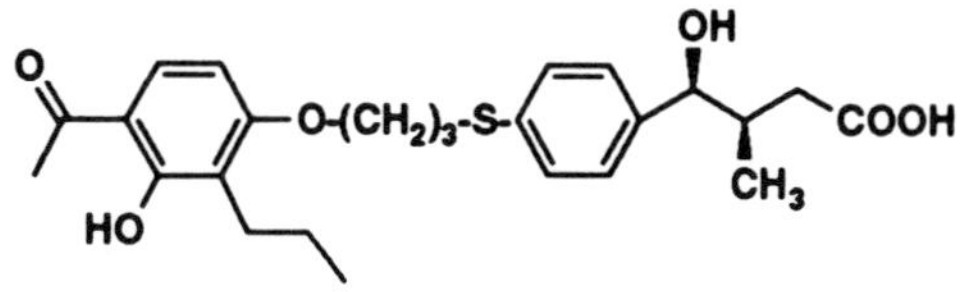

L-649,923

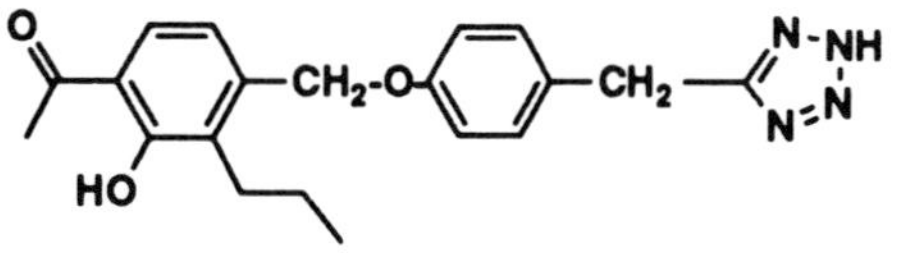

LY 163443

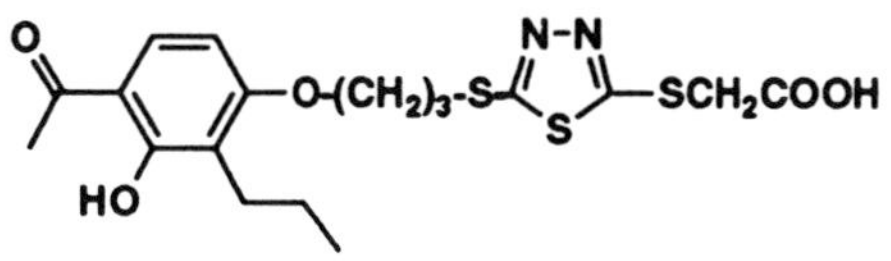

YM-16638

(c)

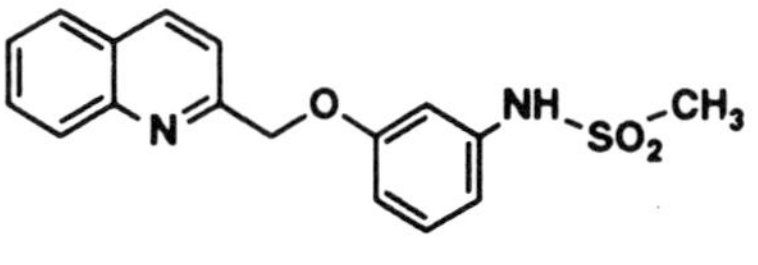

WY-48,252

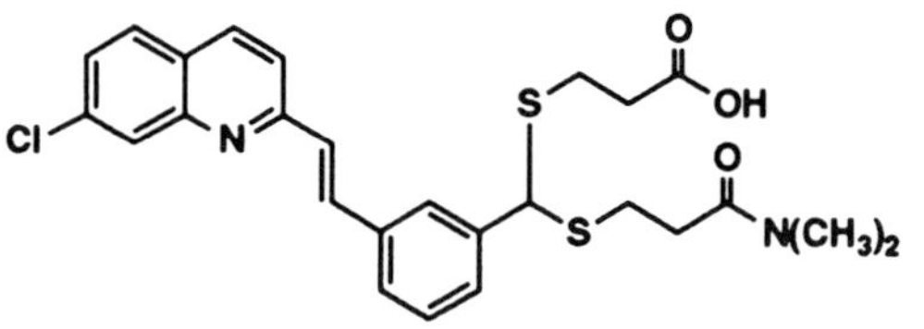

L-660,711 (MK-571)

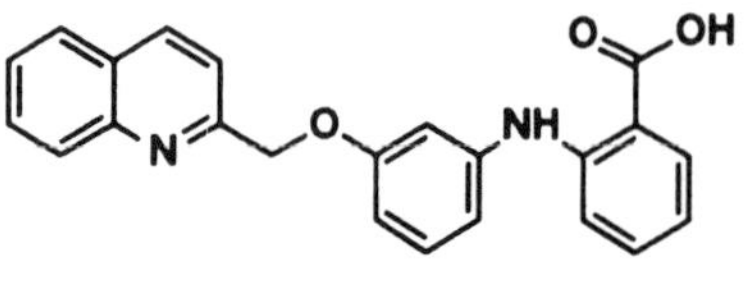

SR 2640

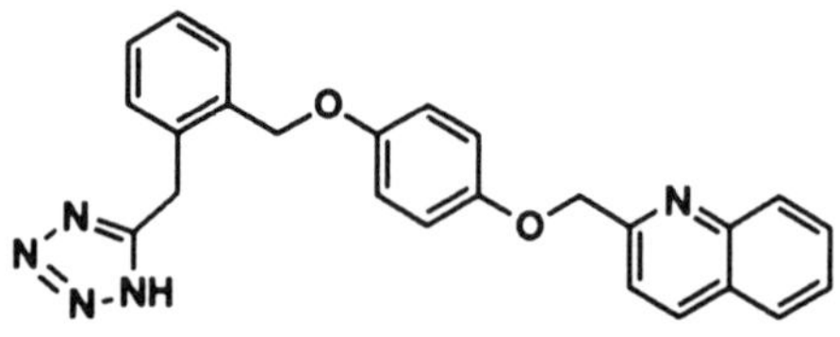

RG 12525

role in disease processes. The disease for which there is most convincing evidence for an involvement of leukotrienes in its aetiology is asthma (Samuelsson, 1983; Drazen and Austen, 1987; Piacentini and Kaliner, 1991). In addition, the leukotrienes, which are released from a variety of cells, have been implicated in a variety of disorders, including bronchitis (Luck *et al.*, 1989), endotoxaemia or septic shock (Ball *et al.*, 1986; Lefer, 1986; Feuerstein and Hallenbeck, 1987), ARDS, cystic fibrosis (Zakrzewski *et al.*, 1984), myocardial ischaemia (Feuerstein, 1984; Lefer, 1986), cardiac anaphylaxis (Levi *et al.*, 1982; Zukowska-Grojec and Feuerstein, 1985) and cerebral vasospasm and ischaemia (Feuerstein, 1984).

In view of the interesting and diverse biological profile of the leukotrienes, and their possible pathophysiological

(b)

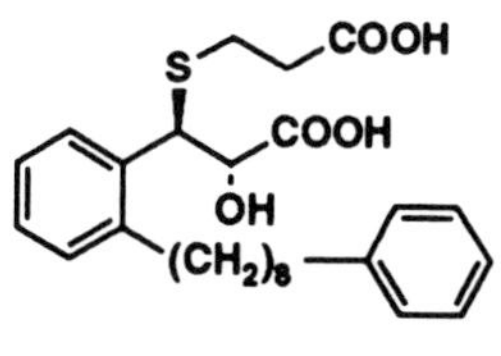

SK&F **104353**

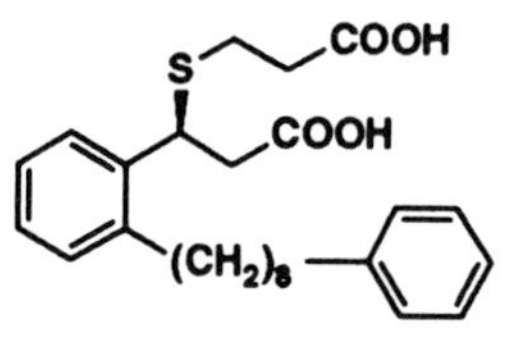

SK&F **106203**

(d)

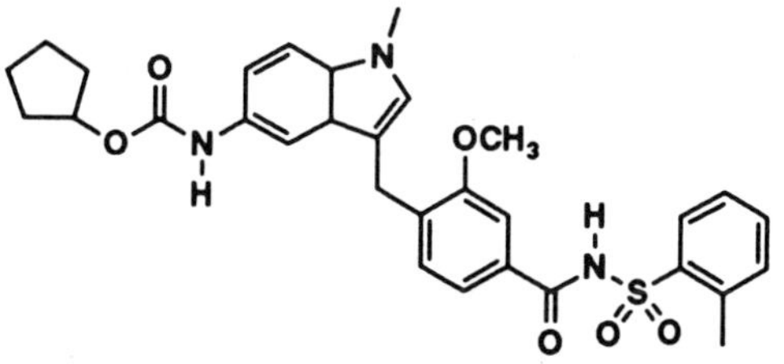

ICI 204219

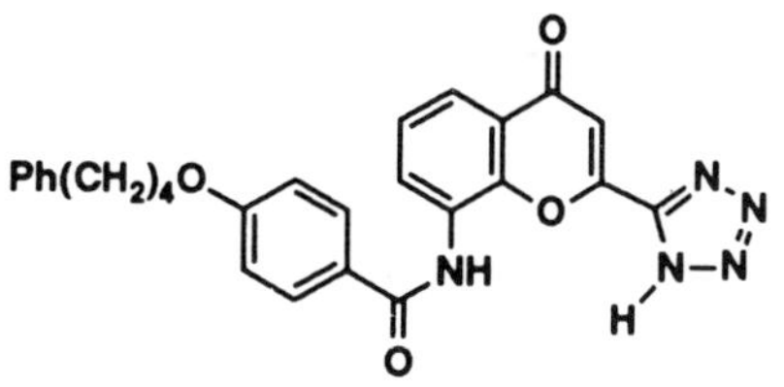

ONO-1078

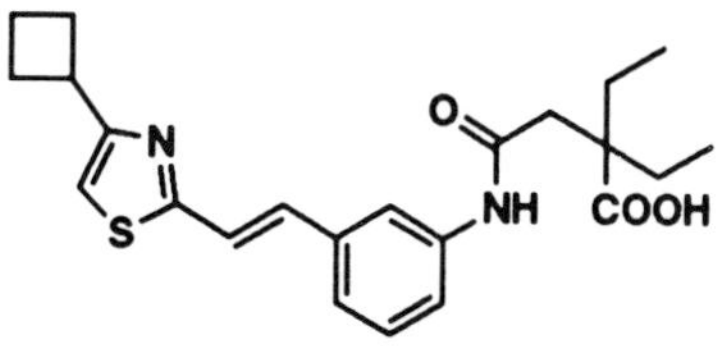

Ro 24-5913

Figure 5.12 Names and chemical structures of representatives of different classes of peptidoleukotriene receptor antagonists. (a) FPL-55712 and analogues (acetophenones). (b) Analogues of the peptidoleukotrienes. (c) Compounds having a quinoline nucleus. (d) Miscellaneous structures.

significance in an array of diseases, there has been a tremendous amount of interest and investment by the pharmaceutical industry in the synthesis and development of potent and selective compounds which will interfere with the formation (5-lipoxygenase inhibitors) and effects (LTD_4 receptor antagonists) of the leukotrienes. The latter effort, which has probably been the most intensive of those directed towards modulation of arachidonic acid metabolites, has been a very successful one, with several structurally distinct compounds currently in clinical trials, primarily for the treatment of asthma. In this section the focus will be on the significant amount of exciting data from clinical trials that have been conducted during the past few years which are providing increasing information to suggest that LTD_4 receptor antagonists may be of therapeutic benefit in the treatment of asthma. For more detailed information on the preclinical pharmacology of the many LTD_4 receptor antagonists, and also information on leukotriene receptors, the reader is guided to two reviews (Snyder and Fleisch, 1989; Krell *et al.*, 1990b).

8.1.2   Chemical/Pharmacological Classes

8.1.2.1   Historical Perspective: FPL-55712

The first leukotriene receptor antagonist identified was FPL-55712 (Fig. 5.12), synthesized by Fisons during their efforts to develop mast cell stabilizers (Augstein *et al.*, 1973). FPL-55712 was classified as an SRS-A (leukotriene) antagonist before the entities responsible for the biological activities of SRS-A had been elucidated (Chand, 1979), and for several years it was utilized effectively as an experimental tool and pharmacological probe to characterize the pharmacology of the leukotrienes. However, several aspects of the pharmacodynamic profile of FPL-55712, outlined below, limit its effectiveness in *in vivo* animal models and precluded significant clinical development of this antagonist.

In vitro studies indicated that FPL-55712 is an antagonist of LTD_4- and LTE_4-induced contraction of isolated guinea-pig trachea (pK_b = 6.0–6.7), but had minimal effect on contractions induced by LTC_4 (Krell *et al.*, 1981; Snyder and Krell, 1984; Weichman and Tucker, 1985). In contrast, in human bronchus, FPL-55712 was equally effective as an antagonist of contractions elicited by the three leukotrienes (pK_bs $\approx$ 6.0) (Buckner *et al.*, 1986). These data provided the first functional evidence for multiple leukotriene receptor subtypes in guinea-pig airways but a homogeneous leukotriene receptor population in human airway smooth muscle. These initial observations, which were supported by subsequent studies using other LTD_4 receptor antagonists (Hay *et al.*, 1987; Snyder *et al.*, 1987; Jones *et al.*, 1989; Buckner *et al.*, 1990), were critical from a drug development standpoint, in that they provided evidence that a single compound would effectively antagonize the bronchoconstrictor effects of LTC_4, LTD_4 and LTE_4 in humans.

Unfortunately, FPL-55712 has poor oral bioavailability and a short plasma half-life (0.6 min in guinea-pigs) when administered intravenously, which limits its usefulness for testing in animal models (Sheard *et al.*, 1977, 1984; Mead *et al.*, 1981). Nevertheless, limited clinical studies were conducted with FPL-55712, although the data obtained were equivocal and not very impressive. Thus, aerosol administration of FPL-55712 abolished the cough response and substantially reduced the bronchoconstriction induced by inhaled LTD_4 in non-asthmatic volunteers (Holroyde *et al.*, 1981). Nebulized FPL-55712, when administered four times a day for 7 days, produced mild bronchodilation in two of four chronic asthmatic patients, although no overall significant effect on pulmonary function was detected (Lee *et al.*, 1981). In another study, FPL-55712, as an aerosol, prevented antigen-induced decreases in tracheal mucus velocity but was without marked effect on antigen-induced bronchoconstriction in ragweed-sensitive patients (Ahmed *et al.*, 1981).

The unimpressive results of clinical studies with FPL-55712 as an aerosol were probably largely linked to its inherent limited potency. The poor oral absorption and inadequate pharmacokinetic profile of FPL-55712 obviously did not permit utilization of the compound via oral routes. In addition, a critical problem which precludes the potential utility of FPL-55712 as an appropriate clinical probe is its lack of selectivity. Thus, FPL-55712 has a variety of actions other than antagonism of the LTD_4 receptor, including: (1) PDE inhibition (Chasin and Scott, 1978); (2) 5-lipoxygenase inhibition (Casey *et al.*, 1983); (3) inhibition of histamine release and thromboxane synthetase (Welton *et al.*, 1981); and (4) non-specific antagonism of contraction produced by various agonists (Sheard *et al.*, 1984). In order to unequivocally determine in a clinical setting the role of any inflammatory mediator in the aetiology of a disease(s) it is imperative that the test compound is both *potent* and *selective* and, furthermore, possesses a reasonable pharmacokinetic profile. Unfortunately, FPL-55712 fits none of these criteria. Nevertheless, FPL-55712 has proven to be a useful template in the search for more appropriate compounds, and it should be reiterated that it has been a very successful experimental tool with which to define the pharmacology of leukotrienes.

8.1.2.2   Newer Leukotriene D_4 Receptor Antagonists

8.1.2.2.1   Introduction

During the past 12 years or so the pharmaceutical industry has utilized several chemical strategies in their quest for more potent and selective leukotriene receptor antagonists than FPL-55712, and which also possess a better pharmacokinetic profile: (1) analogues of FPL-55712 (which has been the most popular strategy); (2) compounds based on the structure of the natural ligands; (3) compounds containing a quinoline group; and (4) miscellaneous compounds which do not fit into any of the above categories. Examples of representatives – generally the most potent ones – of each of these broad categories, in which there is some overlap, are given in Fig. 5.12. An in-depth discussion of the preclinical pharmacology of these numerous and structurally diverse various compounds will not be attempted here. Rather, a broad summary, with some specific examples will be provided. In addition, the reader is directed to two articles for a comprehensive discussion and review of the medicinal chemistry of the leukotriene antagonists (Musser, 1989; Shaw and Krell, 1991).

8.1.2.2.2 Mechanism of action
Generally, these receptor antagonists act competitively.

8.1.2.2.3 Preclinical Studies
The activity of putative leukotriene receptor antagonists is normally determined by examining their ability to antagonize LTD_4-induced contractions of isolated guinea-pig trachea and to inhibit $[^3H]LTD_4$ binding in guinea-pig lung membranes; whenever possible, similar studies are conducted in human isolated airways and human lung membranes. In general, the majority of the leukotriene receptor antagonists, particularly those synthesized in recent years (e.g. ICI 198,615, ICI 204,219; MK-571; SK&F 104353; ONO-1078; SR 2640; RG 12525; LY 163443; LY 171883; LY 170680; Ro 24-5913; WY-48,252; SK&F 106203; CGP 45715A), are very potent compounds in both functional and binding assays, with pA_2 or pK_b values of about 7.5–9.5 and K_i values of about 0.1–10 nM, respectively (Fleisch et al., 1986; Adaikan et al., 1987; Hay et al., 1987, 1991; Obata et al., 1987; Snyder et al., 1987; Ahnfelt-Rønne et al., 1988; Boot et al., 1989; Hand et al., 1989a,b; Jones et al., 1989; Krell et al., 1989, 1990a; Van Inwegen et al., 1989; Ishii et al., 1990; Bray et al., 1991; O'Donnell et al., 1991). Thus, they are much more potent than FPL-55712, in some cases by several orders of magnitude. In addition, and in further contrast to FPL-55712, these newer compounds are generally also selective. However, the acetophenones LY 171883 and LY 163443 are potent inhibitors of phosphodiesterases (Fleisch et al., 1985, 1986).

The inherent *in vitro* activity of the compounds translated into impressive *in vivo* potency and effectiveness against leukotriene- and also antigen-induced broncho-constriction in guinea-pigs. The compounds were active when given via aerosol, oral and/or intravenous routes of administration, although, reflecting differences in oral absorption and pharmacokinetic profile in addition to inherent differences in potency, significant differences were observed in their potencies and durations of action (Fleish et al., 1986; Hay et al., 1987, 1991; Obata et al., 1987; Ahnfelt-Rønne et al., 1988; Hand et al., 1989a,b; Jones et al., 1989; Krell et al., 1989, 1990a; Torphy et al., 1989; Van Inwegen et al., 1989; Ishii et al., 1990; O'Donnell et al., 1991).

Of particular interest and potential relevance is the ability of SK&F 104353 (Hay et al., 1987) or MK-571 (Jones et al., 1989) to inhibit antigen-induced contraction of isolated human bronchus. This supports previous observations that antigen challenge elicited release of leukotrienes in parallel with contraction in lungs of two asthmatic patients natively sensitive to birch pollen (Dahlen et al., 1983). In addition, intravenous administration of SK&F 104353 (in the presence of the antihistamine mepyramine) (Osborn et al., 1992) or oral administration of MK-571 (Jones et al., 1989) inhibited antigen-induced bronchoconstriction in cynomolgus and squirrel monkeys, respectively. Furthermore, oral administration of WY-48,252 (Abraham et al., 1988) or YM-16638 (Tomioka et al., 1989) inhibited both the early and late-phase responses to antigen challenge in allergic sheep. These preclinical studies provided a strong rationale for the potential utility of potent and selective leukotriene receptor antagonists to alleviate antigen-induced bronchoconstriction. A more difficult task will be the assessment of the effects of this class of compounds on other aspects of the overall pathophysiology of asthma.

Several compounds possessed the desired criteria of potency and selectivity to warrant investigation in a clinical setting in diseases in which the leukotrienes have been implicated, notably asthma. Clinical trials with many of these compounds are still in progress.

8.1.2.2.4 Clinical Studies
The strategy employed in the testing of compounds as novel therapeutic agents in asthma has utilized investigation of the effects of the compounds against LTD_4-induced bronchoconstriction, antigen-induced broncho-constriction (early phase and late phase), and/or cold air-, aspirin- or exercise-induced bronchoconstriction.

8.1.2.2.4.1 Leukotriene D_4-induced Bronchoconstriction
The early compounds tested had unimpressive effects on LTD_4-induced bronchoconstriction. Thus, oral administration of LY 171883 (up to 400 mg) or L-649,923 (1000 mg) produced only an approximately three- to five-fold and four-fold shifts in LTD_4 dose–response curves, respectively, in non-asthmatic volunteers (Barnes et al., 1987; Phillips et al., 1988; Fuller et al., 1989). In contrast, the more recent compounds have exhibited substantially more dramatic effects on LTD_4-induced bronchospasm. For example, ICI 204,219 (40 mg p.o.) 2 h after administration produced an approximately 120-fold shift in the dose of LTD_4 aerosol required to reduce sGaw by 35% in non-asthmatic subjects; the compound continued to exhibit activity after 12 h (an approximately nine-fold increase over the LTD_4 dose) and 24 h (an approximately five-fold increase) (Smith et al., 1990). Intravenous infusion of the racemate MK-571 (approximately 28 mg) produced a shift of over 44-fold to the right of the LTD_4 aerosol dose–response curve in asthmatics (Kips et al., 1990); a similar potency was observed in another study with the more active enantiomer MK-0679 (Kips et al., 1991). In a recent publication, RG 12525 (800 mg) produced a 7.5-fold shift to the right in LTD_4 dose–response curves in mild asthmatics (Wahedna et al., 1991). In agreement with data from

preclinical experiments, clinical studies indicate a selectivity of action. Thus, L-649,923 (Barnes *et al.*, 1987) and SK&F 104353 (Dinh Xuan *et al.*, 1990; Evans *et al.*, 1990) were without effect on histamine-induced bronchoconstriction. Inhalation of SK&F 104353 reduced PAF-induced bronchoconstriction by about 30%, but had no effect on the fall in neutrophil count induced by PAF in non-asthmatic volunteers (Spencer *et al.*, 1991a); *in vitro* SK&F 104353 is without effect on PAF-induced contraction in guinea-pig trachea (Hay *et al.*, 1987). These data suggest that PAF-induced bronchoconstriction in humans is mediated in part via the release of leukotrienes, as proposed in other systems (Voelkel *et al.*, 1982; Bonnet *et al.*, 1983).

8.1.2.2.4.2   Baseline Lung Function

Information on the effects of leukotriene receptor antagonists alone on baseline pulmonary lung function is equivocal. ICI 204,219 (40 mg orally) produced a small but significant bronchodilator effect in asthmatic patients, reflected by an 8% increase of FEV_1 between 1 and 4 h after dosing; there appeared to be an additive bronchodilator effect with salbutamol (Hui and Barnes, 1991). In non-asthmatic subjects, ICI 204,219 had no effect on baseline pulmonary function (Smith *et al.*, 1990). In normal patients, aerosol administration of SK&F 104353 was without effect on baseline airway calibre (Evans *et al.*, 1990; Spencer *et al.*, 1991a), whereas in two studies in asthmatics a bronchodilator effect was observed in one study (Joos *et al.*, 1991) but not the other (Dinh Xuan *et al.*, 1990). Similarly, MK-571 (L-660,771), when given via intravenous infusion, did not improve pulmonary lung function in non-asthmatic individuals (Ford-Hutchinson, 1991), but did so in two of three studies in asthmatics (Gaddy *et al.*, 1990; Kips *et al.*, 1990; Manning *et al.*, 1990). In one of the studies the bronchodilator activity of MK-571 was observed for at least 5 h after stopping infusion of the compound (Kips *et al.*, 1990). No improvement in baseline pulmonary function was observed in some reported clinical studies with LY 171883 (Phillips *et al.*, 1988; Fuller *et al.*, 1989), L-649,913 (Barnes *et al.*, 1987) or RG 12525 (Wahedna *et al.*, 1991). Overall, the data suggest a role for the leukotrienes in the maintenance of baseline airway tone in some humans, with the influence more prevalent in asthmatic versus non-asthmatic airways.

8.1.2.2.4.3   Antigen-Induced Responses

Exposure of sensitized, atopic individuals to appropriate antigen results in significant bronchospasm, and this has been utilized extensively as a clinical pharmacological model of asthma. In a significant number (approximately 50–60%) of individuals, antigen provocation elicits a dual response: (1) an EAR which commences very quickly after challenge, reaches a maximum after approximately 30 min and then returns to the baseline after approximately 2 h; (2) an LAR which begins about 6–8 h after antigen exposure, reaches a maximum after about 12 h and then gradually returns to the baseline. The early asthmatic response has been attributed to activation of mast cells and release of bronchoconstrictor mediators, whereas the inflammatory cells and mediators, in addition to the underlying mechanisms responsible for the late-phase asthmatic response, remain uncertain (Larsen, 1985, 1989; Pelikan *et al.*, 1986; Holgate *et al.*, 1987). Nevertheless, the late-phase response has been associated with an increase in airway hyper-reactivity (Cartier *et al.*, 1982), suggesting that antigen challenge studies of this sort may be of relevance to asthma.

In six atopic subjects, oral administration of LY 171883 (400 mg; 2 h pretreatment) inhibited the EAR but had no effect on the LAR elicited by inhaled antigen; the compound had no effect on the wheal or flare response to intradermal antigen (Fuller *et al.*, 1989). Pretreatment for 2 h with ICI 204,219 (40 mg orally) inhibited the EAR and LAR to inhaled antigen in 10 mild atopic individuals and also inhibited the antigen-induced increase in non-specific airway reactivity (as assessed by responsiveness to histamine). However, it was proposed that the effect of ICI 204,219 on PC_{20} histamine may be reflective of a reduction in airway calibre induced by the compound and not an effect on the antigen-induced increase in bronchial responsiveness *per se* (Aalbers and DeMonchy, 1991; Taylor *et al.*, 1991). In another study, ICI 204,219 produced a shift to the right (approximately six-fold) in the antigen dose–response curve (Dahlén *et al.*, 1991). In one of the first studies, oral L-649,923 (1000 mg; 2 h pretreatment) had only a small effect on the EAR to antigen in mild asthmatics; this study was associated with drug-related gastrointestinal side-effects (Britton *et al.*, 1987).

Inhalation of L-648,051 (800 µg) was not effective in reversing antigen-induced bronchospasm. Furthermore, it was without effect, when given as a pretreatment (5 min), on the LAR and caused only a slight reduction in the EAR (Rasmussen *et al.*, 1991). In another study, L-648,051, again given via inhalation, was without effect on the EAR or LAR elicited by antigen challenge and also the associated increase in bronchial hyper-responsiveness (Bel *et al.*, 1990). In contrast, intravenous infusion of MK-571 decreased both the EAR and LAR to inhaled antigen (Hendeles *et al.*, 1990). Antigen-induced bronchoconstriction (EAR) in seven asthmatic subjects was reduced after 7 days of treatment with 150 mg/day of ONO-1078 (Nakagawa *et al.*, 1991). Another study reported a significant inhibitory effect of ONO-1078 on the LAR but not the EAR following exposure to mite antigen in 16 asthmatics (Yamai *et al.*, 1989). In a study using 10 mild asthmatics sensitive to ragweed pollen or cat dander, inhalation of SK&F 104353 (640 µg; 20 min pretreatment) produced a substantial shift to the right in antigen dose–response curves, with a degree of protection similar to that observed with the β_2 adrenoceptor agonist metaproterenol (Creticos *et al.*, 1989). Importantly,

SK&F 104353 abolished, whereas metaproterenol had no effect, on the LAR that was observed in two of the 10 subjects. In another study, in order to elucidate the role of leukotrienes and histamine in mediating antigen-induced bronchoconstriction, the effects of inhaled SK&F 104353 (800–1200 µg) or the antihistamine terfenadine (120 mg p.o.), alone or in combination, was examined in mild asthmatics. The antigen-induced decrease in FEV_1 was inhibited to a similar extent by either SK&F 104353 or terfenadine, and it was observed that the combination was more effective than either agent alone (Eiser *et al.*, 1989). The combination of SK&F 104353 and terfenadine essentially abolished both the EAR and LAR in three of the five subjects who were dual responders; the effects of either agent alone were unfortunately not examined. Caution should be exercised in the interpretation of this interesting latter data, as terfenadine has been shown to inhibit the late-phase response (Hamid *et al.*, 1990) and to inhibit mast cell degranulation (Naclerio *et al.*, 1990). More comprehensive combination studies of this genre with appropriate selective and potent drugs are required to determine with any degree of confidence the relative roles of putative mediators in the EAR and LAR elicited by antigen. More importantly, the extension of these combination studies with different classes of compounds to complex, multifactorial diseases such as asthma in which multiple mediators are likely to play a role is awaited with eager interest.

Overall, these preliminary studies with the peptido-leukotriene receptor antagonists indicate marked effects on the EAR, and in some cases significant effects on the LAR, elicited by antigen provocation. In view of the known *in vitro* effects of peptidoleukotriene receptor antagonists against antigen-induced contraction in isolated human and guinea-pig airways, and also their ability to inhibit bronchospasm produced by antigen challenge in guinea pigs and monkeys *in vivo* (*vide supra*), an appreciable effect of these compounds on the antigen-induced EAR would perhaps have been expected. Of more interest and potential relevance is their inhibitory action against the LAR, and also the associated increased bronchial responsiveness, suggesting a role for leukotrienes in these phenomena. However, more comprehensive studies with various compounds are required.

8.1.2.2.4.4   Exercise-, Cold Air- and Aspirin-Induced Bronchoconstriction

Up to 70–80% of asthmatics are susceptible to exercise-induced asthma which is manifest as acute bronchoconstriction that is normally maximal at 5–10 min after exercise and usually recovers spontaneously to baseline levels within 30–90 min. The underlying mechanism responsible for exercise-induced asthma remains controversial but it has been proposed that it may be related to cooling and/or loss of water from the airway mucosa, which may cause release of bronchoconstrictor mediators; there may be a neurogenic component and an inflamma-

tory response may be involved in late-phase response to exercise (Anderson, 1985; Godfrey, 1986; McFadden, 1987; Lee and O'Hickey, 1989). In a study involving 18 subjects with exercise-induced asthma, administration of nebulized SK&F 104353 (800 µg) 15 min before exercise inhibited the fall in FEV_1 by about 40%; SK&F 104353 was as effective as sodium cromoglycate (20 mg, nebulized aerosol) (Robuschi *et al.*, 1991; Torphy *et al.*, 1991). Similarly, in 12 subjects with stable asthma, intravenous administration of MK-571 (160 mg), 20 min before exercise, produced about 70% inhibition of the decrease in FEV_1 and markedly decreased the mean time to recovery after exercise from 33.4 ± 4.0 to 8.4 ± 2.5 min (Manning *et al.*, 1990). These data suggest a role for leukotrienes in exercise-induced asthma. The mechanism whereby the leukotrienes may be involved is uncertain but may be linked to the recent finding that leukotriene receptor antagonists, including SK&F 104353, reduced C fibre-mediated contractions of guinea-pig airways (Ellis and Undem, 1991).

Treatment for 2 weeks with oral LY 171883, 600 mg twice a day, inhibited cold air-induced bronchoconstriction in asthmatics (Israel *et al.*, 1989).

In five of six aspirin-sensitive asthmatics, inhalation of SK&F 104353 (approximately 900 µg, 30 min pretreatment), substantially inhibited ($56.0 \pm 6.7\%$) the decrease in FEV_1 following aspirin challenge (Christie *et al.*, 1991).

8.1.2.2.4.5   Asthma

A major deficit has been the lack of information on the effects of the leukotriene receptor antagonists in the native disease. In the only published study from a multicentre trial in chronic asthma, the effects of LY 171883, 60 mg twice daily for up to 6 weeks, in 138 subjects was reported. There was a small, gradual improvement in FEV_1 values in patients receiving LY 171883 but not placebo. Furthermore, in a subset of patients who used at least 23 mg/week of the β_2 adrenoceptor agonist metaproterenol there was a decrease in its use in those who received LY 171883 (Cloud *et al.*, 1989). However, overall in this study the effects of LY 171883 were not very impressive, perhaps due to limited inherent potency of the compound. In a recent publication on ONO-1078 it was stated that, to date, studies in mild to moderate asthmatics indicated that "overall efficacy reached more than 80% in these asthmatic patients" (Nakagawa *et al.*, 1991); however, no data were presented.

8.1.3   Summary

In recent years a variety of chemically distinct compounds have been identified as potent and selective LTD_4 receptor antagonists and many are presently in clinical trials, with the primary indication being asthma. These newer compounds are not plagued with the limited potency and lack of selectivity which was associated with the early LTD_4 receptor antagonists and appear to possess an appropriate pharmacokinetic and pharmacodynamic profile with which to unequivocally determine the role of the leukotrienes in the variety of diseases in which they have been implicated. Thus far, the data from clinical trials for asthma are interesting and encouraging. For example, several compounds, in addition to exhibiting the expected activity as potent and effective antagonists of LTD_4-induced bronchoconstriction, have been demonstrated to inhibit antigen-induced bronchoconstriction (the EAR and in some cases the LAR) and also exercise-induced asthma. The results of the pivotal clinical trials in the native population are eagerly awaited. Overall, there is the real potential that, by the latter half of this decade, leukotriene receptor antagonists will be the first novel class of compounds for the treatment of asthma for over a quarter of a century.

Regardless of whether the potential therapeutic utility of leukotriene receptor antagonists is borne out, the next few years promise to be an exciting culmination of the extensive efforts over many decades on the biological effects and pathophysiological role of the peptidoleukotrienes (formerly SRS-A).

8.2   LEUKOTRIENE B_4 RECEPTOR ANTAGONISTS

8.2.1   Introduction

The other prominent product of the 5-lipoxygenase-driven pathway of arachidonic acid metabolism is LTB_4 ($5S,12R$)-6,14-*cis*-8,10-*trans*-eicosatetraenoic acid), which was identified and named by Borgeat and Samuelsson in 1979. The biological effects of LTB_4, which is formed from LTA_4 by the action of LTB_4 synthetase, will be summarized briefly. In view of the numerous effects of LTB_4 that have been identified, particularly its actions on leucocyte function, it is being put forward as a strong candidate to be a pivotal player in immediate hypersensitivity reactions in various inflammatory processes and diseases.

LTB_4 is one of the most potent chemotactic and chemokinetic substances for human neutrophils that has been identified (Ford-Hutchinson *et al.*, 1980; Bray, 1983; Thorsen, 1986; Goldman *et al.*, 1987; Ford-Hutchinson, 1990). It also produces aggregation (Ford-Hutchinson *et al.*, 1980) and degranulation of neutrophils (Bray, 1983), mediates superoxide generation and lysosomal enzyme release (Rae and Smith, 1981; Serhan *et al.*, 1982), stimulates adhesion of leucocytes to the vascular endothelium (Gimbrone *et al.*, 1984; Hoover *et al.*, 1984), elicits bronchoconstriction (Bray, 1983) and increases vascular permeability (Bray *et al.*, 1981; Ford-Hutchinson, 1990). Inhalation of LTB_4 results in a profound infiltration of eosinophils and neutrophils into guinea-pig airways (Silbaugh *et al.*, 1987). *In vivo* in various species, including humans, LTB_4 administration

or exposure elicits the accumulation of PMNs into several sites in the body (Bray, 1983; Thorsen, 1986; Ford-Hutchinson, 1990). In addition, elevated levels of LTB_4 have been detected in several animal models of inflammation (McMillan and Foster, 1988; Ford-Hutchinson, 1990). LTB_4 is released from a variety of inflammatory cells, including human neutrophils, alveolar macrophages, monocytes and lung mast cells (Fels *et al.*, 1982; Ferreri *et al.*, 1986; Williams *et al.*, 1986; Freeland *et al.*, 1988; Ford-Hutchinson, 1990).

The actions of LTB_4 are mediated via specific receptors. In human neutrophils, radioligand-binding experiments have identified two separate, stereospecific populations of LTB_4-binding sites (receptors?): a high-affinity site (K_d = 0.4 nM) which mediates chemotaxis and chemokinesis and aggregation, and a low-affinity site (K_d = 61 nM) which is responsible for degranulation (Goldman *et al.*, 1987; Ford-Hutchinson, 1990).

In relation to a potential significant role for LTB_4 in the pathophysiology of diseases the most convincing evidence is in psoriasis (and other skin disorders), and inflammatory bowel disease, inflammatory disorders which have a significant neutrophil infiltration. For example, elevated levels of LTB_4 were detected in and released from lesional skin of patients with psoriasis (Brain *et al.*, 1984a,b; Grabbe *et al.*, 1984). LTB_4 is a potent stimulator of keratinocyte proliferation in cultured human epidermal cells (Kragballe *et al.*, 1985) and also following topical application in normal and psoriatic patients (Van der Kerkhof *et al.*, 1985). In addition, 5-lipoxygenase activity is increased in epidermis from patients with psoriasis (Ziboh *et al.*, 1984). There is evidence that LTB_4 is present in other inflammatory skin disorders such as allergic contact and atopic dermatitis (Barr *et al.*, 1984; Thorsen *et al.*, 1990), dyshidrotic eczema and bullous pemphigoid (Rosenbach *et al.*, 1985; Kawana *et al.*, 1990) and eosinophilic cellulitis (Wong, 1984).

In IBD, elevated levels or increased production of LTB_4 have been detected in colonic mucosa and there was increased chemotactic activity – attributed to LTB_4 – in homogenates of diseased versus control mucosa (Sharon and Stenson, 1984; Nielsen *et al.*, 1986; Lobos *et al.*, 1987). Furthermore, mucosal LTB_4 levels were increased in patients with gastritis associated with *Campylobacter pylori* (Fukuda *et al.*, 1990) or duodenal ulcers (Ackerman *et al.*, 1990); in the latter case the generation of LTB_4 was significantly reduced following treatment (Ackerman *et al.*, 1990).

Levels of LTB_4, or its metabolite 20-hydryoxy-LTB_4, have been reported to be elevated in body fluids of patients with a variety of diseases, including asthma (O'Driscoll *et al.*, 1984; Lam *et al.*, 1988; Shindo *et al.*, 1990), cystic fibrosis and chronic bronchitis (O'Driscoll *et al.*, 1984), gout (Rae *et al.*, 1982) and rheumatoid arthritis (Klickstein *et al.*, 1980; Davidson *et al.*, 1983) and also following antigen challenge in atopic patients and those with allergic

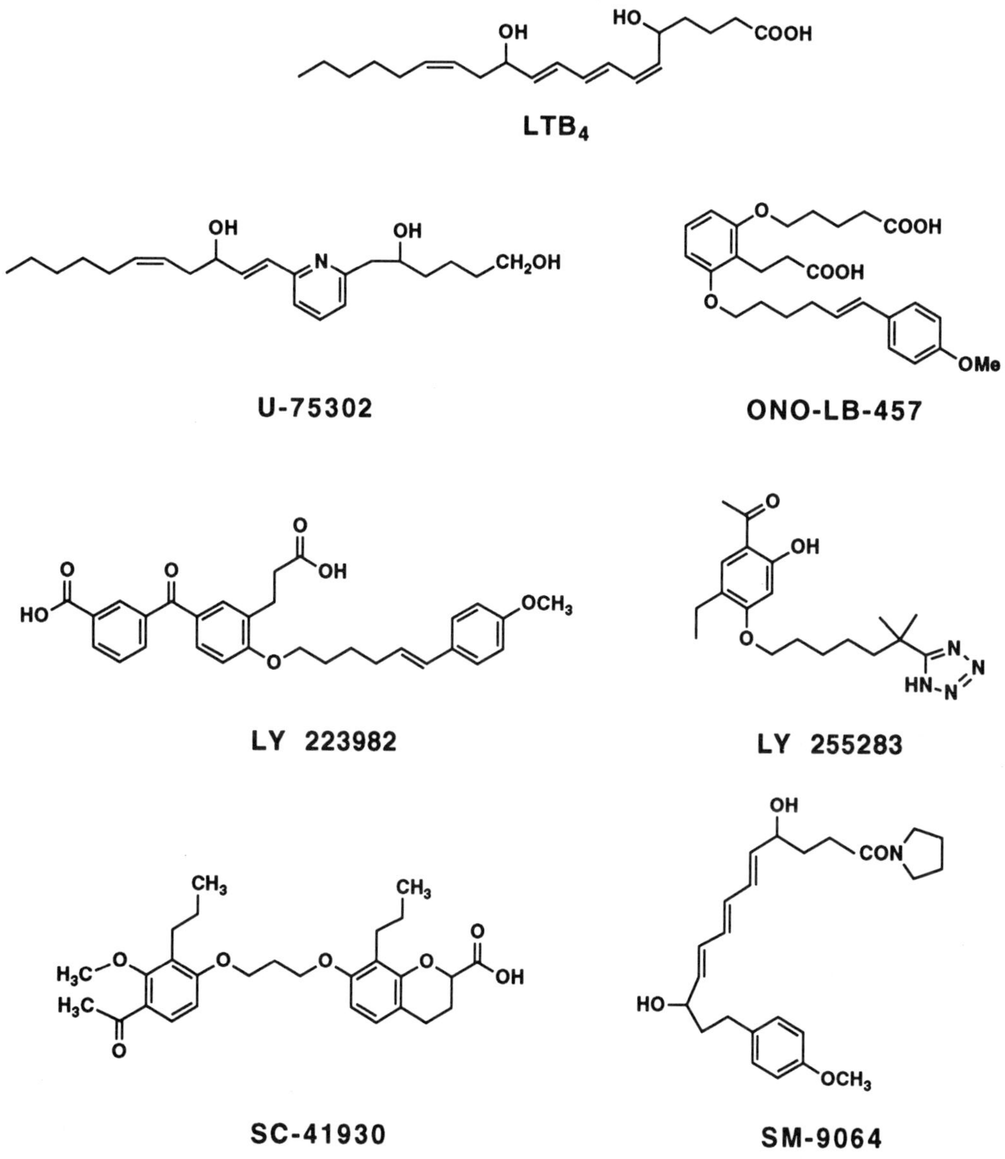

Figure 5.13 Names and chemical structures of LTB$_4$ and representative LTB$_4$ receptor antagonists.

rhinitis (Freeland *et al.*, 1988, 1989). It has also been speculated that LTB$_4$ may be involved in pain (Levine *et al.*, 1984; Bisgaard and Kristensen, 1985; Thorsen, 1986).

8.2.2 Chemical/Pharmacological Classes

One obvious potential therapeutic approach to combat diseases in which LTB$_4$ is implicated is to utilize potent 5-lipoxygenase inhibitors which will attenuate the production of LTB$_4$, in addition to the peptidoleukotrienes; this strategy has been discussed above in Sections 4 and 5. An alternative and more selective mechanism is to develop agents which will antagonize the actions of LTB$_4$ at the receptor level. Another theoretical strategy would

be the development of compounds which inhibit the enzyme LTB$_4$ synthetase, involved in the formation of LTB$_4$; there is no information available on research based on this approach.

The structures of purported LTB$_4$ receptor antagonists for which the most information is available is given in Fig. 5.13; the activity of the individual compounds is summarized separately below.

8.2.2.1 *U-75,302*

Upjohn directed one of the earliest efforts towards the development of LTB$_4$ receptor antagonists. The strategy employed was to synthesize molecules that were based on the structure of the natural ligand LTB$_4$. Specifically, they designed a series of compounds in which they altered the *cis,trans,trans*-triene moiety of LTB_4, by replacing C-7–C-9 with a 2,6-disubstituted pyridine ring (Morris and Wishka, 1988). Several of these structural analogues of LTB_4, including U-75,302, inhibited [^{3}H]LTB_4 binding to human neutrophil membranes, LTB_4-induced neutrophil aggregation and LTB_4-induced contraction of guinea-pig lung parenchyma strips (Lin *et al.*, 1988; Morris and Wishka, 1988). Interestingly, in general, this series of compounds were more potent inhibitors of [^{3}H]LTB_4 binding in isolated membranes versus whole cells (Lin *et al.*, 1988). However, the compounds generally had agonist activity; perhaps in view of their structural similarity to LTB_4 this is not surprising. Nevertheless, U-75,302 effectively antagonized LTB_4-induced contraction of guinea-pig lung parenchyma strips at a concentration (0.3 μM) that was devoid of agonist activity, and several compounds inhibited LTB_4-induced neutrophil aggregation at concentrations much lower ($>$ 100-fold) than those that produced agonist activity (Morris and Wishka, 1988). Selectivity of action of U-75,303 was suggested by its lack of effect against contractions of guinea-pig lung produced by a variety of agonists (Lawson *et al.*, 1989). In addition, U-75,302 inhibited LTB_4-induced release of TXB_2 from lung parenchyma, apparently via an interaction at the LTB_4 receptor level (Lawson *et al.*, 1989). U-75,302 was a weak and not very effective inhibitor of LTB_4-induced chemotaxis of guinea-pig eosinophils (33% inhibition at 10 μM) and also exhibited some chemotactic activity on its own (Taylor *et al.*, 1989). In addition, lack of selectivity of the compound was suggested as it also inhibited chemotaxis induced by zymosan-activated plasma (Taylor *et al.*, 1989).

In vivo, oral administration of U-75,302 (10 and 30 mg/kg; 1 h pretreatment) inhibited antigen-induced influx and adherence of eosinophils (but not neutrophils) in guinea-pig airways (Richards *et al.*, 1989). Similarly, in *Ascaris*-sensitive primates, U-75,302 (30 mg/kg p.o.) inhibited the eosinophilia but not the bronchoconstriction elicited by antigen exposure (Stout and Johnson, 1990). These *in vivo* data suggest that LTB_4 plays a significant chemotactic role in eosinophil influx following antigen challenge, at least in guinea-pigs and monkeys.

8.2.2.2 SC-41930

SC-41930 (7-[3-(4-acetyl-3-methoxy-2-propylphenoxy)propoxy] - 3,4 - dihydro - 8 - propyl - *2H* - 1 - benzopyran - 2-carboxylic acid) from G. D. Searle and Co. has been identified as an orally active, selective LTB_4 receptor antagonist. SC-41930 inhibited [^{3}H]LTB_4 binding to human neutrophils (IC_{50} = 0.3 μM), and inhibited both LTB_4-induced neutrophil degranulation (IC_{50} = 1 μM) and neutrophil chemotaxis (IC_{50} = 2 μM) (Djuric *et al.*, 1989). This suggests that SC-41930 antagonizes the effects of LTB_4 at both the high- and low-affinity PMN receptors. SC-41930 competitively inhibited LTB_4-induced

chemotaxis of human neutrophils (pA_2 = 6.35). Although without effect on C5a-induced chemotaxis, SC-41930 inhibited chemotaxis elicited by FMLP (pA_2 = 5.37). In addition, the compound was about 10-fold more potent as an antagonist of PMN chemotaxis induced by 20 hydroxy-LTB_4 or 12(*R*)-HETE than that elicited by LTB_4 (Tsai *et al.*, 1989). SC-41930 inhibited LTB_4-induced degranulation of PMN, albeit in a non-competitive manner, and was about 10-fold less effective as an antagonist of FMLP- or C5a-induced degranulation (Tsai *et al.*, 1989). SC-41930 was 10 times less potent as an inhibitor of [^{3}H]FMLP binding compared to [^{3}H]LTB_4 binding in human neutrophils (Tsai *et al.*, 1989). With regard to selectivity, SC-41930 did not inhibit bovine heart phosphodiesterase, and at a concentration of 10 μM was without significant effect on [^{3}H]LTD_4 binding to guinea-pig lung homogenate (Djuric *et al.*, 1988; Tsai *et al.*, 1989). In relation to the latter observation, it is of interest that SC-41930 was synthesized by phenolic methylation of a chroman carboxylic acid peptidoleukotriene receptor antagonist related to FPL-55712 (Djuric *et al.*, 1988).

SC-41930 has been shown to be effective in various *in vivo* models of inflammation. For example, when given topically or orally, SC-41930 inhibited PMA- or A23187-induced ear inflammation (Djuric *et al.*, 1988; Fretland *et al.*, 1990b), and following oral administration (30 min pre- and postchallenge) in guinea-pigs it inhibited acetic acid-induced colonic inflammation (a model of IBD) with an ID_{50} of 41 $\pm$ 4.3 μmol/kg (approximately 20 mg/kg) (Djuric *et al.*, 1988; Fretland *et al.*, 1989). In the latter model, oral administration of 40 mg/kg resulted in plasma levels of 7 μg/μl after 1 h and 12 μg/μl after 12 h (Djuric *et al.*, 1989). Rectal administration of SC-41930 inhibited colonic mucosal inflammation induced by acetic acid in rats, guinea-pigs and rabbits, as reflected by attenuation of the increase in myeloperoxidase, vascular permeability, oedema and neutrophil influx that is observed in inflamed tissues: the ID_{50} values were about 20–30 mg/kg (Fretland *et al.*, 1990a). Importantly, SC-41930 has not been reported to possess intrinsic agonist activity.

Interestingly, in a study using rectal biopsies from patients with ulcerative colitis, SC-41930 produced a small (17%) but significant decrease in the musocal release of LTB_4, being as effective as 5-aminosalicylic acid (18% decrease), a standard treatment for this disease (Gertner *et al.*, 1990). This finding suggests that SC-41930 may have some 5-lipoxygenase inhibitory activity. Researchers from Searle recently reported that SC-41930 has a variety of effects other than LTB_4 receptor antagonism; these effects occurred in concentrations not much higher than those required for its recognized action. Thus, SC-41930 inhibited agonist-induced superoxide generation, LTB_4 and 5-HETE production from human PMNs (IC_{50} = 4–12 μM), and LTB_4 and PGE_2 release from HL-60 cells (Villani-Price *et al.*, 1992). In addition, when co-injected intradermally (400 μg/site), SC-41930 attenuated A23187-induced increases in LTB_4 levels in guinea-pig skin.

Higher concentrations of SC-41930 inhibited human synovial PLA_2 (IC_{50} = 72 μM) and rat peritoneal LTA_4 hydrolase (IC_{50} = 20 μM). This profile was postulated to contribute to the anti-inflammatory activity of the compound and it was proposed that SC-41930 produced these effects, in part, by post-receptor inhibition of mediator production by PMNs.

SC-41930 is a novel but not very selective LTB_4 receptor antagonist. It is without agonist activity and has recently been reported to be under evaluation in clinical trials for the treatment of ulcerative colitis (Gaginella, 1990), although no information is yet available on the results. However, the utility and clinical efficacy of SC-41930 may be limited by its inherent low potency (e.g. pA_2 = 6.35 versus LTB_4-induced PMN chemotaxis). A clear elucidation of the role of LTB_4 in inflammatory diseases is likely to require much more potent and selective compounds than SC-41930.

8.2.2.3 *LY 223982 and LY 255283*

Researchers at Eli Lilly have synthesized and examined the activity of a series of lipophilic benzophenone dicarboxylic acid derivatives as LTB_4 receptor antagonists. The most potent compound was LY 223982 ((E)-5-(3-carboxybenzoyl)-2-{[6-(4-methoxyphenyl)-5-hexenyl]oxy}benzene-propanoic acid) which had an IC_{50} of 13.2 ± 2.2 nM for inhibition of [^{3}H]LTB_4 binding to human neutrophils (Gapinski *et al.*, 1990a,b). In addition, modification of the structure of acetophenone peptidoleukotriene receptor antagonists resulted in compounds which were LTB_4 receptor antagonists, the most potent of which was LY 255283 (1-{5-ethyl-2-hydroxy-4-[6-methyl-6-(1H-tetrazol-5-yl)heptyloxy]phenyl}ethanone) (Herron *et al.*, 1988). LY 255283 is approximately sevenfold less potent than LY 223982 as an inhibitor of [^{3}H]LTB_4 binding in human neutrophils (IC_{50} = 87 and 12 nM, respectively) (Jackson *et al.*, 1988a). Both compounds were much less potent inhibitors of [^{35}S] FMLP binding to human neutrophils (IC_{50}s > 1 μM)(Jackson *et al.*, 1988a). LY 223982 (IC_{50} = 0.074 μM) was about twice as potent as LY 255283 (IC_{50} = 0.15 μM) for inhibition of LTB_4-induced aggregation of elicited peritoneal guinea-pig neutrophils. The antagonism produced by LY 223982 was competitive in nature (pA_2 = 8.3), whereas LY 255283 appeared to have a non-competitive component to its action (Jackson *et al.*, 1988b).

LY 255283 and, also, SC-41930 were less potent (about 100-fold) inhibitors of LTB_4-induced human neutrophil adhesion than superoxide production *in vitro* (Schultz *et al.*, 1991). *In vivo*, intravenous administration of LY 223982 (2–10 mg/kg; IC_{50} = 3 mg/kg) or LY 255283 (0.02–2 mg/kg; IC_{50} = 0.2 mg/kg) produced dose-dependent inhibition of LTB_4-induced leucopenia in rabbits, but had no effect on FMLP-induced leucopenia (Jackson *et al.*, 1988a). It is of interest that, despite being inherently less potent *in vitro*, LY 255283 was more

potent than LY 223982 *in vivo* in this study. Intravenous infusion of LY 255283, at a dose which afforded almost complete inhibition of LTB_4-induced leucopenia, hypotension or tachycardia, did not affect the degree of myocardial ischaemia or infarct size, or intensity and size of cardiac arrhythmias following coronary artery occlusion in dogs; there was also no effect of LY 255283 on neutrophil infiltration into the ischaemic myocardium (Hahn *et al.*, 1990).

There is no clinical information on LY 223982 or LY 255283, although LY 223982 has been reported to be in development (Gapinski *et al.*, 1990b). Both compounds appear to be of appropriate potency and selectivity, based on *in vitro* and limited *in vivo* studies, to warrant investigation in humans, and these results should provide important information on the potential role of LTB_4 in inflammatory diseases.

8.2.2.4 *ONO-LB-457*

ONO-LB-457 (5-{2-(2-carboxyethyl)-3-[p-methoxyphenyl)-5E-hexenyl]oxyphenoxy}valeric acid) is the most potent example of a series of β-phenylpropionic acid-containing compounds synthesized by the ONO Pharmaceutical Company that have been proposed to be orally active LTB_4 receptor antagonists (Konno *et al.*, 1991). ONO-LB-457 is a potent inhibitor of [^{3}H]LTB_4 binding in human (K_i = 6 nM) and also guinea-pig (K_i = 18 nM) and rat PMNs (K_i = 86 nM). ONO-LB-457 inhibited LTB_4-induced aggregation (IC_{50} = 2.4 μM) and degranulation (IC_{50} = 3.0 μM) of human PMNs, suggesting that it interacts with both high- and low-affinity LTB_4 receptors, respectively (Kishikawa *et al.*, 1991). In these two assays, ONO-LB-457, up to a concentration of 10 μM, was without agonist activity. The compound also inhibited LTB_4- but not LTC_4-induced contractions of guinea-pig lung parenchymal strips (IC_{50} = 0.7 μM) (Kishikawa *et al.*, 1991). *In vivo* intravenous administration of ONO-LB-457 (0.3–3 mg/kg) produced a dose-related inhibition of LTB_4-induced leucopenia in guinea-pigs. Furthermore, in the same species, oral administration of ONO-LB-457 (3–10 mg/kg) dose-dependently attenuated LTB_4-induced neutrophil accumulation in the skin; the effects of 30 mg/kg of ONO-LB-457 persisted for longer than 6 h (Kishikawa *et al.*, 1991). Thus, the very limited preclinical information that is available suggests that ONO-LB-457 is a potent, orally active and long-acting LTB_4 receptor antagonist.

8.2.2.5 *Miscellaneous Compounds*

SM-9064, a dihydroxy unsaturated fatty acid derivative, was reported to inhibit the chemotaxis of rat PMNs induced by LTB_4 (IC_{50} = 0.13 μM) but had little or no effect on that elicited by other agonists. SM-9064 alone had some chemotactic activity, about 20% of LTB_4. When given orally (5–10 mg/kg) or intraperitoneally (2.5–10 mg/kg), it provided dose-dependent inhibition of Arthus reaction-induced paw oedema in mice, but, even at a dose of 50 mg/kg, oral SM-9064 had no effect on carrageenan-induced paw oedema in rats (Namiki *et al.*, 1986). Some interphenylene derivatives, based on the natural ligand LTB_4, were reported to have mixed LTB_4 agonist/antagonist activity and to have anti-inflammatory activity when administered topically, i.e. inhibition of A23187-induced oedema and cell infiltration in mouse skin. However, the compounds were not very potent (Ekerdt *et al.*, 1991).

A number of structures very closely related to LTB_4, e.g. acetyl-LTB_4 (Goetzl and Pickett, 1981), LTB_4 dimethylamide (Showell *et al.*, 1982) and 5(*S*),12(*S*)-diHETE (Feinmark *et al.*, 1981), have been reported to inhibit LTB_4-induced effects *in vitro*. However, it is not know if these effects were due to classical, competitive antagonism at the LTB_4 receptor or to some other effect, e.g. receptor desensitization.

8.2.3   Mechanism of Action

The compounds considered in this section are competitive and non-competitive receptor antagonists.

8.2.4   Clinical Studies

Other than the preliminary data with SC-41930, there is no information on clinical studies with LTB_4 receptor antagonists.

8.2.5   Summary

To date, research directed towards LTB_4 receptor antagonists has lagged somewhat behind corresponding intensive efforts aimed at developing potent and selective peptidoleukotriene receptor antagonists. However, recently this imbalance has been to some extent redressed and compounds with the appropriate profile with regard to potency and selectivity to warrant investigation in a clinical setting have been identified (e.g. LY 255283, LY 223982 and ONO-LB-457). Accordingly, within the next few years, valuable information should be obtained from clinical studies with LTB_4 receptor antagonists, which ought to clarify the extent of the role that LTB_4 plays in inflammatory diseases in which it has been implicated, particularly psoriasis and other skin disorders, and IBD.

9. *Potential Novel Agents/Future Directions*

9.1   INTRODUCTION

The preceding sections outlined the status of past and present preclinical and clinical studies directed towards delineating the pathophysiological effects of the arachidonic acid-derived mediators, in addition to the potential therapeutic utility of potent and selective agents which interfere with their synthesis and/or effects. Outlined briefly below are some of the other strategies that are currently being employed in addition to our own thoughts as to potential molecular targets and research approaches that may be investigated and utilized in the future. It is recognized that this is an incomplete list and that some of the compounds included have a greater chance of success than others. It should be noted that several of these classes of compounds possess a potential mechanism of action which is directed towards modulation of the release of, rather than the synthesis or actions of, the mediators.

9.2   CHEMICAL/PHARMACOLOGICAL CLASSES

9.2.1   Molecules with Dual Actions

These are similar to compounds such as the combined thromboxane synthetase inhibitors/thromboxane receptor antagonists (*vide supra*) and dual 5-lipoxygenase inhibitors/LTD_4 receptor antagonists (e.g. WY-48,252).

9.2.2   Phosphodiesterase Inhibitors

There is presently considerable interest in the synthesis and development of selective inhibitors of the various PDE isoenzymes that are involved in the metabolism of cAMP and cGMP. With regard to potential novel and anti-inflammatory agents, the focus is on the identification of potent and selective inhibitors of PDE IV (cAMP-specific PDE). This appears to be the predominant, but not the only, PDE isoenzyme present in inflammatory cells, and PDE IV inhibitors have been shown to inhibit inflammatory cell function, including mediator release, from a variety of human cells (Torphy and Undem, 1991). This class of compounds has the potential to be an exciting breakthrough therapy for several inflammatory disorders, including asthma.

9.2.3   Activators of Cyclic Nucleotide Second-Messenger Pathways

Rather than increasing the activity of the second-messenger pathway by inhibiting the metabolism of cAMP or cGMP, i.e. PDE inhibition, another potential mechanism is via stimulation of specific processes, e.g. protein kinases, activated during the modulation of cellular function produced by cAMP and cGMP.

9.2.4   K$^+$ Channel Activators

By virtue of their ability to hyperpolarize cell membranes and relax smooth muscle tissues, K$^+$ channel activators have been developed as potential antihypertensives or bronchodilators. Similarly, if specific K$^+$ channels are identified in relevant inflammatory and other cells, the possibility exists that selective activators of these channels may modulate the function of these cells, including the release of fatty acid-derived mediators (Black and Barnes, 1990).

9.2.5   Coenzyme A-Independent Transacylase Inhibitors

CoA-independent transacylase (CoA-IT) is a recently identified enzyme which has been postulated to selectively transfer polyunsaturated fatty acids between phospholipids (Chilton and Murphy, 1986; MacDonald and Sprecher, 1991; Winkler *et al.*, 1991) and to be involved in the production pathways for leukotrienes, prostaglandins and PAF. Accordingly, specific inhibitors of this enzyme may, like PLA_2 inhibitors, concomitantly regulate the formation and release of several pro-inflammatory mediators.

9.2.6   Leukotriene-A Synthetase Inhibitors

In an analogous manner to 5-lipoxygenase inhibitors, inhibitors of LTA synthetase should theoretically effectively inhibit the formation of both the peptidoleukotrienes and the pro-inflammatory LTB_4; unlike 5-lipoxygenase inhibitors they would have no inhibitory effect on the formation of 5-HETE.

9.2.7   Other Agents

Three other classes of interest are: (1) lipoxin inhibitors or receptor antagonists; (2) inhibitors of the formation and/or actions of the various HETEs; (3) inhibitors of the putative phospholipase-activating protein (PLAP) (Bomalaski *et al.*, 1991a,b; Clark *et al.*, 1991).

10.   Conclusions

During the past 15 years or so considerable progress has been made in our understanding of the biology of the numerous metabolic products of the arachidonic acid cascade and their potential role in the pathophysiology of a variety of diseases, predominantly inflammatory in nature. In parallel with this and stimulated by the elucidation of the structures and synthetic pathways for these fatty acid-derived mediators and the pivotal enzymes involved in their production, the pharmaceutical industry and other bodies have expended a formidable amount of resources towards the synthesis and development of novel chemical entities which will attenuate the effects of these substances. The central tenet of this effort is the hypothesis that various 5-lipoxygenase and cyclooxygenase products of arachidonic acid metabolism are key players in the aetiology of various diseases.

Several strategies have been utilized but, until recently, they have focused primarily on the identification of potent and selective compounds which either antagonize the effects of mediators at the receptor level (receptor antagonists) or specifically inhibit an enzyme involved in their production (enzyme inhibitors). Examples of the former class of compounds are peptidoleukotriene receptor antagonists, LTB_4 receptor antagonists and TXA_2 receptor antagonists, and examples of the latter are cyclooxygenase inhibitors, 5-lipoxygenase inhibitors, combined cyclooxygenase and 5-lipoxygenase inhibitors,

TXA_2 synthetase inhibitors and, more recently, PLA_2 inhibitors. Of these group of compounds, only cyclooxygenase inhibitors are presently marketed, at least in Western countries; there are numerous cyclooxygenase inhibitors, generally over-the-counter products, which are widely utilized. In addition, glucocorticoids continue to be used for an array of inflammatory disorders, although this class of compounds is burdened with significant side-effects. Representatives of the other class of compounds, apart from the PLA_2 inhibitors, have been, or are presently, in clinical trials for a variety of disorders. However, most of these critical studies are at a relatively early stage and, accordingly, it is premature to make definitive conclusions regarding their potential therapeutic utility. Nevertheless, the next few years should be a very informative and exciting period, as the results of the pivotal clinical trials with the appropriate tool compounds will identify which classes of compounds will be of therapeutic benefit in specific disorders. Clinicians are now accumulating a powerful arsenal of potent and specific modulators of the arachidonic acid cascade with which to probe and characterize, with an increasing degree of certainty, the relative roles, if any, of the fatty acid-derived mediators in the aetiology of an array of conditions. Past concerns about the limited potency and specificity of compounds, by and large, do not exist. In general, it appears that drugs affecting the 5-lipoxygenase arm of the cascade are more likely to have significant clinical benefits than agents that control the cyclooxygenase products; for example, clinical trials to date with the thromboxane receptor antagonists and, especially, the thromboxane synthetase inhibitors have proven to be disappointing.

Nevertheless, despite being the cornerstone of drug development efforts, debate persists regarding the likelihood of the clinical success of what can be called the "single mediator antagonist or inhibitor strategy" for complex diseases, such as asthma, in which evidence strongly suggests many mediators and several cells are involved. Accordingly, there has recently been a perceptible shift in the strategies towards identifying novel compounds which control the release and/or production, or alleviate the effects, of several mediators rather than a single mediator. The effort on PLA_2 inhibitors fits into this category with the hope that these compounds will attenuate the deleterious effects of appropriate 5-lipoxygenase and cyclooxygenase products in addition to PAF and exert powerful anti-inflammatory actions; the main issue in this area relates to which is the most relevant form of PLA_2 to target. Another newer effort that is likely to expand significantly is the development of specific inhibitors of the PDE isoenzymes (particularly PDE IV inhibitors) which control the metabolism of the intracellular messengers cAMP and cGMP. PDE IV inhibitors comprise an exciting class of compounds which in pre-clinical experiments exhibit impressive anti-inflammatory activity; for example, they are potent inhibitors of the release, and also the effects, of fatty acid-derived mediators from a variety of inflammatory cells. There may be further interest in compounds which possess more than one activity in an analogous manner to the dual cyclooxygenase and 5-lipoxygenase inhibitors and combined thromboxane receptor antagonists and thromboxane synthetase inhibitors. Other potential novel strategies are CoA-IT inhibitors and K^+ channel activators/antagonists. In addition, as yet unidentified molecular targets and approaches will undoubtedly become apparent in the coming years. Although not discussed in this chapter, recent advances in molecular biology and their application to the area of fatty acid-derived mediators should assist generally in these future endeavours.

In summary, the clinical trials in the next few years will establish whether potent and selective receptor antagonists or enzyme inhibitors of the individual fatty acid-derived mediators of arachidonic acid metabolism exert sufficient clinical efficacy to be considered realistically as novel therapies for diseases in which these substances have been implicated. In essence, this can be considered as the completion of the first phase of the development of novel therapeutic agents based on compounds that influence fatty acid-derived mediators. The next phase is likely to be directed primarily towards the development of potent and selective agents that affect the release, production and/or actions of several mediators rather than one mediator. Drugs emanating from both of these major efforts could herald a new era in the treatment of various inflammatory and other disorders in the 21st century.

11. Acknowledgements

The authors wish to thank Dotti Lavan for her excellent efforts in the typing of the manuscript and Dr Lisa Marshall for helpful discussions.

12. References

Aalbers, R. and De Monchy, J.G.R. (1991). Cysteinyl-leukotriene receptor antagonist, bronchoconstriction, and airway hyperreactivity. Lancet 338, 445.

Aarsman, A.J., de Jong, J.G.H., Arnoldussen, E., Neys, F.W., van Wassenaar, P.D. and Ven den Bosch, H. (1989). Immunoaffinity purification, partial sequence and subcellular localization of rat liver phopholipase A_2. J. Biol. Chem. 264, 10008–10014.

Abraham, W.M., Stevenson, J.S. and Garrido, R. (1988). The effect of an orally active leukotriene (LT)D_4 antagonist WY-48,252 on LTD_4 and antigen-induced bronchoconstrictions in allergic sheep. Prostaglandins 35, 733, 745.

Ackerman, N.R., Batt, D.G., Gans, K.R., Galbraith, W. and Harris, R.R. (1989). In "3rd Interscience World Conf. on Inflammation Antirheumatics, Analgesics, Immunomodulators", p. 159 (Abstr.).

Ackerman, Z., Karmeli, F., Ligumsky, M. and Rachmilewitz, D. (1990). Enhanced gastric and duodenal platelet-activing factor and leukotriene generation in duodenal ulcer patients. Scand. J. Gastroenterol. 25, 925–934.

Adaikan, P.G., Lauc, L.C., Kottegoda, S.R. and Ratnam, S.S. (1987). Effects of two new leukotriene antagonists ONO-RS-347 and ONO-RS-411 (ONO-1078) on the guinea pig and human respiratory and other systems. Adv. Prostaglandin Thromboxane Leukotriene Res. 17, 549–553.

Ahmed, T., Greenblatt, W., Birch, S., Marchette, B. and Wanner, A. (1981). Abnormal mucociliary transport in allergic patients with antigen-induced bronchospasm: Role of slow reacting substance of anaphylaxis. Am. Rev. Respir. Dis. 124, 110–114.

Ahnfelt-Rønne, I. and Arrigoni-Martelli, E. (1982). Multiple effects of a new anti-inflammatory agent, timegadine, on arachidonic acid release and metabolism in neutrophils and platelets. Biochem. Pharmacol. 31, 2619–2624.

Ahnfelt-Rønne, I., Kirstein, D. and Kaergaard-Neilsen, C. (1988). A novel leukotriene D_4/E_4 antagonist, SR2640 (2-[3-(2-quinolylmethoxy)phenylamino]benzoic acid). Eur. J. Pharmacol. 155, 117–128.

Aiken, J.W. (1983). Pharmacology of thromboxane synthetase inhibitors. Adv. Prostaglandin Thromboxane Leukotriene Res. 11, 253–258.

Akbar, H., Mukhopadhyay, A., Anderson, K.S., Navran, S.S., Romstedt, K., Miller, D.D. and Feller, D.R. (1985). Antagonism of prostaglandin-mediated responses in platelets and vascular smooth muscle by 13-azaprostanoic acid analogs. Biochem. Pharmacol. 34, 641–647.

Allen, B.R. and Littlewood, S.M. (1983). The aetiology of psoriasis: clues provided by benoxaprofen. Br. J. Dermatol. 198(Suppl. 25), 126–129.

Ambler, J., Butler, K.D., Ku, E.C., Maguire, E.D., Smith, J.R. and Wallis, R.B. (1985). CGS 12970: a novel, long acting thromboxane synthetase inhibitor. Br. J. Pharmacol. 86, 497–504.

Ambrose, M.P., Bahns, C.-L.C. and Hunninghake, G.W. (1992). Lipocortin I production by human alveolar macrophages. Am. J. Respir. Cell. Mol. Biol. 6, 17–21.

Andersen, N.H., Eggerman, T.L., Harker, L.A., Wilson, C.H. and De, B. (1980). On the multiplicity of platelet prostaglandin receptors. I. Evaluation of competitive antagonism by aggregometry. Prostaglandins 19, 711–715.

Anderson, R. and Jooné, G. (1984). Inhibition of polymorphonuclear leucocyte motility by benoxaprofen related to activation of cellular oxidative metabolism. Int. J. Immunopharmacol. 6, 269–274.

Anderson, S.D. (1985). Exercise-induced asthma. The state of the art. Chest 87, 191S-195S.

Antiplatelet Trialists' Collaboration (1988). Secondary prevention of vascular diseases by prolonged antiplatelet treatment. Br. Med. J. 296, 320–331.

Armstrong, R.A., Jones, R.L., Peesapati, V., Will, S.G. and Wilson, N.H. (1985). Competitive antagonism at thromboxane receptors in human platelets. Br. J. Pharmacol. 84, 595–607.

Arnold, H.L. Jr, Odom, R.B. and James, W.D. (1990). In "Andrews' Diseases of the Skin", 8th edn (eds H.L. Arnold, Jr, R.B. Odom and W.D. James), pp. 194–226. W.B. Saunders, Philadelphia.

Ashida, Y., Matsumoto, T., Kuriki, H., Shiraishi, M., Kato, K. and Terao, S. (1989). A novel anti-asthmatic quinone derivative. AA-2414 with a potent antagonistic activity against a variety of spasmogenic prostanoids. Prostglandins 38, 91–112.

Ashton, J.H., Schmitz, J.M., Campbell, W.B., Ogletree, M.L., Raheja, S., Taylor, A.L., Fitzgerald, C., Buja, L.M. and Willerson, J.T. (1986). Inhibition of cyclic flow variations in stenosed canine coronary arteries by thromboxane A_2/prostaglandin H_2 receptor antagonists. Circ. Res. 59, 568–579.

Augstein, J., Farmer, J.B., Lee, T.B., Sheard, P. and Tattersall, M.L. (1973). Selective inhibition of slow reacting substance of anaphylaxis. Nature (New Biol.) 245, 215–216.

Avery, M.A., Detre, G., Yasuda, D., Chad, W.R., Tanabe, M., Crowe, D., Peters, R. and Chong, W.K.M. (1990). Synthesis and antiinflammatory activity of novel 12β-substituted analogues of betamethasone. J. Med. Chem. 33, 1852–1858.

Aviado, D.M. and Carrillo, L.R. (1970). Antiasthmatic action of corticosteroids: a review of the literature on their mechanism of action. J. Clin. Pharmacol. 10, 3–11.

Axelrod, L. (1989). In "Textbook of Rheumatology" (eds W.N. Kelley, E.D. Harris, Jr, S. Ruddy and C.B. Sledge), pp. 845–861. Harcourt Brace Jovanovich, Philadelphia.

Ball, H.A., Parratt, J.R. and Zeitlin, I.J. (1983). Effect of dazoxiben, a specific inhibitor of thromboxane synthetase, on acute pulmonary responses to *E. coli* endotoxin in anaesthetized cats. Br. J. Clin. Pharmacol. 15, 127S–131S.

Ball, H.A., Cook, J.A., Wise, W.C. and Halushka, P.V. (1986). Role of thromboxane, prostaglandins and leukotrienes in endotoxic and septic shock. Intensive Care Med. 12, 116–126.

Barlow, P.N., Lister, M.D., Sigler, P.B. and Dennis, E.A. (1988a). Probing the role of substrate conformation in phospholipase A_2 action on aggregated phospholipids using constrained phosphatidylcholine analogues. J. Biol. Chem. 263, 12954–12958.

Barlow, P.N., Vidal, J.-C., Lister, M.D., Hancock, A.J. and Sigler, P.B. (1988b). Synthesis and some properties of constrained short-chain phosphatidylcholine analogues: (+)- and (−)-(1,3/2)-1-O-(phosphocholine)2,3-O-dihexanoyl cyclopentane-1,2,3-triol. Chem. Phys. Lipids 46, 157–164.

Barnes, N., Piper, P.J. and Costello, J. (1987). The effect of an oral leukotriene antagonist L-649,923 on histamine and leukotriene D_4-induced bronchoconstriction in normal man. J. Allergy Clin. Immunol. 79, 816–821.

Barnett, A.H., Wakelin, K., Leatherdale, B.A., Britton, J.R., Polak, A., Bennett, J., Toop, M., Rowe, D. and Dallinger, K. (1984). Specific thromboxane synthetase inhibition and albumin excretion rate in insulin-dependent diabetes. Lancet i, 1322–1325.

Barr, R.M., Brain, S., Camp, R.D.R., Cilliers, J., Greaves, M.W., Mallet, A.I. and Misch, K. (1984). Levels of arachidonic acid and its metabolites in the skin in human allergic and irritant contact dermatitis. Br. J. Dermatol. 111, 23–28.

Beasley, R.C.W., Featherstone, R.L., Church, M.K., Rafferty, P., Varley, J.G., Harris, A., Robinson, C. and Holgate, S.T. (1989). Effect of thromboxane receptor antagonist on PGD_2- and allergen-induced bronchoconstriction. J. Appl. Physiol. 66, 1685–1693.

Bel, E.H., Timmers, M.C., Dijkman, J.H., Stahl, E.G. and Sterk, P.K. (1990). The effect of an inhaled leukotriene antagonist, L-648,051, on early and late asthmatic reactions and subsequent increase in airway responsiveness in man. J. Allergy Clin. Immunol. 85, 1067–1075.

Belch, J.J., Cormie, J., Newman, P., McLaren, M., Barbenel, J., Capell, H., Leiberman, P., Forbes, C.D. and Prentice, C.R. (1983). Dazoxiben, a thromboxane synthetase inhibitor in the

treatment of Raynaud's syndrome: a double-blind trial. Br. J. Clin. Pharmacol. 15, 113S–116S.

Bianchi, P.G., Petrillo, M., Ardizzone, S., Caruso, I. and Montrone, F. (1991). Endoscopic assessment of tenoxicam- and piroxicam-induced lesions in osteoarthritic patients. Drug. Invest. 3, 53–56.

Bianco, A., Passacantilli, P., Righi, G., Brufani, M., Cellai, L. and Marchi, E. (1989). Synthesis of 2-hydroxyacetyl-7-acetyl-xanthone, a new xanthone derivative endowed with antianaphylactic, analgesic and antiinflammatory activities. Farmaco 44, 547–554.

Birch, J., Brown, E., Calnan, C., Jessup, C.L., Jessup, R. and Wayne, M. (1988). Studies in the guinea-pig with ICI 185, 282: a thromboxane A_2 receptor antagonist. J. Pharm. Pharmacol. 40, 706–710.

Bird, T.G.C., Bruneau, P., Crawley, G.D. *et al.* (1991). (Methoxyalkyl)thiazoles: a new series of potent, selective and orally active 5-lipoxygenase inhibitors displaying high enantio-selectivity. J. Med. Chem. 34, 2176–2186.

Bisgaard, H. and Kristensen, J.K. (1985). Leukotriene B_4 produces hyperalgesia in humans. Prostaglandins 30, 791–797.

Black, A.K., Camp, R.D.R., Mallet, A.I., Cunningham, F.M., Hofbauer, M. and Greaves, M.W. (1990). Pharmacologic and clinical effects of lonapalene (RS43179), a 5-lipoxygenase inhibitor in psoriasis. J. Invest. Dermatol. 95, 50–54.

Black, J.L. and Barnes, P.J. (1990). Potassium channels and airway function: new therapeutic prospects. Thorax 45, 213–218.

Blackburn, W.D. Jr, Heck, L.W., Loose, L.D., Eskra, J.D. and Carty, T.J. (1991a). Inhibition of 5-lipoxygenase product formation and polymorphonuclear cell degranulation by tenidap sodium in patients with rheumatoid arthritis. Arthritis Rheum. 34, 204–210.

Blackburn, W.D., Jr, Loose, L.D., Heck, L.W. and Chatham, W.W. (1991b). Tenidap, in contrast to several available nonsteroidal antiinflammatory drugs, potently inhibits the release of activated neutrophil collagenase. Arthritis Rheum. 34, 211–216.

Blackham, A., Norris, A.A. and Woods, F.A.M. (1985). Models for evaluating the anti-inflammatory effects of inhibitors or arachidonic acid metabolism. J. Pharm. Pharmacol. 37, 787–793.

Boctor, A.M., Eickholt, M. and Pugsley, T.A. (1986). Meclofenamate sodium is an inhibitor of both the 5-lipoxygenase and cyclooxygenase pathways of the arachidonic acid cascade in vitro. Prostaglandins Leukotrienes Med. 23, 229–238.

Bomalaski, J.S., Ford, T.A., Baker, D. and Clark, M.A. (1991a). PLAP (phospholipase A_2 activating protein) stimulates tumor necrosis factor release. Arthritis Rheum. 34, S153 (Abstr.).

Bomalaski, J.S., Ford, T.A., Simon, P. and Clark, M.A. (1991b). Interleukin 1 stimulates phospholipase A_2 activating protein (PLAP) and PLAP stimulates interleukin 1 release: phospholipase–cytokine circuitry. Arthritis Rheum. 34, S47 (Abstr.).

Bomalaski, J.S., Lawton, P. and Browning, J.L. (1991c). Human extracellular recombinant phospholipase A_2 induces an inflammatory response in rabbit joints. J. Immunol. 146, 3904–3910.

Bonnet, J., Thibaudean, D. and Bessin, P. (1983). Dependency of PAF-acether induced bronchospasm on the lipoxygenase pathway in the guinea pig. Prostaglandins 26, 457–466.

Bonow, R.O., Lipson, L.C., Sheehan, F.H., Capurro, N.L.,

Isner, J.M., Roberts, W.C., Goldstein, R.E. and Epstein, S.L. (1981). Lack of effect of aspirin on myocardial infarct size in the dog. Am. J. Cardiol. 47, 258–264.

Boot, J.R., Bond, A., Gooderham, R., O'Brien, A., Parsons, M. and Thomas, K.H. (1989). The pharmacological evaluation of LY17060, a novel leukotriene D_4 and E_4 antagonist in the guinea pig. Br. J. Pharmacol. 98, 259–267.

Borgeat, B. and Samuelsson, B. (1979). Transformation of arachidonic acid by rabbit polymorphonuclear leukocytes. Formation of a novel dihydroxyeicosatetraenoic acid. J. Biol. Chem. 254, 2643–2646.

Brain, S.D., Camp, R.D.R., Cunningham, F.M., Dowd, P.M., Greaves, M.W. and Kobza Black, A. (1984a). Leukotriene B_4-like material in scale of psoriatic skin lesions. Br. J. Pharmacol. 83, 313–317.

Brain, S., Camp, R., Dowd, P., Black, A.K. and Greaves, M. (1984b). The release of leukotriene B_4-like material in biologically active amounts from the lesional skin of patients with psoriasis. J. Invest. Dermatol. 83, 70–73.

Bramm, E., Binderup, L. and Arrigoni-Martelli, E. (1981). An unusual profile of activity of a new basic anti-inflammatory drug, timegadine. Agents Actions 11, 402–409.

Bray, M., Cunningham, F., Ford-Hutchinson, A. and Smith, M. (1981). Leukotriene B_4: a mediator of vascular permeability. Br. J. Pharmacol. 72, 483–486.

Bray, M.A. (1983). The pharmacology and pathophysiology of leukotriene B_4. Br. Med. Bull. 39, 249–254.

Bray, M.A., Anderson, W.H., Subramanian, N., Niederhauser, U., Kuhn, M., Erard, M. and von Sprecher, A. (1991). CGP 45715: a leukotriene D_4 analogue with potent peptido-LT antagonist activity. Adv. Prostaglandin Thromboxane Leukotriene Res. 21B, 503–507.

Brewster, A.G., Brown, G.R., Jessup, R., Smithers, M.J. and Stocker, A. (1990). In "7th Int. Conf. on Prostaglandins and Related Compounds, Florence", p. 178 (Abstr.)

Brezinski, M.E., Yanagisawa, A., Darius, J. and Lefer, A.M. (1985). Anti-ischemic actions of a new thromboxane receptor antagonist during acute myocardial ischemia in cats. Am. Heart J. 110, 1161–1167.

Brezinski, M.E., Bowker, B. and Lefer, A.M. (1986). Membrane labilizing actions of thromboxane A_2. Pharmacologist 28, 147 (Abstr.)

Brittain, R.T., Coleman, R.A., Collington, E.W., Hallet, P., Humphrey, P.P.A., Kennedy, I., Lumley, P., Sheldrick, R.L.G. and Wallis, C.J. (1984). Br. J. Pharmacol. 83, 377P (Abstr.; no title given).

Brittain, R.T., Boutal, L., Carter, M.C., Coleman, R.A., Collington, E.W., Geisow, H.P., Hallett, P., Hornby, E.J., Humphrey, P.P.A., Jack, D., Kennedy, I., Lumley, P., McCabe, P.J., Skidmore, I.F., Thomas, M. and Wallis, C.J. (1985). AH23848: a thromboxane receptor-blocking drug that can clarify the pathophysiologic role of thromboxane A_2. Circulation 72, 1208–1218.

Britton, J.R., Hanley, S.P. and Tattersfield, A.E. (1987). The effect of an oral leukotriene D_4 antagonist L-649,923 on the response to inhaled antigen in asthma. J. Allergy Clin. Immunol. 79, 811–816.

Brocklehurst, W.E. (1960). The release of histamine and slow reacting substance (SRS-A) during anaphylactic shock. J. Physiol. 151, 416–435.

Brooks, P.M., Buchanan, W.W., Rosenbloom, D. and Bellamy, N. (1985). in "Anti-Inflammatory and Anti-Rheumatic Drugs", Vol III (ed. K.D. Rainsford), pp. 166–204. CRC Boca Raton.

Brooks, P.M., Kean, W.F. and Buchanan, W.A. (1986). In "The Clinical Pharmacology of Anti-Inflammatory Agents" (eds P.M. Brooks, W.F. Kean and W.W. Buchanan), pp. 59–85. Taylor and Francis, London.

Brown, K.A., Ferrie, J., Wilbourn, B. and Dumonde, D.C. (1984). Benoxaprofen, a potent inhibitor of monocyte/endothelial-cell interaction. Lancet ii, 643.

Buckner, C.K., Krell, R.D., Laravuso, R.B., Coursin, D.B., Bernstein, P.R. and Will, J.A. (1986). Pharmacological evidence that human intralobar airways do not contain different receptors that mediate contractions to leukotriene C_4 and leukotriene D_4. J. Pharmacol. Exp. Ther. 237, 558–562.

Buckner, C.K., Fedyna, J.S., Robertson, J.L., Will, J.A., England, D.M., Krell, R.D. and Saban, R. (1990). An examination of the influence of the epithelium on contractile responses to peptidoleukotrienes and blockade by ICI 204,219 in isolated guinea pig trachea and human intralobar airways. J. Pharmacol. Exp. Ther. 252, 77–85.

Burke, S.E., Lefer, A.M., Smith, B.M. and Smith, J.B. (1983a). Prevention of extension of ischemic damage following acute myocardial ischemia by dazoxiben, a new thromboxane synthetase inhibitor. Br. J. Clin. Pharmacol. 15, 97S-101S.

Burke, S.E., Roth, D.M. and Lefer, A.M. (1983b). Antagonism of platelet aggregation by 13-azaprostanoic acid in acute myocardial ischemia and sudden death. Thromb. Res. 29, 473–488.

Burnstein, K.L., Bellingham, D.L., Jewell, C.M., Powell-Oliver, F.E. and Cidlowski, J.A. (1991). Autoregulation of glucocorticoid receptor gene expression. Steroids 56, 52–58.

Bush, L.R. and Smith, S.G. (1987). Comparative and combined effects of U-63557A, a TxA_2 synthetase inhibitor, and BM 13,177, a TxA_2 receptor antagonist, on cyclic flow reductions in stenosed canine coronary arteries. Fed. Proc. 46, 1316 (Abstr.).

Caldwell, A.G., Harris, C.J., Stepney, R. and Whittaker, N. (1979). Hydantoin prostaglandin analogues, potent and selective inhibitors of platelet aggregation. J. Chem. Soc. Chem. Commun. 101, 561–562.

Campbell, M.M., Long-Fox, J., Osuguthorpe, D.J., Sainsbury, M. and Sessions, R.B. (1988). Inhibition of phospholipase A_2; a molecular recognition study. J. Chem. Soc. Chem. Commun. 1560–1562.

Carey, F. and Haworth, D. (1986). Thromboxane synthase inhibition: Implications for prostaglandin endoperoxide metabolism. II. Testing the 'redirection hypothesis' in an acute intravenous challenge model. Prostaglandins 31, 47–59.

Carlson, R.P., O'Neill-Davis, L., Chang, J. and Lewis, A.J. (1985). Modulation of mouse ear edema by cyclooxygenase and lipoxygenase inhibitors and other pharmacologic agents. Agents Actions 17, 197–204.

Carlson, R.P., Glaser, K.B., Kreft, A., Hartman, D.A., Tomchek, L., Lock, Y.W. and Weichman, B.M. (1991). In "Int. Congr. on Inflammation, Rome, Italy", p. 236 (Abstr.).

Carter, G.W., Young, P.R., Albert, D.H., Bouska, J.B., Dyer, R.D., Bell, R.L., Summers, J.B., Brooks, D.W., Gunn, B.P., Rubin, P. and Kesterson, J. (1989). In "Leukotrienes and Prostanoids in Health and Disease", Vol 3 (eds U. Zor, Z. Naor and A. Danon), pp. 50–55. Karger, Basel.

Cartier, A.C., Thompson, N.C., Frith, P.A., Roberts, R. and Hargreave, F.E. (1982). Allergen-induced increase in bronchial responsiveness to histamine: relationship to the late asthmatic response and change in airways calibre. J. Allergy Clin. Immunol. 70, 170–177.

Cartwright, P.H., Ilderton, E., Sowden, J.M. and Yardley, H.J. (1989). Inhibition of normal and psoriatic epidermal phospholipase A_2 by picomolar concentrations of recombinant human lipocortin I. Br. J. Dermatol. 121, 155–160.

Casey, F.B., Applebey, B.J. and Buck, D.C. (1983). Selective inhibition of the lipoxygenase metabolic pathway of arachidonic acid by the SRS-A-antagonist FPL 55712. Prostaglandins 25, 1–11.

Casey, L.C., Fletcher, J.R., Zmudka, M.I. and Ramwell, P.W. (1982). Prevention of endotoxin-induced pulmonary hypertension in primates by the use of a selective thromboxane synthetase inhibitor, OKY-1581. J. Pharmacol. Exp. Ther. 222, 441–446.

Cashman, J.R. (1985). Leukotriene biosynthesis inhibitors. Pharmaceut. Res. 2, 253–261.

Chagnon, M., Gentile, J., Gladu, M. and Sirois, P. (1985). The mechanism of action of leukotrienes A_4, C_4 and D_4 on human lung parenchyma in vitro. Lung 163, 55–62.

Chand, N. (1979). FPL-55712 – an antagonist of slow reacting sustance of anaphylaxis (SRS-A): a review. Agents Actions 9, 133–140.

Chang, J., Musser, J.H. and McGregor, H. (1987a). Phospholipase A_2: function and pharmacological regulation. Biochem. Pharmacol. 36, 2429–2436.

Chang, H.W., Kudo, I., Tomita, M. and Inoue, K. (1987b). Purification and characterization of extracellular phospholipase A_2 from peritoneal cavity of caseinate-treated rat. J. Biochem. 102, 147–154.

Chasin, M. and Scott, C. (1978). Inhibition of cyclic nucelotide phosphodiesterase by FPL 55712, an SRS-A antagonist. Biochem. Pharmacol. 27, 2065–2067.

Check, W.A. and Kaliner, M.A. (1990). Pharmacology and pharmacokinetics of topical corticosteroid derivatives used for asthma therapy. Am. Rev. Respir. Dis. 141, S44–S51.

Cheng, J.B., Pillar, J.S., Conklyn, M.J., Breslow, R., Shirley, J.T. and Showell, H.J. (1990). Evidence that peptidoleukotriene is a prerequisite for antigen-dependent thromboxane synthesis in IgG1-passively sensitized guinea pig lungs. J. Pharmacol. Exp. Ther. 255, 664–667.

Chilton, F.H. and Murphy, R.C. (1986). Remodeling of arachidonate-containing phosphoglycerides within the human neutrophil. J. Biol. Chem. 261, 7771–7777.

Christie, P.E., Smith, C.M. and Lee, T.H. (1991). The potent and selective SRS-A leukotriene antagonist, SKandF 104353, inhibits aspirin-induced asthma. Am. Rev. Respir. Dis. 144, 957–958.

Chung, K.F., Aizawa, H., Becker, A.B., Frick, O., Gold, W.M. and Nadel, J.A. (1986). Inhibition of antigen-induced airway hyperresponsiveness by a thromboxane synthetase inhibitor (OKY-046) in allergic dogs. Am. Rev. Respir. Dis. 134, 258–261.

Cirino, G., Peers, S.H., Flower, R.J., Browning, J.L. and Pepinsky, R.B. (1989). Human recombinant lipocortin 1 has acute local anti-inflammatory properties in the rat paw edema test. Proc. Natl Acad. Sci. USA 86, 3428–3432.

Clark, J.D., Milona, N. and Knopf, J.L. (1990). Purification of a 110-kilodalton cytosolic phopholipase A_2 from the human monocytic cell line U937. Proc. Natl. Acad. Sci. USA 87, 7708–7712.

Clark, M.A., Steiner, M.R., Simon, P.L. and Bomalaski, J.S. (1991). Interleukin-1 induces increased phospholipase A_2 activity, synthesis of phospholipase A_2 activating protein (PLAP), and release of linoleic acid from the murine (EL-4) T helper cell line. J. Cell Biol. 115, 218a (Abstr.)

Cloud, M.L., Enas, G.C., Kemp, J., Platts-Mills, T., Altman, L.C., Townley, R., Tinkelman, D., King, T. Jr, Middleton, E., Sheffer, A.L., McFadden, E.R. Jr. and Farlow, D.S. (1989). A specific LTD_4/LTE_4-receptor antagonist improves pulmonary function in patients with mild, chronic asthma. Am. Rev. Respir. Dis. 140, 1336–1339.

Coffman, T.M., Yarger, W.E., and Klotman, P.E. (1985). Functional role of thromboxane production by acutely rejecting renal allografts in rats. J. Clin. Invest. 75, 1242–1248.

Coffman, T.M., Schwertschlag, U., Yarger, W.E., Pirotsky, E., Benveniste, J. and Klotman, P.E. (1986). Lipid mediators in acute renal allograft rejection. Transplant Proc. XVIII, 94–97.

Coffman, T.M., Ruiz, P., Sanfilippo, F. and Klotman, P.E. (1989). Chronic thromboxane inhibition preserves function of rejecting rat renal allografts. Kidney Int. 35, 24–30.

Coker, S.J. and Parratt, J.R. (1983). Effects of dazoxiben on arrhythmias and ventricular fibrillation induced by coronary artery occlusion and reperfusion in anesthetized greyhounds. Br. J. Clin. Pharmacol. 15, 87S–95S.

Coleman, R.A. (1988). Prostaglandin receptors. Prostaglandin Perspect. 4, 33–35.

Coleman, R.A. and Kennedy, J. (1985). Characterisation of the prostanoid receptors mediating contraction of guinea-pig isolated trachea. Prostaglandins 29, 363–375.

Coleman, R.A., Kennedy, I. and Levy, G.P. (1980). SC-19220, a selective prostanoid receptor antagonist. Br. J. Pharmacol. 69, 266P–267P (Abstr.).

Coleman, R.A., Humphrey, P.P.A., Kennedy, I. and Lumley, P. (1984). Prostanoid receptors – the development of a working classification. Trends Pharmacol. Sci. 5, 303–306.

Coleman, R.A., Humphrey, P.P.A. and Kennedy, I. (1985a). In "Trends in Autonomic Pharmacology", Vol. 3 (ed. S. Kalsner) pp. 35–49. Taylor and Francis, London.

Coleman, R.A., Kennedy, I. and Sheldrick, R.L.G. (1985b). AH6809, a prostanoid EP_1-receptor blocking drug. Br. J. Pharmacol. 85, 273P (Abstr.).

Corey, E.J., Clark, D.A., Goto, G., Marfat, A., Mioskowski, C., Samuelsson, B. and Hammarström, S. (1980). Stereospecific total synthesis of a "slow-reacting substance of anaphylaxis" (SRS-A), leukotriene C-1. J. Am. Chem. Soc. 102, 1436–1439.

Coutts, S.M., Khandwala, A., Van Inwegen, R., Chakraborty, U., Musser, J., Bruens, J., Jariwala, N., Dally-Meade, V., Ingram, R., Pruss, T., Jones, H., Neiss, E. and Weinryb, I. (1985) In "Prostaglandins, Leukotrienes and Lipoxins, Biochemistry, Mechanism of Action, and Clinical Applications" (ed. J. M. Bailey), pp. 627–637. Plenum Press, New York.

Creese, B.R., Bach, M.K., Fitzpatrick, F.A. and Bothwell, W.M. (1984). Leukotriene-induced contraction and thromboxane production in guinea-pig lung parenchymal strips. Eur. J. Pharmacol. 102, 197–204.

Creticos, P.S., Bodenheimer, S., Albright, A., Lichtenstein, L.M. and Norman, P.S. (1989). Effects of an inhaled leukotriene antagonist on bronchial challenge with antigen. J. Allergy Clin. Immunol. 83, 187 (Abstr.)

Dahlén, S.-E., Hansson, G., Hedqvist, P., Björck, t., Granström, E. and Dahlén, B. (1983). Allergen challenge of lung tissue from asthmatics elicits bronchial contraction that correlates with the release of leukotrienes C_4, D_4 and E_4. Proc. Natl Acad. Sci. USA 80, 1712–1716.

Dahlén, S.-E., Dahlén, B., Eliasson, E., Johansson, H., Björck, T., Kumlin, M., Boo, K., Whitney, J., Binks, S., King, B.,

Stark, R. and Zetterström, O. (1991). Adv. Prostaglandin Thromboxane Leukotriene Res. 21A, 461–464.

Dandona, P. and Jeremy, J.Y. (1990). Nonsteroidal anti-inflammatory drug therapy and gastric side effects. Does nabumetone provide a solution? Drugs 40, 16–24.

Danon, A. and Assouline, G. (1978). Inhibition of prostaglandin biosynthesis by corticosteroids requires RNA and protein synthesis. Nature 273, 552–554.

Darius, H. and Lefer, A.M. (1985). Antiaggregatory effects of thromboxane receptor antagonist *in vivo*. Thromb. Res. 40, 663–675.

Darius, H., Smith, J.B. and Lefer, A.M. (1985). Beneficial effects of a new potent and specific thromboxane receptor antagonist (SQ-29,548) *in vitro* and *in vivo*. J. Pharmacol. Exp. Ther. 235, 274–281.

Davidson, E.M., Rae, S.A. and Smith, M.J.H. (1983). Leukotriene B_4, a mediator of inflammation present in synovial fluid in rheumatoid arthritis. Ann. Rheum. 42, 677–679.

Davidson, F.F. and Dennis, E.A. (1989). Biological relevance of lipocortins and related proteins as inhibitors of phospholipase A_2. Biochem. Pharmacol. 38, 3645–3651.

Davidson, J., Flower, R.J., Milton, A.S., Peers, S.H. and Rotondo, D. (1991). Antipyretic actions of human recombinant lipocortin-1. Br. J. Pharmacol. 102, 7–9.

Davis, P.D., Nixon, J.S., Wilkinson, S.E. and Russell, M.G.N. (1988). Inhibition of phospholipase A_2 by some long chain alkylamines. Biochem. Soc. Trans. 16, 816–817.

Dawson, W. (1980). The comparative pharmacology of benoxaprofen. J. Rheumatol. 7(Suppl. 6), 5–11.

Day, R.O., Graham, G.G., Williams, K.M., Champion, G.D. and De Jager, J. (1987). Clinical pharmacology of non-steroidal anti-inflammatory drugs. Pharmacol. Ther. 33, 383–433.

De Carvalho, M.S. and Jacobs, R.S. (1991). Two-step inactivation of bee venom phospholipase A_2 by scalaradial. Biochem. Pharmacol. 42, 1621–1626.

De Clerck, F., Beetens, J., de Chaffoy de Courcelles, D., Freyne, E. and Janssen, P.A.J. (1989). R68070: thromboxane A_2 synthetase inhibition and thromboxane A_2/prostaglandin endoperoxide receptor blockade, combined in one molecule. I. Biochemical profile *in vitro*. Thromb. Haemost. 61, 35–42.

Debono, D.P., Lumley, P., Been, M., Keery, R., Ince, S.E. and Woodings, D.F. (1986). Effect of the specific thromboxane receptor blocking drug AH23848 in patients with angina pectoris. Br. Heart J. 56, 509–517.

Deems, R.A., Eaton, B.R. and Dennis, E.A. (1975). Kinetic analysis of phospholipase A_2 activity toward mixed micelles and its implications for the study of lipolytic enzymes. J. Biol. Chem. 250, 9013–9020.

Deems, R.A., Lombardo, D., Morgan, B.P., Mihelich, E.D. and Dennis, E.A. (1987). Inhibition of phospholipase A_2 by manoalide and manoalide analogues. Biochim. Biophys. Acta 917, 258–268.

Defreyn, G., Deckmyn, H. and Vermylen, J. (1982). A thromboxane synthetase inhibitor reorients endoperoxide metabolism in whole blood towards prostacyclin and prostaglandin E_2. Thromb. Res. 26, 389–400.

Dejong, E.M.G.J., Vanvlijmen, I.M.M.J.M., Scholte, J.C.M., Buntinx, A., Friedman, B., Tanaka, W. and Vanderkerkhof, P.C.M. (1991). Clinical and biochemical effects of an oral leukotriene biosynthesis inhibitor (MK886) in psoriasis. Skin Pharmacol. 4, 278–285.

Dennis, E.A. (1983). In "The Enzymes. Lipid Enzymology", Vol. 16 (ed. P.D. Boyer), pp. 307–353. Academic Press, New York.

Dennis, E.A. (1987). Phospholipase A_2 mechanism: inhibition and role in arachidonic acid release. Drug. Develop Res. 10, 205–220.

Diez, E. and Mong, S. (1990). Purification of a phospholipase A_2 from human monocytic leukemic U937 cells. Calcium-dependent activation and membrane association. J. Biol. Chem. 265, 14654–14661.

Dinh Xuan, A.T., Regnard, J., Similowski, T., Rey, J., Marsac, J. and Lockhart, A. (1990). Effects of SKandF 104353, a leukotriene receptor antagonist, on the bronchial responses to histamine in subjects with asthma: a comparative study with terfenadine. J. Allergy Clin. Immunol. 85, 865–871.

Dixon, R.A.F., Jones, R.E., Diehl, R.E., Bennett, C.D., Kargman, S. and Rouzer, C.A. (1988). Cloning of the cDNA for human 5-lipoxygenase. Proc. Natl Acad. Sci. USA 85, 416–420.

Djuric, S.W., Miyashiro, J.M. and Penning T.D. (1988). A practical enantioselective synthesis of 12-hydroxyeicosatetraenoic acids. Tetrahedron Lett. 29, 3459–3462.

Djuric, S.W., Collins, P.W., Jones, P.H., Shone, R.L., Tsai, B.S., Fretland, D.J., Butchko, G.M., Villani-Price, D., Keith, R.H., Zemaitis, J.M., Metcalf, L. and Bauer, R.F. (1989). 7-[3-(4-Acetyl-3-methoxy-2-propylphenoxy)-propoxy]-3,4-dihydro-8-propyl-2*H*-1-benzopyran-2-carboxylic acid: an orally active selective leukotriene B_4 receptor antagonist. J. Med. Chem. 32, 1145–1147.

Doherty, N.S, (1987). Mediators of the pain of inflammation. Ann. Rep. Med. Chem. 22, 245–252.

Douglas, C.E., Chan, A.C. and Choy, P.C. (1986). Vitamin E inhibits platelet phospholipase A_2. Biochim. Biophys. Acta 876, 639–645.

Drazen, J.M. and Austen, K.F. (1987). Leukotrienes and airway responses. Am. Rev. Respir. Dis. 136, 985–998.

Drvota, V., Vesterqvist, O. and Gréen, K. (1991). Adv. Prostaglandin Thromboxane Leukotriene Res. 21A, 153–156.

Dufton, M.J. and Hider, R.C. (1983). Classification of phospholipase A_2 according to sequence. Evolutionary and pharmacological implications. Eur. J. Biochem. 137, 545–551.

Dufton, M.J., Eaker, D. and Hider, R.C. (1983). Conformational properties of phospholipase A_2. Secondary structure prediction, circular dichroism and relative interface hydrophobicity. Eur. J. Biochem. 137, 537–544.

Dunn, M.J. (1984). Nonsteroidal antiinflammatory drugs and renal function. Ann. Rev. Med. 35, 411–428.

Dusting, G.J., Moncada, S. and Vane, J.R. (1982). In "Prostaglandins and the Cardiovascular System" (ed. J.A. Oates), pp. 59–106. Raven Press, New York.

Eakins, K.E., Rajadhyaksha, V. and Schroer, R. (1976). Prostaglandin antagonism by sodium *p*-benzyl-4-[1-oxo-2-(4-chlorobenzyl)-3-phenylpropyl]phenyl phosphate (N-0164). Br. J. Pharmacol. 58, 333–339.

Egsmose, C., Lund, B. and Andersen, R.B. (1988). Timegadine: more than a non-steroidal for the treatment of rheumatoid arthritis. Scand. J. Rheumatol. 17, 103–111.

Ehrly, A.M. (1983). Influence of a thromboxane synthesis inhibitor on the muscle tissue microcirculation of patients with intermitten claudication. Br. J. Clin. Pharmacol. 15, 117S–118S.

Eiser, N., Hayhurst, M. and Denman, W. (1989). The contribution of histamine and leukotriene to the production of early and late asthmatic responses to antigen. Am. Rev. Respir. Dis. 139, A462 (Abstr.)

Ekerdt, R., Buchmann, B., Frölich, W., Giesen, C., Heindl, J. and Skuballa, W. (1991). The role of leukotriene B_4 as an inflammatory mediator in skin and the functional characterization of LTB_4 receptor antagonists. Adv. Prostaglandin, Thromboxane Leukotriene Res. 21B, 565–568.

Ellis, E.F., Oelz, O., Roberts, L.G., Payne, M.A., Sweetman, B.J., Nies, A.S. and Oates, J.A. (1976). Coronary arterial smooth muscle contraction by a substance released from platelets; evidence that it is thromboxane A_2. Science 193, 1135–1137.

Ellis, J.L. and Undem, B.J. (1991). Role of peptidoleukotrienes in capsaicin-sensitive sensory fibre-mediated responses in guinea-pig airways. J. Physiol. 436, 469–484.

Elwood, P.C. (1983). British studies of aspirin and myocardial infarction. Am. J. Med. 74, 50–54.

Ettinger, W.J., Wise, R.A., Schafhauser, D. and Wigley, F.M. (1984). Controlled double-blind trial of dazoxiben and nifedipine in the treatment of Raynaud's phenomenon. Am. J. Med. 77, 451–456.

Evans, J.M., Piper, P.J. and Costello, J.F. (1990). The pharmacological profile of SKandF 104353-Z_2, a potent, selective inhaled antagonist of cysteinyl leukotrienes, in normal man. Adv. Prostaglandin Thromboxane Leukotriene Res. 21, 469–472.

Evans, J.F., Léveillé, C., Mancini, J.A., Prasit, P., Thérien, M., Zamboni, R., Gauthier, J.Y., Fortin, R., Charleson, P., MacIntyre, D.E., Luell, S., Bach, T.J., Meurer, R., Guay, J., Vickers, P.J., Rouzer, C.A., Gillard, J.W. and Miller, D.L. (1991). 5-Lipoxygenase-activating protein is the target of a quinoline class of leukotriene synthesis inhibitors. Mol. Pharmacol. 40, 22–27.

Fagan, J.M. and Goldberg, A.L. (1986). Inhibitors of protein and RNA synthesis cause a rapid block in prostaglandin production at the prostaglandin synthase step. Proc. Natl Acad. Sci. USA 83, 2771–2775.

Feinmark, S.J., Lindgren, J.A., Claesson, H.-E., Malsten, C. and Samuelsson, B. (1981). Stimulation of human leukocyte degranulation by leukotriene B_4 and its ω-oxidized metabolites. FEBS Lett. 136, 141–144.

Fels, A.O.S., Pawlowski, N.A., Cramer, E.B., King, T.K.C., Cohn, Z.A. and Scott, W.A. (1982). Human alveolar macrophages produce leukotriene B_4. Proc. Natl Acad. Sci USA 79, 7866–7870.

Ferreira, S.H., Moncada, S. and Vane, J.R. (1971). Indomethacin and aspirin abolish prostaglandin release from the spleen. Nature (New Biol.) 231, 237–239.

Ferreri, N.R., Howland, W.C. and Spiegelberg, H.L. (1986). Release of leukotrienes C_4 and B_4 and prostaglandin E_2 from human monocytes stimulated with aggregated IgG, IgA, and IgE. J. Immunol. 136, 4188–4193.

Feuerstein, G. (1984). Leukotrienes and the cardiovascular system. Prostaglandins 27, 781–802.

Feuerstein, G. and Hallenbeck, J.M. (1987). Prostaglandins, leukotrienes and platelet-activating factor in shock. Am. Rev. Pharmacol. Toxicol. 27, 301–313.

Fiddler, G.I. and Lumley, P. (1990). Preliminary clinical studies with thromboxane synthase inhibitors and thromboxane receptor blockers. Circulation 81(Suppl. I), 169–178.

Finnerty, J.P., Twentyman, O.P., Harris, A., Palmer, J.B.D. and Holgate, S.T. (1991). Effect of GR32191, a potent thromboxane receptor antagonist, on exercise induced bronchoconstriction in asthma. Thorax 46, 190–192.

Fischer, S., Struppler, M., Böhlig, B., Bernutz, C., Wober, W. and Weber, P.C. (1983). The influence of selective thromboxane synthetase inhibition with a novel imidazole derivative UK-38,485 on prostanoid formation in man. Circulation 68, 821–826.

Fitzgerald, D.J., Roy, L., Catella, F. and Fitzgerald G.A. (1986). Platelet activation in unstable coronary disease. N. Engl. J. Med. 315, 983–989.

Fitzgerald, D.J., Fragetta, J. and Fitzgerald, G.A. (1988). Prostaglandin endoperoxides modulate the response to thromboxane synthase inhibition during coronary thrombosis. J. Clin. Invest. 82, 1708–1713.

Fitzgerald, G.A., Oates, J.A., Hawiger, J., Maas, R.L. II, Lawson, J.A. and Brash, A.R. (1983a). Endogenous biosynthesis of prostacyclin and thromboxane and platelet function during chronic administration of aspirin in man. J. Clin. Invest. 71, 676–688.

Fitzgerald, G.A., Brash, A.R., Oates, J.A. and Pedersen, A.K. (1983b). Endogenous prostacyclin biosynthesis and platelet function during selective inhibition of thromboxane synthase in man. J. Clin. Invest. 72, 1336–1343.

Fitzgerald, G.A., Reilly, I.A. and Pedersen, A.K. (1985). The biochemical pharmacology of thromboxane synthase inhibition in man. Circulation 72, 1194–1201.

Fitzpatrick, F.A., Bundy, G.L., Gorman, R.R. and Honohan, T. (1978) 9,11-Epoxyiminoprosta-5,13-dienoic acid is a thromboxane A_2 antagonist in human platelets. Nature 275, 764–766.

Fitzsimmons, B.J., Adams, J., Evans, J.F., Leblanc, Y. and Rokach, J. (1985). The lipoxins. Stereochemical identification and determination of their biosynthesis. J. Biol. Chem. 260, 13008–13012.

Fleisch, J.H., Rinkema, L.E., Haish, K.D., Swanson-Bean, D., Goodson, T., Ho, P.P.K. and Marshall, W.S. (1985). LY 171883, 1-[2-Hydroxy-3-propyl-4-[4-(1H-tetrazol-5-yl)butoxy]phenyl]ethanone, an orally active leukotriene D_4 antagonist. J. Pharmacol. Exp. Ther. 233, 148–157.

Fleisch, J.H., Rinkema, L.E., Haisch, K.D., McCullough, D., Carr, F.P. and Dillard, R.D. (1986). Evaluation of LY 163443, 1-]2-hydroxy-3-propyl-4-{[4-(1H-tetrazol-5-ylmethyl)phenoxy]methyl}phenyl]ethanone, as a pharmacologic antagonist of leukotrienes D_4 and E_4. Naunyn-Schmiedebergs Arch. Pharmacol. 333, 70–77.

Fletcher, B.S., Kujubu, D.A., Perrin, D.M. and Herschman, H.R. (1992). Structure of the mitogen-inducible TIS10 gene and demonstration that TIS10-encoded protein is a functional prostaglandin G/H synthase. J. Biol. Chem. 267, 4338–4344.

Flower, R.J. (1988). Lipocortin and the mechanism of action of the glucocorticoids. Br. J. Pharmacol. 94, 987–1015.

Flower, R.J., Moncada, S.and Vane, J.R. (1985). In "The Pharmacological Basis of Therapeutics", 7th edn (eds A.G. Gilman, L.S. Goodman, T.W. Rall and F. Murad), pp. 674–715. Macmillan, New York.

Foegh, M.L., Zmudka, M., Cooley, C., Winchester, J.F., Helfrich, G.B., Ramwell, P.W. and Schreiner, G.E. (1981). Urine i-TXB_2 in renal allograft rejection. Lancet ii, 431–434.

Foegh, M.L., Lim, K., Douglas, F., Turk, J., Helfrich, G.B., Taher, S.A. and Aligani, M.R. (1988). Differential effect of CGS 13080, a thromboxane synthase inhibitor, in suppressing serum and urine immunoreactive thromboxane B_2 in kidney transplant patients. Transplant Proc. 20, 424–427.

Ford-Hutchinson, A.W. (1990). Leukotriene B_4 in inflammation. Crit. Rev. Immunol. 10, 1–12.

Ford-Hutchinson, A.W. (1991). Regulation of the production

and action of leukotrienes by MK-571 and MK-886. Adv. Prostaglandin Thromboxane Leukotriene Res. 21A, 9–16.

Ford-Hutchinson, A.W., Bray, M.A., Doig, M.V., Shipley, M.E. and Smith, M.J.H. (1980). Leukotriene B: a potent chemokinetic and aggregating substance released from polymorphonuclear leukocytes. Nature 286, 264–265.

Forst, J., Weiss, J., Elsbach, P., Maraganore, J.M., Reardon, I. and Heinrikson, R.L. (1986). Structural and functional properties of a phospholipase A_2 purified from an inflammatory exudate. Biochemistry 25, 8381–8385.

Foster, K.A., Buckle, D.R., Crescenzi, K.L., Fenwick, A.E. and Taylor, J.E. (1987). Arachidonic acid analogues as inhibitors of phospholipase A_2 activity. Biochem. Soc. Trans. 15, 418–419.

Freeland, H.S., Schleimer, R.P., Schulman, E.S., Lichtenstein, L.M. and Peters, S.P. (1988). Generation of leukotriene B_4 by human lung fragments and purified human lung mast cells. Am. Rev. Resp. Dis. 138, 389–394.

Freeland, H.S., Pipkorn, U., Schleimer, R.P., Bascom, R., Lichtenstein, L.M., Naclerio, R.M. and Peters, S.P. (1989). Leukotriene B_4 as a mediator of early and late reactions to antigen in humans: the effect of systemic glucocorticoid treatment in vivo. J. Allergy Clin. Immunol. 83, 634–642.

Fretland, D.J., Levin, S., Tsai, B.S., Djuric, S.W., Widomski, D.L., Zemaitis, J.M., Shone, R.L. and Bauer, R.F. (1989). The effect of leukotriene-B_4 receptor antagonist, SC-41930, on acetic acid-induced colonic inflammation. Agents Actions 27, 395–397.

Fretland, D.J., Widomski, D., Tsai, B.-S., Zemaitis, J.M., Levin, S., Djuric, S.W., Shone, R.L. and Gaginella, T.S. (1990a). Effect of leukotriene B_4 receptor antagonist SC-41930 on colonic inflammation in rat, guinea pig and rabbit. J. Pharmacol. Exp. Ther. 255, 572–576.

Fretland, D.J., Widomski, D.L., Zemaitis, J.M., Walsh, R.E., Levin, S., Djuric, S.W., Shone, R.L., Tsai, B.S. and Gaginella, T.S. (1990b). Inflammation of guinea pig dermis. Effect of leukotriene B_4 receptor antagonist. SC-41930. Inflammation 14, 727–739.

Friedhoff, L.T., Manning, J., Funke, P.T., Ivashkiv, E., Tu, J., Cooper, W. and Willard, D.A. (1986). Quantitation of drug levels and platelet receptor blockade caused by a thromboxane antagonist. Clin. Pharmacol. Ther. 40, 634–642.

Fujimura, M., Sasaki, F., Nakatsumi, Y., Takahashi, Y., Hifumi, S., Taga, K., Mifune, J.-I., Tanaka, T. and Matsuda, T. (1986). Effects of a thromboxane synthetase inhibitor (OKY-046) and a lipoxygenase inhibitor (AA-861) on bronchial responsiveness to actylcholine in asthmatic subjects. Thorax 41, 955–959.

Fujimura, M., Sakamoto, S., Saito, M., Miyake, Y. and Matsuda, T. (1991). Effect of a thromboxane A_2 receptor antagonist (AA-2414) on bronchial hyperresponsiveness to methacholine in subjects with asthma. J. Allergy Clin. Immunol. 87, 23–27.

Fukuda, T., Kimura, S., Arakawa, T. and Kobayashi, K. (1990). Possible role of leukotrienes in gastritis associated with *Campylobacter pylori*. J. Clin. Gastroenterol. 12(Suppl. 1), S131–S134.

Fuller, R.W., Black, P.N. and Dollery, C.T. (1989). Effect of the oral leukotriene D_4 antagonist LY171883 on inhaled and intradermal challenge with antigen and leukotriene D_4 in atopic subjects. J. Allergy Clin. Immunol. 83, 939–944.

Gaddy, J., Bush, R.K., Margolskee, D., Williams, U.C. and Busse, E. (1990). The effects of a leukotriene D_4 (LTD_4) antagonist (MK-571) in mild to moderate asthma. J. Allergy Clin. Immunol. 85, A197 (Abstr.)

Gaginella, T.S. (1990). Role of leukotrienes in peptic ulcer and inflammatory bowel diseases. Dig. Dis. Sci. 35, 1029 (Abstr.)

Gapinski, D.M., Mallett, B.E., Froelich, L.L. and Jackson, W.T. (1990a). Benzophenone dicarboxylic acid antagonist of leukotriene B_4. 1. Structure–activity relationships of the benzophenone nucleus. J. Med. Chem. 33, 2798–2807.

Gapinski, D.M., Mallett, B.E., Froelich, L.L. and Jackson, W.T. (1990b). Benzophenone dicarboxylic acid antagonists of leukotriene B_4. 2. Structure–activity relationships of the lipophilic side chain. J. Med. Chem. 33, 2807–2813.

Garcia-Szabo, R., Peterson, M., Watkins, W., Bizios, R., Kong, D. and Malik, A. (1983). Thromboxane generation after thrombin: protective effect of thromboxane synthetase inhibition on lung fluid balance. Circ. Res. 53, 214–222.

Garcia-Szabo, R., Kern, D.F. and Malik, A.B. (1984). Pulmonary vascular response to thrombin: effects of thromboxane synthetase inhibition with OKY-046 and OKY-1581. Prostaglandins 28, 851–866.

Gardiner, P.J. (1990). Classification of prostanoid receptors. Adv. Prostaglandin Thromboxane Leukotriene Res. 20, 110–118.

Gardiner, P.J. and Collier, H.O.J. (1980). Specific receptors for prostaglandins in airways. Prostaglandins 19, 819–841.

Garland, L.G. and Salmon, J.A. (1991). Hydroxamic acids and hydroxyureas as inhibitors of arachidonate 5-lipoxygenase. Drugs of the Future 16, 547–558.

Gattaz, W.F., Köllisch, M., Thuren, T., Virtanen, J.A. and Kinnunen, P.K.J. (1987). Increased plasma phospholipase-A2 activity in schizophrenic patients: reduction after neuroleptic therapy. Biol. Psychiatry 22, 421–426.

Gertner, D.J., Rampton, D.S. and Lennard-Jones, J.E. (1990). *In vitro* leucotriene B_4 production in ulcerative colitis: effects of three potentially efficacious new agents (SC 45662, SC 46193 and misoprostol). Gastroenterol. 98, A450 (Abstr.).

Gialih, L., Noel, J., Loffredo, W., Stable, H.Z. and Tsai, M.-D. (1988). Use of short-chain cyclopentano-phosphatidylcholines to probe the mode of activation of phospholipase A_2 from bovine pancreas and bee venom. J. Biol. Chem. 263, 13208–13214.

Giles, H. and Leff, P. (1988). The biology and pharmacology of PGD_2. Prostaglandins 35, 277–300.

Giles, H., Leff, P., Bologo, M.L., Kelly, M.G. and Robertson, A.D. (1989). The classification of prostaglandin DP-receptors in platelets and vasculature using BW A868C, a novel, selective and potent competitive antagonist. Br. J. Pharmacol. 96, 291–300.

Gillard, J., Ford-Hutchinson, A.W., Chan, C., Charleson, S., Denis, D., Foster, A., Fortin, R., Leger, S., McFarlane, C.S., Morton, H., Piechuta, H., Riendeau, D., Rouzer, C.A., Rokach, J., Young, R., MacIntyre, D.E., Peterson, L., Bach, T., Eiermann, G., Hopple, S., Humes, J., Hupe, L., Luell, S., Metzger, J., Meurer, R., Miller, D.K., Opas, E. and Pacholok, S. (1989). L-663,536 (MK-886) (3-[1-(4-chlorobenzyl)-3-t-butyl-thio-5-isopropylindol-2-yl]-2,2-dimethylpropanoic acid), a novel, orally active leukotriene biosynthesis inhibitor. Can. J. Physiol. Pharmacol. 67, 456–464.

Gimbrone, M.A. Jr, Brock, A.F. and Schafe, A.I. (1984). Leukotriene B_4 stimulates polymorphonuclear leukocyte adhesion to cultured vascular endothelial cells. J. Clin. Invest. 74, 1552–1555.

Glaser, K.B. and Jacobs, R.S. (1986). Molecular pharmacology of manoalide, inactivation of bee venom phospholipase A_2. Biochem. Pharmacol. 35, 449–453.

Godfrey, S. (1986). Controversies in the pathogenesis of exercise-induced asthma. Eur. J. Respir. Dis. 68, 81–88.

Goetzl, E.J. and Pickett, W.C. (1981). Novel structural determinants of the human neutrophil chemotactic activity of leukotriene B. J. Exp. Med. 153, 482–487.

Goldman, D.W., Gifford, L.A., Marotti, T., Koo, C.H. and Goetzl, E.J. (1987). Molecular and cellular properties of human polymorphonuclear leukocyte receptors for leukotriene B_4. Fed. Proc. 46, 200–203.

Gorman, R.R., Bundy, G.L., Pederson, D.C., Sun, F.F., Miller, O.V. and Fitzpatrick, F.A. (1977). Inhibition of human platelet thromboxane synthetase by 9,11-azoprosta-5,13-dienoic acid. Proc. Natl Acad. Sci. USA 74, 4007–4011.

Gorman, R.R., Johnson, R.A., Spilman, C.H. and Aiken, J.W. (1983). Inhibition of platelet thromboxane A_2 synthetase activity by sodium 5-(3′pyridinylmethyl) benzofuran-2-carboxylate. Prostaglandins 26, 325–342.

Grabbe, J., Czarnetzki, B.M., Rosenbach, T. and Mardin, M. (1984). Identification of chemotactic lipoxygenase products of arachidonate metabolism in psoriatic skin. J. Invest. Dermatol. 82, 477–479.

Gresele, P., Deckmyn, H., Arnout, J., Lemmens, J., Janssens, W. and Vermylen, J. (1984). BM13.177, a selective blocker of platelet and vessel wall thromboxane receptors, is active in man. Lancet I, 991–994.

Gresele, P., Arnout, J., Deckmyn, H., Huybrechts, E., Pieters, G. and Vermylen, J. (1987). Role of proaggregatory and antiaggregatory prostaglandins in hemostasis. Studies with combined thromboxane synthase inhibition and thromboxane receptor antagonism. J. Clin. Invest. 80, 1435–1442.

Gresele, P., Deckmyn, H., Arnout, J., Nenci, G.G. and Vermylen, J. (1989). Characterization of N,N'-bis(3-picolyl)-4-methoxy-isophtalamide (Picotamide) as a dual thromboxane synthase inhibitor/thromboxane A_2 receptor antagonist in human platelets. Thromb. Haemost. 61, 479–484.

Gresele, P., Deckmyn, H., Nenci, G.G. and Vermylen, J. (1991). Thromboxane synthase inhibitors, thromboxane receptor antagonists and dual blockers in thrombotic disorders. Trends Pharmacol. Sci. 12, 158–163.

Griswold, D.E., Marshall, P.J., Webb, E.F., Godfrey, R., Newton, J. Jr, DiMartino, M.J., Sarau, H.M., Gleason, J.G., Poste, G. and Hanna, N. (1987). SKandF 86002: a structurally novel anti-inflammatory agent that inhibits lipoxygenase- and cyclooxygenase-mediated metabolism of arachidonic acid. Biochem. Pharmacol. 36, 3463–3470.

Griswold, D.E., Hillegass, L.M., Meunier, P.C. DiMartino, M.J. and Hanni, N. (1988). Effect of inhibitors of eicosanoid metabolism in murine collagen-induced arthritis. Arthritis Rheum. 31, 1406–1412.

Griswold, D.E., Marshall, P.J., Lee, J.C., Webb, E.F., Hillegass, L.M., Wartell, J., Newton, J. Jr and Hanna, N. (1991). Pharmacology of the pyrroloimidazole, SKandF 105809–11. Antiinflammatory activity and inhibition of mediator production *in vivo*. Biochem. Pharmacol. 42, 825–831.

Grone, H.-J., Grippo, R.S., Arendshorst, W.J. and Dunn, M.J. (1986). Role of thromboxane in control of arterial pressure and renal function in young spontaneously hypertensive rats. Am. J. Physiol. 250, F488–F496.

Grover, G.J. and Schumacher, W.A. (1988). Effect of the thromboxane receptor antagonist SQ 29,548 on myocardial infarct size in dogs. J. Cardiovasc. Pharmacol. 11, 29–35.

Grover, G.J. and Schumacher, W.A. (1989). Effect of the thromboxane A_2 receptor antagonist SQ 30,741 on ultimate myocardial infarct size, reperfusion injury and coronary flow reserve. J. Pharmacol. Exp. Ther. 248, 484–491.

Gunther, R.A., Smith, G.J. and Holcroft, J.W. (1984). Pulmonary response to selective inhibition of thromboxane A_2 synthesis during endotoxemia utilizing a unique inhibitor. Surg. Forum 35, 42–44.

Hahn, R.A., MacDonald, B.R., Simpson, P.J., Potts, B.D. and Parli, C.J. (1990). Antagonist of leukotriene B_4 receptors does not limit canine myocardial infarct size. J. Pharmacol. Exp. Ther. 253, 58–66.

Hall, R.A., Gillard, J., Guindon, Y., Letts, G., Champion, E., Elhier, D., Evans, J., Ford-Hutchinson, A.W., Fortin, R., Jones, T.R., Lord, A., Morton, H.E., Rokach, J. and Yoakim, C. (1987). Pharmacology of L655,240 (3-[1-(4-chlorobenzyl)-5-fluoro-3-methyl-indol-2-yl[2,2-dimethyl propanoic acid); a potent selective thromboxane/prostaglandin endoperoxide antagonist. Eur. J. Pharmacol. 135, 193–202.

Halushka, P.V. (1989). Pharmacology of thromboxane A_2 receptor antagonists. Z. Kardiol. 78, 42–47.

Halushka, P.V., Cook, J.A. and Wise, W.C. (1983). Beneficial effects of UK37248, a thromboxane synthetase inhibitor, in experimental endotoxic shock in the rat. Br. J. Clin. Pharmacol. 15, 133S–139S.

Halushka, P.V., Mais, D.E. and Saussy, D.L. Jr (1987). Platelet and vascular smooth muscle thromboxane A_2/prostaglandin H_2 receptors. Fed. Proc. 46, 149–153.

Halushka, P.V., Mais, D.E., Mayeux, P.R. and Morinelli, T.A. (1989a). Thromboxane, prostaglandin and leukotriene receptors. Annu. Rev. Pharmacol. Toxicol. 29, 213–239.

Halushka, P.V., Mais, D.E. and Morinelli, T.A. (1989b). Thromboxane and prostacyclin receptors. Prog. Clin. Biol. Res. 301, 21–28.

Hamberg, M., Hedqvist, P., Strandberg, K., Svensson, J. and Samuelsson, B. (1975a). Prostaglandin endoperoxides IV. Effects on smooth muscle. Life Sci. 16, 451–462.

Hamberg, M., Svensson, J. and Samuelsson, B. (1975b). Thromboxanes: a new group of biologically active compounds derived from prostaglandin endoperoxides. Proc. Natl Acad. Sci. USA 72, 2994–2998.

Hamid, M., Rafferty, P. and Holgate, S.T. (1990). The inhibitory effect of terfanadine and flurbiprofen on early and late-phase bronchoconstriction following allergen challenge in atopic asthma. Clin. Exp. Allergy 20, 261–267.

Hamid-Bloomfield, S. and Whittle, B.J.R. (1986). Prostaglandin D_2 interacts at thromboxane receptor sites on guinea-pig platelets. Br. J. Pharmacol. 88, 931–936.

Hanasaki, K., Nagasaki, T. and Arita, H. (1989). Characterization of platelet thromboxane A_2/prostaglandin H_2 receptor by a novel thromboxane receptor antagonist, (^{3}H)S-145. Biochem. Pharmacol. 38, 2007–2017.

Hand, J.M., Schwalm, S.F. and Lewis, A.J. (1986). antagonism of antigen-induced contraction of isolated guinea-pig trachea by 5-lipoxygenase inhibitors. Int. Arch. Allergy Appl. Immunol. 79, 8–13.

Hand, J.M., Auen, M.A., Chang, J. and Englebach, I.M. (1989a). Prevention and reversal of aerosol LTD_4-induced changes in guinea pig pulmonary mechanics by Wy-48252, an orally active LTD_4/E_4 receptor antagonist. Int. Arch. Allergy Appl. Immunol. 89, 78–82.

Hand, J.M., Schwalm, S.F., Auen, M.A., Kreft, A.F., Musser, J.H. and Chang, J. (1989b). WY-48,252 (1,1,1-trifluoro-N-[3-(2-quinolinylmethoxy)phenyl]methane sulfonamide) an orally active leukotriene D_4 antagonist: Pharmacological characterization *in vitro* and *in vivo* in the guinea pig. Prostaglandins Leukot. Essent. Fatty Acids 37, 97–106.

Hanley, S.P. (1986). Prostaglandins and the lung. Lung 164, 65–77.

Hara, S., Kudo, I., Matsuta, K., Miyamoto, T. and Inoue, K. (1988). Amino acid composition and NH_2-terminal amino acid sequence of human phospholipase A_2 purified from rheumatoid synovial fluid. J. Biochem. 104, 326–328.

Harnett, J.C., Spector, S.L. and Farr, R.S. (1978). In "Principles and Practice" (eds E. Middleton, C.E. Reed and E.F. Ellis), pp 1002–1022. CV Mosby, St Louis.

Hawkey, C.J. and Rampton, D.S. (1983). Benoxaprofen in the treatment of active ulcerative colitis. Prostaglandins Leukotrienes Med. 10, 405–409.

Hay, D.W.P., Muccitelli, R.M., Tucker, S.S., Vickery-Clark, L.M., Wilson, K.A., Gleason, J.G., Hall, R.F., Wasserman, M.A. and Torphy, T.J. (1987). Pharmacologic profile of SKandF 104353: a novel, potent and selective peptidoleukotriene receptor antagonist in guinea pig and human airways. J. Pharmacol. Exp. Ther. 243, 474–481.

Hay, D.W.P., Muccitelli, R.M., Vickery-Clark, L.M., Novak, L.S., Osborn, R.R., Gleason, J.G., Yodis, L.-A.P., Saverino, S.M., Eckardt, R.D., Sarau, H.M., Wasserman, M.A., Torphy, T.J. and Newton, J.F. (1991). Pharmacologic and pharmacokinetic profile of SKandF S-106203, a potent, orally active peptidoleukotriene receptor antagonist, in guinea-pig. Pulm. Pharmacol. 4, 177–189.

Hayakawa, M., Horigome, K., Kudo, I., Tomita, M., Nojima, S. and Inoue, K. (1987). Amino acid composition and NH. terminal amino acid sequence of rat platelet secretory phospholipase A_2. J. Biochem. 101, 1311–1314.

Hayakawa, M., Kudo, I., Tomita, M. and Inoue, K. (1988a). Purification and characterization of a membrane-bound phospholipase A_2 from rat platelets. J. Biochem. 103, 263–266.

Hayakawa, M., Kudo, I., Tomita, M., Nojima, S. and Inoue, K. (1988b). The primary structure of rat platelet phospholipase A_2. J. Biochem. 104, 767–772.

Haynes, R.C. Jr and Murad, F. (1985). In "The Pharmacological Basis of Therapeutics", 7th edn (eds A.G. Gilman, L.S. Goodman, T.W. Rall and F. Murad), pp 1459–1489. MacMillan, New York.

Heinrikson, R.L., Krueger, E.T. and Keim, P.S. (1977). Amino acid sequence of phospholipase A_2-α from the venom of *Crotalus adamanteus*. J. Biol. Chem. 232, 4913–4921.

Hench, P.S., Kendall, E.C., Alocumb, C.H. and Polley, H.F. (1949). Effect of hormone of adrenal cortex (17-hydroxy-11-dehydrocorticosterone: compound E) and of pituitary adrenocorticotropic hormone on rheumatoid arthritis: preliminary report. Proc. Staff Meet. Mayo Clinic 24, 181–197.

Hendeles, L., Davison, D., Blake, K., Harman, E., Cooper, R. and Margolskee, D. (1990). Leukotriene D_4 is an important mediator of antigen-induced bronchoconstriction: attenuation of dual response with MK-571, a specific LTD_4 receptor antagonist. J. Allergy Clin. Immunol. 85, 197A (Abstr.).

Hendra, T., Collins, P., Penny, W. and Sheridan, D. (1983). Dazoxiben in stable angina (letter). Lancet i, 1041.

Herron, D.K., Bollinger, N.G., Swanson-Bean, D., Jackson, W.T., Froelich, L.L. and Goodson, T. (1988). LY255283, a new leukotriene B_4 antagonist. FASEB J. 2, A1110. (Abstr.).

Hibbeln, J.R., Palmer, J.W. and Davis, J.M. (1989). Are disturbances in lipid-protein interactions by phospholipase-A_2

a predisposing factor in affective illness? Biol. Psychiatry 25, 945–961.

Higgs, G.A. and Mugridge, K.G. (1983). Inhibition of mononuclear leukocyte accumulation by the arachidonic acid lipoxygenase inhibitor BW755C. Adv. Prostaglandin Thromboxane Leukotriene Res. 12, 19–23.

Higgs, G.A., Eakins, K.E., Mugridge, K.G., Moncada, S. and Vane, J.R. (1980). The effects of non-steroid anti-inflammatory drugs on leukocyte migration in carrageenan-induced inflammation. Eur. J. Pharmacol. 66, 81–86.

Higgs, G.A., Follenfant, R.L. and Garland, L.G. (1988). Selective inhibition of arachidonate 5-lipoxygenase by novel acetohydroxamic acids: effects on acute inflammatory responses. Br. J. Pharmacol. 94, 547–551.

Hirata, F., Del Carmine, R., Nelson, C.A., Axelrod, J., Schiffmann, E., Warabi, A., De Blas, A.L., Nirenberg, M., Manganiello, V., Vaughan, M., Kumagai, S., Green, I., Decker, J.L. and Steinberg, A.D. (1981). Presence of autoantibody for phospholipase inhibitory protein, lipomodulin, in patients with rheumatic diseases. Proc. Natl Acad. Sci. USA 78, 3190–3194.

Hock, C.E., Phillips, G.R. III and Lefer, A.M. (1985). Protective action of a thromboxane synthetase inhibitor in preventing extension of infarct size in acute myocardial infarction. Prostaglandins Leukotrienes Med. 17, 339–346.

Hoet, B., Arnout, J. and Vermylen, J. (1989). R68070, a combined thromboxane/endoperoxide receptor antagonist (TRA) and thromboxane synthetase inhibitor (TSI), prolongs the bleeding time more than aspirin in man. Prog. Clin. Biol. Res. 301, 573–577.

Hoet, B., Arnout, J., Van Geet, C., Deckmyn, H., Verhaeghe, R. and Vermylen, J. (1990a). Ridogrel, a combined thromboxane synthase inhibitor and receptor blocker, decreases elevated plasma β-thromboglobulin levels in patients with documented peripheral arterial diseases. Thromb. Haemost. 64, 87–90.

Hoet, B., Falcon, C., De Reys, S., Arnout, J., Deckmyer, H. and Vermylen, J. (1990b). R68070, a combined thromboxane/endoperoxide receptor antagonist and thromboxane synthase inhibitor, inhibits human platelet activation *in vitro* and *in vivo*: a comparison with aspirin. Blood 75, 646–653.

Hofling, B., Meier, B., Bulitta, M. and Etti, H. (1988). Sulotroban a thromboxane receptor antagonist in prevention of occlusion after transluminal coronary angioplasty: a placebo controlled, double blind study (abstract). In "Proc. Xth Congr. European Society of Cardiology, Satellite Symposium on New Approaches in the Pharmacology of Ischemic Heart Disease, Vienna."

Holgate, S.T., Twentyman, O.P., Rafferty, P., Beasley, R., Hutson, P.A., Robinson, C. and Church, M.K. (1987). Primary and secondary effector cells in the pathogenesis of bronchial asthma. Int. Arch. Allergy Appl. Immunol. 82, 498–506.

Holroyde, M.C., Cole, M., Altounyan, R.E.C., Dixon, M. and Elliott, E.V. (1981). Bronchoconstriction produced in man by leukotrienes C and D. Lancet I, 17–18.

Hoover, R.L., Kamovsky, M.J., Austen, K.F., Corey, E.J. and Lewis, R.A. (1984). Leukotriene B_4 action on endothelium mediates augmented neutrophil/endothelial adhesion. Proc. Natl Acad. Sci. USA 81, 2191–2193.

Hope, W.C., Patel, B.J. and Wiggan, G.A. (1988). In "Inflammation Research Association Meeting", Therapeutic Control of Inflammatory Diseases, Laurel Resort and Conference Center, White Haven, PA, October 23–27, Abstr. 133.

Hoshino, M. and Fukushima, Y. (1991). Effect of OKY-046 (thromboxane A_2 synthetase inhibitor) on exercise-induced asthma. J. Asthma 28, 19–29.

Hossman, V., Schäfer, H.-J., Auel, H. and Etti, H. (1987). Treatment of advanced stages of peripheral obliterative disease with the thromboxane receptor antagonist BM 13,177: a placebo-controlled double blind study. Thromb. Haemost. 58, 182 (Abstr.).

Huddleston, C.B., Lupinetti, F.M., Laws, K.H., Collins, J.C., Clanton, J.A., Hawinger, J.J., Oates, J.A. and Hammon, J.W. Jr (1983). The effects of Ro-22-4679, a thromboxane synthetase inhibitor, on ventricular fibrillation induced by coronary artery occlusion in conscious dogs. Circ. Res. 52, 608–613.

Hui, K. and Barnes, N.C. (1991). Lung function improvement in asthma with a cysteinyl-leukotriene receptor antagonist. Lancet 337, 1062–1063.

Hui, K.P., Taylor, I.K., Taylor, G.W., Rubin, P., Kesterson, J., Barnes, N.C. and Barnes, P.J. (1991). Effect of a 5-lipoxygenase inhibitor on leukotriene generation and airway responses after allergen challenge in asthmatic patients. Thorax 46, 184–189.

Hume, M. and Eddey, V. (1977). Treatment of chronic airways obstruction with indomethacin. Scand. J. Respir. Dis. 58, 284–286.

Humes, J.L., Sadowski, S., Galavage, M., Goldenberg, M., Subers, E., Kuehl, F.A. Jr and Bonney, R.J. (1983). Pharmacological effects of non-steroidal antiinflammatory agents on prostaglandin and leukotriene synthesis in mouse peritoneal macrophages. Biochem. Pharmacol. 32, 2319–2322.

Huskisson, E.C. and Scott, O. (1979). Treatment of rheumatoid arthritis with a single daily dose of benoxaprofen. J. Rheumatol. Rehab. 18, 110–113.

Hutton, I., Tweddel, A.C., Rankin, A.C., Walker, I.D. Davidson, J.F. (1983). Effects of dazoxiben on transcardiac thromboxane levels and haemodynamics in coronary heart disease. Br. J. Clin. Pharmacol. 15, 79S–82S.

Ishii, A., Nakagawa, T., Nambu, F., Motoishi, M. and Miyamoto, T. (1990). Inhibition of endogenous leukotriene-mediated lung anaphylaxis in guinea-pigs by a novel receptor antagonist ONO-1078. Int. Allergy Appl. Immunol. 92, 404–407.

Israel, E., Juniper, E.F., Callaghan, J.T., Mathur, P.N., Morris, M.M., Dowell, A.R., Enas, G.G., Hargreave, F.E. and Drazen, J.M. (1989). Effect of a leukotriene antagonist, LY 171883, on cold air-induced bronchoconstriction in asthmatics. Am. Rev. Respir. Dis. 140, 1348–1353.

Israel, E., Dermarkarian, R., Rosenberg, M., Sperling, R., Taylor, G., Rubin, P. and Drazen, J.M. (1990). The effects of a 5-lipoxygenase inhibitor on asthma induced by cold, dry air. New Eng. J. Med. 323, 1740–1744.

Iwamoto, I., Ra, C., Sato, T., Tomioka, H. and Yoshida, S. (1988). Thromboxane-A_2 production in allergen-induced immediate and late asthmatic responses. J. Asthma 25, 117–124.

Jackson, W.T., Boyd, R.J., Forelich, L.L., Goodson, T., Bollinger, N.G. Herron, D.K., Mallett, B.E. and Gapinski, D.M. (1988a). Inhibition of LTB_4 binding and aggregation of neutrophils by LY255283 and LY223982. FASEB J. 2, A1110 (abstract).

Jackson, W.T., Froelich, L.L., Goodson, T., Herron, D.K. and Mallett, B.E. (1988b). Inhibition of LTB_4 induced leukopenia by LY255283 and LY223982. Pharmacologist 30, A206 (Abstr.).

Jacobs, R.S., Culver, P., Langdon, R., O'Brien, T. and White, S. (1985). Some pharmacological observations on marine natural products. Tetrahedron 41, 981–984.

Jain, M.K. and Berg, O.B. (1989). The kinetics of interfacial catalysis by phospholipase A_2 and regulation of interfacial activation: hopping versus scooting. Biochim. Biophys. Acta 1002, 127–156.

Johnson, H.G. and Stout, B.K. (1988). Activity of a novel hydroquinone inhibitor of leukotriene synthesis (U-66,858) in the rhesus monkey Ascaris reactor. Int. Arch. Allergy Appl. Immunol. 87, 204–207.

Johnson, L.K., Frank, S., Vadas, P., Pruzanski, W., Lusis, A.J. and Seilhamer, J.J. (1990). In Phospholipase A_2" (eds P.Y.-K. Wong and E.A. Dennis), pp 17–34. Plenum Press, New York.

Jones, R.L., Peesapati, V. and Wilson, N.H. (1982). Antagonism of the thromboxane-sensitive contractile systems of the rabbit aorta, dog, saphenous vein and guinea-pig trachea. Br. J. Pharmacol. 76, 423–438.

Jones, T.R., Zamboni, R., Belley, M., Champion, E., Charette, L., Ford-Hutchinson, A.W., Frenette, R., Gauthier, J.-Y., Leger, S., Masson, P., McFarlane, C.S., Piechuta, H., Rokach, J., Williams, H., Young, R.N., DeHaven, R.N. and Pong, S.S. (1989). Pharmacology of L-660,711 (MK-571): a novel potent and selective leukotriene D_4 receptor antagonist. Can. J. Physiol. Pharmacol. 67, 17–28.

Joos, G.F., Kips, J.C., Pauwels, R.A. and van der Straeten, M.E. (1991). The effect of aerosolized SkandF 104353-Z_2 on the bronchoconstrictor effect of leukotriene D_4 in asthmatics. Pulm. Pharmacol. 4, 37–42.

Kaliner, M. (1985). Mechanisms of glucocorticosteroid action in bronchial asthma. J. Allergy Clin. Immunol. 76, 321–329.

Karbowski, A. (1991). Double-blind, parallel comparison of etodolac and indomethacin in patients with osteoarthritis of the knee. Curr. Med. Res. Opin. 12, 309–317.

Kawana, S., Ueno, A. and Nishiyama, S. (1990). Increased levels of immunoreactive leukotriene B_4 in blister fluids of bullous pemphigoid patients and effects of a selective 5-lipoxygenase inhibitor on experimental skin lesions. Acta Derm. Venereol. (Stockh.) 70, 281–285.

Keen, M., Kelly, E. and MacDermot, J. (1989). Prostaglandin receptors in the cardiovascular system: potential selectivity from receptor subtypes or modified responsiveness. Eicosanoids 2, 193–197.

Keery, R.J. and Lumley, P. (1985). AH 6809, a selective antagonist at the human platelet DP-receptor? Br. J. Pharmacol. 85, 286P (Abstr.).

Kellaway, C.H. and Trethewie, R.E. (1940). The liberation of a slow reacting smooth muscle stimulating substance of anaphylaxis. Q. J. Exp. Physiol. 30, 121–145.

Kennedy, I., Coleman, R.A., Humphrey, P.P.A., Levy, G.P. and Lumley, P. (1982). Studies on the characterization of prostanoid receptors: a proposed classification. Prostaglandins 24, 667–689.

Kiff, P.S., Bergman, C., Atkinson, L., Jewitt, D.E., Westwick, J. and Kakkar, V.V. (1983). Haemodynamic and metabolic effects of dazoxiben at rest and during a trial pacing. Br. J. Clin. Pharmacol. 15, 73S–77S.

Kimura, Y., Kodama, K., Naka, M., Nanto, S., Taniura, K., Kuzuya, T., Hamanaka, Y. and Tada, M. (1986). Antianginal effect of CV-4151, a new potent, selective inhibitor of thromboxane synthetase. J. Am. Coll. Cardiol. 7, 177A (Abstr.).

Kiniwa, M.L., Yamamoto, N., Hashimoto, Y., Miyake, H. and Masuda, H. (1986). Anti-inflammatory effect of THS-201, a new intra-articular steroid. Folia Pharmacol. Japan 87, 89–97.

Kips, J., Joos, G., Pauwels, R., Van Der Straeten, M., Delepeleire, I., Williams, V. and Margolskee, D. (1990). MK-571 (L-660.771): a potent LTD4 antagonist in asthmatic men. Am. Rev. Respir. Dis. 141, A117 (Abstr.).

Kips, J., Margolskee, D., Delepeleire, I., Joos, G., Williams, V., Buntinx, A. and Pauwels, R. (1991). MK-0679 is a potent and selective LTD$_4$ antagonist in asthmatic men. Am. Rev. Resp. Dis. 143, A599 (Abstr.).

Kirby, B. (1989). A review of the rational use of corticosteroids. J. Int Med. Res. 17, 493–505.

Kishikawa, K., Matsunaga, N., Maruyama, T., Seo, R., Toda, M., Miyamoto, T. and Kawasaki, A. (1991). ONO-LB-457: a novel and orally active leukotriene B$_4$ receptor antagonist. Adv. Prostaglandin Thromboxane Leukotriene Res. 21A, 407–410.

Klickstein, L.B., Shapleigh, C. and Goetzl, E.J. (1980). Lipoxygenation of arachidonic acid as a source of polymorphonuclear leukocyte chemotactic factors in synovial fluid and tissue in rheumatoid arthritis and spondyloarthritis. J. Clin. Invest. 66, 1166–1170.

Kondo, K., Seo, R., Kitagawa, T., Omawari, N., Kira, H., Okegawa, T. and Kawasaki, A. (1987). Preventative effect of ONO-3708 on thrombosis and vasospasms *in vitro* and *in vivo*. Adv. Prostaglandin Thromboxane Leukotriene Res. 17, 423–426.

Konno, M., Sakuyama, S., Nakae, T., Hamanaka, N., Miyamoto, T. and Kawasaki, A. (1991). Synthesis and structure-activity relationships of a series of substituted-phenylpropionic acids as a novel class of leukotriene B4 antagonists. Adv. Prostaglandin Thromboxane Leukotriene Res. 21A, 411–414.

Konturek, S.J. and Pawlik, W. (1986). Physiology and pharmacology of prostaglandins. Digest. Dis. Sci. 31, 6S–19S.

Kragballe, K., Desjarlais, L. and Voorhees, J.J. (1985). Leukotrienes B$_4$, C$_4$ and D$_4$ stimulate DNA synthesis in cultured human epidermal keratinocytes. Br. J. Dermatol. 113, 43–52.

Kramer, R.M., Hession, C., Johansen, B., Hayes, G., McGray, P., Chow, E.P., Tizard, R. and Pepinsky, R.B. (1989). Structure and properties of a human non-pancreatic phospholipase A$_2$. J. Biol. Chem. 264, 5768–5775.

Kramer, R.M., Johansen, B., Hession, C. and Pepinsky, R.B. (1990). In "Phospholipase A$_2$" (eds P.Y.-K. Wong and E.A. Dennis), pp 35–53. Plenum Press, New York.

Kramer, R.M., Roberts, E.F., Manetta, J. and Putnam, J.E. (1991). The Ca^{2+}-sensitive cytosolic phospholipase A$_2$ is a 100 k-Da protein in human monoblast U937 cells. J. Biol. Chem. 266, 5268–5272.

Krell, R.D., Osborn, R., Vickery, L., Falcone, K., O'Donnell, M., Gleason, J., Kinzig, C. and Bryan, D. (1981). Contraction of isolated airway smooth muscle by synthetic leukotrienes C$_4$ and D$_4$. Prostaglandins 22, 387–409.

Krell, R.D., Kusner, E.J., Aharony, D. and Giles, R.E. (1989). Biochemical and pharmacological characterization of ICI 198,615: a peptide leukotriene receptor antagonist. Eur. J. Pharmacol. 159, 73–81.

Krell, R.D., Aharony, D., Buckner, C.K., Keith, R.A., Kusner, E.J., Snyder, D.W., Bernstein, P.R., Matassa, V.G., Yee, Y.K., Brown, F.J., Hesp, B. and Giles, R.E. (1990a). The preclinical pharmacology of ICI 204,219. A peptide leukotriene antagonist. Am. Rev. Respir. Dis. 141, 978–987.

Krell, R.D., Aharony, D., Buckner, C.K. and Kusner, E.J. (1990b). Peptide leukotriene receptors and antagonists. Adv. Prostaglandin Thromboxane Leukotriene Res. 20, 119–126.

Kriseman, T., Underwood, D., McNamara, D. and Kadowitz, P. (1987). Inhibition of A23187-induced bronchoconstriction by dazoxiben, a thromboxane synthetase inhibitor. Am. J. Med. Sci. 293, 349–353.

Ku, E.C., McPherson, S.E., Signor, C., Chertock, H. and Cash, W.D. (1983). Characterization of imidazo[1,5-*a*]pyridine-5-hexanoic acid (CGS 13080) as a selective thromboxane synthetase inhibitor using *in vitro* and *in vivo* biochemical models. Biochem. Biophys. Res. Commun. 112, 899–906.

Ku, E.C., Raychaudhuri, A., Ghai, G., Kimble, E.F., Lee, W.H., Colombo, C., Dotson, R., Oglesby, T.D. and Wasley, J.W.F. (1988). Characterization of CGS 8515 as a selective 5-lipoxygenase inhibitor using in vitro and in vivo models. Biochim. Biophys. Acta 959, 332–342.

Kulkarni, P.S. and Eakins, K.E. (1976). N-0164 inhibits generation of thromboxane-A$_2$-like activity from prostaglandin endoperoxides by human platelet microsomes. Prostaglandins 12, 465–469.

Kulmacz, R.J., Ren, Y., Tsai, A.-L. and Palmer, G. (1991). PGH synthase: interaction with hydroperoxides and indomethacin. Adv. Prostaglandin Thromboxane Leukotriene Res. 21A, 137–140.

Laghi Pasini, F., Pasqui, A.L., Ceccatelli, L. and Di Perri, T. (1984). *In vitro* inhibition of granulocyte function by timegadine, a new anti-inflammatory agent. Int. J. Tissue React. VI, 9–15.

Lai, C.-Y. and Wada, K. (1988). Phospholipase A$_2$ from human synovial fluid: purification and structural homology to the placental enzyme. Biochem. Biophys. Res. Commun. 157, 488–493.

Lam, S., Chan, H., LeRiche, J.C., Chan-Yeung, M. and Salari, H. (1988). Release of leukotrienes in patients with bronchial asthma. J. Allergy Clin. Immunol. 81, 711–717.

Lane, I.F., Irwin, J.T.C., Jennings, S.A., Poskitt, K.R., Greenhaulgh, R.M. and McCollum, C.N. (1984). A specific thromboxane A$_2$ antagonist evaluated in vascular graft patients. Br. J. Surg. 71, 903 (Abstr.).

Lanni, C. and Becker, E.L. (1985). Inhibition of neutrophil phospholipase A$_2$ by *p*-bromophenylacyl bromide, nordihydroguaiaretic acid, 5,8,11,14-eicosatetrayenoic acid and quercetin. Int. Arch. Allergy Appl. Immunol. 76, 214–217.

Larochelle, B., Du Souich, P., Bolte, E., Lelorier, J. and Goyer, R. (1983). Tixocortol pivalate, a corticosteroid with no systemic glucocorticoid effect after oral, intrarectal and intranasal application. Clin. Pharmacol. Ther. 33, 343–350.

Larsen, G.L., (1985). Late phase reactions: observations on pathogenesis and prevention. J. Allergy Clin. Immunol. 76, 665–669.

Larsen, G.L. (1989). New concepts in the pathogenesis of asthma. Clin. Immunol. Immunopathol. 53, S107–S118.

Lawson, C.F., Wishka, D.G., Morris, J. and Fitzpatrick, F.A. (1989). Receptor antagonism of leukotriene B$_4$ myotropic activity by the 2,6 disubstituted pyridine analog U-75302: characterization on lung parenchyma strips. J. Lipid Mediators 1, 3–12.

LeBreton, G.C., Venton, D.K., Enke, S.E. and Halushka, P.V. (1979). 13-Azaprostanoid acid: a specific antagonist of the human blood platelet thromboxane/endoperoxide receptor. Proc. Natl Acad. Sci. USA 76, 4097–4101.

LeDuc, L.E., Wyche, A.A., Sprecher, H., Sankarappe, S.K. and Needleman, P. (1981). Analogues of arachidonic acid used to evaluate structural determinants of prostaglandin receptor and enzyme specificities. Mol. Pharmacol. 19, 242–247.

Lee, J.C., Griswold, D.E., Votta, B. and Hanna, N. (1988). Inhibition of monocyte IL-1 production by the anti-inflammatory compound, SKandF 86002. Int. J. Immunopharmacol. 10, 835–843.

Lee, T.H. and O'Hickey, S.P. (1989). Exercise-induced asthma and late phase reactions. Eur. Respir. J. 2, 195–197.

Lee, T.H., Walport, M.J., Wilkinson, A.H., Turner-Warwick, M. and Kay, A.B. (1981). Slow-reacting substance of anaphylaxis antagonist FPL 55712 in chronic asthma. Lancet ii, 304–305.

Leeman, M., Boeynaems, J.-M., Degaute, J.-P., Vincent, J.-L. and Kahn, R.J. (1985). Administration of dazoxiben, a selective thromboxane synthetase inhibitor, in the adult respiratory distress syndrome. Chest 87, 726–730.

Lefer, A.M. (1985). Comparison of the actions of thromboxane receptor antagonists in biological systems. Drugs of Today 21, 283–291.

Lefer, A.M. (1986). Leukotrienes as mediators of ischemia and shock. Biochem. Pharmacol. 35, 123–127.

Lefer, A.M. and Darius, H. (1987). A pharmacological approach to thromboxane receptor antagonism. Fed. Proc. 46, 144–148.

Lefer, A.M., Araki, H., Smith, J.B., Nicolaou, K.C. and Magolda, R.L. (1979). Protective effects of a novel thromboxane analog in lethal traumatic shock. Prostaglandin Med. 3, 139–146.

Lefer, A.M., Smith, E.F. III, Araki, H., Smith, J.B., Aharony, D., Claremon, D.A., Magolda, R.L. and Nicolaou, K.C. (1980). Dissociation of vasoconstrictor and platelet aggregatory activities of thromboxane by carbocyclic thromboxane A_2, a stable analog of thromboxane A_2. Proc. Natl Acad. Sci. USA 77, 1706–1710.

Lefer, A.M., Okamatsu, S., Smith, E.F. III and Smith J.B. (1981). Beneficial effects of a new thromboxane synthetase inhibitor in arachidonate-induced sudden death. Thromb. Res. 23, 265–273.

Lefer, D.J., Mentley, R.K. and Lefer, A.M. (1987). Protective effects of a new specific thromboxane antagonist in arachidonate-induced sudden death. Arch. Int. Pharmacodyn. 287, 89–95.

Leslie, C.C., Voelker, D.R., Channon, J.Y., Wall, M.M. and Zelarney, P.T. (1988). Properties and purification of an arachidonoyl-hydrolyzing phospholipase A_2 from a macrophage cell line. RAW 264.7. Biochem. Biophys. Acta 963, 476–492.

Levi, R., Burke, J.A. and Corey, E.J. (1982). Leukotrienes and immediate hypersensitivity reactions of the heart. Adv. Prostaglandin Thromboxane Leukotriene Res. 9, 215–222.

Levine, J.D., Lau, W., Kwait, G. and Goetzl, E.J. (1984). Leukotriene B_4 produces hyperalgesia that is dependent on polymorphonuclear leukocytes. Science 225, 743–745.

Levine, L. (1983). Inhibition of the A-23187-stimulated leukotriene and prostaglandin biosynthesis of rat basophil leukemia (RBL-1) cells by non-steroidal anti-inflammatory drugs, anti-oxidants, and calcium channel blockers. Biochem. Pharmacol. 32, 3023–3026.

Lewy, R.I., Smith, J.B., Silver, M.J., Saia, J., Walinsky, P. and Wiener, L. (1979). Detection of thromboxane B_2 in peripheral blood of patients with Prinzmetal's angina. Prostaglandins Med. 5, 243–248.

Lin, A.H., Morris, J., Wishka, D.G. and Gorman, R.R. (1988). Novel molecules that antagonize leukotriene B_4 binding to neutrophils. Ann. N.Y. Acad. Sci. 524, 196–200.

Lister, M.D. and Hancock, A.J. (1988). Cyclopentanoid analogs of phosphatidylcholine: susceptibility to phospholipase A_2. J. Lipid Res. 29, 1297–1308.

Lobos, E.A., Sharon, P. and Stenson, W.F. (1987). Chemotactic activity in inflammatory bowel disease. Role of leukotriene B_4. Dig. Dis. Sci. 32, 1380–1388.

Lockey, R.F. (1978). Aspirin improved ASA triad. Hosp. Pract. 13, 129–133.

Loeb, L.A. and Gross, R.W. (1986). Identification and purification of sheep platelet phospholipase A_2 isoforms. J. Biol. Chem. 261, 10467–10470.

Lombardo, D. and Dennis, E.A. (1985). Cobra venom phospholipases A_2 inhibition by manoalide: a novel type of phospholipase inhibitor. J. Biol. Chem. 260, 7234–7240.

Luck, W., Simmet, T., Schmidt, E.W. and Peskar, B.A. (1989). Formation of cysteinyl-leukotrienes by human bronchial mucosa in chronic obstructive bronchitis. Prog. Clin. Biol. Res. 301, 259–265.

Luderer, J.R., Nicholas, G.G., Neumyer, M.M., Riley, D.L., Vary, J.E., Garcia, G. and Schneck, D.W. (1984). Dazoxiben, a thromboxane synthetase inhibitor, in Raynaud's phenomenon. Clin. Pharmacol. Ther. 36, 105–115.

Ludwig, B., Hofling, B., Stablein, A. and von Arnim, T. (1987). Thromboxane-antagonist BM 13.177 compared to aspirin in preventing restenoses after PTCA. Eur. Heart J. 8(Suppl. 2), 225 (Abstr.).

Lumley, P., White, B.P. and Humphrey P.P.A., (1989). GR 32191, a highly potent and specific thromboxane A_2 receptor blocking drug on platelets and vascular and airways smooth muscle *in vitro*. Br. J. Pharmacol. 97, 783–794.

MacDonald, J.I.S. and Sprecher, H. (1991). Phospholipid fatty acid remodelling in mammalian cells. Biochim. Biophys. Acta 1084, 105–121.

MacIntyre, D.E. and Gordon, J.L. (1977). Discrimination between platelet prostaglandin receptors with a specific antagonist of bisenoic prostaglandins. Thromb. Res. 11, 705–713.

MacNab, M.W., Foltz, E.L., Graves, B.S., Rinehart, R.K., Tripp, S.L., Feliciano, N.R. and Sen, S. (1984). The effects of a new thromboxane synthetase inhibitor, CGS-13080, in man. J. Clin. Pharmacol. 24, 76–83.

McFadden, E.R. Jr (1987). Exercise-induced asthma. Assessment of current etiologic concepts. Chest 91, 151S–157S.

McFarlane, C.S., Hamel, R. and Ford-Hutchinson, A.W. (1987). Effects of a 5-lipoxygenase inhibitor (L-651,392) on primary and late pulmonary responses to ascaris antigen in the squirrel monkey. Agents Actions 22, 63–68.

McGibney, D., Menys, V.C., Nelson, G.I.C., Kumar, E.B., Taylor, S.J. and Davies, J.A. (1982). Effects of UK 37248, a thromboxane synthetase inhibitor, on platelet behaviour in patients with coronary heart disease (CHD) undergoing exercise to angina. Clin. Sci. 62, 12P (Abstr.).

McKenniff, M., Rodger, I.W., Norman, P. and Gardiner, P.J. (1988). Characterization of receptors mediating the contractile effects of prostanoids in guinea-pig and human airways. Eur. J. Pharmacol. 153, 149–159.

McKenniff, M.G., Norman, P. Cuthbert, N.J. and Gardiner, P.J. (1991). BAY u3405, a potent and selective thromboxane A_2 receptor antagonist on airway smooth muscle *in vitro*. Br. J. Pharmacol. 104, 585–590.

McMillan, R.M. and Foster, S.J. (1988). Leukotriene B_4 and inflammatory disease. Agents Actions 24, 114–119.

McMillan, R.M., Girodeau, J.-M. and Foster, S.J. (1990). Selective chiral inhibitors of 5-lipoxygenase with anti-inflammatory activity. Br. J. Pharmacol. 101, 501–503.

Mais, D., Dunlap, C., Hamanaka, N. and Halushka, P. (1985a). Further studies on the effects of epimers of 13-azapinane-thromboxane A_2 antagonists on platelets and veins. Eur. J. Pharmacol. 111, 125–128.

Mais, D.E., Saussy, D.L. Jr, Chaikhouni, A., Kochel, P.J., Knapp, D.R., Hamanaka, N. and Halushka, P.V. (1985b). Pharmacologic characterization of human and canine thromboxane A_2/prostaglandin H_2 receptors in platelets and blood vessels: evidence for different receptors. J. Pharmacol. Exp. Ther. 233, 418–424.

Mais, D.E., DeHoll, D., Sightler, H. and Halushka, P. (1988). Different pharmacologic activities for 13-azapinane thromboxane A_2 analogs in platelets and blood vessels. Eur. J. Pharmacol. 148, 309–315.

Manning, P.J., Watson, R.M., Margolskee, D.J., Williams, V.C., Schwartz, J.I. and O'Byrne, P.M. (1990). Inhibition of exercise-induced bronchoconstriction by MK-571, a potent leukotriene D_4-receptor antagonist. N. Engl. J. Med. 323, 1736–1739.

Marnett, L.J., Siedlik, P.H. and Fung, L.W.M. (1982). Oxidation of phenidone and BW755C by prostaglandin endoperoxide synthetase. J. Biol. Chem. 257, 6957–6964.

Marshall, L.A. (1988). Arachidonic acid metabolism of cultured peritoneal rat macrophages and its manipulation by nonsteroidal antiinflammatory agents. Immunopharmacology 15, 177–187.

Marshall, L.A. and Chang, J.Y. (1990). In "Phospholipase A_2" (eds P.H.-K. Wong and E.A. Dennis), pp 169–182. Plenum Press, New York.

Marshall, P.J., Griswold, D.E. Breton, J., Webb, E.F., Hillegass, L.M., Sarau, H.M., Newton, J. Jr., Lee, J.C., Bender, P.E. and Hanna, N. (1991). Pharmacology of the pyrroloimidazole, SKandF 105809–1. Inhibition of inflammatory cytokine production and of 5-lipoxygenase- and cyclooxygenase-mediated metabolism of arachidonic acid. Biochem. Pharmacol. 42, 813–824.

Matsuda, Y., Ogawa, M., Shibata, T., Nakaguchi, K., Nishijima, J., Wakasugi, C. and Mori, T. (1987). Distribution of immunoreactive pancreatic phospholipase A_2 (iPPL-2) in various tissues. Res. Commun. Chem. Pathol. Pharmacol. 58, 281–284.

Matera, G., Cook, J.A., Hennigar, R.A., Tempel, G.E., Wise, W.C., Oglesby, T.D. and Halushka, P.V. (1988). Beneficial effects of a 5-lipoxygenase inhibitor in endotoxic shock in the rat. J. Pharmacol. Exp. Ther. 247, 363–371.

Matsumoto, J. and Saito, K. (1984). The effect of some antibiotics on endogenous phospholipase activity in rat liver. J. Antibiot. (Japan) 37, 910–916.

Mayer, A.M.S., Glaser, K.B. and Jacobs, R.S. (1988). Regulation of eicosanoid biosynthesis *in vitro* and *in vivo* by the marine natural product manoalide: a potent inactivator of venom phospholipases. J. Pharmacol. Exp. Ther. 244, 871–878.

Mead, B., Patterson, L.H. and Smith, D.A. (1981). The disposition of FPL 55712 acid (7-[3(4-acetyl-3-hydroxy-2-propylphenoxy)-2-hydroxypropyl-4*H*-benzopyran-2-carboxylic acid) in rat and dog. J. Pharm. Pharmacol. 33, 682–684.

Meade, E.A., Smith, W.L., and DeWitt, D.L. (1993). Differential inhibition of prostaglandin endoperoxide synthase (cyclooxygenase) isozymes by aspirin and other non-

steroidal antiinflammatory drugs. J. Biol. Chem. 268, 6610–6614.

Melarange, R., Gentry, C., O'Connell, C., Blower, P.R., Neil, C., Kelvin, A.S. and Toseland, C.D.N. (1992). Anti-inflammatory and gastrointestinal effects of nabumetone or its active metabolite, 6MNA (6-methoxy-2-naphthylacetic acid): comparative studies with indomethacin. Dig. Dis. Sci. 37, 1847–1852.

Miele, L., Cordella Miele, E., Facchiano, A. and Mukherjee, A.B. (1988). Novel anti-inflammatory peptides from the region of highest similarity between uteroglobin and lipocortin I. Nature 335, 726–730.

Mihelich, E.D., Morgan, B.P., Ho, P.P.K., Walters, C.P. and Bertsch, B.A. (1988). Inhibition of 5-lipoxygenase and phospholipase A_2 by manoalide analogues. Ann. N.Y. Acad. Sci. 524, 445–457.

Miller, O.V. and Gorman, R.R. (1979). Evidence for distinct PGI_2 and D_2 receptors in human platelets. J. Pharmacol. Exp. Ther. 210, 134–141.

Mobilio, D. and Marshall, L.A. (1989). Recent advances in the design and evaluation of inhibitors of phospholipase A_2. Ann. Rep. Med. Chem. 24, 157–165.

Moilanen, E., Alanko, J., Asmawi, M.Z. and Vapaatalo, H. (1988). CP-66,248, a new anti-inflammatory agent in a potent inhibitor of leukotriene by and prostanoid synthesis in human polymorphonuclear leucocytes in vitro. Eicosanoids 1, 35–39.

Moncada, S., Needleman, P., Bunting, S. and Vane, J.R. (1976). Prostaglandin endoperoxide and thromboxane generating systems and their selective inhibition. Prostaglandins 12, 323–335.

Moncada, S., Bunting, S., Mullane, K., Thorogood, P. and Vane, J.R. (1977). Imidazole: a selective inhibitor of thromboxane synthetase. Prostaglandins 13, 611–618.

Morris, H.G. (1978). In "Allergy Principles and Practice", Vol 1 (eds E. Middleton, Jr, C.E. Reed and E.F. Ellis), pp 464–480. C.V. Mosby, St Louis.

Morris, J. and Wishka, D.G. (1988). Synthesis of novel antagonists of leukotriene B_4. Tetrahedron Lett. 29, 143–146.

Mullane, K., Read, N., Salmon, J.A. and Moncada, S. (1984). Role of leukocytes in acute myocardial infarction in anaesthetized dogs: relationship to myocardial salvage by anti-inflammatory drugs. J. Pharmacol. Exp. Ther. 228, 510–522.

Mullane, K., Hatala, M.A., Kraemer, R., Sessa, W. and Westlin, W. (1987). Myocardial salvage induced by REV-5901: an inhibitor and antagonist of the leukotrienes. J. Cardiovas. Pharmacol. 10, 398–406.

Mullane, K.M. and Fornabaio, D. (1988). Thromboxane synthetase inhibitors reduce infarct size by a platelet-dependent, aspirin-sensitive mechanism. Circ. Res. 62, 668–678.

Murakami, M., Kudo, I., Nakamura, H., Yokoyama, Y., Mori, H. and Inoue, K. (1990). Exacerbation of rat adjuvant arthritis by intradermal injection of purified mammalian 14-kDa group II phospholipase A_2. FEBS Lett. 268, 113–116.

Murphy, R.C., Hammarström, S. and Samuelsson, B. (1979). Leukotriene C: a slow reacting substance from murine mastocytoma cells. Proc. Natl Acad. Sci. USA 76, 4275–4279.

Musser, J.H. (1989). Leukotriene D_4 receptor antagonists: a new approach to antiasthma drug therapy. Drugs, News Perspectives 2, 202–213.

Mustard, J.F., Kinlough-Rathbone, R.L. and Packham, M.A. (1983). Aspirin in the treatment of cardiovascular disease: a review. Am. J. Med. 74, 43–49.

Naclerio, R.M., Kagey-Sobotka, A., Lichtenstein, L.M.,

Freidhoff, L. and Proud, D. (1990). Terfenadine, an H1 antihistamine, inhibits histamine release *in vivo* in the human. Am. Rev. Respir. Dis. 142, 167–171.

Nagai, H., Yakuo, I., Togawa, M., Arimura, A., Matsuura, N., Koda, A., Hamano, S., Ujie, A. and Nakazawa, M. (1987). Effect of OKY-046, a new thromboxane A_2 synthetase inhibitor, on experimental asthma in guinea pigs. Prostaglandins Leukotrienes Med. 30, 111–121.

Nakagawa, T., Mizushima, Y., Ishii, A., Nambu, F., Motoishi, M., Yui, Y., Shida, T. and Miyamoto, T. (1991). Effect of a leukotriene antagonist on experimental and clinical bronchial asthma. Adv. Prostaglandin Thromboxane Leukotriene Res. 21A, 465–468.

Namiki, M., Igarashi, Y., Sakamoto, K., Nakamura, T. and Koga, Y. (1986). Pharmacological profiles of a potential LTB_4-antagonist, SM-9064. Biochem. Biophys. Res. Commun. 138, 540–546.

Needleman, P., Minkes, M. and Raz, A. (1976). Thromboxanes: selective biosynthesis and distinct biological properties. Science 193, 163–165.

Needleman, P., Bryan, B., Wyche, A., Bronson, S.D., Eakins, K., Ferrendelli, J.A. and Minkes, M. (1977). Thromboxane synthetase inhibitors as pharmacological tools: differential biochemical and biological effects on platelet suspensions. Prostaglandins 14, 897–907.

Nicolaou, K.C., Magolda, R.L., Smith, J.B., Aharony, D., Smith, E.F. III and Lefer, A.M. (1979). Synthesis and biological properties of pinane-thromboxane A_2, a selective inhibitor of coronary artery constriction, platelet aggregation, and thromboxane formation. Proc. Natl Acad. Sci. USA 70, 2566–2570.

Nielsen, O.H., Elmgreen, J., Thomsen, B.S., Ahnfelt-Rønne, I. and Wiik, A. (1986). Release of leukotriene B_4 and 5-hydroxyeicosatetraenoic acid during phagocytosis of artificial immune complexes by peripheral neutrophils in chronic inflammatory bowel disease. Clin. Exp. Immunol. 65, 465–471.

Nishikiori, T., Naganawa, H., Muraoka, Y., Aoyagi, T. and Umezawa, H. (1986). Plipstatins: new inhibitors of phospholipase A_2, produced by *Bacillus cereus* BM G302–fF67 III. Structural elucidation of plipstatins. J. Antibiot. (Japan) 39, 755–761.

Norel, X., Labat, C., Gardiner, P.J. and Brink, C. (1991). Inhibitory effects of BAY U3405 on prostanoid-induced contractions in human isolated bronchial and pulmonary arterial muscle preparations. Br. J. Pharmacol. 104, 591–595.

Obata, T., Nambu, F., Kitagawa, T., Terashima, H., Toda, M., Okegawa, T. and Kawasaki, A. (1987). ONO-1078: an antagonist of leukotrienes. Adv. Prostaglandin Thromboxane Leukotriene Res. 17, 540–542.

O'Byrne, P.M. and Fuller, R.W. (1989). The role of thromboxane A_2 in the pathogenesis of airway hyperresponsiveness. Eur. Respir. J. 2, 782–786.

O'Donnell, M., Crowley, H.J., Yaremko, B., O'Neill, N. and Welton, A.F. (1991). Pharmacologic actions of Ro 24–5913, a novel antagonist of leukotriene D_4. J. Pharmacol. Exp. Ther. 259, 751–758.

O'Driscoll, B.R., Cromwell, O. and Kay, A.B. (1984). Sputum leukotrienes in obstructive airways diseases. Clin. Exp. Immunol. 55, 397–404.

Ogletree, M.L. (1987). Overview of physiological and pathophysiological effects of thromboxane A_2. Fed. Proc. 46, 133–138.

Ogletree, M.L. and Lefer, A.M. (1976). Influence of nonsteroidal anti-inflammatory agents on myocardial ischemia in the cat. J. Pharmacol. Exp. Ther. 197, 582–593.

Ogletree, M.L., Harris, D.N., Greenberg, R., Haslanger, M.F. and Nakane, M. (1985a). Pharmacological actions of SQ 29,548, a novel selective thromboxane antagonist. J. Pharmacol. Exp. Ther. 234, 435–441.

Ogletree, M.L., Heran, C.L. and Harris, D.N. (1985b). SQ 28,668, a platelet TxA_2-receptor/PGH_2 antagonist, inhibits thrombotic cyclic reductions in coronary blood flow in anaesthetized pigs. Thromb. Haemost. 54, 280 (Abstr.).

Okimasu, E., Moromizato, Y., Watanabe, S., Sasaki, J., Shiraishi, N., Morimoto, Y.M., Miyahara, M. and Utsumi, K. (1983). Inhibition of phospholipase A_2 and platelet aggregation by glycyrrhizin, an antiinflammation drug. Acta Med. Okayama 37, 385–391.

Ono, T., Tojo, H., Kuramitsu, S., Kagamiyama and Okamoto, M. (1988). Purification and characterization of a membrane-associated phospholipase A_2 from rat spleen. Its comparison with a cytosolic phospholipase A_2 S-1. J. Biol. Chem. 263, 5732–5738.

Osborn, R.R., Hay, D.W.P., Wasserman, M.A. and Torphy, T.J. (1992). SKandF 104353, a selective leukotriene receptor antagonist, inhibits leukotriene D_4- and antigen-induced bronchoconstriction in cynomolgus monkeys. Pulm. Pharmacol. 5, 153–157.

Otterness, I.G., Bliven, M.L., Downs, J.T., Natoli, E.J. and Hanson, D.C. (1991). Inhibition of interleukin 1 synthesis by tenidap: a new drug for arthritis. Cytokine 3, 277–283.

Papatheofanis, F.J. and Lands, W.E.M. (1985). In "Biochemistry of Arachidonic Acid Metabolism" (ed W.E.M. Lands), pp 9–39. Martinus Nihjoff, Boston.

Parks, T.P., Lukas, S. and Hoffman, A.F. (1990). In "Phospholipase A_2" (eds P.Y.-K. Wong and E.A. Dennis), pp 55–81. Plenum Press, New York.

Passeri, M. (1990). Piroxicam- B-cyclodextrin in rheumatology. Drug Invest. 2, 86–88.

Patrignani, P., Filabozzi, P. and Patrono, C. (1982). Selective cumulative inhibition of platelet thromboxane production by low-dose aspirin in healthy subjects. J. Clin. Invest. 69, 1366–1372.

Patrignani, P., Filabozzi, P., Catella, F., Pugliese, F. and Patrono, C. (1984). Differential effects of dazoxiben, a selective thromboxane-synthase inhibitor, on platelet and renal prostaglandin-endoperoxide metabolism. J. Pharmacol. Exp. Ther. 228, 472–477.

Patrono, C. (1989). Aspirin and human platelets: from clinical trials to acetylation of cyclooxygenase and back. Trends Pharmacol. Sci. 10, 453–458.

Patrono, C, (1990). Thromboxane synthesis inhibitors and receptor antagonists. Thromb. Res. 60(Suppl.), 15–23.

Patrono, C., Ciabattoni, G., Filabozzi, P., Catella, F., Forni, L., Segni, M., Patrignani, P., Pugliese, F., Simonetti, B.M. and Pierucci, A. (1985). Drugs, prostaglandin and renal function. Adv. Prostaglandin Thromboxane Leukotriene Res. 13, 131–139.

Patrono, C., Ciabattoni, G., Pugliese, F., Pierucci, A., Blair, I.A. and Fitzgerald, G.A. (1986). Estimated rate of thromboxane secretion into the circulation of normal humans. J. Clin. Invest. 77, 590–594.

Patscheke, H. (1990a). In "Proc. 7th Int. Conf. on Prostaglandins and Related Compounds, Florence, May 1990," pp 32 (Abstr.).

Patscheke, H. (1990b). Thromboxane A_2/prostaglandin H_2

receptor antagonists. A new therapeutic principle. Stroke 21(Suppl. IV), 139–142.

Patscheke, H. and Stegmeier, K. (1985). Effects of the prostaglandin H_2/thromboxane A_2 antagonist BM 13.177 on human platelets. Adv. Prostaglandin Thromboxane Leukotriene Res. 13, 371–373.

Patscheke, H., Staiger, C., Neugebauer, G., Kaufmann, B., Strein, K., Endele, R. and Stegmeier, K. (1986). The pharmacokinetic and pharmacodynamic profiles of the thromboxane A_2 receptor blocker BM 13.177. Clin. Pharmacol. Ther. 39, 145–150.

Paulus, H.E. (1989). In "Textbook of Rheumatology" (eds Kelley, W.N., Harris, E.D. Jr, Ruddy, S. and Sledge, C.B.), pp 765–791. W.B. Saunders, Philadelphia.

Paulus, H.E. and Furst, D.E. (1989). In "Arthritis and Allied Conditions: A Textbook of Rheumatology", 11th edn (ed D.J. McCarty), pp 507–543. Lea and Febiger, Philadelphia.

Payne, A.N., Garland, L.G., Lees, I.W. and Salmon, J.A. (1988). Selective inhibition of arachidonate 5-lipoxygenase by novel acetohydroxamic acids: effects on bronchial anaphylaxis in anaesthetized guinea-pigs. Br. J. Pharmacol. 94, 540–546.

Pelikan, Z., Pelikan, M., Kruis, M. and Berger, M.P.F. (1986). The immediate asthmatic response to allergen challenge. Ann. Allergy 56, 252–260.

Pepinsky, R.B., Sinclair, L.K., Browning, J.L., Mattaliano, R.J., Smart, J.E., Pingchang Chow, E., Falbel, T., Ribolini, A., Garwin, J.L. and Wallner, B.P. (1986). Purification and partial sequence analysis of a 37-kDa protein that inhibits phospholipase A_2 from rat peritoneal exudates. J. Biol. Chem. 261, 4239–4246.

Percival, M.D. (1991). Human 5-lipoxygenase contains an essential iron. J. Biol. Chem. 22, 10058–10061.

Perzborn, E., Fiedler, V.B., Seuter, F., Stasch, J.P., Wever, H., Sander, E., Böshagen, H. and Rosentreter, U. (1990). Characterization of Bay U 3405, a novel thromboxane A_2/endoperoxide receptor antagonist. Stroke 21(Suppl. IV), 143–145.

Phillips, G.D., Rafferty, P., Robinson, C. and Holgate, S.T. (1988). Dose-related antagonism of leukotriene D_4-induced bronchoconstriction by p.o. administration of LY-171883 in nonasthmatic subjects. J. Pharmacol. Exp. Ther. 246, 732–738.

Piacentini, G.L. and Kaliner, M.A. (1991). The potential roles of leukotrienes in bronchial asthma. Am. Rev. Respir. Dis. 143, S96–S99.

Piechuta, H., Ford-Hutchinson, A.W. and Letts, L.G. (1987). Inhibition of allergen-induced bronchoconstriction in hyper-reactive rats as a model for testing 5-lipoxygenase inhibitors and leukotriene D_4 receptor antagonists. Agents Actions 22, 69–74.

Pierucci, A., Simonetti, B.M., Pecci, G., Mavrikakis, G., Feriozzi, S., Cinotti, G.A., Patrignani, P., Ciabattoni, G. and Patrono, C. (1989). Improvement of renal function with selective thromboxane antagonism in lupus nephritis. N. Engl. J. Med. 320, 421–425.

Prié, S. and Sirois, P. (1988). Mechanism of action of leukotrienes on the guinea pig bronchus. Prostaglandins Leukot. Essent. Fatty Acids 34, 19–25.

Pruzanski, W., Vadas, P., Stefanski, E. and Urowitz, M.B. (1985). Phospholipase A_2 activity in sera and synovial fluids in rheumatoid arthritis and osteoarthritis. Its possible role as a proinflammatory enzyme. J. Rheumatol. 12, 211–216.

Pruzanski, W., Keystone, E.C., Sternby, B., Bombardier, C., Snow, K.M. and Vadas, P. (1988). Serum phospholipase A_2 correlates with disease activity in rheumatoid arthritis. J. Rheumatol. 15, 1351–1355.

Pruzanski, W., Vadas, P., Kim, J., Jacobs, J. and Stefanski, E. (1989). Phospholipase A_2 activity associated with synovial fluid cells. J. Rheumatol. 15, 791–794.

Puig-Parellada, P. and Planas, J.A. (1977). Action of selective inhibitor of thromboxane synthetase on experimental thrombosis induced by arachidonic acid in rabbits. Lancet ii, 40.

Quilley, J., McGiff, J.C. and Nasjletti, A. (1989). Role of endoperoxides in arachidonic acid-induced vasoconstriction in the isolated perfused kidney of the rat. Br. J. Pharmacol. 96, 111–116.

Radmark, D., Malmsten, C. and Samuelsson, B. (1980). Leukotriene A_4 enzymatic conversion to leukotriene C_4. Biochem. Biophys. Res. Commun. 96, 1679–1687.

Rae, S.A. and Smith, M.J.H. (1981). The stimulation of lysosomal enzyme secretion from human polymorphonuclear by leukotriene B_4. J. Pharm. Pharmacol. 33, 616–617.

Rae, S.A., Davidson, E.M. and Smith, M.J. (1982). Leukotriene B_4, an inflammatory mediator in gout. Lancet ii, 1122–1124.

Raftery, A.T., Livesey, S. and Forty, J. (1983). Initial impression of dazoxiben in the treatment of the ischaemic limb. Br. J. Clin. Pharmacol. 15, 119S–121S.

Rainsford, K.D. (1985). In "Anti-Inflammatory and Anti-Rheumatic Drugs", Vol 1 (ed. K.D. Rainsford), pp 109–137. CRC Press, Boca Raton.

Rainsford, K.D. (1987). The effect of 5-lipoxygenase inhibitors and leukotriene antagonists on the development of gastric lesions induced by nonsteroidal antiinflammatory drugs in mice. Agents Actions 21, 316–319.

Rainsford, K.D. (1990). NSAID gastropathy. Novel physico-chemical approaches for reducing gastric mucosal injury by drug complexation with cyclodextrins. Drug. Invest. 2, 3–10.

Randall, M.J. and Wilding, R.I.R. (1982). Acute arterial thrombosis in rabbits: reduced platelet accumulation after treatment with thromboxane synthetase inhibitor dazoxiben hydrochloride (UK-37,248–01). Thromb. Res. 28, 607–616.

Randall, M.J. Parry, M.J. Hawkeswood, E., Cross, P.E. and Dickinson, R.P. (1981). UK-37,248, a novel, selective thromboxane synthetase inhibitor with platelet anti-aggregatory and anti-thrombotic activity. Thromb. Res. 23, 145–162.

Rasmussen, J.B., Eriksson, L.-O. and Andersson, K.-E. (1991). Reversal and prevention of airway response to antigen challenge by the inhaled leukotriene D_4 antagonist (L-648,051) in patients with atopic asthma. Allergy 46, 266–273.

Reilly, I.A. and Fitzgerald, G.A. (1987). Inhibition of thromboxane formation *in vivo* and *ex vivo*: Implications for therapy with platelet inhibitory drugs. Blood 69, 180–186.

Reilly, I.A., Doran, J.B., Smith, B. and Fitzgerald, G.A. (1986a). Increased thromboxane biosynthesis in a human preparation of platelet activation: Biochemical and functional consequences of selective inhibition of thromboxane synthase. Circulation 73, 1300–1309.

Reilly, I.A., Roy, F. and Fitzgerald, G.A. (1986b). Biosynthesis of thromboxane in patients with systemic sclerosis and Raynaud's phenomenon. Br. Med. J. 292, 1037–1039.

Reines, H.D., Halushka, P.V., Olanoff, L.S. and Hunt, P.S. (1985). Dazoxiben in human sepsis and adult respiratory distress syndrome. Clin. Pharmacol. Ther. 37, 391–395.

Renetsder, R., Dijkstra, B.W., Huizinga, K., Kalk, K.H. and Drenth, J. (1988). Crystal structure of bovine pancreatic phospholipase A_2 covalently inhibited by p-bromo-phenacyl-bromide. J. Mol. Biol. 200, 181–188.

Reuben, S.R., Kuan, P., Cairns, J. and Gyde, O.H. (1983). Effects of dazoxiben on exercise performance in chronic stable angina. Br. J. Clin. Pharmacol. 15, 83S-86S.

Rich, S., Hart, K., Kieras, K. and Brundage, B.H. (1987). Thromboxane synthetase inhibition in primary pulmonary hypertension. Chest 91, 356–360.

Richards, I.M., Griffin, R.L. Oostveen, J.A., Morris, J., Wishka, D.G. and Dunn, C.J. (1989). Effect of the selective leukotriene B_4 antagonist U-75302 on antigen-induced bronchopulmonary eosinophilia in sensitized guinea pigs. Am. Rev. Respir. Dis. 140, 1712–1716.

Riess, H., Hiller, E., Reinhardt, B. and Bräuning, C. (1984). Effects of BM 13,177, a new antiplatelet drug in patients with atherosclerotic disease. Thromb. Res. 35, 371–378.

Ripka, W.C., Sipio, W.J. and Blaney, J.M. (1987). Molecular modeling and drug design: strategies in the design and synthesis of phospholipase A_2 inhibitors. Lectures Heterocyclic Chem. 9, S95-S104.

Roberts, M.F., Deems, R.A., Mincey, T.C. and Dennis, E.A. (1977). Chemical modification of the histidine residue in phospholipase A_2 (*Naja naja naja*): a case of half-site reactivity. J. Biol. Chem. 252, 2405–2411.

Robuschi, M., Fuccella, L.M., Riva, E., Vida, E., Barnabe, R., Rossi, M., Gambaro, G., Spagnotto, S. and Bianco, S. (1991). Prevention of exercise-induced bronchoconstriction by a leukotriene antagonist (SKandF 104353): a double blind study vs disodium cromoglycate (DSCG) and placebo. Am. Rev. Resp. Dis. 143, A642 (abstract).

Rosenbach, T., Grabbe, J., Möller, A., Schwanitz, H.J. and Czarnetzki, B.M. (1985). Generation of leukotrienes from normal epidermis and their demonstration in cutaneous disease. Br. J. Dermatol. 113(Suppl. 28), 157–167.

Rosenthale, M.E. (1974). In "Antiinflammatory Agents: Chemistry and Pharmacology", Vol 13-II (eds R.A. Scherrer and M.W. Whitehouse), pp 123–192. Academic Press, New York.

Roth, G.J., Stanford, N. and Majerus, P.W. (1975). Acetylation of prostaglandin synthase by aspirin. Proc. Natl. Acad. Sci. USA 72, 3073–3076.

Roth, S.H. (1980). In "New directions in Arthritis Therapy" (ed S.H. Roth), pp 59–98. PSG, Littleton, MA.

Roundtree, L.V. and Calhoun, W. (1991). Zileuton News 1, 1.

Rouzer, C.A., Rands, E., Kargman, S., Jones, R.E., Register, R.B. and Dixon, R.A.F. (1988). Characterization of cloned human leukocyte 5-lipoxygenase expressed in mammalian cells. J. Biol. Chem. 263, 10135–10140.

Rubin, P.D., Kesterson, J.W., Swanson, L., Henson, B., Dube, L.M., Bell, R.L., Albert, D., Braeckman, R.A. and Carter, G.W. (1989). In "3rd Interscience World Conf. on Antirheumatics, Analgesics, Immunomodulators, Monte Carlo, Monaco", p 335 (Abstr.).

Rustin, M.H., Grimes, S.M. Kovacs, I.B., Cooke, E.D., Bowcock, S.S., Sowemimo-Coker, S.P., Turner, P. and Kirby, J.D. (1984). A double-blind trial of UK-38,485, an orally active thromboxane synthetase inhibitor in the treatment of Raynaud's syndrome. Eur. J. Clin. Pharmacol. 27, 61–65.

Salmon, J.A., Simmons, P.M. and Moncada, S. (1983). The effects of BW755C and other anti-inflammatory drugs on eicosanoid concentrations and leukocyte accumulation in experimentally-induced acute inflammation. J. Pharm. Pharmacol. 35, 808–813.

Salmon, J.A., Tilling, L.C. and Moncada, S. (1985). Evaluation of inhibitors of eicosanoid synthesis in leukocytes: possible pitfall of using the calcium ionophore A23187 to stimulate 5-lipoxygenase. Prostaglandins 29, 377–387.

Samter, M. and Beers, R.F. Jr (1968). Intolerance to aspirin. Clinical studies and consideration of its pathogenesis. Ann. Int. Med. 68, 975–983.

Samuelsson, B. (1983). Leukotrienes: mediators of immediate hypersensitivity reactions and inflammation. Science 220, 568–575.

Sanduja, R., Loose-Mitchell, D. and Wu, K.K. (1991). Inhibition of *de novo* synthesis and message expression of prostaglandin H synthase by salicylates. Adv. Prostaglandin Thromboxane Leukotriene Res. 21A, 149–152.

Saris, C.J.M., Tack, B.F., Kristensen, T., Glenney, J.R. and Hunter, T. (1986). The cDNA sequence for the protein-tyrosine kinase substrate p 36 (calpactin I heavy chain) reveals a multidomain protein with internal repeats. Cell 46, 201–212.

Schalkwijk, C.G., Marki, F. and Van den Bosch, H. (1990). Studies on the acyl-chain selectivity of cellular phospholipase A_2. Biochim. Biophys. Acta. 1044, 139–146.

Schane, H.P., Harding, H.R., Bell, M.R., Castracane, V.D., Winneker, R.C. and Snyder, B.W. (1985). Glucocorticoid/antiinflammatory activities of nonsteroidal phenylpyrazoles in the rat. Steroids 45, 171–185.

Scholer, D.W., Ku, E.C., Boettcher, I. and Schweizer, A. (1986). Pharmacology of diclofenac sodium. Am. J. Med. 80, 34–38.

Schrör, K., Smith, E.F. III, Bickerton, M., Smith, J.B., Nicolaou, K.C., Magolda, R. and Lefer, A.M. (1980). Preservation of ischemic myocardium by pinane thromboxane A_2. Am. J. Physiol. 238, H87–H92.

Schultz, R.M., Marder, P., Spaethe, S.M., Herron, D.K. and Sofia, M.J. (1991). Effects of two leukotriene B_4 (LTB$_4$) receptor antagonists (LY 255283 and SC-41930) on LTB$_4$-induced human neutrophil adhesion and superoxide production. Prostaglandins Leukot. Essent. Fatty Acids 43, 267–271.

Schumacher, W.A., Goldenberg, H.J., Harris, D.N. and Ogletree, M.L. (1987). Effect of thromboxane receptor antagonists on renal artery thrombosis in the cynomolgus monkey. J. Pharmacol. Exp. Ther. 243, 460–466.

Scott, D.L., White, S.P., Browning, J.L., Rosa, J.J., Gelb, M.H. and Sigler, P.B. (1991). Structures of free and inhibited human secretory phospholipase A_2 from inflammatory exudate. Science 254, 1007–1010.

Scudeletti, L., Castagnetta, B., Imbimbo, B., Puppo, F., Pierri, I. and Indiveri, F. (1990). New glucocorticoids: mechanisms of immunological activity at the cellular level and in the clinical setting. Ann. N.Y. Acad. Sci. 595, 368–382.

Seilhamer, J.J., Randall, T.L., Yamanaka, M. and Johnson, L.K. (1986). Pancreatic phospholipase A_2: isolation of the human gene and cDNAs from porcine pancreas and human lung. DNA 5, 519–527.

Seilhamer, J.J., Randall, T.L., Johnson, L.K., Heinzmann, C., Klisak, I., Sparkes, R.S. and Lusis, A.J. (1988). Novel gene exon homologous to pancreatic phospholipase A_2: sequence and chromosomal mapping of both human genes. J. Cell. Biochem. 39, 327–337.

Seilhamer, J.J., Plant, S., Pruzanski, W., Schilling, J., Stefanski, E., Vadas, P. and Johnson, L.K. (1989a). Multiple forms of phospholipase A_2 in arthritic synovial fluid. J. Biochem. 106, 38–42.

Seilhamer, J.J., Pruzanski, W., Vadas, P., Plant, S., Miller, J.A., Kloss, J. and Johnson, L.K. (1989b). Cloning and recombinant expression of phospholipase A_2 present in rheumatoid arthritic synovial fluid. J. Biol. Chem. 264, 5335–5338.

Sengupta, Ch., Afèche, P., Meyer-Brunot, H.G. and Rensing, U. (1985). In "Anti-Inflammatory and Anti-Rheumatic Drugs", Vol 2 (ed K.D. Rainsford), pp 49–60. CRC Press, Boca Raton.

Serhan, C.N., Radin, A., Smolen, J.E., Korchak, H., Samuelsson, B. and Weissmann, G. (1982). Leukotriene B_4 is a complete secretagogue in human neutrophils: a kinetic analysis. Biochem. Biophys. Res. Commun. 107, 1006–1012.

Seuter, F., Perzborn, E. and Fiedler, V.-B. (1990). Effect of Bay U 3405, a new thromboxane antagonist, on collagen-induced thromboembolism in rabbits. Stroke 21(Suppl. IV), 146–148.

Sharon, P. and Stenson, W.F. (1984). Enhanced synthesis of leukotriene B_4 by colonic mucosa in inflammatory bowel disease. Gastroenterol. 86, 453–460.

Shaw, A. and Krell, R.D. (1991). Peptide leukotrienes: Current status of research. J. Med. Chem. 34, 1235–1242.

Sheard, P., Lee, T.B. and Tattersall, M.L. (1977). Further studies on the SRS-A antagonist FPL 55712. Monogr. Allergy 12, 245–249.

Sheard, P., Holroyde, M.C., Bantick, J.R. and Lee, T.B. (1984). In "Development of Anti-asthma Drugs" (ed D.R. Buckle), pp 133–158. Butterworth, London.

Shen, T.Y. (1985). In "Anti-inflammatory and Anti-rheumatic Drugs", Vol 1 (ed K.D. Rainsford), pp 149–158. CRC Press, Boca Raton.

Shibouta, Y., Terashita, Z.-N., Inada, Y. and Nishikawa, K. (1985). Delay of the initiation of hypertension in spontaneously hypertensive rats by CV-4151, a specific thromboxane receptor A_2 synthetase inhibitor. Eur. J. Pharmacol. 109, 135–144.

Shindo, K., Matsumoto, Y., Hirai, Y., Sumitomo, M., Amano, T., Miyakawa, K., Matsumura, M. and Mizuno, T. (1990). Measurement of leukotriene B_4 in arterial blood of asthmatic patients during wheezing attacks. J. Intern. Med. 228, 91–96.

Showell, H.J., Otterness, I.G., Marfat, A. and Corey, E.J. (1982). Inhibition of leukotriene B_4-induced neutrophil degranulation by leukotriene B_4-dimethylamide. Biochem. Biophys. Res. Commun. 106, 741–747.

Silbaugh, S.S., Stengel, P.W., Williams, G.D., Herron, D.K., Gallagher, P. and Baker, S.R. (1987). Effects of leukotriene B_4 inhalation. Airway sensitization and lung granulocyte infiltration in the guinea pig. Am. Rev. Respir. Dis. 136, 930–934.

Singh, P.D., Johnson, J.H., Aklonis, C.A., Bush; K., Fisher, S.M. and O'Sullivan, J. (1985). Two new inhibitors of phospholipase A_2 produced by Penicillium chermesinum. J. Antibiot. (Japan) 38, 706–712.

Smith, E.F. III (1989). Thromboxane A_2 in cardiovascular and renal disorders: Is there a defined role of thromboxane receptor antagonists or thromboxane synthase inhibitors? Eicosanoids 2, 199–212.

Smith, E.F. III and McDonald, J. (1988). Protective effects of the thromboxane receptor antagonists BM 13.177 and BM 13.505 against U 46619-induced sudden death in rats. Pharmacology 36, 340–347.

Smith, E.F. III, Lefer, A.M. and Smith, J.B. (1980). Influence of thromboxane inhibition on the severity of myocardial ischemia in cats. Can. J. Physiol. Pharmacol. 58, 294–300.

Smith, E.F. III, Griswold, D.E., Egan, J.W., Hillegass, L.M. and DiMartino, M.J. (1989). Reduction of myocardial damage and polymorphonuclear leukocyte accumulation following coronary artery occlusion and reperfusion by the thromboxane receptor antagonist BM 13.505. J. Cardiovasc. Pharmacol. 13, 715–722.

Smith, J.B. (1987). Pharmacology of thromboxane synthetase inhibitors. Fed. Proc. 46, 139–143.

Smith, J.B. and Willis, A.L. (1971). Aspirin selectively inhibits prostaglandin production in human platelets. Nature (New Biol.) 231, 235–237.

Smith, L.J., Geller, S., Ebright, L., Glass, M. and Thyrum, P.T. (1990). Inhibition of leukotriene D_4-induced bronchoconstriction in normal subjects by the oral LTD_4 receptor antagonist ICI 204,219. Am. Rev. Respir. Dis. 141, 988–992.

Snyder, D.W. and Fleisch, J.H. (1989). Leukotriene receptor antagonists as potential therapeutic agents. Annu. Rev. Pharmacol. Toxicol. 29, 123–143.

Snyder, D.W. and Krell, R.D. (1984). Pharmacological evidence for a distinct leukotriene C_4 receptor in guinea-pig trachea. J. Pharmacol Exp. Ther. 231, 616–622.

Snyder, D.W., Giles, R.E., Keith, R.A., Yee, Y.K. and Krell, R.D. (1987). In vitro pharmacology of ICI 198,615: a novel, potent and selective peptide leukotriene antagonist. J. Pharmacol. Exp. Ther. 243, 548–556.

Spencer, D.A., Evans, J.M., Green, S.E., Piper, P.J. and Costello, J.F. (1991a). Participation of the cysteinyl leukotrienes in the acute bronchoconstrictor response to inhaled platelet factor in man. Thorax 46, 441–445.

Spencer, D.A., Sampson, A.P., Evans, J.M., Garland, L.G., Piper, P.J. and Costello, J.F. (1991b). The effects of a 5-lipoxygenase inhibitor, BW A4C, on the acute response to inhaled PAF in man. Ann. N.Y. Acad. Sci. 629, 430–431.

Staerk, Laursen, L., Naesdal, J., Bukhave, K., Lauritsen, K. and Rask-Madsen, J. (1990). Selective 5-lipoxygenase inhibition in ulcerative colitis. Lancet 335, 683–685.

Stegmeier, K., Pill, J., Müller-Beckmann, B., Schmidt, F.H., Witt, E.C., Wolff, H.P. and Patscheke, H. (1984). The pharmacological profile of the thromboxane A_2 antagonist BM 13,177, a new anti-platelet and anti-thrombotic drug. Thromb. Res. 35, 379–395.

Steinbacher, T.E. and Ogletree, M.L. (1985). Preferential antagonism of airways responses to prostaglandins in anaesthetized guinea pigs. Pharmacologist 27, 214 (Abstr.).

Stier, C.T. Jr, Benter, I.F. and Levine, S. (1988). Thromboxane A_2 in severe hypertension and stroke in stroke-prone spontaneously hypertensive rats. Stroke 19, 1145–1150.

Stinger, R.B., Fitzpatrick, T.M., Corey, E.J., Ramwell, P.W., Rose, J.C. and Kot, P.A. (1982). Selective antagonism of prostaglandin $F_{2\alpha}$-mediated vascular responses by N-dimethylamino substitution of prostaglandin $F_2\alpha$. J. Pharmacol. Exp. Ther. 220, 521–525.

Stout, B.K. and Johnson, H.G. (1990). Inhibition of late phase bronchoconstriction and eosinophilia in Ascaris ova sensitized primates by steroids, an LTB_4 antagonist and a lazaroid. Amer. Rev. Resp. Dis. 141, A473 (Abstr.).

Strandhoy, J.W. (1986). In "Rheumatology" (eds R.A. Turner and C.M. Wise), pp 124–152. Elsevier, New York.

Summers, J.B. (1990). Inhibitors of leukotriene biosynthesis. Drug News Perspectives 3, 517–526.

Summers, J.B., Gunn, B.P., Martin, J.G., Mazdiyasni, H., Stewart, A.O., Young, P.R., Goetze, A.M., Bouska, J.B., Dyer, R.D., Brooks, D.W. and Carter, G.W. (1988). Orally active hydroxamic acid inhibitors of leukotriene biosynthesis. J. Med. Chem. 31, 3–5.

Sun, F.F. (1977). Biosynthesis of thromboxanes in human platelets. I. Characterization and assay of thromboxane synthetase. Biochem. Biophys Res. Commun. 74, 1432–1440.

Svensson, J., Strandberg, K., Tuvemo, T. and Hamberg, M. (1977). Thromboxane A_2: effect on airway and vascular smooth muscle. Prostaglandins 14, 425–436.

Swayne, G.T.G., Maguire, J., Dolan, J., Raval, P., Dane, G., Greener, M. and Owen, D.A.A. (1988). Evidence for homogeneity of thromboxane A_2 receptor using structurally different antagonists. Eur. J. Pharmacol. 152, 311–319.

Swingle, K.F. (1974). In "Antiinflammatory Agents: Chemistry and Pharmacology", Vol 13 (eds R.A. Scherrer and M.W. Whitehouse), pp 34–110. Academic Press, New York.

Szczeklik, A. (1986). Analgesics, allergy and asthma. Drugs 32(Suppl. 4), 148–163.

Szczeklik, A. (1990). The cyclooxygenase theory of aspirin-induced asthma. Eur. Respir. J. 3, 588–593.

Szczeklik, A. and Nizankowska, E. (1983). Asthma improved by aspirin-like drugs. Br. J. Dis. Chest. 77, 153–158.

Szczeklik A., Gryglewski, R.J. and Czerniawska-Mysik, G. (1975). Relationship of inhibition of prostaglandin biosynthesis by analgesics to asthma attacks in aspirin-sensitive patients. Br. Med. J. 11, 67–69.

Tannenbaum, J.S., Anderson, C.B., Sicard, G.A., McKeel, D.W. and Etheredge, E.E. (1984). Prostaglandin synthesis associated with renal allograft rejection in the dog. Transplantation 37, 438–443.

Tateson, J.E., Randall, R.W. Reynolds, C.H., Jackson, W.P., Bhattacherjee, P., Salmon, J.A. and Garland, L.G. (1988). Selective inhibition of arachidonate 5-lipoxygenase by novel acetohydroxamic acids: biochemical assessment *in vitro* and *ex vivo*. Br. J. Pharmacol. 94, 528–539.

Taylor, B.M., Czuk, C.I., Brunden, M.N., Wishka, D.G., Morris, J. and Sun, F.F. (1989). Inhibition of *in vitro* guinea pig eosinophil chemotaxis by leukotriene B_4 antagonist U-75,302. Adv. Prostaglandin Thromboxane Leukotriene Res. 19, 195–198.

Taylor, I.K., O'Shaugnessy, K.M., Fuller, R.W. and Dollery, C.T. (1991). Effect of cysteinyl-leukotriene receptor antagonist ICI 204.219 on allergen-induced bronchoconstriction and airway hyperreactivity in atopic subjects. Lancet 337, 690–694.

Tennant, C.M., Seale, J.P. and Temple, D.M. (1987). Effects of a 5-lipoxygenase inhibitor, REV-5901, on leukotriene and histamine release from human lung tissue in-vitro. J. Pharm. Pharmacol. 39, 309–311.

Terashita, Z., Imura, Y., Tanabe, M., Kawazoe, K., Nishikawa, K., Kato, K. and Terao, S. (1986). CV-4151 – a potent, selective thromboxane A_2 synthetase inhibitor. Thromb. Res. 41, 223–237.

Thaulow, E., Dale, J. and Myhre, E. (1984). Effects of a selective thromboxane synthetase inhibitor, dazoxiben, and of acetylsalicyclic acid on myocardial ischaemia in patients with coronary artery disease. Am. J. Cardiol. 53, 1255–1258.

Thiemermann, C. and Schrör, K. (1985). In "Prostaglandins and Other Eicosanoids in the Cardiovascular System. Proc. of the 2nd Int. Symp. on Prostaglandins" (ed K. Schrör), pp 316–321. Karger, Basel.

Thiemermann, C., Ney, P. and Schrör, K. (1988). The thromboxane receptor antagonist, daltroban, protects the myocardium from ischaemic injury resulting in suppression of leukocytosis. Eur. J. Pharmacol. 155, 57–67.

Thieuleux, F.A., Lablanche, J.M., Lombard, F.A., Nocon, P.E. and Bertrand, M.E. (1986). Prevention of coronary arterial spasm with a thromboxane synthetase inhibitor. J. Am. Coll. Cardiol. 7, 177A (Abstr.).

Thorsen, S. (1986). Leukotriene B_4, a mediator of inflammation. Scand. J. Rheumatol. 15, 225–236.

Thorsen, S., Fogh, K., Broby-Johansen, U. and Søndergaard, J. (1990). Leukotriene B_4 in atopic dermatitis: increased skin levels and altered sensitivity of peripheral blood T-cells. Allergy 45, 457–463.

Tilden, S.J., Underwood, D.C., Cowen, K.H., Wegmann, M.J., Graybar, G.B., Hyman, A.L., McNamara, D.B. and Kadowitz, P.J. (1987). Effects of OKY 1581 on bronchoconstrictor responses to arachidonic acid and PGH_2. J. Appl. Physiol. 62, 2066–2074.

Tojo, H., Ono, T. and Okamoto, M. (1988a). A pancreatic-type phospholipase A_2 in rat gastric mucosa. Biochem. Biophys. Res. Commun. 151, 1188–1193.

Tojo, H., Ono, T., Kuramitsu, S., Kagamiyama, H. and Okamoto, M. (1988b). A phospholipase A_2 in the supernatant fraction of rat spleen. Its similarity to rat pancreatic phospholipase A_2. J. Biol. Chem. 263, 5724–5731.

Toki, Y., Hieda, N., Okumura, K., Hashimoto, H., Ito, T., Ogawa, K. and Satake, T. (1988). Myocardial salvage by a novel thromboxane A_2 synthetase inhibitor in a canine coronary occlusion–reperfusion model. Arzneim-Forsch/Drug Res. 38, 224–227.

Tomioka, K., Garrido, R., Stevenson, J.S. and Abraham, W.M. (1989). The effect of an orally active leukotriene (LT) antagonist YM-16638 on antigen-induced early and late airway responses in allergic sheep. Prostaglandins Leukot. Essent. Fatty Acids 36, 43–47.

Torka, M.C., Hacker, R.W., Yukseltan, I., Pohlmann, V., Meier, P., Zimmermann, T., Etti, H., Bulitta, M. and Kondor, P. (1988). Reduction of the vein graft occlusion rate after coronary artery bypass surgery by treatment with a thromboxane receptor antagonist. Eur. Heart J. 9(Suppl I), 325 (Abstract).

Torphy, T.J. and Undem, B.J. (1991). Phosphodiesterase inhibitors: new opportunities for the treatment of asthma. Thorax 46, 512–523.

Torphy, T.J., Newton, J.F., Wasserman, M.A., Vickery-Clark, L., Osborn, R.R., Bailey, L.S., Yodis, L.A.P., Underwood, D.C. and Hay, D.W.P. (1989). The bronchopulmonary pharmacology of SKandF 104353 in anaesthetized guinea pigs: demonstration of potent and selective antagonism of responses to peptidoleukotrienes. J. Pharmacol Exp. Ther. 249, 430–437.

Torphy, T.J., Faiferman, I., Gleason, J.G., Hall, R.F., Lewis, M.A., Broom, C., Helfrich, H.M., Newton, J.F. and Hay, D.W.P. (1991). The preclinical and clinical pharmacology of SKandF 104353, a potent and selective peptidoleukotriene receptor antagonist. Ann. N.Y. Acad. Sci. 157–167.

Tsai, B.S., Villani-Price, D., Keith, R.H., Zemaitis, J.M., Bauer, R.F., Leonard, R., Djuric, S.W. and Shone, R.L. (1989). SC-41930: an inhibitor of leukotriene B_4-stimulated human neutrophil functions. Prostaglandins 38, 655–674.

Tsuji, M., Saito, S. and Tamura, T. (1985). Efficacy of thromboxane A_2 synthetase inhibitor (OKY-046) in patients with angina pectoris. Jap. Circ. J. 49, 143 (Abstr.).

Tyler, H.M., Saxton, C.A.P.D. and Parry, M.J. (1981). Administration to man of UK 37,248–01, a selective inhibitor of thromboxane synthetase. Lancet i, 629.

Tynan, S.S., Anderson, N.H., Wills, M.T., Harker, L.A. and Hanson, S.R. (1984). On the multiplicity of platelet prostaglandin receptors. II. The use of N-0164 for distinguishing the loci of action for PGI_2, PGD_2 and PGE_2 and hydantoin analogues. Prostaglandins 27, 683–696.

Uderman, H.D., Workman, R.J. and Jackson, E.K. (1982). Attenuation of the development of hypertension in spontaneously hypertensive rats by the thromboxane synthetase inhibitor, 4'-(imidazol-1-yl)acetophone. Prostaglandins 24, 237–244.

Uderman, H.D., Jackson, E.K., Puett, D. and Workman, R.J. (1984). Thromboxane synthetase inhibitor UK 38,485 lowers blood pressure in the adult spontaneously hypertensive rat. J. Cardiovasc. Pharmacol. 6, 969–972.

Ueno, A., Tanaka, K. and Katori, M. (1982). Possible involvement of thromboxane in bronchoconstrictive and hypertensive effects of LTC_4 and LTD_4 in guinea pigs. Prostaglandins 23, 865–880.

Vadas, P. and Hay, J.B. (1983). Involvement of circulating phospholipase A_2 in the pathogenesis of the hemodynamic changes in endotoxin shock. Can. J. Physiol. Pharmacol. 61, 561–566.

Vadas, P. and Pruzanski, W. (1984). In "Advances in Inflammation Research", Vol 7 (eds I. Otterness, R. Capetola and S. Wong), pp 51–59. Raven Press, New York.

Vadas, P. and Pruzanski, W. (1986). Role of secretory phospholipase A_2 in the pathology of disease. Lab. Invest. 55, 391–404.

Vadas, P. and Pruzanski, W. (1990). In "Phospholipase A_2" (eds P.H.-K. Wong and E.A. Dennis), pp 83–101. Plenum Press, New York.

Vadas, P., Stefanski, W. and Pruzanski, W. (1985). Characterization of extracellular phospholipase A_2 in rheumatoid synovial fluid. Life Sci. 36, 579–587.

Vadas, P., Pruzanski, W., Stefanski, E., Sternby, B., Mustard, R., Bohnen, J., Frazer, I., Farewell, V. and Bombardier, C. (1988a). Pathogenesis of hypotension in septic shock: correlation of circulating phospholipase A_2 levels with circulatory collapse. Crit. Care Med. 16, 1–7.

Vadas, P., Pruzanski, W. and Stefanski. (1988b). Extracellular phospholipase A_2: causative agent in circulatory collapse of septic shock? Agents Actions 24, 320–325.

Vadas, P., Pruzanski, W., Kim, J. and Fornasier, V. (1989a). The proinflammatory effect of intra-articular injection of soluble human and venom phospholipase A_2. Am. J. Pathol. 134, 807–811.

Vadas, P., Pruzanski, W., Stefanski, E., Johnson, L., Seilhamer, J., Mustard, R. Jr and Bohnen, J. (1989b). In "Cell Activation and Signal Initiation: Receptor and Phospholipase Control of Inositol Phosphate, PAF and Eicosanoid Production" (ed E. Dennis), pp 155–166. Alan R. Liss, New York.

Vainio, P., Thurén, T., Wichman, K., Luukkainen, T. and Kinnunen, P.K.J. (1985). Hydrolysis of phospholipid mono-layers by human spermatozoa. Inhibition by male contraceptive gossypol. Biochim. Biophys. Acta 814, 405–408.

Van de Kerkhof, P.C.M., Bauer, F.W. and de Grood, R.M. (1985). Leukotriene B_4-induced micropustule formation in psoriasis: interference with epidermal growth-control and therapeutical modulation. J. Invest. Dermatol. 84, 450 (Abstr.).

Vane, J.R. (1971). Inhibition of prostaglandin synthesis as a mechanism of action for aspirin-like drugs. Nature (New Biol.) 231, 232–235.

Van Inwegen, R.G., Nuss, G.W. and Carnathan, G.W. (1989). Antagonism of peptidoleukotrienes and inhibition of systemic anaphylaxis by RG 12525 in guinea pigs. Life Sci. 44, 799–807.

Verger, R., Ferrato, F., Mansback, C.M. and Pieroni, G. (1982). Novel intestinal phospholipase A_2: purification and some molecular characteristics. Biochemistry 21, 6883–6889.

Verheij, H.M., Slotboom, A.J. and de Haas, G.H. (1981). Structure and function of phospholipase A_2. Rev. Physiol. Biochem. Pharmacol. 9, 191–203.

Vermylen, J. and Deckmyn, H. (1983). Reorientation of prostaglandin endoperoxide metabolism by a thromboxane synthetase inhibitor: *in vitro* and clinical observations. Br. J. Clin. Pharmacol. 15, 17S–22S.

Vesterqvist, O. (1986). Rapid recovery of in vivo prostacyclin formation after inhibition by aspirin. Evidence from measurements of the major urinary metabolite of prostacyclin by GC–MS. Eur. J. Clin. Pharmacol. 30, 69–73.

Vesterqvist, O. and Gréen, K. (1984). Development of a GC–MS method for quantitation of 2,3-dinor-6-keto-$PGF_{1\alpha}$ and determination of the urinary excretion rates in healthy humans under normal conditions and following drugs. Prostaglandins 28, 139–154.

Vesterqvist, O., Gréen, K., Henriksson, P., Rasmanis, G. and Edhag, P. (1991). Selective thromboxane synthase inhibition does not increase in vivo synthesis of prostacyclin in healthy humans. Adv. Prostaglandin Thromboxane Leukotriene Res. 21A, 165–168.

Villani-Price, D., Yang, D.C., Walsh, R.E., Fretland, D.J., Keith, R.H., Kocan, G., Kachur, J.F., Gaginella, T.S. and Tsai, B.S. (1992). Multiple actions of the leukotriene B_4 receptor antagonist SC-41930. J. Pharmacol. Exp. Ther. 260, 187–191.

Violi, F., Ghiselli, A., Iuliano, L., Pratico, D., Alessandri, C. and Balsano, F. (1988). Inhibition by picotamide of thromboxane production *in vitro* and *ex vivo*. Eur. J. Clin. Pharmacol. 33, 599–602.

Voelkel, N.F., Worthen, S., Reeves, J.T., Henson, P.M. and Murphy, R.C. (1982). Nonimmunological production of leukotrienes induced by platelet-activating factor. Science 218, 286–288.

Wahedna, I., Wisniewski, A.S. and Tattersfield, A.E. (1991). Effect of RG 12525, an oral leukotriene D_4 antagonist, on the airway response to inhaled leukotriene D_4 in subjects with mild asthma. Br. J. Clin. Pharmacol. 32, 512–515.

Wainwright, C.L. and Parratt, J.R. (1988). The effects of L655,240, a selective thromboxane and prostaglandin endoperoxide antagonist, on ischemia- and reperfusion-induced cardiac arrhythmias. J. Cardiovasc. Pharmacol. 12, 264–271.

Walinsky, P., Smith, J.B., Lefer, A.M., Lebenthal, M., Urban, P., Greenspon, A. and Goldberg, S. (1984). Thromboxane A_2 in acute myocardial infarction. Am. Heart J. 108, 868–872.

Walker, J.R. and Dawson, W. (1979). Inhibition of rabbit PMN lipoxygenase activity by benoxaprofen. J. Pharm. Pharmacol. 31, 778–780.

Wallach, D.P. and Brown, V.J.R. (1981). Studies on the arachidonic acid cascade – I. Inhibition of phospholipase A_2 *in vitro* and *in vivo* by several novel series of inhibitor compounds. Biochem. Pharmacol. 30, 1315–1324.

Wargovich, T.J., Mehta, J., Nichols, W.W., Ward, M.B., Lawson, D., Franzini, D. and Conti, C.R. (1987). Reduction in myocardial neutrophil accumulation and infarct size following administration of thromboxane inhibitor U-63,557A. Am. Heart J. 114, 1078–1085.

Watkins, D., Hüttemeier, P., Kong, D. and Peterson, M. (1982). Thromboxane and pulmonary hypertension following *E. coli* endotoxin infusion in sheep: effect of an imidazole derivative. Prostaglandins 23, 273–285.

Weichman, B.M. and Tucker, S.S. (1985). Differentiation of the mechanisms by which leukotrienes C_4 and D_4 elicit contraction of the guinea pig trachea. Prostaglandins 29, 547–560.

Weichman, B.M., Muccitelli, R.M., Osborn, R.R., Holden, D.A., Gleason, J.G. and Wasserman, M.A. (1982). *In vitro* and *in vivo* mechanisms of leukotriene-mediated broncho-constriction in the guinea pig. J. Pharmacol. Exp. Ther. 222, 202–208.

Weichman, B.M., DeVan, J.F., Muccitelli, R.M., Tucker, S.S., Vickery, L.M. and Wasserman, M.A. (1984a). Analysis of the antagonist profile of SKandF 88046 on guinea pig trachea. Prostaglandins Leukotrienes Med. 15, 167–175.

Weichman, B.M., Wasserman, M.A. and Gleason, J.G. (1984b). SKandF 88046: A unique pharmacologic antagonist of bronchoconstriction induced by leukotriene D_4, thromboxane and prostaglandins $F_2\alpha$ and D_2 *in vitro*. J. Pharmacol. Exp. Ther. 228, 128–132.

Welton, A.F., Hope, W.C., Tobias, L.D. and Hamilton, J.G. (1981). Inhibition of antigen-induced histamine release and thromboxane synthase by FPL 55712, a specific SRS-A antagonist? Biochem. Pharmacol. 30, 1378–1382.

Wery, J.-P., Schevitz, R.W., Clawson, D.K., Bobbitt, J.L., Dow, E.R., Gamboa, G., Goodson, T. Jr, Hermann, R.B., Kramer, R.M., McClure, D.B., Mihelich, E.D., Putnam, J.E., Sharp, J.D. Stark, D.H., Teater, C., Warrick, M.W. and Jones, N.D. (1991). Structure of recombinant human rheumatoid arthritic synovial fluid phospholipase A_2 at 2.2 Å resolution. Nature 352, 79–82.

Whittle, B.J.R., Moncada, S. and Vane, J.R. (1978). Comparison of the effects of prostacyclin (PGI_2), prostaglandin E_1 and D_2 on platelet aggregation in different species. Prostaglandins 16, 373–388.

Wilkerson, W., DeLucca, I., Galbraith, W., Gans, K., Harris, R., Jaffee, B. and Kerr, J. (1991). Antiinflammatory phospholipase-A_2 inhibitors. I. Eur. J. Med. Chem. 26, 667–676.

Williams, J.D., Robin, J.-L., Lewis, R.A., Lee, T.H. and Austen, K.F. (1986). Generation of leukotrienes by human monocytes pretreated with cytochalasin B and stimulated with formyl-methionyl-leucyl-phenylalanine. J. Immunol. 136, 642–648.

Winkler, J.D., Sung, C.-M., Bennett, CF. and Chilton, F.H. (1991). Characterization of CoA-independent transacylase activity in U937 cells. Biochim. Biophys Acta 1081, 339–346.

Winn, R., Harlan, J., Nadir, B., Harker, L. and Hildebrandt, J. (1983). Thromboxane A_2 mediates lung vasoconstriction but not permeability after endotoxin. J. Clin. Invest. 72, 911–918.

Wise, W.C., Cook, J.A., Halushka, P.V. and Knapp, D.R. (1980). Protective effects of thromboxane synthetase inhibitors in rats in endotoxic shock. Circ. Res. 46, 854–859.

Woinar, R.J. and Varma, R.K. (1983). Glucocorticoid receptor binding, anti-proliferative and anti-inflammatory activities of some novel 20-thiasteroids. Pharmacology 25, 236 (Abstr.).

Wong, E., Greaves, M.W. and O'Brien, T. (1984). Increased concentrations of immunoreactive leukotrienes in cutaneous lesions of eosinophilic cellulitis. Br. J. Dermatol. 110, 653–656.

Yamai, T., Watanabe, S., Motojima, S., Fuduka, T. and Makino, S. (1989). The significance of leukotriene in antigen-induced late asthmatic response. Am. Rev. Respir. Dis. 139, A462 (Abstr.).

Yamamoto, S. (1989). Mammalian lipoxygenases: molecular and catalytic properties. Prostaglandins Leukot. Essent. Fatty Acids 35, 219–229.

Yamamura, H., Taira, M., Negi, H., Nanbu, F., Kohno, S.W. and Ohata, K. (1988). Effect of AA-861, a selective 5-lipoxygenase inhibitor, on models of allergy in several species. Jpn. J. Pharmacol. 47, 261–271.

Yoshimoto, T., Yamamoto, S. and Hayaishi, O. (1978). Selective inhibition of prostaglandin endoperoxide thromboxane iso-merase by 1-carboxyalkylimidazoles. Prostaglandins 16, 529–539.

Young, J.M., Befrod, C.T., Murthy, D.V.K., Venati, M.C., Jones, G.H., DeYoung, L., Akers, W.A. and Scholtz, J.R. (1985). The biological activities of the novel topical antipsoriatic agent 6-chloro-1,4-diacetoxy-2,3-dimethoxyna-phthalene (RS-43179). J. Invest. Dermat. 84, 358 (Abstr.).

Yuan, W. and Gelb, M.H. (1988). Phosphate-containing phospholipid analogues of tight-binding inhibitors of phospholipase A_2. J. Am. Chem. Soc. 110. 2665–2666.

Yuan, W., Berman, R.J. and Gelb, M.H. (1987). Synthesis and evaluation of phospholipid analogues as inhibitors of cobra venom phospholipase A_2. J. Am. Chem. Soc. 109, 8071–8081.

Yui, Y., Hattori, R., Takatsu, Y. and Kawai, C. (1986). Selective thromboxane A_2 synthetase inhibition in vasospastic angina pectoris. J. Am. Coll. Cardiol. 7, 25–29.

Zakrzewski, J.F., Barnes, N.C., Piper, P.J. and Costello, J.F. (1984). Sputum leukotrienes and prostanoids – possible synergistic mediators in cystic fibrosis, bronchiectasis or chronic bronchitis. Prostaglandins 28, 641A (Abstr.).

Ziboh, V.A., Casebolt, T.L., Marcelo, C.L. and Voorhees, J.J. (1984). Biosynthesis of lipoxygenase products by enzyme preparations from normal and psoriatic skin. J. Invest. Dermatol. 83, 426–430.

Zipser, R.D., Kronborg, I., Rector, W. Reynolds, T. and Daskalopoulos, G. (1984). Therapeutic trial of thromboxane synthetase inhibition in the hepatorenal syndrome. Gastroenterology 87, 1228–1232.

Zukowska-Grojec, Z. and Feuerstein, G. (1985). In "Leukotrienes in Cardiovascular and Pulmonary Function" (ed M. Gee), pp 101–113. Alan R. Liss, Philadelphia.

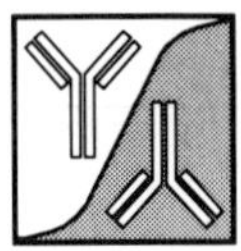

6. Biosynthesis and Catabolism of Platelet-Activating Factor

F. Snyder

1. Introduction

This chapter provides an up-to-date overview of the biosynthetic and catabolic pathways responsible for the metabolism of PAF (1-alkyl-2-acetyl-*sn*-glycero-3-phosphocholine; see Fig. 6.1). A section is also included on the metabolic regulation of PAF levels. Emphasis is on enzymatic studies reported in the literature that have firmly documented how PAF is metabolized.

It is important to note that when PAF is biosynthesized the acyl and alk-1-enyl analogues (i.e. 1-acyl-2-acetyl-*sn*-glycero-3-phosphocholine (ethanolamine) or 1-alk-1-enyl-2-acetyl-*sn*-glycero-3-phosphoethanolamine (choline)) can also both be produced (Fig. 6.1). Occurrence of the acyl analogue of PAF (1-acyl-2-acetyl-*sn*-glycero-3-phosphocholine) has been well documented in both human (Mueller *et al.*, 1984; Satouchi *et al.*, 1987) and rabbit (Mueller *et al.*, 1984; Satouchi *et al.*, 1985) neutrophils, lipid extracts of brain (Tokumura *et al.*, 1987, 1989), rat uterus (Yasuda *et al.*, 1988), HL-60 cells (Suga *et al.*, 1990) and in a variety of human inflammatory cells (Triggiani *et al.*, 1991a). Other reports can also be found in the literature regarding the formation of the acyl

analogue, and in some instances (e.g. in endothelial cells) this analogue of PAF is the predominant form (Garcia *et al.*, 1991; Mueller *et al.*, 1991; Whatley *et al.*, 1992). Tessner and Wykle (1987) have described an ethanolamine-containing plasmalogen analogue of PAF (1-alk-1-enyl-2-acetyl-*sn*-glycero-3-phosphoethanolamine) in human neutrophils, and it has also been shown to be synthesized in differentiated HL-60 cells (Suga *et al.*, 1990). Nakayama *et al.* (1988) have also found the choline plasmalogen analogue of PAF in heart muscle. Little is known about the function or metabolism of these PAF analogues, except that both the acyl analogue of lyso-PAF (Wykle *et al.*, 1980a; Ninio *et al.*, 1983; Lee, 1985) and 1-alk-1-enyl-2-lyso-*sn*-glycero-3-phosphoethanolamine (Tessner and Wykle, 1987) can be acetylated by the acetyl-CoA acetyltransferase found in membrane preparations.

The reader is referred to earlier reviews (Lee and Snyder, 1985, 1989; Snyder, 1989, 1990; Prescott *et al.*, 1990) for the extensive literature available on PAF metabolism in a variety of intact cells, tissues and animals. However, it is noteworthy that the metabolic results obtained in the more complex systems are compatible

Lipid Mediators
ISBN 0–12–198875–9

$$
\begin{array}{l}
\text{H}_2\text{COCH}_2\text{CH}_2\text{R} \\
\quad\ \ \text{O} \ | \\
\text{CH}_3\overset{\|}{\text{C}}\text{OCH} \\
\qquad\quad \text{O} \\
\qquad\quad \overset{\|}{\ } \qquad\qquad + \\
\text{H}_2\text{COPOCH}_2\text{CH}_2\text{N(CH}_3)_3 \\
\mathbf{I} \qquad \text{O-}
\end{array}
$$

PAF (1-alkyl-2-acetyl-<u>sn</u>-glycero-
3-phosphocholine)

$$
\begin{array}{l}
\qquad\quad \text{O} \\
\qquad\quad \overset{\|}{\ } \\
\text{H}_2\text{COCCH}_2\text{R} \\
\quad\ \ \text{O} \ | \\
\text{CH}_3\overset{\|}{\text{C}}\text{OCH} \\
\qquad\quad \text{O} \\
\qquad\quad \overset{\|}{\ } \qquad\qquad + \\
\text{H}_2\text{COPOCH}_2\text{CH}_2\text{N(CH}_3)_3 \\
\mathbf{II} \qquad \text{O-}
\end{array}
$$

Acyl Analog of PAF (1-acyl-2-acetyl-
<u>sn</u>-glycero-3-phosphocholine)

$$
\begin{array}{l}
\text{H}_2\text{COCH=CHR} \\
\quad\ \ \text{O} \ | \\
\text{CH}_3\overset{\|}{\text{C}}\text{OCH} \\
\qquad\quad \text{O} \\
\qquad\quad \overset{\|}{\ } \\
\text{H}_2\text{COPOCH}_2\text{CH}_2\text{NH}_2 \\
\mathbf{III} \qquad \text{O-}
\end{array}
$$

Plasmalogen Analog of PAF
(1-alk-1'-enyl-2-acetyl-<u>sn</u>-
glycero-3-phosphoethanolamine)

$$
\begin{array}{l}
\text{H}_2\text{COR} \\
\quad\ \ \text{O} \ | \\
\text{CH}_3\overset{\|}{\text{C}}\text{OCH} \\
\qquad\quad | \\
\text{H}_2\text{COH} \\
\mathbf{IV}
\end{array}
$$

1-alkyl-2-acetyl-<u>sn</u>-glycerol

Figure 6.1 The chemical structure of PAF and important structurally related analogues.

with the enzymatic pathways described for the isolated membrane systems studied *in vitro*.

2. Biosynthesis of Platelet-Activating Factor

2.1 REMODELLING PATHWAY

Cell stimuli of various sorts (e.g. FMLP, zymosan, thrombin, etc.) can trigger the biosynthesis of PAF via the remodelling pathway in a reaction sequence involving (1) the hydrolysis of the *sn*-2 acyl moiety of a membrane phospholipid precursor (1-alkyl-2-acyl-*sn*-glycero-3-phosphocholine) of PAF to form lyso-PAF and (2) the acetylation of the lyso-PAF intermediate to produce PAF (Fig. 6.2). Either step can be rate-limiting. Since the lyso-PAF intermediate, formed both in the forward reaction and also in the back reaction (catalysed by an acetylhydrolase), can be rapidly re-acylated into the PAF precursor pool of alkylacylglycerophosphocholine by a CoA-independent transacylase that is highly selective for arachidonate and other polyenoates, the remodelling pathway is also referred to as the "PAF cycle."

Studies of human platelets and neutrophils by Sturk *et al.* (1989) have demonstrated that the initial hydrolysis of an acyl group to form lyso-PAF or its acyl analogue does not appear to discriminate between alkylacylglycerophosphocholines and diacylglycerophosphocholines as the precursors, respectively, when PMSF is used to block the degradation of the acetylated product. However, in the absence of the PMSF inhibitor, the acyl analogue is more rapidly degraded and, therefore, the net amount of PAF formed is considerably greater than its acyl counterpart.

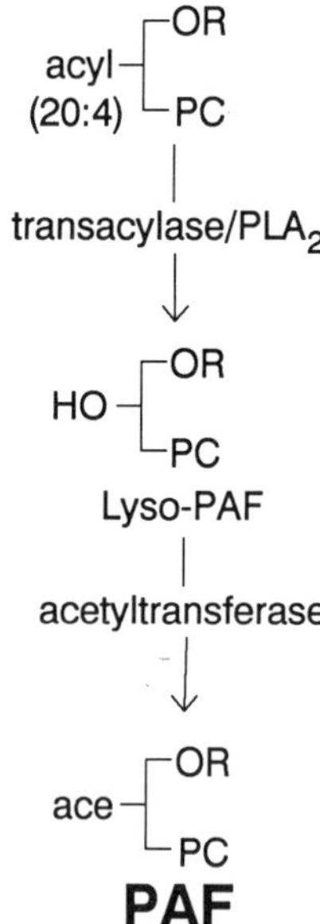

Figure 6.2 A simplified illustration of the remodelling pathway for PAF biosynthesis.

The acetylation step catalysed by acetyl-CoA:lyso-PAF acetyltransferase (Wykle *et al.*, 1980a) has been well documented; it is also known to be activated through phosphorylation via a protein kinase (see Section 4). This acetyltransferase is able to utilize both the acyl analogue (1-acyl-2-lyso-*sn*-glycero-3-phosphocholine) of lyso-PAF (Wykle *et al.*, 1980a; Ninio *et al.*, 1983; Lee, 1985) and the lysoethanolamine plasmalogen (1-alk-1-enyl-2-lyso-*sn*-glycero-3-phosphoethanolamine) analogue of PAF

(Tessner and Wykle, 1987) as substrates. Thus, the acetyltransferase in the remodelling pathway does not appear to discriminate among PAF and its analogues as substrates, albeit the reaction rates are considerably lower for the acyl and alk-1-enyl forms than for lyso-PAF. Acyl-CoAs ranging in size from C_2 to C_6 serve as substrates for the lyso-PAF acetyltransferase; however, the C_4 and C_6 species have considerably lower rates than the C_2 and C_3 species, which have similar V_{max} values (Lee, 1985). Polar head group modifications of lyso-PAF influence activity rates in a decreasing order, with choline $> N',N'$-dimethylethanolamine $>$ monomethylethanolamine $>$ ethanolamine (Lee, 1985). All of the results regarding the substrate specificity of the lyso-PAF acetyltransferase show a good positive correlation between reaction rates and the known biological activities (Satouchi *et al.*, 1981; Tence *et al.*, 1981; Blank *et al.*, 1982; O'Flaherty *et al.*, 1983) of the various PAF analogues tested.

Until recently, little was known about the initial step involving acyl group hydrolysis in the remodelling pathway of PAF biosynthesis. A number of reports have documented that alkylarachidonoylglycerophosphocholine species are selectively utilized as precursors for the production of the lyso-PAF intermediate (Chilton *et al.*, 1984; Ramesha and Pickett, 1986; Suga *et al.*, 1990). In the past the hydrolytic reaction was always depicted as being catalysed by a putative phospholipase A_2; however, there has never been direct evidence for a phospholipase A_2 activity during the period when PAF is formed following cell stimulation with the calcium ionophore A23187 or physiological agonists. Moreover, in stimulated differentiated HL-60 cells it was not possible to show the presence of a radiolabelled lyso-PAF intermediate during PAF production unless unlabelled lyso-PAF was added as a trapping agent (Uemura *et al.*, 1991). The rationale for trapping the labelled lyso-PAF in this manner was to slow down the rapid CoA-independent transacylation re-acylation of lyso-PAF with arachidonate back into the membrane precursor pool (Chilton *et al.*, 1983; Kramer *et al.*, 1984; Robinson *et al.*, 1985; Sugiura *et al.*, 1985, 1987; Sugiura and Waku, 1985; Winkler *et al.*, 1991) (see Fig. 6.2).

Recent studies of the remodelling pathway of PAF biosynthesis (Sugiura *et al.*, 1990; Uemura *et al.*, 1991; Venable *et al.*, 1991) indicate that the lyso-PAF intermediate is generated from alkylarachidonoylglycerophosphocholines by a CoA-independent transacylase activity. This transacylase is thought to possess both phospholipase A_2 and acyltransferase activities. The transfer of arachidonate from alkylarachidonoylglycerophosphocholines to a lysophospholipid acceptor results in the formation of lyso-PAF with no release of free arachidonic acid in this step. The lysophospholipid acceptor molecule required for the transacylase reaction is thought to be generated *in situ* from diacyl-, alkylacyl- or alk-1-enylacyl species of either choline- or ethanolamine- containing glycerophosphatides since only the lyso forms of these

phospholipids serve as acyl acceptor molecules. Other lysophospholipids containing serine/inositol, neutral lipids with free hydroxyl groupings, or cholesterol are not acyl acceptors for the CoA-independent transacylase (Uemura *et al.*, 1991). These studies indicate that the coupled transacylase/phospholipase A_2 and the phospholipase A_2 responsible for generation of the lysophospholipid acceptor can provide the alkyl, acyl and alk-1-enyl forms of precursors required for the biosynthesis of PAF as well as the acyl and plasmalogen analogues of PAF following agonist stimulation. As mentioned earlier, both the acyl (Wykle *et al.*, 1980a; Ninio *et al.*, 1983; Lee, 1985) and alk-1-enyllysoglycerophosphoethanolamine (Tessner and Wykle, 1987) analogues of lyso-PAF can be acetylated by the acetyl-CoA:lyso-PAF acetyltransferase to form the "PAF-like" structure. The plasmalogen analogue of PAF can also be produced via transacetylation of the ethanolamine lysoplasmalogen via transfer of the acetate from PAF in a reaction catalysed by a CoA-independent transacetylase (Lee *et al.*, 1992).

Of great importance is that the lyso-PAF generated by the CoA-independent transacylase has been shown to be directly linked to the synthesis of PAF via the lyso-PAF acetyltransferase since PAF production can be significantly increased simply by adding lysoethanolamine plasmalogen to membrane preparations containing acetyl-CoA (Uemura *et al.*, 1991; Venable *et al.*, 1991). The overall reaction sequence for PAF formation through the combined reactions catalysed by a phospholipase A_2, a transacylase/phospholipase A_2 and the acetyltransferase is illustrated in Fig. 6.3. Whereas the transacylase-induced formation of lyso-PAF introduces a new concept for how PAF can be biosynthesized in the remodelling pathway, the reader is cautioned that these findings do not completely rule out the importance of a single phospholipase A_2 activity in the formation of lyso-PAF at the present time. Complex regulatory control of a phospholipase A_2 by activation/inactivation mechanisms not yet understood could account for the lack of direct experimental proof for the sole involvement of a conventional phospholipase A_2 activity in PAF biosynthesis via the remodelling route.

A novel CoA-independent transacetylase, just recently characterized in HL-60 cells, also appears to play a role in the metabolism of PAF and the formation of other acetylated lipids (Lee *et al.*, 1992). Interestingly, the transacetylase has completely different properties than the CoA-independent transacylase that transfers arachidonate moieties among choline- and ethanolamine-containing phospholipids (see above). For example, the CoA-independent transacetylase can selectively transfer the *sn*-2 acetate from PAF to a wide range of acceptor lysophospholipids possessing choline, ethanolamine, serine, inositol or phosphate groups; moreover, it also utilizes long-chain fatty alcohols as an acetate acceptor. The transacetylase appears to be particularly important in the formation of the acetylated plasmalogen analogue of PAF

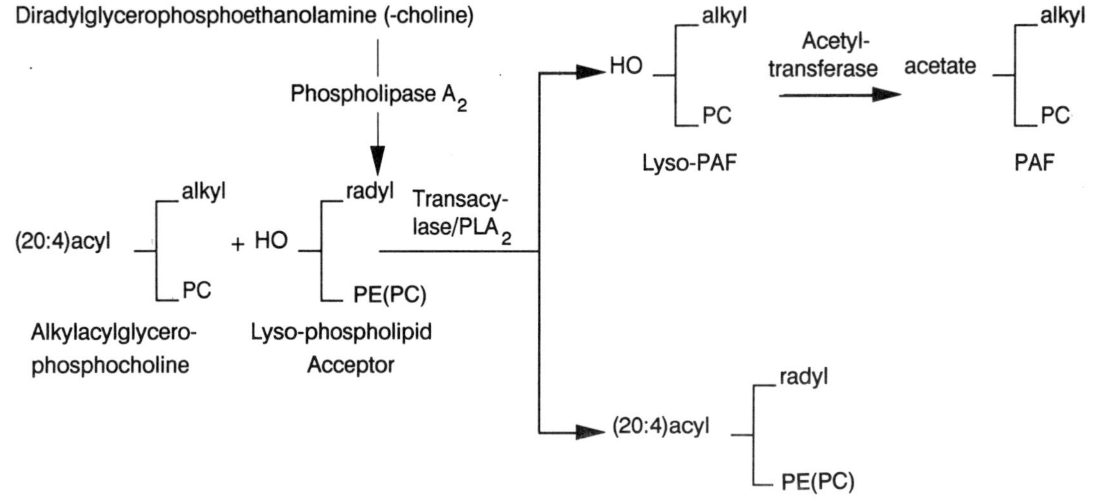

Figure 6.3 The involvement of a phospholipase A₂ and a transacylase/phospholipase A₂ in the generation of lyso-PAF for PAF synthesis in the remodelling pathway.

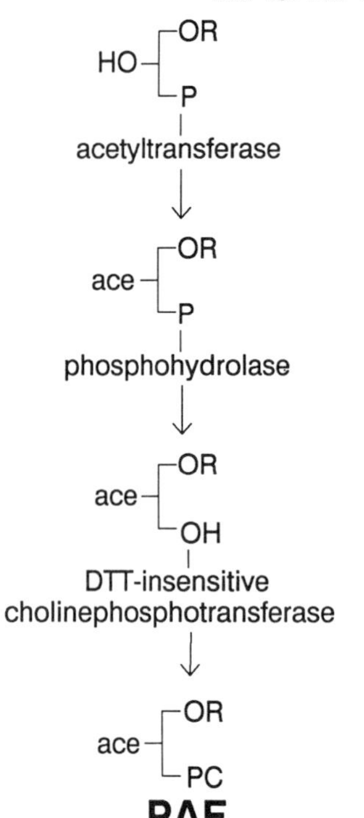

Figure 6.4 Enzymatic steps required for the formation of PAF in the *de novo* pathway.

via the direct transfer of the acetate from PAF to an ethanolamine lysoplasmalogen. In addition to generating a variety of acetylated lipid products, it has been suggested that the transacetylase might be responsible for fine tuning the biological responses induced by PAF (Lee *et al.*, 1992).

2.2   DE NOVO PATHWAY

An alternative, yet prominent, means of PAF biosynthesis occurs in many cells by a *de novo* pathway (Renooij and Snyder, 1981; Lee *et al.*, 1986, 1988; Woodard *et al.*, 1987) rather than the remodelling mechanism. The series of reactions leading to the *de novo* synthesis of PAF includes: (1) the acetylation of 1-alkyl-2-lyso-*sn*-glycero-3-phosphate by an acetyl-CoA acetyltransferase to form 1-alkyl-2-acetyl-*sn*-glycerol-3-phosphate; (2) the dephosphorylation of the 1-alkyl-2-acetyl-*sn*-glycero-3-phosphate by a phosphohydrolase to form 1-alkyl-2-acetyl-*sn*-glycerols; and (3) the transfer of the phosphocholine moiety of CDP-choline to the alkylacetylglycerols by a DTT-insensitive cholinephosphotransferase to produce PAF (Fig. 6.4). All three of the *de novo* enzymes have different properties from other known enzymes in lipid metabolism. Thus, the acetyltransferase that utilizes the alkyllysoglycerophosphate as a substrate differs from the lyso-PAF acetyltransferase in the remodelling pathway (Lee *et al.*, 1986). Moreover, the phosphohydrolase (Lee *et al.*, 1988) and the DTT-insensitive cholinephosphotransferase (Renooij and Snyder, 1981; Woodard *et al.*, 1987) in the *de novo* synthesis of PAF differ from their corresponding catalytic counterparts in the synthesis of phosphatidylcholine. In view of these important differences in enzyme properties, the design and development of specific enzyme inhibitors could provide highly selective tools for the pharmacological intervention in crucial reaction steps involved in the production of excess levels of PAF via the remodelling route during inflammatory responses.

Results obtained by Heller and coworkers (1991) indicate that the acetyltransferase and DTT-insensitive choline phosphotransferase in the *de novo* pathway are activated by protein kinase C. The activation of these enzymes via a phosphorylation step could represent a crucial control mechanism in regulating the *de novo* pathway of PAF biosynthesis.

Substrate specificities of the three enzymes in the *de novo* pathway of PAF biosynthesis have been investigated in some detail. The acetyl-CoA:1-alkyl-2-lyso-*sn*-glycero-3-phosphate acetyltransferase can also catalyse the acetylation of the acyl analogue (1-oleoyl-2-lyso-*sn*-glycero-3-phosphate) but at significantly lower rates than the alkyl precursor (Lee *et al.*, 1991). Both the $C_{16:0}$ and $C_{18:0}$ alkyl forms of the substrate are utilized at identical rates. The *de novo* acetyltransferase exhibits the following order of preference for acyl-CoAs to acylate the *sn*-2 position of 1-alkyl-2-lyso-*sn*-glycero-3-phosphate: $C_{3:0} > C_{2:0} > C_{6:0} = C_{4:0}$. Even arachidonoyl-CoA and linoleoyl-CoA are substrates for the acetyltransferase, but they are utilized at considerably lower rates than acetyl-CoA (Snyder *et al.*, 1992).

The phosphohydrolase, the second enzyme in the *de novo* biosynthetic sequence, exhibits the same hydrolytic activity towards substrates with $C_{16:0}$ and $C_{18:0}$ alkyl chains but somewhat less activity towards the ester-linked analogue (1-palmitoyl-2-acetyl-*sn*-glycero-3-phosphate) (Lee *et al.*, 1988). Substrates with $C_2 - C_6$ acyl chains at the *sn*-2 position exhibit similar dephosphorylation rates. The rate of phosphate hydrolysis for 1-alkyl-2-lyso-*sn*-glycero-3-phosphate is about two-thirds less than when an acetate group is substituted at the *sn*-2 position. Similarly, phosphatidic acid (1,2-dioleoyl-*sn*-glycero-3-phosphate) is a poor substrate for the phosphohydrolase. Substrate specificity and competition experiments, and differences in responses to cations, temperature and detergents, indicate that the alkylacetylglycerophosphate phosphohydrolase is not a non-specific phosphomonoesterase or a typical phosphatidate phosphohydrolase (Lee *et al.*, 1988).

The final enzyme in the *de novo* route, the DTT-insensitive cholinephosphotransferase, is also capable of utilizing the acyl analogue (1-oleoyl-2-acetyl-*sn*-glycerol) at rates comparable to those obtained with 1-alkyl-2-acetyl-*sn*-glycerols. Also, similar activity rates were obtained for substrates having $C_{16:0}$ and $C_{18:1}$ alkyl moieties, whereas the rates were substantially decreased when a $C_{18:0}$ alkyl group occupied the *sn*-1 position (Renooij and Snyder, 1981; Woodard *et al.*, 1987). Rate data obtained with various alkylacetylglycerol analogues possessing modified substituents at the *sn*-2 position were almost identical for the propionyl and acetyl derivatives but about 50% less for the butyryl derivative (Woodard *et al.*, 1987). Replacement of the *sn*-2 acetate moiety of the alkylacetylglycerols with acetamide or methoxy substituents rendered the analogues unsuitable as substrates for the DTT-insensitive cholinephosphotransferase (Woodard *et al.*, 1987).

The alkylacetylglycerols (see Fig. 6.1) produced in the *de novo* route would appear to be more important than just being intermediates in PAF synthesis since they are known to induce cell differentiation (McNamara *et al.*, 1984), attenuate protein kinase C activation (Bass *et al.*, 1989), and can be converted to biologically active lipids such as alkylglycerols, a macrophage activator (Yamamoto *et al.*, 1988) and alkylacetylglycerophosphate, a calcium ionophore (Bussolino *et al.*, 1984). Alkylacetylglycerols have also been shown to aggregate platelets (Satouchi *et al.*, 1984) and to exhibit a hypotensive response when given intravenously to rats (Blank *et al.*, 1984); these biological activities are thought to be due to the fact that the alkylacetylglycerols can be converted to PAF by the DTT-insensitive cholinephosphotransferase (Renooij and Snyder, 1981; Woodard *et al.*, 1987).

Rate-limiting steps in the *de novo* pathway are the initial reaction catalysed by the acetyltransferase (Lee *et al.*, 1986, 1989a) and the reaction responsible for the production of CDP-choline via cytidylyltransferase (Blank *et al.*, 1988; Lee *et al.*, 1990; Vallari *et al.*, 1990). Factors involved in control of these enzymes will be discussed in Section 4.

Unlike the remodelling pathway of PAF biosynthesis, the *de novo* route is not stimulated by inflammatory agents and arachidonic acid, or its metabolites, is not directly involved. It appears that the *de novo* pathway is important in maintaining the physiological levels of PAF found in blood and various tissues. Support for such a concept has been suggested from results obtained in a study of the renal medulla (Lee *et al.*, 1989b), where PAF is produced almost exclusively by the *de novo* enzymes (Woodard *et al.*, 1987). When PAF production in the kidney medulla is reduced by experimentally inhibiting the DTT-insensitive cholinephosphotransferase *in vivo*, the blood levels of PAF are also significantly decreased (Lee *et al.*, 1989b). Moreover, the distribution of the enzymes of the *de novo* pathway of PAF biosynthesis in a wide variety of tissues (Lee *et al.*, 1989a) suggests that the *de novo* synthesis of PAF plays an important role in cellular functions.

An interesting feature of the *de novo* pathway is that alkylacetylacylglycerols (neutral lipid analogues of PAF) can also be produced in addition to PAF (Kawasaki and Snyder, 1988). Formation of the acetylated neutral lipids from 1-alkyl-2-acetyl-*sn*-glycerols is catalysed by an acyl-CoA acyltransferase with properties different than acyltransferases that utilize long-chain diacylglycerols as substrates. Among a variety of acyl-CoAs ($C_{8:0}-C_{20:4}$) tested as substrates in the acylation of 1-hexadecyl-2-acetyl-*sn*-glycerol by the acyltransferase, linoleoyl-CoA was preferred. The physiological function of the alkylacetylacylglycerols is not established but it has been proposed they could represent a precursor reservoir pool of the alkylacetylglycerols that can be converted to PAF and other biologically active products (Kawasaki and Snyder, 1988).

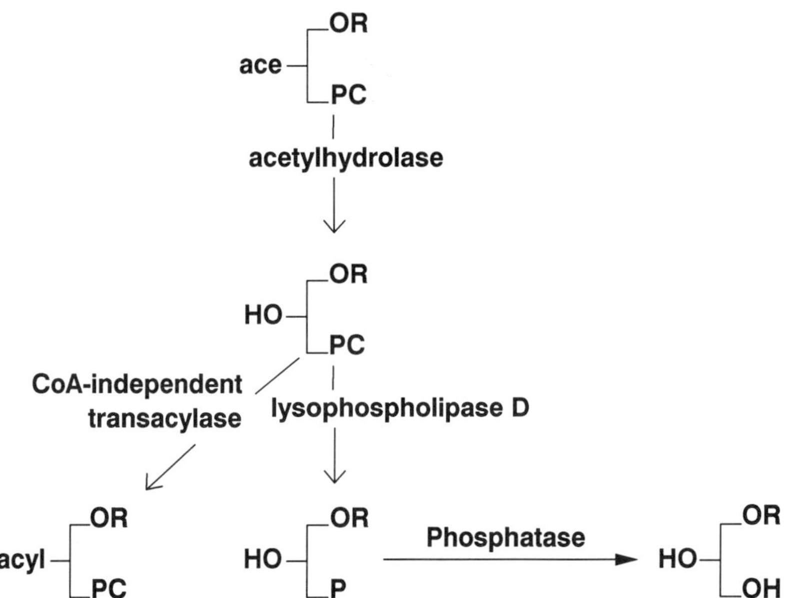

Figure 6.5 Catabolism of PAF and lyso-PAF: involvement of acetylhydrolase, a lysophospholipase D, and a CoA-independent transacylase. The alkyl moiety of the catabolic products (lyso-PAF, 1-alkyl-2-lyso-*sn*-glycero-3-phosphate, and the alkylglycerol) can be cleaved by a Pte-H$_4$-dependent *O*-alkyl monooxygenase (see Fig. 6.6).

3. Catabolism of Platelet-Activating Factor and Related Metabolites

3.1 PLATELET-ACTIVATING FACTOR ACETYLHYDROLASE

PAF is inactivated by an acetylhydrolase (Fig. 6.5), an enzyme that catalyses the hydrolysis of the acetate moiety at the *sn*-2 position (Blank *et al.*, 1981, 1983; Farr *et al.*, 1983; Malone *et al.*, 1985; Stafforini *et al.*, 1987a,b, 1989, 1991). The hydrolytic activity appears to be distributed ubiquitously in mammalian cells and blood (Blank *et al.*, 1981), and has been detected in a wide variety of animal species (Cabot *et al.*, 1984). Acetylhydrolase has distinctively different properties than phospholipase A$_2$ since it is Ca^{2+}-independent and utilizes only relatively short-chain esterified aliphatic chains at the *sn*-2 position of phospholipids as substrates. The enzyme exhibits no selectivity for the groups at the other two positions of the glycerol moiety since both ether- and ester-linked substituents at the *sn*-1 position and either choline or ethanolamine head groups at the *sn*-3 position serve as substrates. The lyso-PAF product can be further metabolized by a Pte-H$_4$-dependent alkyl monooxygenase (Section 3.3), a lysophospholipase D (Section 3.4) or a CoA-independent transacylase (Section 3.5).

Acetylhydrolase exists in both an intracellular (Blank *et al.*, 1981, 1983; Stafforini *et al.*, 1991) and an extracellular (Blank *et al.*, 1983; Farr *et al.*, 1983; Stafforini *et al.*, 1987a,b, 1989, 1991) form. Except for the resistance of the plasma acetylhydrolase to proteases and different molecular weights, the two forms of the acetylhydrolases have identical catalytic properties (Blank *et al.*, 1983; Stafforini *et al.*, 1991). However, only the plasma acetylhydrolase has been purified to near-homogeneity (Stafforini *et al.*, 1987b). The purified enzyme has a molecular mass of about 43 kDa and exhibits a high affinity for both LDL and HDL (Stafforini *et al.*, 1987a, 1989). The kinetic properties of the enzyme associated with either lipoprotein fraction are identical, except that the acetylhydrolase in the LDL fraction is much more active. The origin of acetylhydrolase in plasma is unknown, but it has been shown that macrophages can synthesize and secrete an enzyme having the same properties as the plasma acetylhydrolase (Elstad *et al.*, 1989).

PAF acetylhydrolase also utilizes phospholipid substrates containing oxidized fatty acid fragments at the *sn*-2 position (Steinbrecher and Pritchard, 1989; Stremler *et al.*, 1989, 1991). This finding is of considerable interest since oxidative fragmentation of phospholipids appears to be a critical step in the conversion of LDL to a modified particle that is taken up by macrophages to produce foam cells in the development of atherosclerotic lesions (Brown and Goldstein, 1983). Of additional interest is that the phospholipids with the short-chain oxidized acyl moiety can also activate the PAF receptor in human neutrophils (Smiley *et al.*, 1991).

3.2 1-ALKYL-2-ACETYL-*SN*-GLYCEROL ACETYLHYDROLASE

As mentioned in Section 2.2, on the biosynthesis of PAF via the *de novo* pathway, alkylacetylglycerols are also

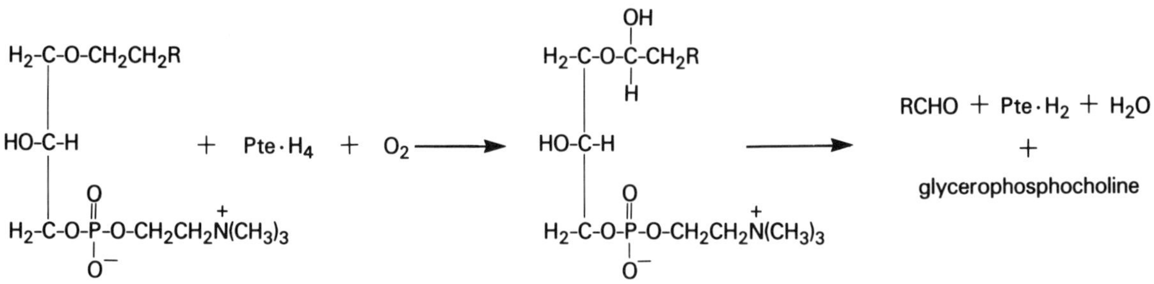

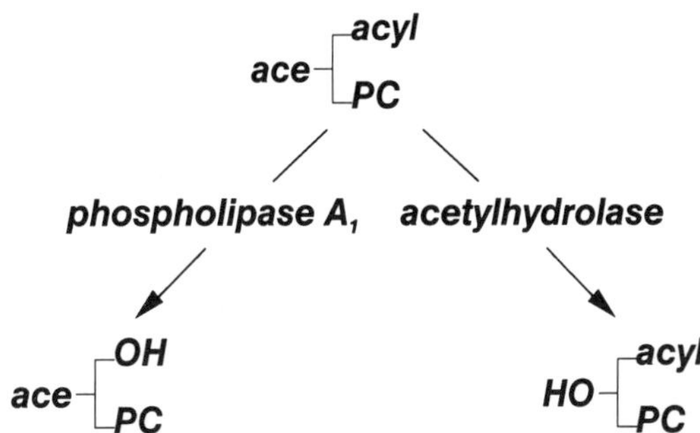

Figure 6.6 Cleavage of the *O*-alkyl moiety of lyso-PAF by the Pte-H$_4$-dependent alkyl cleavage enzyme (an *O*-alkyl monooxygenase).

formed. The enzyme responsible for the catabolism or inactivation of alkylacetylglycerols is an acetylhydrolase that catalyses the hydrolysis of the *sn*-2 acetate group (Blank *et al.*, 1990). This enzyme has completely different properties from PAF acetylhydrolase since it is inhibited by sodium fluoride, exhibits a different substrate specificity, and is membrane bound. Furthermore, alkylacetylglycerol acetylhydrolase does not appear to be a typical diacylglycerol lipase or a non-specific esterase. Interestingly, 1-alkyl-2-acetyl-*sn*-glycero-3-phosphate is not a substrate for this neutral lipase, which further documents the high degree of substrate specificity of this enzyme. It appears that the alkylacetylglycerol acetylhydrolase could be an important regulatory enzyme in controlling the levels of both alkylacetylglycerols and PAF produced by the *de novo* route.

3.3 TETRAHYDROPTERIDINE-DEPENDENT ALKYL CLEAVAGE ENZYME

Cleavage of the alkyl linkage of lyso-PAF (Fig. 6.6) is catalysed by a Pte-H$_4$ *O*-alkyl monooxygenase (Lee *et al.*, 1981). This enzyme activity exhibits identical properties towards lyso-PAF (Lee *et al.*, 1981) as alkylglycerols or other glycerolipids that possess at least one free hydroxyl group (Snyder *et al.*, 1973). PAF is not a substrate for the alkyl cleavage enzyme *per se* (Lee *et al.*, 1981).

3.4 PHOSPHOLIPASES

Phospholipases can also play a role in the metabolism of PAF and its metabolites, since it is known that phospholipase A$_2$ from snake venom is capable of hydrolysing the acetate group of PAF (Renooij and Snyder, 1981) and a phospholipase C inactivation mechanism for PAF has been reported (Nishihira and Ishibashi, 1986; Okayasu *et al.*, 1986). However, the significance of phospholipase C in degrading PAF is questionable since the rate of hydrolysis is extremely low relative to the hydrolytic rates with PAF acetylhydrolase. Lysophospholipase D, an enzyme that selectively hydrolyses the choline or ethanolamine base groups of lyso ether-linked phospholipids (Wykle and Schremmer, 1974; Wykle *et al.*, 1977,

Figure 6.7 Catabolism of the acyl analogue of PAF by a phospholipase A$_1$ or acetylhydrolase.

1980b), has also been investigated with regard to the metabolism of PAF and lyso-PAF (Kawasaki and Snyder, 1987; see Fig. 6.5). Whereas, the importance of lysophospholipase D in PAF metabolism is not fully understood, it could effectively remove the lyso-PAF intermediate from the "PAF cycle" and thereby influence PAF levels.

Recent results reported by Triggiani *et al.* (1991b) for human neutrophils indicate the acyl analogue of PAF is catabolized primarily by an unusual pathway involving a cytosolic phospholipase A$_1$ to produce 1-lyso-2-acetyl-*sn*-glycero-3-phosphocholine and fatty acids (Fig. 6.7). The acetate can then apparently be hydrolysed by an acetylhydrolase in intact neutrophils. Neither the properties of the phospholipase A$_1$ or the acetylhydrolase activities described in this work have been well characterized.

Other studies with suspensions of either human platelets or human leucocytes enriched in neutrophils suggested that the degree of deacylation occurring with the acyl analogues of either lyso-PAF or PAF determines the relative amounts of PAF and the acyl-PAF analogue formed (Sturk *et al.*, 1989; van den Bosch *et al.*, 1991). In fact, the possible importance of lysophospholipases in the catabolism of the acyl analogue of PAF has been further emphasized by results that show purified lysophospholipases I and II from bovine liver can hydrolyse the *sn*-1 acyl moiety of acyl-PAF to form 2-acetyl-*sn*-glycero-3-phosphocholine (Aarsman *et al.*, 1991). Interestingly, the lysophospholipase II from bovine liver can also deacetylate PAF to lyso-PAF; nevertheless, purified lysophospholipases from the cytosol of rat kidney and human platelets do not possess any intrinsic PAF acetylhydrolase activity.

Still further support for the potential role that lysophospholipase plays in regulating the production of acyl-PAF comes from experiments done with intact cells and lysates of alveolar and peritoneal macrophages from guinea-pigs (Nakagawa *et al.*, 1992). These workers found elevated lysophospholipase activity greatly reduced the levels of lysophosphatidylcholine available as a substrate for the acetyl-CoA acetyltransferase in the synthesis of acyl-PAF, thus causing a preferential synthesis of PAF instead of its acyl analogue. However, it should be noted that the lysophospholipases have no direct effect on PAF *per se* since they are unable to hydrolyse any aliphatic moieties in ether linkage with glycerol.

3.5 TRANSACYLASE

As mentioned in Section 2.1 on the remodelling pathway of PAF biosynthesis, a CoA-independent transacylase can play a key role in generating lyso-PAF. Similarly, after inactivation of newly formed PAF by acetylhydrolase, the lyso-PAF intermediate is re-acylated to produce alkylacyl-glycerophosphocholines by a CoA-independent transacylase (see Fig. 6.5). This transacylase activity exhibits a high degree of specificity for arachidonate and other polyunsaturates (Kramer *et al.*, 1984; Robinson, *et al.*, 1985; Sugiura and Waku, 1985; Sugiura *et al.*, 1985, 1987; Lee *et al.*, 1991; Uemura *et al.*, 1991; Winkler *et al.*, 1991). It is unknown at the present time whether the transacylase responsible for catalysing the back re-acylation reaction of lyso-PAF in the PAF cycle is the same transacylase that forms lyso-PAF in the forward reaction during PAF biosynthesis.

4. *Regulation of Platelet-Activating Factor Metabolism*

The regulatory mechanisms involved in PAF metabolism are complex and still poorly understood at the cellular level. Calcium would appear to have an important regulatory role, since it is required for PAF synthesis in the remodelling pathway (Lee *et al.*, 1984; Ludwig *et al.*, 1984) but inhibits all three enzymes in the *de novo* route (Renooij and Snyder, 1981; Lee *et al.*, 1986, 1988; Woodard *et al.*, 1987). Obviously, arachidonic acid is a major factor in PAF production via the remodelling mechanism, but it is not released or known to affect the acetyltransferase, phosphohydrolase, or DTT-insensitive cholinephosphotransferase in the *de novo* pathway of PAF biosynthesis. Arachidonate-depleted cells lack the capacity to produce significant quantities of PAF following agonist stimulation but regain their ability to respond after the cells are supplemented with arachidonic acid (Ramesha and Pickett, 1986; Suga *et al.*, 1990). This effect of

arachidonic acid is thought to be due to the fact that alkylarachidonoylglycerophosphocholines are selectively utilized as precursors of PAF in the remodelling route. Thus, arachidonate would appear to be a regulator of PAF biosynthesis at the substrate level in the remodelling pathway.

Enzyme activation by a protein kinase appears to be important in regulating the activities of lyso-PAF acetyltransferase in the biosynthesis of PAF. The phosphorylated form of lyso-PAF acetyltransferase represents the activated state of this enzyme and it can be inactivated by a phosphatase (Lenihan and Lee, 1984; Gomez-Cambronero *et al.*, 1985, 1987; Domenech *et al.*, 1987; Ninio *et al.*, 1987; Nieto *et al.*, 1988). However, the exact mechanisms and the protein kinases responsible have not been fully elucidated. Both a cAMP-dependent protein kinase (Gomez-Cambronero *et al.*, 1985, 1987; Nieto *et al.*, 1988) and a Ca^{2+}-calmodulin-dependent kinase (Domenech *et al.*, 1987) have been reported to be able to phosphorylate the acetyltransferase in studies of membrane preparations from rat spleens and guinea-pig parotid glands, respectively. Indirect evidence has also suggested that the phospholipase A_2 step in the remodelling pathway requires activation by a protein kinase C (McIntyre *et al.*, 1987).

Studies of the *de novo* pathway of PAF biosynthesis have shown the activation of cytidylyltransferase by fatty acids can greatly stimulate the production of PAF (Blank *et al.*, 1988; Lee *et al.*, 1990; Vallari *et al.*, 1990). The cytidylyltransferase is activated through translocation from the cytosol to membranes, where it catalyses the formation of CDP-choline (Pelech and Vance, 1984). The availability of CDP-choline is rate-limiting in the *de novo* synthesis of PAF (Lee *et al.*, 1990), as it is in the *de novo* synthesis of phosphatidylcholine (Pelech and Vance, 1984). Results obtained on the stimulation of PAF synthesis via the *de novo* route in human endothelial cells by phorbol 12-myristate-13-acetate have suggested that the increased activities of acetyl-CoA:1-alkyl-2-lyso-*sn*-glycero-3-phosphate acetyltransferase and the DTT-insensitive cholinephosphotransferase are due to protein kinase C activation (Heller *et al.*, 1991). Thus, any factor that activates protein kinase C could also stimulate the *de novo* synthesis of PAF by phosphorylating either the acetyltransferase or cholinephosphotransferase in the *de novo* route. Factors such as neurotransmitters also stimulate the *de novo* synthesis of PAF (Bussolino *et al.*, 1986, 1988); however, their exact mode of action is uncertain, except that they are known to stimulate the DTT-insensitive cholinephosphotransferase activity.

In addition, PAF acetylhydrolase, which inactivates PAF by catalysing the hydrolysis of the *sn*-2 acetate moiety, appears to have an important role in the regulation of PAF levels (Snyder, 1990); furthermore, the activity is influenced by hormones. Although the mechanisms involved are still obscure, experiments by Miyaura and coworkers (1991) have shown that the administration of oestrogen (17α-ethynyloestradiol) causes a significant decrease in the acetylhydrolase activity, whereas steroid treatment (dexamethasone) results in a striking increase in the plasma PAF acetylhydrolase activity. On the other hand, testosterone and progesterone were shown to exert no effect on the plasma levels of acetylhydrolase (Miyaura *et al.*, 1991).

Changes in the activity of PAF acetylhydrolase also occur under a variety of pathophysiological conditions. Increased plasma acetylhydrolase activities have been observed during platelet aggregation (Suzuki *et al.*, 1988), responses to stress (Lenihan *et al.*, 1985), ischaemic cerebrovascular disease (Satoh *et al.*, 1988), periods shortly after birth (Maki *et al.*, 1988), insulin-dependent diabetes mellitus (Hofmann *et al.*, 1989), hypertension (Blank *et al.*, 1983; Satoh *et al.*, 1989), and cellular differentiation of human macrophages (Stafforini *et al.*, 1990). In contrast, decreases in the PAF acetylhydrolase activity in plasma are known to occur in mothers during the latter half of pregnancy (Maki *et al.*, 1988; Johnston, 1989) and in children with asthma (Miwa *et al.*, 1988). It is thought the activity of PAF acetylhydrolase is inversely related to the actual levels of PAF in plasma (Snyder, 1990).

5. Acknowledgements

This work was supported by the Office of Energy Research, US Department of Energy (Contract No. DE-AC05-760R00033), The American Cancer Society (Grant BE-26X), the National Heart, Lung, and Blood Institute (Grants 35495-07 and 27109-12), and the National Institute of Diabetes and Digestive and Kidney Diseases (Grant R01 DK42804-03).

6. References

Aarsman, A.J., Neys, F.W. and van den Bosch, H. (1991). Catabolism of platelet-activating factor and its acyl analog. Differentiation of the activities of lysophospholipase and platelet-activating factor acetylhydrolase. J. Biochem. 200, 187–193.

Bass, D.A., McPhail, L.C., Schmitt, J.D., Morris-Natschke, S., McCall, C.E. and Wykle, R.L. (1989). Selective priming of rate and duration of the respiratory burst of neutrophils by 1,2-diacyl and 1-O-alkyl-2-acyl diglycerides. Possible relation to effects on protein kinase C. J. Biol. Chem. 263, 19610–19617.

Blank, M.L., Lee, T-c., Fitzgerald, V. and Snyder, F. (1981). A specific acetylhydrolase for 1-alkyl-2-acetyl-*sn*-glycero-3-phosphocholine (a hypotensive platelet-activating lipid). J. Biol. Chem. 256, 175–178.

Blank, M.L., Cress, E.A., Lee, T-c., Malone, B., Surles, J.R., Piantadosi, C., Hajdu, J. and Snyder, F. (1982). Structural features of platelet activating factor (1-alkyl-2-acetyl-*sn*-glycero-3-phosphocholine) required for hypotensive and platelet serotonin responses. Res. Commun. Chem. Pathol. Pharmacol. 38, 3–20.

Blank, M.L., Hall, N., Cress, E.A. and Snyder, F. (1983). Inactivation of 1-alkyl-2-acetyl-*sn*-glycero-3-phosphocholine by a plasma acetylhydrolase: higher activities in hypertensive rats. Biochem. Biophys. Res. Commun. 113, 666–671.

Blank, M.L., Cress, E.A. and Snyder, F. (1984). A new class of antihypertensive neutral lipids: 1-alkyl-2-acetyl-*sn*-glycerols. Biochem. Biophys. Res. Commun. 118, 344–350.

Blank, M.L., Lee, Y.J., Cress, E.A. and Snyder, F. (1988). Stimulation of the *de novo* pathway for the biosynthesis of platelet activating factor (PAF) via cytidylyltransferase activation in cells with minimal endogenous PAF production. J. Biol. Chem. 263, 5656–5661.

Blank, M.L., Smith, Z.L., Cress, E.A. and Snyder, F. (1990). Characterization of the enzymatic hydrolysis of acetate from alkylacetylglycerols in the *de novo* pathway of PAF biosynthesis. Biochim. Biophys. Acta 1042, 153–158.

Brown, M.S. and Goldstein, J.L. (1983). Lipoprotein metabolism in the macrophage: implications for cholesterol deposition in atherosclerosis. Annu. Rev. Biochem. 52, 223–261.

Bussolino, F., Camussi, G. and Arese, P. (1984). Platelet-activating factor phosphatidate, but not platelet-activating factor, is a powerful calcium ionophore in the human red cell. Cell Calcium 5, 463–473.

Bussolino, F., Gremo, F., Tetta, C., Pescarmona, G. and Camussi, G. (1986). Production of platelet-activating factor by chick retina. J. Biol. Chem. 261, 16502–16506.

Bussolino, F., Pescarmona, G., Camussi, G. and Gremo, F. (1988). Acetylcholine and dopamine promote the production of platelet activating factor in immature cells of chick embryonic retina. J. Neurochem. 51, 1755–1759.

Cabot, M.C., Faulkner, L.A., Lackey, R.J. and Snyder, F. (1984). Vertebrate class distribution of 1-alkyl-2-acetyl-*sn*-glycero-3-phosphocholine acetylhydrolase in serum. Comp. Biochem. Physiol. 78B, 37–40.

Chilton, F.H., O'Flaherty, J.T., Ellis, J.M., Swendsen, C.L. and Wykle, R.L. (1983). Selective acylation of lyso platelet activating factor by arachidonate in human neutrophils. J. Biol. Chem. 258, 7268–7271.

Chilton, F.H., Ellis, J.M., Olson, S.C. and Wykle, R.L. (1984). 1-O-Alkyl-2-arachidonoyl-*sn*-glycero-3-phosphocholine. A common source of platelet-activating factor and arachidonate in human polymorphonuclear leukocytes. J. Biol. Chem. 259, 12014–12019.

Domenech, C.E., Machado-DeDomenech, E. and Soling, H.D. (1987). Regulation of acetyl-CoA: 1-alkyl-*sn*-glycero-3-phosphocholine O-acetyltransferase (lyso-PAF-acetyltransferase) in exocrine glands. Evidence for an activation via phosphorylation by calcium/calmodulin-dependent protein kinase. J. Biol. Chem. 262, 5671–5676.

Elstad, M.R., Stafforini, D.M., McIntyre, T.M., Prescott, S.M. and Zimmerman, G.A. (1989). Platelet-activating factor acetylhydrolase increases during macrophage differentiation. A novel mechanism that regulates accumulation of platelet-activating factor. J. Biol. Chem. 264, 8467–8470.

Farr, R.S., Wardlow, M.L., Cox, C.P., Meng, K.E. and Greene, D.E. (1983). Human serum acid-labile factor is an acylhydrolase that inactivates platelet-activating factor. Fed. Proc. 42, 3120–3122.

Garcia, M.C., Mueller, H.W. and Rosenthal, M.D. (1991). C_{20} polyunsaturated fatty acids and phorbol myristate acetate enhance agonist-stimulated synthesis of 1-radyl-2-acetyl-*sn*-glycero-3-phosphocholine in vascular endothelial cells. Biochim. Biophys. Acta 1083, 37–45.

Gomez-Cambronero, J., Velasco, S., Mato, J.M. and Sanchez-Crespo, M. (1985). Modulation of lyso-platelet activating factor:acetyl-CoA acetyltransferase from rat splenic microsomes. The role of cyclic AMP-dependent protein kinase. Biochim. Biophys. Acta 845, 516–519.

Gomez-Cambronero, J., Mato, J.M., Vivanco, F. and Sanchez-Crespo, M. (1987). Phosphorylation of partially purified 1-O-alkyl-2-lyso-*sn*-glycero-3-phosphocholine:acetyl-CoA acetyltransferase from rat spleen. Biochem. J. 245, 893–898.

Heller, R., Bussolino, F., Ghigo, D., Garbarino, G., Pescarmona, G., Till, U. and Bosia, A. (1991). Stimulation of platelet-activating factor synthesis in human endothelial cells by activation of the de novo pathway. Phorbol 12-myristate 13-acetate activates 1-alkyl-2-lyso-*sn*-glycero-3-phosphate:acetyl-CoA acetyltransferase and dithiothreitol-insensitive 1-alkyl-2-acetyl-*sn*-glycerol:CDP-choline cholinephosphotransferase. J. Biol. Chem; 226, 21358–21361.

Hofmann, B., Ruhling, K., Spangenberg, P. and Ostermann, G. (1989). Enhanced degradation of platelet-activating factor in serum from diabetic patients. Haemostasis 19, 180–184.

Johnston, J.M. (1989). In "Platelet Activating Factor and Diseases" (eds K. Saito and D.J. Hanahan), pp 129–151. International Medical Publishers, Tokyo.

Kawasaki, T. and Snyder, F. (1987). The metabolism of lyso-platelet-activating factor (1-O-alkyl-2-lyso-*sn*-glycero-3-phosphocholine) by a calcium-dependent lysophospholipase D in rabbit kidney medulla. Biochim. Biophys. Acta 920, 85–93.

Kawasaki, T. and Snyder, F. (1988). Synthesis of a novel acetylated neutral lipid related to platelet activating factor by acyl-CoA:1-O-alkyl-2-acetyl-*sn*-glycerol acyltransferase in HL-60 cells. J. Biol. Chem. 263, 2593–2596.

Kramer, R.M., Patton, G.M., Pritzker, C.R. and Deykin, D. (1984). Metabolism of platelet-activating factor in human platelets. Transacylase-mediated synthesis of 1-O-alkyl-2-arachidonoyl-*sn*-glycero-3-phosphocholine. J. Biol. Chem. 259, 13316–13320.

Lee, T-c. (1985). Biosynthesis of platelet activating factor. Substrate specificity of 1-alkyl-2-lyso-*sn*-glycero-3-phosphocholine:acetyl-CoA acetyltransferase in rat spleen microsomes. J. Biol. Chem. 260, 10952–10955.

Lee, T-c. and Snyder, F. (1985). In "Phospholipids and Cellular Regulation" (ed J.F. Kuo), pp 1–39. CRC Press, Boca Raton.

Lee, T-c. and Snyder, F. (1989). In "Overview of PAF Biosynthesis and Catabolism in Platelet Activating Factor and Human Disease" (eds P.S. Barnes, C.P. Page, and P.M. Henson), pp 1–22. Blackwell, Oxford.

Lee, T-c., Blank, M.L., Fitzgerald, V. and Snyder, F. (1981). Substrate specificity in the biocleavage of the O-alkyl bond: 1-alkyl-2-acetyl-*sn*-glycero-3-phosphocholine (a hypotensive and platelet-activating lipid) and its metabolites. Arch. Biochem. Biophys. 208, 353–357.

Lee, T-c., Lenihan, D.J., Malone, B., Roddy, L.L. and Wasserman, S.I. (1984). Increased biosynthesis of platelet-activating factor in activated human eosinophils. J. Biol. Chem. 259, 5526–5530.

Lee, T-c., Malone, B. and Snyder, F. (1986). A new de novo pathway for the formation of 1-alkyl-2-acetyl-*sn*-glycerols, precursors of platelet activating factor. Biochemical characterization of 1-alkyl-2-lyso-*sn*-glycero-3-P:acetyl-CoA acetyltransferase in rat spleen. J. Biol. Chem. 261, 5373–5377.

Lee, T-c., Malone, B. and Snyder, F. (1988). Formation of 1-alkyl-2-acetyl-*sn*-glycerols via *de novo* biosynthetic pathway for platelet activating factor. Characterization of 1-alkyl-2-acetyl-*sn*-glycero-3-phosphate phosphohydrolase in rat spleens. J. Biol. Chem. 263, 1755–1760.

Lee, T-c., Malone, B. and Snyder, F. (1989a). In "Proc. of the Taipei Satellite Symp. of Platelet Activating Factor" (ed N. Hicks), pp 1–4. Excerpta Medica, Hong Kong.

Lee, T-c., Malone, B., Woodard, D. and Snyder, F. (1989b). Renal necrosis and the involvement of a single enzyme of the de novo pathway for the biosynthesis of platelet-activating factor in rat kidney inner medulla. Biochem. Biophys. Res. Commun. 163, 1002–1005.

Lee, T-c., Malone, B., Blank, M.L., Fitzgerald, V. and Snyder, F. (1990). Regulation of the synthesis of platelet activating factor and its inactive storage precursor (1-alkyl-2-acyl-*sn*-glycero-3-phosphocholine) from 1-alkyl-2-acetyl-*sn*-glycerol by rabbit platelets. J. Biol. Chem. 265, 9181–9187.

Lee, T-c., Blank, M.L., Fitzgerald, V. and Snyder, F. (1991). Acylation of alkyllysophospholipids by Fischer sarcoma microsomes. Arch. Biochem. Biophys. 288, 600–608.

Lee, T-c., Uemura, Y. and Snyder, F. (1992). A novel CoA-independent transacetylase produces the ethanolamine plasmalogen and acyl analogs of PAF with PAF as the acetate donor in HL-60 cells. J. Biol. Chem. 267, 19992–20001.

Lenihan, D.J. and Lee, T-c. (1984). Modulation of 1-alkyl-2-lyso-*sn*-glycero-3-phosphocholine:acetyl-CoA acetyltransferase by phosphorylation and dephosphorylation in rat spleen microsomes Biochem. Biophys. Res. Commun. 120, 834–839.

Lenihan, D.J., Greenberg, N. and Lee, T-c. (1985). Involvement of platelet activating factor in physiological stress in the lizard, Anolis carolinensis. Comp. Biochem. Physiol. 81C, 81–86.

Ludwig, J.C., McManus, L.M., Clar, P.O., Hanahan, D.J. and Pinckard, R.N. (1984). Modulation of platelet-activating factor (PAF) synthesis and release from human polymorphonuclear leukocytes (PMN): role of extracellular Ca^{2+}. Arch. Biochem. Biophys. 232, 102–110.

Maki, N., Hoffman, D.R. and Johnston, J.M. (1988). Platelet-activating factor acetylhydrolase activity in maternal, fetal, and newborn rabbit plasma during pregnancy and lactation. Proc. Natl Acad. Sci. USA 85, 728–732.

Malone, B., Lee, T.-c. and Snyder, F. (1985). Inactivation of platelet activating factor (PAF) by rabbit platelets: Lyso-PAF as a key intermediate with phosphatidylcholine as the source of arachidonic acid in its conversion to a tetraeonoic acylated product. J. Biol. Chem. 260, 1531–1534.

McIntyre, T.M., Reinhold, S.L., Prescott, S.M. and Zimmerman, G.A. (1987). Protein kinase C activity appears to be required for the synthesis of platelet-activating factor and leukotriene B_4 by human neutrophils. J. Biol. Chem. 263, 15370–15376.

McNamara, M.J.C., Schmitt, J.D., Wykle, R.L. and Daniel, L.W. (1984). 1-Hexadecyl-2-acetyl-*sn*-glycerol stimulates differentiation of HL-60 human promyelocytic leukemia cells to macrophage-like cells. Biochem. Biophys. Res. Commun. 122, 824–830.

Miwa, M., Miyake, T., Yamanaka, T., Sugatani, J., Suzuki, Y., Sakata, S., Araki, Y. and Matsumoto, M. (1988). Characterization of serum platelet-activating factor (PAF) acetylhydrolase: correlation between deficiency of serum PAF acetylhydrolase and respiratory symptoms in asthmatic children. J. Clin. Invest. 82, 1983–1991.

Miyaura, S., Maki, N., Byrd, W. and Johnston, J.M. (1991). The hormonal regulation of platelet-activating factor acetylhydrolase activity in plasma. Lipids 26, 1015–1020.

Mueller, H.W., O'Flaherty, J.T. and Wykle, R.L. (1984). The molecular species distribution of platelet activating factor synthesized by rabbit and human neutrophils. J. Biol. Chem. 259, 14554–14559.

Mueller, H.W., Nollert, N.U. and Eskin, S.G. (1991). Synthesis of 1-acyl-2-[^{3}H]acetyl-*sn*-glycero-3-phosphocholine, a structural analog of platelet activating factor, by vascular endothelial cells. Biochem. Biophys. Res. Commun. 176, 1557–1564.

Nakagawa, Y., Sugai, M., Karasawa, K., Tokumura, A., Tsukatani, H., Setaka, M. and Nojima, S. (1992). Possible influence of lysophospholipase on the production of 1-acyl-2-acetylglycerophosphocholine in macrophages. Biochim. Biophys. Acta 1126, 277–285.

Nakayama, R., Yasuda, K., Satouchi, K. and Saito, K. (1988). 1-*O*-Hexadec-1'-enyl-2-acetyl-*sn*-glycero-3-phosphocholine and its biological activity. Biochem. Biophys. Res. Commun. 151, 1256–1261.

Nieto, M.L., Velasco, S. and Sanchez-Crespo, M. (1988). Modulation of acetyl-CoA:1-alkyl-2-lyso-*sn*-glycero-3-phosphocholine (lyso-PAF) acetyltransferase in human polymorphonuclears. The role of cyclic AMP-dependent and phospholipid sensitive, calcium-dependent protein kinases. J. Biol. Chem. 263, 4607–4611.

Ninio, E., Mencia-Huerta, J.M. and Benveniste, J. (1983). Biosynthesis of platelet-activating factor (PAF-acether) V. Enhancement of acetyltransferase activity in murine peritoneal cells by calcium ionophore A23187. Biochim. Biophys. Acta 751, 298–304.

Ninio, E., Joly, F., Heiblot, C., Bessou, G., Mencia-Huerta, J.M. and Benveniste, J. (1987). Biosynthesis of PAF-acether. IX. Role for a phosphorylation-dependent activation of acetyltransferase in antigen-stimulated mouse mast cells. J. Immunol. 139, 154–160.

Nishihira, J. and Ishibashi, T. (1986). A phospholipase C with a high specificity for platelet-activating factor in rabbit liver light mitochondria. Lipids 21, 780–785.

O'Flaherty, J.T., Salzer, W.L., Cousart, S., McCall, C.E., Piantadosi, C., Surles, J.R., Hammett, M.J. and Wykle, R.L. (1983). Platelet-activating factor and analogues: comparative studies with human neutrophils and rabbit platelets. Res. Commun. Chem. Pathol. Pharmacol. 39, 291–309.

Okayasu, T., Hoshii, K., Seyama, K., Ishibashi, T. and Iami, Y. (1986). Metabolism of platelet-activating factor in primary cultured adult rat hepatocytes by a new pathway involving phospholipase C and alkyl monooxygenase. Biochim. Biophys. Acta 876, 58–64.

Pelech, S.L. and Vance, D.E. (1984). Regulation of phosphatidylcholine biosynthesis. Biochim. Biophys. Acta 779, 217–251.

Prescott, S.M., Zimmerman, G.A. and McIntyre, T.M. (1990). Platelet-activating factor. J. Biol. Chem. 265, 17381–17384.

Pritchard, P.H. (1987). The degradation of platelet-activating factor by high-density lipoproteins in rat plasma. Effect of ethynyloestradiol administration. Biochem. J. 246, 791–794.

Ramesha, C.S. and Pickett, W.C. (1986). Platelet-activating factor and leukotriene biosynthesis is inhibited in polymorphonuclear leukocytes depleted of arachidonic acid. J. Biol. Chem. 261, 7592–7596.

Renooij, W. and Snyder, F. (1981). Biosynthesis of 1-alkyl-2-acetyl-*sn*-glycero-3-phosphocholine (platelet activating factor and a hypotensive lipid) by cholinephosphotransferase in various rat tissues. Biochim. Biophys. Acta 663, 545–556.

Robinson, M., Blank, M.L. and Snyder, F. (1985). Acylation of lysophospholipids by rabbit alveolar macrophages. Specificities of CoA-dependent and CoA-independent reactions. J. Biol. Chem. 260, 7889–7895.

Satoh, K., Imaizumi, T., Kawamura, Y., Yoshida, H., Takamatsu, S., Mizuno, S., Shoji, B. and Takamatsu, M. (1988). Activity of platelet-activating factor (PAF) acetylhydrolase in plasma from patients with ischemic cerbravascular disease. Prostaglandins 35, 685–689.

Satoh, K., Imaizumi, T., Kawamura, Y., Yoshida, H., Takamatsu, S. and Takamatsu, M. (1989). Increased activity of the platelet-activating factor acetylhydrolase in plasma low density lipoprotein from patients with essential hypertension. Prostaglandins 37, 673–82.

Satouchi, K., Pinckard, R.N. and Hanahan, D.J. (1981). Influence of alkyl ether chain length of acetyl glyceryl ether phosphorylcholine and its ethanolamine analog on biological activity toward rabbit platelets. Arch. Biochem. Biophys. 211, 683–688.

Satouchi, K., Oda, M., Saito, K., and Hanahan, D.J. (1984). Metabolism of 1-*O*-alkyl-2-acetyl-*sn*-glycerol by washed rabbit platelets: formation of platelet activating factor. Arch. Biochem. Biophys. 235, 318–321.

Satouchi, K., Oda, M., Yasunaga, K. and Saito, K. (1985). Evidence for production of 1-acyl-2-acetyl-*sn*-glycerol-3-phosphorylcholine concomitantly with platelet-activating factor. Biochem. Biophys. Res. Commun. 128, 1409–1417.

Satouchi, K., Oda, M. and Saito, K. (1987). 1-Acyl-2-acetyl-*sn*-glycero-3-phosphocholine from stimulated human polymorphonuclear leukocytes. Lipids 22, 285–287.

Smiley, P.L., Stremler, K.E., Prescott, S.M., Zimmerman, G.A. and McIntyre, T.M. (1991). Oxidatively fragmented phosphatidylcholines activate human neutrophils through the receptor for platelet-activating factor. J. Biol. Chem. 266, 11104–11110.

Snyder, F. (1989). Biochemistry of platelet-activating factor: a unique class of biologically active phospholipids. Proc. Soc. Exp. Biol. Med. 190, 125–135.

Snyder, F. (1990). Platelet activating factor and related acetylated lipids as potent biologically active cellular mediators. Am. J. Physiol. 259, C697–C708.

Snyder, F., Malone, B. and Piantadosi, C. (1973). Tetrahydropteridine-dependent cleavage enzyme for *O*-alkyl lipids: substrate specificity. Biochim. Biophys. Acta 316, 259–265.

Snyder, F., Lee, T-c. and Malone, B. (1992). Acetyl-COA 1-Alkyl-2-lyso-*sn*-glycero-3-phosphate acetyltransferase. Methods Enzymol. 209, 407–412.

Stafforini, D.M., McIntyre, T.M., Carter, M.E. and Prescott, S.M. (1987a). Human plasma platelet-activating factor acetylhydrolase. Association with lipoprotein particles and role in the degradation of platelet-activating factor. J. Biol. Chem. 262, 4215–4222.

Stafforini, D.M., Prescott, S.M. and McIntyre, T.M. (1987b). Human plasma platelet-activating factor acetylhydrolase. Purification and properties. J. Biol. Chem. 262, 4223–4230.

Stafforini, D.M., Carter, M.E., Zimmerman, G.A., McIntyre, T.M. and Prescott, S.M. (1989). Lipoproteins alter the catalytic behavior of the platelet-activating factor acetylhydrolase in human plasma. Proc. Natl Acad. Sci. USA 86, 2393–2397.

Stafforini, D.M., Elstad, M.R., McIntyre, T.M., Zimmerman, G.A. and Prescott, S.M. (1990). Human macrophages secrete platelet-activating factor acetylhydrolase. J. Biol. Chem. 265, 9682–9687.

Stafforini, D.M., Prescott, S.M., Zimmerman, G.A. and McIntyre, T. M. (1991). Platelet-activating factor acetylhydrolase activity in human tissues and blood cells. Lipids 26, 979–985.

Steinbrecher, U.P. and Pritchard, P.H. (1989). Hydrolysis of phosphatidylcholine during LDL oxidation is mediated by platelet-activating factor acetylhydrolase. J. Lipid Res. 30, 305–315.

Stremler, K.E., Stafforini, D.M., Prescott, S.M., Zimmerman, G.A. and McIntyre, T.M. (1989). An oxidized derivative of phosphatidylcholine is a substrate for the platelet-activating factor acetylhydrolase from human plasma. J. Biol. Chem. 264, 5331–5334.

Stremler, K.E., Stafforini, D.M., Prescott, S.M. and McIntyre, T.M. (1991). Human plasma platelet-activating factor acetylhydrolase. J. Biol. Chem. 266, 11095–11103.

Sturk, A., Schaap, M.C.L., Prins, A., ten Cate, J.W. and van den Bosch, H. (1989). Synthesis of platelet-activating factor by human blood platelets and leucocytes. Evidence against selective utilization of cellular ether-linked phospholipids. Biochim. Biophys. Acta 993, 148–156.

Suga, K., Kawasaki, T., Blank, M.L. and Snyder, F. (1990). An arachidonoyl (polyenoic)-specific phospholipase A_2 (PLA_2) activity regulates the synthesis of platelet activating factor (PAF) in granulocytic HL-60 cells. J. Biol. Chem. 265, 12363–12371.

Sugiura, T. and Waku, K. (1985). CoA-independent transfer of arachidonic acid from 1,2-diacyl-sn-glycero-3-phosphocholine to 1-O-alkyl-sn-glycero-3-phosphocholine (lyso platelet-activating factor) by macrophage microsomes. Biochem. Biophys. Res. Commun. 127, 384–390.

Sugiura, T., Masuzawa, Y. and Waku, K. (1985). Transacylation 1-O-alkyl-sn-glycero-3-phosphocholine (lyso platelet activating factor) and 1-O-alkenyl-sn-glycero-3-phosphoethanolamine with docosahexaenoic acid. Biochem. Biophys. Res. Commun. 133, 574–580.

Sugiura, T., Masuzawa, Y., Nakagawa, Y. and Waku, K. (1987). Transacylation of lyso platelet-activating factor and other lysophospholipids by macrophage microsomes. Distinct donor and acceptor selectivities. J. Biol. Chem. 262, 1199–1205.

Sugiura, T., Fukuda, T., Masuzawa, Y. and Waku, K. (1990). Ether lysophospholipid-induced production of platelet-activating factor in human polymorphonuclear leukocytes. Biochim. Biophys. Acta 1047, 223–232.

Suzuki, Y., Miwa, M., Harada, M. and Matsumoto, M. (1988). Release of acetylhydrolase from platelets on aggregation with platelet-activating factor. Eur. J. Biochem. 172, 117–120.

Tence, M., Michel, E., Coeffier, E., Polonsky, J., Godfroid, J.J. and Benveniste, J. (1981). Synthesis and biological activity of some structural analogs of platelet-activating factor (PAF-acether). Agents Actions 11, 558–559.

Tessner, T.G. and Wykle, R.L. (1987). Stimulated neutrophils produce an ethanolamine plasmalogen analog of platelet-activating factor. J. Biol. Chem. 262, 12660–12664.

Tokumura, A., Kamiyasu, K., Takauchi, K. and Tsukatani, H. (1987). Evidence for existence of various homologues and analogues of platelet activating factor in a lipid extract of bovine brain. Biochem. Biophys. Res. Commun. 145, 415–425.

Tokumura, A., Takauchi, K., Asai, T., Kamiyasu, K., Ogawa, T. and Tsukatani, H. (1989). Novel molecular analogues of phosphatidylcholines in a lipid extract from bovine brain: 1-long-chain acyl-2-short-chain acyl-sn-glycero-3-phosphocholines J. Lipid Res. 30, 219–224.

Triggiani, M., Schleimer, R.P., Warner, J.A. and Chilton, F.H. (1991a). Differential synthesis of 1-acyl-2-acetyl-sn-glycero-3-phosphocholine and platelet-activating factor by human inflammatory cells. J. Immunol. 147, 660–6.

Triggiani, M., D'Souza, D.M. and Chilton, F.H. (1991b). Metabolism of 1-acyl-2-acetyl-sn-glycero-3-phosphocholine in the human neutrophil. J. Biol. Chem. 266, 6928–6935.

Uemura, Y., Lee, T-c. and Snyder, F. (1991). A CoA-independent transacylase is linked to the formation of platelet activating factor (PAF) by generating the lyso-PAF intermediate in the remodelling pathway. J. Biol. Chem. 266, 8268–8272.

Vallari, D.S., Record, M. and Snyder, F. (1990). Conversion of alkylacetylglycerol to platelet-activating factor in HL-60 cells and subcellular localization of the mediator. Arch. Biochem. Biophys. 276, 538–545.

van den Bosch, H., Sturk, A., ten Cate, J.W. and Aarsman, A.J. (1991). Studies on the selectivity of enzymes involved in platelet-activating factor formation in stimulated cells. Lipids 26, 967–973.

Venable, M.E., Nieto, M.L., Schmitt, J.D. and Wykle, R.L. (1991). Conversion of 1-O-[^{3}H]alkyl-2-arachidonoyl-sn-glycero-3-phosphorylcholine to lyso platelet-activating factor by the CoA-independent transacylase in membrane fractions of human neutrophils. J. Biol. Chem, 266. 18691–18698.

Whatley, R.E., Clay, K.L., Chilton, F.H., Triggiani, M., Zimmerman, G.A., McIntyre, T.M. and Prescott, S.M. (1992). Relative amounts of 1-O-alkyl- and 1-acyl-2-acetyl-sn-glycero-3-phosphocholine in stimulated endothelial cells. Prostaglandins 43, 21–29.

Winkler, J.D., Sung, C.-M., Bennett, C.F. and Chilton, F.H. (1991). Characterization of CoA-independent transacylase activity in U937 cells. Biochim. Biophys. Acta 1081, 339–346.

Woodard, D.S., Lee, T.-c. and Snyder, F. (1987). The final step in the de novo biosynthesis of platelet activating factor. Properties of a unique CDP-choline:1-alkyl-2-acetyl-sn-glycerol cholinephosphotransferase in microsomes from the renal inner medulla of rats. J. Biol. Chem. 262, 2520–2527.

Wykle, R.L. and Schremmer, J.M. (1974). A lysophospholipase D pathway in the metabolism of ether-linked lipids in brain microsomes. J. Biol. Chem. 259, 1742–1746.

Wykle, R.L., Kraemer, W.F. and Schremmer, J.M. (1977). Studies of lysophospholipase D of rat liver and other tissues. Arch. Biochem. Biophys. 184, 149–155.

Wykle, R.L., Malone, B. and Snyder, F. (1980a). Enzymatic synthesis of 1-alkyl-2-acetyl-sn-glycero-3-phosphocholine, a hypotensive and platelet-aggregating lipid. J. Biol. Chem. 255, 10256–10260.

Wykle, R.L., Kraemer, W.F. and Schremmer, J.M. (1980b). Specificity of lysophospholipase D. Biochim. Biophys. Acta 619, 58–67.

Yamamoto, N., St Claire, D.A., Homma, S. and Ngwenya, B.Z. (1988). Activation of mouse macrophages by alkylglycerols, inflammation products of cancerous tissues. Cancer Res. 48, 6044–6049.

Yasuda, K., Satouchi, K. Nakayama, R. and Saito, K. (1988). Acyl type platelet-activating factor in normal rat uterus determined by gas chromatography mass spectometry. Biomed. Environ. Mass Spectrom. 16, 137–141.

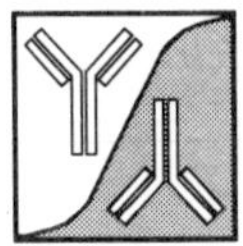

7. Cellular Sources of Platelet-Activating Factor and Related Lipids

D.L. Bratton, K.L. Clay and P.M. Henson

1. Introduction

PAF (also known as PAF-acether and AGEPC) was shown first to be synthesized by the basophil in rabbit blood (Benveniste *et al.*, 1972). Since its chemical identification as 1-*O*-alkyl-2-acetyl-*sn*-glycero-3-phosphocholine (Demopoulos *et al.*, 1979) this and related compounds have been shown to be produced by an ever-expanding variety of cells. This includes cells of ectodermal, endodermal and mesodermal origin, and would certainly imply a general importance for these types of molecules, perhaps extending well beyond their well-studied functions as mediators of cell communication. Significant differences have been noted between mammalian species with regard to the cells that synthesize PAF, but it is unclear how many of these represent fundamental

Lipid Mediators
ISBN 0–12–198875–9

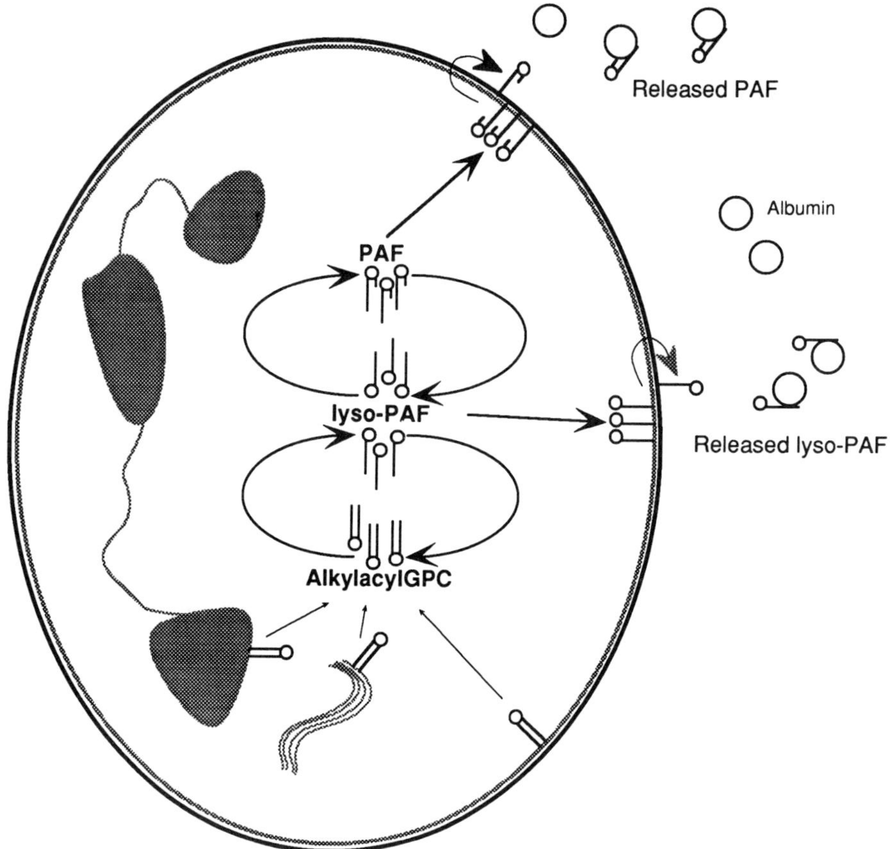

Figure 7.1 Cellular topography of PAF synthesis and release. The remodelling pathway for PAF biosynthesis (see Chapter 6) is likely to occur at intracellular membrane sites. A proportion of the synthesized PAF and lyso-PAF will be transported to the cell plasma membrane inner leaflet, flipped to the outer leaflet, and released to the extracellular milieu complexed with albumin or other binding proteins. Transcellular biosynthesis of PAF may result from the uptake of lyso-PAF by other neighbouring inflammatory cells.

differences or, rather, methodological approaches to the question. In any event, in this chapter we will focus primarily on production of PAF by human cells, in part because most work has been done in this species and in part also because of the interest in these molecules as mediators of allergic and inflammatory disease conditions. However, it is worth noting that, from a teleological perspective, the many cell types that make PAF and the equal diversity of cells that respond to the molecule(s) (including most of the cells that also synthesize it) might suggest fundamental physiological as well as pathological functions.

A review of the cellular origins of PAF must take into account the technical issues of detection and identification of this unique lipid mediator. PAF is not a single molecular species but, rather, represents a family of molecules that classically exhibited an alkyl linkage in the *sn*-1 position, short-chain acyl groups in the *sn*-2 position and phosphocholine as the polar head group. However, it is now apparent that a number of cell types originally thought to produce PAF, produce instead, or in addition,

the *sn*-1 acyl-linked analogue, 1-acyl-2-acetyl-*sn*-glycero-3-phosphocholine, "acyl-PAF" (*vide infra*). Accordingly, several studies of PLA$_2$ activated during the first step of the synthetic pathway demonstrate that either the alkyl or acyl precursor serves as an appropriate substrate (Alonso *et al.*, 1986; Leslie *et al.*, 1988). Acyl-lyso-PAF is then likely to be acetylated in an analogous manner to lyso-PAF (Triggiani *et al.*, 1991a). Evidence of biological activity, a potential role in modulating the response of cells to PAF, and inhibition in the bioassay of PAF by this analogue necessarily require its thorough consideration together with the alkyl-linked, "authentic" PAF (Clay *et al.*, 1991; Columbo *et al.*, 1991; Triggiani *et al.*, 1991b). In this review we will attempt to identify the circumstances where production of the acyl analogue has been documented. It should be noted, however, that for many cells examined, and under many conditions, the relative proportions of authentic PAF and the acyl analogue produced are unknown. In this chapter, the term "PAF" will be used to identify the family of alkyl compounds (which for the most part seem to be comprised of predominantly the 1-O-octadecyl-2 acetyl-*sn*-glycero-3-phosphocholine and 1-O-hexadecyl-2-acetyl-*sn*-glycero-3-phosphocholine species (Pinckard *et al.*, 1984; Mueller *et al.*, 1984; but see also Weintraub *et al.*, 1985; Tessner and Wykle, 1987; Smiley *et al.*, 1991) and acyl-PAF to describe the *sn*-1 acyl-linked analogues.

The reader is alerted to the phrase "PAF release" as opposed to "PAF synthesis" or "PAF production" in reviewing the data presented here and throughout the literature. Data suggest that PAF is synthesized in an intracellular membrane, possibly a light-membrane fraction (Mollinedo *et al.*, 1988), or in the endoplasmic reticulum where its precursor, alkylacyl-glycerophosphocholine GPC, appears in abundance (Saffitz *et al.*, 1986) and the acetyltransferase has been localized (Ribbes *et al.*, 1985) (Fig. 7.1 and see Chapter 6). PAF (and probably lyso-PAF (Vallari *et al.*, 1990)) is then transferred to the plasma membrane (Record and Snyder, 1986; Riches *et al.*, 1990), possibly by specific transfer proteins (Lumb *et al.*, 1983; Banks *et al.*, 1988), where PAF and lyso-PAF become available for release to carrier proteins once they appear in the membrane outer leaflet (Bratton *et al.*, 1991) (see discussion on release below). Because of concurrent catabolism from intracellular and extracellular acetyltransferases, the term "PAF accumulation" (with designation of intracellular and extracellular localization) is the most accurate description. In many studies, particularly in the earlier investigations, only the cell incubation medium (supernatant) was tested for the presence of released PAF. Because both PAF and its acyl analogue remain largely cell associated, they may have intracellular roles in cellular functioning. The lack of released mediator does not constitute a lack of synthesis, nor, as discussed below, is the quantity of released material necessarily reflective of the total amount synthesized. Additionally, these issues have sometimes led to the perception that significant retention of PAF by synthesizing cells means lack of extracellular effect. This is far from the case. The extreme potency of the molecule as a mediator, and the ability of cells to liberate small but significant quantities over time (particularly with the right carriers) would suggest that, whatever the role of intracellular PAF, even the cells that retain it may represent critical reservoirs of locally acting extracellular mediator. The mechanisms of PAF liberation are therefore of considerable importance and will be considered below.

We suggest that identification of the cell types that synthesize (and release) PAF will help our understanding of the role that the mediator plays in physiological and pathological processes. Despite its myriad *potential* biological activities in stimulating cell functions, the real importance of PAF and its acyl analogue remains to be determined. There is a finite risk that when a particular cell type is shown to produce PAF the molecule may be inappropriately tied to pathophysiological processes associated with that cell. There is little doubt that the paradigm of PAF participation in allergic and inflamma-

tory reactions derives in significant measure from the early finding of production by basophils and inflammatory cells, respectively. However, while it may well be true that PAF is a critical mediator in such reactions, definitive demonstration is still lacking. Tissue cells are harder to isolate and study, and may well represent critically important sources of PAF *in vivo*, sources involved with pathophysiological processes that are as yet completely undefined.

2. Detection and Measurement of Platelet-Activating Factor and Related Analogues

2.1 ISOLATION AND SEPARATION OF PLATELET-ACTIVATING FACTOR AND ACYL-PLATELET-ACTIVATING FACTOR

A variety of technical issues are of importance in the consideration of PAF and acyl-PAF synthesis and release. These include methods of detection of the molecules as well as of their isolation before application of the chosen detection system. Generally, isolation can be accomplished by extraction of the lipid from biological samples using a system of organic solvents (Bligh and Dyer, 1959) and separation by one of several chromatographic techniques (Blank *et al.*, 1987). Thin-layer chromatography is perhaps the easiest and most widely used method to isolate PAF, particularly for radiolabelled biological samples, but will not allow adequate resolution of PAF from other structurally similar lipids (phosphatidylinositol, phosphatidylserine, acyl-PAF and sphingomyelin). Normal-phase HPLC is also frequently used, but likewise presents difficulty in detection of PAF except where radiolabelled samples are used. Reverse-phase HPLC has been used to discriminate several species of PAF that differ in the length of the alkyl chain (Mueller *et al.*, 1984; Pinckard *et al.*, 1984). Typically, losses of PAF due to sticking to glass plates, plasticware and columns in other isolation techniques are substantial, making quantification of material from biological samples difficult (Alam *et al.*, 1983a). Losses must be monitored meticulously with known standards to obtain reasonable quantification. As noted above, undegraded acyl-PAF is very difficult to separate adequately from PAF and, furthermore, may interfere with bioassays for PAF. Accordingly, PLA_1 from various sources has been used to degrade acyl-linked PAF species in biological samples before isolation of the resistant alkyl-linked molecules (authentic PAF). However, the success of this approach hinges on exhaustive degradation of the acyl-linked species, which has recently been shown to be unreliable (Mueller *et al.*, 1991). Recent studies utilizing several separatory procedures to discriminate between acyl-PAF and PAF have determined its presence, even predominance, among the acetylated products of some cells. First, the use of normal-phase HPLC can reportedly discriminate the *sn*-1 linkage (Blank and Snyder, 1983). A second approach employs sequential removal of the polar head group by PLC followed by acetylation. In the absence of the polar head group, the subtle differences in polarity of the *sn*-1 linkage can then be exploited to chromatographically separate the acyl and alkyl species (Clay *et al.*, 1991; Mueller *et al.*, 1991). Finally, use of GC–MS procedures has facilitated detailed analysis of the molecular species composition of the acetylated GPC (PAF and analogues) produced in various cells (*vide infra*).

2.2 DETECTION OF PLATELET-ACTIVATING FACTOR AND ACYL-PLATELET-ACTIVATING FACTOR

Until recently, most measurements of PAF have used the same biological activities which initially defined the molecule, platelet aggregation or release reaction. Bioassays, particularly if extended to other cell types in addition to the platelet, continue to be valuable, but there are now compelling reasons to additionally employ analytical methods which can detect and quantify PAF based upon its unique physicochemical properties. It has been shown, for example, that PAF species with different chain lengths in the *sn*-1 position have different potencies in the rabbit platelet aggregation assay (Satouchi *et al.*, 1981; Surles *et al.*, 1985) and that PAF in many cell types is a mixture of species with differing chain lengths (Mueller *et al.*, 1984; Weintraub *et al.*, 1985). Of even more importance, and as indicated above, it is becoming clear that stimulation of cells to produce PAF also leads to acetylation of 1-acyl-lyso-GPC to give 1-acyl-2-acetyl-GPC (acyl-PAF) (Clay *et al.*, 1991; Satouchi *et al.*, 1985). It is very probable that these acyl-PAF species interact with PAF receptors and have some agonist activities which could interfere with bioassays (Columbo *et al.*, 1991; Clay *et al.*, 1991). For example, assay of samples contaminated with large amounts of the acyl-PAF (Satouchi *et al.*, 1985), which does exert activity in the rabbit platelet secretion assay (although approximately 200-fold less than PAF; Demopoulos *et al.*, 1979), could lead to the erroneous assumption of authentic PAF activity. Conversely, and even more probable and important, these PAF analogues could inhibit the responses of PAF in the assay. Several phospholipids from biological sources (Miwa *et al.*, 1987; Nakayama *et al.*, 1987) exhibiting PAF-like chromatographic characteristics have been shown to have inhibitor activity in bioassay systems. Thus, lack of separation of PAF from other lipids, which exert either stimulatory or inhibitory activity in bioassay systems, limits both detection and quantification.

A radioimmunoassay has recently been established and made commercially available, and the advantages of speed and ease of use will probably ensure its ready application. However, the disadvantages for quantitative analysis noted for the bioassays also apply here and the antibody specificity critically determines the spectrum of PAF analogues that can be measured. The value of bioassays and radioimmunoassays as screening procedures for detecting PAF activity should not be underestimated, but quantitative characterization of a biological sample with respect to its PAF and acyl-PAF content requires procedures which can distinguish between those various forms.

Spectroscopic techniques have been widely used to measure many biological substances with improved accuracy and precision. There are no structural features of PAF, however, which allow its measurement at the relevant biological concentrations by the most common spectroscopic techniques such as absorption or emission of light or reaction in an electrochemical cell. As one remedy, Blank *et al.* (1987) have used unsaturated lysophosphatidylcholine detected by UV absorption as a "marker" of PAF which elutes between it and sphingomyelin. Alternatively, the use of benzoate derivatives of PAF allow detection and separation of different diradyl species when present at less than microgram quantities (Blank *et al.*, 1987). The ability to derivatize PAF to an UV-absorbing, fluorescent or electrically active diglyceride would vastly expand the future utilization of HPLC in the detection of PAF (Murphy and Clay, 1987).

Typically, isolated material suspected of being PAF may be further characterized by its pattern of degradation by various phospholipases and acetylhydrolase. The ensuing products of degradation can then be identified physicochemically, or themselves assayed for biological activity. Such degradation schemes include hydrolysis by PLA_2 or PLC (both obtained from a variety of biological sources) which, upon cleavage of the 2-position acetyl group, or polar head, respectively, abolishes bioactivity of the sample. Alternatively, treatment of test samples with acetylhydrolase obtained from serum (Farr *et al.*, 1980; Wardlow *et al.*, 1986; Stafforini *et al.*, 1987) results in the formation of lyso-PAF and likewise in the loss of biological activity. In contrast, treatment of test samples by PLA_1 (also obtained from a variety of biological sources) results in no loss of bioactivity of authentic PAF as the enzyme is unable to cleave to the *sn*-1 ether linkage. The latter has long been routinely used to remove any possible acyl-PAF; however, recent findings suggest that such cleavage by PLA_1 may be far from complete, and has led to substantial confusion as to the products, acyl or alkyl linked, of many cell types. Newer methods for more stringent differentiation of acyl-PAF from PAF are required, and these are outlined below.

The measurement of PAF has been most successfully carried out with mass spectroscopic techniques. Assay of intact PAF with fast atom bombardment mass spectrometry (FABMS) has been demonstrated (Weintraub *et al.*, 1985; Clay *et al.*, 1984a), but has seldom been applied to biological problems because of the requirement for extensive purification prior to analysis. FABMS is a very powerful tool for structural elucidation of PAF and its analogues, but is less useful for quantitative analysis. For quantitative analysis, the method of choice is currently GC–MS, using stable isotope-labelled internal standards for PAF and its analogues. The utility of GC–MS was demonstrated (Satouchi *et al.*, 1983) in a study of neutrophil-derived PAF, and measurement at the subpicogram level by negative ion chemical ionization GC–MS has been described (Ramesha and Pickett, 1986a). In addition to the sensitivity demonstrated in the fore-mentioned study, the value of the use of stable isotopically labelled variants of PAF was illustrated. Use of such internal standards has the dual virtues of compensating for losses incurred during isolation procedures and of being a positive indicator of the success (or failure) of the isolation method. Detection of the internal standard assures that recovery of PAF from the biological sample was successful, and, conversely, failure to detect the ion corresponding to the internal standard indicates that the measurement has failed due to sample degradation or inadequate recovery for other reasons. Stable isotope-labelled forms of most of the compounds of interest are readily available at little cost by means of simple synthetic procedures (Clay, 1990; Pickett and Ramesha, 1990).

The GC–MS procedures in use are all quite similar. Following addition of a stable isotopically labelled internal standard, or standards, a total lipid extract of the biological mixture is purified by silicic acid chromatography to obtain the acetylated GPC species. The procedures have in common as a final step the cleavage of the intact GPC molecules to their diglycerides and the conversion of those diglycerides to molecules amenable to GC–MS by chemical derivatization of the diglyceride alcohol moiety. Various derivatives have been reported, including trimethylsilyl (Clay, 1990), *t*-butyldimethylsilyl (Satouchi *et al.*, 1983), pentafluorbenzoyl esters (Ramesha and Pickett, 1986a; Pickett and Ramesha, 1990) and heptafluorobutyryl esters (Weintraub *et al.*, 1990). The resolution of the capillary gas chromatography columns in use is such that the individual diglycerides of all the acetylated GPC species in biological mixtures are readily separable and thus are individually measurable. In the mass spectrometer, ions characteristic of each species to be measured and its stable isotopically labelled variant are selected, the abundances of the ions are measured and the ratio of the endogeneous to internal standard ion signal is used as a measure of the amount present in the initial sample.

These procedures have been used to demonstrate the presence of PAF in human saliva (Christman and Blair, 1989) and human blood (Yamada *et al.*, 1988) at concentrations too low for satisfactory analysis by other quantitative techniques. In addition, the ability of the procedures to profile mixtures has been used to advantage

in the study of human lung mast cells (Triggiani *et al.*, 1990a), human umbilical vein endothelial cells (Clay *et al.*, 1991), bovine brain (Tokumura *et al.*, 1987a) and human neutrophils (Weintraub *et al.*, 1990) (*vide infra*).

Synthesis of PAF has also been detected by utilizing radiolabelled precursors, either alkyllysophosphatidylcholine or acetate followed by isolation of labelled PAF. However, a shortcoming of this approach is that data generated are only representative of potentially non-equilibrium precursor "pools". Such radiolabelled data estimating the relative amounts of PAF synthesized and secreted may be very misleading if not correlated with quantification of material produced, either by bioassay or, preferably, by GC–MS or FABMS (Chilton and Connell, 1988).

2.3 CATABOLISM

When a certain cell type is assessed for the capability of PAF biosynthesis, catabolism of PAF must always be considered together with synthesis. Acetylhydrolase activity (see Chapter 6) in plasma (Farr *et al.*, 1980; Wardlow *et al.*, 1986; Stafforini *et al.*, 1987), numerous cells and tissues (Blank *et al.*, 1981; Alam *et al.*, 1983b; Averill *et al.*, 1991), and apparently secreted from mast cells (Triggiani and Chilton, 1989; Triggiani *et al.*, 1989), may obscure production of this unique lipid by rapid hydrolysis to the biologically inactive product lyso-PAF. For example, use of a serine-hydrolase inhibitor, PMSF, to inhibit acetylhydrolase degradation of PAF allowed Touqui *et al.* (1985) to detect 10–12 times more PAF synthesis in thrombin-stimulated human platelets. Subsequent studies (Hirafuji *et al.*, 1987; Sturk *et al.*, 1989), however, have suggested that a more predominant effect of PMSF is inhibition of de-acylation of acyl-PAF, which may be mistakenly identified as PAF without specific analysis (see above). Additionally, cellular activation itself can sometimes lead to enhanced activity of catabolic pathways, as has been shown for acetylhydrolase activity in PAF-stimulated human hepatoma cells (Satoh *et al.*, 1991), ionophore- and antigen-stimulated murine mast cells (Triggiani and Chilton, 1989), and ionophore-stimulated human neutrophils and eosinophils (from patients with eosinophilia) (Lee *et al.*, 1982).

2.4 LYSO ANALOGUES OF PLATELET-ACTIVATING FACTOR

The release of lyso-PAF (considered to be both a precursor and a metabolite of PAF (Fig. 7.1)) (Albert and Snyder, 1983; Robinson and Snyder, 1985; Blank *et al.*, 1986; Chignard *et al.*, 1986; Arnoux *et al.*, 1987; Miadonna *et al.*, 1989) and its potential use as a substrate for PAF synthesis by other cells (Coeffier *et al.*, 1986) has made the detection of lyso-PAF synthesis and release of growing importance. To date, four basic techniques for the detection and measurement of lyso-PAF have been devised but have not been widely employed or verified.

Lyso-PAF from biological samples can be acetylated to PAF and identified and quantified by bioassay (Polonsky *et al.*, 1980) or, alternatively, acetylated with radiolabelled acetate and detected after purification (Wientzek, 1985). The use of acetate of known specific activity in the latter method allows quantification of material isolated as PAF. Both methods assume acetylation proceeds reproducibly to a reasonable yield. However, estimates of completeness of acetylation range from 50% (K.L. Clay, unpublished data) to 80% (Michel *et al.*, 1988). Specificity is another shortcoming of both acetylation methods as other molecular species of PAF may be acetylated (i.e. acyl-GPC, acetyl-GPC) and detected with radiolabel or exhibit agonist or antagonist activity in biological assays. More recently, GC–MS and FABMS have been applied to the detection and measurement of lyso-PAF synthesis following stimulation of human neutrophils (Haroldsen *et al.*, 1987) and in nasal lavage fluid following appropriate antigen challenge (Shin *et al.*, 1991). Sensitivity ranged from the picomolar range for GC–MS and the nanomolar range for FABMS.

2.5   PURITY OF CELL POPULATIONS

Additional technical concerns include the potential mis-assignment of a particular cell as the source of PAF in mixed cell populations. An avid producer of PAF could lead to an erroneous conclusion even if it were a quite minor contaminant of the cells under study. This type of problem has led to the continuing question of PAF production by the human basophil (see below), which is very difficult to isolate from other PAF-producing blood cells (Camussi *et al.*, 1977; Betz *et al.*, 1980; Sanchez Crespo *et al.*, 1980; Camussi *et al.*, 1981a; Lynch and Henson, 1986). The difficulty is even greater in cells in tissues, but, if they synthesize and accumulate enough PAF, it may be possible to use immunohistochemical approaches to identify PAF-producing cells.

3. *Synthesis and Release of Platelet-Activating Factor and Related Analogues*

PAF is thought to function as a cell communication molecule or mediator in numerous physiological or pathophysiological processes. To do this it must either be released from the cell of origin or must be expressed on the surface in such a way that it can interact and stimulate the target. PAF has been detected in normal body fluids – urine (Billah and Johnston, 1983; Sanchez Crespo *et al.*, 1983); blood (Caramelo *et al.*, 1984; Yamada *et al.*, 1988); amniotic fluid (Billah and Johnston, 1983) and human saliva (Cox *et al.*, 1981; Christman and Blair, 1989) – suggesting that, at least in some cases, it could exert effects either physiologically (or

pathophysiologically) on relatively distant target cells/tissues. However, for the most part, it is rapidly inactivated in such fluids and in inflammatory exudates by acetylhydrolases, and is therefore suspected of exhibiting local action on its target cells. From the data shown in Table 7.1 it can be seen that a wide variety of cells release PAF where cell-free supernatants have been examined for the presence of PAF (or, as in the older studies, biologically active PAF-like material). It should be noted that cell-associated PAF was not examined or quantified in the majority of these studies, which were designed only to examine release, so that, unless noted by a comment or as a percentage in Table 7.1, the relative proportion of PAF released to that synthesized is unknown. However, when this has been examined it appears that in most cell types only a small proportion of PAF was released. For example, this is seen in the neutrophil (Oda *et al.*, 1985; Lynch and Henson, 1986; Sisson *et al.*, 1987; Worthen *et al.*, 1988), endothelial cell (McIntyre *et al.*, 1986; Lewis *et al.*, 1988), monocyte (Jouvin-Marche *et al.*, 1984) and mast cell (Lichtenstein *et al.*, 1984; Schleimer *et al.*, 1986), regardless of the agonist used (Lynch and Henson, 1986), dose of agonist (Lynch and Henson, 1986; Sisson *et al.*, 1987), cell-isolation technique (Lynch and Henson, 1986), albumin concentration (Oda *et al.*, 1985; Sisson *et al.*, 1987), presence of plasma (Lynch and Henson, 1986), or cell "priming" (Sisson *et al.*, 1987). Furthermore, low amounts of PAF in supernatants cannot be attributed simply to extracellular inactivation (metabolism) of PAF, nor solely to cellular reabsorption of PAF (Lynch and Henson, 1986). The data indicate a dissociation of synthesis from release. Thus, Sisson *et al.* (1987) noted that total synthesis of PAF in adherent PMNs was decreased when compared to that of PMNs in suspension, while the amount released was independent of the state of adherence. Additionally, although neutrophils stimulated with FMLP (and cytochalasin B) produced far less PAF than cells stimulated with calcium ionophore or opsonized zymosan, the relative proportion of PAF released was greatest following FMLP stimulation (Lynch and Henson, 1986). By contrast, the dependence of PAF synthesis and secretion on the presence of extracellular calcium and albumin (Ludwig *et al.*, 1984, 1985) led to the proposal of a synthesis-release coupling mechanism for the human neutrophil in which albumin and extracellular calcium are required for release of PAF, and that release triggers further synthesis. In this way, release of newly synthesized PAF abrogates feedback inhibition of PAF on PAF synthesis. By contrast, Elstad *et al.* (1988) have shown that, while both synthesis and release were dependent on the extracellular albumin concentration, the optimal concentration for release was less than that for optimal synthesis, thus dissociating these phenomena.

The amount of PAF that *is* released appears to be a function of cell type (Elstad *et al.*, 1988) and species (Lynch and Henson, 1986), with further modulation by

Table 7.1 Cellular sources of PAF and acyl-PAF[a]

Cell	Stimulus[b]	Released[c]	Reference
Cells synthesizing predominantly alkyl-linked PAF species			
Neutrophil	A23187	+	Betz and Henson (1980)
		49%	Jouvin-Marche *et al.* (1984)
		10–20%	Lynch and Henson (1986)
		1–10%	Sisson *et al.* 1987
	A23187 + cyto B	+/−	Oda *et al.* (1985)
	Opsonized zymosan	+	Betz and Henson (1980)
		+	Lotner *et al.* (1980)
		2–6%	Lynch and Henson (1986)
		1–10%	Sisson *et al.* (1987)
	Zymosan	+	Sanchez Crespo *et al.* (1980)
	FMLP	1–10%	Sisson *et al.* (1987)
	FMLP + cyto B	+	Betz and Henson (1980)
		33%	Lynch and Henson (1986)
		30–40%	Ludwig *et al.* (1984)
		30–40%	Ludwig *et al.* (1985)
	C5a	+	Betz and Henson (1980)
		+	Camussi *et al.* (1981a)
	C5a des arg	+	Camussi *et al.* (1981a)
	Aggregated IgG	+	Betz and Henson (1980)
		+	Sanchez Crespo *et al.* (1980)
	Immune complexes	+	Sanchez Crespo *et al.* (1980)
		+	Camussi *et al.* (1981a)
	Neutrophil cationic protein	+	Camussi *et al.* (1981a)
	Neutrophil cationic protein des arg	+	Camussi *et al.* (1981a)
	Chymotrypsin	+	Camussi *et al.* (1988)
	Elastase	+	Camussi *et al.* (1988)
	Cathepsin G	+	Camussi *et al.* (1988)
	Complement-activated baker's yeast spores	+	Camussi *et al.* (1981a)
	IgG ± C3 coated RBC	+	Betz and Henson (1980)
	pH 10.6	+	Camussi *et al.* (1981a)
	NaF	+	Betz and Henson (1980)
	PMA	+	Betz and Henson (1980)
		+	Nieto *et al.* (1988)
	LPS	−	Worthen *et al.* (1988)
	IL-1	+	Camussi *et al.* (1988)
	TNF	+	Camussi *et al.* (1988)
	Carbamyl-PAF	NA	Tessner *et al.* (1989)
	PAF	NA	Doebber and Wu (1987)
	GM-CSF	−	Stewart *et al.* (1991)
Eosinophil	A23187	+	Lee *et al.* (1984)
		10%	Burke *et al.* (1990)
		35%	Cromwell *et al.* (1990)
		NA	Triggiani *et al.* (1991c)
	FMLP	+	Lee *et al.* (1984)
		NA	Triggiani *et al.* (1991c)
	ECF-A	+	Lee *et al.* (1984)
	C5a	+	Lee *et al.* (1984)
	IgG–Sepharose beads	40–55%	Cromwell *et al.* (1990)
	Unopsonized zymosan	9%	Burke *et al.* (1990)
Monocyte	A23187	+	Sanchez Crespo *et al.* (1980)
		+	Camussi *et al.* (1981a)
		+	Arnoux *et al.* (1982)
		37%	Jouvin-Marche *et al.* (1984)
		67%	Elstad *et al.* (1988)
	Immunoglobulin aggregates	+	Sanchez Crespo *et al.* (1980)
	Opsonized zymosan	+	Sanchez Crespo *et al.* (1980)

Table 7.1 Continued

Cell	Stimulus[b]	Released[c]	Reference
		+	Jouvin-Marche et al. (1984)
		49%	Elstad et al. (1988)
	Zymosan	+	Sanchez Crespo et al. (1980)
		+	Arnoux et al. (1982)
		+	Jouvin-Marche et al. (1984)
	RSV	NA	Villani et al. (1991)
	PMA	32%	Elstad et al. (1988)
	C3d–baker's yeast spores	+	Camussi et al. (1981a)
		+	Camussi et al. (1983a)
	C3b–baker's yeast spores	+	Camussi et al. (1981a)
		+	Camussi et al. (1983a)
	IgG–baker's yeast spores	+	Camussi et al. (1983a)
	Bacteria	+	Arnoux et al. (1982)
	Immune complexes	+	Sanchez Crespo et al. (1980)
		+	Camussi et al. (1981a)
		+	Arnoux et al. (1982)
	pH 10.6	+	Camussi et al. (1981a)
		+	Camussi et al. (1983a)
	IL-1β		Valone and Epstein (1988)
	TNF } Biphasic release: <15% early; >70% late		Valone and Epstein (1988)
	IFN-γ		Valone and Epstein (1988)
	PAF	NA	Valone (1991)
Alveolar macrophage	A23187	+	Arnoux et al. (1980)
		+	Arnoux et al. (1987)
	Anit-IgE	+	Arnoux et al. (1987)
	Specific antigen[d]	+	Arnoux et al. (1987)
	Opsonized zymosan	NA	Triggiani et al. (1991c)
Monocyte-derived macrophage	A23187	+	Elstad et al. (1988)
	PMA	+	Elstad et al. (1988)
	Opsonized zymosan	+	Elstad et al. (1988)

Cells that synthesize predominantly acyl-PAF

Cell	Stimulus[b]	Released[c]	Reference
Endothelial cell	A23187	+	Camussi et al. (1983b)
		NA	Triggiani et al. (1991c)
	Angiotensin II	+	Camussi et al. (1983b)
	Anti-human factor VIII	+	Camussi et al. (1983b)
	IL-1	25%	Bussolino et al. (1986)
		30%	Bussolino et al. (1986)
	LTC$_4$ and LTD$_4$	2–3%	McIntyre et al. (1986)
	Thrombin	NA	Prescott et al. (1984)
		NA	Triggiani et al. (1991c)
		−	McIntyre et al. (1985)
	H$_2$O$_2$	−	Lewis et al. (1988)
		NA	Whatley et al. (1988)
	Clostridial φ toxin	NA	Whatley et al. (1988)
	Melittin	NA	Whatley et al. (1988)
	Membrane attack complex of complement	NA	Whatley et al. (1988)
	Vasopressin	+	Camussi et al. (1983b)
	Histamine	−	McIntyre et al. (1985)
	Bradykinin	−	McIntyre et al. (1985)
	ATP	−	McIntyre et al. (1985)
	TNF	+	Camussi et al. (1987)
		30%	Bussolino et al. (1988)
	LPS	4–5%	Lynch and Henson (1986)
	Elastase	+	Camussi et al. (1988)
Mast cell (pulmonary)	Anti-IgE	−	Lichtenstein et al. (1984)
		−	Schleimer et al. (1986)

Table 7.1 Continued

Cell	Stimulus[b]	Released[c]	Reference
		NA	Triggiani *et al.* (1991c)
	A23187	NA	Triggiani *et al.* (1991c)
Basophil[e]	A23187	+	Camussi *et al.* (1981a)
		NA	Triggiani *et al.* (1991c)
	Synacthen	+	Camussi *et al.* (1981a)
	Complement-coated baker's yeast spores	+	Camussi *et al.* (1981a)
	C3a	+	Camussi *et al.* (1977)
	C5a	+	Camussi *et al.* (1981a)
	Neutrophil cationic protein	+	Camussi *et al.* (1977)
		+	Camussi *et al.* (1981a)
	DNA[f]	+	Camussi *et al.* (1981b)
	Anti-IgE	+	Camussi *et al.* (1981a)
		NA	Triggiani *et al.* (1991c)
	pH 10.6	+	Camussi *et al.* (1981a)

Cells for which PAF synthesis is attributed, but linkage predominance is unknown

Cell	Stimulus[b]	Released[c]	Reference
Platelet	Thrombin	NA	Touqui *et al.* (1985)
		NA	Alam and Silver (1987)
	A23187	NA	Chignard *et al.* (1979)
		NA	Alam *et al.* (1983b)
Large granular lymphocytes (NK cells)	Immobilized anti-Fc receptor	+	Malavasi *et al.* (1986)
	A23187	+	Malavasi *et al.* (1986)
Fibroblasts (human skin)	A23187	50%	Michel *et al.* (1988)

Transformed or neoplastic cells
HL-60

Cell	Stimulus[b]	Released[c]	Reference
Neutrophil-like[g]	A23187	NA	Billah *et al.* (1986)
	FMLP	NA	Billah *et al.* (1986)
Macrophage-like[h]	pH 9.5	+	Camussi *et al.* (1982)
	C3b–baker's yeast spores	+	Camussi *et al.* (1982)
	C3d–baker's yeast spores	+	Camussi *et al.* (1982)
U937[i]			
Macrophage-like	RSV	+	Villani *et al.* (1991)
Basophilic leukaemia	A23187	+	Lewis *et al.* (1975)
T-cell ALL	A23187 + acetyl-CoA ± PHA	+	Foa *et al.* (1985)
Common ALL	A23187 + acetyl-CoA ± PHA	+	Foa *et al.* (1985)
CLL (B cell)	A23187 + acetyl-CoA ± PHA	+	Foa *et al.* (1985)
CML	A23187 + acetyl-CoA ± PHA	+	Foa *et al.* (1985)
AML	A23187 + acetyl-CoA ± PHA	+	Foa *et al.* (1985)
T cell leukaemia cell lines	A23187 ± acetyl-CoA ± PHA	+	Bussolino *et al.* (1984)
B cell leukaemia cell lines	A23187 ± acetyl-CoA ± PHA	+	Bussolino *et al.* (1984)
EBV transformed lymphoblastoid cell lines	A23187 ± acetyl-CoA ± PHA	+	Bussolino *et al.* (1984)

[a] Table adapted from Bratton and Henson (1989)

[b] Abbreviations: A23187, calcium ionophore; cyto B, cytochalasin B; carbamyl-PAF, 1-*O*-alkyl-2-*N*-methylcarbamyl-*sn*-glycero-3-phosphocholine; RSV, respiratory syncytial virus.

[c] PAF released: +, detectable PAF release; −, no detectable PAF release; NA, PAF synthesized but release not assessed.

[d] Cells from atopic asthmatics or following passive sensitization of cells from normal patients.

[e] Basophil-enriched leucocyte preparations.

[f] In cells from patients with systemic lupus erythematosus.

[g] Differentiated towards neutrophil-like cells with DMSO.

[h] Differentiated towards macrophage-like cells with tetradodecanoylphorbol acetate (TPA).

[i] Human histiocytic lymphoma cell line differentiated towards macrophage-like cells with PMA.

agonist (T-c. Lee *et al.*, 1984; Ludwig *et al.*, 1986; Lynch and Henson, 1986; Elstad *et al.*, 1988), priming (Stewart *et al.*, 1991), state of adherence versus suspension (Sisson *et al.*, 1987; Elstad *et al.*, 1988), "phase of response" (Valone and Epstein, 1988), pH of the medium (Leyravaud and Benveniste, 1989) and presence of antiproteinases (Camussi *et al.*, 1988). Demonstration that this is also affected by cell density

(Betz and Henson, 1980; Betz *et al.*, 1980; Leyravaud and Benveniste, 1989), and more recently that it is markedly enhanced by dynamic removal of albumin-containing medium (Cluzel *et al.*, 1989) or saponin treatment (Vallari *et al.*, 1990), has suggested to us that the molecules must first be expressed on the outer leaflet of the plasma membrane and may then be released into the medium in the presence of appropriate conditions and carriers (*vide infra*).

Despite the potential importance of the released PAF, the large amount of these molecules that can be produced and retained in many cell types has led to hypotheses regarding an intracellular role for PAF in addition to its mediator functions (Henson, 1987; Sisson *et al.*, 1987; Worthen *et al.*, 1988; Tool *et al.*, 1989). Additionally, if external expression of the molecules occurs, cell-to-cell recognition of PAF without actual release of the mediator would allow stimulation of nearby or adherent cells. Such a process may explain the PAF-dependent interaction of neutrophils with appropriately stimulated endothelial cells (McIntyre *et al.*, 1985; Zimmerman *et al.*, 1985; McIntyre *et al.*, 1986). Interestingly, recent evidence suggests that this effect may be mediated largely by acyl-PAF, rather than alkyl-linked species, that is synthesized by the endothelial cells and then expressed in such a manner as to stimulate the neutrophil. In these experiments, treatment of intact endothelial cells with phospholipases has demonstrated that PAF or acyl-PAF located on the cell membrane outer leaflet is accessible to enzymatic cleavage (McIntyre *et al.*, 1985, 1986). Thus, recognition of endothelial cell-associated PAF by neutrophils may cause neutrophil adherence in the absence of detectable release (Zimmerman *et al.*, 1985; McIntyre *et al.*, 1986). In this context, PAF bound (or incorporated) into red cell membranes (Shaw and Henson, 1980) or liposomes (Hayashi *et al.*, 1985; Prescott *et al.*, 1990) can be presented to cells. However, in any of these studies, possible diffusion of the mediator over small distances from cell to cell cannot be excluded.

3.1 MECHANISMS OF RELEASE OF PLATELET-ACTIVATING FACTOR AND RELATED ANALOGUES

The mechanisms by which PAF is released are poorly understood. According to several sources, it appears that release is dependent on the presence of albumin in the extracellular medium (Benveniste *et al.*, 1972; Ludwig *et al.*, 1985; Bratton *et al.*, 1991). The ability of albumin to bind PAF and other phospholipids and lysophospholipids has long been known, and we have suggested that at least four binding sites are involved (Clay *et al.*, 1990). Albumin has been shown to solubilize PAF in aqueous media (Ludwig *et al.*, 1986), and PAF added exogenously to plasma co-elutes with the fraction containing albumin and other plasma proteins (Ludwig *et al.*, 1985). Albumin

appears to compete with cells and cell membranes for PAF as it inhibits binding of PAF to human neutrophil membranes (Valone, 1987) and also inhibits PAF-induced secretion and aggregation of platelets (Hoppens *et al.*, 1983; Tokumura *et al.*, 1987b). Other proteins, both intracellular and extracellular (in plasma), may also bind PAF. A PAF-specific transport protein(s) has been isolated from bovine and rabbit lung macrophages (Banks *et al.*, 1988) and rat lung (Lumb *et al.*, 1983) which can transfer PAF between lipid bilayers. A PAF-binding protein found in human serum has also been described that can apparently bind PAF more tightly than albumin (Matsumoto and Miwa, 1985). From such data it is clear that albumin, and possibly other proteins, bind to PAF that is released from cells, and, indeed, may "extract" PAF and related molecules from the outer leaflet of the membrane.

The mechanism of actual release of PAF is, as yet, unknown but could proceed by (1) transmembrane passage, (2) via an exocytotic process (as in secretion of surfactant phospholipids from alveolar type II cells) or (3) from shed membranes of disrupted or lysed cells. As cells do not appear to make PAF in sufficient quantity to exceed the critical micellar concentration of PAF (around 1 μM) (Kramp *et al.*, 1984; Ludwig *et al.*, 1986), it is unlikely that PAF is released as micelles.

Recent data utilizing the erythrocyte ghost as a model plasma membrane to study the transbilayer movement of PAF inserted in the plasma membrane inner leaflet have led us to suggest the first of these alternatives as the most likely mechanism for PAF release (although similar processes could also contribute to PAF liberation from shed membrane). Enhanced transbilayer movement across the plasma membrane of PAF and related analogues, both lyso-PAF and acyl-PAF, was demonstrated in the setting of calcium-induced membrane perturbation and enhanced (endogenous and exogenously added) phospholipid flipping (Bratton *et al.*, 1991; Fig. 7.2). Subsequent release of PAF from these model membranes was then dependent on the presence of albumin. Similar phospholipid rearrangement and perturbation of the membrane appear to occur in neutrophils following their activation, as shown by fluorescence techniques (McEvoy *et al.*, 1988; Bratton *et al.*, 1992), and is likely to exist in a number of circulating inflammatory cells (Schlegel and Williamson, 1987; McEvoy *et al.*, 1988). Alterations in the cytoskeleton accompanying cellular activation (Verhallen *et al.*, 1987; Jesaitis *et al.*, 1988) probably enhance transbilayer movement (flipping) of phospholipids (Schneider *et al.*, 1986), resulting in enhanced movement of PAF to the outer leaflet, where albumin (Clay *et al.*, 1990) or other appropriate carrier(s) (Matsumoto and Miwa, 1985) can remove it from the membrane (Figs 7.1 and 7.2). The suggested involvement of calcium in the release process (Ludwig *et al.*, 1984) could be explained either as a requirement for cell activation and/or by its involvement

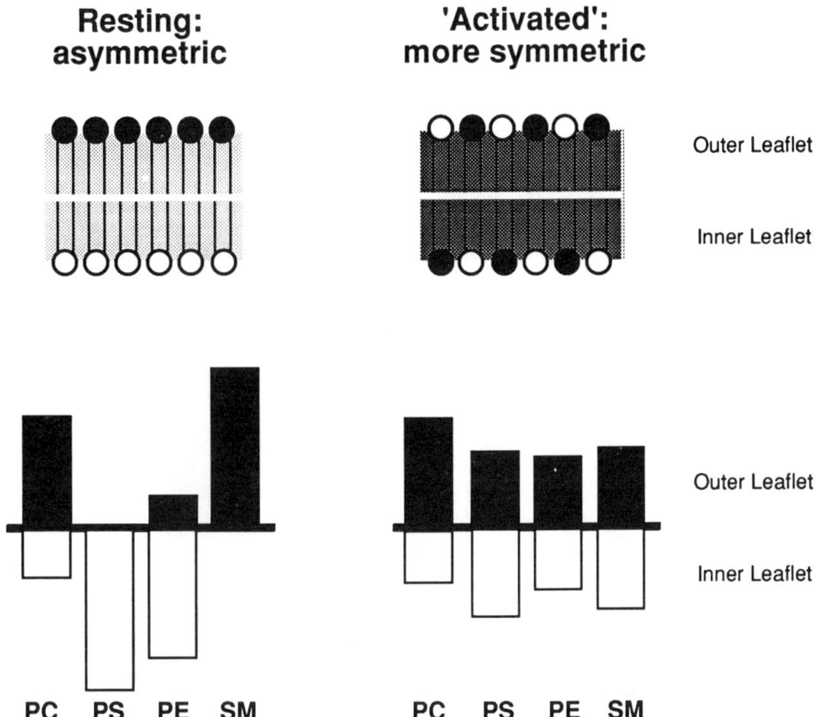

Figure 7.2 The distribution and movement of phospholipid classes in the plasma membrane of the inflammatory cell is likely to govern transbilayer movement and subsequent release of PAF from the cell. An asymmetric distribution of phospholipid classes is seen in the resting plasma membrane. With cellular activation, enhanced transbilayer movement leads to a more random distribution of the phospholipids, including PAF (Bratton *et al.*, 1991). Enhanced movement of PAF to the outer leaflet will result in enhanced release to albumin or other binding proteins. Uptake of released PAF or recycling at the plasma membrane of activated cells may also occur (Bratton *et al.*, 1992). PC, phosphatidylcholine; PS, phosphatidylserine; PE, phosphatidylethanolamine; SM, sphingomyelin.

in membrane lipid reorganization (Bratton *et al.*, 1988). It is proposed then that these plasma membrane physical changes, which are likely to involve heightened hydrophobicity, looser packing of phospholipids and loss of resting plasma membrane phospholipid asymmetry (Williamson *et al.*, 1983), will enhance the movement of PAF that reaches the inner leaflet of the plasma membrane to the outer leaflet.

4. Stimulation of Inflammatory Cells to Produce Platelet-Activating Factor and Related Analogues

Studies indicate that inflammatory cells do not synthesize significant levels of PAF and related analogues in the baseline, non-stimulated state. Upon stimulation, however, large amounts of acetylated phospholipids are synthesized and, as pointed out earlier, there is a growing appreciation of the contribution of not only *sn*-1 alkyl-linked but *sn*-1 acyl-linked species (Fig. 7.3). In the human neutrophil, hexadecyl-PAF has been shown by

GC–MS to be the predominant species (Satouchi *et al.*, 1983; Clay *et al.*, 1984b; Ramesha and Pickett, 1986a) although, as noted above, other analogues, both 1-*O*-alkyl and 1-*O*-acyl, are present in the mixture (Ramesha and Pickett, 1986a; Weintraub *et al.*, 1990). There are even reports of similar species in the phosphatidylethanolamine class (1-*O*-alk-1′-enyl-2-acetyl-*sn*-glycero-3-phosphoethanolamines) although the biological significance of these is unknown (Tessner and Wykle, 1987). In the human lung mast cell and basophil, the 1-*O*-acyl-2-acetyl-GPC species (acyl-PAF) were more abundant than the 1-*O*-alkyl-linked species when IgE and antigen were used as stimuli, but predominance of the acyl species was not seen when the cells were stimulated with a calcium ionophore (Triggiani *et al.*, 1990a). In the endothelial cell, most of the tritiated acetate which is incorporated into GPC upon cellular stimulation is associated with the 1-*O*-acyl species (Clay *et al.*, 1991). Cell stimulation initiates the remodelling pathway for PAF synthesis (discussed in Chapter 6) to generate the large amounts of acetylated lipids. Whether the array of acetylated products simply reflects precursor frequency (Clay *et al.*, 1991) and/or is determined by intensity or

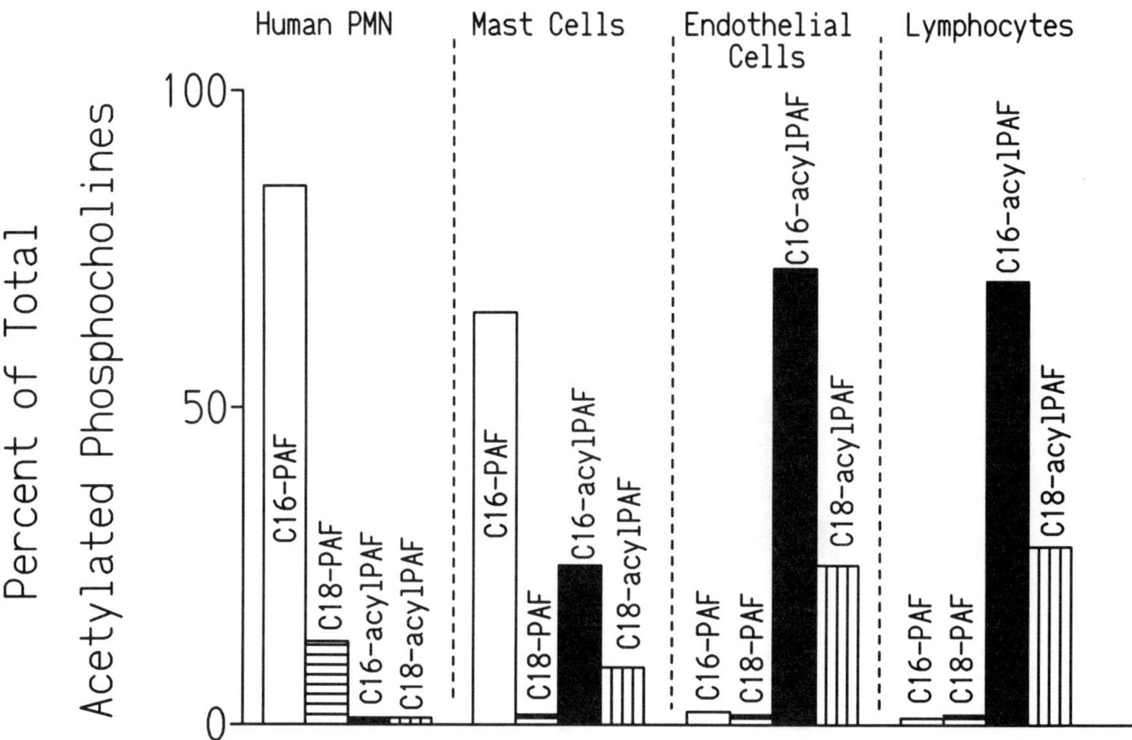

Figure 7.3 Distribution of acetylated phosphocholines amongst various cell types: C16-PAF, 1-*O*-hexadecyl-2-*O*-acetyl-*sn*-glycero-3-phosphocholine; C18-PAF, 1-*O*-octadecyl-2-*O*-acetyl-*sn*-glycero-3-phosphocholine; C16-acyl-PAF, 1-*O*-hexadecanoyl-2-O-acetyl-*sn*-glycero-3-phosphocholine; C18-acylPAF, 1-*O*-octadecanoyl-2-*O*-acetyl-*sn*-glycero-3-phosphocholine. Human neutrophils (PMN) stimulated with opsonized zymosan; murine mast cells stimulated with calcium ionophore, A23187; endothelial cells (human umbilical vein and bovine pulmonary artery) stimulated with thrombin or calcium ionophore; lymphocytes (human B cell line, EBV transformed) not overtly stimulated.

type of stimulus, some as yet unknown specificity of the enzymatic pathways, or other more complicated factors, has yet to be determined. Certainly there is some evidence that the cells that produce predominantly alkyl-PAF exhibit a relative enrichment in their content of ether-linked phospholipids. For example, in human polymorphonuclear leucocytes, more than 50% of choline glycerophospholipids are ether linked in the *sn*-1 position (Sugiura and Waku, 1987). Adequate amounts of ether-linked substrate (alkylacyl-GPC), the precursor for subsequent PAF synthesis, have been documented in neutrophils (Jouvin-Marche *et al.*, 1984; Mueller *et al.*, 1984; Oda *et al.*, 1985; Tence *et al.*, 1985), platelets (Natarajan *et al.*, 1983), human monocytes (Jouvin-Marche *et al.*, 1984) and human fibroblasts (Michel *et al.*, 1988). Additionally, a note of caution should be introduced in evaluating studies of cell lines in culture, since culture conditions and fatty acid supplementation (Suga *et al.*, 1990; Triggiani *et al.*, 1990b; Garcia *et al.*, 1991a; Whellan *et al.*, 1992) may markedly influence the phospholipid species. However, recent evidence suggests that the issue of precursor pools of phospholipid may be much more complicated (Chilton and Murphy, 1986;

Chilton and Connell, 1988) and, in addition, the source of lysophosphocholine for acetylation to acyl-PAF requires detailed investigation. In addition to the remodelling pathway, a *de novo* synthetic pathway for PAF synthesis has been described involving cholinephosphotransferase (Snyder, 1987). Its contribution to the overall levels of PAF within cells remain unclear (see Chapter 6.).

4.1 Human Cells that Synthesize Predominantly Alkyl-Linked Platelet-Activating Factor Species

4.1.1 Neutrophils

Neutrophils remain the paradigm of PAF-producing cells because they synthesize predominantly the alkyl species (Fig. 7.3; Mueller *et al.*, 1984; Pinckard *et al.*, 1984), produce more of the material than any other inflammatory cell and have been most often studied. From 10 to 100 pmol/10^6 cells can be detected following stimulation with calcium ionophores (Jouvin-Marche *et al.*, 1984; Oda *et al.*, 1985; Lynch and Henson, 1986).

Activation of cell surface receptors by various complete or partial secretagogues and phagocytic particles is effective at initiating synthesis of PAF (Table 7.1). However, the cells have also been reported not to make PAF in response to stimulation with such agents as LTB_4, LTC_4, LTD_4, LTE_4, zymosan-activated serum, platelet-derived growth factor, adrenaline, bradykinin, histamine, thrombin, adenosine triphosphate, vasopressin or fibrinogen (Sisson *et al.*, 1987). The response to TNF is controversial (Camussi *et al.*, 1987, 1989; Sisson *et al.*, 1987), as are the effects of phorbol esters and PAF itself (Betz and Henson, 1980; Ludwig *et al.*, 1986; Sisson *et al.*, 1987; Tessner *et al.*, 1989). The issue of neutrophil stimulation is also complicated by the process of priming, whereby exposure of the cell to an agent that in itself does not initiate a response nevertheless causes the cell to become much more responsive to a second stimulus. Thus, prior exposure to bacterial lipopolysaccharide (priming) significantly increased the amounts of PAF that were synthesized by neutrophils stimulated with the tripeptide FMLP (Worthen *et al.*, 1988). Additionally, IFNγ (Geffner *et al.*, 1991) and GM-CSF (Wirthmueller *et al.*, 1990) have been shown to enhance PAF production, the latter probably acting by a process that enhances PLA_2 activity (McColl *et al.*, 1991) and requires protein synthesis (Wirthmueller *et al.*, 1990). Cells that are adherent also show a number of enhanced responses, including that of PAF synthesis (Sisson *et al.*, 1987), and we have earlier suggested that the process of adherence may resemble that of priming. If, in the various investigations of PAF production by neutrophils, the cells had been inadvertently primed during preparation (a number of agents, including PAF itself, are candidates), the subsequent PAF synthesis could have been affected considerably. It should also be noted that many stimuli which can in themselves cause PAF synthesis can prime for other stimuli, when applied at subthreshold concentrations.

Production of PAF in the neutrophil has been dissociated from other responses to stimuli, including secretion of granules (Betz and Henson, 1980; Ludwig *et al.*, 1984, 1985), phagocytosis (Sanchez Crespo *et al.*, 1980), superoxide production (Betz and Henson, 1980), aggregation and increased adhesivity (Sisson *et al.*, 1987). On the other hand, production of LTB_4 appears to be closely associated with that of PAF (Sisson *et al.*, 1987), probably as a function of the coordinate release of the respective precursors of eicosanoids (arachidonate) and PAF (lyso-PAF) from the same phospholipid (Swendsen *et al.*, 1983; Chilton *et al.*, 1984) (see below). Additionally, however, PAF synthesized by the neutrophil may act in an autocrine/paracrine fashion to stimulate, in turn, these same cells to make LTB_4 as a secondary event (Chilton *et al.*, 1982; Lin *et al.*, 1982), thereby complicating analysis of such data. Because of the amount of PAF produced by neutrophils (and monocytes, see below), contamination of other isolated cell preparations,

particularly eosinophils and basophils, with relatively few neutrophils is clearly a problem when determining the capacity of a given cell type to synthesize PAF.

Neutrophils also are capable of making acyl-PAF, but accumulation of this species does not occur because of rapid catabolism (Sturk *et al.*, 1989; Triggiani *et al.*, 1991a). Acyl-PAF catabolism proceeds at roughly twice the rate of that demonstrated for PAF and involves primarily de-acylation of the *sn*-1 group in addition to de-acetylation of the *sn*-2 group (Triggiani *et al.*, 1991a). Pretreatment with PMSF to block de-acylation of acyl-PAF results in a distribution of products reflective of precursor phosphatidylcholines (Sturk *et al.*, 1989). Therefore, the apparent selectivity for ether-linked lipids in this cell, and probably others, may simply reflect differential catabolism of the various PAF species.

4.1.2 Eosinophils

Eosinophils have been shown to produce PAF in response to both calcium ionophores and receptor-mediated stimuli (T.-c. Lee *et al.*, 1984; Cromwell *et al.*, 1990; Triggiani *et al.*, 1991c). The concurrent production of acyl-PAF appears to be minimal (<10% of the acetylated GPC) (Triggiani *et al.*, 1991c). Such studies have generally utilized eosinophils isolated from patients with a variety of disease states, characterized by eosinophilia, in which prior activation (and/or priming) of eosinophils as part of the disease state is quite likely. A recent study has shown that both normal-density and low-density eosinophils have a much greater capacity to synthesize PAF (28 and 23 ng/10^6 cells, respectively) in response to calcium ionophore stimulation than had been noted previously (Cromwell *et al.*, 1990). Additionally, it was found that release of PAF from both types of eosinophils was greater than from neutrophil preparations with comparable synthesis. Heightened activity of several enzymes involved in PAF metabolism has been demonstrated in activated (Lee *et al.*, 1982) and hypodense eosinophils (Cromwell *et al.*, 1990), and heightened catabolism may obscure the potential for PAF production, particularly in hypodense eosinophils (Cromwell *et al.*, 1990).

4.1.3 Monocytes and Macrophages

With the stimuli studied to date (Table 7.1), human monocytes appear to synthesize less PAF than neutrophils – 44 pmol/10^6 cells when stimulated with A23187 (Jouvin-Marche *et al.*, 1984). As was found with neutrophils, the calcium ionophore is the most potent stimulus but phagocytosis of opsonized particles is also an effective inducer of PAF synthesis (Elstad *et al.*, 1988). However, other agents that induce PAF synthesis in neutrophils (C5a, C5a des arg, neutrophil cationic protein or neutrophil cationic protein des arg) (Betz *et al.*, 1980; Camussi *et al.*, 1981a), or PAF analogue synthesis in endothelial cells (rabbit anti-human factor VIII, angiotensin II or vasopressin) (Camussi *et al.*, 1983b), are ineffective for monocytes. As with neutrophils, the state of adherence may modulate the amount of PAF produced by monocytes, although there are conflicting reports as to whether synthesis is increased (Camussi *et al.*, 1983a) or decreased (Elstad *et al.*, 1988), and may signify a state of activation or priming (Sanchez Crespo *et al.*, 1980). Also as seen in the neutrophil, secretion of lysozyme and phagocytosis can be dissociated from the PAF production (Camussi *et al.*, 1983a). It has been suggested that release of PAF may be greater on a per cell basis for the monocyte than the neutrophil (Camussi *et al.*, 1981a; Elstad *et al.*, 1988). Additionally, Valone and Epstein (1988) have reported a novel biphasic PAF biosynthetic response in human monocytes to three cytokines, IL-1β, TNF and IFNγ. These authors noted that while PAF was largely retained intracellularly during the early peak at 1–2 h of stimulation it was secreted (70%) during the late peak at 6–8 h. In light of the hypotheses outlined above to explain the release of PAF, the possibility of significant membrane perturbations and/ or the production of a carrier for PAF would seem fruitful areas for investigation, particularly because the late phase was shown to be dependent on protein synthesis. Furthermore, the late peak induced by IFNγ was inhibited by specific antibodies to TNF or IL-1β, suggesting a secondary role for these monocyte-derived cytokines in stimulated PAF production. In contrast, although a similar delay in PAF synthesis was seen in response to PAF stimulation of human monocytes, protein synthesis inhibition did not inhibit PAF biosynthesis (Valone, 1991). Although these last studies are technically difficult because stimulus added may contribute to assayed product, the notion that PAF may initiate its own synthesis is suggested.

The state of cellular differentiation and the source of mononuclear phagocytes can significantly alter their ability to synthesize PAF. For example, undifferentiated human pro-myelocytic HL-60 cells appear to contain PLA_2 and acetyltransferase, but do not synthesize PAF by the remodelling pathway until differentiation has occurred (Billah *et al.*, 1986). Similarly, respiratory syncytial virus infection stimulated both PAF synthesis and release from the human histiocytic lymphoma cell line (U937 cells), following monocyte-like differentiation by prior treatment with PMA (Villani *et al.*, 1991). In contrast, maturation of monocytic cells to alveolar macrophages may result in reduced PAF production (Elstad *et al.*, 1988). However, studies of macrophages removed from various sites *in vivo* are fraught with the problems inherent in their handling and subsequent culture, often for considerable periods, as well as with the types of stimuli used for their analysis. Triggiani *et al.* (1991c) demonstrated PAF production by lung macrophages in response to both calcium ionophore and opsonized zymosan. These investigators further showed that, in addition to PAF, approximately 20% of the acetylated GPC produced by these cells was acyl-PAF.

Conversely, normal human alveolar macrophages were reported to release PAF in response to A23187 but not to phagocytosis of opsonized zymosan (Arnoux *et al.*, 1980) whereas anti-IgE and specific allergen were effective on cells from asthmatics (Arnoux *et al.*, 1987), suggesting to the authors that the overall state of activation may modulate (production) release. However, these types of observations have not been consistent in either humans or other species. The enzymes involved in synthesis and degradation of PAF appear to vary with the maturational state of the cells, but, in addition, the cells themselves may alter that state upon removal from the *in vivo* environment (Villiani *et al.*, 1991). Additionally, the content of degradative enzymes in murine macrophages has already been shown to determine the levels of PAF detected (Roubin *et al.*, 1986), and it certainly seems reasonable to extrapolate such observations to humans. Suffice it to say that the mononuclear phagocyte can produce PAF and may constitute a significant source in both inflammatory and non-inflammatory processes.

Human alveolar macrophages also produce and release considerable quantities of lyso-PAF following stimulation with ionophores or, in the case of cells from asthmatics, a specific antigen (Arnoux *et al.*, 1987). Release of lyso-PAF, the immediate precursor and catabolic product of PAF, may place cells such as these in a unique position to participate in transcellular biosynthesis of PAF by supplying other cells with the immediate precursor of PAF (*vide infra*).

4.2 HUMAN CELLS THAT SYNTHESIZE PREDOMINANTLY ACYL-PLATELET-ACTIVATING FACTOR

4.2.1 Endothelial Cells

Human endothelial cells obtained from several vascular beds, including the umbilical cord vein (Prescott *et al.*, 1984; McIntyre *et al.*, 1985, 1986), iliac vein (Alam *et al.*, 1986) and microvasculature (Lynch and Henson, 1986), are reported to make large amounts of acetylated phospholipids in response to various stimuli (Table 7.1). Recent studies by several groups now suggest that the bulk of the acetylated phospholipid made by the umbilical vein and other endothelial cell sources (Garcia *et al.*, 1991a; Mueller *et al.*, 1991) is acyl-PAF and not authentic PAF (Clay *et al.*, 1991; Mueller *et al.*, 1991). While most studies have not formally quantified the amounts of acetylated phospholipid produced, available estimates of 50 ng/10^6 cells in response to thrombin (Prescott *et al.*, 1984; Bussolino *et al.*, 1987) suggest a remarkable capacity to synthesize these products. A recent study by Clay *et al.* (1991) using the GC–MS approach outlined above demonstrated approximately 15–20 ng/10^6 cells of acyl-PAF compared to 1 ng/10^6 cells or less of authentic

PAF in response to thrombin or an ionophore. Analysis of the composition of choline-containing phosphoglycerides (PAF precursors) in human umbilical vein endothelial cultures showed that only 2–3% were composed of the *sn*-1 alkyl-linked species (Takamura *et al.*, 1990), which again suggests that the distribution of acetylated phosphocholines that represent the final product closely reflects the original distribution of precursor species (Mueller *et al.*, 1991). It should be noted that, while endothelial cells appear to make little authentic PAF in comparison to acyl-PAF, the alkyl species, because of its great potency, may still be physiologically relevant.

Several unique characteristics of acetylated phospholipid synthesis have been noted for endothelial cells and warrant mention. Unlike the usual response to physiological stimuli which induce a peak of PAF production which then decays within about an hour, PAF (and/or acyl-PAF – the distinction was not drawn) synthesis by endothelial cells in response to stimulation with LTC_4 and LTD_4 was prolonged over 2 h (McIntyre *et al.*, 1986) while that with IL-1 and TNF stimulation was considerably delayed (Bussolino *et al.*, 1988). The onset of PAF synthesis in response to the last two stimuli was at 2 h with a peak production at 8–12 h. Inhibition of this delayed, prolonged response to cytokines by treatment of cells with cycloheximide and actinomycin D led Bussolino *et al.* (1988) to suggest that in endothelial cells *de novo* synthesis of synthetic enzymes, controlling elements or specific activation pathways, may be responsible for the delayed response. Inhibition of cyclooxygenase has been reported to result in increased production of PAF by endothelial cells in response to thrombin (Camussi *et al.*, 1983b), IL-1 and TNF (Bussolino *et al.*, 1988), raising the possiblity of prostanoid regulatory effects on PAF synthesis in these cells.

As mentioned above, data from PAF production by endothelial cells (Prescott *et al.*, 1990; Zimmerman *et al.*, 1990) provide the strongest support for the suggestion that cell-associated PAF/PAF analogues may be exposed in the outer leaflet or "presented" on a surface protein to initiate adherence responses in neutrophils that may come in contact. Thus, PAF (PAF analogues) synthesized by endothelial cells in response to LTC_4, histamine, bradykinin, ATP and thrombin can only be detected in a cell-associated state (Zimmerman *et al.*, 1985, 1990; McIntyre *et al.*, 1986), even in the presence of large amounts of albumin, lipoproteins and liposomes (Prescott *et al.*, 1990). Additionally, following gluteraldehyde fixation, activated endothelial monolayers retain their ability to stimulate neutrophil adhesiveness, a property lost upon treatment of the endothelial cells with either PLA_2 or acetylhydrolase (but not other proteases) (Prescott *et al.*, 1990). Prescott *et al.* (1990) have suggested that the PAF is available on the endothelial cell surface to activate passing leucocytes and initiate their adhesion to the endothelium McIntyre *et al.* (1985, 1986), perhaps in concert with

recognition of the glycoprotein GMP-140 (Prescott *et al.*, 1990). Indeed, release of PAF/PAF analogues from endothelial cells appears to be variable and only occurs to any degree after delayed production as a result of cytokine stimulation (Table 7.1) and, interestingly, under these conditions is not implicated in PMN binding to endothelial cells (Prescott *et al.*, 1990). It is worth noting that there is no suggestion that the PAF molecule binds simultaneously to both the putative binding site in or on the endothelial surface and to the neutrophil receptor for PAF. It is probably too small for this, and the specificity of the receptor certainly suggests recognition of all three domains on the molecule. Accordingly, another way of describing this phenomenon is that the endothelial cell appears to have external binding sites for PAF of higher affinity than those of albumin, only exceeded perhaps by the affinity of the receptors on neutrophils. Alternatively, the neutrophils may have a way of extracting the PAF, which can then act to initiate adhesive responses.

4.2.2   Human Lung Mast Cells

PAF synthesis has been demonstrated in mast cells isolated from animal sources (Mencia-Huerta *et al.*, 1983) and from human lung tissue (Schleimer *et al.*, 1986). In preparations purified to 70% or more, both acyl-PAF and PAF (Triggiani *et al.*, 1990a) were produced in response to anti-IgE and calcium ionophore, although release into the media was undetectable (Lichtenstein *et al.*, 1984) (Table 7.1). Of interest, anti-IgE was shown in this cell type to be a superior stimulus to calcium ionophore in the production of acetylated GPC. Furthermore, the ratio of acyl-PAF to PAF was greater (approximately three times more acyl-PAF than PAF) when stimulated with anti-IgE and was nearly equal when stimulated with calcium ionophore, raising the notion that the type of stimulus used can result in differential output of these acetylated products (Triggiani *et al.*, 1991c). The rates of catabolism of acyl-PAF and PAF were noted to be quite similar following anti-IgE stimulation and could not explain the predominant finding of acyl-PAF as the major acetylated species. Of note was that acyl-PAF was catabolized primarily by removal of the *sn*-1 acyl group in addition to removal of the *sn*-2 acetyl group of PAF (Triggiani *et al.*, 1990a).

4.2.3   Basophils

Rabbit basophils were, in fact, the first cells shown to make PAF and they appear to synthesize and release large quantities of this compound (Betz *et al.*, 1980; Benveniste *et al.*, 1972; Lynch and Henson, 1986). It has been suggested that human basophils isolated in crude buffy coat preparations can make PAF in response to a variety of stimuli (Table 7.1).

Additionally, PAF synthesis was detected from human basophil leukaemia cells in response to A23187 (Lewis *et al.*, 1975). However, the general use of enriched fractions or whole leucocyte preparations for the study of human basophils raises important questions about their ability to synthesize these molecules. In mixed-cell populations, it would be difficult to demonstrate this when the proportion of basophils in the populations is generally so low (often no more than 5–20%). Furthermore, the use of basophil-specific stimuli in mixed-cell populations to implicate the human basophil as a producer of PAF (Camussi et al., 1981a) does not, unfortunately, eliminate the presence of secondary signals and other complex cell–cell interactions which could result in PAF production from contaminating cells. Recently, Triggiani et al. (1991c) have shown in human basophil preparations of 70–80% purity that calcium ionophore stimulation resulted in an approximately six-fold greater production of acetylated GPCs, and that the presence of contaminating neutrophils could not account for the amounts produced in the basophil preparations. Of note was that stimulation with calcium ionophore resulted in enhanced PAF production, while stimulation with anti-IgE resulted in equal production of PAF and acyl-PAF. Such production of acyl-PAF would probably not have been detected by investigators relying on bioassay methods and may explain some of the previous controversy concerning the capability of the human basophil to synthesize PAF. Thus, in basophils, as in mast cells, anti-IgE may cause marked stimulation of acyl-PAF, which again raises the question of the biological importance of this lipid in allergic reactions, either as a mediator itself, or in the modulation of cellular response to PAF. Furthermore, the shift in the relative output of acetylated phospholipids depending on stimulus suggests some discriminate level of cellular control rather than output merely being a reflection of precursor prevalence. Of technical concern is that many of these studies utilize incorporation of tritiated acetate and await confirmation of mass measurements.

4.3   Human Cells for which Platelet-Activating Factor Synthesis is Attributed, but the Linkage Predominance is Unknown

4.3.1   Lymphocytes

Early studies suggested that lymphocytes, including those from humans, produce little, if any, detectable PAF and the cell seemingly lacks acetyltransferase activity (Jouvin-Marche et al., 1984; Garcia et al., 1991b). However, to demonstrate definitively that a cell such as the lymphocyte does not, and cannot, synthesize the molecule it will be necessary to show that it either does not make the enzymes required, does not activate them or lacks the precursor phospholipid. There is good evidence that PAF can activate and modify a variety of lymphocyte responses via a receptor-mediated process(es)

(Rola-Pleszcynski et al., 1987; Delioust et al., 1988; Mazer et al., 1990) and, thus, unlike the cells listed above, which appear to both produce PAF and be a target for PAF, lymphocytes may only respond to this mediator. Of note is that cells isolated from patients with several different types of lymphocytic leukaemias have been shown to produce PAF in response to various stimuli, some only after the addition of acetyl-CoA (Table 7.1; Bussolino et al., 1984; Foa et al., 1985). The ability to produce PAF may therefore be lost with ultimate differentiation along the lymphocytic lineage. Furthermore, recent studies of an EBV-transformed B cell line demonstrated notable acetyltransferase activity and production of acetylated GPCs, predominantly of the acyl linkage (Fig. 7.3).

4.3.2   Platelets

Recent reports by two laboratories report little or no tritiated acetate incorporation into PAF in calcium ionophore-treated human platelets (Sturk et al., 1989; Triggiani et al., 1991c). Thus, whether this cell represents a significant source of either acyl-PAF or PAF during inflammatory or thrombotic events is still unclear. Other investigators, using other means of detection, have demonstrated that platelets produce small quantities of PAF, approximately 2–4 fmol/10^6 cells, in response to thrombin or calcium ionophore stimulation (Touqui et al., 1985; Alam and Silver, 1987). Additionally, thrombin-stimulated platelets produced 10–12 times more acetylated GPC when treated with PMSF (Touqui et al., 1985), although most of this product now seems to be acyl-PAF (Sturk et al., 1989). Once again, the distribution of alkyl and acyl acetylated products from PMSF-treated platelets closely resembled the distribution of precursor species. The fact that this cell seems relatively deficient in acetyltransferase may explain its relatively poor capacity to make either PAF or acyl-PAF. Platelets may, however, release lyso-PAF or, potentially, the lyso precursor of acyl-PAF, which can in turn be utilized by other cells to result in heightened PAF/acyl-PAF production via transcellular biosynthesis (Coeffier et al., 1986) (see below).

4.3.3   Natural Killer Cells

Another cell type that may make PAF is the natural killer cell found in the rather diverse population of circulating cells, known by their phenotypic appearance as large granular lymphocytes. PAF synthesis and release has been reported in response to stimulation by A23187 or the Fc receptor (which in this cell is thought to be similar to that found in the human neutrophil) (Malavasi et al., 1986). However, any role that PAF may play in the cytotoxic repertoire of the natural killer cell is, at this time, speculative.

5. Transcellular Biosynthesis of Platelet-Activating Factor

It is now apparent that complex interactions between different cell types are often involved in the generation of lipid mediators *in vitro*, and almost certainly *in vivo* as well. Transcellular biosynthesis is a term used to represent the release of an intermediate from one cell that is then taken up and converted into the final mediator by another. Transcellular metabolism refers to the further metabolism by one cell of a mediator first produced by another cell and usually results in inactive metabolites. Both of these processes are likely to be important in the production and further handling of PAF in mixed populations of cells.

Thus, lyso-PAF has been shown to be released from a number of different cell types, including alveolar macrophages from humans (Albert and Snyder, 1983; Robinson and Snyder, 1985; Arnoux et al., 1987) and rats (Albert and Snyder, 1983; Robinson and Snyder, 1985), platelets (Chignard et al., 1986) and endothelial cells (Blank et al., 1986), and has been detected in body fluids such as amniotic fluid (Billah and Johnson, 1983) and nasal lavage following allergen challenge (Miadonna et al., 1989; Shin et al., 1991). Whether or not this represents direct release of lyso-PAF itself or of the action of extracellular acetylhydrolases on released PAF, the molecule could nevertheless be taken up into cells and used for synthesis of either precursor phospholipids or directly of PAF, i.e. in a transcellular fashion. Cells such as neutrophils can take up lyso-PAF (Chilton et al., 1984; Jouvin-Marche et al., 1984) in an efficient manner (especially if activated) (Bratton et al., 1992) and can metabolize this into other phospholipids. If the acetyltransferase is active (*vide infra*) it seems reasonable to expect the synthesis of PAF and, in fact, such a transcellular process has been described (Coeffier et al., 1986). Fibroblasts have also been reported to utilize exogenous lyso-PAF to enhance PAF synthesis (Michel et al., 1988), suggesting that other tissues/cells, if supplied with the immediate precursor, may have enhanced biosynthetic capability. Furthermore, in the case of neutrophils and platelets (Jouvin-Marche et al., 1984; Coeffier et al., 1986), the availability of lyso-PAF appears to be a limiting factor in PAF synthesis, perhaps relating to the likelihood that cells have several mechanisms by which to rid themselves of lysophospholipids, since their accumulation would be expected to be toxic. In fact, acetylation to PAF and acyl-PAF may be such a mechanism that only secondarily creates molecules with extracellular mediator function. In contrast, uptake of related but inactive lysophospholipids (both choline- and ethanolamine-containing) also results in enhanced PAF production, in this case probably secondary to increased precursor availability in neutrophils (Sugiura et al., 1990) or by a novel, CoA-independent transacylase activity resulting in lyso-PAF generation in HL-60 cells (Uemura et al., 1991). In support of the presence of such a transacylase, Uemura et al. (1991) demonstrated that exogenously added 1-acyl-2-lyso-sn-glycero-3-phosphoethanolamine, 1-alkyl-2-lyso-sn-glycero-3-phosphoethanolamine, lyso-PAF and 1-acyl-2-lyso-sn-glycero-3-phosphocholine, but not other lysophospholipid classes, were all able to stimulate (enhance) the formation of lyso-PAF from an alkyl-labelled 1-alkyl-2-acyl-GPC precursor by serving as acyl group acceptors. The activity was independent of calcium, EGTA or the presence of bromophenacyl bromide, all known to affect certain PLA_2 activity.

Recent studies utilizing the erythrocyte ghost to examine the transbilayer movement of PAF and related analogues, including acyl-PAF and lyso-PAF, have not only shed light on the process of release of PAF and related analogues, but also on the uptake of these lipids (Bratton et al., 1991). As with release (see above), PAF/PAF analogues and lyso-PAF uptake into the erythrocyte was associated with enhanced flip/flop of the exogenously added analogues and endogenous phospholipids, and was influenced by carrier proteins such as albumin. Studies of activated neutrophils (Bratton et al., 1992) have subsequently shown that enhanced uptake of lyso-PAF (as well as PAF and inactive PAF analogues) did not require the PAF receptor, analogue metabolism, or endocytosis, was non-specific for certain structural features of the analogues, and was accompanied by enhanced flipping of endogenous plasma membrane phospholipids. Taken together these findings suggest that either lyso-PAF itself or other lysophospholipids of the choline and ethanolamine classes liberated from one cell upon activation of PLA_2 could be taken up into an adjacent cell and stimulate PAF production, either by directly supplying precursor (lyso-PAF) in the rate-limiting step of PAF synthesis, or through the generation of lyso-PAF by involvement/activation of a transacylase.

Additionally, other factors in the extracellular milieu that appear to affect PAF synthesis may also affect transcellular biosynthesis of PAF. A synthesis–release coupling mechanism involving serum proteins, particularly albumin, is suggested by several lines of evidence. The presence of albumin is required for PAF biosynthesis in the neutrophil (Ludwig et al., 1985) and rabbit buffy coat leucocytes (Benveniste et al., 1972). Data from Ninio et al. (1988) suggest that the complex interaction of albumin with PAF, lyso-PAF and other phospholipid analogues may influence acetyltransferase activity by removing end-product feedback inhibition. Inhibition of acetyltransferase occurs with high concentrations of lyso-PAF or PAF itself, while low concentrations of lyso-PAF (or lyso-PC) and increased albumin concentrations were protective of enzymatic activity (Ninio et al., 1988).

In the environment of the inflammatory reaction, with multiple cell types in close contact (both inflammatory and tissue cells), exchange of intermediates would not be unexpected. The presence of albumin from the increased vascular permeability seen in inflammation and allergy would only serve to enhance this exchange by stabilizing the lipids during the transfer process.

6. Other Cellular Sources of Platelet-Activating Factor

As has already been mentioned, the finding of PAF in pathological fluids, while important in itself, does not usually help determine the cellular source of the molecule. Nevertheless, PAF and its analogues have been detected in such diverse materials as the following: nasal washings (Friedman *et al.*, 1986; Miadonna *et al.*, 1989; Shin *et al.*, 1991), plasma (Chan-Yeung *et al.*, 1991) and skin blisters (Valone *et al.*, 1987; Shalit *et al.*, 1989) from allergic patients challenged with antigen; sputa from patients with asthma and chronic obstructive pulmonary disease (Grandel *et al.*, 1985a); lung lavage from infants with bronchopulmonary dysplasia (Stenmark *et al.*, 1987); plasma from infants with persistent pulmonary hypertension of the newborn (Caplan *et al.*, 1990a); and pleural fluid from patients with eosinophil or neutrophil infiltrations (Oda *et al.*, 1989). Additionally, blood from patients with cold urticaria undergoing cold challenge (Grandel *et al.*, 1985b), blood and ascitic fluid in cirrhotics (Caramelo *et al.*, 1987), psoriatic scales (Mallet and Cunningham, 1985), and cerebrospinal fluid in bacterial meningitis (Arditi *et al.*, 1990) have been reported as sources of PAF. Stool in infectious diarrhoea (Denizot *et al.*, 1991) and pouchitis (Chaussade *et al.*, 1991), mucosa in inflammatory bowel disease (Eliakim *et al.*, 1988; Kald *et al.*, 1990), and plasma in neonatal necrotizing enterocolitis (Caplan *et al.*, 1990b) are also reported sources. While it is clear that tissue cells could have supplied the PAF in each of these circumstances, inflammatory cells could also have been the source.

Tissue cells that have been reported to make PAF now include mesangial cells (Lianos and Zanglis, 1987; Neuwirth *et al.*, 1988), type II cells (Hoffman *et al.*, 1986; Kumar *et al.*, 1987), keratinocytes (Denizot *et al.*, 1989), hepatocytes (Miwa *et al.*, 1987), fibroblasts (Michel *et al.*, 1988), airway epithelial cells (Holtzman *et al.*, 1991), thymus (Salem *et al.*, 1989) and fetal brain cells (Sogos *et al.*, 1990). In most of these cases, however, the exact composition of the molecular species was not determined. Furthermore, the majority of studies have been carried out using cells from non-human sources. As described above, the reported presence of PAF in body fluids under normal conditions could suggest the ongoing production of the molecule from tissue cells under physiological conditions and would certainly be consistent with a role for PAF in normal body functions. However, even here the presence of a little inflammation could sometimes account for the PAF detected. For example, the PAF found in saliva could perhaps have been derived from neutrophils emigrating through the gingival crevice (McManus *et al.*, 1990). Nevertheless, in general,

it seems reasonable to suggest that most cell types can, under the right conditions of stimulation, produce PAF. In fact, as discussed below, there is a distinct possibility that either-PC turnover, and even perhaps the production of PAF, may play a role in normal cell function, let alone in processes involving whole organs and tissues. Critical to considering PAF as a mediator, therefore, will be the determination of the specific circumstances *in vivo* under which the molecule is synthesized and, just as critically, the conditions leading to its release from the cell of origin.

7. Synthesis of Platelet-Activating Factor and Related Analogues with Other Lipid Mediators

Data from several laboratories suggest that the re-modelling pathway for PAF synthesis (see Chapter 6) preferentially utilizes alkylarachidonoyl-GPC as the preformed precursor of PAF (Chilton *et al.*, 1984; Ramesha and Pickett, 1986b). More recently, similar data suggest that acyl-arachidonyl-GPC may likewise be a preferred substrate for the acyl analogue acyl-PAF (Clay *et al.*, 1991). The finding of preferential use of the *sn*-2 arachidonoyl-substituted GPCs by PLA_2 has led to the hypothesis that the first step of PAF/acyl-PAF synthesis, the cleavage of arachidonate, would in essence liberate two important mediator precursor molecules, the lyso-PAF/acyl-PAF precursor and arachidonic acid which, depending on the cell type (or in transcellular bio-synthesis, on neighbouring cells), could then be metabolized to PAF/acyl-PAF and lipoxygenase and/or cyclooxygenase products (Swendsen *et al.*, 1983; Walsh *et al.*, 1983; Chilton and Murphy, 1986; Ramesha and Pickett, 1986b). To date, perhaps the best studied example of dual synthesis is the concurrent synthesis of PAF and LTB_4 (a lipoxygenase product) in human PMNs. Data linking this type of concurrent synthesis are outlined below.

The production of LTB_4 and PAF by the human neutrophil in response to similar stimuli (Walsh *et al.*, 1981; Chilton *et al.*, 1982; Prescott, 1984) prompted studies to examine whether dual synthesis of these mediators was linked. The tight coupling of the time-course of PAF and LTB_4 biosynthesis for both adherent and suspended PMNs stimulated with calcium ionophore (Sisson *et al.*,1987), or primed with GM-CSF and then stimulated with chemotactic factors (Dahinden *et al.*, 1988; Wirthmueller *et al.*, 1989), support this hypothesis. Furthermore, the demonstration of a common precursor for both mediators, alkylarachidonoyl-GPC (Chilton, 1989), strengthens this linkage. Complicating this schema is the finding that PAF is itself a stimulus for LTB_4 synthesis in the neutrophil (Chilton *et al.*, 1982; Lin *et al.*, 1982), while LTB_4 does not apparently stimulate PAF synthesis (Sisson *et al.*, 1987). In theory, linkage of PAF and eicosanoid biosynthesis would dictate that the inhibition of PAF synthesis by pharmacological agents at the level of the PLA_2 would be accompanied by inhibition of synthesis of eicosanoids. Such circumstantial evidence that the two pathways are linked and can be concomitantly inhibited has been demonstrated in rat neutrophils treated with dexamethasone (Fradin *et al.*, 1988) and in recent studies by Chilton *et al.* (1989), who treated human neutrophils with isoprenaline and were able to inhibit synthesis of both PAF and LTB_4.

Data similar to that described above linking eicosanoid and PAF production in the neutrophil have been obtained in human endothelial cells, which, when stimulated for PAF synthesis (but, in light of more recent data, are most likely synthesizing acyl-PAF (Clay *et al.*, 1991)), also produce PGI_2. Agonists for dual synthesis include LTC_4, LTD_4, histamine (through H_1 receptors), bradykinin (through B_2 receptors) and ATP (McIntyre *et al.*, 1985, 1986). Of the leukotrienes, each appeared to have identical agonist activity in the production of both PAF and PGI_2 with the rank order of potency being LTC_4 >LTD_4>>LTB_4=LTE_4 (McIntyre *et al.*, 1986). Of note is that, while the time-courses of initiation of synthesis of the two mediators appear to be closely linked, the duration of synthesis (7.5–15 min for PGI_2 and > 40 min for PAF) was widely divergent. Also, in contrast were the fates of the two mediators, PGI_2 being almost entirely released, while PAF was not released into the medium even in the presence of high concentrations of albumin and liposomes (McIntyre *et al.*, 1985, 1986; Zimmerman *et al.*, 1985). Now that findings suggest that the major acetylated product of endothelial cells is actually acyl-PAF, the importance of this analogue in the inflammatory site during PGI_2-enhanced blood flow and inflammatory cell delivery would seem to be underscored. Recent studies suggest that acyl-PAF may have an antagonistic role in neutrophil activation (Triggiani *et al.*, 1991b), or possibly this analogue stimulates neutrophil adhesion (the first step in egress into inflamed tissues) without full activation within the intravascular space.

Since the liberation and reincorporation of arachidonic acid has been closely linked to PAF synthesis and catabolism (Chilton *et al.*, 1984; Chilton and Murphy, 1986; Swendsen *et al.*, 1987), depletion of the potential common precursor, alkylarachidonoyl-GPC, has been studied to further define the relationship between the two mediator pathways. Such limitation of substrate availability may serve as a clinically relevant mechanism whereby both eicosanoids and PAF synthesis may be inhibited. Inhibition of both PAF and LTB_4 synthesis was demonstrated in arachidonic acid-deficient rat polymorphonuclear leucocytes and supported the possibility of modifying the activity of both pathways by depletion of the common precursor molecule, alkylarachidonoyl-GPC, through dietary manipulation (Ramesha and Pickett, 1986b). Most attempts to date utilize the supplementation of marine fish oils to inflam-

matory cells in culture, or to whole animals, to increase incorporation of the supplied fatty acid, decrease arachidonate release, and/or shift the yield of leukotrienes and prostaglandins to less potent analogues upon inflammatory cell stimulation (T.H. Lee *et al.*, 1984; Lee *et al.*, 1985; Prescott, 1984; Strasser *et al.*, 1985). Studying the biosynthesis of PAF, several laboratories have demonstrated decreased synthesis following arachidonic acid depletion. Codde *et al.* (1987) found a decrease in whole blood PAF and lyso-PAF in human subjects fed a diet enriched in fish oils. Similarly, Sperling *et al.* (1987) found a decrease in PAF production in ionophore-stimulated human monocytes following fish oil supplementation either *in vivo* or *in vitro*. Kawasaki *et al.* (1988) demonstrated decreased PAF production in arachidonic acid-deficient HL-60 cells differentiated to granulocytic-type cells and stimulated with calcium ionophore, while synthesis was restored following supplementation with arachidonic acid or its precursor. Conversely, no decrease in PAF synthesis was demonstrated in ionophore-stimulated neutrophils from human subjects fed a fish oil supplemented diet (Triggiani *et al.*, 1990b). Further, these investigators demonstrated avid incorporation of both arachidonic acid and eicosapentaenoic acid into endogenous PAF precursors and exogenously added lyso-PAF. Subsequent stimulus-induced liberation suggests that PLA_2 can utilize either *sn*-2 substituted precursor. Finally, enrichment of cultured murine mast cells (Triggiani *et al.*, 1990b) or endothelial cells (Garcia *et al.*, 1991a) with unsaturated marine fatty acids (or arachidonic acid) actually resulted in enhanced PAF (or acyl-PAF) production. In much of this preliminary work directed at decreasing substrate availability, it is unclear whether arachidonate depletion (or substitution) to any degree, as measured for whole cells or tissues, is reflective of limiting substrate in specific precursor pools utilized in PAF biosynthesis. The concept that substrate pools and subcellular locations may restrict substrate availability is unproven but attractive. As more is learned of the intracellular topography of these events, perhaps the means to modulate PAF synthesis will be identified.

In conclusion, while data linking tandem PAF and eicosanoid biosynthesis in several cell types are available, pinpointing the actual source of these mediators from a common precursor molecule is yet to be firmly established. Conclusive proof of such tandem biosynthesis and its relative contribution to cellular biosynthesis of these mediators awaits refinement in detection methods and effective dual labelling of precursor pools to a high degree of specific activity.

8. *Conclusions*

In summary, PAF is synthesized and, under certain conditions, released in response to a variety of stimuli from numerous cells, particularly those involved in inflammation.

New information has come to light that suggests release of PAF to the extracellular milieu is in part governed by bulk transbilayer movement of phospholipids in the activated plasma membrane. The precise role of PAF as a released mediator in either pathological processes or in homeostatic mechanisms is far from clear, and proof of such a role requires detection of PAF *in vivo* and abrogation of its effects with specific blockade by antagonists or specific modulation of its synthesis/release. Alternatively, the retention of PAF by many cells suggests that it may have its effects in intimate cell–cell interactions, and/or in many settings which have as yet unknown intracellular functions.

New evidence that the PAF precursor and metabolite lyso-PAF is similarly released and taken up by cells opens the door for new transcellular mechanisms of cooperation in the biosynthesis of PAF. Likewise, the emerging prominence of acyl-PAF as an acetylated analogue, synthesized under many of the same circumstances as PAF, leaves many unanswered questions. Whether it will have similar pro-inflammatory activity or, conversely, different attributes or antagonist roles to play is yet to be determined.

9. References

Alam, I. and Silver, M.J. (1987). Metabolism of 1-alkyl-2-acyl-GPC in human platelets in response to stimulation by thrombin. Thromb. Res. 45, 311–322.

Alam, I., Smith, J.B. and Silver, M.J. (1983a). Human and rabbit platelets form platelet-activating factor in response to calcium ionophore. Thromb. Res. 30, 71–79.

Alam, I., Smith, J.B. and Silver, M.J. (1983b). Metabolism of platelet-activating factor by blood platelets and plasma. Lipids 18, 534–538.

Alam, I., Silver, M.J., Mueller, S.N. and Levine, E.M. (1986). In "Proc. 2nd Int. Conf. on Platelet-Activating Factor and Structurally Related Alkyl Ether Lipids, Gatlinburg, Tennessee", p. 58.

Albert, D.H. and Snyder, H.F. (1983). Biosynthesis of 1-alkyl-2-acetyl-*sn*-glycero-3-phosphocholine (platelet-activating factor) from 1-alkyl-2-acyl-*sn*-glycero-3-phosphocholine by rat alveolar macrophages. J. Biol. Chem. 258, 97–102.

Alonso, F., Henson, P.M. and Leslie, C.C. (1986). A cytosolic phospholipase in human neutrophils that hydrolyzes arachidonoyl-containing phosphatidylcholine. Biochim. Biophys. Acta 878, 273–280.

Arditi, M., Manogue, K.R., Caplan, M. and Yogev, R. (1990). Cerebrospinal fluid cachectin/tumor necrosis factor-a and platelet-activating factor concentrations and severity of bacterial meningitis in children. J. Infect. Dis. 162, 139–147.

Arnoux, B., Duval, D. and Benveniste, J. (1980). Release of platelet-activating factor (PAF-acether) from alveolar macrophages by the calcium ionophore A23187 and phagocytosis. Eur. J. Clin. Invest. 10, 437–441.

Arnoux, B., Jouvin-Marche, E., Arnoux, A. and Benveniste, J. (1982). Release of PAF-acether from human blood monocytes. Agents Actions 12, 713–716.

Arnoux, B., Joseph, M., Simoes, M.H., Tonnel, A.B., Duroux, P., Capron, A. and Benveniste, J. (1987). Antigenic release of PAF-acether and b-glucuronidase from alveolar macrophages of asthmatics. Bull. Eur. Physiopathol. Respir. 23, 119–124.

Averill, E., Shin, M.H., Hubbard, W.C., Chilton, F., Nacleno, R. and Liu, M.C. (1991). Metabolism of platelet activating factor (PAF) in nasal and lung fluids following antigen challenge. J. Allergy Clin. Immunol. 87, 247.

Banks, J.B., Wykle, R.L., O'Flaherty, J.T. and Lumb, R.H. (1988). Evidence for protein-catalyzed transfer of platelet activating factor by macrophage cytosol. Biochim. Biophys. Acta 961, 48–52.

Benveniste, J., Henson, P.M. and Cochrane, C.G. (1972). Leukocyte-dependent histamine release from rabbit platelets – the role of IgE, basophils and a platelet-activating factor. J. Exp. Med. 136, 1356–1377.

Betz, S.J. and Henson, P.M. (1980). Production and release of platelet-activating factor (PAF); dissociation from degranulation and superoxide production in the human neutrophil. J. Immunol. 125, 2756–2763.

Betz, S.J., Lotner, G.Z. and Henson, P.M. (1980). Generation and release of platelet-activating factor (PAF) from enriched preparations of rabbit basophils; failure of human basophils to release PAF. J. Immunol. 125, 2749–2755.

Billah, M.M. and Johnston, J.M. (1983). Identification of phospholipid platelet-activating factor (1-*O*-alkyl-2-acetyl-*sn*-3-phosphocholine) in human amniotic fluid and urine. Biochem. Biophys. Res. Commun. 113, 51–58.

Billah, M.M., Eckel, S., Myers, R.F. and Siegel, M.I. (1986). Metabolism of platelet-activating factor (1-*O*-alkyl-acetyl-*sn*-glycero-3-phosphocholine) by human promyelocytic leukemic HL60 cells. J. Biol. Chem. 261, 5824–5831.

Blank, M.L. and Snyder, F. (1983). Improved high-performance liquid chromatographic method for isolation of platelet-activating factor from other phospholipids. J. Chromatography 273, 415–420.

Blank, M.L., Lee, T.-C., Fitzgerald, V. and Snyder, F. (1981). A specific acetylhydrolase for 1-alkyl-2-acetyl-*sn*-glycero-3-phosphocholine (a hypotensive and platelet-activating lipid). J. Biol. Chem., 256, 175–178.

Blank, M.L., Spector, A.A., Kaduce, T.L., Lee, T.-c. and Snyder, F. (1986). Metabolism of platelet activating factor (1-alkyl-2-acetyl-*sn*-glycero-3-phosphocholine) and 1-alkyl-2-acetyl-*sn*-glycerol by human endothelial cells. Biochim. Biophys. Acta 876, 373–378.

Blank, M.L., Robinson, M. and Snyder, F. (1987). In "Platelet-Activating Factor and Related Lipid Mediators" (ed F. Snyder), pp 33–52. Plenum Press, New York.

Bligh, E.G. and Dyer, W.J.A. (1959). A rapid method of total lipid extraction and purification. Can. J. Biochem. Physiol. 37, 911–917.

Bratton, D. and Henson, P.M. (1989). In "Platelet Activating Factor and Human Disease" (eds P.J. Barnes, C.P. Page, and P.M. Henson), pp 23–57. Blackwell, Oxford.

Bratton, D.L., Harris, R.A., Clay, K.L. and Henson, P.M. (1988). Effects of platelet activating factor on calcium–lipid interactions and lateral phase separations in phospholipid vesicles. Biochim. Biophys. Acta 943, 211–219.

Bratton, D.L., Kailey, J.M., Clay, K.L. and Henson, P.M. (1991). A model for the extracellular release of PAF: the influence of plasma membrane phospholipid asymmetry. Biochim. Biophys. Acta 1062, 24–34.

Bratton, D.L., Dreyer, E., Kailey, J.M., Fadok, V.A., Clay, K.L.

and Henson, P.M. (1992). The mechanism of internalization of PAF in activated human neutrophils: enhanced transbilayer movement across the plasma membrane. J. Immunol. (in press).

Burke, L.A., Crea, A.E.G., Wilkinson, J.R.W., Arm, J.P., Spur, B.W. and Lee, T.H. (1990). Comparison of the generation of platelet-activating factor and leukotriene C_4 in human eosinophils stimulated by unopsonized zymosan and by the calcium ionophore A23187: the effects of nedocromil sodium. J. Allergy Clin. Immunol. 85, 26–35.

Bussolino, F., Foa, R., Malavasi, F., Ferrando, M.L. and Camussi, G. (1984). Release of platelet-activating factor (PAF)-like material from human lymphoid cell lines. Exp. Hematol. 12, 688–693.

Bussolino, F., Breviario, F., Tetta, C., Aglietta, M., Mantovani, A. and Dejana, E. (1986). Interleukin-1 stimulates platelet-activating factor production in cultured human endothelial cells. J. Clin. Invest. 77, 2027–2033.

Bussolino, F., Breviario, F., Aglietta, M., Sanavio, F., Bosia, A. and Dejana, E. (1987). Studies on the mechanism of interleukin-1 stimulation of platelet activating factor synthesis in human endothelial cells in culture. Biochim. Biophys. Acta 927, 43–54.

Bussolino, F., Camussi, G. and Baglioni, C. (1988). Synthesis and release of platelet-activating factor by human vascular endothelial cells treated with tumor necrosis factor or interleukin-1α. J. Biol. Chem. 263, 11856–11861.

Camussi, G., Mencia-Huerta, J.M. and Benveniste, J. (1977). Release of platelet-activating factor and histamine. I. Effect of immune complexes, complement and neutrophils on human and rabbit mastocytes and basophils. Immunology 33, 532–532.

Camussi, G., Aglietta, R., Coda, T., Bussolino, F., Piacibello, W. and Tetta, C. (1981a). Release of platelet-activating factor (PAF) and histamine. II. The cellular origin of human PAF: monocytes, polymorphonuclear neutrophils and basophils. Immunology 42, 191–199.

Camussi, G., Tetta, C., Coda, R. and Benveniste, J. (1981b). Release of platelet-activating factor in human pathology. I. Evidence for the occurrence of basophil degranulation and release of platelet-activating factor in systemic lupus erythematosus. Lab. Invest. 44. 241–251.

Camussi, G., Bussolino, F., Ghezzo, F. and Pegoraro, L. (1982). Release of platelet-activating factor from HL-60 human leukemic cells following macrophage-like differentiation. Blood 59, 16–22.

Camussi, G., Bussolino, F., Tetta, C., Piacibello, W. and Aglietta, M. (1983a). Biosynthesis and release of platelet-activating factor from human monocytes. Int. Arch. Allergy Appl. Immunol. 70, 245–251.

Camussi, G., Aglietta, M., Malavasi, F., Tetta, C., Piacibello, W., Sanavio, F. and Bussolino, F. (1983b). The release of platelet-activating factor from human endothelial cells in culture. J. Immunol. 131, 2397–2403.

Camussi, G., Bussolino, F., Salvidio, G. and Baglioni, C. (1987). Tumor necrosis factor/cachectin stimulates peritoneal macrophages, polymorphonuclear neutrophils, and vascular endothelial cells to synthesize and release platelet-activating factor. J. Exp. Med. 166, 1390–1404.

Camussi, G., Tetta, C., Bussolino, F. and Baglioni, C. (1988). Synthesis and release of platelet-activating factor is inhibited by plasma α₁-proteinase inhibitor or α₁-antichymotrypsin and is stimulated by proteinases. J. Exp. Med. 168, 1293–1306.

Camussi, G., Tetta, C., Bussolino, F. and Baglioni, C. (1989). Tumor necrosis factor stimulates human neutrophils to release leukotriene B_4 and platelet-activating factor. Induction of phospholipase A_2 and acetyl-CoA:1-alkyl-sn-glycero-3-phosphocholine O_2-acetyltransferase activity and inhibition by antiproteinase. Eur. J. Biochem. 182, 661–666.

Caplan, M.S., Hsueh, W., Sun, X.-M., Gidding, S.S. and Hageman, J.R. (1990a). Circulating plasma platelet activating factor in persistent pulmonary hypertension of the newborn. Am. Rev. Respir. Dis. 142, 1258–1262.

Caplan, M.S., Sun, X.-M., Hsueh, W. and Hageman, J.R. (1990b). Role of platelet activating factor and tumor necrosis factor-alpha in neonatal necrotizing enterocolitis. J. Pediatr. 116, 960–964.

Caramelo, C., Fernandez-Gallardo, S. and Marin-Cao, D. (1984). Presence of platelet-activating factor in blood from humans and experimental animals; its absence in anephric individuals. Biochem. Biophys. Res. Commun. 120, 789–796.

Caramelo, C., Fernandez-Gallardo, S., Santos, J.C. et al. (1987). Increased levels of platelet-activating factor in blood from patients with cirrhosis of the liver. Eur. J. Clin. Invest. 17, 7–11.

Chan-Yeung, M., Lam, S., Chan, H., Tse, K.S. and Salari, H. (1991). The release of platelet-activating factor into plasma during allergen-induced bronchoconstriction. J. Allergy Clin. Immunol. 87, 667–673.

Chaussade, S., Denizot, Y., Valleur, P., Nicoli, J., Raibaud, P., Guerre, J., Hautefeuille, P., Couturier, D. and Benveniste, J. (1991). Presence of PAF-acether in stool of patients with pouch ileoanal anastomosis and pouchitis. Gasteroenterology 100, 1509–1514.

Chignard, M., Le Couedic, J.P., Tence, M., Vargaftig, B.B. and Benveniste, J. (1979). The role of platelet-activating factor in platelet aggregation. Nature 279, 799–800.

Chignard, M., Coeffier, E., LeCouedic, J.P. and Benveniste, J. (1986). In "Proc. 2nd Int. Conf. on Platelet-Activating Factor and Structurally Related Alkyl Ether Lipids, Gatlinburg, Tennessee", p 70.

Chilton, F.H. (1989). Potential phospholipid source(s) of arachidonate used for the synthesis of leukotrienes by the human neutrophil. Biochem. J. 258, 327–333.

Chilton, F.H. and Connell, T.R. (1988). Major endogenous sources of arachidonate in the human neutrophil. J. Biol. Chem. 263, 5260–5265.

Chilton, F.H. and Murphy, R.C. (1986). Remodelling of arachidonate-containing phosphoglycerides within the human neutrophil. J. Biol. Chem. 261, 7771–7777.

Chilton, F.H., O'Flaherty, J.T., Walsh, C.E. et al. (1982). Platelet activating factor-stimulation of the lipoxygenase pathway in polymorphonuclear leukocytes by 1-O-alkyl-2-O-acetyl-sn-glycero-3-phosphocholine. J. Biol. Chem. 257, 5402–5407.

Chilton, F.H., Ellis, J.M., Olson, S.C. and Wykle, R.L. (1984). 1-O-Alkyl-2-arachidonoyl-sn-glycero-3-phosphocholine, a common source of platelet-activating factor and arachidonate in human polymorphonuclear leukocytes. J. Biol. Chem. 259, 12014–12019.

Chilton, F.H., Schmidt, D., Torphy, T., Goldman, D. and Undem, B. (1989). cAMP inhibits platelet activating factor (PAF) biosynthesis in the human neutrophil. FASEB J. 3, A308 (Abstr.).

Christman, B.W. and Blair, I.A. (1989). Analysis of platelet activating factor in human saliva by gas chromatography/mass spectrometry. Biomed. Environ. Mass Spectrom. 18, 258–264.

Clay, K.L. (1990) Quantitation of platelet activating factor by gas chromatography–mass spectrometry. Methods Enzymol. 187, 134–142.

Clay, K.L., Stene, D.O. and Murphy, R.C. (1984a). Quantitative analysis of platelet activating factor (AGEPC) by fast atom bombardment mass spectrometry. Biomed. Mass Spectrom. 11, 47–49.

Clay, K.L., Murphy, R.C., Andres, J.L., Lynch, J. and Henson, P.M. (1984b). Structure elucidation of platelet-activating factor derived from human neutrophils. Biochem. Biophys. Res. Commun. 121, 815–825.

Clay, K.L., Johnson, C. and Henson, P. (1990). Binding of platelet activating factor to albumin. Biochim. Biophys. Acta 1046, 309–314.

Clay, K.L., Johnson, C. and Worthen, G.S. (1991). Biosynthesis of platelet activating factor and 1-O-acyl analogues by endothelial cells. Biochim. Biophys. Acta 1094, 43–50.

Cluzel, M., Undem, B.J. and Chilton, F.H. (1989). Release of platelet-activating factor and the metabolism of leukotriene B_4 by the human neutrophil when studied in a cell superfusion model. J. Immunol. 143, 3659–3665.

Codde, J.P., Vandongen, R., Mori, T.A., Beilin, L.J. and Hill, K.J. (1987). Can the synthesis of platelet-activating factor, a potent vasodilator and pro-aggregatory agent, be altered by dietary marine oils? Clin. Exp. Pharmacol. Physiol. 14, 197–202.

Coeffier, E., Delautier, D., Le Couedic, J.-P., Chignard, M. and Benveniste, J. (1986). In "Proc. 2nd Int. Conf. Platelet-Activating Factor and Structurally Related Alkyl Ether Lipids, Gatlinburg, Tennessee", p 124.

Columbo, M., Horowitz, E.M., Patella, V., Kagey-Soborka, A., MacGlashan, D.W. Jr, Chilton, F.H. and Lichtenstein, L.M. (1991). The effects of 1-alkyl-2-acetyl-sn-glycero-3-phospho-choline (AAPC) on human basophil mediator release. J. Allergy Clin. Immunol. 87, 241.

Cox, C.P., Wardlow, M.L., Jorgensen, R. and Farr, R.S. (1981). The presence of platelet-activating factor (PAF) in normal human mixed saliva. J. Immunol. 127, 46–50.

Cromwell, O., Wardlaw, A.J., Champion, A., Moqbel, R., Osel, D. and Kay, A.B. (1990). IgG-dependent generation of platelet-activating factor by normal and low density human eosinophils. J. Immunol. 145, 3862–3868.

Dahinden, C.A., Zingg, J., Maly, F.E. and de Weck, A.I. (1988). Leukotriene production in human neutrophils primed by recombinant human granulocyte/macrophage colony-stimulating factor and stimulated with the complement component C5A and FMLP as second signals. J. Exp. Med. 167, 1281–1295.

Delioust, A., Vivier, E., Salem, P., Benveniste, J. and Thomas, Y. (1988). Immunoregulatory functions of PAF-acether. I. Effect of PAF-acether on cell proliferation. J. Immunol. 140, 240–245.

Demopoulos, C.A., Pinckard, R.N. and Hanahan, D.J. (1979). Platelet-activating factor. Evidence for 1-O-alkyl-2-acetyl-sn-glyceryl-3-phosphorylcholine as the active component (a new class of lipid mediators). J. Biol. Chem. 254, 9355–9358.

Denizot, Y., Michel, L., Thomas, Y., Benveniste, J. and Dubertret, L. (1989). Human epidermal cells produce PAF-acether in vitro. FASEB J. 3, A609 (Abstr.).

Denizot, Y., Chaussade, S., Benveniste, J. and Couturier, D. (1991). Presence of PAF-acether in stool of patients with infectious diarrhea. J. Infect. Dis. 163, 1168.

Doebber, T.W. and Wu, M.S. (1987). Platelet-activatig factor (PAF) stimulates the PAF-synthesizing enzyme acetyl-CoA:1-alkyl-sn-glycero-3-phosphocholine-O-acetyl-transferase and PAF synthesis in neutrophils. Proc. Natl Acad. Sci. USA 84, 7557–7561.

Eliakim, R., Karmeli, F., Razin, E. and Rachmilewitz, D. (1988). Role of platelet-activating factor in ulcerative colitis. Gastroenterology 95, 1167–1172.

Elstad, M.R., Prescott, S.M., McIntyre, T.M. and Zimmerman, G.A. (1988). Synthesis and release of platelet-activating factor by stimulated human mononuclear phagocytes. J. Immunol. 140, 1618–1624.

Farr, R.S., Cox, C.P., Wardlow, M.L. and Jorgensen, R. (1980). Preliminary studies of an acid-labile factor (ALF) in human seras that inactivates platelet-activating factor (PAF). Clin. Immunol. Immunopathol. 15, 318–330.

Foa, R., Bussolino, F., Ferrando, M.L. et al. (1985). Release of platelet-activating factor in human leukemia. Cancer Res. 45, 4483–4485.

Fradin, A., Tothhut, B., Poincelot-Canton, B., Errasfa, M. and Russo-Marie, F. (1988). Inhibition of eicosanoid and PAF formation by dexamethasone in rat inflammatory polymorphonuclear neutrophils may implicate lipocortins. Biochim. Biophys. Acta 963, 248–257.

Friedman, B.F., Grandel, K.E. and Farr, R.S. (1986). Platelet activating factor-like lipid in the nasal washings of allergic individuals following specific nasal antigen challenge. Clin. Res. 34, 625A (Abstr.).

Garcia, M.C., Mueller, H.W. and Rosenthal, M.D. (1991a). C_{20} polyunsaturated fatty acids and phorbol myristate acetate enhance agonist-stimulated synthesis of 1-radyl-2-acetyl-sn-glycero-3-phosphocholine in vascular endothelial cells. Biochim. Biophys. Acta 1083, 37–45.

Garcia, M.d.C., Garcia, C., Gijon, M.A., Fernandez-Gallardo, S., Mollinedo, F. and Sanchez Crespo, M. (1991b). Metabolism of platelet-activating factor in human haemato-poietic cell lines. Biochem. J. 273, 573–578.

Geffner, J.R., Schattner, M.A., Lazzari, M.A. and Isturiz, M.A. (1991). Interferon-gamma enhances PAF-acether production by stimulated human polymorphonuclear leukocytes. Scand. J. Immunol. 33, 575–578.

Grandel, K.E., Wardlow, M.L. and Farr, R.S. (1985a). Platelet activating factor in sputum of patients with asthma and COPD. J. Allergy Clin. Immunol., 75(Suppl. 2), 184 (Abstr.).

Grandel, K.E., Farr, R.S., Wanderer, A.A., Eisenstadt, T.C. and Wasserman, S.I. (1985b). Association of platelet-activating factor with primary acquired cold urticaria. N. Engl. J. Med. 313, 405–409.

Haroldsen, P.E., Clay, K.L. and Murphy, R.C. (1987). Quantitation of lyso-platelet activating factor molecular species from human neutrophils by mass spectrometry. J. Lipid Res. 28, 42–49.

Hayashi, H., Kudo, I., Inoue, K., Nomura, H. and Nojima, S. (1985). Macrophage activation by PAF incorporated into dipalmitoyl-phosphatidylcholine-cholesterol liposomes. J. Biochem. 97, 1255–1258.

Henson, P.M. (1987). In "Platelet-Activating Factor and Related Lipid Mediators" (ed. F. Snyder), pp 255–271. Plenum Press, New York.

Hirafuji, M., Mencia-Huerta, J.M. and Benveniste, J. (1987). Regulation of PAF-acether (platelet-activating factor) bio-synthesis in cultured human vascular endothelial cells stimulated with thrombin. Biochim. Biophys. Acta 930, 359–369.

Hoffman, D.R., Truong, C.T. and Johnston, J.M. (1986). Metabolism and function of platelet-activating factor in fetal rabbit lung development. Biochim. Biophys. Acta 879, 88–96.

Holtzman, M.J., Ferdman, B., Bohrer, A. and Turk, J. (1991). Synthesis of the 1-O-hexadecyl molecular species of platelet-activating factor by airway epithelial and vascular endothelial cells. Biochem. Biophys. Res. Commun. 177, 357–364.

Hoppens, C.M., Ludwig, J.C., Castillo, R., McManus, L.M. and Pinckard, R.N. (1983). Albumin modulation of rabbit platelet activation by 1-O-hexadecyl-2-acetyl-sn-glyceryl-3-phosphorylcholine (AGEPC). Fed. Proc. 42, 693 (Abstr.).

Jesaitis, A.J., Bokoch, G.M., Tolley, J.O. and Allen, R.A. (1988). Lateral segregation of neutrophil chemotactic receptors into actin- and foldrin-rich plasma membrane microdomains depleted in guanyl nucleotide regulatory proteins. J. Cell. Biol. 107, 921–928.

Jouvin-Marche, E., Ninio, E., Beaurain, G., Tence, M., Niaudet, P. and Benveniste, J. (1984). Biosynthesis of PAF-acether (platelet-activating factor). VII. Precursors of PAF-acether and acetyl-transferase activity in human leukocytes. J. Immunol. 133, 892–898.

Kald, B., Olaison, G., Sjödahl, R. and Tagesson, C. (1990). Novel aspect of Crohn's disease: increased content of platelet-activating factor in ileal and colonic mucosa. Digestion 46, 199–204.

Kawasaki, T., Blank, M.L. and Snyder, F. (1988). Arachidonic acid regulates platelet activating factor (PAF) synthesis in HL-60 cells. FASEB J. 2, A1376 (Abstr.).

Kramp, W., Pieroni, G., Pinckard, R.N. and Hanahan, D.J. (1984). Observations on the critical micellar concentration of 1-alkyl-2-acetyl-sn-glycero-3-phosphocholine and a series of its homologs and analogs. Chem. Phys. Lipids 35, 49–62.

Kumar, R., King, R.J., Martin, H.M. and Hanahan, D.J. (1987). Metabolism of platelet-activating factor (alkylacetylphosphocholine) by type-II epithelial cells and fibroblasts from rat lungs. Biochim. Biophys. Acta 917, 33–41.

Lee, T.-c., Malone, B., Wasserman, S.I., Fitzgerald, V. and Snyder, F. (1982). Activities of enzymes that metabolize platelet-activating factor (1-O-alkyl-2-acetyl-sn-glycero-3-phosphocholine) in neutrophils and eosinophils from humans and the effect of a calcium ionophore. Biochem. Biophys. Res. Commun. 105, 1303–1308.

Lee, T.-c., Lenihan, D.J., Malone, B., Roddy, L. and Wasserman, S.L. (1984). Increased biosynthesis of platelet-activating factor in activated human eosinophils. J. Biol. Chem. 259, 5526–5530.

Lee, T.H., Mencia-Huerta, J.-M., Shih, C., Corey, E.J., Lewis, R.A. and Austen, K.F. (1984). Characterization and biologic properties of 5,12-dihydroxy derivatives of eicosapentaenoic acid, including leukotriene B_5 and the double lipoxygenase product. J. Biol. Chem. 259, 2383–2389.

Lee, T.H., Hoover, R.L., Williams, J.D., Sperling, R.I., Ravalese, J. III, Spur, B.W., Robinson, D.R., Corey, E.J., Lewis, R.A. and Austen, K.F. (1985). Effect of dietary enrichment with eicosapentaenoic and docosahexaenoic acids on in vitro neutrophil and monocyte leukotriene generation and neutrophil function. New Engl. J. Med. 312, 1217–1224.

Leslie, C.C., Voelker, D.R., Channon, J.Y., Wall, M.M. and Zelarney, P.T. (1988). Properties of purification of an arachidonoyl-hydrolyzing phospholipase A_2 from a macrophage cell line, RAW 264.7 Biochim. Biophys. Acta 963, 476–492.

Lewis, M.S., Whatley, R.E., Cain, P., McIntyre, T.M. and Zimmerman, G.A. (1988). Hydrogen peroxide stimulates the synthesis of platelet-activating factor by endothelium and induces endothelial cell-dependent neutrophil adhesion. J. Clin. Invest. 82, 2045–2055.

Lewis, R.A., Goetzl, E.J., Wasserman, S.I., Valone, F.H., Rubin, R.H. and Austen, K.F. (1975). The release of four mediators of immediate hypersensitivity from human leukemic basophils. J. Immunol. 114, 87–92.

Leyravaud, S. and Benveniste, J. (1989). Regulation of cellular retention of paf-acether by extracellular pH and cell concentration. Biochim. Biophys. Acta 1005, 192–195.

Lianos, E.A. and Zanglis, A. (1987). Biosynthesis and metabolism of 1-O-alkyl-2-acetyl-sn-glyceryl-3-phosphorylcholine in rat glomerular mesangial cells. J. Biol. Chem. 262, 8990–8993.

Lichtenstein, L.M., Schleimer, R.P., MacGlashan, D.W. et al. (1984). In "Asthma: Physiology, Immunopharmacology, and Treatment. Third International Symposium," (eds A.B. Kay, K.F. Austen and L.M. Lichtenstein), p. 1–18. Academic Press, London.

Lin, A.H., Morton, D.R. and Gorman, R.R. (1982). Acetyl glyceryl ether phosphorylcholine stimulates leukotriene B_4 synthesis in human polymorphonuclear leukocytes. J. Clin. Invest. 70, 1058–1065.

Lotner, G.Z., Lynch, J.M., Betz, S.J. and Henson, P.M. (1980). Human neutrophil-derived platelet activating factor. J. Immunol. 124, 676–684.

Ludwig, J.C., McManus, L.M., Clark, P.O., Janahan, D.J. and Pinckard, R.N. (1984). Modulation of platelet-activating factor (PAF) synthesis and release from human polymorphonuclear leukocytes (PMN): role of extracellular Ca^{2+}. Arch. Biochem. Biophys. 232, 101–110.

Ludwig, J.C., Hoppens, C.L., McManus, L.M., Mott, G.E. and Pinckard, R.N. (1985). Modulation of platelet-activating factor (PAF) synthesis and release from human polymorphonuclear leukocytes (PMN): role of extracellular albumin. Arch. Biochem. Biophys. 241, 337–347.

Ludwig, J.C., McManus, L.M. and Pinckard, R.N. (1986). Synthesis-release coupling of platelet activating factors (PAF) from stimulated human neutrophils. Adv. Inflammation Res. 11, 111–125.

Lumb, T.H., Pool, G.L., Bubacz, D.G., Blank, M.L. and Snyder, F. (1983). Spontaneous and protein-catalyzed transfer of 1-alkyl-2-sn-glycero-3-phosphocholine (platelet-activating factor) between phospholipid bilayers. Biochim. Biophys. Acta 750, 217-222.

Lynch, J.M. and Henson, P.M. (1986). The intracellular retention of newly synthesized platelet-activating factor. J. Immunol. 137, 2653–2661.

McColl, S.R., Krump, E., Naccache, P.H., Poubelle, P.E., Braquet, P., Braquet, M. and Borgeat, P. (1991). Granulocyte-macrophage colony-stimulating factor increases the synthesis of leukotriene B_4 by human neutrophils in response to platelet-activating factor. J. Immunol. 146, 1204–1211.

McEvoy, L., Schlegel, R.A. Williamson, P. and Del Buono, B.J. (1988). Merocyanine 540 as a flow cytometric probe of membrane lipid organization in leukocytes. J. Leukoc. Biol. 44, 337–344.

McIntyre, T.M., Zimmerman, G.A., Satoh, K. and Prescott, S.M. (1985). Cultured endothelial cells synthesize both platelet-activating factor and prostacyclin in response to histamine, bradykinin, and adenosine triphosphate. J. Clin. Invest. 76, 271–280.

McIntyre, T.M., Zimmerman, G.A. and Prescott, S.M. (1986). Leukotrienes C_4 and D_4 stimulate human endothelial cells to synthesize platelet-activating factor and bind neutrophils. Proc. Natl Acad. Sci. USA 83, 2204–2208.

McManus, L.M., Marze, B.T. and Schiess, A.V. (1990). Deficiency of salivary PAF in edentulous individuals. J. Periodont. Res. 25, 347–351.

Malavasi, F., Tetta, C., Funaro, A., Bellone, G., Ferrero, E., Franzone, A.C., Dellabona, P., Rusci, R., Matera, L., Camussi, G. and Caligaris-Cappio, F. (1986). Fc receptor triggering induces expression of surface activation antigens and release of platelet-activating factor in large granular lymphocytes. Proc. Natl Acad. Sci. USA 83, 2443–2447.

Mallet, A.I. and Cunningham, F.M. (1985). Structural identification of platelet activating factor in psoriatic scale. Biochem. Biophys. Res. Commun. 126, 192–198.

Matsumoto, M. and Miwa, M. (1985). Platelet-activating factor-binding protein in human serum. Adv. Prostaglandin Thromboxane Leukotriene Res. 15, 705–706.

Mazer, B., Clay, K.L., Renz, H. and Gelfand, E.W. (1990). Platelet-activating factor enhances Ig production in B lymphoblastoid cell lines. J. Immunol. 145, 2602–2607.

Mencia-Huerta, J.-M., Lewis, R.A., Razin, E. and Austen, K.F. (1983). Antigen-initiated release of platelet-activating factor (PAF-acether) from mouse bone marrow-derived mast cells sensitized with monoclonal IgE. J. Immunol. 131, 2958–2964.

Miadonna, A., Tedeschi, A., Arnoux, B., Sala, A., Zanussi, C. and Benveniste, J. (1989). Evidence of PAF-acether metabolic pathway activation in antigen challenge of upper respiratory airways. Am. Rev. Respir. Dis. 140, 142-147.

Michel, L., Denizot, Y., Thomas, Y., Jean-Louis, F., Pitton, C., Benveniste, J. and Dubertret, L. (1988). Biosynthesis of PAF-acether factor-acether by human skin fibroblasts *in vitro*. J. Immunol. 141, 948–953.

Miwa, M., Hill, C., Kumar, R., Sagatani, J., Olson, M.S. and Hanahan, D.J. (1987). Occurrence of an endogenous inhibitor of platelet-activating factor in rat liver. J. Biol. Chem., 262, 527–530.

Mollinedo, F., Gomez-Cambronero, J., Cano, E. and Sanchez Crespo, M. (1988). Intracellular localization of platelet-activating factor synthesis in human neutrophils. Biochem. Biophys. Res. Commun. 154, 1232–1239.

Mueller, H.W.W., O'Flaherty, J.T. and Wykle, R.L. (1984). The molecular species distribution of platelet-activating factor synthesized by rabbit and human neutrophils. J. Biol. Chem. 259, 554–559.

Mueller, H.W., Nollert, M.U. and Eskin, S.G. (1991). Synthesis of 1-acyl-2-[^{3}H]acetyl-*sn*-glycero-3-phosphocholine, a structural analog of platelet activating factor by vascular endothelial cells. Biochem. Biophys. Res. Commun. 176, 1557–1564.

Murphy, R.C. and Clay, K.L. (1987). In "Platelet-Activating Factor and Related Lipid Mediators" (ed F. Snyder), pp 9–31. Plenum Press, New York.

Nakayama, R., Yasuda, K. and Saito, K. (1987). Existence of endogenous inhibitors of platelet-activating factor (PAF) with PAF in rat uterus. J. Biol. Chem. 262, 13174–13179.

Natarajan, V., Zuzarte-Augstin, M., Schmid, H.H.O. and Graff, G. (1983). The alkylacyl and alkenylacyl glycerophospholipids of human platelets. Thromb. Res. 30, 119–125.

Neuwirth, R., Singhal, P., Diamond, B., Hays, R.M., Lobmeyer, L., Clay, K. and Schlondorff, D. (1988). Evidence for immunoglobulin Fc receptor-mediated prostaglandin $F_{2\alpha}$ and platelet-activating factor formation by cultured rat mesangial cells. J. Clin. Invest. 82, 936–944.

Nieto, M.L., Velasco, S. and Sanchez Crespo, M. (1988). Biosynthesis of platelet-activating factor in human polymorphonuclear leukocytes. J. Biol. Chem. 263, 2217–2222.

Ninio, E., Joly, F. and Bessou, G. (1988). Biosynthesis of PAF-acether. XI. Regulation of acetyltransferase by enzyme-substrate imbalance. Biochim. Biophys. Acta 963, 288–294.

Oda, M., Satouchi, K., Yasunaga, K. and Saito, K. (1985). Molecular species of platelet-activating factor generated by human neutrophils challenged with ionophore A23187. J. Immunol. 134, 1090–1093.

Oda, M., Satouchi, K., Ikeda, I., Sakakura, M., Yasunaga, K. and Saito, K. (1989). The presence of platlet-activating factor associated with eosinophil and/or neutrophil accumulations in the pleural fluids. Am. Rev. Respir. Dis. 141, 1469–1473.

Pickett, W.C. and Ramesha, C.S. (1990). Quantitative analysis of platelet-activating factor by gas chromatography–negative ion chemical ionization mass spectrometry. Methods Enzymol. 187, 142–152.

Pinckard, R.N., Jackson, E.M., Hoppens, C., Weintraub, S.T., Ludwig, J.C., McManus, L.M. and Mott, G.E. (1984). Molecular heterogeneity of platelet-activating factor produced by stimulated human polymorphonuclear leukocytes. Biochem. Biophys. Res. Commun. 122, 325–332.

Polonsky, J., Tence, M., Varenne, P., Das, B.C., Lunel, J. and Benveniste, J. (1980). Release of 1-*O*-alkylglyceryl-3-phosphorylcholine, *O*-deacetyl platelet-activating factor, from leukocytes: chemical ionization mass spectrometry of phospholipids. Proc. Natl Acad. Sci. USA 77, 7019–7023.

Prescott, S.M. (1984). The effect of eicosapentaenoic acid on leukotriene B production by human neutrophils. J. Biol. Chem. 259, 7615–7621.

Prescott, S.M., Zimmerman, G.A. and McIntyre, T.M. (1984). Human endothelial cells in culture produce platelet-activating factor (1-alkyl-2-acetyl-*sn*-glycero-3-phosphocholine) when stimulated with thrombin. Proc. Natl Acad. Sci. USA 81, 3534–3538.

Prescott, S.M., McIntyre, T.M. and Zimmerman, G.A. (1990). The role of platelet-activating factor in endothelial cells. Thromb. Haemost. 64, 99–103.

Ramesha, C.S. and Pickett, W.C. (1986a). Measurement of subpicogram quantities of platelet activating factor (AGEPC) by gas chromatography negative ion chemical ionization mass spectrometry. Biomed. Environ. Mass Spectrom. 13, 107–112.

Ramesha, C.S. and Pickett, W.C. (1986b). Platelet-activating factor and leukotriene biosynthesis is inhibited in polymorphonuclear leukocytes depleted of arachidonic acid. J. Biol. Chem. 251, 7592–7595.

Record, M. and Snyder, F. (1986). Biosynthesis of platelet activating factor (PAF) via alternate pathways: Subcellular distribution of products in HL-60 cells. Fed. Proc. 45, 1529 (Abstract).

Ribbes, G., Ninio, G., Fontan, P., Record, M., Chap, H., Benveniste, J. and Louis, D.-B. (1985). Evidence that biosynthesis of platelet-activating factor (PAF-acether) by human neutrophils occurs in an intracellular membrane. FEBS Lett. 191, 195–199.

Riches, D.W.H., Young, S.K., Seccombe, J.F., Henson, J.E., Clay, K.L. and Henson, P.M. (1990). The subcellular distribution of platelet-activating factor in stimulated human neutrophils. J. Immunol. 145, 3062–3070.

Robinson, M. and Snyder, F. (1985). Metabolism of platelet-activating factor by rat alveolar macrophages: lyso-PAF as an obligatory intermediate in the formation of alkylarachidonoyl glycerophosphocholine species. Biochim. Biophys. Acta 837, 52–56.

Rola-Pleszcynski, M., Pignol, B., Pouliot, C. and Braquet, P. (1987). Inhibition of human lymphocyte proliferation and interleukin-2 production by platelet activating factor (PAF-acether): reversal by a specific antagonist, BN 52021. Biochem. Biophys. Res. Commun. 142, 754–760.

Roubin, R., Dulioust, A., Haye-Legrand, I., Ninio, E. and Benveniste, J. (1986). Biosynthesis of PAF-acether. VII. Impairment of PAF-acether production in activated macrophages does not depend upon acetyltransferase activity. J. Immunol. 136, 1796–1802.

Saffitz, J.E., Krueger, C.M., Loeb, L.A. and Gross, R.W. (1986). Subcellular localization of 1-alkyl-sn-glycero-3-phosphocholine (LPAF) metabolites in human neutrophils. Fed. Proc. 45, 226.

Salem, P., Denizot, Y., Pitton, C., Dulioust, A., Bossant, M.-J., Benveniste, J. and Thomas, Y. (1989). Presence of paf-acether in human thymus. FEBS Lett. 257, 49–51.

Sanchez Crespo, M., Alonso, F. and Egido, J. (1980). Platelet-activating factor in anaphylaxis and phagocytosis. I. Release from human peripheral polymorphonuclears and monocytes during the stimulation by ionophore A23187 and phagocytosis but not from degranulating basophils. Immunology 40, 645–655.

Sanchez Crespo, M., Inarrea, P., Alvarez, V., Alonso, F., Egido, J. and Hernando, L. (1983). Presence in normal human urine of a hypotensive and platelet-activating phospholipid. Am. J. Physiol. 244, F706–F711.

Satoh, K., Imaizumi, T.-A., Kawamura, Y., Yoshida, H., Hiramoto, M., Takamatsu, S. and Takamatsu, M. (1991). Platelet-activating factor (PAF) stimulates the production of PAF acetylhydrolase by the human hepatoma cell line, HepG2. J. Clin. Invest. 87, 476–481.

Satouchi, K., Pinckard, R.N. and Hanahan, D.J. (1981). Influence of alkyl ether chain length of acetyl glyceryl ether phosphorylcholine and its ethanolamine analog on biological activity toward rabbit platelets. Arch. Biochem. Biophys. 211, 683–688.

Satouchi, K., Oda, M., Yasunga, K. and Saito, K. (1983). Application of selected ion monitoring to determination of platelet activating factor. J. Biochem. 94, 2067–2070.

Satouchi, K., Oda, M., Yasunaga, K. and Saito, K. (1985). Evidence for production of 1-acyl-2-acetyl-sn-glyceryl-3-phosphorylcholine concomitantly with platelet-activating factor. Biochem. Biophys. Res. Commun. 128, 1409–1417.

Schlegel, R.A. and Williamson, P. (1987). Membrane phospholipid organization as a determinant of blood cell–reticuloendothelial cell interaction. J. Cell. Physiol. 132, 381–384.

Schleimer, R.P., Macglashan, D.W. Jr, Peters, S.P., Pinckard, R.N., Adkinson, N.F. and Lichtenstein, L.M. (1986). Characterization of inflammatory mediator release from purified human lung mast cells. Am. Rev. Resp. Dis. 133, 614–617.

Schneider, E., Haest, C.W.M., Plasa, G. and Deuticke, B. (1986). Bacterial cytotoxins, amphotericin B and local anaesthetics enhance transbilayer mobility of phospholipids in erythrocyte membranes. Biochim. Biophys. Acta 855, 325–336.

Shalit, M., Valone, F.H., Atkins, P.C., Ratnoff, W.D., Goetzl, E.J. and Zweiman, B. (1989). Late appearance of phospholipid platelet-activating factor and leukotriene B4 in human skin after repeated antigen challenge. J. Allergy Clin. Immunol. 83, 691–696.

Shaw, J.O. and Henson, P.M. (1980). The binding of rabbit basophil-derived platelet-activating factor to rabbit platelets. Am. J. Pathol. 98, 791–810.

Shin, M.H., Averill, F., Hubbard, W.C., Chilton, F.H., Liu, M.C. and Naclerio, R.M. (1991). Detection of platelet activating factor (PAF) and its 2 lyso metabolite in allergic rhinitics following antigen challenge. J. Allergy Clin. Immunol. 87, 145 (Abstr.).

Sisson, J.H., Prescott, S.M., McIntyre, T.M. and Zimmerman, G.A. (1987). Production of platelet-activating factor by stimulated human polymorphonuclear leukocytes. J. Immunol. 138, 3918–3926.

Smiley, P.L., Stremler, K.E., Prescott, S.M., Zimmerman, G.A, and McIntyre, T.M. (1991). Oxidatively fragmented phosphatidylcholines activate human neutrophils through the receptor for platelet-activating factor. J. Biol. Chem. 266, 11104–11110.

Snyder, F. (1987). In "Platelet-Activating Factor and Related Lipid Mediators" (ed F. Snyder), pp 89–113. Plenum Press, New York.

Sogos, V., Bussolino, F., Pilia, E., Torelli, S. and Gremo, F. (1990). Acetylcholine-induced production of platelet-activating factor by human fetal brain cells in culture. J. Neurosci. Res. 27, 706–711.

Sperling, R.I., Robin, J.-L., Kylander, K.A., Lee, T.H., Lewis, R.A. and Austen, K.F. (1987). The effects of N-3 polyunsaturated fatty acids on the generation of platelet-activating factor-acether by human monocytes. J. Immunol. 139, 4186–4191.

Stafforini, D.M., Prescott, S.M. and McIntyre, T.M. (1987). Human plasma platelet-activating factor acetylhydrolase. J. Biol. Chem. 262, 4223–4230.

Stenmark, K.R., Eyzaguirre, M., Westcott, J.Y., Henson, P.M. and Murphy, R.C. (1987). Potential role of eicosanoids and PAF in the pathophysiology of bronchopulmonary dysplasia. Am. Rev. Resp. Dis. 136, 770–772.

Stewart, A.G., Harris, T., De Nichilo, M. and Lopez, A.F. (1991). Involvement of leukotriene B4 and platelet-activating factor in cytokine priming of human polymorphonuclear leucocytes. Immunology 72, 206–212.

Strasser, T., Fischer, S. and Weber, P.C. (1985). Leukotriene B5 is formed in human neutrophils after dietary supplementation with icosapentaenoic acid. Proc. Natl Acad. Sci. USA 82, 1540–1543.

Sturk, A., Schaap, M.C.L., Prins, A., Wouterten Cate, J. and van den Bosch, H. (1989). Synthesis of platelet-activating factor by blood platelets and leucocytes. Evidence against selective utilization of cellular ether-linked phospholipids. Biochim. Biophys. Acta 993, 148–156.

Suga, K., Kawasaki, T., Blank, M.L. and Snyder, F. (1990). An arachidonoyl (polyenoic)-specific phospholipase A2 activity regulates the synthesis of platelet-activating factor in granulocytic HL-60 cells. J. Biol. Chem. 265, 12363-12371.

Sugiura, T. and Waku, K. (1987). In "Platelet-Activating Factor and Related Lipid Mediators" (ed F. Snyder), p. 55–85. Plenum Press, New York.

Sugiura, T., Fukuda, T., Masuzawa, Y. and Waku, K. (1990). Ether lysophospholipid-induced production of platelet-activating factor in human polymorphonuclear leukocytes. Biochim. Biophys. Acta 1047, 223–232.

Surles, J.R., Wykle, R.L., O'Flaherty, J.T., Salzer, W.L.,

Thomas, M.J., Snyder, F. and Piantadosi, C. (1985). Facile synthesis of platelet-activating factor and racemic analogues containing unsaturation in the *sn*-1-alkyl chain. J. Med. Chem. 28, 73–78.

Swendsen, C.L., Ellis, J.M., Chilton, F.H., O'Flaherty, J.T. and Wykle, R.L. (1983). 1-*O*-Alkyl-2-arachidonoyl-*sn*-glycero-3-phosphocholine: a novel source of arachidonic acid in neutrophils stimulated by the calcium ionophore A23187. Biochem. Biophys. Res. Commun. 113, 72–79.

Swendsen, C.L., Chilton, F.H., O'Flaherty, J.T., Surles, J.R., Piantadosi, C., Waite, M. and Wykle, R.L. (1987). Human neutrophils incorporate arachidonic acid and saturated fatty acids into separate molecular species of phospholipids. Biochim. Biophys. Acta 919, 79–89.

Takamura, H., Kasai, H., Arita, H. and Kito, M. (1990). Phospholipid molecular species in human umbilical artery and vein endothelial cells. J. Lipid Res. 31, 709–717.

Tence, M., Jouvin-Marche, E., Bessou, G., Record, M. and Benveniste, J. (1985). Ether-phospholipids composition in neutrophils and platelets. Thromb. Res. 38, 207–214.

Tessner, T.G. and Wykle, R.L. (1987). Stimulated neutrophils produce an ethanolamine plasmalogen analog of platelet-activating factor. J. Biol. Chem. 262, 12660–12664.

Tessner, T.G., O'Flaherty, J.T. and Wykle, R.L. (1989). Stimulation of platelet-activating factor synthesis by a non-metabolizable bioactive analog of platelet-activating factor and influence of arachidonic acid metabolites. J. Biol. Chem. 264, 4794–4799.

Tokumura, A., Kamiyasu, K., Takauchi, K. and Tsukatani, H. (1987a). Evidence for existence of various homologues and analogues of platelet activating factor in a lipid extract of bovine brain. Biochem. Biophys. Res. Commun. 121, 815–825.

Tokumura, A., Yoshida, J.-I., Maruyama, T., Fukuzawa, K. and Tsukatani, H. (1987b). Platelet aggregation induced by ether-linked phospholipids. I. Inhibitory actions of bovine serum albumin and structural analogues of platelet activating factor. Thromb. Res. 47, 51–63.

Tool, A.T.J., Verhoeven, A.J., Roos. D. and Koenderman, L. (1989). Platelet-activating factor (PAF) acts as an intracellular messenger in the changes of cytosolic free Ca^{2+} in human neutrophils induced by opsonized particles. FEBS Lett. 259, 209–212.

Touqui, L., Hatmi, M. and Vargaftig, B.B. (1985). Human platelets stimulated by thrombin produce platelet-activating factor (1-*O*-alkyl-2-acetyl-*sn*-glycero-3-phosphocholine) when the degrading enzyme acetylhydrolase is blocked. Biochem. J. 229, 811–816.

Triggiani, M. and Chilton, F.H. (1989). Influence of immuno-logic activation and cellular fatty acid levels on the catabol-ism of platelet-activating factor within the murine mast cell (PT-18). Biochim. Biophys. Acta 1006, 41–51.

Triggiani, M., Chilton, F.H. and Lichtenstein, L.M. (1989). Metabolism of platelet activating factor (PAF) in human and murine mast cells. FASEB J. 3, A1277 (Abstr.).

Triggiani, M., Hubbard, W.C. and Chilton, F.H. (1990a). Synthesis of 1-acyl-2-acetyl-*sn*-glycero-3-phosphocholine by an enriched preparation of the human lung mast cell. J. Immunol. 144, 4773–4780.

Triggiani, M., Connell, T.R. and Chilton, F.H. (1990b). Evidence that increasing the cellular content of eicosapentae-noic acid does not reduce the biosynthesis of platelet-activating factor. J. Immunol. 145, 2241–2248.

Triggiani, M., Schleimer, R.P., Warner, J.A. and Chilton, F.H. (1991a). Differential synthesis of 1-acyl-2-acetyl-*sn*-glycero-3-phosphocholine and platelet-activating factor by human inflammatory cells. J. Immunol. 147, 660–666.

Triggiani, M., Goldman, D.W. and Chilton, F.H. (1991b). Biological effects of 1-acyl-2-acetyl-*sn*-glycero-3-phosphocholine in the human neutrophil. Biochim. Biophys. Acta 1084, 41–47.

Triggiani, M., D'Souza, D.M. and Chilton, F.H. (1991c). Metabolism of 1-acyl-2-acetyl-*sn*-glycero-3-phosphocholine in the human neutrophil. J. Biol. Chem. 266, 6928–6935.

Uemura, Y., Lee, T.-c. and Snyder, F. (1991). A coenzyme A-independent transacylase is linked to the formation of platelet-activating factor (PAF) by generating the lyso-PAF intermedi-ate in the remodeling pathway. J. Biol. Chem. 266, 8268–8272.

Vallari, D.S., Record, M. and Snyder, F. (1990). Conversion of alkylacetylglycerol to platelet-activating factor in HL-60 cells and subcellular localization of the mediator. Arch. Biochem. Biophys. 276, 538–545.

Valone, F.H. (1987). In "Platelet-Activating Factor and Related Lipid Mediators" (ed. F. Snyder), pp 137–151. Plenum Press, New York.

Valone, F.H. (1991). Synthesis of platelet-activating factor by human monocytes stimulated by platelet-activating factor. J. Allergy Clin. Immunol. 87, 715–720.

Valone, F.H. and Epstein, L.B. (1988). Biphasic platelet-activating factor synthesis by human monocytes stimulated with IL-1β, tumor necrosis factor, or IFN-γ. J. Immunol. 141, 3945–3950.

Valone, F., Shalit, M., Atkins, P., Goetzl, E. and Zweiman, B. (1987). Platelet activating factor release in allergic skin sites in humans. J. Allergy Clin. Immunol. 79, 248A (Abstr.).

Verhallen, P.F.J., Bevers, E.M., Comfurius, P. and Zwaal, R.F.A. (1987). Correlation between calpain-mediated cyto-skeletal degradation and expression of platelet procoagulant activity. A role for the platelet membrane-skeleton in the regulation of membrane lipid asymmetry? Biochim. Biophys. Acta 903, 206–217.

Villani, A., Cirino, N.M., Baldi, E., Kester, M., McFadden, E.R. Jr and Panuska, J.R. (1991). Respiratory syncytial virus infection of human mononuclear phagocytes stimulates syn-thesis of platelet-activating factor. J. Biol. Chem. 266, 5472–5479.

Walsh, C.E., Waite, B.M., Thomas, M.J. and DeChatelet, L.R. (1981). Release and metabolism of arachidonic acid in human neutrophils. J. Biol. Chem. 256, 7228–7234.

Walsh, C.E., Dechatelet, L.R., Chilton, F.H., Wykle, R.L. and Waite, M. (1983). Mechanism of arachidonic acid release in human polymorphonuclear leukocytes. Biochim. Biophys. Acta 750, 32–40.

Wardlow, M.L., Cox, C.P., Meng, K.E., Greene, D.E. and Farr, R.S. (1986). Substrate specificity and partial characterization of the PAF-acylhydrolase in human serum that rapidly inactivates platelet-activating factor. J. Immunol. 136, 3441–3446.

Weintraub, S.T., Ludwig, J.C., Mott, G.E., McManus, L.M., Lear, C. and Pinckard, R.N. (1985). Fast atom bombardment mass spectrometric identification of molecular species of platelet activating factor produced by stimulated human polymorphonuclear leukocytes. Biochem. Biophys. Res. Commun. 129, 868–876.

Weintraub, S.T., Lear, C.S. and Pinckard, R.N. (1990). Analysis of platelet activating factor by GC–MS after direct derivatization with pentafluorobenzoyl chloride and heptafluorobutyric anhydride. J. Lipid Res. 31, 719–725.

Whatley, R.E., Lewis, M.S., Zimmerman, G.A. and McIntyre, T.M. (1988). The regulation of synthesis of platelet-activating factor by endothelial cells. Chest 93, 110S–111S.

Whellan, P., Zirrolli, J.A. and Clay, K. (1992). Analysis of glycerophosphocholine molecular species as derivatives of 7[(chlorocarbonyl)-methoxyl]-4-methylcoumarin. J. Lipid Res. 33, 111–121.

Wientzek, M., Arthur, G., Man, R.Y.K. and Choy, P.C. (1985). A sensitive method for the quantitation of lysophosphatidylcholine in canine heart. J. Lipid Res. 26, 1166–1169.

Williamson, P., Mattocks, K. and Schlegel, R.A. (1983). Merocyanine 540, a fluorescent probe sensitive to lipid packing. Biochim. Biophys. Acta 732, 387–393.

Wirthmueller, U., De Weck, A.L. and Dahinden, C.A. (1989). Platelet-activating factor production in human neutrophils by sequential stimulation with granulocyte-macrophage colony-stimulating factor and the chemotactic factors C5A or formyl-methionyl-leucyl-phenylalanine. J. Immunol. 142, 3213–3218.

Wirthmueller, U., de Weck, A.L. and Dahinden, C.A. (1990). Studies on the mechanism of platelet-activating factor production in GM-CFS primed neutrophils: involvement of protein synthesis and phospholipase A_2 activation. Biochem. Biophys. Res. Commun. 170, 556–562.

Worthen, G.S., Seccombe, J.F., Clay, K.L., Guthrie, L.A. and Johnston, R.B. Jr (1988). The priming of neutrophils by lipopolysaccharide for production of intracellular platelet-activating factor. J. Immunol. 140, 3553–3559.

Yamada, K., Asano, O., Yoshimura, T. and Katayama, K. (1988). Highly sensitive gas chromatographic–mass spectrometric method for determination of platelet-activating factor in human blood. J. Chromatogr. 433, 243–247.

Zimmerman, G.A., McIntyre, T.M. and Prescott, S.M. (1985). Thrombin stimulates the adherence of neutrophils to human endothelial cells *in vitro*. J. Clin. Invest. 76, 2235–2246.

Zimmerman, G.A., McIntyre, T.M., Mehra, M. and Prescott, S.M. (1990). Endothelial cell-associated platelet-activating factor: a novel mechanism for signaling intercellular adhesion. J. Cell. Biol. 110, 529–540.

8. Biological Properties of Platelet-Activating Factor

A.G. Stewart

Lipid Mediators
ISBN 0–12–198875–9

1. Introduction

The structure elucidation of PAF as a unique, biologically active phospholipid (Benveniste *et al.*, 1979; Demopoulos *et al.*, 1979; Blank *et al.*, 1979) held promise for the discovery of some unusual properties of this mediator. The ensuing 12 years have partially fulfilled that promise with the identification of a bewildering array of actions in diverse cell types and of its production by both inflammatory and non-inflammatory cell types. The range of biological activity of PAF extends over 9 log units from femtomolar potency in lymphocytes and macrophages (Braquet *et al.*, 1989b) to micromolar potency in induction of eosinophil O_2^- (Kroegel *et al.*, 1989c; Rouis *et al.*, 1988). The actions of PAF at these extreme ends of the concentration range are, nonetheless, apparently mediated by specific, and perhaps even similar, receptors. The phospholipid nature of PAF has evoked novel suggestions regarding the form of the molecule that is active at PAF receptors, including plasma membrane complexes (Zimmerman *et al.*, 1990), albumin complexes (Clay *et al.*, 1990) and liposomal forms (Hayashi *et al.*, 1985a) in addition to its free monomeric form (Grigoriadis and Stewart, 1992). Retention of PAF by some cell types has led to speculation on potential intracellular roles (Lynch and Henson, 1986; Bratton *et al.*, 1988b; Stewart and Phillips, 1989; Stewart *et al.*, 1990b) in addition to the accepted function as an intercellular mediator of allergy and inflammation.

The purpose of this review is to describe the effects of PAF on isolated cells, tissues and in intact organisms and to relate these effects to potential involvement in (patho)physiological processes, particularly in humans. The broad scope of this review and the ever-growing relevant literature enforce a restricted bibliography. Wherever possible, relevant, comprehensive reviews of specific aspects of the actions of PAF are cited.

1.1 PHYSICOCHEMICAL PROPERTIES

The unique structure of PAF provides some necessity for a careful interpretation of data obtained from isolated cells, tissues and, indeed, whole animals exposed to high concentrations/doses of this mediator. Most studies cited in this review have used concentrations of PAF of less than 1 μM. However, in those that have not, the possibility that PAF is acting in a micellar form at concentrations below the nominal concentration, or exerts a non-specific detergent-like effect (Sawyer and Andersen, 1989), must be considered, since the CMC of C_{16}-PAF is reported to be 1.1 μM (Kramp *et al.*, 1984). The impact of the inclusion of BSA in the suspending medium on the CMC has not been reported.

It is known that PAF binds with high affinity to BSA and that such binding profoundly influences the potency of PAF (Clay *et al.*, 1990). It has been suggested that the high degree of albumin binding is consistent with the albumin complex being the active form, on the basis that the concentration of free PAF is negligible in the presence of the standard concentration of BSA used as a suspending agent (0.25%) (Clay *et al.*, 1990). However, cellular responses to PAF are competitively inhibited by increasing concentrations of BSA, consistent with the notion that albumin reduces the free concentration of PAF available for receptor interaction (Grigoriadis and Stewart, 1992). The reversal of PAF-induced receptor desensitization by BSA (Gerkens, 1990; Tokumura *et al.*, 1988) may simply represent the removal of PAF sequestered in cell membranes, rather than a requirement for albumin for biological activity.

Table 8.1   Cellular responses to PAF

Response	Platelet	Macrophage	Neutrophil	Eosinophil
Aggregation	+	+(> 1 μM)	+	?
Adhesion	+	?	+	+
Arachidonate mobilization	+	+	+	+
CD11/CD18 expression		?	+	+
Ca^{2+}_i	+	+	+	+
Chemotaxis		+	+	+
Degranulation	+		+/−	+
Priming		+	+	+
O_2^- generation	?	+	+/−	+

Consideration of the effects of high concentrations of PAF with respect to intracellular or intramembranous actions may not be subject to the same caveat, since it has been found that high local concentrations do indeed occur (Riches *et al.*, 1990). In this respect it is interesting that 19–380 μM PAF activates a protein kinase in plants (Martiny-Baron and Scherer, 1989).

The physicochemical properties of PAF have provided evidence consistent with a receptor-independent role in the fusion of membranes (Bratton *et al.*, 1988a,b), but the further evaluation of this proposal awaits the availability of specific inhibitors of PAF biosynthesis.

2. Cellular Actions

The cellular actions of PAF are summarized in Table 8.1.

2.1   Platelets

The discovery of PAF as a factor causing platelet aggregation ensured that this aspect of the inflammatory mediator has received considerable attention. Activation of platelets by PAF occurs in the picomolar concentration range and its specific inhibition by PAF receptor antagonists enables the use of platelets as a bioassay for PAF (Bossant *et al.*, 1990; Moon *et al.*, 1990; Sugatani *et al.*, 1990). In addition, the platelet has been used as the primary screen for the development of a wide range of PAF receptor antagonists (Braquet *et al.*, 1987; Saunders and Handley, 1987; Casals Stenzel and Heuer, 1988) and much is known about the signal transduction cascade initiated by PAF binding (Salari *et al.*, 1990).

2.1.1   Platelet-Activating Factor Receptors on Platelets

Studies by Henson and coworkers on partially purified PAF established that it induced rabbit platelet aggregation and degranulation (Henson and Landes, 1976) by activation of serine proteases (Henson and Oades, 1976). PAF-induced human platelet aggregation *in vitro* could be distinguished from immune complex activation by its independence from ADP (Camussi *et al.*, 1978b).

Furthermore, PAF-induced activation of serotonin release by human platelets was not altered by pretreatment with indomethacin (O'Donnell *et al.*, 1978). The high potency of PAF, the importance of the short side-chain on C-2, the alkyl-ether substituent on C-1 and agonist-specific desensitization (Keraly and Benveniste, 1982; Keraly *et al.*, 1983) all suggested that the actions of PAF on platelets were mediated by activation of specific receptors. This conclusion was greatly strengthened by the radioligand-binding studies of Hwang *et al.* (1983) demonstrating saturable, high-affinity and stereospecific binding sites on rabbit platelet membranes. Moreover, binding of [^{3}H]PAF to human platelets correlated with functional responses, including exposure of fibrinogen-binding sites, aggregation and increased PI turnover (Kloprogge and Akkerman, 1984). However, the strongest evidence that PAF activates platelets via activation of specific receptors is provided by the now enormous and diverse range of compounds that specifically antagonize PAF responses (Kornecki *et al.*, 1984; Hosford *et al.*, 1989).

2.1.2   Third Pathway of Platelet Aggregation

Chignard *et al.* (1980, 1984) demonstrated that rabbit platelets produced PAF in response to A23187 and thrombin and suggested that PAF may mediate the ADP and TXA$_2$-independent platelet activation by A23187 or thrombin. It was shown by a combination of HPLC and bioassay techniques that human platelets produce PAF in response to A23187 (Alam *et al.*, 1983). Other workers, however, reported that neither thrombin nor A23187 were stimuli for PAF formation by human platelets (Marcus *et al.*, 1981). The variation in the literature may relate to the high activity of the catabolic enzyme acetylhydrolase, since inhibition of this enzyme with the protease inhibitor PMSF resulted in a 10–12-fold increase in PAF levels of thrombin-stimulated platelets (Touqui *et al.*, 1985). The observation of a thrombin-mediated increase in acetyltransferase activity in human platelets also supports the capacity of this cellular element to produce PAF (Alam and Silver, 1986; Coeffier *et al.*, 1986).

The failure of PAF desensitization to influence aggregation induced by thrombin or collagen provided evidence against PAF as a mediator of the third pathway (Hornby and Perry, 1983; Kloprogge *et al.*, 1983; Nunn, 1983). Perhaps the best evidence against a role for PAF formation in thrombin-induced activation of rabbit platelets was provided by the failure of 3-deazaadenosine, which inhibited thrombin-induced PAF formation, to inhibit thrombin-induced aggregation (Touqui *et al.*, 1984). Furthermore, a number of chemically distinct PAF receptor antagonists failed to inhibit thrombin-induced aggregation of platelets treated with aspirin and ADP scavengers (Adnot *et al.*, 1987). A study of thrombin-induced platelet aggregation in platelets from Zellweger's syndrome patients with a deficiency in plasmalogens and defective PAF formation revealed no diminution in the aggregatory response to thrombin, indicating that PAF is unlikely to be the mediator of the third pathway of platelet aggregation (Sturk *et al.*, 1987).

2.1.3 Physiological Regulators of Platelet-Activating Factor-Induced Aggregation

Physiological modulation of PAF-induced platelet activation was suggested by potent inhibition by PGI_2 ($IC_{50} \approx 5$ nM) (Chesney *et al.*, 1982). A potential negative feedback on PAF-induced platelet activation is indicated by its ability to elicit PGI_2 release from vascular rings obtained from humans, rats and rabbits (Acharya and MacIntyre, 1983). A comparison of VIP and PGI_2 indicated that these agents have a similar inhibitory potency against PAF-induced rabbit platelet activation ($IC_{50} \approx 1$–4 µM) accompanied by accumulation of cAMP (Cox *et al.*, 1984), but the potency of PGI_2 was very much less than that reported for inhibition of PAF-induced human platelet aggregation (Chesney *et al.*, 1982). PGD_2 has also been recognized as a modulator of PAF-induced aggregation of human platelets, with an IC_{50} of about 3 nM, by a mechanism involving elevation of cAMP (Bushfield *et al.*, 1985). PAF, thrombin and ADP are able to elicit the release of acetylhydrolase from bovine platelets, suggesting a potential negative-feedback mechanism on PAF accumulation in thrombosis (Suzuki *et al.*, 1988).

Adrenaline synergizes with PAF in inducing full aggregation of plasma-free human platelet suspensions whereas, in the absence of adrenaline, PAF induces only a partial aggregatory response. It has been suggested that this interaction may have pathophysiological relevance to thrombosis during stress or exercise (Vargaftig *et al.*, 1982).

2.1.4 Species Variation in Platelet Responses

The species variation in responses of isolated platelets to PAF has been subjected to extensive investigation with the goals of identifying the most appropriate source of platelets for bioassay and to facilitate analysis of *in vivo* PAF actions in relation to platelet dependence. Studies of platelet response to PAF indicate that, though widespread, the *in vitro* platelet activation by PAF is not

universal, being absent in rats and certain non-human primates (Cargill *et al.*, 1983). Absence of PAF responsiveness in rat platelets correlates with the absence of detectable PAF-binding sites (Inarrea *et al.*, 1984). Comparison of structural features for activation of rabbit and human platelets indicates that, although the latter are 30–100-fold less sensitive, relative potencies were maintained, suggesting the presence of similar, if not identical, receptors (Hadvary and Baumgartner, 1983). However, radioligand-binding studies establishing rank orders of potency of antagonists provide evidence for potential subtle differences between receptors of these species (Hwang and Lam, 1986).

2.1.5 Role of Cyclooxygenase Products in Platelet Activation

Some of the confusion in the literature on the importance of cyclooxygenase products for PAF-induced human platelet aggregation was resolved by the recognition that only low doses of PAF were inhibited by *in vitro* treatment with NSAIDs (Fouque and Vargaftig, 1984; Cattaneo *et al.*, 1985). Although there is continuing debate as to the importance of cyclooxygenase products in aggregation of human platelets by PAF, there is little doubt that PAF is a stimulus for TXA_2 generation by platelets (Ostermann *et al.*, 1984a,b; Rao *et al.*, 1984; Greco *et al.*, 1985; Lauri *et al.*, 1985; Altman *et al.*, 1986; McCulloch and Vandongen, 1990). Nevertheless, some studies have failed to detect TXA_2 formation (Dahl, 1985). Variable conclusions regarding the importance of TXA_2 for PAF-induced aggregation may have resulted from differences in the concentration of PAF and the use of either *in vivo* or *in vitro* treatment with aspirin (McCulloch *et al.*, 1989). In gel-filtered human platelets, indomethacin or aspirin pretreatment inhibited both the release reaction and the aggregation response, supporting the involvement of cyclooxygenase products (Chesney *et al.*, 1982). Aspirin was reported to inhibit PAF-induced release of 5-HT, β-thromboglobulin and platelet factor 4, but not aggregation responses in human PRP (Marcus *et al.*, 1981). On the other hand, indomethacin was reported to have no effect on 5-HT release from rabbit platelets (O'Donnell *et al.*, 1978).

TXA_2 or cyclic endoperoxide synthesis in response to low concentrations of PAF contributed to the second phase of aggregation and ingestion of aspirin or *in vitro* treatment resulted in a conversion of responses to low concentrations of PAF from irreversible to reversible aggregation (Lauri *et al.*, 1985; Sturk *et al.*, 1985). Furthermore, it has been suggested that the ability of PAF to synergize with a range of other stimuli, including collagen, thrombin or adrenaline is dependent on ADP and cyclooxygenase products (Sturk *et al.*, 1985; Lauri *et al.*, 1986). However, Vargaftig and coworkers have repeatedly and convincingly shown that synergism between adrenaline and PAF is independent of cyclooxygenase products. PAF-induced generation of TXA_2 and release of ADP appear to mediate the appearance of high- and low-affinity fibrinogen-binding sites, respectively (Kloprogge and Akkerman, 1986).

Thus, the balance of evidence indicates that PAF-induced aggregation and the release reaction do not have an absolute requirement for ADP or TXA$_2$. However, these products do reinforce platelet stimulation by low concentrations of PAF in particular.

2.1.6   Thrombocytopenia

Intravenous administration of PAF results in a reversible thrombocytopenia in guinea-pigs (Demopoulos et al., 1981; Lefort et al., 1982), rabbits (McManus et al., 1980; Okamoto et al., 1986) and sheep (Burhop et al., 1986b; Toyofuku et al., 1988), indicating a potential contribution to thrombosis and thrombocytopenia. However, there is little information on PAF levels in patients with thrombocytopenia (for a review, see Meade et al., 1991). In a patient with DIC, PAF levels ($\approx$ 0.4 nM) in the range capable of activating human platelets have been observed to coincide with decreased platelet counts (Meade et al., 1991).

Interestingly, intravenous administration of PAF to rats induced a "DIC-like state", despite the insensitivity of the rat platelets to direct (*in vitro*) activation by PAF (Imura et al., 1986). Nevertheless, there was no thrombocytopenic response to intravenous PAF in the rat or in the non-human primate *Cebus apella* in which platelets are also insensitive to stimulation by PAF (Handley et al., 1986b). Endotoxin treatment of rats caused DIC involving thrombocytopenia and fibrinogen consumption (Imura et al., 1986). Hence, it was of interest that the PAF receptor antagonist CV 3988 appeared to inhibit endotoxin-induced DIC despite the insensitivity of rat platelets to PAF. Intraplantar PAF in the rat (4–16 µg/paw) did induce a thrombocytopenic response, suggesting the involvement of a locally produced mediator (Martins et al., 1987). Indeed, high doses of PAF (6 µg/kg intravenously) did induce a thrombocytopenia in the rat by a lipoxygenase-dependent mechanism, although the aggregates were only observed in blood diluted in formalin-containing saline, indicating a relatively weak and indirect aggregation response (Martins et al., 1988). In addition, immune complex-induced alveolitis in the rat is accompanied by a thrombocytopenia that is blocked by pretreatment with antagonists of either PAF or TXA$_2$ receptors (Tavares de Lima et al., 1989).

Endotoxin shock and systemic anaphylaxis are conditions that have been documented to involve PAF formation and detection in plasma and are also associated with thrombocytopenia. Evidence for a significant role for PAF in these thrombocytopenic responses has been produced by the use of specific PAF receptor antagonists, although only a partial protective effect has been observed in some studies (Okamoto et al., 1986; Touvay et al., 1986; Yue et al., 1990), and in active anaphylaxis PAF antagonists were ineffective (Darius et al., 1986b;

Pretolani et al., 1987; Lohman and Halonen, 1990). Considering the multitude of mediators released during these phenomena, it is hardly surprising that PAF antagonists are at best only weakly active.

Reversals of the fall in platelet count in a single patient with ITP treated with WEB 2086 (Lohmann et al., 1988) and the absence of any effect of WEB 2086 on platelet counts in healthy subjects (Brecht et al., 1991) led to a clinical trial of this compound in ITP in adults. The PAF receptor antagonist failed to show any beneficial effects (Giers et al., 1990). Nevertheless, there are no clinical data as yet on the potential therapeutic effects of PAF antagonists on other forms of thrombocytopenia (Meade et al., 1991).

It is of interest that PAF formation by embryos was first discovered by its ability to account for pregnancy-induced thrombocytopenia in mice (O'Neill, 1985). The normally high plasma acetylhydrolase level decreases in maternal circulation at term (Maki et al., 1988) and may predispose to thrombosis and thrombocytopenia at this time. Thus, treatment of platelet-related disorders in pregnancy with PAF receptor antagonists may have beneficial effects (Meade et al., 1991). In this respect, platelets from pregnant subjects (22–30 weeks) showed diminished responsiveness to PAF but not to ADP, which may indicate, albeit indirectly, that PAF release is elevated in this condition (Kornecki and Ehrlich, 1990).

2.1.7   Platelet Accumulation *in vivo*

Page and Morley have developed a technique to monitor organ-specific accumulation of platelets *in vivo* (Page et al., 1982; Smith et al., 1989). Homologous platelets are labelled *ex vivo* with indium-111 oxine and, following washing, are reinjected into anaesthetized animals. The radioactivity is determined by γ counting over the region of interest versus a reference region, which should itself be insensitive to the effects of the stimulus and therefore control for changes in circulating platelet numbers. Tissue localization of platelets is then determined by changes in the ratio of radioactivity in the region of interest (commonly thoracic) to that in the reference region (commonly abdominal). Using this methodology, PAF has been shown to increase platelet accumulation in the thorax of the anaesthetized guinea-pig (Page et al., 1982) by a mechanism that is inhibited by high (supra-anticoagulant) doses of heparin (Barrett et al., 1984). A component of the PAF-induced intrathoracic platelet accumulation is due to intravascular mixed aggregates of platelets and neutrophils (Dewar et al., 1984). Inhalation of endotoxin by conscious guinea-pigs induces an accumulation of platelets in the thoracic region which can be prevented by pretreatment with a prostacyclin analogue, heparin or PAF receptor antagonists, whereas cyclo-oxygenase inhibition was ineffective (Beijer et al., 1987).

This method has recently been applied to recording of cranial platelet accumulation (May et al., 1990). Following intracarotid administration of PAF and other platelet stimulants, there is a prolonged accumulation of platelets in the cranial region, indicating that this method may be suitable for the treatment of thromboembolic disorders in the cerebral circulation. In anaesthetized rabbits, the *in vivo* platelet aggregation in response to PAF and other agents is inhibited by endothelin-1 through a mechanism which is likely to involve PGI_2 (Thiemermann *et al.*, 1990).

An alternative model for examining thrombotic responses to PAF has been developed which measures thrombus formation optically (Bourgain *et al.*, 1985). Superfusion of PAF on to the guinea-pig mesenteric circulation results in a thrombus that may grow to occlude the arteriolar lumen (Bourgain *et al.*, 1987) and comprises leucocytes, lymphocytes, monocytes and eosinophils (Maes *et al.*, 1986; Vargaftig and Bourgain, 1989). Electrical challenge induced ultrastructural changes to endothelium and synergized with PAF in inducing thrombus formation. The thrombus could be embolized by superfusion with PGI_2 or PAF receptor antagonists but, upon cessation of inhibitor superfusion, the thrombus would redevelop without a further PAF superfusion (Bourgain *et al.*, 1985). This ongoing, PAF-initiated thrombosis suggested the possibility of autogeneration, and shows similarity to the prolonged effects of PAF on the coronary vasculature of buffer-perfused guinea-pig hearts (Piper and Stewart, 1987b). PAF has been shown to stimulate its own formation in neutrophils (Doebber and Wu, 1987) and it is possible that endothelial cells may also respond with autogeneration. Interestingly, thrombus formation in response to the electrical stimulation alone was inhibited by PAF antagonists (Vargaftig and Bourgain, 1989), possibly by antagonizing the effects of PAF release from the damaged endothelium or the degranulating platelets and leucocytes in the growing thrombus.

2.1.8 Platelet Responsiveness to Platelet-Activating Factor in Disease

A major obstacle to establishing the importance of PAF in human disease has been the difficulty encountered in developing suitable assays and sampling conditions to detect PAF in plasma and other relevant biological fluids. One approach, first used in studies of rabbit anaphylaxis (Henson and Pinckard, 1977), is to monitor the sensitivity of platelets to *ex vivo* PAF activation – subsensitivity being indicative of PAF release *in vivo*. Several problems with this approach are apparent: a genetically based absence of PAF-induced platelet aggregation has been described (Pelczar Wissner *et al.*, 1984); inhalation of PAF at a dose adequate to elicit bronchospasm in human volunteers had no effect on platelet responsiveness to PAF or platelet counts, indicating that *ex vivo* platelet responsiveness may not be a sufficiently sensitive approach (Kioumis *et al.*, 1988), especially when PAF release is restricted to a tissue compartment; changes in PAF-specific signal transduction mechanisms could occur in

the absence of any PAF release. Nevertheless, decreased responsiveness of *ex vivo* platelets to PAF following allergen-induced bronchospasm (Thompson *et al.*, 1984) and endotoxaemia (Meade *et al.*, 1991) has been documented.

A modification of this approach has used radioligand-binding studies. In patients with sepsis, the number of PAF-binding sites decreased to approximately 20% of control and high (> 1 nM) levels of PAF were measured in the blood of these individuals (Lopez Diez *et al.*, 1989).

In atopic subjects, an increased bleeding time has been observed, but the platelet aggregation response to PAF was normal, and responses to arachidonate, collagen and ADP were depressed (Szczeklik *et al.*, 1987). In patients with myeloproliferative disorders, the irreversible phase of PAF-induced platelet aggregation is absent (Viero *et al.*, 1986).

Specific desensitization of platelets to activation by PAF following IgE-mediated anaphylaxis in rabbits provided indirect evidence for the involvement of PAF in anaphylaxis (Henson and Pinckard, 1977). Further studies established that depletion of platelets attenuated IgE-mediated anaphylaxis in the rabbit (Pinckard *et al.*, 1977).

An increase in platelet responsiveness to PAF has been reported in a number of conditions, including acute experimental pancreatitis in dogs (Lukaszyk *et al.*, 1989) and type I diabetes (Greco *et al.*, 1985). Other workers reported that treatment with insulin either *in vivo* or *in vitro* decreased platelet responsiveness to PAF, collagen, ADP and adrenaline, suggesting that hypoinsulinaemia may be responsible for increased platelet responsiveness in diabetes (Trovati *et al.*, 1988).

β Thalassaemia is associated with an increased platelet responsiveness to PAF, whereas ADP responses were normal or decreased (Demopoulos *et al.*, 1988).

A further consideration in analysing the sensitivity of human PRP to PAF is the level of activity of plasma acetylhydrolase. Several conditions and disease states have been associated with increases in plasma acetylhydrolase levels, including diabetes (Spangenberg *et al.*, 1989), essential hypertension (Satoh *et al.*, 1989) and asthma (Miwa *et al.*, 1988), and with decreases in near-term pregnancy. Such changes in acetylhydrolase activity may well confound interpretation of aggregation studies unless these are carried out in washed platelets.

2.2 MONOCYTES/MACROPHAGES

The macrophage produces large amounts of PAF and is, itself, highly sensitive to stimulation by PAF. Cytokine production in response to PAF may contribute to some of its persistent inflammatory effects.

2.2.1 Platelet-Activating Factor Formation and Release

Benveniste and colleagues identified PAF synthesis and release by rat and mouse peritoneal (Mencia Huerta and

Benveniste, 1979) and rat, rabbit and human alveolar (Arnoux *et al.*, 1980, 1981) macrophages following exposure to A23187 and zymosan. The release of PAF by monocytes/macrophages (Arnoux *et al.*, 1980, 1981; Camussi *et al.*, 1981a; Mencia Huerta and Benveniste, 1981; Arnoux *et al.*, 1982) and its pronounced broncho-constrictor actions in baboons suggested a role in asthma (Denjean *et al.*, 1981). Specific antigen elicited the generation and release of PAF from alveolar macrophages isolated from untreated allergic asthmatic subjects (Tonnel *et al.*, 1986), but not from those receiving theophylline with or without corticosteroid (Arnoux *et al.*, 1987). Alveolar hypoxia in rats caused PAF release into rat pulmonary surfactant from alveolar macrophages (Prevost *et al.*, 1988). The range of stimuli documented to elicit PAF synthesis by macrophages includes TNF-α (Camussi *et al.*, 1987a), particulate stimuli such as zymosan (Mencia Huerta *et al.*, 1981) and possibly silica (Davenas *et al.*, 1987) in addition to the chemotactic peptide FMLP (Stewart and Phillips, 1989).

2.2.2 Respiratory Burst

PAF stimulates the respiratory burst in guinea-pig macrophages (Hartung *et al.*, 1982) and aggregation of human peripheral blood monocytes (Yasaka *et al.*, 1982; Boxer *et al.*, 1983). PAF, but not the anticancer ether lipid 1-octadecyl-2-methylglycerophosphocholine, was reported to stimulate chemiluminescence in bone marrow-derived mouse macrophages (Storch *et al.*, 1988). However, some agonist activity of the C_{18}-methyl-PAF was suggested by its ability to desensitize PAF responses and prime responses to PMA. C_{18}-Methyl-PAF is a full agonist at PAF receptors on guinea-pig macrophages linked to O_2^- generation and on platelet PAF receptors, albeit at about 1% of the potency of C_{18}-PAF (Stewart and Grigoriadis, 1991). Elicitation of macrophages to the peritoneal cavity of guinea-pigs using the inflammatory stimulus *Corynebacterium parvum* markedly increased the macrophage capacity to generate oxygen-derived free radicals in response to PAF (Hartung *et al.*, 1983).

It is of considerable interest that while PAF does not stimulate O_2^- generation in murine resident peritoneal macrophages, there is convincing evidence for the presence of PAF receptors linked to PLC activation on these cells (Parnham *et al.*, 1984; Prpic *et al.*, 1988). In contrast, Huang *et al.* reported that in bone marrow-derived macrophages of this species, PAF elicits both a respiratory burst and PLC activation (Huang *et al.*, 1988).

In human monocyte-derived macrophages, PAF stimulates generation of oxygen-derived free radicals at high concentrations with an EC_{50} value of 5 μM and a maximum response at 13 μM (Rouis *et al.*, 1988). Antagonism of the respiratory burst by the PAF analogue 1-*O*-hexadecyl-2-acetyl-*sn*-glycero-3-phospho(*N, N, N*-tri-methyl)hexanolamine suggested the involvement of a specific PAF receptor. However, a partial agonist activity of hexanolamine PAF in guinea-pig macrophages and rabbit platelets has been described, indicating that its inhibitory effect may have been via receptor desensitiza-tion rather than antagonism (Grigoriadis and Stewart, 1991). The high levels of PAF required to stimulate O_2^- generation in human macrophages are known to disturb membrane permeability to direct physicochemical inter-actions. Nevertheless, similar concentrations were re-quired for the stimulation of O_2^- generation by eosino-phils, and these actions were blocked by a modest (10 μM) concentration of the specific PAF receptor antagonist WEB 2086 (Kroegel *et al.*, 1989c). It seems likely that effects requiring a minimum of micromolar concentra-tions of PAF are mediated by a PAF receptor that is atypical, either in its binding affinity or in its location, and therefore accessibility, to exogenous PAF. The more important question is whether PAF ever achieves these concentrations *in vivo* or whether PAF in another form (e.g. liposomes (Hayashi *et al.*, 1985a) or lipoproteins (Imaizumi *et al.*, 1991)) is a more potent and therefore more likely pathophysiological stimulus.

2.2.3 Chemotaxis and Adhesion

PAF stimulates chemotaxis of monocytes of healthy donors and elicits greater responses from cells of patients with psoriasis and atopic eczema, despite high-dose therapy with oral steroids. This supersensitivity subsided when the skin lesions healed (Czarnetzki, 1983). PAF is reported to induce chemotaxis of thioglycollate-elicted murine macrophages (Prpic *et al.*, 1988).

Thrombin-stimulated endothelial cells display an in-creased adhesiveness to human monocytes and the monocyte cell line U937 (Dicorleto and de la Motte, 1989) by a mechanism which appears to be independent of PAF production and requires 4 h to reach a maximal effect, in contrast to the rapid adherence of human neutrophils to endothelial cells that has been ascribed to endothelial membrane expression of PAF (Zimmerman *et al.*, 1990b).

2.2.4 Macrophage Activation

Using glucose metabolism over a 72 h period as an index of macrophage activation, Hayashi *et al.* (1985a) reported that free PAF was a relatively weak stimulant, but that upon incorporation into liposomes its potency was greatly increased. Since non-metabolizable PAF analogues were more active than PAF in activating macrophages, it was suggested that the inactivity of PAF may have been the result of metabolism (Hayashi *et al.*, 1985b). Other explanations of the higher potency of liposomal PAF include the possibility that the liposomes fuse with the macrophage plasma membrane and deliver PAF to an intracellular or intramembranous biophase not readily accessible to PAF in its monomeric form. In guinea-pig bone marrow macrophages, but not in resident peritoneal macrophages, PAF induces an increased microbicidal activity through activation of specific receptors (Hayashi *et al.*, 1988).

In contrast to weak activation of peritoneal macrophage glucose consumption, PAF is a potent stimulus for O_2^- generation by guinea-pig alveolar (Maridonneau Parini *et al.*, 1985; Desquand *et al.*, 1986) or peritoneal macrophages (Lambrecht and Parnham, 1986; Parnham and Bittner, 1986; Stewart, 1989) by activation of specific receptors.

2.2.5 Cellular Responses

PAF induces a number of responses in isolated macrophages, including mobilization of arachidonic acid from its membrane phospholipid stores (Bachelet *et al.*, 1986) and an increase in intracellular calcium, which is mainly due to activation of receptor-operated Ca^{2+} channels (Conrad and Rink, 1986). PAF induces a rapid increase in IP_3 and IP_4 in murine bone marrow-derived macrophages consistent with activation of PLC (Stephens *et al.*, 1988). Studies of K^+ flux in mouse peritoneal macrophages suggest that PAF activates a quinidine-sensitive K^+ channel leading to membrane hyperpolarization (Garay and Braquet, 1986). PAF lowers basal cAMP levels. Moreover, increases in cAMP levels induced by either salbutamol or PGE_2 are reduced by PAF (Bachelet *et al.*, 1988). These effects may increase the capacity of the macrophage to generate O_2^-.

Aggregation of both human and rat macrophages has been observed in response to PAF stimulation, but responses in the rat require enormous concentrations, and lyso-PAF is equi-active in inducing these effects (Rabovsky *et al.*, 1990), indicating a non-specific cytotoxic effect that is unlikely to be relevant *in vivo*. It has been suggested that PAF regulates the production and secretion of elastase in activated macrophages, but this action is also observed only at high concentrations (> 25 µM) (Rouis *et al.*, 1990).

2.2.6 Eicosanoids

There is overwhelming evidence in macrophages and other cell types that PAF stimulates the mobilization of arachidonic acid and the production of eicosanoids. On the other hand, direct addition of PAF and several anticancer ether lipids to bone marrow macrophage homegenates inhibits re-acylation of arachidonic acid (Herrmann *et al.*, 1986). Thus, exogenous PAF may stimulate eicosanoid generation by receptor-mediated PLA_2 activation, whereas endogenous PAF may elevate arachidonic acid by preventing its reincorporation into membrane phospholipids. PAF (10–1000 nM) induces the synthesis and release of both LTB_4 and LTC_4 from human monocytes by a receptor-dependent mechanism (Rubin *et al.*, 1987). In contrast to observations in guinea-pig macrophages in which PAF stimulates both PGI_2 (Stewart and Dusting, 1988) and TXA_2 synthesis (Hartung *et al.*, 1983), equine macrophages respond with a selective increase in TXA_2 synthesis which is not prevented by the phospholipid-type PAF receptor antagonist SRI 63-441 (Morris and Moore, 1989).

2.2.7 Platelet-Activating Factor Receptors

Several cell lines having some of the characteristics of macrophages, including U937 cells and J774 cells, express PAF receptors that are linked to the elevation of intracellular Ca^{2+} (Maudsley and Morris, 1987; Ward and Westwick, 1988). The murine macrophage cell line P388D1 expresses high-affinity PAF-binding sites ($K_d \approx$ 0.08 nM) and responds to low concentrations of PAF (1–100 pM) with an increase in Ca^{2+}_i (Valone, 1988). These findings contrast with the relatively poor potency of PAF in stimulating O_2^- generation in murine bone marrow macrophages (Storch *et al.*, 1988) and a lack of effect in murine resident peritoneal macrophages (Parnham *et al.*, 1984). Thus, elevation of Ca^{2+}_i may be an insufficient stimulus for a complete respiratory burst.

Lambrecht and Parnham (1986) suggested that PAF receptors on macrophages differ from those on platelets on the basis of a relatively higher affinity of kadsurenone for PAF receptors on elicited macrophages. Subsequent studies indicated that this apparent difference in receptors could not be explained by the activation status of the macrophages (Parnham *et al.*, 1989). Antagonism of PGI_2 and O_2^- generation in response to PAF has also revealed differences in the potencies and characteristics of agonists and antagonists between resident macrophages and platelets (Stewart and Dusting, 1988; Stewart, 1989; Stewart and Grigoriadis, 1991), confirming that the expression of PAF receptor subtype on macrophages is not a result of macrophage activation. These observations indicate the potential for selective modulation of PAF-induced activation of macrophages in relevant diseases.

2.2.8 Immunoregulation

The immunoregulatory actions of PAF are highly complex and involve interactions with a number of different cell types, including macrophages. Concentrations of PAF ranging from 1 pM to 100 nM were reported to have no direct effect on IL-1α synthesis by rat spleen adherent monocytes. However, picomolar concentrations of PAF increase LPS-induced IL-1α synthesis and release and 100 nM PAF had an inhibitory effect by a BN 52021-sensitive mechanism (Pignol *et al.*, 1987). Interestingly, LPS is itself a stimulus for PAF release by alveolar macrophages (Rylander and Beijer, 1987). In human monocytes, PAF (0.001–5 µM) induced a weak stimulation of IL-1, but had marked effects in cells pretreated with muramyl dipeptide (MDP) (Salem *et al.*, 1990). In addition, MDP directly stimulated IL-1 production and stimulated PAF synthesis and retention with a time-course preceding that of IL-1 production. However, L-652,731 did not inhibit the MDP-induced IL-1 synthesis, suggesting that release of endogenous PAF did not contribute to the MDP effect. It remains possible that L-652,731 failed to achieve a high enough concentration to antagonize the actions of PAF in the intracellular compartment (Salem *et al.*, 1990). PAF alone, and in combination with IFNγ, stimulated the production of IL-1 by human monocytes at concentrations between 0.1 pM and 1 nM (Barthelson and Valone, 1990), providing a potential explanation of some of the more protracted responses to *in vivo* PAF treatment. PAF had no effect on the production of TNF-α by rat alveolar macrophages, but in combination with muramyl dipeptide PAF (100 pM) increased the levels up to three-fold (Dubois *et al.*, 1989).

In human monocytes, low concentrations of PAF (1 pM to 1 nM) increased cytotoxicity for WEHI 164 cells by activating specific PAF receptors, leading to an increase in TNF-α secretion (Valone *et al.*, 1988). In contrast, PAF stimulation of human peripheral blood monocytes was observed to induce only a transient cytotoxicity, and low levels of TNF-α were secreted (Bonavida *et al.*, 1989a).

Regulation of macrophage function by the PAF antagonists BN 52021 and SRI 63-441 was suggested to explain the inhibitory effect of these compounds on the growth of Ehrlich ascites tumour cells inoculated into the mouse peritoneal cavity, since neither compound influenced the proliferation of the tumour cells in *in vitro* cultures (Fecchio *et al.*, 1990).

2.2.9   Monocytes and Disease

PAF levels in the first exchange dialysate of continuous ambulatory peritoneal dialysis patients with infectious peritonitis were approximately 7 nM and there was a linear correlation of the PAF levels with leucocyte numbers and albumin (Montrucchio *et al.*, 1989b). PAF was not observed in dialysate following recovery from infection, suggesting that PAF may contribute to the pathogenesis of the accompanying inflammation. It is also useful to consider the level of PAF in the exudate compared with the concentrations required *in vitro* to elicit certain of the effects of PAF on human monocytes/macrophages. In addition, high levels of albumin in the exudate would lower the effective concentration of PAF. Macrophage responses requiring micromolar concentrations of PAF are therefore unlikely to be evoked, even in such a florid inflammatory reaction as that occurring in peritonitis.

2.3   NEUTROPHILS

The participation of neutrophils in the pulmonary, vascular and inflammatory actions of PAF have been established in a large number of studies. Apart from being an important target for the actions of PAF, neutrophils also generate and release PAF. The regulation of PAF production and its release has received increasing attention, since the modulation of these events may provide an additional therapeutic approach to certain inflammatory and ischaemic diseases.

2.3.1   Platelet-Activating Factor Formation and Release

Betz and Henson demonstrated that human neutrophils obtained from healthy and CGD subjects produced and released PAF upon stimulation with A23187 or the DAG analogue, PMA (Betz *et al.*, 1980). This study provided evidence that PAF production was not dependent on the respiratory burst which is absent in the CGD patients. The stimuli for PAF synthesis by neutrophils was extended by Camussi *et al.* (1981a,b) to include immune complexes, C5a, cationic proteins and phagocytic stimuli.

Regulation of PAF release from human neutrophils by the rise in cAMP levels in response to zymosan was suggested by the inhibitory effects of preincubation of neutrophils in phosphodiesterase inhibitors (Alonso *et al.*, 1982). The stimulation of PAF biosynthesis by A23187 was reduced by treatment with the Ca^{2+} channel antagonist, nifedipine (Jouvin Marche *et al.*, 1983) or with calmodulin antagonists (Billah and Siegel, 1984; Sanchez Crespo *et al.*, 1985). The synthesis of PAF by FMLP-stimulated neutrophils is largely blocked by the absence of external Ca^{2+} in the medium (Ludwig *et al.*, 1984).

Relatively few studies have examined physiological stimuli for PAF synthesis in human neutrophils. However, the putative mediator of endotoxaemia, TNF-α, is a stimulus for PAF synthesis and release in rat neutrophils (Camussi *et al.*, 1987a) and PAF synthesis in human neutrophils (DeNichilo *et al.*, 1991).

Further analyses of the regulation of acetyltransferase in neutrophils revealed that acetyltransferase is stimulated by PKA and that the duration of the stimulation is regulated by alkaline phosphatases (Nieto *et al.*, 1988). Despite the fact that PKC activation stimulates PAF formation, stimulants of this kinase were reported to reduce acetyltransferase activity (Nieto *et al.*, 1988). However, the metabolically unstable PKC stimulant OAG was reported to increase acetyltransferase activity (Leyravaud *et al.*, 1989).

2.3.2   Molecular Heterogeneity

Multiple molecular forms of PAF may be significant for autocrine actions, since the nature of the alkyl chain is a significant determinant of chemotactic activity (Carolan and Casale, 1990) and platelet aggregation (Hanahan *et al.*, 1981). Nevertheless, C_{16}- and C_{18}-PAF are reported to be equipotent in inducing weal and flare responses in the skin of human volunteers (Archer *et al.*, 1988).

The extent of the molecular heterogeneity of PAF produced by neutrophils is controversial: Clay *et al.* (1984) suggested that C_{16}-PAF was the exclusive product; Satouchi *et al.* (1983) suggested that up to 30% of neutrophil PAF is C_{18}; Pinckard and coworkers reported considerable molecular heterogeneity of neutrophil PAF with five classes being isolated (Weintraub *et al.*, 1985); Bossant *et al.* (1989) observed acyl-PAF, in addition to C_{16}/C_{18}-alkyl-PAF.

2.3.3   Platelet-Activating Factor Release

There is considerable variance in the literature regarding the relative proportion of PAF released from stimulated neutrophils, and several studies have investigated the factors that regulate PAF release. Intracellular calcium increases are considered to be necessary, but not sufficient, for PAF release (Gomez Cambronero *et al.*, 1989). However, the experiments on which this conclusion was based did not appear to discriminate between released and retained PAF. Benveniste and colleagues have suggested that PAF release is inversely correlated with cell number by a mechanism relating to increased retention of PAF by neutrophils in an acidic environment (Leyravaud and Benveniste, 1989). On the other hand, Cluzel *et al.* (1989) suggested that static neutrophil suspensions release a lesser proportion of the total than those that are continually perfused with new buffer. A number of cytokines, including TNF-α (Stewart *et al.*, 1991), GM-CSF (Wirthmueller *et al.*, 1989; DeNichilo *et al.*, 1991) and endotoxin (Worthen *et al.*, 1988; Stewart and Harris, 1991), cause marked increases not only in PAF synthesis in response to FMLP, but also in its release. The mechanism of the amplification of PAF synthesis has not been elucidated, but evidence of PAF-induced PAF formation (Doebber and Wu, 1987) from human neutrophils provides the potential for a positive feedback.

The complexity of regulation of neutrophil PAF synthesis was shown by the observation by Ludwig *et al.* (1985) that human serum albumin increased not only the release of PAF, but also the total amounts formed following stimulation with FMLP. A non-albumin serum protein with a high affinity for PAF was also shown to enhance PAF release from stimulated neutrophils (Matsumoto and Miwa, 1985). In contrast, Lynch and Henson (1986) reported that bovine albumin had little influence on the release of PAF from human neutrophils.

The mechanism for the export of PAF from stimulated neutrophils remains to be established, but studies by Ribbes *et al.* (1985) indicate that acetyltransferase activity is localized to intracellular membranes, necessitating transport to the cytosol prior to transfer across the plasma membrane. Such cytosolic transfer proteins have been identified in macrophages (Banks *et al.*, 1988).

The generation and release of PAF in neutrophils exposed to a variety of stimuli using radiochemical detection of acetate incorporation into PAF have been characterized (Sisson *et al.*, 1987). Most of the neutrophil PAF was retained, irrespective of the stimulus or whether the cells had been primed by endotoxin exposure, leading to the conclusion that PAF may have an intracellular role. In contrast to several other studies (Sanchez Crespo and Nieto, 1989; DeNichilo *et al.*, 1991), PMA was reported to be an ineffective stimulus for PAF formation, even though several PKC inhibitors reduced A23187-induced PAF synthesis (McIntyre *et al.*, 1987). However, the acetate labelling method may not detect PAF formed by the choline phosphotransferase pathway that mediates

PAF formation in response to PMA (Nieto *et al.*, 1988).

2.3.4   Aggregation

PAF induces aggregation of neutrophils (Camussi *et al.*, 1980). Parallel inhibition by PGI_2 of PAF release and immune complex-induced aggregation suggested that PAF may be the mediator of immune complex-mediated neutropenia in rabbits (Camussi *et al.*, 1981b,c). The aggregation of human neutrophils by PAF is reported to be accompanied by the generation of TXA_2 and PGE_2, but these eicosanoids do not regulate the aggregation response (Dahl, 1985). PAF- and LTB_4-induced aggregation of human neutrophils are potentiated by 5-HPETE and 5-HETE, which have little direct aggregatory effect (O'Flaherty *et al.*, 1984). The aggregation response of isolated neutrophils has been used as a bioassay for PAF, but this response is less sensitive than platelet aggregation.

2.3.5   Chemotaxis

PAF induces chemokinesis and chemotaxis of human neutrophils (Czarnetzki and Benveniste, 1981a,b), but the concentrations required are much greater than those for eosinophil migration (Sigal *et al.*, 1987; Morita *et al.*, 1989). In contrast to their relative potency in inducing platelet aggregation, C_{18}-PAF was more potent than the more abundant C_{16}-PAF in inducing neutrophil chemotaxis (Archer *et al.*, 1988). At the cellular level, PAF elicits so-called oscillatory motion in neutrophils, which is considered to reflect changes in the cytoskeleton associated with migration (Wymann *et al.*, 1987).

2.3.6   Degranulation

PAF was reported to have no direct effect on the degranulation of neutrophils, but enhanced responses to zymosan or A23187 by a mechanism that appeared to involve lipoxygenase products (Czarnetzki, 1982). In the presence of cytochalasin B, PAF alone induced degranulation (Ingraham *et al.*, 1982). However, Dewald and Baggiolini (1986) suggested that PAF-induced degranulation (assessed by gelatinase release) did not require cytochalasin B.

Studies using desensitization to PAF, LTB_4 and 5-HETE indicated that these lipids interact to induce degranulation upon stimulation with A23187, but not C5a (O'Flaherty, 1985). Mutual interactions between PAF and the lipoxygenase pathway have been observed in A23187-stimulated neutrophils in which PAF formation is reduced by lipoxygenase inhibitors (Billah *et al.*, 1985), whereas other studies have shown that exogenous PAF stimulates LTB_4 production (Hopkins *et al.*, 1983).

2.3.7   Superoxide Anion Generation

PAF was reported to be an ineffective stimulus for the respiratory burst unless neutrophils were first treated with cytochalasin B (Ingraham *et al.*, 1982). Of greater importance was the observation that PAF pretreatment of normal neutrophils caused a three-fold increase in O_2^- generation in response to an optimal concentration of FMLP (Ingraham *et al.*, 1982). However, Benveniste and colleagues suggested that PAF could elicit a direct increase in O_2^- generation and degranulation by neutrophils not primed with cytochalasin B (Jouvin Marche *et al.*, 1982).

The requirement for prior treatment of human neutrophils to induce responsiveness to PAF is controversial. Several considerations apply to the variance in the results: the concentration of PAF; the time of incubation; the detection methods; and the possibility of priming by inadvertent exposure to endotoxin (Haslett *et al.*, 1985). In those studies that do describe a direct effect, the O_2^- stimulation is weak relative to that induced by FMLP or PMA (Dewald and Baggiolini, 1986; Gay *et al.*, 1986). On the other hand, the ability of PAF to prime neutrophils for stimulation by powerful agonists such as FMLP has been documented repeatedly by many groups and without exception in human cells (Dewald and Baggiolini, 1985; Gay *et al.*, 1986; Baggiolini and Dewald, 1986; Vercellotti *et al.*, 1988, 1989; Stewart *et al.*, 1991).

Gay and coworkers (1986) described a priming effect of PAF on neutrophil responses to FMLP and the DAG analogue PMA. The priming by PAF was independent of extracellular Ca^{2+} and persisted for up to 60 min. PAF was reported to induce luminol-dependent chemiluminescence (O_2^-) in human neutrophils at high concentrations ($0.1–10$ μM) and to enhance zymosan-induced chemiluminescence at lower concentrations (0.1 μM) (Poitevin *et al.*, 1984). Although earlier studies reported a priming effect of PAF on FMLP-induced respiratory burst of neutrophils, a priming action at concentrations of PAF as low as 0.1 nM was subsequently identified (Dewald and Baggiolini, 1985; Baggiolini and Dewald, 1986). Furthermore, in neutrophils pretreated with FMLP, normally ineffective concentrations of PAF were found to stimulate O_2^- generation, indicating a bidirectional priming effect. The priming effects of PAF were inhibited by several PAF receptor antagonists, including BN 52021, L-652,731 and kadsurenone (Vercellotti *et al.*, 1988), and WEB 2086 (Stewart *et al.*, 1991). Although priming is best established with soluble PAF, a study by Vercellotti *et al.* (1989) indicated that PAF produced by cultured endothelial cells in response to thrombin could enhance the respiratory burst by cocultured neutrophils. *In vivo* studies in a rat model of immune complex-induced vasculitis suggest that picomolar concentrations of PAF can prime neutrophils to exacerbate the vascular injury (Warren *et al.*, 1989).

The synergism between PAF and FMLP for O_2^- generation also extends to arachidonic acid mobilization and eicosanoid formation by human neutrophils (Tanizawa and Tai, 1989).

The ability of PAF to prime neutrophil O_2^- generation and the observation that endotoxin and a number of cytokines prime for O_2^- generation and show a marked

increase in PAF release raised the possibility that the enhanced release of PAF contributed to the cytokine/endotoxin priming by an extracellular autocrine mechanism. Indeed, it was shown that the second phase of Ca^{2+}_i rise in response to zymosan was mediated by PAF release and autocrine stimulation (Tool *et al.*, 1989). Nevertheless, no evidence for an extracellular action of PAF in cytokine or endotoxin priming was obtained from studies examining the effects of PAF receptor antagonists (Stewart and Harris, 1991; Stewart *et al.*, 1991). A role for PAF in intracellular signal transduction in the priming phenomenon was not excluded by these studies.

The priming of neutrophils by PAF may be an important mechanism for rendering the quiescent circulating neutrophil sensitive to stimuli encountered at a site of inflammation following its transmigration through the vascular endothelium. A number of chemotactic substances elicit PAF synthesis by the endothelium, and in some cases these stimuli also elicit PAF release from inflammatory cells. Thus, both soluble and membrane-bound PAF may play a role in the priming phenomenon (Fig. 8.1).

2.3.8 Adherence

PAF superfused in a concentration of 20 nM over the hamster cheek pouch preparation elicited adherence of circulating neutrophils, and the accompanying increase in vascular permeability by this, but not lower, concentrations (5 nM), was reduced in animals made neutropenic (Bjork and Smedegard, 1983). In rabbit skin, the PAF-induced oedema was not affected by neutrophil depletion (Wedmore and Williams, 1981b). Thus, PAF-induced increases in vascular permeability appeared to arise through both neutrophil-dependent and -independent mechanisms. PAF, in common with LTB_4, has the capacity to increase the expression of the complement receptors (CR3 and CR1) by an action at specific PAF receptors, suggesting a regulatory role in adhesion and phagocytic responses (Shalit *et al.*, 1988). PAF increases the expression of the CD11/CD18 adhesion complex (Vercellotti *et al.*, 1988).

The adherence of neutrophils to cultured endothelial cells has received considerable attention as a model for neutrophil transmigration. In particular, studies by Prescott and coworkers have demonstrated that PAF is synthesized by endothelial cells in response to a wide range of inflammatory mediators, including thrombin and cysteinyl leukotrienes, and is expressed on the cell surface (Zimmerman *et al.*, 1985a,b; McIntyre *et al.*, 1986; Lewis *et al.*, 1988). This PAF, which appears to be firmly anchored in the membrane, can activate adherence of contacting neutrophils by actions at a specific PAF receptor located on the neutrophil (Zimmerman *et al.*, 1990a). The adherence is rapid, independent of protein synthesis, is not inhibited by fixation of the endothelium, but is prevented by fixation of the neutrophils. Initially, it was reported that LTC_4 stimulated the accumulation of PAF with a time-course and concentration–response

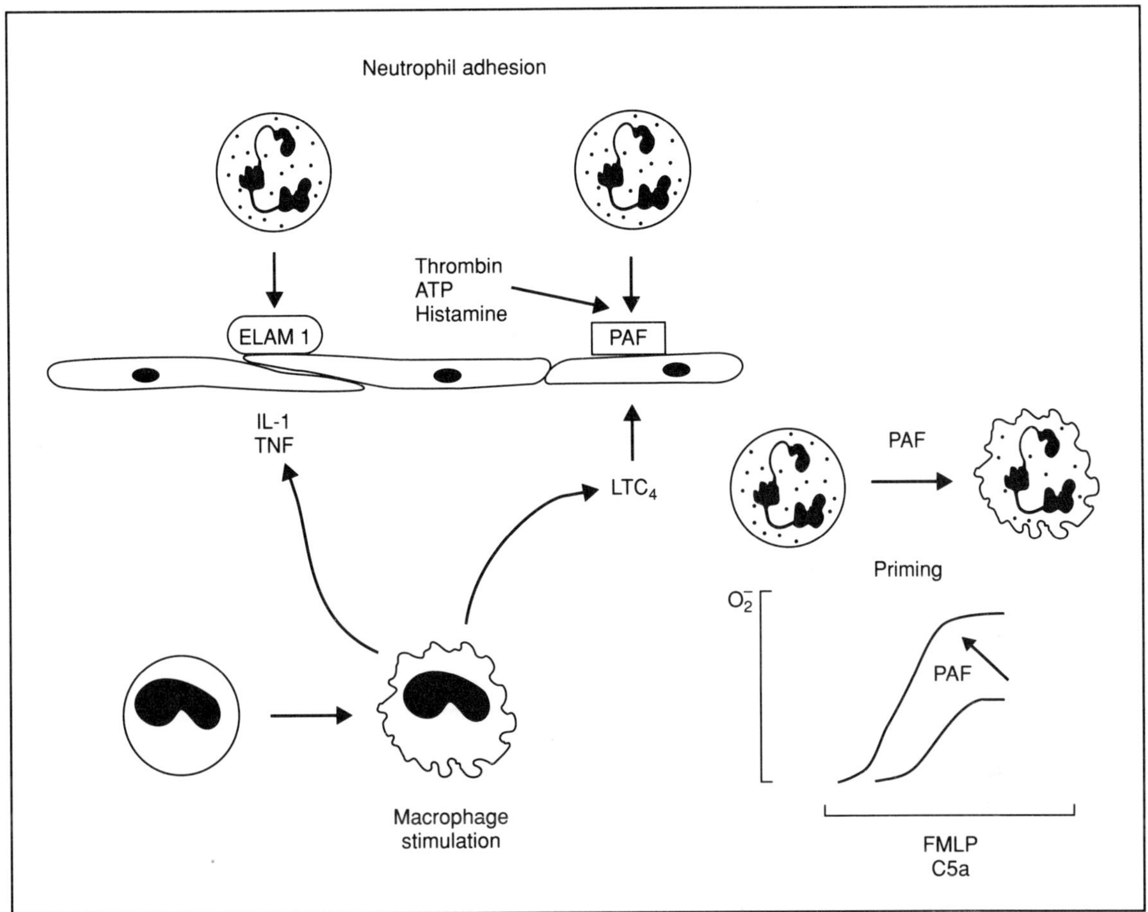

Figure 8.1 The role of PAF in the adhesion of neutrophils is of less significance for cytokines than for low molecular-weight mediators such as LTC₄. In addition, both soluble PAF and that expressed in the plasma membrane induce priming of neutrophils, rendering them highly responsive to activation of the respiratory burst and degranulation by FMLP and C5a.

relationship similar to that for neutrophil adherence to cultured endothelial monolayers (McIntyre *et al.*, 1986). Further evidence of a role for PAF was provided by specific desensitization by PAF and the ability of exogenous PAF to promote neutrophil adhesion to human endothelial cells (Zimmerman *et al.*, 1985a). However, given that all aspects of the structure of PAF are considered important for its binding to receptors, activation of PAF receptors by PAF anchored in a lipid membrane seems unlikely. Nevertheless, activation of macrophage PAF receptors by liposome-incorporated PAF (Hayashi *et al.*, 1985a) is compatible with the endothelial mechanism. Alternatively, upon close opposition of the lipid membranes, PAF may diffuse from the endothelial cell to the neutrophil PAF receptor without ever being "released" into the extracellular medium (Zimmerman *et al.*, 1990). Evidence that the action of endothelial PAF is a result of plasma membrane expression was provided by experiments showing that acetyl-

hydrolase could reduce neutrophil adherence, whereas endothelial cell fixation had no effect (Zimmerman *et al.*, 1990). The neutrophil adherence to thrombin-stimulated endothelium has been shown to be blocked by PAF receptor antagonists, including BN 52021 (Vercellotti *et al.*, 1989; Zimmerman *et al.*, 1990). More recent studies by Zimmerman and colleagues indicate that the coordinate expression of PAF and GMP-140 is required for maximal adhesion of neutrophils and there is cooperation between these endothelial products (Lorant *et al.*, 1991). GMP-140 also appears to be important for the rapidly developing adherence of neutrophils to endothelium (Toothill *et al.*, 1990).

However, it appears that cytokine-mediated adherence, which is associated with a delayed and long-lasting PAF synthesis, occurs independently of PAF, since it is only partially sensitive to BN 52021 and correlates closely with upregulation of ELAM-1 (Breviario *et al.*, 1988). Thus, it was suggested that IL-1α, which induces PAF synthesis by a mechanism that has a lag time of 2–4 h and requires protein synthesis, could stimulate PAF synthesis in a different cellular compartment to thrombin and other mediators. Furthermore, the partial inhibition of PMN adhesion by PAF antagonists may result from an inhibition of PMN-derived PAF rather than endothelial PAF (Breviario *et al.*, 1988). The function of IL-1α-induced PAF production by endothelial cells remains unclear, but it may contribute to intracellular signalling processes.

2.3.9   Neutrophil Accumulation

In cultured endothelial cells grown on nucleopore filters, both PAF and LTB$_4$ induce neutrophil transmigration by a cAMP-sensitive mechanism, and the response to PAF appeared to be mediated by LTB$_4$ (Hopkins *et al.*, 1984). This observation is consistent with the heterologous desensitization of chemotactic responses between PAF and LTB$_4$ *in vivo* (Colditz and Movat, 1984). The absence of a transmigration response to either LTC$_4$ or LTD$_4$, which are stimuli for PAP production by endothelium suggests that PAF-dependent neutrophil adherence is insufficient to induce transmigration.

Intratracheal instillation of PAF induced pathological responses similar to those induced immunologically, including platelet and neutrophil accumulation in the alveolar spaces (Camussi *et al.*, 1983d). In addition, challenge of rabbits with anti-ACE antibodies elicited intravascular PAF release and isolated native endothelial cells released PAF (Camussi *et al.*, 1983c), indicating that PAF may play a role in the pulmonary oedema accompanying the immunological challenge.

PAF induced neutrophil influx in rabbit skin with a time-course similar to that of FMLP (Vemulapalli *et al.*, 1984) and showed homologous desensitization (Colditz and Movat, 1984). In guinea-pig skin, the injection of PAF elicited a neutrophil influx with a similar time of onset (2 h) to that in rabbit skin, but with a more protracted duration (Dewar *et al.*, 1984). Intradermal PAF in humans also induced acute accumulation of neutrophils (4 h) with a later mixed infiltrate containing monocytes and neutrophils (24 h), leading to the speculation that PAF could be an important mediator of acute and chronic inflammatory skin disorders (Archer *et al.*, 1985b,d,e).

2.3.10   Neutrophils and Disease

Neutrophils from the blood of patients undergoing cardiopulmonary bypass showed increased reactivity 24 h after surgery as measured by the generation of O$_2^-$ and the release of PAF and β-glucuronidase (Hoshikawa Fujimura *et al.*, 1989). Aggregation of plasma-free human neutrophils has been observed in a model of dialysis and correlated with PAF production when regenerated cellulose membranes were used (Tetta *et al.*, 1989). PAF release has been documented in an experimental model of dialysis in which neutrophil aggregates were formed. These data suggest a role for PAF in the adverse cardiovascular consequences of haemodialysis (Tetta *et al.*, 1985).

Indirect evidence for priming of neutrophils in atopic subjects is provided by studies showing that neutrophils from asthmatic subjects are more sensitive to the chemotactic actions of PAF (Rabier *et al.*, 1989) and release greater amounts of the granule marker lactoferrin than those of non-atopic subjects, following *in vitro* PAF stimulation (Taylor *et al.*, 1990).

2.3.11   Fish Oil and Platelet-Activating Factor Production/Actions

Many epidemiological and experimental studies have suggested a bene fish oil effect of diets rich in EPA in cardiovascular diseases, particularly thrombosis and hypertension. More recent studies have provided some evidence that EPA may be useful in treating inflammatory disorders such as rheumatoid arthritis (Sperling *et al.*, 1987b) and asthma (Arm and Lee, 1989). Whilst the therapeutic effect of EPA is usually ascribed to effects on plasma membrane ion channel activity through modification of membrane fluidity or alterations in the profile of eicosanoid generation, several studies in rats indicate that EPA feeding may also reduce PAF biosynthesis (Sperling *et al.*, 1987a,b; Yeo *et al.*, 1989c).

Neutrophils from subjects with a fish oil-enriched diet incorporated EPA into the PAF precursor alkyl ether phospholipids, and subsequent stimulation with the non-physiological stimulus ionophore A23187 indicated that the increased EPA content had no effect on PAF biosynthesis (Triggiani *et al.*, 1990a). EPA was also ineffective in altering PAF formation by rat-elicited neutrophils (Pickett *et al.*, 1986). Hence, neutrophil PAF formation may not be a target for the effects of EPA supplements.

2.3.12   Cell Interactions

Co-incubates of rabbit neutrophils and platelets stimulated with FMLP release a platelet aggregating factor that has the characteristics of PAF (Coeffier *et al.*, 1987). However, similar experiments with human cells indicate that a protease is responsible for the platelet activation (Chignard *et al.*, 1986) and the activity appears to be a combination of cathepsin G and elastase (Ferrer Lopez *et al.*, 1990).

The transfer of arachidonic acid between cell types is a well-established mechanism for the generation of increased amounts of eicosanoids. A recent study suggests that PAF biosynthesis may also occur through transcellular movement of precursor lyso-PAF from platelets to neutrophils, providing an amplification of the aggregation signal in a growing thrombus (Coeffier *et al.*, 1990).

2.3.13   Pharmacological Modulation

It has been suggested that some of the therapeutic benefits of diflunisal and salicylate may derive from their inhibition of neutrophil acetyltransferase at clinically relevant concentrations (White and Faison, 1988). Nedocromil sodium, a relatively new prophylactic anti-asthma drug, inhibits both PAF and ZAS-induced chemotactic responses of human isolated neutrophils (Bruijnzeel *et al.*, 1989b).

PAF-induced neutrophil chemotaxis is reported to be selectively inhibited by BN 52021 and resistant to dexamethasone and nedocromil sodium (Kurihara *et al.*, 1989), in contrast to the findings of Bruijnzeel *et al.* (1990), who reported that nedocromil sodium inhibited responses to PAF. The PAF antagonist BN 52021 prevented PAF-induced and attenuated ZAS-induced Ca^{2+}_i increases (Bruijnzeel *et al.*, 1989a), whereas nedocromil sodium was completely ineffective against the Ca^{2+}_i rise, suggesting that the increase in Ca^{2+}_i is not a target for modulation of ZAS-induced neutrophil migration by this drug. PAF-induced degranulation of human neutrophils is attenuated by PGE_2 and PGD_2 by actions at distinct prostaglandin receptors (Rossi and O'Flaherty, 1989) and by the prostaglandin-elevating agent defibrotide (Schror *et al.*, 1989).

2.3.14 Platelet-Activating Factor Receptor Antagonists

The correlation between affinity constants for WEB 2086 obtained from radioligand-binding studies and antagonism of PAF-induced degranulation indicates the utility of WEB 2086 in defining the actions of PAF at its neutrophil receptors (Dent *et al.*, 1989a) and gives further support to the receptor dependence of most, if not all, of the effects of PAF on neutrophils.

Most studies of the influence of PAF on neutrophil function have focused on receptor-mediated effects. However, Henson and colleagues recognized that the retention of PAF by neutrophils could lead to high concentrations within the membrane, which may exert physicochemical effects independent of receptor activation. It was shown that PAF could attenuate the Ca^{2+}-induced increase in rigidity of artificial membranes, possibly by altering phospholipid head packing (Bratton *et al.*, 1988b). Furthermore, PAF and related lipids were shown to reduce the phase transition temperature for dipalmitoylphosphatidylcholine membranes (Bratton *et al.*, 1988a). Interestingly, an ester long-chain substituent on C-1 was found to be more potent than the ether linkage present in native PAF, and this acyl form of PAF has been reported to occur in human neutrophils (Mueller *et al.*, 1984). These properties of PAF have led to the proposal that it may play a role in the fusion of membranes during exocytosis (Bratton *et al.*, 1988a). This proposal will be possible to evaluate with the availability of selective inhibitors of PAF synthesis.

2.4 EOSINOPHILS

The importance of the eosinophil in the pathogenesis of asthma has been firmly established (Evans *et al.*, 1988;

Kay *et al.*, 1989). Pulmonary eosinophilia has been associated with the late phase of allergen-induced asthma and is believed to contribute to the bronchial hyper-reactivity that is a hallmark of this disease.

2.4.1 Platelet-Activating Factor Production

Lee *et al.* (1984) established that eosinophils generate and release PAF in response to FMLP, C5a or A23187 and that the amounts generated by cells from patients with eosinophilia are greater than those in cells from healthy subjects (Lee *et al.*, 1984; Chamone *et al.*, 1986).

2.4.2 Cellular Responses

Functional responses to PAF include degranulation (Kroegel *et al.*, 1989b) and the induction of hypodense eosinophils (Yukawa *et al.*, 1989). It is now established that PAF induces degranulation of eosinophils via a WEB 2086-sensitive receptor, causing release of eosinophil peroxidase and β-glucuronidase (Kroegel *et al.*, 1988, 1989c). In addition, PAF elicits an increase in intracellular Ca^{2+} at nanomolar concentrations and a stimulation of O_2^- generation at micromolar concentrations (Kroegel *et al.*, 1989a). Eosinophils generate LTC_4 in response to PAF (Bruynzeel *et al.*, 1986, 1987) and PAF primes for increased O_2^- generation at concentrations well below those required to elicit a direct effect (Bruynzeel *et al.*, 1986). In guinea-pig eosinophils, PAF stimulated TXA_2 generation (Hirata *et al.*, 1989; Giembycz *et al.*, 1990). The LTC_4 synthesis in response to PAF was confirmed in a study that also demonstrated a priming effect of PAF on A23187-induced LTC_4 production (Tamura *et al.*, 1988). Thus, while these overt effects of PAF on eosinophils require relatively high concentrations, pathophysiological levels of PAF may well induce priming of this cell type, enabling responses to be elicited by the prevailing concentrations of other simultaneously generated inflammatory mediators. Furthermore, a number of the bronchopulmonary effects of PAF may be secondary to the release of eicosanoids from PAF-stimulated eosinophils.

PAF or LTB_4 induces an increase in IgE-binding capacity of normodense human eosinophils (Walsh *et al.*, 1989), and this correlates with an increased cytotoxicity of eosinophils for *Schistosoma mansoni* (Moqbel *et al.*, 1990).

2.4.3 Chemotaxis

Perhaps the most important functional responses of eosinophils to PAF are the chemotaxis and chemokinesis which sets PAF apart from other potential chemical mediators of asthma (Colditz and Movat, 1984; Wardlaw *et al.*, 1986), since the effects of PAF are relatively selective for eosinophils (Hakansson *et al.*, 1987; Sigal *et al.*, 1987; Czarnetzki and Csato, 1989) and the maximum response to PAF is greater than that to LTB_4 or FMLP (Morita *et al.*, 1989).

PAF increased adherence of human eosinophils to

HUVECs by an action on eosinophils to increase CD11/CD18 expression (Kimani *et al.*, 1988; Lamas *et al.*, 1988). Townley and coworkers established that PAF was a potent chemotactic agent for human normodense eosinophils and its actions could be blocked by either desensitization or by BN 52021 (Tamura *et al.*, 1987).

Eosinophil migration into lung tissue has been observed following either aerosol or intravenous PAF challenge and has been related to the ensuing increase in bronchial reactivity (Lellouch Tubiana *et al.*, 1987), through the release of the epithelial toxins major basic protein (Grundel *et al.*, 1991) and other less specific inflammatory mediators, including eicosanoids and oxygen-derived free radicals. Indeed, coculture of PAF-activated eosinophils with guinea-pig trachea resulted in exfoliation of the ciliated epithelium and a decrease in ciliary beating frequency (Yukawa *et al.*, 1990). Bronchoconstrictor doses of PAF (60 pmol/kg) elicited increases in BAL eosinophils that were blocked by CV 3988, and the eosinophil infiltration was associated with damage to ciliated bronchial epithelial cells (Takizawa *et al.*, 1988). In addition, the pulmonary eosinophilia induced by aerosol challenge of PAF is enhanced by the cytokine, GM-CSF (Sanjar *et al.*, 1990b).

2.4.4 Pharmacological Modulation of Chemotaxis

Both PAF- and antigen-induced eosinophil accumulation in the lung were inhibited by antiplatelet antiserum or PGI_2 infusion, indicating a partial dependence on platelets (Lellouch Tubiana *et al.*, 1988). The recently developed anti-asthma drug, nedocromil sodium inhibited PAF- and zymosan-activated serum-induced PMN migration *in vitro* and reduced LTC_4 but not LTB_4 production, indicating a selective effect on the eosinophil leukotriene pathway (Bruijnzeel *et al.*, 1989b). On the other hand, nedocromil sodium had no effect on PAF- or zymosan-activated plasma-induced increases in Ca^{2+}_i levels (Bruijnzeel *et al.*, 1989a). In partial agreement with these findings, it was reported that both nedocromil sodium and DSCG reduced eosinophil cytotoxicity and LTC_4 production (Kurihara *et al.*, 1989), but had no effect on chemotaxis (Moqbel *et al.*, 1989).

2.5 Vascular Endothelium

2.5.1 Platelet-Activating Factor Formation and Release

Camussi and coworkers (1983a) first demonstrated the synthesis and release of PAF from endothelial cells in response to the non-physiological stimulus, A23187 and antibody to the endothelial-specific cell surface marker factor VIII. Additional *in vitro* stimuli included AII and AVP, and thrombin was effective when the cyclooxygenase pathway was blocked. It was suggested that PAF released from endothelium could influence the functional properties of local smooth muscle cells and inflammatory cells.

In contrast to the findings of Camussi *et al.* (1983a), thrombin was found to be a stimulant for PAF production by cultured HUVECs without the need for indomethacin treatment, although indomethacin did increase the amount of PAF produced by endothelial cells (Prescott *et al.*, 1984; Zimmerman *et al.*, 1985b). Since thrombin is a stimulus for endothelial PGI_2 production, and exogenous PGI_2 inhibits endothelial PAF production, the enhancing effect of indomethacin is likely to result from inhibition of PGI_2 synthesis (Zimmerman *et al.*, 1985b).

Subsequently, a wide range of pro-inflammatory mediators have been identified as stimuli for PAF production by endothelial cells, including bradykinin, histamine and ATP (McIntyre *et al.*, 1985), IL-1α (Bussolino *et al.*, 1986b,c, 1987a; Dejana *et al.*, 1987), TNF-α (Camussi *et al.*, 1987a; Bussolino *et al.*, 1988a) and hydrogen peroxide (Lewis *et al.*, 1988). These inflammatory mediators are also characterized by their ability to stimulate the vasodilator and antiaggregatory products, PGI_2 and nitric oxide (McIntyre *et al.*, 1985; Moncada *et al.*, 1988).

2.5.2 Cellular Responses

PAF induces a rise in Ca^{2+}_i levels in cultured cells via activation of specific membrane receptors that are linked to receptor-operated Ca^{2+} channels and to PLC and consequent IP_3-mediated mobilization of Ca^{2+} from intracellular stores (Brock and Gimbrone, 1986; Hirafuji *et al.*, 1988). In addition, high concentrations (1–10 μM) of PAF were reported to induce membrane depolarization (Lerner *et al.*, 1988) and to increase endothelial conversion of AI to AII with a maximum effect at 10 μM (Kawaguchi and Yasuda, 1987). This increase in ACE activity appeared to be related to PAF-induced PLC activation and increases in intracellular Ca^{2+} (Kawaguchi *et al.*, 1990).

2.5.3 Neutrophil Adhesion

Studies by Prescott's group led to the proposal that PAF is retained by vascular endothelial cells and its function may be restricted to regulating the adherence and reactivity of contacting cells, particularly neutrophils (Zimmerman *et al.*, 1990) (see Section 2.3.8).

2.5.4 Endothelial Thrombogenicity

Exogenous PAF makes the endothelium thrombogenic (Bourgain *et al.*, 1985; Maes *et al.*, 1986), despite the reported ability of this mediator to stimulate the production of the antiaggregatory mediator PGI_2 by cultured HUVECs (D'Humieres *et al.*, 1986) and by bovine pulmonary aortic endothelial cells (Grigorian and Ryan, 1987). However, several studies indicate that PAF is not a stimulus for either PGI_2 or nitric oxide production by endothelial cells cultured from arteries (Gryglewski *et al.*, 1986; Stewart *et al.*, 1990). Whether PAF is a stimulus for PGI_2 production by endothelial cells in the microcirculation remains to be established, but its capacity to

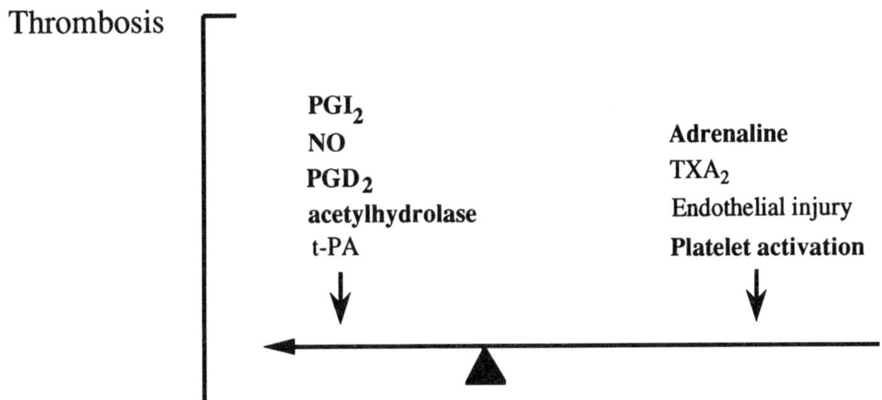

Figure 8.2 The thrombogenic actions of PAF arise from its direct activation of platelets, its ability to cause damage to the endothelial surface and synergism with TXA$_2$ or adrenaline. Counterbalancing these actions are the synergistic inhibitory effects of PGI$_2$ and nitric oxide (NO), inhibition by PGD$_2$, the presence of high acetylhydrolase activity in plasma and activation of fibrinolysis through the release of tPA from the endothelium.

release PGI$_2$ from perfused hearts (Piper and Stewart, 1986b; Stewart and Piper, 1986) and lungs (Hamasaki *et al.*, 1984) is suggestive of such an action.

Direct superfusion of PAF over the mesenteric vasculature induces a protracted thrombogenesis (Bourgain *et al.*, 1986a,b). Nevertheless, PAF is also reported to release tissue-type plasminogen activator from the perfused pig ear (Hofmann *et al.*, 1988b) and the buffer-perfused rat hindlimb preparation by a Ca^{2+}- and PLA$_2$-dependent mechanism (Emeis and Kluft, 1985; Tranquille and Emeis, 1990) via specific PAF receptors (Hofmann *et al.*, 1988a). This fibrinolytic action may serve as a negative feedback on the pronounced prothrombogenic actions of PAF. The balance between the thrombotic and fibrinolytic actions of PAF may depend on the amounts released and whether its source is intravascular or local within a tissue (Fig. 8.2).

2.5.5 Structural Damage and Permeability Changes

Fluorescence, light and electron microscopy studies reveal ultrastructural changes in the cytoskeleton of PAF-stimulated endothelial cells (Bussolino *et al.*, 1987b) and gross morphological abnormalities have been observed by light microscopy (Grigorian and Ryan, 1987). These changes included blebbing, vacuole formation, retraction from neighbouring cells and the assumption of a spindle, rather than a polygonal shape. Such structural changes may be important for thrombogenesis and increased permeability. In addition, PAF induced similar ultrastructural changes in the endothelium *in vivo* (Lewis *et al.*, 1983).

PAF induced a non-selective increase in vascular permeability to plasma macromolecules, including albumin and lipoproteins (Handley *et al.*, 1984; Burhop *et al.*, 1986a). The permeability of confluent monolayers of cultured vascular endothelial cells to macromolecules is not always increased by PAF (Stoll and Spector, 1989),

despite an invariable effect on vascular permeability *in vivo*. Thus, the response to PAF may be indirect, even though it does not appear to involve platelets. Alternatively, the arterial cultured endothelial cells commonly used in *in vitro* studies may be inappropriate for investigating the effects of PAF on vascular permeability.

Although there have been a number of studies confirming the ability of endothelial PAF to stimulate neutrophil adherence, there are no corresponding studies on the ability of inflammatory mediators to induce activation of contacting platelets. Importantly, it is evident that cysteinyl leukotrienes, bradykinin and histamine do not induce neutropenia, nor are they considered to be thrombogenic. Thus, while the elegant studies of Prescott and colleagues have firmly established a novel mechanism for neutrophil/endothelial plasma membrane interactions in extensive *in vitro* studies using cultured endothelial cells, it is as yet unclear as to whether such a mechanism contribute to PMN/endothelial interactions *in vivo* (Faulk *et al.*, 1989). There is, however, unequivocal evidence indicating that PAF induces the transmigration of neutrophils *in vivo* and also impairs the integrity of the endothelial barrier, resulting in increased vascular permeability.

2.6 MESANGIAL CELLS

Early studies on PAF established that it evoked renal dysfunction in various forms of cardiovascular shock (Camussi *et al.*, 1985; Plante *et al.*, 1986). The mesangial cell is thought to play a role in the regulation of renal blood flow and the surface area available for filtration. Thus, there has been great interest in the role of PAF in regulating the function of mesangial cells.

2.6.1 Eicosanoid Synthesis and Mesangial Cell Contraction

In cultured rat mesangial cells, PAF (0.1–100 nM) elicited an increase in PGE$_2$ synthesis and the release of preincorporated [^{14}C]arachidonic acid. Over the same concentration range, PAF induced a contraction of mesangial cells that was enhanced by indomethacin and attenuated by exogenous PGE$_2$, suggesting an important negative-feedback role for this eicosanoid (Schlondorff *et al.*, 1984). Similarly, mesangial cell contraction induced by AII or AVP was found to be modulated by PGE$_2$ or PGI$_2$ (Ardaillou *et al.*, 1985). In confirmation of these findings in cultured mesangial cells, it has been demonstrated that PAF, in common with other mediators, including AII, AVP and cysteinyl leukotrienes, reduces the planar surface area of rat isolated glomeruli (Barnett *et al.*, 1986).

PAF increases Ca^{2+}$_i$ level in cultured mesangial cells via activation of a specific PAF receptor linked to PLC activation (Kester *et al.*, 1987) and by activation of Ca^{2+} influx (Bonventre *et al.*, 1988). PAF is also able to elicit an increase in hydrogen peroxide production by mesangial cells that is inhibited by dexamethasone pretreatment (Baud *et al.*, 1986). BN 52021 and SRI 63-072 specifically antagonized the PAF-induced increases in PGE$_2$ levels in rat cultured mesangial cells, since responses to AII or A23187 were unaffected (Neuwirth *et al.*, 1987). Thus, endogenous PAF is unlikely to mediate the increased eicosanoid synthesis in response to these agents. Likewise, these PAF antagonists specifically inhibited PAF-induced mesangial cell contraction, but another antagonist, kadsurenone, was found not to be specific for PAF-induced effects (Neuwirth *et al.*, 1987).

2.6.2 Platelet-Activating Factor Formation and Release

Rat cultured mesangial cells generate PAF constitutively, the level of which can be stimulated to increase about eight-fold by A23187 (Schlondorff *et al.*, 1986; Lianos and Zanglis, 1987). Endotoxin was shown to cause the release of PAF and an IL-1-like material from both human and rat isolated glomeruli (Morell *et al.*, 1988). Rat cultured mesangial cells produced PAF in response to endotoxin (Wang *et al.*, 1988), suggesting that this cell type was the source of the PAF identified in rat glomeruli. AII induced PAF formation in mesangial cells, and, interestingly, about 90% remained cell associated (Neuwirth *et al.*, 1989), whereas endotoxin released the majority of the newly synthesized PAF (Wang *et al.*, 1988).

Nephrotoxicity is a significant problem with the clinical use of CSA. In mesangial cells, CSA reduced the planar surface area by a mechanism that was sensitive to inhibition by alprazolam or BN 52021. In addition, CSA increased the mesangial PAF levels more than two-fold (Rodriguez Puyol *et al.*, 1989). Similarly, TNF-α contracted human cultured mesangial cells and elicited an increase in PAF synthesis. The contractile response was inhibited by PAF synthesis inhibitors and receptor antagonists, including BN 52021 and SRI 63-072 (Camussi *et al.*, 1990). These observations suggest an

important role for PAF in the renal effects of CSA and TNF-α, and possibly other mediators. It is not yet clear whether the mechanism involves autocrine actions of PAF or an intracellular action.

3. *Organ Systems*
3.1 CARDIOVASCULAR ACTIONS
3.1.1 Systemic Blood Pressure
The intravenous administration of PAF elicits a dose-related decrease in arterial blood pressure that results from a complex series of events involving decreases in cardiac output and peripheral vascular resistance (Feuerstein, 1989). The relative importance of these effects depends on the dose of PAF. All species examined to date have shown a depressor response to PAF, but the dose eliciting this effect is dependent on the species being examined.

3.1.2 Peripheral Vascular Resistance
In dogs, the peak femoral vasodilator response was unaffected by inhibition of cyclooxygenase/lipoxygenase pathways. However, the systemic hypotension induced by PAF was attenuated by either indomethacin or BW-755C (Sybertz *et al.*, 1985). In another study in anaesthetized dogs, it was reported that PAF-induced femoral vasodilatation was prolonged by indomethacin, whereas responses in the mesenteric and gastric circulations were curtailed (Chu *et al.*, 1988). Candidate vasodepressor mediators include PGI$_2$ from the endothelium and leukotrienes from circulating inflammatory cells. Furthermore, there is considerable circumstantial evidence that EDRF/nitric oxide may mediate some of the vasodilator response to PAF. Studies with isolated arterial preparations have failed to reveal any direct vasomotor actions, but PAF may have direct effects on the microvasculature. However, C$_{18}$-PAF induced relaxation of the rat isolated thoracic aorta at relatively high concentrations (> 100 nM) that was dependent on an intact endothelium (Kamitani *et al.*, 1984). In addition to releasing a number of vasodilator substances, PAF releases 5-HT and TXA$_2$ from platelets, and LTC$_4$ and its active metabolites are known to have vasoconstrictor actions in certain vascular beds. Local blood flow may also be influenced by oedema and leucocyte aggregation induced by PAF (Whittle *et al.*, 1986) (Fig. 8.3).

In vivo studies in the rat established that PAF decreased blood pressure without influencing cardiac output (Sanchez Crespo *et al.*, 1981, 1982), suggesting that hypotension resulted from regional decreases in vascular resistance. A more recent study in conscious rats indicates that while hypotensive responses to lower doses of PAF can occur in the absence of a change in cardiac output, higher doses (3 nmol/kg) do cause a profound fall in cardiac output, and these are accompanied by a more prolonged hypotensive response (Siren and Feuerstein, 1989).

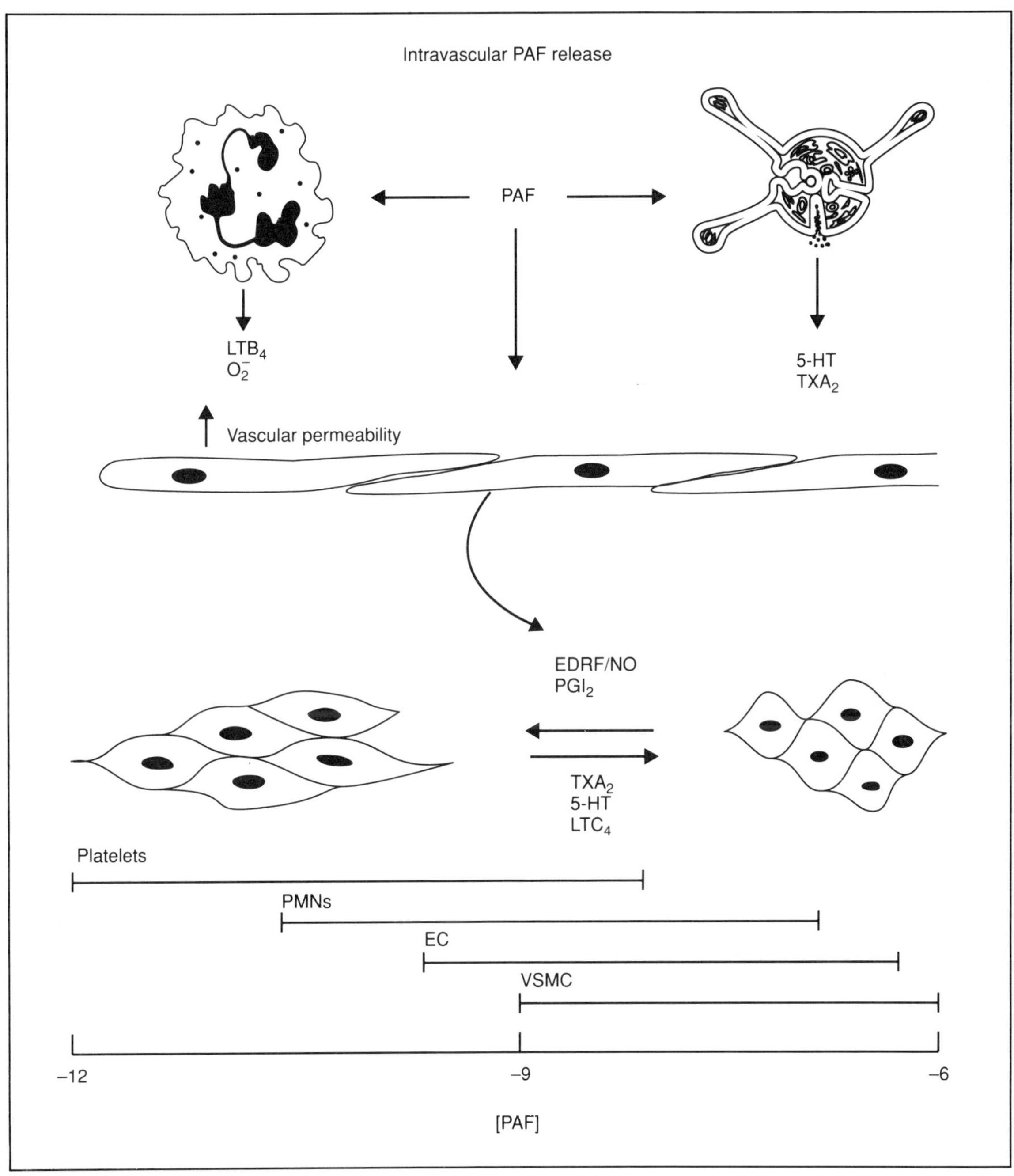

Figure 8.3 The vasoactive properties of PAF are the result of relatively weak direct actions and a number of indirect actions via activation of platelets, neutrophils and endothelium. The absence of any pronounced direct effect of PAF on large blood vessels at biologically relevant concentrations suggests that the microvascular smooth muscle may respond differently or that most of the effect of PAF is secondary to interaction with the depicted, and possibly other unidentified, cell types.

Studies in dogs (Kenzora *et al.*, 1984), pigs (Feuerstein *et al.*, 1984) and rabbits (Montrucchio *et al.*, 1987) have clearly established that the systemic and regional effects of PAF occur in phases, and that pharmacological treatments have different effects on these phases. However, the relationship between the phases of hypotension and secondary mediators is not consistent between species, possibly because of differences in dose.

3.1.3   Cardiac Actions

PAF reduces cardiac contractility (Benveniste *et al.*, 1983; Levi *et al.*, 1984; Saeki *et al.*, 1985; Piper and Stewart, 1987) in Langendorff-perfused guinea-pig hearts. This effect was present, but less pronounced, in Krebs-perfused rat hearts (Piper and Stewart, 1986b; Stewart and Piper, 1986). *In vivo* studies confirm that PAF reduces cardiac output in dogs (Kenzora *et al.*, 1984; Vemulapalli *et al.*, 1984; Sybertz *et al.*, 1985), pigs (Feuerstein *et al.*, 1984; Ezra *et al.*, 1987; Feuerstein, 1989; Laurindo *et al.*, 1989) and rabbits (Montrucchio *et al.*, 1987). Thus, despite species variability in responsiveness of platelets to PAF, this mediator consistently produces a negative inotropic effect.

3.1.4   Coronary Vasomotor Actions

Although the vasodilator actions of PAF were recognized in early studies (Sanchez Crespo *et al.*, 1981, 1982; Vemulapalli *et al.*, 1984; Bergelson *et al.*, 1985), it soon became clear that, in some vascular beds, PAF had pronounced vasoconstrictor effects (Voelkel *et al.*, 1982; Benveniste *et al.*, 1983; Saeki *et al.*, 1985; Piper and Stewart, 1986b; Schlondorff and Neuwirth, 1986).

In isolated perfused guinea-pig hearts, a bolus injection of PAF elicited a coronary vasoconstriction within 10 s, which peaked after 2–5 min, and subsided over the next 10 min to a level of about 50% of the peak response (Piper and Stewart, 1987). The mechanism of this increase in perfusion pressure remains controversial. However, it is generally accepted that the rise in perfusion pressure results from vasoconstriction rather than oedema or changes in extramural pressure secondary to changes in cardiac contractility or rate.

The protracted nature of the increase in coronary perfusion pressure would be consistent with a role for increased vascular permeability induced by PAF, as demonstrated in *in vivo* studies in the rabbit (Camussi *et al.*, 1985). However, several observations argue against such an explanation: most studies of the cardiac actions of PAF have noted a variable but significant degree of spontaneous reversibility within minutes of a bolus dose (Piper and Stewart, 1986, 1987b; Piper *et al.*, 1987); rapid reversal of PAF-induced vasoconstriction can be achieved either by infusion of PAF receptor antagonists (Piper and Stewart, 1987; Felix *et al.*, 1990b) or by administration of vasodilators such as bradykinin and acetylcholine, which are believed to release the vascular smooth muscle relaxant nitric oxide from the endothelium lining arterioles (Stewart and Piper, 1988a).

It is noteworthy that isolated preparations of large coronary arteries have little response to relevant (up to 1 µM) concentrations of PAF (Lefer, 1987). Thus, large coronary arteries may not express receptors for PAF and may therefore be unable to respond directly. In contrast, it is clear from *in vivo* studies that large arteries are a target for secondary mediators released by PAF from platelets or other primary targets (Fiedler *et al.*, 1987).

The second controversial aspect of the coronary actions of PAF relates to whether the vasoconstrictor actions of PAF are exerted by interaction with receptors on microvascular smooth muscle or whether metabolites released from other cellular targets play a role. Involvement of muscarinic receptors was excluded by the failure of atropine to modify the cardiac actions of PAF (Benveniste *et al.*, 1983). Studies by Levi *et al.* (1984) suggested that neither cyclooxygenase nor lipoxygenase metabolites contributed to the coronary vasoconstrictor action of PAF. On the other hand, isolated perfused rat hearts underwent a coronary vasoconstrictor response to PAF that was accompanied by the release of cysteinyl leukotrienes and TXB_2 into the coronary effluent, and inhibitors of the lipoxygenase and cyclooxygenase pathways attenuated the ensuing vasoconstriction (Piper and Stewart, 1986b; Stewart and Piper, 1986; Mest *et al.*, 1987; Stahl and Lefer, 1987a; Hu *et al.*, 1991). These observations were consistent with the actions of PAF in the pulmonary circulation of the rat (Voelkel *et al.*, 1982), and suggested that species differences may underlie the differing observations. However, further studies indicated that PAF released leukotrienes and TXB_2 from guinea-pig hearts and that these products contributed to the functional effects of PAF (Piper and Stewart, 1987; Stewart and Piper, 1987; Viossat *et al.*, 1989; Felix *et al.*, 1990b), albeit to a lesser extent than in rat hearts. The release of eicosanoids from guinea-pig hearts was transient, whereas the actions of PAF were maintained in excess of 2 h (Piper and Stewart, 1987; Stewart and Piper, 1987). Thus, in the study of Levi *et al.* (1984) the use of infusions of PAF may have obscured a contribution of eicosanoids to the early phase of the response.

Of the eicosanoids released by PAF-challenged hearts, LTC_4 and TXA_2 are the likely candidates for the vasoconstrictor actions, since each of these substances has pronounced vasoconstrictor activity. In rat hearts, the release of 6-oxo-$PGF_{1\alpha}$ indicated that the vasodilator cyclooxygenase product PGI_2 may modulate the vasoconstrictor actions of PAF. In contrast, in guinea-pig hearts (Piper and Stewart, 1987), there was no increase in PGI_2 formation, suggesting that vascular endothelial cells were neither a direct, nor a secondary target for the actions of PAF.

3.1.5   *In vivo* Studies

There have been several *in vivo* studies of the coronary vasomotor actions of PAF. In the pig, PAF elicits a decrease in coronary blood flow that is dependent on TXA_2 formation (Ezra *et al.*, 1987) and, like the *in vitro* responses of guinea-pig hearts, is not accompanied by PGI_2 release. In the dog, PAF has complex effects on the coronary vasculature. The studies of Sybertz *et al.* (1985) suggested that intracoronary PAF (0.2–6.4 nmol) induced a decrease in coronary flow that resulted partly from a direct effect on the coronary circulation (i.e.

independent of changes in systemic blood pressure) and was mediated by both cyclooxygenase and lipoxygenase products (Sybertz *et al.*, 1985). However, Jackson *et al.* (1986) reported that intracoronary PAF at a level of 0.5–2 nmol increased coronary flow by a platelet-dependent mechanism, whereas higher doses induced a biphasic effect with a brief increase followed by a decrease in flow.

In vivo and *in vitro* studies indicate that the coronary constrictor response to PAF results largely from the secondary release of TXA_2 and partly from the release of LTC_4 and an eicosanoid-independent (direct?) action.

3.1.6 Myocardial Depression

PAF invariably exerts a negative inotropic action, but the mechanism has been difficult to elucidate. Systemic administration of PAF at doses causing a profound hypotension are associated with a depression of cardiac performance, as determined by reductions in cardiac output and left ventricular pressure development (Alloatti *et al.*, 1987; Montrucchio *et al.*, 1987; Feuerstein, 1989). In a number of studies, the impairment of cardiac contractility has been accompanied by conduction abnormalities. The PAF-induced cardiac dysfunction could result from the precipitous fall in blood pressure, the coronary vasoconstriction, and/or a negative inotropic effect through PAF itself or secondary mediators such as LTC_4.

3.1.7 Is Cardiodepression Secondary to Coronary Flow Reduction?

In those studies that describe a reduction in coronary flow in response to PAF, a decrease in cardiac contractility has been observed. In a study reporting a sustained increase in circumflex coronary artery blood flow in response to intracoronary injection of PAF in anaesthetized dogs, there was no inotropic effect (Jackson *et al.*, 1986). Stahl *et al.* (1987) have suggested that although PAF may have a small direct negative inotropic effect, doses of PAF that reduce contractility of isolated perfused hearts invariably decrease coronary flow, and higher concentrations are required to decrease contractility of isolated tissues. Notwithstanding these arguments, PAF decreases left ventricular contractility in isolated Langendorff guinea-pig and rat hearts perfused at constant flow, suggesting that the negative inotropic action was unlikely to be secondary to changes in the coronary circulation (Levi *et al.*, 1984; Piper and Stewart, 1986b; Stewart and Piper, 1986; Piper and Stewart, 1987; Stahl and Lefer, 1987a; Stahl *et al.*, 1987; Felix *et al.*, 1990a). Pharmacological dissociation of the coronary and myocardial effects of PAF was evident in the guinea-pig, since the cardiodepressant response to PAF was not modified by the cysteinyl leukotriene receptor antagonist FPL 55712 (Piper and Stewart, 1987), but this compound attenuated the increase in coronary perfusion pressure.

The results of the foregoing studies in isolated hearts perfused at constant flow are consistent with a direct negative inotropic effect of PAF. However, the mainte-nance of global coronary flow may not preclude PAF-induced local ischaemia. Thus, local increases in microvascular tone and regional shunting could impair contractility, even under conditions of constant total flow.

3.1.8 Isolated Cardiac Tissues

A number of studies have investigated the effect of PAF on isolated cardiac tissue preparations to exclude the confounding influence of changes in coronary flow on myocardial contractility.

In guinea-pig papillary muscles, PAF (0.1 nM) induces a biphasic response comprising an initial positive and subsequent negative inotropic effect (Camussi *et al.*, 1984). Biphasic changes in action potential duration with an increase followed by a decrease have been observed at the same concentration of PAF (Alloatti *et al.*, 1987). These observations suggest an involvement of the L-type Ca^{2+} channel, since the duration of the action potential plateau is controlled largely by the magnitude of the inward calcium current. However, only the negative inotropic phase was suppressed by the Ca^{2+} channel blocker verapamil (Camussi *et al.*, 1984). This result is difficult to interpret, since the Ca^{2+} channel antagonists bind to PAF-binding sites with a high affinity and may therefore be unsuitable for evaluating the role of L-type Ca^{2+} channels in the inotropic actions of PAF (Wade *et al.*, 1986; Filep and Foldes-Filep, 1990).

A further study of the inotropic actions of PAF in papillary muscle indicated that 10–100 pM PAF had a positive inotropic effect, whereas higher concentrations (1–100 nM) had a negative inotropic effect (Tamargo *et al.*, 1985). Robertson *et al.* (1988) reported that PAF (1–1000 nM) induced a negative inotropic effect, shortened the action potential duration and reduced intracellular Na^+ levels. They suggested that the Na^+/Ca^{2+} exchanger may reduce calcium levels as a result of the sodium influx in response to the lowered intracellular Na^+ level (Robertson *et al.*, 1988).

Similar biphasic responses have been reported in studies of guinea-pig atrial preparations, with a positive inotropic response to PAF at a concentration of less than 1 nM, and a negative inotropic effect at higher concentrations (Diez *et al.*, 1990). In the study by Diez *et al.* (1990), verapamil blocked the positive inotropic effect but did not influence the action potential configuration. In addition, PAF increased Ca^{2+} uptake at the lower concentration range only. The failure to observe a reduction in Ca^{2+} uptake at the higher concentration range is not consistent with the involvement of either Na^+/Ca^{2+} exchange or Ca^{2+} L channels (Diez *et al.*, 1990). Other studies of atrial responses have provided contradictory results: PAF (10 pM to 10 µM) reduced the contractility of isolated left atria (Levi *et al.*, 1984); PAF at 0.1–100 µM was reported to have no effect on contractility (Kamitani *et al.*, 1984).

In rat atria, PAF increased contractility with no evidence of a depressor effect (Cervoni *et al.*, 1983;

Kamitani *et al.*, 1984). The inhibitory effect of propranolol suggested the involvement of locally released catecholamines (Cervoni *et al.*, 1983). In dog Purkinje fibres, PAF in the concentration range 1–10 μM elicited a transient positive inotropic effect followed by a maintained decrease in twitch amplitude, but only a negative inotropic effect was observed at higher concentrations. The action potential amplitude, upstroke velocity and resting membrane potential were all decreased by PAF, and these electrophysiological and functional effects were ascribed to non-specific amphiphilic effects, since similar effects were observed with other phospholipids known to be inactive at PAF receptors (Nakaya and Tohse, 1986).

Studies of isolated human cardiac tissues also provide evidence for direct inotropic actions. PAF (100 pM to 1 μM) elicited a biphasic response from papillary muscle fragments comprising an initial increase in contractility that was blocked by propranolol, followed by a negative inotropic effect that was only partially suppressed by indomethacin (Alloatti *et al.*, 1986). Over the same concentration range, PAF elicited only a negative inotropic response in human atrial pectinate muscle (Robertson *et al.*, 1987). This response was unaffected by either indomethacin or FPL 55712, but was blocked by PAF antagonists.

These studies in isolated tissues strongly support the contention that PAF exerts a direct negative inotropic action, at least at concentrations above 1 nM. The local release of PAF in cardiac anaphylaxis or during ischaemia may produce such concentrations. The ultimate evaluation of the mechanism of the inotropic actions of PAF will require the demonstration of PAF receptors on cardiomyocytes using both functional and binding assays.

3.1.9   Cardiac Anaphylaxis

Levi *et al.* (1984) noted the similarity between the effects of PAF and some of the pathophysiological responses of guinea-pig isolated perfused hearts during cardiac anaphylaxis. In the same study, the release of biologically active amounts of PAF by antigen challenge was demonstrated. The ability of antigen-challenge to release PAF was confirmed (Stewart and Piper, 1988b), and several PAF receptor antagonists were found to inhibit both the coronary vasoconstriction and the impairment of cardiac contractility that accompany the anaphylactic response (Koltai *et al.*, 1986; Piper and Stewart, 1986a). Despite the evidence from a number of previous studies suggesting an important role for cyclooxygenase products in anaphylaxis, PAF receptor antagonists did not modify the antigen-induced release of either TXB_2 or 6-oxo-$PGF_{1\alpha}$, indicating that PAF may have an independent role in cardiac anaphylaxis.

3.2   BRONCHOPULMONARY ACTIONS

The participation of PAF in systemic anaphylaxis (Pinckard *et al.*, 1977) signalled a possible involvement in the respiratory manifestations of this disorder. It is now established that PAF is an extremely potent bronchoconstrictor in a number of species including humans, and perhaps more importantly induces a long-lasting inflammatory response that may be associated with increases in bronchial reactivity (Fig. 8.4).

3.2.1   Airways Smooth Muscle

In general, the lack of direct effect of PAF on airways smooth muscle is consistent with a similar lack of activity on vascular smooth muscle. Nevertheless, PAF has extremely potent bronchoconstrictor actions *in vivo* that clearly result, at least in part, from airways smooth muscle constriction. These findings imply that the majority of the bronchoconstrictor effect of PAF may result from an interaction with non-airways smooth muscle cells with the release of secondary mediators of bronchoconstriction.

3.2.2   Parenchymal Lung Strips

The parenchymal lung strip, a preparation containing a number of cell types, has been used extensively to probe the mechanism of the bronchoconstrictor effects of PAF. Early studies suggested that PAF constricted rabbit isolated lung strips by a mechanism involving lipoxygenase products (Camussi *et al.*, 1983b). However, McManus and coworkers found that similarly prepared rabbit lung strips failed to respond to PAF (Halonen *et al.*, 1990). In contrast, guinea-pig lung parenchyma strips have invariably responded to PAF. The reason for the difference between guinea-pig and rabbit lung parenchymal responsiveness to PAF has recently been shown to result from the existence of a layer of smooth muscle in the pleura of the former species, but not the latter, removal of which precludes a contractile response to PAF (Halonen *et al.*, 1990).

In contrast to studies with rabbit lung strips, PAF-induced contractions of guinea-pig lung strips were reported to be resistant to lipoxygenase inhibition and to the leukotriene receptor antagonist FPL 55712 (Stimler and O'Flaherty, 1983). Furthermore, these responses were not due to the release of histamine, cyclooxygenase products or products derived from PAF activation of either platelets or neutrophils (Stimler and O'Flaherty, 1983). Several lines of evidence implicated postganglionic cholinergic nerves in the guinea-pig parenchyma response, including partial inhibition by hexamethonium and tetrodotoxin, enhancement by physostigmine and a lack of effect of atropine (Stimler-Gerard, 1986). However, the direct bronchomotor effects of PAF *in vivo* do not appear to be dependent on cholinergic nerves.

Although it was established that PAF-induced contractions of guinea-pig lung parenchyma were accompanied by the generation of TXA_2, neither thromboxane synthetase inhibitors nor a TXA_2 antagonist inhibited the contractile response (Yaghi *et al.*, 1989). Moreover, repeated exposure to PAF led to tachyphylaxis to the contractile effect, but not to the PAF-induced increase in

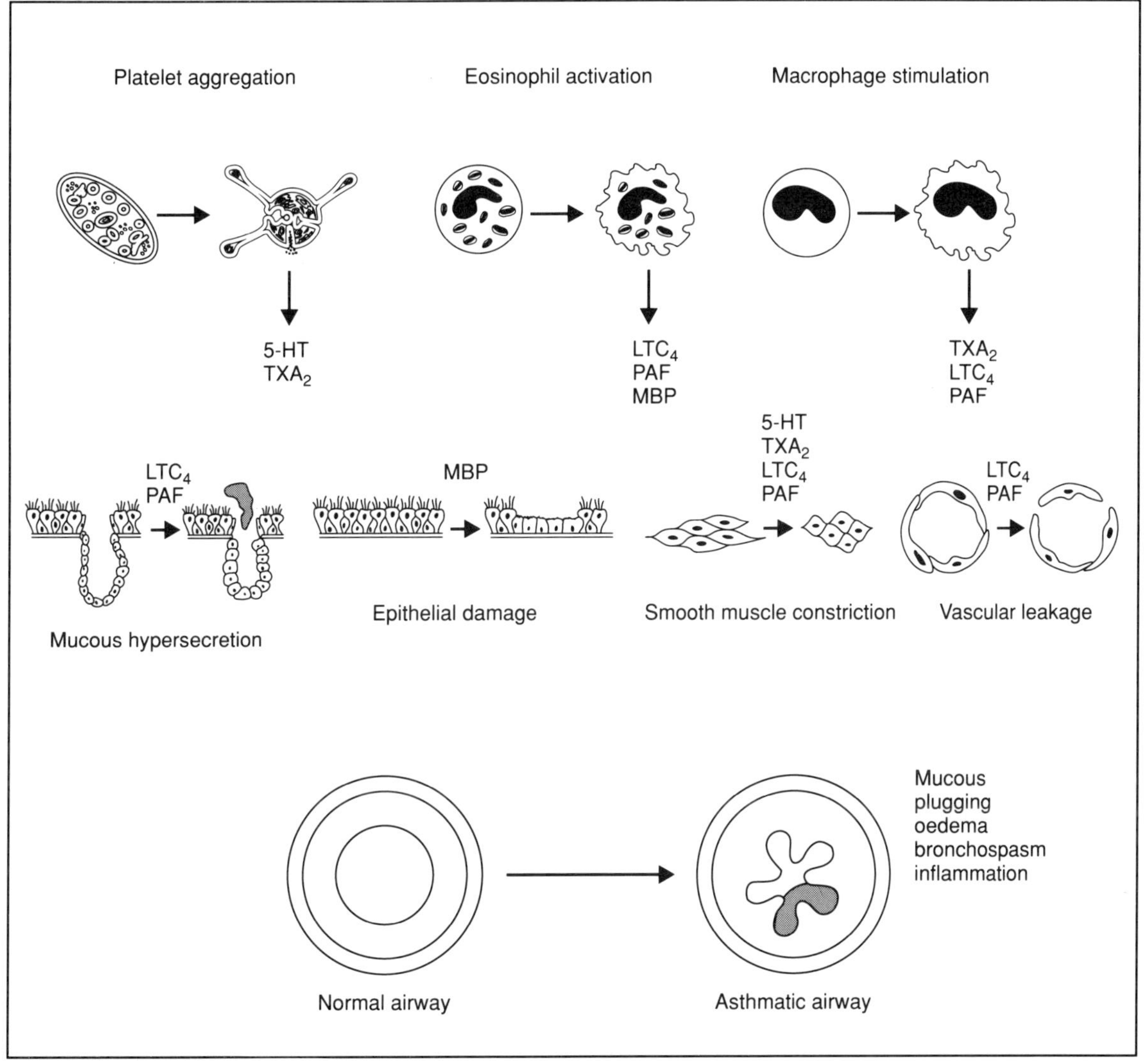

Figure 8.4 Immunological stimulation in the airways is documented to activate macrophages and to elicit the accumulation and activation of platelets and eosinophils. PAF is capable of activating these cells and is also released upon their activation. The cascade of mediators released in response to PAF and/or immunological activation act in concert to induce mucus plugging, oedema, bronchospasm and inflammation, the hallmarks of an asthmatic airway.

TXB_2 levels. In contrast to the failure of FPL 55712 to inhibit contraction, inhibitors of cyclooxygenase and lipoxygenase reduced contractile responses, suggesting the involvement of lipoxygenase products (Jancar *et al.*, 1988). PAF-induced contractile responses of guinea-pig lung strips were attenuated by gossypol, a flavanoid that inhibits eicosanoid formation (Touvay *et al.*, 1987), consistent with the suggestion of a lipoxygenase product (LTB_4?) being the final mediator.

3.2.3 Conducting Airways Smooth Muscle

Tracheas from sensitized guinea-pigs contracted in response to antigen by a mechanism that appeared to depend on both cysteinyl leukotrienes and PAF, since a combina-

tion of FPL 55712 and L-652,731 were additive in their inhibitory effects (Burka *et al.*, 1989). However, in contrast to PAF-induced contractions in guinea-pig lung strips, PAF was observed to cause an aspirin-insensitive relaxation of isolated guinea-pig trachea (Prancan *et al.*, 1982). This relaxant response has recently been shown to be mediated by the release of prostaglandins from the epithelium (Brunelleschi *et al.*, 1987, 1989; Conroy *et al.*, 1990). Thus, it appears that endogenous PAF production induced by antigen challenge has the opposite effect to that of exogenous PAF. On the other hand, several studies have described contractile responses of isolated guinea-pig trachea to PAF (Jancar *et al.*, 1987; Malo *et al.*, 1987; Chand *et al.*, 1988; Medeiros *et al.*

1989). One explanation for the discrepancy between these studies and those reporting relaxant responses may be variation in the integrity of the epithelium, which appears to be essential for relaxant responses. The latter are mediated by PGE_2 and therefore may be distinct from the EDRF.

In human isolated airways, PAF (0.07–0.7 μM) is reported to cause a single contraction which varies in magnitude on different tissues from the same lung and between different donors (Johnson et al., 1990). The contractile response was blocked by WEB 2086, and the PAF exposure also resulted in an increased responsiveness to histamine, suggesting that PAF-induced bronchoconstrictor responses and hyper-responsiveness could result from direct effects on airways smooth muscle. The variability in these responses did not appear to correlate with the presence of inflammatory cells within the preparation (Johnson et al., 1990).

3.2.4   Inflammatory Cells, Platelets and Bronchoconstriction

3.2.4.1   Platelets

Intrathoracic accumulation of platelets in PAF-challenged guinea-pigs is consistent with the suggestion that they play an important role in bronchoconstriction (Page et al., 1984). Moreover, platelets migrating through the alveolar wall and into the lumen were observed to precede the migration of polymorphonuclear leucocytes (Lellouch Tubiana et al., 1985). The lack of evidence for participation of TXA_2 in the bronchoconstrictor effects of intravenous PAF in the anaesthetized guinea-pig and the dependence of this response on platelets indicated the involvement of a non-cyclooxygenase platelet product (Lefort et al., 1984). In contrast to its effect in the rabbit (Camussi et al., 1983e), PGI_2 did not influence the bronchoconstrictor effects of PAF in the guinea-pig, but it did inhibit the platelet extravasation (Lellouch Tubiana et al., 1985). The platelet-independent bronchospastic effects of PAF aerosols indicated a distinct cellular target for PAF administered by this route. Vargaftig and coworkers suggested that alveolar macrophages may play a role since, upon exposure to PAF in vivo, these cells were desensitized when examined in vitro for PAF-induced O_2^- generation (Maridonneau Parini et al., 1985).

There have been few studies of PAF-induced bronchomotor actions in the rat, since this species is usually relatively unresponsive to the bronchomotor actions of mediators of anaphylaxis and its platelets are insensitive to activation by PAF. However, it was reported that PAF aerosols induced a bronchoconstriction with about 10 times the potency of acetylcholine and that no synergistic interaction between PAF and acetylcholine or histamine was observed (Misawa and Takata, 1988).

The bronchoconstrictor actions of PAF have been examined in the anaesthetized dog, in a preparation which allows simultaneous recording of tracheal smooth muscle tone and airways resistance. Local injection of PAF into the arterial circulation of the tracheal smooth muscle segment elicited a contractile response that was blocked by atropine, but not hexamethonium, indicating a stimulatory effect on postganglionic cholinergic nerves (Leff et al., 1987). Later studies showed that lower doses of PAF could sensitize the tracheal segment to parasympathetic stimulation (Bethel et al., 1989). PAF was inactive on isolated canine tracheal smooth muscle but, in the presence of platelets, a serotonin-dependent contraction was observed (Popovich et al., 1988). In vivo experiments revealed a correlation between arterial/venous differences in serotonin levels and the tracheal response to intra-arterial PAF (Murphy et al., 1989). However, the failure of atropine to block the constrictor effect of serotonin suggests that another platelet product may be involved in the interaction of PAF with cholinergic nerves. In vitro experiments in dogs suggest that TXA_2 is a candidate mediator of the enhancement of cholinergic neurotransmission by PAF in vivo (Serio and Daniel, 1988).

3.2.4.2   Neutrophils and Macrophages

Intravenous administration of PAF in rabbits induced thrombocytopenia and neutropenia associated with a selective sequestration of neutrophil aggregates in the pulmonary microcirculation, and these effects were abrogated by infusion of PGI_2 (Camussi et al., 1983e). Intratracheal instillation of PAF in rabbits elicited a similar pathological response and also caused degenerative changes in the alveolar epithelium with accumulation of alveolar macrophages (Camussi et al., 1983d). Intratracheal administration of PAF in baboons evoked a bronchoconstrictor response that was associated with neutropenia and thrombocytopenia and was independent of the formation of cyclooxygenase products (Denjean et al., 1983).

3.2.4.3   Eosinophils

In passively sensitized guinea-pigs, BN 52021 and WEB 2086 prevent both PAF and antigen-induced lung eosinophil infiltration, whereas PGI_2 infusion and platelet depletion were only partially effective, suggesting that platelets contributed to the PAF-dependent eosinophil recruitment in anaphylaxis (Lellouch Tubiana et al., 1988). PAF induced an early, but not a late, bronchoconstrictor response in guinea-pigs, associated with mucus hypersecretion, eosinophil infiltration and epithelial cell damage, all of which were prevented in animals pretreated with CV 3988 (Takizawa et al., 1988).

The early (6 h) and late (24 h to 5 days) eosinophil infiltration following antigen challenge of actively sensitized guinea-pigs has been shown to be blocked by CV 6209 and CSA, respectively, suggesting the involvement of PAF in the early response and a T cell factor in the late response (Fukuda et al., 1990). An obvious candidate for the latter would be the eosinophil activator IL-5 (for a review, see Lopez *et al.*, 1992), although a number of other cytokines are also known to induce eosinophilia (Kings *et al.*, 1990).

A single intraperitoneal injection of PAF induced an increase in eosinophil numbers in BAL fluid which could be prevented by 7 days of pretreatment with a number of prophylactic anti-asthma agents, including dexamethasone, aminophylline, DSCG or ketotifen, suggesting that inhibition of eosinophilia may comprise part of their mechanism of action (Sanjar *et al.*, 1989a). In guinea-pigs, it has recently been shown that PAF aerosol-induced pulmonary eosinophilia is blocked by a number of prophylactic anti-asthma drugs, including aminophylline, DSCG, ketotifen and dexamethasone, but not by indomethacin, salbutamol or the histamine H_1 receptor antagonist, mepyramine (Sanjar *et al.*, 1990a).

In PAF-challenged baboons, the eosinophil but not the platelet recruitment in the airways, as assessed by BAL, was inhibited by the anti-allergic, antihistamine ketotifen but not by the H_1 histamine receptor blocker pyribenzamine, indicating a potential contribution of PAF to the eosinophilia in asthma and an explanation of the therapeutic efficacy of ketotifen in asthma (Arnoux *et al.*, 1988).

3.2.5 Platelet-Activating Factor Formation and Release in the Lung

PAF was shown to be released from sensitized rabbit lungs (Kravis and Henson, 1975). Subsequently, PAF release was observed from rabbit, rat and human alveolar macrophages stimulated with A23187 (Arnoux *et al.*, 1980) and from human monocytes stimulated with various agents, including zymosan, immune complexes and A23187 (Arnoux *et al.*, 1982). Antigen challenge of isolated lungs from guinea-pigs actively sensitized to ovalbumin has been shown to release lyso-PAF and PAF (Fitzgerald *et al.*, 1986), whereas prolonged exposure of guinea-pigs to hyperbaric oxygen increased lyso-PAF, but not PAF release (Parente *et al.*, 1985). In addition, a human lung epithelial cell line (ATC-CCL-185) generates and releases PAF upon stimulation with A23187 or PMA, but physiological stimuli were not identified (Salari and Wong, 1990).

3.2.6 Eicosanoids and Airway Responses

The observation that PAF elicits leukotriene release from isolated buffer-perfused rat lungs indicated a close link between the bronchopulmonary effects of PAF and eicosanoid release (Voelkel *et al.*, 1982). Similarly, PAF induced TXB_2 release from guinea-pig lungs (Hamasaki *et al.*, 1984). The vasoconstrictor and oedema responses were largely mediated by leukotrienes and TXA_2, respectively. In contrast to observations in rat lung, PAF appears to induce the release of LTB_4 rather than the cysteinyl leukotrienes from guinea-pig lung, as assessed by bioassay and by the use of cyclooxygenase and lipoxygenase blockers (Jancar *et al.*, 1989).

The cellular targets for PAF-stimulated eicosanoid release in the airways are yet to be identified. Collagenase-dispersed human lung cells synthesized PGE_2 and TXA_2 in response to a relatively high concentration of PAF (0.5 μM) (Robidoux *et al.*, 1988). In addition, PAF has been shown to stimulate the release of TXB_2 from eosinophils isolated from guinea-pig lung (Hirata *et al.*, 1989). Antigen (anti-IgE) stimulation of an enriched preparation of human lung mast cells elicited the production of PAF and LTC_4 and the release of histamine (Triggiani *et al.*, 1990b). Antigen-stimulated release of histamine and TXA_2 from guinea-pig lungs was reduced by BN 52021, implying a role for PAF in activation of the mediator cascade (Berti *et al.*, 1988), confirming results of a similar study (Harczy *et al.*, 1986).

3.2.7 Anti-Asthma Drugs and Bronchomotor Responses

The potential involvement of PAF in asthma prompted an examination of the modulatory effects of anti-asthma drugs, but DSCG, ipratropium bromide (a muscarinic receptor antagonist) and ketotifen (a non-competitive antihistamine with the capacity to inhibit mediator release) did not influence PAF-induced bronchoconstriction (Lewis *et al.*, 1984). The TXA_2 receptor antagonist ICI 185,282 inhibited bronchospasm induced by PAF in anaesthetized guinea-pigs without affecting responses to histamine (Birch *et al.*, 1988).

3.2.8 Microvascular Leakage

Since it is believed that inflammation is a key feature of BHR and asthma, increasing attention has been paid to the regulation of inflammatory oedema in the airways (Chung *et al.*, 1990). Microvascular leakage induced by intravenous PAF is observed in the larynx, bronchi and intrapulmonary airways (Evans *et al.*, 1989), is independent of platelet activation (O'Donnell and Barnett, 1987) and is inhibited by adrenaline (via α adrenoceptors) and verapamil, but not by theophylline or the β adrenoceptor agonist salbutamol (Boschetto *et al.*, 1989). The inhibitory effects of a cyclooxygenase inhibitor, a TXA_2 synthetase inhibitor and a TXA_2 receptor antagonist on the increase in vascular permeability to PAF (intravenous) in rat airways suggested that TXA_2 made a significant contribution even though the TXA_2 mimetic, U44069 had no effect on permeability (Sirois *et al.*, 1990). Sirois *et al.* (1990) suggested that the vasoconstrictor actions of TXA_2 may be responsible, by increasing hydrostatic pressure. However, the leakage takes place in the postcapillary venules, and arteriole constriction would therefore decrease hydrostatic pressure. Indeed, the latter effect has been proposed to explain the adrenaline (at α adrenoceptors) inhibition of microvascular leakage. Hence, only a venoconstrictor action of TXA_2 would be compatible with an explanation of its role based on hydrostatic pressures. The increase in microvascular permeability of guinea-pig airways in response to bradykinin was found to be mediated largely by PAF, and, in agreement with findings in the rat, cyclooxygenase inhibition also reduced the response (Rogers *et al.*, 1990). A role for endogenous catecholamines was suggested by Rogers *et al.* (1990), who showed that treatment of guinea-pigs with antagonists of both α and β adrenoceptors enhanced leakage, presumably by dilating arterioles and increasing local blood flow. Plasma exudation is also observed when PAF is administered to the mucosal surface, and the response is inhibited by WEB 2086 or enprofylline pretreatment (O'Donnell *et al.*, 1990).

Aerosols of PAF were shown to induce pulmonary extravasation of Evans blue in amounts that correlated with the increase in airways resistance. Matching bronchoconstrictor doses of histamine induced considerably less Evans blue extravasation (Tokuyama *et al.*, 1991). Hyperinflation of the lungs (which reverses increases in airways resistance due to airways collapse as a result of smooth muscle constriction) had a greater effect on histamine than on PAF-induced bronchoconstriction. These findings provide strong indirect evidence that at least part of the PAF-induced bronchoconstriction results from bronchial oedema (Tokuyama *et al.*, 1991).

3.2.9 Mucus Secretion

Using the "tantalum hillock" technique for examining mucus secretion, Steiger *et al.* (1987) reported that PAF stimulated fluid secretion from porcine isolated trachea via activation of specific PAF receptors by a mechanism independent of acetylcholine, histamine or cysteinyl leukotrienes. In addition, PAF induced mucus secretion in ferrets by a CV 3988-sensitive mechanism (Lang *et al.*, 1987). Isolated human airways secrete mucus in response to PAF (10–200 nM) via specific PAF receptors linked to the production of cysteinyl leukotrienes, but independent of cholinergic mechanisms (Goswami *et al.*, 1989). In a study of cultured tracheal explants, PAF stimulated mucin production and the generation of cysteinyl leukotrienes from epithelial cells (Adler *et al.*, 1987). The lipoxygenase/cyclooxygenase inhibitor NDGA, but not the LTD$_4$ receptor antagonist FPL 55712, reduced the response to PAF.

3.2.10 Bronchial Anaphylaxis

The advent of specific receptor antagonists greatly facilitated investigations of the role of PAF in pulmonary pathology. BN 52021 proved to be a selective and complete inhibitor of cardiopulmonary responses to PAF (Berti *et al.*, 1986; Desquand *et al.*, 1986). A number of studies provided evidence of an inhibitory effect of PAF receptor antagonists on the respiratory manifestations of systemic anaphylaxis (Berti *et al.*, 1986; Darius *et al.*, 1986a; Touvay *et al.*, 1986). Similar inhibitory effects on both PAF and antigen-induced bronchoconstriction have been observed with the structurally distinct PAF receptor antagonist WEB 2086 (Casals Stenzel, 1987b; Casals Stenzel *et al.*, 1987). In guinea-pigs pretreated with H$_1$ histamine receptor antagonists, WEB 2086 inhibited the pulmonary anaphylaxis, irrespective of the route of antigen administration or the mode (active or passive) of sensitization (Heuer and Casals Stenzel, 1988). It is of interest that the non-histamine anaphylactic bronchoconstriction is also blocked by leukotriene synthesis inhibitors and receptor antagonists (Anderson *et al.*, 1983), further suggesting that both the production and actions of these mediators in immunological reactions are intrinsically linked.

The PAF receptor antagonist Ro-19,3704 blocked anaphylactic bronchoconstriction in passively, but not in actively, sensitized guinea-pigs (Lagente *et al.*, 1988a). However, this compound also suppressed FMLP-induced O$_2^-$ generation in alveolar macrophages (Lagente *et al.*, 1988a) by an action that was later ascribed to inhibition of PLA$_2$ (Gilfillan *et al.*, 1990).

Dogs natively allergic to *Ascaris suum* showed tachyphylaxis to PAF-induced increases in bronchial tension following allergen challenge, suggesting that endogenous PAF may have been released and contributed to the allergen-induced constriction (Munoz *et al.*, 1989). In contrast, in rhesus monkeys with *A. suum* sensitivity, the combination of effective doses of an LTD$_4$ and a PAF receptor antagonist failed to inhibit antigen-induced airways responses (Patterson *et al.*, 1989).

On the basis of the inhibitory effects of WEB 2086 on IgE-mediated anaphylaxis in the rabbit, it has been concluded that PAF contributes to the lethality and the bronchoconstriction, but is not the major mediator of the thrombocytopenia, leucopenia or hypotension (Lohman and Halonen, 1990).

3.2.11 Platelet-Activating Factor Receptors in the Lung

PAF receptors in lung tissue have been demonstrated by radioligand-binding studies using [^{3}H]PAF (Hwang *et al.*, 1983) and, more recently, [^{3}H]WEB 2086 (Dent *et al.*, 1989b). In addition, a PAF receptor from guinea-pig lung has been cloned (Honda *et al.*, 1991), but it is not yet firmly established whether this receptor is the same as that on platelets. The generation of a monoclonal anti-idiotypic, anti-PAF antibody holds promise for new insights into the PAF receptor (Wang and Tai, 1991), and may facilitate autoradiographic studies of PAF receptors in lung and other tissues which have hitherto been precluded by high non-specific binding and low specific activity of the available ligands.

3.2.12 Human Airways Responses

In a study of a mixed group of subjects (three healthy, four mildly asthmatic), the acute bronchoconstrictor effect of PAF inhalation was attenuated by the histamine H$_1$ receptor blocker chlorpheniramine, unaffected by indomethacin and, surprisingly, enhanced by atropine (Smith *et al.*, 1988).

BN 52063, a mixture of ginkgolides that contains BN

52021, inhibited the bronchoconstrictor effects of PAF in healthy human volunteers (Roberts *et al.*, 1988). In addition, BN 52063 blunted the early bronchoconstrictor response to allergen challenge in eight asthmatic subjects (Guinot *et al.*, 1987a). Earlier studies in human skin suggested that both cellular infiltration and weal and flare responses to PAF could be inhibited by the anti-asthma drug DSCG (Basran *et al.*, 1982). In addition, both the acute bronchoconstrictor and the BHR-inducing effects of PAF in guinea-pigs were attenuated by DSCG and other prophylactic anti-asthma drugs (Page *et al.*, 1985b; Morley *et al.*, 1988; Sanjar *et al.*, 1989b). However, ketotifen had no effect on PAF-induced bronchoconstriction, neutropenia or BHR in healthy human subjects, whereas its inhibitory effects on the cutaneous actions of PAF were confirmed (Chung *et al.*, 1988). Neither the acute bronchoconstriction nor the two-fold increase in responsiveness to methacholine induced by PAF inhalation in healthy subjects was affected by pretreatment with a single oral dose of theophylline (Chung *et al.*, 1989b). Thus, it appears that species differences are an important consideration in the mechanism of BHR and, hence, its susceptibility to modulation by anti-asthma drugs.

Inhalation of PGI_2 failed to inhibit the acute bronchoconstrictor effect of PAF in healthy humans, nor did it influence the neutropenia, despite inhibiting *ex vivo* platelet aggregation responses to PAF (Lammers *et al.*, 1990). Thus, it appears that acute bronchoconstriction induced by inhalation of PAF in either guinea-pigs or humans may be independent of platelet activation, especially since platelet numbers and *ex vivo* sensitivity to PAF were unaffected by PAF aerosols (Kioumis *et al.*, 1988). In contrast, blood neutrophil levels decreased, and increased numbers were recovered from BAL fluid after 4 h without an increase in eosinophil numbers, suggesting a possible role for the former cell type (Wardlaw *et al.*, 1990).

Investigation of the effects of nasal aerosols of PAF in healthy subjects and those with allergic rhinitis suggested that it had little direct effect on nasal patency and induced only a minor exacerbation of the nasal response to allergen challenge (Andersson and Pipkorn, 1988).

3.2.13 Pulmonary Circulation

PAF induces the release of TXA_2 and LTC_4 from buffer-perfused rat lungs and increases perfusion pressure and oedema formation by an eicosanoid-dependent mechanism (Voelkel *et al.*, 1982). Similarly, guinea-pig lung perfusion pressure and oedema formation were increased by 2 nmol of PAF via the formation of TXA_2, but higher doses (6 and 20 nmol) were only partly sensitive to inhibition with indomethacin (Hamasaki *et al.*, 1984). In isolated perfused rat lung, PAF-induced hypertension and oedema were prevented by BN 52021 or L-652,731, suggesting the involvement of a specific PAF receptor (Imai *et al.*, 1988).

Infusion of high concentrations of PAF (2 μM) decrease

vascular reactivity to AII but not potassium chloride in buffer-perfused rat lungs, and PAF infusion reduced pulmonary perfusion pressure in lungs perfused with AII (Gillespie and Bowdy, 1986). Several subsequent studies support the role of EDRF/nitric oxide in these vasodilator responses to PAF. It was found that low doses of PAF elicited vasodilatation *in vivo* (McMurtry and Morris, 1986; Voelkel *et al.*, 1986) and isolated human pulmonary arteries showed an endothelium-dependent relaxation to 100 nM PAF (Ono *et al.*, 1989).

Infusion of human neutrophils into isolated lungs of guinea-pigs treated *in vivo* with TNF-α elicited oedema, an increase in pulmonary capillary pressure and an increase in TXB_2 levels (Hocking *et al.*, 1990). The TXA_2 synthetase inhibitor Dazoxiben inhibited oedema formation and WEB 2086 inhibited both the oedema and the rise in capillary pressure (Hocking *et al.*, 1990). It is not clear whether guinea-pig lungs or human neutrophils were the source of PAF in these experiments.

PAF-induced bronchoconstrictor and pulmonary hypertensive effects in unanaesthetized sheep were unaffected by granulocyte depletion and only modest inhibition was observed with platelet depletion (Christman *et al.*, 1988), possibly suggesting a non-platelet source for the TXB_2 measured in lung lymph. Studies of the interaction between protamine, a cationic substance that induces endothelial injury, and PAF in inducing pulmonary-oedema suggest that the effects of PAF were dependent on a venoconstriction rather than an increase in vascular permeability alone (Chen *et al.*, 1990). Marked changes in the ultrastructure of rabbit pulmonary vascular endothelium following acute or chronic PAF administration indicated that PAF may contribute to the adult respiratory distress syndrome (Lewis *et al.*, 1983).

3.3 THE CENTRAL NERVOUS SYSTEM

The observation that the triazolobenzodiazepines were antagonists of PAF receptors on platelets indicated a potential link with CNS function (Kornecki *et al.*, 1984). However, it was soon discovered that the PAF antagonist properties of this class of compounds could be dissociated from their actions at the benzodiazepine site on the GABA receptor complex (Casals Stenzel and Weber, 1987).

3.3.1 Platelet-Activating Factor Formation

PAF production by bovine brain tissue has been demonstrated, but the acyl forms of PAF were more abundant than alkyl forms (Tokumura *et al.*, 1987). The formation of PAF by acetylcholine- or dopamine-stimulated chick retinal cells in culture gave support to a role for PAF in vision or in the development of the visual system (Bussolino *et al.*, 1986d, 1988b). Interestingly, this PAF was formed by the choline phosphotransferase pathway (Bussolino *et al.*, 1986d) and thus represents one of the few findings indicating that the activity of this pathway can be regulated. In a study of PAF production by rat brain, it was reported that the convulsants bicuculline or picrotoxin could elicit 20–40-fold increases in the PAF level (Kumar *et al.*, 1988b). Moreover, experiments with brains perfused with physiological buffer suggested a neuronal origin for the PAF. PAF was reported to stimulate PI turnover and Na^+/Ca^{2+} exchange in synaptosomes and to increase the permeability of the blood–brain barrier following intravenous administration (Kumar *et al.*, 1988b).

3.3.2 Physiological Roles?

The presence of functional PAF receptors on rod cells was indicated by an inhibitory effect of PAF on the amplitude of the β wave of the electroretinogram, which was inhibited by BN 52021 (Doly *et al.*, 1987). In addition, rat retinal membranes express binding sites for PAF that have similar characteristics to those on platelets, with a K_d of approximately 3 nM (Thierry *et al.*, 1989). Caution in interpretation of PAF antagonist-binding sites as evidence for PAF receptors was advocated by Laduron and colleagues, who reported that 52770 RP bound to guinea-pig cortical membranes with high affinity, but that the binding was not displaced by an excess of PAF (Levy *et al.*, 1989).

It has been suggested that PAF may play a role in regulation of the long-term phenotypic changes in the CNS, since PAF causes a receptor-mediated coordinate activation of the proto-oncogenes, c-*fos* and c-*jun* in SH-SY5Y neuroblastoma cells (Squinto *et al.*, 1989). The oncogene proteins form a heterodimer that binds to the AP-1 regulatory site that is found in a number of important neuronal genes, including proenkephalin and tyrosine hydroxylase (Squinto *et al.*, 1989).

PAF receptors have been identified in the CNS of gerbils (Domingo *et al.*, 1988), and functional evidence for PAF receptors on the neuronal cell lines, NG108-15 and PC12 has been reported (Kornecki and Ehrlich, 1988). In the latter study, 2–8 μM PAF induced receptor-mediated increases in Ca^{2+}_i that were suggested to be involved in differentiation, whereas slightly higher concentrations (4–10 μM) of PAF were neurotoxic over a period of 3–4 days (Kornecki and Ehrlich, 1988). The significance of these effects at very high concentrations of PAF is unclear, even though the antagonist CV 3988 blocked the increase in Ca^{2+}_i, a concentration of 350 μM was used.

Extensive studies have implicated PAF as a mediator of gastric ulceration in animal models (for reviews, see Esplugues and Whittle, 1989b; Wallace, 1990). However, the intracerebroventricular administration of PAF decreased gastric acid output and reduced intravenous pentagastrin-induced acid secretion (Cucala *et al.*, 1989a,b). Whether CNS PAF is involved in the physiological regulation of gastric acid secretion awaits studies with centrally administered PAF receptor antagonists.

3.3.3 Neuroendocrine Actions

Several studies have provided evidence that PAF stimulates the hypothalamic–pituitary–adrenal (HPA) axis. PAF, in common with a wide range of inflammatory mediators, including IL-1, IL-2, TNF-α and $PGF_{2\alpha}$, stimulates hypothalamic CRH secretion (for a review, see Calogero *et al.*, 1988). At low femtomolar concentrations, PAF inhibits the release of LHRH and somatostatin from the median eminence of the rat brain by a mechanism which may involve inhibition of Ca^{2+}-dependent processes (Junier *et al.*, 1988). This effect of PAF correlated with the presence of high-affinity ($K_d \approx 2$ nM) binding sites in hypothalamic membranes, but not in those from the pituitary, in which PAF had no effect on the release of luteinizing hormone or growth hormone (Junier *et al.*, 1988). However, *in vivo* administration of PAF (intravenous) stimulated both ACTH and corticosterone secretion via receptors that were blocked by intraperitoneal BN 52021 (Bernardini *et al.*, 1989). This finding appears to conflict with the absence of PAF-binding sites in the pituitary (Junier *et al.*, 1988), but may be explained by the ability of PAF to stimulate CRH secretion from explanted rat hypothalami, suggesting an indirect effect on pituitary ACTH release (Bernardini *et al.*, 1989).

Despite these studies on the production and effects of PAF in the CNS, it remains uncertain as to whether PAF functions as a neurotransmitter. On the other hand, a neuromodulatory role is well established. Most attention on the CNS actions of PAF has been directed towards its potential contribution to ischaemic brain injury (see 4.7.2).

3.4   THE IMMUNE SYSTEM

3.4.1 Lymphocytes

The importance of PAF in non-specific immunity following its release from leucocytes is well established. The first appreciation of a role in specific immunity came from studies indicating that lymphocytes had specific receptors that bind PAF (Shaw and Henson, 1980). Subsequently, it was reported that lymphocytes could produce PAF in response to stimulation by A23187, albeit in amounts about two orders of magnitude less than that produced by neutrophils (Jouvin Marche *et al.*, 1984).

3.4.2 Lymphocyte Accumulation

The observation that intradermal injection of PAF could induce lymphocyte infiltration in guinea-pig skin prompted the suggestion that PAF was a mediator of persistent inflammation (Archer *et al.*, 1985d). Intradermal injection of PAF elicited lymphocyte infiltration in human skin after 24 h (Archer *et al.*, 1985b). Other studies have confirmed the ability of PAF to induce lymphocyte infiltration in the rat paw (Silva *et al.*, 1986). PAF aerosols failed to elicit lymphocyte infiltration of the airways in guinea-pigs (Coyle *et al.*, 1988), whereas in human asthmatic subjects a correlation was reported between PAF levels and lymphocyte number in BAL fluid (Stenton *et al.*, 1990a).

The significance of the effect of PAF on lymphocyte traffic has been evaluated in a model of heart transplantation in which the PAF receptor antagonist BN 52021 failed to prevent the lymphocyte infiltration despite prolonging the survival of the cardiac allograft.

3.4.3 Lymphocyte Proliferation

The influence of PAF on lymphocyte proliferation has been examined in detail (for a review, see Braquet and Rola Pleszynski, 1987). *In vitro*, high concentrations of PAF (0.01–1 μM) inhibit PHA-stimulated proliferation of lymphocytes in a mixed culture of peripheral blood mononuclear leucocytes (Rola Pleszczynski *et al.*, 1987). The inhibition could be prevented by BN 52021 or the cyclooxygenase inhibitor indomethacin, consistent with the involvement of an inhibitory prostaglandin released by stimulation of a specific PAF receptor. The inhibition of lymphocyte proliferation was paralleled by a reduction in IL-2 levels (Rola Pleszczynski *et al.*, 1987). A subsequent study showed that BN 52021 by itself could enhance the generation of effector cells independent of the presence of IL-2 in the culture medium (Gebhardt *et al.*, 1988). A more detailed investigation of the effects of PAF on lymphocyte proliferation in mixed cultures revealed that the indomethacin-sensitive component of the suppression was dependent on the presence of monocytes in the culture (Rola Pleszczynski *et al.*, 1988). In the absence of monocytes, PAF induced a net helper effect with a more significant induction of $CD4^+$ helper cell function than $CD8^+$ suppressor cell function. These findings were further complicated by the observation that PAF decreased the number of $CD4^+$ and increased that of $CD8^+$ T cells (Rola Pleszczynski *et al.*, 1988). In contrast to the above studies showing that PAF decreased IL-2 production *in vitro*, chronic (7 days) *in vivo* PAF administration to rats resulted in a receptor-mediated increase in IL-2 production (Misawa and Takata, 1988). This report is consistent with the finding that PAF enhances the release of IL-1 from rat spleen monocytes (Pignol *et al.*, 1987) and human macrophages (Barret *et al.*, 1987).

In contrast to studies describing a decrease in lymphocyte proliferation by PAF, Patrignani *et al.* (1987) reported that PAF had no effect up to 10 μM. However, PAF was able to partially reverse the inhibitory effect of the PAF receptor antagonist L-652,731 on [^{3}H]thymidine incorporation (Patrignani *et al.*, 1987). The effect of L-652,731 could not be explained by inhibition of IL-2 production or receptor expression. Hence, it was proposed that endogenous PAF may have a role in the proliferative response. Given that PAF itself did not stimulate proliferation, the role of PAF could well be via an intracellular action not accessible to exogenous PAF. Patrignani *et al.* (1987) reported an increase in PGE_2

production by PAF-stimulated PBML cells in agreement with Rola Pleszynski *et al.* (1988). In a more complex culture system involving PBML cells incubated with endothelium, it was shown that the cytokines, IL-1 and TNF-α could induce suppression of concanavalin A mitogenesis by a mechanism that was inhibited by BN 52021 and depended on the presence of monocytes (Lacasse and Rola Pleszynski, 1991). In these experiments, the participation of PGE_2 was excluded by the failure of indomethacin to inhibit the suppressor activity.

In monocyte-depleted lymphocyte cultures, low concentrations of exogenous PAF (0.3 nM) were reported to enhance proliferation induced by suboptimal concentrations of IL-2 (Ward *et al.*, 1987). Ward *et al.* (1987) also showed an inhibitory effect of either L-652,731 or CV 3988, but not BN 52021, on [^{3}H]thymidine incorporation in IL-2-stimulated lymphocytes. These studies imply that lymphocytes produce PAF which contributes to the proliferation signal induced by IL-2 or PHA (Rola Pleszczynski *et al.*, 1987).

Thus, the complex influence of PAF on lymphocyte proliferation (Fig. 8.5) has not been clearly elucidated despite several extensive *in vitro* studies. The presence of monocytes in the culture, the concentration range of PAF and the type of PAF antagonists used all appear to be important variables in the conclusions reached. In addition, the use of [^{3}H]thymidine to estimate cell proliferation may provide misleading results if any of the PAF analogues or antagonists have effects on the uptake rather than the incorporation of [^{3}H]thymidine into DNA. The latter possibility may explain the effects of L-652,731, which appeared to decrease the uptake of [^{3}H]thymidine into the myeloid cell lines, WEHI-3B and FDCP mix A4 (Bazill and Dexter, 1989a). Nevertheless, L-652,731 had an unequivocal cytostatic effect on these two cell lines. The observation of the cytostatic activity of the PAF antagonist L-652,731 is of interest, since early studies established cytotoxic activity of alkyl ether phospholipids and, particularly, octadecylmethoxy-PAF for some tumour cells (Section 4.1).

A recent study indicated that the direct cytostatic effect of PAF was at least partly mediated via binding to a PAF receptor, since octadecylmethoxy-PAF-induced cytotoxicity for the IL-3-dependent WEHI-3B cell line was attenuated by L-652,731 (Bazill and Dexter, 1989b). It was suggested (Bazill and Dexter, 1989b) that PAF may be involved in the signalling induced by IL-3 in WEHI-3B cells, but did not provide evidence that PAF was produced in response to IL-3. Nevertheless, there is considerable evidence that other growth factors such as GM-CSF stimulate the production of PAF (Stewart *et al.*, 1991; DeNichilo *et al.*, 1991), raising the possibility that PAF may be involved in signalling mechanisms induced by a wider range of growth factors.

Less attention has been directed towards the influence of PAF on humoral immunity. A recent study indicates that PAF (0.1 pM to 1 nM) increased the secretion of IgE

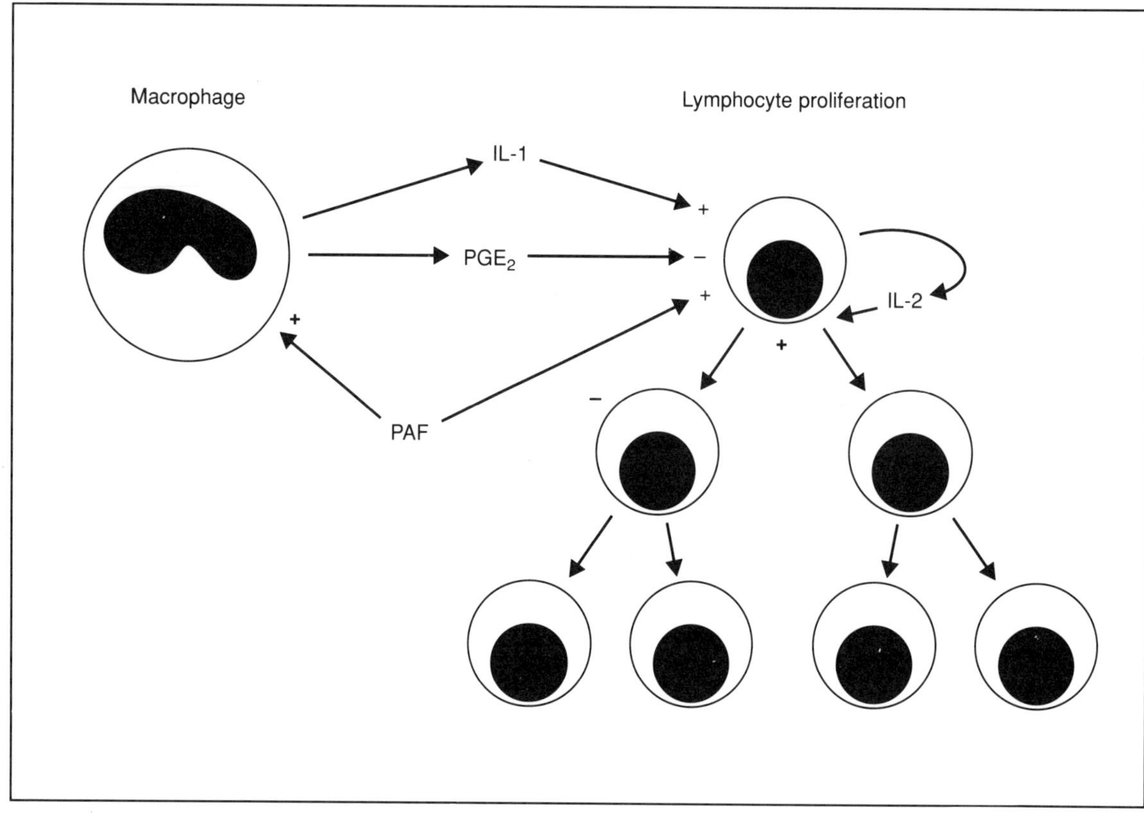

Figure 8.5 Two major targets for the immunoregulatory actions of PAF are the macrophage and the lymphocyte. PAF elicits a suppressor activity from the macrophage that is likely to be PGE₂.
However, *in vitro* studies suggest a direct stimulatory action via enhancing IL-2-dependent proliferation, and the release of IL-1 provides a further stimulus. The regulation of *in vivo* proliferation responses is certain to be more complex.

and IgG by B lymphoblastoid cells by an action that was independent of cell proliferation (Mazer *et al.*, 1990).

A role for PAF in NK cell cytotoxicity has been suggested, since PAF exerted a receptor-dependent cytotoxic action on the target cell line K562, and BN 52021 reduced NK cytotoxicity for K562 cells (Mandi *et al.*, 1989a,b). However, the effect of PAF generated by NK cells (Mandi *et al.*, 1989a) appears to be on the target cells rather than on the NK cells. The production of PAF by lymphocytes is further supported by a report that the thymus contains PAF and that A23187 can elicit PAF synthesis by isolated thymic cells (Salem *et al.*, 1989). A physiological stimulus for PAF production by these cells has yet to be identified.

3.5   RENAL ACTIONS OF PLATELET-ACTIVATING FACTOR

Prior to its structural characterization, PAF was implicated in poststreptococcal glomerulonephritis, since PAF levels in blood were reduced in association with a depletion of basophils. In addition, basophils from patients recovering from glomerulonephritis released PAF in response to streptococcal antigens (Camussi *et al.*, 1978a). The availability of synthetic PAF has revealed a variety of renal actions that are consistent with an important role for this mediator in renal pathology (Perico and Remuzzi, 1990).

3.5.1   Renal Vascular Actions

PAF induced a fall in renal blood flow in the rat, suggesting a platelet-independent effect (Sanchez Crespo *et al.*, 1981, 1982). Detailed studies in anaesthetized dogs indicated that intravenous infusion of PAF significantly reduced renal function (Vemulapalli *et al.*, 1984). The primary effect appeared to be on renal blood flow, which was decreased by arterial hypotension due to diminished cardiac output, and a generalized fall in peripheral vascular resistance coupled with renal vasoconstriction. Consequently, GFR, urine production and Na⁺ and K⁺ output were decreased (Vemulapalli *et al.*, 1984).

PAF-induced renal vasoconstriction was blocked by the AII antagonist saralasin, whereas the cyclooxygenase inhibitor indomethacin blocked the systemic, but not the local, vascular actions of PAF (Plante *et al.*, 1986; Hebert *et al.*, 1987). It was found that PAF decreased Na^+ excretion by a mechanism independent of changes in systemic blood pressure (Plante *et al.*, 1986; Hebert *et al.*, 1987). Infusion of PAF into the renal circulation elicited a similar profile of renovascular changes without affecting the systemic circulation of anaesthetized dogs (Scherf *et al.*, 1986; Santos *et al.*, 1988). Pretreatment with indomethacin increased the renal dysfunction induced by a local infusion of PAF (Scherf *et al.*, 1986). A more extensive investigation of the intra-arterial effects of PAF (40–160 pmol) in anaesthetized dogs excluded the involvement of adrenoceptor activation, AII and serotonin, but confirmed a net protective effect of cyclooxygenase metabolites (Baer and Cagen, 1987). A specific receptor appeared to be involved, since lyso-PAF (2 nmol) was inactive and the effects of PAF were selectively inhibited by the PAF receptor antagonists alprazolam (Baer and Cagen, 1987) or BN 52021 (Hebert *et al.*, 1987). In the anaesthetized dog, the TXA_2 receptor antagonist L-655,240 inhibited the renal effects of PAF, including the decreased output of urine and Na^+ (Hebert *et al.*, 1989). Studies in isolated buffer-perfused rat kidney also supported a role for TXA_2 in the reduced GFR following PAF infusion and indicated that, in the presence of a TXA_2 receptor antagonist, PAF increased renal plasma flow via the formation of a vasodilator prostaglandin (Badr *et al.*, 1989). The vasoconstrictor effect of PAF appears to be selective for the microcirculation, since isolated cat renal arteries failed to respond to high doses of PAF (up to 40 nmol) (Stahl and Lefer, 1987b). *In vivo* studies in the rat indicate that the PAF-induced decrease in GFR and proteinuria is secondary to the generation of TXA_2 (Badr *et al.*, 1989; Yoo *et al.*, 1990). Interestingly, low doses of PAF (0.1–3 nmol/kg intravenously) administered to conscious rats resulted in a receptor-dependent increase, rather than a decrease, in renal blood flow (Siren and Feuerstein, 1989), but the consequences for other renal actions of PAF were not elucidated. These findings were confirmed in anaesthetized rats in which PAF (1–20 pmol), administered as a bolus into the renal artery, increased renal blood flow before inducing any systemic effect (Handa *et al.*, 1990). The vasodilator response was independent of eicosanoid synthesis or neuronal mechanisms, but was blocked by a specific PAF antagonist.

3.5.2   Platelet-Activating Factor Formation and Release

The ionophore A23187 induces the release of PAF from isolated perfused rat kidneys (Pirotzky *et al.*, 1980). A23187 is also reported to induce PAF formation by freshly isolated glomerular and medullary cells but not by tubular cells (Pirotzky *et al.*, 1984b). Antigen challenge of kidneys isolated from sensitized rats elicits the release of PAF (Pirotzky *et al.*, 1984a, 1987). PAF production by glomeruli can also be stimulated by ischaemia (Lopez

Farre *et al.*, 1988) and by endotoxin (Morell *et al.*, 1988), and the latter stimulus elicits PAF formation by cultured rat mesangial cells (Wang *et al.*, 1988). Cultured rat mesangial cells produce PAF in response to AII, suggesting that part of the beneficial action of ACE inhibitors in models of renal disease may be secondary to inhibition of PAF formation (Neuwirth *et al.*, 1989). Rats fed diets containing fish oil had reduced renal capacity to generate PAF, suggesting a possible beneficial effect of fish oil feeding in renal disease (Yeo *et al.*, 1989a,c). Hyperacute renal allograft rejection is associated with the release of PAF into the venous blood, and PAF has been suggested as a mediator of the vasoconstriction, thrombosis and cellular injury that accompany this reaction (Ito *et al.*, 1984).

The presence of PAF in the urine of normotensive humans (Sanchez Crespo *et al.*, 1983) and mice (Benfenati *et al.*, 1989) has been reported, and the likely renal origin of this PAF may provide an index of tissue damage in renal disease (Benfenati *et al.*, 1989). In isolated perfused rat kidney preparations, infusion of PAF did not result in significant PAF levels in the urine (Noris *et al.*, 1990), further supporting a renal origin for urinary PAF.

There is no clear physiological role for PAF in the kidney, but its effects on GFR, permeability and blood flow make it a strong candidate as a mediator in a variety of renal disorders.

3.6   REPRODUCTION

The role of PAF in general physiological processes remains ill-defined, but there have been substantial advances in the area of reproductive physiology (for a review, see Harper, 1989).

3.6.1   Ovulation

The association between follicle rupture during ovulation and the inflammatory response led Braquet and colleagues to investigate the role of PAF in ovulation (Abisogun *et al.*, 1989). BN 52021 attenuated follicle rupture in rats stimulated to ovulate with human chorionic gonadotrophin, raising the possibility for a further site of contraceptive action of PAF antagonists (Abisogun *et al.*, 1989).

3.6.2   Embryo-Derived Platelet-Activating Factor

It has been established by O'Neill and colleagues that, in mice, a factor produced early in pregnancy elicits thrombocytopenia and is closely related to PAF (O'Neill, 1985, 1987; Collier *et al.*, 1988), but shows chromatographic differences to PAF on TLC, HPLC or GC–MS (Kodama *et al.*, 1989). The production of this so-called embryo-derived PAF by human cells appears to correlate with embryo quality and viability (O'Neill *et al.*, 1987). Significantly higher levels of PAF have been observed in culture media from embryos that achieved pregnancy on transfer to patients (≈ 300 nM) compared to those which failed to establish pregnancy (≈ 75 nM) (Collier *et al.*, 1990). Supplementation of embryo culture medium with PAF ($0.2–1.5$ µM) resulted in a significant increase in the pregnancy rate in women taking part in an *in vitro* fertilization programme (O'Neill *et al.*, 1989). Ongoing PAF production has been reported in postimplantation rat embryos subsequently explanted in culture (Geissler *et al.*, 1989), but the relationship of this PAF production to the maintenance of pregnancy is not known.

More recently, it has been reported that a range of chemically distinct PAF receptor antagonists reduce implantation of mouse embryos (Spinks and O'Neill, 1988; Spinks *et al.*, 1990). Nevertheless, some workers have been unable to detect PAF release from cultured human embryos (Amiel *et al.*, 1989) and one study reports that several PAF antagonists were unable to block embryo implantation in mice (Milligan and Finn, 1990). The failure to detect PAF production at a level greater than 1 pg per embryo (mouse) using a sensitive immunoassay (Smal *et al.*, 1990) indicates that the previously observed material shows biological PAF activity, but appears to differ immunologically and therefore may not be identical to authentic PAF.

The mechanism of the involvement of PAF in embryo implantation awaits complete elucidation. It has been proposed that PAF may initiate uterine decidualization and interfere with immunologically based rejection of the implanting embryo (Adamson *et al.*, 1987). In pseudo-pregnant rats, PAF induces a decidua-like reaction by activating specific PAF receptors that are linked to the generation of cyclooxygenase products (Acker *et al.*, 1989). This activity may be species-specific, since PAF was reported not to induce decidualization in mice (Milligan and Finn, 1990). Alternatively, embryo-derived PAF may be an autocrine growth factor, since its addition to culture media stimulates the embryo utilization of energy substrates (Ryan *et al.*, 1989).

3.6.3 Spermatozoa

There is now evidence for PAF production by human, mouse (Kuzan *et al.*, 1990), rabbit (Kumar *et al.*, 1988a) and bovine (Parks *et al.*, 1990) spermatozoa. Exogenous PAF increases sperm motility and induces the acrosome reaction (Ricker *et al.*, 1989), but the PAF synthesized by spermatozoa is retained intracellularly, implying an intracellular role which is more difficult to elucidate. One interesting proposal is that the high levels of the PAF precursor, alkylacylglycerylphosphocholine present in sperm may provide the embryo with the capacity to produce sufficient PAF to act in an autocrine or paracrine manner (Parks *et al.*, 1990). The PAF antagonist WEB 2086 inhibits *in vitro* fertilization of mouse oocytes, whereas addition of PAF ($10–1000$ nM) to the culture medium increased fertilization rates, suggesting that PAF derived from spermatozoa may play a role in fertilization

(Kuzan *et al.*, 1990). PAF has been reported to promote the fusion of membranes, albeit by a receptor-independent mechanism (Bratton *et al.*, 1988a,b), indicating the potential for a receptor-independent role in gamete fusion. PAF receptor antagonists have been reported to have spermicidal actions, but these actions were manifest only at high concentrations and did not appear to be related to antagonism of PAF receptors, since compounds that were not structurally related to PAF did not have a similar effect (Harper *et al.*, 1989). It will be of considerable interest to determine the stability of PAF in seminal fluid and to assess the effects of PAF antagonists, administered to males, on coital fertilization rates. Pre- or postovulation intravenous administration of SRI 63-441 in rabbits had no effect on fertilization rate, but instillation into the vagina 2 min before insemination reduced fertilization, consistent with a receptor-independent action of high local concentrations of this compound (Harper *et al.*, 1989).

3.6.4 Parturition

Evidence for a physiological role of PAF at both ends of pregnancy is accumulating. Johnston and colleagues (Hoffman *et al.*, 1986b, 1988a) have described the synthesis of PAF by the developing fetal lung, and have suggested that release of PAF may initiate parturition (Hoffman *et al.*, 1986a). Treatment of rabbits with a PAF receptor antagonist is reported to prolong delivery time (Johnston, 1991). There is a temporal correlation in the maturation of type II pneumocytes, the decrease in glycogen content and the increase in PAF content in the rabbit fetal lung (Hoffman *et al.*, 1986b), and injection of PAF induces glycogen breakdown in the rabbit fetal lung by activation of a specific PAF receptor (Hoffman *et al.*, 1988b). PAF receptors have been detected by radioligand-binding studies on uterine membrane preparations obtained from pregnant rabbits (Kudolo and Harper, 1989) and on human myometrium (Johnston, 1991). Binding sites were located predominantly on the myometrium and, surprisingly, lyso-PAF appeared to displace the binding of [^{3}H]PAF. These binding sites may represent an important target for fetal lung-derived PAF. Alternatively, since PAF is known to be produced by the stromal cells of the endometrium of the rat (Yasuda *et al.*, 1986; Nakayama *et al.*, 1987), rabbit (Angle *et al.*, 1988) and human (Alecozay *et al.*, 1989), it is possible that locally generated PAF could be the source of the physiological stimulus for myometrial receptors. PAF contracts the myometrium of a number of species, including the rat (Medeiros and Calixto, 1989), guinea-pig (Montrucchio *et al.*, 1986a) and human (Tetta *et al.*, 1986). The PAF-induced myometrial contraction is blocked by inhibitors of lipoxygenase and cyclooxygenase and also by the PAF antagonist CV 3988, indicating a role for the receptor-mediated release of eicosanoids (Tetta *et al.*, 1986). Cultured human myometrial smooth muscle cells show increased intracellular calcium and myosin light-chain phosphorylation in response to an exceedingly low concentration (10 fM) of PAF (Johnston, 1991). Thus, the late gestational increase in amniotic PAF levels elicits both fetal and maternal responses. The latter are reinforced by a precipitous fall in maternal plasma acetylhydrolase levels, allowing greater levels of PAF to reach the myometrium (Maki *et al.*, 1988).

Amniotic fluid obtained at term from women in labour contains a large amount of PAF, of which approximately half is associated with a lamellar body-enriched fraction, suggesting that type II pneumocytes may be the source (Billah and Johnston, 1983). Moreover, amniotic fluid PAF levels are elevated in women with preterm labour compared with those at term and either in, or not in labour (Hoffman *et al.*, 1990). Interestingly, maternal acetylhydrolase activities may fail to decrease in the serum of preeclamptic patients, and the maintained catabolism of PAF could contribute to the increased vascular reactivity in these patients (Benedetto *et al.*, 1989).

Thus, there is extensive evidence in support of a physiological role for PAF in all stages of the reproductive processes. The implications for the therapeutic use of PAF receptor antagonists are two-fold: first, these compounds offer potential as novel contraceptive agents and, secondly, it follows necessarily that PAF antagonists may produce the highly undesirable side-effect of infertility.

3.7   SKIN

3.7.1   Responses in Animal Skin

The pro-inflammatory actions of PAF are exemplified by the sequelae of effects seen on intradermal administration. In rabbit skin, intradermal injection of PAF induced an increase in permeability as measured by the accumulation of ^{125}I-labelled albumin (Wedmore and Williams, 1981b). In a series of studies, Morley's group (Morley *et al.*, 1983; Page *et al.*, 1983; Archer *et al.*, 1984a) established that PAF induced a plasma protein extravasation in guinea-pig skin that was accompanied by accumulation of platelets and red blood cells. The magnitude and duration of the response were both related to the dose, and responses were enhanced by simultaneous injection of PGE_2 (Archer *et al.*, 1984a), suggesting that local blood flow may limit the response under normal conditions, according to the two-mediator hypothesis of inflammatory oedema (Wedmore and Williams, 1981a).

PAF induced inflammatory responses in the skin of rhesus monkeys that appeared to be mediated by specific receptors (Patterson *et al.*, 1986). PAF was also shown to be a potent (0.04 pmol to 4.5 nmol) inducer of plasma exudation in rat skin, and local platelet accumulation was associated with the higher doses (Pirotzky *et al.*, 1984c). Inhibitors of platelet degranulation products and depletion of neutrophils did not interfere with the oedema response. In a subsequent study, the oedema-producing activity of PAF in rat skin was confirmed, as was its independence of neutrophil accumulation (Gerdin *et al.*, 1985).

3.7.2   Human Skin Responses

PAF induced an early weal and flare reaction followed, in 60% of human volunteers, by a late (3–6 h)-onset erythema (Archer *et al.*, 1984b, 1985e); PGE_2 was also able to enhance PAF-induced oedema in human skin by a mechanism that was synergistic rather than additive (McGivern and Basran, 1984). In a double-blind study, PAF was shown to be more potent than PGE_2 or histamine in inducing weal responses, but did not induce hyperalgesia (Sciberras *et al.*, 1987).

In normal subjects, skin responses were accompanied by perivascular infiltration with neutrophils (2–8 h) and a mixed infiltrate, consisting primarily of mononuclear cells, at 24 h (Archer *et al.*, 1985b). Using a technique involving dermal abrasion and the application of inflammatory mediators in solution to a skin chamber, it was shown that PAF induced neutrophil migration only at a concentration of about 1 μM. Nevertheless, PGE_2 enhanced the cell migration response and protein leakage to both PAF and LTB_4 (Michel *et al.*, 1987). Intradermal injection of PAF (800 pmol) in human skin caused vessel destruction with gross endothelial swelling and a mixed cellular infiltrate. Direct effects of PAF on resident cells have been demonstrated: PAF-induced activation of keratinocytes, resulting in arachidonic acid mobilization and prostaglandin formation (Fisher *et al.*, 1989). Such observations have led to the suggestion that PAF may be a potential mediator of psoriasis and other inflammatory skin disorders (Archer *et al.*, 1985b,d; Michel and Dubertret, 1985; Pirotzky *et al.*, 1985b).

The similarity in the dual response to PAF with that seen in response to allergen given intradermally prompted consideration of a mediator role for PAF in atopy (Basran *et al.*, 1984). In atopic individuals, PAF and FMLP, like allergen, caused eosinophil infiltration in atopic, but not normal, subjects (Henocq and Vargaftig, 1986, 1988). No difference in the magnitude of the immediate weal and flare responses was observed in the two groups.

In the context of high levels of PAF precursors in neoplastic cells and the cytotoxic effects of certain alkyl ether lipids, it is interesting that decreased responses to intradermal PAF have been observed in cancer patients (Burtin *et al.*, 1990). The biological significance of this finding awaits further studies.

3.7.3   Pharmacological Modulation

Despite convincing evidence that PGE_2 and PGI_2 enhanced PAF-induced skin oedema and the ability of PAF to stimulate eicosanoid production by a range of cell types, the cyclooxygenase inhibitor indomethacin had no effect on PAF responses in guinea-pig or human skin (Archer *et al.*, 1985a). Similarly, PAF responses in rat skin were unaffected by indomethacin, but those in guinea-pig skin were attenuated (Hwang *et al.*, 1985), in contrast to the findings of Archer *et al.* (1985a). Other vasodilators, such as the neuropeptide calcitonin gene-related peptide, enhanced the microvascular permeability response to a wide range of mediators, including PAF (Brain and Williams, 1985).

Antihistamines were found to have a modest inhibitory effect, whereas serum albumin potentiated responses (Archer *et al.*, 1985a). A number of other drugs have been tested for modulatory activity against PAF-induced skin responses. The anti-allergic, anti-asthma drug DSCG attenuated PAF responses (Archer *et al.*, 1985c). It was suggested that dermal mast cells were the prime target for intradermal PAF, since histamine H_1 receptor antagonism was found to inhibit flare and itching responses of human skin (Fjellner and Hagermark, 1985). These findings suggest distinct mechanisms for the flare and itching, as opposed to the oedema responses to PAF that were only weakly inhibited by histamine antagonism (Archer *et al.*, 1985a). Another anti-allergic compound, ketotifen, was reported to inhibit the flare and weal responses to PAF as well as those to histamine (Chung *et al.*, 1988). However, it is difficult to draw a conclusion from this study on the histamine-dependence of PAF actions in human skin, since ketotifen has a number of other properties apart from being a non-competitive histamine H_1 receptor antagonist.

The use of pharmacological inhibitors revealed species differences between the rat and guinea-pig: oedema responses in the guinea-pig were attenuated by cyclo-oxygenase inhibitors by a mechanism that could be reversed by PGE_1 locally, whereas in rats these compounds had no effect, but PGE_1 was an effective enhancing agent (Hwang *et al.*, 1985). Oedema responses of both species were resistant to histamine H_1 receptor blockade and, interestingly, the PAF antagonist kadsurenone also inhibited histamine and bradykinin induced permeability responses in rat skin, suggesting an involvement of endogenous PAF (Hwang *et al.*, 1985). This suggestion is consistent with the ability of histamine and bradykinin to stimulate PAF synthesis by vascular endothelial cells (McIntyre *et al.*, 1985), and the inhibitory effect of various PAF antagonists on bradykinin-induced PGI_2 generation (Stewart *et al.*, 1989). However, a phospholipid-type PAF antagonist, CV 3988, was reported to be ineffective against bradykinin and histamine in the rat (Hayashi *et al.*, 1987a), as were other antagonists in other species (Hellewell and Williams, 1986; Issekutz and Szpejda, 1986).

Passive cutaneous anaphylactic reactions and those induced by PAF in rats were inhibited by the β adrenoceptor stimulant, isoprenaline (Inagaki *et al.*, 1989). Since the vasodilation would be expected to enhance the oedema response, another target for the inhibition by isoprenaline, probably an inflammatory cell, requires identification.

3.7.4 Inflammatory Oedema

Hellewell and Williams (1986) investigated the role of PAF in a model of inflammatory oedema in rabbit skin using the reversed passive Arthus reaction, in which immune complexes initiate the reaction. Oedema formation was inhibited by L-652,731 which had no effect on responses to histamine, bradykinin or preformed immune complexes. The phospholipid-type PAF antagonist, CV 3988 inhibited inflammatory oedema induced by zymosan, zymosan-activated plasma, endotoxin or the reversed passive Arthus reaction, but not that induced by bradykinin or histamine (Issekutz and Szpejda, 1986), confirming earlier observations (Hellewell and Williams, 1986).

Contact and, to a lesser extent, irritant dermatitis was attenuated by oral treatment of mice with BN 52021, suggesting a role for PAF in these conditions (Csato and Czarnetzki, 1988; Kemeny *et al.*, 1989). Dithranol-induced dermatitis was inhibited by topical application of an ointment containing 0.45% BN 52021 to the skin of human volunteers, as measured by increase in skin temperature and swelling (Kemeny *et al.*, 1990).

3.7.5 Platelet-Activating Factor Formation

PAF has been isolated from human psoriatic skin scales using HPLC (Mallet *et al.*, 1984) and its identity established by platelet aggregation bioassay and GC–MS (Mallet and Cunningham, 1985). A PAF-like substance has been described in the blister fluid from patients with the autoimmune disease, bullous pemphigoid (Okubo *et al.*, 1989). Cold challenges, carried out by immersion of the arm in a water-bath (4–8°C) in patients with acquired cold urticaria, resulted in the release of PAF into the venous forearm blood in four of six patients and none of five healthy control subjects (Grandel *et al.*, 1985).

The release of PAF and lyso-PAF during allergen-induced skin reactions strengthens the suggestion that PAF contributes to inflammatory skin diseases (Michel *et al.*, 1988b). Antigen challenge resulted in the appearance of LTB_4 and PAF in skin chamber fluid between 3 and 9 h postchallenge, indicating a persistent production of these inflammatory mediators (Shalit *et al.*, 1989). Allergen induced the accumulation of PAF in skin chamber fluid 6 h after allergen challenge and this response was blocked by the histamine H_1 receptor antagonist cetirizine, which also blocked eosinophil recruitment (Michel *et al.*, 1988a; Fadel *et al.*, 1990).

Further evidence for local PAF production was provided by an investigation of PAF biosynthesis by human cultured skin fibroblasts which were shown to produce PAF in response to A23187 (Michel *et al.*, 1988c). Human cultured epidermal cells have also been found to produce PAF in response to A23187, with approximately half being released into the culture medium (Michel *et al.*, 1990). However, it will be important to characterize other, more physiological stimuli for PAF production and release by fibroblasts and epidermal cells. In addition, the means of obtaining skin samples or superfusing skin chambers may induce sufficient damage to elicit PAF synthesis (Cunningham, 1989).

PAF production has also been observed in the skin of guinea-pigs irradiated by UV light (Calignano *et al.*, 1988), and in the skin of rats subjected to moxibustion (Nakayama *et al.*, 1985). The relationship of PAF formation to the therapeutic effect of moxibustion was not established in the latter study.

3.7.6 Platelet-Activating Factor Receptor Antagonists

Although several studies report that the acute skin response to PAF is blocked by BN 52063, this compound does not appear to inhibit the later cellular recruitment to the site of injection (Markey *et al.*, 1990). The complex pattern of inhibition of PAF-induced inflammatory oedema by PAF antagonists observed in rabbits led Hellewell and Williams (1989) to suggest caution in interpreting inhibition of skin responses by PAF antagonists as evidence for an involvement of PAF.

In humans, the ginkgolide mixture BN 52063 given orally produced a dose-related inhibition of the weal and flare reactions to intradermal PAF, suggesting the involvement of specific PAF receptors in these responses (Chung *et al.*, 1987; Guinot *et al.*, 1987b), a finding of particular importance when high doses are injected in relatively small volumes and may therefore contain very high concentrations of PAF (e.g. 400 pmol in 50 µl is equivalent to 8 µM).

3.7.7 Sensory Nerve Fibres

Neurogenic oedema is considered to be an important feature of bronchial oedema in asthma and may be involved in inflammatory skin lesions. Evidence for the independence of PAF-induced skin oedema from neurogenic mechanisms was provided by the observation that ruthenium red inhibited the oedema response to capsaicin, but not that to PAF, histamine, CGRP, bradykinin or FMLP in rabbits (Buckley *et al.*, 1990). However, topical capsaicin pretreatment, using a protocol to deplete sensory neuropeptides, in either healthy or atopic volunteers reduced the immediate flare response to histamine or PAF, but not that to antigen (McCusker *et al.*, 1989). It is not yet clear whether these observations indicate a differential involvement of neurogenic processes in flare and oedema responses, or whether the differences are dependent on species or the agents used to investigate neuropeptide involvement.

The acute and persistent effects of PAF in the skin are now well characterized and there is evidence of PAF release in some inflammatory skin disorders. The potential therapeutic role for PAF receptor antagonists in the treatment of inflammatory skin disorders awaits clinical trials.

4. *Pathophysiology*

4.1 ANTICANCER ETHER PHOSPHOLIPIDS

The recognition that certain ether phospholipids exerted cytotoxic actions on tumour growth predated the elucidation of the structure of PAF as an alkyl ether lipid by a number of years. Further interest in this field was stimulated by the demonstration that neoplastic cells contained large proportions of alkyl ether lipids (Andreesen, 1988) and have the capacity to generate PAF (Camussi *et al.*, 1982; Bussolino *et al.*, 1984; Blank *et al.*, 1988). However, lymphoblastic leukaemia cells were found to produce PAF in response to A23187 in five of six patients whereas myeloid cells were found to be producers less frequently (two of ten patients), suggesting that the production of PAF may be related to the stage of differentiation rather than to neoplasia (Foa *et al.*, 1985). This possibility is strengthened by the requirement for differentiation of HL-60 cells to the macrophage phenotype for PAF production in response to A23187 (Camussi *et al.*, 1982). The significance of the high concentration of alkyl ether lipids in neoplastic cells (Record *et al.*, 1984) has not been fully elucidated, but may have some importance for the selectivity of the cytotoxic effects of exogenous alkyl ether lipids in these cells.

The primary question has been whether the anticancer activity of the alkyl ether phospholipids occurs directly via a cytotoxic effect or indirectly by inducing antitumour activity in macrophages. There is evidence for both of these actions (Berdel, 1987; Andreesen, 1988).

4.1.1 Cytotoxicity

Ether lipids, including PAF, exert a selective cytotoxic effect on tumour cells *in vitro* (Hoffman *et al.*, 1984; Berdel *et al.*, 1987; Noseda *et al.*, 1987; Schick *et al.*, 1987; Ukawa *et al.*, 1989; Gati *et al.*, 1991). These direct cytotoxic effects may be the result of activation of specific PAF receptors, since the cytotoxicity of the octadecylmethoxy-PAF analogue on the WEHI-3B cell line was prevented by L-652,731 and the phospholipid-type antagonist, SRI 62-834 (Bazill and Dexter, 1989b). However, a more recent study using the potent, non-phospholipid PAF antagonist WEB 2086 failed to show any inhibition of the cytotoxicity induced by PAF, octadecylmethoxy-PAF or SRI 62-834 on HL-60 cells (Workman *et al.*, 1991). Further confusion is provided by the observation that L-652,731 inhibits thymidine transport, as opposed to DNA synthesis, in WEHI-3B and FDCP mix A4 cells, confounding interpretation of studies using this technique to estimate cell proliferation (Bazill and Dexter, 1989a). The potencies of octadecyl-methoxy-PAF, SRI 63-154 and BM 41.440 did not correlate with their ability to displace [^{3}H]PAF from platelet PAF receptors, providing additional indirect evidence for a PAF receptor-independent cytotoxic effect of PAF-related ether lipids (Berdel *et al.*, 1987). Some stereospecificity in the cytotoxic activity of octadecylmethoxy-PAF has been observed, but no clear structure–activity relationship has emerged from studies of cytotoxic alkyl ether lipids (Danhauser *et al.*, 1987).

The antitumour ether lipids appear to act by physicochemical interaction with plasma membranes (Noseda *et al.*, 1989) to increase intracellular calcium levels (Lazenby *et al.*, 1990), which has been recognized as an important cytotoxic mechanism (Orrenius *et al.*, 1989). In addition, it is possible that these compounds act by inhibition of PKC (Kramer *et al.*, 1989; Zheng *et al.*, 1990). These physicochemical mechanisms are more likely to explain the direct cytotoxic effects of PAF than actions at a specific receptor.

4.1.2   Macrophage-Dependent Actions

In contrast to the direct cytotoxic effects of PAF, activation of macrophage functions such as stimulation of O_2^- generation and induction of microbicidal activity appear to be the result of activation of specific receptors (Lambrecht and Parnham, 1986; Hayashi *et al.*, 1988; Parnham, 1988; Stewart and Dusting, 1988; Stewart, 1989).

Longer-term effects of PAF include its capacity to synergize with other immunoregulators to induce Il-1α (Barret *et al.*, 1987; Pignol *et al.*, 1987, 1990) or TNF-α (Bonavida *et al.*, 1989b; Hayashi *et al.*, 1989) release by a receptor-dependent mechanism. Nevertheless, it remains possible that certain aspects of macrophage activation induced by PAF and related ether lipids, particularly those that occur above the critical micellar concentration (1 μM; Kramp *et al.*, 1984) also occur via a receptor-independent interaction with the membrane (Storch *et al.*, 1988).

4.2   ASTHMA

Elucidation of the bronchopulmonary actions of PAF raised interest about the potential role of this mediator in the pathogenesis of asthma. The extent of the importance of PAF is currently being tested in clinical trials of PAF receptor antagonists. It is pertinent that only one receptor antagonist, ipratropium bromide (a muscarinic receptor blocker), is effective in this disease and that all other effective pharmacological treatments have activity against a multitude of mediators through either functional antagonism or inhibition of mediator release. Thus, failure of PAF receptor antagonists to produce a beneficial effect will not exclude an important role for this mediator, but may indicate the value of assessing combined treatment with antagonists of other mediators that are likely to be quantitatively important such as LTC_4 or TXA_2.

4.2.1   Platelet-Activating Factor Release in Asthma

In asthmatics undergoing allergen challenge and in those with acute asthma requiring emergency treatment in hospital, there was a significant reduction in platelet responsiveness to PAF tested *ex vivo*, indicative of PAF release *in vivo* (Thompson *et al.*, 1984). Allergen challenge of asthmatic subjects had no effect on circulating platelet numbers, plasma concentrations of the markers of platelet activation, platelet factor 4 or β-thromboglobulin, or *ex vivo* aggregation responses to PAF, arachidonic acid or adrenaline, leading to the conclusion that PAF does not have an important direct role in asthma (Hemmendinger *et al.*, 1989). However, given that aerosols of PAF fail to desensitize PAF-mediated platelet aggregation (Kioumis *et al.*, 1988), the study by Hemmendinger *et al.* (1989) merely suggests that significant blood levels of PAF are not reached following allergen inhalation. The variation between these findings may relate to the severity of the asthma.

Bioassay of BAL supernatants for PAF showed that eight of 28 asthmatic subjects and no controls had detectable levels and, in the positive subgroup, PAF levels correlated with low neutrophil and high lymphocyte numbers and with macrophage activation (Stenton *et al.*, 1990a). The amount of PAF acetylhydrolase in the serum of asthmatic children has been positively correlated with the severity of asthma (Miwa *et al.*, 1988) and the differences do not appear to be affected by theophylline or steroids (Miyake and Miwa, 1989). The biological significance of these potentially important findings and their extension to other asthmatic populations requires further studies, particularly in view of the role of acetylhydrolase in regulating the levels of oxidized phosphatidylcholines (Smiley *et al.*, 1991).

Incubation of human lung membranes, with PAF, LTB_4 or LTC_4 reduced β adrenoceptor binding, and these mediators also decreased lymphocyte cAMP levels, consistent with changes noted in patients with chronic obstructive airways disease (Raaijmakers *et al.*, 1989).

4.2.2   Platelet-Activating Factor Antagonists in Asthma

Following successful bioavailability studies in human volunteers (Adamus *et al.*, 1988), it was established that WEB 2086 (APAFANT) at a dose of 40 mg prevented the bronchoconstriction and attenuated the cardiovascular responses to inhaled PAF (Adamus *et al.*, 1990).

The ginkgolide mixture BN 52063 inhibited the late but not the early bronchoconstriction induced by exercise challenge in asthmatic subjects and also decreased the elevation of platelet factor 4 and β-thromboglobulin (Wilkens *et al.*, 1990). In addition, inhaled BN 52021 attenuated allergen-induced bronchoconstriction in asthmatic children (Hsieh, 1991). Interestingly, in the latter study it was reported that BN 52021 prevented the

Table 8.2 Comparison of the properties of putative mediators of asthma

Effect	Mediator						
	ACh	HA	$PGF_{2\alpha}$	TXA_2	LTB_4	$LTC_4/LTD_4/LTE_4$	PAF
Airways smooth muscle constriction							
In vitro	+	+	+	+	−	+	+/−
In vivo	+	+	+	+	+	+	+
Bronchial oedema		+				+	+
BHR			+	+		+(E_4)	+
Cytokine release					+		+
Epithelial toxicity							+
Mucus hypersecretion	+	+	+			+	+
Eosinophilia							+[a]
Neutrophil activation					+		+
Platelet activation				+			+

Modified from Page and Morley (1986).

[a] Not established in human airways.

PAF-induced decrease in circulating neutrophils and eosinophils in healthy, but not in asthmatic, children. This latter finding may be analogous to the resistance of PAF responses in immunologically sensitized guinea-pig lungs to inhibition by PAF antagonists (Pretolani *et al.*, 1989).

4.2.3 Bronchial Hyper-Reactivity

Asthma is characterized by a non-specific increase in bronchial reactivity. Currently, it is believed that BHR is fundamental to asthma (Chung, 1990; Meade and Heuer, 1991) and there is evidence that disease severity correlates with the degree of BHR (Sotomayor *et al.*, 1984). There have been a multitude of hypotheses put forward to explain the phenomenon of BHR, including the following: diminished responsiveness to β adrenoceptor stimulation; overactivity of bronchoconstrictor cholinergic nerves; enhanced permeability and other functional deficits of the epithelial layer; intrinsic defects in bronchial smooth muscle.

Earlier work on BHR focused on interactions between bronchoactive substances released during asthmatic bronchospasm, including prostaglandins, leukotrienes and histamine (Fennessy *et al.*, 1987). However, several studies provided compelling evidence for a link between ongoing inflammation and the degree of BHR (Sotomayor *et al.*, 1984; Dutoit *et al.*, 1987). The protracted increase in BHR observed following antigen challenge in atopic asthmatics further supports the view that persistent and, possibly, low-grade bronchial inflammation is a key feature of BHR (Cockcroft *et al.*, 1977).

Clearly, antigen-induced increases in BHR are the result of the release of mediator(s) from immuno-competent cells in the airway. The emerging concept of the importance of inflammation in asthma (Kay, 1983; Kay *et al.*, 1989) provided Page and Morley (Page and Morley, 1986; Morley, 1987; Page and Robertson, 1987)

with a set of criteria to assess the likely chemical mediators of asthma and BHR (Table 8.2). On this basis it has been proposed that PAF has the relevant biological actions in the airways to be an important mediator of asthma (Page *et al.*, 1983, 1985a; Roberts *et al.*, 1988; Barnes and Chung, 1989; Page and Coyle, 1989) and of BHR (Morley *et al.*, 1985; Page, 1987; Page and Robertson, 1988; Barnes, 1989; Chung, 1990).

Perhaps the most important feature of the potential for PAF to contribute to BHR is the persistence of its pro-inflammatory effects. Investigations in humans suggested that BHR was increased up to 4 weeks after a single aerosol exposure (Cuss *et al.*, 1986), and increases have been observed up to 24 h in dogs (Chung *et al.*, 1986). The mechanism of the PAF-induced increase in BHR has been subject to extensive investigation.

4.2.4 Mechanism of Platelet-Activating Factor-Induced Bronchial Hyper-Reactivity

PAF infusions in anaesthetized guinea-pigs, in a protocol known to elicit a non-specific increase in responsiveness, had no effect on pulmonary β adrenoceptor density or on isoprenaline-induced relaxation of tracheas isolated from PAF-treated animals (Barnes *et al.*, 1987). Despite the lack of detectable changes in β adrenoceptor function in isolated airways, PAF treatment resulted in an impairment of isoprenaline-induced reversal of bombesin-induced bronchospasm *in vivo*, leading the authors to suggest an involvement of β adrenoceptors on non-spasmogenic elements in the airways that may contribute to the bronchoconstrictor response (Barnes *et al.*, 1987). Agrawal and Townley (1987) reported that a relatively high concentration of PAF (0.1 μM) decreased the number of β adrenoceptor-binding sites in human lung slices and also decreased relaxant responses of airway smooth muscle to the non-selective β agonist isoprenaline. However, PAF aerosol treatment in healthy human volunteers had no effect on responsiveness to isoprenaline (Hopp *et al.*, 1990).

In the rabbit, aerosols of PAF resulted in BHR at 24 h that was associated with inflammatory cell recruitment. Isolated intrapulmonary airways from animals with PAF-induced BHR did not exhibit any hyper-responsiveness to histamine or acetylcholine, nor was there any reduction in the relaxant effects of isoprenaline (Coyle *et al.*, 1990). The lack of effect of PAF on isolated airways response to isoprenaline is in agreement with observations from similar studies in guinea-pigs (Robertson *et al.*, 1988). In rabbits pretreated with a regimen of capsaicin that induced desensitization to rechallenge, but did not deplete substance P or CGRP, PAF aerosols failed to induce BHR. Thus, the inhibitory effects of capsaicin do not appear to be related to depletion of sensory neuropeptides and there was no inhibition of pulmonary eosinophilia (Spina *et al.*, 1991). Characterization of the BHR response to a bolus dose of PAF in the anaesthetized guinea-pig indicated that cholinergic nerves were not involved, since both bilateral vagotomy and atropine had no effect (Anderson *et al.*, 1988). In addition, the BHR was blocked by SRI 63-441 and compounds that block leukotriene generation, but not by cyclooxygenase inhibitors (Anderson and Fennessy, 1988).

Intravenous PAF induces BHR in conscious sheep, as measured by histamine-induced decreases in dynamic compliance 4 h after the dose of PAF, at which time its bronchoconstrictor effects had subsided (Christman *et al.*, 1987).

Administration of PAF aerosol to conscious guinea-pigs caused selective recruitment of eosinophils to the lung, as assessed by BAL (Coyle *et al.*, 1988). In addition, inhibition by the PAF receptor antagonist BN 52021 of increases in BHR induced by either antigen or by PAF and concurrent reductions in eosinophil accumulation provides circumstantial evidence that antigen-induced BHR resulted from PAF release and consequent eosinophil infiltration of the airways (Coyle *et al.*, 1988). However, in a similar study the inhibition of the antigen-induced airway hyper-responsiveness was confirmed with WEB 2086 or SDZ 64-412 pretreatment, but these PAF antagonists had no effect on eosinophil accumulation (Havill *et al.*, 1990). In a distinct model involving chronic aerosol challenge of conscious guinea-pigs with ovalbumin, the PAF receptor antagonist SDZ 64-412 reduced the BHR, but had no effect on either the hypoxia or the pulmonary accumulation of eosinophils (Ishida *et al.*, 1990). The latter findings question the importance of the eosinophil in BHR, but it remains possible that the PAF antagonists modify the function of eosinophils, in addition to their accumulation in the lung. Indeed, the observation that cytokines and lymphokines, including GM-CSF, IL-3 and TNF-α, induce pulmonary eosinophilia without increasing BHR in guinea-pigs, indicates

that the presence of this cell type in the airways is not sufficient to evoke BHR (Kings *et al.*, 1990).

In a longer term study, PAF infusion over 15 days was achieved by the use of Alzet osmotic minipumps (Mencia Huerta *et al.*, 1989). PAF treatment induced BHR to histamine and was accompanied by eosinophil infiltration, an increase in peribronchial mast cells, evidence of airways smooth muscle contraction and hyperplasia of the bronchial wall. Thus, PAF infusions induce histological changes that resemble those observed in asthmatic patients (Mencia Huerta *et al.*, 1989).

In guinea-pigs that have been sensitized to ovalbumin by a regimen that is selective for raising homocytotropic antibodies, isolated lungs show a marked increase in responsiveness to PAF, comprising both an increase in smooth muscle constriction and in release of histamine and eicosanoids (Pretolani *et al.*, 1988). The immunization-induced hyper-responsiveness in guinea-pigs was accompanied by an influx of eosinophils in the airways which could be reduced by pretreatment with nedocromil sodium for 7 days, but not by a single administration (Pretolani *et al.*, 1990). This chronic treatment also reduced PAF and antigen-induced mediator release. *In vivo* studies revealed that guinea-pigs sensitized by this regimen and challenged with intratracheal ovalbumin had an anaphylactic response that was insensitive to either BN 52021 (Desquand *et al.*, 1991) or WEB 2086 (Desquand *et al.*, 1990).

A study of an *in situ* preparation of canine airways showed that infusion of PAF into the tracheal circulation increased cholinergic responses of the trachea, but not of more distal airways, indicating a local effect of PAF on the tracheal muscle or the parasympathetic nerves supplying it (Bethel *et al.*, 1989). Further, evidence suggested the involvement of TXA$_2$ in the development of BHR both in dogs (Chung *et al.*, 1986) and in guinea-pigs (Takehana *et al.*, 1990), and it has been shown that TXA$_2$ induces increased responsiveness to cholinergic bronchoconstriction (Serio and Daniel, 1988).

4.2.5   Platelet-Activating Factor Receptor Antagonists in Animal Models of Bronchial Hyper-Reactivity

In the anaesthetized guinea-pig, PAF receptor antagonists in doses adequate to block the BHR induced by PAF (1.1 nmol kg^{-1} h^{-1}), had no effect on BHR induced by infusion of either indomethacin or propranolol (Dixon *et al.*, 1989). However, BHR induced by ovalbumin aerosols in sensitized animals was blocked by BN 52021 (Coyle *et al.*, 1988).

A recent study indicated that pretreatment (30 min), but not post-treatment (4 h) of sheep with WEB 2086 inhibited both the development of BHR following antigen challenge, as measured by enhancement of carbachol-induced increases in airways resistance, and the increase in BAL neutrophils (Soler *et al.*, 1989). This group also showed that another PAF receptor antagonist,

YM461, blocked both the antigen-induced late response (Abraham *et al.*, 1989) and BHR (Tomioka *et al.*, 1989), and that BHR induced by intratracheal instillation of PAF in allergic sheep was prevented by the anti-asthma drug nedocromil sodium (Soler *et al.*, 1990). Thus, the lack of a late response to PAF does not preclude a role for this mediator in the allergen-induced late response.

4.2.6   Bronchial Hyper-Reactivity in Humans

Following the report by Cuss *et al.* (1986) showing that PAF induced BHR in healthy subjects, great interest was stimulated in the role of PAF in BHR in asthma. This group was able to reproduce the BHR in subsequent studies (Chung *et al.*, 1988, 1989a; Wardlaw *et al.*, 1990) in which they showed that the airways hyper-responsiveness was associated with a neutropenia and was not inhibited by ketotifen. However, a number of similar studies have failed to observe an increase in BHR following PAF inhalation (Hopp *et al.*, 1989; Lai *et al.*, 1990; Spencer *et al.*, 1990; Stenton *et al.*, 1990b). Thus, Lai and Holgate (1990) have concluded that the relatively small increases in BHR in healthy subjects are not sufficiently reproducible to provide a reliable model of BHR in asthma. PAF does not cause a persistent increase in BHR in asthmatic subjects (Rubin *et al.*, 1987), but was reported to transiently (2 h) increase methacholine responsiveness (Hopp *et al.*, 1990).

A number of explanations for the inconsistent effects of PAF-induced BHR have been discussed (Lai and Holgate, 1990), but it would be premature to discount the potential contribution of PAF to BHR on the basis of its lack of effect on BHR in asthmatics and its inconsistent actions in healthy subjects. It may not be possible to mimic the actions of endogenously released PAF with aerosol administration of PAF, particularly in studies in humans which have entailed a single exposure to PAF. To resolve adequately the possibility that PAF is a key mediator of BHR in asthma, a pharmacological approach is required using long-term treatment with effective doses of PAF receptor antagonists and specific PAF biosynthesis inhibitors, should they become available. Interestingly, in one study, the PAF receptor antagonist mixture BN 52063 inhibited allergen-induced bronchospasm, but had no effect on responses to acetylcholine (Guinot *et al.*, 1987a). It will be especially important to examine the effects of long-term administration of PAF receptor antagonists on BHR, since anti-asthma drugs which do modulate BHR (e.g. beclomethasone dipropionate) require repeated doses to achieve a maximal effect. This slow onset of action is suggestive of a reversal of chronic inflammation (Sotomayor *et al.*, 1984; Dutoit *et al.*, 1987) or, possibly, structural changes in the airways.

4.3   ENDOTOXAEMIA

4.3.1   Platelet-Activating Factor Receptor Antagonists

The cardiovascular and haematological effects of intravenous PAF are similar to those observed in Gram-negative sepsis and in experimental models involving endotoxin administration. Doebber *et al.* (1985) provided convincing evidence that PAF was implicated in endotoxin shock by demonstrating an increase in PAF levels in the blood and a partial reversal of the endotoxin-induced hypotension by kadsurenone. In addition, Gram-negative sepsis induced by intraperitoneal injection of *Escherichia coli* in rats was associated with significant PAF synthesis in peritoneal exudate cells and in spleen (Inarrea *et al.*, 1985). The case for the involvement of PAF in endotoxin shock was greatly strengthened by the demonstration that CV 3988 was an effective treatment of the fall in blood pressure given either pre- or post-endotoxin infusion (Terashita *et al.*, 1985). In addition, CV 3988 significantly increased survival rates, up to 30 h post-endotoxin infusion when given simultaneously with the endotoxin.

The reduction in endotoxin-induced lethality has now been confirmed with a large number of PAF antagonists, including BN 52021 (Etienne *et al.*, 1986a), SRI 63-072 (Handley *et al.*, 1986a), CV 3988 (Toth and Mikulaschek, 1986), ONO-6240 (Toyofuku *et al.*, 1986), L-652,731 (Wu *et al.*, 1986), WEB 2086 (Casals Stenzel, 1987a), SRI 63-441 (Handley *et al.*, 1987), PCA 4248 (Fernandez Gallardo *et al.*, 1990) and BN 50739 (Yue *et al.*, 1990).

4.3.2   Relationship to Other Mediators

A great number of mediators has been implicated in the pathogenesis of endotoxin shock, including both lipoxygenase (Feuerstein and Hallenbeck, 1987; Feuerstein and Siren, 1988) and cyclooxygenase products (Feuerstein and Hallenbeck, 1987; Feuerstein and Siren, 1988) and, more recently, TNF-α (Ferguson Chanowitz *et al.*, 1990; Rabinovici *et al.*, 1990; Sun *et al.*, 1990) and nitric oxide (Salvemini *et al.*, 1990; Thiemermann and Vane, 1990). Thus, several studies have attempted to identify the relative roles of PAF and other mediators of shock in this phenomenon.

The finding that lipoxygenase inhibitors had little effect on endotoxin-induced lethality, whereas indomethacin could synergize with BN 52021, suggested a predominant role for PAF and cyclooxygenase products, but not lipoxygenase products (Etienne *et al.*, 1986b). The inhibitory effect of indomethacin implies a net deleterious effect of cyclooxygenase products, but PGE_2 abrogates a number of features of endotoxin shock in rats, including the haemoconcentration and increase in vascular permeability (Ibbotson and Wallace, 1989). Since PAF is clearly implicated in these phenomena, the inhibitory effects of PGE_2 pretreatment on endotoxin-induced PAF formation by stomach, duodenum or lung tissues were investigated. Although PGE_2 had no effect on these PAF levels (Ibbotson and Wallace, 1989), it remains possible that leucocytes are the major source of PAF in endotoxaemia and may be sensitive to inhibition by PGE_2. Endotoxin-induced PAF formation in rats was implicated in the elevation of plasma TXB_2 and PGI_2 levels, since BN 52021 attenuated the increase in eicosanoids, the fall in systemic blood pressure and increased the survival rate (Fletcher et al., 1990). In addition, in endotoxin-challenged pigs, SRI 63-675 attenuated the pulmonary vasoconstriction, broncho-constriction, hypokalaemia, permeability changes and the increase in plasma levels of 5-HETE, 12-HETE, TXB_2 and 6-oxo-$PGF_{1\alpha}$. Direct inhibition of the biosynthetic enzymes for these eicosanoids was excluded as an explanation because SRI 63-675 had no effect on the production of these monoHETEs or cyclooxygenase products by A23187-stimulated ex vivo whole blood (Olson et al., 1990a,b).

It has been suggested that PAF is involved in a neutrophil-independent oxidative stress induced by endotoxin in rats, as shown by the ability of PAF antagonists to inhibit the increase in plasma and tissue GSSG levels (Chang et al., 1988).

4.3.3 Interactions with Tumour Necrosis Factor-α

Although there is clear evidence that both PAF and eicosanoids contribute to endotoxin shock, it is equally clear that TNF-α and possibly other cytokines play an important role. Heuer et al. (1990) have shown that PAF and TNF-α synergize in increasing mortality in mice. Furthermore, TNF-α is known to be a stimulant of PAF production by endothelial cells (Camussi et al., 1987a), macrophages/monocytes (Camussi et al., 1987a), neutro-phils (Camussi et al., 1987a; Stewart et al., 1991), mesangial cells (Camussi et al., 1990) and bowel tissue (Sun and Hsueh, 1988). Conversely, PAF is reported to stimulate TNF-α production by human monocytes and alveolar macrophages (Bonavida et al., 1989a; Dubois et al., 1989). Thus, the potential exists for independent and interdependent roles for PAF and TNF-α in the pathogenesis of endotoxin shock. Recent studies with a new PAF receptor antagonist, BN 50739, indicate a role for PAF in both TNF-α and endotoxin-induced increases in eicosanoid formation, but not the thrombocytopenia or leucopenia induced by this cytokine (Rabinovici et al., 1990). In contrast, Chang et al. (1989) did not obtain any evidence for an involvement of PAF in the pulmonary and vascular responses of rats to TNF-α, including the increases in plasma levels of TXB_2 and 6-oxo-$PGF_{1\alpha}$. In a mouse model of endotoxin shock, the inhibition of endotoxin lethality by BN 52021 was confirmed, but in animals treated with a combination of low doses of endotoxin and TNF-α, BN 52021 had no effect on the lethality (Myers et al., 1990). This study indicates that

PAF and TNF-α may have separate roles in endotoxaemia. However, the involvement of PAF in TNF-α-induced shock warrants further investigation since, in a guinea-pig model involving TNF-α challenge in vivo and perfusion of lungs isolated from challenged animals with human neutrophils, PAF was found to contribute to the increase in permeability and the pulmonary capillary pressure (Hocking et al., 1990).

Plasma protease activity increases following endotoxin challenge of rats and is inhibited by BN 52021 and BN 52063 (Etienne et al., 1987). The source of the protease was not identified but may be from leucocytes, suggesting that the lack of effect of PAF antagonists on endotoxin-induced leucopenia does not indicate a complete lack of effect on the participation of those cell types in the progression of shock.

4.3.4 Haematological Effects of Endotoxaemia

In guinea-pigs, a biphasic hypotensive response was observed upon endotoxin challenge comprising an initial rapid drop in blood pressure (≈ 10 min) followed by partial recovery and a further fall commencing at 30 min (Adnot et al., 1986a,b). In this model, BN 52021 inhibited the early and late phases of hypotension and the thrombocytopenia, but not the leucopenia, that accompanied the initial hypotension. It was proposed that BN 52021 may act by a platelet-independent mechanism, since platelet depletion abrogated the early phase of hypotension, but BN 52021 had a beneficial effect on the persisting second phase of hypotension (Adnot et al., 1986a,b).

In sheep, endotoxin induced thrombocytopenia and leucopenia, but the PAF receptor antagonist FR-900452 inhibited only the thrombocytopenia (Okamoto et al., 1986). Contrasting observations have been made with ONO-6240, which attenuated endotoxin-mediated leucopenia in the sheep (Toyofuku et al., 1986).

4.3.5 Pulmonary Consequences of Endotoxin Shock

A PAF-dependent pulmonary platelet accumulation was observed in guinea-pigs exposed to an aerosol of endotoxin that was inhibited by CV 3988, brotizolam or BN 52021 and a combination of heparin and PGI_2, but not by indomethacin (Beijer et al., 1987), confirming a role for PAF in endotoxin-mediated platelet activation. A further study identified the alveolar macrophage as a potential target for endotoxin aerosol-induced PAF synthesis (Rylander and Beijer, 1987). Lung parenchymal strips from endotoxaemic guinea-pigs showed a reduced sensitivity to the relaxant actions of isoprenaline that could be prevented by in vivo treatment with BN 52021, indicating the PAF may induce down-regulation of β adrenoceptors (Touvay et al., 1988). These observations signal the need to consider PAF as a mediator of lung disease associated with inhalation of endotoxin-containing dusts.

Intraperitoneal endotoxin challenge of rats increased blood levels of PAF and induced haemodynamic changes, including hypotension and increased pulmonary microvascular permeability, that were attenuated by pretreatment with SRI 63-441 or CV 3988 (Chang *et al.*, 1987). In contrast to observations made with BN 52021, endotoxin-induced pulmonary recruitment of platelets was not reduced by the PAF receptor antagonist 48740 RP, whereas vascular permeability and neutrophil influx were attenuated (Rylander *et al.*, 1988).

Endotoxin induced an early (0.5 h) and a late (5 h) increase in pulmonary vascular resistance and a decrease in p_aO_2, but only the early effects were inhibited by SRI 63-441. Furthermore, SRI 63-441 attenuated the endotoxin-mediated increase in lung lymph flow, suggesting an inhibition of microvascular permeability increases, but had no effect on the gradual fall in systemic blood pressure, in contrast to the protective effect noted in many other studies (Sessler *et al.*, 1988).

4.3.6 Gastrointestinal Tract

Endotoxin-induced increases in microvascular permeability of the mucosa of the gastrointestinal tract were inhibited by CV 3988, BN 52021 or Ro-19,3704, indicating a contribution of PAF to endotoxin-induced gastric ulceration (Wallace *et al.*, 1987). In addition, endotoxin challenge of rats induced a 20-fold increase in jejunum PAF levels which correlated with hyperaemia and haemorrhage (Whittle *et al.*, 1987). The participation of PAF and TXA_2 in endotoxin-induced jejunal damage was assessed with a range of pharmacological inhibitors, establishing that TXA_2 synthetase, cyclooxygenase or lipoxygenase products had no influence on PAF formation by the jejunum (Boughton Smith *et al.*, 1989). It remains possible, however, that endotoxin-induced TXB_2 formation could be secondary to PAF formation, since the effects of PAF antagonists on TXB_2 were not evaluated. The persistence of some jejunal damage by endotoxin in rats treated with TXA_2 synthase blockers indicates that PAF and TXA_2 are, to a certain extent, independent mediators of this damage (Boughton Smith *et al.*, 1989).

Endotoxin induces an increase in gastric motility in rats, as does PAF, and both of these reponses are inhibited by L-652,731 (Esplugues and Whittle, 1989a). Studies of the motility of the duodenum assessed by monitoring electrical activity revealed that intravenous endotoxin- or PAF-induced increases in motility were attenuated by BN 52021 or by indomethacin (Pons *et al.*, 1989).

4.3.7 Kidney

A low dose of endotoxin (150 µg/kg) in the rat induced a renal dysfunction without altering systemic blood pressure, in which the decreases in GFR, renal plasma flow and filtration fraction were inhibited by either a pretreatment or a post-treatment with L-652,731 but not by the cyclooxygenase inhibitor meclofenamate (Wang

and Dunn, 1987). Similar results were obtained using BN 52021 and SRI 63-675 (Tolins *et al.*, 1989). A potential source of PAF in the renal response to endotoxin is the mesangial cell, since endotoxin evoked the synthesis and release of PAF from cultured rat glomerular mesangial cells (Wang *et al.*, 1988).

4.3.8 Vasculature

PAF and endotoxin cause similar increases in arteriolar diameter of mesenteric vessels exteriorized for video microscopy, and the vascular effects of both stimuli were blocked by PAF receptor antagonists (Lagente *et al.*, 1988b). Endotoxin induces an increase in ocular vascular permeability which is attenuated by SRI 63-441 and prevented with the addition of indomethacin or a corticosteroid to the therapeutic regimen (Rubin *et al.*, 1988).

In a complex investigation of the potential role of macrophages in endotoxin-induced cardiac dysfunction, it was reported that rat peritoneal macrophages adhered to glass beads and, when stimulated with endotoxin, released a material which decreased coronary flow and cardiac contractility by a PAF antagonist-sensitive, but leukotriene- and prostaglandin-resistant, mechanism (Salari and Walker, 1989).

4.4 GASTRIC ULCERATION AND ISCHAEMIC BOWEL DISEASE

4.4.1 Ulcerogenic Actions

PAF is regarded as one of the most potent ulcerogens and its involvement in a range of pathological conditions that are frequently associated with gastric ulceration provided a stimulus for the considerable effort directed towards defining its role in this condition (Esplugues and Whittle, 1989b; Wallace, 1990).

Intravenous administration of PAF at 20–200 pmol kg^{-1} min^{-1} caused extensive haemorrhagic erosions in the gastric mucosa that were not mimicked by matching the hypotensive response to PAF with infusions of PGI_2, isoprenaline or nitroprusside (Rosam *et al.*, 1986). Furthermore, close intra-arterial infusion of PAF into the rat stomach *in vivo* produced gastric damage that was independent of changes in systemic arterial blood pressure (Esplugues and Whittle, 1988). These observations indicate that the PAF-induced gastric damage was not solely due to systemic hypotension, even though the latter action would be expected to exacerbate the ischaemia induced by the systemic release of PAF accompanying anaphylaxis or septicaemia. In the gastric mucosal microcirculation, in contrast to most vascular beds, PAF-induced decreases in blood flow appear to result from vascular stasis rather than changes in vessel diameter (Whittle *et al.*, 1986; Sato *et al.*, 1988). The stasis may be a consequence of haemoconcentration or leucocyte aggregation, leading to a change in the "stiffness" of leucocytes via changes in the cytoskeleton entailing decreased ability to pass through the capillaries.

PAF-induced gastric damage in the rat was not affected by platelet depletion, histamine H_1 and H_2 receptor antagonists, α adrenoceptor antagonism or inhibition of cyclooxygenase (Rosam *et al.*, 1986). In contrast, Etienne *et al.* (1988) reported that cimetidine attenuated PAF-induced gastric damage, and suggested that the administration of cimetidine by the oral route and at a higher dose may explain the difference between their observations and those of Rosam *et al.* (1986). In addition, parasympathetic activation and neutrophils were implicated in the complex sequence of events involved in PAF-induced gastric ulceration, since atropine and neutrophil depletion inhibited ulcer formation (Etienne *et al.*, 1988).

A number of studies indicate that eicosanoids, particularly leukotrienes, are pro-ulcerogenic (Rainsford, 1986; Konturek *et al.*, 1988). PAF-induced gastric damage is not affected by the cyclooxygenase inhibitor indomethacin, whereas corticosteroids and the dual cyclooxygenase/lipoxygenase inhibitor BW-755C reduced damage without influencing PAF-induced leucopenia, suggesting a role for leukotrienes (Wallace and Whittle, 1986b, 1988). Further evidence of the involvement of lipoxygenase products in PAF-induced gastric ulceration was provided by the observation that the cysteinyl leukotriene antagonist FPL 55712 inhibited PAF-induced augmentation of ethanol-induced lesions (Konturek *et al.*, 1988). The leukotriene-dependence of the gastric ulcerogenic actions of PAF is supported by the ability of PAF to augment LTC_4 generation in response to ethanol via a specific PAF receptor (Konturek *et al.*, 1989). In addition, infusion of PAF in the rat increased the formation of LTB_4 throughout the gastrointestinal tract, whereas LTC_4 synthesis was detected only in the duodenum (Wallace and MacNaughton, 1988). Infusion of leukotrienes failed to reproduce the damage induced by lower doses of PAF, suggesting that leukotrienes were unlikely to be major mediators of PAF-induced gastrointestinal tract damage (Wallace and MacNaughton, 1988). However, the failure of leukotrienes to produce overt gastric damage does not, of itself, preclude a significant contribution to the complex sequence of events initiated by PAF infusion. Biopsy specimens from duodenal ulcer patients showed two- to three-fold higher levels of PAF and LTB_4 and LTC_4 generation, and this increased biosynthetic capacity was reduced upon healing of the ulcer (Ackerman *et al.*, 1990). The restricted distribution of LTC_4 synthesis observed in the rat was not seen in the human biopsy specimens in which LTC_4 synthesis was reported to occur in both antral and fundic regions, but was unchanged in patients with duodenal ulcers (Ackerman *et al.*, 1990).

Indirect evidence for a role for lipid peroxidation in gastric damage was provided by the observation that PAF induced an increase in thiobarbituric acid reactants (Binnaka *et al.*, 1989). PAF-induced gastric ulceration was reduced by a combination of catalase and superoxide dismutase, whereas the decreased gastric mucosal blood flow was unaffected (Yoshida *et al.*, 1989). The free radical scavenger allopurinol gave an almost complete suppression of PAF-induced gastric damage. The abrogation of damage by neutrophil depletion suggested that this cell type was the major source of oxygen-derived free radicals (Etienne *et al.*, 1988).

Enhancement of PAF-induced gastric ulceration by morphine or neonatal capsaicin treatment has been attributed to the existence of sensory afferent neurones that promote local mucosal blood flow in response to noxious stimuli (Pique *et al.*, 1990). Since mast cells may interact with sensory fibres in the reflex increase in mucosal blood flow, it is of interest to note that a relationship of PAF-induced gastric ulceration to local mast cell degranulation has been reported (Escolar *et al.*, 1989).

4.4.2 Haemorrhage

In a model of gastric haemorrhage in rats, exogenous PAF decreased bleeding but a PAF antagonist, SRI 63-675, had no effect, suggesting that endogenous PAF plays no role in haemostasis in this model (Berstad *et al.*, 1988). However, PAF increased bleeding from lesions induced by 0.15 N hydrochloric acid in anaesthetized rats by a mechanism that involved an increase in mucosal permeability that was blocked by CV 3988 (Hatakeyama *et al.*, 1989). Further studies in species with platelets that are responsive to PAF seem warranted before extending this conclusion.

4.4.3 Involvement of Platelet-Activating Factor in Animal Models of Ulceration

Although PAF induces hyperaemia, vascular stasis and gastric lesions, it does not appear to contribute to the non-inflammatory gastric ulceration induced by aspirin, or ethanol with hydrochloric acid (Rainsford, 1986). The gastric damage induced by a single dose of PAF in the conscious rat was inhibited by pretreatment with WEB 2086 (Brambilla *et al.*, 1987), whereas PGE_2 and analogues had no inhibitory effects despite their effectiveness in irritant-induced ulceration (Steel *et al.*, 1987) as a result of maintaining the gastric microcirculation (Whittle, 1989). Nevertheless, non-ulcerogenic doses of PAF predispose the rat gastric mucosa to damage induced by the topically applied irritants, 20% ethanol or 2 mM sodium taurocholate (Wallace and Whittle, 1986a).

Neither PAF nor leukotrienes appear to contribute to NSAID-induced gastric ulceration, whereas reduction of circulating neutrophil numbers using antineutrophil antiserum or methotrexate decreased indomethacin-induced gastric damage (Wallace *et al.*, 1990b).

In rats treated singly with LPS or PAF, ischaemic intestinal necrosis was observed, whereas combined administration resulted in necrotizing lesions of the gastrointestinal tract that showed morphological similarities to necrotizing enterocolitis (Gonzalez Crussi and

Hsueh, 1983). This finding indicated a synergistic action of these stimuli consistent with the previous report that picomolar doses of PAF enhance the ulcerogenic actions of topically applied irritants (Wallace and Whittle, 1986a). In addition, endotoxin-induced necrosis and vascular congestion in the stomach and small intestine were inhibited by CV 3988, BN 52021 and Ro-19,3704 (Wallace *et al.*, 1987). The level of PAF formation by the jejunum correlated with the damage induced by endotoxin administration in rats (Whittle *et al.*, 1987). In a complex study of the interrelationship between eicosanoids and PAF in endotoxin-induced gastric ulceration, it was observed that inhibition of either cyclooxygenase or lipoxygenase enzymes had no effect on PAF formation by the jejunum, but the inhibitors reduced damage. It remains to be seen whether PAF receptor antagonists inhibit prostaglandin and leukotriene production in this model (Boughton-Smith *et al.*, 1989). Nevertheless, establishing whether a hierarchy of mediator production and release is triggered by endotoxin in gut tissue may have important implications for the treatment of ulceration with compounds that modulate these lipid mediators.

BN 52021 was completely effective against PAF-induced ulceration and decreased the damage observed in endotoxin or cold restraint stress-induced gastrointestinal lesions, suggesting a contribution of PAF to the ulceration observed in the latter models (Etienne *et al.*, 1989). The water immersion stress model of gastric ulceration was accompanied by a rise in the serum acetylhydrolase level, which may serve as a negative feedback or protect the systemic circulation from locally released PAF (Fujimura *et al.*, 1989). In addition, PAF levels in the stomach, particularly the antral region, decrease in this model of ulceration (Sugatani *et al.*, 1989). The role of this constitutive PAF is not known and it is not clear whether the decrease in endogenous PAF levels results from PAF release.

PAF appears to play a role in a wide range of models of gastric ulceration, including that induced by haemorrhagic shock in rats in which WEB 2086 and BN 52021 pretreatment show protective effects, such as a reduction in lumenal haemoglobin (Wallace *et al.*, 1990a).

4.5 HYPERTENSION AND ATHEROSCLEROSIS

4.5.1 Pulmonary

PAF has potent pulmonary vasoconstrictor activities (Voelkel *et al.*, 1982; Hamasaki *et al.*, 1984) that have prompted the suggestion that it may play a role in pulmonary hypertension (McManus and Deavers, 1989).

4.5.2 Endotoxaemia

In endotoxin shock, in which pulmonary hypertension is a common finding, there is some variation in the literature regarding the effectiveness of PAF antagonists: ONO-6240 had no effect in unanaesthetized sheep (Toyofuku *et al.*, 1986); SRI 63-441 blocked the early but not the late endotoxin-induced pulmonary hypertension (Sessler *et al.*, 1988); group B streptococcus-induced pulmonary hypertension was resistant to SRI 63-441 or SRI 63-072 (Pinheiro *et al.*, 1989); SRI 63-441 attenuated endotoxin-induced pulmonary hypertension, but WEB 2086 had little effect (Christman *et al.*, 1990); in pigs, SRI 63-675 was effective against pulmonary hypertension (Olson *et al.*, 1990b). Thus, while there is clear evidence of PAF release in endotoxaemia, it seems likely that other mediators such as TXA_2 make a major contribution to the pulmonary hypertension.

4.5.3 Hypoxia

PAF has received some consideration as a mediator of hypoxic pulmonary vasoconstriction. PAF levels have been shown to be elevated in plasma of neonates with pulmonary hypertension (Caplan *et al.*, 1990). However, in rat lungs the constrictor response to ventilation with 3% oxygen was attenuated with a high concentration of PAF (≈ 2 μM), making it an unlikely candidate (Gillespie and Bowdy, 1986). In a more definitive study, anaesthetized, mechanically ventilated pigs were ventilated with 10% oxygen to evaluate the potential protective effects of WEB 2086 against the hypoxic pulmonary vasoconstriction. The dose of WEB 2086 was effective against PAF-induced pulmonary hypertension, but had no effect on that induced by hypoxia (McCormack *et al.*, 1990).

The role of PAF has also been evaluated in pulmonary hypertension following protamine neutralization of heparin (Habazettl *et al.*, 1990) and complement-mediated hypertension in sheep (Smallbone *et al.*, 1987). Neither study indicated a major role for PAF. Despite the apparent failure of PAF antagonists to modify TNF-α-induced lung injury in rats *in vivo* (Chang *et al.*, 1989), WEB 2086 attenuated TNF-α-induced oedema and vasoconstrictor responses of guinea-pig isolated perfused lungs (Hocking *et al.*, 1990). Bleomycin-induced pulmonary oedema in rats was attenuated by pretreatment with BN 52021, whereas platelet depletion was ineffective (Evans *et al.*, 1990).

Thus, several forms of acute pulmonary hypertension do not appear to involve PAF as the major mediator. Involvement in chronic pulmonary hypertension remains possible, particularly in view of the stimulation by PAF of pulmonary vascular smooth muscle proliferation (Stoll and Spector, 1989).

4.5.4 Systemic Blood Pressure

The simultaneous discovery of PAF as an antihypertensive polar renal-medullary lipid (APRL) released from the kidney to lower systemic blood pressure by Muirhead and colleagues (Blank *et al.*, 1979) raised the possibility that altered PAF production may contribute to some forms of hypertension (Blank *et al.*, 1984). The concept that

PAF functions as the vasodepressor renal hormone was proposed by Muirhead and colleagues, but their own and other studies (*vide infra*) indicate that PAF is not the renomedullary hypotensive lipid now called medullipin, but not yet structurally identified (Muirhead, 1990).

Croft and colleagues (Croft *et al.*, 1986; Codde *et al.*, 1987) found that SHR lyso-PAF levels were less than those in Wistar Kyoto rats and that the cardioprotective lipid EPA ($C_{20:5}$) lowered lyso-PAF levels in the plasma of normotensive rats. Unclipping of one-kidney, one-clip hypertensive rats led to a significant increase in whole-blood PAF levels and a decrease in plasma lyso-PAF levels in association with the reversal of the hypertension (McGowan *et al.*, 1986, 1988). Chemically induced necrosis specific to the renal medulla caused a lowering of PAF levels to 50% of the control level, and a selective decrease in the activity of choline phosphotransferase in the inner medulla, indicating that this *de novo* pathway is responsible for the release of PAF (Lee *et al.*, 1989). However, a recent study indicates that WEB 2086 has no effect on the extent of the fall in blood pressure upon unclipping in this model, providing strong evidence against a role for PAF in this hypotensive response (Cotter *et al.*, 1990).

Percutaneous transluminal renal angioplasty reduces blood pressure in patients with renovascular hypertension, and there is an associated release of PAF and a phospholipid antagonist of PAF into the renal venous blood (Masugi *et al.*, 1988a,b). In addition, plasma PAF acetylhydrolase levels are reported to be increased in essential hypertension (Satoh *et al.*, 1989).

Infusion of 14,15-epoxyeicosatrienoic acid resulted in hypotensive responses in spontaneously hypertensive and normotensive rats that could be blocked with BN 52021, but not indomethacin (Lin *et al.*, 1990). It remains to be established whether the epoxide arachidonate metabolite stimulates PAF synthesis, as suggested by this finding.

It seems intrinsically unlikely that a substance with the thrombotic and inflammatory potential of PAF could function as a circulating hormone. However, it is possible that PAF circulates in a form that is able to lower blood pressure without activating inflammatory cells or platelets. One such form may be lipoproteins (Benveniste *et al.*, 1988). Although PAF receptor antagonists have little acute effect on systemic arterial blood pressure, it remains to be determined whether PAF antagonists have any effect on blood pressure when given chronically to hypertensive animals.

4.5.5   Atherosclerosis

Activation of macrophages, the capacity to induce endothelial damage and proliferation of vascular smooth muscle, together with platelet activation, are properties consistent with a role for PAF in atherosclerosis (Fig. 8.6). In rabbits fed diets enriched in cholesterol, the development of atherosclerotic lesions and the vascular wall content of cholesterol ester was reduced by pretreat-ment with BN 52021 (Feliste *et al.*, 1989). It is now of importance to determine whether other PAF antagonists have similar activities in this and other models of atherosclerosis that are not dependent on cholesterol feeding.

Lyso-PAF levels were reported to be unaltered in cigarette smokers or in patients with atherosclerosis, but platelet aggregation in response to ADP and collagen was increased (Sturm *et al.*, 1989). A recent study suggests that cigarette smoking acutely elevates PAF levels in plasma and, more particularly, in LDL and HDL, indicating a potential contribution to atherosclerosis (Imaizumi *et al.*, 1991). Whether this material is PAF or one of the recently described oxidized phosphatidyl-choline derivatives with PAF-like biological activity (Stremler *et al.*, 1989) is yet to be established. It is possibly significant that acetylhydrolase is also associated with HDL and LDL particles (Stafforini *et al.*, 1987) and that this enzyme degraded an oxidized phospholipid (Stremler *et al.*, 1989, 1991) that appears to have activity at PAF receptors on human neutrophils (Smiley *et al.*, 1991). Thus, should lipoprotein-associated PAF escape catabolism by acetylhydrolase, the uptake of lipoproteins by monocytes/macrophages within the atherosclerotic plaque may provide access of PAF to intracellular/intramembranous sites of action inferred by previous studies (Stewart and Phillips, 1989; Stewart *et al.*, 1989, 1990) and documented in neutrophils (Hwang, 1988) and cerebellar neurones (Marcheselli *et al.*, 1990).

4.6   INFLAMMATION

The pro-inflammatory actions of PAF have been studied in a multitude of models in different organ systems. The following sections describe the involvement of PAF in inflammation not covered in sections on the major organ systems or specific diseases.

4.6.1   Rat Paw Oedema

Injection of PAF into the rat paw induces oedema and desensitization to subsequent injections of PAF, but neither desensitization nor BN 52021 pretreatment inhibits carrageenan-induced oedema (Cordeiro *et al.*, 1986). PAF-induced rat paw oedema was inhibited by the dual cyclooxygenase/lipoxygenase inhibitor, BW-755C and by the histamine/serotonin receptor antagonist cyproheptadine (Calhoun *et al.*, 1987). Involvement of lipoxygenase products in PAF-induced hyperalgesia of the rat hindpaw was demonstrated by the inhibitory effect of a lipoxygenase inhibitor, L-615,919, whereas indomethacin had no effect (Dallob *et al.*, 1987). It is likely that LTB_4 was the target leukotriene, since the cysteinyl leukotrienes are reported to have analgesic effects (Schweizer *et al.*, 1984).

The dose–response relationship for PAF-induced rat paw oedema has a bell shape with a leucopenia accompanying the ascending phase, and thrombocytopenia,

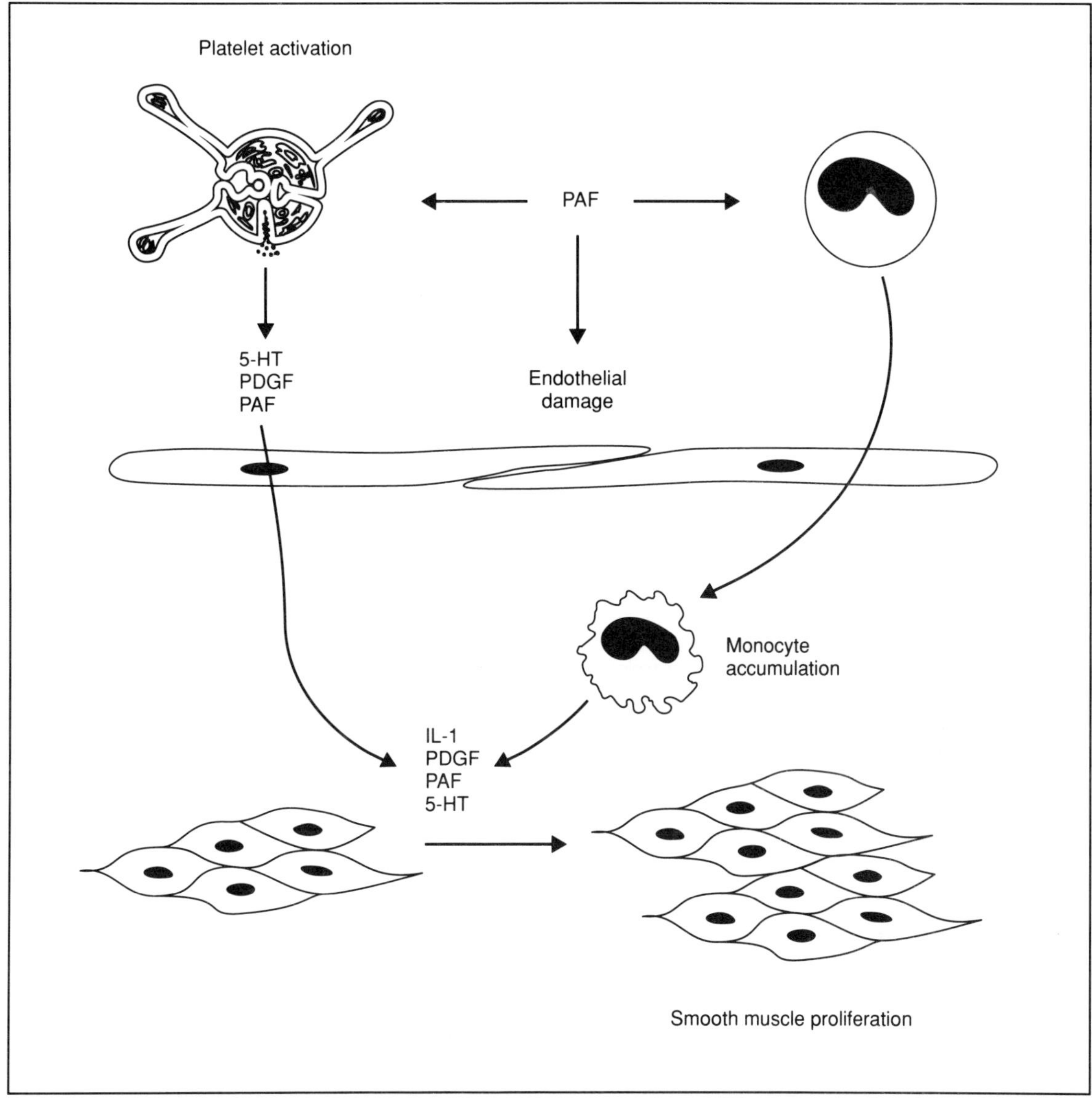

Figure 8.6 Potential contribution of PAF to atherosclerosis. The role of macrophages and platelets in the atherosclerotic process is well established, but the initiating stimuli are yet to be clearly elucidated. Several observations are compatible with a role for PAF in atherosclerosis, including platelet activation, monocyte activation and chemotaxis, and the recently recognized ability to induce proliferation of vascular smooth muscle cells.

haemoconcentration and leucocytosis in the descending phase (Martins *et al.*, 1987). Although RP 48740 blocked the thrombocytopenia and leucocytosis, it did not block the oedema, suggesting that the last effect is local and independent of platelet and leucocyte activation (Martins *et al.*, 1987).

4.6.2   Animal Models of Arthritis

In the acute, but not the chronic, stages of antigen-induced arthritis in the rabbit, detectable levels of PAF were observed in the synovial fluid. However, intra-articular injection of PAF at up to 100 times the detected level failed to induce swelling or leucocyte accumulation (Pettipher *et al.*, 1987). These observations suggest an involvement of PAF only in the acute stage of lesion development.

Intraperitoneal injection of monosodium urate crystals induces plasma protein extravasation and lyso-PAF was detectable in the exudate, but WEB 2086 was ineffective in reducing the permeability change, suggesting that the lyso-PAF was not indicative of PAF involvement (Griswold *et al.*, 1989).

Thus, while neither of the above studies gives support for an important role for PAF in arthritis, this disease is difficult to model in experimental animals. The activation of macrophages by PAF and its production by cells involved in the pathology of arthritis indicate that PAF antagonists may produce some improvement in this condition.

4.6.3 Pleurisy

Intrapleural PAF injection in rats elicited a biphasic change in leucocyte numbers, with an initial fall (0.5 h) followed by a rise accompanied by the formation of an exudate which peaked at 0.5–1 h (Tarayre et al., 1986). PAF-induced pleurisy was reported to be inhibited by CV 3988, but not by either indomethacin or dexamethasone, excluding a role for eicosanoids in this response (Ohishi et al., 1986). In the pleural exudate of rats treated with PMA, PAF was detected by bioassay following HPLC, and pleural cells from untreated rats produced PAF upon stimulation with PMA or A23187 (Hayashi et al., 1987b). In the pleural exudate, approximately two-thirds of the PAF was present in the cells, whereas an even greater proportion was retained by pleural cells stimulated in vitro with either PMA or A23187. Furthermore, the PAF receptor antagonist CV 3988, reduced the pleural fluid accumulation following intrapleural PMA, but CV 3988 was without effect on exudation induced by intrapleural carrageenan (Hayashi et al., 1987b).

PAF desensitization or WEB 2086 pretreatment of rats reduced pleural exudates induced by zymosan, but not those in response to carrageenan (Martins et al., 1989). In contrast, mouse paw oedema and pleurisy induced by carrageenan were sensitive to PAF desensitization and to WEB 2170, whereas the related antagonist WEB 2086 blocked only the first phase of the carrageenan-induced oedema, and failed to block either pleurisy or paw oedema induced by PAF in this species (Henriques et al., 1990).

4.6.4 Ocular Inflammation

PAF (20–2000 pmol) application to the rat conjunctiva induced an increase in Evans blue extravasation together with an increase in weight that could be blocked with L-652,731, but not with inhibitors of cyclooxygenase or lipoxygenase (Gautheron et al., 1987). This finding was confirmed in a model of ocular inflammation using endotoxin challenge, in which SRI 63-441 inhibited vascular leakage of fluoresceinated dextran without affecting aqueous humour PGE_2 levels and had additive effects with either indomethacin or corticosteroid (Rubin et al., 1988). Similarly, topical application of PAF on to guinea-pig conjunctiva elicited an increase in vascular permeability as determined by Evans blue extravasation, indicating its potential as a mediator of allergic conjunctivitis (Stock et al., 1990).

Alkali burn injury of the rabbit cornea in vivo is followed by the accumulation of prostaglandins, 5-HETE and 12-HETE in the anterior chamber. Topical application of BN 52021 inhibited accumulation of the eicosanoids when applied in the first 4 h after injury, but it had no effect on the accumulation of protein in the aqueous humour (Bazan et al., 1987).

4.7 ISCHAEMIA

Braquet et al. (1989a) reviewed the evidence supporting a therapeutic role for PAF antagonists in the treatment of ischaemic conditions and suggested that, in combination with other compounds, PAF antagonists may prove useful in cerebral, renal and myocardial ischaemia.

4.7.1 Myocardial Ischaemia

The prolonged increase in coronary resistance and the electrophysiological actions of PAF make it a likely candidate as a mediator of angina. Montrucchio et al. (1986b) provided direct evidence for the involvement of PAF in angina in a study which showed that PAF was released into the coronary sinus blood of patients with angina during atrial pacing challenges. A more recent study by this group described PAF release from buffer-perfused rabbit hearts following ischaemia and reperfusion, indicating a cardiac origin for at least some of the PAF released during ischaemia (Montrucchio et al., 1989a). Nevertheless, other cellular sources are likely to contribute to PAF production in vivo, particularly neutrophils which migrate into the ischaemic tissue soon after reperfusion.

Infusion of PAF in anaesthetized cats at a dose with no detectable direct effects, evoked the release of creatine phosphokinase in animals previously subjected to 30 min of coronary artery ligation. This aggravation of ischaemia was attenuated with a TXA_2 synthetase inhibitor (Lepran and Lefer, 1985). In anaesthetized rats, coronary artery ligation-induced loss of myocardial amino nitrogen and cathepsin D was inhibited by post-treatment with CV 6209, and in vitro studies revealed an inhibition of global ischaemia-induced increases in cardiac permeability in rat hearts treated with CV 6209 (Stahl et al., 1988).

In isolated perfused rat hearts, the rhythm disturbances induced by 30 min of coronary artery ligation were attenuated by BN 52021 through a mechanism proposed to involve antagonism of the increase in slow inward calcium current (Koltai et al., 1989). Inhibition of PAF formation during myocardial ischaemia is a potential mechanism for the beneficial effects of cardioprotective fish oils ($C_{20:5}$, $C_{22:6}$), since these are incorporated into PAF precursors in cardiac tissues of rats fed fish oil (Yeo et al., 1989b).

4.7.2 Cerebral Ischaemia

Pretreatment of anaesthetized dogs with kadsurenone 60 min before induction of multifocal ischaemia enhanced recovery of the cortical somatosensory evoked response, but had no effect on the accumulation of indium-labelled platelets, suggesting a platelet-independent action (Kochanek *et al.*, 1987). In gerbils subjected to bilateral common carotid occlusion, BN 52021, brotizolam or kadsurenone inhibited the ischaemia-induced cerebral impairment, even when administered during the postischaemia period, indicating that this class of antagonist may be useful in the treatment of ischaemic brain injury (Spinnewyn *et al.*, 1987). Intracarotid infusion of PAF in anaesthetized rats caused a reduction in cerebral blood flow associated with an increase in cerebral oxidative metabolism, and it was suggested that these effects may underlie its contribution to the pathogenesis of brain ischaemic injury (Kochanek *et al.*, 1988).

It has been established that the release of free fatty acids as a consequence of PLA_2 and PLC activation accompanies cerebral ischaemia and reperfusion injury. Pretreatment of gerbils with BN 52021 prior to bilateral carotid occlusion reduced the ensuing cerebral injury and the accumulation of free fatty acids and DAG, suggesting an inhibitory effect on the activation of PLA_2 and PLC (Panetta *et al.*, 1989). Similar observations were made on the accumulation of free fatty acids in mouse brain following either ischaemia or electroconvulsive shock (Birkle *et al.*, 1988).

In a study of laser-induced multifocal cortical lesions in rats, the secondary brain damage, including reduced cortical blood flow, oedema and dysfunction of the blood–brain barrier, were reduced by pretreatment with BN 50739, indicating a therapeutic potential for PAF antagonists in the prevention of brain damage that follows neurological trauma (Frerichs *et al.*, 1990). Constriction of cortical blood vessels by PAF (Armstead *et al.*, 1988) could serve to increase the original focal damage in this model. Furthermore, the progressive neuronal cell degeneration that follows such injury was significantly reduced by BN 50739 (Frerichs *et al.*, 1990). PAF receptors have also been observed on astrocytes maintained in primary culture, in which picomolar but not nanomolar concentrations of PAF increase PI turnover, suggesting that these cells may also be a target for PAF release during ischaemia (Murphy and Welk, 1990).

The foregoing studies give strong support to the notion that PAF is a key mediator of neuronal death following ischaemic injury, since a wide range of structurally distinct PAF antagonists have shown activity and similar results have been reported in mice, dogs, rats and gerbils in models using global ischaemia or multifocal injury.

4.7.3 Ischaemia in Other Vascular Beds

The importance of neutrophil adhesion to endothelium and accumulation in ischaemic tissues is well recognized. In a study of ischaemia and reperfusion of the cat mesenteric circulation, pretreatment with either BN 52021 or WEB 2086 was found to inhibit both neutrophil adhesion and extravasation (Kubes *et al.*, 1990).

Ischaemia-induced damage in several organs, including the gastrointestinal tract (Hsueh *et al.*, 1986), the

mesentery (Filep *et al.*, 1989; Mozes *et al.*, 1989) and the kidney (Lopez Farre *et al.*, 1988; Braquet *et al.*, 1989a) appears to involve the release of PAF.

It is less clear whether PAF receptor antagonists will prove useful in the clinical treatment of ischaemia-related disorders since, in most of the reviewed studies, PAF antagonists have been administered either prior to ischaemia or postischaemia, but prior to reperfusion.

4.8 RENAL PATHOLOGY

4.8.1 Proteinuria

In isolated perfused rat kidneys, PAF induces proteinuria, indicating its potential as a mediator of renal disease (Pirotzky *et al.*, 1985a; Schwertschlag *et al.*, 1988). In an isolated buffer-perfused rat kidney preparation, it was shown that the proteinuria induced by PAF is receptor mediated and does not appear to involve either cyclooxygenase or lipoxygenase products or a change in the electrostatic barrier of the glomerulus (Perico *et al.*, 1988; Schwertschlag *et al.*, 1988). This property, in addition to known platelet and leucocyte activation, led to the suggestion that PAF may be involved in microangiopathy (Bussolino *et al.*, 1986a). Proteinuria and ultrastructural changes associated with adriamycin-induced nephropathy in rats were attenuated with BN 52021 and WEB 2086 pretreatment (Egido *et al.*, 1987).

4.8.2 Hydronephrotic Kidney

In the hydronephrotic kidney, PAF is a powerful stimulant of eicosanoid (PGE_2, PGI_2 and TXA_2) generation in comparison to its effects in normal, isolated perfused kidneys (Weisman *et al.*, 1985). In the same model, it was observed that TXB_2 release in response to AII or bradykinin was inhibited by kadsurenone, CV 3988 or triazolam, implicating endogenous PAF in the stimulation of eicosanoid synthesis; arachidonate-induced TXB_2 was unaffected by these PAF receptor antagonists (Weisman *et al.*, 1988). However, the failure of a PAF antagonist (L-659,989) to inhibit the inflammatory cell infiltration, induction of microsomal cyclooxygenase activity or increased eicosanoid generation in a rabbit kidney made hydronephrotic by unilateral obstruction suggests that PAF is not the primary mediator in this renal disease model (Spaethe *et al.*, 1989). Similarly, PAF appears to play a minor role in the acute renal haemodynamic response to unilateral obstruction (Felsen *et al.*, 1990).

In addition to the release of cyclooxygenase products, PAF is also reported to release histamine (Alessandri *et al.*, 1988) and N-acetyl-β-glucosaminidase (Giovannini *et al.*, 1988), but the contribution of these products to PAF-induced renal damage is not established.

4.8.3 Immune Complex Deposition –
Glomerulonephritis

In rabbits treated with PAF, immune complexes formed by administration of BSA and anti-BSA antibodies were selectively deposited in heart, lung and kidney (Camussi *et al.*, 1985). In a model of glomerulonephritis in the rat, the PAF receptor antagonist SRI 63-072 inhibited the proteinuria and gross morphological abnormalities, but did not influence the deposition of immune complexes (Camussi *et al.*, 1987c). Thus, PAF formation in response to immune complex deposition appears to be an important contributor to the renal dysfunction in this model. Induction of nephrotoxic nephritis in rabbits by administration of antiglomerular basement protein antiserum increases TXA_2 synthesis by a PAF-dependent mechanism, and a component of the beneficial effect of PAF receptor antagonism in this model is likely to result from reduced TXA_2 synthesis (Macconi *et al.*, 1989).

4.9   TRANSPLANTATION

The vasoactive, immunoregulatory and haemostatic (Faulk *et al.*, 1989) actions of PAF make it a candidate mediator for participation in the transplant rejection process.

4.9.1   Graft Survival and Platelet-Activating Factor Antagonists

In rats, cardiac allograft survival was improved by combined treatment with BN 52021 and azathioprine; a 5-lipoxygenase inhibitor had similar synergistic activity to that of BN 52021 (Foegh *et al.*, 1986). In a dog-to-pig renal xenograft, treatment with SRI 63-441 alone did not increase survival or improve function, but SRI 63-441 showed a synergistic interaction with PGI_2 or PGE_1 infusion, with six- to nine-fold increases in survival and three- to 20-fold increases in urine output (Makowka *et al.*, 1987). BN 52021 increased murine tail skin graft survival and concomitantly reduced the TXB_2 content of the skin at and around the transplantation site, indicating that TXA_2 may contribute to the ischaemia component of the rejection process (Becker *et al.*, 1988, 1989). The similarity in the protective actions of BN 52021 and TXA_2 receptor antagonists on cardiac allograft rejection gives additional support to this suggestion (Foegh *et al.*, 1985). In addition, TXA_2 antagonists, but not PAF antagonists, inhibit platelet deposition in allografts, whereas neither class of compound influences lymphocyte accumulation (Khirabadi *et al.*, 1987). It remains possible that the PAF receptor antagonists modulate the function rather than the accumulation or proliferation of these cells. It has become apparent from the work of Foegh *et al.* (1987) that there is considerable overlap in the protective effects of corticosteroids and those of PAF antagonists or inhibitors and blockers of the prorejection eicosanoids TXA_2, LTB_4, LTC_4 and LTD_4, suggesting that the corticosteroids act by inhibition of the latter substances.

In Lewis rats pre-immunized against Brown Norway (BN) rat antigens and bilaterally nephrectomized before transplantation of kidneys from BN rats, survival was prolonged by RP 48740 (Freiche *et al.*, 1990). In non-immunized recipients, RP 48740 prolonged survival to a lesser extent. The absence of the expected synergism of RP 48740 with CSA was ascribed to an effect of the PAF antagonist on the bioavailability of CSA (Freiche *et al.*, 1990).

In models of cardiac xenografts, SRI 63-441 is effective in prolonging survival, suggesting that PAF receptor antagonists may be useful alone or in combination with CSA in rejection processes involving immune complexes and complement activation (Makowka *et al.*, 1990).

4.9.2   Platelet-Activating Factor Release During Graft Rejection

In rabbits presensitized to donor antigens by multiple skin grafts, renal allografts were hyperacutely rejected (Ito *et al.*, 1984). The deposition of immune complexes and fixation of complement on the endothelium of the graft were accompanied by an early accumulation of platelets and by detectable PAF release within 5 min of revascularization. At 60 min the microvasculature was obliterated with PMNs (Ito *et al.*, 1984). Further evidence for the involvement of PAF in the hyperacute rejection was provided by the observation of PAF release from rabbit hearts perfused with buffer containing transplantation alloantibodies and fresh serum. The challenged hearts displayed conduction disturbances and decreased coronary flow and stopped beating within 30 min (Camussi *et al.*, 1987b). The PAF receptor antagonist SRI 63-072 partially reversed the reduction in coronary flow, decreased the rhythm disturbances and delayed cessation of heart beating until 50 min (Camussi *et al.*, 1987b).

Thus, a potentially important role of this mediator in transplant rejection is indicated by the actions of exogenous PAF, evidence of PAF release during the rejection process and the well-established protective effects of PAF receptor antagonists.

5. *Perspectives*

There is little doubt that PAF is released in and contributes to a number of life-threatening conditions, including cardiovascular shock, septicaemia and various forms of ischaemia. Similarly, there is considerable evidence for the participation of PAF in chronic inflammatory conditions, such as asthma. The presence of endogenous inhibitors of PAF in plasma and tissues/cells, high-affinity binding of PAF to albumin and other plasma proteins, and the relatively small amounts released suggest that, in less severe reactions, the actions of PAF will be highly localized. Ultimately, the evaluation of the importance of PAF in such conditions will be determined by clinical trials, many of which are currently underway.

The enormous effort to establish the actions of PAF has generated detailed knowledge of the respiratory, cardiovascular and renal actions, and the responses of isolated inflammatory cells, platelets and endothelial cells are well characterized. On the other hand, while it seems likely that PAF has a number of physiological roles, their elucidation has proved more difficult. Nevertheless, partial retention of PAF by all cell types capable of its synthesis, together with constitutive PAF synthesis by the gastrointestinal tract and uterine tissues, gives strong indications of a physiological role for this cell-associated PAF. In addition, the presence of PAF in lipoproteins in plasma and in lamellar bodies secreted from type II pneumocytes indicates the need to consider alternative biologically active forms of this unusual phospholipid mediator. The number and range of PAF receptor antagonists has proved particularly useful in studies examining pathophysiological roles of endogenously released PAF. However, PAF receptor antagonists may not be the most appropriate tool to investigate the function of cell-associated PAF, since the retained PAF may act in a receptor-independent manner or at specific sites not available to, or blocked by, existing PAF receptor antagonists. The availability of a selective inhibitor of PAF synthesis would greatly facilitate the definition of the role of endogenous PAF in both pathological and physiological processes.

6. *Acknowledgements*

The author's work cited in this review has been supported by grants from the National Health and Medical Research Council (Australia) and by the National Heart Foundation (Australia). I thank my colleagues, Dr A. Lopata, Lea Delbridge and George Grigoriadis, for their helpful comments on this manuscript.

7. *References*

Abisogun, A.O., Braquet, P. and Tsafriri, A. (1989). The involvement of platelet activating factor in ovulation. Science 243, 381–383.

Abraham, W.M., Stevenson, J.S. and Garrido, R. (1989). A possible role for PAF in allergen-induced late responses: Modification by a selective antagonist. J. Appl. Physiol. 66, 2351–2357.

Acharya, S.B. and MacIntyre, D.E. (1983). Platelet products and vascular PGI_2 production. Prostaglandins, Leukotrienes Med. 10, 73–81.

Acker, G., Braquet, P. and Mencia Huerta, J.M. (1989). Role of platelet-activating factor (PAF) in the inititation of the decidual reaction in the rat. J. Reprod. Fertil. 85, 623–629.

Ackerman, Z., Karmeli, F., Ligumsky, M. and Rachmilewitz, D. (1990). Enhanced gastric and duodenal platelet-activating factor and leukotriene generation in duodenal ulcer patients. Scand. J. Gastroenterol. 25, 925–934.

Adamson, L.M., Smart, Y.C., Stanger, J.D., Murdoch, R.N. and Roberts, T.K. (1987). Mechanistic studies of early pregnancy associated thrombocytopenia (EPAT) in the mouse. Am. J. Reprod. Immunol. Microbiol. 13, 117–120.

Adamus, W.S., Heuer, H., Meade, C.J., Frey, G. and Brecht, H.M. (1988). Inhibitory effect of oral WEB 2086, a novel selective PAF-acether antagonist, on *ex vivo* platelet aggregation. Eur. J. Clin. Pharmacol. 35, 237–240.

Adamus, W.S., Heuer, H.O., Meade, C.J. and Schilling, J.C. (1990). Inhibitory effects of the new PAF acether antagonist WEB-2086 on pharmacologic changes induced by PAF inhalation in human beings. Clin. Pharmacol. Ther. 47, 456–462.

Adler, K.B., Schwarz, J.E., Anderson, W.H. and Welton, A.F. (1987). Platelet activating factor stimulates secretion of mucin by explants of rodent airways in organ cultured. Exp. Lung Res. 13, 25–43.

Adnot, S., Lefort, J., Braquet, P. and Vargaftig, B.B. (1986a). Interference of the PAF-acether antagonist BN 52021 with endotoxin-induced hypotension in the guinea-pig. Prostaglandins 32, 791–802.

Adnot, S., Lefort, J., Lagente, V., Braquet, P. and Vargaftig, B.B. (1986b). Interference of BN 52021, a PAF-acether antagonist, with endotoxin-induced hypotension in the guinea-pig. Pharmacol. Res. Commun. 18 (Suppl.), 197–200.

Adnot, S., Joseph, D. and Vargaftig, B.B. (1987). Antagonists of PAF-acether do not suppress thrombin-induced aggregation of ADP-deprived and aspirin-treated human platelets. Agents Actions 21, 195–202.

Agrawal, D.K. and Townley, R.G. (1987). Effect of platelet activating factor on beta adrenoceptors in human being. Biochem. Biophys. Commun. 143, 1–6.

Alam, I. and Silver, M.J. (1986). Acetyltransferase activity in human platelet microsomes and washed platelets. Biochim. Biophys. Acta 884, 67–72.

Alam, I., Smith, J.B. and Silver, M.J. (1983). Human and rabbit platelets form platelet-activating factor in response to calcium ionophore. Thromb. Res. 30, 71–79.

Alecozay, A.A., Casslén, B.G., Riehl, R.M. *et al.* (1989). Platelet-activating factor in human luteal phase endometrium. Biol. Reprod. 41, 578–586.

Alessandri, M.G., Giovanni, L., Mian, M. *et al.*, (1988). PAF-induced histamine release in the isolated perfused rat kidney. Int. J. Tissue React. 10, 33–38.

Alloatti, G., Montrucchio, G., Mariano, F. *et al.*, (1986). Effect of platelet-activating factor (PAF) on human cardiac muscle. Int. Arch. Allergy Appl. Immunol. 79, 108–112.

Alloatti, G., Montrucchio, G., Mariano, F. Tetta, C., Emanuelli, G. and Camussi, G. (1987). Protective effect of verapamil on the cardiac and circulatory alterations induced by platelet-activating factor. J. Cardiovasc. Pharmacol. 9, 181–186.

Alonso, F., Sanchez Crespo, M. and Mato, J.M. (1982). Modulatory role of cyclic AMP in the release of platelet-activating factor from human polymorphonuclear leucocytes. Immunology 45, 493–500.

Altman, R., Scazziota, A., Rouvier, J. and Cacchione, R. (1986). Synergistic actions of PAF-acether and sodium arachidonate in human platelet aggregation. 1. Studies in normal human platelet rich plasma. Thromb. Res. 43, 103–111.

Amiel, M.L., Duquenne, C., Benveniste, J. and Testart, J. (1989). Platelet aggregating activity in human embryo culture media free of PAF-acether. Hum. Reprod. 4, 327–330.

Anderson, G.P. and Fennessy, M.R. (1988). Lipoxygenase metabolites mediate increased airways responsiveness to histamine after acute platelet activating factor exposure in the guinea pig. Agents Actions 24, 8–19.

Anderson, G.P., White, H.L. and Fennessy, M.R. (1988). Increased airways responsiveness to histamine induced by platelet activating factor in the guinea-pig: possible role of lipoxygenase metabolites. Agents Actions 24, 1–7.

Anderson, V.H., O'Donnell, M., Simko, B. and Welton, A.F. (1983). An *in vivo* model for measuring antigen-induced SRS-A mediated bronchoconstriction and plasma SRS-A levels in the guinea-pig. Br. J. Pharmacol. 78, 67–74.

Andersson, M. and Pipkorn, U. (1988). The effect of platelet activating factor on nasal hypersensitivity. Eur. J. Clin. Pharmacol. 35, 231–235.

Andressen, R. (1988). In "Progress in Biochemical Pharmacology", Vol 22 (ed. P. Braquet, H.K. Mangold and B.B. Vargaftig), pp. 118–131. Karger, Basel.

Angle, M.J., Jones, M.A., McManus, L.M., Pinckard, R.N. and Harper, M.J. (1988). Platelet-activating factor in the rabbit uterus during early pregnancy. J. Reprod. Fertil. 83, 711–722.

Archer, C.B., Frohlich, W., Page, C.W., Paul, W., Morley, J. and MacDonald, D.M. (1984a). Synergistic interaction between prostaglandins and PAF-acether in experimental animals and man. Prostaglandins 27, 495–501.

Archer, C.B., Page, C.P., Paul, W., Morley, J. and MacDonald, D.M. (1984b). Inflammatory characteristics of platelet activating factor (PAF-acether) in human skin. Br. J. Dermatol. 110, 45–50.

Archer, C.B., MacDonald, D.M., Morley, J., Page, C.P., Paul, W. and Sanjar, S. (1985a). Effects of serum albumin, indomethacin and histamine H_1-antagonists on PAF-acether-induced inflammatory responses in the skin of experimental animals and man. Br. J. Pharmacol. 85, 109–113.

Archer, C.B., Page, C.P., Morley, J. and MacDonald, D.M. (1985b). Accumulation of inflammatory cells in response to intracutaneous platelet activating factor (PAF-acether) in man. Br. J. Dermatol. 112, 285–290.

Archer, C.B., Page, C.B., Paul, W., Morley, J. and MacDonald, D.M. (1985c). Actions of disodium cromoglycate (DSCG) on human skin responses to histamine, codeine and Paf-acether. Agents Actions 16, 6–8.

Archer, C.B., Page, C.P., Paul, W., Morley, J. and MacDonald, D.M. (1985d). Inflammatory cell accumulation in response to intracutaneous Paf-acether: a mediator of acute and persistent inflammation? Br. J. Dermatol. 113 (Suppl. 28), 133–135.

Archer, C.B., Page, C.P., Paul, W., Morley, J. and MacDonald, D.M. (1985e). Inflammatory characteristics of PAF-acether in the skin of experimental animals and man. Int. J. Tissue React. 7, 363–365.

Archer, C.B., Cunningham, F.M. and Greaves, M.W. (1988). Actions of platelet activating factor (PAF) homologues and their combinations on neutrophil chemokinesis and cutaneous inflammatory responses in man. J. Invest. Dermatol. 91, 82–85.

Ardaillou, N., Hagege, J., Nivez. M.P., Ardaillou, R. and Schlondorff, D. (1985). Vasoconstrictor-evoked prostaglandin synthesis in cultured human mesangial cells. Am. J. Physiol. 248, F240–F246.

Arm, J.P. and Lee, T.H. (1989). The use of fish oil in bronchial asthma. Allergy Proc. 10, 185–187.

Armstead, W.M., Pourcyrous, M., Mirro, R., Leffler, C.W. and Busija, D.W. (1988). Platelet activating factor: a potent constrictor of cerebral arterioles in newborn pigs. Circ. Res. 62, 1–7.

Arnoux, B., Duval, D. and Benveniste, J. (1980). Release of platelet-activating factor (PAF-acether) from alveolar macrophages by the calcium ionophore A23187 and phagocytosis. Eur. J. Clin. Invest. 10, 437–441.

Arnoux, B., Durand, J., Rigaud, M., Vargaftig, B.B. and Benveniste, J. (1981). Release of platelet-activating factor (PAF-acether) and arachidonic acid metabolites from alveolar macrophages. Agents Actions 11, 555–556.

Arnoux, B., Jouvin Marche, E., Arnoux, A. and Benveniste, J. (1982). Release of PAF-acether from human blood monocytes. Agents Actions 12, 713–716.

Arnoux, B., Joseph, M., Simoes, M.H. *et al.* (1987). Antigenic release of paf-acether and beta-glucuronidase from alveolar macrophages of asthmatics. Bull. Eur. Physiopathol. Respir. 23, 119–124.

Arnoux, B., Denjean, A., Page, C.P., Nolibe, D., Morley, J. and Benveniste, J. (1988). Accumulation of platelets and eosinophils in baboon lung after paf-acether challenge. Inhibition by ketotifen. Am. Rev. Respir. Dis. 137, 855–860.

Bachelet, M., Masliah, J., Vargaftig, B.B., Bereziat, G. and Colard, O. (1986). Changes induced by PAF-acether in diacyl and ether phospholipids from guinea-pig alveolar macrophages. Biochim. Biophys. Acta 878, 177–183.

Bachelet, M., Adlofs, M.J., Masliah, J., Bereziat, G., Vargaftig, B.B. and Bonta, I.L. (1988). Interaction between PAF-acether and drugs that stimulate cyclic AMP in guinea-pig alveolar macrophages. Eur. J. Pharmacol. 149, 73–78.

Badr, K.F., DeBoer, D.K., Takahashi, K., Harris, R.C., Fogo, A. and Jacobson, H.R. (1989). Glomerular responses to platelet-activating factor in the rat: role of thromboxane A_2. Am. J. Physiol. 256, F35-F43.

Baer, P.G., and Cagen, L.M. (1987). Platelet activating factor vasoconstriction of dog kidney. Inhibition by alprazolam. Hypertension 9, 253–260.

Baggiolini, M. and Dewald, B. (1986). Stimulus amplification by PAF and LTB_4 in human neutrophils. Pharmacol. Res. Commun. 18 (Suppl.), 51–59.

Banks, J.B., Wykle, R.L., O'Flaherty, J.T. and Lumb, R.H. (1988). Evidence for protein-catalyzed transfer of platelet activating factor by macrophage cytosol. Biochim. Biophys. Acta 961, 48–52.

Barnes, P.J., (1989). New concepts in the pathogenesis of bronchial hyperresponsiveness and asthma. J. Allergy Clin. Immunol. 83, 1013–1026.

Barnes, P.J. and Chung, K.F. (1989). PAF antagonists in asthma. Lancet i, 1071–1072.

Barnes, P.J., Grandordy, B.M., Page, C.P., Rhoden, K.J. and Robertson, D.N. (1987). The effect of platelet activating factor on pulmonary beta-adrenoceptors. Br. J. Pharmacol. 90, 709–715.

Barnett, R., Goldwasser, P., Scharschmidt, L.A. and Schlondorff, D. (1986). Effects of leukotrienes on isolated rat glomeruli and cultured mesangial cells. Am. J. Physiol. 250, F838–F844.

Barret, M.L., Lewis, G.P., Ward, S. and Westwick, J. (1987). Platelet-activating factor induces interleukin-1 production from human adherent macrophages. Br. J. Pharmacol. 90, 113P.

Barrett, P.A., Butler, K.D., Morley, J., Page, C.P., Paul, W. and White, A.M. (1984). Inhibition by heparin of platelet accumulation *in vivo*. Thromb. Haemost. 51, 366–370.

Barthelson, R. and Valone, F. (1990). Interaction of platelet-activating factor with interferon-gamma in the stimulation of interleukin-1 production by human monocytes. J. Allergy Clin. Immunol. 86, 193–201.

Basran, G.S., Page, C.P., Paul, W. and Morley, J. (1982). Cromoglycate (DSCG) inhibits responses to platelet-activating factor (PAF-acether) in man: an alternative mode of action for DSCG in asthma? Eur. J. Pharmacol. 86, 143–144.

Basran, G.S., Page, C.P., Paul, W. and Morley, J. (1984). Platelet-activating factor: a possible mediator of the dual response to allergen? Clin. Allergy 14, 75–79.

Baud, L., Perez, J. and Ardaillou, R. (1986). Dexamethasone and hydrogen peroxide production by mesangial cells during phagocytosis. Am. J. Physiol. 250, F596-F604.

Bazan, H.E., Braquet, P., Reddy, S.. and Bazan, N.G. (1987). Inhibition of the alkali burn-induced lipoxygenation of arachidonic acid in the rabbit cornea *in vivo* by a platelet activating factor antagonist. J. Ocul. Pharmacol. 3, 357–365.

Bazill, G.W. and Dexter, T.M. (1989a). Platelet-activating factor antagonist L-652,731 inhibits thymidine transport. Biochem. Pharmacol. 38, 373–374.

Bazill, G.W. and Dexter, T.M. (1989b). An antagonist to platelet-activating factor counteracts the tumoricidal action of alkyl lysophospholipids. Biochem. Pharmacol. 38, 374–377.

Becker, K., Braquet, P. and Forster, W. (1988). Influence of the specific antagonist of PAF-acether, BN 52021, and of Ginkgo extract on the rejection of murine tail skin allografts and the PAF-acether mortality in mice in particular consideration of the role of TXB2. Biomed. Biochim. Acta 47, S165–S168.

Becker, K., Lueddeckens, G., Braquet, P. and Forster, W. (1989). Influence of the specific antagonist of PAF-acether, BN 52021, on the rejection of murine tail skin allografts and the PAF-acether mortality in mice in particular consideration of the role of TXB$_2$. Prog. Clin. Biol. Res. 301, 499–503.

Beijer, L., Botting, J., Crook, P., Oyekan, A.O., Page, C.P. and Rylander, R. (1987). The involvement of platelet-activating factor in endotoxin-induced pulmonary platelet recruitment in the guinea-pig. Br. J. Pharmacol. 92, 803–808.

Benedetto, C., Massobrio, M., Bertini, E., Abbondanza, M., Enrien, H. and Tetta, C. (1989). Reduced serum inhibition of platelet-activating factor in preeclampsia. Am. J. Obstet. Gynecol. 160, 100–104.

Benfenati, E., Macconi, D., Noris, M. *et al.* (1989). Development of a mass spectrometric method to quantitate platelet activating factor in mouse urine. J. Lipid Res. 30, 1977–1981.

Benveniste, J., Tence, M., Varenne, P., Bidault, J., Boullet, C. and Polonsky, J. (1979). [Semi-synthesis and proposed structure of platelet-activating factor (P.A.F.): PAF-acether an alkyl ether analog of lysophosphatidycholine.] C.R. Seances Acad. Sci. D 289, 1037–1040.

Benveniste, J., Boullet, C., Brink, C. and Labat, C. (1983). The actions of Paf-acether (platelet-activating factor) on guinea-pig isolated heart preparations. Br. J. Pharmacol. 80, 81–83.

Beneviste, J., Nunez, D., Duriez, P., Korth, R., Bidault, J. and Fruchart, J.C. (1988). Preformed PAF-acether and lyso PAF-acether are bound to blood lipoproteins. FEBS Lett. 226, 371–376.

Berdel, W.E. (1987). Ether lipids and analogs in experimental cancer therapy. A brief review of the Munich experience. Lipids 22, 970–973.

Berdel, W.E., Korth, R., Reichert, A. *et al.* (1987). Lack of correlation between cytotoxicity of agonists and antagonists of platelet activating factor in neoplastic cells and modulation of-paf-acether binding to platelets from humans *in vitro*. Anticancer Res. 7, 1181–1187.

Bergelson, L.D., Kulikov, V.I. and Muzia, G.I. (1985). Influence of platelet activation factor and prostaglandins on cholesterol esterification in human plasma. FEBS Lett. 190, 305–306.

Bernardini, R., Calogero, A.E., Ehrlich, Y.H., Brucke, T., Chrousos, G.P. and Gold, P.W. (1989). The alkyl-ether phospholipid platelet-activating factor is a stimulator of the hypothalamic-pituitary-adrenal axis in the rat. Endocrinology 125, 1067–1073.

Berstad, A., Almodovar, K., Weatherstone, R.G. and Hirschowitz, B.I. (1988). Effect of platelet-activating factor and its receptor antagonist, SRI 63–675, on gastric bleeding in rats. Scand. J. Gastroenterol 23, 738–742.

Berti, F., Omini, C., Rossoni, G. and Braquet, P. (1986). Protection by two ginkgolides, BN-52020 and BN-52021, against guinea-pig lung anaphylaxis. Pharmacol. Res. Commun. 18, 775–793.

Berti, F., Galli, G., Magni, F. *et al.* (1988). The PAF-acether receptor antagonist BN-52021 inhibits mediator release during guinea-pig active lung anaphylaxis. Pharmacol. Res. Commun. 20, 1047–1059.

Bethel, R.A., Curtis, S.P., Lien, D.C. *et al.* (1989). Effect of PAF on parasympathetic contraction of canine airways. J. Appl. Physiol. 66, 2629–2634.

Betz, S.J., Lotner, G.Z. and Henson, P.M. (1980). Generation and release of platelet-activating factor (PAF) from enriched preparations of rabbit basophils; failure of human basophils to release PAF. J. Immunol. 125, 2749–2755.

Billah, M.M. and Johnston, J.M. (1983). Identification of phospholipid platelet-activating factor (1-*O*-alkyl-2-acetyl-*sn*-glycero-3-phosphocholine) in human amniotic fluid and urine. Biochem. Biophys. Res. Commun. 113, 51–58.

Billah, M.M. and Siegel, M.I. (1984). Calmodulin antagonists inhibit formation of platelet-activating factor in stimulated human neutrophils. Biochem. Biophys. Res. Commun. 118, 629–635.

Billah, M.M., Bryant, R.W. and Siegel, M.I. (1985). Lipoxygenase products of arachidonic acid modulate biosynthesis of platelet-activating factor (1-*O*-alkyl-2-acetyl-*sn*-glycero-3-phosphocholine) by human neutrophils via phospholipase A$_2$. J. Biol. Chem. 6899–6906.

Binnaka, T., Yamaguchi, T., Hirohara, J., Hiramatsu, A., Mizuno, T. and Sameshima, Y. (1989). Gastric mucosal damage induced in rats by intravenous administration of platelet-activating factor. Scand. J. Gastroenterol. Suppl. 162, 67–70.

Birch, J., Brown, E., Calnan, C., Jessup, C.L., Jessup, R. and Wayne, M. (1988). Studies in the guinea-pig with ICI 185,282: a thromboxane A$_2$ receptor antagonist. J. Pharm. Pharmacol. 40, 706–710.

Birkle, D.L., Kurian, P., Braquet, P. and Bazan, N.G. (1988). Platelet-activating factor antagonist BN52021 decreases accumulation of free polyunsaturated fatty acid in mouse brain during ischemia and electroconvulsive shock. J. Neurochem. 51, 1900–1905.

Bjork, J. and Smedegard, G. (1983). Acute microvascular effects of PAF-acether, as studied by intravital microscopy. Eur. J. Pharmacol. 96, 87–94.

Blank, M.L., Snyder, F., Byers, L.W., Brooks, B. and Muirhead, E.E. (1979). Antihypertensive activity of an alkyl ether analog of phosphatidylcholine. Biochem. Biophys. Res. Commun. 90, 1194–1200.

Blank, M.L., Cress, E.A. and Snyder, F. (1984). A new class of antihypertensive neutral lipid: 1-alkyl-2-acetyl-*sn*-glycerols, a precursor of platelet activating factor. Biochem. Biophys. Res. Commun. 118, 344–350.

Blank, M.L., Lee, Y.J., Cress, E.A. and Snyder, F. (1988). Stimulation of the de novo pathway for the biosynthesis of platelet-activating factor (PAF) via cytidylyltransferase activation in cells with minimal endogenous PAF production. J. Biol. Chem. 263, 5656–5661.

Bonavida, B., Mencia Huerta, J.M. and Braquet, P. (1989a). Effect of platelet-activating factor on monocyte activation and production of tumor necrosis factor. Int. Arch. Allergy Appl. Immunol. 88, 157–160.

Bonavida, B., Mencia-Huerta, J.-M. and Braquet, P. (1989b). Effect of platelet-activating factor on monocyte activation and production of tumor necrosis factor. Int. Arch. Allergy Appl. Immunol. 88, 157–160.

Bonventre, J.V., Weber, P.C. and Gronich, J.H. (1988). PAF and PDGF increase cytosolic and phospholipase activity in mesangial cells. Am. J. Physiol. 254, F87–F94.

Boschetto, P., Roberts, N.M., Rogers, D.F. and Barnes, P.J. (1989). Effect of antiasthma drugs on microvascular leakage in guinea pig airways. Am. Rev. Respir. Dis. 139, 416–421.

Bossant, M.J., Farinotti, R., De Maack, F., Mahuzier, G., Benveniste, J. and Ninio, E. (1989). Capillary gas chromatography and tandem mass spectrometry of paf-acether and analogs: absence of 1-O-alkyl-2-propionyl-sn-glycero-3-phosphocholine in human polymorphonuclear neutrophils. Lipids 24, 121–124.

Bossant, M.J., Ninio, E., Delautier, D. and Benveniste, J. (1990). Bioassay of paf-acether by rabbit platelet aggregation. Methods Enzymol. 187, 125–130.

Boughton-Smith, N.K., Hutcheson, I. and Whittle, B.J. (1989). Relationship between PAF-acether and thromboxane A_2 biosynthesis in endotoxin-induced intestinal damage in the rat. Prostaglandins 38, 319–333.

Bourgain, R.H., Maes, L., Braquet, P., Andries, R., Touqui, L. and Braquet, M. (1985). The effect of 1-O-alkyl-2-acetyl-sn-glycero-3-phosphocholine (paf-acether) on the arterial wall. Prostaglandins 30, 185–197.

Bourgain, R.H., Maes, L., Andries, R. and Braquet, P. (1986a). Thrombus induction by endogenic paf-acether and its inhibition by *Ginkgo biloba* extracts in the guinea pig. Prostaglandins 32, 142–144.

Bourgain, R.H., Maes, L., Andries, R. and Braquet, P. (1986b). *In vivo* effects of PAF-acether and arterial thrombosis. Adv. Exp. Med. Biol. 200, 439–447.

Bourgain, R.H., Andries, R., Bourgain, C. and Braquet, P. (1987). Arterial obstruction induced by PAF-acether (1-O-alkyl-sn-glycero-3-phosphocholine). Adv. Exp. Med. Biol. 215, 209–214.

Boxer, L.A., Yasaka, T., Butterick, C.J., Tzeng, D.Y. and Baehner, R.L. (1983). Comparative studies of functional characteristics of mononuclear cell subsets and granulocytes. Am. J. Pediatr. Hematol. Oncol. 5, 181–188.

Brambilla, A., Ghiorzi, A. and Giachetti, A. (1987). WEB 2086. A potent PAF antagonist exerts protective effect toward PAF-induced gastric damage. Pharmacol. Res. Commun. 19, 147–151.

Braquet, P. and Rola-Pleszynski, M. (1987). The role of Paf in immunological responses: a review. Prostaglandins 34, 143–147.

Braquet, P., Touqui, L., Shen, T.Y. and Vargaftig, B.B. (1987). Perspectives in platelet-activating factor research. Pharmacol. Rev. 39, 97–145.

Braquet, P., Paubert Braquet, M., Koltai, M., Bourgain, R., Bussolino, F. and Hosford, D. (1989a). Is there a case for PAF antagonists in the treatment of ischemic states? Trends. Pharmacol. Sci. 10, 23–30.

Braquet, P., Paubert-Braquet, M., Bourgain, R.H., Bussolino, F. and Hosford, D. (1989b). PAF/cytokine auto-generated feedback networks in microvascular immune injury: consequences in shock, ischaemia and graft rejection. J. Lipid Mediators 1, 75–112.

Bratton, D.L., Harris, R.A., Clay, K.L. and Henson, P.M. (1988a). Effects of platelet activating factor and related lipids on phase transition of dipalmitoylphosphatidylcholine. Biochim. Biophys. Acta 941, 76–82.

Bratton, D.L., Harris, R.A., Clay, K.L. and Henson, P.M. (1988b). Effects of platelet activating factor on calcium-lipid interactions and lateral phase separations in phospholipid vesicles. Biochim. Biophys. Acta 943, 211–219.

Brecht, H.M., Adamus, W.S., Heuer, H.O., Birke, F.W. and Kempe, E.R. (1991). Pharmacodynamics, pharmacokinetics and safety profile of the new platelet-activating factor antagonist Apafant in man. Arzneimittelforschung 41, 51–59.

Breviario, F., Bertocchi, F., Dejana, E. and Bussolino, F. (1988). IL-1-induced adhesion of polymorphonuclear leukocytes to cultured human endothelial cells. Role of platelet-activating factor. J. Immunol. 141, 3391–3397.

Brock, T.A. and Gimbrone, M.A.J. (1986). Platelet activating factors alters calcium homeostasis in cultured vascular endothelial cells. Am. J. Physiol. 250, H1086-H1092.

Bruijnzeel, P.L., Kok, P.T., Hamelink, M.L., Kijne, A.M. and Verhagen, J. (1987). Platelet-activating factor induces leukotriene C_4 synthesis by purified human eosinophils. Prostaglandins 34, 205–214.

Bruijnzeel, P.L., Warringa, R.A. and Kok. P.T. (1989a). Inhibition of platelet-activating factor- and zymosan-activated serum-induced chemotaxis of human neutrophils by nedocromil sodium, BN 52021 and sodium cromoglycate. Br. J. Pharmacol. 97, 1251–1257.

Bruijnzeel, P.L., Warringa, R.A., Kok, P.T., Hamelink, M.L. and Kreukniet, J. (1989b). Inhibitory effects of nedocromil sodium on the *in vitro* induced migration and leukotriene formation of human granulocytes. Drugs 37(Suppl. 1), 9–18.

Bruijnzeel, P.L., Warringa, R.A., Kok, P.T. and Kreukniet, J. (1990). Inhibition of neutrophil and eosinophil induced chemotaxis by nedocromil sodium and sodium cromoglycate. Br. J. Pharmacol. 99, 798–802.

Brunelleschi, S., Haye Legrand, I., Labat, C., Norel, X., Benveniste, J. and Brink, C. (1987). Platelet-activating factor-acether-induced relaxation of guinea pig airway muscle: role of prostaglandin E_2 and the epithelium. J. Pharmacol. Exp. Ther. 243, 356–363.

Brunelleschi, S., Renzi, D., Ledda, F. *et al.* (1989). Interference of WEB 2086 and BN 52021 with Paf-induced effects on guinea-pig trachea. Br. J. Pharmacol. 97, 469–474.

Bruynzeel, P.L., Koenderman, L., Kok, P.T., Hameling, M.L. and Verhagen, J. (1986). Platelet-activating factor (PAF-acether) induced leukotriene C_4 formation and luminol dependent chemiluminescence by human eosinophils. Pharmacol. Res. Commun. 18(Suppl.), 61–69.

Buckley, T.L., Brain, S.D. and Williams, T.J. (1990). Ruthenium red selectively inhibits oedema formation and increased blood flow induced by capsaicin in rabbit skin. Br. J. Pharmacol. 99, 7–8.

Burhop, K.E., Garcia, J.G., Selig, W.M. *et al.* (1986a). Platelet-activating factor increases lung vascular permeability to protein. J. Appl. Physiol. 61, 2210–2217.

Burhop, K.E., Van Der Zee, H., Bizios, R., Kaplan, J.E. and Malik, A.B. (1986b). Pulmonary vascular response to platelet-activating factor in awake sheep and the role of cyclooxygenase metabolites. Am. Rev. Respir. Dis. 134, 548–554.

Burka, J.F., Briand, H., Scott Savage, P. and Pasutto, F.M. (1989). Leukotriene D_4 and platelet-activating factor-acether antagonists on allergic and arachidonic acid-induced reactions in guinea pig airways. Can. J. Physiol. Pharmacol. 67, 483–490.

Burtin, C., Noirot, C., Scheinmann, P., Paupe, J. and Benveniste, J. (1990). Decreased skin response to intradermal platelet-activating factor (PAF-acether) in cancer patients. J. Allergy. Clin. Immunol. 86, 418–419.

Bushfield, M., McNicol, A. and MacIntyre, D.E. (1985). Inhibition of platelet-activating-factor-induced human platelet activation by prostaglandin D_2. Differential sensitivity of platelet transduction processes and functional responses to inhibition by cyclic AMP. Biochem. J. 232, 267–271.

Bussolino, F., Foa, R., Malavasi, F., Ferrando, M.L. and Camussi, G. (1984). Release of platelet-activating factor (PAF)-like material from human lymphoid cell lines. Exp. Hematol. 12, 688–693.

Bussolino, F., Biffignandi, P. and Arese, P. (1986a). Platelet-activating factor – a powerful lipid autacoid possibly involved in microangiopathy. Acta Haematol. Basel 75, 129–140.

Bussolino, F., Brevario, F., Tetta, C., Aglietta, M., Mantovani, A. and Dejana, E. (1986b). Interleukin 1 stimulates platelet-activating factor production in cultured human endothelial cells. J. Clin. Invest. 77, 2027–2033.

Bussolino, F., Breviario, F., Tetta, C. et al. (1986c). Interleukin 1 stimulates platelet activating factor production in cultured human endothelial cells. Pharmacol. Res. Commun. 18(Suppl.), 133–137.

Bussolino, F., Gremo, F., Tetta, C., Pescarmona, G.P. and Camussi, G. (1986d). Production of platelet-activating factor by chick retina. J. Biol. Chem. 261, 16502–16508.

Bussolino, F., Breviario, F., Aglietta, M., Sanavio, F., Bosia, A. and Dejana, E. (1987a). Studies on the mechanism of interleukin 1 stimulation of platelet activating factor synthesis in human endothelial cells in culture. Biochim. Biophys. Acta 927, 43–54.

Bussolino, F., Camussi, G., Aglietta, M. et al. (1987b). Human endothelial cells are target for platelet-activating factor. I. Platelet-activating factor induces changes in cytoskeleton structures. J. Immunol. 139, 2439–2446.

Bussolino, F., Camussi, G. and Baglioni, C. (1988a). Synthesis and release of platelet-activating factor by human vascular endothelial cells treated with tumor necrosis factor or interleukin 1α. J. Biol. Chem. 263, 11856–11861.

Bussolino, F., Pescarmona, G., Camussi, G. and Gremo, F. (1988b). Acetylcholine and dopamine promote the production of platelet activating factor in immature cells of chick embryonic retina. J. Neurochem. 51, 1755–1759.

Calhoun, W., Chang, J. and Carlson, R.P. (1987). Effect of selected antiinflammatory agents and other drugs on zymosan, arachidonic acid, PAF and carrageenan induced paw edema in the mouse. Agents Actions 21, 306–309.

Calignano, A., Cirino, G., Meli, R. and Persico, P. (1988). Isolation and identification of platelet-activating factor in UV-irradiated guinea pig skin. J. Pharmacol. Methods 19, 89–91.

Calogero, A.E., Bernardini, R., Gold, P.W. and Chrousos, G.P. (1988). Regulation of rat hypothalmic corticotropin-releasing hormone secretion in vitro: potential clinical implications. Adv. Exp. Med. Biol. 245, 167–181.

Camussi, G., Bosio, D., Segoloni, G. Tetta, C. and Vercellone, A. (1978a). Evidence for the involvement of the IgE-basophil-mastocyte system in human acute post-streptococcal glomerulonephritis. Ric. Clin. Lab. 8, 56–64.

Camussi, G., Tetta, C., Segoloni, G. and Vercellone, A. (1978b). Immunological mechanisms of human platelet involvement. Ric. Clin. Lab. 8, 262–272.

Camussi, G., Bussolino, F., Tetta, C., Brusca, R. and Ragni, R. (1980). The binding of platelet-activating factor (PAF) to polymorphonuclear neutrophils (PMN) as a trigger for the immune-induced PMN aggregation. Panminerva Med. 22, 1–5.

Camussi, G., Aglietta, M., Coda, R., Bussolino, F., Piacibello, W. and Tetta, C. (1981a). Release of platelet-activating factor (PAF) and histamine. II. The cellular origin of human PAF: monocytes, polymorphonuclear neutrophils and basophils. Immunology 42, 191–199.

Camussi, G., Tetta, C., Bussolino, F. et al. (1981b). Mediators of immune-complex-induced aggregation of polymorphonuclear neutrophils. II. Platelet-activating factor as the effector substance of immune-induced aggregation. Int. Arch. Allergy Appl. Immunol. 64, 25–41.

Camussi, G., Tetta, C., Segoloni, G., Chiara Deregibus, M. and Bussolino, F. (1981c). Neutropenia induced by platelet-activating factor (PAF-acether) released from neutrophils: the inhibitory effect of prostacyclin (PGI$_2$). Agents Actions 11, 550–553.

Camussi, G., Bussolino, F., Ghezzo, F. and Pegoraro, L. (1982). Release of platelet-activating factor from HL-60 human leukemic cells following macrophage-like differentiation. Blood 59, 16–22.

Camussi, G., Aglietta, M., Malavasi, F. et al. (1983a). The release of platelet-activating factor from human endothelial cells in culture. J. Immunol. 131, 2397–2403.

Camussi, G., Montrucchio, G., Antro, C., Bussolino, F., Tetta, C. and Emanuelli, G. (1983b). Platelet-activating factor-mediated contraction of rabbit lung strips: pharmacologic modulation. Immunopharmacology 6, 87–96.

Camussi, G., Pawlowksi, I., Bussolino, F., Caldwell, P.R., Brentjens, J. and Andres, G. (1983c). Release of platelet activating factor in rabbits with antibody-mediated injury of the lung: the role of leukocytes and of pulmonary endothelial cells. J. Immunol. 131, 1802–1807.

Camussi, G., Pawlowski, I., Tetta, C. et al. (1983d). Acute lung inflammation induced in the rabbit by local instillation of 1-O-octadecyl-2-acetyl-sn-glyceryl-3-phosphorylcholine or of native platelet-activating factor. Am. J. Pathol. 112, 78–88.

Camussi, G., Tetta, C. and Bussolino, F. (1983e). Inhibitory effect of prostacyclin (PGI$_2$) on neutropenia induced by intravenous injection of platelet-activating-factor (PAF) in the rabbit. Prostaglandins 25, 343–351.

Camussi, G., Alloatti, G., Montrucchio, G., Meda, M. and Emanuelli, G. (1984). Effect of platelet activating factor on guinea-pig papillary muscle. Experientia 40, 697–699.

Camussi, G., Tetta, C., Alberton, M. et al. (1985). The role of platelet-activating factor in experimental immune complex pathology. Int. J. Tissue. React. 7, 355–362.

Camussi, G., Bussolino, F., Salvidio, G. and Baglioni, C. (1987a). Tumor necrosis factor/cachectin stimulates peritoneal macrophages, polymorphonuclear neutrophils, and vascular endothelial cells to synthesize and release platelet-activating factor. J. Exp. Med. 166, 1390–1404.

Camussi, G., Niesen, N., Tetta, C., Saunders, R.N. and Milgrom, F. (1987b). Release of platelet-activating factor from rabbit heart perfused in vitro by sera with transplantation alloantibodies. Transplantation 44, 113–118.

Camussi, G., Pawlowski, I., Saunders, R., Brentjens, J. and Andres, G. (1987c). Receptor antagonist of platelet activating factor inhibits inflammatory injury induced by *in situ* formation of immune complexes in renal glomeruli and in the skin. J. Lab. Clin. Med. 110, 196–206.

Camussi, G., Turello, E., Tetta, C., Bussolino, F. and Baglioni, C. (1990). Tumor necrosis factor induces contraction of mesangial cells and alters their cytoskeletons. Kidney Int. 38, 795–802.

Caplan, M.S., Hsueh, W., Sun, X.M., Gidding, S.S. and Hageman, J.R. (1990). Circulating plasma platelet activating factor in persistent pulmonary hypertension of the newborn. Am. Rev. Respir. Dis. 142, 1258–1262.

Cargill, D.I., Cohen, D.S., Van Valen, R.G., Klimek, J.J. and Levin, R.P. (1983). Aggregation, release and desensitization induced in platelets from five species by platelet activating factor (PAF). Thromb. Haemost. 49, 204–207.

Carolan, E.J. and Casale, T.B. (1990). Degree of platelet activating factor-induced neutrophil migration is dependent upon the molecular species. J. Immunol. 145, 2561–2565.

Casals Stenzel, J. (1987a). Protective effect of WEB 2086, a novel antagonist of platelet activating factor, in endotoxin shock. Eur. J. Pharmacol. 135, 117–122.

Casals Stenzel, J. (1987b). Effects of WEB 2086, a novel antagonist of platelet activating factor, in active and passive anaphylaxis. Immunopharmacology 13, 117–124.

Casals Stenzel, J. and Heuer, H. (1988). Pharmacology of PAF antagonists. Prog. Biochem. Pharmacol. 22, 58–65.

Casals Stenzel, J. and Weber, K.H. (1987). Triazolodiazepines: dissociation of their Paf (platelet activating factor) antagonistic and CNS activity. Br. J. Pharmacol. 90, 139–146.

Casals Stenzel, J., Franke, J., Friedrich, T. and Lichey, J. (1987). Bronchial and vascular effects of Paf in the rat isolated lung are completely blocked by WEB 2086, a novel specific Paf antagonist. Br. J. Pharmacol. 91, 799–802.

Cattaneo, M., Canciani, M.T. and Mannucci, P.M. (1985). Human platelet aggregation and release reaction induced by platelet activating factor (PAF-acether) – effects of acetylsalicylic acid and external ionized calcium. Thromb. Haemost. 53, 221–224.

Cervoni, P., Herzlinger, H.E., Lai, F.M. and Tanikella, T.K. (1983). Aortic vascular and atrial responses to (+/−)-1-*O*-octadecyl-2-acetyl-glyceryl-3-phosphorylcholine. Br. J. Pharmacol. 79, 667–671.

Chamone, D.F., Fujimura, A.Y. and Jamra, M. (1986). Increased levels of eosinophil-derived PAF-acether activity in a patient with idiopathic hypereosinophilic syndrome. Braz. J. Med. Biol. Res. 19, 135–136.

Chand, N., Diamantis, W. and Sofia, R.D. (1988). Effect of paf-acether on isoprenaline-induced relaxation in isolated tracheal segments of rats and guinea pigs. Eur. J. Pharmacol. 158, 135–137.

Chang, S.W., Feddersen, C.O., Henson, P.M. and Voelkel, N.F. (1987). Platelet-activating factor mediates hemodynamic changes and lung injury in endotoxin-treated rats. J. Clin. Invest. 79, 1498–1509.

Chang, S.W., Lauterburg, B.H. and Voelkel, N.F. (1988). Endotoxin causes neutrophil-independent oxidative stress in rats. J. Appl. Physiol. 65, 358–367.

Chang, S.W., Ohara, N., Kuo, G. and Voelkel. N.F. (1989). Tumor necrosis factor-induced lung injury is not mediated by platelet-activating factor. Am. J. Physiol. 257, L232–L239.

Chen, C.R., Voelkel, N.F. and Chang, S.W. (1990). PAF potentiates protamine-induced lung edema: role of pulmonary venoconstriction. J. Appl. Physiol. 68, 1059–1068.

Chesney, C.M., Pifer, D.D., Byers, L.W. and Muirhead, E.E. (1982). Effect of platelet-activating factor (PAF) on human platelets. Blood 59, 582–585.

Chignard, M., Selak, M.A. and Smith, J.B. (1986). Direct evidence for the existence of a neutrophil-derived platelet activator (neutrophilin). Proc. Natl Acad. Sci. USA 83, 8609–8613.

Christman, B.W., Lefferts, P.L. and Snapper, J.R. (1987). Effect of platelet-activating factor on aerosol histamine responsiveness in awake sheep. Am. Rev. Respir. Dis. 135, 1267–1270.

Christman, B.W., Lefferts, P.L., King, G.A. and Snapper, J.R. (1988). Role of circulating platelets and granulocytes in PAF-induced pulmonary dysfunction in awake sheep. J. Appl. Physiol. 64, 2033–2041.

Christman, B.W., Lefferts, P.L., Blair, I.A. and Snapper, J.R. (1990). Effect of platelet-activating factor receptor antagonism on endotoxin-induced lung dysfunction in awake sheep. Am. Rev. Respir. Dis. 142, 1272–1278.

Chu, K.M., Gerber, J.G. and Nies, A.S. (1988). Local vasodilator effect of platelet activating factor in the gastric, mesenteric and femoral arteries of the dog. J. Pharmacol. Exp. Ther. 246, 996–1000.

Chung, K.F. (1990). Mediators of bronchial hyperresponsiveness. Clin. Exp. Allergy 20, 453–458.

Chung, K.F., Aizawa, H., Leikauf, G.D., Ueki, I.F., Evans, T.W. and Nadel, J.A. (1986). Airway hyperresponsiveness induced by platelet-activating factor: role of thromboxane generation. J. Pharmacol. Exp. Ther. 236, 580–584.

Chung, K.F., Dent, G., McCusker, M., Guinot, P., Page, C.P. and Barnes, P.J. (1987). Effect of a ginkgolide mixture (BN 52063) in antagonising skin and platelet responses to platelet activating factor in man. Lancet i, 248–251.

Chung, K.F., Minette, P., McCusker, M. and Barnes, P.J. (1988). Ketotifen inhibits the cutaneous but not the airway responses to platelet-activating factor in man. J. Allergy Clin. Immunol. 81, 1192–1198.

Chung, K.F., Cuss, F.M. and Barnes, P.J. (1989a). Platelet activating factor: effects on bronchomotor tone and bronchial responsiveness in human beings. Allergy Proc. 10, 333–337.

Chung, K.F., Lammers, J.W., McCusker, M., Roberts, N.M., Nichol, G.M. and Barnes, P.J. (1989b). Effect of theophylline on airway responses to inhaled platelet-activating factor in man. Eur. Respir. J. 2, 763–768.

Chung, K.F., Rogers, D.F., Barnes, P.J. and Evans, T.W. (1990). The role of increased airway microvascular permeability and plasma exudation in asthma. Eur. Respir. J. 3, 329–337.

Clay, K.L., Murphy, R.C., Andres, J.L., Lynch, J. and Henson, P.M. (1984). Structure elucidation of platelet activating factor derived from human neutrophils. Biochem. Biophys. Res. Commun. 121, 815–825.

Clay, K.L., Johnson, C. and Henson, P. (1990). Binding of platelet activating factor to albumin. Biochim. Biophys. Acta 1046, 309–314.

Cluzel, M., Undem, B.J. and Chilton, F.H. (1989). Release of platelet-activating factor and the metabolism of leukotriene B_4 by the human neutrophil when studied in a cell superfusion model. J. Immunol. 143, 3659–3665.

Cockcroft, D.W., Ruffin, R.E., Dolovich, J. and Hargreave, F.E. (1977). Allergen-induced increase in non-allergic bronchial reactivity. Clin. Allergy 7, 503–513.

Codde, J.P., Beilin, L.J., Croft, K.D. and Vandongen, R. (1987). Effects of altered prostanoid and lyso-PAF synthesis by marine oil diets on blood pressure of salt loaded spontaneously hypertensive rats. Agents Actions Suppl. 22, 101–109.

Coeffier, E., Ninio, E., Le Couedic, J.P. and Chignard, M. (1986). Transient activation of the acetyltransferase necessary for paf-acether biosynthesis in thrombin-activated platelets. Br. J. Haematol. 62, 641–651.

Coeffier, E., Joseph, D., Prevost, M.C. and Vargaftig, B.B. (1987). Platelet–leukocyte interaction: activation of rabbit platelets by FMLP-stimulated neutrophils. Br. J. Pharmacol. 92, 393–406.

Coeffier, E., Delautier, D., Le Couedic, J.P., Chignard, M., Denizot, Y. and Benveniste, J. (1990). Cooperation between platelets and neutrophils for paf-acether (platelet-activating factor) formation. J. Leukocyte Biol. 47, 234–243.

Colditz, I.G. and Movat, H.Z. (1984). Kinetics of neutrophil accumulation in acute inflammatory lesions induced by chemotaxins and chemotaxinigens. J. Immunol. 133, 2169–2173.

Collier, M., O'Neill, C., Ammit, A.J. and Saunders, D.M. (1988). Biochemical and pharmacological characterization of human embryo-derived platelet activating factor. Hum. Reprod. 3, 993–998.

Collier, M., O'Neill, C., Ammit, A.J. and Saunders, D.M. (1990). Measurement of human embryo-derived platelet-activating factor (PAF) using a quantitative bioassay of platelet aggregation. Hum. Reprod. 5, 323–328.

Conrad, G.W. and Rink, T.J. (1986). Platelet activating factor raises intracellular calcium ion concentration in macrophages. J. Cell Biol. 103, 439–450.

Conroy, D.M., Samhoun, M.N. and Piper, P.J. (1990). Relaxations of guinea-pig isolated trachea induced by platelet-activating factor are epithelial-dependent and are antagonized by WEB 2086. Eur. J. Pharmacol. 186, 315–318.

Cordeiro, R.S., Martins, M.A., Silva, P.M., Faria Neto, H.C., Castanheira, J.R. and Vargaftig, B.B. (1986). Desensitization to PAF-induced rat paw oedema by repeated intraplantar injections. Life Sci. 39, 1871–1878.

Cotter, J.L., Vandongen, R., Burton, D.L. and Sturm, M.J. (1990). Platelet-activating factor and one-kidney, one clip hypertension. Hypertension 15, 628–632.

Cox, C.P., Linden, J. and Said, S.I. (1984). VIP elevates platelet cyclic AMP (cAMP) levels and inhibits in vitro platelet activation induced by platelet-activating factor (PAF). Peptides 5, 325–328.

Coyle, A.J., Urwin, S.C., Page, C.P., Touvay, C., Villain, B. and Braquet, P. (1988). The effect of the selective PAF antagonist BN 52021 on PAF- and antigen-induced bronchial hyper-reactivity and eosinophil accumulation. Eur. J. Pharmacol. 148, 51–58.

Coyle, A.J., Spina, D. and Page, C.P. (1990). Paf-induced bronchial hyperresponsiveness in the rabbit: contribution of platelets and airways smooth muscle. Br. J. Pharmacol. 101, 31–38.

Croft, K.D., Sturm, M.J., Codde, J.P., Vandongen, R. and Beilin, L.J. (1986). Dietary fish oils reduce plasma levels of platelet activating factor precursor (lyso-PAF) in rats. Life Sci. 38, 1875–1882.

Csato, M. and Czarnetzki, B.M. (1988). Effect of BN 52021, a platelet activating factor antagonist, on experimental murine contact dermatitis. Br. J. Dermatol. 118, 475–479.

Cucala, M., Wallace, J.L., Salas, A., Guarner, F., Rodriguez, R. and Malagelada, J.R. (1989a). Central regulation of gastric acid secretion by platelet-activating factor in anesthesized rats. Prostaglandins 37, 275–285.

Cucala, M., Wallace, J.L., Salas, A., Guarner, F., Rodriguez, R. and Malagelada, J.R. (1989b). CNS effects of platelet-activating factor on the rat stomach. Methods Find. Exp. Clin. Pharmacol. 11(Suppl. 1), 67–71.

Cunningham, F.M. (1989). In "Platelet-Activating Factor and Human Disease" (ed P.J. Barnes, C.P. Page and P.M. Henson), pp 231–249. Blackwell, Oxford.

Cuss, F.M., Dixon, C.M. and Barnes, P.J. (1986). Effects of inhaled platelet activating factor on pulmonary function and bronchial responsiveness in man. Lancet ii, 189–192.

Czarneztki, B.M. (1982). Effect of platelet activating factor on leukocytes. II. Enhancement of eosinophil chemotactic factor and beta-glucuronidase release. Chem. Phys. Lipids 31, 205–211.

Czarnetzki, B. (1983). Increased monocyte chemotaxis towards leukotriene B_4 and platelet activating factor in patients with inflammatory dermatoses. Clin. Exp. Immunol. 54, 486–492.

Czarnetzki, B.M. and Benveniste, J. (1981a). Effect of synthetic PAF-acether on human neutrophil function. Agents Actions 11, 549–550.

Czarnetzki, B.M. and Benveniste, J. (1981b). Effect of 1-O-octadecyl-2-O-acetyl-sn-glycero-3-phosphocholine (PAF-acether) on leukocytes I. Analysis of the in vitro migration of human neutrophils. Chem. Phys. Lipids 29, 317–326.

Czarnetzki, B.M. and Csato, M. (1989). Comparative studies of human eosinophil migration towards platelet-activating factor and leukotriene B_4. Int. Arch. Allergy Appl. Immunol. 88, 191–193.

D'Humieres, S., Russo Marie, F. and Vargaftig, B.B. (1986). PAF-acether-induced synthesis of prostacyclin by human endothelial cells. Eur. J. Pharmacol. 131, 13–19.

Dahl, M.L. (1985). Aggregating and prostanoid-releasing effects of platelet-activating factor and leukotrienes on human polymorphonuclear leukocytes and platelets. Int. Arch. Allergy Appl. Immunol. 76, 145–150.

Dallob, A., Guindon, Y. and Goldenberg, M.M. (1987). Pharmacological evidence for a role of lipoxygenase products in platelet-activating factor (PAF)-induced hyperalgesia. Biochem. Pharmacol. 36, 3201–3204.

Danhauser, S., Berdel, W.E., Schick, H.D. et al. (1987). Structure–cytotoxicity studies on alkyl lysophospholipids and some analogs in leukemic blasts of human origin in vitro. Lipids 22, 911–915.

Darius, H., Lefer, D.J., Smith, J.B. and Lefer, A.M. (1986a). Role of platelet-activating factor-acether in mediating guinea pig anaphylaxis. Science 232, 58–60.

Darius, H., Smith, J.B. and Lefer, A.M. (1986b). Inhibition of the platelet activating factor mediated component of guinea pig anaphylaxis by receptor antagonists. Int. Arch. Allergy Appl. Immunol. 80, 369–375.

Davenas, E., Poitevin. B. and Benveniste, J. (1987). Effect on mouse peritoneal macrophages of orally administered very high dilutions of silica. Eur. J. Pharmacol. 135, 313–319.

Dejana, E., Breviario, F., Erroi, A. et al. (1987). Modulation of endothelial cell functions by different molecular species of interleukin 1. Blood 69, 695–699.

Demopoulos, C.A., Pinckard, R.N. and Hanahan, D.J. (1979). Platelet-activating factor. Evidence for 1-O-alkyl-2-acetyl-sn-glyceryl-3-phosphorylcholine as the active component (a new class of lipid chemical mediators). J. Biol. Chem. 254, 9355–9358.

Demopoulos, C.A., Tsabikakis, G.E. and Kapoulas, V.M. (1981). Intravascular pathobiology of acetyl glyceryl ether phosphorylcholine (AGEPC), a synthetic platelet-activating factor (PAF). I. Intravenous infusion in guinea pigs. Immunol. Lett. 3, 133–135.

Demopoulos, C.A., Andrikopoulos, N.K. and Stathopoulou Sparou, R. (1988). Abnormal platelet response to PAF and ADP in beta-thalassaemia. Int. J. Biochem. 20, 599–604.

DeNichilo, M.O., Stewart, A.G., Vadas, M.A. and Lopez, A.F. (1991). Granulocyte–macrophage colony-stimulating factor is a stimulant of platelet-activating factor and superoxide anion generation by human neutrophils. J. Biol. Chem. 266, 4896–4902.

Denjean, A., Arnoux, B., Benveniste, J., Lockhart, A. and Masse, R. (1981). Bronchoconstriction induced by intratracheal administration of platelet-activating factor (PAF-acether) in baboons. Agents Actions 11, 567–568.

Denjean, A., Arnoux, B., Masse, R., Lockhart, A. and Benveniste, J. (1983). Acute effects of intratracheal administration of platelet-activating factor in baboons. J. Appl. Physiol. 55, 799–804.

Dent, G., Ukena, D., Chanez, P., Sybrecht, G. and Barnes, P. (1989a). Characterization of PAF receptors on human neutrophils using the specific antagonist, WEB 2086: correlation between receptor binding and function. FEBS Lett. 244, 365–368.

Dent, G., Ukena, D., Sybrecht, G.W. and Barnes, P.J. (1989b). [^{3}H]WEB 2086 labels platelet activating factor receptors in guinea pig and human lung. Eur. J. Pharmacol. 169, 313–316.

Desquand, S., Touvay, C., Randon, J. et al. (1986). Interference of BN 52021 (ginkgolide B) with the bronchopulmonary effects of PAF-acether in the guinea-pig. Eur. J. Pharmacol. 127, 83–95.

Desquand, S., Lefort, J., Dumarey, C. and Vargaftig, B.B. (1990). The booster injection of antigen during active sensitization of guinea-pig modifies the anti-anaphylactic activity of the Paf antagonist WEB 2086. Br. J. Pharmacol. 100, 217–222.

Desquand, S., Lefort, J., Dumarey, C. and Vargaftig, B.B. (1991). Interference of BN 52021, an antagonist of PAF, with different forms of active anaphylaxis in the guinea-pig: importance of the booster injection. Br. J. Pharmacol. 102, 687–695.

Dewald, B. and Baggiolini, M. (1985). Activation of NADPH oxidase in human neutrophils. Synergism between fMLP and the neutrophil products PAF and LTB4. Biochem. Biophys. Res. Commun. 128, 297-304.

Dewald, B. and Baggiolini, M. (1986). Platelet-activating factor as a stimulus of exocytosis in human neutrophils. Biochim. Biophys. Acta 888, 42–48.

Dewar, A., Archer, C.B., Paul, W., Page, C.P., MacDonald, D.M. and Morley, J. (1984). Cutaneous and pulmonary histopathological responses to platelet activating factor (Paf-acether) in the guinea-pig. J. Pathol. 144, 25–34.

Dicorleto, P.E. and de la Motte, C.A. (1989). Thrombin causes increased monocytic-cell adhesion to endothelial cells through a protein kinase C-dependent pathway. Biochem. J. 264, 71–77.

Diez, J., Delpon, E. and Tamargo, J. (1990). Effects of platelet-activating factor on contractile force. Br. J. Pharmacol. 100, 305–311.

Dixon, E.J., Wilsoncroft, P., Robertson, D.N. and Page, C.P. (1989). The effect of Paf antagonists on bronchial hyper-responsiveness induced by Paf, propranolol or indomethacin. Br. J. Pharmacol. 97, 717–722.

Doebber, T.W. and Wu, M.S. (1987). Platelet-activating factor (PAF) stimulates the PAF-synthesizing enzyme acetyl-CoA:1-alkyl-sn-glycero-3-phosphocholine O$_2$-acetyltransferase and PAF synthesis in neutrophils. Proc. Natl Acad. Sci. USA 84, 7557–7561.

Doebber, T.W., Wu, M.S., Robbins, J.C., Choy, B.M., Chang, M.N. and Shen, T.Y. (1985). Platelet activating factor (PAF) involvement in endotoxin-induced hypotension in rats. Studies with PAF-receptor antagonist kadsurenone. Biochem. Biophys. Res. Commun. 127, 799–808.

Doly, M., Braquet, P., Bonhomme, B. and Meyniel. G. (1987). Effects of PAF-acether on electrophysiological response of isolated retina. Int. J. Tissue React. 9, 33–37.

Domingo, M.T., Spinnewyn, B., Chabrier, P.E. and Braquet, P. (1988). Presence of specific binding sites for platelet-activating factor (PAF) in brain. Biochem. Biophys. Res. Commun. 151, 730–736.

Dubois, C., Bissonnette, E. and Rola Pleszczynski, M. (1989). Platelet-activating factor (PAF) enhances tumor necrosis factor production by alveolar macrophages. Prevention by PAF receptor antagonists and lipoxygenase inhibitors. J. Immunol. 143, 964–970.

Dutoit, J., Salome, C.M. and Woolcock, A.J. (1987). Inhaled corticosteroids reduce the severity of bronchial hyperresponsiveness in asthma but oral theophylline does not. Am. Rev. Respir. Dis. 136, 1174–1178.

Egido, J., Robles, A., Ortiz, A. et al. (1987). Role of platelet-activating factor in adriamycin-induced nephropathy in rats. Eur. J. Pharmacol. 138, 119–123.

Emeis, J.J. and Kluft, C. (1985). PAF-acether-induced release of tissue-type plasminogen activator from vessel walls. Blood 66, 86–91.

Escolar, G., Navarro, C., Galmes, J.L., Casanovas, L.I. and Bulbena, O. (1989). Zinc acexamate reduces gastric damage induced by platelet-activating factor. Prostaglandins Leukot. Essent. Fatty Acids 38, 49–53.

Esplugues, J.V. and Whittle, B.J. (1988). Gastric mucosal damage induced by local intra-arterial administration of Paf in the rat. Br. J. Pharmacol. 93, 222–228.

Esplugues, J.V. and Whittle, B.J. (1989a). Mechanisms contributing to gastric motility changes induced by PAF-acether and endotoxin in rats. Am. J. Physiol. 256, G275–G282.

Esplugues, J.V. and Whittle, B.J. (1989b). Gastric effects of PAF. Methods Find. Exp. Clin. Pharmacol. 11(Suppl.1), 61–66.

Etienne, A., Hecquet, F., Soulard, C., Spinnewyn, B., Clostre, F. and Braquet, P. (1986a). In vivo inhibition of plasma protein leakage and Salmonella enteridis-induced mortality in the rat by a specific paf-acether antagonist: BN 52021. Agents Actions 17, 368–370.

Etienne, A., Hecquet, F., Soulard, C., Touvay, C., Clostre, F. and Braquet, P. (1986b). The relative role of PAF-acether and icosanoids in septic shock. Pharmacol. Res. Commun. 18(Suppl.), 71–79.

Etienne, A., Hecquet, F., Guilmard, C., Soulard, C. and Braquet, P. (1987). Inhibition of rat endotoxin-induced lethality by BN 52021 and BN 52063, compounds with PAF-acether antagonistic effect and protease-inhibitory activity. Int. J. Tissue React. 9, 19–26.

Etienne, A., Thonier, F., Hecquet, F. and Braquet, P. (1988). Role of neutrophils in gastric damage induced by platelet activating factor. Naunyn-Schmiedebergs Arch. Pharmacol. 338, 422–425.

Etienne, A., Thonier, F. and Braquet, P. (1989). Protective effect of the PAF-antagonist BN 52021 on several models of gastrointestinal mucosal damage in rats. Int. J. Tissue React. 11, 59–64.

Evans, T.W., Rogers, D.F., Aursudkij, B., Chung, K.F. and Barnes, P.J. (1988). Inflammatory mediators involved in antigen-induced airway microvascular leakage in guinea pigs. Am. Rev. Respir. Dis. 138, 395–399.

Evans, T.W., Rogers, D.F., Aursudkij, B., Chung, K.F. and Barnes, P.J. (1989). Regional and time-dependent effects of inflammatory mediators on airway microvascular permeability in the guinea pig. Clin. Sci. 76, 479–485.

Evans, T.W., McAnulty, R.J., Rogers, D.F., Chung, K.F., Barnes, P.J. and Laurent, G.J. (1990). Bleomycin-induced lung injury in the rat: effects of the platelet-activating factor (PAF) receptor antagonist BN 52021 and platelet depletion. Environ. Health Perspect. 85, 65–69.

Ezra, D., Laurindo, F.R., Czaja, J.F., Snyder, F., Goldstein, R.E. and Feuerstein, G. (1987). Cardiac and coronary consequences of intracoronary platelet activating factor infusion in the domestic pig. Prostaglandins 34, 41–57.

Fadel, R., David, B., Herpin Richard, N., Borgnon, A., Rassemont, R. and Rihoux, J.P. (1990). In vivo effects of cetirizine on cutaneous reactivity and eosinophil migration induced by platelet-activating factor (PAF-acether) in man. J. Allergy Clin. Immunol. 86, 314–320.

Faulk, W.P., Gargiulo, P., McIntyre, J.A. and Bang, N.U. (1989). Hemostasis and fibrinolysis in renal transplantation. Semin. Thromb. Hemost. 15, 88–98.

Fecchio, D., Russo, M., Sirois, P., Braquet, P. and Jancar, S. (1990). Inhibition of Ehrlich ascites tumor in vivo by PAF-antagonists. Int. J. Immunopharmacol. 12, 57–65.

Feliste, R., Perret, B., Braquet, P. and Chap. H. (1989). Protective effect of BN 52021, a specific antagonist of platelet-activating factor (PAF-acether) against diet-induced cholesteryl ester deposition in rabbit aorta. Atherosclerosis 78, 151–158.

Felix, S.B., Baumann, G., Ahmad, Z., Hashemi, T., Niemczyk, M. and Berdel, W.E. (1990a). Effects of platelet-activating factor on myocardial contraction and myocardial relaxation of isolated, perfused guinea-pig hearts. J. Cardiovasc. Pharmacol. 16, 750–756.

Felix, S.B., Steger, A., Baumann, G., Busch, R., Ochsenfeld, G. and Berdel, W.E. (1990b). Platelet-activating factor-induced coronary constriction in the isolated perfused guinea-pig heart and antagonistic effects of the Paf antagonist WEB 2086. J. Lipid Mediators 2, 9–20.

Felsen, D., Loo, M.H., Marion, D.N. and Vaughan, E.D.J. (1990). Involvement of platelet activating factor and thromboxane A_2 in the renal response to unilateral ureteral obstruction. J. Urol. 144, 141–145.

Fennessy, M., Anderson, G.P. and Stewart. A.G. (1987). Lipid mediators of anaphylaxis and increased airways reactivity. Clin. Exp. Pharmac. Physiol. 14, 393–399.

Ferguson Chanowitz, K.M., Katocs, A.S.J., Pickett, W.C. et al. (1990). Platelet-activating factor or a platelet-activating factor antagonist decreases tumor necrosis factor-alpha in the plasma of mice treated with endotoxin. J. Infect. Dis. 162, 1081–1086.

Fernandez Gallardo, S., Ortega, M.P., Priego, J.G., de Casa Juana, M.F., Sunkel, C. and Sanchez Crespo, M. (1990). Pharmacological actions of PCA 4248, a new platelet-activating factor receptor antagonist: in vivo studies. J. Pharmacol. Exp. Ther. 255, 34–39.

Ferrer Lopez, P., Renesto, P., Schattner, M., Bassot, S., Laurent, P. and Chignard, M. (1990). Activation of human platelets by C5a-stimulated neutrophils: a role for cathepsin G. Am. J. Physiol. 258, C1100–C1107.

Feuerstein, G. (1989). In "Platelet Activating Factor and Human Disease" (ed C.P. Page, P.J. Barnes and P.M. Henson), pp 138–157. Blackwell, Oxford.

Feuerstein, G. and Hallenbeck, J.M. (1987). Prostaglandins, leukotrienes, and platelet-activating factor in shock. Annu. Rev. Pharmacol. Toxicol. 27, 301–313.

Feuerstein, G. and Siren, A.L. (1988). Platelet-activating factor and shock. Prog. Biochem. Pharmacol. 22, 181–190.

Feuerstein, G., Boyd, L.M. and Goldstein, R.E. (1984). Effect of platelet-activating factor on the coronary circulation of the domestic pig. Am. J. Physiol. 246, 466–471.

Fiedler, V.B., Mardin, M. and Abram, T.S. (1987). Comparison of cardiac and hemodynamic effects of platelet-activating factor-acether and leukotriene D4 in anesthetized dogs. Basic Res. Cardiol. 82, 197–208.

Filep, J., Herman, F., Braquet, P. and Mozes, T. (1989). Increased levels of platelet-activating factor in blood following intestinal ischemia in the dog. Biochem. Biophys. Res. Commun. 158, 353–359.

Filep, J.G. and Foldes-Filep, E. (1990). Inhibition by calcium channel blockers of the binding of platelet-activating factor to human neutrophil granulocytes. Eur. J. Pharmacol. 190, 67–73.

Fisher, G.J., Talwar, H.S., Ryder, N.S. and Voorhees, J.J. (1989). Differential activation of human skin cells by platelet activating factor: stimulation of phosphoinositide turnover and arachidonic acid mobilization in keratinocytes but not in fibroblasts. Biochem. Biophys. Res. Commun, 163, 1344–1350.

Fitzgerald, M.F., Moncada, S. and Parente, L. (1986). The anaphylactic release of platelet-activating factor from perfused guinea-pig lungs. Br. J. Pharmacol. 88, 149–153.

Fjellner, B. and Hagermark, O. (1985). Experimental pruritus evoked by platelet activating factor (PAF-acether) in human skin. Acta. Derm. Venereol. (Stockh.) 65, 409–412.

Fletcher, J.R., DiSimone, A.G. and Earnest, M.A. (1990). Platelet activating factor receptor antagonist improves survival and attenuates eicosanoid release in severe endotoxemia. Ann. Surg. 211, 312–316.

Foa, R., Bussolino, F., Ferrando, M.L. et al. (1985). Release of platelet-activating factor in human leukemia. Cancer Res. 45, 4483–4485.

Foegh, M.L., Alijani, M.R., Helfrich, G.B., Khirabadi, B.S. and Ramwell, P.W. (1985). Prolongation of cardiac allograft survival with the PAF antagonist BN-52021 and the thromboxane receptor antagonists L640035 and L636499. Adv. Prostaglandin Thromboxane Leukotriene Res. 15, 381–384.

Foegh, M.L., Khirabadi, B.S., Rowles, J.R., Braquet, P. and Ramwell, P.W. (1986). The causal role of PAF and leukotrienes in acute cardiac allograft rejection in rats. Pharmacol. Res. Commun. 18 Suppl. 127–132.

Foegh, M.L., Hartmann, D.P., Rowles, J.R. et al. (1987). Leukotrienes, thromboxane, and platelet activating factor in organ transplantation. Adv. Prostaglandin Thromboxane Leukotriene Res. 17A, 140–146.

Fouque, F. and Vargaftig, B.B. (1984). Triggering by Paf-acether and adrenaline of cyclo-oxygenase-independent platelet aggregation. Br. J. Pharmacol. 83, 625–633.

Freiche, J.C., Lang, J., Sedivy, P. and Touraine, J.L. (1990).

Prolonged survival of renal transplants in nonimmunized and hyperimmunized rats receiving a platelet-activating factor antagonist. Transplantation 50, 8–13.

Frerichs, K.U., Lindsberg, P.J., Hallenbeck, J.M. and Feuerstein, G.Z. (1990). Platelet-activating factor and progressive brain damage following focal brain injury. J. Neurosurg. 73, 223–233.

Fujimura, K., Sugatani, J., Miwa, M., Mizuno, T., Sameshima, Y. and Saito, K. (1989). Serum platelet-activating factor acetylhydrolase activity in rats with gastric ulcers induced by water-immersion stress. Scand. J. Gastroenterol. Suppl. 162, 59–62.

Fukuda, T., Akutsu, I., Amagai, M., Numao, T., Motojima, S. and Makino, S. (1990). Antigen-induced biphasic eosinophil infiltration in the airways of actively sensitized guinea pigs and its inhibition by PAF antagonist and cyclosporin A. Arerugi 39, 548–552.

Garay, R. and Braquet, P. (1986). Involvement of K^+ movements in the membrane signal induced by PAF-acether. Biochem. Pharmacol. 35, 2811–2815.

Gati, I., Bergstrom, M., Muhr, C. and Carlsson, J. (1991). Effects of Paf analog and antagonist CV 6209 on cultured human glioma cell lines. Prostaglandins Leukot. Essent. Fatty Acids 43, 103–110.

Gautheron, P.D., Coulbault, L. and Sugrue, M.F. (1987). A study of PAF-induced ocular inflammation in the rat and its inhibition by the PAF antagonist, L-652,731. J. Pharm. Pharmacol. 39, 857–859.

Gay, J.C., Beckman, J.K., Zaboy, K.A. and Lukens, J.N. (1986). Modulation of neutrophil oxidative responses to soluble stimuli by platelet-activating factor. Blood 67, 931–936.

Gebhardt, B.M., Braquet, P., Bazan, H. and Bazan, N. (1988). Modulation of in vitro immune reactions by platelet activating factor and a platelet activating factor antagonist. Immunopharmacology 15, 11–19.

Geissler, F.T., Kuzan, F.B., Faustman, E.M. and Henderson, W.R.J. (1989). Lipid mediator production by post-implantation rat embryos in vitro. Prostaglandins 38, 145–155.

Gerdin, B., Lundberg, C. and Smedegard, G. (1985). Permeability-increasing ability of PAF-acether in rat skin. Inflammation 9, 107–112.

Gerkens, J.F. (1990). Reproducible vasodilation by platelet-activating factor in blood- and Krebs-perfused rat kidneys is albumin-dependent. Eur. J. Pharmacol. 177, 119–126.

Giembycz, M.A., Kroegel, C. and Barnes, P.J. (1990). Platelet activating factor stimulates cyclo-oxygenase activity in guinea pig eosinophils. Concerted biosynthesis of thromboxane A_2 and E-series prostaglandins. J. Immunol. 144, 3489–3497.

Giers, G., Janzarik, H., Kempe, E.R. and Mueller-Eckhardt, C. (1990). Failure of platelet-activating factor antagonist WEB 2086 BS in treatment of chronic autoimmune thrombocytopaenia. Blut 61, 21–24.

Gilfillan, A.M., Wiggan, G.A., Hope, W.C., Patel, B.J. and Welton, A.F. (1990). Ro 19–3704 directly inhibits immunoglobulin E-dependent mediator release by a mechanism independent of its platelet-activating factor antagonist properties. Eur. J. Pharmacol. 176, 255–262.

Gillespie, M.N. and Bowdy, B.D. (1986). Impact of platelet activating factor on vascular responsiveness in isolated rat lungs. J. Pharmacol. Exp. Ther. 236, 396–402.

Giovannini, L., Mian, M., Alessandri, M.G., Galligani, P. and Bertelli, A.A. (1988). N-Acetyl-beta-glucosaminidase (NAG) and alpha-glycosidase released by PAF in isolated perfused rat kidney. Int. J. Tissue. React. 10, 39–43.

Gomez Cambronero, J., Durstin, M., Molski, T.F., Naccache, P.H. and Sha'afi, R.I. (1989). Calcium is necessary but not sufficient for the platelet-activating factor release in human neutrophils stimulated by physiological stimuli. Role of G-proteins. J. Biol. Chem. 264, 12699–12704.

Gonzalez Crussi, F. and Hsueh, W. (1983). Experimental model of ischemic bowel necrosis. The role of platelet-activating factor and endotoxin. Am. J. Pathol. 112, 127–135.

Goswami, S.K., Ohashi, M., Stathas, P. and Marom, Z.M. (1989). Platelet-activating factor stimulates secretion of respiratory glycoconjugate from human airways in culture. J. Allergy Clin. Immunol. 84, 726–734.

Grandel, K.E., Farr, R.S., Wanderer, A.A., Eisenstadt, T.C. and Wasserman, S.I. (1985). Association of platelet-activating factor with primary acquired cold urticaria. N. Engl. J. Med. 313, 405–409.

Greco, N.J., Arnold, J.H., O'Dorisio, T.M., Cataland, S. and Panganamala, R.V. (1985). Action of platelet-activating factor on type 1 diabetic human platelets. J. Lab. Clin. Med. 105, 410–416.

Grigoriadis, G. and Stewart, A.G. (1991). 1-O-Hexadecyl-2-acetyl-sn-glycero-3-phospho (N,N,N trimethyl) hexanolamine: an analogue of platelet-activating factor with partial agonist activity. Br. J. Pharmacol. 104, 171–177.

Grigoriadis, G. and Stewart, A.G. (1992). Albumin inhibits platelet-activating factor (PAF)-induced responses in platelets and macrophages: implications for the biologically active form of PAF. Br. J. Pharmacol. 107, 73–77.

Grigorian, G.Y. and Ryan, U.S. (1987). Platelet-activating factor effects on bovine pulmonary artery endothelial cells. Circ. Res. 61, 389–395.

Griswold, D.E., Hillegass, L., Rotilio, D., Mong, S. and Hanna, N. (1989). Absence of a lyso-PAF relationship with PAF (platelet activating factor) in monosodium urate crystal-induced inflammatory exudates. Eicosanoids 2, 151–156.

Grundel, R.H., Letts, L.G. and Gleich, G.J. (1991). Human eosinophil major basic protein induces airway constriction and airway hyperresponsiveness. J. Clin. Invest. 87, 1470–1473.

Gryglewski, R.J., Moncada, S. and Palmer, R.M. (1986). Bioassay of prostacyclin and endothelium-derived relaxing factor (EDRF) from porcine aortic endothelial cells. Br. J. Pharmacol. 87, 685–694.

Guinot, P., Brambilla, C., Duchier, J., Braquet, P., Bonvoisin, B. and Cournot, A. (1987a). Effect of BN 52063, a specific PAF-acether antagonist, on bronchial provocation test to allergens in asthmatic patients. A preliminary study. Prostaglandins 34, 723–731.

Guinot, P., Braquet, P., Duchier, J. and Cournot, A. (1987b). Effects of BN 52063 on PAF-acether induced weal and flare in man. Agents Actions Suppl. 21, 239–243.

Habazettl, H., Conzen, P.F., Vollmar, B. et al. (1990). Pulmonary hypertension after heparin-protamine: roles of left-sided infusion, histamine, and platelet-activating factor. Anesth. Analg. 71, 637–644.

Hadvary, P. and Baumgartner, H.R. (1983). Activation of human and rabbit blood platelets by synthetic structural analogs of platelet activating factor. Thromb. Res. 30, 143–156.

Hakansson, L., Westerlund, D. and Venge, P. (1987). New method for the measurement of eosinophil migration. J. Leukocyte Biol. 42, 689–696.

Halonen, M., Dunn, A.M., Palmer, J.D. and McManus, L.M. (1990). Anatomic basis for species differences in peripheral lung strip contraction to PAF. Am. J. Physiol. 259, L81–L86.

Hamasaki, Y., Mojarad, M., Saga, T., Tai, H.H. and Said, S.I. (1984). Platelet-activating factor raises airway and vascular pressures and induces edema in lungs perfused with platelet-free solution. Am. Rev. Respir. Dis. 129, 742–746.

Hanahan, D.J., Munder, P.G., Satouchi, K., McManus, L. and Pinckard, R.N. (1981). Potent platelet stimulating activity of enantiomers of acetyl glyceryl ether phosphorylcholine and its methoxy analogues. Biochem. Biophys. Res. Commun. 99, 183–188.

Handa, R.K., Strandhoy, J.W. and Buckalew, V.M.J. (1990). Platelet-activating factor is a renal vasodilator in the anesthetized rat. Am. J. Physiol. 258, F1504–F1509.

Handley, D.A., Arbeeny, C.M., Lee, M.L., Van Valen, R.G. and Saunders, R.N. (1984). Effect of platelet activating factor on endothelial permeability to plasma macromolecules. Immunopharmacology 8, 137–142.

Handley, D.A., Van Valen, R.G., Melden, M.K., Flury, S., Lee, M.L. and Saunders, R.N. (1986a). Inhibition and reversal of endotoxin-, aggregated IgG- and paf-induced hypotension in the rat by SRI 63–072, a paf receptor antagonist. Immunopharmacology 12, 11–16.

Handley, D.A., Van Valen, R.G. and Saunders, R.N. (1986b). Vascular responses of platelet-activating factor in the Cebus apella primate and inhibitory profiles of antagonists SRI 63-072 and SRI 63-119. Immunopharmacology 11, 175–182.

Handley, D.A., Van Valen, R.G., Tomesch, J.C. et al. (1987). Biological properties of the antagonist SRI 63–441 in the PAF and endotoxin models of hypotension in the rat and dog. Immunopharmacology 13, 125–132.

Harczy, M., Maclouf, J., Pradelles, P., Braquet, P., Borgeat, P. and Sirois, P. (1986). Inhibitory effects of a novel platelet activating factor (PAF) antagonist (BN 52021) on antigen-induced prostaglandin and thromboxane formation by the guinea pig lung. Pharmacol. Res. Commun. 18(Suppl.), 111–117.

Harper, M.J. (1989). Platelet-activating factor: a paracrine factor in preimplantation stages of reproduction? Biol. Reprod. 40, 907–913.

Harper, M.J.K., Woodard, D.S. and Norris, C.J. (1989). Spermicidal effect of antagonists of platelet-activating factor. Fertil. Steril. 51, 890–895.

Hartung, H.P., Parnham, M.J., Winkelmann, J., Rasokat, H. and Hadding, U. (1982). Stimulation of the oxidative burst in macrophages with platelet activating factor (PAF-acether). Agents Actions Suppl. 11, 139–146.

Hartung, H.P., Parnham, M.J., Winkelmann, J., Englberger, W. and Hadding, U. (1983). Platelet activating factor (PAF) induces the oxidative burst in macrophages. Int. J. Immunopharmacol. 5, 115–121.

Haslett, C., Guthrie, L.A., Kopaniak, M.A., Johnston, R.B. Jr and Henson, P.M. (1985). Modulation of multiple neutrophil functions by preparative methods or trace concentrations of bacterial lipopolysaccharide. Am. J. Pathol. 119, 101–110.

Hatakeyama, K., Yano, S. and Watanabe, K. (1989). Platelet activating factor induced gastric damage in rats: participation of increased gastric vascular permeability. Scand. J. Gastroenterol. Suppl. 162, 71–74.

Havill, A.M., Van Valen, R.G. and Handley, D.A. (1990). Prevention of non-specific airway hyperreactivity after allergen challenge in guinea-pigs by the PAF receptor antagonist SDZ 64–412. Br. J. Pharmacol. 99, 396–400.

Hayashi, H., Kudo, I., Inoue, K., Nomura, H. and Nojima, S. (1985a). Macrophage activation by PAF incorporated into dipalmitoylphosphatidylcholine-cholesterol liposomes. J. Biochem. Tokyo 97, 1255–1258.

Hayashi, H., Kudo, I., Inoue, K. et al. (1985b). Activation of guinea pig peritoneal macrophages by platelet activating factor (PAF) and its agonists. J. Biochem. Tokyo 97, 1737–1745.

Hayashi, H., Kudo, I., Kato, T., Nozawa, R., Nojima, S. and Inoue, K. (1988). A novel bioaction of PAF: induction of microbicidal activity in guinea pig bone marrow cells. Lipids 23, 1119–1124.

Hayashi, H., Kudo, I., Kato, T. and Inoue, K. (1989). In "Platelet-activating Factor and Diseases." (ed K. Saito and D.J. Hanahan), pp 51–68. International Medical Publishers, Tokyo.

Hayashi, M., Kimura, J., Oh ishi, S., Tsushima, S. and Nomura, H. (1987a). Characterization of the activity of a platelet activating factor antagonist, CV-3988. Jpn. J. Pharmacol. 44, 127–134.

Hayahsi, M., Kimura, J., Yamaki, K. et al. (1987b). Detection of platelet-activating factor in exudates of rats with phorbol myristate acetate-induced pleurisy. Thromb. Res. 48, 299–310.

Hebert, R.L., Sirois, P., Braquet, P. and Plante, G.E. (1987). Hemodynamic effects of PAF-acether on the dog kidney. Prostaglandins Leukotrienes Med. 26, 189–202.

Hebert, R.L., Sirois, P. and Plante, G.E. (1989). Inhibition of platelet-activating factor induced renal hemodynamic and tubular dysfunctions with L-655,240, a new thromboxane–prostaglandin endoperoxide antagonist. Can. J. Physiol. Pharmacol. 67, 304–308.

Hellewell, P.G. and Williams, T.J. (1986). A specific antagonist of platelet-activating factor suppresses oedema formation in an Arthus reaction but not oedema induced by leukocyte chemoattractants in rabbit skin. J. Immunol. 137, 302–307.

Hellewell, P.G. and Williams, T.J. (1989). Antagonism of Paf-induced oedema formation in rabbit skin: a comparison of different antagonists. Br. J. Pharmacol. 97, 171–180.

Hemmendinger, S., Pauli, G., Tenabene, A. et al. (1989). Platelet function; aggregation by PAF or sequestration in lung is not modified during immediate or late allergen-induced bronchospasm in man. J. Allergy. Clin. Immunol. 83, 990–996.

Henocq, E. and Vargaftig, B.B. (1986). Accumulation of eosinophils in response to intracutaneous PAF-acether and allergens in man. Lancet i, 1378–1379.

Henocq, E. and Vargaftig, B.B. (1988) Skin eosinophilia in atopic patients. J. Allergy Clin. Immunol. 81, 691–695.

Henriques, M.G., Weg, V.B., Martins, M.A. et al. (1990). Differential inhibition by two hetrazepine PAF antagonists of acute inflammation in the mouse. Br. J. Pharmacol. 99, 164–168.

Henson, P.M. and Landes, R.R. (1976). Activation of platelets by platelet activating factor (PAF) derived from IgE-sensitized basophils. IV. PAF does not activate platelet factor 3 (PF3). Br. J. Haematol. 34, 269–282.

Henson, P.M. and Oades, Z.G. (1976). Activation of platelets by platelet-activating factor (PAF) derived from IgE-sensitized basophils. II. The role of serine proteases, cyclic nucleotides, and contractile elements in PAF-induced secretion. J. Exp. Med. 143, 953–968.

Henson, P.M. and Pinckard, R.N. (1977). Basophil-derived platelet-activating factor (PAF) as an in vivo mediator of acute allergic reactions: demonstration of specific desensitization of platelets to PAF during IgE-induced anaphylaxis in the rabbit. J. Immunol. 119, 2179–2184.

Herrmann, D.B., Ferber, E. and Munder, P.G. (1986). Ether phospholipids as inhibitors of the arachidonoyl-CoA:1-acyl-*sn*-glycero-3-phosphocholine acyltransferase in macrophages. Biochim. Biophys. Acta 876, 28–35.

Heuer, H. and Casals Stenzel, J. (1988). Effect of the PAF-antagonist WEB 2086 on anaphylactic lung reaction: comparison of inhalative and intravenous challenge. Agents Actions Suppl. 23, 207–215.

Heuer, H.O., Letts, G. and Meade, C.J. (1990). Tumour necrosis factor (TNF) and endotoxin prime effects of PAF *in vivo*. J. Lipid Mediators 2, s101–s108.

Hirafuji, M., Maeyama, K., Watanabe, T. and Ogura, Y. (1988). Transient increase of cytosolic free calcium in cultured human vascular endothelial cells by platelet-activating factor. Biochem. Biophys. Res. Commun. 154, 910–917.

Hirata, K., Pele, J.P., Robidoux, C. and Sirois, P. (1989). Guinea pig lung eosinophil: purification and prostaglandin production. J. Leukocyte Biol. 45, 523–528.

Hocking, D.C., Phillips, P.G., Ferro, T.J. and Johnson, A. (1990). Mechanisms of pulmonary oedema induced by tumor necrosis factor-alpha. Circ. Res. 67, 68–77.

Hoffman, D.R., Hajdu, J. and Snyder, F. (1984). Cytotoxicity of platelet activating factor and related alkyl-phospholipid analogs in human leukemia cells, polymorphonuclear neutrophils, and skin fibroblasts. Blood 63, 545–552.

Hoffman, D.R., Truong, C.T. and Johnston, J.M. (1986a). The role of platelet-activating factor in human fetal lung maturation. Am. J. Obstet. Gynecol. 155, 70–75.

Hoffman, D.R., Truong, C.T. and Johnston, J.M. (1986b). Metabolism and function of platelet-activating factor in fetal rabbit lung development. Biochim. Biophys. Acta 879, 88–96.

Hoffman, D.R., Bateman, M.K. and Johnston, J.M. (1988a). Synthesis of platelet activating factor by cholinephosphotransferase in developing fetal rabbit lung. Lipids 23, 96–100.

Hoffman, D.R., White, R.G., Angle, M.J., Maki, N. and Johnston, J.M. (1988b). Platelet-activating factor induces glycogen degradation in fetal rabbit lung *in utero*. J. Biol. Chem. 263, 9316–9319.

Hoffman, D.R., Romero, R. and Johnston, J.M. (1990). Detection of platelet-activating factor in amniotic fluid of complicated pregnancies. Am. J. Obstet. Gynecol. 162, 525–528.

Hofmann, B., Meisgeier, U., Kertscher, P. and Ostermann, G. (1988a). Effects of synthetic analogues of platelet-activating factor (PAF) on fibrinolytic activity in the rat. Biomed. Biochim. Acta 47, S161–S164.

Hofmann, B., Ostermann, G., Hoffmann, A., Klocking, H.P. and Kertscher, H.P. (1988b). Effect of specific antagonists on PAF-induced platelet aggregation and release of plasminogen activator. Biomed. Biochim. Acta 47, S157–S160.

Honda, Z.-I., Nakamura, M., Miki, I. *et al.* (1991). Cloning by functional expression of platelet-activating factor receptor from guinea-pig lung. Nature 349, 342–346.

Hopkins, N.K., Lin, A.H. and Gorman, R.R. (1983). Evidence for mediation of acetyl glyceryl ether phosphorylcholine stimulation of adenosine 3′,5′-(cyclic)monophosphate levels in human polymorphonuclear leukocytes by leukotriene B4. Biochim. Biophys. Acta 763, 276–283.

Hopkins, N.K., Schaub, R.G. and Gorman, R.R. (1984). Acetyl glyceryl ether phosphorylcholine (PAF-acether) and leukotriene B₄-mediated neutrophil chemotaxis through an intact endothelial cell monolayer. Biochim. Biophys. Acta 805, 30–36.

Hopp, R.J., Bewtra, A.K., Agrawal, D.K. and Townley, R.G. (1989). Effect of platelet-activating factor inhalation on nonspecific bronchial reactivity in man. Chest 96, 1070–1072.

Hopp, R.J., Bewtra, A.K., Nabe, M., Agrawal, D.K. and Townley, R.G. (1990). Effect of PAF-acether inhalation on nonspecific bronchial reactivity and adrenergic response in normal and asthmatic subjects. Chest 98, 936–941.

Hornby, E.J. and Perry, C.R. (1983). 1-*O*-hexadecyl-2-acetyl-*sn*-3-glycerophosphorylcholine (PAF): some effects on the aggregation of human platelets by thrombin or collagen. Thromb. Haemost. 50, 586–587.

Hosford, D., Page, C.P., Barnes, P.J. and Braquet, P. (1989). In "Platelet Activating Factor and Human Disease", (ed P.J. Barnes, C.P. Page and P. Henson), pp 82–116. Blackwell, Oxford.

Hoshikawa Fujimura, A.Y., Auler Junior, J.O., Da Rocha, T.R. *et al.* (1989). PAF-acether, superoxide anion and beta-glucuronidase as parameters of polymorphonuclear cell activation associated with cardiac surgery and cardiopulmonary bypass. Braz. J. Med. Biol. Res. 22, 1077–1082.

Hsieh, K.-H. (1991). Effects of Paf antagonist, BN 52021, on the Paf-, methacholine-, and allergen-induced bronchoconstriction in asthmatic children. Chest 99, 877–882.

Hsueh, W., Gonzalez Crussi, F. and Arroyave, J.L. (1986). Platelet-activating factor-induced ischemic bowel necrosis. An investigation of secondary mediators in its pathogenesis. Am. J. Pathol. 122, 231–239.

Hu, W., Kinnaird, A.A.A and Man, R.Y.K. (1991). Mechanisms of the coronary vascular effects of platelet-activating factor in the rat perfused heart. Br. J. Pharmacol. 103, 1097–1102.

Huang, S.J., Monk, P.N., Downes, C.P. and Whetton, A.D. (1988). Platelet-activating factor-induced hydrolysis of phosphatidylinositol 4,5-bisphosphate stimulates the production of reactive oxygen intermediates in macrophages. Biochem. J. 249, 839–845.

Hwang, S.B. (1988). Identification of a second putative receptor of platelet-activating factor from human polymorphonuclear leukocytes. J. Biol. Chem. 263, 3225–3233.

Hwang, S.B. and Lam, M.H. (1986). Species difference in the specific receptors of platelet activating factor. Biochem. Pharmacol. 35, 4511–4518.

Hwang, S.B., Lee, C.S., Cheah, M.J. and Shen, T.Y. (1983). Specific receptor sites for 1-*O*-alkyl-2-*O*-acetyl-*sn*-glycero-3-phosphocholine (platelet activating factor) on rabbit platelet and guinea pig smooth muscle membranes. Biochemistry 22, 4756–4763.

Hwang, S.B., Li, C.L., Lam, M.H. and Shen, T.Y. (1985). Characterization of cutaneous vascular permeability induced by platelet-activating factor in guinea pigs and rats and its inhibition by a platelet-activating factor receptor antagonist. Lab. Invest. 52, 617–630.

Ibbotson, G.C. and Wallace, J.L. (1989). Beneficial effects of prostaglandin E₂ in endotoxic shock are unrelated to effects on PAF-acether synthesis. Prostaglandins 37, 237–250.

Imai, T., Vercellotti, G.M., Moldow, C.F., Jacob, H.S. and Weir, E.K. (1988). Pulmonary hypertension and edema induced by platelet-activating factor in isolated, perfused rat lungs are blocked by BN52021. J. Lab. Clin. Med. 111, 211–217.

Imaizumi, T.-A., Satoh, K., Yoshida, H., Kawamura, Y., Hiramoto, M. and Takamatsu, S. (1991). Effect of cigarette smoking on the levels of platelet-activating factor-like lipids in plasma lipoproteins. Atherosclerosis 87, 47–55.

Imura, Y., Terashita, Z. and Nishikawa, K. (1986). Possible role of platelet activating factor (PAF) in disseminated intravascular coagulation (DIC), evidenced by use of a PAF antagonist, CV-3988. Life Sci. 39, 111–117.

Inagaki, N., Miura, T., Daikoku, M., Nagai, H. and Koda, A. (1989). Inhibitory effects of beta-adrenergic stimulants on increased vascular permeability caused by passive cutaneous anaphylaxis, allergic mediators, and mediator releasers in rats. Pharmacology 39, 19–27.

Inarrea, P., Gomez-Cambronero, J., Nieto, M. and Crespo, M.S. (1984). Characteristics of the binding of platelet-activating factor to platelets of different animal species. Eur. J. Pharmacol. 105, 309–315.

Inarrea, P., Gomez Cambronero, J., Pascual, J., Ponte, M.C., Hernando, L. and Sanchez Crespo, M. (1985). Synthesis of PAF-acether and blood volume changes in gram-negative sepsis. Immunopharmacology 9, 45–52.

Ingraham, L.M., Coates, T.D., Allen, J.M., Higgins, C.P., Baehner, R.L. and Boxer, L.A. (1982). Metabolic, membrane, and functional responses of human polymorphonuclear leukocytes to platelet-activating factor. Blood 59, 1259–1266.

Ishida, K., Thomson, R.J., Beatti, L.L., Wiggs, B. and Schellenberg, R.R. (1990). Inhibition of antigen-induced airway hyperresponsiveness, but not acute hypoxia nor airway eosinophilia, by an antagonist of platelet-activating factor. J. Immunol. 144, 3907–3911.

Issekutz, A.C. and Szpejda, M. (1986). Evidence that platelet activating factor may mediate some acute inflammatory responses. Studies with the platelet-activating factor antagonist, CV3988. Lab. Invest. 54, 275–281.

Ito, S., Camussi, G., Tetta, C., Milgrom, F. and Andres, G. (1984). Hyperacute renal allograft rejection in the rabbit. The role of platelet-activating factor and of cationic proteins derived from polymorphonuclear leukocytes and from platelets. Lab. Invest. 51, 148–161.

Jackson, C.V., Schumaker, W.A., Kunkel, S.L., Driscoll, E.M. and Lucchesi, B.R. (1986). Platelet-activating factor and the release of a platelet-derived coronary artery vasodilator substance in the canine. Circ. Res. 58, 218–229.

Jancar, S., Theriault, P., Braquet, P. and Sirois, P. (1987). Comparative effects of platelet activating factor, leukotriene D4 and histamine on guinea pig trachea, bronchus and lung parenchyma. Prostaglandins 33, 199–208.

Jancar, S., Theriault, P., Provencal, B., Cloutier, S. and Sirois, P. (1988). Mechanism of action of platelet-activating factor on guinea-pig lung parenchyma strips. Can. J. Physiol. Pharmacol. 66, 1187–1191.

Jancar, S., Theriault, P., Lauziere, M., Braquet, P. and Sirois, P. (1989). Paf-induced release of spasmogens from guinea-pig lungs. Br. J. Pharmacol. 96, 153–162.

Johnson, P.R., Armour, C.L. and Black, J.L. (1990a). The action of platelet activating factor and its antagonism by WEB 2086 on human isolated airways. Eur. Respir. J. 3, 55–60.

Johnson, P.R.A., Black, J.L. and Armour, C.L. (1990b). Is the platelet activating factor (PAF) induced contraction of human isolated airway tissue related to the presence of inflammatory cells within the tissue? Clin. Exp. Pharmacol. Physiol. Suppl. 17, 37P.

Johnston, J.M. (1991). In "Platelet-Activating Factor and Diseases" (ed. K. Saito and D.J. Hanahan), pp 129–151. International Medical Publishers, Tokyo.

Jouvin Marche, E., Poitevin, B. and Benveniste, J. (1982). Platelet-activating factor (PAF-acether), an activator of neutrophil functions. Agents Actions 12, 716–720.

Jouvin Marche, E., Cerrina, J., Coeffier, E., Duroux, P. and Benveniste, J. (1983). Effect of the Ca^{2+} antagonist nifedipine on the release of platelet-activating factor (PAF-acether), slow-reacting substance and beta-glucuronidase from human neutrophils. Eur. J. Pharmacol. 89, 19–26.

Jouvin Marche, E., Ninio, E., Beaurain, G., Tence, M., Niaudet, P. and Benveniste, J. (1984). Biosynthesis of Paf-acether (platelet-activating factor). VII. Precursors of Paf-acether and acetyl-transferase activity in human leukocytes. J. Immunol. 133, 892–898.

Junier, M.P., Tiberghien, C., Rougeot, C., Fafeur, V. and Dray, F. (1988). Inhibitory effect of platelet-activating factor (PAF) on luteinizing hormone-releasing and hormone somatostatin release from rat median eminence in vitro correlated with the characterization of specific PAF receptor sites in rat hypothalamus. Endocrinology 123, 72–80.

Kamitani, T., Katamoto, M., Tatsumi, M. et al. (1984). Mechanisms of the hypotensive effect synthetic 1-O-octadecyl-2-O-acetyl-glycero-3-phosphorylcholine. Eur. J. Pharmacol. 98, 357–366.

Kawaguchi, H. and Yasuda, H. (1987). Effects of platelet-activating factor on conversion of angiotensin I to II. FEBS Lett. 221, 305–308.

Kawaguchi, H., Sawa, H. and Yasuda, H. (1990). Mechanism of increased angiotensin-converting enzyme activity stimulated by platelet-activating factor. Biochim. Biophys. Acta 1052, 503–508.

Kay, A.B. (1983). Mediators of hypersensitivity and inflammatory cells in the pathogenesis of bronchial asthma. Eur. J. Respir. Dis. Suppl. 129, 1–44.

Kay, A.B., Frew, A.J., Moqbel, R. et al. (1989). The activated eosinophil in allergy and asthma. Prog. Clin. Biol. Res. 297, 183–196.

Kemeny, L., Csato, M. and Dobozy, A. (1989). Pharmacological studies on dithranol-induced irritative dermatitis in mice. Arch. Dermatol. Res. 281, 362–365.

Kemeny, L., Csato, M., Braquet, P. and Dobozy, A. (1990). Effect of BN 52021, a platelet activating factor antagonist, on dithranol-induced inflammation. Br. J. Dermatol. 122, 539–544.

Kenzora, J.L., Perez, J.E., Bergman, S.R. and Lange, L.G. (1984). Effects of acetyl glyceryl ether phosphorylcholine (platelet-activating factor) on ventricular preload, afterload and contractility in dogs. J. Clin. Invest, 74, 1193–1202.

Keraly, C.L. and Benveniste, J. (1982). Specific desensitization of rabbit platelets by platelet-activating factor (PAF-acether) and derivatives. Br. J. Haematol. 51, 313–322.

Keraly, C.L., Coeffier, E., Tence, M., Borrel, M.C. and Benveniste, J. (1983). Effect of structural analogues of PAF-acether on platelet desensitization. Br. J. Haematol. 53, 513–521.

Kester, M., Mene, P., Dubyak, G.R. and Dunn, M.J. (1987). Elevation of cytosolic free calcium by platelet-activating factor in cultured rat mesangial cells. FASEB J. 1, 215–219.

Khirabadi, B.S., Foegh, M.L., Goldstein, H.A. and Ramwell, P.W. (1987). The effect of prednisolone, thromboxane, and platelet-activating factor receptor antagonists on lymphocyte and platelet migration in experimental cardiac transplantation. Transplantation 43, 626–630.

Kimani, G., Tonnesen, M.G. and Henson, P.M. (1988). Stimulation of eosinophil adherence to human vascular endothelial cells in vitro by platelet-activating factor. J. Immunol. 140, 3161–3166.

Kings, M.A., Chapman, I., Kristersson, A., Sanjar, S. and Morley, J. (1990). Human recombinant lymphokines and cytokines induce pulmonary eosinophilia in the guinea pig which is inhibited by ketotifen and AH 21–132. Int. Arch. Allergy. Appl. Immunol. 91, 354–361.

Kioumis, I., Lammers, J.W., Dent, G., Chung, K.F. and Barnes, P.J. (1988). Effect of inhaled platelet-activating factor on circulating neutrophils and platelets *in vivo* and *ex vivo* in man. Prostaglandins 36, 343–354.

Kloprogge, E. and Akkerman, J.W. (1984). Binding kinetics of PAF-acether (1-*O*-alkyl-2-acetyl-*sn*-glycero-3-phosphocholine) to intact human platelets. Biochem. J. 223, 901–909.

Kloprogge, E. and Akkerman, J.W. (1986). Platelet-activating factor (PAF-acether) induces high- and low-affinity binding of fibrinogen to human platelets via independent mechanisms. Biochem. J. 240, 403–412.

Kloprogge, E., De Haas, G.H., Gorter, G. and Akkerman, J.W. (1983). Stimulus–response coupling in human platelets. Evidence against a role of PAF-acether in the "third pathway". Thromb. Res. 30, 107–112.

Kochanek, P.M., Dutka, A.J., Kumaroo, K.K. and Hallenbeck, J.M. (1987). Platelet activating factor receptor blockade enhances recovery after multifocal brain ischemia. Life Sci. 41, 2639–2644.

Kochanek, P.M., Nemoto, E.M., Melick, J.A., Evans, R.W. and Burke, D.F. (1988). Cerebrovascular and cerebrometabolic effects of intracarotid infused platelet-activating factor in rats. J. Cereb. Blood Flow. Metab. 8, 546–551.

Kodama, H., Muto, H. and Maki, M. (1989). Isolation and identification of embryo-derived platelet-activating factor in mice. Nippon Sanka Fujinka Gakkai Zasshi. 41, 899–906.

Koltai, M., Lepran, I., Szekeres, L., Viossat, I., Chabrier, E. and Braquet, P. (1986). Effect of BN 52021, a specific PAF-acether antagonist, on cardiac anaphylaxis in Langendorff hearts isolated from passively sensitized guinea-pigs. Eur. J. Pharmacol. 130, 133–136.

Koltai, M., Tosaki, A., Hosford, D. and Braquet, P. (1989). Ginkgolide B protects isolated hearts against arrhythmias induced by ischemia but not reperfusion. Eur. J. Pharmacol. 164, 293–302.

Konturek, S.J., Brzozowski, T., Drozdowicz, D. and Beck, G. (1988). Role of leukotrienes in acute gastric lesions induced by ethanol, taurocholate, aspirin, platelet-activating factor and stress in rats. Dig. Dis. Sci. 33, 806–813.

Konturek, S.J., Brzozowski, T., Drozdowicz, D., Garlicki, J. and Beck, G. (1989). Role of leukotrienes and platelet activating factor in acute gastric mucosal lesions in rats. Eur. J. Pharmacol. 164, 285–292.

Kornecki, E. and Ehrlich, Y.H. (1988). Neuroregulatory and neuropathological actions of the ether-phospholipid platelet-activating factor. Science 240, 1792–1794.

Kornecki, E. and Ehrlich, Y.H. (1990). Diminished responsiveness of human platelets to platelet-activating factor during pregnancy. Am. J. Physiol. 259, H766–H771.

Kornecki, E., Ehrlich, Y.H. and Lenox, R.H. (1984). Platelet-activating factor-induced aggregation of human platelets specifically inhibited by triazolobenzodiazepines. Science 226, 1454–1456.

Kramer, I.M., Van der Bend, R.L., Tool, A.T.J., Van Blitterswijk, W.J., Roos, D. and Verhoeven, A.J. (1989). 1-*O*-Hexadecyl-2-*O*-methylglycerol, a novel inhibitor of protein kinase C, inhibits the respiratory burst in human neutrophils. J. Biol. Chem. 264, 5876–5884.

Kramp, W., Pieroni, G., Pinckard, R.N. and Hanahan, D.J. (1984). Observations on the critical micellar concentration of 1-*O*-alkyl-2-acetyl-*sn*-glycero-3-phosphocholine and a series of its homologs and analogs. Chem. Phys. Lipids 35, 49–62.

Kravis, T.C. and Henson, P.M. (1975). IgE-induced release of a platelet-activating factor from rabbit lung. J. Immunol. 115, 1677–1681.

Kroegel, C., Yukawa, T., Dent, G., Chanez, P., Chung, K.F. and Barnes, P.J. (1988). Platelet-activating factor induces eosinophil peroxidase release from purified human eosinophils. Immunology 64, 559–561.

Kroegel, C., Pleass, R., Yukawa, T., Chung, K.F., Westwick, J. and Barnes, P.J. (1989a). Characterization of platelet-activating factor-induced elevation of cytosolic free calcium concentration in eosinophils. FEBS Lett. 243, 41–46.

Kroegel, C., Yukawa, T., Dent, G., Venge, P., Chung, K.F. and Barnes, P.J. (1989b). Stimulation of degranulation from human eosinophils by platelet-activating factor. J. Immunol. 142, 3518–3526.

Kroegel, C., Yukawa, T., Westwick, J. and Barnes, P.J. (1989c). Evidence for two platelet activating factor receptors on eosinophils: dissociation between PAF-induced intracellular calcium mobilization degranulation and superoxides anion generation in eosinophils. Biochem. Biophys. Res. Commun. 162, 511–521.

Kubes, P., Ibbotson, G., Russell, J., Wallace, J.L. and Granger, D.N. (1990). Role of platelet-activating factor in ischemia/reperfusion-induced leukocyte adherence. Am. J. Physiol. 259, G300–G305.

Kudolo, G.B. and Harper, M.J.K. (1989). Characterisation of platelet-activating factor binding sites on uterine membranes from pregnant rabbits. Biol. Reprod. 41, 587–603.

Kumar, R., Harper, M.J. and Hanahan, D.J. (1988a). Occurrence of platelet-activating factor in rabbit spermatozoa. Arch. Biochem. Biophys. 260, 497–502.

Kumar, R., Harvey, S.A., Kester, M., Hanahan, D.J. and Olson, M.S. (1988b). Production and effects of platelet-activating factor in the rat brain. Biochim. Biophys. Acta 963, 375–383.

Kurihara, K., Wardlaw, A.J., Moqbel, R. and Kay, A.B. (1989). Inhibition of platelet-activating factor (PAF)-induced chemotaxis and PAF binding to human eosinophils and neutrophils by the specific ginkgolide-derived PAF antagonist, BN 52021. J. Allergy. Clin. Immunol. 83, 83–90.

Kuzan, F.B., Geissler, F.T. and Henderson, W.R.J. (1990). Role of spermatozoal platelet-activating factor in fertilization. Prostaglandins 39, 61–74.

Lacasse, C. and Rola-Pleszynski, M. (1991). Immune regulation by platelet-activating factor: II Mediation of suppression by cytokine-stimulated endothelial cells *in vitro*. J. Leukocyte Biol. 49, 245–252.

Lagente, V., Desquand, S., Hadvary, P. *et al.* (1988a). Intereference of the Paf antagonist Ro 19-3704 with Paf and antigen-induced bronchoconstriction in the guinea-pig. Br. J. Pharmacol. 94, 27–36.

Lagente, V., Fortes, Z.B., Garcia Leme, J. and Vargaftig, B.B. (1988b). PAF-acether and endotoxin display similar effects on rat mesenteric microvessels: inhibition by specific antagonists. J. Pharmacol. Exp. Ther. 247, 254–261.

Lai, C.K. and Holgate, S.T. (1990). Does inhaled PAF cause airway hyperresponsiveness in humans? Clin. Exp. Allergy 20, 449–452.

Lai, C.K., Jenkins, J.R., Polosa, R. and Holgate, S.T. (1990). Inhaled PAF fails to induce airway hyperresponsiveness to methacholine in normal human subjects. J. Appl. Physiol. 68, 919–926.

Lamas, A.M., Mulroney, C.M. and Schleimer, R.P. (1988). Studies on the adhesive interaction between purified human eosinophils and cultured vascular endothelial cells. J. Immunol. 140, 1500–1505.

Lambrecht, G. and Parnham, M.J. (1986). Kadsurenone distinguishes between different platelet activating factor receptor subtypes on macrophages and polymorphonuclear leucocytes. Br. J. Pharmacol. 87, 287–289.

Lammers, J.-W.J., Kioumis, I., McCusker, M., Nichol. G.M., Barnes, P.J. and Chung, K.F. (1990). Effects of prostacyclin on bronchoconstriction and neutropenia induced by inhaled platelet-activating factor in man. J. Allergy Clin. Immunol. 85, 763–769.

Lang, M., Hansen, D. and Hahn, H.L. (1987). Effects of the PAF-antagonist CV-3988 on PAF-induced changes in mucus secretion and in respiratory and circulatory variables in ferrets. Agents Actions Suppl. 21, 245–252.

Lauri, D., Cerletti, C. and De Gaetano, G. (1985). Amplification of primary response of human platelets to platelet-activating factor: aspirin-sensitive and aspirin-insensitive pathways. J. Lab. Clin. Med. 105, 653–658.

Lauri, D., Cerletti, C. and De Gaetano, G. (1986). Contribution of ADP to the amplification of primary platelet aggregation by platelet activating factor (PAF): modulatory role of aspirin. Agents Actions 17, 506–511.

Laurindo, F.R., Goldstein, R.E., Davenport, N.J., Ezra, D. and Feuerstein, G.Z. (1989). Mechanisms of hypotension produced by platelet-activating factor. J. Appl. Physiol. 66, 2681–2690.

Lazenby, C.M., Dive, C., Thompson, M.G. and Hickman, J.A. (1990). Elevation of leukemic cell intracellular calcium by ether lipid SRI 62–834. Cancer Res. 50, 3327–3330.

Lee, T., Lenihan, D.J., Malone, B., Roddy, L.L. and Wasserman, S.I. (1984). Increased biosynthesis of platelet-activating factor in activated human eosinophils. J. Biol. Chem. 259, 5526–5530.

Lee, T.C., Malone, B., Woodard, D. and Snyder, F. (1989). Renal necrosis and the involvement of a single enzyme of the de novo pathway for the biosynthesis of platelet-activating factor in the rat kidney inner medulla. Biochem. Biophys. Res. Commun. 163, 1002–1005.

Lefer, A.M. (1987). Interaction between myocardial depressant factor and vasoactive mediators with ischemia and shock. Am. J. Physiol. 252, R193–R205.

Leff, A.R., White, S.R., Munoz, N.M., Popovich, K.J., Shioya, T. and Stimler Gerard, N.P. (1987). Parasympathetic involvement in PAF-induced contraction in canine trachealis in vivo. J. Appl. Physiol. 62, 599–605.

Lefort, J., Wal, F., Chignard, M., Medeiros, M.C. and Vargaftig, B.B. (1982). Pharmacological properties of PAF-acether in the guinea-pig: platelet-dependent and independent reactions. Agents Actions 12, 723–725.

Lefort, J., Rotilio, D. and Vargaftig, B.B. (1984). The platelet-independent release of thromboxane A$_2$ by Paf-acether from guinea-pig lungs involves mechanisms distinct from those for leukotriene. Br. J. Pharmacol. 82, 565–575.

Lellouch, Tubiana, A., Lefort, J. Pirotzky, E., Vargaftig, B.B. and Pfister, A. (1985). Ultrastructural evidence for extravascular platelet recruitment in the lung upon intravenous injection of platelet-activating factor (PAF-acether) to guinea-pigs. Br. J. Exp. Pathol. 66, 345–355.

Lellouch, Tubiana, A., Lefort, J., Pfister, A. and Vargaftig, B.B. (1987). Interactions between granulocytes and platelets with the guinea-pig lung in passive anaphylactic shock. Correlations with PAF-acether-induced lesion. Int. Arch. Allergy Appl. Immunol. 83, 198–205.

Lellouch Tubiana, A., Lefort, J., Simon, M.T., Pfister, A. and Vargaftig, B.B. (1988). Eosinophil recruitment into guinea pig lungs after PAF-acether and allergen administration. Modulation by prostacyclin, platelet depletion, and selective antagonists. Am. Rev. Respir. Dis. 137, 948–954.

Lepran, I. and Lefer, A.M. (1985). Ischemia aggravating effects of platelet-activating factor in acute myocardial ischemia. Basic. Res. Cardiol. 80, 135–141.

Lerner, R., Lindstrom, P. and Palmblad, J. (1988). Platelet activating factor and leukotriene B$_4$ induce hyperpolarization of human endothelial cells but depolarization of neutrophils. Biochem. Biophys. Res. Commun. 153, 805–810.

Levi, R., Burke, J.A., Guo, Z.-G., Hattori, Y., Hoppens, C.M., McManus, L.M., Hanahan, D.J. and Pinckard, R.N. (1984). Acetyl glyceryl ether phosphorylcholine (AGEPC): a putative mediator of cardiac anaphylaxis in the guinea-pig. Circ. Res. 54, 117–124.

Levy, C., Baudouy, N., Brun, V. et al. (1989). High affinity acceptor sites in guinea pig brain for 52770 RP, a PAF antagonist. Fundam. Clin. Pharmacol. 3, 37–46.

Lewis, A.J., Dervinis, A. and Chang, J. (1984). The effects of antiallergic and bronchodilator drugs on platelet-activating factor (PAF-acether) induced bronchospasm and platelet aggregation. Agents Actions 15, 636–642.

Lewis, J.C., O'Flaherty, J.T., McCall, C.E., Wykle, R.L. and Bond, M.G. (1983). Platelet-activating factor effects on pulmonary ultrastructure in rabbits. Exp. Mol. Pathol. 38, 100–108.

Lewis, M.S,, Whatley, R.E., Cain, P., McIntyre, T.M., Prescott, S.M. and Zimmerman, G.A. (1988). Hydrogen peroxide stimulates the synthesis of platelet-activating factor by endothelium and induces endothelial cell-dependent neutrophil adhesion. J. Clin. Invest. 82, 2045–2055.

Leyravaud, S. and Benveniste, J. (1989). Regulation of cellular retention of paf-acether by extracellular pH and cell concentration. Biochim. Biophys. Acta 1005, 192–195.

Leyravaud, S., Bossant, M.J., Joly, F., Bessou, G., Benveniste, J. and Ninio, E. (1989). Biosynthesis of paf-acether. X. Phorbol myristate acetate-induced paf-acether biosynthesis and acetyltransferase activation in human neutrophils. J. Immunol. 143, 245–249.

Lianos, E.A. and Zanglis, A. (1987). Biosynthesis and metabolism of 1-O-alkyl-2-acetyl-sn-glycero-3-phosphocholine in rat glomerular mesangial cells. J. Biol. Chem. 262, 8990–8993.

Lin, W.K., Falck, J.R. and Wong, P.Y. (1990). Effect of 14,15-epoxyeicosatrien ic acid infusion on blood pressure in normal and hypertensive rats. Biochem. Biophys. Res. Commun. 167, 977–981.

Lohman, I. and Halonen, M. (1990). Effects of the PAF antagonist WEB 2086 on PAF-induced physiologic alterations and on IgE anaphylaxis in the rabbit. Am. Rev. Respir. Dis. 142, 390–397.

Lohmann, H.F., Adamus, W.S. and Meade, C.J. (1988). Idiopathic thrombocytopenia treated with Paf-acether antagonist WEB 2086. Lancet ii, 1147.

Lopez Diez, F., Nieto, M.L., Fernandez Gallardo, S., Gijon, M.A. and Sanchez Crespo, M. (1989). Occupancy of platelet receptors for platelet-activating factor in patients with septicemia. J. Clin. Invest. 83, 1733–1740.

Lopez Farre, A., Torralbo, M. and Lopez Novoa, J.M. (1988). Glomeruli from ischemic rat kidneys produce increased amounts of platelet activating factor. Biochem. Biophys. Res. Commun. 152, 129–135.

Lopez, A.F., Woodcock, J., Gillis, D., Stewart, A.G. and Vadas, M. (1992). In "Allergy and Immunity to Helminth Parasites" (ed R. Moqbel), pp. 205–227. Taylor and Francis, London.

Lorant, D.E., Patel, K.D., McIntyre, T.M., McEver, R.P., Prescott, S.M. and Zimmerman, G.A. (1991). Coexpression of GMP-140 and PAF by endothelium stimulated by histamine or thrombin: a juxtacrine system for adhesion and activation of neutrophils. J. Cell. Biol. 115, 223–234.

Ludwig, J.C., McManus, L.M., Clark, P.O., Hanahan, D.J. and Pinckard, R.N. (1984). Modulation of platelet-activating factor (PAF) synthesis and release from human polymorphonuclear leukocytes (PMN): role of extracellular Ca^{2+}. Arch. Biochem. Biophys. 232, 102–110.

Ludwig, J.C., Hoppens, C.L., McManus, L.M., Mott, G.E. and Pinckard, R.N. (1985). Modulation of platelet-activating factor (PAF) synthesis and release from human polymorphonuclear leukocytes (PMN): role of extracellular albumin. Arch. Biochem. Biophys. 241, 337–347.

Lukaszyk, A., Bodzenta Lukaszyk, A., Gabryelewicz, A. and Bielawiec, M. (1989). Does acute experimental pancreatitis affect blood platelet function? Thromb. Res. 53, 319–325.

Lynch, J.M. and Henson, P.M. (1986). The intracellular retention of newly synthesized platelet-activating factor. J. Immunol. 137, 2653–2661.

Macconi, D., Benigni, A., Morigi, M. et al. (1989). Enhanced glomerular thromboxane A_2 mediates some pathophysiologic effect of platelet-activating factor in rabbit nephrotoxic nephritis: evidence from biochemical measurements and inhibitor trials. J. Lab. Clin. Med. 113, 549–560.

McCormack, D.G., Barnes, P.J. and Evans, T.W. (1990). Platelet-activating factor: evidence against a role in hypoxic pulmonary vasoconstriction. Crit. Care Med. 18, 1398–1402.

McCulloch, R.K. and Vandongen, R. (1990). Mechanisms of platelet activating factor-induced aggregation and secretion in human platelets. Prostaglandins 39, 13–21.

McCulloch, R.K., Summers, J. Vandongen, R. and Rouse I.L. (1989). Role of thromboxane A_2 as a mediator of platelet-activating-factor-induced aggregation of human platelets. Clin. Sci. 77, 99–103.

McCusker, M.T., Chung, K.F., Roberts, N.M. and Barnes, P.J. (1989). Effect of topical capsaicin on the cutaneous responses to inflammatory mediators and to antigen in man. J. Allergy Clin. Immunol. 83, 1118–1124.

McGivern, D.V. and Basran, G.S. (1984). Synergism between platelet-activating factor (PAF-acether) and prostaglandin E2 in man. Eur. J. Pharmacol. 102, 183–185.

McGowan, H.M., Vandongen, R., Codde, J.P. and Croft, K.D. (1986). Increased aortic PGI_2 and plasma lyso-PAF in the unclipped one-kidney hypertensive rat. Am. J. Physiol. 251, H1361–H1364.

McGowan, H.M., Vandongen, R., Kelly, L.D. and Hill, K.J. (1988). Increased levels of platelet-activating factor (1-O-alkyl-2-acetylglycerophosphocholine) in blood after reversal of renal clip hypertension in the rat. Clin. Sci. 74, 393–396.

McIntyre T.M., Zimmerman, G.A., Satoh, K. and Prescott, S.M. (1985). Cultured endothelial cells synthesize both platelet-activating factor and prostacyclin in response to histamine,

bradykinin, and adenosine triphosphate. J. Clin. Invest. 76, 271–280.

McIntyre, T.M., Zimmerman, G.A. and Prescott, S.M. (1986). Leukotrienes C_4 and D_4 stimulate human endothelial cells to synthesize platelet-activating factor and bind neutrophils. Proc. Natl Acad. Sci USA 83, 2204–2208.

McIntyre, T.M., Reinhold, S.L., Prescott, S.M. and Zimmerman, G.A. (1987). Protein kinase C activity appears to be required for the synthesis of platelet-activating factor and leukotriene B_4 by human neutrophils. J. Biol. Chem. 262, 15370–15376.

McManus, L.M. and Deavers, S.I. (1989). Platelet activating factor in pulmonary pathobiology. Clin. Chest. Med. 10, 107–118.

McManus, L.M., Hanahan, D.J., Demopoulos, C.A. and Pinckard, R.N. (1980). Pathobiology of the intravenous infusion of acetyl glyceryl ether phosphorylcholine (AGEPC), a synthetic platelet-activating factor (PAF), in the rabbit. J. Immunol. 124, 2919–2924.

McMurtry, I.F. and Morris, K.G. (1986). Platelet-activating factor causes pulmonary vasodilation in the rat. Am. Rev. Respir. Dis. 134, 757–762.

Maes, L., Andries, R., Bourgain, R.H. and Braquet, P. (1986). Endogenic PAF-acether production by guinea pig endothelial cells in experimental arterial thrombosis. Pharmacol. Res. Commun. 18 Suppl, 81–89.

Maki, N., Hoffman, D.R. and Johnston, J.M. (1988). Platelet-activating factor acetylhydrolsase activity in maternal, fetal, and newborn rabbit plasma during pregnancy and lactation. Proc. Natl Acad. Sci. USA 85, 728–732.

Makowka, L., Miller, C., Chapchap. P. et al. (1987). Prolongation of pig-to-dog renal xenograft survival by modification of the inflammatory mediator response. Ann. Surg. 206, 482–495.

Makowka, L. Chapman, F.A., Cramer, D.V., Qian, S.G., Sun, H. and Starzl, T.E. (1990). Platelet-activating factor and hyperacute rejection. The effect of a platelet-activating factor antagonist, SRI 63–441, on rejection of xenografts and allografts in sensitized hosts. Transplantation 50, 359–365.

Mallet, A.I. and Cunningham, F.M. (1985). Structural identification of platelet-activating factor in psoriatic scale. Biochem. Biophys. Res. Commun. 126, 192–198.

Mallet, A.I., Cunningham, F.M. and Daniel, R. (1984). Rapid isocratic high-performance liquid chromatographic purification of platelet activating factor (PAF) and lyso-PAF from human skin. J. Chromatogr. 309, 160–164.

Malo, P.E., Wasserman, M.A. and Pfeiffer, D.F. (1987). Enhancement of leukotriene D_4-induced contraction of guinea-pig isolated trachea by platelet activating factor. Prostaglandins 33, 209–225.

Mandi, Y., Farkas, G., Koltai, M., Beladi, I. and Braquet, P. (1989a). Effect of the platelet-activating factor antagonist BN 52021 on human and natural killer cell cytotoxicity. Int. Arch. Allergy Appl. Immunol. 88, 222–224.

Mandi, Y., Farkas, G., Koltai, M., Beladi, I., Mencia Huerta, J.M. and Braquet, P. (1989b). The effect of the platelet-activating factor antagonist, BN 52021 on human natural killer cell-mediated cytotoxicity. Immunology 67, 370–374.

Marcheselli, V.L., Rossowska, M.J., Domingo, M.-T., Braquet, P. and Bazan, N.G. (1990). Distinct platelet-activatng factor binding sites in synaptic endings and in intracellular membranes of rat cerebral cortex. J. Biol. Chem. 265, 9140–9145.

Marcus, A.J., Safier, L.B., Ullman, H.L. et al. (1981). Effects of acetyl glyceryl ether phosphorylcholine on human platelet function in vitro. Blood 58, 1027–1031.

Maridonneau Parini, I., Lagente, V., Lefort, J., Randon, J., Russo Marie, F. and Vargaftig, B.B. (1985). Desensitization to PAF-induced bronchoconstriction and to activation of alveolar macrophages by repeated inhalations of PAF in the guinea pig. Biochem. Biophys. Res. Commun. 131, 42–49.

Markey, A.C., Barker, J.N., Archer, C.B., Guinot, P., Lee, T.H. and MacDonald, D.M. (1990). Platelet activating factor-induced clinical and histopathologic responses in atopic skin and their modification by the platelet activating factor antagonist BN52063. J. Am. Acad. Dermatol. 23, 263–268.

Martins, M.A, Silva, P.M., Castro, H.C. et al. (1987). Interactions between local inflammatory and systemic haematological effects of PAF-acether in the rat. Eur. J. Pharmacol. 136, 353–360.

Martins, M.A., Martins, P.M., Faria Neto, H.C. et al. (1988). Intravenous injections of PAF-acether induce platelet aggregation in rats. Eur. J. Pharmacol. 149, 89–96.

Martins, M.A., Silva, P.M., Faria Neto, H.C. et al. (1989). Pharmacological modulation of Paf-induced rat pleurisy and its role in inflammation by zymosan. Br. J. Pharmacol. 96, 363–371.

Martiny-Baron, G. and Scherer, G.F. (1989). Phospholipid-stimulated protein kinase in plants. J. Biol. Chem. 264, 18052–18059.

Masugi, F., Ogihara, T., Saeki, S., Sakaguchi, K. and Kumahara, Y. (1988a). Platelet-activating factor and anti-platelet-aggregating factor in acute reduction of blood pressure following percutaneous transluminal renal angioplasty in patients with renovascular hypertension. J. Hum. Hypertens. 2, 111–116.

Masugi, F., Ogihara, T., Saeki, S. et al. (1988b). Endogenous platelet-activating factor and anti-platelet-activating factor in patients with renovascular hypertension. Life. Sci. 42, 455–460.

Matsumoto, M. and Miwa, M. (1985). Platelet-activating factor-binding protein in human serum. Adv. Prostaglandin Thromboxane Leukotriene Res. 15, 705–706.

Maudsley, D.J. and Morris, A.G. (1987). Rapid intracellular calcium changes in U937 monocyte cell line: transient increases in response to platelet-activating factor and chemotactic peptide but not interferon-gamma or lipopolysaccharide. Immunology 61, 189–194.

May, G.R., Herd, C.M., Butler, K.D. and Page, C.P. (1990). Radioisotopic model for investigating thromboembolism in the rabbit. J. Pharmacol. Methods 24, 19–35.

Mazer, B., Clay, K.L., Renz, H. and Gelfand, E.W. (1990). Platelet-activating factor enhances Ig production in B lymphoblastoid cell lines. J. Immunol. 145, 2602–2607.

Meade, C.J. and Heuer, H. (1991). PAF antagonism as an approach to the treatment of airway hyperreactivity. Am. Rev. Respir. Dis. 143, s79–s82.

Meade, C.J., Heuer, H. and Kempe, R., (1991). Biochemical pharmacology of platelet-activating factor (and Paf antagonists) in relation to clinical and experimental thrombocytopaenia. Biochem Pharmacol. 41, 657–668.

Medeiros, Y.S. and Calixto, J.B. (1989). Effect of PAF-acether on the reactivity of the isolated rat myometrium. Braz. J. Med. Biol. Res. 22, 1131–1135.

Medeiros, Y.S., Torres, R.C. and Calixto, J.B. (1989). Tracheal responsiveness to histamine, PAF-acether and acetylcholine in normal and actively ovalbumin-sensitized guinea pigs. Braz. J. Med. Biol. Res. 22, 97–101.

Mencia Huerta, J.M. and Benveniste, J. (1979). Platelet-activating factor and macrophages. I. Evidence for the release from rat and mouse peritoneal macrophages and not from mastocytes. Eur. J. Immunol. 9, 409–415.

Mencia Huerta, J.M. and Benveniste, J. (1981). Platelet-activating factor (PAF-acether) and macrophages. II. Phagocytosis-associated release of PAF-acether from rat peritoneal macrophages. Cell Immunol. 57, 281–292.

Mencia Huerta, J.M., Ninio, E., Roubin, R. and Benveniste, J. (1981). Is platelet-activating factor (PAF-acether) synthesis by murine peritoneal cells (PC) a two-step process? Agents Actions 11, 556–558.

Mencia Huerta, J.M., Touvay, C., Pfister, A. and Braquet, P. (1989). Effect of long-term administration of platelet-activating factor on pulmonary responsiveness and morphology in the guinea pig. Int. Arch. Allergy Appl. Immunol. 88, 154–156.

Mest, H.J., Schneider, S. and Riedel, A. (1987). Relevance of eicosanoids for biochemical regulation of cardiac rhythm disturbances. Biomed. Biochim. Acta 46, S534–S538.

Michel, L. and Dubertret, L. (1985). A simple method for studying chemotaxis, vascular permeability and histological modifications induced by mediators of inflammation in vivo in man. Br. J. Dermatol. 113(Suppl. 28), 61–66.

Michel, L., Mencia Huerta, J.M., Benveniste, J. and Dubertret, L. (1987). Biologic properties of LTB$_4$ and paf-acether in vivo in human skin. J. Invest. Dermatol. 88, 675–681.

Michel, L., De Vos, C., Rihoux, J.P., Burtin, C., Benveniste, J. and Dubertret, L. (1988a). Inhibitory effect of oral cetirizine on in vivo antigen-induced histamine and PAF-acether release and eosinophil recruitment in human skin. J. Allergy Clin. Immunol. 82, 101–109.

Michel, L., Denizot, Y., Thomas, Y., Benveniste, J. and Dubertret, L. (1988b). Release of PAF-acether and precursors during allergic cutaneous reactions. Lancet ii, 404.

Michel, L., Denizot, Y., Thomas, Y. et al. (1988c). Biosynthesis of paf-acether factor-acether by human skin fibroblasts in vitro. J. Immunol. 141, 948–953.

Michel, L., Denizot, Y., Thomas, Y. et al. (1990). Production of paf-acether by human epidermal cells. J. Invest. Dermatol. 95, 576–581.

Milligan, S.R. and Finn, C.A. (1990). Failure of platelet-activating factor (PAF-acether) to induced decidualization in mice and failure of antagonists of PAF to inhibit implantation. J. Reprod. Fertil. 88, 105–112.

Misawa, M. and Takata, T. (1988). Effects of platelet-activating factor on rat airways. Jpn. J. Pharmacol. 48, 7–13.

Miwa, M., Miyake, T., Yamanaka, T. et al. (1988). Characterization of serum platelet-activating factor (PAF) acetylhydrolase. Correlation between deficiency of serum PAF acetylhydrolase and respiratory symptoms in asthmatic children. J. Clin. Invest. 82, 1983–1991.

Miyake, T. and Miwa, M. (1989). Influence of theophylline and corticosteroid preparations on serum PAF acetylhydrolase activity in asthmatic children. Arerugi 38, 423–427.

Moncada, S., Radomski, M.W. and Palmer, R.M.J. (1988). Endothelium-derived relaxing factor: identification as nitric oxide and role in the control of vascular tone and platelet function. Biochem. Pharmacol. 37, 2495–2501.

Montrucchio, G., Alloatti, G., Tetta, C., Roffinello, C., Emanuelli, G. and Camussi, G. (1986a). In vitro contractile effect of platelet-activating factor on guinea-pig myometrium. Prostaglandins 32, 539–554.

Montrucchio, G., Camussi, G., Tetta, C. et al. (1986b).

Intravascular release of platelet-activating factor during atrial pacing. Lancet ii, 293.

Montrucchio, G., Alloatti, G., Mariano, F. et al. (1987). The pattern of cardiovascular alterations induced by infusion of platelet-activating factor in rabbit is modified by pretreatment with H_1–H_2 receptor antagonists but not by cyclooxygenase inhibition. Agents Actions 21, 72–78.

Montrucchio, G., Alloatti, G., Tetta, C. et al. (1989a). Release of platelet-activating factor from ischemic-reperfused rabbit heart. Am. J. Physiol. 256, H1236–H1246.

Montrucchio, G., Mariano, F., Cavalli, P.L. et al. (1989b). Platelet-activating factor is produced during infectious peritonitis in CAPD patients. Kidney Int. 36, 1029–1036.

Moon, D.G. Van Der Zee, H., Morton, K.D., Krasodomski, J.A., Kaplan, J.E. and Fenton, J.W. (1990). Platelet activating factor and sheep platelets: a sensitive new bioassay. Thromb. Res. 57, 551–564.

Moqbel, R, Cromwell, O. and Kay, A.B. (1989). The effect of nedocromil sodium on human eosinophil activation. Drugs 37(Suppl.1), 19–22.

Moqbel, R., Walsh, G.M., Nagakura, T. et al. (1990). The effect of platelet-activating factor on IgE binding to, and IgE-dependent biological properties of, human eosinophils. Immunology 70, 251–257.

Morell, G.P., Pirotzky, E., Erard, D. et al. (1988). Paf-acether (platelet-activating factor) and interleukin-1-like cytokine production by lipopolysaccharide-stimulated glomeruli. Clin. Immunol. Immunopathol. 46, 396–405.

Morita, E., Schroder, J.M. and Christophers, E. (1989). Differential sensitivities of purified human eosinophils and neutrophils to defined chemotaxins. Scand. J. Immunol. 29, 709–716.

Morley, J. (1987). PAF and airway hyperreactivity; prospects for novel prophylactic anti-asthma drugs. Agents Actions Suppl. 21, 87–95.

Morley, J., Page, C.P. and Paul, W. (1983). Inflammatory actions of platelet activating factor (Paf-acether) in guinea-pig skin. Br. J. Pharmacol. 80, 503–509.

Morley, J., Page, C.P. and Sanjar, S. (1985). PAF and airway hyperreactivity. Lancet ii, 451.

Morley, J., Sanjar, S., Boubekeur, K., Aoki, S. and Kristersson, A. (1988). Pharmacological evaluation of prophylactic anti-asthma drugs by reference to the pathological sequelae of exposure to allergen or platelet activating factor. Agents Actions Suppl. 23, 187–194.

Morris, D.D. and Moore, J.N. (1989). Equine peritoneal macrophage production of thromboxane and prostacyclin in response to platelet activating factor and its receptor antagonist SRI 63–441. Circ. Shock. 28, 149–158.

Mozes, T. Braquet, P. and Filep, J. (1989). Platelet-activating factor: an endogenous mediator of mesenteric ischemia-reperfusion-induced shock. Am. J. Physiol. 257, R872–R877.

Mueller, H.W., O'Flaherty, J.T. and Wykle, R.L. (1984). The molecular species distribution of platelet-activating factor synthesized by rabbit and human neutrophils. J. Biol. Chem. 259, 14554–14559.

Muirhead, E.E. (1990). Discovery of the renomedullary system of blood pressure control and its hormones. Hypertension 15, 114–116.

Munoz, N.M., Stimler Gerard, N.P., Mack, M.M. et al. (1989). Evidence for platelet-activating factor secretion during immune activation in dogs. J. Appl. Physiol. 66, 1860–1866.

Murphy, S. and Welk, G. (1990). Hydrolysis of polyphosphoinositides in astrocytes by platelet-activating factor. Eur. J. Pharmacol. 188, 399–401.

Murphy, T.M., Munoz, N.M., Moss, J., Blake, J.S., Mack, M.M. and Leff, A.R. (1989). PAF-induced contraction of canine trachea mediated by 5-hydroxytryptamine in vivo. J. Appl. Physiol. 66, 638–643.

Myers, A.K., Robey, J.W. and Price, R.M. (1990). Relationships between tumour necrosis factor, eicosanoids and platelet-activating factor as mediators of endotoxin-induced shock in mice. Br. J. Pharmacol. 99, 499–502.

Nakaya, H. and Tohse, N. (1986). Electrophysiological effects of acetyl glyceryl ether phosphorylcholine on cardiac tissues: comparison with lyso-phosphatidyl choline and long chain acyl-carnitine. Br. J. Pharmacol. 89, 749–757.

Nakayama, R., Oda, M., Satouchi, K. and Saito, K. (1985). Generation of acetyl glyceryl ether phosphoryl choline from the rat skin and muscle tissues stimulated by moxibustion. Biochem. Biophys. Res. Commun. 127, 629–634.

Nakayama, R., Yasuda, K. and Saito, K. (1987). Existence of endogenous inhibitors of platelet-activating factor (PAF) with PAF in rat uterus. J. Biol. Chem. 262, 13174–13179.

Neuwirth, R., Singhal, P., Satriano, J.A., Braquet, P. and Schlondorff, D. (1987). Effect of platelet activating factor antagonists on cultured rat mesangial cells. J. Pharmacol. Exp. Ther. 243, 409–414.

Neuwirth, R., Satriano, J.A., DeCandido, S., Clay, K. and Schlondorff, D. (1989). Angiotensin II causes formation of platelet activating factor in cultured rat mesangial cells. Circ. Res. 64, 1224–1229.

Nieto, M.L., Velasco, S. and Sanchez Crespo, M. (1988). Biosynthesis of platelet-activating factor in human polymorphonuclear leukocytes. Involvement of the cholinephosphotransferase pathway in response to the phorbol esters. J. Biol. Chem. 263, 2217–2222.

Noris, M., Perico, N., Macconi, D. et al. (1990). Renal metabolism and urinary excretion of platelet-activating factor in the rat. J. Biol. Chem. 265, 19414–19419.

Noseda, A., Berens, M.E., Piantadosi, C. and Modest, E.J. (1987). Neoplastic cell inhibition with new ether lipid analogs. Lipids 22, 878–883.

Noseda, A., White, J.G., Godwin, P.L., Jerome, W.G. and Modest, E.J. (1989). Membrane damage in leukemic cells induced by ether and ester lipids: an electron microscopic study. Exp. Mol. Pathol. 50, 69–83.

Nunn, B. (1983). Total and selective desensitization of human platelets to synthetic platelet activating factor (PAF): evidence that extracellular PAF does not mediate collagen-induced aggregation. Thromb. Res. 31, 657–663.

O'Donnell, M.C., Henson, P.M. and Fiedel, B.A. (1978). Activation of human platelets by platelet activating factor (PAF) derived from sensitized rabbit basophils. Immunology 35, 953–958.

O'Donnell, S.R. and Barnett, C.J. (1987). Microvascular leakage to platelet activating factor in guinea-pig trachea and bronchi. Eur. J. Pharmacol. 138, 385–396.

O'Donnell, S.R., Erjefält, I. and Persson, C.G.A. (1990). Early and late tracheobronchial plasma exudation by platelet-activating factor administered to the airway mucosal surface in guinea pigs: Effects of WEB 2086 and enprofylline. J. Pharmacol. Exp. Ther. 254, 65–70.

O'Flaherty, J.T., (1985). Neutrophil degranulation: evidence pertaining to its mediation by the combined effects of leukotriene B_4, platelet-activating factor, and 5-HETE. J. Cell. Physiol. 122, 229–239.

O'Flaherty, J.T., Wykle, R.L., Thomas, M.J. and McCall, C.E. (1984). Neutrophil degranulation responses to combinations of arachidonate metabolites and platelet-activating factor. Res. Commun. Chem. Pathol. Pharmacol. 43, 3–23.

O'Neill, C., (1985). Partial characterization of the embryo-derived platelet-activating factor in mice. J. Reprod. Fertil. 75, 375–380.

O'Neill, C., Gidley Baird, A.A., Pike, I.L. and Saunders, D.M. (1987). Use of a bioassay for embryo-derived platelet-activating factor as a means of assessing quality and pregnancy potential of human embryos. Fertil. Steril. 47, 969–975.

O'Neill, C., Ryan, J.P., Collier, M., Saunders, D.M., Ammit, A.J. and Pike, I.L. (1989). Supplementation of in-vitro fertilization culture medium with platelet activating factor. Lancet ii, 769–772.

Oh-ishi, S., Hayashi, M. and Yamaki, K. (1986). Inflammatory effects of acetylglyceryl ether phosphorylcholine: vascular permeability increase and induction of pleurisy in rats. Prostaglandins Leukotrienes Med. 22, 21–33.

Okamoto, M., Yoshida, K., Nishikawa, M., Kohsaka, M. and Aoki, H. (1986). Platelet activating factor (PAF) involvement in endotoxin-induced thrombocytopenia in rabbits: studies with FR-900452, a specific inhibitor of PAF. Thromb. Res. 42, 661–671.

Okubo, K., Karashima, T., Hachisuka, H. and Sasai, Y. (1989). The existence of a platelet-activating factor (PAF-acether)-like substance in blister fluid derived from patients with bullous pemphigoid as demonstrated by human platelet aggregation techniques. Kurume Med. J. 36, 173–179.

Olson, N.C., Joyce, B.B. and Fleisher, L.N. (1990a). Mono-hydroxyeicosatetraenoic acids during porcine endotoxemia. Effect of a platelet-activating factor receptor antagonist. Lab. Invest. 63, 221–232.

Olson, N.C., Joyce, P.B. and Fleisher, L.N. (1990b). Role of platelet-activating factor and eicosanoids during endotoxin-induced lung injury in pigs. Am. J. Physiol. 258, H1674–H1686.

Ono, S., Koike, K., Tanita, T. et al. (1989). Possible involvement of vascular endothelium-derived relaxing factor induced by platelet-activating factor (PAF) on human pulmonary artery after contraction in vitro. Nippon Kyobu Shikkan Gakkai Zasshi. 27, 910–916.

Ostermann, G., Block, H.U. and Till, U. (1984a). The influence of PAF on the metabolization of arachidonic acid by human blood platelets. Biomed. Biochim. Acta 43, S323–S326.

Ostermann, G., Till, U. and Thielmann, K. (1984b). Cooperative effects of 1-O-alkyl-2-O-acetyl-sn-glycero-3-phosphocholine (PAF-acether) and exogenous arachidonic acid in stimulation of human blood platelets. Thromb. Res. 33, 409–418.

Page, C.P. (1987). A role for platelet activating factor and platelets in the induction of bronchial hyperreactivity. Int. J. Tissue React. 9, 27–32.

Page, C.P. and Coyle, A.J. (1989). The interaction between PAF, platelets and eosinophils in bronchial asthma. Eur. Respir. J. Suppl. 2, 483s–487s.

Page, C.P. and Morley, J. (1986). Evidence favouring PAF rather than leukotrienes in the pathogenesis of asthma. Pharmacol. Res. Commun. 18(Suppl.), 217–237.

Page, C.P. and Robertson, D.N. (1987). Properties of PAF relevant to asthma. Agents Actions Suppl. 21, 67–76.

Page, C.P. and Robertson, D.N. (1988). The role of PAF in altered airway responsiveness. Prog. Clin. Biol. Res. 263, 71–80.

Page, C.P., Paul, W. and Morley, J. (1982). An in vivo model for studying platelet aggregation and disaggregation. Thromb. Haemost. 47, 210–213.

Page, C.P., Paul, W., Dewar, A., Wood, L., Basran, G.S. and Morley, J. (1983). PAF-acether: a putative mediator of asthma and inflammation. Agents Actions Suppl. 13, 177–183.

Page, C.P., Paul, W. and Morley, J. (1984). Platelets and bronchospasm. Int. Arch. Allergy Appl. Immunol. 74, 347–350.

Page, C.P., Archer, C.B., Guerreiro, D. and Morley, J. (1985a). Properties of PAF-acether appropriate to a mediator of the inflammatory aspects of asthma. Int. J. Tissue React. 7, 351–354.

Page, C.P., Tomiak, R.H., Sanjar, S. and Morley, J. (1985b). Suppression of PAF-acether responses: an anti-inflammatory effect of anti-asthma drugs. Agents Actions 16, 33–35.

Panetta, T., Marcheselli, V.L. Braquet, P. and Bazan, N.G. (1989). Arachidonic acid metabolism and cerebral blood flow in the normal, ischemic, and reperfused gerbil brain. Inhibition of ischemia-reperfusion-induced cerebral injury by a platelet-activating factor antagonist (BN 52021). Ann. N.Y. Acad. Sci. 559, 340–351.

Parente, L., Fitzgerald, M., De Nucci, G. and Moncada, S. (1985). The release of lyso-PAF from guinea-pig lungs. Eur. J. Pharmacol. 112, 281–284.

Parks, J.E, Hough, S. and Elrod, C. (1990). Platelet-activating activity in the phospholipids of bovine spermatozoa. Biol. Reprod. 43, 806–811.

Parnham, M.J. (1988). PAF receptors. Prog. Biochem. Pharmacol. 22, 17–24.

Parnham, M.J. and Bittner, C. (1986). Pharmacological analysis of guinea-pig macrophage chemiluminescence responses to platelet activating factor and opsonized zymosan. Int. J. Immunopharmacol. 8, 951–959.

Parnham, M.J., Bittner, C. and Lambrecht, G. (1989). Antagonism of platelet activating factor-induced chemiluminescence in guinea-pig peritoneal macrophages in differing states of activation. Br. J. Pharmacol. 98, 574–580.

Parnham, M.J., Winkelman, J., Hartung, H.P. and Hadding, U. (1984). Regulation of the oxidative burst of macrophages by lipid mediators. Agents Actions Suppl. 14, 215–226.

Patrignani, P., Valitutti, S., Aiello, F. and Musiani, P. (1987). Platelet-activating factor (PAF) receptor antagonists inhibit mitogen-induced human peripheral blood T-cell proliferation. Biochem. Biophys. Res. Commun. 148, 802–810.

Patterson, R., Harris, K.E., Lee, M.L. and Houlihan, W.J. (1986). Inhibition of rhesus monkey airway and cutaneous responses to platelet-activating factor (PAF) (AGEPC) with the anti-PAF agent SRI 63–072. Int. Arch. Allergy Appl. Immunol. 81, 265–268.

Patterson, R., Harris, K.E., Bernstein, P.R., Krell, R.D., Handley, D.A. and Saunders, R.N. (1989). Effects of combined receptor antagonists of leukotriene D_4 (LTD_4) and platelet-activating factor (PAF) on rhesus airway responses to LTD4, PAF and antigen. Int. Arch. Allergy Appl. Immunol. 88, 462–270.

Pelczar Wissner, C.J., McDonald, E.G. and Sussman, I.I. (1984). Absence of platelet activating factor (PAF) mediated platelet aggregation: a new platelet defect. Am. J. Hematol. 16, 419–422.

Perico, N. and Remuzzi, G. (1990). Role of platelet-activating factor in renal immune injury and proteinuria. Am. J. Nephrol. 10(Suppl. 1), 98–104.

Perico, N., Delani, F., Tagliaferri, M. *et al.* (1988). Effect of platelet-activating factor and its specific receptor antagonist on glomerular permeability to proteins in isolated perfused rat kidney. Lab. Invest. 58, 163–171.

Pettipher, E.R., Higgs, G.A. and Henderson, B. (1987). PAF-acether in chronic arthritis. Agents Actions 21, 98–103.

Pickett, W.C., Nytko, D., Dondero Zahn, C. and Ramesha, C. (1986). The effect of endogenous eicosapentaenoic acid on PMN leukotriene and PAF biosynthesis. Prostaglandins Leukotrienes Med. 23, 135–140.

Pignol, B., Henane, S., Mencia Huerta, J.M., Rola Pleszczynski, M. and Braquet, P. (1987). Effect of platelet-activating factor (PAF-acether) and its specific receptor antagonist, BN 52021, on interleukin 1 (IL1) release and synthesis by rat spleen adherent monocytes. Prostaglandins 33, 931–939.

Pignol, B., Hénane, S., Sorlin, B., Rola-Pleszczynski, M., Mencia-Huerta, J.-M. and Braquet, P. (1990). Effect of long-term treatment with platelet-activating factor on IL-1 and IL-2 production by rat spleen cells. J. Immunol. 145, 980–984.

Pinckard, R.N., Halonen, M., Palmer, J.D., Butler, C., Shaw, J.O. and Henson, P.M. (1977). Intravascular aggregation and pulmonary sequestration of platelets during IgE-induced systemic anaphylaxis in the rabbit: abrogation of lethal anaphylactic shock by platelet depletion. J. Immunol. 119, 2185–2193.

Piper, P.J. and Stewart, A.G. (1986a). Evidence of a role for platelet-activating factor in antigen-induced coronary vasoconstriction in guinea-pig perfused hearts. Br. J. Pharmacol. 88, 238P.

Piper, P.J. and Stewart, A.G. (1986b). Coronary vasoconstriction in the rat, isolated perfused heart induced by platelet-activating factor is mediated by leukotriene C_4. Br. J. Pharmacol. 88, 595–605.

Piper, P.J. and Stewart, A.G. (1987a). Antagonism of vasoconstriction induced by platelet-activating factor in guinea-pig perfused hearts by selective platelet-activating factor receptor antagonists. Br. J. Pharmacol. 90, 771–783.

Piper, P.J., Stanton, A.W. and Stewart, A.G. (1987). Mechanisms of vascular actions of PAF. Int. J. Tissue. React. 9, 15–17.

Pique, J.M., Esplugues, J.V. and Whittle, B.J. (1990). Influence of morphine or capsaicin pretreatment on rat gastric microcirculatory response to PAF. Am. J. Physiol. 258, G352–G357.

Pirotzky, E., Misumi, J., Boullet, C. and Benveniste, J. (1980). [Release of platelet activating factor (P.A.F.-acether) by isolated perfused rat kidney]. C.R. Seances Acad. Sci. D. 290, 1079–1082.

Pirotzky, E., Bidault, J., Burtin, C., Gubler, M.C. and Benveniste, J. (1984a). Release of platelet-activating factor, slow-reacting substance, and vasoactive amines from isolated rat kidneys. Kidney Int. 25, 404–410.

Pirotzky, E., Ninio, E., Bidault, J. Pfister, A. and Benveniste, J. (1984b). Biosynthesis of platelet-activating factor. VI. Precursor of platelet-activating factor and acetyltransferase activity in isolated rat kidney cells. Lab. Invest. 51, 567–572.

Pirotzky, E., Page, C.P., Roubin, R. *et al.* (1984c). PAF-acether-induced plasma exudation in rat skin is independent of platelets and neutrophils. Microcirc. Endothelium Lymphatics 1, 107–122.

Pirotzky, E., Page, C., Morley, J., Bidault, J. and Benveniste, J. (1985a). Vascular permeability induced by Paf-acether (platelet-activating factor) in the isolated perfused rat kidney. Agents Actions 16, 17–18.

Pirotzky, E. Pfister, A. and Benveniste, J. (1985b). A role for Paf-acether (platelet-activating factor) in acute skin inflammation? Br. J. Dermatol. 113(Suppl. 28), 91–94.

Pirotzky, E., Pintos Morell, G. Burtin, C., Boudet, J., Bidault, J. and Benveniste, J. (1987). Renal anaphylaxis, I. Antigen-initiated responses from isolated perfused rat kidney. Kidney Int. 32, 233–237.

Plante, G.E., Hebert, R.L., Lamoureux, C., Braquet, P. and Sirois, P. (1986). Hemodynamic effects of PAF-acether. Pharmacol. Res. Commun. 18(Suppl.), 173–179.

Poitevin, B., Roubin, R. and Benveniste, J. (1984). Paf-acether generates chemiluminescence in human neutrophils in the absence of cytochalasin B. Immunopharmacology 7, 135–144.

Pons, L. Droy Lefaix, M.T., Braquet, P. and Bueno, L. (1989). Involvement of platelet-activating factor (PAF) in endotoxin-induced intestinal motor disturbances in rats. Life Sci. 45, 533–541.

Popovich, K.J., Sheldon, G., Mack, M. *et al.* (1988). Role of platelets in contraction of canine tracheal muscle elicited by PAF *in vitro*. J. Appl. Physiol. 65, 914–920.

Prancan, A., Lefort, J., Barton, M. and Vargaftig, B.B. (1982). Relaxation of the guinea-pig trachea induced by platelet-activating factor and by serotonin. Eur. J. Pharmacol. 80, 29–35.

Prescott, S.M., Zimmerman, G.A. and McIntyre, T.M. (1984). Human endothelial cells in culture produce platelet-activating factor (1-alkyl-2-acetyl-*sn*-glycero-3-phosphocholine) when stimulated with thrombin. Proc. Natl Acad. Sci. USA 81, 3535-3538.

Pretolani, M., Lefort, J., Malanchere, E. and Vargaftig, B.B. (1987). Interference by the novel PAF-acether antagonist WEB 2086 with the bronchopulmonary responses to PAF-acether and to active and passive anaphylactic shock in guinea-pigs. Eur. J. Pharmacol. 140, 311–321.

Pretolani, M., Lefort, J. and Vargaftig, B.B. (1988). Active immunization induces lung hyperresponsivieness in the guinea pig. Pharmacologic modulation and triggering role of the booster injection. Am. Rev. Respir. Dis. 138, 1572–1578.

Pretolani, M., Lefort, J. and Vargaftig, B.B. (1989). Limited interference of specific Paf antagonists with hyper-responsiveness to Paf itself of lungs from actively sensitized guinea-pigs. Br. J. Pharmacol. 97, 433–442.

Pretolani, M., Lefort, J., Silva P. *et al.* (1990). Protection by nedocromil sodium of active immunization-induced bronchopulmonary alterations in the guinea pig. Am. Rev. Respir. Dis. 141, 1259–1265.

Prevost, M.C., Cariven, C. and Chap, H. (1988). Possible origins of PAF-acether and lyso-PAF-acether in rat lung alveoli secondary to hypoxia. Biochim. Biophys. Acta 962, 354–361.

Prpic, V., Uhing, R.J., Weiel, J.E. *et al.* (1988). Biochemical and functional responses stimulated by platelet-activating factor in murine peritoneal macrophages. J. Cell. Biol. 107, 363–372.

Raaijmakers, J.A., Beneker, C., van Geffen, E.C., Meisters, T.M. and Pover, p. (1989). Inflammtory mediators and beta-adrenoceptor function. Agents Actions 26, 45–57.

Rabier, M., Damon, M., Chanez, P. *et al.* (1989). Inhibition by histamine of platelet-activating-factor-induced neutrophil chemotaxis in bronchial asthma. Int. Arch. Allergy Appl. Immunol. 89, 314–317.

Rabinovici, R., Yue, T.L., Farhat, M. *et al.* (1990). Platelet activating factor (PAF) and tumor necrosis factor-alpha (TNF alpha) interactions in endotoxemic shock: studies with BN

50739, a novel PAF antagonist. J. Pharmacol. Exp. Ther. 255, 256–263.

Rabovsky, J. Goddard, M., Pailes, W.H., Judy, D.J. and Castranova, V. (1990). Platelet-activating factor-induced aggregation of rat alveolar macrophages. Res. Commun. Chem. Pathol. Pharmacol. 69, 163–172.

Rainsford, K.D. (1986). Relative roles of leukotrienes and platelet activating factor in experimentally-induced gastric ulceration. Pharmacol. Res. Commun. 18(Suppl.), 209–215.

Ramesha, C.S. and Pickett, W.C. (1987). Human neutrophil platelet-activating factor: molecular heterogeneity in unstimulated and ionophore-stimulated cells. Biochim. Biophys. Acta 921, 60–66.

Rao, A.K., Willis, J., Hassell, B., Dangelamier, C., Holmsen, H. and Smith, J.B. (1984). Platelet-activating factor is a weak platelet agonist: evidence from normal human platelets and platelets with congenital secretion defects. Am. J. Hematol. 17, 153–165.

Record, M., El Tamer, A., Chap, H. and Douste Blazy, L. (1984). Evidence for a highly asymmetric arrangement of ether- and diacyl-phospholipid subclasses in the plasma membrane of Krebs II ascites cells. Biochim. Biophys. Acta 778, 449–456.

Ribbes, G., Ninio, E., Fontan, P. et al. (1985). Evidence that biosynthesis of platelet-activating factor (PAF-acether) by human neutrophils occurs in an intracellular membrane. FEBS Lett. 191, 195–199.

Riches, D.W., Young, S.K., Seccombe, J.F., Henson, J.E., Clay, K.L. and Henson, P.M. (1990). The subcellular distribution of platelet-activating factor in stimulated human neutrophils. J. Immunol. 145, 3062–3070.

Ricker, D.D., Robertson, J.L., Minhas, B.S., Dodson, M.G. and Kumar, R. (1989). The effects of platelet-activating factor on the motility of human spermatozoa. Fertil. Steril. 52, 655–658.

Roberts, N.M., McCusker, M., Chung, K.F. and Barnes, P.J. (1988). Effect of a PAF antagonist, BN 52063, on PAF-induced bronchoconstriction in normal subjects. Br. J. Clin. Pharmacol. 26, 65–72.

Robertson, D.A., Genovese, A. and Levi, R. (1987). Negative inotropic effect of platelet-activating factor on human myocardium: a pharmacological study. J. Pharmacol. Exp. Ther. 243, 834–839.

Robertson, D.A, Wang, D.Y., Lee, C.O. and Levi, R. (1988). Negative inotropic effect of platelet-activating factor: association with a decrease in intracellular sodium activity. J. Pharmacol. Exp. Ther. 245, 124–128.

Robertson, D.N., Coyle, A.J., Rhoden, K.J., Grandordy, B., Page, C.P. and Barnes, P.J. (1988). The effect of platelet-activating factor on histamine and muscarinic receptor function in guinea pig airways. Am. Rev. Respir. Dis. 137, 1317–1322.

Robidoux, C., Maclouf, J., Pradelles, P. and Sirois, P. (1988). Release of prostaglandin E_2 and thromboxane B_2 by mixed isolated human lung cells. Prostaglandins Leukot. Essent. Fatty Acids 32, 23–28.

Rodriguez Puyol, D., Lamas, S., Olivera, A. et al. (1989). Actions of cyclosporin A on cultured rat mesangial cells. Kidney Int. 35, 632–637.

Rogers, D.F., Dijk, S. and Barnes, P.J. (1990). Bradykinin-induced plasma exudation in guinea-pig airways: involvement of platelet-activating factor. Br. J. Pharmacol. 101, 739–745.

Rola Pleszczynski, M., Pignol, B., Pouliot, C. and Braquet, P. (1987). Inhibition of human lymphocyte proliferation and interleukin 2 production by platelet activating factor (PAF-acether): reversal by a specific antagonist, BN 52021. Biochem. Biophys. Res. Commun. 142, 754–760.

Rola Pleszczynski, M., Pouliot, C., Turcotte, S., Pignol. B. Braquet, P. and Bouvrette, L. (1988). Immune regulation by platelet-activating factor. I. Induction of suppressor cell activity in human monocytes and CD8[+] T cells and of helper cell activity in CD4[+] T cells. J. Immunol. 140, 3547–3552.

Rosam, A.C., Wallace, J.L. and Whittle, B.J. (1986). Potent ulcerogenic actions of platelet-activating factor on the stomach. Nature 319, 54–56.

Rossi, A.G. and O'Flaherty, J.T. (1989). Prostaglandin binding sites in human polymorphonuclear neutrophils. Prostaglandins 37, 641–653.

Rouis, M., Nigon, F. and Chapman, M.J. (1988). Platelet activating factor is a potent stimulant of the production of active oxygen species by human monocyte-derived macrophages. Biochem. Biophys. Res. Commun, 156, 1293–1301.

Rouis, M., Nigon, F., Lafuma, C., Hornebeck, W. and Chapman, M.J. (1990). Expression of elastase activity by human monocyte-macrophages is modulated by cellular cholesterol content, inflammatory mediators, and phorbol myristate acetate. Arteriosclerosis 10, 246–255.

Rubin, A.H., Smith, L.J. and Patterson, R. (1987). The bronchoconstrictor properties of platelet-activating factor in humans. Am. Rev. Respir. Dis. 136, 1145–1151.

Rubin, R.M., Samples, J.R. and Rosenbaum, J.T. (1988). Prostaglandin-independent inhibition of ocular vascular permeability by a platelet-activating factor antagonist. Arch Ophthalmol. 106, 1116–1120.

Ryan, J.P., Spinks, N.R., O'Neill, C., Ammit, A.J. and Wales, R.G. (1989). Platelet activating factor (PAF) production by mouse embryos in vitro and its effect on embryonic metabolism. J. Cell. Biochem. 40, 387–395.

Rylander, R. and Beijer, L. (1987). Inhalation of endotoxin stimulates alveolar macrophage production of platelet-activating factor. Am. Rev. Respir. Dis. 135, 83–86.

Rylander, R., Beijer, L., Lantz, R.C. Burrell, R. and Sedivy, P. (1988). Modulation of pulmonary inflammation after endotoxin inhalation with a platelet-activating factor antagonist (48740 RP). Int. Arch. Allergy Appl. Immunol. 86, 303–307.

Saeki, S., Masugi, F., Ogihara, T. et al. (1985). Effects of 1-O-alkyl-2-acetyl-sn-glycero-3-phosphocholine (platelet activating factor) on cardiac funtion in perfused guinea-pig heart. Life Sci. 37, 325–329.

Salari, H. and Walker, M.J. (1989). Cardiac dysfunction caused by factors released from endotoxin-activated macrophages. Circ. Shock 27, 263–272.

Salari, H. and Wong, A. (1990). Generation of platelet activating factor (PAF) by a human lung epithelial cell line. Eur. J. Pharmacol. 175, 253–259.

Salari, H., Duronio, V., Howard, S.L. et al. (1990). Erbstatin blocks platelet activating factor-induced protein-tyrosine phosphorylation, polyphosphoinositide hydrolysis, protein kinase C activation, serotonin secretion and aggregation of rabbit platelets. FEBS Lett. 263, 104–108.

Salem, P., Denizot, Y., Pitton, C. et al. (1989). Presence of paf-acether in human thymus. FEBS Lett. 257, 49–51.

Salem, P., Deryckx, S., Dulioust, A. et al. (1990). Immunoregulatory functions of Paf-acether. IV. Enhancement of IL-1 production by muramyl dipeptide-stimulated monocytes. J. Immunol. 144, 1338–1344.

Salvemini, D., Korbut, R., Änggård, E. and Vane, J. (1990). Immediate release of a nitric oxide-like factor from bovine aortic edothelial cells by *Escherichia coli* lipopolysaccharide. Proc. Natl Acad. Sci. USA 87, 2593–2597.

Sanchez Crespo, M. and Nieto, M.L. (1989). Modulation of PAF biosynthesis in human polymorphonuclear leukocytes. The role of phorbol esters and protein kinases. Agents Actions 26, 111–112.

Sanchez-Crespo, M., Alonso, F., Inarrea, P. and Egido, J. (1981). Non-platelet-mediated vascular actions of 1-*O*-alkyl-2-acetyl-*sn*-3-glyceryl phosphorycholine (a synthetic PAF). Agents Actions 11, 565–566.

Sanchez Crespo, M., Alonso, F., Inarrea, P. Alvarez, V. and Egido, J. (1982). Vascular actions of synthetic PAF-acether (a synthetic platelet-activating factor) in the rat: evidence for a platelet independent mechanism. Immunopharmacology 4, 173–185.

Sanchez Crespo, M. Inarrea, P., Alvarez, V., Alonso, F., Egido, J. and Hernando, L. (1983). Presence in normal human urine of a hypotensive and platelet-activating phospholipid. Am. J. Physiol. 244, F706–F711.

Sanchez Crespo, M., Alonso, F., Garcia Gill, M., Gomez Cambronero, J. and Nieto, M.L. (1985). Synthesis of platelet-activating factor from human polymorphonuclear leukocytes: regulation and pharmacological approaches. Int. J. Tissue. React. 7, 345–349.

Sanjar, S., Aoki, S., Boubekeur, K. *et al.* (1989a). Inhibition of PAF-induced eosinophil accumulation in pulmonary airways of guinea pigs by anti-asthma drugs. Jpn. J. Pharmacol. 51, 167–172.

Sanjar, S., Smith, D., Schaeublin, E. *et al.* (1989b). The effect of prophylactic anti-asthma drugs on PAF-induced airway hyperreactivity. Jpn. J. Pharmacol. 51, 151–160.

Sanjar, S., Aoki, S., Boubekeur, K. *et al.* (1990a). Eosinophil accumulation in pulmonary airways of guinea-pigs induced by exposure to an aerosol of platelet-activating factor: effect of anti-asthma drugs. Br. J. Pharmacol. 99, 267–272.

Sanjar, S., Smith, D., Kings, M.A. and Morley, J. (1990b). Pretreatment with rh-GMCSF, but not rh-IL3, enhances PAF-induced eosinophil accumulation in guinea-pig airways. Br. J. Pharmacol. 100, 399–400.

Santos, J.C., Sanz, E. Caramelo, C. and Lopez Novoa, J.M. (1988). Effect of intrarenal infusion of synthetic PAF-acether in dogs. Rev. Esp. Fisiol. 44, 273–277.

Sato, N., Kawano, S., Tsuji, S. and Kamada, T. (1988). Microvascular basis of gastric mucosal protection. J. Clin. Gastroenterol. 10(Suppl.1), S13–S18.

Satoh, K., Imaizumi, T., Kawamura, Y., Yoshida, H., Takamatsu, S. and Takamatsu, M. (1989). Increased activity of the platelet-activating factor acetylhydrolase in plasma low density lipoprotein from patients with essential hypertension. Prostaglandins 37, 673–682.

Satouchi, K., Oda, M., Yasunaga, K. and Saito, K. (1983). Application of selected ion monitoring to determination of platelet-activating factor. J. Biochem. Tokyo 94, 2067–2070.

Saunders, R.N. and Handley, D.A. (1987). Platelet-activating factor antagonists. Annu. Rev. Pharmacol. Toxicol. 27, 237–255.

Sawyer, D.B. and Andersen, O.S. (1989). Platelet-activating factor is a general membrane perturbant. Biochim. Biophys. Acta 987, 129–132.

Scherf, H., Nies, A.S., Schwertschlag, U., Hughes, M. and Gerber, J.G. (1986). Hemodynamic effects of platelet activating factor in the dog kidney *in vivo*. Hypertension 8, 737–741.

Schick, H.D., Berdel, W.E., Fromm, M. *et al.* (1987). Cytotoxic effects of ether lipids and derivatives in human nonneoplastic bone marrow cells and leukemic cells in vitro. Lipids 22, 904–910.

Schlondorff, D. and Neuwirth, R. (1986). Platelet-activating factor and the kidney. Am. J. Physiol. 251, F1–11.

Schlondorff, D., Satriano, J.A., Hagege, J., Perez, J. and Baud, L. (1984). Effect of platelet-activating factor and serum-treated zymosan on prostaglandin E$_2$ synthesis, arachidonic acid release, and contraction of cultured rat mesangial cells. J. Clin. Invest. 73, 1227–1231.

Schlondorff, D., Goldwasser, P., Neuwirth, R., Satriano, J.A. and Clay, K.L. (1986). Production of platelet-activating factor in glomeruli and cultured glomerular mesangial cells. Am. J. Physiol. 250, F1123–F1127.

Schror, K., Ackermann, G., Hohlfeld, T. *et al.* (1989). Endothelial protection by defibrotide – a new strategy for treatment of myocardial infarction? Z. Kardiol. 78(Suppl.6), 35–41.

Schweizer, A., Brom, R., Glatt, M. and Bray, M. (1984). Leukotrienes reduce nociceptive responses to bradykinin. Eur. J. Pharmacol. 105, 105–112.

Schwertschlag, U.S., Dennis, V.W., Tucker, J.A. and Camussi, G. (1988). Nonimmunological alterations of glomerular filtration by s-PAF in the rat kidney. Kidney Int. 34, 779–785.

Sciberras, D.G., Goldenberg, M.M., Bolognese, J.A., James, I. and Baber, N.S. (1987). Inflammatory responses to intra-dermal injection of platelet activating factor, histamine and prostaglandin E$_2$ in healthy volunteers: a double blind investigation. Br. J. Clin. Pharmacol. 24, 753–761.

Serio, R. and Daniel, E.E. (1988). Thromboxane effects on canine trachealis neuromuscular function. J. Appl. Physiol. 64, 1979–1988.

Sessler, C.N., Glauser, F.L., Davis, D. and Fowler, A.A. (1988). Effects of platelet-activating factor antagonist SRI 63-441 on endotoxemia in sheep. J. Appl. Physiol. 65, 2624–2631.

Shalit, M., von Allmen, C., Atkins, P.C. and Zweiman, B. (1988). Platelet activating factor increases expression of complement receptors on human neutrophils. J. Leukocyte Biol. 44, 212–217.

Shaw, J.O. and Henson, P.M. (1980). The binding of rabbit basophil-derived platelet-activating factor to rabbit platelets. Am. J. Pathol. 98, 791–810.

Sigal, C.E., Valone, F.H., Holtzman, M.J. and Goetzl, E.J. (1987). Preferential human eosinophil chemotactic activity of the platelet-activating factor (PAF) 1-*O*-hexadecyl-2-acetyl-*sn*-glyceryl-3-phosphocholine (AGEPC). J. Clin. Immunol. 7, 179–184.

Silva, P.M., Cordeiro, R.S., Martins, M.A., Henriques, M.G. and Vargaftig, B.B. (1986). Platelet involvement in rat paw oedema induced by 2-methoxy-PAF. Inflammation 10, 393–401.

Siren, A.L. and Feuerstein, G. (1989). Effects of PAF and BN 52021 on cardiac function and regional blood flow on conscious rats. Am. J. Physiol. 257, H25–H32.

Sirois, M.G., Plante, G.E., Braquet, P. and Sirois, P. (1990). Role of eicosanoids in PAF-induced increases of the vascular permeability in rat airways. Br. J. Pharmacol. 101, 896–900.

Sisson, J.H., Prescott, S.M., McIntyre, T.M. and Zimmerman, G.A. (1987). Production of platelet-activating factor by stimulated human polymorphonuclear leukocytes. Correlation of synthesis with release, functional events, and leukotriene B4 metabolism. J. Immunol. 138, 3918–3926.

Smal, M.A,., Dziadek, M., Cooney, S.J., Attard, M. and Baldo, B.A. (1990). Examination for platelet-activating factor production by preimplantation mouse embryos using a specific radioimmunoassay. J. Reprod. Fertil. 90, 419–425.

Smallbone, B.W., Taylor, N.E. and McDonald, J.W. (1987). Effects of L-652,731, a platelet-activating factor (PAF) receptor antagonist, on PAF- and complement-induced pulmonary hypertension in sheep. J. Pharmacol. Exp. Ther. 242, 1035–1040.

Smiley, P.L., Stremler, K.E., Prescott, S.M., Zimmerman, G.A. and McIntrye, T.M. (1991). Oxidatively fragmented phosphatidylcholines activate human neutrophils through the receptor for platelet-activating factor. J. Biol. Chem. 266, 11104–11110.

Smith, D., Sanjar, S., Herd, C. and Morley, J. (1989). In vivo method for the assessment of platelet accumulation. J. Pharmacol. Methods 21, 45–59.

Smith, L.J., Rubin, A.H. and Patterson, R. (1988). Mechanism of platelet activating factor-induced bronchoconstriction in humans. Am. Rev. Respir. Dis. 137, 1015–1019.

Soler, M., Sielczak, M.W. and Abraham, W.M., (1989). A PAF antagonist blocks antigen-induced airway hyperresponsiveness and inflammation in sheep. J. Appl. Physiol. 67, 406–413.

Soler, M., Mansour, E., Fernandez, A., D'Brot, J., Ahmed, T. and Abraham, W.M. (1990). PAF-induced airway responses in sheep: effects of a PAF antagonist and nedocromil sodium. J. Allergy Clin. Immunol. 85, 661–668.

Sotomayor, H., Badier, M. Vervloet, D. and Orehek, J. (1984). Seasonal increase in carbachol airway responsiveness in patients allergic to grass pollen. Am. Rev. Respir. Dis. 130, 56–58.

Spaethe, S.M., Hsueh, W. and Needleman, P. (1989). Pharmacological manipulation of inflammation in rabbit hydronephrosis: effect of a combined cyclooxygenase/lipoxygenase inhibitor ethoxyquin, a thromboxane synthase inhibitor RS–5186 and a PAF antagonist L-659,989. J. Pharmacol. Exp. Ther. 248, 1308–1316.

Spangenberg, P., Schymik, C., Hoffman, B., Ostermann, G., Ruhling, K. and Till, U. (1989). Blood platelet behaviour in patients with a type I diabetes mellitus. Exp. Clin. Endocrinol. 94, 329–337.

Spencer, D.A., Green, S.E., Evans, J.M., Piper, P.J. and Costello, J.F. (1990). Platelet activating factor does not cause a reproducible increase in bronchial responsiveness in normal man. Clin. Exp. Allergy 20, 525–532.

Sperling, R.I., Robin, J.L., Kylander, K.A., Lee, T.H., Lewis, R.A. and Austen, K.F. (1987a). The effects of N-3 polyunsaturated fatty acids on the generation of platelet-activating factor-acether by human monocytes. J. Immunol. 139, 4186–4191.

Sperling, R.I., Weinblatt, M., Robin, J.-L. et al. (1987b). Effects of dietary supplementation with marine fish oil on leukocyte mediator generation and function in rheumatoid arthritis. Arthritis Rheum. 30, 988–997.

Spina, D., McKenniff, M.G., Coyle, A.J. et al. (1991). Effect of capsaicin on PAF-induced bronchial hyperresponsiveness and pulmonary cell accumulation in the rabbit. Br. J. Pharmacol. 103, 1268–1274.

Spinks, N.R. and O'Neill, C. Antagonists of embryo-derived platelet-activating factor prevent implantation of mouse embryos. J. Reprod. Fertil. 84, 89–98.

Spinks, N.R., Ryan, J.P. and O'Neill, C. (1990). Antagonists of embryo-derived platelet-activating factor act by inhibiting the ability of the mouse embryo to implant. J. Reprod. Fertil. 88, 241–248.

Spinnewyn, B., Blavet, N., Clostre, F., Bazan, N. and Braquet, P. (1987). Involvement of platelet-activating factor (PAF) in cerebral post-ischemic phase in Mongolian gerbils. Prostaglandins 34, 337–349.

Squinto, S.P., Block, A.L., Braquet, P. and Bazan, N.G. (1989). Platelet-activating factor stimulates a Fos/Jun/AP-1 transcriptional system in human neuroblastoma cells. J. Neurosci. Res. 24, 558–566.

Stafforini, D.M., McIntyre, T.M., Carter, M.E. and Prescott, S.M. (1987). Human plasma platelet-activating factor acetylhydrolase. Association with lipoprotein particles and role in the degradation of platelet-activating factor. J. Biol. Chem. 262, 4215–4222.

Stahl, G.L. and Lefer, A.M. (1987a). Mechanisms of platelet-activating factor-induced cardiac depression in the isolated perfused rat heart. Circ. Shock. 23, 165–177.

Stahl, G.L. and Lefer, A.M. (1987b). Heterogeneity of vascular smooth muscle responsiveness to lipid vasoactive mediators. Blood Vessels 24, 24–30.

Stahl, G.L., Lefer, D.J. and Lefer, A.M. (1987). PAF-acether induced cardiac dysfunction in the isolated perfused guinea pig heart. Naunyn-Schmiedebergs Arch. Pharmacol. 336, 459–463.

Stahl, G.L., Terashita, Z. and Lefer, A.M. (1988). Role of platelet activating factor in propagation of cardiac damage during myocardial ischemia. J. Pharmacol. Exp. Ther. 244, 898–904.

Steel, G., Wallace, J.L. and Whittle, B.J. (1987). Failure of prostaglandin E_2 and its 16,16-dimethyl analogue to prevent the gastric mucosal damage induced by PAF. Br. J. Pharmacol. 90, 365–371.

Steiger, J., Bray, M.A. and Subramanian, N. (1987). Platelet activating factor (PAF) is a potent stimulator of porcine tracheal fluid secretion in vitro. Eur. J. Pharmacol. 142, 367–372.

Stenton, S.C., Court, E.N., Kingston, W.P. et al. (1990a). Platelet-activating factor in bronchoalveolar lavage fluid from asthmatic subjects. Eur. Respir. J. 3, 408–413.

Stenton, S.C., Ward, C., Duddridge, M. et al. (1990b). The actions of GR32191B, a thromboxane receptor antagonist, on the effects of inhaled PAF on human airways. Clin. Exp. Allergy 20, 311–317.

Stephens, L., Hawkins, P.T., Carter, N. et al. (1988). L-Myo-inositol 1,4,5,6-tetrakisphosphate is present in both mammalian and avian cells. Biochem. J. 249, 271–282.

Stewart, A.G. (1989). CV 6209 is a non-competitive antagonist of platelet-activating factor receptors on guinea-pig resident peritoneal macrophages. Clin. Exp. Pharmacol. Physiol. 16, 813–820.

Stewart, A.G. and Dusting, G.J. (1988). Characterization of receptors for platelet-activating factor on platelets, polymorphonuclear leukocytes and macrophages. Br. J. Pharmacol. 94, 1225–1233.

Stewart, A.G. and Grigoriadis, G. (1991). Structure-activity relationships for platelet-activating factor (Paf) and analogues reveal differences between Paf receptors on platelets and macrophages. J. Lipid Mediators 4, 299–308.

Stewart, A.G. and Harris, T. (1991). Involvement of platelet-activating factor in endotoxin-induced priming of rabbit polymorphonuclear leukocytes. J. Lipid Mediators, 3, 125–138.

Stewart, A.G. and Phillips, W.A. (1989). Intracellular platelet-activating factor regulates eicosanoid generation in guinea-pig resident peritoneal macrophages. Br. J. Pharmacol. 98, 141–148.

Stewart, A.G. and Piper, P.J. (1986). Platelet-activating factor-induced vasoconstriction in rat isolated, perfused hearts: contribution of cyclo-oxygenase and lipoxygenase arachidonic acid metabolites. Pharmacol. Res. Commun. 18(Suppl.), 163–172.

Stewart, A.G. and Piper, P.J. (1987). In "Prostaglandins in Clinical Research" (ed H. Sinzinger), pp 229–234. Alan Liss, New York.

Stewart, A.G. and Piper, P.J. (1988a). Vasodilator actions of acetylcholine, A23187 and bradykinin in the guinea-pig isolated perfused heart are independent of prostacyclin. Br. J. Pharmacol. 95, 379–384.

Stewart, A.G. and Piper, P.J. (1988b). In "Biologically Active Ether Lipids" (ed P. Braquet, H.K. Mangold and B.B. Vargaftig), pp 132–140. Karger, Basel.

Stewart, A.G., Dubbin, P.N., Harris, T. and Dusting, G.J. (1989). Evidence for an intracellular action of platelet-activating factor in bovine cultured aortic endothelial cells. Br. J. Pharmacol. 96, 503–505.

Stewart, A.G. Dubbin, P.N., Harris, T. and Dusting, G.J. (1990). Platelet-activating factor may act as a second messenger in the release of icosanoids and superoxide anions from leukocytes and endothelial cells. Proc. Natl Acad. Sci. USA 87, 3215–3219.

Stewart, A.G., Harris, T., DeNichilo, M. and Lopez, A.F. (1991). Involvement of leukotriene B_4 and platelet-activating factor in cytokine priming of human polymorphonuclear leukocytes. Immunology 72, 206–212.

Stimler Gerard, N.P. (1986). Parasympathetic stimulation as a mechanism for platelet-activating factor-induced contractile responses in the lung. J. Pharmacol. Exp. Ther. 237, 209–213.

Stimler, N.P. and O'Flaherty, J.T. (1983). Spasomogenic properties of platelet-activating factor: evidence for a direct mechanism in the contractile response of pulmonary tissues. Am. J. Pathol. 113, 75–84.

Stock, E.L., Roth, S.I., Kim, E.D., Walsh, M.K. and Thamman, R. (1990). The effect of platelet-activating factor (PAF), histamine, and ethanol on vascular permeability of the guinea pig conjunctiva. Invest. Ophthalmol. Vis. Sci. 31, 987–992.

Stoll, L.L. and Spector, A.A. (1989). Interaction of platelet-activating factor with endothelial and vascular smooth muscle cells in coculture. J. Cell. Physiol. 139, 253–261.

Storch, J., Ferber, E. and Munder, P.G. (1988). Influence of platelet activating factor and a nonmetabolizable analogue on superoxide production by bone marrow derived macrophages. J. Leukocyte Biol. 44, 385–390.

Stremler, K.E., Stafforini, D.M., Prescott, S.M., Zimmerman, G.A. and McIntyre, T.M. (1989). An oxidized derivative of phosphatidylcholine is a substrate for the platelet-activating factor acetyl-hydrolase from human plasma. J. Biol. Chem. 264, 5331–5334.

Stremler, K.E., Stafforini, D.M., Prescott, S.M. and McIntyre, T.M. (1991). Human plasma platelet-activating factor acetyl-hydrolase: oxidatively fragmented phospholipids as substrates. J. Biol. Chem. 266, 11095–11103.

Sturk, A., Asyee, G.M., Schaap, M.C., van Maanen, M. and Ten Cate, J.W. (1985). Synergistic effects of platelet-activating factor and other platelet agonists in human platelet aggregation and release: the role of ADP and products of the cyclooxygenase pathway. Thromb. Res. 40, 359–372.

Sturk, A., Schaap, M.C., Ten Cate, J.W. et al. (1987). Platelet-activating factor: mediator of the third pathway of platelet aggregation? A study in three patients with deficient platelet-activating factor synthesis. J. Clin. Invest. 79, 344–350.

Sturm, M.J., Strophair, J.M., Kendrew, P.J., Vandongen, R., Beilin, L.J. and Taylor, R.R. (1989). Whole blood aggregation and plasma lyso-PAF related to smoking and atherosclerosis. Clin. Exp. Pharmacol. Physiol. 16, 597–605.

Sugatani, J., Fujimura, K., Miwa, M., Mizuno, T., Sameshima, Y. and Saito, K. (1989). Occurrence of platelet-activating factor (PAF) in normal rat stomach and alteration of PAF level by water immersion stress. FASEB. J. 3, 65–70.

Sugatani, J., Lee, D.Y., Hughes, K.T. and Saito, K. (1990). Development of novel scintillation proximity radioimmunoassay for platelet-activating factor measurement: comparison with bioassay and GC/MS techniques. Life Sci. 46, 1443–1450.

Sun, X.M. and Hsueh, W. (1988). Bowel necrosis induced by tumor necrosis factor in rats is mediated by platelet-activating factor. J. Clin. Invest. 81, 1328–1331.

Sun, X.M., Hsueh, W. and Torre Amione, G. (1990). Effects of in vivo "priming" on endotoxin-induced hypotension and tissue injury. The role of PAF and tumor necrosis factor. Am. J. Pathol. 136, 949–956.

Suzuki, Y., Miwa, M., Harada, M. and Matsumoto, M. (1988). Release of acetylhydrolase from platelets on aggregation with platelet-activating factor. Eur. J. Biochem. 172, 117–120.

Sybertz, E.J., Watkins, R.W., Baum, T., Pula, K. and Rivelli, M. (1985). Cardiac, coronary and peripheral vascular effects of acetyl glyceryl ether phosphoryl choline in the anaesthetized dog. J. Exp. Pharmacol. Ther. 232, 156–162.

Szczeklik, A., Krzanowski, M., Nizankowska, E. and Musial, J. (1987). Bleeding time and PAF-acether-induced platelet aggregation in atopy. Agents Actions Suppl. 21, 145–150.

Takehana, Y., Hamano, S., Kikuchi, S., Komatsu, H., Okegawa, T. and Ikeda, S. (1990). Inhibitory action of OKY-046.HCl, a specific TXA_2 synthetase inhibitor, on platelet activating factor (PAF)-induced airway hyperresponsiveness of guinea pigs: role of TXA_2 in development of PAF-induced nonspecific airway hyperresponsiveness. Jpn. J. Pharmacol. 52, 621–630.

Takizawa, H., Ishii, A., Suzuki, S., Shiga, J. and Miyamoto, T. (1988). Bronchoconstriction induced by platelet-activating factor in the guinea pig and its inhibition by CV-3988, a PAF antagonist: serial changes in findings of lung histology and bronchoalveolar lavage cell population. Int. Arch. Allergy Appl. Immunol. 86, 375–382.

Tamargo, J. Tejerina, T., Delgado, C. and Barrigon, S. (1985). Electrophysiological effects of platelet-activating factor (PAF-acether) in guinea-pig papillary muscles. Eur. J. Pharmacol. 109, 219–227.

Tamura, N., Agrawal, D.K., Suliaman, F.A. and Townley, R.G. (1987). Effects of platelet activating factor on the chemotaxis of normodense eosinophils from normal subjects. Biochem. Biophys. Res. Commun. 142, 638–644.

Tamura, N., Agrawal, D.K. and Townley, R.G. (1988). Leukotriene C_4 production from human eosinophils in vitro. Role of eosinophil chemotactic factors on eosinophil activation. J. Immunol. 141, 4291–4297.

Tanizawa, H. and Tai, H.H. (1989). Synergism between chemotactic peptide and platelet-activating factor in stimulating thromboxane B_2 and leukotriene B_4 biosynthesis in human neutrophils. Biochem. Pharmacol. 38, 2559–2563.

Tarayre, J.P., Aliaga, M., Barbara, M., Caillol, V. and Tisne

Versailles, J. (1986). Inflammatory action of PAF-acether in the rat pleural cavity. Agents Actions 17, 397–398.

Tavares de Lima, W., Teixeira, C.F., Sirois, P. and Jancar, S. (1989). Involvement of eicosanoids and PAF in immune-complex alveolitis. Braz. J. Med. Biol. Res. 22, 745–748.

Taylor, M.B., Zweiman, B., Moskovitz, A.R., von Allmen, C. and Atkins, P.C. (1990). Platelet-activating factor- and leukotriene B_4-induced release of lactoferrin from blood neutrophils of atopic and nonatopic individuals. J. Allergy Clin. Immunol. 86, 740–748.

Terashita, Z., Imura, Y., Nishikawa, K. and Sumida, S. (1985). Is platelet activating factor (PAF) a mediator of endotoxin shock? Eur. J. Pharmacol. 109, 257–261.

Tetta, C., Jeantet, A., Camussi, G. et al. (1985). Direct interaction between polymorphonuclear neutrophils and cuprophan membranes in a plasma-free model of dialysis. Proc. Eur. Dial. Transplant Assoc. Eur. Ren. Assoc. 21, 150–155.

Tetta, C., Montrucchio, G., Alloatti, G. et al. (1986). Platelet-activating factor contracts human myometrium in vitro. Proc. Soc. Exp. Biol. Med. 183, 376–381.

Tetta, C., Segoloni, G., Pacitti, A. et al. (1989). The production of platelet-activating factor during hemodialysis. Int. J. Artif. Organs 12, 766–772.

Thiemermann, C. and Vane, J. (1990). Inhibition of nitric oxide synthesis reduces the hypotension induced by bacterial lipopolysaccharides. Eur. J. Pharmacol. 182, 591–595.

Thiemermann, C., May, G.R., Page, C.P. and Vane, J.R. (1990). Endothelin-1 inhibits platelet aggregation in vivo: a study with ^{111}indium-labelled platelets. Br. J. Pharmacol. 99, 303–308.

Thierry, A., Doly, M., Braquet, P., Cluzel, J. and Meyniel, G. (1989). Presence of specific platelet-activating factor binding sites in the rat retina. Eur. J. Pharmacol. 163, 97–101.

Thompson, P.J., Hanson, J.M., Bilani, H., Turner-Warwick, M. and Morley, J. (1984). Platelets, platelet-activating factor and asthma. Am. Rev. Respir. Dis. 129, A3.

Tokumura, A., Kamiyasu, K., Takauchi, K. and Tsukatani, H. (1987). Evidence for existence of various homologues and analogues of platelet activating factor in a lipid extract of bovine brain. Biochem. Biophys. Res. Commun. 145, 415–425.

Tokumara, A., Tereao, M., Okamoto, M., Yoshida, K. and Tsukatani, H. (1988). Study of platelet-activating factor and its antagonists on rat colon strip with a new method avoiding tachyphylaxis. Eur. J. Pharmacol. 148, 353–360.

Tokuyama, K., Lotvall, J.O., Barnes, P.J. and Chung, K.F. (1991). Airway narrowing caused by inhaled platelet-activating: role of airway microvascular leakage. Am. Rev. Respir. Dis. 143, 1345–1349.

Tolins, J.P., Vercellotti, G.M., Wilkowske, M., Ha, B., Jacob, H.S. and Raij, L. (1989). Role of platelet activating factor in endotoxemic acute renal failure in the male rat. J. Lab. Clin. Med. 113, 316–324.

Tonnel, A.B., Gossett, P., Joseph, M., Lassalle, P., Dessant, J.P. and Capron, A. (1986). Alveolar macrophage and its participation in the inflammatory processes of allergic asthma. Bull. Eur. Physiopathol. Respir. 22(Suppl. 7), 70–77.

Tool, A.T.J., Verhoeven, A.J., Roos, D. and Koenderman, L. (1989). Platelet-activating factor (PAF) acts as an intercellular messenger in the changes of cytosolic free Ca^{2+} in human neutrophils induced by opsonized particles. FEBS Lett. 259, 209–212.

Toothill, V.J., Van Mourik, J.A., Niewenhuis, H.K., Metzelaar, M.J. and Pearson, J.D. (1990). Characterization of the enhanced adhesion of neutrophil leukocytes to thrombin-stimulated endothelial cells. J. Immunol. 145, 283–291.

Toth, P.D. and Mikulaschek, A.W. (1986). Effects of a platelet-activating factor antagonist, CV-3988, on different shock models in the rat. Circ. Shock 20, 193–203.

Touqui, L., Chignard, M., Jacquemin, C. Wal, F. and Vargaftig, B.B. (1984). Dissociation between inhibition of phospholipid methylation and production of PAF-acether by rabbit platelets. Experientia 40, 374–377.

Touqui, L., Hatmi, M. and Vargaftig, B.B. (1985). Human platelets stimulated by thrombin produced platelet-activating factor (1-O-alkyl-2-acetyl-sn-glycero-3-phosphocholine) when the degrading enzyme acetyl hydrolase is blocked. Biochem. J. 229, 811–816.

Touvay, C., Etienne, A. and Braquet, P. (1986). Inhibition of antigen-induced lung anaphylaxis in the guinea-pig by BN 52021 a new specific paf-acether receptor antgonist isolated from Ginkgo biloba. Agents Actions 17, 371–372.

Touvay, C., Vilain, B., Sirois, P. Soufir, M. and Braquet, P. (1987). Gossypol: a potent inhibitor of PAF-acether- and leukotriene-induced contractions of guinea-pig lung parenchyma strips. J. Pharm. Pharmacol. 39, 454–458.

Touvay, C., Vilain, B. Carre, C., Mencia Huerta, J.M. and Braquet, P. (1988). Role of platelet-activating factor (PAF) in the bronchopulmonary alterations and beta-adrenoceptor function induced by endotoxin. Biochem. Biophys. Res. Commun. 152, 527–533.

Toyofuku, T., Kubo, K., Kobayashi, T. and Kusama, S. (1986). Effects of ONO-6240, a platelet-activating factor antagonist, on endotoxin shock in unanaesthetized sheep. Prostaglandins 31, 271–281.

Toyofuku, T., Kobayashi, T., Koyama, S. and Kusama, S. (1988). Pulmonary vascular response to platelet-activating factor in conscious sheep. Am. J. Physiol. 255, H434–H440.

Tranquille, N. and Emeis, J.J. (1990). The simultaneous acute release of tissue-type plasminogen activator and von Willebrand factor in the perfused rat hindleg region. Thromb. Haemost, 63, 454–458.

Triggiani, M., Connell, T.R. and Chilton, F.H. (1990a). Evidence that increasing the cellular content of eicosapentaenoic acid does not reduce the biosynthesis of platelet-activating factor. J. Immunol. 145, 2241–2248.

Triggiani, M., Hubbard, W.C. and Chilton, F.H. (1990b). Synthesis of 1-acetyl-sn-glycero-3-phosphocholine by an enriched preparation of the human lung mast cell. J. Immunol. 144, 4773–4780.

Trovati, M., Anfossi, G., Cavalot, F., Massucco, P., Mularoni, E. and Emanuelli, G. (1988). Insulin directly reduces platelet sensitivity to aggregating agents. Studies in vitro and in vivo. Diabetes 37, 780–786.

Ukawa, K., Imamiya, E., Yamomoto, H. et al. (1989). Synthesis and antitumor activity of new amphiphilic alkylglycerolipids substituted with a polar head group, 2-(2-trimethylammonio-ethoxy)ethyl or a congeneric oligo (ethyleneoxy)ethyl group. Chem. Pharm. Bull. Tokyo 37, 3277–3285.

Valone, F.H. (1988). Identification of platelet-activating factor receptors in P388D1 murine macrophages. J. Immunol. 140, 2389–2394.

Valone, F.H., Philip, R. and Debs, R.J. (1988). Enhanced human monocyte cytotoxicity by platelet activating factor. Immunology 64, 715–718.

Vargaftig, B.B. and Bourgain, R. (1989). In "Platelet-Activating

Factor and Human Disease" (ed P.J. Barnes, C.P. Page and P.M. Henson) pp 220–230. Blackwell, Oxford.

Vargaftig, B.B., Fouque, F., Benveniste, J. and Odiot, J. (1982). Adrenaline and PAF-acether synergize to trigger cyclooxygenase-independent activation of plasma-free human platelets. Thromb. Res. 28, 557–573.

Vemulapalli, S., Chiu, P.J. and Barnett, A. (1984). Cardiovascular and renal action of platelet-activating factor in anaesthetized dogs. Hypertension 6, 489–493.

Vercellotti, G.M., Yin, H.Q., Gustafson, K.S,. Nelson, R.D. and Jacob, H.S. (1988). Platelet-activating factor primes neutrophil responses to agonists: role in promoting neutrophil-mediated endothelial damage. Blood 71, 1100–1107.

Vercellotti, G.M., Wickham, N.W., Gustafson, K.S., Yin, H.Q., Hebert, M. and Jacob, H.S. (1989). Thrombin-treated endothelium primes neutrophil functions: inhibition by platelet-activating factor receptor antagonists. J. Leukocyte Biol. 45, 483–490.

Viero, P., Cortelazzo, S., Barbui, T. and De Gaetano, G. (1986). Defective platelet aggregation induced by platelet activating factor in myeloproliferative disorders: deficiency of an aspirin-independent mechanism? Haemostasis 16, 27–33.

Viossat, I., Chapelat, M., Chabrier, P.E. and Braquet, P. (1989). Effects of platelet-activating factor (PAF) and its receptor antagonist BN 52021 on isolated guinea-pig heart. Prostaglandins Leukot. Essent. Fatty Acids 38, 189–194.

Voelkel, N.F., Worthen, S., Reeves, J.T., Henson, P.M. and Murphy, R.C. (1982). Nonimmunological production of leukotrienes induced by platelet-activating factor. Science 218, 286–288.

Voelkel, N.F., Chang, S.W., Pfeffer, K.D., Worthen, S.G., McMurtry, I.F. and Henson, P.M. (1986). PAF antagonists: different effects on platelets, neutrophils, guinea pig ileum and PAF-induced vasodilation in isolated rat lung. Prostaglandins 32, 359–372.

Wade, P.J., Lunt, D.O., Lad, N., Tuffin, D.P. and McCullagh, K.G. (1986). Effect of calcium and calcium antagonists on PAF-acether binding to washed human platelets. Thromb. Res. 41, 251–262.

Wallace, J.L. (1990). Lipid mediators of inflammation in gastric ulcer. Am. J. Physiol. 258, G1–G11.

Wallace, J.L. and MacNaughton, W.K. (1988). Gastrointestinal damage induced by platelet-activating factor: role of leukotrienes. Eur. J. Pharmacol. 151, 43–50.

Wallace, J.L. and Whittle, B.J. (1986a). Picomole doses of platelet-activating factor predispose the gastric mucosa to damage by topical irritants. Prostaglandins 31, 989–998.

Wallace, J.L. and Whittle, B.J. (1986b). Effects of inhibitors of arachidonic acid metabolism on Paf-induced gastric mucosal necrosis and haemoconcentration. Br. J. Pharmacol. 89, 415–422.

Wallace, J.L. and Whittle, B.J. (1988). Gastrointestinal damage induced by platelet-activating factor. Inhibition by the corticoid, dexamethasone. Dig. Dis. Sci. 33, 225–232.

Wallace, J.L., Steel, G., Whittle, B.J., Lagente, V. and Vargaftig, B. (1987). Evidence for platelet-activating factor as a mediator of endotoxin-induced gastrointestinal damage in the rat. Effects of three platelet-activating factor antagonists. Gastroenterology 93, 765–773.

Wallace, J.L., Hogaboam, C.M. and McKnight, G.W. (1990a). Platelet-activating factor mediates gastric damage induced by hemorrhagic shock. Am. J. Physiol. 259, G140–G146.

Wallace, J.L., Keenan, C.M. and Granger, D.N. (1990b). Gastric

ulceration induced by nonsteroidal anti-inflammatory drugs in a neutrophil-dependent process. Am. J. Physiol. 259, G462–G467.

Walsh, G.M., Nagakura, T. and Iikura, Y. (1989). Flow-cytometric analysis of increased IgE uptake by normal eosinophils following activation with PAF-acether and other inflammatory mediators. Int. Arch. Allergy Appl. Immunol. 88, 194–196.

Wang, C.-J. and Tai, H.-H. (1991). Monoclonal anti-idiotypic antibodies to platelet-activating factor (PAF) and their interaction with PAF receptors. J. Biol. Chem. 266, 12372–12378.

Wang, J. and Dunn, M.J. (1987). Platelet-activating factor mediates endotoxin-induced acute renal insufficiency in rats. Am. J. Physiol. 253, F1283–F1289.

Wang, J., Kester, M. and Dunn, M.J. (1988). The effects of endotoxin on platelet-activating factor synthesis in cultured rat glomerular mesangial cells. Biochim. Biophys. Acta 969, 217–224.

Ward, S.G. and Westwick, J. (1988). Antagonism of the platelet factor-induced rise of the intracellular calcium ion concentration of U937 cells. Br. J. Pharmacol. 93, 769–774.

Ward, S.G., Lewis, G.P. and Westwick, J. (1987). In "Lipid Mediators in Immunology of Shock" (ed M. Paubert-Braquet, P. Braquet, B. Demling, J.R. Fletcher and M. Foegh), pp 483–493. Plenum Press, New York.

Wardlaw, A.J., Moqbel, R., Cromwell, O. and Kay, A.B. (1986). Platelet-activating factor. A potent chemotactic and chemokinetic factor for human eosinophils. J. Clin. Invest. 78, 1701–1706.

Wardlaw, A.J., Chung, K.F., Moqbel, R. et al. (1990). Effects of inhaled PAF in humans on circulating and bronchoalveolar lavage fluid neutrophils. Relationship to bronchoconstriction and changes in airway responsiveness. Am. Rev. Respir. Dis. 141, 386–392.

Warren, J.S., Mandel, D.M., Johnson, K.J. and Ward, P.A. (1989). Evidence for the role of platelet-activating factor in immune complex vasculitis in the rat. J. Clin. Invest. 83, 669–678.

Wedmore, C.V. and Williams, T.J. (1981a). Control of vascular permeability by polymorphonuclear leukocytes in inflammation. Nature 289, 646–650.

Wedmore, C.V. and Williams, T.J. (1981b). Platelet-activating factor (PAF), a secretory product of polymorphonuclear leukocytes that increases vascular permeability in rabbit skin. Br. J. Pharmacol. 74, 91P.

Weintraub, S.T., Ludwig, J.C., Mott, G.E., McManus, L.M., Lear, C. and Pinckard, R.N. (1985). Fast atom bombardment-mass spectrometric identification of molecular species of platelet-activating factor produced by stimulated human polymorphonuclear leukocytes. Biochem. Biophys. Res. Commun. 129, 868–876.

Weisman, S.M., Felsen, D. and Vaughan, E.D.J. (1985). Platelet-activating factor is a potent stimulus for renal prostaglandin synthesis: possible significance in unilateral ureteral obstruction. J. Pharmacol. Exp. Ther. 235, 10–15.

Weisman, S.M., Freund, R.M., Felsen, D. and Vaughan, E.D.J. (1988). Differential effect of platelet-activating factor (PAF) receptor antagonists on peptide and PAF-stimulated prostaglandin release in unilateral ureteral obstruction. Biochem. Pharmacol. 37, 2927–2932.

White, H.L. and Faison, L.D. (1988). Inhibition of lyso-PAF: acetyl-CoA acetyltransferase by salicylates and other compounds. Prostaglandins 35, 939–944.

Whittle, B.J. (1989). The defensive role played by the gastric microcirculation. Methods Find. Exp. Clin. Pharmacol. 11(Suppl.1), 35–43.

Whittle, B.J., Morishita, T., Ohya, Y., Leung, F.W. and Guth, P.H. (1986). Microvascular actions of platelet-activating factor on rat gastric mucosa and submucosa. Am. J. Physiol. 251, G772–G778.

Whittle, B.J., Boughton Smith, N.K., Hutcheson, I.R., Esplugues, J.V. and Wallace, J.L. (1987). Increased intestinal formation of PAF in endotoxin-induced damage in the rat. Br. J. Pharmacol. 92, 3–4.

Wilkens, J.H., Wilkens, H., Uffmann, J., Bovers, J., Fabel, H. and Frolich, J.C. (1990). Effects of a PAF-antagonist (BN 52036) on bronchoconstriction and platelet activation during exercise induced asthma. Br. J. Clin. Pharmacol. 29, 85–91.

Wirthmueller, U., De Weck, A.L. and Dahinden, C.A. (1989). Platelet-activating factor production in human neutrophils by sequential stimulation with granulocyte–macrophage colony-stimulating factor and the chemotactic factors C5A or formyl-methionyl-leucyl-phenylalanine. J. Immunol. 142, 3213–3218.

Workman, P., Donaldson, J. and Lohmeyer, M. (1991). Platelet-activating factor (PAF) antagonist WEB 2086 does not modulate the cytotoxicity of PAF or antitumour alkyl lysophospholipids ET-18-O-methyl and SRI 62-834 in HL-60 promyelocytic leukemia cells. Biochem. Pharmacol. 41, 319–322.

Worthen, G., Seccombe, J.F., Clay, K.L., Guthrie, L.A. and Johnston, R.B. (1988). The priming of neutrophils by lipopolysaccharide for production of intracelluular platelet-activating factor. J. Immunol. 140, 3553–3559.

Wu, M.S., Biftu, T. and Doebber, T.W. (1986). Inhibition of the platelet activating factor (PAF)-induced *in vivo* responses in rats by *trans*-2,5-(3,4,5-trimethoxyphenyl)tetrahydrofuran (L-652,731), a PAF receptor antagonist. J. Pharmacol. Exp. Ther. 841–845.

Wymann, M.P., Kerren, P., Deranleau, D.A., Dewald, B., von Tscharner, V. and Baggiolini, M. (1987). Oscillatory motion in human neutrophils responding to chemotactic stimuli. Biochem. Biophys. Res. Commun. 147, 361–368.

Yaghi, A., Hamilton, J.T. and Paterson, N.A. (1989). Influence of platelet-activating factor on leukotriene D_4-induced contractions of the guinea pig parenchymal strip. Can. J. Physiol. Pharmacol. 67, 315–321.

Yasaka, T., Boxer, L.A. and Baehner, R.L. (1982). Monocyte aggregation and superoxide anion release in response to formyl-methionyl-leucyl-phenylalanine (FMLP) and platelet-activating factor (PAF). J. Immunol. 128, 1939–1944.

Yasuda, K., Satouchi, K. and Saito, K. (1986). Platelet-activating factor in normal rat uterus. Biochem. Biophys. Res. Commun. 138, 1231–1236.

Yeo, Y.K., Philbrick, D.J. and Holub, B.J. (1989a). Effects of dietary *n*-3 fatty acids on mass changes and glycerol incorporation in various glycerolipid classes of rat kidney *in vivo*. Biochim. Biophys. Acta 1006, 9–14.

Yeo, Y.K., Philbrick, D.J. and Holub, B.J. (1989b). Altered acyl chain compositions of alkylacyl, alkenylacyl, and diacyl subclasses of choline and ethanolamine glycerophospholipids in rat heart by dietary fish oil. Biochim. Biophys. Acta 1001, 25–30.

Yeo, Y.K., Philbrick, D.J. and Holub, B.J. (1989c). The effect of long-term consumption of fish oil on platelet-activating factor synthesis in rat renal microsomes. Biochem. Biophys. Res. Commun. 160, 1238–1242.

Yoo, J., Schlondorff, D. and Neugarten, J. (1990). Thromboxane mediates the renal hemodynamic effects of platelet activating factor. J. Pharmacol. Exp. Ther. 253, 743–748.

Yoshida, N., Yoshikawa, T., Ando, T. *et al*. (1989). Pathogenesis of platelet-activating factor-induced gastric mucosal damage in rats. Scand. J. Gastroenterol. Suppl. 162, 210–214.

Yue, T.L., Farhat, M., Rabinovici, R., Perara, P.Y., Vogel, S.N. and Feuerstein, G. (1990). Protective effect of BN 50739, a new platelet-activating factor antagonist, in endotoxin-treated rabbits. J. Pharmacol. Exp. Ther. 254, 976–981.

Yukawa, T., Kroegel, C., Evans, P., Fukuda, T., Chung, K.F. and Barnes, P.J. (1989). Density heterogeneity of eosinophil leucocytes: induction of hypodense eosinophils by platelet-activating factor. Immunology 68, 140–143.

Yukawa, T., Read, R.C., Kroegel, C. *et al*. (1990). The effects of activated eosinophils and neutrophils on guinea pig airway epithelium *in vitro*. Am. J. Respir. Cell. Mol. Biol. 2, 341–353.

Zheng, B., Oishi, K., Shoji, M. *et al*. (1990). Inhibition of protein kinase C, (sodium plus potassium)-activated adenosine triphosphatase, and sodium pump by synthetic phospholipid analogues. Cancer Res. 50, 3025–3031.

Zimmerman, G.A., McIntyre, T.M. and Prescott, S.M. (1985a). Thrombin stimulates the adherence of neutrophils to human endothelial cells *in vitro*. J. Clin. Invest. 76, 2235–2246.

Zimmerman, G.A., McIntyre, T.M. and Prescott, S.M. (1985b). Production of platelet-activating factor by human vascular endothelial cells: evidence for a requirement for specific agonists and modulation by prostacyclin. Circulation 72, 718–727.

Zimmerman, G.A., McIntyre, T.M. and Prescott, S.M. (1986). Thrombin stimulates neutrophil adherence by an endothelial cell-dependent mechanism: characterization of the response and relationship to platelet-activating factor synthesis. Ann. N.Y. Acad. Sci. 485, 349–368.

Zimmerman, G.A., McIntyre, T.M., Mehra, M. and Prescott, S.M. (1990). Endothelial cell-associated platelet-activating factor: a novel mechanism for signalling intercellular adhesion. J. Cell Biol. 110, 529–540.

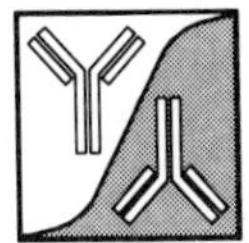

9. Platelet-Activating Factor: Receptors and Receptor Antagonists

S.-B. Hwang

1. Introduction

PAF, identified as 1-O-alkyl-2-acetyl-*sn*-glycero-3-phosphocholine (Fig. 9.1) (Benveniste *et al.*, 1979; Blank *et al.*, 1979; Demopoulos *et al.*, 1979) is a potent lipid mediator. PAF can induce the aggregation of rabbit platelets with the simultaneous release of granular constituents that include histamine. Subsequent studies have shown that PAF likewise activates PMNs, eosinophils, basophils, lymphocytes, monocytes, vascular endothelium, macrophages, exocrine pancreas, parotid acini, neurocytes, smooth muscle, and cells from skin, heart, liver and kidney (O'Flaherty, 1982; Camussi, 1986; Hanahan, 1986; Schlondorf and Neuwirth, 1986; Braquet *et al.*, 1987; Pinckard *et al.*, 1988; Snyder, 1989; Prescott *et al.*, 1990). PAF was initially identified as a platelet-aggregating soluble product released by IgE-sensitized rabbit basophils (Benveniste *et al.*, 1972). It is now clear that PAF can be synthesized by a wide variety of cells and tissues, including platelets, macrophages, neutrophils, endothelial cells, lung tissue, uterus, brain and kidney (Chignard *et al.*, 1980; Zimmerman *et al.*, 1985;

Lipid Mediators
ISBN 0–12–198875–9

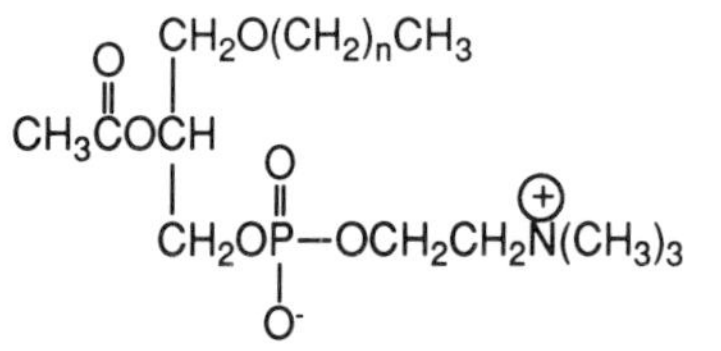

$n = 15$ C$_{16}$ PAF (1)
$n = 17$ C$_{18}$ PAF (2)

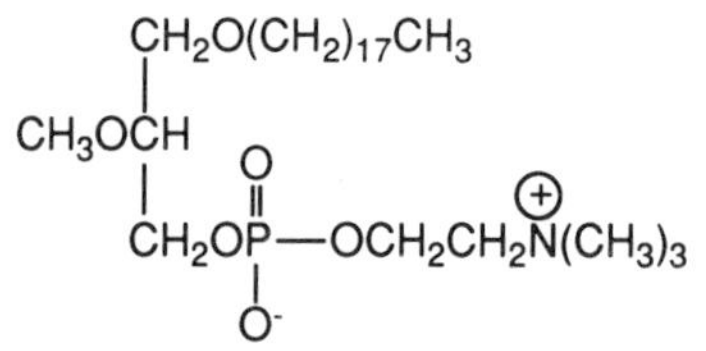

ET-18-O-methyl (3)

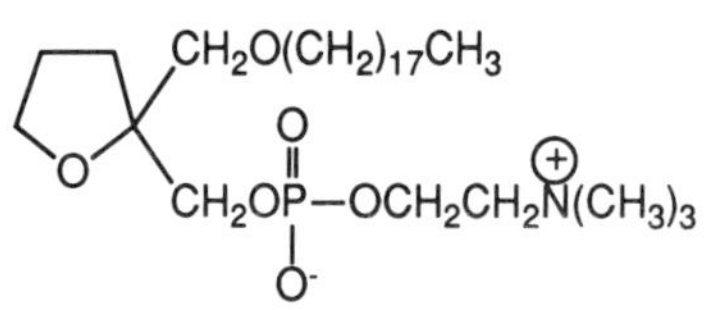

SRI 62-834 (4)

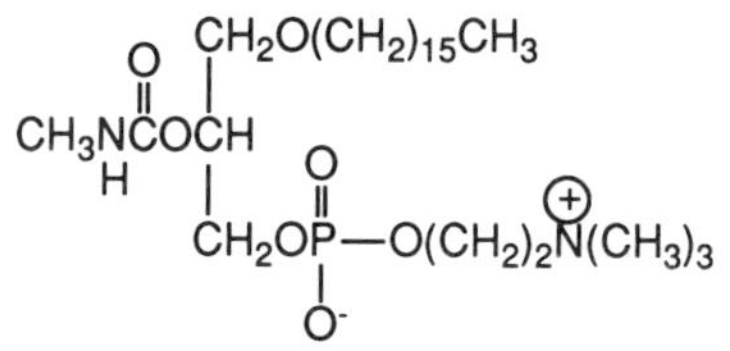

N-Methylcarbamyl PAF (5)

Figure 9.1 Structures of PAF and PAF analogues.

Snyder, 1985; Hoffman *et al.*, 1986a; Lee, 1987; Angle *et al.*, 1988; Kumar *et al.*, 1988; Camussi *et al.*, 1989b).

PAF has been considered, along with other lipid mediators, to be involved in allergic and inflammatory reactions (Humphrey *et al.*, 1982; Braquet *et al.*, 1987) and, more recently, in shock, asthma, pulmonary hypertension, brain injury, and transplant rejection (Braquet *et al.*, 1987; Feuerstein and Hallenbeck, 1987; Vargaftig and Braquet, 1987; Vargaftig, 1987; Foegh, 1988; Page, 1990). In addition to its involvement in allergy and inflammation, pathology and disease, the role of PAF in normal physiological processes is now becoming better understood. Such physiological processes include pre-implantation stages of reproduction (Harper, 1989), parturition (Billah and Johnston, 1983; Billah *et al.*, 1985; Johnston and Miyaura, 1990), and fetal lung maturity (Hoffman *et al.*, 1986a,b; Johnston and Miyaura, 1990). Development of PAF-specific receptor antagonists may prove to be useful for further investigation of the role of PAF in human diseases and normal physiological processes.

Several PAF analogues (Fig. 9.1), such as 1-O-octadecyl-2-O-methyl-*rac*-3-phosphocholine (ET-18-O-methyl) and (±)-2-{hydroxy[tetrahydro-2-(octadecyloxy)-methylfuran-2-yl]methoxyl}phosphinyloxy-*N,N,N*-trimethylethaniminium hydroxide (SRI 62-834) exhibit antitumour activity (Berdel *et al.*, 1981; Berdel and Munder, 1987; Houlihan *et al.*, 1987). These ether lipids cause immune effects such as macrophage activation as well as direct cytotoxic effects against tumour cells *in vitro*. Compared to the extremely potent effects of PAF, much higher concentrations of ether lipids, i.e. micromolar, are required for the pharmacological and cytotoxic actions seen in *in vitro* cell lines. Although ET-18-O-methyl does

exhibit weak PAF agonistic activity in rabbit platelets, the direct cytotoxicity of this and related ether lipids against HL-60 human promyelocytic leukaemia cells and erythrocytes does not exhibit PAF receptor-like stereospecificity (Kudo *et al.*, 1987). A lack of correlation between cytotoxicity of ether lipids in neoplastic cells and modulation of [^{3}H]PAF binding in human platelets was also reported (Berdel *et al.*, 1987). Furthermore, the PAF receptor antagonists L-652,731 (Lazenby *et al.*, 1990) and WEB 2086 (Workman *et al.*, 1990) were unable to counteract the *in vitro* cytotoxicity of these ether lipids. Therefore, the molecular nature of these ether lipids as antitumour agents may not act through PAF receptors. The antitumour activity of these ether lipids will not be included in this chapter.

2. Existence of Platelet-Activating Factor Receptors

Subsequent to the determination of the chemical structure of PAF (Demopoulos *et al.*, 1979; Benveniste *et al.*, 1979; Blank *et al.*, 1979), the structural requirements for activity had been determined using a large number of synthetic analogues and have been comprehensively reviewed (Shen *et al.*, 1987a; Vargaftig *et al.*, 1989). At the 1 position, the length of the straight-chain alkyl ether moiety affects function. Compounds lacking the ether linkage are less active in stimulating various cell types. Unsaturation of the alkyl ether residue produces little change in potency. At the second position, acetic or propionic acid esters give a very potent substance, but the activity decreases as the length of the fatty acid ester increases. Substitution of the 2 position with a variety of other groups results in little or no activity. At the 3 position, compounds lacking one, two or three choline N-methyl residues show progressive loss in potency, and absence of a choline or phosphocholine group yields little, if any activity. Only the stereoisomer of PAF (the (R) form but not the (S) form), exhibits activity. The results obtained using these analogues indicate that PAF acts by a stereospecific and structurally specific mechanism. Very low concentrations (usually less than 1 nM) are necessary to trigger biological effects. Also, specific desensitization takes place after exposure of cells or tissues to PAF. A cell surface receptor that specifically binds PAF is therefore generally believed to initiate various responses to PAF both *in vitro* and *in vivo*.

The existence of specific PAF receptors has been further confirmed by binding experiments using [³H]PAF. The specific binding of [³H]PAF to a variety of cells and tissue homogenates was found to have high affinity with a finite number of specific binding sites (Hwang, 1990). High-affinity binding sites have been demonstrated in human (Valone *et al.*, 1982; Chesney *et al.*, 1983; Kloprogge and Akkerman, 1984; Inarrea *et al.*, 1984; Hwang and Lam, 1986), rabbit (Hwang *et al.*, 1983; Homma *et al.*, 1987) and canine platelets (Janero *et al.*, 1988a,b; Tahraoui *et al.*, 1988), human PMNs (Valone and Goetzl, 1983; O'Flaherty *et al.*, 1986; Hwang, 1988), human mononuclear leucocytes (Ng and Wong, 1988a; Hwang *et al.*, 1991), macrophages (Prpic *et al.*, 1988; Valone, 1988), rat Kupffer cells (Chao *et al.*, 1989), and plasma membranes derived from several tissue types (Hwang *et al.*, 1983, 1985a; Hwang, 1987; Domingo *et al.*, 1988, 1989; Thierry *et al.*, 1989).

The specific binding of [³H]PAF was reversible, since [³H]PAF bound specifically to cells (Homma *et al.*, 1987; Ng and Wong, 1988a; Chao *et al.*, 1989) or isolated membranes (Hwang *et al.*, 1983) and could be dissociated effectively by an excess of unlabelled PAF. The PAF receptor was functionally active. The relative potencies of PAF and several structurally related analogues both *in vitro* and *in vivo* (Hwang *et al.*, 1985b, 1986a) correlate very well with their relative inhibition of [³H]PAF binding to specific receptor sites. Rat platelets do not aggregate in the presence of PAF. Red blood cells were found to be unresponsive to PAF. These results suggest that these cells do not have high-affinity PAF receptors (Hwang *et al.*, 1983; Inarrea *et al.*, 1984; Hwang and Lam, 1986).

Several PAF receptor antagonists have been identified in the past few years (for reviews, see Saunders and Handley, 1987; Shen *et al.*, 1987a; Weber and Heuer, 1989; Hwang, 1990). These receptor antagonists include natural products and synthetic compounds. All are reported to compete with the binding of [³H]PAF to rabbit (Hwang *et al.*, 1985b,c; Hwang and Lam, 1986), guinea-pig (Terashita *et al.*, 1985a) and human platelets (Terashita *et al.*, 1985a; Hwang and Lam, 1986), as well as binding to human PMNs (Hwang, 1988), human lung

tissues (Hwang *et al.*, 1985a; Hwang and Lam, 1986) and rat liver tissues (Hwang, 1987). These receptor antagonists are also capable of inhibiting the function induced by PAF. An excellent structure–activity relationship for these receptor antagonists in inhibiting either [³H]PAF binding or induction of PAF action has been well established (Hwang *et al.*, 1985b; Ponpipom *et al.*, 1987; Weber and Heuer, 1989). Furthermore, tritium-labelled PAF receptor antagonists were demonstrated to share a common binding site with PAF (Hwang *et al.*, 1985b,c, 1986b, 1989; Robaut *et al.*, 1987; Marquis *et al.*, 1988).

These results strongly suggest that binding sites for [³H]PAF truly represent the pharmacologically relevant receptor sites. An "earmuffs" model, a central hydrophobic pocket embedded below two opposing zones of positive charge, has been proposed for the binding site of PAF, PAF analogues and receptor antagonists to the PAF receptor (Dive *et al.*, 1989). The PAF receptor of guinea-pig lung has now been cloned and expressed in *Xenopus* oocytes (Honda *et al.*, 1991), and has a calculated molecular weight of 38 982. Further work is needed to determine the structural features of the PAF receptor that allow it to bind the ligand and trigger the post-receptor signal transduction processes.

3. Platelet-Activating Factor Receptor Antagonists

The discovery of PAF antagonists took place relatively soon after the chemical structure of PAF was firmly established. Good progress has been made in three directions: phospholipid analogues of PAF, natural products, and synthetic compounds from *in vitro* screening efforts. In general, these PAF antagonists are relatively potent inhibitors of [³H]PAF binding to its receptors *in vitro*. Most of them are highly specific to PAF. The availability of a variety of PAF receptor antagonists with different chemical and physical characteristics provides valuable research tools to delineate the pathophysiological roles of PAF as well as an opportunity to investigate their potential therapeutic applications in various cardiovascular, pulmonary, ophthalmic and dermatological disorders.

A wide variety of *in vitro* and *in vivo* assays have been used to evaluate the PAF antagonism of putative PAF antagonists as well as to demonstrate the involvement of PAF in various pathological conditions. Several more commonly used assays will be described in this section.

Two *in vitro* assays have been adapted for the primary screening of PAF receptor antagonists. One is the inhibition of PAF-induced aggregation of either rabbit platelets or human platelets. The other is the binding inhibition of [³H]PAF to either isolated intact cells or fractionated membrane fragments from cells or homogenized tissues. Platelet aggregation is easy to perform but is limited to the finite survival time of platelets. Also, inhibition of the platelet aggregation may not necessarily be due to receptor antagonism, but may be related to post-receptor-binding steps involved in platelet aggregation. Therefore, direct receptor-binding inhibition is always required to make certain that the inhibitors are truly receptor antagonists.

The receptor antagonistic activity of a selected compound depends on the PAF concentration used to induce the platelet aggregation as well as the concentration of $[^3H]PAF$ used in the receptor-binding assay. Furthermore, the receptor antagonistic activity in the receptor-binding assay depends on the ionic conditions of the assay (Hwang *et al.*, 1985c, 1989). These contributing factors make it difficult to compare the anti-PAF activity of a compound evaluated in one assay with that run in another type of assay. In this chapter, the equilibrium inhibition constant (K_i) derived from the Cheng and Prusoff equation (Cheng and Prusoff, 1973) will be given if it is available. Otherwise the ED_{50} value will be given to represent the antagonistic potency.

A useful PAF response for the study of the effectiveness of PAF receptor antagonists is PAF-induced hypotension in rats (Blank *et al.*, 1979). Most of the published PAF receptor antagonists very effectively inhibit and/or reverse PAF-induced hypotension. There is a good correlation between the amount of each PAF receptor antagonist needed to inhibit and/or reverse PAF-induced hypotension and the quantity that blocks the binding of $[^3H]PAF$. Also, as rats are small animals, a small quantity of the drug is usually sufficient to show the effectiveness of PAF receptor antagonists. This *in vivo* model of PAF-induced hypotension has been invaluable for the evaluation of the oral activity of antagonists in rats.

PAF may play a major role in asthma, especially since it can cause bronchial hyper-reactivity in humans (Cuss *et al.*, 1986; Barnes *et al.*, 1988), which is a common characteristic of this disease (Boushey *et al.*, 1980). Although bronchoconstriction may not be directly correlated with hyper-reactivity, inhibition of PAF-induced bronchoconstriction in guinea-pigs has been commonly used to evaluate the *in vivo* activity of the PAF receptor antagonist.

Probably the most important scientific utility of PAF receptor antagonists will be to delineate the role of PAF in different disease processes. Many published studies have provided circumstantial evidence for the involvement of PAF in human disease models by demonstrating that PAF is produced or increased in the pathological state and that PAF elicits the same response as the natural stimulus. Both endotoxin in animal shock models (Doebber *et al.*, 1985; Terashita *et al.*, 1985b; Chang *et al.*, 1987) and animal models of anaphylaxis induced by either antigen (Halonen *et al.*, 1980; Touvay *et al.*, 1985) or soluble immune complexes (Doebber *et al.*, 1986) have been extensively used to evaluate the effectiveness of PAF receptor antagonists in blocking the action of endogenous

PAF. Rats have also been used successfully to prove that PAF has a pathological role in anaphylaxis and endotoxin shock (Shen *et al.*, 1987a).

3.1   PHOSPHOLIPID ANALOGUES

Terashita *et al.* (1983) reported the first PAF receptor antagonist, CV-3988 (Fig. 9.2), a phospholipid derivative originally developed as a cytotoxic and antifungal agent. The incorporation of a carbamate linkage at the C-1 position and replacement of the trimethylammonium polar head with a thiazolium group yielded a potent receptor antagonist, with very weak agonistic activity demonstrable only at high doses. CV-3988 inhibited $[^3H]PAF$ binding with a K_i value of about 1 μM in rabbit and human platelet (Hwang and Lam, 1986), human PMN (Hwang, 1988) and human lung membranes (Hwang and Lam, 1986). This compound was further elaborated to a more potent substance, CV-6209 (Terashita *et al.*, 1987) (Fig. 9.2). CV-6209 is about 100-fold more potent than CV-3988 in inhibiting the specific $[^3H]PAF$ binding to human platelet membranes. CV-6209 potently inhibited the binding of $[^3H]C_{18}$-PAF to human platelet, PMN and mononuclear leucocyte membranes with K_i values of 5.37 ± 1.93 ($n = 4$), 7.26 ± 2.93 ($n = 6$) and 24.54 ± 5.36 nM ($n = 4$), respectively (Hwang *et al.*, 1991). The K_i values showed no significant difference between human platelet and PMN membranes, whereas the K_i value for human mononuclear leucocyte membranes was signficantly different from those for human platelet and human PMN membranes. Differences in PAF receptors between human platelets and human PMNs have been demonstrated (Hwang, 1988). It will be further emphasized below that PAF receptors in platelets, PMNs and monocytes are different, based on the differences in the ionic modulation of $[^3H]PAF$ binding as well as the differences in the affinity constants for CV-6209 and other PAF receptor antagonists, 52770 RP, ONO-6240 and FR-72112 (Hwang, 1988, 1990; Hwang *et al.*, 1991). The compound CV-6209 is not well absorbed when given orally. CV-6209 showed *in vivo* activity only when it was intravenously administered (Terashita *et al.*, 1987), even though CV-6209 has a potency at a subnanomolar concentration in inhibiting $[^3H]PAF$ binding to human platelet, human PMN and human monocyte membranes (Hwang *et al.*, 1991), as well as in inhibiting the platelet aggregation induced by PAF (Terashita *et al.*, 1987). It protected mice and rats from PAF-induced lethality with ED_{50} values of 0.009 and 0.014 mg/kg, respectively, when the compound was administered intravenously (Terashita *et al.*, 1987).

A series of PAF analogues was developed by the Upjohn Company in which the methylene chains between the phosphate and the choline nitrogen were increased from three to 10. The ability to induce platelet aggregation decreased with the increased chain length. U 66985 and U 66982 (Fig. 9.2) were essentially inactive as agonists

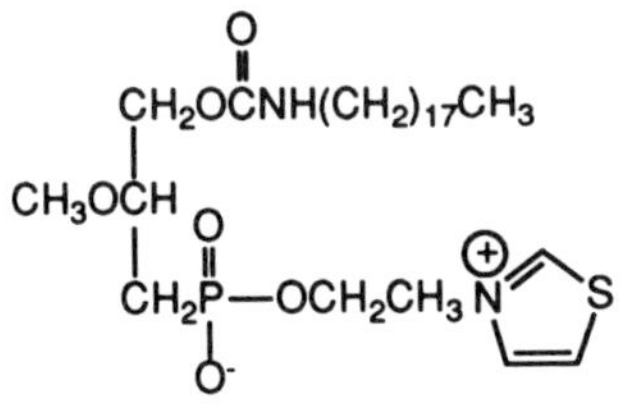
CV-3988 (6)

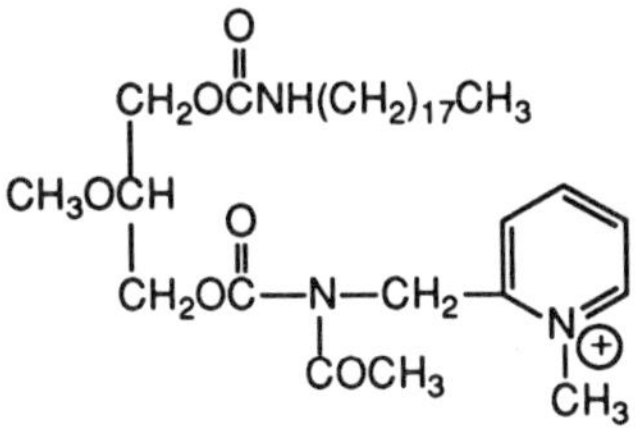
CV-6209 (7)

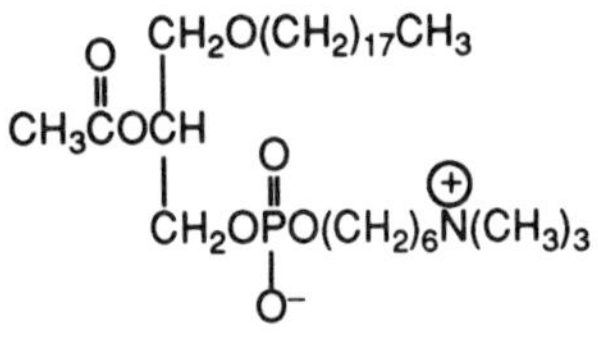
U-66985 (8)

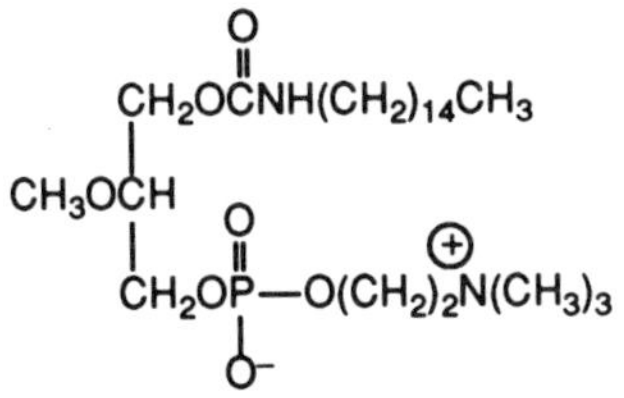
U-68043 (9)

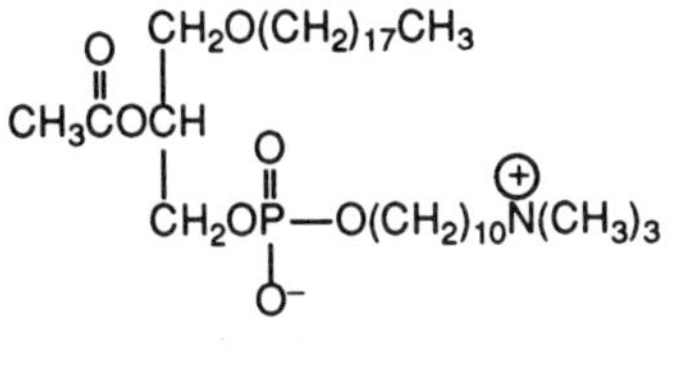
U-66982 (10)

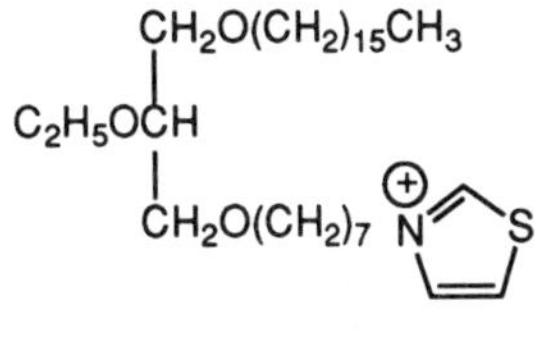
Ono-6240 (11)

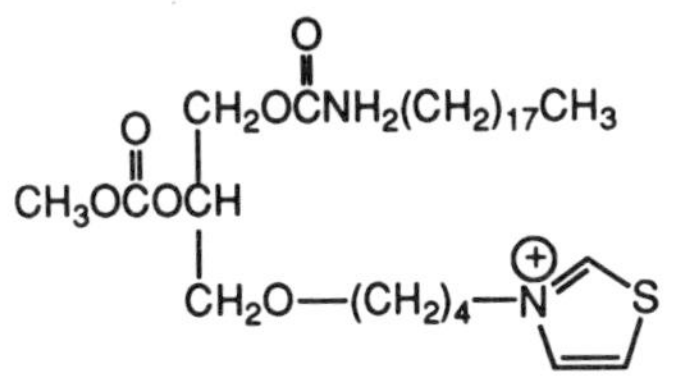
Ro 19-3704 (12)

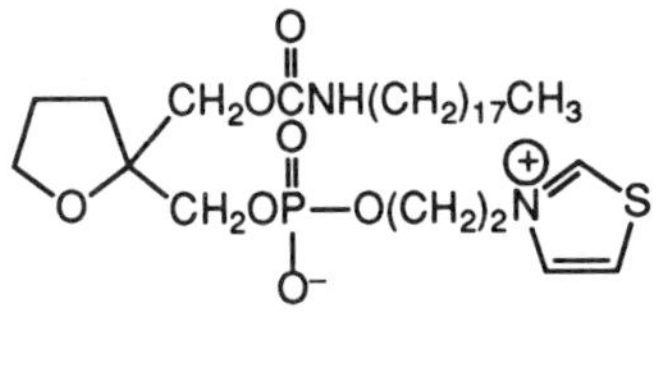
SDZ 63-072 (13)

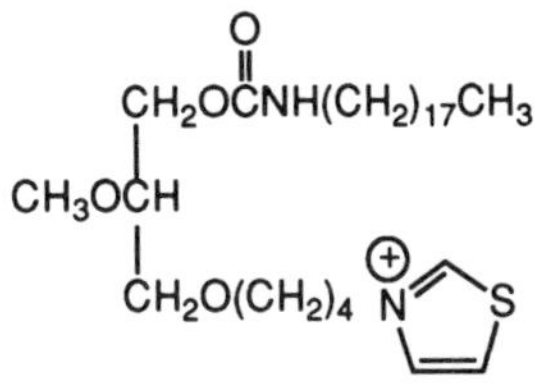
SDZ 63-019 (14)

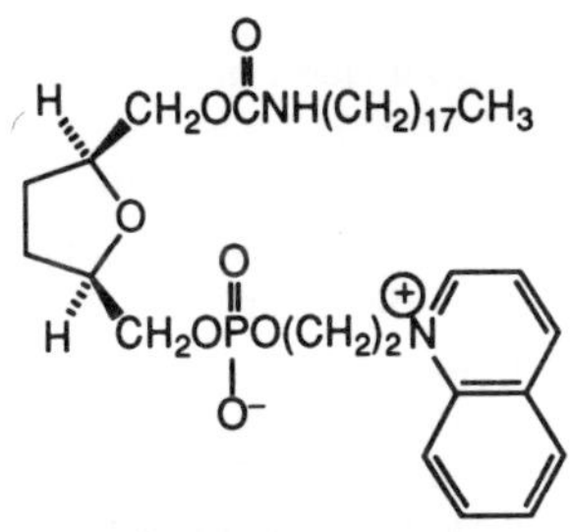
SDZ 63-675 (15)

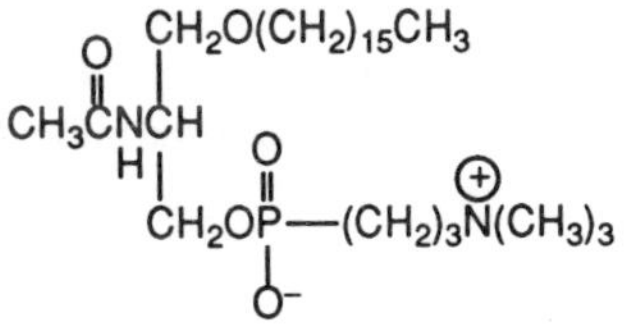
PAF-AP (16)

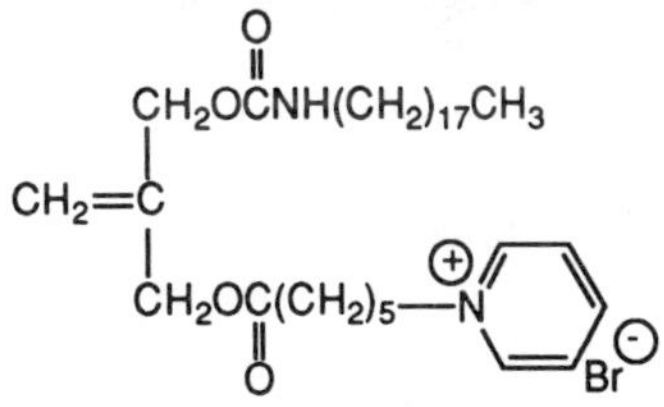
GS-1065-180 (17)

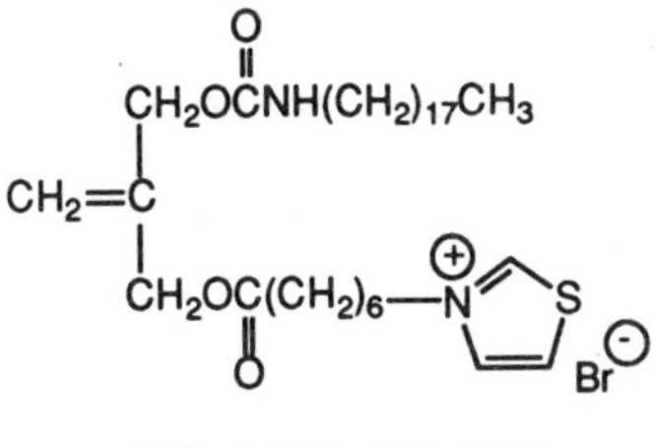
GS-1160-180 (18)

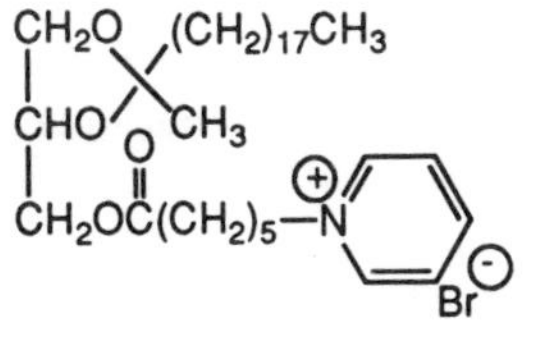
BN 52111 (19)

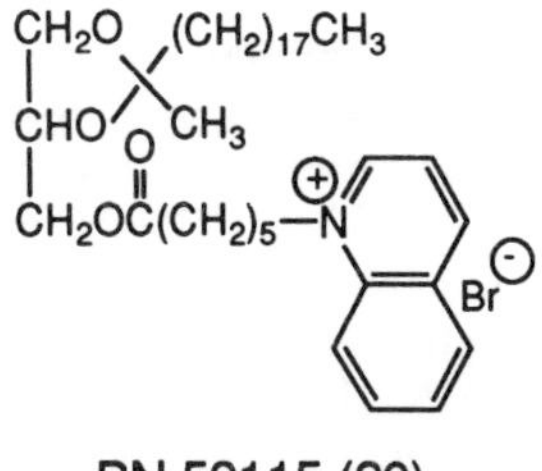
BN 52115 (20)

Figure 9.2 Structures of phospholipid derivatives as PAF receptor antagonists.

(Tokumura *et al.*, 1985). However, U 66985 and U 66982 showed antagonistic activity on PAF-induced platelet aggregation. U 66985 is the most potent inhibitor in this series. It inhibited [^{3}H]PAF binding to its receptors (Homma *et al.*, 1987) and PAF-induced aggregation of rabbit platelets (Tokumura *et al.*, 1985) with an ED_{50} value nearly 20 times less than that for CV-3988. U 68043 (Fig. 9.2), which is an analogue of CV-3988 but with no thiazolioethyl phosphate group, is not active at concentrations in the micromolar range in inhibiting PAF binding to rabbit platelets (Homma *et al.*, 1987). Therefore, the thiazolioethyl phosphate group in CV-3988 plays an important role in the antagonistic activity.

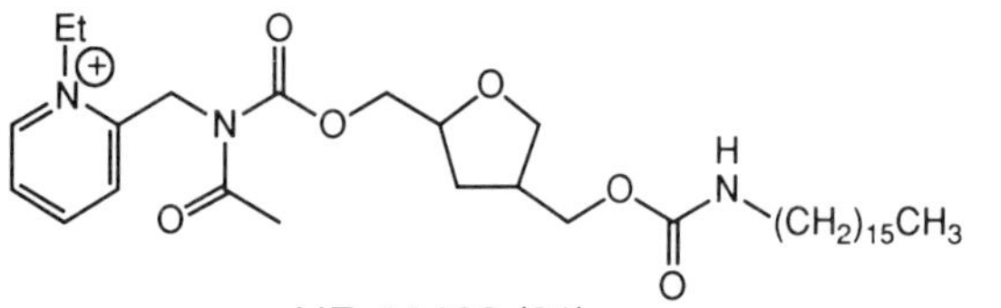

UR-11402 (21)

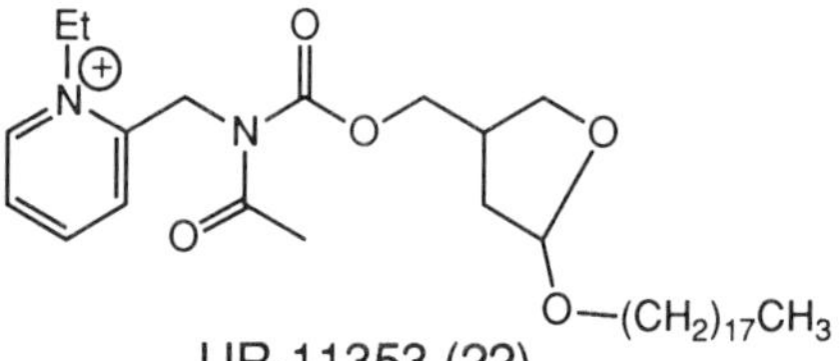

UR-11353 (22)

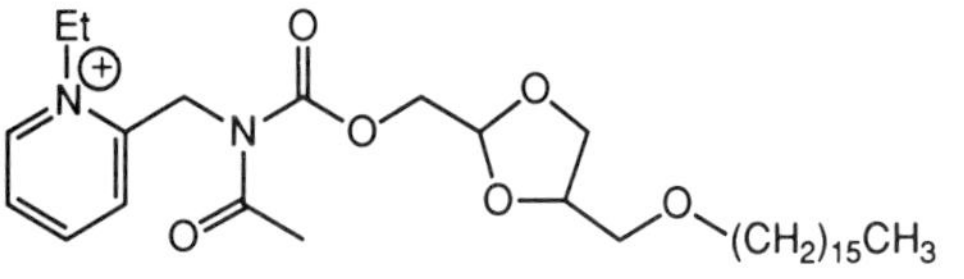

UR-10197 (23)

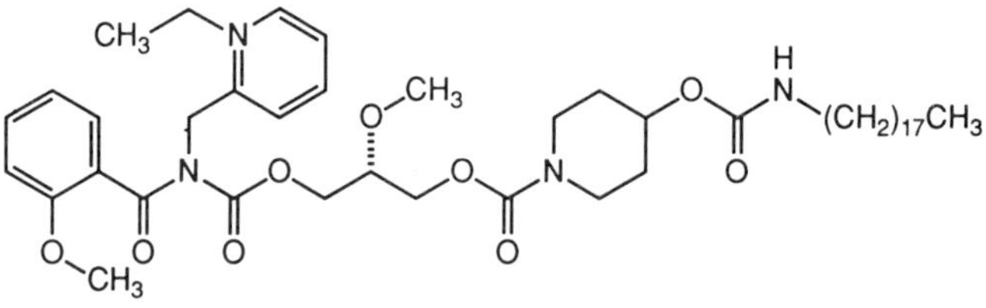

E-5880 (24)

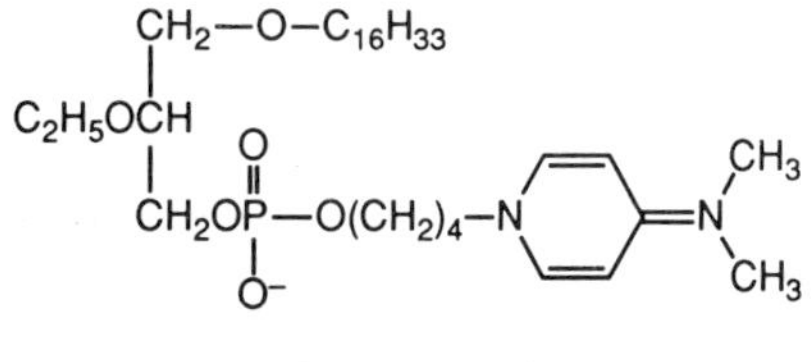

KO-286011 (25)

Figure 9.3 Structures of phospholipid derivatives as PAF receptor antagonists (continued from Fig. 9.2).

Another series of antagonists was obtained by replacing the phosphocholine moiety in PAF with a hepta-methylene thiazolium side-chain. The most potent compound in this series is Ono-6240 (Fig. 9.2) (Toyofuku *et al.*, 1986). Ono-6240 inhibited [³H]PAF binding to human platelet and human lung membranes with K_i values of 4.86 ($\pm$ 1.44) $\times$ 10^{-8} and 8.9 ($\pm$ 6.7) $\times$ 10^{-8} M, respectively, which is about 10 times more potent than CV-3988 (Hwang and Lam, 1986). Therefore, substitution of the phosphate group in CV-3988 by a non-polar hydrocarbon fragment improved the PAF antagonistic activity of Ono-6240. Ono-6240 also inhibited [³H]PAF binding to human PMN membranes but at a lower potency. This compound led us to identify a second type of PAF receptor present in human PMNs (Hwang, 1988). *In vivo*, Ono-6240 (3 mg/ kg i.v.) significantly prevented the decreases in systemic arterial pressure, left atrial pressure and cardiac output during endotoxin shock (1 µg/kg over 30 min) in unanaesthetized sheep.

Hoffman-La Roche's group (Burri *et al.*, 1985) adopted a similar approach and developed new potent PAF antagonists. These antagonists have an ED_{50} value against PAF (4 nM)-induced aggregation of rabbit platelets in plasma that varies between 10^{-8} and 10^{-6} M (Hadvary and Baumgarter, 1985). Ro 19-3704 (Fig. 9.2) is the most efficient antagonist. *In vivo*, Ro 19-3704

inhibited PAF-induced bronchoconstriction, hypotension and thrombocytopenia, and increased vasopermeability in the guinea-pig (Hadvary and Baumgartner, 1985).

Sandoz's group (Saunders and Handley, 1987; Handley, 1988) also developed several phospholipid-related antagonists (Fig. 9.2). SDZ 63-019 is a non-phosphorus thiazolium PAF receptor antagonist. SDZ 63-072, SDZ 63-441 and SDZ 63-675 are tetrahydro-furan analogues with a thiazolium (SDZ 63-072) or a quinolinium polar head group (SDZ 63-441 and SDZ 63-675). SDZ 63-675 is the most potent, and this analogue inhibited all major PAF responses in the rat, guinea-pig, dog and primate at doses between 10 and 30 µg/kg i.v. (Handley, 1988).

A novel amidophosphophonate analogue of PAF (PAF-AP) (Fig. 9.2) has been described by Steiner *et al.* (1985) as a potent inhibitor of platelet aggregation induced by PAF. However, this inhibitor was not specific, since it also inhibited platelet aggregation induced by arachidonic acid, A 23187, ADP and thrombin. This inhibitor may also function as a partial agonist that can activate the PAF receptor and receptor-mediated signal transduction in human monocytic leukaemic U-937 cells (Barzaghi *et al.*, 1989). Inhibition of either phospholipase A_2 or C activity was suggested from suppression of release of [¹⁴C] arachidonic acid from prelabelled platelet glycerophospholipids (Steiner *et al.*, 1985).

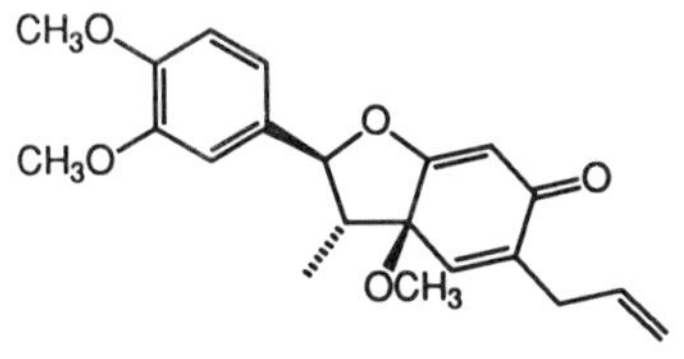

Kadsurenone (26)

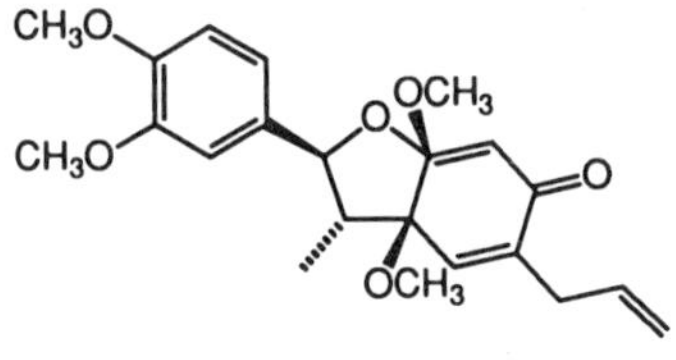

Piperenone (27)

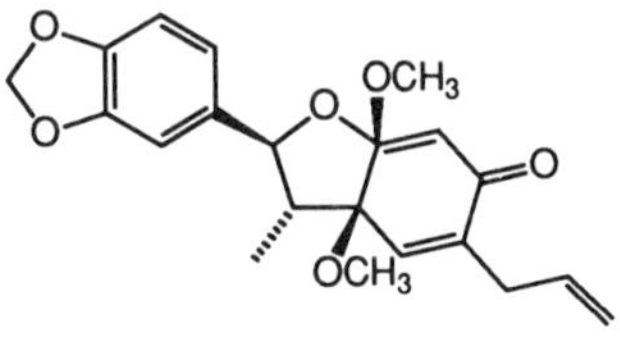

Kadsurin A (28)

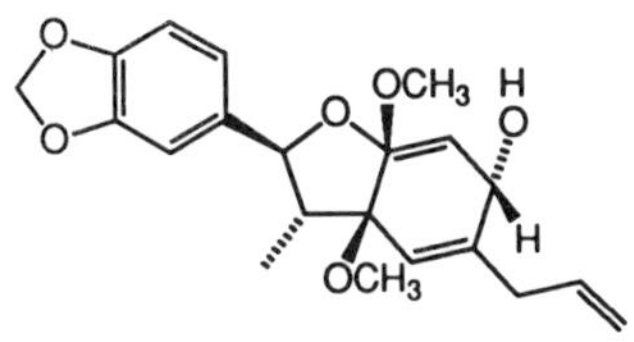

Kadsurin B (29)

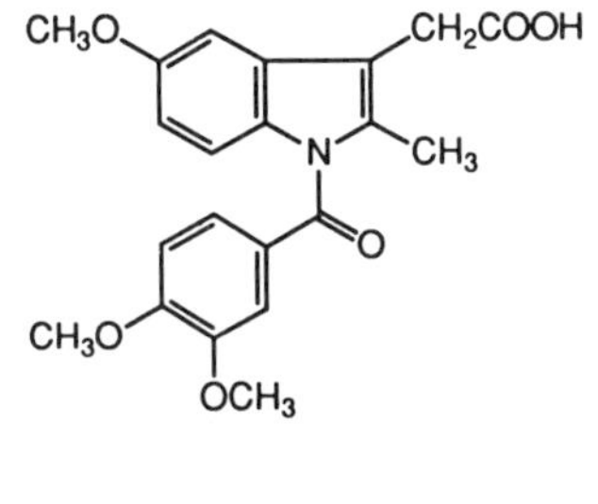

Indomethacin **analogue (30)**

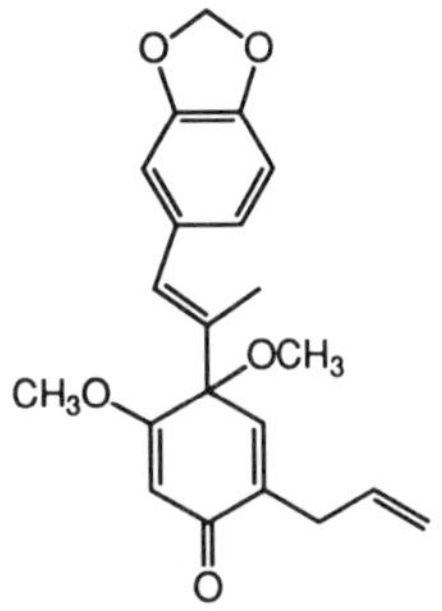

Futoquinol (31)

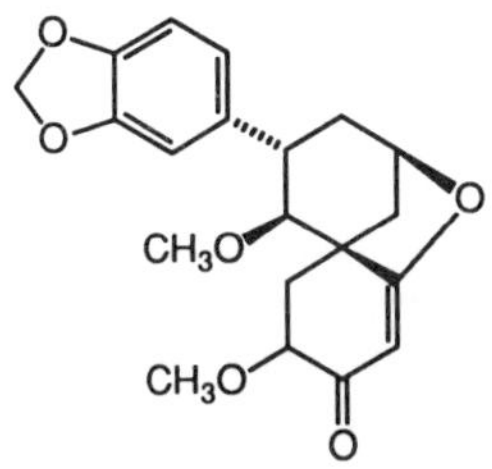

Futoenone (32)

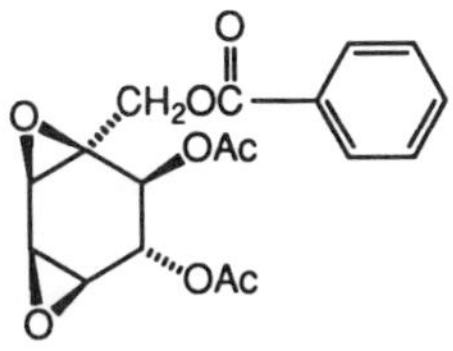

Futoxide (33)

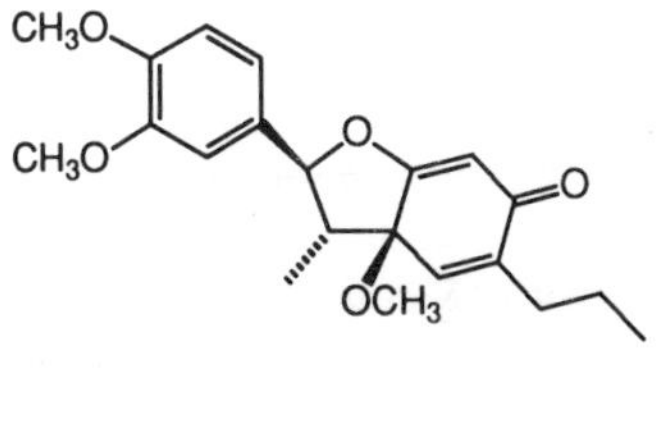

Dihydrokadsurenone (34)

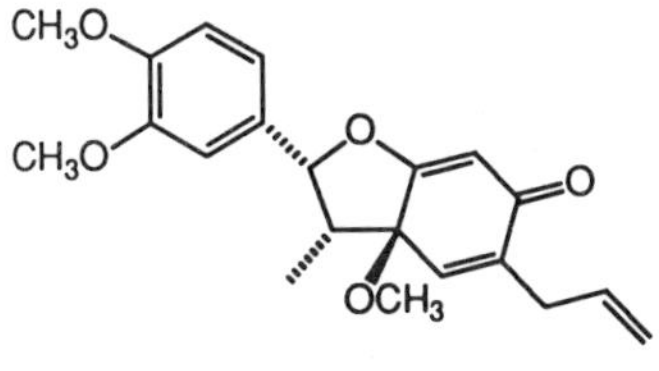

2 epimer of kadsurenone (35)

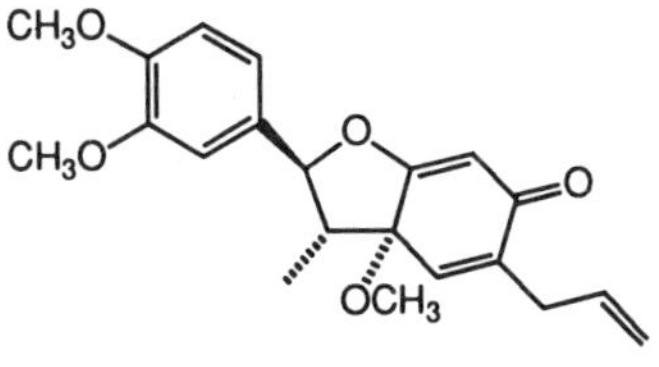

Denudatin B (36)

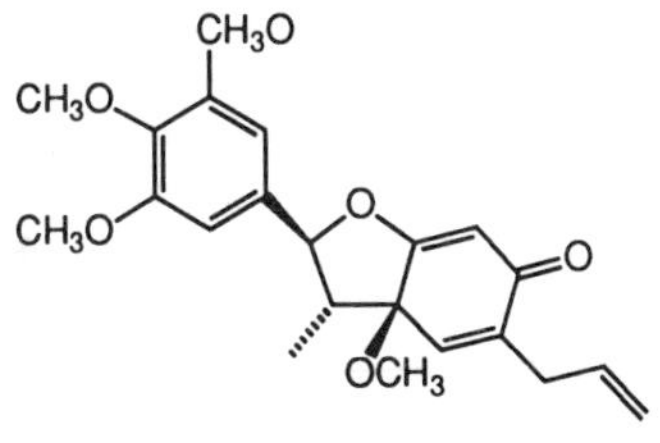

Mirandin A (37)

Figure 9.4 Structures of kadsurenone and its structural analogues.

Two new derivatives of 2-methylenepropane-1,3-diol (GS-1065-180) and (GS-1160-180) (Fig. 9.2) were reported to competitively antagonize PAF-induced aggregation in platelet-rich plasma from rabbits with ED_{50} values of 0.5 and 0.25 µM, respectively (Grue-Sorensen *et al.*, 1988). In the same experiment, CV-3988 was about 80 times less potent as an antagonist than GS-1160-180. At concentrations less than 10^{-4} M, neither GS-1065-180 nor GS-1160-180 caused platelet aggregation, nor did they inhibit platelet aggregation induced by collagen or ADP. *In vivo*, both compounds inhibited PAF-induced bronchoconstriction in guinea-

pigs. GS-1160-180 (ED_{50} = 0.1 mg/kg i.v.) was twice as active as GS-1065-180 and was 35 times more active than CV-3988. Both compounds were equipotent in inhibiting PAF-induced hypotension in rats (ED_{50} = 0.5 mg/kg i.v.), and were twice as active as CV-3988.

Broquet *et al.* (1990) from the Henri Beaufour Institute recently reported two new PAF-related antagonists but with constrained frameworks. They are BN 52111 and BN 52115 (Fig. 9.2) (Braquet *et al.*, 1990). Both compounds were about equipotent (ED_{50} = 0.3 and 0.4 µM for BN 52115 and BN 52111, respectively) as BN 52021, ginkgolide B, a terpenoid PAF antagonist (see below). *In vivo*, BN 52111 was active at 0.1 mg/kg i.v. in inhibiting PAF-induced bronchoconstriction and thrombocytopenia in the guinea-pig.

Following the strategy of Sandoz, the replacement of the PAF glycerol backbone with a disubstituted five-membered ring containing one (tetrahydrofuran) or two oxygen atoms (dioxolane) and the optimization of these two substituents has led to a new series of compounds showing high PAF antagonist activity (Bartroli *et al.*, 1991). Several of these compounds, such as UR-11353, UR-11402 and UR-10197 (Fig. 9.3) with a 2-pyridinium salt exhibited more potent activity than CV-6209 in inhibiting PAF-induced aggregation of rabbit platelets in plasma. A similar *N*-ethylpyridinium phospholipid analogue (E-5880) (Fig. 9.3) was also reported in Paris at a IUPHAR satellite symposium held in June 1990 (Sanchez Crespo, 1990). This compound competitively inhibited [^{3}H]PAF binding to PAF receptors of human platelets and inhibited PAF-induced hypotension and lethality. It was also active in some endotoxin-induced disturbances, including disseminated intravascular coagulation.

A 4-(*N,N*-dimethylamino)pyridium butyl ester analogue, KO-286011 (Fig. 9.3), was reported recently to have anti-PAF activity (Ostermann *et al.*, 1990). It inhibited [^{3}H]PAF binding to human platelets with a potency about three-fold higher than BN 52021. It also inhibited PAF-induced aggregation of human platelets in plasma with a K_b value of 3.5 μM, as derived from a Schild analysis, which is about 12-fold more potent than BN 52021.

3.2 LIGNANS AND NEOLIGNANS

Haifenteng is a Chinese herbal preparation of *Piper futokadsurae* for the general relief of asthma and the stiffness, inflammation and pain of rheumatic conditions. A neolignan, kadsurenone (Fig. 9.4), isolated from the stem of *P. futokadsurae*, was the first natural product to be discovered as a potent inhibitor of the binding of [^{3}H]PAF to rabbit platelet membrane preparation (Chang *et al.*, 1985; Hwang *et al.*, 1985b; Shen *et al.*, 1985). Kadsurenone is a competitive inhibitor of specific [^{3}H]PAF binding to its receptor preparations with an equilibrium dissociation constant of 86 nM (Hwang *et al.*, 1985b, 1986c). At the cellular level, it potently inhibited the PAF-induced aggregation of rabbit platelets (Hwang *et al.*, 1985b) and degranulation of human neutrophils (Shen *et al.*, 1985). *In vivo*, kadsurenone inhibited PAF-induced cutaneous vascular permeability in the guinea-pig (Hwang *et al.*, 1985b; Shen *et al.*, 1985) and in the rat (Hwang *et al.*, 1985b), as well as haematocrit changes (Shen *et al.*, 1985), foot oedema (Hwang *et al.*, 1986a) and hypotension in the rat (Shen *et al.*, 1985). Finally, kadsurenone partially antagonized endotoxic shock in rats (Doebber *et al.*, 1986).

Other related structures from *P. futokadsurae* (Hwang

et al., 1986c) and another related medicinal herb, *Piper hancei maxim* (Han *et al.*, 1986), also showed PAF antagonistic activity but at a lower potency. The data revealed a structural specificity of kadsurenone in the inhibition of PAF receptor binding (Hwang *et al.*, 1986c). Piperenone (Fig. 9.4), a 7α-methoxy analogue of kadsurenone, had an ED$_{50}$ value of 2.51 μM, which is about 25 times less potent than kadsurenone (ED$_{50}$ = 0.1 μM). The other two analogues with 2α(3′,4′-methylenedioxylphenyl) groups, kadsurin A and kadsurin B (Fig. 9.4), are even less potent, with ED$_{50}$ values of 10 and 40 μM, respectively, in inhibiting specific [^{3}H]PAF binding to rabbit platelet membranes. Clearly, the 3′,4′-dimethoxy moiety in the 2α-phenyl group is important in blocking PAF receptor binding. A similar structure–activity relationship was also found with indomethacin analogues (Hwang *et al.*, 1984). The 3′,4′-dimethoxy analogue of indomethacin (Fig. 9.4) was found to be the most potent one in that series.

Several other lignans, i.e. futoquinol, futoenone and futoxide (Fig. 9.4), have recently been isolated from samples of material from *P. futokadsurae* and found to possess comparable PAF antagonist activities in the platelet aggregation and receptor-binding assays (Shen and Hussaini, 1990).

A more detailed structure–activity relationship was not done until total synthesis of kadsurenone became possible (Ponpipom *et al.*, 1986, 1987). Synthetic *rac*-kadsurenone had an ED$_{50}$ value of 0.24 μM, which is about half the activity of the natural product. Dihydrokadsurenone (Fig. 9.4) (Hwang *et al.*, 1986b) prepared by hydrogenation of the alkyl side-chain of kadsurenone was slightly more active than kadsurenone with an ED$_{50}$ value of 0.07 μM. The 2 epimer of kadsurenone (Fig. 9.4) was much less active (16% inhibition at 1 μM). The 3*a* epimer of kadsurenone (denudatin B) (Fig. 9.4) was also weakly active (31% at 1 μM). The inversion of the configuration at either the 2 or 3*a* position appears to diminish the receptor inhibitory activities. Racemic mirandin A (Fig. 9.4) was a poor PAF antagonist. Interestingly, desallylkadsurenone was also weakly active, indicating the contribution to receptor binding of a lipophilic allyl or propyl side-chain at the 5 position (Ponpipom *et al.*, 1987). Curiously, denudatin B, although showing poor PAF receptor antagonist activity, has been reported to inhibit the aggregation of rabbit platelets caused by PAF, inhibit phosphoinositide breakdown caused by thrombin and collagen, and possibly inhibit the Ca^{2+} influx through voltage-gated and receptor-operated calcium channels (Yu *et al.*, 1990).

The flower buds of *Magnolia biondii* have been widely used as a folk medicine in China for the treatment of nasal congestion, headache, sinusitis and allergic activity. Six lignans isolated from the flower buds of *M. biondii* showed anti-PAF activities (Pan *et al.*, 1987). They are liroresinol-B dimethyl ether, magnolin, pinoresinol dimethyl ether, fargesin, demethoxyaschantin and aschantin (Fig. 9.5).

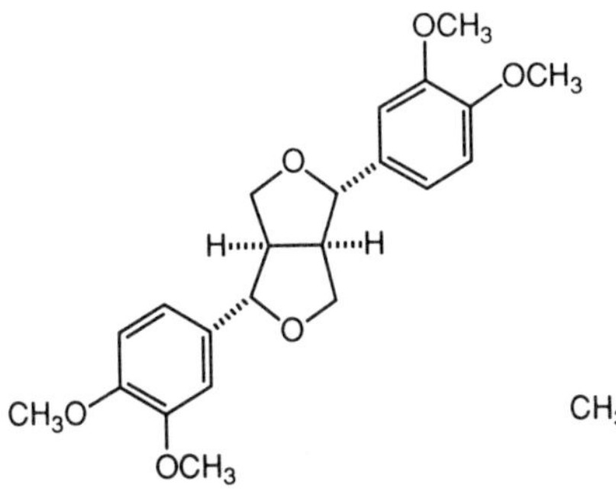

Pinoresinol dimethylether (38)

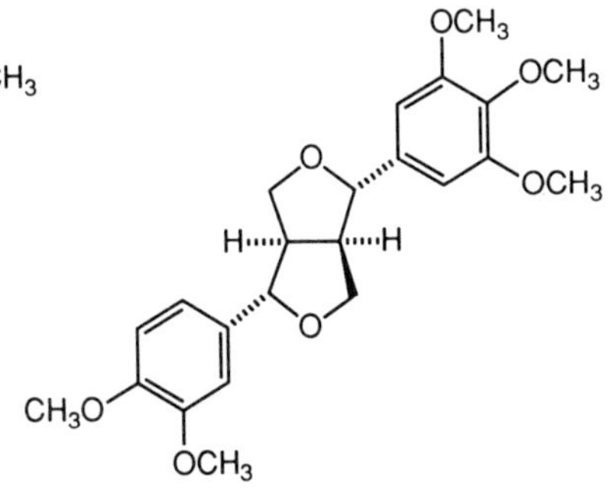

Magnolia (39)

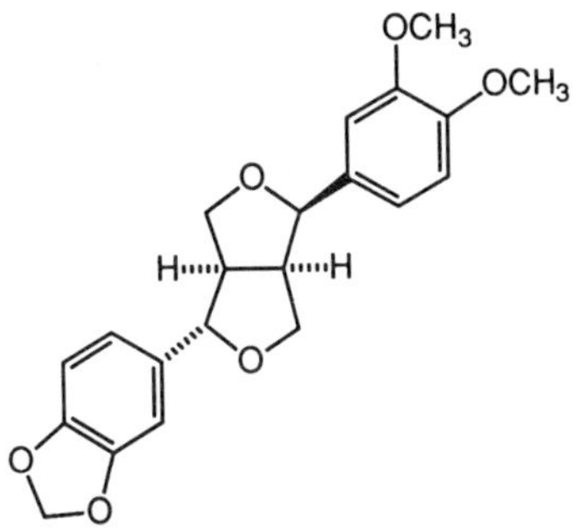

Fargesin (41)

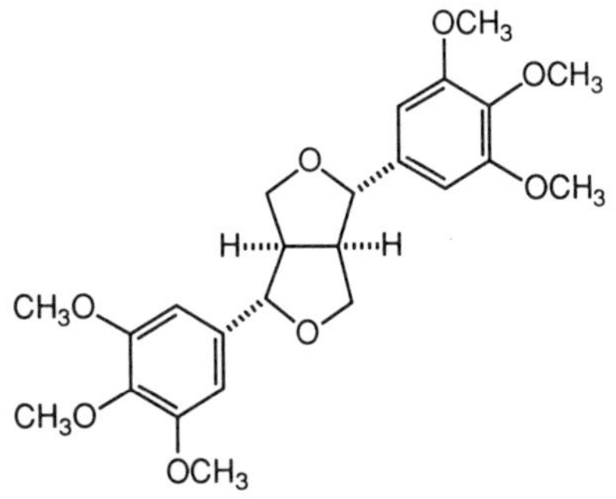

Liroresinol-B dimethyl ether (40)

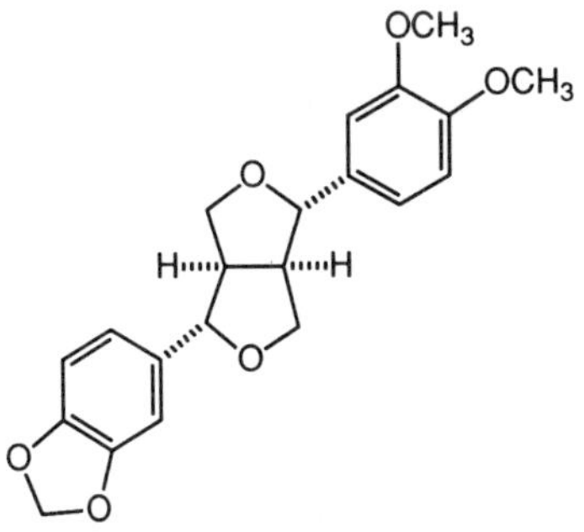

Demethoxyaschantin (42)

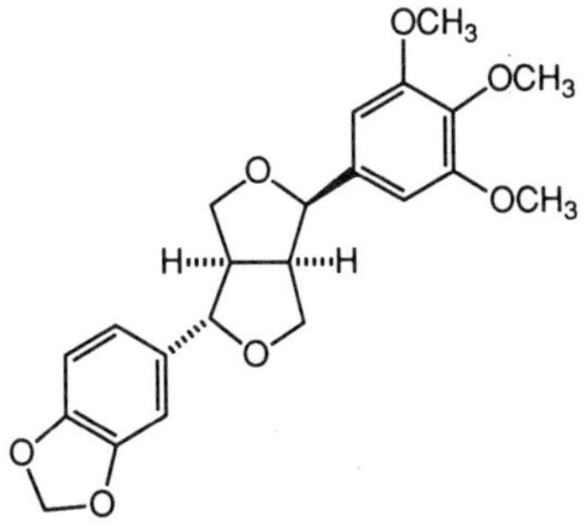

Aschantin (43)

Figure 9.5 Six active components isolated from *Magnolia biondii* as PAF receptor antagonists.

Fargesin and pinoresinol dimethyl ether showed roughly identical potencies in inhibiting the specific [³H]PAF binding to rabbit platelet membranes, with ED_{50} values of 1.3 and 1.7 μM, respectively, which are about 10 times less potent than kadsurenone. The other four lignans were less potent and followed the order demethyoxyaschantin > magnolin > liroresinol-B dimethyl ether > aschantin (Pan *et al.*, 1987). Similar structures and analogues were also isolated from the fruits of *Forsythia suspensa* (Oleaceae), the seeds of *Arctium lappa* (Compositae) and the herb *Centipeda minima* (Compositae) (Iwakami *et al.*, 1990). The presence of at least one 3,4-dimethoxyphenyl or 3,4,5-trimethoxyphenyl group is essential for high PAF antagonist activity. However the presence of hydrophilic groups, such as glucosyl, hydroxyl and phenolic moieties, lowers the antagonistic activity. No significant difference was observed between *trans* and *cis* isomers of bistetrahydrofuran lignans.

Several butanolide-type lignans were also reported to have PAF antagonistic activity (Iwakami *et al.*, 1990).

Table 9.1 Structure–activity relationship of butanolide

Compound	R¹	R²	R³	ED_{50} (μM)[a]
(43)	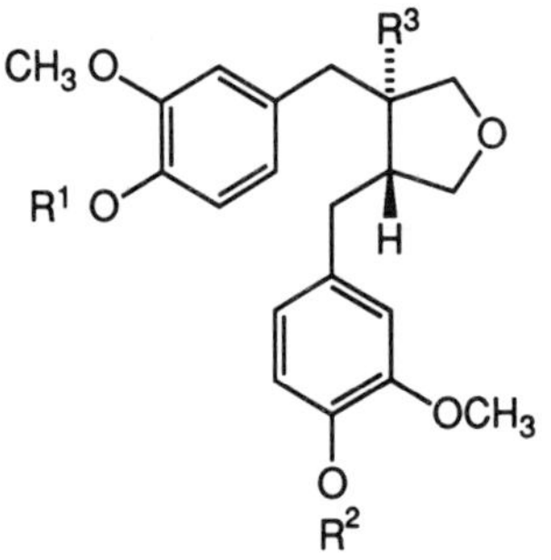 H	H	H	<30% (10 μM)
(44)	H	CH_3	H	2.9
(45)	CH_3	CH_3	H	0.56
(46)	H	CH_3	OH	57% (10 μM)
(47)	CH_3	CH_3	OH	60% (10 μM)

[a] Inhibition of PAF-induced rabbit platelet aggregation.

All cis Compound (48)

Veraguensin (49)

Saucerneol (50)

Galbelgin (51)

(52)

Galbelgin (53)

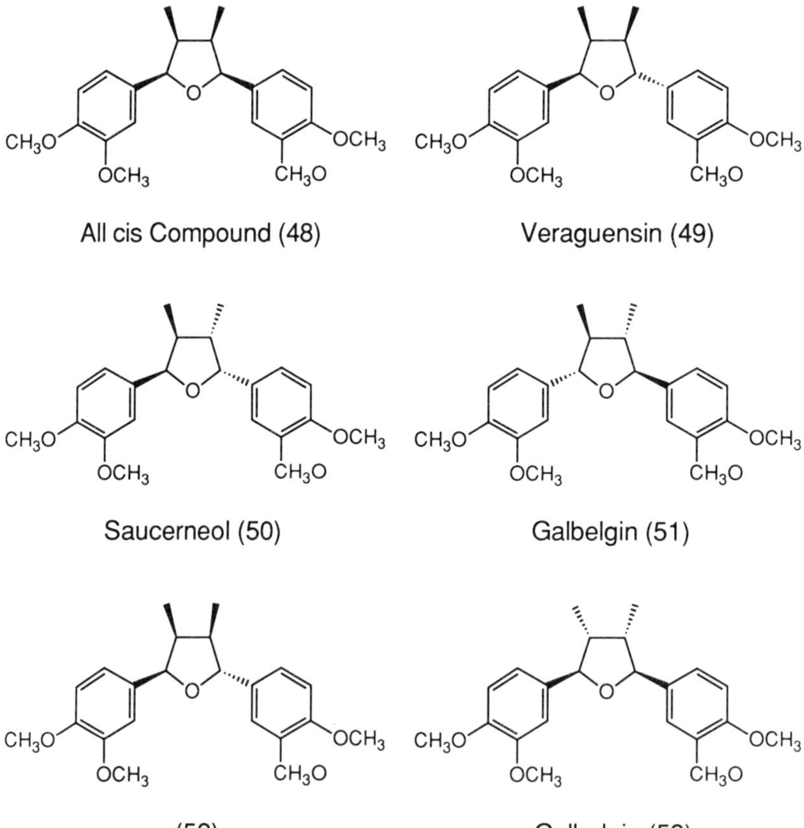

Figure 9.6 Six stereoisomers of veraguensin as PAF receptor antagonists.

The presence of the 3,4-dimethoxyphenyl group is also required for high PAF antagonistic activity in this series (Table 9.1). Subsequent demethylation of the 3,4-dimethyphenyl group resulted in progressive loss of PAF antagonist activity. Substitution of hydrogen by a hydroxyl group at the R^3 position also causes loss of anti-PAF activity (Iawakami *et al.*, 1990). Burseran, isolated from *Bursera microphylla* A (Burseraceae), showed moderate activity in inhibiting PAF-induced rabbit platelet aggregation (Braquet and Godfroid, 1987).

Members of the 2,5-bisaryltetrahydrofuran class of lignans, veraguensin, galbegin and galgravin, isolated

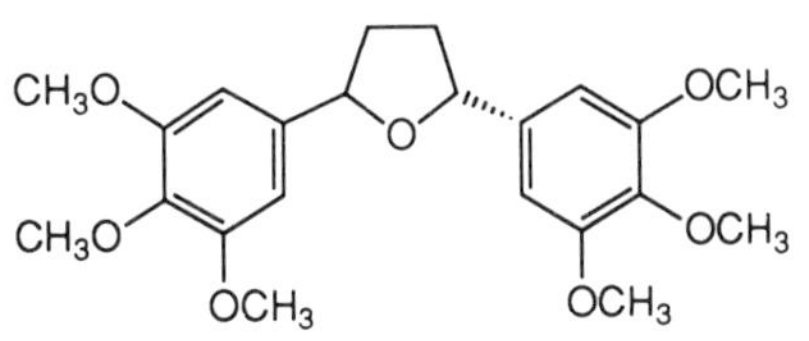

L-652,731 (54)

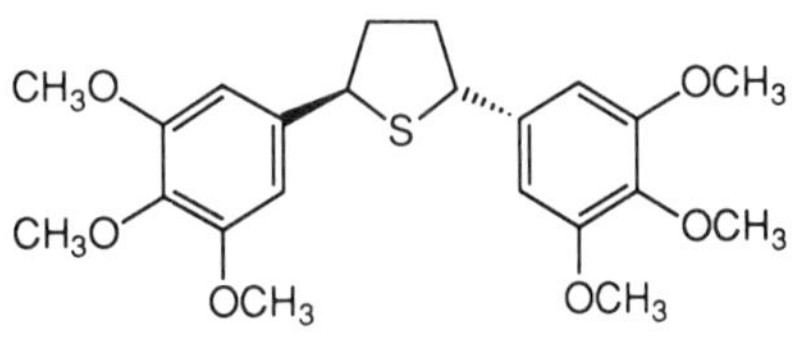

L-653,150 (55)

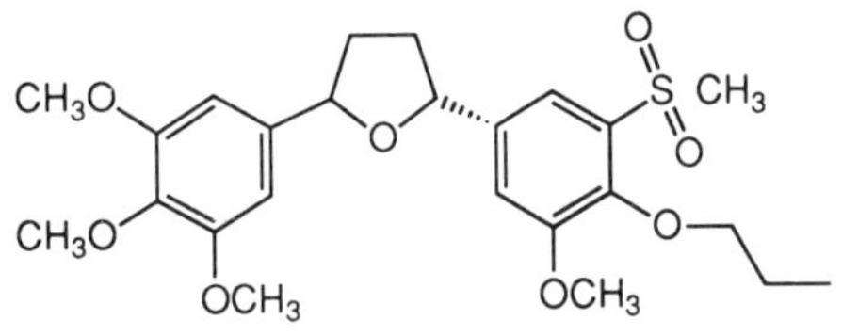

L-659,989 (56)

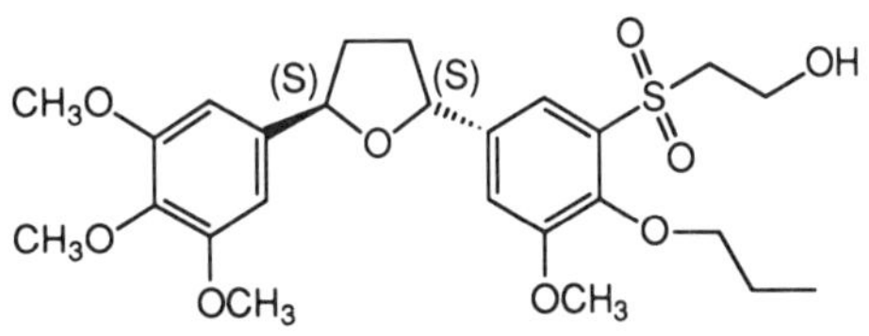

L-680,573 or MK 287 (57)

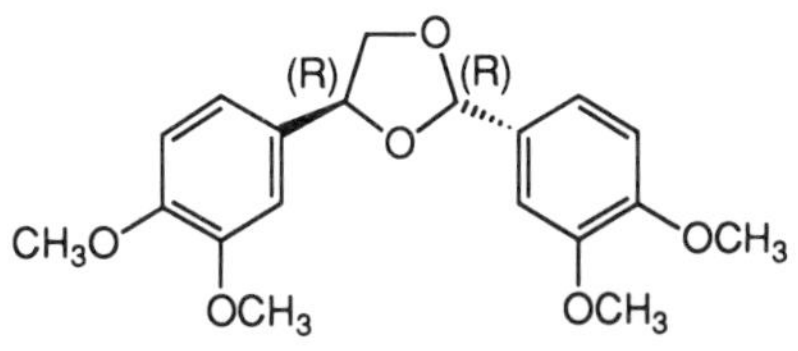

(84)

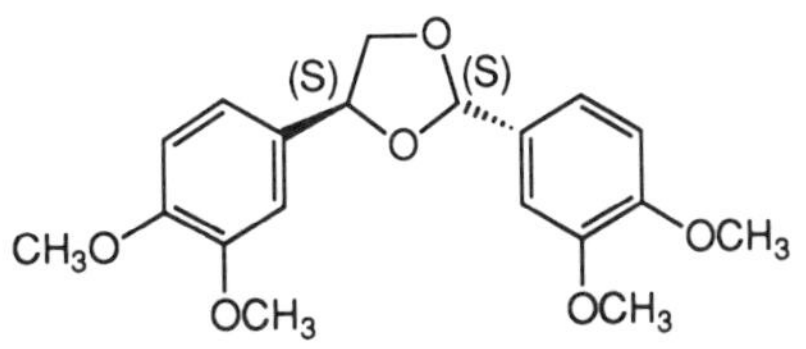

(85)

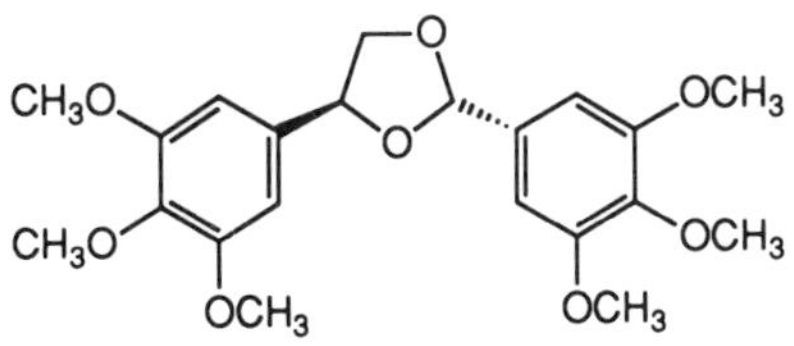

(86)

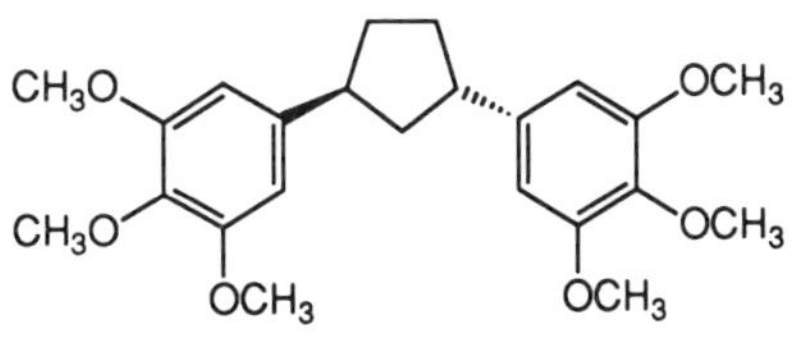

L-658,994 (88)

Figure 9.7 Structures of tetrahydrofuran analogues as PAF receptor antagonists.

from *Magnolia acuminata* (cucumber tree) and *Himantandra belgraveana*, are relatively potent specific inhibitors of the binding of [^{3}H]PAF to isolated rabbit platelet membranes and PAF-induced rabbit platelet aggregation (Biftu *et al.*, 1986a). Six stereoisomers of these tetrahydrofuran lignans (Fig. 9.6) were synthesized. The inhibition of PAF binding to its receptors by the isomers is stereodependent. The lowest-energy conformation of 2,5-diaryltetrahydrofuran has been obtained by using the molecular mechanism progam MM2 (Biftu *et al.*, 1986a). The spatial arrangements of the Ar–C–O–C–Ar fragment have three distinct shapes that resemble the letters W, S and U. As the shape changes from W to S to U, the activity falls from an ED_{50} value of 0.2 μM for the all-*cis* isomer to 4.5 μM for galgravin in inhibiting [^{3}H]PAF (1 nM) binding to rabbit platelet membranes in the presence of 150 mM sodium chloride.

Using these structure–activity relationships of kadsurenone and veraguensen, L-652,731 (Fig. 9.7), a trimethoxy analogue of tetrahydrofuran, was synthesized (Hwang *et al.*, 1985c). L-652,731 showed no agonist activities at the cellular and tissue levels. It specifically and competitively inhibited [^{3}H]PAF (1 nM) binding to its receptors on the following membranes: rabbit (Hwang *et al.*, 1985c) and human platelet membranes (Hwang and Lam, 1986); rabbit (Paulson *et al.*, 1990), human (Hwang, 1988) and rat PMN membranes (Hwang, 1991a); human (Hwang *et al.*, 1985a) and guinea-pig lung membranes (Gomez *et al.*, 1990); and rat liver membranes (Hwang, 1987). L-652,731 showed different

Table 9.2 Structure–activity relationship of L-652,731 analogues in inhibiting [^{3}H]PAF binding to rabbit platelet membranes

Compound	X	ED_{50} (nM)	L number
(54)	OCH$_3$	20	L-652,731
(58)	OCH$_2$CH(CH$_3$)$_2$	8	L-659,600
(59)	Cl	50	L-658,451
(60)	OCH$_2$CH=CH$_2$	58	L-659,509
(61)	I	100	L-658,996
(62)	OCH$_2$Ph	120	L-659,368
(63)	CH$_3$	130	L-657,760
(64)	COOH	170	L-658,353
(65)	COOCH$_2$CH$_3$	200	L-658,351
(66)	CH$_2$CH$_3$	280	L-658,324
(67)	SCH$_3$	500	L-658,426
(68)	H	>5000	L-656,783
(69)	(CH$_2$)$_{15}$CH$_3$	>5000	L-658,375

Table 9.3 Structure–activity relationship of L-659,989 analogues in inhibiting [^{3}H]PAF binding to rabbit platelet membranes

Compound	X	ED_{50} (nM)	L number
(56)	SO$_2$CH$_3$	1	L-659,989
(70)	CH$_2$SCH$_3$	280	L-661,101
(71)	CH$_2$NH$_2$	210	L-660,621
(72)	CH$_2$NHAc	180	L-661,059
(73)	CH$_2$OAc	45	L-661,060
(74)	NH$_2$Ac	33	L-659,660
(75)	NH$_2$	30	L-659,659
(76)	CH$_2$OH	25	L-661,090
(77)	CH$_2$CN	25	L-662,036
(78)	CH$_2$CH$_2$CH$_3$	25	L-659,927
(79)	CONH$_2$	23	L-659,891
(80)	CN	9	L-659,915
(81)	SCH$_3$	7	L-659,967
(82)	NO$_2$	5	L-659,169
(83)	SO$_2$CH$_2$CH$_2$CH$_3$	0.2	L-668,446

potency among species as well as cell types in inhibiting the specific [^{3}H]PAF binding to isolated membranes (Hwang, 1987, 1991a). L-652,731, as well as other tetrahydrofuran analogues including L-653,150, L-659,989 and MK 287, showed differences in antagonist potency between rabbit and human membranes (Hwang and Lam, 1986; Hwang *et al.*, 1988, 1990), being about 10 times more potent in rabbits than in humans in inhibiting [^{3}H]PAF binding to isolated membranes, as well as in inhibiting platelet aggregation induced by PAF (Hwang, 1988). L-652,731 was also about 10 times more potent in inhibiting the [^{3}H]PAF binding in rat liver tissues than in rat PMNs (Hwang, 1991a). A difference in PAF receptors between rat PMNs and liver can be accepted if one adopts the guidelines suggested by Furchgott (1972) that affinity constants for antagonists differing by more than three-fold should be considered as evidence for differences in receptors. Therefore, PAF receptors in rabbits and humans are

different. PAF receptors are also different between PMNs and liver tissues. The K_b value of L-652,731 to rabbit platelet membranes obtained from a Schild plot by performing a Scatchard analysis of the receptor-binding inhibition is 0.27 μM. The K_b value obtained from a Schild plot by comparing the dose–response curves of platelet aggregation in the presence and in the absence of L-652,731 is 0.21 μM (Hwang *et al.*, 1985c). The close agreement of K_b values determined either from receptor inhibition or from platelet aggregation inhibition indicated that L-652,731 is a specific and pure PAF receptor antagonist and that L-652,731 should exhibit no inhibitory effect on other enzymes involved in platelet aggregation following binding to the receptors. L-652,731 showed no inhibition of platelet aggregation stimulated by other stimuli, including collagen, arachidonic acid, ADP, thrombin, adrenaline and A 23187.

For L-652,731, a structural specificity was clearly demonstrated by a group of structural analogues. The *cis* isomer of L-652,731 was 1000 times less potent (Hwang *et al.*, 1985c). Aromatic ring substitutions were also relatively specific, with the decrease in methoxy substitution generally leading to a fall in activity. The positions of trimethoxy groups in the phenyl rings were important. *trans*-2,5-bis(2,3,4-trimethoxyphenyl)tetrahydrofuran was about 500-fold less potent than L-652,731. Similarly, the substitution of methoxy groups by isopropyloxy moieties significantly reduced the PAF antagonistic activity (Hwang *et al.*, 1985c).

Asymmetric substitutions in the phenyl rings of tetrahydrofuran were further explored. This led to the identification of L-659,989 (Hwang *et al.*, 1988) and L-690,573 (MK 287) (Hwang *et al.*, 1990) (Fig. 9.7). Maintaining the 3,4,5-trimethoxyphenyl group on one side and the 3-methoxyphenyl group on the other side of the tetrahydrofuran ring seemed to be essential for PAF antagonist activity. Tables 9.2 and 9.3 demonstrate that the PAF antagonist activities of 2,5-diaryltetrahydrofurans varied with the electron density of the 3-aryl substituents and the chain length of the 4-aryl substituents. Again, the *trans* isomer of L-659,989 (Hwang *et al.*, 1988) or MK 287 (Hwang *et al.*, 1990) was about 1000 times more potent than the corresponding *cis* isomer in inhibiting the specific PAF binding to either rabbit or human platelet membranes. In addition, the laevorotatory, *trans-(S,S)* isomer was about 30 times more potent than the corresponding dextrorotatory, *trans-(R,R)* isomer (Hwang *et al.*, 1988, 1990). Interestingly, *trans*-2,5-(*R,R*)-bis(4,5-dimethoxyphenyl)dioxolane (compound (84)) and its *trans-(S,S)* enantiomer (compound (85)) showed roughly the same potency in inhibiting PAF-induced rabbit platelet aggregation, with K_i values of 0.43 and 0.37 μM, respectively (Corey *et al.*, 1988a).

L-659,989 showed species differences. It inhibited [^{3}H]PAF binding to either rabbit platelet or rabbit PMN membranes, with a K_i value of 1.1 nM; whereas, in human platelet, human PMN or human lung membranes, L-659,989 was about 10 times less potent, with a K_i value of 14.3 nM (Hwang *et al.*, 1988). Such variation in potency between species seems to be a general feature of tetrahydrofuran analogues. Similar differences in potency between humans and rabbits were also observed with other tetrahydrofuran analogues, such as L-652,731, L-653,150 (Hwang and Lam, 1986) and MK 287 (Hwang *et al.*, 1990).

Kadsurenone and the tetrahydrofuran analogues are competitive receptor antagonists. They may share a common binding site with PAF in the receptor. In the receptor-binding assays, kadsurenone (Hwang *et al.*, 1985b, 1986c), L-652,731 (Hwang *et al.*, 1985c), L-659,989 (Hwang *et al.*, 1988) and MK 287 (Hwang *et al.*, 1990) altered only the apparent dissociation constant (K_d') of [^{3}H]PAF binding to its receptors, but showed no effect on the maximal detected number of receptor sites. In Schild plots, by plotting the logarithm of the mean of $((K_d'/K_d)-1)$ versus the logarithm of the antagonist concentration, a unit slope was obtained. L-659,989, like PAF, PAF analogues and several selected PAF receptor antagonists, can fully displace the specific binding of [^{3}H]dihydrokadsurenone to rabbit platelet membranes (Hwang *et al.*, 1986b, 1988). Similarly, PAF, PAF analogues, dihydrokadsurenone and several selected PAF receptor antagonists can also fully displace the specific binding of [^{3}H]L-659,989 (Hwang *et al.*, 1989). With [^{3}H]52770 RP (see below), Robaut *et al.* (1987) also obtained a good correlation between the antagonist potencies of several structural analogues of 52770 RP in inhibiting the specific binding of [^{3}H]52770 RP to rabbit platelet membranes as well as in inhibiting PAF-induced aggregation of rabbit platelets. These results strongly suggested that L-652,731, L-659,989, MK 287, kadsurenone and 52770 RP share a common binding site with PAF.

In vivo, L-652,731 (Hwang *et al.*, 1985c; Doebber *et al.*, 1986), L-659,989 (Hwang *et al.*, 1988; Ponpipom *et al.*, 1988) and MK 287 (Hwang *et al.*, 1990) are orally active. They inhibited guinea-pig bronchoconstriction induced by PAF and showed protective effects in several disease models in animals. L-659,989 was also shown to prolong parturition in the pregnant rat (Zhu *et al.*, 1991). Administration by mouth with three concentrations (1.6, 16 and 48 mg/kg/day)) of L-659,989 did not alter the gestation period; however, the duration of parturition was increased from two-fold to five-fold by such treatment. No toxicity of the compound was apparent and fetal mortality was not significantly altered by the treatment of L-659,989. These results suggest that PAF may be involved in the process of parturition.

The bioavailability of L-659,989 in rhesus monkeys was low (< 10%). The compound was metabolized predominantly at the C-4' propoxy side-chain. Two major metabolites in the excreta of rhesus monkeys were identified as the (±)-*trans*/(±)-*cis* mixture of the 4'-hydroxy species and the glucuronide conjugate of the *trans*-4'-hydroxy compound (Thompson *et al.*, 1991). In rats, L-659,989 exhibited gender-dependent pharmacological activity, with the compound showing good activity in female rats but no activity in males, whereas, in monkeys, both genders demonstrated similar absorption, bioavailability, metabolism and excretion patterns (Thompson *et al.*, 1991).

Modification of the tetrahydrofuran ring into dioxolane retains the anti-PAF activity. *trans*-2,5-bis(3,4,5-Trimethoxyphenyl)dioxolane (compound (86), Fig. 9.7) also showed good competitive PAF antagonism in the rabbit washed platelet assay (K_i = 0.3 μM). The *cis* isomer was also less active (Corey *et al.*, 1988a). *trans*-2,5-bis(3,4,5-Trimethoxyphenyl)dioxolane inhibited specific binding of [^{3}H]C$_{18}$-PAF to human platelet and PMN membranes with ED$_{50}$ values of 0.75 and 0.6 μM, respectively (Fig. 9.8), roughly the same potency as L-652,731.

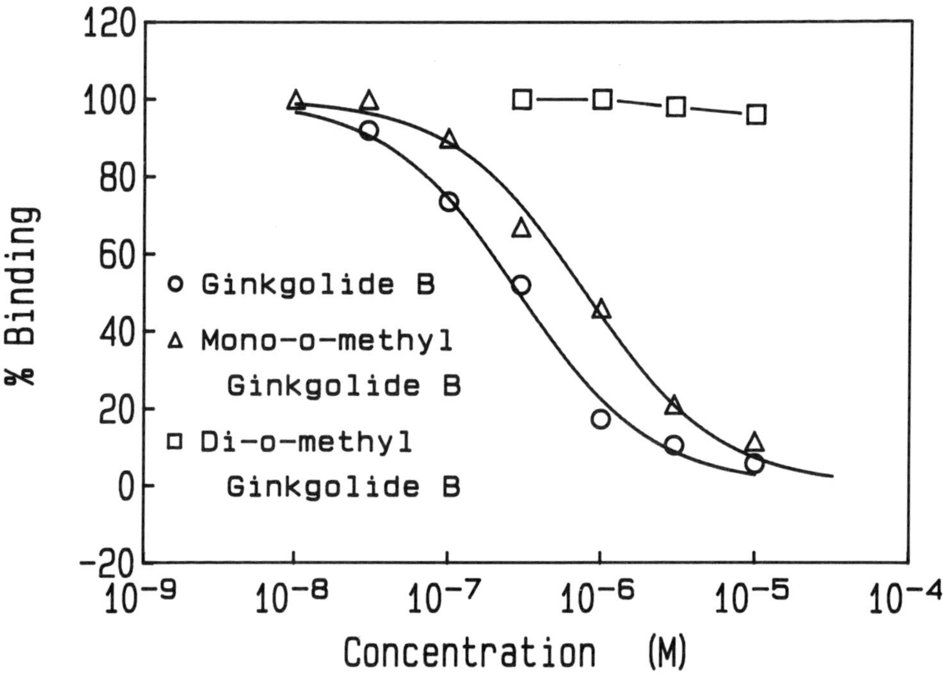

Figure 9.8 Inhibition of [³H]C₁₈-PAF binding to human platelet and PMN membranes by *trans*-2,5-bis(3,4,5-trimethyoxyphenyl)dioxolan. Membrane protein (100 µg) was added to an incubation mixture (1 ml) containing 0.3 nM [³H]C₁₈-PAF and a known amount of inhibitor in a medium of 10 mM magnesium chloride, 10 mM Tris and 0.25% BSA at pH 7.0. The compound was dissolved in DMSO. The data points and error bars are the means and the standard deviations of three independent experiments. In each experiment, triplicate samples were prepared.

Table 9.4 Structure–activity relationship of cyclopentane analogues related to L-652,731 in inhibiting [³H]PAF binding to rabbit platelet membranes

Compound	X	ED_{50} (nM)	L number
(87)	$=O$	440	L-658,927
(88)	H	20	L-658,994
(89)	OH	84% (5 µM)	L-658,996
(90)	OCH_3	8.3	L-659,248
(91)	$OCH_2CH_2CH_3$	30	L-660,317
(92)	OCH_2CH_3	30	L-660,318
(93)	$OCH_2CH=CH_2$	4.8	L-660,319
(94)	$=NOH$	5.8	L-660,320
(95)	$=NOCH_3$	5.0	L-660,321
(96)	OCH_2COOH	16% (300 nM)	L-661,000
(97)	$OCH_2C\equiv CH$	1.0	L-661,002
(98)	$OCH_2COOCH_2CH_3$	3.6	L-661,003
(99)	OCH_2Ph	48	L-661,004
(100)	OCH	19	L-661,005
(101)	$OCH_2CH_2NCH_3$	110	L-662,042
(102)	$OCH_2CON(CH_3)_2$	23% (300 nM)	L-662,389
(103)	$OCH_2CH_2N(CH_3)_2$	100	L-662,390

1,3-Diarylcyclopentanes, carbocylic analogues of 2,5-diaryltetrahydrofuran, also have potent PAF receptor antagonist activity (Graham *et al.*, 1989). The activities in inhibiting the PAF receptor binding for the cyclopentane analogues are given in Tables 9.4 and 9.5. With respect to the symmetrical analogues (Table 9.4), the unadorned cyclopentane (L-659,994) is as active as the corresponding tetrahydrofuran (L-652,731) towards rabbit platelet receptors (ED_{50} = 20 nM). Although the oximes (L-660,321 and L-660,320) showed excellent

Table 9.5 Structure–activity relationship of cyclopentane analogues related to L-659,989 in inhibiting [³H]PAF binding to rabbit and human platelet membranes

Compound	X	ED_{50} (nM)		L number
		Rabbit	Human	
(104)	OCH_3	4.3		L-662,427
(105)	OCH_3	3.8	17	L-662,429
(106)	$OCH_2CH{=}CH_2$		19% (30 nM)	L-662,910
(107)	$OCH_2CH{=}CH_2$		4.1	L-662,911
(108)	$OCH_2C{\equiv}CH$		60	L-665,911
(109)	$OCH_2C{\equiv}CH$		3.6	L-667,130
(56)		3.0	30	L-659,989

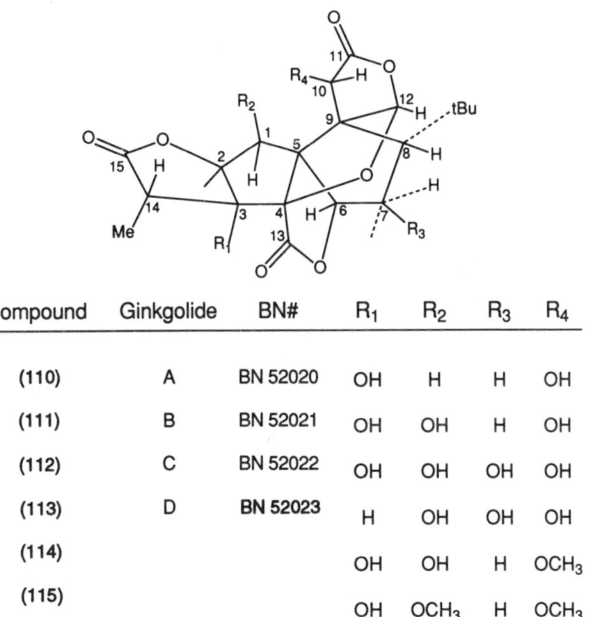

Compound	Ginkgolide	BN#	R_1	R_2	R_3	R_4
(110)	A	BN 52020	OH	H	H	OH
(111)	B	BN 52021	OH	OH	H	OH
(112)	C	BN 52022	OH	OH	OH	OH
(113)	D	BN 52023	H	OH	OH	OH
(114)			OH	OH	H	OCH_3
(115)			OH	OCH_3	H	OCH_3

Figure 9.9 Structures of ginkgolides A, B, C and D, and the monomethoxy- and dimethoxy-substituted structural analogues.

activity, other small polar substituents such as the alcohol (L-648,996), the ketone (L-658,927) and the acetate (L-659,247) are not very active. Ethers are the most active class of substituents. From the representative examples in Table 9.4, it can be seen that the methyl moiety (L-659,248) and small unsaturated groups (L-661,002, L-661,003 and L-660,319) are most active, while saturated alkyl (L-660,318 and L-660,317), aromatic (L-661,004) and alkyl groups containing hetero atoms (L-662,042 and L-662,390), acidic (L-661,000) or basic (L-662,390) groups are much less active.

The activities of analogues with the best ether substituents on the asymmetric cyclopentanes related to L-659,989 are given in Table 9.5. The α- and β-methoxy isomers (L-662,427 and L-662,429) are nearly equipotent, while the β-allyl and propargyl analogues (L-662,911 and L-667,130) are much more active than the corresponding α substituents (L-662,910 and L-665,911) against the human receptor. The same relative order of potency (methyl < allyl < propargyl) seen in the symmetrical series (see Table 9.4) is maintained with the L-659,989-related compounds (Table 9.5). L-662,911 and L-667,130 are two of the most potent PAF receptor antagonists in this cyclopentane series.

3.3   GINKGOLIDE B AND OTHER TERPENOID-RELATED COMPOUNDS

Ginkgolides A, B and C (BN 52020, BN 52021 and BN 52022, respectively) (Fig. 9.9) are terpenoids isolated

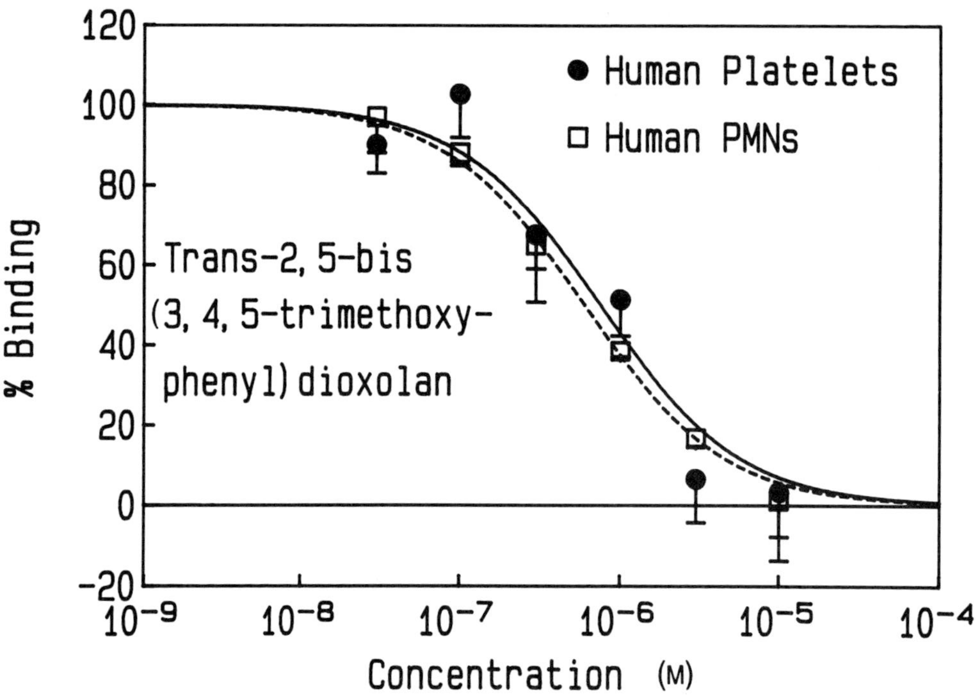

Figure 9.10 Inhibition of $[^3H]C_{18}$-PAF binding to rabbit platelet membranes by ginkgolide B and di-*o*-methyl-ginkgolide B. Membrane protein (100 µg) was added to an incubation mixture (1 ml) containing 1 nM $[^3H]C_{16}$-PAF and a known amount of inhibitor in a medium of 150 mM sodium chloride, 10 mM Tris and 0.25% BSA at pH 7.0. The compounds were dissolved in DMSO. The data points are the means of three independent experiments. In each experiment, triplicate samples were prepared.

from the Chinese tree *Ginkgo biloba* L. These, together with BN 52063, a mixture of the three above-mentioned ginkgolides in a ratio of 2:2:1, constitute another family of natural PAF antagonists (Nunez *et al.*, 1986; Braquet, 1987). *G. biloba* extracts are used as anti-hypertensives in Chinese herbal medicine under the name *ba-guo*. Similar extracts are widely used in Europe under such trade names as Tanakan (France) and Tebonin (Germany) to treat peripheral and cerebral ischaemic diseases.

Ginkgolide B (BN 52021) is the most potent ingredient as a PAF receptor antagonist in *G. biloba* extracts. It inhibited the binding of $[^3H]$PAF to rabbit and human platelet membranes, human PMN membranes and human lung membranes with a similar potency ($K_i = 2$ µM) (Hwang *et al.*, 1985a; Hwang and Lam, 1986; Hwang, 1988). A high degree of structure specificity was again evident. The closely related ginkgolides A, C and D, which differ only in the number and position of the hydroxyl groups attached to a polycyclic backbone, are less active (Braquet, 1987). The potency in inhibiting $[^3H]$PAF binding to rabbit platelet membranes was also progressively lost by substituting the hydroxy group(s) at C-10 (mono-*o*-methyl-ginkgolide B; compound (114)) or at both C-1 and C-10 with methoxy group(s) (di-*o*-methyl-ginkgolide B; compound (115)). Mono-*o*-methyl-ginkgolide B is about three-fold less potent than ginkgolide B, whereas no PAF antagonist activity was found for di-*o*-methyl-ginkgolide B, even at a concentration of 10 µM (Fig. 9.10).

BN 52021 is the most evaluated PAF antagonist to date (Braquet, 1987), and two monographs on the chemistry, biology and pharmacology of ginkgolides have been published. BN 52021 inhibited the formation of thrombi induced by PAF in guinea-pigs, in addition to inducing a rapid curative thrombolysis (Bourgain *et al.*, 1988). BN 52021 and other ginkgolides (BN 52063 in particular) showed a marked protective activity in pulmonary anaphylaxis and hyper-reactivity of the respiratory tract. BN 52063 inhibited not only the bronchoconstriction and airway hyper-reactivity caused by PAF in animals and humans, but also several models of immune anaphylaxis in animals and the antigen-induced bronchial provocation test in asthmatic patients (Braquet *et al.*, 1987). BN 52021 also considerably reduced the fall in contracting force and the increase in perfusion pressure due to antigen challenge in the heart of passively sensitized guinea-pigs (Koltai *et al.*, 1986). BN 52021 prevented in a dose-dependent manner the endotoxic shock caused by *Salmonella enteritidis* in rats (Etienne *et al.*, 1985). BN 52021 prevented the transitory increase in intraocular pressure induced by laser iridal burns (Verby *et al.*, 1989). BN 52021 inhibited stomach lesions induced by endotoxin, alcohol and stress in which the participation of PAF has already been confirmed (Wallace *et al.*, 1987). BN 52021

Table 9.6 Inhibitory activity of ginkgolide B and three synthetic ginkgolide analogues on PAF-induced rabbit platelet aggregation

Compound		R^1	R^2	ED_{50} (μM)
(116)		CH_2OCH_3	CH_2OCH_3	0.3
(117)		Cl	H	0.4
(118)		Cl	CH_2OCH_3	0.2
(111)	Ginkgolide B			0.6

alone or in combination with either azathioprine or cyclosporin significantly delayed graft rejection in rats (Foegh *et al.*, 1986). Intraperitoneal injection of BN 52021 at a dose of 25 mg/kg suppressed PAF-induced eosinophil accumulation in bronchoalveolar lavage fluid and bronchial hyper-reactivity in guinea-pigs challenged with either PAF or antigen (Coyle *et al.*, 1988). The ginkgolides could also antagonize the increase in vascular permeability in rats and guinea-pigs and the eosinophil infiltration in atopic patients induced by intracutaneous PAF injection (Cordeiro *et al.*, 1988; Vargaftig and Braquet, 1987). BN 52021 has also been shown to antagonize the suppressor action of PAF on T lymphocyte proliferation and cytokine production (Rola-Pleszczynski *et al.*, 1986). Egido *et al.* (1987) reported that BN 52021 inhibited the urinary protein excretion in adriamycin-induced nephropathy in rats. Finally, BN 52021 at a dose of 20 mg/kg p.o. per day for 1 month showed protective effects against diet-induced cholesteryl ester deposition in rabbit aorta, without changing the plasma cholesterol levels (Feliste *et al.*, 1989).

Total synthesis of ginkgolide B has been achieved (Corey, 1988; Corey and Ghosh, 1988; Corey *et al.*, 1988b). Three synthetic ginkgolide analogues (Table 9.6) were reported to be somewhat more active than ginkgolide B as antagonists of PAF in inhibiting PAF-induced aggregation of rabbit platelets. Ginkgolide B has an ED_{50} value of 0.6 μM, as measured in the same assay (Corey and Gavai, 1990).

The ether extract from the seeds of *Swietenia mahagoni* Jacq. (Meliaceae), a medicinal plant from Indonesia, inhibited PAF-induced aggregation of rabbit platelets (Kadota *et al.*, 1989). Systematic separation of this extract led to the isolation of 28 tetranortriterpenoids related to swietenine and swietenolide (Ekimoto *et al.*, 1991). Seven of them, namely swietemahonins A, D, E and G, and 3-*O*-acetylswietenolide and 6-*O*-acetylswietenolide (Fig. 9.11), showed anti-PAF activity. The inhibitory activities are weaker than that of kadsurenone, but are roughly comparable with that of verapamil (Kadota *et al.*, 1989; Ekimoto *et al.*, 1991).

Manoalide (Fig. 9.12), a sesterpenoid first isolated from the marine sponge *Luffariela variabilis* (De Silva and Scheuer, 1980), was also found to inhibit platelet aggregation induced by PAF (Freitas *et al.*, 1988). However, recent research suggests that anti-PAF activity may be due to its inhibitory effect on the phosphoinositide-specific phospholipase C activity rather than its binding to the receptor site of PAF (Bennet *et al.*, 1987; Burtin *et al.*, 1987; Barzaghi *et al.*, 1989).

Several sesquiterpene lactones, isolated from the herb *Centipeda minima*, showed PAF antagonist activity (Iwakami *et al.*, 1990). High activity was observed only in plenolin esters having a double bond in the ester groups, i.e. 6-*O*-angeloylphenolin (compound (127)) and 6-*O*-senecioylphenolin (compound (128)) (Fig. 9.12). These two plenolin esters showed comparable activity to that of CV-3988 in inhibiting PAF-induced aggregation of rabbit platelets, with ED_{50} values of 0.30 and 0.25 μM, respectively. The size of the acryl group also seemed to be important for activity, although the effects were much lower when compared to those of unsaturated acryl groups (Iwakami *et al.*, 1990).

A sesquiterpene lactone, scandenolide (Fig. 9.12), isolated from the Philippine medicinal plant *Mikania cordata*, at a dose of 100 μM completely inhibited whole-blood chemiluminescence in response to the activator PMA and zymosan (Ysrael and Croft, 1990). The compound is interesting, because it is the first reported non-phospholipid compound which suppresses PAF production in isolated rat leucocytes with an ED_{50} value of less than 20 μM. However, scandenolide may not be a specific inhibitor of PAF synthesis. It also potently inhibited both LTB_4 and 5-HETE production with ED_{50} values of 15 and 30 μM, respectively, while the formation of the cyclooxygenase product TXB_2 was not inhibited by 10–200 μM scandenolide. This could indicate that scandenolide is not a phospholipase A_2 inhibitor.

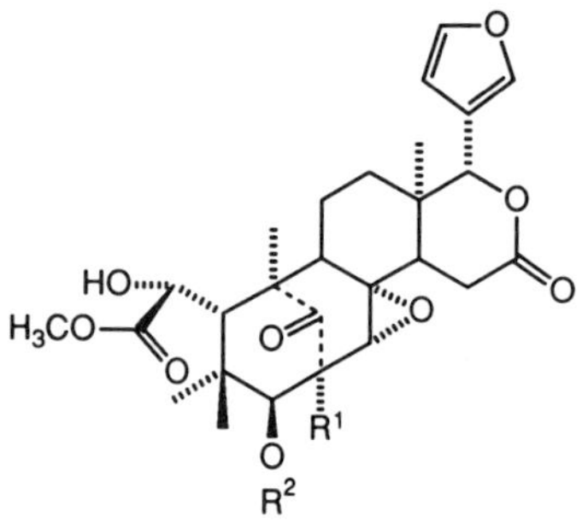

			R^1	R^2
(119)	Swietemahonin	A	H	COC_2H_5
(120)		D	H	$COCH_3$
(121)		E	H	$COC(CH_3)=CHCH_3$
(122)		G	OH	$COC(CH_3)=CH(CH_3)$

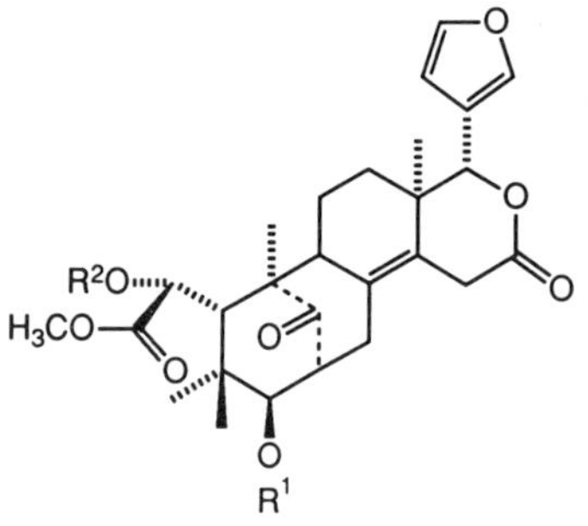

		R^1	R^2
(123)	3-*O*-Acetylswietenolide	$COCH_3$	H
(124)	6-*O*-Acetylswietenolide	H	$COCH_3$
(125)	3,6-*O,O*-Diacetylswietenolide	$COCH_3$	$COCH_3$

Figure 9.11 Structures of swietenine and swietenolide analogues.

3.4   CALCIUM CHANNEL BLOCKERS

Some class II (gallopamil and verapamil) (Fig. 9.13) and class III (diltiazem) (Fig. 9.13) calcium channel blockers are potent PAF antagonists (Wade *et al.*, 1986). The order of potency against receptor binding ((±)-gallopamil > (±)-*cis*-diltiazem > (±)-verapamil > nifedipine) closely resembled the order of potency observed against PAF-induced aggregation of human platelets in plasma. This inhibition appears to be stereospecific since the (±)-*cis*-diltiazem is 4.3-fold more potent than (−)-*cis*-diltiazem. Both (±)-gallopamil and (+)-*cis*-diltiazem caused competitive inhibition of [³H]PAF binding. In contrast, class I calcium blockers

(1,4-dihydropyridines) (Fig. 9.13) are only weak inhibitors. A new series of 4-alkyl-1,4-dihydropyridines were recently reported to be potent and specific PAF antagonists with no significant cardiovascular activity related to calcium antagonist properties (Sunkel *et al.*, 1990). The most potent compound in this series is PCA-4248 (Fig. 9.13), with a K_i value of 15 nM in inhibiting the specific [³H]PAF binding to rabbit platelets. Incorporation of a thioaryl group into the chain linked to C-3 of the dihydropyridine ring leads to an increased affinity for the PAF receptor and, consistently, PAF antagonist activity. On the other hand, incorporation of a bulky substituent at C-5 of the dihydropyridine ring

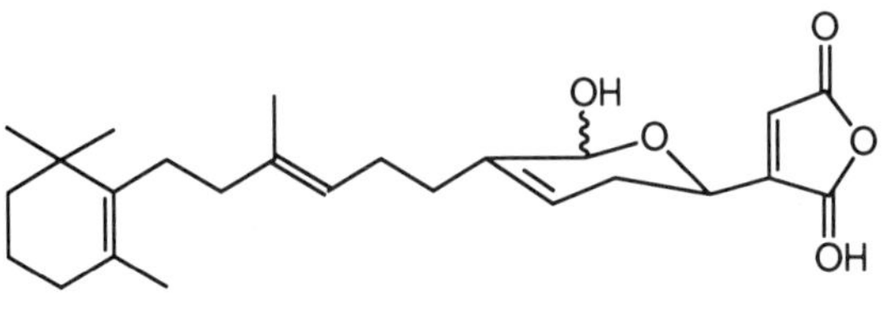

Manoalide (126)

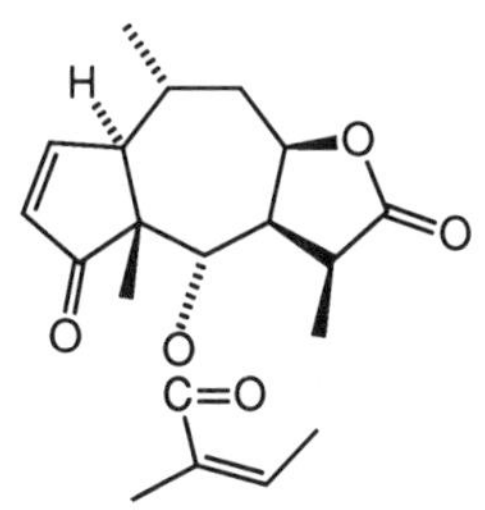

(127)

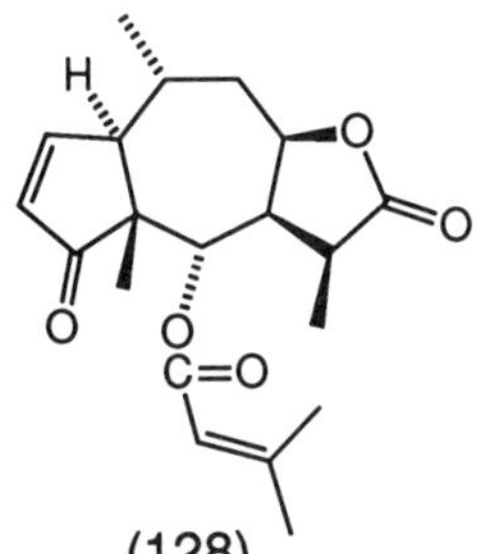

(128)

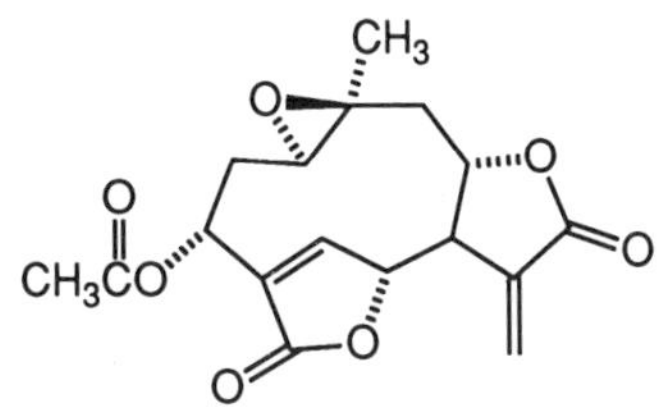

Scandenolide (129)

Figure 9.12 Structures of manoalide, two sesterpenoids and scandenolide.

results in loss of activity. PCA-4233 (Fig. 9.13), produced by substituting a methyl by an ethyl group at C-5 in PCA-4248, was about 100-fold less potent than PCA-4248 in inhibiting specific [^{3}H]PAF binding to either rabbit (Sunkel *et al.*, 1990) or human platelets (Ortega *et al.*, 1990). Nitrendipine, in the same assay, is not active, with a K_i value of 7.3 M. At the cellular level, these compounds inhibited PAF-induced aggregation of and ATP release from rabbit platelets, showing a very good correlation between antiaggregatory activity and K_i value for binding of [^{3}H]PAF to rabbit platelets (Sunkel *et al.*, 1990). *In vivo*, PCA-4248 was able to reverse the PAF-induced hypotension in rats (ED$_{50}$ = 0.45 mg/kg i.v., against PAF 0.33 µg/kg i.v.), and to inhibit the PAF-induced plasma extravasation in rats (ED$_{50}$ = 0.36 mg/ kg i.v., against PAF 1 µg/kg i.v.). Pretreatment of mice with an oral dose of 30 mg/kg of PCA-4248, 5 min before PAF challenge, increased the survival rate from 16 to 68% (Fernandez-Gallardo *et al.*, 1990).

Another type of 1,4-dihydropyridine (UK-74,505) (Fig. 9.13) was also recently reported by Pfizer Inc. (Parry *et al.*, 1990). UK-74,505 inhibited PAF-induced aggregation of rabbit platelets in plasma and washed rabbit platelets with ED$_{50}$ values of 14.7 and 4.4 nM, respectively, and showed inhibition of rabbit platelet aggregation induced by ADP, collagen or thrombin only at high concentrations (ED$_{50}$ > 100 µM). UK-74,505 antagonized the [^{3}H]PAF binding to rabbit platelets (ED$_{50}$ = 15 nM) but competed with the binding of [^{3}H]nitrendipine, binding only at a 1000-fold higher concentration (ED$_{50}$ = 7 µM). *In vivo*, UK-74,505 showed long-lasting effects. It totally blocked PAF-

induced aggregation of dog platelets *ex vivo* at 8 h when administered at a dose of 0.075 mg/kg p.o. It effectively protected the PAF-induced lethality in mice with an ED$_{50}$ value of 0.11 mg/kg p.o. It inhibited guinea-pig bronchoconstriction induced by PAF aerosol (ED$_{50}$ = 0.054 mg/kg i.v.) or by antigen aerosol (ED$_{50}$ = 2 mg/ kg i.v.).

3.5   TRIAZOLOBENZODIAZEPINES

Triazolobenzodiazepines (triazolam and alprazolam) (Fig. 9.14) (Kornecki *et al.*, 1984) and brotizolam (Fig. 9.14) (Casals-Stenzel, 1987) are classical psycho-tropic agents. Some of these are used clinically as hypnotics. Alprozolam and triazolam inhibited PAF-induced aggregation of human platelets at about 5 µM (Kornecki *et al.*, 1984). Brotizolam was 14 times as potent as triazolam, and active at 1–10 mg/kg p.o. (Casals-Stenzel, 1987). Flunitrazepam and diazepam, two benzodiazepines lacking the triazole ring, were consider-ably less potent in inhibiting PAF-induced platelet aggregation. Both triazolam and brotizolam showed little or no effect on the aggregation induced by ADP, adrenaline, serotonin, collagen and arachidonic acid, suggesting a specific action of these drugs against PAF.

A successful separation of hypnotic and anti-PAF activities was first accomplished by the synthesis of WEB 2086, a thienotriazolodiazepine carrying a hydrophilic side-chain which prevents its penetration into the brain (Weber and Heuer, 1989). Modification of the side-chain at position 2 on the thiophene ring was further extended at the Boehringer Ingelheim by ring closure to give a new

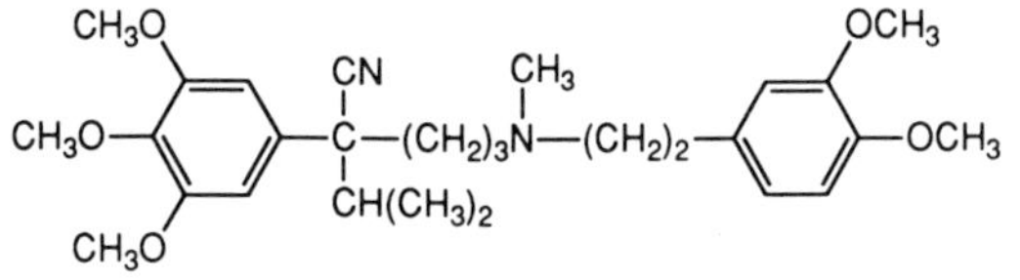

Gallopamil (130)

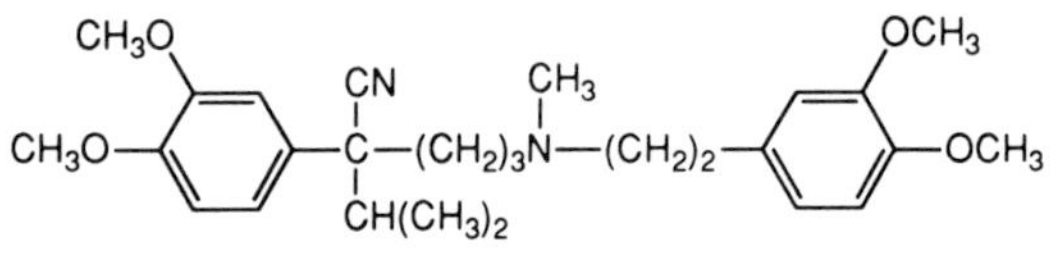

Verapamil (131)

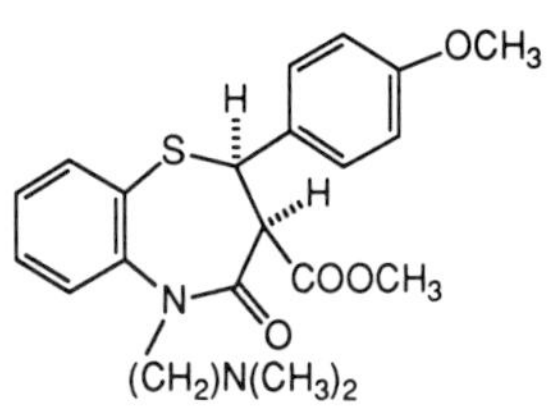

Diltiazem (132)

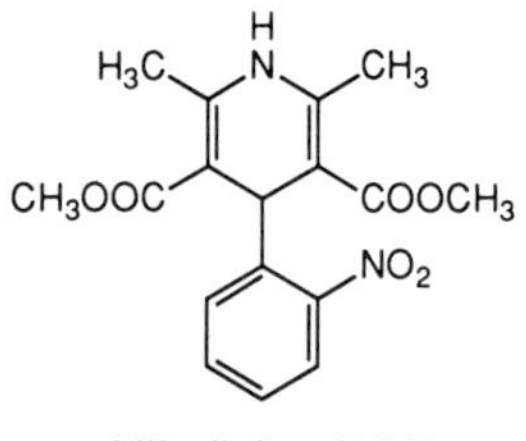

Nifedipine (133)

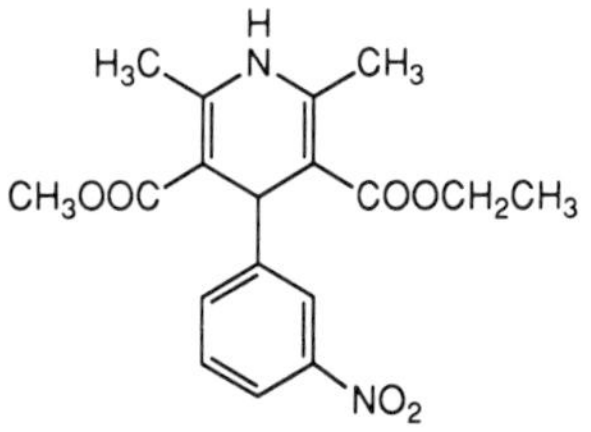

Nitrendipine (134)

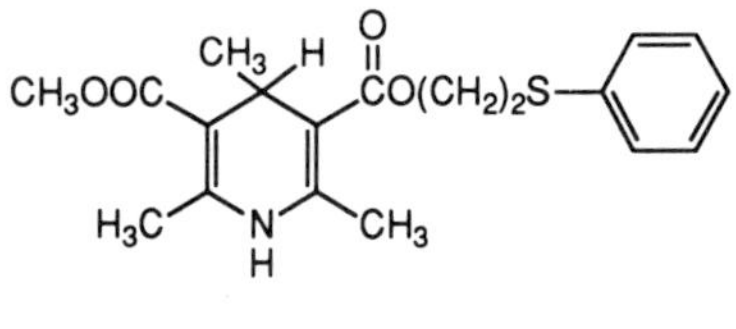

PCA-4248 (135)

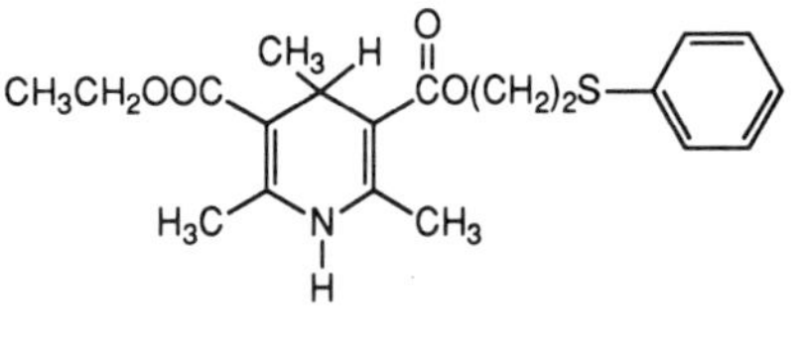

PCA-4233 (136)

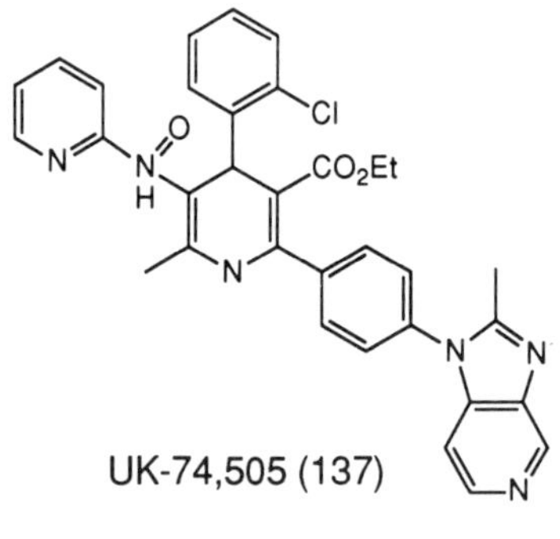

UK-74,505 (137)

Figure 9.13 Structures of calcium channel blockers as PAF receptor antagonists.

series of tetracyclic compounds that had both potent PAF antagonist activity and low affinity for the benzodiazepine receptor. WEB 2170, WEB 2347 and STY 2108 are examples of this series (Weber and Heuer, 1989). Yoshitomi Pharmaceutical Industries, Ltd, Hoffman-La Roche Inc., the Henri Beaufour Institute and Tsukuba Research Laboratories, Eisai Co., Ltd also followed the same trend and successfully developed Y-24180, Ro 24-4736, BN 50739 and E-6123, respectively, as potent PAF receptor antagonists. WEB 2086, WEB 2170, STY 2108 and Y-24180 had been assayed in the same assay for inhibition of $[^3H]C_{18}$-PAF binding in isolated human platelet membranes. Y-24180 is the most potent antagonist

with an ED_{50} value of 0.89 nM in inhibiting specific $[^3H]C_{18}$-PAF (0.3 nM) binding to human platelet membranes, followed by STY 2108 (ED_{50} = 25.2 nM), WEB 2170 (117.6 nM) and WEB 2086 (117.6 nM) (Fig. 9.15). These types of compounds showed no differences in potency in inhibiting PAF binding to human platelet, PMN or lung membranes. Y-24180 had K_i values of 0.77 and 0.89 nM in human platelet and human PMN membranes, respectively (Figs 9.15 and 9.16); WEB 2086 also showed similar K_i values in human platelet (94 nM), human PMN (103 nM) and human lung membranes (84.1 nM) (Fig. 9.16). In inhibiting specific $[^3H]N$-methylcarbamyl PAF binding to the receptor in human

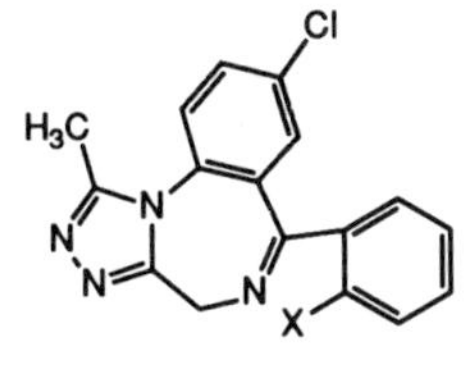

X = Cl, Triazolam (138)
X = H, Alprozolam (139)

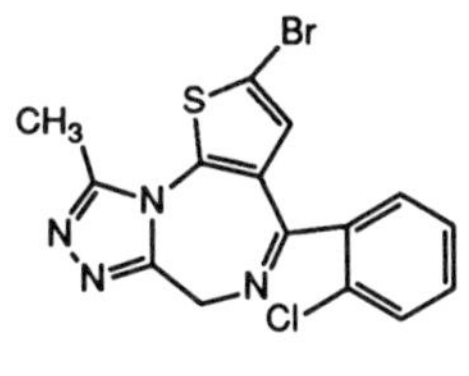

Brotizolam (140)

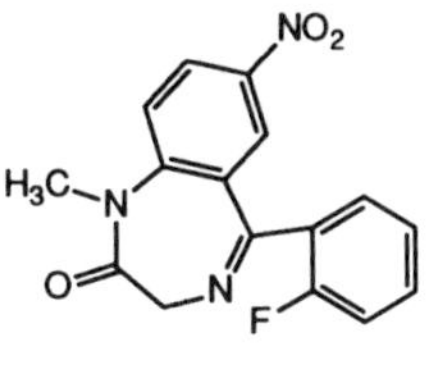

Flunitrazepam (141)

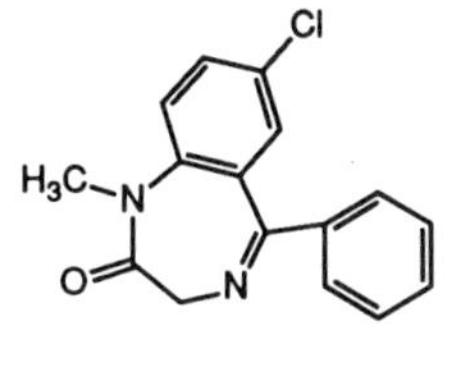

Diazepam (142)

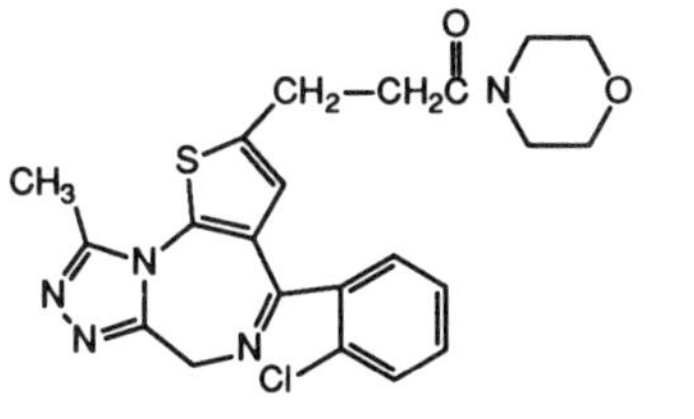

WEB 2086 (143)

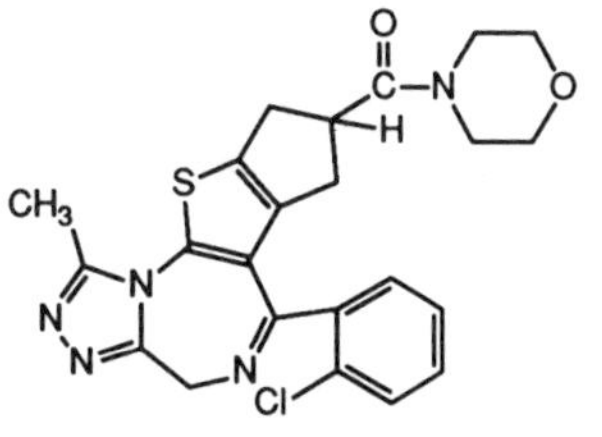

WEB 2170 (144)

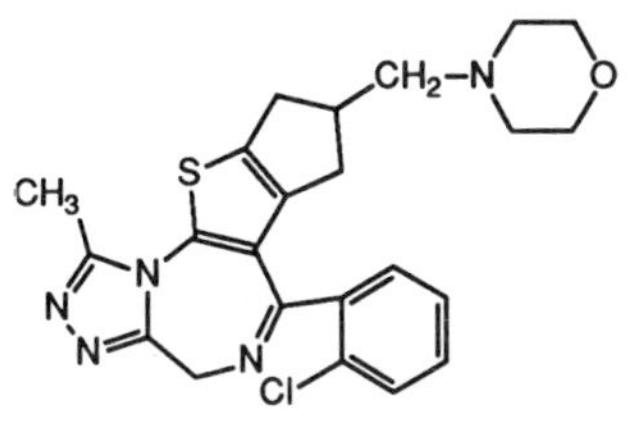

STY 2108 (145)

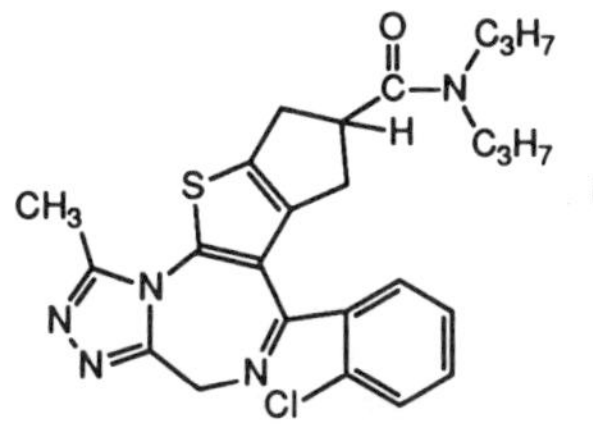

WEB 2347 (146)

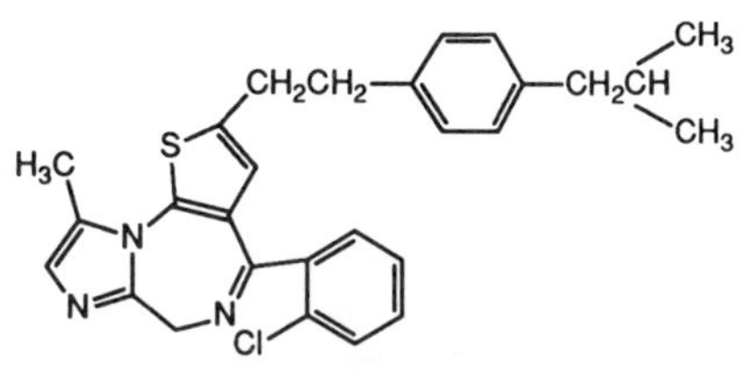

Y-24180 (147)

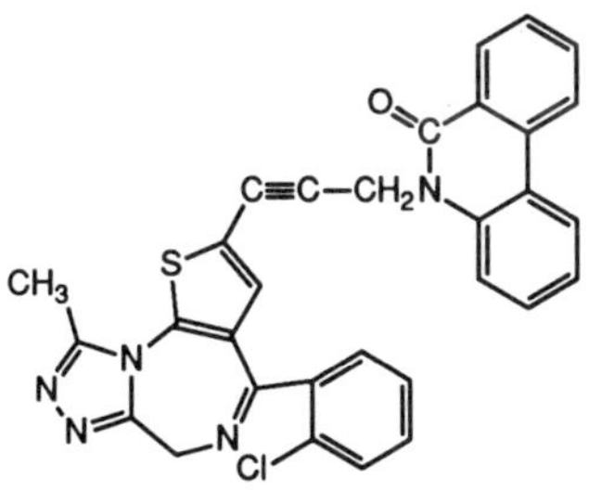

Ro 24-4736 (148)

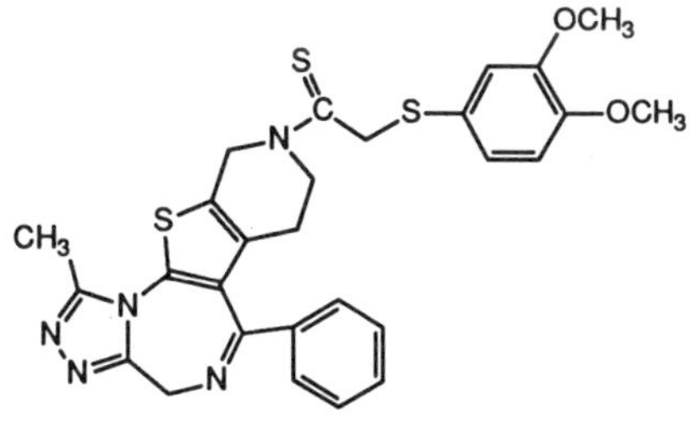

BN 50739 (149)

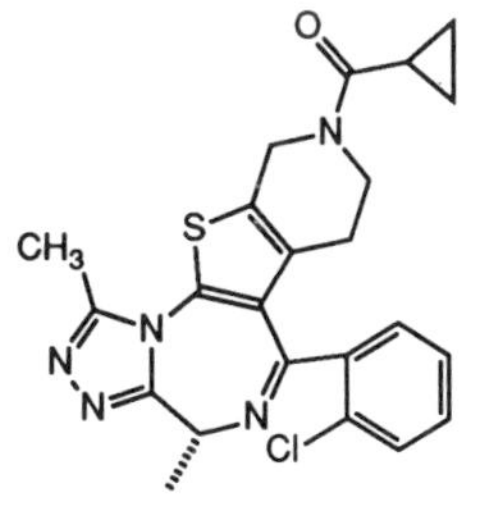

E-6123 (150)

Figure 9.14 Molecular structures of hetrazepine compounds that inhibit PAF actions.

platelet and PMN membranes, WEB 2086 showed similar K_i values (83 and 77 nM in human platelet and PMN membranes, respectively). [³H]WEB 2086 is commercially available from Du Pont-NEN (Boston, MA). So far, no high-affinity binding of [³H]WEB 2086 to either rabbit, human platelet or human PMN membranes with concentrations of [³H]WEB 2086 up to 10 nM has been found. These results are in good agreement with the low affinity of WEB 2086 for the PAF receptor in these membrane systems (Figs 9.15 and 9.16). However, a high-affinity binding of [³H]WEB 2086 has been reported in human platelets (K_d = 6.1 nM) (Ukena *et al.*, 1988), guinea-pig (K_d = 16.8 nM) and human lung membranes (K_d = 22.6 nM) (Dent *et al.*, 1989b), human

neutrophils (K_d = 18.9 nM) (Dent *et al.*, 1989a) and guinea-pig eosinophils (K_d = 16.1 nM) (Ukena *et al.*, 1989). A K_d value of 40 nM for [³H]WEB 2086 was recently reported in guinea-pig lung membranes by another group (Gomez *et al.*, 1990). The reason for this controversy in the determination of the affinity of WEB 2086 is not known.

Despite the contradiction in the affinity measurements for WEB 2086, WEB 2086 has been one of the most popular PAF antagonists tested in a variety of model systems *in vitro* and *in vivo*. WEB 2086 inhibited PAF-induced aggregation of human platelets and neutrophils *in vitro* (ED_{50} = 0.17 and 0.36 μM, respectively), but had little or no effect on the action of other

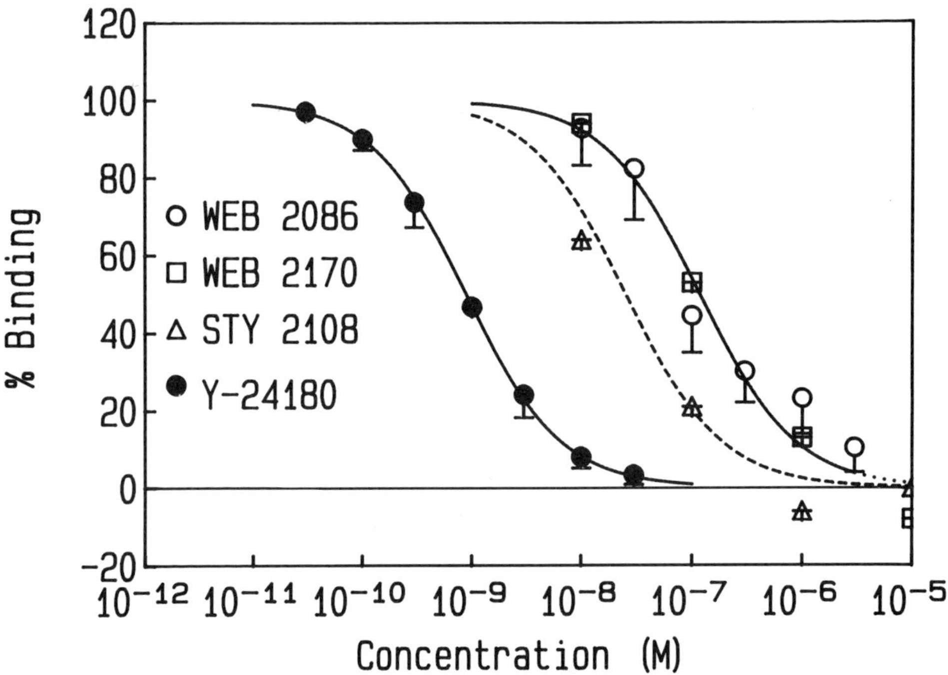

Figure 9.15 Inhibition of [³H]C₁₈-PAF binding to human platelet membranes by WEB 2086, WEB 2170, STY 2108 and Y-24180. Membrane protein (100 μg) was added to an incubation mixture (1 ml) containing 0.3 nM [³H]C₁₈-PAF and a known amount of inhibitor in a medium of 10 mM magnesium chloride, 10 mM Tris and 0.25% BSA at pH 7.0. The compounds were dissolved in DMSO. The data points and error bars are the means and the standard deviations of three independent experiments. In each experiment, triplicate samples were prepared.

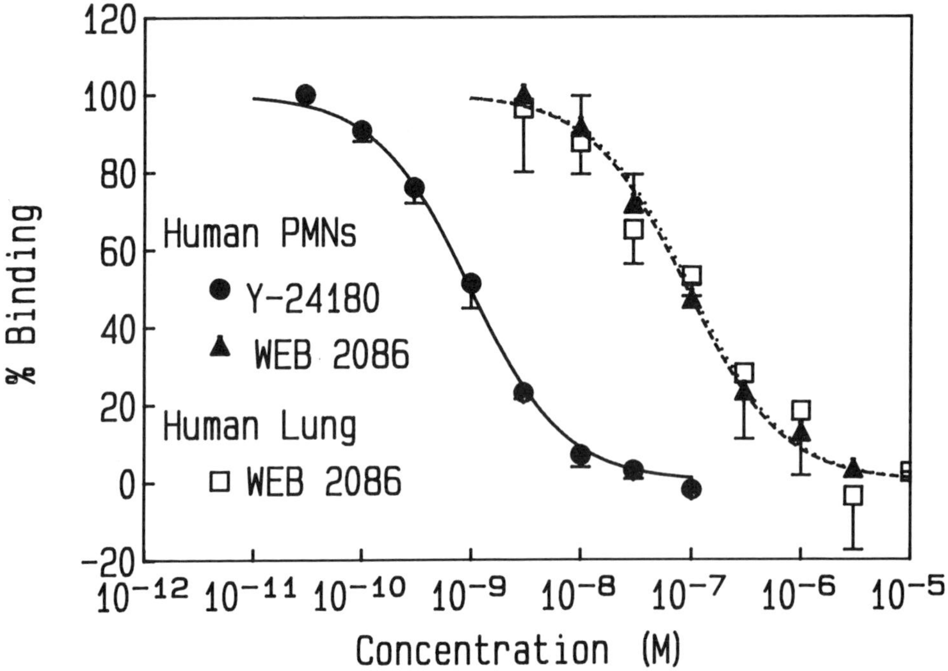

Figure 9.16 Inhibition of [³H]C₁₈-PAF binding to human PMN and human lung membranes by WEB 2086 and Y-24180. Membrane protein (100 μg) was added to an incubation mixture (1 ml) containing 0.3 nM [³H]C₁₈-PAF and a known amount of inhibitor in a medium of 10 mM magnesium chloride, 10 mM Tris and 0.25% BSA at pH 7.0. The compounds were dissolved in DMSO. The data points and error bars are the means and the standard deviations of three to four independent experiments. In each experiment, triplicate samples were prepared.

platelet-aggregating agents (Casals-Stenzel *et al.*, 1987). It is interesting to note that the potencies of WEB 2086 in inhibiting PAF-induced aggregation of human platelets and human neutrophils reported by Casals-Stenzel *et al.* (1987) are similar to those in inhibiting the specific binding of [^{3}H]PAF in human platelet or PMN membranes. Due to its hydrophilic property, it is likely that WEB 2086 was able to cross cell membranes and blocked intracellular PAF receptors. Inhibition of PGI_2 production by WEB 2086 as well as by a structurally unrelated PAF antagonist, CV-6209, in endothelial cells stimulated by bradykinin (Stewart *et al.*, 1989), or in macrophages with FMLP (Stewart and Phillips, 1989), has been used as evidence in suggesting that intracellular PAF has a messenger role in signal transduction. This can only be explained if WEB 2086 also blocks intracellular receptors. *In vivo*, WEB 2086 inhibited PAF-induced bronchoconstriction and several other features of clinical asthma, such as eosinophil influx and bronchial hyper-responsiveness in guinea-pigs when given either intravenously (Lellouch-Tubiana *et al.*, 1988) or orally, with an ED_{50} value of 0.058 mg/kg p.o. (Weber and Heuer, 1989; Meade and Heuer, 1990). Orally, WEB 2086 also potently inhibited puromycin amino-nucleoside-induced nephropathy at 5 mg/kg in the rat (Yamada *et al.*, 1991).

PAF antagonist activities of a number of thienotriazolo-diazepines with a substituted ethynyl group at the 2 position, and the corresponding *cis*-olefins and fully saturated analogues, were evaluated both *in vitro* by inhibiting [^{3}H]PAF binding to washed dog platelets and *in vivo* by inhibiting the PAF-induced bronchocon-striction in the guinea-pig (Walser *et al.*, 1991). The compounds with a triple bond were generally less active than those with a double or single bond. The rigidity and excessive bulk of the 2 position substituent was presumably responsible for the lower affinity and potency of the ethynyl compounds as compared to the corresponding (*Z*)-ethenyl and ethyl derivatives.

WEB 2170 (bepafant) showed comparable potency to WEB 2086 both *in vitro* and *in vivo* (Heuer *et al.*, 1990). WEB 2170 inhibited PAF-induced platelet and neutro-phil aggregation *in vitro* (ED_{50} = 0.3 and 0.83 μM, respectively) but showed little or no inhibitory action against aggregation induced by other agonists. WEB 2170 is about five to 40 times more potent against exogenous PAF-induced effects than WEB 2086 *in vivo*. In guinea-pigs, oral (0.005–0.5 mg/kg) as well as intravenous (0.005–0.05 mg/kg) treatment with WEB 2170 abrogated the intrathoracic accumulation of ^{111}In-labelled platelets, the bronchoconstriction and the hypo-tension as well as death in a dose-dependent fashion. In anaesthetized rats, intravenous (0.001–0.1 mg/kg) and oral (0.05–1 mg/kg) WEB 2170 inhibited PAF induced hypotension in a dose-related manner.

Y-24180 inhibited PAF-induced rabbit platelet aggregation *in vitro* (ED_{50} = 3.84 nM), but had little effect on ADP- or arachidonic acid-induced aggregation. In rabbits, pretreatment with Y-24180 antagonized PAF-induced aggregation *ex vivo*. The significant inhibitory effect of Y-24180 (1 mg/kg p.o.) lasted 72 h after oral administration of a single dose (Takehara *et al.*, 1990). *In vivo*, Y-24180 (0.0003–0.003 mg/kg i.v.) dose-dependently inhibited PAF induced bronchoconstriction in guinea-pigs, and even at a high dosage of 10 mg/kg i.v. it was either inactive or weakly active against the bronchoconstriction induced by histamine, serotonin, acetylcholine, arachidonic acid, bradykinin or LTD_4. Oral doses (0.003–0.1 mg/kg) of Y-24180 also prevented haemoconcentration due to PAF in a dose-dependent manner and produced a parallel shift of the PAF dose–response curve in conscious guinea-pigs. In mice, PAF-induced lethality was inhibited by Y-24180, with an ED_{50} value of 0.022 mg/kg p.o. This protective effect of Y-24180 given orally persisted for at least 6 h. In actively sensitized mice, Y-24180 protected against lethal anaphylac-tic shock, with an ED_{50} value of 0.095 mg/kg (Terasawa *et al.*, 1990). In general, Y-24180 is about 100 times more potent than WEB 2086 both *in vitro* and *in vivo*.

Ro 24-4736 is also a potent and competitive PAF receptor antagonist. *In vitro*, Ro 24-4736 competed with [^{3}H]PAF for its receptor sites on dog platelets with an ED_{50} value of 13 nM. Ro 24-4736 also inhibited PAF-induced aggregation of human platelets in a concentration-dependent manner with an ED_{50} value of 200 nM. *In vivo*, Ro 24-4736 inhibited PAF-induced guinea-pig bronchoconstriction with an ED_{50} value of 0.006 mg/kg p.o. The hypotension caused by endotoxin in rats was also prevented by intravenous (0.001–1.0 mg/kg) and oral (0.1–10 mg/kg) doses of Ro 24-4736 (O'Donnell *et al.*, 1990).

No receptor data were available for BN 50739. However, it has been shown to block PAF-induced aggregation of rabbit platelets by 60–100% at 0.2×10^{-7} to 1×10^{-7} M (Yue *et al.*, 1990). Under *in vitro* conditions, BN 50739 exerted dose-dependent protection against ventricular tachycardia and fibrillation in isolated working rat hearts (Koltai *et al.*, 1991). *In vivo*, it showed beneficial effects on endotoxin-induced hypotension, haemoconcentration, TXB_2 and TNF production, and it also improved survival of rats exposed to lethal endotoxic shock (Karasawa *et al.*, 1990; Robinovici *et al.*, 1990; Yue *et al.*, 1990). In addition to its anti-PAF activity, BN 50739 also showed an inhibitory effect on free radical-scavenging activity (Castaner *et al.*, 1991). PAF and oxygen free radicals are endogenous mediators of tissue damage under cerebral and myocardial ischaemia and shock conditions (Gilboe *et al.*, 1991).

Tsunoda *et al.* (1990) recently reported E-6123 as a potent PAF antagonist. E-6123 inhibited [^{3}H]PAF binding to human and guinea-pig platelets with ED_{50} values of 2.7 and 3.0 nM, respectively. The ED_{50} values of E-6123 in inhibiting PAF-induced aggregation in platelet-rich plasma of humans, guinea-pigs, and beagle

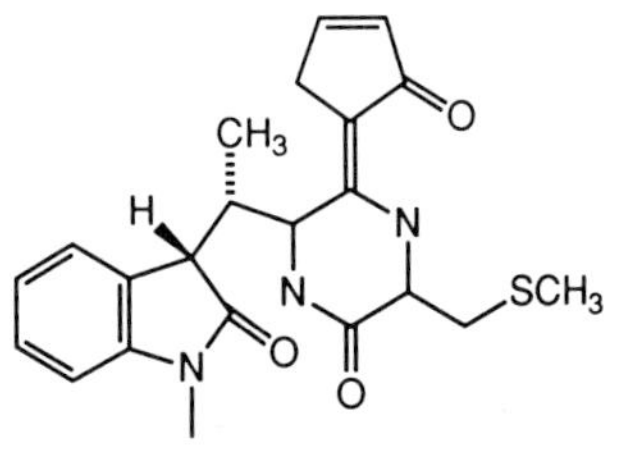

FR 900495 (151) 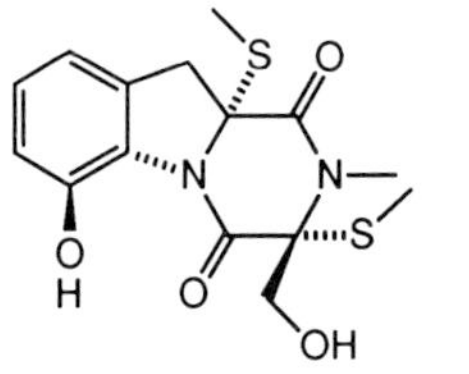   FR 49175 (152) 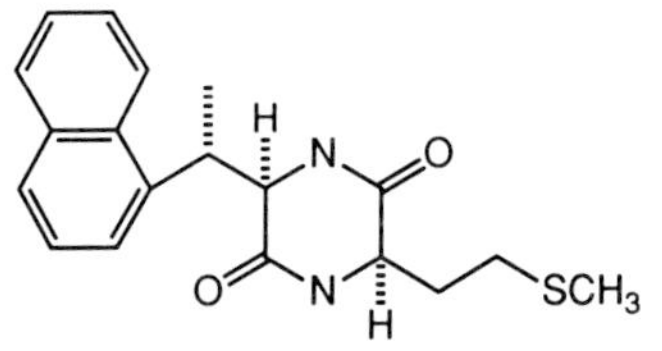   FR 72112 (153)

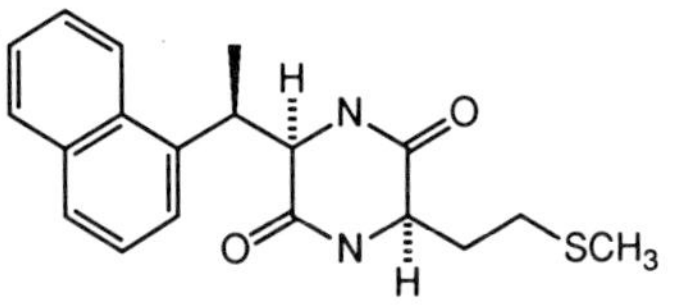

(154) 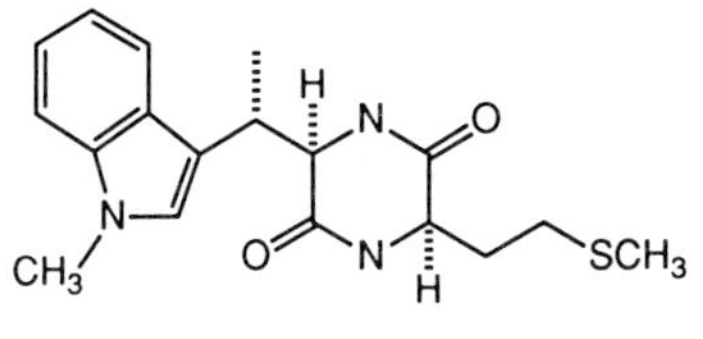   (155) 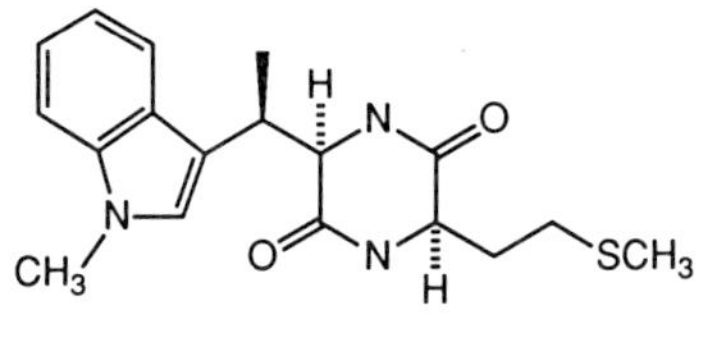   (156)

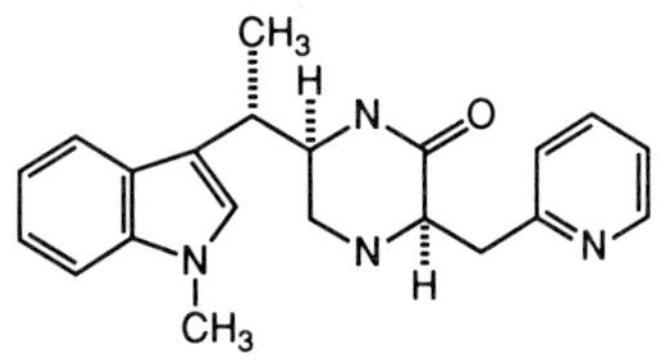

FR 76600 (157) 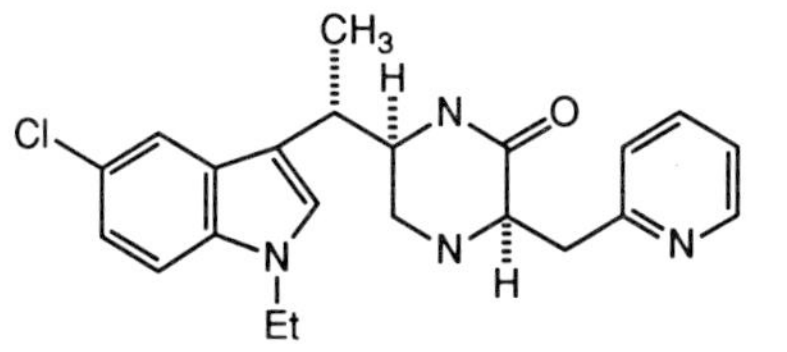   FR 103409 (158)

Figure 9.17 Fermentation products and diketopiperazines as PAF receptor antagonists.

dogs were 10.1, 14.7 and 16 nM, respectively. *In vivo*, E-6123 had a long duration of action. E-6123 (3 µg/kg p.o.) inhibited PAF-induced bronchoconstriction in guinea-pigs by more than 90% for up to 8 h. The inhibition was then gradually decreased to a low level at 36 h after oral administration. Haemoconcentration induced by PAF injection in guinea-pigs was inhibited by oral administration of E-6123 at 10 µg/kg. E-6123 protected mice from PAF-induced death in a dose-dependent manner. E-6123 showed activity against PAF- or antigen-induced bronchial hyper-reactivity when given 3 h before the challenge in guinea-pigs. The compound was also effective in suppressing eosinophil accumulation in the bronchoalveolar lavage fluid and tissue after challenge by PAF or antigen inhalation in guinea-pigs (Tsundo *et al.*, 1991; Clark, 1991). In safety studies (Clark, 1991), E-6123 showed no mutagenicity by the Ames method. The affinity of E-6123 for the central diazepine receptor in cortical P2 homogenate was less than 2/10 000 that for the PAF receptor. The maximal tolerated single dose of E-6123 was 500 mg/kg p.o. in rats and 100 mg/kg p.o. in dogs. E-6123 showed no toxic effects in rats or dogs when given at a dose of 10 mg/kg in 4 week oral toxicity studies.

3.6   GLIOTOXIN AND DIKETOPIPERAZINES

Some antagonists were recently isolated by the Fujisawa Pharmaceutical Company from the fermentation products of several fungal and microbial strains such as *Streptomyces phacofaciens* (Okamoto *et al.*, 1986a), *Gliocladium delquescens* (Okamoto *et al.*, 1986b) and *Penicillium terlikowskii* (Okamoto *et al.*, 1986b,c). The most potent antagonists in this series are FR-900452 (Okamoto *et al.*, 1986a) and FR-49175 (Okamoto *et al.*, 1986b) (Fig. 9.17), with ED_{50} values of 0.37 and 8 µM, respectively, in inhibiting PAF-induced aggregation of rabbit platelets. In receptor-binding assays, FR-49175 inhibited specific PAF binding to human platelet and PMN membranes with ED_{50} values of 5.8 and 4.2 µM respectively (Fig. 9.18). The structure of FR-900452 is unique in its incorporation of an unusual vinylogous diketopiperazine moiety, 5-(2-oxocyclopent-3-en-1-ylidene)-2-oxopiperazine. FR-49175 also possesses a diketopiperazine backbone. Total synthesis of the diketopiperazine analogues by replacing the oxindole moiety of FR-900452 with the structurally simpler *N*-methylindole and α-naphthalene led them to select a PAF-specific antagonist, a naphthalene derivative of diketopiperazine containing the 7(R)-methyl group (FR-72112). FR-72112 showed high potency, with an ED_{50} value of 0.18 µM, which is about twice as

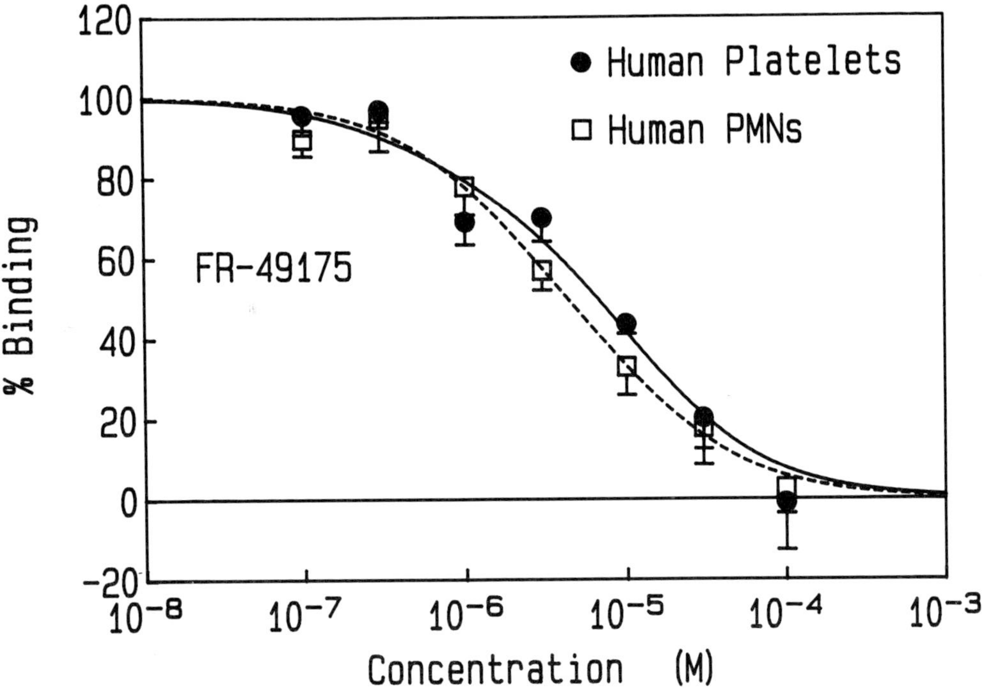

Figure 9.18 Inhibition of [³H]C₁₈-PAF binding to human platelet and PMN membranes by FR-49175. Membrane protein (100 μg) was added to an incubation mixture (1 ml) containing 0.3 nM [³H]C₁₈-PAF and a known amount of inhibitor in a medium of 10 mM magnesium chloride, 10 mM Tris and 0.25% BSA at pH 7.0. The compound was dissolved in DMSO. The data points and error bars are the means and the standard deviations of three independent experiments. In each experiment, triplicate samples were prepared.

potent as the natural product, FR-900452, in inhibiting the PAF-induced rabbit platelet aggregation (Shimazaki *et al.*, 1987). FR-72112 showed specificity in different receptor subtypes. It is more potent in human PMN membranes (K_i = 0.32 μM) than in human platelet membranes (K_i = 0.95 μM) in inhibiting the specific [³H]PAF binding to its receptors (Hwang, 1990). The biological activity of this series is also affected by the stereochemistry of the tertiary carbon (C-7) bonded to the aromatic nucleus. The 7(*S*)-methyl compounds were considerably less potent than the corresponding 7(*R*)-methyl compounds. The 7(*S*)-methyl compound (compound (154)) has an ED₅₀ value of 1.9 μM in inhibiting the PAF-induced aggregation of rabbit platelets, which is about 10 times less potent than the corresponding derivative FR-72112 (Shimazaki *et al.*, 1987). Similarly in the *N*-methylindole series, the 7(*R*)-methyl compound (compound (155)), with an ED₅₀ value of 0.7 μM in inhibiting PAF-induced rabbit platelet aggregation, is about five times more potent than the 7(*S*)-methyl compound (compound (156)) (ED₅₀ = 3.6 μM) (Shimazaki *et al.*, 1987). Unlike FR-900452, FR-72112 showed no inhibitory activity in platelet aggregation induced by arachidonic acid and ADP, suggesting that the compound is a PAF-specific inhibitor.

In the Third International Conference on Platelet-Activating Factor and Structurally Related Alkyl Ether

Lipids in Tokyo, a diketopiperazine analogue (FR-76600) was reported (Shimazaki *et al.*, 1989). It has an ED₅₀ value of 0.44 μM in inhibiting PAF-induced aggregation of rabbit platelets. FR-76600 is orally active. It inhibited PAF-induced bronchoconstriction in guinea-pigs with an ED₅₀ value of 2.5 mg/kg p.o. FR-76600 also reversed PAF-induced hypotension in rats with an ED₅₀ value of 4.6 mg/kg i.v. In the same report, FR-103409 was shown to be about 10 times more potent than FR-76600 in inhibiting PAF-induced aggregation of rabbit platelets with an ED₅₀ value of 36 nM. However, no *in vivo* data were available.

3.7 SYNTHETIC PLATELET-ACTIVATING FACTOR ANTAGONISTS

A heterocyclic compound, 48740 RP (Fig. 9.19), originally synthesized as an anti-allergic agent, was the first synthetic compound identified as a PAF receptor antagonist (Sedivy *et al.*, 1985). 48740 RP inhibited PAF-induced human and rabbit platelet aggregation (ED₅₀ = 69 and 3.3 μM, respectively) (Sedivy *et al.*, 1986). It has a K_i value of 2 μM for the PAF receptor. However, the inhibition was not PAF-specific, since 48740 RP interfered with the aggregating response to arachidonic acid, U 46619, collagen and thrombin. *In*

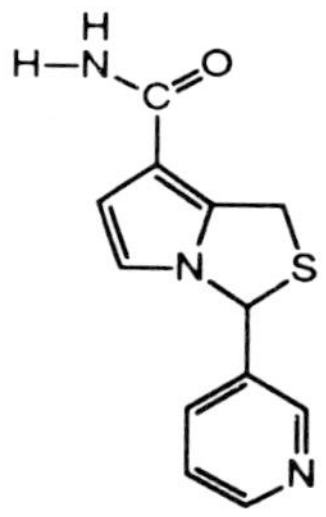

48740 RP(159)

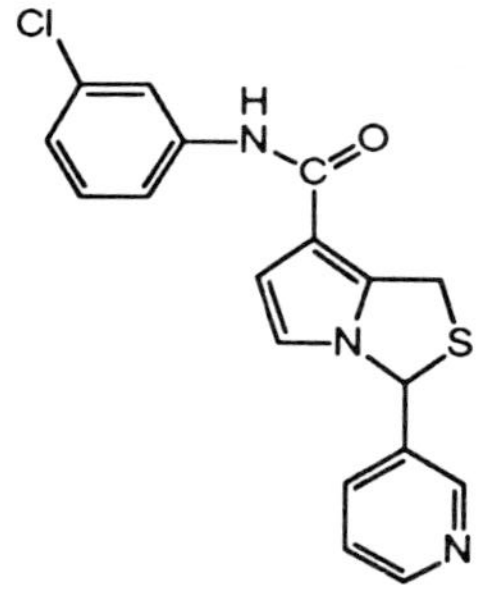

52770 RP(160)

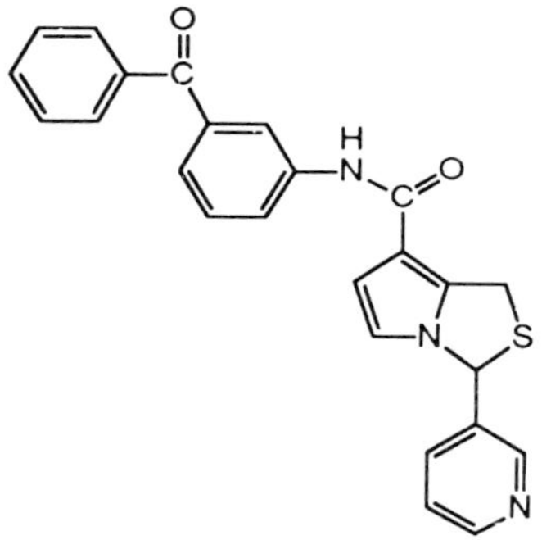

59227 RP(161)

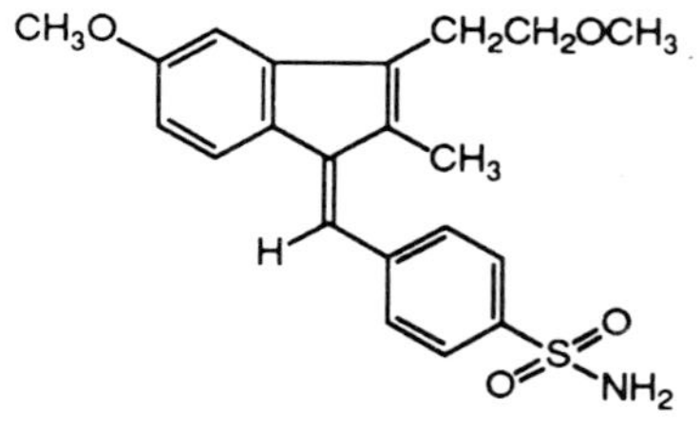

L-651,142 (162)

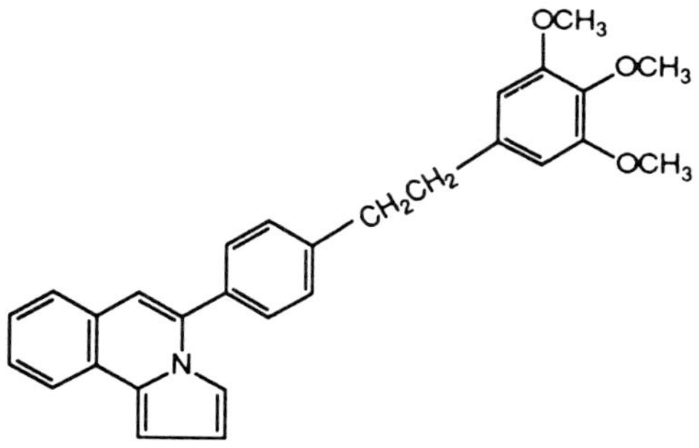

SDZ 64-412 (163)

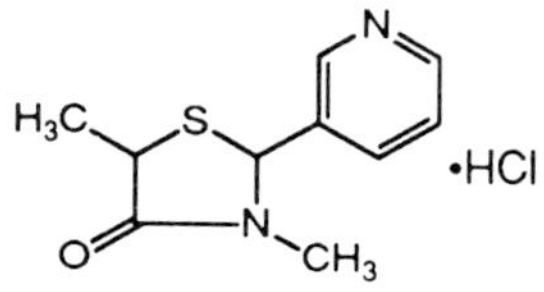

SM-10661 (164)

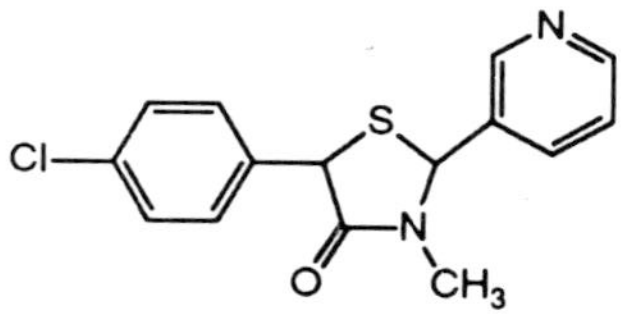

(165)

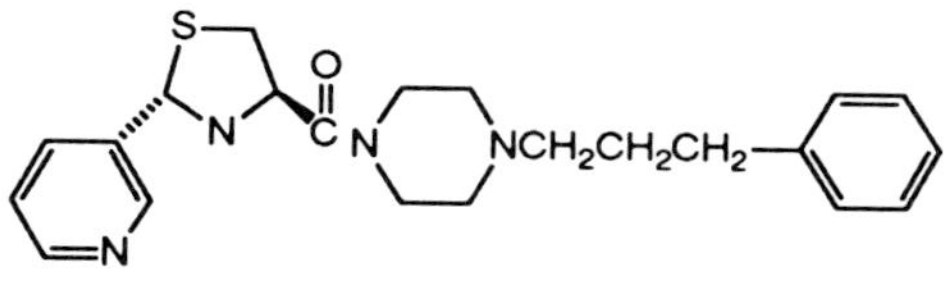

YM461 (172)

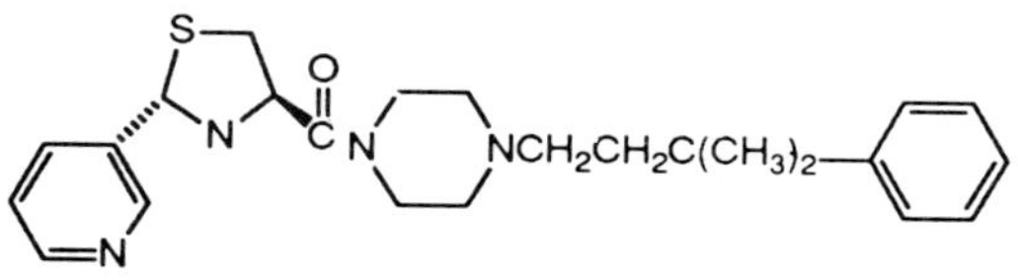

YM264 (173)

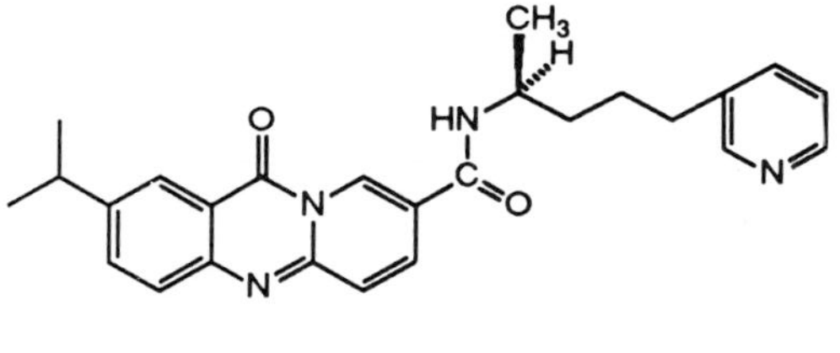

(174)

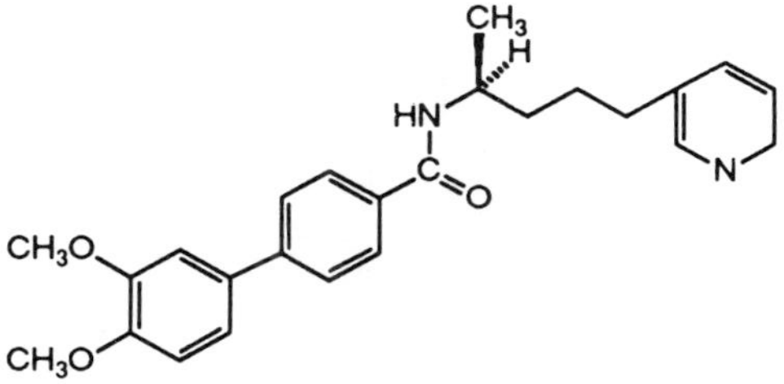

(175)

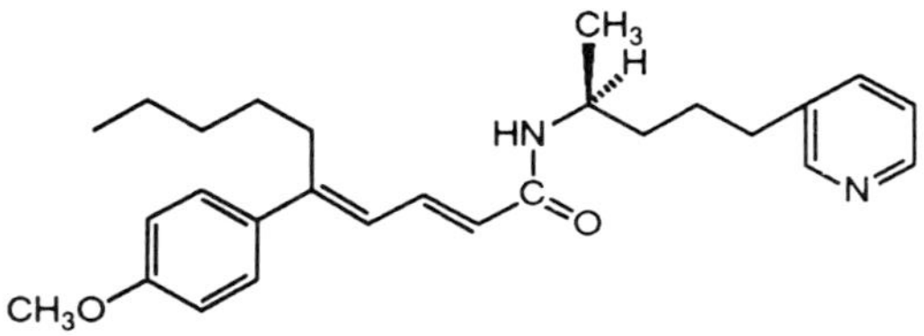

Ro 24-0238 (176)

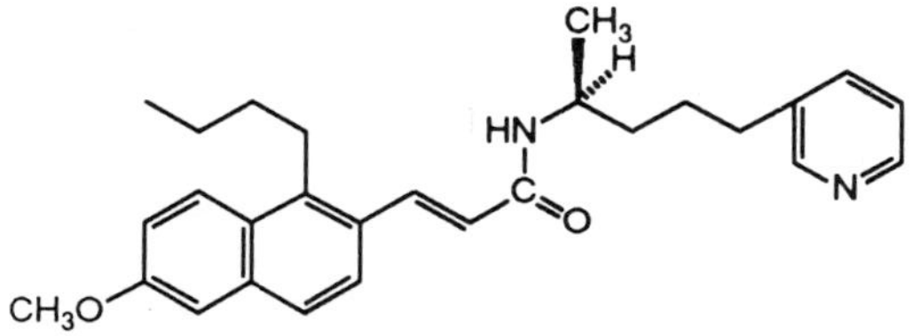

(177)

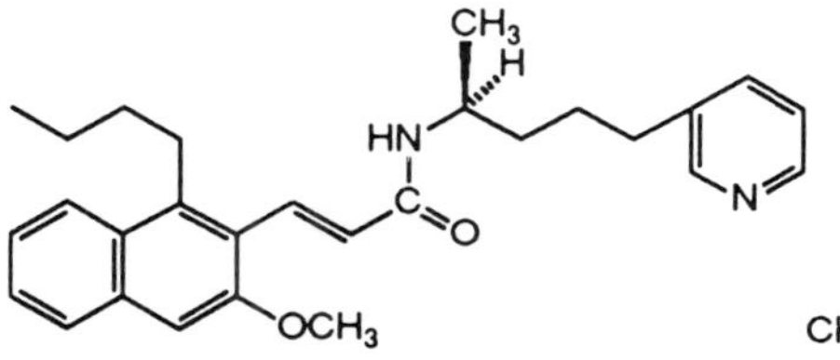

(178)

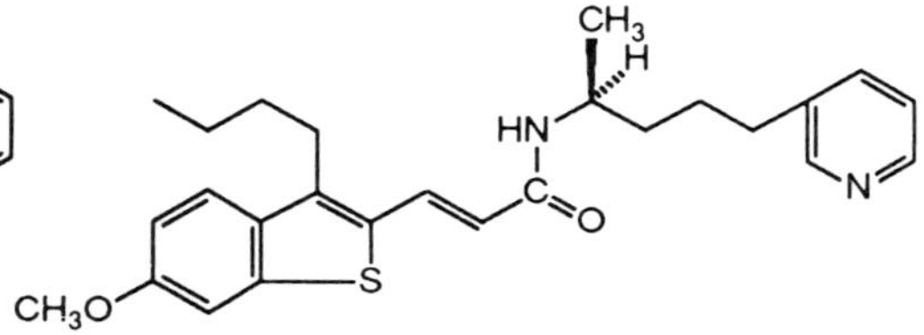

(179)

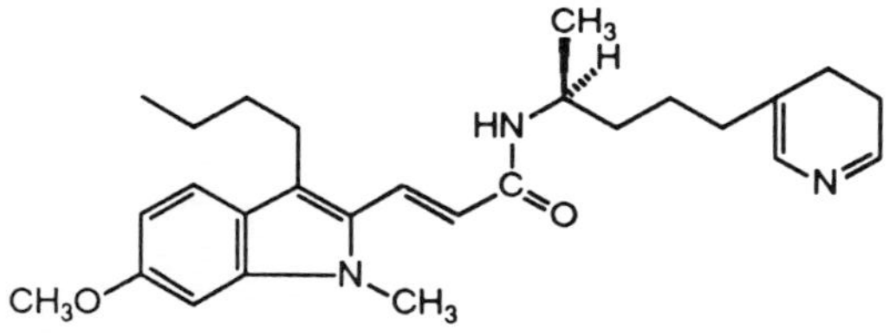

(180)

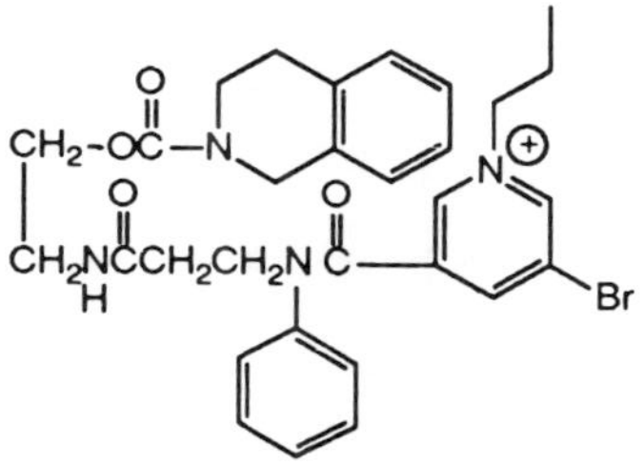

TCV-309 (181)

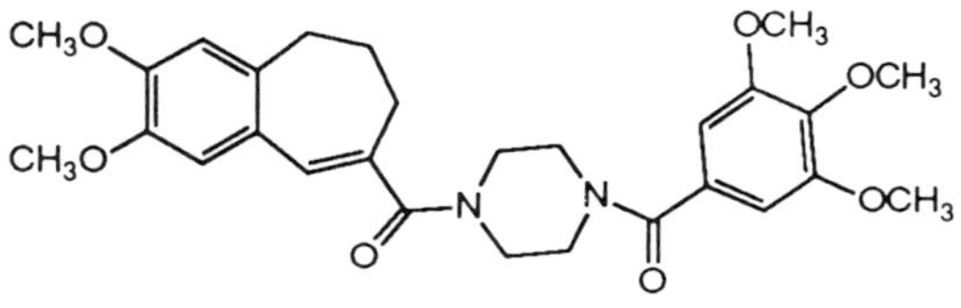

(182)

Figure 9.19 Structures of synthetic PAF receptor antagonists.

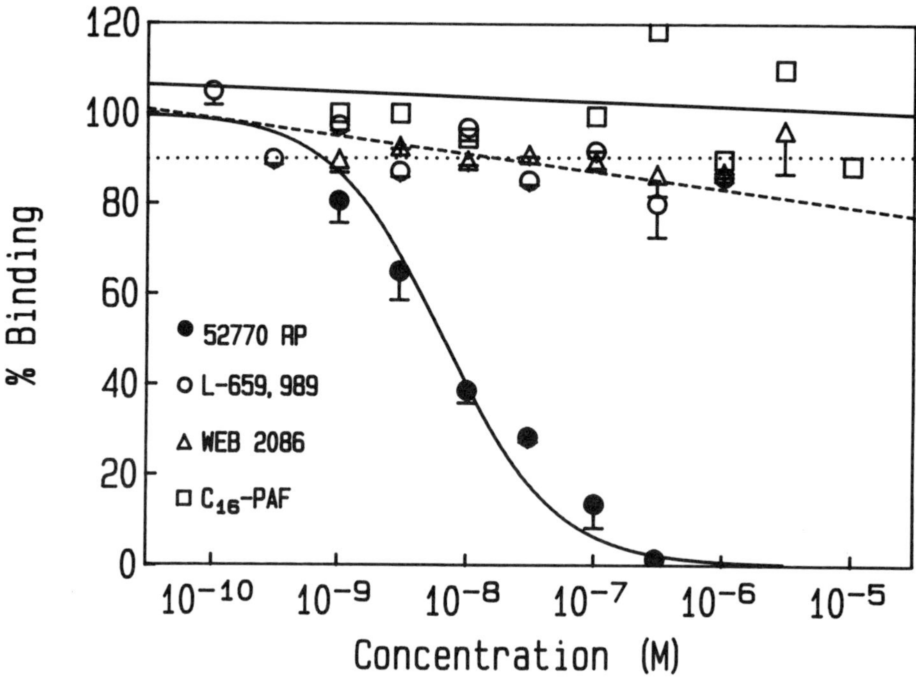

Figure 9.20 Inhibition of [^{3}H]52770 RP binding to human platelet membranes by 52770 RP, L-659,989, WEB 2086 and C$_{16}$-PAF. Membrane protein (100 µg) was added to an incubation mixture (1 ml) containing 0.3 nM [^{3}H]C$_{18}$-PAF and a known amount of inhibitor in a medium of 10 mM magnesium chloride, 10 mM Tris and 0.25% BSA at pH 7.0. C$_{16}$-PAF was dissolved in 10 mM Tris and 0.25% BSA, and the antagonists in DMSO. The data points and error bars are the means and the standard deviations of three independent experiments. In each experiment, triplicate samples were prepared.

vivo, it showed protective effects against PAF-induced hypotension, thrombocytopenia, haemoconcentration and hyperfibrinogenolysis in the rabbit and rat, and PAF-induced bronchospasm in the guinea-pig (Sedivy *et al.*, 1986). However, 48740 RP failed to inhibit passive homologous bronchospasm and *Ascaris suum*-induced allergic reactions in conscious dogs (Stenzel *et al.*, 1986).

Modification of the pyrrolodihydrothiozole, 48740 RP, led to the more potent series of *N*-aryl amides (Lave *et al.*, 1989). 52770 RP, the *N*-(3-chlorophenyl) derivative, displayed highly potent and specific PAF antagonist activity, some four- and 200-fold more potent than L-652,731 and BN 52021, respectively. The inhibition of the specific [^{3}H]PAF binding by 52770 RP to human monocyte membranes is quite different from either human platelet or human PMN membranes (Hwang *et al.*, 1991). 52770 RP showed roughly similar potencies in human platelet and PMN membranes, with K_i values of 24.08 ± 8.35 and 18.58 ± 6.91 nM, respectively (Hwang *et al.*, 1991). However, 52770 RP is about four times more potent in human monocyte (K_i = 5.25 ± 1.78 nM) than in human platelet and human PMN membranes in inhibiting specific [^{3}H]PAF binding (Hwang *et al.*, 1991). As described above, anti-PAF receptor-binding activity of CV 6209 is about three-fold less potent in human monocyte membranes than in human platelet and in human PMN membranes (Hwang *et al.*, 1991). The differences in

potencies of 52770 RP further confirmed the difference of PAF receptors within those three cell types.

[^{3}H]52770 RP has been used to characterize the PAF-binding sites on rabbit platelets (Robaut *et al.*, 1987) and human PMNs (Marquis *et al.*, 1988). However, it showed marked discrepancies in competition with either PAF or other PAF receptor antagonists. In rabbit intact platelets and crude platelet membranes, the values for [^{3}H]52770 RP to its receptors were 8.5 and 7.6 nM, respectively (Robaut *et al.*, 1987). Similar K_d values were also obtained for [^{3}H]52770 RP in intact humans PMNs (4.2 ± 0.3 nM, n = 8) and in human PMN membranes (2.8 nM) (Marquis *et al.*, 1988). The (+) enantiomer of 52770 RP (56972 RP) was 300- to 600-fold more potent than the (−) enantiomer in either rabbit platelet or human PMN membranes. In rabbit platelets and human PMNs, PAF and several representative PAF receptor antagonists were able to fully displace the specific [^{3}H]52770 RP binding to its receptors. Furthermore, the potency of each compound in inhibiting either [^{3}H]PAF or [^{3}H]52770 RP specific binding was of the same order of magnitude (the slope of regression straight line was not significantly different from unity) (Marquis *et al.*, 1988). However, it was also reported that neither PAF nor WEB 2086 was able to displace specific [^{3}H]52770 RP binding to either human platelets (Ukena *et al.*, 1988) or guinea-pig lung membranes (Gomez *et al.*, 1990).

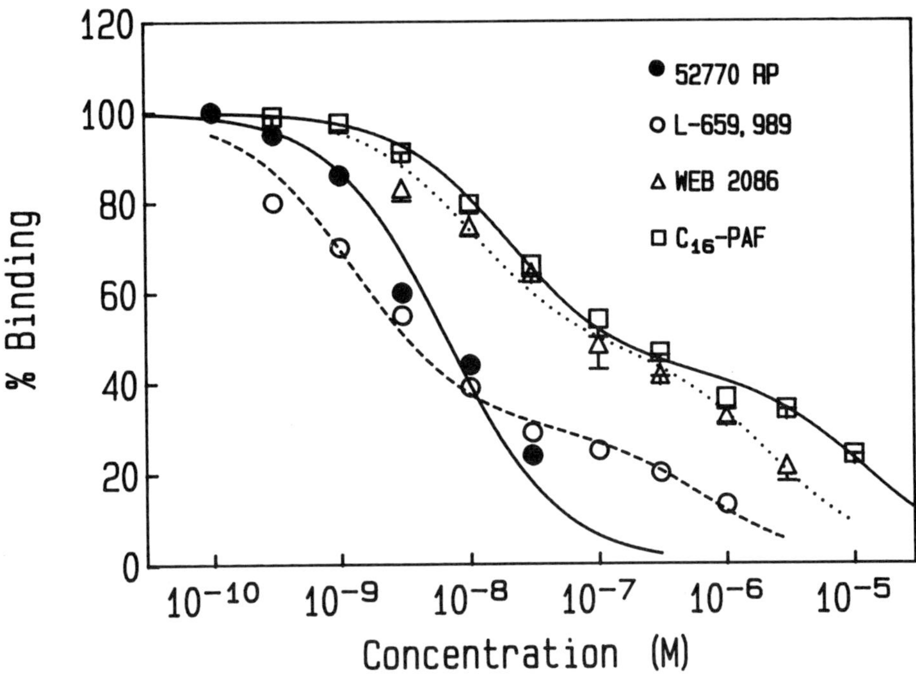

Figure 9.21 Inhibition of [³H]52770 RP binding to human PMN membranes by 52770 RP, L-659,989, WEB 2086 and C₁₆-PAF. Membrane protein (100 μg) was added to an incubation mixture (1 ml) containing 0.3 nM [³H]C₁₈-PAF and a known amount of inhibitor in a medium of 10 mM magnesium chloride, 10 mM Tris and 0.25% BSA at pH 7.0. C₁₆-PAF was dissolved in 10 mM Tris and 0.25% BSA, and the antagonists were dissolved in DMSO. The data points and error bars are the means and the standard deviations of three independent experiments. In each experiment, triplicate samples were prepared. The curves were plotted by fitting the competitive binding data using a non-linear regression computer program, GraphPAD, version 3.0 (GraphPAD Institute Software for Science, San Diego, CA).

[³H]52770 RP is commercially available from Du Pont-NEN (Boston, MA). Binding of [³H]52770 RP to either isolated human platelet or PMN membranes was saturable (S.-B. Hwang and M.-H. Lam, unpublished results). A linear plot was obtained in a Scatchard analysis, suggesting the existence of a single population of non-cooperative binding sites. The K_d values were 13.79 ± 3.56 nM ($n = 7$) and 3.5 ± 0.28 nM ($n = 6$) in human platelet and PMN membranes, respectively. The difference in the K_d values for [³H]52770 RP between receptors in human platelet and PMN membranes is statistically significant ($p < 0.0001$). The K_d value (3.5 nM) obtained for [³H]52770 RP in human PMN membranes is not different from that (2.8 nM) reported by Marquis *et al.* (1988). Unlabelled 52770 RP competitively inhibited the binding of [³H]52770 RP (1 nM) to human platelet membranes with an ED₅₀ value of 6.6 nM (or a K_i value of 6.2 nM), whereas C₁₆-PAF, WEB 2086 and L-659,989 were unable to compete with the binding of [³H]52770 RP (Fig. 9.20). In human PMN membranes, the unlabelled 52770 RP could also displace the binding of [³H]52770 RP with an ED₅₀ value of 6.5 nM, which corresponds to a K_i value of 5.0 nM (Fig. 9.21). The K_d value measured directly from the binding of [³H]52770 RP is similar to the K_i value in

inhibiting the specific [³H]C₁₈-PAF binding, but is about twice the K_i value for the unlabelled 52770 RP in inhibiting the specific [³H]52770 RP binding to human platelet membranes. However, in human PMN membranes, the K_d value of [³H]52770 RP is similar to the K_i value in inhibiting the specific [³H]52770 RP binding, but is only about one-half of the K_i value measured from the inhibition of [³H]C₁₈-PAF binding to human PMN membranes. The reason for this is not known. As shown in Fig. 9.21, C₁₆-PAF, WEB 2086 and L-659,989 could also displace the specific binding of [³H]52770 RP to human PMN membranes. However, the displacement curves were shallow and statistically better described by a two-site rather than a one-site model (Fig. 9.21). The order of potency (L-659,989 > WEB 2086 > C₁₆-PAF) in inhibiting the binding of [³H]52770 RP as shown in Fig. 9.21 is quite different from the one in inhibiting the binding of [³H]PAF to human PMN membranes (C₁₆-PAF > L-659,989 > WEB 2086). These results further indicate the complexity of [³H]52770 RP binding *in vitro*.

To study in further detail the mechanism of inhibition of 52770 RP binding by C₁₆-PAF, saturation binding studies were performed with C₁₆-PAF and the data subjected to a Scatchard analysis. Figure 9.22 shows one

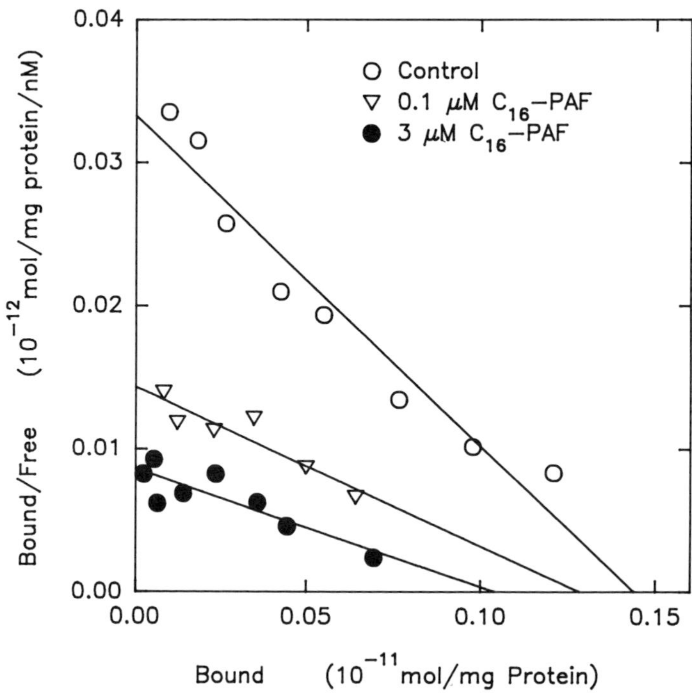

Figure 9.22 Scatchard plots of [³H]52770 RP binding to human PMN membranes in the absence and in the presence of C₁₆-PAF at a concentration of 0.1 and 3 μM. The assay medium was 10 mM magnesium chloride, 10 mM Tris and 0.25% BSA at pH 7.0. Straight lines were drawn from the linear regression using the program Sigmaplot, version 4.0 (Jandel Scientific, Corte Madera, CA). The data points were the means of triplicate determinations.

example of such a study. Control membranes display a K_d value of 4.32 nM and a maximal level of detectable binding sites (B_{max}) of 1.44 pmol/mg protein for [³H]52770 RP. When the experiment was performed in the presence of 0.1 and 3 μM C₁₆-PAF, the K_d value of 52770 RP was shifted to 8.95 and 12.81 nM, respectively. Besides the alteration of the K_d value in the presence of PAF, the B_{max} value changed from 1.44 pmol/mg protein in the absence of PAF to 1.04 and 0.89 pmol/mg protein, respectively, in the presence of 0.1 and 3 μM C₁₆-PAF (Fig. 9.22), indicating that binding inhibition of [³H]52770 RP by C₁₆-PAF in human PMN membranes is not competitive. Both the affinity and the B_{max} value of [³H]52770 RP in human PMN membranes decreased progressively in the presence of C₁₆-PAF. Therefore, the inhibition of [³H]52770 RP by PAF is not due to direct competition between agents at the same site but to a separate site which allosterically modulates 52770 RP binding. The reason for this contradiction between our results and those reported by Marquis *et al.* (1988) is not known.

The 3-benzoyl derivative, RP 58467, was also found to be very potent, with an ED₅₀ value of 0.2 μM in inhibiting PAF-induced aggregation of rabbit platelets in plasma and a K_i value of 1.5 nM in antagonizing [³H]PAF binding to washed rabbit platelets (Lave *et al.*, 1989). The dextrorotatory isomer, RP 59227, was approximately 300 times more potent than the laevorotatory

enantiomer, RP 59228 (Robaut *et al.*, 1988; Lave *et al.*, 1989). RP 59227 did not interfere with the aggregation induced by other aggregating agents, indicating that it has a high specificity for PAF receptor sites (Robaut *et al.*, 1988). *In vivo*, 59227 RP inhibited PAF-induced hypotension in rats at 10 mg/kg p.o. Pretreatment of rats with 59227 RP prevented the hypotension and haemoconcentration due to endotoxin (5 mg/kg i.v.). Furthermore, 59227 RP specifically antagonized PAF-induced bronchoconstriction with an ED₅₀ value of 20 μg/kg i.v. (Robaut *et al.*, 1988).

In a comparative study of hydrophobic non-steroidal anti-inflammatory agents, while some analogues of indomethacin and sulindac interfered with the binding to the PAF receptor at micromolar levels, no correlation with their anti-inflammatory activity was observed (Hwang *et al.*, 1984). Further structure modification of some sulindac analogues uncovered several more potent PAF receptor antagonists. The stereochemistry and substituents optimal for PAF antagonism are distinctly different from those required for cyclooxygenase inhibition. A *trans*-conformation with a *p*-sulphonyl group and a neutral side-chain is apparently preferred (Shen *et al.*, 1987b). The most potent compound in this series is L-651,142. It inhibited [³H]PAF binding to human platelet and human PMN membranes with ED₅₀ values of 6.9 and 6.3 μM, respectively (Hwang, 1988). However, the

Table 9.7 Anti-PAF activities of optical active forms of two 5-substituted 3-methyl-2-(3-pyridyl)thiazolidin-4-ones

Compound	Configuration	Anti-PAF activity (ED_{50}, µM)[a]
(164)	(±)-*cis*	2.7
(166)	(2R,5S)-(+)-*cis*	0.98
(167)	(2S,5R)-(−)-*cis*	145.00
(168)	(2R,5R)-(+)-*trans*	9.26
(169)	(2S,5S)-(−)-*trans*	42.0
(170)	(2R,5R)-(+)-*trans*	1.5
(171)	(2S,5S)-(−)-*trans*	30.0
(165)	(±)-*cis*	>30

[a] The anti-PAF activity of these compounds is expressed as the ED_{50} value to inhibit PAF-induced aggregation of rabbit platelets.

inhibitory potency of L-651,142 in rat peritoneal PMN membranes was about five-fold less potent than in human PMN membranes (Hwang, 1991a). The ED_{50} value of L-651,142 in rat PMN membranes was 30 µM. Therefore, there seem to be clear differences between PAF receptors on human and rat PMNs (Hwang, 1991a). *In vivo*, L-651, 142 was orally active. It inhibited PAF-induced cutaneous vascular permeability in guinea-pigs with an ED_{50} value of 15 mg/kg p.o. (S.-B. Hwang, unpublished results).

Sandoz recently reported a newly developed PAF antagonist designed by using the PAF molecule as the template (Houlihan *et al.*, 1988). The compound (SDZ 64-412) inhibited PAF-induced aggregation of human platelets and specific [3H]PAF binding with an ED_{50} value of about 60 nM. *In vivo*, SDZ 64-412 exhibited potent oral activity in a broad range of species and demonstrated dose-dependent antagonism of PAF-induced effects of hypotension, bronchoconstriction and haemoconcentration, as well as protection against PAF or endotoxin-induced systemic lethal shock (Handley *et al.*, 1988). Protection against non-specific airways hyper-reactivity by SDZ 64-412 when given orally as a single treatment 2 h before an aerosol allergen challenge in sensitized guinea-pigs was also recently reported (Havill *et al.*, 1990). SDZ 64-412 shares this protective effect with two other clinically used drugs, methylprednisolone and ketotifen. The compound had no effect on pulmonary eosinophilia.

A soluble thiazolidine derivative, SM-10661 ((±)-*cis*-3,5-dimethyl-2-(3-pyridyl)thiazolidin-4-one), was also reported to be a potent and selective antagonist (Komuro *et al.*, 1990). SM-10661 antagonized [3H]PAF binding to rabbit platelets competitively with an ED_{50} value of 1.0 µM. PAF-induced aggregation of rabbit and human platelets in plasma was inhibited by SM-10661 dose-dependently with an ED_{50} value of about 3 µM. Platelet aggregation induced either by ADP, collagen, thrombin, arachidonic acid, A 23187 or U 46619 were not inhibited by SM-10661 even at concentrations up to 0.4 mM. The *cis* isomer of SM-10661 showed stronger anti-PAF activity than the *trans* isomer. *In vivo*, SM-10661 at 100 mg/kg p.o. inhibited PAF-induced pathology, including hypotension in rats, haemoconcentration in guinea-pigs and bronchoconstriction in guinea-pigs (Morooka *et al.*, 1988).

The 5-(4-chlorophenyl) analogue of SM-10661 was reported to have roughly the same anti-PAF activity (Tanabe *et al.*, 1991a). Although the anti-PAF activity of the *cis* isomer of SM-10661 was more potent than its *trans* isomer, *trans* isomers of other 5-substituted (3-pyridyl)thiazolidin-4-ones generally showed higher activities than the *cis* isomers (Tanabe *et al.*, 1991a). The position of the nitrogen in these thiazolidin-4-ones is important since 4-pyridyl analogues were generally found to show about one-third lower anti-PAF activity when compared to the corresponding 3-pyridyl ones. The

2-pyridyl analogues showed almost no activity (Tanabe *et al.*, 1991a). The stereospecificity of the anti-PAF action of SM-10661 and its 4-chlorophenyl analogues was also reported by Tanabe *et al.* (1991b), as summarized in Table 9.7. There is a distinct difference in activities between enantiomers. Clearly, (2*R*) stereoisomers of the thiazolidin-4-ones (compounds (166), (168) and (170)) showed much higher activities when compared with the corresponding (2*S*) enantiomers (compounds (167), (169) and (171)) in inhibiting either the PAF binding or PAF-induced aggregation of rabbit platelets.

Yamada *et al.* (1989) reported the pharmacological properties of the piperazine derivative YM-461 at the Third International Conference on Platelet-Activating Factor and Structurally Related Alkyl Ether Lipids in Tokyo. YM-461 inhibited [^{3}H]PAF binding to rabbit platelet membranes with a K_i value of 1.3 nM. It inhibited PAF-induced aggregation of rabbit and human platelets with pA$_2$ values of 7.52 and 7.29, respectively. Here, pA$_2$ is defined as the negative logarithm of the molar concentration of antagonist that produces a two-fold shift to the right in the agonist dose–response curve. *In vivo*, YM-461 dose-dependently inhibited PAF-induced death in mice (ED$_{50}$ = 0.35 mg/kg p.o.) and PAF-induced haemoconcentration in rats (ED$_{50}$ = 0.16 mg/kg p.o.). In anaesthetized guinea-pigs, YM-461 inhibited the bronchoconstriction induced by PAF and antigen-induced anaphylactic bronchoconstriction with ED$_{50}$ values of 1.2 and 3 mg/kg p.o., respectively. YM-461 at 3 and 10 mg/kg p.o. also inhibited the antigen-induced late asthmatic responses by 66 and 82%, respectively, and the associated airways responsiveness in allergic sheep at 24 h (Tomioka *et al.*, 1989). Modification of the 3-phenylpropyl moiety in YM-461 into 3-methyl-3-phenylbutyl (YM-264) improved the anti-PAF activity both *in vitro* and *in vivo*. YM-264 was twice as active as YM-461 in inhibiting [^{3}H]PAF binding to rabbit platelet membranes as well as in inhibiting PAF-induced aggregation of rabbit platelets (Yamada *et al.*, 1990). In anaesthetized rats, YM-264 inhibited PAF-induced hypotension with an ED$_{50}$ value of 0.005 mg/kg i.v. Orally, YM-264 also inhibited PAF-induced death in mice, haemoconcentration in rats and increased vascular permeability in guinea-pigs with ED$_{50}$ values of 0.19, 0.30 and 0.49 mg/kg p.o., respectively (Yamada *et al.*, 1990) and inhibited puromycin-induced proteinuria, hypercholesterolaemia and hypoalbuminaemia at doses of 0.3 and 3.0 mg/kg p.o. in rats (Yamada *et al.*, 1991).

In the last few years, Hoffman-La Roche has reported a series of synthetic compounds with PAF antagonist activities exemplified by pyridoquinazoline (Tilley *et al.*, 1988) and biphenyl carboxamide (Tilley *et al.*, 1989), (*E,E*)-5-phenyl-2,4-pentadienamide (Guthrie *et al.*, 1989) and 5-arylpentadienecarboxamide (Guthrie *et al.*, 1990). (*R*)-2-(1-Methylethyl)-*N*-[1-methyl-4-(3-pyridinyl)butyl]-11-oxo-11-*H*-pyrido[2,1-*b*]quinazoline-8-carboxamide (compound (174)) was the most potent

agonist in the pyrido[2,1-*b*]quinazoline-8-carboxamide series in inhibiting [^{3}H]PAF binding to dog platelets with an ED$_{50}$ value of 0.25 μM (Tilley *et al.*, 1988). The pyridoquinazolines in which the key aromatic ring is part of a planar heteroaromatic ring are generally less potent PAF antagonists than the biphenylcarboxamides (compound (175)) (Tilley *et al.*, 1989) or the pentadienamides (Ro 24-0238) (Guthrie *et al.*, 1989), in which rotation of the corresponding aromatic ring out of conjugation with the remainder of the system is possible. Ro 24-0238 with an ED$_{50}$ value of 40 nM in inhibiting the [^{3}H]PAF binding to dog platelets was selected for further clinical evaluation (Guthrie *et al.*, 1989). Ro 24-0238 shows a dual activity, combining potent PAF antagonism with thromboxane synthase activity (Welton and O'Donnell, 1989). Results from propenamide derivatives with the constraining aromatic ring led to the conclusion that the aromatic ring present in the 5-arylpentadienecarboxamides is preferentially coplanar with the diene system for good anti-PAF activity (Guthrie *et al.*, 1990). Several compounds, including compounds (177), (178), (179) and (180), showed potent activity in inhibiting [^{3}H]PAF binding to canine platelets with an ED$_{50}$ value in the nanomolar concentration range (Guthrie *et al.*, 1990). However, oral activity in inhibiting PAF-induced bronchoconstriction in guinea-pigs is highly sensitive to the position of the methoxy group. Compounds (177), (178) and (180) inhibited PAF-induced bronchoconstriction in guinea-pigs with ED$_{50}$ values of 3.0–5.4 mg/kg p.o. (Guthrie *et al.*, 1990).

As described above, CV-6209 showed poor oral activity. To minimize this disadvantage, Takatani *et al.* (1990) carried out modifications of CV-6209 which led to a nicotinamide derivative, TCV-309, which was twice as potent in inhibiting platelet aggregation induced by PAF than CV-6209. TCV-309 was orally active with an ED$_{50}$ value of 0.3 mg/kg p.o. in inhibiting PAF-induced hypotension in rats. The inhibitory effects continued for more than 8 h at a dose of 10 μg/kg i.v. or 1 mg/kg p.o. Furthermore, TCV-309 did not exhibit the undesirable side-effects associated with the lipid-type analogues. It did not show haemolytic action at a concentration of 3 mM and showed no vascular damage at the site of injection at a dose of 10 mg/kg in the rabbit (Takatani *et al.*, 1990).

Finally, a new PAF antagonist containing the 1,4-disubstituted piperazine moiety (compound (182)) was patented by Takeda Chemical Industries (Hirosada *et al.*, 1988). This compound potently inhibited PAF-induced aggregation of rabbit platelets. *In vivo*, the compound was orally active, with an ED$_{50}$ value of less than 3 mg/kg in inhibiting both PAF- and endotoxin-induced hypotension in the rat.

3.8   Dual Inhibitors

Given the pharmacological profile of PAF antagonists, and in view of the involvement and interplay of multiple

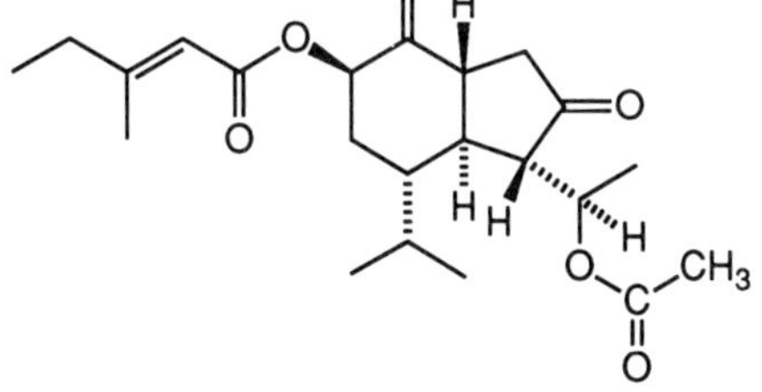

L-652,469 (183)

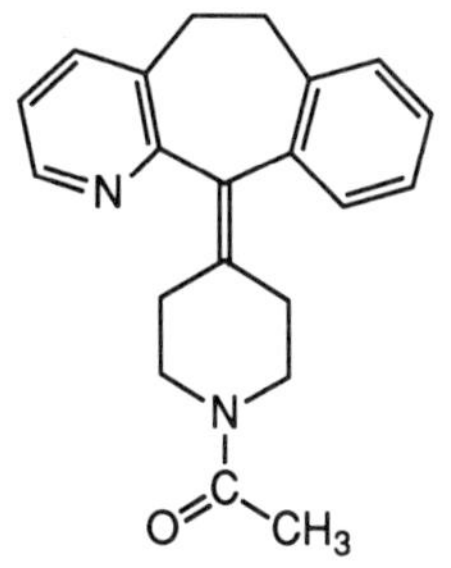

SCH 37370 (184)

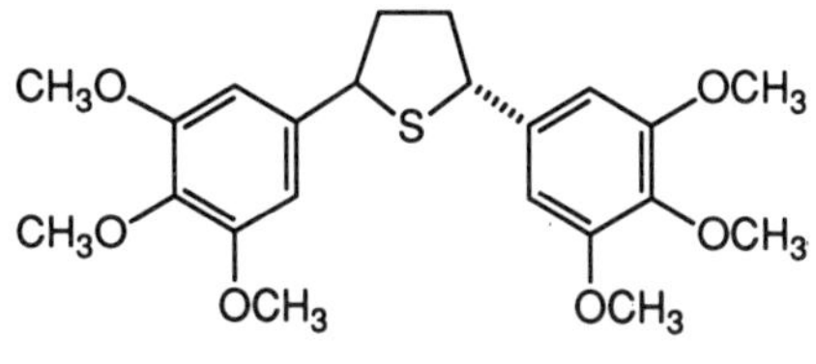

L-653,150 (55)

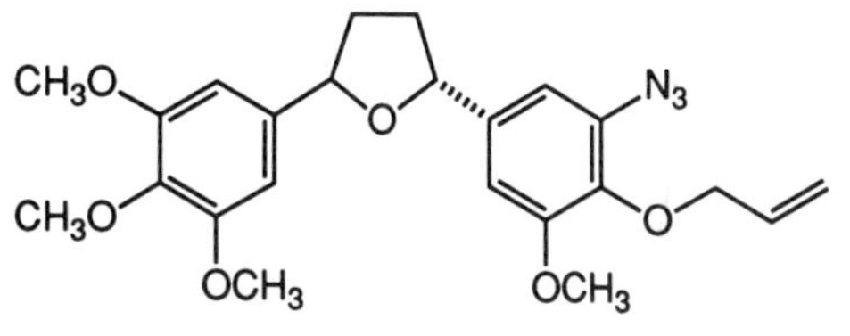

L-662,025 (185)

Figure 9.23 Structures of dual and irreversible PAF receptor antagonists.

mediators in many inflammatory processes, an antagonist with a broader spectrum of action, affecting the function or synthesis of other mediators, might have great efficacy in some clinical disorders. Those compounds listed in Fig. 9.23 showed dual inhibitory activities.

L-652,469 (Fig. 9.23), isolated from the Chinese medicinal herb *Tussilago farfara*, was identified to be an inhibitor for both PAF and calcium entry blocker binding to membrane vesicles. It inhibited [^{3}H]PAF binding to rabbit platelet membranes with a K_i value of 4.0 μM in the presence of 10 mM magnesium chloride (Hwang *et al.*, 1987). Although the potency of L-652,469 is low as a PAF receptor antagonist, it competitively inhibited the specific binding of Ca^{2+} channel blockers (e.g. [^{3}H]nitrendipine; K_i = 1.2 μM) in cardiac sarcolemmal vesicles. At a concentration of 10^{-5} M, L-652,469 caused a 60% relaxation of Ca^{2+}-induced contraction of rat thoracic aorta strips. As a dual PAF antagonist and calcium channel blocker, it showed a beneficial effect orally in inhibiting PAF-induced rat foot oedema and the first phase of carregeenan-induced rat hindpaw oedema (Hwang *et al.*, 1987).

Based on the strategy that PAF and histamine may complement each other during an inflammatory or allergic response, an agent (SCH 37370) (Fig. 9.23) which inhibits the actions of histamine in addition to those of PAF was identified (Billah *et al.*, 1990). SCH 37370 selectively inhibited PAF-induced aggregation of human platelets (IC$_{50}$ = 0.6 μM) and also competed with PAF binding to specific sites in membrane preparations from human lung with a potency similar to that of L-652,731 (IC$_{50}$ = 1.2 μM). As an indication of its H$_1$

antihistamine activity, SCH 37370 blocked pyrilamine binding to rat brain membranes with a K_i value of 310 nM. In guinea-pig ileum, SCH 37370 effectively inhibited H$_1$ histamine-induced contraction; however, it exhibited no H$_2$ antihistamine activity as measured by contraction in guinea-pig atrial tissue.

The anti-PAF activity appears to be optimal for acetamide. Removing a methylene group from the acetamide moiety, or adding one to it, reduced the activity (Piwinski *et al.*, 1991). Analogues with large amide groups were essentially inactive. Placement of a chlorine atom at C-8 appears to be essential. C-9 chloro or C-9 and C-10 dichloro analogues were less active. Conformational rigidity is also important for optimal anti-PAF activity. The less constrained analogues with a broken piperidine ring or with a removed ethylene bridge are weak antagonists of PAF. Furthermore, anti-PAF activity requires both nitrogen atoms. The bisaryl derivative, which lacks the pyridine nitrogen, was a weak antagonist, while the compound which lacked nitrogen in the piperidine ring was completely inactive at a dose of 50 μM.

When administered intravenously to guinea-pigs, SCH 37370 is an equipotent antagonist of PAF and histamine-induced bronchospasm (ED$_{50}$ = 0.5 mg/kg). Orally, SCH 37370 blocked hypotension in the rat and a cutaneous reaction in monkeys induced by PAF or by histamine, as well as PAF-induced lethality in mice and allergic bronchospasm in guinea-pigs (ED$_{50}$ = 4.4 mg/kg).

L-653,150 (see Fig. 9.23), a tetrahydrothiophene analogue of L-652,731, showed roughly the same potency as L-652,731 in inhibiting the specific binding of [³H]PAF binding to rabbit platelet (Biftu *et al.*, 1986b), human platelet (Hwang and Lam, 1986) and human PMN membranes (Hwang, 1988). However, L-652,150 showed dual activities. It also inhibited the RBL-5-lipoxygenase-catalysed transformation of [¹⁴C]arachidonic acid to 5-HETE with an ED_{50} value of 5 μM (Biftu *et al.*, 1986b). The higher efficacy and longer duration of action of L-653,150, when compared to L-652,731, were demonstrated in a variety of models, including PAF-induced platelet aggregation, PAF-induced human neutrophil degranulation and aggregation *in vitro* and PAF-induced enzyme release, vascular permeability, and hypotension *in vivo* (Biftu, 1985).

As described above, BN 50739 and Ro 24-0238 were also demonstrated to have dual inhibitory activity. BN 50739 inhibits both PAF- and oxygen free radical-induced functions (Castaner *et al.*, 1991). Ro 24-0238 was found to inhibit both PAF and thromboxane synthase activity (Welton and O'Donnell, 1989).

3.9   IRREVERSIBLE PLATELET-ACTIVATING FACTOR ANTAGONISTS

L-662,025 (Fig. 9.23), a tetrahydrofuran analogue, is a photolabile, irreversible PAF antagonist (Hussaini and Shen, 1989; Shen and Hussaini, 1990). Before irradiation, L-662,025 inhibited the binding of [³H]PAF in a competitive and reversible manner at a dose of 1–5 μM. Upon incubation of rabbit platelets with L-652,025 for 2 min before addition of PAF, L-652,025 produces a dose-related inhibition of platelet aggregation with a parallel shift of the concentration–response curve to the right without significant changes in the maximal response. Photoactivation of L-662,025 at wavelengths greater than 320 nm produced an irreversible inhibition of platelet aggregation and PAF receptor binding. Following photoactivation of human platelets with L-662,025, the maximal aggregation of the dose–response curve decreased with the increase of antagonist concentration. The antagonist effect cannot be reversed by increasing the concentration of PAF or by repeatedly washing the platelets, suggesting irreversible inhibition. Photolysis of L-652,025 did not affect collagen or ADP-induced aggregation of human platelets, thus indicating its specificity for the PAF receptor. Futoxide, isolated from a batch of *haifenteng (Piper futokadsurae)* from Taiwan, is another irreversible receptor antagonist (Shen *et al.*, 1989). Futoxide inhibited specific binding of [³H]PAF to human platelet and leucocyte membranes at a dose of 0.1 μM. It could irreversibly inactivate the receptors in a concentration- and time-dependent manner. It has no effect on platelet aggregation induced by collagen and ADP.

4. *Multiple Conformational States of Platelet-Activating Factor Receptors*

In isolated rabbit platelet membranes, ions and GTP regulated the specific [³H]PAF binding. Na⁺ specifically inhibited the binding with an ED_{50} value of 6 mM (Hwang *et al.*, 1986d). Li⁺ also inhibited the binding but at a much higher concentration (ED_{50} = 150 mM) (Hwang *et al.*, 1986d). A similar ionic effect on the specific [³H]PAF binding has also been reported in human platelet membranes (Valone and Ruis, 1986; Hwang, 1988). In contrast, K⁺, Cs⁺ and Rb⁺, as well as Mg²⁺, Ca²⁺ and Mn²⁺, enhance [³H]PAF binding to rabbit platelet membranes (Hwang *et al.*, 1986d). From both Scatchard and Klotz analysis, the inhibitory effect of Na⁺ is apparently due to an increase in the K_d value of PAF binding to its receptors. In contrast, the Mg²⁺-induced enhancement of specific PAF binding arises from both an increased affinity to the receptor and an increased number of receptor sites (Hwang *et al.*, 1986d). The inhibitory effect of Na⁺ is generally attributed to the direct linkage between receptors and adenylate cyclases through an inhibitory G protein (Hwang, 1990). Haslam and Vanderwel (1982) also reported that PAF inhibited basal, PGE₁-stimulated and fluoride-stimulated adenylate cyclase activities. A metal ion site that binds Na⁺ or Li⁺ ions has been proposed (Rodbell, 1983). A membrane component of 168 kDa, different from G proteins, was also identified in the opioid receptor complex to act as an allosteric inhibitor that mediates the effect of sodium on the receptor (Ott *et al.*, 1988). This Na⁺ regulation site appears to be different from the divalent ion regulatory site(s) (Hwang *et al.*, 1986d; Hwang, 1990). Similar to the α-adrenergic receptor system of human platelets (Motulsky and Insel, 1983), this Na⁺-binding site which modulates the affinity of PAF to specific PAF receptors was also found to be located on the cytoplasmic side of the plasma membranes (Hwang and Lam, 1989, 1991). The mechanism of the inhibitory effects by Li⁺ on specific binding is not well characterized. However, the Li⁺-binding site may be on the G protein since Li⁺ has been shown to attenuate the ADP ribosylation of the G_i protein of the cell membrane (Kawamoto *et al.*, 1991). The mechanism for potentiation by K⁺, Mg²⁺, Ca²⁺ and Mn²⁺ is not known, but it may also be attributed to the interaction between PAF receptors and other closely coupled regulatory proteins.

In contrast, Na⁺ has quite different effects on the specific binding of [2,5-³H]L-659,989, an orally active, PAF specific and competitive receptor antagonist (Hwang *et al.*, 1988; Ponpipom *et al.*, 1988). The specific [³H]L-659,989 binding to rabbit platelet membranes is not only potentiated by K⁺, Mg²⁺, Ca²⁺ and Mn²⁺ but is also equally potentiated by Na⁺ and Li⁺ (Hwang *et al.*, 1989). In the presence of either 150 mM sodium chloride or 10 mM magnesium chloride, the PAF receptors show identical K_d values and similar B_{max} values for L-659,989

(Hwang *et al.*, 1989). In contrast, in the absence of cations, the detectable receptor number for [³H]L-659,989 is only about 50% of that in the presence of either 10 mM magnesium chloride or 150 mM sodium chloride (Hwang *et al.*, 1989). Ni²⁺ inhibits specific [³H]L-659,989 binding (Hwang *et al.*, 1989), even though Ni²⁺ has no effect on specific [³H]C₁₆-PAF binding to rabbit platelet membranes (Hwang and Lam, 1987).

GTP was also found to specifically inhibit PAF binding (Hwang *et al.*, 1986d), but showed no effect on the binding of [³H]L-659,989 (Hwang *et al.*, 1989). Therefore, there appears to be differences in the modulation by ions and by GTP on the binding of [³H]PAF and [³H]L-659,989 to rabbit platelet membranes. Since both PAF and L-659,989 bind to the same receptor and share a common binding site (Hwang *et al.*, 1988, 1989), the variation in the detectable receptor number under different ionic conditions could be due to coexistence of several conformational states of the PAF receptors. Since receptors that cannot be detected by [³H]PAF in the presence of 150 mM sodium chloride are detectable with [³H]L-659,989, the lower detectable receptor number for [³H]PAF is not due to the receptor being either embedded in the membrane or translocated into the inner membrane surface, but rather due to the transformation of the receptor into the low affinity state(s) for [³H]PAF with the K_d value(s) possibly in the micromolar range, which cannot be detected in Scatchard plots with the radioligand in the nanomolar concentration range. Such clear differences in the binding of an agonist and antagonist caused by changing ionic conditions strongly suggest that the variation in the B_{max} value under varying ionic conditions results from a direct receptor effect rather than from an alteration in membrane structure. Hence, the modulation of [³H]PAF and/or [³H]L-659,989 binding by ions and/or GTP clearly indicates that ions and GTP can modulate the conformation of the receptor, and thus multiple conformational states can exist in a single type of PAF receptor.

The existence of multiple conformational states of the PAF receptor can be further confirmed by the competitive binding studies of [³H]L-659,989 by PAF under different ionic conditions, either in the presence or absence of GTP. Since GTP shows no effect on the binding of [³H]L-659,989 to rabbit platelet membranes and since the PAF receptor exhibits identical affinity to [³H]L-659,989 in both 10 mM magnesium chloride and 150 mM sodium chloride (Hwang *et al.*, 1989), it becomes possible to compare and to calculate the affinity of PAF to PAF receptors under different ionic conditions. In the presence of 10 mM magnesium chloride, the ED₅₀ value for C₁₆-PAF inhibition of specific [³H]L-659,989 binding is 1.4 nM whereas, in the presence of 150 mM sodium chloride, the ED₅₀ value is shifted to about 1 μM, which is about 1000-fold higher than that in 10 mM magnesium chloride. GTP at a concentration of 1 mM

also significantly shifts the C₁₆-PAF competition curves to the right, from an ED₅₀ value of 1.4 nM to 10 nM in the presence of 10 mM magnesium chloride and from 1 μM to 2 μM in the presence of 150 mM sodium chloride. The competition curves of C₁₆-PAF in inhibiting [³H]L-659,989 binding to rabbit platelet membranes are "shallow", with a cooperativity index (R_s) greater than 81 (Hwang *et al.*, 1989), and can be best fitted with two sites rather than with one site (Hwang and Lam, 1991), indicating that multiple orders of binding sites for C₁₆-PAF are present in the competition binding. Such a biphasic displacement curve in rabbit platelet membranes for C₁₆-PAF becomes more evident in the presence of 150 mM sodium chloride. However, in the presence of 150 mM sodium chloride and 1 mM GTP, a normal steepness competition curve was observed having a pseudo-Hill coefficient equal to -1. A single competitive antagonism by C₁₆-PAF of the specific binding of [³H]L-659,989 is therefore expected under these experimental ionic conditions. The receptor is in a single conformational state and the K_b value can thus be obtained from a Schild plot based on the competitive inhibition of [³H]L-659,989 binding by C₁₆-PAF. The K_b value for PAF to its receptors in the presence of 150 mM sodium chloride and 1 mM GTP is 0.93 μM, a 1000-fold higher value than the K_d value obtained from the Scatchard plot of the saturation binding of [³H]C₁₆-PAF to rabbit platelet membranes in the presence of 10 mM magnesium chloride ($K_d = 0.53$ nM) (Hwang *et al.*, 1986d). Under these ionic conditions, the PAF receptor probably has the lowest affinity for PAF. PAF receptors therefore exist in several distinct conformational forms with different affinities either for PAF or for a PAF receptor antagonist, L-659,989. These multiple conformational states could possibly regulate and mediate the function of PAF.

5. *Endogenous Modulator(s) of Platelet-Activing Factor Receptors*

Specific binding of [³H]PAF or [³H]*N*-methylcarbamyl-PAF to human platelet membranes was potentiated specifically by WGA or erythroagglutinin (PHA-E) (Hwang and Wang, 1991). WGA and PHA-E at concentrations about four times higher than those used to initiate the agglutination of red blood cells significantly potentiated the specific binding of [³H]*N*-methylcarbamyl-PAF to human platelet membranes, whereas concanavalin A from *Canavalia ensiformia*, the lectin from *Ulex europaeus* (Gorse), and the lectin from *Bandeiraea simplicifolia* (BS-I) at similar dose ratios showed no effects on the binding (Hwang and Wang, 1989, 1991). The potentiation effect, as demonstrated in Scatchard plots, was due to an increase in the detectable number of receptor sites but not to an alteration in the K_d value of the radioligand from receptors. The WGA-potentiated specific binding of [³H]*N*-methylcarbamyl-PAF was specifically inhibited by N-acetylglucosamine with an ED_{50} value of 7 mM (Hwang and Wang, 1989, 1991). D-Glucosamine and N-acetylgalactosamine were less potent, while L-fucose at similar concentrations did not inhibit WGA-potentiated binding of [³H]N-methylcarbamyl-PAF to human platelet membranes.

Binding of tritium-labelled PAF receptor antagonists, including MK 287 and 52770 RP, was also potentiated by WGA at similar concentrations. The potentiating effect was again due to an increase in the B_{max} value but not to an alteration of the K_d value of the receptor antagonist. Enantio-C_{16}-PAF, the biologically inactive enantiomer of C_{16}-PAF, and L-659,989 fully displaced the binding of N-methylcarbamyl-PAF to human platelet membranes with identical K_i values either in the presence or absence of 30 µg WGA/ml (Hwang and Wang, 1991). The WGA-potentiated binding was also inhibited by Na^+ in a manner similar to that in the absence of WGA (Hwang and Wang, 1991). These results suggest that the PAF receptors exposed by WGA are identical to the PAF receptors present in the absence of WGA.

The potentiation of the specific binding by WGA is due to the existence of an endogenous modulator that is either a glycoprotein or a glycolipid which appears to modulate the conformation of the PAF receptor. WGA could bind to the carbohydrate portion of the inhibitor and prevent it from interacting with the receptor, since removal of the carboyhdrate portion of the inhibitor with neuraminidase and β-N-acetylglucosaminidase increased the PAF receptor number to roughly the same as that in the presence of WGA. Therefore, the binding of WGA to the saccharide substituent of such an endogenous glycolipid or glycoprotein modulator could induce either dissociation from, or association with, the PAF receptor, and the endogenous modulator could change the receptor conformation from a low-affinity to a high-affinity state, a change which can be detected in the conventional receptor-binding assay at nanomolar concentrations.

The existence of endogenous PAF inhibitors has also been reported in liver (Miwa et $al.$, 1987) and in uterus (Nakayama et $al.$, 1987). The endogenous PAF inhibitor in freeze-clamped perfused liver was subsequently identified as a mixture of free fatty acids. Among the free fatty acids present, oleic acid at a concentration of between 10 and 100 µM was found to be a potent inhibitor of platelet aggregation and serotonin release (Nunez et $al.$, 1990). Oleic acid did not block [³H]PAF binding to platelets, and it inhibited platelet aggregation in a non-competitive manner, indicating that inhibition of the PAF response by this acid may be at one of the steps in the signal transduction. Oleic acid induced a dose-dependent decrease in the levels of [³²P]PIP and [³²P]PIP₂, but did not induce the formation of IP₃, suggesting that the decrease in [³²P]PIP and [³²P]PIP₂ was not due to a stimulation of phospholipase C. In contrast to the effect of oleic acid, PAF induced a dose-dependent increase in the [³²P]PIP level. Oleic acid and PAF, when added together to platelets, interacted by affecting the level of PIP and PIP₂ in opposite ways. The decrease in the level of PIP and PIP₂ induced by oleic acid was partially reversed by an excess of PAF. Therefore, the inhibition of PAF-induced platelet aggregation by oleic acid is associated with a decrease in the level of PIP and PIP₂.

Two kinds of phospholipids in normal uterus were found to inhibit PAF-induced aggregation of rabbit platelets (Nakayama et $al.$, 1987). Inhibitor I was a mixture of 1-acyl($C_{16:0}$, $C_{18:0}$, $C_{18:1}$, $C_{18:2}$ and $C_{20:4}$)-2-lyso-sn-glycero-3-phosphocholine and 1-alkyl($C_{16:0}$, $C_{18:0}$ and $C_{18:1}$)-2-lyso-sn-glycero-3-phosphocholine (alkyl lyso–PC). Their ED_{50} values were in the range 1×10^{-5} to 4×10^{-5} M against PAF (0.1 nM)-induced aggregation, which is about 100 times weaker than CV-3988. The other inhibitor (inhibitor II) was a mixture of N-acyl($C_{18:1}$/$C_{18:0}$, $C_{18:1}$/$C_{20:0}$, $C_{18:1}$/$C_{24:0}$ and $C_{18:1}$/$C_{24:3}$)sphingo-4-enylphosphocholines; and their ED_{50} values were in the range 4×10^{-5} to 5×10^{-5} M in inhibiting the PAF-induced aggregation of rabbit platelets. The content of these inhibitors was found to be 10^5 times that of the PAF precursor, 1-alkyl-2-acylphosphocholine in the normal uterus. The function of these inhibitors needs to be further elucidated.

C-reactive protein is a prototypical acute-phase plasma protein. Serum C-reactive protein concentrations rise as much as 1000-fold during acute inflammation and tissue destruction (Anderson and McCarty, 1950; Claus et $al.$, 1976). C-reactive protein has been shown to inhibit PAF-induced aggregation of rabbit (Vigo, 1985) and human platelets (Filep et $al.$, 1991) as well as PAF- and FMLP-induced degranulation and superoxide production of human PMNs (Filep and Foldes-Filep, 1989). C-reactive protein was also able to compete and displace the specific [³H]PAF binding to human platelets (Filep et $al.$, 1991) as well as the specific [³H]PAF and [³H]FMLP binding to human PMNs (Filep and Foldes-Filep, 1989). Since antiserum to C-reactive protein blocked the inhibitory effects of C-reactive protein, the inhibitory effects of C-reactive protein on PAF- and/or FMLP-induced functions may not be due to its ability to bind PAF and FMLP (Filep and Foldes-Filep, 1989) or its membrane-stabilizing effects (Vigo, 1985), but rather due to its direct interaction with membrane receptors. Elevated plasma and tissue levels of PAF have been detected in association with infection or tissue injury (Inarrea et $al.$, 1985; Doebber et $al.$, 1985; Filep et $al.$, 1989). Since C-reactive protein appears in serum under the same pathological conditions (Claus et $al.$, 1976; Pepys, 1981; Morley and Kushner, 1982), this protein might act as an early protective recognition mechanism in acute inflammatory states.

6. Existence of Intracellular Platelet-Activing Factor Receptors

Synthesis of PAF occurs concomitantly with cellular activation. The majority of PAF produced intracellularly is retained inside the cell. This is evident in neutrophils (Ludwig *et al.*, 1984; Lynch and Henson, 1985), mononuclear phagocytes (Benveniste *et al.*, 1982; Lynch and Henson, 1985), mast cells (Triggiani *et al.*, 1990) and endothelial cells (Prescott *et al.*, 1984; Lynch and Henson, 1985). Such retention led to the speculation that PAF could act as both an intracellular and extracellular mediator (Henson, 1987). Specific receptors of PAF in plasma membranes mediate the function of PAF extracellularly, as discussed above. However, the function of the intracellularly retained PAF needs to be further elucidated. Due to the high concentration of PAF produced in the phagolysosome fraction after subcellular fractionation of phagocytosing neutrophils (Riches *et al.*, 1985) and the specific increase in the fluidity of artificial dipalmitoyl phosphatidylcholine membranes by PAF (Bratton *et al.*, 1988a), intracellularly retained PAF may act as a fusogen (Bratton *et al.*, 1988b). However, a high molar ratio of PAF to synthetic phospholipid is normally required in order to effect a significant change in the phase transition temperature or to broaden the transition width. In addition, there is no correlation between the effect of PAF or PAF analogues on activation of platelets and the effect on the alteration of physical properties of pure dipalmitoylphosphocholine bilayers determined by the differential scanning calorimetry (Hwang *et al.*, 1983). Here, we will present the evidence that functions of PAF produced and retained intracellularly may also act through PAF specific receptors inside the cells.

In the preparation of human platelet membranes, lysed membranes were separated into two fractions, membrane fractions A and B collected from the interfaces between 0.25 and 1.03 M and between 1.03 and 1.5 M sucrose, respectively (Hwang and Lam, 1986). Membrane fraction B is the major portion of membrane preparations and contains more PAF receptor sites than membrane fraction A (Hwang and Lam, 1986); membrane fraction A has a higher alkaline phosphatase activity (Hwang and Lam, 1986; Hwang *et al.*, 1986b). WGA and PHA-E potentiate the binding of tritium-labelled PAF agonists or antagonists to isolated human platelet membranes (Hwang and Wang, 1991); this potentiation by WGA or PHA-E is preferentially located in membrane fraction A (Hwang and Wang, 1991). Membrane fraction A may thus be enriched with either glycoprotein and/or glycolipid and, therefore, membrane fraction A is likely to be a plasma membrane-enriched fraction. Membrane fraction B contains more receptors for IP_3 than membrane fraction A (Hwang, 1991b) and a higher activity of antimycin-insensitive NADH–cytochrome-*c* reductase than membrane fraction A. Membrane fraction B is therefore enriched with the dense tubular system. Membrane fraction B consistently contains more PAF receptor sites than membrane fraction A (Hwang and Lam, 1986). Therefore, the intracellular membranes from human platelets contain PAF receptors which may initiate the cellular function induced by PAF synthesized intracellularly.

So far, no difference in human platelets between the intracellular and extracellular receptors for PAF have been detected. The K_d values obtained from binding studies of tritium-labelled ligands (Hwang, 1991c) and the K_i values of several selected PAF agonists and antagonists in inhibiting [³H]PAF binding to either membrane fraction A or B (Hwang and Lam, 1986; Hwang and Wang, 1991) are identical.

Intracellular PAF receptors were also demonstrated in rat cerebral cortex (Marcheselli *et al.*, 1990). Using [³H]PAF binding and several receptor antagonists, three distinct sites were identified. Two of them were in intracellular membranes. The microsomal membrane sites may be involved, at least in part, in intracellular events such as gene expression (see below).

7. Possible Functions of Intracellular Platelet-Activating Factors Receptors

Lectins such as WGA and PHA and that from *Ricinus communis* are strong inducers of both aggregation and release reactions for platelets (Majerus and Brodie, 1972; Greenberg and Jamieson, 1974). The exact mechanism for lectin-induced platelet activation is still unknown. However, intracellular PAF receptors mediating the function of intracellularly synthesized PAF may be involved in the platelet activation induced by WGA. WGA induces aggregation of gel-filtered human platelets in the presence of a cyclooxygenase inhibitor and ADP scavengers (creatine phosphate/creatine-phosphate kinase); the aggregation can be blocked by a specific PAF synthesis inhibitor, L-648,611 (3-(*N*-palmitoylamino)propylphosphocholine), and by a specific PAF receptor antagonist, CV-6209 (Hwang and Wang, 1991). Activation of PAF acetyltransferase activity can be observed even at concentrations of WGA below those needed for the induction of platelet aggregation. Most importantly, synthesis of PAF induced by WGA preceded the WGA-induced aggregation of human platelets and PAF synthesis can be blocked by the PAF synthesis inhibitor (Hwang and Wang, 1991). That a long incubation time with platelets is needed for L-648,611 and CV-6209 before the stimulation of WGA, and that CV-6209, but not L-659,989, inhibits the WGA-induced aggregation suggest that the inhibitory effect by receptor antagonists depends on the membrane permeability of the compounds and hence the ability to get into the cells, and that PAF could act as an intracellular mediator. The intracellular receptor may in turn mediate the actions of PAF synthesized and retained inside the platelets.

The possibility that PAF may act as an intracellular mediator has also been demonstrated in human (Tool *et al.*, 1989) and rabbit neutrophils (Stewart *et al.*, 1990), guinea-pig peritoneal macrophages (Stewart and Phillips, 1989) and bovine cultured aortic endothelial cells (Stewart *et al.*, 1989, 1990). The PAF receptor antagonists WEB

2086 and CV-6209 inhibited bradykinin-, A 23187- and ATP-induced PGI_2 generation, but had no effect on basal PGI_2 production. The lack of inhibition of basal PGI_2 generation by WEB 2086 and CV-6209 suggests that the inhibitory effect of PAF receptor antagonists is not directly on the enzymes involved in PGI_2 synthesis but, rather, on the action of PAF synthesized and retained intracellularly. Also, PAF formation preceded or accompanied eicosanoid generation (Stewart *et al.*, 1990) and exogenous PAF does not stimulate PGI_2 generation in bovine aortic endothelial cells in culture (Stewart *et al.* 1989). PAF synthesized intracellularly may therefore be responsible for eicosanoid generation.

PAF binds to the intracellular receptor and causes the stimulation of cells; it is tempting to speculate that it is possibly linked, at least in part, to gene expression. Indeed, PAF stimulated the production and secretion of PAF acetylhydrolase by the human hepatoma cell line $HepG_2$ (Satoh *et al.*, 1991). PAF also leads to the transcriptional activation of the proto-oncogenes c-*fos* and c-*jun* (Squinto *et al.*, 1989) and to the accumulation of calcyclin mRNA in human neuroblastoma cells (Allan and Bazan, 1989), possibly directly through the activation of the gene promotor (Squinto *et al.*, 1990). Proto-oncogenes are genes that normally serve the body well but cause cancer when they undergo a certain mutation. Both the *jun* and *fos* gene products are gene regulatory proteins. The *jun* protein could combine with the *fos* protein to form a dimer which is required for DNA binding to regulate gene expression. These effects on the induction of the transcriptional activation by PAF were inhibited by pretreatment of the cells with the PAF antagonist BN 52021, confirming the involvement of the PAF-specific receptor (Squinto *et al.*, 1989, 1990).

8. *Heterogeneity of Platelet-Activating Factor Receptors*

8.1   RECEPTOR HETEROGENEITY FOR PLATELET-ACTIVATING FACTOR IN THE SAME CELL

The possible existence of PAF receptor subtypes within the same type of cell was first proposed by Naccache *et al.* (1985) in rabbit neutrophils. Pertussis toxin treatment of the neutrophils essentially abolishes the turnover of polyphosphoinositides and the degranulation stimulated by PAF. On the other hand, the rise in intracellular Ca^{2+} caused by PAF is insensitive to treatment with pertussis toxin. Therefore, two separate sets of PAF-binding sites are proposed to mediate the effects of PAF.

The existence of two PAF receptors on guinea-pig eosinophils has been demonstrated (Kroegel *et al.*, 1989). Different dose–response curves to PAF were observed for intracellular calcium mobilization, degranulation and superoxide generation. A PAF receptor antagonist, WEB

2086, also shows differences in potency in inhibiting PAF-induced eosinophil functions with pA_2 values of 8.5 for PAF-induced peroxidase release, 8.3 for the rise in intracellular Ca^{2+} and 5.8 for superoxide anion production. In addition, PAF-induced degranulation and elevation of intracellular Ca^{2+} concentration were dependent on extracellular Ca^{2+} whereas PAF-stimulated superoxide anion generation was dependent on the presence of extracellular Mg^{2+} ions.

Three different types of PAF-binding sites were reported in rat cerebral cortex subcellular fractions (Marcheselli *et al.*, 1990). Two of the sites are located in purified microsomal fractions and one in synaptic plasma membranes. However, unlike PAF receptors in human platelets, the intracellular PAF receptors are distinctly different from those in the synaptic plasma membranes. A site displaying the highest affinity reported to date in all cells and membranes ($K_d = 0.022$ nM) was found in the microsomal fractions. There was also a second binding site in microsomal fractions ($K_d = 25$ nM). The binding site in synaptic plasma membranes displays relatively low specific binding with a K_d value of 1.2 nM. Several PAF receptor antagonists show differences in potencies in inhibiting [^{3}H]PAF binding to microsomal membranes and to synaptic plasma membranes. Two of the Takeda compounds, CV-3988 and CV-6209, are more potent than the BN compounds, BN-50727 and BN-52021, in intracellular membranes, whereas BN compounds are more potent than CV compounds in synaptic plasma membranes. Furthermore, specific [^{3}H]PAF binding to microsomal membranes is inhibited by Ca^{2+}. This is the only PAF receptor so far not potentiated by Ca^{2+} in the specific binding of [^{3}H]PAF. These results indicated that PAF binding sites are heterogeneous in rat cerebral cortex. Furthermore, CV-3988 appeared to be more potent than L-652,731 and CV-6209 in rat cerebral cortex (Marcheselli *et al.*, 1990), even though L-652,731 has been demonstrated to be a much more potent antagonist than CV-3988 in other membrane systems (Hwang and Lam, 1986; Hwang, 1988, 1991a). PAF receptors in the central nervous system may be quite different from other sources.

8.2   RECEPTOR HETEROGENEITY FOR PLATELET-ACTIVATING FACTOR BETWEEN DIFFERENT CELL TYPES

There is considerable variation between different cell types in their sensitivity to PAF. Femtomolar concentrations are normally required to significantly enhance IL-1 production in lymphocytes (Braquet and Rola-Pleszczynski, 1987), whereas stimulation of eosinophil or neutrophil superoxide generation requires micromolar concentrations (Doebber and Wu, 1987; Kroegel *et al.*, 1989). Differences in sensitivity in the same cell type were also noticed. Activation of acetyltransferase and PAF synthesis in

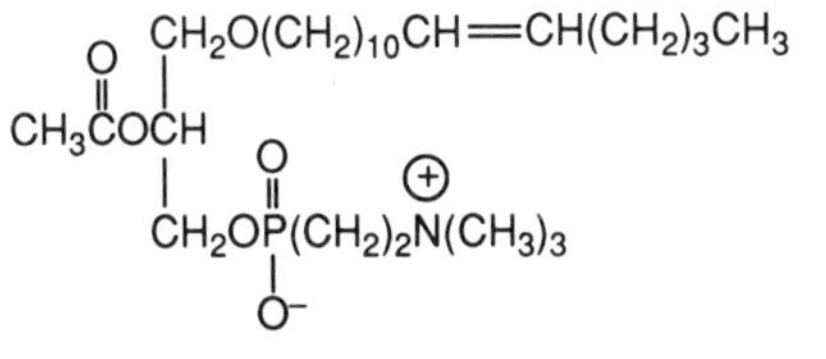

$CH_3\overset{O}{\overset{\|}{C}}OCH$ — $CH_2O(CH_2)_{10}CH{=}CH(CH_2)_3CH_3$ / $CH_2O\overset{O}{\overset{\|}{P}}(CH_2)_2\overset{+}{N}(CH_3)_3$ / O^-

$C_{16,11}$ PAF (186)

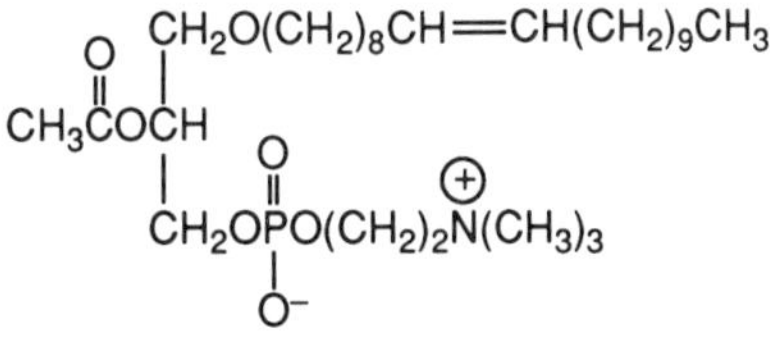

$CH_3\overset{O}{\overset{\|}{C}}OCH$ — $CH_2O(CH_2)_8CH{=}CH(CH_2)_9CH_3$ / $CH_2O\overset{O}{\overset{\|}{P}}O(CH_2)_2\overset{+}{N}(CH_3)_3$ / O^-

$C_{18,9}$ PAF (187)

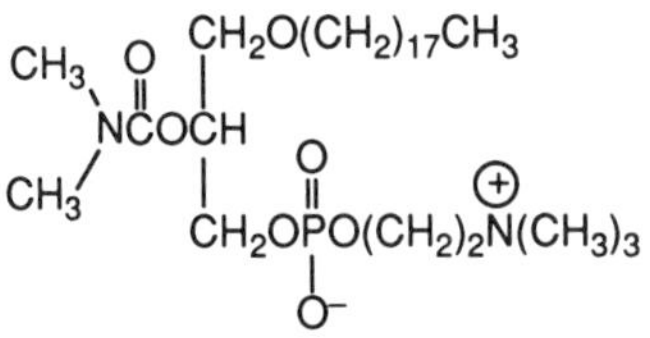

$(CH_3)_2N\overset{O}{\overset{\|}{C}}OCH$ — $CH_2O(CH_2)_{17}CH_3$ / $CH_2O\overset{O}{\overset{\|}{P}}O(CH_2)_2\overset{+}{N}(CH_3)_3$ / O^-

(188)

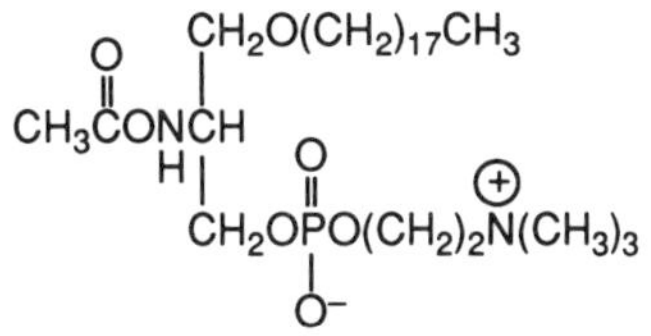

$CH_3\overset{O}{\overset{\|}{C}}O\underset{H}{N}CH$ — $CH_2O(CH_2)_{17}CH_3$ / $CH_2O\overset{O}{\overset{\|}{P}}O(CH_2)_2\overset{+}{N}(CH_3)_3$ / O^-

(189)

CH_3OCH — $CH_2O(CH_2)_{17}CH_3$ / $CH_2O\overset{O}{\overset{\|}{P}}O(CH_2)_2\overset{+}{N}(CH_3)_3$ / O^-

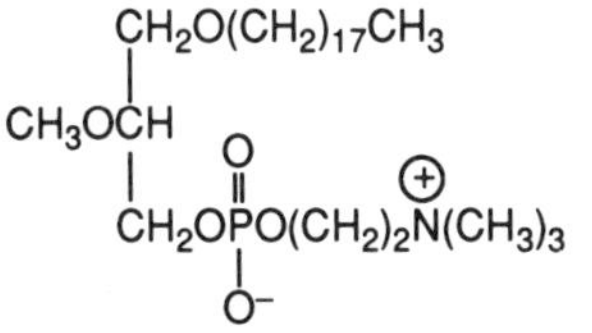

(3)

Figure 9.24 Structures of PAF analogues.

neutrophils was 10–30 times more sensitive to activation by PAF than was degranulation (Doebber and Wu, 1987). Multiple molecular species of PAF are produced as a result of the inflammatory processes (Ramesha and Pickett, 1987; Pinckard *et al.*, 1988). The PAF species produced varies with both cell of origin and stimulus (Ramesha and Pickett, 1987; Pinckard *et al.*, 1988). Moreover, identical cells from different animal species produced different spectra of PAF molecules (Ramesha and Pickett, 1987; Pinckard *et al.*, 1988). These results suggest the possible existence of PAF receptor heterogeneity and that different subtypes of the PAF receptor may induce different biological functions.

Using isolated membranes, a single type of PAF-specific receptor for [³H]C₁₆-PAF has been observed in rabbit (Hwang *et al.*, 1983) and human platelets (Hwang and Lam, 1986), human PMNs (Hwang, 1988), human lung tissue (Hwang *et al.*, 1985a), rat liver tissue (Hwang, 1987) and rat peritoneal PMNs (Hwang, 1991) with a K_d value of 0.5 nM under identical ionic conditions (Hwang, 1990). [³H]C₁₈-PAF showed no significant difference in the K_d values in human platelet, PMN and monocyte membranes (Hwang *et al.*, 1991). C₁₆-PAF, 1-*O*-hexadecyl-11-enyl-2-acetyl-*sn*-glycero-3-phosphocholine (C₁₆,₁₁-PAF (Fig. 9.24)) and 1-*O*-octadecyl-9-enyl-2-acetyl-*sn*-glycero-3-phosophocholine (C₁₈,₉-PAF) (Fig. 9.24) showed roughly the same potency in inhibit-

Table 9.8 Inhibition of [³H]C₁₈-PAF binding to human platelet and PMN membranes by C₁₆-PAF, C₁₆,₁₁-PAF and C₁₈,₉-PAF

Compound	K_i (nM)[a]	
	Platelets	PMNs
C₁₆-PAF	0.92 ± 0.17	1.80 ± 0.37
C₁₆,₁₁-PAF	1.76 ± 0.31	4.11 ± 0.20
C₁₈,₉-PAF	2.05 ± 0.19	4.68 ± 0.48

[a] K_d values of [³H]C₁₈-PAF in human platelets and PMN membranes used in the calculations are 2.05 and 3.15 nM, respectively.

ing specific [³H]C₁₈-PAF binding to human platelet and human PMN membranes (Table 9.8). In a comparison of the oral potency of several leading PAF antagonists, Saunders and Handley (1987) reported that the order of potency for four orally active antagonists (L-652,731 > BN 52021 > 48740 RP > SDZ 63-072) observed for inhibition of PAF-induced hypotension in the rat was similar to that for inhibition of PAF-induced haemoconcentration and bronchoconstriction in the guinea-pig. Therefore, they appear to have the same PAF receptor.

The possible existence of PAF receptor subtypes was also proposed by Lambrecht and Parnham (1986) on the basis of the different potencies of kadsurenone in inhibiting PAF-induced chemiluminescence of guinea-pig peritoneal macrophages and aggregation of pig peripheral blood leucocytes. Previously, species differences in PAF receptors between rabbit and human platelets (Hwang and Lam, 1986) and between rat and human PMNs (Hwang, 1991a) have been shown. The tetrahydrofuran derivatives of PAF receptor antagonists, including L-652,731, L-653,150 and L-659,989, are about 10 times more potent in rabbit platelets than in human platelets in inhibiting either the specific $[^3H]PAF$ binding (Hwang and Lam, 1986; Hwang et al., 1988) or PAF-induced platelet aggregation (Hwang, 1988). Differences between the potencies of several PAF agonists and L-651,142 in inhibiting the binding of $[^3H]PAF$ to rat and human PMN membranes was also observed (Hwang, 1991a). Enantio-C_{16}-PAF is only about 1500-fold less potent than C_{16}-PAF in inhibiting PAF binding to rat PMNs as well as in inhibiting PAF-induced rat cutaneous vascular permeability in vivo (Hwang et al., 1985b). The inactive enantiomer in human PMNs is about 30 000 times less potent than C_{16}-PAF (Hwang, 1988). Moreover, L-651,142 was about six times less potent in inhibiting the binding of $[^3H]PAF$ to rat PMN membranes than to human PMN membranes. This seems to indicate clear differences in PAF receptors between species. Whether the differences in potencies of kadsurenone between the two cell types is due to the species differences of the PAF receptor needs to be further confirmed. Dihydrokadsurenone showed no differences in potencies in inhibiting the specific $[^3H]C_{18}$-PAF binding to either human PMN or human mononuclear leucocyte membranes (Hwang et al., 1991). The difference in potency shown by kadsurenone may indeed be due to the difference in receptors between guinea-pigs and pigs.

In contrast, in studies of the specific binding of a $[^3H]N$-methylcarbamyl-PAF, the K_d value is significantly different for human platelet and human PMN membranes. The K_d value for $[^3H]N$-methylcarbamyl-PAF in human platelet membranes is 2.18 ± 0.48 nM (mean $\pm$ SD; $n = 8$), which is significantly different ($p<0.0001$) from the K_d value in human PMN membranes (5.18 ± 0.54 nM, $n = 8$) (Hwang, 1990).

As described above, sodium specifically inhibited the binding of $[^3H]PAF$ to isolated rabbit (Hwang et al., 1986d) or human platelet membranes (Hwang, 1988); lithium is about 10–20 times less potent than sodium. On the other hand, K^+, Mg^{2+} and Ca^{2+} potentiated the specific $[^3H]PAF$ binding to platelet membranes (Hwang et al., 1986d; Hwang, 1988). From both Scatchard and Klotz analyses, the inhibitory effect of Na^+ is apparently due to an increase in the K_d value of PAF binding to its receptors. In contrast, the Mg^{2+}-induced enhancement of specific PAF binding arises from both an increased affinity of the receptor and an increased number of receptor sites (Hwang et al., 1986d). The variation in the detectable receptor number under different ionic conditions has been

attributed to the coexistence of several conformational states of PAF receptors (Hwang et al., 1989; Hwang and Lam, 1991). In human PMN membranes, K^+, Mg^{2+} and Ca^{2+} potentiated the specific $[^3H]PAF$ binding, whereas Na^+ and Li^+ showed no effects on the specific $[^3H]PAF$ binding even at concentrations up to 300 mM (Hwang, 1988). In human monocyte membranes, as in human platelet membranes, both Na^+ and Li^+ inhibited the binding of $[^3H]PAF$ but, unlike in human platelet membranes, Na^+ and Li^+ showed identical potency in inhibiting the specific $[^3H]PAF$ binding (Hwang et al., 1991). Although both K^+ and Mg^{2+} potentiated the specific $[^3H]PAF$ binding to all three cell types, the potentiated effects of K^+ and Mg^{2+} on the $[^3H]PAF$ binding are not synergistic. The effects of K^+ and Mg^{2+}-potentiated $[^3H]PAF$ binding are also different in the above three cell types. K^+ showed no effects on the Mg^{2+}-potentiated $[^3H]PAF$ binding to human platelet and monocyte membranes, whereas K^+ inhibited the Mg^{2+}-potentiated $[^3H]PAF$ binding to human PMN membranes (Hwang, 1988; Hwang et al., 1991). Since ionic modulation of the specific $[^3H]PAF$ binding is presumably due to the binding of ions to the nearby regulatory protein, differences in the regulation of $[^3H]PAF$ binding may indicate differences in the signal transduction mechanism of PAF receptors from different sources.

The relative potencies of several selected PAF agonists and PAF receptor antagonists for human platelet, human PMN, and human monocyte membranes were also compared. Two receptor antagonists (Ono-6240 and FR 72112) were found to have significant differences in potency in the binding assay as well as in functional studies (Hwang, 1988, 1990). Ono-6240 is about six times more potent in human platelet ($K_i = 4.6 \times 10^{-8}$ M) than in human PMN membranes ($K_i = 8.0 \times 10^{-7}$ M). In contrast, FR-72112 is more potent in human PMN membranes than in human platelet membranes in inhibiting the specific $[^3H]C_{16}$-PAF binding. In indirect Hill plots, FR 72112 has an IC_{50} value of 1 μM ($K_i = 0.32$ μM) in inhibiting the $[^3H]C_{16}$-PAF (1 nM) binding to human PMN membranes, a value about four times lower than that in human platelet membranes ($IC_{50} = 3.6$ μM, $K_i = 9.51 \times 10^{-7}$ M). At the cellular level, Ono-6240 is also found to be less potent in inhibiting the PAF-induced aggregation of human PMNs than human platelets (Hwang, 1988). A linear Schild plot with a unit slope was observed for the inhibition of PAF-induced aggregation of rabbit platelets, human platelets, and human PMNs by Ono-6240. While the inhibitory potency of Ono-6240 towards PAF-induced aggregation of gel-filtered human platelets was identical to that of rabbit platelets ($K_b = 3.4 \times 10^{-7}$ M), Ono-6240 is significantly less potent in inhibiting PAF-induced human PMN aggregation. The K_b value of Ono-6240 for inhibition of PAF-induced PMN aggregation was 2×10^{-6} M, which is about six-fold higher than the corresponding value for platelets (Hwang, 1988). BN 52111 and BN 52115 were also reported to be more active on human PMNs than on platelets (Hosford *et al.*, 1990). Recently, we found two compounds showing similar potencies in human platelet and human PMN membranes, but they had different potencies in human monocyte membranes (Hwang *et al.*, 1991). CV-6209 is less potent but 52770 RP is more potent in inhibiting the specific [^{3}H]PAF binding in human monocyte membranes than in human platelet and human PMN membranes. Zinc chloride has also been demonstrated to specifically inhibit the binding of [^{3}H]PAF to human platelets and PAF-induced aggregation in human platelets (Nunez *et al.*, 1989). However, it showed no significant inhibition of the specific [^{3}H]C$_{18}$-PAF binding to either human PMN or human monocyte membranes at those zinc chloride concentrations where a significant inhibition can be demonstrated in human platelet membranes (Hwang and Wang, 1990; Hwang *et al.*, 1991). Receptor heterogeneity is therefore clearly demonstrated by the differences in (1) the K_d values of [^{3}H]N-methylcarbamyl-PAF, (2) the K_i values of the PAF receptor antagonists Ono-6240 and FR 72112 between human platelet and human PMN membranes, and the K_i values of CV-6209 and 52770 RP between human monocyte membranes and membranes from either human PMNs or human platelets, and (3) the ionic modulation of the specific [^{3}H]PAF binding to PAF receptors in human platelet, PMN and monocyte membranes. These results further confirm the differences of PAF receptors within these three cell types.

A difference in the potency of L-652,731 in inhibiting specific [^{3}H]PAF binding was also found between rat PMNs and liver (Hwang, 1991a). L-652,731 was about 10 times more potent in rat liver than in PMNs. A difference in PAF receptors between rat PMNs and liver tissues can be expected if one adopts the guidelines suggested by Furchgott (1972) that affinity constants for antagonists differing by more than three-fold should be considered as evidence for differences in receptors.

Numao *et al.* (1989) presented evidence for the possible existence of PAF receptor subtypes in human eosinophils and neutrophils. CV-6209 inhibited PAF-induced chemotaxis of eosinophils and neutrophils in a dose-dependent manner with IC$_{50}$ values of 1.1 ($\pm$ 0.4) $\times$ 10^{-7} M and 1.9 ($\pm$ 0.3) $\times$ 10^{-6} M, respectively. The IC$_{50}$ values for WEB 2086 were 6.6 ($\pm$ 0.3) $\times$ 10^{-7} M for eosinophils and 5.0 ($\pm$ 1.0) $\times$ 10^{-8} M for neutrophils. Both results support the existence of PAF receptor subtypes in human eosinophils and neutrophils. Since WEB 2086 shows no difference in inhibiting [^{3}H]C$_{16}$-PAF binding to either human platelet or human neutrophil membranes (Hwang, 1988), the PAF receptor in human platelets may also differ from that in human eosinophils.

Differences in rank orders of potency of PAF and PAF structural analogues in different cell types from the same species have been reported by Hayashi *et al.* (1985) and Levi *et al.* (1989). Three PAF analogues, 1-*O*-octadecyl-2 - *O* - (*N*,*N*-dimethylcarbamyl) - *sn* - glycero - 3 - phosphocholine (compound (188)), 1-*O*-octadecyl-2-acetamido-2-deoxy-*sn*-glycero-3-phosphocholine (compound (189)) and 1-*O*-octadecyl-2-*O*-methyl-*sn*-glycero-3-phosphocholine (compound (3)), showed higher activity in stimulating macrophage function than PAF. However, PAF is more potent than these three PAF analogues in stimulating platelet function (Hayashi *et al.*, 1985). On the isolated guinea-pig heart, the rank order of potency for the negative inotropic effect was C$_{16}$-PAF > C$_{18,1}$-PAF > semisynthetic beef heart-derived PAF > C$_{15}$-PAF > C$_{18}$-PAF > C$_{14}$-PAF, whereas the rank order for the coronary-vasoconstricting effect was C$_{16}$-PAF = C$_{18,1}$-PAF = semisynthetic beef heart-derived PAF > C$_{15}$-PAF > C$_{18}$-PAF = C$_{14}$-PAF (Levi *et al.*, 1989). With the exception of C$_{16}$-PAF and C$_{18,1}$-PAF, the relative potencies for the cardiac and coronary effects of these PAF analogues did not correlate well with their relative potencies in stimulating rabbit platelets and human neutrophils (Levi *et al.*, 1989). These results further demonstrated the possible existence of PAF receptor subtypes.

The existence of PAF receptor subtypes as described above is indicated by observed potency differences of either agonists or receptor antagonists in different tissues. Now that the PAF receptor from guinea-pig lung has been cloned (Honda *et al.*, 1991), the subtypes present in each tissue can be identified. The genes can presumably be transfected into immortal cell lines which previously lacked any member of the receptor family; thus, the properties of the individual subtypes of PAF receptors can be investigated in isolation. Characterization of individual receptor subtypes may need to wait until each individual receptor subtype gene is cloned.

9. Solubilization and Identification of Platelet-Activating Factor Receptors

The PAF receptor of rabbit platelet membranes is heat-labile and sensitive to protease (Hwang *et al.*, 1983). In addition, there is no correlation between the effect of PAF on platelet aggregation and several physical properties of pure dipalmitoylphosphatidycholine bilayers as detected by differential scanning calorimetry (Hwang *et al.*, 1983). These results suggest that PAF-binding sites are not phospholipids but are likely to be membrane-bound proteins.

A PAF receptor complex has been solubilized with 1% digitonin from rabbit platelet membranes with an apparent molecular mass of 220 kDa (Chau and Jii, 1988). Binding of [^{3}H]PAF to the solubilized receptor complex can be fully blocked by unlabelled PAF or by a PAF receptor antagonist, L-652,731. Dissociation of [^{3}H]PAF from the radioligand receptor complex was facilitated by Na$^+$ and Li$^+$, whereas K$^+$ and Cs$^+$ were ineffective. GTP was also found to synergize with Na$^+$-induced dissociation of [³H]PAF from the solubilized PAF receptor complex. These observations led to the speculation that G protein may coexist with the PAF receptor within this solubilized PAF receptor complex. Using a photoaffinity probe, 1-*O*-(4-azido-2-hydroxy-3-iodobenzamido)undecyl-2-*O*-acetyl-*sn*-glycero-3-phosphocholine, Chau *et al.* (1989) further demonstrated that the PAF receptor has a molecular mass of 52 kDa. However, no binding affinity for PAF in the solubilized receptor complex has been reported. In the past few years, I have also screened through a variety of detergent systems, and found that PAF receptors solubilized with the detergent lauryl dodecylamine oxide retained their high affinity for PAF. Binding of [³H]C$_{18}$-PAF to solubilized PAF receptors, as analysed in a Scatchard plot, is linear. The K_d value of solubilized receptor for PAF is roughly the same as that in isolated membranes. From several experiments, the solubilized receptor from human platelet membranes had a K_d value of 2.02 ± 0.48 nM, which is not significantly different from the K_d value of PAF receptor binding in isolated membranes for [³H]C$_{18}$-PAF(2.32 ± 0.27 nM) (Hwang, 1993).

Two auto-anti-idiotypic monoclonal antibodies to PAF were generated by immunizing mice with an aldehydic PAF analogue–thyroglobulin conjugate and by a hybridoma technique (Wang and Tai, 1991). These monoclonal antibodies were able to cause the dissociation of the idiotypic antibodies–[³H]PAF complex. The monoclonal antibodies could also compete with [³H]PAF binding to PAF receptors in rabbit platelet membranes. They also mimic PAF agonistic activity. Both could stimulate the aggregation of rabbit platelets in a dose-dependent manner and this aggregation could be inhibited by PAF receptor antagonists such as SDZ 63-441. These antibodies could serve as immunoreagents to purify PAF receptors.

10. *The Cloned Platelet-Activating Factor Receptor*

Researchers at the University of Tokyo in Japan have cloned the PAF receptor of guinea-pig lung using the gene expression system in *Xenopus* oocytes (Honda *et al.*, 1991). Murphy *et al.* (1990) also applied the same method and demonstrated PAF receptor activity in *Xenopus* oocytes injected with mRNA from promyelocytic leukaemia cell line HL-60 differentiated with dibutyryl-cAMP. However, no amino acid sequences were reported. In the guinea-pig, the cloned PAF receptor cDNA contains 3020 nucleotide base pairs. The open-reading frame (1026 base pairs) codes for a protein of 342 amino acids. The calculated molecular mass based on the deduced amino acid sequences is 38 982, which is smaller than that detected by SDS–PAGE in rabbit platelets (52 000) (Chau *et al.*, 1989) or in human platelets (65 000) (Shen *et al.*, 1989). This could be due to a

difference in subtype, the presence of the carbohydrate residues in the native receptor or a combination of both. Northern blot analysis showed that leucocytes, lung, spleen and kidney were rich in the receptor mRNA.

Hydropathicity analysis of the cloned primary structure revealed the existence of seven hydrophobic, putative transmembrane α helices characteristic of G protein-coupled receptors. Several amino acids are highly conserved. The aspartic acid residue of the second transmembrane helix, the cysteine residues on the second and third extracellular loops, and the three proline residues in the sixth and the seventh transmembrane helices, may contribute to the ligand-building pocket, which has been described for β adrenoceptors (Strader *et al.*, 1989). Two cysteine residues are in the second and third extracellular loops, which may make a disulphide bond, and three proline residues are in the sixth and seventh transmembrane segments. This conservative intramolecular disulphide bond may make the PAF receptor, similar to other G protein-coupled receptors (Malbon *et al.*, 1987), sensitive to thiol compounds (Sugatani *et al.*, 1987; Ng and Wong, 1989). As in the rhodospin (Nathans and Hogness, 1984), substance K (Masu *et al.*, 1991) and TXA$_2$ receptors (Hirata *et al.*, 1991), there is no aspartic acid residue in the third transmembrane domain of the PAF receptor, a residue which is presumed to bind the amino group of the ligand in adrenergic receptors (Strader *et al.*, 1989). The cytoplasmic tail of the PAF receptor contains four serine residues and five threonine residues as possible phosphate acceptors. Phosphorylation of these residues could result in rapid receptor desensitization (Dohlman *et al.*, 1987). The PAF receptor is strongly desensitized by the application of the ligand (see below) and in oocytes injected with poly(A)$^+$ RNA (Honda *et al.*, 1991).

11. *Signal Transduction Mechanisms*

G proteins are a family of proteins involved in a variety of functions at the cell surface membranes (Stryer and Bourne, 1986; Gilman, 1987). A number of G proteins have been identified and purified. These include: G$_s$ and G$_i$, the G proteins associated with the stimulation and inhibition, respectively, of adenylate cyclase; transducin, the G protein coupling rhodopsin to cGMP phosphodiesterase in rod photoreceptors; and a number of G proteins, including G$_k$, the protein that stimulates K$^+$ channels in a variety of cells and G$_o$, a G protein that is highly abundant in brain. G proteins also participate in other signal transduction pathways, notably those involving phosphoinositide breakdown and de-acylation of phospholipids. They are all heterotrimeric in nature, consisting of α, β and γ subunits. G proteins undergo a cyclic dissociation and reassociation which is superimposed upon a cycle of guanine nucleotide exchange and hydrolysis. In the basal state, G protein is a heterotrimer with GDP bound to the α subunit. Receptors coupled to G proteins show high affinity to ligands. Dissociation of GDP from the α subunit is enhanced by the interaction of the oligomer with an agonist-bound receptor. Subsequent binding of GTP to the α subunit causes a conformational change of the polypeptide that results in βγ subunit dissociation. The G protein-dissociated receptor exhibits low affinity for the ligand. The GTP-bound α subunit then interacts with effectors until the system is deactivated by hydrolysis of GTP to GDP. The α subunit is unique for each G protein and may determine the specificity of a given G protein for receptor and effector coupling. The size of the α subunit varies from 39 to 52 kDa. The α subunit contains the GTP-binding site, where GTP is hydrolysed by terminal phosphatase activity. The α subunit also contains the acceptor sites for ADP ribosylation by cholera toxin and pertussin toxin. Many G protein α subunits are myristoylated (Buss *et al.*, 1987; Schultz *et al.*, 1987; Jones *et al.*, 1990b; Mumby *et al.*, 1990a). Myristoylation increases the affinity of α subunits for βγ and may thus facilitate formation of the heterotrimer and the localization of α subunits to the plasma membranes (Linder *et al.*, 1991). The β subunit (35 kDa) is very similar for all of the G proteins. Phosphorylation of a G protein β subunit in yeasts has been shown to be associated with an adaptive response to mating pheromone (Cole and Reed, 1991). The γ subunit (5–7 kDa) appears to be very tightly associated with the β subunit and can be separated only under denaturating conditions. The γ subunit is prenylated (Mumby *et al.*, 1990b; Yamane *et al.*, 1990; Fukada *et al.*, 1990) and may serve as the membrane anchor for the G proteins. Ligands bound to receptors cause the dissociation of α subunits from βγ subunits.

The signal transduction mechanism induced by PAF appears to be mediated by G proteins. GTP modulates the binding of PAF to rabbit platelet membranes (Hwang *et al.*, 1986d), human PMN membranes (Ng and Wong, 1986) as well as guinea-pig lung membranes (Votta and Mong, 1990). PAF stimulated GTPase activity in human (Avdonin *et al.*, 1985; Hwang, 1988) and rabbit platelets (Hwang *et al.*, 1986d) and in human PMNs (Hwang, 1988) with an ED_{50} value near 1 nM, whereas the biologically inactive enantiomer of PAF showed no stimulatory effect even at micromolar concentrations. The activated GTPase activity can be specifically inhibited by the PAF receptor antagonist kadsurenone. However, the inactive analogue kadsurin B showed no inhibitory effect at concentrations where kadsurenone showed significant inhibition (Hwang *et al.*, 1986d). These results suggest that PAF-induced GTPase activity is a receptor-mediated process. The primary sequence of cloned PAF receptors also has a structure consisting of seven transmembrane segments with an extracellular amino terminus and a cytoplasmic carboxyl terminus (Honda *et al.*, 1991), a specific feature for those receptors coupled to G proteins. This further confirms that PAF receptors belong to the family of receptors coupled to G proteins.

Table 9.9 Amino acid sequences of the cytoplasmic loop between transmembrane segment I and II

Cloned receptor	Amino acid sequences
PAF receptor	Y P S K K L N E I K I F
TXA$_2$ receptor	Q G G S H T R S S F L
β$_1$ adrenoreceptor	G R T Q R L Q T L T N
Porcine M$_1$	K V N T E L K T V N N
β$_2$ adrenoreceptor	A K F E R L Q T V T N
Bovine rhodopsin	A T T K S L R T P A N
Drosophila opsin	A T T K S L R T P A N
Human colour pigment opsin	X X X K K L R X P L N
α$_2$ adrenoreceptor	A L R A P Q N
Substance P	H K R M R T V T N

The cytoplasmic loop between transmembrane segments I and II has been suggested as a possible site for interaction with G proteins due to the similarity of the eight amino acids on the C-terminal side of this loop in the cloned β-adrenergic receptor, the muscarinic acetylcholine receptor, and rhodopsin (Hall, 1987). Asparagine (N) and leucine (L) are identical in all sequences and threonine (T) is retained in all but the human colour pigment opsins. Even though leucine is retained in the cytoplasmic loop between segments I and II of the cloned PAF receptor, asparagine and threonine are not present (Table 9.9). Instead Lys-Lys-Leu residues are homologous to the rhodopsin molecule and the human colour pigment opsin. Recent studies from *in vitro* site-specific mutagenesis (Dohlman *et al.*, 1987; Franke *et al.*, 1988; Strader *et al.*, 1989) and from chimeric receptors (Wong *et al.*, 1990) suggest that the third cytoplasmic loop between segments V and VI may be involved in the regulation of G proteins. The length and amino acid sequence of this loop is not well conserved in the family of homologous receptors that couple to G proteins. Proteolytic digestion of the large third intracellular loop of rhodopsin was first demonstrated to abolish its interaction with transducin (Findlay and Pappin, 1986). Deletion mutagenesis analysis of the hamster β adrenoreceptor further demonstrated that regions at both the amino and carboxyl ends of the putative third intracellular loop of the protein were critical for the coupling of β adrenoreceptors to the G$_s$ protein and the subsequent activation of adenylate cyclase (Dixon *et al.*, 1987). The use of synthetic peptides from the α$_2$-adrenergic receptor to probe the functional regions of that receptor provides further evidence that the distal portion of the third cytoplasmic loop of the α$_2$-adrenergic receptor is important for coupling to the G$_i$ protein (Dalman and Neubig, 1991). The cloned PAF receptor contains only 20 amino acids at the third intracellular loop. Whether this short third intracellular sequence is sufficient to interact with the G protein is still an open question.

G proteins coupled to PAF receptors are not well characterized. However, with the help of cholera toxin

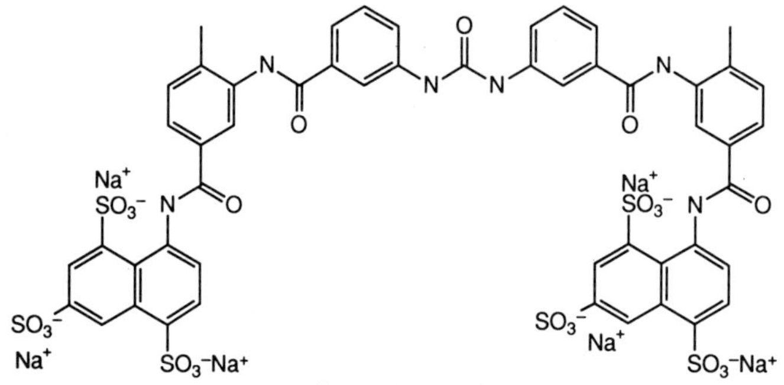

Suramin (190)

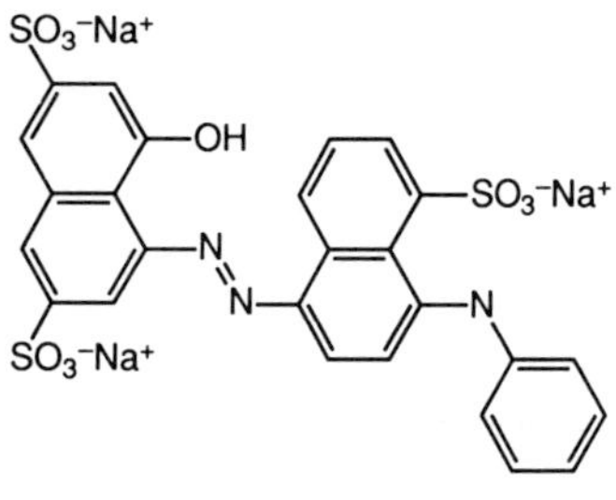

L-451,167 (191)

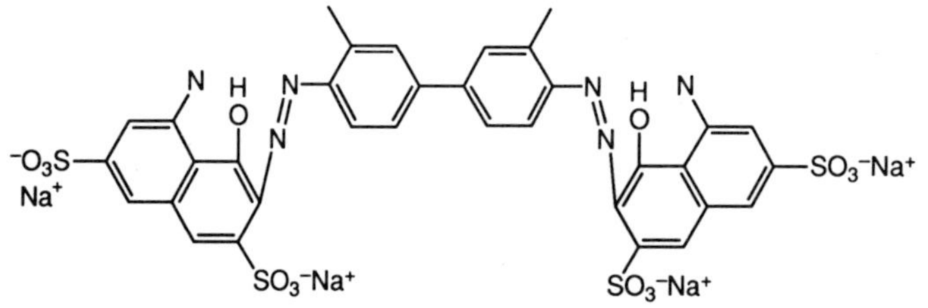

Trypan Blue (192)

Figure 9.25 Structures of polyanions, suramin, trypan blue and L-451,167.

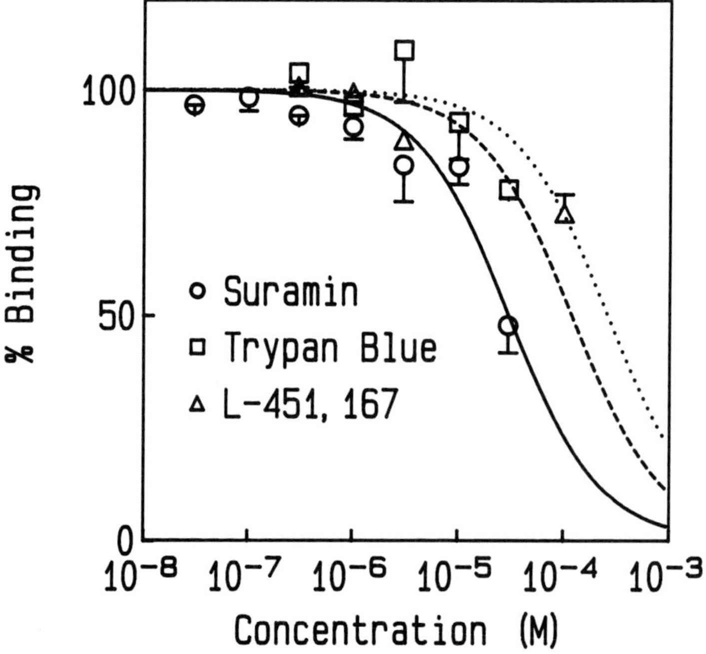

Figure 9.26 Inhibition of [³H]C₁₈-PAF binding to human platelet membranes by suramin, trypan blue and L-451,167. Membrane protein (100 μg) was added to an incubation mixture (1 ml) containing 0.3 nM [³H]C₁₈-PAF and a known amount of inhibitor in a medium of 10 mM magnesium chloride, 10 mM Tris and 0.25% BSA at pH 7.0. The compounds were dissolved in DMSO. The data points and error bars are the means and the standard deviations of three independent experiments. In each experiment, triplicate samples were prepared.

and pertussin toxin, we have demonstrated that the G-protein coupled to the PAF receptor in human platelets is different from that in human PMNs (Hwang, 1988). Pertussis toxin and cholera toxin totally abolished PAF-stimulated GTPase activity in human neutrophils, whereas neither toxin has an effect on human platelets. Similar results on the sensitivity of G proteins coupled to PAF receptors in human platelets and PMNs had been reported previously (Lad *et al.*, 1985; Houslay *et al.*, 1986).

A series of polyanionic compounds, including suramin, trypan blue and L-451,167 (Fig. 9.25), have been shown to inhibit the coupling of the α_2-adrenergic receptors and the β_2-adrenergic receptors to G proteins at submicromolar concentrations (Huang *et al.*, 1989). However, much higher concentrations are required to inhibit the binding of [^{3}H]C_{18}-PAF to human platelet membranes for suramin, trypan blue and L-451,167 with ED_{50} values of 30, 120 and 261 μM, respectively (Fig. 9.26). These values are about 40-fold higher than those seen in the [^{3}H]UK 14304 binding to the α-adrenergic receptors (Huang *et al.*, 1989) or on the [^{3}H]IP$_3$ binding to human platelet membranes (Hwang, 1991b). These results suggest that G proteins coupled to PAF receptors may be different from those coupled to either α_2- or β_2-adrenergic receptors.

PAF induced transient elevation of intracellular Ca^{2+} in several cell types, including rabbit and human platelets (Avdonin *et al.*, 1987; Valone and Johnson, 1985; Yamada *et al.*, 1988), human and rabbit neutrophils (Naccache *et al.*, 1985; Molski *et al.*, 1988; O'Flaherty and Rossi, 1989), human peripheral blood mononuclear leucocytes (Ng and Wong, 1989), neural cells (Kornecki and Ehrlich, 1988) and human vascular endothelial cells (Hirafuji *et al.*, 1988). The PAF-induced calcium responses appear to be involved in both intracellular calcium mobilization and extracellular calcium entry. Since PAF-induced elevation of intracellular calcium can be blocked by PAF-specific receptor antagonists (Kornecki and Ehrlich, 1988; Ng and Wong, 1989), the transient elevation of intracellular Ca^{2+} is a receptor-mediated process. Kinetic studies of changes in the elevation of intracellular Ca^{2+} in human platelets have indicated that PAF evoked a rise in intracellular Ca^{2+} which was delayed in onset by 200–400 ms (Sage and Rink, 1987). These delays are sufficient for one or more biochemical steps to intercede between ligand receptor binding and calcium influx generation (Sage and Rink, 1987). It is now thought that activation of phospholipase C upon stimulation of cells with ligands is responsible for this receptor-mediated elevation of Ca^{2+} (Putney, 1986). The phosphodiesteratic cleavage of PIP_2 yields the two second-messenger molecules IP_3 and DAG (Berridge and Irvine, 1984). IP_3 has been demonstrated in many cell types to release intracellular Ca^{2+} stores (Berridge and Irvine, 1984; O'Rouke *et al.*, 1985). PAF has been reported to stimulate phosphatidylinositol turnover in

horse platelets (Lapetina, 1982; Billah and Lapetina, 1983), rabbit platelets (Morrison and Shukla, 1988, 1989), human platelets (MacIntyre and Pollock, 1983; Mauco *et al.*, 1983; Shukla, 1985; Kloprogge *et al.*, 1986), human PMNs (Verghese *et al.*, 1987; O'Flaherty and Rossi, 1989), rat hepatocytes (Shukla *et al.*, 1983; Fisher *et al.*, 1984) and primary cultures of rat liver Kupffer cells (Gandhi *et al.*, 1990). Specific IP_3 receptor-binding sites have been identified in guinea-pig hepatocytes (Spat *et al.*, 1986), rabbit neutrophils (Spat *et al.*, 1986) and human platelets (Hwang, 1991b). The released IP_3 could therefore bind to the specific receptor and induce the intracellular mobilization of Ca^{2+}. The exact mechanism regulating the external Ca^{2+} entry is still obscure; however, a receptor-mediated calcium channel (Avodonin *et al.*, 1987; Sage *et al.*, 1989) or a direct pathway for external Ca^{2+} between the plasma membrane and endoplasmic reticulum (Merritt and Rink, 1987) has been proposed to regulate the external calcium entry.

PAF (Doebber and Wu, 1987) and *N*-methylcarbamyl-PAF (Tessner *et al.*, 1989) stimulated the synthesis of PAF in human PMNs. Release of [^{3}H]arachidonic acid from human PMNs in response to PAF stimulation has also been reported and arachidonic acid release can be dose-dependently blocked by a PAF receptor antagonist, Ono-6240 (Nakashima *et al.*, 1989). Intracellular calcium mobilization in the same cells was stimulated by PAF at concentrations as low as 80 pM. Therefore, PAF receptors in PMNs may be coupled to a calcium-dependent phospholipase A_2. The G protein involved in the activation of phospholipase A_2 is pertussis toxin-sensitive (Nakashima *et al.*, 1989). On the other hand, PAF was unable to stimulate phospholipase A_2 in human platelets (Crouch and Lapetina, 1988). Therefore the signal transduction mechanisms appear to be different between human platelets and neutrophils, in addition to the differences in PAF receptors and G proteins coupled to PAF receptors as discussed above. The PAF receptors in human platelets are coupled to both adenylate cyclase and phospholipase C but not phospholipase A_2. On the other hand, PAF receptors in human neutrophils may be coupled to phospholipase C and phospholipase A_2, which has been demonstrated to be activated by phospholipase A_2-specific G protein (Axelrod *et al.*, 1988).

The signal transduction mechanism in macrophages may also be unique. In human peritoneal macrophages, PAF stimulated adenylate cyclase, which could be reversed by pretreatment with BN 52021 (Adolfs *et al.*, 1989). In guinea-pig alveolar macrophages, PAF increased the liberation of arachidonic acid, which could be counteracted by compounds that augment intracellular cAMP (Bachelet *et al.*, 1988). Therefore, the cAMP levels modulated the PAF-induced release of arachidonic acid. However, a recent report by Beusenberg *et al.* (1991) concluded that both the cyclooxygenase and lipoxygenase products of the released arachidonic acid may modulate the cAMP levels in guinea-pig alveolar macrophages.

More work is needed to clarify the signal transduction mechanism in macrophages.

In primary cultures of rat Kupffer cells, two distinct pathways of PAF-induced hydrolysis of phosphoinositides were also demonstrated (Gandhi et al., 1990). One was phospholipase C-mediated hydrolysis of phosphoinositides, which was not dependent on extracellular Ca^{2+}, but was inhibited by the intracellular Ca^{2+} antagonist 3,4,5-trimethoxybenxoic acid 8-(diethylamino)octylester (TMB-8). The other was characterized by de-acylation of phosphatidylinositol, leading to the accumulation of glycerophosphoinositol. PAF-induced synthesis of glycerophosphoinositol was not inhibited by TMB-8. Both cholera toxin and pertussis toxin inhibited PAF-induced phospholipase C activity. Pertussis toxin also inhibited PAF-stimulated de-acylation. However, cholera toxin itself stimulated release of glycerophosphoinositol. Addition of PAF to the cholera toxin-treated cells caused a further increase in glycerophosphoinositol release. Therefore, PAF, like other stimuli such as thrombin (Grandt et al., 1986; Houslay et al., 1986) or serotonin (Ashkenazi et al., 1987), may stimulate two or more distinct G proteins.

Cells are highly regulated in responding to their immediate intracellular environment as well as to extra-cellular stimuli. Most regulation is mediated either directly or indirectly by conformational changes in protein. Interchanges between active and inactive conformational states can be altered by both allosteric and covalent mechanisms. One of the most common covalent means for the regulation of protein activity is by phosophorylation. Protein phosphorylation may also be involved in the early events in the signal transduction pathway for the stimulation of cells by PAF. Pretreatment of ^{32}P-labelled rabbit platelets with PAF caused a time- and dose-dependent phosphorylation of several proteins, including 20, 35, 40, 50, 60 and 150 kDa proteins (Ieyasu et al., 1982; Lapetina and Siegel, 1983; Kloprogge et al., 1986b; Sugatani and Hanahan, 1986; Shukla et al., 1989; Dhar et al., 1990; Siess and Lapetina, 1990). The level of ^{32}P incorporation into these proteins showed a maximum at 10 s and decreased thereafter to an elevated basal level by 5 min, indicating the occurrence of a phosphorylation and dephosphorylation process. PAF receptor antagonists, including CV-3988, CV-6209, SDZ 63-441 and SDZ 63-675, dramatically reduced the PAF-stimulated phosphorylation of the 20, 40, 60 and 150 kDa proteins (Shukla et al., 1989). These four PAF receptor antagonists have been observed to inhibit PAF-stimulated aggregation of platelets and specific [^{3}H]PAF binding (Saunders and Handley, 1987; Terashita et al., 1987). Antagonists that inhibited PAF receptor binding also inhibited phosphorylation. Phosphorylation of proteins by PAF is therefore mediated through the PAF receptor (Sugatani and Hanahan, 1986; Shukla et al., 1989).

The 20 kDa protein, myosin light chain, is the substrate of myosin light-chain kinase (Nishizuka, 1984). The 40 kDa protein is the substrate of protein kinase C (Nishizuka, 1984). The phosphorylation of this 40 kDa protein is insensitive to changes in the intracellular Ca^{2+} concentration, whereas an elevation of the intracellular Ca^{2+} is essential for the 20 kDa protein phosphorylation (Lyons et al., 1975; Yamanishi et al., 1983; Sugatani and Hanahan, 1986). Earlier studies by Nishizuka (1984) demonstrated that the phosphorylation of both the 40 kDa protein and the 20 kDa protein is an essential step in degranulation. Indeed, in PAF-stimulated rabbit platelets, the inhibition of protein phosphorylation was closely correlated with the inhibition of serotonin release by K-252a, an inhibitor of protein kinase C (Yamada et al., 1988).

The 50 kDa protein had been demonstrated to be tyrosine phosphorylated in rabbit platelets stimulated with PAF (Dhar et al., 1990). However, phosphorylation of this 50 kDa protein was cAMP-dependent in human platelets stimulated with thrombin (Siess and Lapetina, 1990). One protein with a molecular mass of 50 kDa has been purified from platelet membranes (Halbrugge and Walter, 1989). This protein could be phosphorylated in vitro by cAMP and cGMP (Waldmann et al., 1984), and the phosphorylation was reversible (Halbrugge et al., 1990). Identification of this 50 kDa protein is required to further elucidate its role in the mechanism of signal transduction.

The 60 kDa protein is the gene product p60^{c-src}. The translocation of activated p60^{c-src} to the plasma membrane may be involved in the signal transduction pathway of cells. Association of p60^{src} with the plasma membrane is thought to be mediated by a membrane-bound receptor protein (Resh, 1989; Goddard et al., 1989). Interaction of p60^{src} with the receptor is dependent on amino-terminal p60^{src} sequences. It is also dependent upon N-myristoylation. A 32 kDa membrane protein has been identified as a component of the receptor for p60^{src} (Resh and Ling, 1990). The transforming active src gene product has been shown to be associated with the Triton X-100-resistant cytoskeletal structure in chicken fibro-blasts, while the inactive src proteins, including p60^{c-src}, were only found in soluble fractions (Umezawa et al., 1986). Therefore, p60^{src} tyrosine kinase activity may be involved in the mechanism of cytoskeletal organization (Umezawa et al., 1991).

The 60 kDa protein p60^{src} is tyrosine phosphorylated upon stimulation with PAF in rabbit platelets (Dhar et al., 1990). p60^{src}, together with another 50 kDa protein, reacted with the monoclonal antibody of phosphotyrosine residues. Genistein, a specific inhibitor for tyrosine-specific protein kinase (Akiyama et al., 1987), inhibited PAF-induced aggregation of rabbit platelets and PAF-stimulated phosphorylation of tyrosine residues (Dhar et al., 1990). Monoclonal antibody for phosphotyrosine residues did not react with the 20 and 40 kDa proteins, because the tyrosine residues of these two proteins were not phosphorylated (Ganguly et al., 1985; Nishizuka, 1984). However, genistein did decrease the PAF-stimulated phosphorylation of the 20, 40 and 50 kDa proteins (Dhar *et al.*, 1990). These results suggest that the activation of tyrosine kinase induced by PAF is an earlier event, in comparison to those steps involved in the phosphorylation of the 20 and 40 kDa proteins. PAF alone caused stimulation of phospholipase C activity in rabbit platelets, whereas pretreatment with genistein diminished PAF-stimulated phospholipase C activity to basal levels. Genistein did not inhibit the binding of [^{3}H]PAF to platelets. The inhibition of phospholipase C activity by genistein raised the possibility of a tyrosine kinase involvement in PAF receptor-coupled phospholipase C activation (Dhar *et al.*, 1990). Tyrosine kinase-dependent activation of phospholipase C has also been demonstrated to be involved in signal transduction by receptors with tyrosine kinase activity (Ullrich and Schlessinger, 1990; Finkel *et al.*, 1991). Phospholipase C_γ is an excellent substrate for those receptors with tyrosine kinase activity. Tyrosine phosphorylation in intact cells, or *in vitro*, increased the catalytic activity of phospholipase C_γ (Nishibe *et al.*, 1990). Tyrosine phosphorylation presumably acts as a switch to induce the binding of phospholipase C to tyrosine phosphorylated proteins (Koch *et al.*, 1991). The formation of these heteromeric protein complexes at or near the plasma membrane is likely to control the activation of signal transduction pathways by tyrosine kinases (Koch *et al.*, 1991). p60^{c-src} is abundantly available in platelets (Golden *et al.*, 1986). Whether p60^{c-src} is involved in the activation of phospholipase C in the PAF-induced signal transduction pathway requires further confirmation.

PAF-induced tyrosine phosphorylation has also been demonstrated in human neutrophils (Gomez-Cambronero *et al.*, 1991). The effects of PAF on the phosphorylation of several proteins were dose-dependent, could be seen at concentrations as low as 1 nM, and were prevented by the PAF receptor antagonist BN 52021. Thus, PAF-induced tyrosine phosphorylation is mediated through its receptors. Neutrophils treated with pertussis toxin showed a significant reduction in the tyrosine phosphorylation of several high molecular weight proteins (66, 104 and 116 kDa) but showed no effect on the phosphorylation of a 41 kDa protein. The phosphorylation of the high but not the low molecular-weight proteins is therefore mediated, either directly or indirectly, through a pertussis toxin-sensitive G-protein. The phosphorylation of the high and low molecular-weight proteins is probably mediated by two different sets of kinases and/ or phosphatases.

PAF at subnanomolar concentrations will induce platelet aggregation. However, the major links between PAF binding to its receptor at the platelet surface and platelet aggregation are still missing. It is now recognized that platelet aggregation is the result of fibrinogen binding to its specific binding sites that are not accessible in the resting platelets. Upon activation of platelets with PAF, two different classes of fibrinogen-binding sites are

exposed. One has a high affinity with a K_d value of 72 nM, and the other shows a low affinity with a K_d value of 0.59 μM (Kloprogge and Akkerman, 1986). Pretreatment of platelets with inhibitors of cyclooxygenase or thromboxane synthetase completely abolished high-affinity binding, leaving low-affinity binding unchanged. In contrast, ADP scavengers completely prevented low-affinity binding, leaving high-affinity binding unaltered (Kloprogge and Akkerman, 1986). Therefore, the interaction of PAF and the PAF receptor in platelets somehow triggers exposure of high-affinity fibrinogen binding via arachidonate metabolites, and low-affinity binding via secreted ADP. The velocity with which PAF occupies a certain number of PAF receptors rather than the number of occupied receptors controlled the exposure of fibrinogen-binding sites (Kloprogge *et al.*, 1986c). In the absence of fibrinogen, exposed fibrinogen-binding sites gradually disappear (Kloprogge *et al.*, 1986a). This disappearance is enhanced by metabolic inhibitors. Thus, metabolic energy appears to be required for mechanisms that regulate the exposure of the binding sites. On the other hand, the normal maintenance of the fibrinogen-binding site complex appears to be energy-independent (Kloprogge *et al.*, 1986a).

12. *Priming Effects of Platelet-Activating Factor*

Inflammation is the primary physiological response to tissue injury. Recent research suggests that complex interactions between PAF and various cytokines, such as TNF, interleukins and growth factors, may be involved in the immune and inflammatory responses. At extremely low concentrations, these mediators appear to be able to prime cells to respond in an exaggerated manner to subsequent stimuli, a mechanism which confers great sensitivity to the immuno-inflammatory processes. Studies with PMNs have shown that exposure to substimulatory concentrations of PAF induces a "primed" state such that exposure to a second agonist elicits markedly enhanced responses. PAF by itself induced very little respiratory burst activity in PMNs at concentrations less than 100 nM (Gay *et al.*, 1986). In contrast, PMNs incubated with as little as 0.01 nM PAF exhibited marked enhancement of superoxide anion production and oxygen consumption when stimulated with FMLP (Gay *et al.*, 1986; Dewald and Baggiolini, 1985; Ingraham *et al.*, 1982; Vercellotti *et al.*, 1988). The PAF-primed response could be elicited almost immediately after PAF was added to the cells; this response persisted for at least 60 min and was not reversed by cell washing. PAF also primed PMN for enhanced superoxide production in response to PMA but did not affect the response to opsonized zymosan (Gay *et al.*, 1986). The PAF-primed response in PMNs is not limited to respiratory burst activity. Lysosomal enzyme release and aggregation stimulated by

FMLP or PMA are increased in PAF-primed cells. Priming does not occur with lyso-PAF or with PAF in the presence of PAF receptor antagonists, indicating that the priming effect requires specific binding of PAF to membrane receptors (Vercellotti *et al.*, 1988). Priming is not observed if the order of presentation of stimuli is reversed. The intracellular changes induced by priming stimuli that allow for an augmented response have not been clearly defined.

Because physiological primers operate through membrane receptors, it is reasonable to hypothesize that priming is mediated through the subcellular pathways involved in stimulus–response coupling. Indeed, O'Flaherty *et al.* (1990a,b) have demonstrated that FMLP, LTB$_4$ and PAF stimulate PMNs to translocate protein kinase C from the cytosol to the plasmalemma, as judged by their ability to increase PMN binding of, and receptor numbers for, [^{3}H]phorbol dibutyrate ([^{3}H]PDB), which is independent of the intracellular Ca^{2+} transient mechanism. PAF stimulated increases in the reversible and saturable binding of [^{3}H]PDB. Lyso-PAF produced little change in [^{3}H]PDB binding. PMNs incubated with 1–100 nM PAF for 5–40 min had markedly enhanced [^{3}H]PDB binding responses to FMLP. Furthermore, a PAF receptor antagonist, WEB 2086, blocked responses to PAF without altering those to FMLP. PAF thus acted through its receptors to stimulate and prime protein kinase C translocation. PAF, FMLP and LTB$_4$ may stimulate different pools of protein kinase C. PMNs thus treated with PAF became desensitized to PAF while retaining sensitivity to FMLP. PAF has also been demonstrated to induce the translocation of the complement receptors CR1 and CR3 from the intracellular pools to the cell surface at concentrations as low as 10^{-12} M (Shalit *et al.*, 1988). Increased cell surface expression of CR1 and CR3 may play important roles in PMN phagocytosis and adherence, respectively (Medof *et al.*, 1982; Gallin, 1985). Increased expression of complement receptors may also play an important role in priming PMN oxidative metabolism.

The capacity for PAF to prime cell types other than PMNs has not been studied as extensively. Baggiolini *et al.* (1988) have reported that monocytes/macrophages are primed by PAF for enhanced superoxide release in response to FMLP. Picomolar concentrations of PAF prime alveolar macrophages for a two- to three-fold increase in stimulated TNF-α and IL-1 production (Dubois *et al.*, 1989).

Although FMLP showed no priming effect on PAF stimulation (Vercellotti *et al.*, 1988), other synthetic agonists do prime the effect of PAF. WGA potentiated the aggregation of rabbit and human platelets induced by PAF (Hwang and Wang, 1991). PAF induced aggregation of human and rabbit platelets with ED$_{50}$ values of around 4×10^{-9} and 9×10^{-10} M, respectively. PAF-induced aggregation was potentiated with a concentration of WGA below the threshold of platelet

aggregation (< 2 µg/ml in human platelets and < 50 µg/ml in rabbit platelets). The potentiation effect of WGA on PAF-induced aggregation was partly due to an increase in PAF potency (left-shifting of the dose–response curves) and partly due to an increase in the maximal aggregation responses induced by PAF. In the presence of WGA, the PAF concentrations needed to reach an equivalent aggregation of human and rabbit platelets were shifted to 2×10^{-9} and 2×10^{-10} M, respectively. The maximal aggregation responses were increased from 10 and 56% to 22 and 67% aggregation for human and rabbit platelets, respectively. WGA increases in the maximal aggregation responses of platelets induced by PAF indicated an increase in the number of functional PAF receptors. An increase in the PAF receptor number in the presence of WGA can also be demonstrated in the isolated membrane fragment (Hwang and Wang, 1991). A shift of the PAF dose–response curve to the left may correspond to PAF synthesis in platelets. Indeed, WGA activated the acetyltransferase activity and induced PAF synthesis in human platelets (Hwang and Wang, 1991). The synthesized PAF retained inside the cells may prime the subsequent PAF stimulation.

TNF and GM-CSF enhanced stimulatory effects of PAF and other inflammatory mediators toward neutrophils (Weisbart *et al.*, 1987; Naccache *et al.*, 1988; McColl *et al.*, 1991; Stewart *et al.*, 1991). Human neutrophils stimulated with PAF in the range 10^{-7} to 10^{-5} M released small quantities of 5-lipoxygenase products. Preincubation of neutrophils with GM-CSF enhanced the ability of PAF to stimulate leukotriene synthesis by increasing both arachidonic acid availability and 5-lipoxygenase activation in response to PAF (McColl *et al.*, 1991). FMLP is an ineffective stimulus for LTB$_4$ generation unless exogenous arachidonic acid is also present (Palmer and Salmon, 1983). TNF and GM-CSF prime human neutrophils for FMLP-induced increased superoxide generation and for increased PAF synthesis and LTB$_4$ release (Camussi *et al.*, 1989; Stewart *et al.*, 1991). This result suggests that GM-CSF overcomes the requirement for the exogenous arachidonic acid. Indeed, the biochemical response to TNF and GM-CSF is the mobilization of arachidonic acid (Atkinson *et al.*, 1990; McColl *et al.*, 1991). GM-CSF-stimulated PAF synthesis was dependent on a pertussis toxin-sensitive G protein (DeNichilo *et al.*, 1991). Activation of phospholipase A$_2$ is likely to explain the enormous increase in PAF levels in cytokine-primed neutrophils, since this enzyme is the initial step in the remodelling pathway of PAF biosynthesis (Snyder, 1989). In functional studies, GM-CSF stimulated superoxide generation from neutrophils with a time and dose relationship that paralleled PAF synthesis (DeNichilo *et al.*, 1991). The synthesized PAF and LTB$_4$ appear to be candidate mediators for the enhanced superoxide production in cytokine primed neutrophils in response to PAF or other inflammatory stimuli (Stewart *et al.*, 1991).

Bacterial LPS primes neutrophils, resulting in an enhanced release of superoxide ions upon subsequent stimulation with fixed immune complexes, the chemotactic peptide, FMLP or PMA (Guthrie *et al.*, 1984). LPS itself did not appear to act as a stimulus for the release of superoxide ions. Preincubation of cells and LPS at 37°C for at least 30 min was required to achieve a significant enhancement of stimulated superoxide anion release. The priming effect of LPS was inhibited at 0°C and did not require protein synthesis. Incubation of neutrophils with 100 ng/ml of LPS for 60 min resulted in a small but significant increase in intracellular PAF (Worthen *et al.*, 1988). The synthesized PAF was retained inside the cells (Lynch and Henson, 1986; Worthen *et al.*, 1988). The further stimulation of primed neutrophils with FMLP resulted in a marked increase in neutrophil PAF compared with non-LPS-treated cells (Worthen *et al.*, 1988). As discussed above, PAF enhanced the neutrophil oxidative response to other soluble stimuli (Dewald and Baggiolini, 1985; Gay *et al.*, 1986; Ingraham *et al.*, 1982). Therefore, LPS might induce PAF accumulation within the cell, which might enhance subsequent neutrophil responses.

LPS also primed P388D$_1$ macrophage-like cells to enhance PAF-stimulated release of arachidonic acid metabolites (Glaser *et al.*, 1990). LPS itself showed no ability to stimulate arachidonic acid metabolism. LPS priming is inhibited by actinomycin D, while primed PAF-stimulation of arachidonic acid metabolism is blocked by cycloheximide. These results suggest that some transcriptional events are involved in LPS priming whereas some translationally related protein turnover is involved in PAF stimulation. LPS-primed PAF stimulation was also inhibited by the phospholipase A$_2$ inhibitor analogue. This leads to the suggestion that activation of phospholipase A$_2$ is the event which is being stimulated by PAF in these LPS-primed cells, resulting in increased arachidonic acid availability.

Fibronectin molecules are adhesive proteins which act as biological organizers by holding cells in position and guiding their migration (Hynes, 1986). Specific fibronectin receptors are present on the plasma membrane of human PMNs (Pommier *et al.*, 1984) and in specific granules (Singer *et al.*, 1989). Treatment of PMNs with high concentrations of fibronectin does not appear to stimulate chemotactic activity or enzyme release. Fibronectin (up to 100 µg/ml) did not directly stimulate the PMN respiratory burst, suggesting that fibronectin is not a stimulus of PMN function (Stanislawski *et al.*, 1990). Low fibronectin concentrations (up to 25 µg/ml) caused a dose-dependent enhancement of the chemiluminescence induced by PAF and FMLP and also by PMA. Potentiation persisted after cell washing but disappeared when fibronectin was degraded with trypsin (Stanislawski *et al.*, 1990). The enhanced respiratory burst may therefore result from the interaction of intact fibronectin with PMNs, most likely through fibronectin

receptors. These results suggest that fibronectin can serve as a priming agent for the biological function of PMNs induced by soluble inflammatory stimuli, such as PAF and FMLP, even though the mechanism for this priming effect remains to be elucidated.

13. Desensitization

Desensitization is an effect opposite to the priming effects. Pretreatment of cells with an agonist that activates G protein-linked receptors usually results in a reduction in the sensitivity of the second response for that agonist. The receptor desensitization effect provides a mechanism of adaptation and limits the response of the cell in situations where high signal concentrations are present for long periods. The exact mechanism of receptor desensitization is not well determined but may be involved directly in receptor phosphorylation (Sibley *et al.*, 1987), receptor down-regulation (Scarpace and Abrass, 1982), G protein down-regulation (Milligan and Green, 1991) or in inactivation of the signal transduction mechanism (Galizzi *et al.*, 1987; Smith *et al.*, 1987).

Pretreatment of cells with PAF completely abolished any further response to PAF, a process called homologous desensitization, or response to any other stimuli, called heterologous desensitization (Benveniste *et al.*, 1972; Henson, 1976; O'Flaherty *et al.*, 1981; Schwertschlag and Whorton, 1988). The cytoplasmic tail of the recently cloned PAF receptor contains four serine residues and five threonine residues which could possibly serve as phosphate acceptors (Honda *et al.*, 1991). Phosphorylation of the PAF receptor could abolish the subsequent binding of PAF to its receptors and thus desensitize the subsequent responses. However, the whole question of direct PAF effects on platelet PAF receptors is controversial. Prior exposure of human platelets to PAF, followed by washing, decreases PAF binding to its receptors by more than 90% (Kloprogge and Akkerman, 1984; Valone *et al.*, 1982). Deactivation has been attributed either to diminished affinity (Chesney *et al.*, 1986) or to receptor number (Kloprogge and Akkerman, 1984). However, in rabbit platelets desensitized to PAF, no decrease in receptor numbers was described (Homma *et al.*, 1987) and the desensitization was presumed to result from uncoupling of the PAF receptor from GTP-binding protein rather than from changes in PAF receptor binding (Homma and Hanahan, 1988). This apparent discrepancy between human and rabbit platelets may arise from species differences, or result from persistent PAF binding to its receptor due to technical difficulties in those studies of human platelets.

The cAMP-dependent down-regulation of PAF receptors in Kupffer cells reported by Chao *et al.* (1990) seems to be convincing. The β-adrenergic agonist isoproterenol inhibited the cellular responses to PAF, including arachidonic acid release and eicosanoid production in rat Kupffer cells (Chao *et al.*, 1990). The inhibitory effect of isoprenaline on PAF-induced arachidonic acid release appears to be mediated by β_2-adrenergic subtype-specific receptors. Isoprenaline markedly enhanced intracellular cAMP levels. Most importantly, the inhibitory effect of the β-adrenergic agonist can be mimicked by a cAMP analogue, dibutyryl-cAMP. These results suggest that a cAMP-dependent mechanism coupled with β_2-adrenergic receptors may also have important regulatory effects on the PAF receptor and postreceptor signal transduction mechanisms for PAF in Kupffer cells. cAMP analogues (dibutyryl-cAMP and 8-bromo-cAMP) and forskolin induce a long-term down-regulation of the PAF receptor. Prolonged treatment of Chinese hamster fibroblasts with dibutyryl-cAMP caused a down-regulation of the β_2-adrenergic receptor. The mechanism for cAMP-mediated down-regulation of the β_2-adrenergic receptor had been shown to involve the phosphorylation of the receptor and the regulation of its mRNA level (Benovic *et al.*, 1986; Bouvier *et al.*, 1988, 1989). The activation of cAMP-dependent protein kinases may play an important role in down-regulating the PAF receptor and thus desensitizing the second PAF responses in Kupffer cells.

PAF binds to membrane-specific receptors, which then activate phospholipase C and cleave PIP_2 into IP_3 and DAG in several cell types. Inactivation of phospholipase C has been shown to be involved in the desensitization effect in rabbit platelets. The desensitization of rabbit platelets to PAF is specific to PAF since rabbit platelets pretreated with PAF and then washed become unresponsive to further PAF stimulation of either phosphoinositide-specific phospholipase C (Tokumura *et al.*, 1987; Morrison and Shukla, 1988) or protein phosphorylation (Shukla *et al.*, 1989). However, their ability to respond to other agents such as thrombin, ADP and collagen is retained. Conversely, platelets pretreated with thrombin were desensitized to both thrombin and PAF (Morrison and Shukla, 1988; Shukla *et al.*, 1989). Therefore, desensitization by PAF in platelets was homologous in nature whereas that by thrombin was heterologous. PAF apparently induces both homologous and heterologous desensitization in cultured vascular smooth muscle cells since pretreatment of vascular smooth muscle cells with PAF attenuated both PAF and angiotensin II-stimulated Ca^{2+} mobilization (Schwertschlag and Whorton, 1988). Down-regulation in the density of β-adrenergic receptors by PAF was observed in homogenates of rat cerebellum (Braquet *et al.*, 1985) and in human lung slices (Agrawal and Townley, 1987). However, no change in β-adrenergic receptors or binding affinity was reported in lungs removed from PAF-treated guinea-pigs or after direct incubation with PAF *in vitro* (Barnes *et al.*, 1987).

Protein kinase C may also play a role in PAF-induced desensitization effects. IP_3 raises cytoplasmic Ca^{2+} levels by releasing the cation from intracellular storage pools, while DAG remains in the plasma membrane where it binds with and activates protein kinase C through a Ca^{2+}-enhanced reaction (Berridge and Irvine, 1984; Nishizuka, 1984). Ca^{2+} and DAG then cooperate synergistically in moving protein kinase C from the cytosol to the plasma membrane. The membrane-adherent, DAG-bound protein kinase C proceeds to phosphorylate elements of the response mechanism (Nishizuka, 1984), such as a major 40 kDa protein in platelets (Imaoka *et al.*, 1983; Sano *et al.*, 1983; Lapetina, 1986). This enzyme can also be activated by the tumour-promoting phorbol ester. When neutrophils are pretreated either with PAF (O'Flaherty *et al.*, 1981, 1989; O'Flaherty, 1982; Zimmerman *et al.*, 1985; Naccache *et al.*, 1986) or with protein kinase C activators (Naccache *et al.*, 1986; O'Flaherty *et al.*, 1989), the cells rapidly lose their responsiveness to a second PAF challenge. Similarly, PAF-induced Ca^{2+} mobilization either in endothelial cells (Brock *et al.*, 1987) or in vascular smooth muscle cells (Schwertschlag and Whorton, 1988) is blocked by pretreatment of the cells with the activators of protein kinase C. Protein kinase C activators stimulated human neutrophils and reduced the availability of high-affinity receptors for PAF (O'Flaherty *et al.*, 1990b). These effects were concentration-dependent, irreversible, temperature-sensitive, and antagonized by protein kinase C blockers. Thus, activated protein kinase C down-regulated high-affinity PAF receptors and desensitized the second PAF responses.

14. Conclusions

Specific ligands must first bind to a receptor with a high affinity and stereochemical specificity. Second, the receptor must respond to the binding of its ligand by a conformational change that triggers a transmembrane signal. The signal-generating conformational change is caused by active agonists, but not by pure antagonists, which can bind to the receptor but do not activate the system. It is now recognized that for pharmacological receptors it is not the ligand but the receptor *per se* that possesses in its structure the key elements that generate a cellular response. Now that the PAF receptor has been cloned, this should open a new vista to determine the structural features of the PAF receptor that allows it to (1) bind its ligand and (2) trigger a transmembrane signal.

There exist both extracellular and intracellular PAF receptors. Extracellular PAF receptors mediate the functions of PAF synthesized in a stimulated cell and then released. It is being found increasingly in *in vitro* studies that much of the PAF synthesized by stimulated cells is retained within the cells and only small amounts are released. The intracellular receptors may be responsible for the function of PAF synthesized and retained inside the cells. Several PAF receptor antagonists, including WEB 2086 and MK 287, have been investigated in clinical trials for various human diseases. The results have not been promising. A membrane-permeable PAF synthesis inhibitor or receptor antagonist may be required to block the synthesis or inhibit the function of PAF inside the cells. The existing PAF receptor antagonists have been designed to block the actions of extracellular PAF and are not, or only slightly, permeable to the plasma membranes. They either exhibit no inhibitory effect on the function of PAF produced and retained inside the cells or inhibit only at high concentrations (Hwang and Wang, 1991). Furthermore, the intracellular and extracellular PAF receptors could be identical or different, depending on the source of cells. PAF also appears to interact with arachidonate metabolites in mediating the inflammatory responses (Pelow and Mikhailidis, 1990). This could further complicate the issue.

Receptor subtypes do exist in the PAF systems. The existing PAF receptor antagonists show small or no differentiation among receptor subtypes. Development of PAF receptor antagonists specific to each subtype is required to understand the differentiated role of each individual receptor subtype. Cloning of each individual subtype of receptor could provide important information to improve our designs of specific receptor antagonists for each individual subtype.

Multiple mediators may be involved in human diseases. One mediator may stimulate the synthesis of another mediator and multiple mediators could synergize with each other in isolated organs and cell preparations *in vitro* and in whole animals to amplify the production of autotoxins to enhance inflammatory injury. The mechanism of priming has not been elucidated, despite its intensive investigation. Clarification of the biochemical mechanisms involved in the priming and amplification processes may provide important information regarding the pathogenesis of inflammation.

15. Acknowledgements

I wish to thank Dr M.M. Ponpipom for synthesizing 52770 RP, $C_{16,11}$-PAF and $C_{18,9}$-PAF, Dr D.W. Graham for the synthesis of Y-24180 and My-Hanh Lam for technical assistance.

16. References

Adolfs, M.J.P., Beusenberg, F.D. and Bonta, I.L. (1989). PAF-acether modifies the cAMP levels in human peritoneal macrophages in a biphasic fashion. Agents Actions 26, 119–120.

Agrawal, D.K. and Townley, R.G. (1987). Effect of platelet-activating factor on beta-adrenoceptors in human lung. Biochem. Biophys. Res. Commun. 143, 1–6.

Akiyama, T., Ishida, J., Nakagawa, S., Ogawara, H., Watanabe, S.-I., Itoh, N., Shibuya, M. and Fukami, Y. (1987). Genistein, a specific inhibitor of tyrosine-specific protein kinases. J. Biol. Chem. 262, 5592–5595.

Allan, G. and Bazan, N.G. (1989). Differential increase in levels of c-fos and calcyclin, but not heat-shock protein, mRNAs in response to platelet-activating factor in a human. neuroblastoma cell line. Soc. Neurosci. 15, 842 (Abstr.).

Anderson, H.C. and McCarty, M. (1950). Determination of C-reactive protein in the blood as a measure of the activity of the disease process in acute rheumatic fever. Am. J. Med. 8, 445–455.

Angle, M.J., Jones, M.A., McManus, L.M. and Harper, M.J.K. (1988). Platelet-activating factor (PAF) in the rabbit uterus during early pregnancy. J. Reprod. Fert. 83, 711–722.

Ashkenazi, A., Winslow, J.W. G., Peralta, G.L. Peterson, M.I. Schimerlik, Capon, D.J. and Ramachandran, J. (1987). An M_2 muscarinic receptor subtype coupled to both adenyl cyclase and phosphoinositide turnover. Science 238, 672–675.

Atkinson, Y.H., Murray, A.W., Krillis, S., Vadas, M.A. and Lopez, A.F. (1990). Human tumor necrosis factor-alpha directly stimulates arachidonic acid release in human neutrophils. Immunology 70, 82–87.

Avdonin, P.V., Svitina-Ulitina I.V. and Kulikov, V.I. (1985). Stimulation of high affinity hormone-sensitive GTPase of human platelets by 1-O-alkyl-2-acetyl-*sn*-glyceryl-3-phosphocholine (platelet-activating factor). Biochem. Biophys. Res. Commun. 131, 307–313.

Advonin, P.V., Chelakov, I.V., Boogry, E.M., Svitina-Ulitina, I.V., Mazaev, A.V. and Tkachuk, V.A. (1987). Evidence for the receptor-operated calcium channels in human platelet plasma membranes. Thromb. Res. 46, 29–37.

Axelrod, J., Burch, R.M. and Jelsema, C.L. (1988). Receptor-mediated activation of phospholipase A_2 via GTP-binding proteins: arachidonic acid and its metabolites as second messangers. Trends Neurosci. 11, 117–123.

Bachelet, M., Adolf, M.J.P., Masliah, J., Bereziat, G., Vargaftig, B.B. and Bonta, I.L. (1988). Interaction between PAF-acether and drugs that stimulate cyclic AMP in guinea pig alveolar macrophages. Eur. J. Pharmacol. 149, 73–78.

Baggiolini, M., Dewald, B. and Thelen, M. (1988). Effects of PAF on neutrophils and mononuclear phagocytes. Prog. Biochem. Pharmacol. 22, 90–105.

Barnes, P.J., Grandordy, B.M., Page, C.P., Rhoden, K.J. and Robertson, D.N. (1987). The effect of platelet activating factor on pulmonary β-adrenoceptors. Br. J. Pharmacol. 90, 709–715.

Barnes, P.J., Chung, K.F. and Page, C.P. (1988). Inflammatory mediators and asthma. Pharmacol. Rev. 40, 49–84.

Bartroli, J., Carceller, E., Merlos, M., Garcia-Rafanell, J. and Form, J. (1991). Disubstituted tetrahydrofurans and dioxolanes as PAF antagonists. J. Med. Chem. 34, 373–386.

Barzaghi, G., Sarau, H.M. and Mong, S. (1989). Platelet-activating factor-induced phosphoinositide metabolism in differentiated U-937 cells in culture. J. Pharmacol. Exp. Ther. 248, 559–566.

Bennet, C.F., Mong, S., Wu, H.-L.W., Clark, M.A., Wheeler, L. and Crooke, S.T. (1987). Inhibition of phosphoinositide-specific phospholipase C by manoalide. Mol. Pharmacol. 32, 587–593.

Benovic, J.L., Strasser, R.H., Benovic, J.L., Daniel, K. and Lefkowitz, R.J. (1986). β-adrenergic receptor kinase: identification of a novel protein kinase that phorphorylates the agonist-occupied form of the receptor. Proc. Natl Acad. Sci. USA 83, 2797–2801.

Benveniste, J., Henson, P.M. and Cochane, C.G. (1972). Leukocyte-dependent histamine release from rabbit platelets: the role of IgE, basophils, and a platelet-activating factor. J. Exp. Med. 136, 1356–1377.

Benveniste, J., Tence, M., Bidault, J., Boullet, C. and Polonsky, J. (1979). Semisynthese et structure proposee du facteur activant les plaquettes (PAF). PAF acether, un alkyl ether analog de la lyso-phosphatidylcholine, C.R. Acad. Sci. Paris D289, 1037–1040.

Benveniste, J., Roubin, R., Chignard, M., Jouvin-Marche, E. and LeCouedic, P. (1982). Release of platelet-activating factor (PAF-acether) and 2-lysoPAF-acether from three cell types. Agents Actions 12, 711–713.

Berdel, W.E. and Munder, P.G. (1987). In "Platelet-Activating Factor and Related Lipid Mediators" (ed F. Snyder), pp 449–467. Plenum Press, New York.

Berdel, W.E., Bausert, W.R.R., Fink, U., Rastetler, J. and Munder, P.G. (1981). Antitumor action of alkyl-lysophospholipids (review). Anticancer Res. 1, 345–352.

Berdel, W.E., Korth, R., Reichert, A., Houlihan, W.J., Bicker, U., Nomura, H., Vogler, W.R., Benveniste, J. and Rastetter, J. (1987). Lack of correlation between cytotoxicity of agonists and antagonists of platelet activating factor (PAF-acether) in neoplastic cells and modulation of [^{3}H]PAF-acether binding to platelets from humans in vitro. Anticancer Res. 7, 1181–1188.

Berridge, M.J. and Irvine, R.F. (1984). Inositol trisphosphate, a novel second messenger in cellular signal transduction. Nature 312, 315–321.

Beusenberg, F.D., van Schaik, A., van Amsterdam, J.G.C. and Bonta, I.L. (1991). Involvement of eicosanoids in platelet-activating factor-induced modulation of adenyl cyclase activity in alveolar macrophages. J. Lipid Mediators 3, 301–310.

Biftu, T. (1985). 2, 5-diacyl tetrahydrothiophenes and analogs thereof as PAF antagonists. European Patent Appl. EP 154887.

Biftu, T., Gamble, N.F., Doebber, T.W., Hwang, S.-B., Shen, T.Y., Snyder, J., Springer, J.P. and Stevenson, R. (1986a). Conformation and activity of tetrahydrofuran lignans and analogues as specific platelet activating factor antagonists. J. Med. Chem. 29, 1917–1921.

Biftu, T., Gamble, N.F., Hwang, S.-B., Chabala, J.C., Doebber, T.W., Dougherty, H.W. and Shen, T.Y. (1986b). In: "6th Int. Conf. on Prostaglandins and Related Compounds, Florence Italy, 3–6 June" (Abstr.).

Billah, M.M. and Johnston, J.M. (1983). Identification of phospholipid platelet-activating factor (1-O-alkyl-2-acetyl-sn-glycero-3-phosphocholine) in human amniotic fluid and urine. Biochem. Biophys. Res. Commun. 113, 51–58.

Billah, M.M. and Lapetina, E.G. (1983). Platelet-activating factor stimulates metabolism of phosphoinositides in horse platelets: possible relationship to Ca^{2+} mobilization during stimulation. Proc. Natl Acad. Sci. USA 80, 965–968.

Billah, M.M., Di Renzo, G.C., Ban, C., Truong, C.T., Hoffman, D.R., Anceschi, M.M., Bleasdale, J.E. and Johnston, J.M. (1985). Platelet-activating factor metabolism in human amnion and the responses of this tissue to extracellular platelet-activating factor. Prostaglandins 30, 841–850.

Billah, M.M., Chapman, D., Egan, R.W., Gilchrest, H., Piwinski, J.J., Sherwood, J., Siegel, M.I. West, R.E. Jr. and Kreutner, W. (1990). Sch 37370: a potent, orally active, dual antagonist of platelet-activating factor (PAF) and histamine. J. Pharmacol. Exp. Ther. 252, 1090–1096.

Blank, N.L., Snyder, F., Byer, L.W., Brooks B. and Muirhead, E.E. (1979). Antihypertensive activity of an alkyl ether analog of phosphatidylcholine, Biochem. Biophys. Res. Commun. 90, 1194–1200.

Bourgain, R.H., Andries, R. and Bourgain, C. (1988). In "Ginkgolides – Chemistry, Biology, Pharmacology and Clinical Perspectives", Vol. 1, pp 137–149. J. R. Prous Science Publishers, Barcelona.

Boushey, H.A., Holtzman M.J., Sheller, J.R. and Nadel, J.A. (1980). Bronchial hyperreactivity. Am. Rev. Respir. Dis. 121, 389–413.

Bouvier, M.W., Hausdorff, A., DeBlasi, A., O'Dowd, B.F., Kobilka, B.K., Caron, M.G. and Lefkowitz, R.J. (1988). Removal of phosphorylation sites from the β-adrenergic receptor delays the onset of agonist-promoted desensitization. Nature 333, 370–373.

Bouvier, M., Collins, S., O'Dowd, B.F., Campbell, P.T., de Blasi, A., Kobilka, B.K., MacGregor, C., Irons, G.P., Caron, M.G. and Lefkowitz, R.J. (1989). Two distinct pathways for cAMP-mediated down-regulation of the β-adrenergic receptor. Phosphorylation of the receptor and regulation of its mRNA level. J. Biol. Chem. 264, 16786–16792.

Braquet, P. (1987). The ginkolides: potent platelet-activating factor antagonists isolated from Ginko bilboa: Chemistry, pharmacology and clinical applications. Drugs Future 12, 643–699.

Braquet, P. (1988). Ginkgolides – chemistry, biology, pharmacology, and Clinical Perspectives, Vols 1 and 2. J. R. Prous Science Publishers: Barcelona.

Braquet, P. and Godfroid, J.J. (1987). In "Platelet-Activating Factor and Related Lipid Mediators" (ed F. Snyder), pp 191–235. Plenum Press, New York.

Braquet, P. and Rola-Pleszczynski, M. (1987). The role of PAF in immunological responses: a review. Prostaglandins 34, 143–147.

Braquet, P., Etienne, A. and Clostre, F. (1985). Down-regulation of β$_2$-adrenergic receptors by PAF-acether and its inhibition by the PAF-acether antagonist BN 52021. Prostaglandins 30, 721.

Braquet, P., Touqui, L., Shen, T.Y. and Vargaftig, B.B. (1987). Perspectives in platelet-activating factor research. Pharmacol. Rev. 39, 97–145.

Bratton, D.L., Harris, R.A., Clay, K.L. and Henson, P.M. (1988a). Effects of platelet-activating factor and related lipids on phase transition of dipalmitoylphosphatidylcholine. Biochim. Biophys. Acta 941, 76–82.

Bratton, D.L., Harris, R.A., Clay, K.L. and Henson, P.M. (1988b). Effect of platelet-activating factor on calcium-lipid interactions and lateral phase separations in phospholipid vesicles. Biochim. Biophys. Acta 943, 211–219.

Brock, T.A., Griendling, K.K. and Gimbrone, M.A. (1987). In "New Horizons in Platelet-Activating Factor Research" (ed C.M. Winslow and M.L. Lee), pp 197–202. Wiley, New York.

Broquet, C., Auclair, E., Blavet, N., Touvay, C. and Braquet, P. (1990). Aminoacylates and aminocarbamates of 2-substituted 4-hydroxymethyl-1,3-dioxolans as ammonium salts. A new series of PAF antagonists. Eur. J. Med. Chem. 25, 235–240.

Burri, K., Barner, R., Cassal, J.M., Hadvary, P., Hirth, G. and Muller, K. (1985). PAF: from agonists to antagonists by synthesis. Prostaglandins 30, 691.

Burtin, C., Noirot, C., Scheinmann, P., Paupe, J., Bennet, C.F., Mong, S., Clarke, M.A., Kruse, L.I. and Crooke, S.T. (1987). Differential effects of manoalide on secreted and intracellular phospholipases. Biochem. Pharmacol. 36, 733–740.

Buss, J.E., Mumby, S.M., Casey, P.J., Gilman, A.G., and Sefton,

B.M. (1987). Myristoylated alpha subunits of guanine nucleotide-binding regulatory proteins. Proc. Natl Acad. Sci. USA 84, 7493–7497.

Camussi, G. (1986). Potential role of platelet-activating factor in renal pathophysiology. Kidney Int. 29, 469–477.

Camussi, G., Bussolino, F. and Tetta, C. (1989a). In "Platelet-Activating Factor and Diseases" (eds K. Saito and D.J. Hanahan), pp 113–128. International Medical Publishers, Tokyo.

Camussi, G., Tetta, C., Bussolino, F. and Baglioni, C. (1989b). Tumor necrosis factor stimulates human neutrophils to release leukotriene B$_4$ and platelet-activating factor. Eur. J. Biochem. 182, 661–666.

Casals-Stenzel, J. (1987). In "New Horizons in Platelet-Activating Factor Research" (eds C.M. Winslow and M.L. Lee), pp 277–284. Wiley, New York.

Casals-Stenzel, J., Muacevic, G. and Weber, K.H. (1987). Pharmacological actions of WEB 2086, a new specific antagonist of platelet-activating factor. J. Pharmacol. Exp. Ther. 241, 974–981.

Castaner, J., Koltai, M., Spinnewyn, B., Duverger, D., Pirotzky, E., Esanu, A. and Braquet, P. (1991). BN 50739. Drugs Future 16, 413–419.

Chang, M.N., Han, G.Q., Alison, B.H., Springer, J.P., Hwang, S.-B. and Shen, T.Y. (1985). Neolignans from *Piper futokadsurae*. Phytochemistry 24, 2079–2082.

Chang, S.W., Fedderson, C.O., Henson, P.M. and Voelkel, N.F. (1987). Platelet-activating factor mediates hemodynamic changes and lung injury in endotoxin-treated rats. J. Clin. Invest. 79, 1498–1509.

Chao, W., Liu, H., Debuysere, M., Hanahan, D.J. and Olson, M.S. (1989). Identification of receptors for platelet-activating factor in rat kupffer cells. J. Biol. Chem. 264, 13591–13598.

Chao, W., Liu, H. Zhou, W., Hanahan, D.J. and Olson, M.S. (1990). Regulation of platelet-activating factor receptor and platelet-activating factor receptor-mediated biological responses by cAMP in rat Kupffer cell. J. Biol. Chem. 265, 17576–17583.

Chau, L.-Y. and Jii, Y.-J. (1988). Characterization of ^{3}H-labelled platelet-activating factor receptor complex solubilized from rabbit platelet membranes. Biochim. Biophys. Acta 970, 103–112.

Chau, L.-Y., Tsai, Y.-M. and Cheng, J.-R. (1989). Photoaffinity labelling of platelet-activating factor binding sites in rabbit platelet membranes. Biochem. Biophys. Res. Commun. 161, 1070–1076.

Cheng, Y. and Prusoff, W.H. (1973). Relationship between the inhibition constant (K_i) and the concentration of inhibitor which cause 50% inhibition (I_{50}) of an enzyme reaction. Biochem. Pharmacol. 22, 3099–3108.

Chesney, C.M., Pifer, D.D. and Huch, K.M. (1983). In "Platelet-Activating Factor and Structurally Related Ether Lipids" (eds J. Benveniste and B. Arnoux), pp 177–184. Elsevier, Amsterdam.

Chesney, C.M., Pifer, D.D. and Huch, K.M. (1985). Densensitization of human platelet by platelet-activating factor. Biochem. Biophys. Res. Commun. 127, 24–30.

Chesney, C.M., Pifer, D.D. and Cagen, L.M. (1987). Triazolobenzodiazepines competitively inhibit the binding of platelet-activating factor (PAF) to human platelets. Biochem. Biophys. Res. Commun. 144, 359–366.

Chignard, M., Le Couedic, J.P., Vargaftig, B.B. and Benveniste, J. (1980). Platelet-activating factor (PAF-acether) secretion from platelets: effect of aggregating agents. Br. J. Haematol. 46, 455–464.

Clark, R.S.J. (1991). E6123. Drugs Future 16, 310–312.

Claus, D., Osmand, A.P. and Gewurz, H. (1976) Radioimmunoassay of human C-reactive protein and levels in normal sera. J. Lab. Clin. Med. 87, 120–128.

Cole, G.M. and Reed, S. (1991). Pheromone-induced phosphorylation of a G protein beta subunit in *S. cerevisiae* is associated with an adaptive response to mating pheromone. Cell 64, 703–716.

Cordeiro, R.S.B., Martins, M.A., Silva, P.M.R., Henriques, M.G.M.O., Weg, V.B., Faria Neto, H.C.C. and Lima, M.C.R. (1988). In "Ginkgolides – Chemistry, Biology, Pharmacology and Clinical Perspectives", Vol 1 (ed P. Braquet), pp 261–278. J.R. Prous, Barcelona.

Corey, E.J. (1988). Retrosynthetic thinking – essentials and examples. Chem. Soc. Rev. 17, 111–133.

Corey, E.J. and Gavai, A.V. (1990). Simple analogs of ginkgolide B which are highly active antagonists of platelet activating factor. Tetrahedron Lett. 30, 6959–6962.

Corey, E.J. and Ghosh, A.K. (1988). Total synthesis of ginkgolide A. Tetrahedron Lett. 29, 3205–3206.

Corey, E.J., Chen, C.-P. and Parry, M.J. (1988a). Dual binding modes to the receptor for platelet activating factor (PAF) of anti-PAF *trans*-2,5-diarylfurans. Tetrahedron Lett. 29, 2899–2902.

Corey, E.J., Kang, M.-c., Desai, M.C., Ghosh, A.K. and Houpis, I.N. (1988b). Total synthesis of ($\pm$)-ginkgolide B. J. Am. Chem. Soc. 110, 649–651.

Coyle, A.J., Urwin, S.C., Page, C.P., Touvay, C., Villain, B. and Braquet, P. (1988). The effect of the selective PAF antagonist BN 52021 on PAF- and antigen-induced bronchial hyper-reactivity and eosinophil accumulation. Eur. J. Pharmacol. 148, 51–58.

Crouch, M.F. and Lapetina, E.G. (1988). No direct correlation between Ca^{2+} mobilization and dissociation of G_i during platelet phospholipase A$_2$ activation. Biochem. Biophys. Res. Commun. 153, 21–30.

Cuss, F.M., Dixon, C.M.S. and Barnes, P.J. (1986). Effects of inhaled platelet-activating factor on pulmonary function and bronchial responsiveness in man. Lancet ii, 189–192.

Dalman, H.M. and Neubig, R.R. (1991). Two peptides from the α_{2A}-adrenergic receptor alter receptor G protein coupling by distinct mechanisms. J. Biol. Chem. 266, 11025–11029.

Demopoulos, A., Pinckard R.N. and Hanahan, D.J. (1979). Platelet-activating factor. Evidence for 1-O-alkyl-2-acetyl-*sn*-glycerol-3-phosphocholine as the active component (a new class of lipid chemical mediators). J. Biol. Chem. 254, 9355–9358.

DeNichilo, M.O., Stewart, A.G., Vadas, M.A. and Lopez, A.F. (1991). Granulocyte–macrophage colony-stimulatory factor is a stimulant of platelet-activating factor and superoxide anion generation by human neutrophils. J. Biol. Chem. 266, 4896–4902.

Dent, G., Ukena, D., Chanez, P., Sybrecht, G. and Barnes, P. (1989a). Characterization of PAF receptors on human neutrophils using the specific antagonist, WEB 2086. FEBS Lett. 244, 365–368.

Dent, G., Ukena, D., Sybrecht, G. and Barnes, P. (1989b). [^{3}H]WEB 2086 labels platelet activating factor receptors in guinea pig and human lung. Eur. J. Pharmacol. 169, 313–316.

De Silva, E.D. and Scheuer, P.J. (1980). Manoalide, an antibiotic sesterterpenoid from the marine sponge *Luffariella variabilis* (Polejaeff). Tetrahedron Lett. 21, 1611–1614.

Dewald, B. and M. Baggiolini (1985). Activation of NADPH oxidase in human neutrophils. Synergism between fMPL and the neutrophil products PAF and LTB$_4$. Biochem. Biophys. Res. Commun. 128, 297–304.

Dhar, A., Paul, A.K. and Shukla, S.D. (1990). Platelet-activating factor stimulation of tyrosine kinase and its relationship to phospholipase C in rabbit platelets: studies with genistein and monoclonal antibody to phosphotyrosine. Mol. Pharmacol. 37, 519–525.

Dive, G., Godfroid, J.-J., Lamotte-Brasseur, J., Batt, J.-P., Heymans, F., Dupont, L. and Braquet, P. (1989). PAF receptor: 1. cache-oreilles effect of selected high-potency platelet-activating factor (PAF) antagonists. J. Lipid Mediators 1, 201–215.

Dixon, R.A.F., Sigal, I.S., Rands, E., Register, R.B., Candelore, M.R., Blahe, A.D. and Strader, C.D. (1987). Ligand binding to the β-adrenergic receptor involves its rhodopsin-like core. Nature 326, 73–77.

Doebber, T.W. and Wu, M.S. (1987). Platelet-activating factor (PAF) stimulates the PAF-synthesizing enzyme acetyl-CoA: 1-alkyl-*sn*-glycero-3-phosphocholine O$_2$-acetyltransferase and PAF synthesis in neutrophils. Proc. Natl Acad. Sci. USA 84, 7557–7561.

Doebber, T.W., Wu, M.S., Robbins, J.C., Choy, B., Chang, M.N. and Shen, T.Y. (1985). Platelet-activating factor (PAF) involvement in endotoxin induced hypotension in rats, studies with PAF-receptor antagonist kadsurenone. Biochem. Biophys. Res. Commun. 127, 799–808.

Doebber, T.W., Wu, M.S. and Biftu, T. (1986) Platelet-activating factor (PAF) mediation of rat anaphylatic responses to soluble immune complexes. Studies with PAF-receptor antagonist L-652,731. J. Immunol. 136, 4659–4668.

Dohlman, H.G., Caron, M.G. and Lefkowitz, R.J. (1987). A family of receptors coupled to guanine nucleotide regulatory proteins. Biochemistry 26, 2657–2664.

Domingo, M.T., Spinnewyn, B., Chabrier, P.E. and Braquet, P. (1988). Presence of specific binding sites for platelet-activating factor (PAF) in the brain. Biochem. Biophys. Res. Commun. 151, 730–736.

Domingo, M.T., Chabrier, P.E., VanDelft, J.L., Verbeij, N.L., Van Haeringen, N.J. and Braquet, P. (1989). Characterization of specific binding sites for PAF in the iris and ciliary body of rabbit. Biochem. Biophys. Res. Commun. 160, 250–256.

Dubois, C., Bissonnette, E. and Rola-Pleszczynski, M. (1989). Platelet-activating factor (PAF) enhances tumor necrosis factor production by alveolar macrophages. Inhibition by PAF receptor antagonists and lipoxygenase inhibitors. J. Immunol. 143, 964–970.

Egido, J., Robles, A., Ortiz, A., Ramirez, F., Gonzalez, E., Mampaso, F., Sanchez Crespo, M., Braquet, P. and Herando, L. (1987). Role of platelet-activating factor in adriamycin-induced nephropathy in rats. Eur. J. Pharmacol. 138, 119–123.

Ekimoto, H., Irie, Y., Araki, Y., Han, G.-Q., Kadota, S. and Kikuchi, T. (1991). Platelet aggregation inhibitors from *Swietenia mahagoni*: inhibition of *in vitro* and *in vivo* platelet-activating factor-induced effects of tetranortriterpenoids related to swietenine and swietenolide. Planta Med. 57, 56–58.

Etienne, A., Hecquet, F., Souland, C., Spinnewyn, B., Closter, F. and Braquet, P. (1985). *In vivo* inhibition of plasma protein leakage and *Salmonella enteritidis* induced mortality in the rat by a specific PAF-acether antagonist, BN 52021. Agents Actions 17, 368–370.

Fadel, R., David, B., Herpin-Richard, N., Borgonon, A., Rassemont, R. and Rihoux, J.-.P. (1990). *In vivo* effects of cetirizine on cutaneous reactivity and eosinophil migration induced by platelet-activating factor (PAF-acether) in man. J. Allergy Clin. Immunol. 1990, 314–320.

Feliste, R., Perret, B., Braquet, P. and Chap, H. (1989). Protective effect of BN 52021, a specific antagonist of platelet-activating factor (PAF-acether) against diet-induced cholesteryl ester deposition in rabbit aorta. Atherosclerosis 78, 151–158.

Feogh, M.L. (1988). Eicosanoids and platelet activating factor mechanisms in organ rejection. Transplant Proc. 20, 1260–1263.

Feogh, M.L., Kheirabadi, B.S., Rowles, J.R., Braquet, P. and Ramwell, P.W. (1986). Prolongation of cardiac allograft survival with BN 52021, a specific antagonist of platelet-activating factor. Transplantation 42, 86–88.

Fernandez-Gallardo, S., Ortega, M.D.P., Priego, J.G., Casa-Juana, M.F., Sunkel, C. and Sanchez Crespo, M. (1990). Pharmacological actions of PCA 4248, a new platelet-activating factor receptor antagonist: *in vivo* studies. J. Pharmacol. Exp. Ther. 255, 34–39.

Feuerstein, G. and Hallenbeck, J.M. (1987). Prostaglandins, leukotrienes and platelet-activating factor in shock. Annu. Rev. Pharmacol. Toxicol. 27, 301–313.

Filep, J.G. and Foldes-Felip, E. (1989). Effects of C-reactive protein on human neutrophil granulocytes challenged with *N*-formyl-methionyl-leucyl-phenylalanine and platelet-activating factor. Life Sci. 44, 517–524.

Filep, J.G., Herman, F., Braquet, P. and Mozes, T. (1989). Increased levels of platelet-activating factor in blood following intestinal ischemia in the dog. Biochem. Biophys. Res. Commun. 158, 353–359.

Filep, J.G., Herman, F., Kelemen, E. and Foldes-Felip, E. (1991). C-reactive protein inhibits binding of platelet-activating factor to human platelets. Thromb. Res. 61, 411–421.

Findlay, J.B.C. and Pappin, D.J.C. (1986). The opsin family of proteins. Biochem. J. 238, 625–642.

Finkel, T.H., Kubo, R.T. and Cambier, J.C. (1991). T-cell development and transmembrane signalling: changing biological responses through an unchanging receptor. Immunol. Today 12, 79–85.

Fisher, R.A., Shukla, S.D., Debuysere, M.S., Hanahan, D.J. and Olson, M.S. (1984). The effect of acetylglyceryl ether phosphorylcholine on glycogeneolysis and phosphatidylinositol 4,5-bisphosphate metabolism in rat hepatocytes. J. Biol. Chem. 259, 8685–8688.

Franke, R.R., Sakmar, T.P., Oprian, D.D. and Khoranan, H.G. (1988). A single amino acid substitution in rhodopsin (lysine 248→ leucine) prevents activation of transducin. J. Biol. Chem. 263, 2119–2122.

Freitas, J.C., Rodriques, F., Lavaud, P., Mencia-Huerta, J.M. and Braquet, P. (1988). Inhibition of platelet aggregation by manoalide: preliminary results. Braz. J. Med. Biol. Res. 21, 337–340.

Fukada, Y., Takao, T., Ohguro, H., Yoshizawa, T., Akino, T. and Shimonishi, Y. (1990). Farnesylated γ-subunit of photoreceptor G protein indispensible for GTP binding. Nature 346, 658–660.

Furchgott, R.F. (1972). In "Handbook of Experimental Pharmacology: Catecholamines", Vol 33 (eds H. Blaschko and E. Muscholl), pp 283–335. Springer-Verlag, New York.

Galizzi, J.P., Quar, J., Fosset, M., Van Renterghem, C. and

Lazdunski, D. (1987). Regulation of calcium channels in aortic muscle cells by protein kinase C activators (Diacylglycerol and phorbol esters) and by peptides (Vasopressin and Bombesin) that stimulate phosphoinositide breakdown. J. Biol. Chem. 262, 6947–6950.

Gallin, J.I. (1985). Leukocyte adherence – related glycoproteins LFA-1, Mol and p150,95: a new group of monoclonal antibodies, a new disease, and a possible opportunity to understand the molecular basis of leukocyte adherence. J. Infect. Dis. 152, 661–667.

Gandhi, A.R., Hanahan, D.J. and Olson, M.S. (1990). Two distinct pathways of platelet-activating factor-induced hydrolysis of phosphoinositides in primary cultures of rat kupffer cells. J. Biol. Chem. 265, 18234–18241.

Ganguly, C.L., Chelladurai, M. and Ganguly, P. (1985). Protein phosphorylation and activation of platelets by wheat germ agglutinin. Biochem. Biophys. Res. Commun. 132, 313–319.

Gay, J.C., Beckman, J.K. Zaboy, K.A. and Lukens, J.N. (1986). Modulation of neutrophil oxidative response to soluble stimuli by platelet-activating factor. Blood 67, 931–936.

Giers, G., Janzarik, H., Kempe, E.R. and Mueller-Eckhardt, C. (1990). Failure of the platelet-activating factor antagonist WEB 2086 BS for treatment of chronic autoimmune thrombocytopenia. Blut 61, 21–24.

Gilboe, D.D., Kintner, D., Fitzpatrick, H.J., Emoto, E., Esanu, A., Braquet, P. and Bazan, N.G. (1991). Recovery of postischemic brain metabolism and function following treatment of a free radical scavanger and platelet-activating factor antagonists. J. Neurochem. 56, 113–118.

Gilman, A.G. (1987). G-proteins: Transducers of receptor-generated signals. Annu. Rev. Biochem. 56, 615–649.

Glaser, K.B., Asmis, R. and Dennis, E.A. (1990). Bacterial lipopolysaccharide priming of P388D$_1$ macrophage-like cells for enhanced arachidonic acid metabolism. Platelet-activating factor receptor activation and regulation of phospholipase A$_2$. J. Biol. Chem. 265, 8658–8664.

Goddard, C., Arnold, S.T. and Felsted, R.L. (1989). High affinity binding of an N-terminal myristoylated p60^{src} peptide. J. Biol. Chem. 264, 15173–15176.

Golden, A., Nemeth, S.P. and Brugge, J.S. (1986). Blood platelets express high levels of the pp60^{c-src}-specific tyrosine kinase activity. Proc. Natl Acad. Sci. USA 83, 852–856.

Gomez, J., Bloom, J.W., Yamamura, H.I. and Halonen, M. (1990). Characterization of receptors for platelet-activating factor in guinea pig lung membranes. Am. J. Respir. Cell Mol. Biol. 3, 259–264.

Gomez-Cambronero, J., Wang, E., Johnson, G., Huang, C.-K. and Sha'afi, R.I. (1991). Platelet-activating factor induces tyrosine phosphorylation in human neutrophils. J. Biol. Chem. 266, 6240–6245.

Graham, D.W., Chiang, P., Yang, S.S., Thompson, K.L., Chang, M.N., Doebber, T.W., Hwang, S.-B., Lam, M.-H., Wu, M.S., Alberts, A.W. and Chabala, J.C. (1989). In "197th ACS National Meeting, Dallas, Texas, 9–14 April 1989, Division of Medicinal Chemistry", Poster 25 (Abstract).

Grandt, R., Aktories, K. and Jakobs, K.H. (1986). Evidence for two GTPases activated by thrombin in membranes of human platelets. Biochem. J. 237, 669–674.

Greenberg, J.H. and Jamieson, G.A. (1974). The effects of various lectins on platelet aggregation and release. Biochim. Biophys. Acta 345, 231–242.

Grue-Sorensen, G., Nielsen, I.M. and Neilsen, C.K. (1988).

Derivatives of 2-methylenepropane-1,3-diol as new antagonists of platelet activating factor. J. Med. Chem. 31, 1174–1178.

Grynkiewicz, G., Poenie, M. and Tsien, R.Y. (1985). A new generation of Ca^{2+} indicators with greatly improved fluorescence properties. J. Biol. Chem. 260, 3440–3450.

Guthrie, L.A., McPhail, L.C., Henson, P.M. and Johnston, R.B. Jr (1984). Priming of neutrophils for enhanced release of oxygen metabolites by bacterial lipopolysaccharide. Evidence for increased activity of the superoxide-producing enzyme. J. Exp. Med. 160, 1656–1671.

Guthrie, R.W., Kaplan, G.L., Mennona, F.A., Tilley, J.W., Kierstead, R.W., Mullin, J.G., LeMahieu, R.A., Zawoiski, S., O'Donnell, M., Crowley, H., Yaremko, B. and Welton, A.F. (1989). Pentadienyl carboxamide derivatives as antagonists of platelet activating factor. J. Med. Chem. 33, 1820–1835.

Guthrie, R.W., Kaplan, G.L., Mennona, F.A., Tilley, J.W., Kierstead, R.W., O'Donnell, M., Crowley, H., Yaremko, B. and Welton, A.F. (1990). Propenyl carboxamide derivatives as antagonists of platelet activating factor. J. Med. Chem. 33, 2856–2864.

Hadvary, P. and Baumgartner, H.R. (1985). Interference of PAF-acether antagonists with platelet aggregation and with the formation of platelet thrombi. Prostaglandins 30, 694 (Abstract).

Halbrugge, M. and Walter, U. (1989). Purification of a vasodilator-regulated phosphoprotein from human platelets. Eur. J. Biochem. 185, 41–50.

Halbrugge, M., Friedrich, C., Eigenthaler, M., Schonzenbacher, P. and Walter, U. (1990). Stoichiometric and reversible phosphorylation of a 46-kDa protein in human platelets in response to cGMP- and cAMP-elevating vasodilators. J. Biol. Chem. 265, 3088–3093.

Hall, Z.W. (1987). Three of a kind: the β-adrenergic receptor, the muscarinic acetylcholine receptor, and rhodopsin. Trends in Neurosci. 10, 99–101.

Halonen, M., Palmer, J.D., Lohman, I.C., McManus, L.M. and Pinckard, R.N. (1980). Respiratory and circulatory alterations induced by acetyl glyceryl ether phosphorylcholine, a mediator of IgE anaphylaxis in the rabbit. Am. Rev. Respir. Dis. 128, 915–924.

Han, G.-Q., Ma, Y., Li, S.M., Li, C.-L., Springer, J.P., Hwang, S.-B. and Chang, M.N. (1986). Neolignans from *Piper hanci* Maxim. Acta Pharmaceut. Sinica 21, 361–365.

Hanahan, D.J. (1986). Platelet-activating factor: a biologically active phosphoglyceride. Annu. Rev. Biochem. 55, 483–509.

Handley, D.A. (1988). Development and therapeutic indications for PAF receptor antagonists. Drugs Future 13, 137–152.

Handley, D.A., Van Valen, R.G., Melden, M.K., Houlihan, W.J. and Saunders, R.N. (1988). Biological effects of the orally active platelet activating factor receptor antagonist SDZ 64-412. J. Pharmacol. Exp. Ther. 247, 617–623.

Harper, M.J.K. (1989). Platelet-activating factor: a paracrine factor in preplantation stages of reproduction? Biol. Reprod. 40, 907–913.

Haslam, R.J. and Vanderwel, M. (1982). Inhibition of platelet adenylate cyclase by 1-O-alkyl-2-O-acetyl-sn-glyceryl-3-phosphorylcholine (platelet-activating factor). J. Biol. Chem. 257, 6879–6885.

Havill, A.M., Van Valen, D.G. and Handley, D.A. (1990). Prevention of non-specific airway hyperreactivity after allergen challenge in guinea pigs by the PAF receptor antagonist SDZ 64-412. Br. J. Pharmacol. 99, 396–400.

Hayashi, H., Kudo, I., Inoue, K., Onozaki, K., Tsushima, S., Nomura, H. and Nojima, S. (1985). Activation of guinea pig peritoneal macrophages by platelet-activating factor (PAF) and its agonists. J. Biochem. (Tokyo) 97, 1737–1745.

Henson, P.M. (1976). Activation and desensitization of platelets by platelet-activating factor (PAF) derived from IgE-sensitized basophils. 1. Characteristics of the secretory response. J. Exp. Med. 143, 937–952.

Henson, P.M. (1987). In "New Horizons in Platelet Activating Factor Research" (eds C.M. Winslow and M.L. Lee), pp 3–10. Wiley, New York.

Heuer, H.O., Casals-Stenzel, J., Muacevic, G. and Weber, K.H. (1990). Pharmacologic activity of bepafant (WEB 2170), a new and selective hetrazepinoic antagonist of platelet-activating factor. J. Pharmacol. Exp. Ther. 255, 962–968.

Hirafuji, M., Maeyama, K., Watanabe, T. and Ogura, Y. (1988). Transient increase of cytosolic free calcium in cultured human vascular endothelial cells by platelet-activating factor. Biochem. Biophys. Res. Commun. 154, 910–917.

Hirata, M., Hayashi, Y., Ushikubi, F., Yokota, Y., Kageyama, R., Nakanishi, S. and Narumiya, S. (1991). Cloning and expression of cDNA for a human thromboxane A2 receptor. Nature 349, 617–620.

Hirosada, S., Katsumi, I. and Kohel, N. (1988). European Patent O 284 359.

Hoffman, D.R., Truong, T.C. and Johnston, J.M. (1986a). The role of platelet-activating factor in human fetal lung maturation. Am. J. Obstet. Gynecol. 155, 70–75.

Hoffman, D.R., Truong, T.C. and Johnston, J.M. (1986b). Metabolism and function of platelet-activating factor in rabbit fetal lung development. Biochim. Biophys. Acta 879, 88–96.

Homma, H. and Hanahan, D.J. (1988). Attenuation of platelet activating factor (PAF)-induced stimulation of rabbit platelet GTPase by phorbol ester, dibutyryl cAMP, and desensitization: concomitant effects on PAF receptor binding characteristics. Arch. Biochem. Biophys. 262, 32–39.

Homma, H., Tokumura, A. and Hanahan, D.J. (1987). Binding and internalization of platelet-activating factor 1-O-alkyl-2-acetyl-sn-glycero-3-phosphocholine in washed rabbit platelets. J. Biol. Chem. 262, 10582–10587.

Honda, Z.-I., Nakamura, M., Miki, I., Minami, M., Watanabe, T., Seyama, Y., Okado, H., Toh, H., Ito, K., Miyamoto, T. and Shimizu, T. (1991). Cloning by functional expression of platelet-activating factor receptor from guinea-pig lung. Nature 349, 342–346.

Hosford, D., Mencia-Huerta, J.M. and Braquet, P. (1990). Platelet-activating factor (PAF) and PAF antagonists in asthma. Crit. Rev. Ther. Drug 7, 261–273.

Houlihan, W.J., Lee, M., Munder, P.G., Menecek, G.M., Handley, D.A., Winslow, C.M., Happy, J. and Jaeggi, C. (1987). Antitumor activity of SRI 62–834, a cyclic ether analog of ET-18-OCH₃. Lipids 22, 884–890.

Houlihan, W.J., Cheon, S.H., Handley, D.A., Larson, D., Parrino, V.A., Reitter, B., Schmitt, B. and Winslow, C.M. (1988). 5-aryl-2,3-dihydroimidazo[2,1-a]isoquinolines. A novel class of platelet activating factor (PAF) molecule. Prostaglandins 35, 848.

Houslay, M.D., Bojanic, D., Gawler, D., O'Hagan, S. and Wilson, A. (1986). Thrombin, unlike vasopressin, appears to stimulate two distinct guanine nucleotide regulatory proteins in human platelets. Biochem. J. 238, 109–113.

Huang, R.-R.C., DeHaven, R.N., Cheung, A.H., Diehl, R.E., Dixon, R.A.F. and Strader, C.D. (1989). Identification of allosteric antagonists of receptor-guanine nucleotide-binding protein interactions. Mol. Pharmacol. 37, 304–310.

Humphrey, D.M., Hanahan, D.J. and Pinckard, R.N. (1982). Induction of leukocyte infiltrates in rabbit skin by acetyl glyceryl ether phosphorylcholine. Lab. Invest. 47, 227–234.

Hussaini, I. and Shen, T.Y. (1989). A specific, photolabile and irreversible antagonist (L-662,025) of the PAF-receptor. Biochem. Biophys. Res. Commun. 161, 23–30.

Hwang, S.-B. (1987). Specific receptor sites for platelet activating factor on rat liver plasma membranes. Arch. Biochem. Biophys. 257, 339–344.

Hwang, S.-B. (1988). Identification of a second putative receptor of platelet-activating factor from human polymorphonuclear leukocytes. J. Biol. Chem. 263, 3225–3233.

Hwang, S.-B. (1990). Specific receptors of platelet-activating factor, receptor heterogeneity, and signal transduction mechanisms, J. Lipid Mediators 2, 123–158.

Hwang, S.-B. (1991a). High affinity receptor binding of platelet-activating factor in rat peritoneal polymorphonuclear leukocytes. Eur. J. Pharmacol. 196, 169–175.

Hwang, S.-B. (1991b). Specific binding of tritium-labelled inositol-1,4,5-trisphosphate to human platelet membranes: ionic and GTP regulation. Biochim. Biophys. Acta 1064, 351–359.

Hwang, S.-B. (1991c). Function and regulation of extracellular and intracellular receptors of platelet-activating factor, Ann. N.Y. Acad. Sci. 629, 217–226.

Hwang, S.-B. (1993). In "PAF Receptor Signal Transduction" (ed S.D. Shulka), Plenum Press, New York, pp 8–18.

Hwang, S.-B. and Lam, M.-H. (1986). Species difference in the specific receptor of platelet activating factor. Biochem. Pharmacol. 35, 4511–4518.

Hwang, S.-B. and Lam, M.-H. (1987). In "New Horizons in Platelet Activating Factor Research" (eds C.W. Winslow and M.L. Lee), pp 165–172. Wiley, New York.

Hwang, S.-B. and Lam, M.-H. (1989). In "3rd Int. Conf. on Platelet-Activating Factor and Structurally Related Alkyl Ether Lipids, Tokyo Japan 8–12 May, p 32 (Abstr.).

Hwang, S.-B. and Lam, M.-H. (1991). L-659,989: a useful probe for the detection of multiple conformational states of PAF receptors. Lipids 26, 1148–1153.

Hwang, S.-B. and Wang, S. (1989). In "Platelet-Activating Factor in Immune Responses and Renal Diseases" (eds P. Braquet, K.H. Hsieh, E. Pirotzky and J.M. Mencia-Huerta), pp 9–15. Excerta Medica Asia, Hong Kong.

Hwang, S.-B. and Wang, S. (1990). In "PAF and Antagonists, New Developments and Clinical Applications" (eds J.T. O'Flaherty and P. Ramwell), pp 13–30. Portfolio, The Woodlands, TX.

Hwang, S.-B. and Wang, S. (1991). Wheat germ agglutinin potentiated the specific binding of platelet-activating factor to isolated human platelet membranes and induced the synthesis of platelet-activating factor in human platelets. Mol. Pharmacol. 39, 788–797.

Hwang, S.-B., Lee, C.S.C., Cheah, M.J. and Shen, T.Y. (1983). Specific receptor sites for 1-O-alkyl-2-O-acetyl-sn-glycero-3-phosphocholine (platelet activating factor) on rabbit platelet and guinea pig smooth muscle membranes. Biochemistry 22, 4756–4763.

Hwang, S.-B., Cheah, M.J., Lee, C.-S.C. and Shen, T.Y. (1984). Effects of nonsteroid antiinflammatory drugs on the specific binding of platelet activating factor to membrane preparations of rabbit platelets. Thromb. Res. 34, 519–531.

Hwang, S.-B., Lam, M.-H. and Shen, T.Y. (1985a). Specific sites for platelet activating factor in human lung tissues. Biochem. Biophys. Res. Commun. 128, 972–979.

Hwang, S.-B., Li, C.-L., Lam, M.-H. and Shen, T.Y. (1985b). Characterization of cutaneous vascular permeability induced by platelet activating factor in guinea pigs and rats and its inhibition by a platelet-activating factor receptor antagonist. Lab. Invest. 52, 617–630.

Hwang, S.-B., Lam, M.-H., Biftu, T., Beattie, T.R. and Shen, T.Y. (1985c). 2,5-bis(3,4,5-trimethoxyphenyl)tetrahydrofuran. An orally active specific and competitive receptor antagonist of platelet activating factor. J. Biol. Chem. 260, 15639–15645.

Hwang, S.-B., Lam, M.-H., Li, C.L. and Shen, T.Y. (1986a). Release of platelet activating factor and its involvement in the first phase of carrageenan-induced rat foot edema. Eur. J. Pharmacol. 120, 33–41.

Hwang, S.-B., Lam, M.-H. and Chang, M.N. (1986b). Specific binding of [^{3}H]dihydrokadsurenone to rabbit platelet membranes and its inhibition by the receptor agonists and antagonists of platelet-activating factor. J. Biol. Chem. 261, 13720–13726.

Hwang, S.-B., Lam, M.-H. and Shen, T.Y. (1986c). Membrane receptors for platelet-activating factor (PAF) and a competitive specific PAF-receptor antagonist, kadsurenone. Adv. Inflam. Res. 11,83–95.

Hwang, S.-B., Lam, M.-H. and Pong, S.S. (1986d). Ionic and GTP regulation of binding of platelet-activating factor to receptors and platelet-activating factor-induced activation of GTPase in rabbit platelet membranes. J. Biol. Chem. 261, 532–537.

Hwang, S.-B., Chang, M.N., Garcia, M.L., Han, Q.Q., Huang, L., King, V.F., Kaczorowski, G.J. and Winquist, R.J. (1987). L-652,469-A dual receptor antagonist of platelet activating factor and dihydropyridines from *Tussilago farfara* L. Eur. J. Pharmacol. 141, 269–281.

Hwang, S.-B., Lam, M.-H., Alberts, A.W., Bugianesi, R.L., Chabala, J.C. and Ponpipom, M.M. (1988). Biochemical and pharmacological characterization of L-659,989: an extremely potent, selective and competitive receptor antagonist of platelet-activating factor. J. Pharmacol. Exp. Ther. 246, 534–541.

Hwang, S.-B., Lam, M.-H. and Hsu, A.H.-M. (1989). Characterization of platelet-activating factor (PAF) receptor by specific binding of [^{3}H]L-659,989, a PAF receptor antagonist, to rabbit platelet membranes: possible multiple conformational states of a single type of PAF receptors. Mol. Pharmacol. 35, 48–58.

Hwang, S.-B., McIntyre, D.E., Lam, M.-H., Szalkowski, D.M., Bach, T., Luell, S., Meuer, T., Sahoo, S.P., Graham, D.W., Acton, J., Biftu, T., Bugianesi, R.L., Chabala, J.C., Girotra, N.N., Kuo, H.C.H., Ponpipom, M.M., Davies, P. and Alberts, A.W. (1990). L-680,573: a potent specific and orally active PAF receptor antagonist. FASEB J. 4, A347 (Abstr.).

Hwang, S.-B., Lam, M.-H. and Wu, K. (1991). In "Prostaglandins, Leukotrienes, Lipoxins and PAF" (ed J.M. Bailey), Plenum Press, New York, pp 209–294.

Hynes, R.O. (1986). Fibronectins. Sci. Am. 254, 42–51.

Ieyasu, H., Takai, Y., Kaibuchi, K., Sawamura, M. and Nishizuka, Y. (1982). A role of calcium-activated, phospholipid-dependent protein kinase in platelet-activating factor-induced serotonin release from rabbit platelets. Biochem. Biophys. Res. Commun. 108, 1701–1708.

Imaoka, T., Lynham, J.A. and Haslam, R.J. (1983). Purification and characterization of the 47,000-dalton protein phosphorylated during degranulation of human platelets. J. Biol. Chem. 258, 11404–11414.

Inarrea, P., Gomez-Cambronero, J., Nieto, M. and Sanchez Crespo, M. (1984). Characteristics of the binding of platelet-activating factor to platelets of different animal species. Eur. J. Pharmacol. 105, 309–315.

Inarrea, P., Gomez-Cambronero, J., Pascual, J., Del Carmenponte, M., Hernando, L. and Sanchez Crespo, M. (1985). Synthesis of PAF-acether and blood volume changes in Gram-negative sepsis. Immunopharmacology 9, 45–52.

Ingraham, L.M., Coates, T.D., Allen, J.M., Higgins, C.P., Baehner, R.L. and Boxer, L.A. (1982). Metabolic, membrane, and functional responses of human polymorphonuclear leukocytes to platelet-activating factor. Blood 59, 1259–1266.

Iwakami, S., Ebizuka, Y. and Sankawa, U. (1990). Lignans and sesquiterpenoids as PAF antagonists. Heterocycles 30, 795–799.

Janero, D.R., Burghardt, B. and Burghardt, C. (1988a). Radioligand competitive binding methodology for the evaluation of platelet-activating factor (PAF) and PAF receptor antagonism using intact canine platelets. J. Pharmacol. Meth. 20, 237–253.

Janero, D.R., Burghardt, B. and Burghardt, C. (1988b). Specific binding of 1-O-alkyl-2-acetyl-*sn*-glycero-3-phosphocholine (platelet-activating factor) to the intact canine platelet. Thromb. Res. 50, 789–802.

Johnston, J.M. and Miyaura, S. (1990). In "Platelet-Activating Factor Antagonists: New Developments and Clinical Application" (eds J.T. O'Flaherty and P.W. Ramwell), pp 139–160. Portfolio, The Woodlands, TX.

Jones, T.L.X., Simonds, W.F., Merendino, J.J. Jr, Brann, M.R., and Spiegel, A.M. (1990). Myristoylation of an inhibitory GTP-binding protein alpha subunit is essential for its membrane attachment. Proc. Natl Acad. Sci. USA 87, 568–572.

June, C.H., Fletcher, M.C., Ledbetter, J.A., Schieven, G.L., Siegel, J.N., Phillips, A.F. and Samelson, L.E. (1990). Inhibition of tyrosine phosphorylation prevents T cell receptor-mediated signal transduction. Proc. Natl Acad. Sci. USA 87, 7722–7726.

Kadota, S., Marpaung, L., Kikuchi, T. and Ekimoto, H. (1989). Antagonists of platelet activating factor from *Swietenia mahogani* (L.) Jacq. Tetrahedron Lett. 30, 1111–1114.

Karasawa, A., Rochester, J.A. and Lefter, A.M. (1990). Beneficial actions of BN 50739, a new PAF receptor antagonist, in murine traumatic shock. Meth. Find. Exp. Clin. Pharmacol. 12, 231–237.

Kawamoto, H., Watanabe, Y., Imaizumi, T., Iwasaki, T. and Yoshida, H. (1991). Effects of lithium ion on ADP ribosylation of inhibitory GTP-binding protein by pertussis toxin, islet-activating protein. Eur. J. Pharmacol. 206, 33–37.

Kloprogge, E. and Akkerman, J.W.N. (1984). Binding kinetics of PAF-acether (1-O-alkyl-2-acetyl-*sn*-glycero-3-phosphocholine) to intact human platelets. Biochem. J. 223, 901–909.

Kloprogge, E. and Akkerman, J.W.N. (1986). Platelet-activating factor (PAF-acether) induces high- and low-affinity binding of fibrinogen to human platelets via independent mechanisms. Biochem. J. 240, 403–412.

Kloprogge, E., Hasselaar, P. and Akkerman, J.W.N. (1986a). PAF-acether (1-O-hexadecyl/octadecyl-2-acetyl-*sn*-glycero-3-phosphocholine)-induced fibrinogen binding to platelets depends on metabolic energy. Biochem. J. 238, 885–891.

Kloprogge, E., Hasselaar, P., Gorter, G. and Akkerman, J.W.N. (1986b). Stimulus–aggregation coupling in platelets activated with PAF-acether. Biochim. Biophys. Acta 883, 127–137.

Kloprogge, E., Mommersteeg, M. and Akkerman, J.W.N. (1986c). Kinetics of platelet-activating factor 1-O-alkyl-2-acetyl-sn-glycero-3-phosphocholine-induced fibrinogen binding to human platelets. J. Biol. Chem. 261, 11071–11076.

Koch, C.A., Anderson, D., Moran, M.F., Ellis, C. and Pawson, T. (1991). SH2 and SH3 domains: elements that control interactions of cytoplasmic signalling proteins. Science 252, 668–674.

Koltai, M., Lepran, I., Szekeres, L., Viossat, I., Chabrier, E. and Braquet, P. (1986). Effect of BN 52021, a specific PAF-acether antagonist, on cardiac anaphylaxis in Langendorff hearts isolated from passively sensitized guinea pigs. Eur. J. Pharmacol. 130, 133–136.

Koltai, M., Tosaki, A., Hosford, D., Esanu, A. and Braquet, P. (1991). Effect of BN 50739, a new platelet activating factor antagonist, on ischaemia induced ventricular arrhythmias in isolated working rat hearts. Cardiovas. Res. 391–397.

Komuro, Y., Imanishi, N., Uchida, M. and Morooka, S. (1990). Biological effect of orally active platelet-activating factor receptor antagonist SM-10661. Mol. Pharmacol. 38, 378–384.

Kornecki, E. and Ehrlich, Y.H. (1988). Neuroregulatory and neuropathological actions of the ether phospholipid platelet-activating factor. Science 240, 1792–1794.

Kornecki, E., Ehrlich, Y.H. and Lenox, R.H. (1984). Platelet-activating factor-induced aggregation of human platelets specifically inhibited by triazolobenzodiazepines. Science 226, 1454–1456.

Kroegel, C., Yakawa, T., Westwick, J. and Barnes, P.J. (1989). Evidence for two platelet activating factor receptors on eosinophils: dissociation between PAF-induced intracellular calcium mobilization and degranulation and superoxide anion generation in eosinophils. Biochem. Biophys. Res. Commun. 162, 511–521.

Kudo, I., Nojima, S., Chang, H.W., Yanoshita, R., Hayashi, H., Kondo, E., Nomura, H. and Inoue, K. (1987). Antitumour activity of synthetic alkylphospholipids with or without PAF activity. Lipids 22, 862–867.

Kumar, R., Harvey, S.A.K., Kester, M., Hanahan, D.J. and Olson, M.S. (1988). Production and effects of platelet-activating factor in the rat brain. Biochim. Biophys. Acta 963, 375–383.

Lad, P.M., Olson, C.V. and Grewal, I.S. (1985). Platelet activating factor mediated effects on human neutrophil function are inhibited by pertussis toxin. Biochem. Biophys. Res. Commun. 129, 632–638.

Lambrecht, G. and Parnham, M. (1986). Kadsurenone distinguishes between different platelet activating factor receptor subtypes on macrophages and polymorphonuclear leukocytes. Br. J. Pharmacol. 87, 287–289.

Lapetina, E.G. (1982). Platelet-activating factor stimulates the phosphatidylinositol cycle. Appearance of phosphatidic acid is associated with the release of serotonin in horse platelets. J. Biol. Chem. 257, 7314–7317.

Lapetina, E.G. (1986). In "Phosphoinositides and Receptor Mechanisms" (ed J.W. Putney), pp 271–286. Alan R. Liss, New York.

Lapetina, E.G. and Siegel, E.L. (1983). Shape change induced in human platelets by platelet-activating factor. Correlation with the formation of phosphatidic acid and phosphorylation of a 40,000-dalton protein. J. Biol. Chem. 258, 7241–7245.

Lave, D., James, C., Rajoharison, H., Bost, P.E. and Cavero, I. (1989). Pyrrolo[1,2-c]thiazole derivatives: potent PAF receptor antagonists. Drugs Future 14, 891–898.

Lazenby, C.M., Dive, C., Thompson, M.G. and Hickman, J.A. (1990). The ether lipid SRI 62–834 does not mimic the action of platelet activating factor. Proc. Am. Assoc. Cancer Res. 31, 352.

Lee, T.-C. (1987). In "Platelet-Activating Factor and Related Lipid Mediators" (ed F. Snyder), pp 115–133. Plenum Press, New York.

Lellouch-Tubiana, A., Lefort, J., Simon, M.-T., Pfister, A. and Vargafig, B.B. (1988). Eosinophil recruitment into guinea pig lungs after PAF-acether and allergen administration. Modulation by procyclin, platelet depletion, and selective antagonists. Am. Rev. Respir. Dis. 137, 948–954.

Levi, R., Genovese, A. and Pinckard, R.N. (1989). Alkyl chain homologs of platelet-activating factor and their effects on the mammalian heart. Biochem. Biophys. Res. Commun. 161, 1341–1347.

Linder, M.E., Pang, I.-H., Duronio, R.J., Gordon, J.I., Sternweis, P.C. and Gilman, A.G. (1991). Lipid modifications of G protein subunits, myristoylation of G_0 increases its affinity for beta gamma. J. Biol. Chem. 266, 4654–4659.

Ludwig, J.C., McManus, L.M., Clark, P.D., Hanahan, D.J. and Pinckard, R.N. (1984). Modulation of platelet-activating factor (PAF) synthesis and release from human polymorphonuclear leukocytes (PMN): role of extracellular albumin. Arch. Biochem. Biophys. 241, 337–347.

Lynch, J.M. and Henson, P.M. (1986). The intracellular retention of newly synthesized platelet-activating factor. J. Immunol. 137, 2653–2661.

Lyons, R.M., Stanford, N. and Majerus, P.W. (1975). Thrombin-induced protein phosphorylation in human platelets. J. Clin. Invest. 56, 924–936.

McColl, S.R., Krump, E., Naccache, P.H., Poubelle, P.E., Braquet, P., Braquet, M. and Borgeat, P. (1991). Granulocyte-macrophage colony-stimulatory factor increases the synthesis of leukotriene B_4 by human neutrophils in response to platelet-activating factor: enhancement of both arachidonic acid availability and 5-lipoxygenase activation. J. Immunol. 146, 1204–1211.

MacIntyre, D.E. and Pollack, W.K. (1983). Platelet activating factor stimulates phosphatidylinositol turnover in human platelets. Biochem J. 212, 433–437.

Malbon, C.C., George, S.T. and Moxham, C.P. (1987). Intramolecular disulfide bridges: avenues to receptor activation? Trends Biochem. Sci. 12, 172–175.

Majerus, P.W. and Brodie, G.N. (1972). The binding of phytohemagglutinins to human platelet plasma membranes. J. Biol. Chem. 247, 4253–4257.

Marcheselli, V.L., Rossowska, M.J., Domingo, M.-T., Braquet, P. and Bazan, N.G. (1990). Distinct platelet-activating factor binding sites in synaptic endings and in intracellular membranes of rat cerebral cortex. J. Biol. Chem. 265, 9140–9145.

Marquis, O., Robaut, C. and Cavero, I. (1988) [^{3}H]52770RP, a platelet-activating factor receptor antagonist, and tritiated platelet-activating factor label a common specific binding site in human polymorphonuclear leukocytes. J. Pharmacol. Exp. Ther. 244, 709–715.

Masu, Y., Nakayama, K., Tamaki, H., Harada, Y., Kuno, M. and Nakanishi, S. (1991). cDNA cloning of bovine substance-K receptor through oocyte expression system. Nature 329, 836–838.

Mauco, G., Chap, H. and Douste-Blazy, L. (1983). Platelet activating factor (PAF-acether) promotes an early degradation of phosphatidylinositol-4,5-biphosphate in rabbit platelets. FEBS Lett. 153, 361–365.

Meade, C.J. and Heuer, H.O. (1990). In "PAF and Antagonists, New Developments and Clinical Applications" (eds J.T. O'Flaherty and P. Ramwell), pp 47–80. Portfolio, The Woodlands, TX.

Medof, M.E., Iida, K., Mold, C. and Nussenzweig, V. (1982). Unique role of the complement receptor CR1 in the degradation of C3b associated with immune complexes. J. Exp. Med. 156, 1739–1754.

Merritt, J.E. and Rink, T.J. (1987). Regulation of cytosolic free calcium in fura-2-loaded rat parotid acinar cells. J. Biol. Chem. 262, 17362–17369.

Milligan, G. and Green, A. (1991). Agonist control of G-protein levels. Trends Pharmacol. Sci. 12, 207–209.

Miwa, M., Hill, C., Kumar, R., Sugatani, J., Olson, M.S. and Hanahan, D.J. (1987). Occurrence of an endogenous inhibitor of platelet-activating factor in rat liver. J. Biol. Chem. 262, 527–530.

Molski, T.F.P., Tao, W., Becker, E.L. and Sha'afi, R.I. (1988). Intracellular calcium rise produced by platelet-activating factor is deactivated by fMet-Leu-Phe and this requires uninterrupted activation sequence: role of protein kinase C. Biochem. Biophys. Res. Commun. 151, 836–843.

Morley, J.J. and Kushner, I. (1982). Serum C-reactive protein levels in disease. Ann. N. Y. Acad. Sci. 389, 406–418.

Morooka, S., Komuro, Y., Ando, R. and Uchida, M. (1988). *In vitro* and *in vivo* effects of SM-10661, a newly synthetized platelet-activating factor (PAF) antagonist. Prostaglandins 35, 840–840.

Morrison, W.J. and Shukla, S.D. (1988). Desensitization of receptor-coupled activation of phosphoinositide-specific phospholipase C in platelets: evidence for distinct mechanisms for platelet-activating factor and thrombin. Mol. Pharmacol. 33, 58–63.

Morrison, W.J. and Shukla, S.D. (1989). Antagonism of platelet activating factor receptor binding and stimulated phosphoinositide-specific phospholipase C. J. Pharmacol. Exp. Ther. 250, 831–835.

Motulsky, H.J. and Insel, P.A. (1983). Influence of sodium on the α-adrenergic receptor system of human platelets. Role of intraplatelet sodium in receptor binding. J. Biol. Chem. 258, 3913–3919.

Mumby, S.M., Heuckeroth, R.O., Gordon, J.I. and Gilman, A.G. (1990a). G.-protein α-subunit expression, myristoylation, and membrane association in COS cells. Proc. Natl Acad. Sci. USA 87, 728–732.

Mumby, S.M., Casey, P.J., Gilman, A.G., Gutowski, S. and Sternweis, P.C. (1990b). G protein gamma subunits contain a 20-carbon isoprenoid. Proc. Natl Acad. Sci. USA 87, 5873–5877.

Murphy, P.M., Gallin, E.K. and Tiffany, H.E. (1990). Characterization of human phagocytic cell receptors for C5a and platelet activating factor expressed in *Xenopus* oocytes. J. Immunol. 145, 2227–2234.

Naccache, P.H., Molski, M.M., Volpi, M., Becker, E.L. and Sha'afi, R.I. (1985). Unique inhibitory profile of platelet activating factor induced calcium mobilization, polyphosphoinositide turnover and granule enzyme secretion in rabbit neutrophils towards pertussis toxin and phorbol ester. Biochem. Biophys. Res. Commun. 130, 677–684.

Naccache, P.H., Molski, M.M., Volpi, M., Shefcyk, J., Molski, T.P.F., Loew, L., Becker, E.L. and Sha'afi, R.I. (1986). Biochemical events associated with the stimulation of rabbit neutrophils by platelet-activating factor. J. Leukocyte Biol. 40, 533–548.

Naccache, P.H., Faucher, N., Borgeat, P., Gasson, J.C. and DiPersio, J.F. (1988). Granulocyte–macrophage colony-stimulating factor modulates the excitation–response coupling sequence in human neutrophils. J. Immunol. 140, 3541–3546.

Nakashima, S., Suganuma, A., Sato, M., Tohmatsu, T. and Nozawa, Y. (1989). Mechanism of arachidonic acid liberation in platelet-activating factor-stimulated human polymorphonuclear neutrophils. J. Immunol. 143, 1295–1302.

Nakayama, R., Yasuda, K. and Saito, K. (1987). Existence of endogenous inhibitors of platelet-activating factor (PAF) with PAF in rat uterus. J. Biol. Chem. 262, 13174–13179.

Nathans, J. and Hogness, D.S. (1984). Isolation and nucleotide sequence of the gene encoding human rhodopsin. Proc. Natl Acad. Sci. USA 81, 4851–4855.

Ng, D.S. and Wong, K. (1986). GTP regulation of platelet-activating factor binding to human neutrophil membranes. Biochem. Biophys. Res. Commun. 141, 353–359.

Ng, D.S. and Wong, K. (1988a). Specific binding of platelet-activating factor (PAF) by human peripheral blood mononuclear leukocytes. Biochem. Biophys. Res. Commun. 155, 311–316.

Ng, D.S. and Wong, K. (1988b). Effect of sulfhydryl reagents on PAF binding to human neutrophils and platelets. Eur. J. Pharmacol. 154, 47–52.

Ng, D.S. and Wong, K. (1989). Effect of platelet-activating factor (PAF) on cytosolic free calcium in human peripheral blood mononuclear leukocytes. Res. Commun. Chem. Pathol. Pharmacol. 64, 351–354.

Nishibe, S., Wahl, M.I., Tonks, N.K., Rhee, S.G. and Carpenter, G. (1990). Tyrosine phosphorylation increases the catalytic activity of phospholipase C-gamma. Science 250, 1253–1256.

Nishizuka, Y. (1984). The role of protein kinase C in cell surface signal transduction and tumor promotion. Nature 308, 693–698.

Numao, T., Fukuda, T., Akutsu, I. and Makino, S. (1989). In "3rd Int. Conf. on Platelet-Activating Factor and Structurally Related Alkyl Ether Lipids, Tokyo, Japan, 8–12 May, p 140 (Abstr.).

Nunez, D., Chignard, M., North, R., Le Conedic, J.P., Norel, X., Spinnewyn, B., Braquet, P. and Benveniste, J. (1986). Specific inhibition of PAF-catheter-induced platelet activation by BN 52021 and comparison with PAF-acether inhibitors Kadsurenone and CV 3988. Eur. J. Pharmacol. 123, 197–205.

Nunez, D., Kumar, R. and Hanahan, D.J. (1989). Inhibition of [3H]platelet activating factor (PAF) by Zn^{2+}: a possible explanation for its specific PAF antiaggregating effects in human platelets. Arch. Biochem. Biophys. 272, 466–475.

Nunez, D., Randon, J., Gandhi, C., Siafaka-Kapadai, A., Olson, M.S. and Hanahan, D.J. (1990). The inhibition of platelet-activating factor-induced platelet activation by oleic acid is associated with a decrease in phosphoinositide metabolism. J. Biol. Chem. 265, 18330–18338.

O'Donnell, M., Crowley, H.J., Yaremko, B., O'Nell, N., Burghardt, C. and Welton, A.F. (1990). In "An Int. Meeting of Adv. in the Understanding and Treatment of Asthma, London, UK 1–3 Oct. (Abstr.).

O'Flaherty, J.T. (1982). Lipid mediators of inflammation and allergy. Lab. Invest. 47, 314–329.

O'Flaherty, J.T. and Rossi, A.G. (1989). In "3rd Int. Conf. on Platelet-Activating Factor and Structurally Related Alkyl Ether Lipids", 32 (Abstr.).

O'Flaherty, J.T., Lees, C.J., Miller, C.H., McCall, C.E., Lewis, J.C., Love, S.H. and Wykle, R.L. (1981). Selective desensitization of neutrophils: further studies with 1-O-alkyl-*sn*-glycero-3-phosphocholine analogs. J. Immunol. 127, 731–737.

O'Flaherty, J.T., Surles, J.R., Redman, J., Jacobson, D., Piantadosi, C. and Wykle, R.L. (1986). Binding and metabolism of platelet-activating factor by human neutrophils. J. Clin. Invest. 78, 381–388.

O'Flaherty, J.T., Jacobson, D.P. and Reman, J.F. (1989). Bidirectional effects of protein kinase C activators. Studies with human neutrophils and platelet-activating factor. J. Biol. Chem. 264, 6836–6843.

O'Flaherty, J.T., Jacobson, D.P., Redman, J.F. and Rossi, A.G. (1990a). Translocation of protein kinase C in human polymorphonuclear neutrophils. Regulation by cytosolic Ca^{2+}-independent and Ca^{2+}-dependent mechanisms. J. Biol. Chem. 265, 9146–9152.

O'Flaherty, J.T., Redman, J.F., Jacobson, D.P. and Rossi, A.G. (1990b). Stimulation and priming of protein kinase C translocation by a Ca^{2+} transient independent mechanism. Studies in human neutrophils challenged with platelet-activating factor and other receptor agonists. J. Biol. Chem. 265, 21619–21623.

Okamoto, M., Yoshida, K., Nishikawa, M., Ando, T., Iwami, M., Kohsaka, M. and Aoki, H. (1986a). FR-900452, a specific antagonist of platelet-activating factor (PAF) produced by Streptomyces phaeofaciens. I. Taxonomy, fermentation, isolation and physio-chemical and biological characteristics. J. Antibiot. 1, 141–147.

Okamoto, M., Yoshida, K., Uchida, I., Kohsaka, M. and Aoki, H. (1986b). Studies of platelet-activating factor (PAF) antagonists from microbial products, II. Pharmacological studies of FR-49175 in animal model. Chem. Pharm. Bull. 34, 345–348.

Okamoto, M., Yoshida, K., Uchida, I., Nishikawa, M., Kohsaka, M. and Aoki, H. (1986c). Studies of platelet-activating factor (PAF) antagonists from microbial products (1) Bisdethiobis (methylthio)gliotoxin and its derivatives. Chem. Pharm. Bull. 34, 340–344.

O'Rourke, F.A., Halenda, S.P., Zavoico, G.B. and Feinstein, M.B. (1985). Inositol 1,4,5-trisphosphate releases Ca^{2+} from a Ca^{2+}-transporting membrane vesicle fraction derived from human platelets. J. Biol. Chem. 260, 956–962.

Ortega, M.P., Garcia, M.D.C., Gijon, M.A., Casa-Juana, M.F.D., Priego, J.G., Sanchez Crespo, M. and Sunkel, C. (1990). 1,4-dihydropyridines, a new class of platelet-activating factor receptor antagonists: *in vitro* pharmacologic studies. J. Pharmacol. Exp. Ther. 255, 28–39.

Ostermann, G., Hofmann, B., Kertscher, H.-P. and Till, U. (1990). Platelet-activating factor (PAF) inhibitory profile of KO-286011 on blood platelets *in vitro* and *in vivo*. Naunyn-Schmiedeberg's Arch. Pharmacol. 342, 713–718.

Ott, S., Costa, T. and Herz, A. (1988). Sodium modulates opioid receptors through a membrane component different from G-proteins: demonstration by target size analysis. J. Biol. Chem. 263, 10524–10533.

Page, C.P. (1990). The role of PAF in allergic respiratory disease. Drug News Perspective 3, 389–395.

Palmer, R.M.J. and Salmon, J.A. (1983). Release of leukotriene B_4 from human neutrophils and its relationship to degranulation induced by N-formyl-methionyl-leucyl-phenylalanine, serum-treated zymosan and the ionophore A23187. Immunology 50, 65–73.

Pan, J.-X., Hensens, O.D., Zink, D.L., Chang, M.N. and Hwang, S.-B. (1987). Lignans with platelet-activating factor antagonist activity from *Magnolia biondii*. Phytochemistry 26, 1377–1379.

Parry, M.J., Alabaster, V.A., Cheeseman, H.E., de Souza, R.N. and Keir, R.F. (1990). In "IUPHAR Satellite Symposium, Paris, France, 28–29 June" (Abstr. 13).

Paulson, S.K., Wolf, J.L., Novotney-Barry, A. and Cox, C.P. (1990). Pharmacologic characterization of the rabbit neutrophil receptor for platelet-activating factor. Proc. Soc. Exp. Biol. Med. 195, 247–254.

Peplow, P.V. and Mikhailidis, D.P. (1990). Platelet-activating factor (PAF) and its relation to prostaglandins, leukotrienes, and other aspects of arachidonate metabolism. Prostaglandins Leukot. Essent. Fatty Acids 41, 71–82.

Pepys, M.B. (1981). C-reactive protein fifty years on. Lancet i, 653–656.

Pinckard, R.N., Ludwig, J.C. and McManus, L.M. (1988). In "Inflammation: Basic Principles and Clinical Correlates" (eds J.I. Gallin, I.M. Goldstein and R. Snyderman), pp 139–167. Raven Press, New York.

Piwinski, J.J., Wong, J.K., Green, M.J., Ganguly, A.K., Billah, M.M., West, R.E. and Kreutner, W. (1991). Dual antagonists of platelet activating factor and histamine. Identification of structural requirements for dual activity of N-acyl-4-(5,6-dihydro-11*H*-benzo[5,6]cyclohepta[1,2-*b*]pyridin-11-ylidene)piperidines. J. Med. Chem. 34, 457–461.

Pommier, C.G., O'Shea, J., Chused, T., Yancey, K., Franck, M.M., Takahashi, T. and Brown, E. (1984). Studies on the fibronectin receptors of human peripheral leukocytes. J. Exp. Med. 151, 602–613.

Ponpipom, M.M., Yue, B.-Z., Bugianesi, R.L., Brooker, D.R., Chang, M.N. and Shen, T.Y. (1986). Total synthesis of kadsurenone and its analogs. Tetrahedron Lett. 27, 309–312.

Ponpipom, M.M., Bugianesi, R.L., Brooker, D.R., Yue, B.-Z., Hwang, S.-B. and Shen, T.Y. (1987). Structure-activity relationships of kadsurenone analogues. J. Med. Chem. 30, 136–142.

Ponpipom, M.M., Hwang, S.-B., Doebber, T.W., Acton, J.J., Alberts, A.W., Biftu, T., Brooker, D.R., Bugianesi, R.L., Chabala, J.D., Gamble, N.L., Graham, D.W., Lam, M.-H. and Wu, M.S. (1988). (±)-*trans*-2(3-Methoxy-5-methylsulfonyl-4-propoxyphenyl)-5-(3,4,5-trimethoxyphenyl)tetrahydrofuran (L-659,989), a novel, potent PAF receptor antagonist. Biochem. Biophys. Res. Commun. 150, 1213–1220.

Prescott, S.M., Zimmerman, G.A. and McIntyre, T.M. (1984). Human endothelial cells in culture produce platelet-activating factor (1-alkyl-2-acetyl-*sn*-glycero-3-phosphocholine) when stimulated with thrombin. Proc. Natl Acad. Sci. USA 81, 3534–3538.

Prescott, S.M., Zimmerman, G.A. and McIntyre, T.M. (1990). Platelet-activating factor. J. Biol. Chem. 265, 17381–17384.

Prpic, V., Uhing, R.J., Weiel, J.E., Jakoi, L., Gawdi, G., Herman, B. and Adams, D.O. (1988). Biochemical and functional responses stimulated by platelet-activating factor in murine peritoneal macrophages. J. Cell Biol. 107, 363–372.

Putney, J.W. Jr (ed) (1986). "Phosphoinositides and Receptor Mechanisms". Alan R. Liss, New York.

Ramesha, C.S. and Pickett, W.C. (1987). Species-specific

variations in the molecular heterogeneity of platelet-activating factor. J. Immunol. 138, 1559–1563.

Resh, M.D. (1989). Specific and saturable binding of pp60[v-src] to plasma membranes: evidence for a myristyl-src receptor. Cell 58, 281–286.

Resh, M.D. and Ling, H.-P. (1990). Identification of a 32K plasma membrane protein that binds to the myristylated amino-terminal sequence of p60[v-src]. Nature 346, 84–86.

Riches, D.W.H., Young, S.K., Seccombe, J.F., Lynch, J.M. and Henson, P.M. (1985). The subcellular distribution of platelet-activating factor (PAF) in phagocytosing human neutrophils. Fed. Proc. 44, 737 (Abstr.).

Robaut, C., Durand, G., James, C., Lave, D., Sedivy, P., Floch, A., Mondot, S., Pacot, D., Cavero, I. and LeFur, G. (1987). PAF binding sites: characterization by [^{3}H]52770RP, a pyrrolo[1,2c]thiazole derivative, in rabbit platelets. Biochem. Pharmacol. 36, 3221–3229.

Robaut, C., Mondot, S., Floch, A., Tahraoui, L. and Cavero, I. (1988). Pharmacological profile of a novel potent and specific PAF receptor antagonist, the 59227 RP. Prostaglandins 35, 838–838.

Robinovici, R., Yue, T.-.L., Farhat, M., Smith, E.F., Esser, K.M., Slivjak, M. and Feuerstein, G. (1990). Platelet activating factor (PAF) and tumor necrosis factor-alpha (TNFalpha) interactions in endotoxemic shock: studies with BN 50739, a noval PAF antagonist. J. Pharmacol. Exp. Ther. 255, 256–263.

Rodbell, M. (1983). In "Cell Surface Receptors" (ed P.G. Strange), pp 227–239. Ellis Horwood, Chichester.

Rola-Pleszczynski, M., Pignol, B., Pouliot, C. and Braquet, P. (1986). Inhibition of human lymphocyte proliferation and interleukin 2 production by platelet activating factor (PAF-acether): reversal by a specific antagonist, BN 52021. Biochem. Biophys. Res. Commun. 142, 754–760.

Sage, S.O. and Rink, T.J. (1987). The kinetics of changes in intracellular calcium concentration in fura-2-loaded human platelets. J. Biol. Chem. 262, 16364–16369.

Sage, S.O., Merritt, J.E., Hallam, T.J. and Rink, T.J. (1989). Receptor-mediated calcium entry in fura-2-loaded human platelets stimulated with ADP and thrombin. Dual-wavelengths studies with Mn^{2+}. Biochem. J. 258, 923–926.

Sanchez Crespo, M. (1990). Advances in platelet-activating factor research. Drugs News Perspective 3, 573–576.

Sano, K., Takai, Y., Yamanishi, J. and Nishizuka, Y. (1983). A role of calcium-activated phospholipid-dependent protein kinase in human platelet activation. Comparison of thrombin and collagen actions. J. Biol. Chem. 258, 2010–2013.

Satoh, K., Imaizumi, T.-A., Kawamura, Y., Yoshida, H., Hiramoto, M., Takamatsu, S. and Takamatsu, M. (1991). Platelet-activating factor (PAF) stimulates the production of PAF acetylhydrolase by the human hepatoma cell line, HepG2. J. Clin. Invest. 87, 476–481.

Saunders, R.N. and Handley, D.A. (1987). Platelet-activating factor antagonists. Annu. Rev. Pharmacol. Toxicol. 27, 237–255.

Scarpace, P.J. and Abrass, I.B. (1982). Desensitization of adenylate cyclase and down regulation of beta adrenergic receptors after *in vivo* administration of beta agonists. J. Pharmacol. Exp. Ther. 223, 327–331.

Schlondorf, D. and Neuwirth, R. (1986). Platelet-activating factor and the kidney. Am. J. Physiol. 251, F1–F11.

Schultz, A.M., Tsai, S.-C., Kung, H.-F., Oroszlan, S., Moss, J., and Vaughan, M. (1987). Hydroxylamine-stable covalent linkage of myristic acid in G_0, a guanine nucleotide-binding protein of bovine brain. Biochem. Biophys. Res. Commun. 146, 1234–1239.

Schwertschlag, U.S. and Whorton, A.R. (1988). Platelet-activating factor-induced homologous and heterologous desensitization in cultured vascular smooth muscle cells. J. Biol. Chem. 263, 13791–13796.

Sedivy, P., Caillard, C.G., Floch, A., Folliard, F., Mondot, S., Robaut, C. and Terlain, B. (1985). 48740 RP: a specific PAF-acether antagonist. Prostaglandins 30, 688 (Abstr.).

Sedivy, P., Caillard, C.-G., Carruette, A., Deregnaucourt, J. and Mondot, S. (1986). 48 740 RP: selective anti-PAF agent. Adv. Inflam. Res. 10, 171–173.

Shalit, M., Allmen, C.V., Atkin, P.C. and Zweiman, B. (1988). Platelet-activating increases expression of complement receptors on human neutrophils. J. Leukocyte Biol. 44, 212–217.

Shen, T.Y. and Hussaini, I.M. (1990). Kadsurenone and other related lignans as antagonists of platelet-activating factor receptor. Meth. Enzymol. 187, 446–454.

Shen, T.Y., Hwang, S.-B., Chang, M.N., Doebber, T.W., Lam, M.H., Wu, M.S., Wang, X., Han, G.Q. and Li, R.Z. (1985). Characterization of a platelet-activating factor receptor antagonist isolated from haifentent (*Piper futokadsura*): specific inhibition of *in vitro* and *in vivo* platelet-activating factor-induced effects. Proc. Natl Acad. Sci. USA 82, 672–676.

Shen, T.Y., Hwang, S.-B., Doebber, T.W. and Robbins, J.C. (1987a). In "Platelet-Activating Factor and Related Lipid Mediators" (ed F. Snyder), pp 153–190. Plenum Press, New York.

Shen, T.Y., Yang, S.S. and Hwang, S.-B. (1987b). Indene derivatives useful as PAF antagonists. US Patent No. 4,656, 190.

Shen, T.Y., Hussaini, I., Hwang, S.-B. and Chang, M.N. (1989). Recent development of platelet-activating factor antagonists. Adv. Prostaglandin Thromboxane Leukotriene Res. 19, 359–362.

Shimazaki, N., Shima, I., Hemmi, K. and Hishimoto, M. (1987). Diketopiperazines as a new class of platelet-activating factor inhibitors. J. Med. Chem. 30, 1706–1709.

Shimazaki, N., Shima, I., Okamoto, M., Yoshida, K., Hemmi, K. and Hishimoto, M. (1989). In "3rd Int. Conf. on Platelet-Activating Factor and Structurally Related Alkyl Ether Lipids", 12 (Abstr.).

Shukla, S.D. (1985). Platelet-activating factor-stimulated formation of inositol triphosphate in platelets and its regulation by various agents including Ca^{2+}, indomethacin, CV-3988 and forskolin. Arch. Biochem. Biophys. 240, 674–681.

Shukla, S.D., Buxton, D.B., Olson, M.S. and Hanahan, D.J. (1983). Acetylglyceryl ether phosphorylcholine, a potent activator of hepatic phosphoinositide metabolism and glycogenolysis. J. Biol. Chem. 258, 10212–10214.

Shukla, S.D., Morrison, W.J. and Dhar, A. (1989). Desensitization of platelet-activating factor-stimulated protein phosphorylation in platelets. Mol. Pharmacol. 35, 409–413.

Sibley, D.R., Benovic, J.L., Caron, M.G. and Lefkowitz, R.L. (1987). Regulation of transmembrane signalling by receptor phosphorylation. Cell 48, 913–922.

Siess, W. and Lapetina, E.G. (1990). Functional relationship between cyclic AMP-dependent phosphorylation and platelet inhibition. Biochem. J. 271, 815–819.

Singer, I.I., Scott, S., Kawaka, D.K. and Kazazis, D.M. (1989). Adhesomes: specific granules containing receptors for laminin,

C3Bi/fibrinogen, fibronectin, and vitronectin in human polymorphonuclear leukocytes and monocytes. J. Cell Biol. 109, 3169–3183.

Smith, C.D., Uhing, R.J. and Snyderman, R. (1987). Nucleotide regulatory protein-mediated activation of phospholipase C in human polymorphonuclear leukocytes is disrupted by phorbol esters. J. Biol. Chem. 262, 6121–6127.

Snyder, F. (1989). Biochemistry of platelet-activating factor: a unique class of biologically active phospholipids. Proc. Soc. Exp. Biol. Med. 190, 125–135.

Spat, A., Brodford, P.G., Mekinney, J.S., Rubin, R.P. and Putney, J.W. Jr (1986). A saturable receptor for ^{32}P-inositol-1,4,5-trisphosphate in hepatocytes and neutrophils. Nature 319, 514–516.

Squinto, S.P., Block, A.L., Braquet, P. and Bazan, N.G. (1989). Platelet-activating factor stimulates a fos/jun/AP-1 transcriptional signalling in human neuroblastoma cells. J. Neurosci. Res. 24, 558–566.

Squinto, S.P., Braquet, P. Block, A.L. and Bazan, N.G. (1990). Platelet-activating factor activates HIV promotor in transfected SH-Sy5Y neuroblastoma cells and MOLT-4 T-lymphocytes. J. Mol. Neurosci. 2, 79–84.

Stanislawski, L., Huu, T.P. and Perianin, A. (1990). Priming effect of fibronectin on respiratory burst of human neutrophils induced by formyl peptides and platelet-activating factor. Inflammation 14, 523–530.

Steiner, M., Landolfi, R., Motola, N.C. and Turcotte, J.G. (1985). Biological activity of platelet activating factor-amidophosphonate (PAF-AP), a novel phosphonolipid selective inhibitor of platelet activating factor (PAF). Biochem. Biophys. Res. Commun., 851–855.

Stenzel, H., Sannwald, U. and Hahn, H.L. (1986). In "6th Int. Conf. on Prostaglandins and Related Compounds, Florence, Italy, 3–6 June" (Abstr.).

Stewart, A.G. and Phillips, W.A. (1989). Intracellular platelet-activating factor regulates eicosanoid generation in guinea-pig resident peritoneal macrophages. Br. J. Pharmacol. 98, 141–148.

Stewart, A.G., Dubbin, P.N., Harris, T. and Dusting, G.J. (1989). Evidence for an intracellular action of platelet-activating factor in bovine cultured aortic endothelial cells. Br. J. Pharmacol. 96, 503–505.

Stewart, A.G., Dubbin, P.N., Harris, T. and Dusting, G.J. (1990). Platelet-activating factor may act as a second messenger in the release of eicosanoids and superoxide anions from leukocytes and endothelial cells. Proc. Natl Acad. Sci. USA 87, 3215–3219.

Stewart, A.G., Harris, T., De Nichilo, M. and Lopez, A.F. (1991). Involvement of leukotriene B$_4$ and platelet-activating factor in cytokine priming of human polymorphonuclear leukocytes. Immunology 72, 206–212.

Strader, C.D., Sigal, I.S., Candelore, M.R., Rand, E., Hill, W.S. and Dixon, R.A.F. (1987). Conserved aspartic acid residues 79 and 113 of the β-adrenergic receptor have different roles in receptor function. J. Biol. Chem. 263, 10267–10271.

Strader, C.D., Sigal, I.S. and Dixon, R.A.F. (1989). Structural basis of β-adrenergic receptor function. FASEB J. 3, 1825–1832.

Stryer, L. and Bourne, H.R. (1986). G-protein: a family of signal transducers. Annu. Rev. Cell Biol. 2, 391–419.

Sugatani, J. and Hanahan, D.J. (1986). Characterization of 1-O-alkyl-2-acetyl-sn-glycero-3-phosphocholine (AGEPC)-induced protein phosphorylation in rabbit platelets: Inhibitory effect of AGEPC analogs. Arch. Biochem. Biophys. 246, 855–864.

Sugatani, J., Steinhelper, M.E., Saito, K., Olson, M.S. and Hanahan, D.J. (1987). Potential involvement of vicinal sulfhydryls in stimulus-induced rabbit platelet activation. J. Biol. Chem. 262, 16995–17001.

Sunkel, C.E., Casa-Juana, M.F., Santos, L., Gomez, M.M., Villarroya, M., Gonzalez-Morales, M.A., Priego, J.G. and Ortega, M.P. (1990). 4-Alkyl-1,4-dihydropyridines derivatives as specific PAF-acether antagonists. J. Med. Chem. 33, 3205–3210.

Tahraoui, L., Floch, A., Mondot, S. and Cavero, I. (1988). High affinity specific binding sites for tritiated platelet–activating factor in canine platelet membranes: counterparts of platelet-activating factor receptors mediating platelet aggregation. Mol. Pharmacol. 34, 145–151.

Takehara, S., Mikashima, H., Muramoto, Y., Terasawa, M., Setoguchi, M. and Tahara, T. (1990). Pharmacological actions of Y-24180, a new specific antagonist of platelet activating factor (PAF): II. Interactions with PAF and benzodiazepine receptors. Prostaglandins 40, 571–583.

Takatani, M., Maezaki, N., Imura, Y., Terashita, Z.-I., Nishikawa, K. and Susumu, T. (1990). Platelet-activating factor (PAF) antagonists: development of a highly potent PAF antagonist, TCV-309. Adv. Prostaglandin Thromboxane Leukotriene Res. 21, 943–946.

Tanabe, Y., Kubota, Y.-N., Sanemitsu, Y., Itaya, N. and Suzukamo, G. (1991a). Stereoselective synthesis of anti-PAF active thiazolidin-4-ones via cyclo-condensation of alkyl alpha-mercaptocarboxylates with arylimines. Tetrahedron Lett. 32, 383–386.

Tanabe, Y., Suzukamo, G., Komuro, Y., Imanishi, N., Morooka, S., Enomoto, M., Kojima, A., Sanemitsu, Y. and Mizutani, M. (1991b). Structure-activity relationship of optically active 2-(3-pyridyl)thiazolidin-4-ones as a PAF antagonist. Tetrahedron Lett. 32, 379–382.

Terasawa, M., Aratani, H., Setogushi, M. and Tahara, T. (1990). Pharmacological actions of Y-24180: I. A potent and specific antagonist of platelet-activating factor. Prostaglandins 40, 553–569.

Terashita, Z., Tsushima, S., Yoshioka, Y., Nomura, H., Inada, Y. and Nishikawa, K. (1983). CV-3988-A specific antagonist of platelet-activating factor (PAF). Life Sci. 32, 1975–1982.

Terashita, Z., Imura, Y. and Nishikawa, K. (1985a). Inhibition by CV-3988 of the binding of [^{3}H]platelet-activating factor (PAF) to the platelet. Biochem. Pharmacol. 43, 1491–1495.

Terashita, Z., Imura, Y., Nishikawa, K. and Sumida, S. (1985b). Is platelet-activating factor (PAF) a mediator of endotoxin shock. Eur. J. Pharmacol. 109, 257–261.

Terashita, Z., Imura, Y., Takatani, M., Tsushima, S. and Nishikawa, K. (1987). CV-6209-A highly potent platelet-activating factor (PAF) antagonist in in vitro and in vivo. J. Pharmacol. Exp. Ther. 242, 263–268.

Tessner, T.G., O'Flaherty, J.T. and Wykle, R.L. (1989). Stimulation of platelet-activating factor synthesis by a non-metabolizeable bioactive analog of platelet-activating factor and influence of arachidonic acid metabolites. J. Biol. Chem. 264, 4794–4799.

Thierry, A., Doly, M., Braquet, P., Cluzel, J. and Meyniel, G. (1989). Presence of specific platelet-activating factor binding sites in the rat retina. Eur. J. Pharmacol. 163, 97–101.

Thompson, K.L., Chang, M.N., Bugianesi, R.L., Ponpipom, M.M., Arison, B.H., Hucker, H.B., Sweeney, B.M., White, S.D. and Chabala, J.C. (1991). Metabolism of the platelet-activating factor (±)-trans-2-(3'-methoxy-5'-methylsulphonyl-

4'propoxyphenyl)-5-(3″,4″,5″-trimethoxyphenyl)tetrahydrofuran (L-659,989) in rhesus monkeys. Xenobiotica 21, 613–625.

Tilley, J.W., Burghardt, B., Burghardt, C., Mowles, T.F., Leiweber, F.-J., Klevans, L., Young, R., Hirkaler, G., Fahrenholtz, K., Zawoiski, S. and Todaro, L.J. (1988). Pyrido[2,1-*b*]quinazolinecarboxamide derivatives as platelet activating factor antagonists. J. Med. Chem. 31, 466–472.

Tilley, J.W., Clader, J.W., Zawoiski, S., Wirkus, M., LeMahieu, R.A., O'Donnell, M., Crowley, H. and Welton, A.F. (1989). Biphenylcarboxamide derivatives as antagonists of platelet-activating factor. J. Med. Chem. 32, 1814–1820.

Tokumura, A., Homma, H. and Hanahan, D.J. (1985). Structural analogs of alkylacetyl glycerophosphocholine inhibitory behaviour on platelet activation. J. Biol. Chem. 260, 12710–12714.

Tokumura, A., Yoshida, J.-I., Okasaka, N., Fukuzawa, K. and Tsukatani, H. (1987). Platelet-aggregation induced by ether-linked phospholipids. 2. Mechanism of desensitization of rabbit platelets by platelet-activating factor and reversibility of inhibitory actions of its antagonists. Thromb. Res. 46, 153–161.

Tomioka, K., Garrido, R., Ahmed, A., Stevenson, J.S. and Abraham, W.M. (1989). YM461, a PAF antagonist, blocks antigen-induced late airway responses and airway hyperresponsiveness in allergic sheep. Eur. J. Pharmacol. 170, 209–215.

Tool, A.T.J., Verhoeven, A.J., Roos, D. and Koenderman, L. (1989). Platelet-activating factor (PAF) acts as an intracellular messenger in the changes of cytosolic free Ca^{2+} in human neutrophils induced by opsonized particles. FEBS Lett. 259, 209–212.

Touvay, C., Vilain, B., Etienne, A., Clostre, F., Drieu, K. and Braquet, P. (1985). Proof of the involvement of PAF-acether in pulmonary complex immune systems using a specific PAF-acether receptor antagonist: BN 52021. Int. J. Immunopharmacol. 7, 385 (Abstr.).

Toyofuku, T., Kubo, K., Kobayashi, T. and Kusama, S. (1986). Effects of Ono-6240, a PAF antagonist, on endotoxin shock in unanaesthetized sheep. Prostaglandins 31, 271–281.

Triggiani, M., Hubbard, W.C. and Chilton, F.H. (1990). Synthesis of 1-acyl-2-acetyl-*sn*-3-phosphocholine by an enriched preparation of the human lung mast cells. J. Immunol. 144, 4773–4780.

Tsunoda, H., Sakuma, Y., Harada, K., Muramoto, K., Katayama, S., Horie, T., Shimomura, N., Clark, R., Miyazawa, S., Okano, K., Machida, Y., Katayama, K. and Yamatsu, I. (1990). Pharmacological activities of a novel thienodiazepine derivative as a platelet-activating factor antagonist. Arzeim.-Forsch./Drug Res. 40, 1201–1205.

Tsundo, H., Sakuma, Y., Shirato, M., Obaishi, H., Harada, K., Yamada, K., Shimomura, N., Machida, Y., Yamatsu, I. and Katayama, K. (1991). Activity of a novel thienodiazepine derivative as a platelet-activating factor antagonist in guinea pig lungs. Arzeim.-Forsch./Drug Res. 41, 224–227.

Tuffin, D.P., Davey, P., Dyer, R.L., Lunt, D.O. and Wade, P.J. (1985). Specific binding of [³H]1-*O*-octadecyl PAF-acether to washed human platelets. Adv. Exp. Med. Biol. 192, 83–96.

Ukena, D., Dent, G., Birke, B.W., Robaut, C., Sybrecht, G.W. and Barnes, P.J. (1988). Radioligand binding of antagonists of platelet-activating factor to intact human platelets. FEBS Lett. 228, 285–289.

Ukena, D., Krogel, C., Dent, G., Yukawa, T., Sybrecht, G. and Barnes, P.J. (1989). PAF-receptors on eosinophils: Identification with a novel ligand, [³H]WEB-2086. Biochem. Pharmacol. 38, 1702–1705.

Ullrich, A. and Schlessinger, J. (1990). Signal transduction by receptors with tyrosine kinase activity. Cell 61, 203–212.

Umezawa, K., Imoto, M., Sawa, T., Isshiki, K., Mutsuda, N., Uchida, T., Iinuma, H., Hamada, M. and Takeuchi, T. (1986). Studies on a new epidermal growth factor-receptor kinase inhibitor, erbstatin, produced by MH435-LF3. J. Antibiot. 39, 170–173.

Umezawa, K., Tanaka, K., Hori, T., Abe, S., Sekizawa, R. and Imoto, M. (1991). Induction of morphological change by tyrosine kinase inhibitors in Rous sarcoma virus-transformed rat kidney cells. FEBS Lett. 279, 132–136.

Valone, F.H. (1988). Identification of platelet-activating factor receptors in P388D₁ murine macrophages. J. Immunol. 140, 2389–2394.

Valone, F.H. and Goetzl, E.J. (1983). Specific binding by human polymorphonuclear leukocytes of the immunological mediator 1-*O*-hexadecyl/octadecyl-2-acetyl-*sn*-glycero-3-phosphorylcholine. Immunology 48, 141–149.

Valone, F.H. and Johnson, B. (1985). Modulation of cytoplasmic calcium in human platelets by the phospholipid platelet-activating factor, 1-*O*-alkyl-2-acetyl-*sn*-glycero-3-phosphorylcholine. J. Immunol. 134, 1120–1124.

Valone, F.H. and Ruis, N.M. (1986). Platelet-activating factor binding to human platelet membranes. Biotechnol. Appl. Biochem. 8, 465–470.

Valone, F.H., Cole, E., Reinhold, V.R. and Goetzl, E.J. (1982). Specific binding of phospholipid platelet-activating factor by human platelets. J. Immunol. 129, 1637–1641.

Vargaftig, B.B. (1987). In "Platelet-Activating Factor and Related Lipid Mediators" (ed F. Snyder), pp 341–353. Plenum Press, New York.

Vargaftig, B.B. and Braquet, P.G. (1987). PAF-acether today — relevance for acute experimental anaphylaxis. Br. Med. Bull. 43, 312–335.

Vargaftig, B.B., Pretolani, M., Coeffier, E. and Chignard, M. (1989). In "Handbook of Inflammation", Vol. 6 (eds P.M. Henson and R.C. Murphy), pp 113–146. Elsevier, Amsterdam.

Verby, N.L., Van Delft, J.L., Van Haeringen, N.J. and Braquet, P. (1989). Platelet-activating factor and laser trauma of the iris. Invest. Ophthalmol. Vis. Sci. 30, 1101–1103.

Vercelloti, G.M., Yin, H.Q., Gustafson, K.S., Nelson, R.D. and Jacob, H.S. (1988). Platelet-activating factor primes neutrophil responses to agonists: role in promoting neutrophil-mediated endothelial damage. Blood 71, 1100–1107.

Verghese, M.W., Charles, L., Jakoi, L., Dillon, S.B. and Snyderman, R. (1987). Role of a guanine nucleotide regulatory protein in the activation of phospholipase C by different chemoattractants. J. Immunol. 138, 4374–4380.

Vigo, C. (1985). Effect of C-reactive protein on platelet-activating factor-induced platelet aggregation and membrane stabilization. J. Biol. Chem. 260, 3418–3422.

Votta, B. and Mong, S. (1990). Binding of radiolabelled antagonist to platelet-activating factor in human lung membranes: Shifting of agonist binding affinity states by cations and guanine nucleotide. Life Sci. 46, 309–313.

Wade, P.J., Lad, N. and Tuffin, D.P. (1986). Interaction of the human platelet PAF-acether binding site with calcium and calcium channel antagonists. Prog. Lipid Res. 25, 163–165.

Waldman, R., Bauer, S., Gobel, C., Hoffman, F., Jakobs, K.H. and Walter, U. (1986). Demonstration of cGMP-dependent protein kinase and cGMP-dependent phosphorylation in cell free extracts of platelets. Eur. J. Biochem. 158, 203–210.

Wallace, J.L., Steel, G., Whittle, B.J.R., Lagente, V. and Vargaftig, B. (1987). Evidence for platelet-activating factor as a mediator of endotoxin-induced gastrointestinal damage in the rat. Effects of three platelet-activating factor antagonists. Gasteroenterology 93, 765–773.

Walser, A., Flynn, T., Mason, C., Crowley, H., Maresca, C. and O'Donnell, M. (1991). Thienotriazolodiazepines as platelet-activating factor antagonists. Steric limitations for the substituent in position 2. J. Med. Chem. 34, 1440–1446.

Wang, C.J. and Tai, H.H. (1991). Monoclonal anti-idiotypic antibodies to platelet activating factor (PAF) and their interaction with PAF receptors. J. Biol. Chem. 266, 12372–12378.

Weber, K.H. and Heuer, H.O. (1989). Hetrazepines as antagonists of platelet activating factor. Med. Res. Rev. 9, 181–218.

Weisbart, R.H., Kwan, L., Golde, D.W. and Gasson, J.C. (1987). Human GM-CSF primes neutrophils for enhanced oxidative metabolism in response to the major chemattractants. Blood 69, 18–21.

Welton, A.F. and O'Donnell, M. (1989). New pharmacological agents which antagonize leukotriene D_4 and platelet-activating factor. NATO ASI Ser. A 177, 83–104.

Wong, S.K.-F., Parker, E.M. and Ross, E.M. (1990). Chimeric muscarinic cholinergic: β-adrenergic receptors that activate G_s in response to muscarinic agonists. J. Biol. Chem. 265, 6219–6224.

Workman, P., Donaldson, J. and Lohmeyer, M. (1990). Platelet-activating factor (PAF) antagonist WEB 2086 does not modulate the cytotoxicity of PAF or antitumour alkyl lysophospholipids ET-18-O-methyl and SRI 62–834 in HL-60 promyelocytic leukaemia cells. Biochem. Pharmacol. 41, 319–322.

Worthen, G.S., Seccombe, J.F., Clay, K.L., Guthrie, L.A. and Johnston, J.B. (1988). The priming of neutrophils by lipopolysaccharide for production of intracellular platelet-activating factor. Potential role in mediation of enhanced superoxide secretion. J. Immunol. 140, 3553–3559.

Yamada, K., Iwahashi, K. and Kase, H. (1988). Parallel inhibition of platelet-activating factor-induced protein phosphorylation and serotonin release by K-252a, a new inhibitor of protein kinases, in rabbit platelets. Biochem. Pharmacol. 37, 1161–1166.

Yamada, T., Tomioka, K., Saito, M., Mase, T., Hara, H., Nagaoka, H. and Murase, K. (1989). In "3rd Int. Conf. on Platelet-Activating Factor and Structurally Related Alkyl Ether Lipids", p 14 (Abstr.).

Yamada, T., Tomioka, K., Saito, M., Horie, M., Mase, T., Hara, H. and Nagaoka, H. (1990). Pharmacological properties of YM264, a potent and orally active antagonist of platelet-activating factor. Arch. Int. Pharmacodyn. 308, 123–136.

Yamada, T., Tomioka, K., Horie, M., Sakurai, Y., Nagaoka, H. and Mase, T. (1991). Effects of YM264, a novel PAF antagonist, on puromycin aminonucleoside-induced nephropathy in the rat. Biochem. Biophys. Res. Commun. 176, 781–785.

Yamane, H.K., Farnsworth, C.C., Xie, H., Howald, A., Fung, B.K.-K., Clarke, S., Gelb, M.H. and Glomset, J.A. (1990). Brain G protein and subunits contain an all-*trans*-geranylgeranylcysteine methyl ester at their carboxyl termini. Proc. Natl Acad. Sci. USA 87, 5868–5872.

Yamanishi, J., Takai, Y., Kaibuchi, K., Sano, K., Castagna, M. and Nishizuka, Y. (1983). Synergistic functions of phorbol ester and calcium in serotonin release from human platelets. Biochem. Biophys. Res. Commun. 112, 778–786.

Ysrael, M.C. and Croft, K.D. (1990). Inhibition of leukotriene and platelet-activating factor synthesis in leukocytes by the sysquiterpene lactone scandenolide. Planta Med. 56, 268–270.

Yu, S.-M., Chen, C.-C., Huang, Y.-L., Tsai, C.-W., Lin, C.-H., Huang, T.-F. and Teng, C.-M. (1990). Vasorelaxing effect in rat thoracic aorta caused by denudatin B, isolated from the Chinese herb, *Magnolia fargessi*. Eur. J. Pharmacol. 187, 39–47.

Yue, T.-L., Farhat, M., Rabinovici, R., Perera, P.Y., Vogel, S.N. and Feuerstein, G. (1990). Protective effect of BN 50739, a new platelet-activating antagonist, in endotoxin-treated rabbits. J. Pharmacol. Exp. Ther. 254, 976–981.

Zhu, Y.P., Hoffman, D.R., Hwang, S.-B., Miyaura, S. and Johnston, J.M. (1991). Prolongation of parturition in the pregnant rat following treatment with a platelet activating factor receptor antagonist. Biol. Reprod. 44, 39–42.

Zimmerman, G.A., McIntyre, T.M. and Prescott, S.M. (1985). Production of platelet-activating factor by human vascular endothelium cells: evidence for a requirement for specific agonists and modulation by prostacyclin. Circulation 72, 718–727.

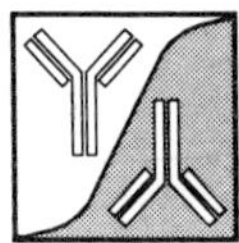

Glossary

A Absorbance

Å Angstrom

AA Arachidonic acid

AAb Autoantibody

Ab Antibody

Abcc Antibody dependent cytotoxic activity

ABA-L-GAT Arsanilicacid conjugated with the synthetic polypeptide L-GAT

AC Adenylate cyclase

ACAT Acyl co-enzyme A acyltransferase

ACE Angiotensin-converting enzyme

ACh Acetylcholine

α_1-ACT α_1-antichymotrypsin

ACTH Adrenocorticotrophin hormone

ADCC Antibody-dependent cell-mediated cytotoxicity

Ado Adenosine

ADP Adenosine diphosphate

AES Anti-eosinophil serum

Ag Antigen

AGE Advanced glycosylation end-product

AGEPC 1-*O*-Alkyl-2-acetyl-*sn*-glyceryl-3-phosphocholine

AI Angiotensin I

AII Angiotensin II

AID Autoimmune disease

AIDS Acquired immune deficiency syndrome

A/J A Jackson inbred mouse strain

cAMP Cyclic adenosine monophosphate (adenosine 3', 5'-phosphate)

AM Alveolar macrophage

AML Acute myelogenous leukaemia

AMP Adenosine monophosphate

ANAb Anti-nuclear antibodies

ANCA Anti-neutrophil cytoplasmic auto antibodies

cANCA Cytoplasmic ANCA

pANCA Perinuclear ANCA

AND Anaphylactic degranulation

ANF Atrial natriuretic factor

ANP Atrial natriuretic peptide

anti-Ig Antibody against an immunoglobulin

anti-RTE Anti-tubular epithelium

APA 13-azaprostanoic acid

APAS Antiplatelet antiserum

APC Antigen-presenting cell

APD Action potential duration

ApO-B Apolipoprotein B

ARDS Acute respiratory distress syndrome

AS Ankylosing spondylitis

4–ASA 4–Aminosalicylic acid

5–ASA 5–Aminosalicylic acid

ASA Acetylsalicylic acid (aspirin)

ATHERO-ELAM A monocyte adhesion molecule

ATL Adult T cell leukaemia

ATP Adenosine triphosphate

AUC Area under curve

AVP Arginine vasopressin

B$_2$ (CD18) A leukocyte integrin

β_2M β_2-microglobulin

BAF Basophil-activating factor

BAL Bronchoalveolar lavage

BALF Bronchoalveolar lavage fluid

BALT Bronchus-associated lymphoid tissue

B cell Bone marrow-derived lymphocyte

BCF Basophil chemotactic factor

BCG Bacillus Calmette-Guérin

bFGF Basic fibroblast growth factor

BG Birbeck granules

BHR Bronchial hyperresponsiveness

BI-CFC Blast colony-forming cells

Bk Bradykinin

BM Bone marrow

BMCMC Bone marrow cultured mast cell

BMMC Bone marrow mast cell

BOC-FMLP Butoxycarbonyl-FMLP

bp Base pair

BPB Para-bromophenacyl bromide

BPI Bacterial permeability-increasing protein

BSA Bovine serum albumin

β-TG β-thromboglobulin

CatG Cathepsin G

C1 The first component of complement

C1 inhibitor A serine protease inhibitor which inactivates C1r/C1s

C1q Complement fragment 1q (anaphylatoxin)

C1qR Receptor for C1w; facilitates attachment of immune complexes to mononuclear leucocytes and endothelium

C2 The second component of complement

C3 The third component of complement

C3a Complement fragment 3a (anaphylatoxin)

C3a$_{72-77}$ A synthetic carboxyterminal peptide C3a analogue

C3aR Receptor for anaphylatoxins, C3a, C4a, C5a

C4 The fourth component of complement

C4b Complement fragment 4b (anaphylatoxin)

C4BP C4 binding protein; plasma protein which acts as co-factor to factor I inactivate C3 convertase

C5 The fifth component of complement

C5a Complement fragment 5a (anaphylatoxin)

C5aR Receptor for anaphylatoxins C3a, C4a and C5a

C5b Complement fragment 5b (anaphylatoxin)

C6 The sixth component of complement

C7 The seventh component of complement

C8 The eighth component of complement

C9 The ninth component of complement

CAH Chronic active hepatitis

CALT Conjunctival associated lymphoid tissue

cAMP Cyclic adenosine monophosphate (adenosine 3', 5'- phosphate)

CAM Cell adhesion molecule

CBH Cutaneous basophil hypersensitivity

CBP Cromolyn binding protein

CCR Creatinine clearance rate

CD Cluster of differentiation (a system of nomenclature for surface molecules on cells of the immune system); cluster determinant

CD$_2$ Present on T cells and involved in antigen non-specific cell activation

CD$_3$ Present on T cells associated with the antigen receptor and involved in antigen-specific cell activation

CD$_4$ Present primarily on helper T cells and involved in class II restricted interactions

CD$_8$ Present primarily on cytotoxic T cells and involved in class I restricted interactions

CD11a α chain of LFA-1 (leucocyte function antigen-1) present on several types of leucocyte and which mediates adhesion

CD11b α chain of CR3 (complement receptor type 3) present on several types of leucocyte and which mediates adhesion

CD11c Cluster of differentiation 11c. Complement receptor 4α chain

CD18 The common β chain of the CD11 family of molecules

CD$_{20}$ Pan B cell

CD31 Platelets, monocytes, macrophages, granulocytes and B-cells

CD34$^-$ Stem cell marker

CD33$^+$ Monocyte and stem cell marker

CD$_{45}$ Cluster of differentiation 45 (Pan leucocyte)

CD59 Low molecular weight HRf; many haemotopolatic and non-haemotopolatic cells

CD62 Activated platelets, endothelial cells

CDC Complement-dependent cytotoxicity

C$_ε$2 Heavy chain of immunoglobulin E: domains 2

C$_ε$3 Heavy chain of immunoglobulin E: domain 3

C$_ε$4 Heavy chain of immunoglobulin E: domain 4

cDNA Complementary DNA

CDP Choline diphosphate

CDR Complementary-determining region

CD$_{xx}$ Common determinant *xx*

CEA Carcinoembryonic antigen

CETAF Corneal epithelial T cell activating factor

CF Cystic fibrosis

Cf Cationized ferritin

CFA Complete Freund's adjuvant

CFC Colony-forming cell

CFU Colony-forming unit

CFU-Eo/B Eosinophil/basophil colony-forming cell

CFU-GM Granulocyte-macrophage colony-forming cell

CFU-S Colony-forming unit, spleen

CGD Chronic granulomatous disease

cGMP Cyclic guanosine monophosphate (guanosine 3′, 5′-phosphate)

CGRP Calcitonin gene-related peptide

CHO Chinese hamster ovary

CI Chemical ionization

CIBD Chronic inflammatory bowel disease

CK Creatine phosphokinase

CKMB The myocardial-specific isoenzyme of creatine phosphokinase

CL Chemiluminescent

CL18/6 Anti-ICAM-1 monoclonal antibody

CLC Charcot-Leyden crystal

CMC Critical micellar concentration

CMI Cell mediated immunity

CML Chronic myeloid leukaemia

CMV Cytomegalovirus

CNS Central nervous system

CO Cyclooxygenase

CoA Coenzyme A

CoA-IT Coenzyme A-independent Transacylase

Con A Concanavalin A

COPD Chronic obstructive pulmonary disease

CoVF Cobra venom

CP Creatine phosphate

CPJ Cartilage/pannus junction

CR Complement receptor

CR1 Complement receptor type 1

CR2 Complement receptor type 2

CR3 Complement receptor type 3

CR3–α Complement receptor type α

CRF Corticotrophin-releasing factor

CRI Cross-reactive idiotype

CRP C-reactive protein

CSA Cyclosporin A

CSF Colony-stimulating factor

CSS Churg-Strauss syndrome

CTAP-III Connective tissue-activating peptide

CTD Connective tissue diseases

CThp Cytotoxic T lymphocyte precursors

C terminus Carboxy terminus of peptide

CTL Cytotoxic T lymphocyte

CTMC Connective tissue mast cell

ct.min^{-1} Counts per minute

Da Dalton (the unit of relative molecular mass)

DAF Decay accelerating factor

DAG Diacylglycerol

DAO Diamine oxidase

D-Arg D-Arginine

DC Dendritic cell

DCF Oxidised DCFH

DCFH 2′,7′-dichlorofluorescin

DEC Diethylcarbamazine

DFMO α-Difluoromethyl ornithine

DFP Diisopropyl fluorophosphate

DGLA Dihomo-γ-linolenic acid

DH Delayed hypersensitivity

DHR Delayed hypersensitivity reaction

DIC Disseminated intravascular coagulation

DGW2 HLA phenotype

DLE *Discoid lupus erythematosus*

DMARD Disease-modifying anti-rheumatic drug

DMF *N,N*-Dimethylformamide

DMSO Dimethylsulphoxide

DNA Deoxyribonucleic acid

D-NAME D-Nitroarginine methyl ester

DNase Deoxyribonuclease

DNCB Dinitrochlorobenzene

DNP Dinitrophenol

Dpt4 *Dermatophagoides pteronyssinus* allergen 4

DREG-2 Murine IgG$_1$ monoclonal antibody against L-selectin

ds Double-stranded

DSCG Disodium cromoglycate

DST Donor-specific transfusion

DTH Delayed-type hypersensitivity

DTPA Diethylenetriamine pentaacetate

DTT Dithiothreitol

dv/dt Rate of change of voltage within time

ε Molar absorption coefficient

EA Egg albumin

EAE Experimental autoimmune encephalomyelitis

EAF Eosinophil-activating factor

EAR Early phase asthmatic reaction

EAT Experimental autoimmune thyroiditis

EBV Epstein–Barr virus

EC Electron capture

ECD Electron capture detector

ECE Endothelin converting enzyme

E-CEF Eosinophil cytotoxicity enhancing factor

ECF-A Eosinophil chemotactic factor of anaphylaxis

ECG Electrocardiogram

ECGF Endothelial cell growth factor

ECGS Endothelial cell growth supplement

E. coli Escherichia coli

ECP Eosinophil cationic protein

ED$_{35}$ Effective dose producing 35% maximum response

ED$_{50}$ Effective dose producing 50% maximum response

EDF Eosinophil differentiation factor

EDN Eosinophil-derived neurotoxin

EDRF Endothelium-derived relaxant factor

EDTA Ethylene diamine tetraacetic acid (etidronic acid)

EE Eosinophilic eosinophils

EEG Electroencephalogram

EET Epoxyeicosatrienoic acid

EFA Essential fatty acid

EFS Electrical field stimulation

EGF Epidermal growth factor

EGTA Ethylene glycol-bis(β-aminoethyl ether) *N,N,N′N′*-tetraacetic acid

EI Electron impact

ELAM Endothelial leucocyte adhesion molecule

ELAM-1 Endothelial leucocyte adhesion molecule-1

ELF Respiratory epithelium lung fluid

ELISA Enzyme-linked immunosorbent assay

EMS Eosinophilia-myalgia syndrome

ENS Enteric nervous system

EO Eosinophil

EOR Early onset reaction

EPA Eicosapentaenoic acid

EPO Eosinophil peroxidase

EpDIF Epithelial-derived inhibitory factor

EpDRF Epithelium-derived relaxant factor

EPX Eosinophil protein X

ER Endoplasmic reticulum

ESP Eosinophil stimulation promoter

ESR Erythrocyte sedimentation rate
ET Endothelin
ETYA Eicosatetraynoic acid

FA Fatty acid
FAB Fast-electron bombardment
factor B Serine protease in the C3 converting enzyme of the alternative pathway
factor D Serine protease which cleaves factor B
factor H Plasma protein which acts as a co-factor to factor I
factor I Hydrolyses C3 converting enzymes with the help of factor H
FBR Fluorescence photobleaching recovery
Fc Portion of immunoglobulin molecule
F$_c$R Receptor for F$_c$ region of antibody
Fc$_e$RI High affinity receptor for IgE
Fc$_e$RII Low affinity receptor for IgE
FCS Fetal calf (bovine) serum
FEV$_1$ Forced expiratory volume in 1 second
FGF Fibroblast growth factor
FID Flame ionization detector
FITC Fluorescein isothiocyanate
FKBP FK506–binding protein
FLAP 5–Lipoxygenase-activating protein
FMLP *N*-Formyl-methionyl-leucyl-phenylalanine
FNLP Formyl-norleucyl-leucyl-phenylalanine
FSG Focal sequential glomerulosclerosis
FTS Facteur thymic serique
5–FU 5–Fluorouracil

G6PD Glucose 6–phosphate dehydrogenase
GABA γ-Aminobutyric acid
GAG Glycosaminoglycan
GALT Gut-associated lymphoid tissue
GBM Glomerular basement membrane
GC Guanylate cyclase
GC-MS Gas chromatography mass spectroscopy
G-CSF Granulocyte colony-stimulating factor
GDP Guanosine 5′-diphosphate
GEC Glomerular epithelial cells
GF-1 Insulin-like growth factor
GFR Glomerular filtration rate
GH Growth hormone
GH-RF Growth hormone releasing factor
GI Gastrointestinal
GIP Granulocyte inhibitory protein
GMC Gastric mast cell
GM-CSF Granulocyte-macrophage colony-stimulating factor
GMP Guanosine monophosphate (guanosine 5′-phosphate)
GMP-140 Granule-associated membrane protein-140
GP Glycoprotein

GPIIb-IIIa Glycoprotein IIb-IIIa – a platelet membrane antigen
GSH Glutathione (reduced)
GSSG Glutathione (oxidized)
GTP Guanosine triphosphate
GTP-γ-S Guanarine 5′O-(3–thiotriphosphate)
GTPase Guanidine triphosphatase
GVHD Graft versus host disease
GVHR Graft versus host reaction

H Haemagglutinin
H$_1$ Histamine receptor type 1
H$_2$ Histamine receptor type 2
H$_2$O$_2$ Chemical symbol for hydrogen peroxide
H$_3$ Histamine receptor type 3
HA Histamine
H & E Haematoxylin and eosin
hIL Human interleukin
Hb Haemoglobin
HBBS Hank's balanced salt solution
HDC Histidine decarboxylase
HDL High-density lipoprotein
HEL Hen egg white lysozyme
HEPE Hydroxyeicosapentanoic acid
HEPES N-2–hydroxylethylpiperazine-N′-2–ethane sulphonic acid
HES Hypereosinophilic syndrome
HETE 5,8,9,11 and 15 Hydroxyeicosatetraenoic acid
HETrE Hydroxyeicosatrienoic acid
HEV High endothelial venule
HFN Human fibronectin
HGF Hepatocyte growth factor
HHT 12–hydroxy-5, 8, 10–heptadecatrienoic acid
HHTrE 12(*S*)-Hydroxy-5,8,10–heptadecatrienoic acid
HIV Human immunodeficiency virus
HLA Human leucocyte antigen
HMG CoA Hydroxyl methyl glutaryl Co-enzyme A
HMW High molecular weight
HMT Histidine methyltransferase
HMVEC Human microvascular endothelial cells
HNC Human neutrophil collagenase (MMP-8)
HNE Human neutrophil elastase
HNG Human neutrophil gelatinase (MMP-9)
HODE Hydroxyoctadecanoic acid
HPETE Hydroperoxyeicosatetraenoic acid
HPETrE Hydroperoxytrienoic acid
HPODE Hydroperoxyoctadecanoic acid
HPLC High-performance liquid chromatography
HRA Histamine-releasing activity
HRAN Neutrophil-derived histamine-releasing activity
HRf Homologous-restriction factor
HRF Histamine-releasing factor
HRP Horseradish peroxidase

HSA Human serum albumin
HSP Heat-shock protein
HS-PG Heparan sulphate proteoglycan
HSV Herpes simplex virus
^{3}HTdR Tritiated thymidine
5–HT 5–Hydroxytryptamine *also known as* Serotonin
HUVEC Human umbilical vein endothelial cell

Ia Immune reaction-associated antigen
Ia+ Murine class II major histocompatibility complex antigen
I$_{sc}$ Short-circuit current
IB$_4$ Anti-CD18 monoclonal antibody
IBD Inflammatory bowel disease
IBMX Isobutylmethylxanthine
IBS Inflammatory bowel syndrome
IC$_{50}$ Concentration producing 50% inhibition
ICAM Intercellular adhesion molecules
ICAM-1 Intercellular adhesion molecule-1
ICAM-2 Intercellular adhesion molecule-2
ICE IL-1β–converting enzyme
IDC Interdigitating cell
IDD Insulin-dependent (type 1) diabetes
IEL Intraepithelial leucocytes
IFA Incomplete Freund's Adjuvant
IFN Interferon
IFNα Interferon α
IFNβ Interferon β
IFNγ Interferon γ
Ig Immunoglobulin
IgA Imunoglobulin A
IgE Immunoglobulin E
IgG Immunoglobulin G
IgG1 Immunoglobulin G class 1
IgG$_{2a}$ Immunoglobulin G class 2a
IgM Immunoglobulin M
IGF-1 Insulin-like growth factor
IHC Immunohistochemistry
IHES Idiopathic hypereosinophilic syndrome
IL Interleukin
IL-1 Interleukin-1
IL-2 Interleukin 2
IL-3 Interleukin-3
IL-5 Interleukin-5
IL-6 Interleukin-6
IL-5R Interleukin-5–receptor
IL-8 Interleukin-8
IL-1α Interleukin-1α
IL-1β Interleukin-1β
IL-1Ra Interleukin-1 receptor antagonist
IL-2R Interleukin-2 receptor
IL-3R Interleukin-3R
ILR Interleukin receptor
IMMC Intestinal mucosal mast cell
INCAM Inducible cell adhesion molecule
INCAM110 Inducible cell adhesion molecule 110
i.p. Intraperitoneally
IP$_3$ Inositol triphosphate
IP$_4$ Inositol tetrakisphosphate

IPO Intestinal peroxidase
IpOCOCq Isopropylidene OCOCq
I/R Ischaemia-reperfusion
IRAP IL-1 receptor antagonist protein
ISCOM Immune-stimulating complexes
ISGF3 Interferon-stimulated gene factor 3
iSRS Immunoreactive slow-reacting substance
IT Immunotherapy
ITP Idiopathic thrombocytopenic purpura
i.v. Intravenous

K_a Association constant
kb Kilobase
20KDHRF A homologous restriction factor; binds to C8
65KDHRF A homologous restriction factor, also known as C8 binding protein; interferes with cell membrane pore-formation by C5b-C8 complex
K_d Equilibrium dissociation constant
kD Kilo Dalton
K_D Dissociation constant
KD Kallidin
Ki Antagonist binding affinity
Ki67 Nuclear membrane antigen
KLH Keyhole limpet haemocyanin
KOS KOS strain of herpes simplex virus

λ_{max} Wavelength of maximum absorbance
LAD Leucocyte adhesion deficiency
LAK Lymphocyte-activated killer (cell)
LAM Leucocyte adhesion molecule
LAM-1 Leucocyte adhesion molecule-1
LAR Late-phase asthmatic reaction
L-Arg L-Arginine
LBP LPS binding protein
LC Langerhans cell
LCF Lymphocyte chemoattractant factor
LCR Locus control region
LDH Lactate dehydrogenase
LDL Low-density lipoprotein
LDV Laser Doppler velocimetry
LECAM Lectin adhesion molecule
LECAM-1 Lection adhesion molecule-1
LFA Leucocyte function-associated antigen
LFA-1 Leucocyte function-associated antigen-1
LG β-Lactoglobulin
LHRH Luteinizing hormone-releasing hormone
LI Labelling index
LIS Lateral intercellular spaces
LMP Low molecular mass polypeptide
LMW Low molecular weight
L-NOARG L-Nitroarginine
LP(a) Lipoprotein a
LPS Lipopolysaccharide
LT Leukotriene
LTA$_4$ Leukotriene A$_4$
LTB$_4$ Leukotriene B$_4$
LTC$_4$ Leukotriene C$_4$

LTD$_4$ Leukotriene D$_4$
LTE$_4$ Leukotriene E$_4$
L$_y$-1$^+$ (Cell line)
LXA$_4$ Lipoxin A$_4$
LXB$_4$ Lipoxin B$_4$
LXC$_4$ Lipoxin C$_4$
LXD$_4$ Lipoxin D$_4$
LXE$_4$ Lipoxin E$_4$

M-540 Merocyanine-540
α_2-M α_2-macroglobulin
mAb Monoclonal antibody
mAB IB4 Monoclonal antibody IB4
mAB PB1.3 Monoclonal antibody PB1.3
mAB R 3.1 Monoclonal antibody R 3.1
mAB R 3.3 Monoclonal antibody R 3.3
mAB 6.5 Monoclonal antibody 6.5
mAB 60.3 Monoclonal antibody 60.3
MAC Membrane attack molecule
Mac Macrophage *also abbreviated to* MO
Mac-1 Macrophage-1
MAF Macrophage-activating factor
MAO Monoamine oxidase
MAP(s) Monophasic action potential(s)
MBP Major basic protein
MBSA Methylated bovine serum albumin
MC Mesangial cells
M cell Microfold or membranous cell of Peyer's patch epithelium
MCP Membrane co-factor protein
M-CSF Monocyte colony-stimulating factor
MC$_T$ Tryptase-containing mast cell
MC$_{TC}$ Tryptase- and chymase-containing mast cell
MDA Malondialdehyde
MDGF Macrophage-derived growth factor
MDP Muramyl dipeptide
MEA Mast cell growth-enhancing activity
MEL Metabolic equivalent level
MEM Minimal essential medium
MG *Myasthenia gravis*
MHC Major histocompatibility complex
MI Myocardial ischaemia
MIF Migration inhibition factor
mIL Mouse interleukin
MI/R Myocardial ischaemia/reperfusion
MIRL Membrane inhibitor of reactive lysis
MLC Mixed lymphocyte culture
MLR Mixed lymphocyte reaction
MMC Mucosal mast cell
MMCP Mouse mast cell protease
MMP Matrix metalloproteinase
MMP1 Matrix metalloproteinase 1
mNA 6-methoxy-2-napthylacetic acid
MNC Mononuclear cells
MO Macrophage *also abbreviated to* Mac
MPO Myeloperoxidase
MRI Magnetic resonance imaging
mRNA Messenger ribonucleic acid
MS Mass spectrometry
MSS Methylprednisoline sodium succinate

MT Malignant tumour
MW Molecular weight

NA Noradrenaline
NAAb Natural autoantibody
NAb Natural antibody
NADH Reduced nicotinamide adenine dinucleotide
NADP Nicotinamide adenine diphosphate
NADPH Reduced nicotinamide adenine dinucleotide phosphate
L-NAME L-Nitroarginine methyl ester
NANC Nonadrenergic, non-cholinergic
NAP Neutrophil-activating peptide
NAP-1 Neutrophil-activating peptide-1
NAP-2 Neutrophil-activating peptide-2
NC1 Non-collagen 1
N-CAM Neural cell adhesion molecule
NCEH Neutral cholesteryl ester hydrolase
NCF Neutrophil chemotactic factor
NDGA Nordihydroguaretic acid
Neca 5'-(N-ethyl carboxamido)-adenosine
NED Nedocromil sodium
NEP Neutral endopeptidase (EC 3.4.24.11)
NF-AT Nuclear factor of activated T lymphocytes
NF-κB Nuclear factor-κB
NGF Nerve growth factor
NGPS Normal guinea-pig serum
NIMA Non-inherited maternal antigens
Nk Neurokinin
NK Natural killer
NkA Neurokinin A
NkB Neurokinin B
L-NMMA L-Nitromonomethyl arginine
NMR Nuclear magnetic resonance
NO Chemical symbol for nitric oxide
NPY Neuropeptide Y
NRS Normal rabbit serum
NSAID Non-steroidal anti-inflammatory drug
NSE Nerve-specific enolase
NT Neurotensin
N terminus Amino terminus of peptide

O_2^- Oxygen free radical
OA Osteoarthritis
OD Optical density
ODC Ornithine decarboxylase
ODS Octadecylsilyl
·OH Chemical symbol for hydroxyl radical
OVA Ovalbumin
ox-LDL Oxidized low-density lipoprotein

p Probability
P Phosphate
P$_a$O$_2$ Arterial oxygen pressure
P$_i$ Inorganic phosphate
α_{1-PI} α_1-proteinase inhibitor
P150,95 A leucocyte integrin
PA Phosphatidic acid

pA₂ Negative logarithm of the antagonist dissociation constant
PADGEM Platelet-activation dependent granule external membrane
PAF Platelet-activating factor
PAGE Polyacrylamide gel electrophoresis
PAI Plasminogen activator inhibitor
PAM Pulmonary alveolar macrophages
PAS Periodic acid-Schiff reagent
PBA Polyclonal B cell activators
PBC Primary biliary cirrhosis
PBL Peripheral blood lymphocytes
PBMC Peripheral blood mononuclear cells
PBS Phosphate-buffered saline
PC Phosphatidylcholine
PCA Passive cutaneous anaphylaxis
PCNA Proliferating cell nuclear antigen
PCR Polymerase chain reaction
p.d. Potential difference
PDBu 4a-phorbol 12,13–dibutyrate
PDE Phosphodiesterase
PDGF Platelet-derived growth factor
PE Phosphatidylethanolamine
PECAM Platelet endothelial cell adhesion molecule
PEG Polyethylene glycol
PET Positron emission tomography
PEt Phosphatidylethanol
PF₄ Platelet factor 4
PG Prostaglandin
PGA Polyglandular autoimmune syndrome
PGE₂ Prostaglandin E₂
PGF Prostaglandin F
PGI Prostacyclin
PGI₂ More standard abbreviation of prostacyclin
PGD₂ Prostaglandin D₂
PGE₁ Prostaglandin E₁
PGF₂α Prostaglandin F₂α
PGF₂ Prostaglandin F₂
PGG₂ Prostaglandin G₂
PGH Prostaglandin H
PₐO₂ Arterial oxygen pressure
PGH₂ Prostaglandin H₂
PGI₂ Prostaglandin I₂
PGP Protein gene-related peptide
Ph¹ Philadelphia (chromosome)
PHA Phytohaemagglutinin
PHI Peptide histidine isoleucine
PHM Peptide histidine-methionine
PI Phosphatidyl inositol
Pᵢ Inorganic phosphate
PIP Phosphatidylinositol monophosphate
PIP₂ Phosphatidylinositol biphosphate
PK Protein kinase
PKA Protein kinase A
PKC Protein kinase C
PL Phospholipase
PLA Phospholipase A
PLA₂ Phospholipase A₂
PLAP Putative phospholipase activating protein
PLC Phospholipase C
PLD Phospholipase D

PLP Proteolipid protein
PLT Primed lymphocyte typing
PMA Phorbol myristate acetate
PMC Peritoneal mast cell
PMD Piecemeal degranulation
PML Polymorphonuclear leucocyte
PMN Polymorphonuclear neutrophil
PMSF Phenylethylsulphonyl fluoride
PNU Protein nitrogen unit
p.o. *Per os* (by mouth)
PPD Purified protein derivative
PRA Percentage reactive activity
PRD Positive regulatory domain
PR3 Proteinase-3
proET-1 Proendothelin-1
PS Phosphatidylserine
PTA₂ Pinane thromboxane A₂
PTCA Percutaneous transluminal coronary angioplasty
PTCR Percutaneous transluminal coronary recanalization
Pte-H₄ Tetrahydropteridine
PtX Pertussis toxin
PUFA Polyunsaturated fatty acid
PUMP-1 Punctuated metalloproteinase
PWM Pokeweed mitogen
PYY Peptide YY

q.i.d. *Quater in die* (four times a day)
QRS Segment of electrocardiogram

·R Free radical
R15.7 Anti-CD18 monoclonal antibody
RA Rheumatoid arthritis
RANTES Regulated on activation, normal T expressed and secreted
RAST Radioallergosorbent test
RBC Red blood cell
RBF Renal blood flow
RBL Rat basophilic leukaemia
RE RE strain of herpes simplex virus type 1
REA Reactive arthritis
REM Relative electrophoretic mobility
RER Rough endoplasmic reticulum
RF Rheumatoid factor
RFL-6 Rat fetal lung-6
RFLP Restriction fragment length polymorphism
rh- (prefix) recombinant human (usually refers to peptide)
RIA Radioimmunoassay
RMCP Rat mast cell protease
RMCPII Rat mast cell protease II
RNA Ribonucleic acid
RNase Ribonuclease
RNHCl N-Chloramine
RNL Regional lymph nodes
ROM Reactive oxygen metabolites
ROS Reactive oxygen species
R-PIA *R-NG*-(1–methyl-1–phenyltheyl)-adenosine
RPMI 1640 Roswell Park Memorial Institute 1640 medium
RS Reiter's syndrome
RSV Rous sarcoma virus

RTE Rabbit tubular epithelium
RW Ragweed

S Svedberg (unit of sedimentation density)
SALT Skin associated lymphoid tissue
SAZ Sulphasalazine
SC Secretory component
SCF Stem cell factor
SCFA Short chain fatty acid
SCG Sodium cromoglycate
SCID Severe combined immunodeficiency sydrome
sCR₁ Soluble type-1 complement receptors
SCW Streptococcal cell wall
SD Standard deviation
SDS Sodium dodecyl sulphate
SDS-PAGE Sodium dodecyl sulphate-polyacrylamide gel electrophoresis
SEM Standard error of the mean
SGAW Specific airway conductance
SIRS Soluble immune response suppressor
SK Streptokinase
Sl Murine Steel mutation
SLE Systemic lupus erythematosus
SLeˣ Sialyl Lewis X antigen
SLO Streptolysin-O
SLPI Secretory leucocyte protease inhibitor
SM Sphingomyelin
SNAP *S*-Nitroso-*N*-acetylpenicillamine
SNP Sodium nitroprusside
SOD Superoxide dismutase
SOM Somatostatin
SOZ Serum-opsonized zymosan
SP Sulphapyridine
S Protein vitronectin
SR Systemic reaction
SRBC Sheep red blood cells
SRIF Somatotrophin release-inhibiting factor (somatostatin)
SRS Slow-reacting substance
SRS-A Slow-reacting substance of anaphylaxis
Sub P Substance P

t₁/₂ Half-life
T84 Human intestinal epithelial cell line
TauNHCl Taurine monochloramine
TBM Tubular basement membrane
TCA Trichloroacetic acid
T cell Thymus-derived lymphocyte
TCR T cell receptor
TDI Toluene diisocyanate
TDID₅₀ Tissue culture infectious dose – 50%
TEC Tubular epithelial cells
TF Tissue factor
Tg Thyroglobulin
TGF Transforming growth factor
TGFβ Transforming growth factor β
TGFβ₁ Transforming growth factor-β₁
Tʜ T helper cells
Tʜo T Helper o

T_Hp T helper precursor
T_H0, T_H1, T_H2 Subsets of helper
T cells
Thy 1+ Murine T cell antigen
t.i.d. *Ter in die* (three times a day)
TIL Tumor-infiltrating lymphocytes
TIMP Tissue inhibitors of
metalloproteinase
TIMP-1 Tissue inhibitor of
metalloproteinases 1
Tla Thymus leukaemia antigen
TLC Thin-layer chromatography
TLP Tumour-like proliferation
Tm T memory
TNF Tumour necrosis factor
TNF-α Tumour necrosis factor-α
tPA Tissue-type plasminogen activator
r-tPA Recombinant tissue-type
plasminogen activator
TPA 12–*o*-Tetradeconylphorbol-13-
acetate

TPK Tyrosine protein kinases
TPP Transpulmonary pressure
Tris Tris(hydroxymethyl)aminomethane
TSH Thyroid-stimulating hormone
TTX Tetrodotoxin
TX Thromboxane
TXA₂ Thromboxane A_2
TXB₂ Thromboxane B_2
Tyk2 Tyrosine kinase

UC Ulcerative colitis
UDP Uridine diphosphate
UPA Urokinase-type plasminogen
activator
UV Ultraviolet

VC Veiled cells
VCAM Vascular cell adhesion
molecule
VCAM-1 Vascular cell adhesion
molecule-1

VF Ventricular fibrillation
VIP Vasoactive intestinal peptide
VLA Very late antigen
VLA-4 Very late activating antigen-4
VLDL Very low-density lipoprotein
V_{max} maximal velocity
vp Viral protein
VP Vasopressin
VPB Ventricular premature beat
VT Ventricular tachycardia

W Murine dominant white spotting
mutation
WBC White blood cell
WGA Wheat germ agglutinin

XO Xanthine oxidase

ZA Zonulae adherens
ZAS Zymosan-activated serum
ZO Zonulae occludentes

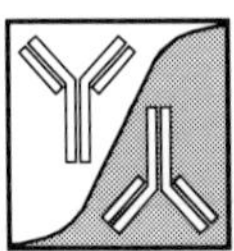

Key to Illustrations

Helper lymphocyte

Suppressor lymphocyte

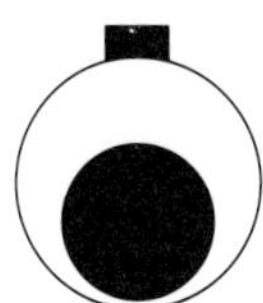

Killer lymphocyte

Plasma cell

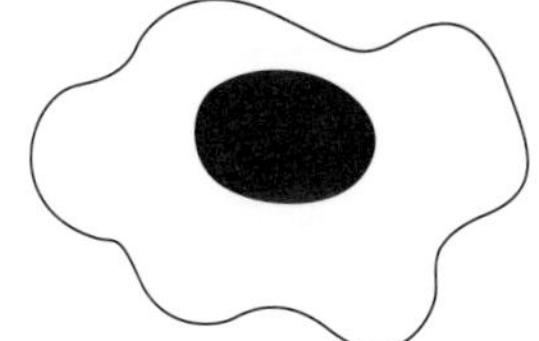

Bacterial or Tumour cell

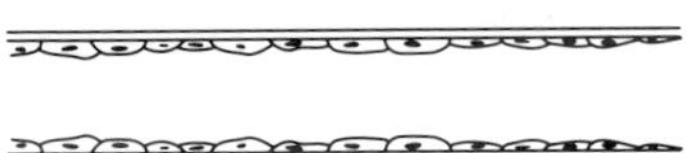

Blood vessel lumen

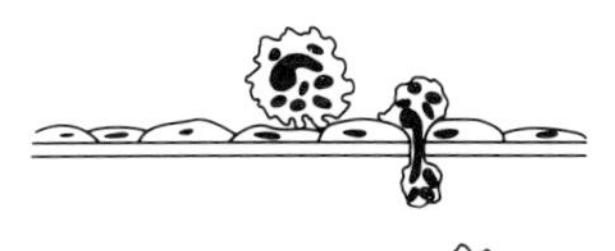

Eosinophil passing through vessel wall

Neutrophil passing through vessel wall

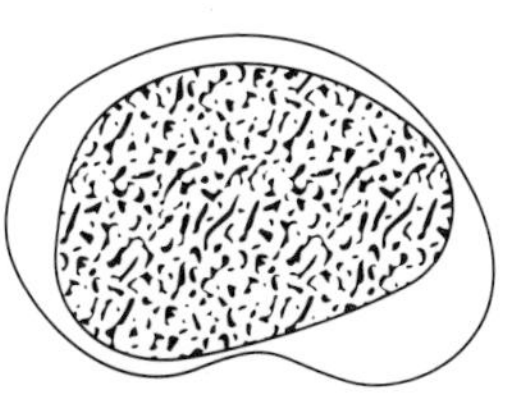

Fibroblast

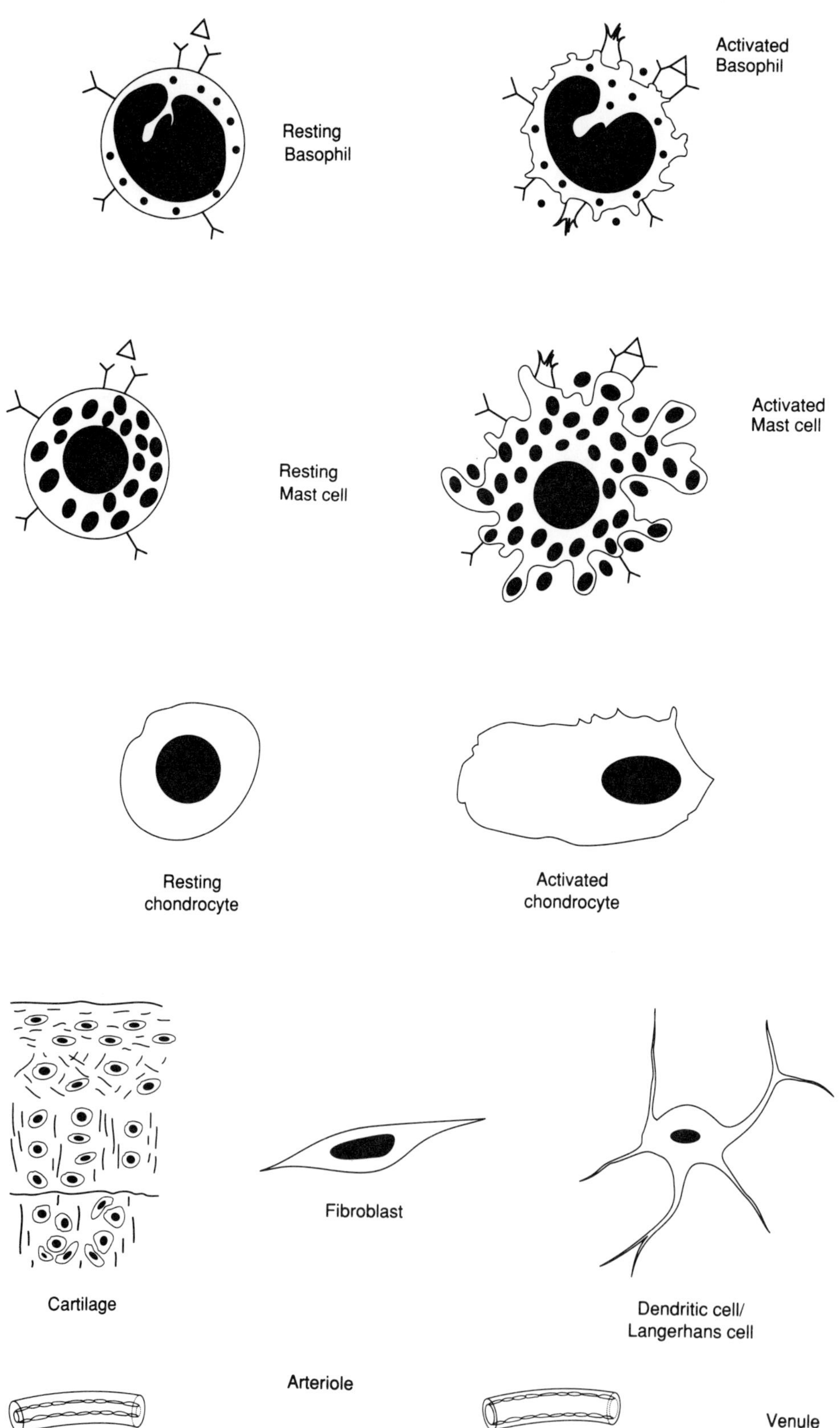
Resting
Basophil
Activated
Basophil
Resting
Mast cell
Activated
Mast cell
Resting
chondrocyte
Activated
chondrocyte
Cartilage
Fibroblast
Dendritic cell/
Langerhans cell
Arteriole
Venule

Resting neutrophil

Activated neutrophil

Resting eosinophil

Activated eosinophil

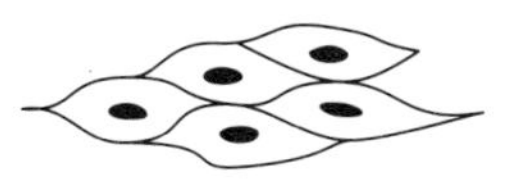

Smooth muscle

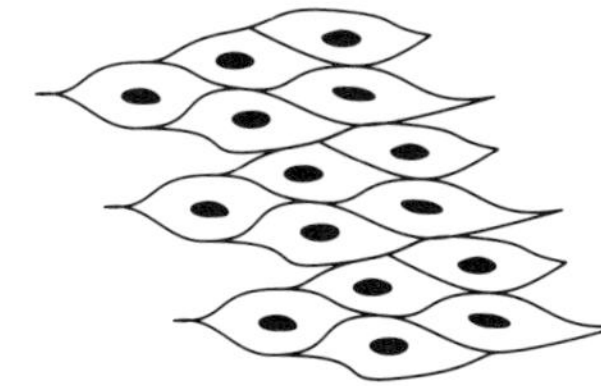

Smooth muscle thickening

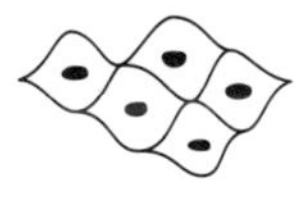

Smooth muscle contraction

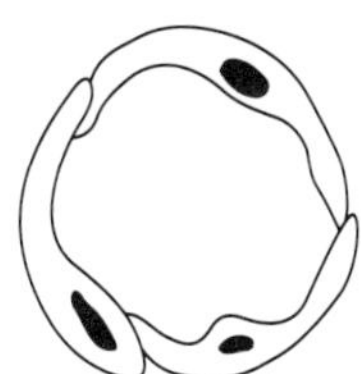

Normal blood vessel

Endothelial cell permiability

Resting macrophage

Activated macrophage

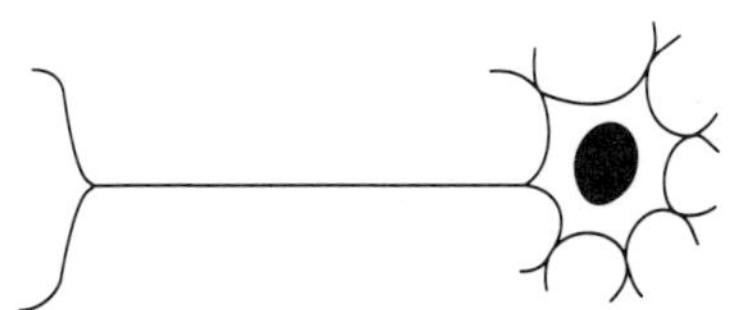

Nerve

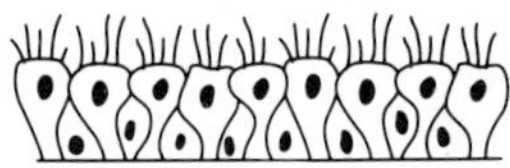

Intact
epithelium

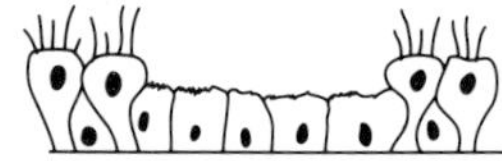

Damaged
epithelium

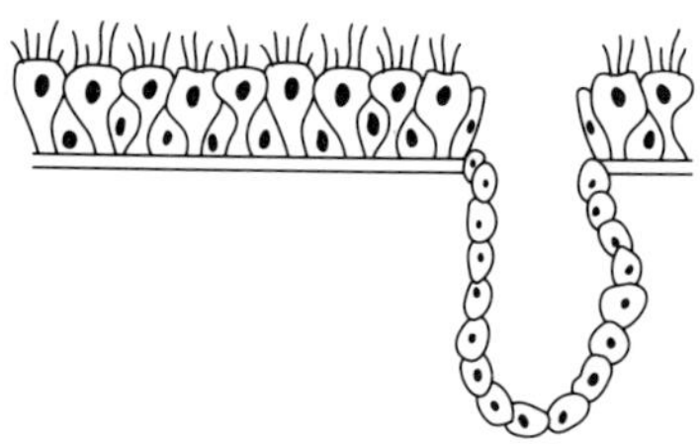

Intact
epithelium
with
submucosal
gland

Normal
submucosal
gland

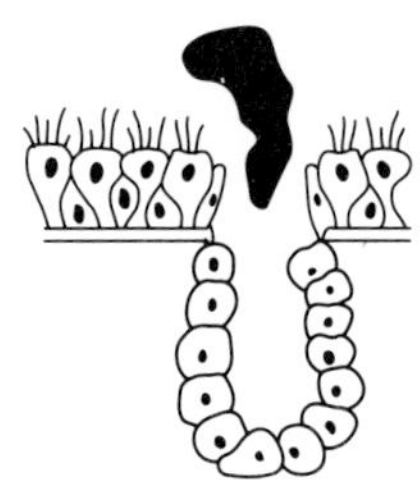

Hypersecreting
submucosal
gland

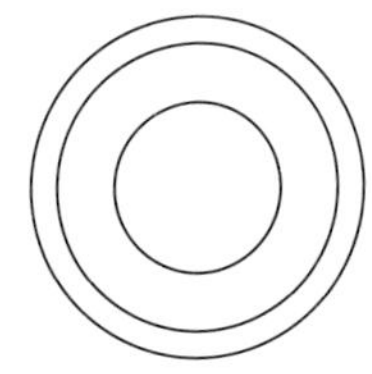

Normal
airway

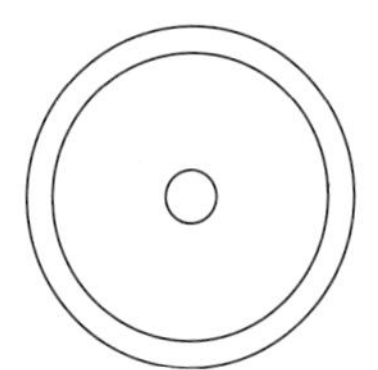

Oedema

Bronchospasm

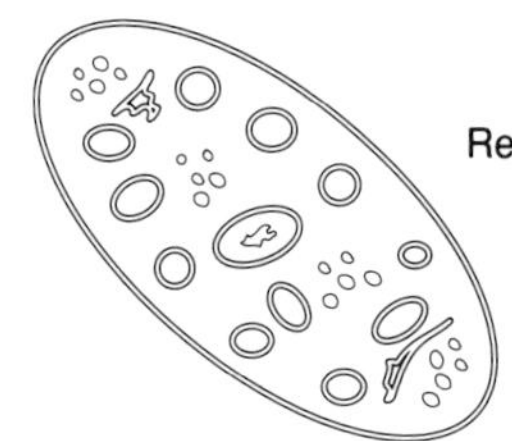

Resting platelet

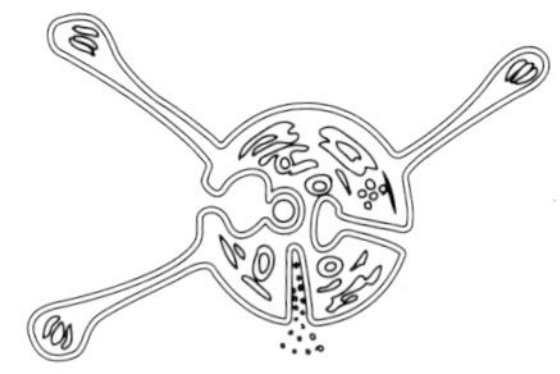

Activated
platelet

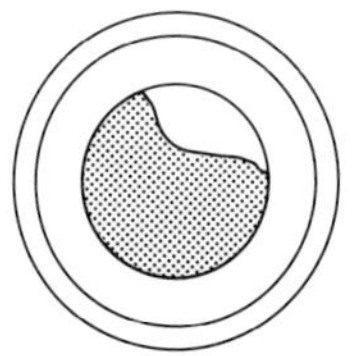

Airway
hypersecreting
mucus

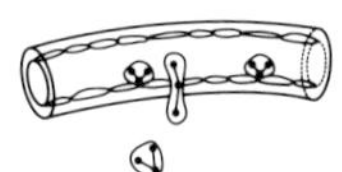

Inflamed
venule

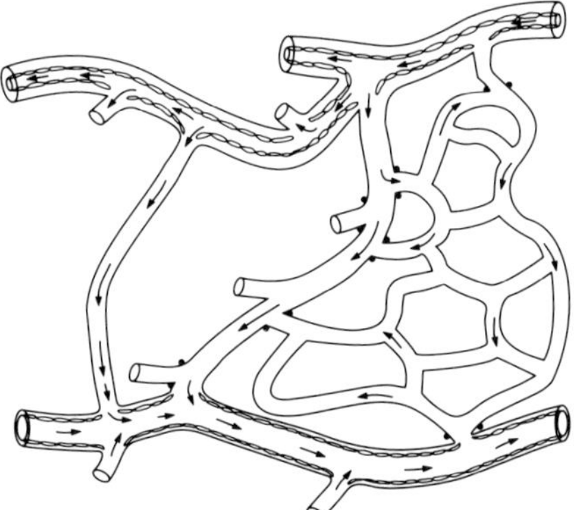

Microcirculatory
system

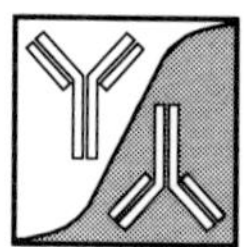

Index